HANDBOOK OF
AIR CONDITIONING
SYSTEM DESIGN

OTHER McGRAW-HILL HANDBOOKS OF INTEREST

HANDBOOK OF
AIR CONDITIONING
SYSTEM DESIGN

Carrier Air Conditioning Company

McGRAW-HILL BOOK COMPANY

New York   San Francisco   Toronto   London   Sydney

HANDBOOK OF AIR CONDITIONING SYSTEM DESIGN

Copyright © 1965 by McGraw-Hill, Inc. All Rights Reserved.

© 1960, 1963, 1964, 1965 by Carrier Corporation. Printed in the United States of America. This book, or parts thereof, may not be reproduced in any form without permission of the publishers. *Library of Congress Catalog Card Number* 65–17650.

ISBN 07-010090-X

17 18 HDHD 8543210

PREFACE

The *Handbook of Air Conditioning System Design* is the first complete practical guide to the design of air conditioning systems. It embodies all the knowledge and experience gained over the past fifty years by the pioneer in the field, Carrier Air Conditioning Company.

This handbook is tailored to the specific needs of the man responsible for the details of design, and, therefore, the foremost consideration has been the requirements of the consulting engineer. In fact, many of the concepts embody the up-to-date thinking of consulting engineers.

If any one word best describes this work, it is the word "practical."

- It is usable at all educational levels.
- It provides practical data for professional designers who need optimum solutions on a day-to-day basis.
- It bridges the gap between air conditioning texts and manufacturers' product catalogs.
- It provides proved system design techniques and assures quality of application with minimum service requirements.
- It provides guidance in simplified form.
- It provides a reference source employing the best techniques of indexing and format.

This *Handbook of Air Conditioning System Design* is a companion piece to manufacturers' product literature. Together the handbook and product literature make up a complete engineer's manual.

Those using this book for study will benefit from clear applicable examples presented in each of the engineering sections.

In summary, this *Handbook of Air Conditioning System Design* is a quick reference for those actively engaged in designing air conditioning systems, a teaching work for those studying air conditioning system design, and a refresher for those engineers with wide experience in the field.

*   *   *

Grateful appreciation is hereby extended to those hundreds of Carrier engineers who generously contributed to the total body of knowledge herein, and to those consulting engineers, mechanical contractors, and architects who so willingly and enthusiastically contributed their experience to this project.

Carrier Air Conditioning Company

CONTENTS

HANDBOOK OF
AIR CONDITIONING
SYSTEM DESIGN

Part 1
LOAD ESTIMATING
CHAPTER 1. BUILDING SURVEY AND LOAD ESTIMATE

The primary function of air conditioning is to maintain conditions that are (1) conducive to human comfort, or (2) required by a product, or process within a space. To perform this function, equipment of the proper capacity must be installed and controlled throughout the year. The equipment capacity is determined by the *actual* instantaneous peak load requirements; type of control is determined by the conditions to be maintained during peak and partial load. Generally, it is impossible to measure either the actual peak or the partial load in any given space; these loads must be estimated. It is for this purpose that the data contained in Part 1 has been compiled.

Before the load can be estimated, it is imperative that a comprehensive survey be made to assure accurate evaluation of the load components. If the building facilities and the actual instantaneous load within a given mass of the building are carefully studied, an economical equipment selection and system design can result, and smooth, trouble free performance is then possible.

The heat gain or loss is the amount of heat instantaneously coming into or going out of the space. *The actual load is defined as that amount of heat which is instantaneously added or removed by the equipment.* The instantaneous heat gain and the actual load on the equipment will rarely be equal, because of the thermal inertia or storage effect of the building structures surrounding a conditioned space.

Chapters 2, 4, 5, 6, and 7 contain the data from which the instantaneous heat gain or loss is estimated. *Chapter 3* provides the data and procedure for applying storage factors to the appropriate heat gains to result in the actual load. *Chapter 8* provides the bridge between the load estimate and the equipment selection. It furnishes the procedure for establishing the criteria to fulfill the conditions required by a given project.

The basis of the data and its use, with examples, are included in each chapter with the tables and charts; also an explanation of how each of the heat gains and the loads manifest themselves.

BUILDING SURVEY
SPACE CHARACTERISTICS AND HEAT LOAD SOURCES

An accurate survey of the load components of the space to be air conditioned is a basic requirement for a realistic estimate of cooling and heating loads. *The completeness and accuracy of this survey is the very foundation of the estimate, and its importance can not be overemphasized.* Mechanical and architectural drawings, complete field sketches and, in some cases, photographs of important aspects are part of a good survey. The following physical aspects must be considered:

1. *Orientation of building* — Location of the space to be air conditioned with respect to:
    a) Compass points — sun and wind effects.
    b) Nearby permanent structures — shading effects.
    c) Reflective surfaces — water, sand, parking lots, etc.
2. *Use of space(s)* — Office, hospital, department store, specialty shop, machine shop, factory, assembly plant, etc.
3. *Physical dimensions of space(s)* — Length, width, and height.
4. *Ceiling height* — Floor to floor height, floor to ceiling, clearance between suspended ceiling and beams.
5. *Columns and beams* — Size, depth, also knee braces.
6. *Construction materials* — Materials and thickness of walls, roof, ceiling, floors and partitions, and their relative position in the structure.
7. *Surrounding conditions* — Exterior color of walls and roof, shaded by adjacent building or sunlit. Attic spaces — unvented or vented, gravity or forced ventilation. Surrounding spaces conditioned or unconditioned — temperature of non-conditioned adjacent spaces, such as furnace or boiler room, and kitchens. Floor on ground, crawl space, basement.
8. *Windows* — Size and location, wood or metal

sash, single or double hung. Type of glass — single or multipane. Type of shading device. Dimensions of reveals and overhangs.

9. *Doors* — Location, type, size, and frequency of use.

10. *Stairways, elevators, and escalators* — Location, temperature of space if open to unconditioned area. Horsepower of machinery, ventilated or not.

11. *People* — Number, duration of occupancy, nature of activity, any special concentration. At times, it is required to estimate the number of people on the basis of square feet per person, or on average traffic.

12. *Lighting* — Wattage at peak. Type — incandescent, fluorescent, recessed, exposed. If the lights are recessed, the type of air flow over the lights, exhaust, return or supply, should be anticipated. At times, it is required to estimate the wattage on a basis of watts per sq ft, due to lack of exact information.

13. *Motors* — Location, nameplate and brake horsepower, and usage. The latter is of great significance and should be carefully evaluated.

   The power input to electric motors is not necessarily equal to the rated horsepower divided by the motor efficiency. Frequently these motors may be operating under a continuous overload, or may be operating at less than rated capacity. It is always advisable to measure the power input wherever possible. This is especially important in estimates for industrial installations where the motor machine load is normally a major portion of the cooling load.

14. *Appliances, business machines, electronic equipment* — Location, rated wattage, steam or gas consumption, hooded or unhooded, exhaust air quantity installed or required, and usage.

   Greater accuracy may be obtained by measuring the power or gas input during times of peak loading. The regular service meters may often be used for this purpose, provided power or gas consumption not contributing to the room heat gain can be segregated.

   Avoid pyramiding the heat gains from various appliances and business machines. For example, a toaster or a waffle iron may not be used during the evening, or the fry kettle may not be used during morning, or not all business

machines in a given space may be used at the same time.

Electronic equipment often requires individual air conditioning. The manufacturer's recommendation for temperature and humidity variation must be followed, and these requirements are often quite stringent.

15. *Ventilation* — Cfm per person, cfm per sq ft, scheduled ventilation (agreement with purchaser), see *Chapter 6*. Excessive smoking or odors, code requirements. Exhaust fans — type, size, speed, cfm delivery.

16. *Thermal storage* — Includes system operating schedule (12, 16 or 24 hours per day) specifically during peak outdoor conditions, permissible temperature swing in space during a design day, rugs on floor, nature of surface materials enclosing the space (see *Chapter 3*).

17. *Continuous or intermittent operation* — Whether system be required to operate every business day during cooling season, or only occasionally, such as churches and ballrooms. If intermittent operation, determine duration of time available for precooling or pulldown.

LOCATION OF EQUIPMENT AND SERVICES

The building survey should also include information which enables the engineer to select equipment location, and plan the air and water distribution systems. The following is a guide to obtaining this information:

1. *Available spaces* — Location of all stairwells, elevator shafts, abandoned smokestacks, pipe shafts, dumbwaiter shafts, etc., and spaces for air handling apparatus, refrigeration machines, cooling towers, pumps, and services (also see *Item 5*).

2. *Possible obstructions* — Locations of all electrical conduits, piping lines, and other obstructions or interferences that may be in the way of the duct system.

3. *Location of all fire walls and partitions* — Requiring fire dampers (also see *Item 16*).

4. *Location of outdoor air intakes* — In reference to street, other buildings, wind direction, dirt, and short-circuiting of unwanted contaminants.

5. *Power service* — Location, capacity, current limitations, voltage, phases and cycle, 3 or 4 wire; how additional power (if required) may be brought in and where.

6. *Water service* — Location, size of lines, ca-

pacity, pressure, maximum temperature.

7. *Steam service* — Location, size, capacity, temperature, pressure, type of return system.

8. *Refrigeration, brine or chilled water (if furnished by customer)* — Type of system, capacity, temperature, gpm, pressure.

9. *Architectural characteristics of space* — For selection of outlets that will blend into the space design.

10. *Existing air conveying equipment and ducts* — For possible reuse.

11. *Drains* — Location and capacity, sewage disposal.

12. *Control facilities* — Compressed air source and pressure, electrical.

13. *Foundation and support* — Requirements and facilities, strength of building.

14. *Sound and vibration control requirements* — Relation of refrigeration and air handling apparatus location to critical areas.

15. *Accessibility for moving equipment to the final location* — Elevators, stairways, doors, accessibility from street.

16. *Codes, local and national* — Governing wiring, drainage, water supply, venting of refrigeration, construction of refrigeration and air handling apparatus rooms, ductwork, fire dampers, and ventilation of buildings in general and apparatus rooms in particular.

AIR CONDITIONING LOAD ESTIMATE

The air conditioning load is estimated to provide the basis for selecting the conditioning equipment. It must take into account the heat coming into the space from outdoors on a design day, as well as the heat being generated within the space. A design day is defined as:

1. A day on which the dry- and wet-bulb temperatures are peaking simultaneously (*Chapter 2, "Design Conditions"*).

2. A day when there is little or no haze in the air to reduce the solar heat (*Chapter 4, "Solar Heat Gain Thru Glass"*).

3. All of the internal loads are normal (*Chapter 7, "Internal and System Heat Gain"*).

The time of peak load can usually be established by inspection, although, in some cases, estimates must be made for several different times of the day.

Actually, the situation of having all of the loads peaking at the same time will very rarely occur. To be realistic, various diversity factors must be applied to some of the load components; refer to *Chapter 3, "Heat Storage, Diversity, and Stratification."*

The infiltration and ventilation air quantities are estimated as described in *Chapter 6*.

Fig. 1 illustrates an air conditioning load estimate form and is designed to permit systematic load evaluation. This form contains the references identified to the particular chapters of data and tables required to estimate the various load components.

OUTDOOR LOADS

The loads from outdoors consist of:

1. *The sun rays entering windows* — *Table 15, pages 44-49*, and *Table 16, page 52*, provide data from which the solar heat gain through glass is estimated.

   The solar heat gain is usually reduced by means of shading devices on the inside or outside of the windows; factors are contained in *Table 16*. In addition to this reduction, all or part of the window may be shaded by reveals, overhangs, and by adjacent buildings. *Chart 1, page 57*, and *Table 18, page 58*, provide an easy means of determining how much the window is shaded at a given time.

   A large portion of the solar heat gain is radiant and will be partially stored as described in *Chapter 3. Tables 7 thru 11, pages 30-34*, provide the storage factors to be applied to solar heat gains in order to arrive at the actual cooling load imposed on the air conditioning equipment. These storage factors are applied to *peak solar heat gains* obtained from *Table 6, page 29*, with overall factors from *Table 16, page 52*.

2. *The sun rays striking the walls and roof* — These, in conjunction with the high outdoor air temperature, cause heat to flow into the space. *Tables 19 and 20, pages 62 and 63*, provide equivalent temperature differences for sunlit and shaded walls and roofs. *Tables 21, 22, 23, 24, 25, 27, and 28, pages 66-72*, provide the transmission coefficients or rates of heat flow for a variety of roof and wall constructions.

3. *The air temperature outside the conditioned space* — A higher ambient temperature causes heat to flow thru the windows, partitions, and floors. *Tables 25 and 26, pages 69 and 70*, and *Tables 29 and 30, pages 73 and 74*, provide the transmission coefficients. The temperature differences used to estimate the heat flow thru these structures are contained in the notes after each table.

CHAP REF	TABLE REFERENCES	CHAP REF	TABLE REFERENCES

Left panel — TABLE REFERENCES

ITEM	AREA OR QUANTITY	SUN GAIN OR TEMP. DIFF.	FACTOR

SOLAR GAIN — GLASS

3 & 4	GLASS — WITH STORAGE	SQ FT ×	TBLS 6&7,8 9,10 or 11 PP 29-34	× TBLS 16,17 PP 52-54
	GLASS — WITH STORAGE	SQ FT ×		
	GLASS — WITHOUT STORAGE	SQ FT ×	TBL 15 PP 44-49	TBL 15 CORR PP 44-49
	GLASS — WITHOUT STORAGE	SQ FT ×		
	SKYLIGHT	SQ FT ×		

SOLAR & TRANS. GAIN — WALLS & ROOF

5	WALL	SQ FT ×	TBL 19 P 62	× TBLS 21,22, 23,24 or 25 PP 66-69
	WALL	SQ FT ×		
	WALL	SQ FT ×		
	WALL	SQ FT ×		
	ROOF—SUN	SQ FT ×	TBL 20 P 63	× TBLS 27,28 71,72
	ROOF—SHADED	SQ FT ×		

TRANS. GAIN — EXCEPT WALLS & ROOF

	ALL GLASS	SQ FT ×	NOTE 1	× TBL 33 P 76
	PARTITION	SQ FT ×	NOTES, PP 69,70	× TBLS 25,26 PP 69,70
	CEILING	SQ FT ×	NOTES	× TBL 29 or 30 PP 73,74
	FLOOR	SQ FT ×		× PP 73,74
6	INFILTRATION NOTE 4	CFM ×	NOTE 1	× 1.08

INTERNAL HEAT

3 & 7	PEOPLE	PEOPLE ×	TBLS 14,48 PP 38,100
	POWER	HP OR KW ×	TBL 53 P 105
	LIGHTS	WATTS × 3.4	TBLS 12,14,49 PP 35,38,101
	APPLIANCES, ETC. TBLS 50-52 PP 101-103		× CORR BELOW TBLS 50-52
	ADDITIONAL HEAT GAINS TBLS 54-57 PP 107-109	×	

2 & 3	STORAGE	SQ FT × TBL 14 P 38	SUB TOTAL	× (TEMP SWING − TBL 13 P 37)
			SUB TOTAL	
	SAFETY FACTOR P 113	%		

ROOM SENSIBLE HEAT ■

7	SUPPLY CHART 3 DUCT P 110 HEAT GAIN %	SUPPLY DUCT P 110 + LEAK. LOSS %	FAN P 111 + H.P. %	TBL 59
8	OUTDOOR AIR NOTE 3 CFM × NOTE 1 F × P 121 BF × 1.08			

EFFECTIVE ROOM SENSIBLE HEAT ■

LATENT HEAT

6	INFILTRATION NOTE 4	CFM ×	NOTE 2	GR/LB × 0.68
3 & 7	PEOPLE	PEOPLE ×	TBLS 14,48; PP 38,100	
	STEAM P 107	LB/HR × 1050		
	APPLIANCES, ETC. TBLS 50-52 PP 101-103	× CORR BELOW TBLS		
	ADDITIONAL HEAT GAINS TBL 58 P 109			
5	VAPOR TRANS.	SQ FT × 1/100 × NOTE 2 GR/LB × TBL 40 P 84		
			SUB TOTAL	
	SAFETY FACTOR P 113	%		

ROOM LATENT HEAT

7	SUPPLY DUCT LEAKAGE LOSS P 110	%
	OUTDOOR AIR NOTE 3 CFM × NOTE 2 GR/LB × P 121 BF × 0.68	

EFFECTIVE ROOM LATENT HEAT

EFFECTIVE ROOM TOTAL HEAT ■

OUTDOOR AIR HEAT

	SENSIBLE: NOTE 3 CFM × NOTE 1 F × (1 — P 121 BF) × 1.08
	LATENT: NOTE 3 CFM × NOTE 2 GR/LB × (1 — P 121 BF) × 0.68

7	RETURN CHART 3 DUCT P 110 HEAT GAIN %	RETURN DUCT P 112 + LEAK. GAIN %	TBL 60 HP P 113 + PUMP %	SUB TOTAL DEHUM. & P 113 PIPE LOSS %

GRAND TOTAL HEAT ■

Right panel — TABLE REFERENCES

ESTIMATE FOR	LOCAL TIME SUN TIME	PEAK LOAD	LOCAL TIME SUN TIME
HOURS OF OPERATION			

CONDITIONS	DB	WB	% RH	DP	GR/LB
OUTDOOR (OA) TBLS 1-3 PP 10-19					
ROOM (RM) TBLS 4,5 PP 20,22,23					
DIFFERENCE	X X X	X X X		X X X	

(CHAP REF 2)

OUTDOOR AIR (CHAP REF 6)

VENTILATION	PEOPLE × [TBL 45 P 97] CFM/PERSON = _____
	SQ FT × CFM/SQ FT = _____
	CFM VENTILATION ■

INFILTRATION	SWINGING REVOLVING DOORS PEOPLE × [TBL 41 P 90] CFM/PERSON = _____
	OPEN DOORS DOORS × CFM/DOOR = _____
	EXHAUST FAN TBLS 46,47; P 98
	CRACK _____ FEET × TBL 44 P 95 CFM/FT = _____
	CFM INFILTRATION ■ TBL 42 P 92

CFM OUTDOOR AIR THRU APPARATUS ■ NOTE 3 _____ CFM_OA

APPARATUS DEWPOINT (CHAP REF 8)

ESHF	EFFECTIVE SENS HEAT FACTOR = $\dfrac{\text{EFFECTIVE ROOM SENS. HEAT}}{\text{EFFECTIVE ROOM TOTAL HEAT}}$ = _____
	TBL 65 P 145, OR PSYCH CHART, FIG 33 P 116
ADP	INDICATED ADP = _____ F SELECTED ADP = _____ F

DEHUMIDIFIED AIR QUANTITY

TEMP. RISE	$(1 - \underset{\text{P 121}}{\text{BF}}) \times (T_{RM}\_\_\_ F - T_{ADP}\_\_\_ F) = \_\_\_\_\_ F$
DEHUM. CFM	$\dfrac{\text{EFFECTIVE ROOM SENS. HEAT}}{1.08 \times \_\_\_\_ F \text{ TEMP. RISE}} = \_\_\_\_\_ CFM_{DA}$
OUTLET TEMP. DIFF.	$\dfrac{\text{ROOM SENS. HEAT}}{1.08 \times \_\_\_\_ CFM_{DA}} = \_\_\_\_\_ F \text{ (RM−OUTLET AIR)}*$

SUPPLY AIR QUANTITY

SUPPLY CFM	$\dfrac{\text{ROOM SENS. HEAT}}{1.08 \times \_\_\_\_ F \text{ DESIRED DIFF}} = \_\_\_\_\_ CFM_{SA}$
BYPASS CFM	_____ CFM_SA − _____ CFM_DA = _____ CFM_BA

RESULTING ENT & LVG CONDITIONS AT APPARATUS

EDB	$T_{RM}\_\_\_ F + \dfrac{CFM_{OA}}{\underset{\text{P 125}}{CFM\dagger}} \times (T_{OA}\_\_\_ F - T_{RM}\_\_\_ F) = T_{EDB}\_\_\_ F$
LDB	$T_{ADP}\_\_\_ F + \underset{\text{P 121}}{BF} \times (T_{EDB}\_\_\_ F - T_{ADP}\_\_\_ F) = T_{LDB}\_\_\_ F$
	FROM PSYCH. CHART: $T_{EWB}\_\_\_ F, T_{LWB}\_\_\_ F$

NOTES

1. USE DRY-BULB (DB) TEMPERATURE DIFFERENCE FROM TOP OF ESTIMATE FORM.

2. USE MOISTURE CONTENT (GR/LB) DIFFERENCE FROM TOP OF ESTIMATE FORM.

3. NORMALLY, USE "CFM VENTILATION" FOR "CFM OUTDOOR AIR." HOWEVER, WHEN INFILTRATION IS TO BE OFFSET, REFER TO PAGE 92 TO DETERMINE "CFM OUTDOOR AIR."

4. WHEN INFILTRATION IS NOT TO BE OFFSET, AND "CFM VENTILATION" IS LESS THAN "CFM INFILTRATION," THEN THE EXCESS INFILTRATION IS ACCOUNTED FOR HERE.

*IF THIS ΔT IS TOO HIGH, DETERMINE SUPPLY CFM FOR *DESIRED DIFFERENCE* BY SUPPLY AIR QUANTITY FORMULA.
†WHEN BYPASSING A MIXTURE OF OUTDOOR AND RETURN AIR, USE SUPPLY CFM. WHEN BYPASSING RETURN AIR ONLY, USE DEHUMIFIED CFM.

With Carrier Masthead Form E20. Without Carrier Masthead Form E5024.

FIG. 1 — AIR CONDITIONING LOAD ESTIMATE

4. *The air vapor pressure* — A higher vapor pressure surrounding conditioned space causes water vapor to flow thru the building materials. This load is significant only in low dew-point applications. The data required to estimate this load is contained in *Table 40, page 84*. In comfort applications, this load is neglected.

5. *The wind blowing against a side of the building*—Wind causes the outdoor air that is higher in temperature and moisture content to infiltrate thru the cracks around the doors and windows, resulting in localized sensible and latent heat gains. All or part of this infiltration may be offset by air being introduced thru the apparatus for ventilation purposes. *Chapter 6* contains the estimating data.

6. *Outdoor air usually required for ventilation purposes* — Outdoor air is usually necessary to flush out the space and keep the odor level down. This ventilation air imposes a cooling and dehumidifying load on the apparatus because the heat and/or moisture must be removed. Most air conditioning equipment permits some outdoor air to bypass the cooling surface (see *Chapter 8*). This bypassed outdoor air becomes a load within the conditioned space, similar to infiltration; instead of coming thru a crack around the window, it enters the room thru the supply air duct. The amount of bypassed outdoor air depends on the type of equipment used as outlined in *Chapter 8*. *Table 45, page 97,* provides the data from which the ventilation requirements for most comfort applications can be estimated.

The foregoing is that portion of the load on the air conditioning equipment that originates outside the space and is common to all applications.

INTERNAL LOADS

Chapter 7 contains the data required to estimate the heat gain from most items that generate heat within the conditioned space. The internal load, or heat generated within the space, depends on the character of the application. Proper diversity and usage factor should be applied to all internal loads. As with the solar heat gain, some of the internal gains consist of radiant heat which is partially stored (as described in *Chapter 3*), thus reducing the load to be impressed on the air conditioning equipment.

Generally, *internal heat gains* consist of some or all of the following items:

1. *People* — The human body thru metabolism generates heat within itself and releases it by radiation, convection, and evaporation from the surface, and by convection and evaporation in the respiratory tract. The amount of heat generated and released depends on surrounding temperature and on the activity level of the person, as listed in *Table 48, page 100*.

2. *Lights* — Illuminants convert electrical power into light and heat (refer to *Chapter 7*). Some of the heat is radiant and is partially stored (see *Chapter 3*).

3. *Appliances* — Restaurants, hospitals, laboratories, and some specialty shops (beauty shops) have electrical, gas, or steam appliances which release heat into the space. *Tables 50 thru 52, pages 101-103,* list the recommended heat gain values for most appliances when not hooded. If a positive exhaust hood is used with the appliances, the heat gain is reduced.

4. *Electric calculating machines* — Refer to manufacturer's data to evaluate the heat gain from electric calculating machines. Normally, not all of the machines would be in use simultaneously, and, therefore, a usage or diversity factor should be applied to the full load heat gain. The machines may also be hooded, or partially cooled internally, to reduce the load on the air conditioning system.

5. *Electric motors* — Electric motors are a significant load in industrial applications and should be thoroughly analyzed with respect to operating time and capacity before estimating the load (see *Item 13* under *"Space Characteristics and Heat Load Sources"*). It is frequently possible to actually measure this load in existing applications, and should be so done where possible. *Table 53, page 105,* provides data for estimating the heat gain from electric motors.

6. *Hot pipes and tanks* — Steam or hot water pipes running thru the air conditioned space, or hot water tanks in the space, add heat. In many industrial applications, tanks are open to the air, causing water to evaporate into the space. *Tables 54 thru 58, pages 107-109* provide data for estimating the heat gain from these sources.

7. *Miscellaneous sources* — There may be other sources of heat and moisture gain within a space, such as escaping steam (industrial cleaning devices, pressing machines, etc.), absorption of water by hygroscopic materials (paper, textiles, etc.); see *Chapter 7*.

In addition to the heat gains from the indoor and outdoor sources, the air conditioning equipment and duct system gain or lose heat. The fans and pumps required to distribute the air or water thru the system add heat; heat is also added to supply and return air ducts running thru warmer or hot spaces; cold air may leak out of the supply duct and hot air may leak into the return duct. The procedure for estimating the heat gains from these sources in percentage of room sensible load, room latent load, and grand total heat load is contained in *Chart 3, page 110,* and *Tables 59 and 60, pages 111-113.*

HEATING LOAD ESTIMATE

The heating load evaluation is the foundation for selecting the heating equipment. Normally, the heating load is estimated for the winter design temperatures (*Chapter 2*) usually occurring at night; therefore, no credit is taken for the heat given off by internal sources (people, lights, etc.). This estimate must take into account the heat loss thru the building structure surrounding the spaces and the heat required to offset the outdoor air which may infiltrate and/or may be required for ventilation. *Chapter 5* contains the transmission coefficients and procedures for determining heat loss. *Chapter 6* contains the data for estimating the infiltration air quantities. *Fig. 2* illustrates a heating estimate form for calculating the heat loss in a building structure.

Another factor that may be considered in the evaluation of the heating load is temperature swing. Capacity requirements may be reduced when the temperature within the space is allowed to drop a few degrees during periods of design load. This, of course, applies to continuous operation only. *Table 4, page 20,* provides recommended inside design conditions for various applications, and *Table 13, page 37,* contains the data for estimating the possible capacity reduction when operating in this manner.

The practice of drastically lowering the temperature to 50 F db or 55 F db when the building is unoccupied precludes the selection of equipment based on such capacity reduction. Although this type of operation may be effective in realizing fuel economy, *additional* equipment capacity is required for

pickup. In fact, it may be desirable to provide the additional capacity, even if continuous operation is contemplated, because of pickup required after forced shutdown. It is, therefore, evident that the use of storage in reducing the heating load for the purpose of equipment selection should be applied with care.

HIGH ALTITUDE LOAD CALCULATIONS

Since air conditioning load calculations are based on pounds of air necessary to handle a load, a decrease in density means an increase in cfm required to satisfy the given sensible load. The weight of air required to meet the latent load is decreased because of the higher latent load capacity of the air at higher altitudes (greater gr per lb per degree difference in dewpoint temperature). For the same dry-bulb and percent relative humidity, the wet-bulb temperature decreases (except at saturation) as the elevation above sea level increases.

The following adjustments are required for high altitude load calculations (see *Chapter 8, Table 66, page 148*):

1. Design room air moisture content must be adjusted to the required elevation.
2. Standard load estimating methods and forms are used for load calculations, except that the factors affecting the calculations of volume and sensible and latent heat of air must be multiplied by the relative density at the particular elevation.
3. Because of the increased moisture content of the air, the effective sensible heat factor must be corrected.

EQUIPMENT SELECTION

After the load is evaluated, the equipment must be selected with capacity sufficient to offset this load. The air supplied to the space must be of the proper conditions to satisfy both the sensible and latent loads estimated. *Chapter 8, "Applied Psychrometrics,"* provides procedures and examples for determining the criteria from which the air conditioning equipment is selected (air quantity, apparatus dewpoint, etc.).

HEATING CONDITIONS		
TEMPERATURE OF AIR ENTERING UNIT		

HEATING CONDITIONS

ROOM TBL 4 P 20 D.B._____ W.B._____ % R.H._____ GR/LB_____

OUTSIDE TBL 1 PP 10-17 D.B._____ W.B._____ % R.H._____ GR/LB_____

DIFF._____°F   DIFF._____GR/LB

TEMPERATURE OF AIR ENTERING UNIT

_____% OUTSIDE AIR   × _____°F = _____°F

_____% RECIRCULATED AIR   × _____°F* = _____°F

TOTAL (AVERAGE ENTERING AIR TEMP.) = _____°F

*ROOM TEMP. PLUS TEMP. CORRECTION FOR HT. OF UNIT

SURFACE	SQ FT	TRANS. FACT.	BTU PER HR PER °F DIFF.	TOTAL BTU PER HR °F DIFF.	TEMP. DIFF.	TOTAL · BTU PER HOUR

CHAPTER 5, TBLS 21-34 PP 66-80

INFILTRATION | **TOTAL TRANSMISSION LOSS**

	CRACK METHOD	AREA METHOD	CFM
WINDOW	___ LIN. FT × ___	___ SQ FT × 0.8	
SKYLIGHT	___ LIN. FT × ___	___ SQ FT × 0.8	
DOOR	___ LIN. FT × ___	___ SQ FT × ___	
DOOR USAGE	___ SQ. FT × ___	___ SQ FT × ___	
		TOTAL	

CHAPTER 6   TBL 44 P 95   TBL 43 P 94

× TEMP. GRADIENT FACTOR_____

OUTSIDE AIR TBL 45 P 97 _____CFM × _____°F × 1.08

INFILTRATION_____CFM × _____°F × 1.08

SUBTOTAL

SAFETY FACTOR _____%

GRAND TOTAL HEAT LOSS ███

AIR CHANGE METHOD	CFM	HUMIDIFICATION	WATER EVAPORATED
___ CU FT × { MAX. ___ / MIN. ___ } CHANGES/HR (60)	MAX._____ MIN._____	(CFM INFIL. + CFM OA (GR/LB DIFF.) 1580	_____ LB/H

EQUIPMENT CHOICE

QUAN.	UNIT SIZE	RPM	CFM	FIN. TEMP. F.	BASIC RATING	BTU CONSTANT	INSTALLED CAPACITY	STEAM COND'S. LB/HR	HOT WATER G P M	CU FT GAS/HR

FORM E10

FIG. 2 — HEATING LOAD ESTIMATE

CHAPTER 2. DESIGN CONDITIONS

This chapter presents the data from which the outdoor design conditions are established for various localities and inside design conditions for various applications. The design conditions established determine the heat content of the air, both outdoor and inside. They directly affect the load on the air conditioning equipment by influencing the transmission of heat across the exterior structure and the difference in heat content between the outdoor and inside air. For further details, refer to *Chapters 5 and 6.*

OUTDOOR DESIGN CONDITIONS — SUMMER AND WINTER

The outdoor design conditions listed in *Table 1* are the industry accepted design conditions as published in ARI Std. 530-56 and the 1958 ASHAE Guide. The conditions, as listed, permit a choice of outdoor dry-bulb and wet-bulb temperatures for different types of applications as outlined below.

NORMAL DESIGN CONDITIONS — SUMMER

Normal design conditions are recommended for use with *comfort and industrial cooling applications* where it is occasionally permissible to exceed the design room conditions. These outdoor design conditions are the *simultaneously occurring dry-bulb and wet-bulb temperatures and moisture content,* which can be expected to be exceeded a few times a year for short periods. The dry-bulb is exceeded more frequently than the wet-bulb temperature, and usually when the wet-bulb is lower than design.

When cooling and dehumidification (dehydration) are performed separately with these types of applications, use the normal design dry-bulb temperature for selecting the sensible cooling apparatus; use a moisture content corresponding to the normal design wet-bulb temperature and 80% rh for selecting the dehumidifier (dehydrator).

Daily range is the average difference between the high and low dry-bulb temperatures for a 24-hr period on a design day. This range varies with local climate conditions.

MAXIMUM DESIGN CONDITIONS — SUMMER

Maximum summer design conditions are recommended for *laboratories and industrial applications* where exceeding the room design conditions for even short periods of time can be detrimental to a product or process.

The maximum design dry-bulb and wet-bulb temperatures are simultaneous peaks (not individual peaks). The moisture content is an individual peak, and is listed only for use in the selection of separate cooling and dehumidifying systems for closely controlled spaces. Each of these conditions can be expected to be exceeded no more than 3 hours in a normal summer.

NORMAL DESIGN CONDITIONS — WINTER

Normal winter design conditions are recommended for use with all *comfort and industrial heating applications.* The outdoor dry-bulb temperature can be expected to go below the listed temperatures a few times a year, normally during the early morning hours. The annual degree days listed are the sum of all the days in the year on which the daily mean temperature falls below 65 F db, times the number of degrees between 65 F db and the daily mean temperature.

TABLE 1—OUTDOOR DESIGN CONDITIONS—SUMMER AND WINTER

STATE AND CITY	NORMAL DESIGN COND.—SUMMER July at 3:00 PM			AVG. DAILY RANGE	MAXIMUM DESIGN COND.—SUMMER July at 3:00 PM			NORMAL DESIGN COND. WINTER		WIND DATA Avg. Velocity and Prevailing Direction		Elevation Above Sea Level (ft)	Latitude (deg)
	Dry-Bulb (F)	Wet-Bulb (F)	Moisture Content* (gr/lb of dry air)	Dry-Bulb (F)	Dry-Bulb (F)	Wet-Bulb (F)	Moisture Content† (gr/lb of dry air)	Dry-Bulb (F)	Annual Degree Days	Summer	Winter		
ALABAMA													
Anniston	95	78	117.5	19				5	2806			733	34
Birmingham	95	78	117.5	19	99	82		10	2611	5.0 S	8.0 N	694	34
Mobile	95	80	131	12	95	82	155.6	15	1566	9.0 SW	9.9 N	10	31
Montgomery	95	78	117.5	15				10	2071		7.5 NW	293	32
ARIZONA													
Flagstaff	90	65	81	26	90			−10	7242		7.7 SW	6,894	35
Phoenix	105	76	94	30	113	78	126.9	25	1441	5.0 W	5.4 E	1,108	33
Tucson	105	72	77	30				25		5.0 W	5.2 NW	2,376	32
Winslow	100	70	85					−10				4,853	35
Yuma	110	78	93	30				30	1036		6.7 N	146	33
ARKANSAS													
Fort Smith	95	76	104.5	16	103			10	3226	7.0 E	8.3 E	448	35
Little Rock	95	78	117.5	16	103	83	145.5	5	3009	6.0 NW	8.3 NW	324	35
CALIFORNIA													
Bakersfield	105	70	54	25				25				499	35
El Centro	110	78	94									43	33
Eureka	90	65	52					30	4758	7.0 N	7.3	132	41
Fresno	105	74	76	35	110	75	95.9	25	2403	8.0 NW	5.4 NW	287	37
Laguna Beach				9	82	70	103.0					10	34
Long Beach	90	70	78	14								47	34
Los Angeles	90	70	78	14	94			35	1391	6.0 SW	6.4 NE	261	34
Oakland	85	65	60	17	94	68	99.3	30				17	38
Montague								0				2,635	42
Pasadena	95	70	70										34
Red Bluff	100	70	62									305	40
Sacramento	100	72	73	18				30	2680		7.2 SE	116	39
San Bernadino	105	72	65					35					34
San Diego	85	68	75	10	88	74	78.4	35	1596	7.0 W	6.3 NW	26	33
San Francisco	85	65	60	17				35	3137	12.0 W	7.5 N	17	38
San Jose	91	70	76.5					25	2823			100	37
Williams				40	110	80	74.4					86	39
COLORADO													
Denver	95	64	60	25	99	68	89.4	−10	5839	7.0 S	7.5 S	5,221	40
Durango	95	65	70									6,558	37
Fort Collins								−30					41
Grand Junction	95	65	62	24	102	68	86.2	−15	5613	6.0 SE	4.4 NW	4,587	39
Pueblo	95	65	63	25				−20	5558		7.9 NW	4,770	38
CONNECTICUT													
Bridgeport	95	75	99	14				0				9	41
Hartford	93	75	102	16	94	82		0	6113	7.0 S	8.7 NW	58	42
New Haven	95	75	99	14	95			0	5880	7.0 S	9.4 N	23	41
Waterbury								−15					42
DELAWARE													
Wilmington	95	78	117.5	15				0		10.0 SW	NW	134	40
DIST. OF COLUMBIA													
Washington	95	78	117.5	18	99	84	155.6	0	4561	5.0 S	7.8 NW	72	39
FLORIDA													
Apalachicola	95	80	131					25	1252	5.0 SW	8.4	23	30
Jacksonville	95	78	117.5	17	99	82	150.5	25	1185	8.0 SW	9.0 NE	18	30
Key West	98	78	112.5					45	59	9.0 SE	10.6 NE	23	25
Miami	91	79	131	12	92	81	150.5	35	185	7.0 SE	10.1 E	11	26
Pensacola	95	78	117.5	12				20	1281		10.9 N	408	31
Tampa	95	78	117.5	14	95			30	571	6.0 NE	8.6 NE	25	28
Tallahassee								25	1463		N	68	30

*Corresponds to dry-bulb and wet-bulb temperatures listed, and is corrected for altitude of city.

†Corresponds to peak dewpoint temperature, corrected for altitude.

TABLE 1—OUTDOOR DESIGN CONDITIONS—SUMMER AND WINTER (Contd)

STATE AND CITY	NORMAL DESIGN COND.—SUMMER July at 3:00 PM			AVG. DAILY RANGE	MAXIMUM DESIGN COND.—SUMMER July at 3:00 PM			NORMAL DESIGN COND. WINTER		WIND DATA Avg. Velocity and Prevailing Direction		Eleva-tion Above Sea Level (ft)	Lati-tude (deg)
	Dry-Bulb (F)	Wet-Bulb (F)	Moisture Content* (gr/lb of dry air)	Dry-Bulb (F)	Dry-Bulb (F)	Wet-Bulb (F)	Moisture Content† (gr/lb of dry air)	Dry-Bulb (F)	Annual Degree Days	Summer	Winter		
GEORGIA													
Atlanta	95	76	109.5	18	101	82	150.5	10	2985	7.0 NW	11.7 NW	975	34
Augusta	98	76	100	18				10	2306		6.5 NW	195	34
Brunswick	95	78	117.5										31
Columbus	98	76	100										33
Macon	95	78	117.5	18				15	2338	5.0 S	6.7 NW	408	33
Savannah	95	78	117.5	17	99			20	1635	8.0 SW	9.5 NW	42	32
IDAHO													
Boise	95	65	54.5	31	109	71	92.6	−10	5678	5.0 NW	9.1 SE	2,705	44
Lewiston	95	65	44	28				5	5109		4.1 E	763	46
Pocatello	95	65	61	28	100			−5	6741		8.9 SE	4,468	43
Twin Falls								−10			W		42
ILLINOIS													
Cairo	98	78	112.5					0	3957		9.8	319	37
Chicago	95	75	99	19	104	80	140.6	−10	6282	10.0 NE	12.0 SW	594	42
Danville								−5			NW		40
Moline	96	76	103	22	103	83	155.6	−10				594	41
Peoria	96	76	103	20	100			−10	6004	8.0 S	8.3 S	602	41
Springfield	98	77	106	20				−10	5446		11.9 NW	603	40
INDIANA													
Evansville	95	78	117.5	19	102	82	150.5	0	4410	7.0 SW	9.7 S	388	38
Fort Wayne	95	75	99	20	100			−10	6232	8.0 SW	10.4 SW	777	41
Indianapolis	95	76	104.5	18	99			−10	5458	9.0 SW	11.3 S	715	40
South Bend								−5			SW	773	42
Terre Haute	95	78	124									1,146	40
IOWA													
Cedar Rapids								−5					42
Davenport	95	78	117.5	18				−15	6252		10.5 NW	648	42
Des Moines	95	78	123	18	102			−15	6375	6.0 SW	10.1 NW	800	42
Dubuque	95	78	117.5					−20	6820		7.1	740	43
Fort Dodge								−20					42
Keokuk	95	78	117.5					−10	5663		8.2 SW	637	41
Sioux City	95	78	124	19	102			−20	6905	10.0 S	11.5 NW	1,111	43
Waterloo								−15					43
KANSAS													
Concordia	95	78	125	20				−10	5425		7.7 S	1,425	39
Dodge City	95	78	132	21	106			−10	5069		10.6	2,522	38
Salina					111			−15			NW	1,226	39
Topeka	100	78	109.5	19				−10	5075	10.0 S	9.2 S	991	39
Wichita	100	75	98	21	110	79	126.9	−10	4644	11.0 S	12.4 S	1,300	38
KENTUCKY													
Lexington								0	4792		13.3 SW	989	38
Louisville	95	78	117.5	22	99			0	4417	7.0 SW	9.8 SW	459	38
LOUISIANA													
Alexandria								20			N	89	32
New Orleans	95	80	131	13	95	83	161.2	20	1203	6.0 SW	8.6 N	9	30
Shreveport	100	78	109.5	15	102	83	150.5	20	2132	5.0 S	8.8 SE	197	33
MAINE													
Augusta	90	73	95	13								362	45
Bangor	90	73	95	13									45
Bar Harbor								−15			NW		44
Belfast								−5					44
Eastport	90	70	78	13				−10	8445	7.0 S	12.6 W	100	45
Millinocket								−15					46
Presque Isle									9644		NW		47
Portland	90	73	95	13	93			−5	7377	7.0 S	10.4 NW	47	44
Rumford								−20					44

*Corresponds to dry-bulb and wet-bulb temperatures listed, and is corrected for altitude of city.
†Corresponds to peak dewpoint temperature, corrected for altitude.

TABLE 1—OUTDOOR DESIGN CONDITIONS—SUMMER AND WINTER (CONT.)

STATE AND CITY	NORMAL DESIGN COND.—SUMMER July at 3:00 PM			AVG. DAILY RANGE	MAXIMUM DESIGN COND.—SUMMER July at 3:00 PM			NORMAL DESIGN COND. WINTER		WIND DATA Avg. Velocity and Prevailing Direction		Elevation Above Sea Level (ft)	Latitude (deg)
	Dry-Bulb (F)	Wet-Bulb (F)	Moisture Content* (gr/lb of dry air)	Dry-Bulb (F)	Dry-Bulb (F)	Wet-Bulb (F)	Moisture Content† (gr/lb of dry air)	Dry-Bulb (F)	Annual Degree Days	Summer	Winter		
MARYLAND													
Baltimore	95	78	117.5	18	99			0	4487	6.0 SW	8.2 NW	14	39
Cambridge								5			NW		39
Cumberland	95	75	99	18									39
Frederick								−5			NW		40
Frostburg								−5			W		40
Salisbury								10			NW		40
MASSACHUSETTS													
Amherst								−10			NW		42
Boston	92	75	104	13	96	78	135.9	0	5936	9.0 SW	12.4 W	14	42
Fall River								−10					42
Fitchburg	93	75	102	17				−10	6743	W	NW	402	43
Lowell								−15					43
Nantucket	95	75	99					0			14.8	45	41
New Bedford								0					42
Plymouth								−5			W		42
Springfield	93	75	102	17				−10		9.0 SW		199	42
Worcester	93	75	102	17				0				625	42
MICHIGAN													
Alpena	95	75	99					−10	8278		11.0 SW	615	45
Big Rapids								−15			NW		43
Detroit	95	75	99	19	101	79	135.9	−10	6560	10.0 SW	12.0 SW	619	42
Escanaba								−15	8777		9.5 NW		46
Flint	95	75	99	20				−10		W	W	766	43
Grand Rapids	95	75	99	20	98			−10	6702	8.0 W	12.1 NW	638	43
Kalamazoo								−5			W		42
Lansing	95	75	104	20				−10	7149		9.8 SW	861	43
Ludington								−10	7458		11.9 W		44
Marquette	93	73	90	20	96			−10	8745		10.6 NW	652	47
Saginaw	95	75	99									601	43
Sault Ste Marie								−20	9307		8.9 SE	724	47
MINNESOTA													
Alexandria								−25			NW		47
Duluth	93	73	96	19				−25	9723	13.4 SW	13.4 SW	1,128	47
Minneapolis	95	75	103	17	102			−20	7966	10.0 S	11.3 NW	839	45
St. Cloud								−25					46
St. Paul	95	75	99	17	103	79	131.1	−20	7975	8.0 SE	9.5 NW	719	45
MISSISSIPPI													
Jackson				21	103	83	155.6	15		5.0 SW	7.7 SE	316	32
Meridian	95	79	124	21				10	2330	4.0 SW	6.3 N	410	32
Vicksburg	95	78	117.5	21	96			10	2069	6.0 SW	8.3	226	32
MISSOURI													
Columbia	100	78	109.5	19				−10	5070		8.9 SW	739	39
Kansas City	100	76	106.5	19	109	79	135.9	−10	4962	9.0 S	10.3 NW	741	39
Kirksville				19	108	82	150.5				SW	969	40
St. Louis	95	78	117.5	20	108	81	135.9	0	4596	9.0 S	11.8 S	465	39
St. Joseph								−10	5596		9.3 NW	817	40
Springfield				18	98	79	135.9	−10	4569	8.0 S	10.9 SE	1,301	37
MONTANA													
Billings	90	66	70	20	104			−25	7213		12.4 W	3,119	46
Butte								−20			NW	5,538	46
Great Falls								−20			SW	3,687	48
Havre	95	70	82	20				−30	8416	7.0 E	9.4 SW	2,498	49
Helena	95	67	71	20	97	70	77.4	−20	7930	7.0 SW	7.4 SW	4,090	47
Kalispell	95	65	56					−20	8032		5.2	3,004	48
Miles City								−35	7591		5.6 S	2,629	47
Missoula	95	66	49	20				−20	7604		E	3,205	47

*Corresponds to dry-bulb and wet-bulb temperatures listed, and is corrected for altitude of city.

†Corresponds to peak dewpoint temperature, corrected for altitude.

TABLE 1—OUTDOOR DESIGN CONDITIONS—SUMMER AND WINTER (CONT.)

STATE AND CITY	NORMAL DESIGN COND.—SUMMER July at 3:00 PM			AVG. DAILY RANGE	MAXIMUM DESIGN COND.—SUMMER July at 3:00 PM			NORMAL DESIGN COND. WINTER		WIND DATA Avg. Velocity and Prevailing Direction		Elevation Above Sea Level (ft)	Latitud (deg)
	Dry-Bulb (F)	Wet-Bulb (F)	Moisture Content* (gr/lb of dry air)	Dry-Bulb (F)	Dry-Bulb (F)	Wet-Bulb (F)	Moisture Content† (gr/lb of dry air)	Dry-Bulb (F)	Annual Degree Days	Summer	Winter		
NEBRASKA													
Grand Island								−20				1,856	41
Lincoln	95	78	124	20	106			−10	5980	9.0 S	10.6 S	1,180	41
Norfolk								−15		NW	NW		42
North Platte	95	78	135	26	104	76	74.4	−20	6384	6.0 S	7.9 W	2,805	41
Omaha	95	78	123	20	108	80	131.1	−10	6095	8.0 S	9.7 NW	978	41
Valentine	95	78	135	20				−25	7197		9.2 NW	2,627	43
York								−15					
NEVADA													
Las Vegas	115	75	76	40				20			S	1,882	36
Reno	95	65	62	41	102	66	66.9	−5	5621	7.0 SW	6.0 W	4,493	40
Tonopah								5	5812		9.9 SE	5,421	38
Winnemucca	95	65	62	40				−15	6357	7.0 SW	8.1 NE	4,293	42
NEW HAMPSHIRE													
Berlin								−25					45
Concord	90	73	95	14				−15	7400	5.0 NW	6.2 NW	289	43
Keene								−20			NW		43
Manchester	90	73	95	14	92							171	43
Portsmouth	90	73	95	14									43
NEW JERSEY													
Atlantic City	95	78	117.5	14				5	5015	13.0 SW	15.8 NW	8	39
Bloomfield	95	75	99	14								125	41
Camden				14	102	82	145.5	0		10.0 SW		30	40
East Orange	95	75	99	14								173	41
Jersey City	95	75	99					0			NW		41
Newark	95	75	99	14	99	81	140.6	0	5500	13.0 SW	17.1 NW	10	41
Paterson	95	75	99	14	95					13.0 SW		10	41
Sandy Hook								0	5369		16.1		41
Trenton	95	78	117.5	14	96			0	5256	9.0 SW	10.9 NW	56	40
NEW MEXICO													
Albuquerque	95	70	94.5	26	98	68	95.9	0	4517	8.0 SW	7.3 N	5,101	35
Roswell	95	70	87	25				−10	3578	6.0 S	7.1 S	3,643	32
Santa Fe	90	65	80	30	90			0	6123	6.0 SE	7.1 NE	7,000	36
NEW YORK													
Albany	93	75	102	18	97	78	131.1	−10	6648	7.0 S	10.5 S	19	43
Binghamton	95	75	103.5					−10	6818		6.8 NW	915	42
Buffalo	93	73	90	18	93	77	126.9	−5	6925	12.0 SW	17.1 W	604	43
Canton	90	73	95					−25	8305	8.0	10.5	458	43
Cortland								−10			NW		43
Glens Falls								−15			W		43
Ithaca								−15	6914		11.3 NW		42
Jamestown								−10			SW		42
Lake Placid								−20			W		44
New York City	95	75	99	14	100	81	145.5	0	5280	13.0 S	16.8 NW	10	41
Ogdensburg								−20			SW		45
Oneonta								−15			SW		43
Oswego	93	73	90					−10	7186		12.1 S	363	43
Rochester	95	75	102	18	95			−5	6772	8.0 SW	9.6 W	543	43
Schenectady	93	75	102	18								235	43
Syracuse	93	75	102	18	96			−10	6899	9.0 S	11.2 S	400	43
Watertown								−15			SW		44
NORTH CAROLINA													
Asheville	93	75	114.5	19	93			0	4236	6.0 NW	9.5 NW	2,192	36
Charlotte	95	78	117.5	16				10	3224	5.0 SW	7.3 SW	809	35
Greensboro	95	78	123.5	15				10	3849		7.9 SW	896	37
Raleigh	95	78	117.5	15	98	82	155.6	10	3275	6.0 SW	7.9 SW	345	36
Wilmington	95	78	117.5	15	95	81	150.5	15	2420	7.0 SW	9.4 SW	6	34

*Corresponds to dry-bulb and wet-bulb temperatures listed, and is corrected for altitude of city.

†Corresponds to peak dewpoint temperature, corrected for altitude.

TABLE 1—OUTDOOR DESIGN CONDITIONS—SUMMER AND WINTER (CONT.)

STATE AND CITY	NORMAL DESIGN COND.—SUMMER July at 3:00 PM			AVG. DAILY RANGE	MAXIMUM DESIGN COND.—SUMMER July at 3:00 PM			NORMAL DESIGN COND. WINTER		WIND DATA Avg. Velocity and Prevailing Direction		Elevation Above Sea Level (ft)	Latitude (deg)
	Dry-Bulb (F)	Wet-Bulb (F)	Moisture Content* (gr/lb of dry air)	Dry-Bulb (F)	Dry-Bulb (F)	Wet-Bulb (F)	Moisture Content† (gr/lb of dry air)	Dry-Bulb (F)	Annual Degree Days	Summer	Winter		
NORTH DAKOTA													
Bismarck	95	73	95.5	19	103			−30	8937	9.0 NW	9.1 NW	1,670	47
Devils Lake	95	70	77					−30	10104		10.1 W	1,481	48
Fargo	95	75	104.5	19				−25			10.9 NW	900	47
Grand Forks								−25	9871		NW	832	48
Williston	95	73	96.5					−35	9301	8.0 SE	8.6 W	1,919	48
OHIO													
Akron	95	75	99	19				−5				104	41
Cincinnati	95	78	117.5	22	106	81	145.5	0	4990	7.0 SW	8.5 SW	553	39
Cleveland	95	75	99	19	101	79	135.9	0	6144	11.0 S	14.7 SW	651	42
Columbus	95	76	104.5	23	95			−10	5506	9.0 SW	11.6 SW	724	40
Dayton	95	78	123	23	99			0	5412	8.0 SW	11.1 SW	900	40
Lima								−5					41
Sandusky	95	75	99					0	6095		11.0	608	42
Toledo	95	75	99	19	99			−10	6269	10.0 SW	12.1 SW	589	42
Youngstown	95	75	99	19								1,186	41
OKLAHOMA													
Ardmore								10			N	762	34
Bartlesville								−10			N		37
Oklahoma City	101	77	108	21	104			0	3670	10.0 S	11.5 S	1,254	35
Tulsa	101	77	101.5		106	79	140.6	0		10.0 S	N	804	36
OREGON													
Baker	90	66	71	19				−5	7197		5.6 SE	3,501	44
Eugene	90	68	67	19				−15				366	44
Medford	95	70	76	19								1,428	42
Pendleton								−15			W	1,494	46
Portland	90	68	67	19	99	70	103.0	10	4353	6.0 NW	7.3 S	30	46
Roseburg	90	66	57	19						4.0 N		523	42
Wamic								0			W		45
PENNSYLVANIA													
Altoona	95	75	99	14				−5				1,469	40
Bethlehem								−5					41
Erie	93	75	102	18				−5	6363	9.0 S	13.6 SW	670	42
Harrisburg	95	75	99	14				0	5412		7.6 NW	339	40
New Castle								0			NW		41
Oil City	95	75	99	18									42
Philadelphia	95	78	117.5	14	97			0	4739	10.0 SW	11.0 NW	26	40
Pittsburgh	95	75	105	14	98	79	126.9	0	5430	9.0 SW	11.6 W	1,248	40
Reading	95	75	99					0	5232		9.0	311	40
Scranton	95	75	99	14	95			−5	6218	6.0 SW	7.6 SW	746	41
Warren								−15			NW		41
Williamsport								−5			NW	525	42
RHODE ISLAND													
Block Island	95	75	99						5897		20.6 NW	46	41
Pawtucket	93	75	102	14									41
Providence	93	75	102	14				0	5984	10.0 NW	12.1 NW	8	42
SOUTH CAROLINA													
Charleston	95	78	117.5	17	98	82	155.6	15	1866	10.0 SW	10.5 SW	9	33
Columbia	95	75	99	17				10	2488		8.0 SW	401	34
Greenville	95	76	104.5	17				10	3059	7.0 NE	8.4	982	35
SOUTH DAKOTA													
Huron	95	75	106	19	106	76	126.9	−20	7940	10.0 SE	10.7 NW	1,282	44
Rapid City	95	70	85	22	103	71	95.9	−20	7197	7.0 W	8.0 W	3,231	44
Sioux Falls	95	75	99	20				−20			NW	1,427	43

*Corresponds to dry-bulb and wet-bulb temperatures listed, and is corrected for altitude of city.
†Corresponds to peak dewpoint temperature, corrected for altitude.

TABLE 1—OUTDOOR DESIGN CONDITIONS—SUMMER AND WINTER (CONT.)

STATE AND CITY	NORMAL DESIGN COND.—SUMMER July at 3:00 PM			AVG. DAILY RANGE	MAXIMUM DESIGN COND.—SUMMER July at 3:00 PM			NORMAL DESIGN COND. WINTER		WIND DATA Avg. Velocity and Prevailing Direction		Elevation Above Sea Level	Latitude
	Dry-Bulb (F)	Wet-Bulb (F)	Moisture Content* (gr/lb of dry air)	Dry-Bulb (F)	Dry-Bulb (F)	Wet-Bulb (F)	Moisture Content† (gr/lb of dry air)	Dry-Bulb (F)	Annual Degree Days	Summer	Winter	(ft)	(deg)
TENNESSEE													
Chattanooga	95	76	104.5	18	98			10	3238	6.0 SW	7.7 NW	689	35
Johnson City								0			W		36
Knoxville	95	75	103.5	17	100	79	135.9	0	3658	6.0 SW	7.2 SW	921	36
Memphis	95	78	117.5	18	103	83	155.6	0	3090	7.0 SW	9.3 W	271	35
Nashville	95	78	117.5	17	98			0	3613	8.0 W	9.8 NW	485	36
TEXAS													
Abilene	100	74	93					15	2573	9.0 S	10.1 S	1,748	32
Amarillo	100	72	91.6	22	101	75	110.4	—10	4196	11.0 S	12.1 SW	3,657	35
Austin	100	78	109.5	19				20	1679		8.3 N	625	31
Brownsville	95	80	131	20	96	80	150.5	30	628	9.0 SE	10.4 SE	35	26
Corpus Christi	95	80	131					20	965	13.0 SE	11.0 SE	21	28
Dallas	100	78	109.5	21	105	80	135.9	0	2367	8.0 S	10.6 NW	460	33
Del Rio	100	78	115					15	1501	10.0 SE	8.0 SE	1,020	29
El Paso	100	69	73	23	101	72	106.6	10	2532	9.0 E	9.0 NW	3,720	32
Fort Worth	100	78	109.5	21				10	2355	10.0	10.5 NW	708	33
Galveston	95	80	131	14				20	1174	9.0 S	11.2 SE	6	29
Houston	95	80	131	14	100	81	150.5	20	1315	8.0 S	10.5 SE	52	30
Palestine	100	78	109.5					15	2068		8.0	555	32
Port Arthur	95	79	124					20	1532		10.7	64	30
San Antonio	100	78	109.5	19	102	83	166.4	20	1435	7.0 SE	8.3 NE	646	29
UTAH													
Modena	95	65	66	25	97	66	80.3	—15	6598	11.0 SW	9.0	5,479	38
Logan								—15					42
Ogden								—10			S	4,446	41
Salt Lake City	95	65	61	25	102	68	89.4	—10	5650	7.0 S	7.8 SE	4,222	41
VERMONT													
Bennington								—10					43
Burlington	90	73	95	17	91			—10	8051	8.0 S	11.6 S	308	44
Rutland	90	73	95	17				—20					43
VIRGINIA													
Cape Henry	95	78	117.5					10	3538		14.0	24	37
Lynchburg	95	75	99	16	99			5	4068		8.1	386	37
Norfolk	95	78	117.5	16	95			15	3364	11.0 S	12.1 N	11	37
Richmond	95	78	117.5	16	98			15	3922	6.0 SW	8.1 SW	162	38
Roanoke	95	76	111.5	16				0	4075		8.2 W	1,194	38
WASHINGTON													
North Head	85	65	60					20	5367		16.1	199	
Seattle	85	65	60	17	86	70	99.3	15	4815	7.0 N	9.8 SE	14	48
Spokane	93	65	54.5	28	106	68	71.9	—15	6318	7.0 SW	6.2 SW	1,879	48
Tacoma	85	64	55.5	17				15	5039		8.0	279	47
Tatoosh Island								15	5857		18.9	110	48
Walla Walla	95	65	47.5	28	105			—10	4910		5.4 S	952	46
Wenatchee	90	65	52	20									48
Yakima	95	65	48	20				5	5585		4.1	1,160	47
WEST VIRGINIA													
Bluefield	95	75	99	16									37
Charleston	95	75	99	16	102			0		4.0 SW	W	603	38
Elkins								—10	5800		6.2 W	2,006	39
Huntington	95	76	104.5	16				—5			W		38
Martinsburg								—5				540	39
Parkersburg	95	75	99	16	98			—10	4928	4.0 SE	7.2 SW	615	39
Wheeling	95	75	99	14				—5					40

*Corresponds to dry-bulb and wet-bulb temperatures listed, and is corrected for altitude of city.

†Corresponds to peak dewpoint temperature, corrected for altitude.

TABLE 1—OUTDOOR DESIGN CONDITIONS—SUMMER AND WINTER (CONT.)

STATE AND CITY	NORMAL DESIGN COND.—SUMMER July at 3:00 PM			AVG. DAILY RANGE	MAXIMUM DESIGN COND.—SUMMER July at 3:00 PM			NORMAL DESIGN COND. WINTER		WIND DATA Avg. Velocity and Prevailing Direction		Elevation Above Sea Level	Latitude
	Dry-Bulb (F)	Wet-Bulb (F)	Moisture Content* (gr/lb of dry air)	Dry-Bulb (F)	Dry-Bulb (F)	Wet-Bulb (F)	Moisture Content† (gr/lb of dry air)	Dry-Bulb (F)	Annual Degree Days	Summer	Winter	(ft)	(deg)
WISCONSIN													
Ashland								−20			SW	885	42
Eau Claire								−20			NW		45
Green Bay	95	75	99	14	99	79	131.1	−20	7931	8.0 S	10.5 SW	589	45
La Crosse	95	75	99	17	100	83	161.2	−25	7421	6.0 S	9.3 S	673	44
Madison	95	75	103.5	18	96			−15	7405	8.0 SW	10.1 NW	938	43
Milwaukee	95	75	99	14	99			−15	7079	9.0 SW	12.1 W	619	43
WYOMING													
Casper								−20			SW	5,321	43
Cheyenne	95	65	68.5	28				−15	7536	9.0 S	13.3 NW	6,139	42
Lander	95	65	66	28				−18	8243	5.0 SW	3.9	5,448	44
Sheridan					102			−30	7239	5.0 NW	4.9 NW	3,773	45
CANADA													
PROVINCE AND CITY													
ALBERTA													
Calgary	90	66	71					−29	9520	9.7	10.1	3,540	51
Edmonton	90	68	77					−33	10320	8.9	7.6	2,219	54
Grand Prairie								−39			7.9	2,190	55
Lethbridge								−32	8650		15.0	3,018	50
McMurray								−42				1,216	57
Medicine Hat	90	65						−35	8650	9.1	9.0	2,365	50
BRITISH COLUMBIA													
Estevan Point								17			9.9	20	49
Fort Nelson								−38			3.7	1,230	59
Penticton								−6				1,121	50
Prince George								−32	9500		7.2	2,218	54
Prince Rupert								8	6910		8.0	170	54
Vancouver	80	67	78					11	5230		7.7	22	49
Victoria								15	5410		12.3	228	48
MANITOBA													
Brandon								−32	10930			1,200	50
Churchill								−42	16810		14.7	115	59
The Pas								−39			6.4	894	54
Winnipeg	90	71	83.5					−29	10630	11.5	12.0	786	50
NEW BRUNSWICK													
Campbellton								−11				42	48
Fredericton	90	75	107					−6	8830		9.2	164	46
Moncton								−8	8700		14.9	248	46
Saint John								−3	8380	7.9	13.8	119	45
NEWFOUNDLAND													
Corner Brook								−1	9210			40	49
Gander								−3	9440		17.2	482	49
Goose Bay								−26	12140		10.3	144	53
Saint Johns								1	8780		19.3	463	48
NORTHWEST TERRITORIES													
Aklavik								−46	17870			30	68
Fort Norman								−42	16020			300	65
Frobisher								−47				68	
Resolute								−42			9.2	56	
Yellowknife								−47				682	62

*Corresponds to dry-bulb and wet-bulb temperatures listed, and is corrected for altitude of city.
†Corresponds to peak dewpoint temperature, corrected for altitude.

TABLE 1—OUTDOOR DESIGN CONDITIONS—SUMMER AND WINTER (CONT.)

CANADA PROVINCE AND CITY	NORMAL DESIGN COND.—SUMMER July at 3:00 PM			AVG. DAILY RANGE	MAXIMUM DESIGN COND.—SUMMER July at 3:00 PM			NORMAL DESIGN COND. WINTER		WIND DATA Avg. Velocity and Prevailing Direction		Elevation Above Sea Level	Latitude
	Dry-Bulb (F)	Wet-Bulb (F)	Moisture Content* (gr/lb of dry air)	Dry-Bulb (F)	Dry-Bulb (F)	Wet-Bulb (F)	Moisture Content† (gr/lb of dry air)	Dry-Bulb (F)	Annual Degree Days	Summer	Winter	(ft)	(deg)
NOVA SCOTIA													
Halifax	90	75	107					4	7570	6.6	9.6	83	45
Sydney								1	8220	9.9	13.1	197	46
Yarmouth								7	7520		13.5	136	44
ONTARIO													
Fort William								−24	10350	8.4	9.6	644	48
Hamilton								0	6890			303	43
Kapuskasing								−30	11790		10.0	752	49
Kingston								−11	7810		10.8	340	44
Kitchener								−3	7380			1,100	43
London								−1			11.9	912	43
North Bay								−20		9.6	11.3	1,210	46
Ottawa	90	75	107					−15	8830	8.9	11.1	339	45
Peterborough								−11				648	44
Souix Lookout								−33			8.5	1,227	50
Sudbury								−17				837	47
Timmins								−26				1,100	48
Toronto	93	75	102					0	7020	8.1	14.1	379	43
Windsor								3			12.3	637	42
Sault Ste. Marie	93	75	102									635	47
PRINCE EDWARD ISLAND													
Charlottetown								−3	8380	8.7	11.3	74	46
QUEBEC													
Arvida								−19	10440		8.2	375	
Knob Lake								−40				1,605	55
Mont Joli								−11			13.3	150	48
Montreal	90	75	107					−9	8130	9.9	12.3	187	46
Port Harrison								−39			13.4	66	58
Quebec City	90	75	107					−12	9070	9.0	12.4	296	47
Seven Islands								−20				190	50
Sherbrooke								−12	8610		8.2	620	45
Three Rivers								−13				50	46
SASKATCHEWAN													
Prince Albert								−41	11430		4.9	1,414	53
Regina	90	71	92.5					−34	10770	12.4	12.1	1,884	50
Saskatoon	90	70	81					−37	10960	10.7	9.7	1,645	52
Swift Current								−33	9660		14.6	2,677	50
YUKON TERRITORY													
Dawson								−56	15040			1,062	64
Whitehorse								−43			8.7	2,289	61

*Corresponds to dry-bulb and wet-bulb temperatures listed, and is corrected for altitude of city.

†Corresponds to peak dewpoint temperature, corrected for altitude.

CORRECTIONS TO OUTDOOR DESIGN CONDITIONS FOR TIME OF DAY AND TIME OF YEAR

The normal design conditions for summer, listed in *Table 1*, are applicable to the month of July at about 3:00 P.M. Frequently, the design conditions at other times of the day and other months of the year must be known.

Table 2 lists the approximate corrections on the dry-bulb and wet-bulb temperatures from 8 a.m. to 12 p.m. based on the average daily range. The dry-bulb corrections are based on analysis of weather data, and the wet-bulb corrections assume a relatively constant dewpoint throughout the 24-hr period.

Table 3 lists the approximate corrections of the dry-bulb and wet-bulb temperatures from March to November, based on the yearly range in dry-bulb temperature (summer normal design dry-bulb minus winter normal design dry-bulb temperature). These corrections are based on analysis of weather data and are applicable only to the cooling load estimate.

Example 1 — Corrections to Design Conditions
Given:
A comfort application in New York City.

Find:
The approximate dry-bulb and wet-bulb temperatures at 12:00 noon in October.

Solution:
Normal design conditions for New York in July at 3:00 p.m. are 95 F db, 75 F wb (*Table 1*).

Daily range in New York City is 14 F db.

Yearly range in New York City = 95 − 0 = 95 F db.

Correction for time of day (12 noon) from *Table 2*:
Dry-bulb = −5 F
Wet-bulb = −1 F

Correction for time of year (October) from *Table 3*:
Dry-bulb = −16 F
Wet-bulb = − 8 F

Design conditions at 12 noon in October (approximate):
Dry-bulb = 95 − 5 − 16 = 74 F
Wet-bulb = 75 − 1 − 8 = 66 F

INSIDE COMFORT DESIGN CONDITIONS — SUMMER

The inside design conditions listed in *Table 4* are recommended for types of applications listed. These conditions are based on experience gathered from many applications, substantiated by ASHAE tests.

The optimum or deluxe conditions are chosen where costs are not of prime importance and for comfort applications in localities having summer outdoor design dry-bulb temperatures of 90 F or less. Since all of the loads (sun, lights, people, outdoor air, etc.) do not peak simultaneously for any prolonged periods, it may be uneconomical to design for the optimum conditions.

TABLE 2—CORRECTIONS IN OUTDOOR DESIGN TEMPERATURES FOR TIME OF DAY
(For Cooling Load Estimates)

DAILY RANGE OF TEMPERATURE* (F)	DRY- OR WET-BULB	SUN TIME									
		AM			PM						
		8	10	12	2	3	4	6	8	10	12
10	Dry-Bulb	−9	−7	−5	−1	0	−1	−2	−5	−8	−9
	Wet-Bulb	−2	−2	−1	0	0	0	−1	−1	−2	−2
15	Dry-Bulb	−12	−9	−5	−1	0	−1	−2	−6	−10	−14
	Wet-Bulb	−3	−2	−1	0	0	0	−1	−1	−3	−4
20	Dry-Bulb	−14	−10	−5	−1	0	−1	−3	−7	−11	−16
	Wet-Bulb	−4	−3	−1	0	0	0	−1	−2	−3	−4
25	Dry-Bulb	−16	−10	−5	−1	0	−1	−3	−8	−13	−18
	Wet-Bulb	−4	−3	−1	0	0	0	−1	−2	−3	−5
30	Dry-Bulb	−18	−12	−6	−1	0	−1	−4	−10	−15	−21
	Wet-Bulb	−5	−3	−1	0	0	0	−1	−3	−4	−6
35	Dry-Bulb	−21	−14	−7	−1	0	−1	−6	−12	−18	−24
	Wet-Bulb	−6	−4	−2	0	0	0	−1	−3	−5	−7
40	Dry-Bulb	−24	−16	−8	−1	0	−1	−7	−14	−21	−28
	Wet-Bulb	−7	−4	−2	0	0	0	−2	−4	−6	−9
45	Dry-Bulb	−26	−17	−8	−2	0	−2	−8	−16	−24	−31
	Wet-Bulb	−7	−5	−2	0	0	−1	−2	−4	−8	−10

*The daily range of dry-bulb temperature is the difference between the highest and lowest dry-bulb temperature during a 24-hour period on a typical design day. (See *Table 1* for the value of daily range for a particular city).

Equation: Outdoor design temperature at any time = Outdoor design temperature from *Table 1* + Correction from above table.

TABLE 3—CORRECTIONS IN OUTDOOR DESIGN CONDITIONS FOR TIME OF YEAR
(For Cooling Load Estimates)

YEARLY RANGE OF TEMPERATURE(F)*	DRY- OR WET-BULB	TIME OF YEAR								
		March	April	May	June	July	August	Sept.	Oct.	Nov.
120	Dry-Bulb	−39	−22	−11	−4	0	0	−9	−24	−44
	Wet-Bulb	−23	−12	−5	−2	0	0	−4	−13	−27
115	Dry-Bulb	−33	−22	−11	−4	0	0	−8	−20	−36
	Wet-Bulb	−18	−11	−5	−2	0	0	−4	−10	−21
110	Dry-Bulb	−30	−20	−11	−4	0	0	−6	−17	−31
	Wet-Bulb	−15	−10	−5	−2	0	0	−3	−8	−16
105	Dry-Bulb	−30	−20	−11	−4	0	0	−6	−17	−29
	Wet-Bulb	−15	−10	−5	−2	0	0	−3	−8	−14
100	Dry-Bulb	−29	−19	−10	−3	0	0	−6	−16	−27
	Wet-Bulb	−14	−10	−5	−2	0	0	−3	−8	−14
95	Dry-Bulb	−29	−19	−10	−3	0	0	−6	−16	−27
	Wet-Bulb	−14	−10	−5	−2	0	0	−3	−8	−14
90	Dry-Bulb	−29	−19	−10	−3	0	0	−6	−16	−26
	Wet-Bulb	−14	−10	−5	−2	0	0	−3	−8	−14
85	Dry-Bulb	−29	−19	−9	−3	0	0	−5	−16	−25
	Wet-Bulb	−14	−10	−5	−2	0	0	−3	−8	−14
80	Dry-Bulb	−24	−16	−8	−3	0	0	−4	−12	−20
	Wet-Bulb	−13	−9	−4	−2	0	0	−2	−6	−11
75	Dry-Bulb	−14	−9	−4	−1	0	0	−3	−7	−15
	Wet-Bulb	−7	−5	−2	0	0	0	−2	−4	−8
70	Dry-Bulb	−13	−9	−4	−1	0	0	−2	−7	−14
	Wet-Bulb	−6	−4	−2	0	0	0	−1	−4	−6
65	Dry-Bulb	−11	−8	−4	−1	0	0	−2	−6	−12
	Wet-Bulb	−6	−4	−2	0	0	0	−1	−3	−6
60	Dry-Bulb	−9	−7	−3	−1	0	0	−2	−5	−10
	Wet-Bulb	−4	−3	−2	0	0	0	−1	−3	−5
55	Dry-Bulb	−6	−5	−3	−1	0	0	−2	−4	−8
	Wet-Bulb	−3	−3	−2	0	0	0	−1	−2	−4
50	Dry-Bulb	−5	−4	−3	−1	0	0	−2	−4	−7
	Wet-Bulb	−3	−2	−1	0	0	0	−1	−2	−3

*Yearly range of temperature is the difference between the summer and winter normal design dry-bulb temperatures (*Table 1*).

Equation: Outdoor design temperature = Outdoor design temperature from *Table 1* + Corrections from above table.

The commercial inside design conditions are recommended for general comfort air conditioning applications. Since a majority of people are comfortable at 75 F or 76 F db and around 45% to 50% rh, the thermostat is set to these temperatures, and these conditions are maintained under partial loads. As the peak loading occurs (outdoor peak dry-bulb and wet-bulb temperatures, 100% sun, all people and lights, etc.), the temperature in the space rises to the design point, usually 78 F db.

If the temperature in the conditioned space is forced to rise, heat will be stored in the building mass. Refer to *Chapter 3, "Heat Storage, Diversity and Stratification,"* for a more complete discussion of heat storage. With summer cooling, the temperature swing used in the calculation of storage is the difference between the design temperature and the normal thermostat setting.

The range of summer inside design conditions is provided to allow for the most economical selection of equipment. Applications of inherently high sensible heat factor (relatively small latent load) usually result in the most economical equipment selection if the higher dry-bulb temperatures and lower relative humidities are used. Applications with low sensible heat factors (high latent load) usually result in more economical equipment selection if the lower dry-bulb temperatures and higher relative humidities are used.

INSIDE COMFORT DESIGN CONDITIONS — WINTER

For winter season operation, the inside design conditions listed in *Table 4* are recommended for general heating applications. With heating, the temperature swing (variation) is below the comfort condition at the time of peak heating load (no people, lights, or solar gain, and with the minimum outdoor temperature). Heat stored in the building structure during partial load (day) operation reduces the required equipment capacity for peak load operation in the same manner as it does with cooling.

TABLE 4—RECOMMENDED INSIDE DESIGN CONDITIONS*—SUMMER AND WINTER

TYPE OF APPLICATION	SUMMER					WINTER				
	Deluxe		Commercial Practice			With Humidification			Without Humidification	
	Dry-Bulb (F)	Rel. Hum. (%)	Dry-Bulb (F)	Rel. Hum. (%)	Temp. Swing† (F)	Dry-Bulb (F)	Rel. Hum. (%)	Temp. Swing‡ (F)	Dry-Bulb (F)	Temp. Swing‡ (F)
GENERAL COMFORT Apt., House, Hotel, Office Hospital, School, etc.	74-76	50-45	77-79	50-45	2 to 4	74-76	35-30	−3 to −4	75-77	−4
RETAIL SHOPS (Short term occupancy) Bank, Barber or Beauty Shop, Dept. Store, Supermarket, etc.	76-78	50-45	78-80	50-45	2 to 4	72-74	35-30**	−3 to −4	73-75	−4
LOW SENSIBLE HEAT FACTOR APPLICATIONS (High Latent Load) Auditorium, Church, Bar, Restaurant, Kitchen, etc.	76-78	55-50	78-80	60-50	1 to 2	72-74	40-35	−2 to −3	74-76	−4
FACTORY COMFORT Assembly Areas, Machining Rooms, etc.	77-80	55-45	80-85	60-50	3 to 6	68-72	35-30	−4 to −6	70-74	−6

*The room design dry-bulb temperature should be reduced when hot radiant panels are adjacent to the occupant and increased when cold panels are adjacent, to compensate for the increase or decrease in radiant heat exchange from the body. A hot or cold panel may be unshaded glass or glass block windows (hot in summer, cold in winter) and thin partitions with hot or cold spaces adjacent. An unheated slab floor on the ground or walls below the ground level are cold panels during the winter and frequently during the summer also. Hot tanks, furnaces or machines are hot panels.

†Temperature swing is above the thermostat setting at peak summer load conditions.

‡Temperature swing is below the thermostat setting at peak winter load conditions (no lights, people or solar heat gain).

**Winter humidification in retail clothing shops is recommended to maintain the quality texture of goods.

INSIDE INDUSTRIAL DESIGN CONDITIONS

Table 5 lists typical temperatures and relative humidities used in preparing, processing, and manufacturing various products, and for storing both raw and finished goods. These conditions are only typical of what has been used, and may vary with applications. They may also vary as changes occur in processes, products, and knowledge of the effect of temperature and humidity. In all cases, the temperature and humidity conditions and the permissible limits of variations on these conditions should be established by common agreement with the customer.

Some of the conditions listed have no effect on the product or process other than to increase the efficiency of the employee by maintaining comfort conditions. This normally improves workmanship and uniformity, thus reducing rejects and production cost. In some cases, it may be advisable to compromise between the required conditions and comfort conditions to maintain high quality commensurate with low production cost.

Generally, specific inside design conditions are required in industrial applications for one or more of the following reasons:

1. A constant temperature level is required for close tolerance measuring, gaging, machining, or grinding operations, to prevent expansion and contraction of the machine parts, machined products and measuring devices.

Normally, a constant temperature is more important than the temperature level. A constant relative humidity is secondary in nature but should not go over 45% to minimize formation of heavier surface moisture film.

Non-hygroscopic materials such as metals, glass, plastics, etc., have a property of capturing water molecules within the microscopic surface crevices, forming an invisible, non-continuous surface film. The density of this film increases when relative humidity increases. Hence, this film must, in many instances, be held below a critical point at which metals may etch, or the electric resistance of insulating materials is significantly decreased.

2. Where highly polished surfaces are manufactured or stored, a constant relative humidity and temperature is maintained, to minimize increase in surface moisture film. The temperature and humidity should be at, or a little

below, the comfort conditions to minimize perspiration of the operator. Constant temperature and humidity may also be required in machine rooms to prevent etching or corrosion of the parts of the machines. With applications of this type, if the conditions are not maintained 24 hours a day, the starting of air conditioning after any prolonged shutdown should be done carefully: (1) During the summer, the moisture accumulation in the space should be reduced before the temperature is reduced; (2) During the winter, the moisture should not be introduced before the materials have a chance to warm up if they are cooled during shutdown periods.

3. Control of relative humidity is required to maintain the strength, pliability, and regain of hydroscopic materials, such as textiles and paper. The humidity must also be controlled in some applications to reduce the effect of static electricity. Development of static electric charges is minimized at relative humidities of 55% or higher.

4. The temperature and relative humidity control are required to regulate the rate of chemical or biochemical reactions, such as drying of varnishes or sugar coatings, preparation of synthetic fibers or chemical compounds, fermentation of yeast, etc. Generally, high temperatures with low humidities increase drying rates; high temperatures increase the rate of chemical reaction, and high temperatures and relative humidities increase such processes as yeast fermentations.

5. Laboratories require precise control of both temperature and relative humidity or either. Both testing and quality control laboratories are frequently designed to maintain the ASTM Standard Conditions* of 73.4 F db and 50% rh.

6. With some industrial applications where the load is excessive and the machines or materials do not benefit from controlled conditions, it may be advisable to apply spot cooling for the relief of the workers. Generally, the conditions to be maintained by this means will be above normal comfort.

*Published in ASTM pamphlet dated 9-29-48. These conditions have also been approved by the Technical Committee on Standard Temperature and Relative Humidity Conditions of the FSB (Federal Specifications Board) with one variation: FSB permits ±4%, whereas ASTM requires ±2% permissable humidity tolerance.

TABLE 5—TYPICAL INSIDE DESIGN CONDITIONS—INDUSTRIAL

(Listed conditions are only typical; final design conditions are established by customer requirements)

INDUSTRY	PROCESS	DRY-BULB (F)	RH (%)
ABRASIVE	Manufacture	75-80	45-50
BAKERY	Dough Mixer	75-80	40-50
	Fermenting	75-82	70-75
	Proof Box	92-96	80-85
	Bread Cooler	70-80	80-85
	Cold Room	40-45	—
	Make-up Rm.	78-82	65-70
	Cake Mixing	95-105	—
	Crackers & Biscuits	60-65	50
	Wrapping	60-65	60-65
	Storage—		
	Dried Ingred.	70	55-65
	Fresh Ingred.	30-45	80-85
	Flour	70-75	50-65
	Shortening	45-70	55-60
	Sugar	80	35
	Water	32-35	—
	Wax Paper	70-80	40-50
BREWERY	Storage—		
	Hops	30-32	55-60
	Grain	80	60
	Liquid Yeast	32-34	75
	Lager	32-35	75
	Ale	40-45	75
	Fermenting Cellar—		
	Lager	40-45	75
	Ale	55	75
	Racking Cellar	32-35	75
CANDY—CHOCOLATE	Candy Centers	80-85	40-50
	Hand Dipping Rm.	60-65	50-55
	Enrobing Rm.	75-80	55-60
	Enrobing—		
	Loading End	80	50
	Enrober	90	13
	Stringing	70	40-50
	Tunnel	40-45	DP — 40
	Packing	65	55
	Pan Specialty Rm.	70-75	45
	General Storage	65-70	40-50
CANDY—HARD	Mfg.	75-80	30-40
	Mixing & Cooling	75-80	40-45
	Tunnel	55	DP — 55
	Packing	65-75	40-45
	Storage	65-75	45-50
	Drying—Jellies, Gums	120-150	15
	Cold Rm.—		
	Marshmallow	75-80	45-50
CHEWING GUM	Mfg.	77	33
	Rolling	68	63
	Stripping	72	53
	Breaking	74	47
	Wrapping	74	58

INDUSTRY	PROCESS	DRY-BULB (F)	RH (%)
CERAMICS	Refractory	110-150	50-90
	Molding Rm.	80	60-70
	Clay Storage	60-80	35-65
	Decal & Decorating	75-80	45-50
CEREAL	Packaging	75-80	45-50
COSMETICS	Mfg.	65-70	—
DISTILLING	Storage—		
	Grain	60	35-40
	Liquid Yeast	32-34	
	Mfg.	60-75	45-60
	Aging	65-72	50-60
ELECTRICAL PRODUCTS	Electronic & X-ray Coils & Trans. Winding	72	15
	Tube Assem.	68	40
	Electrical Inst. Mfg. & Lab.	70	50-55
	Thermostat Assem. & Calib.	76	50-55
	Humidistat Assem. & Calib.	76	50-55
	Close Tol. Assem.	72	40-45
	Meter Assem. Test	74-76	60-63
	Switchgear—		
	Fuse & Cut-Out Assem.	73	50
	Cap. Winding	73	50
	Paper Storage	73	50
	Conductor Wrapping	75	65-70
	Lightning Arrestor	68	20-40
	Circuit Brkr. Assem. & Test	76	30-60
	Rectifiers—		
	Process Selenium & Copper Oxide Plates	74	30-40
FURS	Drying	110	—
	Shock Treatment	18-20	—
	Storage	40-50	55-65
GLASS	Cutting	Comfort	
	Vinyl Lam. Rm.	55	15
LEATHER	Drying—		
	Veg. Tanned	70	75
	Chrome Tanned	120	75
	Storage	50-60	40-60
LENSES—OPTICAL	Fusing	Comfort	
	Grinding	80	50
MATCHES	Mfg.	72-74	50
	Drying	70-75	40
	Storage	60-62	50
MUNITIONS	Metal Percussion Elements—		
	Drying Parts	190	—
	Drying Paints	110	—
	Black Powder Drying	125	—
	Condition & Load Powder Type Fuse	70	40
	Load Tracer Pellets	80	40

TABLE 5—TYPICAL INSIDE CONDITIONS—INDUSTRIAL (Contd)

(Listed conditions are only typical; final design conditions are established by customer requirements)

INDUSTRY	PROCESS	DRY-BULB (F)	RH (%)	INDUSTRY	PROCESS	DRY-BULB (F)	RH (%)
PHARMACEU-TICAL	Powder Storage			TEXTILES (cont.)	Cotton, cont.		
	Before Mfg.	70-80	30-35		Ring Spinning		
	After Mfg.	75-80	15-35		Conventional	80-85	60-70
	Milling Rm.	80	35		Long Draft	80-85	
	Tablet Compressing	70-80	40		Frame Spinning	80-85	55-60
	Tablet Coating	80	35		Spooling, Warping	78-80	60-65
	Effervescent—				Weaving	78-80	70-85
	Tablet & Powder	90	15		Cloth Room	75	65-70
	Hypodermic Tablet	75-80	30		Combing	75	55-65
	Colloids	70	30-50		Linen		
	Cough Syrup	80	40		Carding, Spinning	75-80	60
	Glandular Prod.	78-80	5-10		Weaving	80	80
	Ampule Mfg.	80	35		Woolens		
	Gelatin Capsule	78	40-50		Pickers	80-85	60
	Capsule Storage	75	35-40		Carding	80-85	65-70
	Microanalysis	Comfort			Spinning	80-85	50-60
	Biological Mfg.	80	35		Dressing	75-80	60
	Liver Extract	70-80	20-30		Weaving—		
	Serums	Comfort			Light Goods	80-85	55-70
	Animal Rm.	Comfort			Heavy	80-85	60-65
PHOTO MATERIAL	Drying	20-125	40-80		Drawing	75	50-60
	Cutting & Packing	65-75	40-70		Worsteds		
	Storage—				Carding, Combing, & Gilling	80-85	60-70
	Film Base, Film Paper, Coated Paper	70-75	40-65		Storage	70-85	75-80
	Safety Film	60-80	45-50		Drawing	80-85	50-70
	Nitrate Film	40-50	40-50		Cap Spinning	80-85	50-55
PLASTIC	Mfg.—				Spooling, Winding	75-80	55-60
	Thermo Setting				Weaving	80	50-60
	Compounds	80	25-30		Finishing	75-80	60
	Cellophane	75-80	45-65		Silk		
PLYWOOD	Hot Press—Resin	90	60		Prep. & Dressing	80	60-65
	Cold Press	90	15-25		Weaving & Spinning	80	65-70
PRECISION MACHINING	Spectrographic Anal.	Comfort			Throwing	80	60
	Gear Matching & Assem.	75-80	35-40		Rayon		
	Storage—				Spinning	80-90	50-60
	Gasket	100	50		Throwing	80	55-60
	Cement & Glue	65	40		Weaving		
	Machinings Gaging, Assem.	Comfort			Regenerated	80	50-60
	Adjusting Precision Parts				Acetate	80	55-60
	Honing	75-80	35-45		Spun Rayon	80	80
PRINTING	Multicolor Litho.				Picking	75-80	50-60
	Pressroom	75-80	46-48		Carding, Roving, Drawing	80-90	50-60
	Stockroom	73-80	49-51		Knitting		
	Sheet & Web Print.	Comfort			Viscose or Cuprammonium	80-85	65
	Storage, Folding, etc.	Comfort			Synthetic Fiber Prep. & Weaving		
REFRIGERATION EQUIPMENT	Valve Mfg.	75	40		Viscose	80	60
	Compressor Assem.	70-76	30-45		Celanese	80	70
	Refrigerator Assem.	Comfort			Nylon	80	50-60
	Testing	65-82	47	TOBACCO	Cigar & Cigarette Mfg.	70-75	55-65
RUBBER DIPPED GOODS	Mfg.	90	—		Softening	90	85-88
	Cementing	80	25-30		Stemming & Stripping	75-85	75
	Surgical Articles	75-90	25-30		Storage & Prep.	78	70
	Storage Before Mfg.	60-75	40-50		Conditioning	75	75
	Lab. (ASTM Std.)	73.4	50		Packing & Shipping	75	60
TEXTILES	Cotton						
	Opening & Picking	70-75	55-70				
	Carding	83-87	50-55				
	Drawing & Roving	80	55-60				

CHAPTER 3. HEAT STORAGE, DIVERSITY AND STRATIFICATION

The normal load estimating procedure has been to evaluate the instantaneous heat gain to a space and to assume that the equipment will remove the heat at this rate. Generally, it was found that the equipment selected on this basis was oversized and therefore capable of maintaining much lower room conditions than the original design. Extensive analysis, research and testing have shown that the reasons for this are:

1. Storage of heat in the building structure.
2. Non-simultaneous occurrence of the peak of the individual loads (diversity).
3. Stratification of heat, in some cases.

This chapter contains the data and procedures for determining the load the equipment is actually picking up at any one time (actual cooling load), taking into account the above factors. Application of these data to the appropriate individual heat gains results in the actual cooling load.

The actual cooling load is generally considerably below the peak total instantaneous heat gain, thus requiring smaller equipment to perform a specific job. In addition, the air quantities and/or water quantities are reduced, resulting in a smaller overall system. Also, as brought out in the tables, if the equipment is operated somewhat longer during the peak load periods, and/or the temperature in the space is allowed to rise a few degrees at the peak periods during cooling operation, a further reduction in required capacity results. The smaller system operating for longer periods at times of peak load will produce a lower first cost to the customer with commensurate lower demand charges and lower operating costs. It is a well-known fact that equipment sized to more nearly meet the requirements results in a more efficient, better operating system. Also, if a smaller system is selected, and is based on extended periods of operation at the peak load, it results in a more economical and efficient system at a partially loaded condition.

Since, in most cases, the equipment installed to perform a specific function is smaller, there is less margin for error. This requires more exacting engineering including air distribution design and system balancing.

With multi-story, multi-room application, it is usually desirable to provide some flexibility in the air side or room load to allow for individual room control, load pickup, etc. Generally, it is recommended that the full reduction from storage and diversity be taken on the overall refrigeration or building load, with some degree of conservatism on the air side or room loads. This degree should be determined by the engineer from project requirements and customer desires. A system so designed, full reduction on refrigeration load and less than full reduction on air side or room load, meets all of the flexibility requirements, except at time of peak load. In addition, such a system has a low owning and operating cost.

STORAGE OF HEAT IN BUILDING STRUCTURES

The instantaneous heat gain in a typical comfort application consists of sun, lights, people, transmission thru walls, roof and glass, infiltration and ventilation air and, in some cases, machinery, appliances, electric calculating machines, etc. A large portion of this instantaneous heat gain is radiant heat which does not become an instantaneous load on the equipment, because it must strike a solid surface and be absorbed by this surface before becoming a load on the equipment. The breakdown on the various instantaneous heat gains into radiant heat and convected heat is approximately as follows:

HEAT GAIN SOURCE	RADIANT HEAT	CONVECTIVE HEAT
Solar, without inside blinds	100%	—
Solar, with inside blinds	58%	42%
Fluorescent Lights	50%	50%
Incandescent Lights	80%	20%
People*	40%	20%
Transmission†	60%	40%
Infiltration and Ventilation	—	100%
Machinery or Appliances‡	20-80%	80-20%

*The remaining 40% is dissipated as latent load.

†Transmission load is considered to be 100% convective load. This load is normally a relatively small part of the total load, and for simplicity is considered to be the instantaneous load on the equipment.

‡The load from machinery or appliances varies, depending upon the temperature of the surface. The higher the surface temperature, the greater the radiant heat load.

CONSTANT SPACE TEMPERATURE AND EQUIPMENT OPERATING PERIODS

As the radiant heat from sources shown in the above table strikes a solid surface (walls, floor, ceiling, etc.), it is absorbed, raising the temperature at the surface of the material above that inside the material and the air adjacent to the surface. This temperature difference causes heat flow into the material by conduction and into the air by convection. The heat conducted away from the surface is stored, and the heat convected from the surface becomes an instantaneous cooling load. The portion of radiant heat being stored depends on the ratio of the resistance to heat flow into the material and the resistance to heat flow into the air film. With most construction materials, the resistance to heat flow into the material is much lower than the air resistance; therefore, most of the radiant heat will be stored. However, as this process of absorbing radiant heat continues, the material becomes warmer and less capable of storing more heat.

The highly varying and relatively sharp peak of the instantaneous solar heat gain results in a large part of it being stored at the time of peak solar heat gain, as illustrated in *Fig. 3*.

The upper curve in *Fig. 3* is typical of the *solar heat gain* for a west exposure, and the lower curve

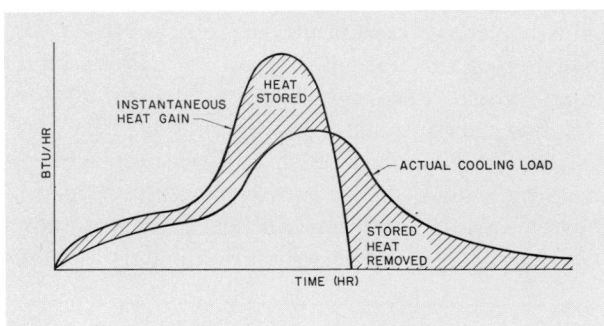

FIG. 3 — ACTUAL COOLING LOAD, SOLAR HEAT GAIN, WEST EXPOSURE, AVERAGE CONSTRUCTION

is the actual cooling load that results in an average construction application with the space temperature held constant. The reduction in the peak heat gain is approximately 40% and the peak load lags the peak heat gain by approximately 1 hour. The cross-hatched areas *(Fig. 3)* represent the Heat Stored and the Stored Heat Removed from the construction. Since all of the heat coming into a space must be removed, these two areas are equal.

The relatively constant light load results in a large portion being stored just after the lights are turned on, with a decreasing amount being stored the longer the lights are on, as illustrated in *Fig. 4*.

The upper and lower curves represent the instantaneous heat gain and actual cooling load from *fluorescent lights* with a constant space temperature. The cross-hatched areas are the Heat Stored and the Stored Heat Removed from the construction. The dotted line indicates the actual cooling load for the first day if the lights are on longer than the period shown.

Figs. 3 and 4 illustrate the relationship between the instantaneous heat gain and the actual cooling load in average construction spaces. With light construction, less heat is stored at the peak (less storage capacity available), and with heavy construction, more heat is stored at the peak (more storage capacity available), as shown in *Fig. 5*. This aspect affects the extent of zoning required in the design of a system for a given building; the lighter the building construction, the more attention should be given to zoning.

The upper curve of *Fig. 5* is the instantaneous solar heat gain while the three lower curves are the actual cooling load for *light, medium and heavy construction* respectively, with a constant temperature in the space.

One more item that significantly affects the storage of heat is the operating period of the air conditioning equipment. All of the curves shown in

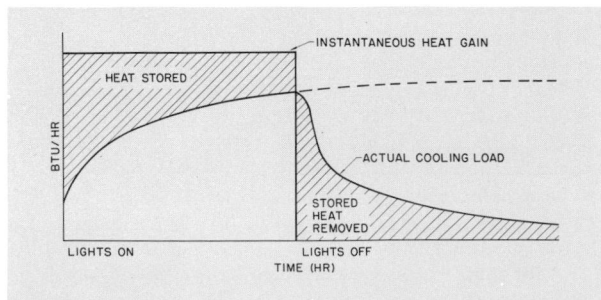

FIG. 4 — ACTUAL COOLING LOAD FROM FLUORESCENT LIGHTS, AVERAGE CONSTRUCTION

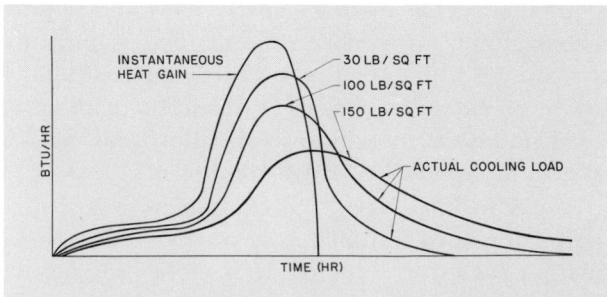

FIG. 5 — ACTUAL COOLING LOAD, SOLAR HEAT GAIN, LIGHT, MEDIUM AND HEAVY CONSTRUCTION

Figs. 3, 4, and 5 illustrate the actual cooling load for 24-hour operation. If the equipment is shut down after 16 hours of operation, some of the stored heat remains in the building construction. This heat must be removed (heat in must equal heat out) and will appear as a pulldown load when the equipment is turned on the next day, as illustrated in *Fig. 6.*

Adding the pulldown load to the cooling load for that day results in the actual cooling load for *16-hour operation,* as illustrated in *Fig. 7.*

The upper curve represents the instantaneous heat gain and the lower curve the *actual cooling load* for that day with a constant temperature maintained within the space during the operating period of the equipment. The dotted line represents the additional cooling load from the heat left in the building construction. The temperature in the space rises during the shutdown period from the nighttime transmission load and the stored heat, and is brought back to the control point during the pulldown period.

Shorter periods of operation increase the pulldown load because more stored heat is left in the building construction when the equipment is shut off. *Fig. 8* illustrates the *pulldown load for 12-hour operation.*

Adding this pulldown load to the cooling load for that day results in the actual cooling load for *12-hour operation,* as illustrated in *Fig. 9.*

The upper and lower solid curves are the instantaneous heat gain and the actual cooling load in average construction space with a constant temperature maintained during the operating period. The cross-hatched areas again represent the Heat Stored and the Stored Heat Removed from the construction.

The *light load (fluorescent)* is shown in *Fig. 10* for *12- and 16-hour operation* with a constant space temperature (assuming 10-hour operation of lights).

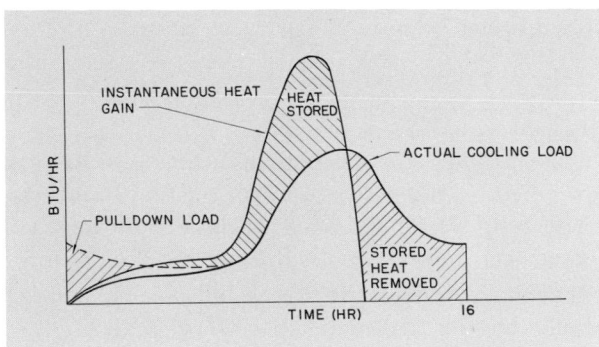

FIG. 7 — ACTUAL COOLING LOAD, SOLAR HEAT GAIN, WEST EXPOSURE, 16-HOUR OPERATION

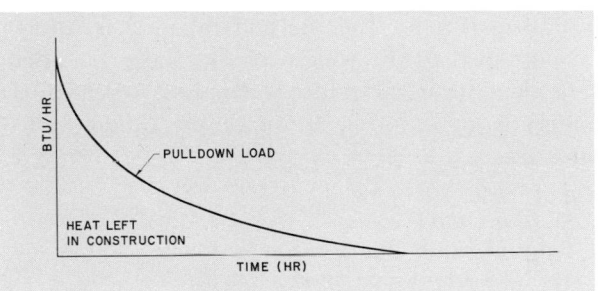

FIG. 8 — PULLDOWN LOAD, SOLAR HEAT GAIN, WEST EXPOSURE, 12-HOUR OPERATION

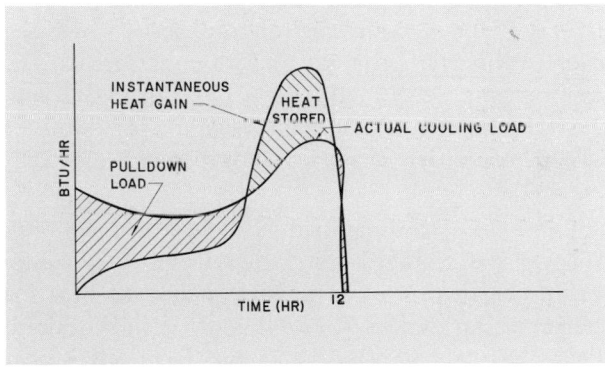

FIG. 9 — ACTUAL COOLING LOAD, SOLAR HEAT GAIN, WEST EXPOSURE, 12-HOUR OPERATION

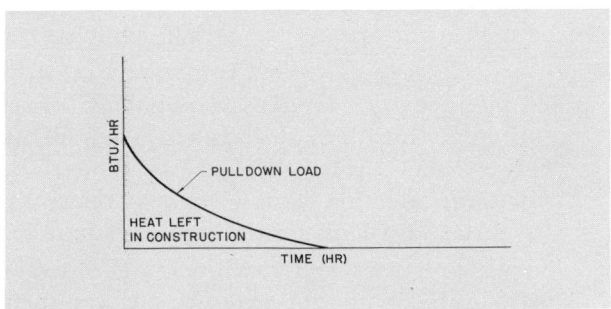

FIG. 6 — PULLDOWN LOAD, SOLAR HEAT GAIN, WEST EXPOSURE, 16-HOUR OPERATION

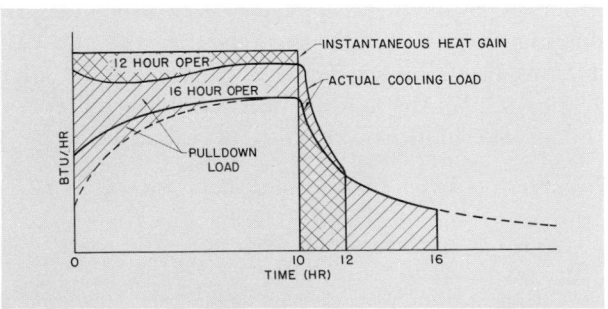

FIG. 10 — ACTUAL COOLING LOAD FROM FLUORESCENT LIGHTS, 12- AND 16-HOUR OPERATION

Basis of Tables 7 thru 12
Storage Load Factors,
Solar and Light Heat Gain
12-, 16-, and 24-hour Operation,
Constant Space Temperature

These tables are calculated, using a procedure developed from a series of tests in actual buildings. These tests were conducted in office buildings, supermarkets, and residences throughout this country.

The magnitude of the storage effect is determined largely by the thermal capacity or heat holding capacity of the materials surrounding the space. The thermal capacity of a material is the weight times the specific heat of the material. Since the specific heat of most construction material is approximately 0.20 Btu/(lb)(F), the thermal capacity is directly proportional to the weight of the material. Therefore, the data in the tables is based on weight of the materials surrounding the space, per square foot of floor area.

Use of Tables 7 thru 12
Storage Load Factors,
Solar and Light Heat Gain
12-, 16-, and 24-hour Operation,
Constant Space Temperature

Tables 7 thru 11 are used to determine the actual cooling load from the solar heat gain with a constant temperature maintained within the space for different types of construction and periods of operation. With both the 12- and 16-hour factors, the starting time is assumed to be 6 a.m. suntime (7 a.m. Daylight Saving Time). The weight per sq ft of types of construction are listed in *Tables 21 thru 33, pages 66-76.*

The actual cooling load is determined by multiplying the storage load factor from these tables for any or all times by the peak solar heat gain for the particular exposure, month and latitude desired. *Table 6* is a compilation of the peak solar heat gains for each exposure, month and latitude. These values are extracted from *Table 15, page 44*. The peak solar heat gain is also to be multiplied by either or both the applicable over-all factor for shading devices *(Table 16, page 52)* and the corrections listed under *Table 6*. Reduction in solar heat gain from the shading of the window by reveals and/or overhang should also be utilized.

Example 1 — Actual Cooling Load, Solar Heat Gain
Given:
A 20 ft × 20 ft × 8 ft outside office room with 6-inch sand aggregate concrete floor, with a floor tile finish, 2½-inch solid sand plaster partitions, no suspended ceiling, and a 12-inch common brick outside wall with ⅝-inch sand aggregate plaster finish on inside surface. A 16 ft × 5 ft steel sash window with a white venetian blind is in the outside wall and the wall faces west.

Find:
A. The actual cooling load from the solar heat gain in July at 4 p.m., 40° North latitude with the air conditioning equipment operating 24 hours during the peak load periods and a constant temperature maintained within the room.

B. The cooling load at 8 p.m. for the same conditions.

Solution:
The weight per sq ft of floor area of this room (values obtained from *Chapter 5*) is:

$$\text{Outside wall} = \frac{(20 \times 8) - (16 \times 5)}{20 \times 20} \times 126 \text{ lb/sq ft}$$
(Table 21, page 66)
$$= 25.2 \text{ lb/sq ft floor area}$$

$$\text{Partitions} = \frac{1}{2} \times \frac{20 \times 8 \times 3}{20 \times 20} \times 22 \text{ lb/sq ft}$$
(Table 26, page 70)
$$= 13.2 \text{ lb/sq ft floor area}$$

$$\text{Floor} = \frac{1}{2} \times \frac{20 \times 20}{20 \times 20} \times 59 \text{ lb/sq ft}$$
(Table 29, page 73)
$$= 29.5 \text{ lb/sq ft floor area}$$

$$\text{Ceiling} = \frac{1}{2} \times \frac{20 \times 20}{20 \times 20} \times 59 \text{ lb/sq ft}$$
(Table 29, page 73)
$$= 29.5 \text{ lb/sq ft floor area}$$

NOTE: One-half of the partition, floor and ceiling thickness is used, assuming that the spaces above and below are conditioned and are utilizing the other halves for storage of heat.

Total weight per sq ft of floor area
$$= 25.2 + 13.2 + 29.5 + 29.5 = 97.4 \text{ lb/sq ft.}$$

The overall factor for the window with the white venetian blind is 0.56 *(Table 16, page 52)* and the correction for steel sash = 1/.85.

A. Storage factor, 4 p.m. = 0.66 *(Table 7)*
The peak solar heat gain for a west exposure in July at 40° North latitude = 164 Btu/(hr)(sq ft), *(Table 6)*.

Actual cooling load
$$= \left(5 \times 16 \times 164 \times .56 \times \frac{1}{.85} \right) \times 0.66 = 5700 \text{ Btu/hr}$$

B. Storage factor, 8 p.m. = .20 *(Table 7)*

Actual cooling load
$$= \left(5 \times 16 \times 164 \times .56 \times \frac{1}{.85} \right) \times .20 = 1730 \text{ Btu/hr}$$

Table 12 is used to determine the actual cooling load from the heat gain from lights. These data may also be used to determine the actual cooling load from:

1. People — except in densely populated areas such as auditoriums, theaters, etc. The radiant heat exchange from the body is reduced in situations like this because there is relatively less surface available for the body to radiate to.

2. Some appliances and machines that operate periodically, with hot exterior surfaces such as ovens, dryers, hot tanks, etc.

NOTE: For Items 1 and 2 above, use values listed for fluorescent exposed lights.

Example 2 — Actual Cooling Load, Lights and People

Given:

The same room as in *Example 1* with a light heat gain of 3 watts per sq ft of floor area not including ballast, exposed fluorescent lights and 4 people. The room temperature to be maintained at 78 F db with 24-hour operation during the peak load periods.

Find:

The actual cooling load at 4 p.m. (with the lights turned on as the people arrive at 8 a.m.).

Solution:

The time elapsed after the lights are turned on is 8 hours (8 a.m. to 4 p.m.).

Storage load factor = .87 *(Table 12)*.

Sensible heat gain from people = 215 Btu/hr

(Table 48, page 100)

Actual cooling load
$$= [(3 \times 3.4 \times 1.25 \times 20 \times 20) + (4 \times 215)] \times .87$$
$$= 5190 \text{ Btu/hr}.$$

TABLE 6—PEAK SOLAR HEAT GAIN THRU ORDINARY GLASS*
Btu/(hr)(sq ft)

NORTH LAT.	MONTH	EXPOSURE NORTH LATITUDE									MONTH	SOUTH LAT.
		N†	NE	E	SE	S	SW	W	NW	Horiz		
0°	June	59	156	147	42	14	42	147	156	226	Dec	0°
	July & May	48	153	152	52	14	52	152	153	233	Nov & Jan	
	Aug & April	25	141	163	79	14	79	163	141	245	Oct & Feb	
	Sept & March	10	118	167	118	14	118	167	118	250	Sept & March	
	Oct & Feb	10	79	163	141	34	141	163	79	245	Aug & April	
	Nov & Jan	10	52	152	153	67	153	152	52	233	July & May	
	Dec	10	42	147	156	82	156	147	42	226	June	
10°	June	40	153	155	55	14	55	155	153	243	Dec	10°
	July & May	30	148	158	66	14	66	158	148	247	Nov & Jan	
	Aug & April	13	130	163	94	14	94	163	130	250	Oct & Feb	
	Sept & March	10	103	164	127	28	127	164	103	247	Sept & March	
	Oct & Feb	10	66	155	149	73	149	155	66	230	Aug & April	
	Nov & Jan	9	37	143	161	106	101	143	37	210	July & May	
	Dec	9	28	137	163	120	163	137	28	202	June	
20°	June	26	154	160	73	14	73	160	154	250	Dec	20°
	July & May	19	138	163	85	14	85	163	138	251	Nov & Jan	
	Aug & April	11	118	165	113	26	113	165	118	247	Oct & Feb	
	Sept & March	10	87	163	140	65	140	163	87	233	Sept & March	
	Oct & Feb	9	52	147	160	111	160	147	52	208	Aug & April	
	Nov & Jan	8	26	128	164	141	164	128	26	1,80	July & May	
	Dec	8	18	121	167	149	167	121	18	170	June	
30°	June	20	139	161	90	21	90	161	139	250	Dec	30°
	July & May	16	131	164	100	30	100	164	131	246	Nov & Jan	
	Aug & April	11	108	165	129	63	129	165	108	235	Oct & Feb	
	Sept & March	9	90	158	152	105	152	158	90	212	Sept & March	
	Oct & Feb	8	39	135	163	145	163	135	39	179	Aug & April	
	Nov & Jan	7	16	116	162	159	162	116	16	145	July & May	
	Dec	6	12	105	162	163	162	105	12	131	June	
40°	June	17	133	162	111	54	111	162	133	237	Dec	40°
	July & May	15	127	164	125	69	125	164	127	233	Nov & Jan	
	Aug & April	11	102	162	146	102	146	162	102	214	Oct & Feb	
	Sept & March	9	58	149	162	140	162	149	58	183	Sept & March	
	Oct & Feb	7	35	122	163	162	163	122	35	129	Aug & April	
	Nov & Jan	5	12	100	156	166	156	100	12	103	July & May	
	Dec	5	10	86	148	165	148	86	10	85	June	
50°	June	16	126	164	135	93	135	164	126	220	Dec	50°
	July & May	14	117	163	143	106	143	163	117	211	Nov & Jan	
	Aug & April	11	94	158	157	138	157	158	94	185	Oct & Feb	
	Sept & March	8	58	138	163	158	163	138	58	148	Sept & March	
	Oct & Feb	5	29	105	157	167	157	105	29	94	Aug & April	
	Nov & Jan	4	9	64	127	153	127	64	9	53	July & May	
	Dec	3	7	47	116	141	116	47	7	40	June	
		S	SE	E	NE	N	NW	W	SW	Horiz		
		EXPOSURE SOUTH LATITUDE										

Solar Gain Correction	Steel Sash or No Sash ×1/.85 or 1.17	Haze −15% (Max)	Altitude +0.7% per 1000 ft	Dewpoint Above 67 F −7% per 10 F	Dewpoint Below 67 F +7% per 10 F	South Lat Dec or Jan +7%

*Abstracted from *Table 15, page 43.*

†Solar heat gain on North exposure (in North latitudes) or on South exposure (in South latitudes) consists primarily of diffuse radiation which is essentially constant throughout the day. The solar heat gain values for this exposure are the average for the 12 hr period (6 a.m. to 6 p.m.). The storage factors in *Tables 7 thru 11* assume that the solar heat gain on the North (or South) exposure is constant.

TABLE 7—STORAGE LOAD FACTORS, SOLAR HEAT GAIN THRU GLASS

WITH INTERNAL SHADE*

24 Hour Operation, Constant Space Temperature†

EXPOSURE (North Lat)	WEIGHTS§ (lb per sq ft of floor area)	AM 6	7	8	9	10	11	12	PM 1	2	3	4	5	6	7	8	9	10	11	12	AM 1	2	3	4	5	EXPOSURE (South Lat)
Northeast	150 & over	.47	.58	.54	.42	.27	.21	.20	.19	.18	.17	.16	.14	.12	.09	.08	.07	.06	.06	.05	.05	.04	.04	.04	.03	Southeast
	100	.48	.60	.57	.46	.30	.24	.20	.19	.17	.16	.15	.13	.11	.08	.07	.06	.05	.05	.04	.04	.03	.03	.02	.02	
	30	.55	.76	.73	.58	.36	.24	.19	.17	.15	.13	.12	.11	.07	.04	.02	.02	.01	.01	0	0	0	0	0	0	
East	150 & over	.39	.56	.62	.59	.49	.33	.23	.21	.20	.18	.17	.15	.12	.10	.09	.08	.08	.07	.06	.05	.05	.05	.04	.04	East
	100	.40	.58	.65	.63	.52	.35	.24	.22	.20	.18	.16	.14	.12	.09	.08	.07	.06	.05	.05	.04	.04	.03	.03	.02	
	30	.46	.70	.80	.79	.64	.42	.25	.19	.16	.14	.11	.09	.07	.04	.02	.02	.01	.01	0	0	0	0	0	0	
Southeast	150 & over	.04	.28	.47	.59	.64	.62	.53	.41	.27	.24	.21	.19	.16	.14	.12	.11	.10	.09	.08	.07	.06	.06	.05	.05	Northeast
	100	.03	.28	.47	.61	.67	.65	.57	.44	.29	.24	.21	.18	.15	.12	.10	.09	.08	.07	.06	.05	.05	.04	.04	.03	
	30	0	.30	.57	.75	.84	.81	.69	.50	.30	.20	.17	.13	.09	.05	.04	.03	.02	.01	0	0	0	0	0	0	
South	150 & over	.06	.06	.23	.38	.51	.60	.66	.67	.64	.59	.42	.24	.22	.19	.17	.15	.13	.12	.11	.10	.09	.08	.07	.07	North
	100	.04	.04	.22	.38	.52	.63	.70	.71	.69	.59	.45	.26	.22	.18	.16	.13	.12	.10	.09	.08	.07	.06	.06	.05	
	30	.10	.21	.43	.63	.77	.86	.88	.82	.56	.50	.24	.16	.11	.08	.05	.04	.02	.02	.01	.01	0	0	0	0	
Southwest	150 & over	.08	.08	.09	.10	.11	.24	.39	.53	.63	.66	.61	.47	.23	.19	.18	.16	.14	.13	.11	.10	.09	.08	.08	.07	Northwest
	100	.07	.08	.08	.08	.10	.24	.40	.55	.66	.70	.64	.50	.26	.20	.17	.15	.13	.11	.10	.09	.08	.07	.06	.05	
	30	.03	.04	.06	.07	.09	.23	.47	.67	.81	.86	.79	.60	.26	.17	.12	.08	.05	.04	.03	.02	.01	.01	0	0	
West	150 & over	.08	.09	.09	.10	.10	.10	.10	.18	.36	.52	.63	.65	.55	.22	.19	.17	.15	.14	.12	.11	.10	.09	.08	.07	West
	100	.07	.08	.08	.09	.09	.09	.09	.18	.36	.54	.66	.68	.60	.25	.20	.17	.15	.13	.11	.10	.08	.07	.06	.05	
	30	.03	.04	.06	.07	.08	.08	.08	.19	.42	.65	.81	.85	.74	.30	.19	.13	.09	.06	.05	.03	.02	.02	.01	0	
Northwest	150 & over	.08	.09	.10	.10	.10	.10	.10	.10	.16	.33	.49	.61	.60	.19	.17	.15	.13	.12	.10	.09	.08	.08	.07	.06	Southwest
	100	.07	.08	.09	.09	.10	.10	.10	.10	.16	.34	.52	.65	.64	.23	.18	.15	.12	.11	.09	.08	.07	.06	.06	.05	
	30	.03	.05	.07	.08	.09	.09	.10	.10	.17	.39	.63	.80	.79	.28	.18	.12	.09	.06	.04	.03	.02	.02	.01	0	
North and Shade	150 & over	.08	.37	.67	.71	.74	.76	.79	.81	.83	.84	.86	.87	.88	.29	.26	.23	.20	.19	.17	15	.14	.12	.11	.10	South and Shade
	100	.06	.31	.67	.72	.76	.79	.81	.83	.85	.87	.88	.90	.91	.30	.26	.22	.19	.16	.15	.13	.12	.10	.09	.08	
	30	0	.25	.74	.83	.88	.91	.94	.96	.96	.98	.98	.99	.99	.26	.17	.12	.08	.05	.04	.03	.02	.01	.01	.01	

Equation: Cooling Load, Btu/hr = [Peak solar heat gain, Btu/(hr) (sq ft), (Table 6)]
× [Window area, sq ft]
× [Shade factor, Haze factor, etc., (Chapter 4)]
× [Storage factor, (above Table at desired time)]

*Internal shading device is any type of shade located on the inside of the glass.

†These factors apply when maintaining a CONSTANT TEMPERATURE in the space during the operating period. Where the temperature is allowed to swing, additional storage will result during peak load periods. Refer to *Table 13* for applicable storage factors.

§Weight per sq ft of floor—

Room on Bldg Exterior (One or more outside walls) = $\dfrac{\text{(Weight of Outside Walls, lb)} + \frac{1}{2}\,\text{(Weight of Partitions, Floor and Ceiling, lb)}}{\text{Floor Area in Room, sq ft}}$

Room in Bldg Interior (No outside walls) = $\dfrac{\frac{1}{2}\,\text{(Weight of Partitions, Floor and Ceiling, lb)}}{\text{Floor Area in Room, sq ft}}$

Basement Room (Floor on ground) = $\dfrac{\text{(Weight of Outside Walls, lb)} + \text{(Weight of Floor, lb)} + \frac{1}{2}\,\text{(Weight of Partitions and Ceiling, lb)}}{\text{Floor Area in Room, sq ft}}$

Entire Building or Zone = $\dfrac{\text{(Weight of Outside Wall, Partitions, Floors, Ceilings, Structural Members and Supports, lb)}}{\text{Air Conditioned Floor Area, sq ft}}$

With rug on floor—Weight of floor should be multiplied by 0.50 to compensate for insulating effect of rug.

Weights per sq ft of common types of construction are contained in *Tables 21 thru 33, pages 66 thru 76.*

TABLE 8—STORAGE LOAD FACTORS, SOLAR HEAT GAIN THRU GLASS

WITH BARE GLASS OR WITH EXTERNAL SHADE‡

24 Hour Operation, Constant Space Temperature†

EXPOSURE (North Lat)	WEIGHT§ (lb per sq ft of floor area)	AM 6	7	8	9	10	11	12	PM 1	2	3	4	5	6	7	8	9	10	11	12	AM 1	2	3	4	5	EXPOSURE (South Lat)
Northeast	150 & over	.17	.27	.33	.33	.31	.29	.27	.25	.23	.22	.20	.19	.17	.15	.14	.12	.11	.10	.09	.08	.07	.07	.06	.06	Southeast
	100	.19	.31	.38	.39	.36	.34	.27	.24	.22	.21	.19	.17	.16	.14	.12	.10	.07	.08	.07	.06	.05	.05	.04	.03	
	30	.31	.56	.65	.61	.46	.33	.26	.21	.18	.16	.14	.12	.09	.06	.04	.03	.02	.01	.01	.01	0	0	0	0	
East	150 & over	.16	.26	.34	.39	.40	.38	.34	.30	.28	.26	.23	.22	.20	.18	.16	.14	.13	.12	.10	.09	.08	.08	.07	.06	East
	100	.16	.29	.40	.46	.46	.42	.36	.31	.28	.25	.23	.20	.18	.15	.14	.12	.11	.09	.08	.08	.06	.06	.05	.04	
	30	.27	.50	.67	.73	.68	.53	.38	.27	.22	.18	.15	.12	.09	.06	.04	.03	.02	.01	.01	.01	.01	0	0	.01	
Southeast	150 & over	.08	.14	.22	.31	.38	.43	.44	.43	.39	.35	.32	.29	.26	.23	.21	.19	.16	.15	.13	.12	.11	.10	.09	.08	Northeast
	100	.05	.12	.23	.35	.44	.49	.51	.47	.41	.36	.31	.27	.24	.21	.18	.16	.14	.12	.10	.09	.08	.08	.06	.06	
	30	0	.18	.40	.59	.72	.77	.72	.60	.44	.32	.23	.18	.14	.09	.07	.05	.03	.02	.01	.01	.01	0	0	0	
South	150 & over	.10	.10	.13	.20	.28	.35	.42	.48	.51	.51	.48	.42	.37	.33	.29	.26	.23	.21	.19	.17	.15	.14	.13	.12	North
	100	.07	.06	.12	.20	.30	.39	.48	.54	.58	.57	.53	.45	.37	.31	.27	.23	.20	.18	.16	.14	.12	.11	.10	.08	
	30	0	0	.12	.29	.48	.64	.75	.82	.81	.75	.61	.42	.28	.19	.13	.09	.06	.04	.03	.02	.01	.01	0	0	
Southwest	150 & over	.11	.10	.10	.10	.10	.14	.21	.29	.36	.43	.47	.46	.40	.34	.30	.27	.24	.22	.20	.18	.16	.14	.13	.12	Northwest
	100	.09	.09	.08	.09	.09	.14	.22	.31	.42	.50	.53	.51	.44	.35	.29	.26	.22	.19	.17	.15	.13	.12	.11	.09	
	30	.02	.03	.05	.06	.08	.12	.34	.53	.68	.78	.78	.68	.46	.29	.20	.14	.09	.07	.05	.03	.02	.02	.01	.01	
West	150 & over	.12	.11	.11	.10	.10	.10	.10	.13	.19	.27	.36	.42	.44	.38	.33	.29	.26	.23	.21	.18	.16	.15	.13	.12	West
	100	.09	.09	.09	.09	.09	.09	.10	.12	.19	.30	.40	.48	.51	.42	.35	.30	.25	.22	.19	.16	.14	.13	.11	.09	
	30	.02	.03	.05	.06	.07	.07	.08	.14	.29	.49	.67	.76	.75	.53	.33	.22	.15	.11	.08	.05	.04	.03	.02	.01	
Northwest	150 & over	.10	.10	.10	.10	.10	.10	.10	.10	.12	.17	.25	.34	.39	.34	.29	.26	.23	.20	.18	.16	.14	.13	.12	.10	Southwest
	100	.08	.09	.09	.09	.09	.09	.09	.09	.11	.19	.29	.40	.46	.40	.32	.26	.22	.19	.16	.14	.13	.11	.10	.08	
	30	.02	.04	.05	.07	.08	.09	.10	.10	.13	.27	.48	.65	.73	.49	.31	.21	.16	.10	.07	.05	.04	.03	.02	.01	
North and Shade	150 & over	.16	.23	.33	.41	.47	.52	.57	.61	.66	.69	.72	.74	.59	.52	.46	.42	.37	.34	.31	.27	.25	.23	.21	.17	South and Shade
	100	.11	.33	.44	.51	.57	.62	.66	.70	.74	.76	.79	.80	.60	.51	.44	.37	.32	.29	.27	.23	.21	.18	.16	.13	
	30	0	.48	.66	.76	.82	.87	.91	.93	.95	.97	.98	.98	.52	.34	.24	.16	.11	.07	.05	.04	.02	.02	.01	.01	

Equation: Cooling Load, Btu/hr = [Peak solar heat gain, Btu/(hr) (sq ft), (Table 6)]

× [Window area, sq ft]

× [Shade factor, Haze factor, etc., (Chapter 4)]

× [Storage factor, (above Table at desired time)]

‡Bare glass — Any window with no inside shading device. Windows with shading devices on the outside or shaded by external projections are considered bare glass.

†These factors apply when maintaining a CONSTANT TEMPERATURE in the space during the operating period. Where the temperature is allowed to swing, additional storage will result during peak load periods. Refer to Table 13 for applicable storage factors.

§Weight per sq ft of floor—

Room on Bldg Exterior (One or more outside walls) = $\dfrac{\text{(Weight of Outside Walls, lb)} + \frac{1}{2}\text{(Weight of Partitions, Floor and Ceiling, lb)}}{\text{Floor Area in Room, sq ft}}$

Room in Bldg Interior (No outside walls) = $\dfrac{\frac{1}{2}\text{(Weight of Partitions, Floor and Ceiling, lb)}}{\text{Floor Area in Room, sq ft}}$

Basement Room (Floor on ground) = $\dfrac{\text{(Weight of Outside Walls, lb)} + \text{(Weight of Floor, lb)} + \frac{1}{2}\text{(Weight of Partitions and Ceiling, lb)}}{\text{Floor Area in Room, sq ft}}$

Entire Building or Zone = $\dfrac{\text{(Weight of Outside Wall, Partitions, Floors, Ceilings, Structural Members and Supports, lb)}}{\text{Air Conditioned Floor Area, sq ft}}$

With rug on floor—Weight of floor should be multiplied by 0.50 to compensate for insulating effect of rug.

Weights per sq ft of common types of construction are contained in Tables 21 thru 33, pages 66 thru 76.

TABLE 9—STORAGE LOAD FACTORS, SOLAR HEAT GAIN THRU GLASS

WITH INTERNAL SHADING DEVICE*

16 Hour Operation, Constant Space Temperature†

EXPOSURE (North Lat)	WEIGHTS§ (lb per sq ft of floor area)	6	7	8	9	10	11	12	1	2	3	4	5	6	7	8	9	EXPOSURE (South Lat)
		AM							PM									
Northeast	150 & over	.53	.64	.59	.47	.31	.25	.24	.22	.18	.17	.16	.14	.12	.09	.08	.07	Southeast
	100	.53	.65	.61	.50	.33	.27	.22	.21	.17	.16	.15	.13	.11	.08	.07	.06	
	30	.56	.77	.73	.58	.36	.24	.19	.17	.15	.13	.12	.11	.07	.04	.02	.02	
East	150 & over	.47	.63	.68	.64	.54	.38	.27	.25	.20	.18	.17	.15	.12	.10	.09	.08	East
	100	.46	.63	.70	.67	.56	.38	.27	.24	.20	.18	.16	.14	.12	.09	.08	.07	
	30	.47	.71	.80	.79	.64	.42	.25	.19	.16	.14	.11	.09	.07	.04	.02	.02	
Southeast	150 & over	.14	.37	.55	.66	.70	.68	.58	.46	.27	.24	.21	.19	.16	.14	.12	.11	Northeast
	100	.11	.35	.53	.66	.72	.69	.61	.47	.29	.24	.21	.18	.15	.12	.10	.09	
	30	.02	.31	.57	.75	.84	.81	.69	.50	.30	.20	.17	.13	.09	.05	.04	.03	
South	150 & over	.19	.18	.34	.48	.60	.68	.73	.74	.64	.59	.42	.24	.22	.19	.17	.15	North
	100	.16	.14	.31	.46	.59	.69	.76	.70	.69	.59	.45	.26	.22	.18	.16	.13	
	30	.12	.23	.44	.64	.77	.86	.88	.82	.56	.50	.24	.16	.11	.08	.05	.04	
Southwest	150 & over	.22	.21	.20	.20	.20	.32	.47	.60	.63	.66	.61	.47	.23	.19	.18	.16	Northwest
	100	.20	.19	.18	.17	.18	.31	.46	.60	.66	.70	.64	.50	.26	.20	.17	.15	
	30	.08	.08	.09	.09	.10	.24	.47	.67	.81	.86	.79	.60	.26	.17	.12	.08	
West	150 & over	.23	.23	.21	.21	.20	.19	.18	.25	.36	.52	.63	.65	.55	.22	.19	.17	West
	100	.22	.21	.19	.19	.17	.16	.15	.23	.36	.54	.66	.68	.60	.25	.20	.17	
	30	.12	.10	.10	.10	.10	.10	.09	.19	.42	.65	.81	.85	.74	.30	.19	.13	
Northwest	150 & over	.21	.21	.20	.19	.18	.18	.17	.16	.16	.33	.49	.61	.60	.19	.17	.15	Southwest
	100	.19	.19	.18	.17	.17	.16	.16	.15	.16	.34	.52	.65	.23	.18	.15	.12	
	30	.12	.11	.11	.11	.11	.11	.11	.10	.17	.39	.63	.80	.79	.28	.18	.12	
North and Shade	150 & over	.23	.58	.75	.79	.80	.80	.81	.82	.83	.84	.86	.87	.88	.39	.35	.31	South and Shade
	100	.25	.46	.73	.78	.82	.82	.83	.84	.85	.87	.88	.89	.90	.40	.34	.29	
	30	.07	.22	.69	.80	.86	.93	.94	.95	.97	.98	.98	.99	.99	.35	.23	.16	

Equation: Cooling Load, Btu/hr = [Peak solar heat gain, Btu/(hr) (sq ft), (Table 6)]
                      × [Window area, sq ft]
                      × [Shade factor, Haze factor, etc., (Chapter 4)]
                      × [Storage factor, (above Table at desired time)]

*Internal shading device is any type of shade located on the inside of the glass.

†These factors apply when maintaining a CONSTANT TEMPERATURE in the space during the operating period. Where the temperature is allowed to swing, additional storage will result during peak load periods. Refer to *Table 13* for applicable storage factors.

§Weight per sq ft of floor—

Room on Bldg Exterior (One or more outside walls) = $\dfrac{\text{(Weight of Outside Walls, lb)} + \frac{1}{2}\text{(Weight of Partitions, Floor and Ceiling, lb)}}{\text{Floor Area in Room, sq ft}}$

Room in Bldg Interior (No outside walls) = $\dfrac{\frac{1}{2}\text{(Weight of Partitions, Floor and Ceiling, lb)}}{\text{Floor Area in Room, sq ft}}$

Basement Room (Floor on ground) = $\dfrac{\text{(Weight of Outside Walls, lb)} + \text{(Weight of Floor, lb)} + \frac{1}{2}\text{(Weight of Partitions and Ceiling, lb)}}{\text{Floor Area in Room, sq ft}}$

Entire Building or Zone = $\dfrac{\text{(Weight of Outside Wall, Partitions, Floors, Ceilings, Structural Members and Supports, lb)}}{\text{Air Conditioned Floor Area, sq ft}}$

With rug on floor—Weight of floor should be multiplied by 0.50 to compensate for insulating effect of rug.

Weights per sq ft of common types of construction are contained in *Tables 21 thru 33, pages 66 thru 76*.

TABLE 10—STORAGE LOAD FACTORS, SOLAR HEAT GAIN THRU GLASS
WITH BARE GLASS OR WITH EXTERNAL SHADE‡
16 Hour Operation, Constant Space Temperature†

| EXPOSURE (North Lat) | WEIGHTS§ (lb per sq ft of floor area) | SUN TIME | | | | | | | | | | | | | | | | EXPOSURE (South Lat) |
| | | AM | | | | | | | PM | | | | | | | | | |
		6	7	8	9	10	11	12	1	2	3	4	5	6	7	8	9	
Northeast	150 & over	.28	.37	.42	.41	.38	.36	.33	.31	.23	.22	.20	.19	.17	.15	.14	.12	Southeast
	100	.28	.39	.45	.45	.41	.39	.31	.27	.22	.21	.19	.17	.16	.14	.12	.10	
	30	.33	.57	.66	.62	.46	.33	.26	.21	.18	.16	.14	.12	.09	.06	.04	.03	
East	150 & over	.29	.38	.44	.48	.48	.46	.41	.36	.28	.26	.23	.22	.20	.18	.16	.14	East
	100	.27	.38	.48	.54	.52	.48	.41	.35	.28	.25	.23	.20	.18	.15	.14	.12	
	30	.29	.51	.68	.74	.69	.53	.38	.27	.22	.18	.15	.12	.09	.06	.04	.03	
Southeast	150 & over	.24	.29	.35	.43	.49	.53	.53	.51	.39	.35	.32	.29	.26	.23	.21	.19	Northeast
	100	.19	.24	.33	.44	.52	.57	.57	.53	.41	.36	.31	.27	.24	.21	.18	.16	
	30	.03	.20	.41	.60	.73	.77	.72	.60	.44	.32	.23	.18	.14	.09	.07	.05	
South	150 & over	.33	.31	.32	.37	.43	.49	.55	.60	.57	.51	.48	.42	.37	.33	.29	.26	North
	100	.27	.24	.28	.34	.42	.50	.58	.60	.60	.57	.53	.45	.37	.31	.27	.23	
	30	.06	.04	.15	.31	.49	.65	.75	.82	.81	.75	.61	.42	.28	.19	.13	.09	
Southwest	150 & over	.35	.32	.30	.28	.26	.28	.30	.37	.43	.47	.46	.40	.34	.30	.27	.24	Northwest
	100	.31	.28	.25	.24	.22	.26	.33	.40	.46	.50	.53	.51	.44	.35	.29	.26	
	30	.11	.10	.10	.09	.10	.14	.35	.54	.68	.78	.78	.68	.46	.29	.20	.14	
West	150 & over	.38	.34	.32	.28	.26	.25	.23	.25	.26	.27	.36	.42	.44	.38	.33	.29	West
	100	.34	.31	.28	.25	.23	.22	.21	.21	.23	.30	.40	.48	.51	.43	.35	.30	
	30	.17	.14	.13	.11	.11	.10	.10	.15	.29	.49	.67	.76	.75	.53	.33	.22	
Northwest	150 & over	.33	.30	.28	.26	.24	.23	.22	.20	.18	.17	.25	.34	.39	.34	.29	.26	Southwest
	100	.30	.28	.25	.23	.22	.20	.19	.17	.17	.19	.29	.40	.46	.40	.32	.26	
	30	.18	.14	.12	.12	.12	.12	.12	.11	.13	.27	.48	.65	.73	.49	.31	.21	
North and Shade	150 & over	.31	.57	.64	.68	.72	.73	.73	.74	.74	.75	.76	.78	.78	.59	.52	.46	South and Shade
	100	.30	.47	.60	.67	.72	.74	.77	.78	.79	.80	.81	.82	.83	.60	.51	.44	
	30	.04	.07	.53	.70	.78	.84	.88	.91	.93	.95	.97	.98	.99	.62	.34	.24	

Equation: Cooling Load, Btu/hr = [Peak solar heat gain, Btu/(hr) (sq ft), (Table 6)]
    × [Window area, sq ft]
    × [Shade factor, Haze factor, etc., (Chapter 4)]
    × [Storage factor, (above Table at desired time)]

‡Bare glass — Any window with no inside shading device. Windows with shading devices on the outside or shaded by external projections are considered bare glass.

†These factors apply when maintaining a CONSTANT TEMPERATURE in the space during the operating period. Where the temperature is allowed to swing, additional storage will result during peak load periods. Refer to *Table 13* for applicable storage factors.

§Weight per sq ft of floor—

Room on Bldg Exterior (One or more outside walls) = $\dfrac{\text{(Weight of Outside Walls, lb)} + \frac{1}{2}\text{(Weight of Partitions, Floor and Ceiling, lb)}}{\text{Floor Area in Room, sq ft}}$

Room in Bldg Interior (No outside walls) = $\dfrac{\frac{1}{2}\text{(Weight of Partitions, Floor and Ceiling, lb)}}{\text{Floor Area in Room, sq ft}}$

Basement Room (Floor on ground) = $\dfrac{\text{(Weight of Outside Walls, lb)} + \text{(Weight of Floor, lb)} + \frac{1}{2}\text{(Weight of Partitions and Ceiling, lb)}}{\text{Floor Area in Room, sq ft}}$

Entire Building or Zone = $\dfrac{\text{(Weight of Outside Wall, Partitions, Floors, Ceilings, Structural Members and Supports, lb)}}{\text{Air Conditioned Floor Area, sq ft}}$

With rug on floor—Weight of floor should be multiplied by 0.50 to compensate for insulating effect of rug.

Weights per sq ft of common types of construction are contained in *Tables 21 thru 33, pages 66 thru 76.*

TABLE 11—STORAGE LOAD FACTORS, SOLAR HEAT GAIN THRU GLASS

12 Hour Operation, Constant Space Temperature†

EXPOSURE (North Lat)	WEIGHTS§ (lb per sq ft of floor area)	INTERNAL SHADE* SUN TIME AM							PM					BARE GLASS OR EXTERNAL SHADE‡ AM							PM					EXPOSURE (South Lat)
		6	7	8	9	10	11	12	1	2	3	4	5	6	7	8	9	10	11	12	1	2	3	4	5	
Northeast	150 & over	.59	.67	.62	.49	.33	.27	.25	.24	.22	.21	.20	.17	.34	.42	.47	.45	.42	.39	.36	.33	.30	.29	.26	.25	Southeast
	100	.59	.68	.64	.52	.35	.29	.24	.23	.20	.19	.17	.15	.35	.45	.50	.49	.45	.42	.34	.30	.27	.26	.23	.20	
	30	.62	.80	.75	.60	.37	.25	.19	.17	.15	.13	.12	.11	.40	.62	.69	.64	.48	.34	.27	.22	.18	.16	.14	.12	
East	150 & over	.51	.66	.71	.67	.57	.40	.29	.26	.25	.23	.21	.19	.36	.44	.50	.53	.53	.50	.44	.39	.36	.34	.30	.28	East
	100	.52	.67	.73	.70	.58	.40	.29	.26	.24	.21	.19	.16	.34	.44	.54	.58	.57	.51	.44	.39	.34	.31	.28	.24	
	30	.53	.74	.82	.81	.65	.43	.25	.19	.16	.14	.11	.09	.36	.56	.71	.76	.70	.54	.39	.28	.23	.18	.15	.12	
Southeast	150 & over	.20	.42	.59	.70	.74	.71	.61	.48	.33	.30	.26	.24	.34	.37	.43	.50	.54	.58	.57	.55	.50	.45	.41	.37	Northeast
	100	.18	.40	.57	.70	.75	.72	.63	.49	.34	.28	.25	.21	.29	.33	.41	.51	.58	.61	.61	.56	.49	.44	.37	.33	
	30	.09	.35	.61	.78	.86	.82	.69	.50	.30	.20	.17	.13	.14	.27	.47	.64	.75	.79	.73	.61	.45	.32	.23	.18	
South	150 & over	.28	.25	.40	.53	.64	.72	.77	.77	.73	.67	.49	.31	.47	.43	.42	.46	.51	.56	.61	.65	.66	.65	.61	.54	North
	100	.26	.22	.38	.51	.64	.73	.79	.79	.77	.65	.51	.31	.44	.37	.39	.43	.50	.57	.64	.68	.70	.68	.63	.53	
	30	.21	.29	.48	.67	.79	.88	.89	.83	.56	.50	.24	.16	.28	.19	.25	.38	.54	.68	.78	.84	.82	.76	.61	.42	
Southwest	150 & over	.31	.27	.27	.26	.25	.27	.50	.63	.72	.74	.69	.54	.51	.44	.40	.37	.34	.36	.41	.47	.54	.57	.60	.58	Northwest
	100	.33	.28	.25	.23	.23	.35	.50	.64	.74	.77	.70	.55	.53	.44	.37	.35	.31	.33	.39	.46	.55	.62	.64	.60	
	30	.29	.21	.18	.15	.14	.27	.50	.69	.82	.87	.79	.60	.48	.32	.25	.20	.17	.19	.39	.56	.70	.80	.79	.69	
West	150 & over	.63	.31	.28	.27	.25	.24	.22	.29	.46	.61	.71	.72	.56	.49	.44	.39	.36	.33	.31	.31	.35	.42	.49	.54	West
	100	.67	.33	.28	.26	.24	.22	.20	.28	.44	.61	.72	.73	.60	.52	.44	.39	.34	.31	.29	.28	.33	.43	.51	.57	
	30	.77	.34	.25	.20	.17	.14	.13	.22	.44	.67	.82	.85	.77	.56	.38	.28	.22	.18	.16	.19	.33	.52	.69	.77	
Northwest	150 & over	.68	.28	.27	.25	.23	.22	.20	.19	.24	.41	.56	.67	.49	.44	.39	.36	.33	.30	.28	.26	.26	.30	.37	.44	Southwest
	100	.71	.31	.27	.24	.22	.21	.19	.18	.23	.40	.58	.70	.54	.49	.41	.35	.31	.28	.25	.23	.24	.30	.39	.48	
	30	.82	.33	.25	.20	.18	.15	.14	.13	.19	.41	.64	.80	.75	.53	.36	.28	.24	.19	.17	.15	.17	.30	.50	.66	
North and Shade	150 & over	.96	.96	.96	.96	.96	.96	.96	.96	.96	.96	.96	.96	.75	.75	.79	.83	.84	.86	.88	.88	.91	.92	.93	.93	South and Shade
	100	.98	.98	.98	.98	.98	.98	.98	.98	.98	.98	.98	.98	.81	.84	.86	.89	.91	.93	.93	.94	.94	.95	.95	.95	
	30	←————————1.00————————→												←————————1.00————————→												

Equation: Cooling Load, Btu/hr = [Peak solar heat gain, Btu/(hr) (sq ft), (Table 6)]
         × [Window area, sq ft]
         × [Shade factor, Haze factor, etc., (Chapter 4)]
         × [Storage factor, (above Table at desired time)]

*Internal shading device is any type of shade located on the inside of the glass.

‡Bare glass — Any window with no inside shading device. Windows with shading devices on the outside or shaded by external projections are to considered bare glass.

†These factors apply when maintaining a CONSTANT TEMPERATURE in the space during the operating period. Where the temperature is ed allowed to swing, additional storage will result during peak load periods. Refer to *Table 13* for applicable storage factors.

§Weight per sq ft of floor—

Room on Bldg Exterior (One or more outside walls) = $\dfrac{\text{(Weight of Outside Walls, lb)} + \frac{1}{2}\,\text{(Weight of Partitions, Floor and Ceiling, lb)}}{\text{Floor Area in Room, sq ft}}$

Room in Bldg Interior (No outside walls) = $\dfrac{\frac{1}{2}\,\text{(Weight of Partitions, Floor and Ceiling, lb)}}{\text{Floor Area in Room, sq ft}}$

Basement Room (Floor on ground) = $\dfrac{\text{(Weight of Outside Walls, lb)} + \text{(Weight of Floor, lb)} + \frac{1}{2}\,\text{(Weight of Partitions and Ceiling, lb)}}{\text{Floor Area in Room, sq ft}}$

Entire Building or Zone = $\dfrac{\text{(Weight of Outside Wall, Partitions, Floors, Ceilings, Structural Members and Supports, lb)}}{\text{Air Conditioned Floor Area, sq ft}}$

With rug on floor—Weight of floor should be multiplied by 0.50 to compensate for insulating effect of rug.

Weights per sq ft of common types of construction are contained in *Tables 21 thru 33, pages 66 thru 76.*

TABLE 12—STORAGE LOAD FACTORS, HEAT GAIN—LIGHTS*

Lights On 10 Hours† with Equipment Operating 12, 16 and 24 Hours, Constant Space Temperature

	EQUIP. OPER-ATION Hours	WEIGHT§ (lb per sq ft of floor area)	0	1	2	3	4	5	6	7	8	9	10	11	12	13	14	15	16	17	18	19	20	21	22	23
									NUMBER OF HOURS AFTER LIGHTS ARE TURNED ON																	
Fluorescent Lights Exposed	24	150 & over	.37	.67	.71	.74	.76	.79	.81	.83	.84	.86	.87	.29	.26	.23	.20	.19	.17	.15	.14	.12	.11	.10	.09	.08
		100	.31	.67	.72	.76	.79	.81	.83	.85	.87	.88	.90	.30	.26	.22	.19	.16	.15	.13	.12	.10	.09	.08	.07	.06
		30	.25	.74	.83	.88	.91	.94	.96	.96	.98	.98	.99	.26	.17	.12	.08	.05	.04	.03	.02	.01	.01	.01	0	0
	16	150 & over	.60	.82	.83	.84	.84	.84	.85	.85	.86	.88	.90	.32	.28	.25	.23	.19								
		100	.46	.79	.84	.86	.87	.88	.88	.89	.89	.90	.90	.30	.26	.22	.19	.16								
		30	.29	.77	.85	.89	.92	.95	.96	.96	.98	.98	.99	.26	.17	.12	.08	.05								
	12	150 & over	.63	.90	.91	.93	.93	.94	.95	.95	.95	.96	.96	.37												
		100	.57	.89	.91	.92	.94	.94	.95	.95	.96	.96	.97	.36												
		30	.42	.86	.91	.93	.95	.97	.98	.98	.99	.99	.99	.26												
Fluorescent Lights Recessed in Susp. Ceiling or Exposed Incandescent Lights.	24	150 & over	.34	.55	.61	.65	.68	.71	.74	.77	.79	.81	.83	.39	.35	.31	.28	.25	.23	.20	.18	.16	.15	.14	.12	.11
		100	.24	.56	.63	.68	.72	.75	.78	.80	.82	.84	.86	.40	.34	.29	.25	.20	.18	.17	.15	.14	.12	.10	.09	.08
		30	.17	.65	.77	.84	.88	.92	.94	.95	.97	.98	.98	.35	.23	.16	.11	.07	.05	.04	.03	.02	.01	.01	0	0
	16	150 & over	.58	.75	.79	.80	.80	.81	.82	.83	.84	.86	.87	.39	.35	.31	.28	.25								
		100	.46	.73	.78	.82	.82	.82	.83	.84	.85	.87	.88	.40	.34	.29	.25	.20								
		30	.22	.69	.80	.86	.89	.93	.94	.95	.97	.98	.98	.35	.23	.16	.11	.07								
	12	150 & over	.69	.86	.89	.90	.91	.91	.92	.93	.94	.95	.95	.50												
		100	.58	.85	.88	.88	.90	.92	.93	.94	.94	.94	.95	.48												
		30	.40	.81	.88	.91	.93	.96	.97	.97	.98	.99	.99	.35												
Fluorescent or Incandescent Lights Recessed in Susp. Ceiling and Ceiling Plenum Return System.	24	150 & over	.23	.33	.41	.47	.52	.57	.61	.66	.69	.72	.74	.59	.52	.46	.42	.37	.34	.31	.27	.25	.23	.21	.18	.16
		100	.17	.33	.44	.52	.56	.61	.66	.69	.74	.77	.79	.60	.51	.44	.37	.32	.30	.27	.23	.20	.18	.16	.14	.12
		30	0	.48	.66	.76	.82	.87	.91	.93	.95	.97	.98	.52	.34	.24	.16	.11	.07	.05	.04	.02	.02	.01	0	0
	16	150 & over	.57	.64	.68	.72	.73	.73	.74	.74	.75	.76	.78	.59	.52	.46	.42	.37								
		100	.47	.60	.67	.72	.74	.77	.78	.79	.80	.81	.82	.60	.51	.44	.37	.32								
		30	.07	.53	.70	.78	.84	.88	.91	.93	.95	.97	.98	.52	.34	.24	.16	.11								
	12	150 & over	.75	.79	.83	.84	.86	.88	.89	.91	.91	.93	.93	.75												
		100	.68	.77	.81	.84	.86	.88	.89	.89	.92	.93	.93	.72												
		30	.34	.72	.82	.87	.89	.92	.95	.95	.97	.98	.98	.52												

*These factors apply when maintaining a CONSTANT TEMPERATURE in the space during the operating period. Where the temperature is allowed to swing, additional storage will result during peak load periods. Refer to *Table 13* for applicable storage factors.

With lights operating the *same* number of hours as the time of equipment operation, use a load factor of 1.00.

†**Lights On for Shorter or Longer Period than 10 Hours**

Occasionally adjustments may be required to take account of lights operating less or more than the 10 hours on which the table is based. The following is the procedure to adjust the load factors:

A—WITH LIGHTS IN OPERATION FOR SHORTER PERIOD THAN 10 HOURS and the equipment operating 12, 16 or 24 hours at the time of the overall peak load, extrapolate load factors as follows:

1. Equipment operating for 24 hours:
   a. Use the storage load factors as listed up to the time the lights are turned off.
   b. Shift the load factors beyond the 10th hour (on the right of heavy line) to the left to the hour the lights are turned off. This leaves last few hours of equipment operation without designated load factors.
   c. Extrapolate the last few hours at the same rate of reduction as the end hours in the table.

2. Equipment operating for 16 hours:
   a. Follow the procedure in Step 1, using the storage load factor values in 24-hour equipment operation table.
   b. Now construct a new set of load factors by adding the new values for the 16th hour to that denoted 0, 17th hour to the 1st hour, etc.
   c. The load factors for the hours succeeding the switching-off the lights are as in Steps 1b and 1c.

3. Equipment operating for 12 hours:
   Follow procedure in Step 2, except in Step 2b add values of 12th hour to that designated 0, 13th hour to the 1st hour, etc.

B—WITH LIGHTS IN OPERATION FOR LONGER PERIOD THAN 10 HOURS and the equipment operating 12, 16 or 24 hours at the time of the overall peak load, extrapolate load factors as follows:

1. Equipment operating for 24 hours:
   a. Use the load factors as listed through 10th hour and extrapolate beyond the 10th hour at the rate of the last 4 hours.
   b. Follow the same procedure as in Step 1b of "A" except shift load factors beyond 10th hour now to the right, dropping off the last few hours.

2. Equipment operating for 16 hours or 12 hours:
   a. Use the load factors in 24-hour equipment operation table as listed through 10th hour and extrapolate beyond the 10th hour at the rate of the last 4 hours.
   b. Follow the procedure in Step 1b of "A" except shift the load factors beyond 10th hour now to the right.
   c. For 16-hour equipment operation, follow the procedure in Steps 2b and 2c of "A".
   d. For 12-hour equipment operation, follow the procedure in Step 3 of "A".

Example

Adjust values for 24-hour equipment operation and derive new values for 16-hour equipment operation for fluorescent lights in operation 8 and 13 hours, and an enclosure of 150 lb/sq ft of floor.

EQUIP OPERATION Hours	WEIGHT§ (lb per sq ft of floor area)	\multicolumn{24}{c}{NUMBER OF HOURS AFTER LIGHTS ARE TURNED ON}																							LIGHTS ON Hours
		0	1	2	3	4	5	6	7	8	9	10	11	12	13	14	15	16	17	18	19	20	21	22	23	
24	150	.37	.67	.71	.74	.76	.79	.81	.83	.84	.86	.87	.89	.90	.92	.29	.26	.23	.20	.19	.17	.15	.14	.12	.11	13
		.37	.67	.71	.74	.76	.79	.81	.83	.84	.29	.26	.23	.20	.19	.17	.15	.14	.12	.11	.10	.09	.08	.07	.06	8
		.37	.67	.71	.74	.76	.79	.81	.83	.84	.86	.87	.29	.26	.23	.20	.19	.17	.15	.14	.12	.11	.10	.09	.08	10
16	150	.60	.87	.90	.91	.91	.93	.93	.94	.94	.95	.95	.96	.96	.97	.29	.26									13
		.51	.79	.82	.84	.85	.87	.88	.89	.90	.29	.26	.23	.20	.19	.17	.15									8
		.60	.82	.83	.84	.84	.84	.85	.85	.86	.88	.90	.32	.28	.25	.23	.19									10

§Weight per sq ft of floor—

Room on Bldg Exterior (One or more outside walls) $= \dfrac{\text{(Weight of Outside Walls, lb)} + \tfrac{1}{2}\,\text{(Weight of Partitions, Floor and Ceiling, lb)}}{\text{Floor Area in Room, sq ft}}$

Room in Bldg Interior (No outside walls) $= \dfrac{\tfrac{1}{2}\,\text{(Weight of Partitions, Floor and Ceiling, lb)}}{\text{Floor Area in Room, sq ft}}$

Basement Room (Floor on ground) $= \dfrac{\text{(Weight of Outside Walls, lb)} + \text{(Weight of Floor, lb)} + \tfrac{1}{2}\,\text{(Weight of Partitions and Ceiling, lb)}}{\text{Floor Area in Room, sq ft}}$

Entire Building or Zone $= \dfrac{\text{(Weight of Outside Wall, Partitions, Floors, Ceilings, Structural Members and Supports, lb)}}{\text{Air Conditioned Floor Area, sq ft}}$

With rug on floor—Weight of floor should be multiplied by 0.50 to compensate for insulating effect of rug.

Weights per sq ft of common types of construction are contained in *Tables 21 thru 33*, pages 66 thru 76.

SPACE TEMPERATURE SWING

In addition to the storage of radiant heat with a constant room temperature, heat is stored in the building structure when the space temperature is forced to swing. If the cooling capacity supplied to the space matches the cooling load, the temperature in the space remains constant throughout the operating period. On the other hand, if the cooling capacity supplied to the space is lower than the actual cooling load at any point, the temperature in the space will rise. As the space temperature increases, less heat is convected from the surface and more radiant heat is stored in the structure. This process of storing additional heat is illustrated in *Fig. 11*.

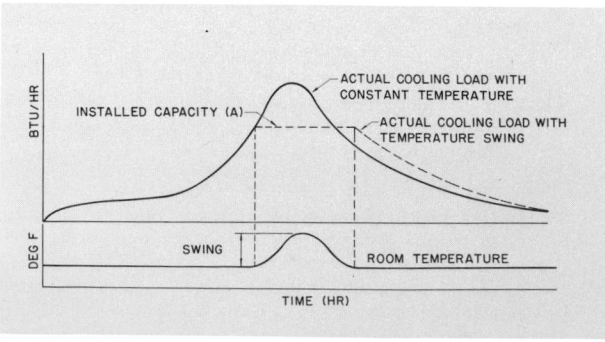

Fig. 11 — Actual Cooling Load With Varying Room Temperature

The solid curve is the actual cooling load from the solar heat gain on a west exposure with a constant space temperature, 24-hour operation. Assume that the maximum cooling capacity available is represented by *A,* and that the capacity is controlled to maintain a constant temperature at partial load. When the actual cooling load exceeds the available cooling capacity, the temperature will swing as shown in the lower curve. The actual cooling load with temperature swing is shown by the dotted line. This operates in a similar manner with different periods of operation and with different types of construction.

NOTE: When a system is designed for a temperature swing, the maximum swing occurs only at the peak on design days, which are defined as those days when all loads simultaneously peak. Under normal operating conditions, the temperature remains constant or close to constant.

Basis of Table 13
— Storage Factors,
Space Temperature Swing

The storage factors in *Table 13* were computed using essentially the same procedure as *Tables 7 thru 12* with the exception that the equipment capacity available was limited and the swing in room temperature computed.

The magnitude of the storage effect is determined largely by the thermal capacity or heat holding

TABLE 13—STORAGE FACTORS, SPACE TEMPERATURE SWING

Btu/(hr) (deg F swing) (sq ft of floor area)

NOTE: This reduction is to be taken at the time of peak load only.

TYPE APPLICATION — Load Pattern	Bldg Type	WEIGHT (lb/sq ft floor area)	GLASS RATIO‡ (%)	24 — 1-2	3-4	5-6	16 — 1-2	3-4	5-6	12 — 1-2	3-4	5-6
VARIABLE INTERMITTENT 24-HOUR PERIOD	Office Bldg Periphery, Except North Side	150 and Over	75	1.90	1.80	1.65	1.80	1.70	1.55	1.60	1.50	1.40
			50	1.70	1.60	1.45	1.60	1.50	1.35	1.50	1.35	1.25
			25	1.50	1.40	—	1.40	1.30	—	1.30	1.20	—
		100	75	1.70	1.60	1.45	1.50	1.45	1.35	1.40	1.35	1.30
			50	1.50	1.40	1.30	1.35	1.30	1.20	1.30	1.25	1.10
			25	1.35	1.25	1.20	1.25	1.00	.90	1.20	.95	.70
		30	75	1.40	1.25	1.00	1.20	1.10	.95	1.00	.95	.88
			50	1.20	.95	.80	1.10	.90	.80	.90	.85	.80
			25	.90	.80	.70	.85	.75	.60	.80	.70	.55
CONSTANT INTERMITTENT 24-HOUR PERIOD	Interior Zones† Department Stores, Factories	150 and Over	—	1.60	1.55	1.50	1.50	1.45	—	1.35	—	—
		100	—	1.40	1.38	1.36	1.30	1.28	1.25	1.25	1.20	—
		30	—	.95	.92	.90	.90	.88	.85	.85	.80	—
VARIABLE CONTINUOUS 24-HOUR PERIOD	Apartment Houses, Hotels, Hospitals Residences	150 and Over	75	1.85	1.75	1.40	—	—	—	—	—	—
			50	1.65	1.50	—	—	—	—	—	—	—
			25	1.45	—	—	—	—	—	—	—	—
		100	75	1.55	1.45	1.40	—	—	—	—	—	—
			50	1.40	1.35	—	—	—	—	—	—	—
			25	1.30	—	—	—	—	—	—	—	—
		30	75	1.20	1.10	.95	—	—	—	—	—	—
			50	1.10	.90	.80	—	—	—	—	—	—
			25	.85	.70	—	—	—	—	—	—	—

Equation: Reduction in Peak Cooling Load, Btu/hr = (Floor Area, sq ft) × (Desired Temp Swing, *Table 4, page 20*) × (Storage Factor, above table)

*Weight per sq ft of floor may be obtained from equation on *page 30.*

† For 12-hour operation, use a 2 degree max temp swing.

‡ Glass ratio is the percent of glass area to the total wall area.

capacity of the materials surrounding the space. It is limited by the amount of heat available for storage. Load patterns for different applications vary approximately as shown in the first column of *Table 13.* For instance, an office building has a rather large varying load with a high peak that occurs intermittently. An interior zone has an intermittent peak but the load pattern is relatively constant. A hospital, on the other hand, has a constant base load which is present for 24 hours with an additional intermittent load occurring during daylight hours. The thermal capacity of a material is the weight times the specific heat of the material. Since the specific heat of most construction material is approximately 0.20 Btu/(lb)(F), the thermal capacity is directly proportional to the weight of the material. Therefore, the data in the tables is based on weight of the materials surrounding the space, per square foot of floor area.

Use of Table 13
— Storage Factors,
  Space Temperature Swing

Table 13 is used to determine the reduction in cooling load when the space temperature is forced to swing by reducing the equipment capacity below that required to maintain the temperature constant. This reduction is to be subtracted from the room sensible heat.

NOTE: This reduction is only taken at the time of *peak cooling load.*

Example 3 — Space Temperature Swing

Given:

The same room as in *Example 1, page 28.*

Find:

The actual cooling load at 4 p.m. from sun, lights, and people with 3 F temperature swing in the space.

Solution:

The peak sensible cooling load in this room from the sun, lights, and people (neglecting transmission infiltration, ventilation and other internal heat gain) is

$$5700 + 5190 = 10,890 \text{ Btu/hr. } (Examples\ 1\ and\ 2.)$$

NOTE: The peak cooling load in this room occurs at approximately 4 p.m. The solar and light loads are almost at their peak at 4 p.m. Although the transmission across the large glass window peaks at about 3 p.m., the peak infiltration and ventilation load also occurs at 3 p.m. and the relatively small transmission load across the wall peaks much later at about 12 midnight. The sum of these loads results in the peak cooling load occurring at about 4 p.m. in the spaces with this exposure.

The weight of the materials surrounding the room in *Example 1* is 97.4 lb/sq ft of floor area.

Reduction in cooling load for a 3 F swing *(Table 13)*
= 20 × 20 × 1.4 × 3 = 1680 Btu/hr.
Cooling load = 10,890 − 1680 = 9210 Btu/hr.

> (For comparison purposes, the instantaneous heat gain from sun, lights, and people in this particular room is 14,610 Btu/hr.)

Since the normal thermostat setting is about 75 F or 76 F db, the design temperature (78 F = 75 F thermostat setting + 3 F swing) occurs only on design peak days at the time of peak load. Under partial load operation, the room temperature is between 75 F db and 78 F db, or at the thermostat setting (75 F), depending on the load.

PRECOOLING AS A MEANS OF INCREASING STORAGE

Precooling a space below the temperature normally desired *increases the storage of heat* at the time of peak load, only when the precooling temperature is maintained as the control point. This is because the potential temperature swing is increased, thus adding to the amount of heat stored at the time of peak load. Where the space is precooled to a lower temperature and the control point is reset upward to a comfortable condition when the occupants arrive, no additional storage occurs. In this situation, the cooling unit shuts off and there is no cooling during the period of warming up. When the cooling unit begins to supply cooling again, the cooling load is approximately up to the point it would have been without any precooling.

Precooling is very useful in reducing the cooling load in applications such as churches, supermarkets, theaters, etc., where the precooled temperature can be maintained as the control point and the temperature swing increased to 8 F or 10 F.

DIVERSITY OF COOLING LOADS

Diversity of cooling load results from the probable non-occurrence of part of the cooling load on a design day. Diversity factors are applied to the refrigeration capacity in large air conditioning systems. These factors vary with location, type and size of the application, and are based entirely on the judgment of the engineer.

Generally, diversity factors can be applied to people and light loads in large multi-story office, hotel or apartment buildings. The possibility of having all of the people present in the building and all of the lights operating at the time of peak load are slight. Normally, in large office buildings,

some people will be away from the office on other business. Also, the lighting arrangement will frequently be such that the lights in the vacant offices will not be on. In addition to lights being off because the people are not present, the normal maintenance procedure in large office buildings usually results in some lights being inoperative. Therefore, a diversity factor on the people and light loads should be applied for selecting the proper size refrigeration equipment.

The size of the diversity factor depends on the size of the building and the engineer's judgment of the circumstances involved. For example, the diversity factor on a single small office with 1 or 2 people is 1.0 or no reduction. Expanding this to one floor of a building with 50 to 100 people, 5% to 10% may be absent at the time of peak load, and expanding to a 20, 30 or 40-story building, 10% to 20% may be absent during the peak. A building with predominantly sales offices would have many people out in the normal course of business.

This same concept applies to apartments and hotels. Normally, very few people are present at the time the solar and transmission loads are peaking, and the lights are normally turned on only after sundown. Therefore, in apartments and hotels, the diversity factor can be much greater than with office buildings.

These reductions in cooling load are real and should be made where applicable. *Table 14* lists some typical diversity factors, based on judgment and experience.

TABLE 14—TYPICAL DIVERSITY FACTORS FOR LARGE BUILDINGS

(Apply to Refrigeration Capacity)

TYPE OF APPLICATION	DIVERSITY FACTOR	
	People	Lights
Office	.75 to .90	.70 to .85
Apartment, Hotel	.40 to .60	.30 to .50
Department Store	.80 to .90	.90 to 1.0
Industrial*	.85 to .95	.80 to .90

Equation:
Cooling Load (for people and lights), Btu/hr
= (Heat Gain, Btu/hr, Chapter 7)
× (Storage Factor, Table 12) × (Diversity Factor, above table)
*A diversity factor should also be applied to the machinery load. Refer to Chapter 7.

Use of Table 14
— Typical Diversity Factors for Large Buildings

The diversity factors listed in *Table 14* are to be used as a guide in determining a diversity factor for any particular application. The final factor must necessarily be based on judgment of the effect of the many variables involved.

STRATIFICATION OF HEAT

There are generally two situations where heat is stratified and will reduce the cooling load on the air conditioning equipment:

1. Heat may be stratified in rooms with high ceilings where air is exhausted through the roof or ceiling.
2. Heat may be contained above suspended ceilings with recessed lighting and/or ceiling plenum return systems.

The first situation generally applies to industrial applications, churches, auditoriums, and the like. The second situation applies to applications such as office buildings, hotels, and apartments. With both cases, the basic fact that hot air tends to rise makes it possible to stratify loads such as convection from the roof, convection from lights, and convection from the upper part of the walls. The convective portion of the roof load is about 25% (the rest is radiation); the light load is about 50% with fluorescent (20% with incandescent), and the wall transmission load about 40%.

In any room with a high ceiling, a large part of the convection load being released above the supply air stream will stratify at the ceiling or roof level. Some will be induced into the supply air stream. Normally, about 80% is stratified and 20% induced in the supply air. If air is exhausted through the ceiling or roof, this convection load released above the supply air may be subtracted from the air conditioning load. This results in a large reduction in load if the air is to be exhausted. It is not normally practical to exhaust more air than necessary, as it must be made up by bringing outdoor air through the apparatus. This usually results in a larger increase in load than the reduction realized by exhausting air.

Nominally, about a 10 F to 20 F rise in exhaust air temperature may be figured as load reduction if there is enough heat released by convection above the supply air stream.

Hot air stratifies at the ceiling even with no exhaust but rapidly builds up in temperature, and no reduction in load should be taken where air is not exhausted through the ceiling or roof.

With suspended ceilings, some of the convective heat from recessed lights flows into the plenum space. Also, the radiant heat within the room (sun, lights, people, etc.) striking the ceiling warms it up and causes heat to flow into the plenum space. These sources of heat increase the temperature of air in the plenum space which causes heat to flow into the underside of the floor structure above. When the ceiling plenum is used as a return air system, some of the return air flows through and over the light fixture, carrying more of the convective heat into the plenum space.

Containing heat within the ceiling plenum space tends to "flatten" both the room and equipment load. The storage factors for estimating the load with the above conditions are contained in *Table 12.*

CHAPTER 4. SOLAR HEAT GAIN THRU GLASS

SOLAR HEAT — DIRECT AND DIFFUSE

The solar heat on the outer edge of the earth's atmosphere is about 445 Btu/(hr)(sq ft) on December 21 when the sun is closest to the earth, and about 415 Btu/(hr)(sq ft) on June 21 when it is farthest away. The amount of solar heat outside the earth's atmosphere varies between these limits throughout the year.

The solar heat reaching the earth's surface is reduced considerably below these figures because a large part of it is scattered, reflected back out into space, and absorbed by the atmosphere. The scattered radiation is termed *diffuse* or *sky radiation*, and is more or less evenly distributed over the earth's surface because it is nothing more than a reflection from dust particles, water vapor and ozone in the atmosphere. The solar heat that comes directly through the atmosphere is termed *direct radiation*. The relationship between the total and the direct and diffuse radiation at any point on earth is dependent on the following two factors:

1. The distance traveled through the atmosphere to reach the point on the earth.
2. The amount of haze in the air.

As the distance traveled or the amount of haze increases, the diffuse radiation component increases but the direct component decreases. As either or both of these factors increase, the overall effect is to reduce the total quantity of heat reaching the earth's surface.

ORDINARY GLASS

Ordinary glass is specified as crystal glass of single thickness and single or double strength. The solar heat gain through ordinary glass depends on its location on the earth's surface (latitude), time of day, time of year, and facing direction of the window. The direct radiation component results in a heat gain to the conditioned space only when the window is in the direct rays of the sun, whereas the diffuse radiation component results in a heat gain, even when the window is not facing the sun.

Ordinary glass absorbs a small portion of the solar heat (5% to 6%) and reflects or transmits the rest. The amount reflected or transmitted depends on the angle of incidence. (The angle of incidence is the angle between the perpendicular to the window surface and the sun's rays, *Fig. 18, page 55.*) At low angles of incidence, about 86% or 87% is transmitted and 8% or 9% is reflected, as shown in *Fig. 12.* As the angle of incidence increases, more solar heat is reflected and less is transmitted, as shown in *Fig. 13.* The total solar heat gain to the conditioned space consists of the transmitted heat plus about 40% of the heat that is absorbed in the glass.

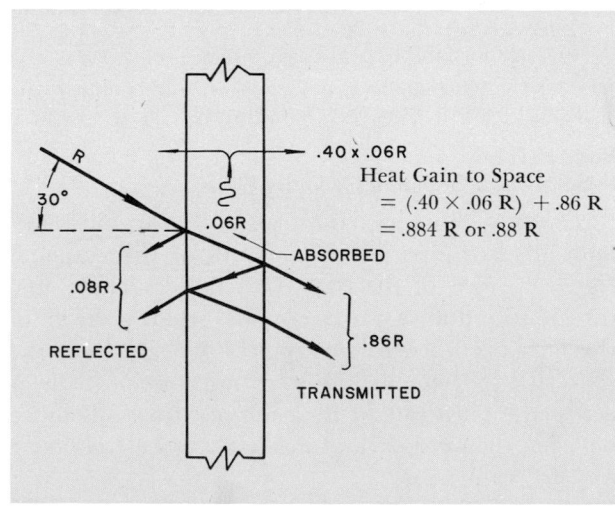

FIG. 12 — REACTION ON SOLAR HEAT (R), ORDINARY GLASS, 30° ANGLE OF INCIDENCE

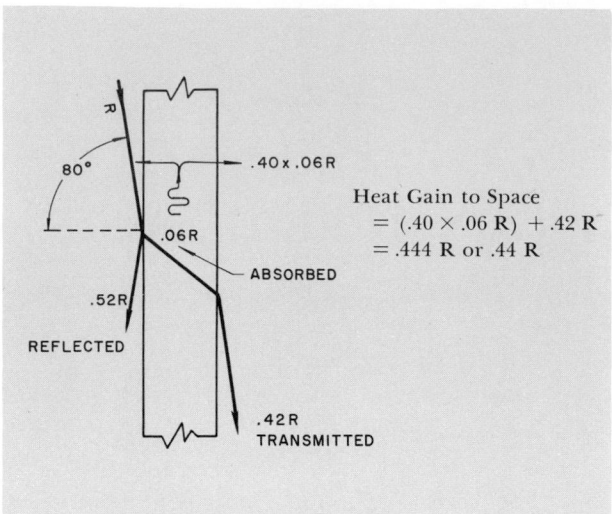

FIG. 13 — REACTION ON SOLAR HEAT (R), ORDINARY
GLASS, 80° ANGLE OF INCIDENCE

NOTE: The 40% of the absorbed solar heat going into the
space is derived from the following reasoning:

1. The outdoor film coefficient is approximately 2.8 Btu/(hr)
(sq ft) (deg F) with a 5 mph wind velocity during the
summer.

2. The inside film coefficient is approximately 1.8 Btu/(hr)
(sq ft) (deg F) because, in the average system design, air
velocities across the glass are approximately 100-200 fpm.

3. If outdoor temperature is equal to room temperature, the
glass temperature is above both. Therefore absorbed heat

flowing in $= \dfrac{1.8 \times 100}{1.8 + 2.8} = 39.2\%$, or 40%.

Absorbed heat flowing out $= \dfrac{2.8 \times 100}{1.8 + 2.8} = 60.8\%$, or 60%.

4. As the outdoor temperature rises, the glass temperature
also rises, causing more of the absorbed heat to flow into
the space. This can be accounted for by adding the transmission of heat across the glass (caused by temperature
difference between inside and outdoors) to the constant
40% of the absorbed heat going inside.

5. This reasoning applies equally well when the outdoor temperature is below the room temperature.

Basis of Table 15
— Solar Heat Gain thru Ordinary Glass

Table 15 provides data for 0°, 10°, 20°, 30°, 40°,
and 50° latitudes, for each month of the year and
for each hour of the day. This table includes the
direct and diffuse radiation and that portion of
the heat absorbed in the glass which gets into the
space. It *does not* include the transmission of heat
across the glass caused by a temperature difference
between the outdoor and inside air. (See *Chapter 5*
for "U" values.)

The data in *Table 15* is based on the following
conditions:—

1. A glass area equal to 85% of the sash area.

This is typical for wood sash windows. For
metal sash windows, the glass area is assumed
equal to 100% of the sash area because the
conductivity of the metal sash is very high
and the solar heat absorbed in the sash is
transmitted almost instantaneously.

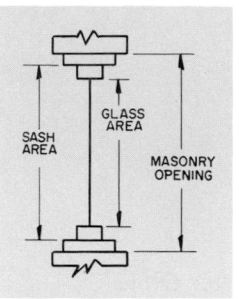

FIG. 14
WINDOW AREAS

NOTE: The sash area equals
approximately 85% of the
masonry opening (or frame
opening with frame walls)
with wood sash windows,
90% of masonry opening
with double hung metal sash
windows, and 100% of ma-
sonry opening with casement
windows.

2. No haze in the air.

3. Sea level elevation.

4. A sea level dewpoint temperature of 66.8 F
(95 F db, 75 F wb) which approximately corresponds to 4 centimeters of precipitable water
vapor. Precipitable water vapor is all of the
water vapor in a column of air from sea level
to the outer edge of the atmosphere.

If these conditions do not apply, use the correction factors at the bottom of each page of *Table 15*.

Use of Table 15
— Solar Heat Gain thru Ordinary Glass

*The bold face values in Table 15 indicate the
maximum solar heat gain for the month for each
exposure. The bold face values that are boxed indicate the yearly maximums for each exposure.*

Table 15 is used to determine the solar heat gain
thru ordinary glass at any time, in any space, zone
or building.

To determine the actual cooling load *due to the
solar heat gain*, refer to *Chapter 3*, *"Heat Storage,
Diversity and Stratification."*

CAUTION — *Where Estimating Multi-Exposure Rooms
or Buildings*

If a haze factor is used on one exposure to determine the peak room or building load, the diffuse
component listed for the other exposures must be
divided by the haze factor to result in the actual
room or building peak load. This is because the
diffuse component increases with increasing haze, as
explained on *page 41*.

Example 1 — Peak Solar Heat Gain (2 Exposures)

Since the time at which the peak solar load occurs in a space with 2 exposures is not always apparent, the solar heat gain is generally calculated at more than one time to determine its peak.

Given:

A room with equal glass areas on the West and South at 40° North latitude.

Find:

Peak solar heat gain.

Solution:

From *Table 15*

Solar heat gain —

September 22	2:00	3:00	4:00 p.m.
West	99	139	149
South	110	81	44
Total	209	220	193

Solar heat gain —

October 23	2:00	3:00	4:00 p.m.
West	88	122	117
South	137	104	59
Total	225	226	176

Solar heat gain —

November 21	2:00	3:00	4:00 p.m.
West	74	100	91
South	139	104	59
Total	213	204	150

The peak solar heat gain to this room occurs at 3:00 p.m. on October 23. The peak room cooling load does not necessarily occur at the same time as the peak solar heat gain, because the peak transmission load, people load, etc., may occur at some other time.

Example 2 — Solar Gain Correction Factors (Bottom Table 15)

The conditions on which *Table 15* is based do not apply to all locations, since many cities are above sea level, and many have different design dew points and some haze in their atmosphere.

Given:

A west exposure with steel casement windows
Location — Topeka, Kansas
Altitude — 991 ft
Design dewpoint — 69.8 F
39° North latitude

Find:

Peak solar heat gain

Solution:

By inspection of *Table 15* the boxed boldface values for peak solar heat gain, occurring at 4:00 p.m. on July 23
= 164 Btu/(hr)(sq ft)

Assume a somewhat hazy condition.
Altitude correction = 1.007 (bottom *Table 15*)
Dewpoint difference = 69.8 − 66.8 = 3 F
Dewpoint correction = 1 − (3/10 × .07) = .979
(bottom *Table 15*)
Haze correction = 1 − .10 = .90 (bottom *Table 15*)
Steel sash correction = 1/.85 (bottom *Table 15*)

Solar heat gain at 4:00 p.m., July 23
= 164 × 1.007 × .979 × .90 × 1/.85
= 171 Btu/(hr)(sq ft)

TABLE 15—SOLAR HEAT GAIN THRU ORDINARY GLASS

Btu/(hr) (sq ft sash area)

0°   **0°**

0° NORTH LATITUDE		AM						SUN TIME				PM			0° SOUTH LATITUDE	
Time of Year	Exposure	6	7	8	9	10	11	Noon	1	2	3	4	5	6	Exposure	Time of Year
JUNE 21	North	0	45	65	74	78	80	82	80	78	74	65	45	0	South	DEC 22
	Northeast	0	119	156	154	133	95	53	20	14	13	11	6	0	Southeast	
	East	0	116	147	135	93	43	14	14	14	13	11	6	0	East	
	Southeast	0	37	42	27	15	14	14	14	14	13	11	6	0	Northeast	
	South	0	6	11	13	14	14	14	14	14	13	11	6	0	North	
	Southwest	0	6	11	13	14	14	14	14	15	27	42	37	0	Northwest	
	West	0	6	11	13	14	14	14	43	93	135	147	116	0	West	
	Northwest	0	6	11	13	14	20	53	95	133	154	156	119	0	Southwest	
	Horizontal	0	28	87	147	191	217	226	217	191	147	87	28	0	Horizontal	
JULY 23 & MAY 21	North	0	37	54	61	65	66	67	66	65	61	54	37	0	South	JAN 21 & NOV 21
	Northeast	0	118	153	150	124	86	43	16	14	13	11	6	0	Southeast	
	East	0	121	152	139	96	43	14	14	14	13	11	6	0	East	
	Southeast	0	46	52	36	18	14	14	14	14	13	11	6	0	Northeast	
	South	0	6	11	13	14	14	14	14	14	13	11	6	0	North	
	Southwest	0	6	11	13	14	14	14	14	18	36	52	46	0	Northwest	
	West	0	6	11	13	14	14	14	43	96	139	152	121	0	West	
	Northwest	0	6	11	13	14	16	43	86	124	150	153	118	0	Southwest	
	Horizontal	0	29	91	151	195	223	233	223	195	151	91	29	0	Horizontal	
AUG 24 & APR 20	North	0	17	28	31	33	34	34	34	33	31	28	17	0	South	FEB 20 & OCT 23
	Northeast	0	110	141	133	102	61	24	14	14	13	12	6	0	Southeast	
	East	0	129	163	148	103	46	14	14	14	13	12	6	0	East	
	Southeast	0	67	79	65	35	15	14	14	14	13	12	6	0	Northeast	
	South	0	6	12	13	14	14	14	14	14	13	12	6	0	North	
	Southwest	0	6	12	13	14	14	14	15	35	65	79	67	0	Northwest	
	West	0	6	12	13	14	14	14	46	103	148	163	129	0	West	
	Northwest	0	6	12	13	14	14	24	61	102	133	141	110	0	Southwest	
	Horizontal	0	31	97	150	206	234	245	234	206	150	97	31	0	Horizontal	
SEPT 22 & MAR 22	North	0	6	12	13	14	14	14	14	14	13	12	6	0	South	MAR 22 & SEPT 22
	Northeast	0	95	118	101	68	31	14	14	14	13	12	6	0	Southeast	
	East	0	134	167	151	107	47	14	14	14	13	12	6	0	East	
	Southeast	0	95	118	101	68	31	14	14	14	13	12	6	0	Northeast	
	South	0	6	12	13	14	14	14	14	14	13	12	6	0	North	
	Southwest	0	6	12	13	14	14	14	31	68	101	118	95	0	Northwest	
	West	0	6	12	13	14	14	14	47	107	151	167	134	0	West	
	Northwest	0	6	12	13	14	14	14	31	68	101	118	95	0	Southwest	
	Horizontal	0	32	100	163	210	240	250	240	210	163	100	32	0	Horizontal	
OCT 23 & FEB 20	North	0	6	12	13	14	14	14	14	14	13	12	6	0	South	APR 20 & AUG 24
	Northeast	0	67	79	65	35	15	14	14	14	13	12	6	0	Southeast	
	East	0	129	163	148	103	46	14	14	14	13	12	6	0	East	
	Southeast	0	110	141	133	102	61	24	14	14	13	12	6	0	Northeast	
	South	0	17	28	31	33	34	34	34	33	31	28	17	0	North	
	Southwest	0	6	12	13	14	14	24	61	102	133	141	110	0	Northwest	
	West	0	6	12	13	14	14	14	46	103	148	163	129	0	West	
	Northwest	0	6	12	13	14	14	14	15	35	65	79	67	0	Southwest	
	Horizontal	0	31	97	150	206	234	245	234	206	150	97	31	0	Horizontal	
NOV 21 & JAN 21	North	0	6	11	13	14	14	14	14	14	13	11	6	0	South	MAY 21 & JULY 23
	Northeast	0	46	52	36	18	14	14	14	14	13	11	6	0	Southeast	
	East	0	121	152	139	96	43	14	14	14	13	11	6	0	East	
	Southeast	0	118	153	150	124	86	43	16	14	13	11	6	0	Northeast	
	South	0	37	54	61	65	66	67	66	65	61	54	37	0	North	
	Southwest	0	6	11	13	14	16	43	86	124	150	153	118	0	Northwest	
	West	0	6	11	13	14	14	14	43	96	139	152	121	0	West	
	Northwest	0	6	11	13	14	14	14	14	18	36	52	46	0	Southwest	
	Horizontal	0	29	91	151	195	223	233	223	195	151	91	29	0	Horizontal	
DEC 22	North	0	6	11	13	14	14	14	14	14	13	11	6	0	South	JUNE 21
	Northeast	0	37	42	27	15	14	14	14	14	13	11	6	0	Southeast	
	East	0	116	147	135	93	43	14	14	14	13	11	6	0	East	
	Southeast	0	119	156	154	133	95	53	20	14	13	11	6	0	Northeast	
	South	0	45	65	74	78	80	82	80	78	74	65	45	0	North	
	Southwest	0	6	11	13	14	20	53	95	133	154	156	119	0	Northwest	
	West	0	6	11	13	14	14	14	43	93	135	147	116	0	West	
	Northwest	0	6	11	13	14	14	14	14	15	27	42	37	0	Southwest	
	Horizontal	0	28	87	147	191	217	226	217	191	147	87	28	0	Horizontal	

Solar Gain Correction	Steel Sash, or No Sash × 1/.85 or 1.17	Haze −15% (Max.)	Altitude +0.7% per 1000 Ft	Dewpoint Decrease From 67 F +7% per 10 F	Dewpoint Increase From 67 F −7% per 10 F	South Lat. Dec. or Jan. +7%

Bold Face Values — Monthly Maximums   Boxed Values — Yearly maximums

TABLE 15—SOLAR HEAT GAIN THRU ORDINARY GLASS (Contd)

Btu/(hr) (sq ft sash area)

10° 10°

10° NORTH LATITUDE		AM				SUN TIME			PM					10° SOUTH LATITUDE	
Time of Year	Exposure	6	7	8	9	10	11	Noon	1	2	3	4	5	6	Exposure	Time of Year
JUNE 21	North	19	44	50	45	44	43	41	43	44	45	50	44	2	South	DEC 22
	Northeast	55	131	153	140	106	65	28	14	14	13	11	8	2	Southeast	
	East	54	134	155	139	98	41	14	14	14	13	11	8	2	East	
	Southeast	18	49	55	43	25	14	14	14	14	13	11	8	2	Northeast	
	South	2	8	11	13	14	14	14	14	14	13	11	8	2	North	
	Southwest	2	8	8	13	14	14	14	14	25	43	55	49	18	Northwest	
	West	2	8	8	13	14	14	14	41	98	139	155	134	54	West	
	Northwest	2	8	8	13	14	18	28	65	106	140	153	131	55	Southwest	
	Horizontal	4	44	107	166	205	233	243	233	205	166	107	44	4	Horizontal	
JULY 23 & MAY 21	North	5	34	39	35	33	31	30	31	33	35	39	34	5	South	JAN 21 & NOV 21
	Northeast	42	127	148	133	109	56	22	14	14	13	11	7	1	Southeast	
	East	50	135	158	142	98	43	14	14	14	13	11	7	1	East	
	Southeast	26	57	66	56	32	14	14	14	14	13	11	7	1	Northeast	
	South	1	7	11	13	14	14	14	14	14	13	11	7	1	North	
	Southwest	1	7	11	13	14	14	14	14	32	56	66	57	26	Northwest	
	West	1	7	11	13	14	14	14	43	98	142	158	135	50	West	
	Northwest	1	7	11	13	14	14	22	56	109	133	148	127	42	Southwest	
	Horizontal	3	42	107	166	210	236	247	236	210	166	107	42	3	Horizontal	
AUG 24 & APR 20	North	1	15	16	15	15	14	14	14	15	15	16	15	1	South	FEB 20 & OCT 23
	Northeast	17	113	130	111	80	34	14	14	14	13	11	7	1	Southeast	
	East	25	138	163	149	104	46	14	14	14	13	11	7	1	East	
	Southeast	18	79	94	85	60	27	14	14	14	13	11	7	1	Northeast	
	South	1	7	11	13	14	14	14	14	14	13	11	7	1	North	
	Southwest	1	7	11	13	14	14	14	27	60	85	94	79	18	Northwest	
	West	1	7	11	13	14	14	14	46	80	149	163	138	25	West	
	Northwest	1	7	11	13	14	14	14	34	65	111	130	113	17	Southwest	
	Horizontal	2	38	105	167	213	242	250	242	213	167	105	38	2	Horizontal	
SEPT 22 & MAR 22	North	1	6	11	13	14	14	14	14	14	13	11	6	1	South	MAR 22 & SEPT 22
	Northeast	1	89	103	80	45	17	14	14	14	13	11	6	1	Southeast	
	East	1	130	164	151	106	47	14	14	14	13	11	6	1	East	
	Southeast	1	97	127	122	94	56	21	14	14	13	11	6	1	Northeast	
	South	1	6	13	19	24	27	28	27	24	19	13	6	1	North	
	Southwest	1	6	11	13	14	14	21	56	94	122	127	97	1	Northwest	
	West	1	6	11	13	14	14	14	47	106	151	164	130	1	West	
	Northwest	1	6	11	13	14	14	14	17	45	80	103	89	1	Southwest	
	Horizontal	1	31	97	160	207	235	247	235	207	160	97	31	1	Horizontal	
OCT 23 & FEB 20	North	0	5	10	13	14	14	14	14	14	13	10	5	0	South	APR 20 & AUG 24
	Northeast	0	58	66	44	28	14	14	14	14	13	10	5	0	Southeast	
	East	0	118	155	145	100	40	14	14	14	13	10	5	0	East	
	Southeast	0	103	147	149	123	81	46	18	14	13	10	5	0	Northeast	
	South	0	18	40	55	65	71	73	71	65	55	40	18	0	North	
	Southwest	0	5	10	13	14	18	46	81	123	149	147	103	0	Northwest	
	West	0	5	10	13	14	14	14	40	100	145	155	118	0	West	
	Northwest	0	5	10	13	14	14	14	14	28	44	66	58	0	Southwest	
	Horizontal	0	22	85	139	193	220	230	220	193	139	85	22	0	Horizontal	
NOV 21 & JAN 21	North	0	4	9	12	13	14	14	14	13	12	9	4	0	South	MAY 21 & JULY 23
	Northeast	0	27	37	17	13	14	14	14	13	12	9	4	0	Southeast	
	East	0	99	143	132	93	39	14	14	13	12	9	4	0	East	
	Southeast	0	99	153	161	146	109	70	31	17	12	9	4	0	Northeast	
	South	0	35	65	91	96	104	106	104	96	91	65	35	0	North	
	Southwest	0	4	9	12	17	31	70	109	146	161	153	99	0	Northwest	
	West	0	4	9	12	13	14	14	39	93	132	143	99	0	West	
	Northwest	0	4	9	12	13	14	14	14	13	17	37	27	0	Southwest	
	Horizontal	0	17	62	131	175	202	210	202	175	131	62	17	0	Horizontal	
DEC 22	North	0	4	9	12	13	14	14	14	13	12	9	4	0	South	JUNE 21
	Northeast	0	15	28	17	13	14	14	14	13	12	9	4	0	Southeast	
	East	0	86	137	130	91	42	14	14	13	12	9	4	0	East	
	Southeast	0	99	154	163	149	121	79	36	23	12	9	4	0	Northeast	
	South	0	50	74	94	109	116	120	116	109	94	74	50	0	North	
	Southwest	0	4	9	12	23	36	79	121	149	163	154	99	0	Northwest	
	West	0	4	9	12	13	14	14	42	91	130	137	86	0	West	
	Northwest	0	4	9	12	13	14	14	14	13	17	28	15	0	Southwest	
	Horizontal	0	14	66	120	167	193	202	193	167	120	66	14	0	Horizontal	

Solar Gain Correction	Steel Sash, or No Sash × 1/.85 or 1.17	Haze −15% (Max.)	Altitude +0.7% per 1000 Ft	Dewpoint Decrease From 67 F +7% per 10 F	Dewpoint Increase From 67 F −7% per 10 F	South Lat. Dec. or Jan. +7%

Bold Face Values — Monthly Maximums   Boxed Values — Yearly maximums

TABLE 15—SOLAR HEAT GAIN THRU ORDINARY GLASS (Contd)

20°   Btu/(hr) (sq ft sash area)   **20°**

20° NORTH LAT — Time of Year	Exposure	6	7	8	9	10	11	Noon	1	2	3	4	5	6	Exposure	20° SOUTH LAT — Time of Year
JUNE 21	North	28	41	33	25	19	17	15	17	19	25	33	41	28	South	DEC 22
	Northeast	81	154	144	122	83	38	15	14	14	14	12	9	3	Southeast	
	East	81	148	160	143	96	41	14	14	14	14	12	9	3	East	
	Southeast	28	62	73	66	44	21	14	14	14	14	12	9	3	Northeast	
	South	3	9	12	14	14	14	14	14	14	14	12	9	3	North	
	Southwest	3	9	12	14	14	14	14	21	44	66	73	62	28	Northwest	
	West	3	9	12	14	14	14	14	41	96	143	160	148	81	West	
	Northwest	3	9	12	14	14	14	15	38	83	122	144	154	81	Southwest	
	Horizontal	11	60	121	176	216	232	250	232	216	176	121	60	11	Horizontal	
JULY 23 & MAY 21	North	20	28	23	17	15	14	14	14	15	17	23	28	20	South	JAN 21 & NOV 21
	Northeast	71	132	138	111	73	31	14	14	14	13	12	8	3	Southeast	
	East	75	148	163	145	99	46	14	14	14	13	12	8	3	East	
	Southeast	31	70	85	79	57	29	14	14	14	13	12	8	3	Northeast	
	South	3	8	12	13	14	14	14	14	14	13	12	8	3	North	
	Southwest	3	8	12	13	14	14	14	29	57	79	85	70	31	Northwest	
	West	3	8	12	13	14	14	14	46	99	145	163	148	75	West	
	Northwest	3	8	12	13	14	14	14	31	73	111	138	132	71	Southwest	
	Horizontal	8	55	118	175	216	240	251	240	216	175	118	55	8	Horizontal	
AUG 24 & APR 20	North	6	10	11	13	14	14	14	14	14	13	11	10	6	South	FEB 20 & OCT 23
	Northeast	45	111	118	89	50	18	14	14	14	13	11	7	2	Southeast	
	East	53	142	165	149	106	51	14	14	14	13	11	7	2	East	
	Southeast	29	89	113	108	98	55	20	14	14	13	11	7	2	Northeast	
	South	2	7	11	14	20	24	26	24	20	14	11	7	2	North	
	Southwest	2	7	11	13	14	14	20	55	98	108	113	89	29	Northwest	
	West	2	7	11	13	14	14	14	51	106	149	165	142	53	West	
	Northwest	2	7	11	13	14	14	14	18	50	89	118	111	45	Southwest	
	Horizontal	5	48	107	167	210	235	247	235	210	167	107	48	5	Horizontal	
SEPT 22 & MAR 22	North	0	6	11	13	14	14	14	14	14	13	11	6	0	South	MAR 22 & SEPT 22
	Northeast	0	83	87	59	22	14	14	14	14	13	11	6	0	Southeast	
	East	0	130	163	149	104	45	14	14	14	13	11	6	0	East	
	Southeast	0	99	136	140	120	84	41	15	14	13	11	6	0	Northeast	
	South	0	8	22	38	52	63	65	63	52	38	22	8	0	North	
	Southwest	0	6	11	13	14	15	41	84	120	140	136	99	0	Northwest	
	West	0	6	11	13	14	14	14	45	104	149	163	130	0	West	
	Northwest	0	6	11	13	14	14	14	14	22	59	87	83	0	Southwest	
	Horizontal	0	30	93	153	198	225	233	225	198	153	93	30	0	Horizontal	
OCT 23 & FEB 20	North	0	4	9	12	13	14	14	14	13	12	9	4	0	South	APR 20 & AUG 24
	Northeast	0	44	52	29	13	14	14	14	13	12	9	4	0	Southeast	
	East	0	99	147	141	100	49	14	14	13	12	9	4	0	East	
	Southeast	0	91	146	160	149	119	74	27	13	12	9	4	0	Northeast	
	South	0	21	50	76	93	106	111	106	93	76	50	21	0	North	
	Southwest	0	4	9	12	13	27	74	119	149	160	146	91	0	Northwest	
	West	0	4	9	12	13	14	14	49	100	141	147	99	0	West	
	Northwest	0	4	9	12	13	14	14	14	13	29	52	44	0	Southwest	
	Horizontal	0	18	68	127	171	196	208	196	171	127	68	18	0	Horizontal	
NOV 21 & JAN 21	North	0	3	8	11	13	13	13	13	13	11	8	3	0	South	MAY 21 & JULY 23
	Northeast	0	24	26	14	13	13	13	13	13	11	8	3	0	Southeast	
	East	0	71	128	127	91	43	13	13	13	11	8	3	0	East	
	Southeast	0	73	144	164	158	135	91	46	16	11	8	3	0	Northeast	
	South	0	28	69	100	123	136	141	136	123	100	69	28	0	North	
	Southwest	0	3	8	11	16	46	91	135	158	164	144	73	0	Northwest	
	West	0	3	8	11	12	13	13	43	91	127	128	71	0	West	
	Northwest	0	3	8	11	12	13	13	13	14	26	24	0	0	Southwest	
	Horizontal	0	5	48	101	146	172	180	172	146	101	48	5	0	Horizontal	
DEC 22	North	0	2	7	11	12	13	13	13	12	11	7	2	0	South	JUNE 21
	Northeast	0	14	18	12	12	13	13	13	12	11	7	2	0	Southeast	
	East	0	56	118	121	85	34	13	13	12	11	7	2	0	East	
	Southeast	0	59	139	167	159	134	97	60	20	11	7	2	0	Northeast	
	South	0	25	74	111	132	146	149	146	132	111	74	25	0	North	
	Southwest	0	2	7	11	20	60	97	134	159	167	139	59	0	Northwest	
	West	0	2	7	11	12	13	13	34	85	121	118	56	0	West	
	Northwest	0	2	7	11	12	13	13	13	12	12	18	14	0	Southwest	
	Horizontal	0	4	36	92	135	161	170	161	135	92	36	4	0	Horizontal	

Solar Gain Correction	Steel Sash, or No Sash × 1/.85 or 1.17	Haze −15% (Max.)	Altitude +0.7% per 1000 Ft	Dewpoint Decrease From 67 F +7% per 10 F	Dewpoint Increase From 67 F −7% per 10 F	South Lat. Dec. or Jan. +7%

Bold Face Values — Monthly Maximums   Boxed Values — Yearly maximums

TABLE 15—SOLAR HEAT GAIN THRU ORDINARY GLASS (Contd)

30° Btu/(hr) (sq ft sash area) **30°**

Time of Year (N)	Exposure	6	7	8	9	10	11	Noon	1	2	3	4	5	6	Exposure (S)	Time of Year (S)
JUNE 21	North	33	29	18	14	14	14	14	14	14	14	18	29	33	South	DEC 22
	Northeast	105	139	130	97	55	19	14	14	14	14	12	10	5	Southeast	
	East	108	156	161	143	98	44	14	14	14	14	12	10	5	East	
	Southeast	42	75	90	90	73	44	17	14	14	14	12	10	5	Northeast	
	South	5	10	12	14	15	19	21	19	15	14	12	10	5	North	
	Southwest	5	10	12	14	14	14	17	44	73	90	90	75	42	Northwest	
	West	5	10	12	14	14	14	14	44	98	143	161	156	108	West	
	Northwest	5	10	12	14	14	14	14	19	55	97	130	139	105	Southwest	
	Horizontal	19	61	131	180	217	240	250	240	217	180	131	61	19	Horizontal	
JULY 23 & MAY 21	North	22	20	14	13	14	14	14	14	14	13	14	20	22	South	JAN 21 & NOV 21
	Northeast	93	131	123	89	46	16	14	14	14	13	12	9	4	Southeast	
	East	100	155	164	145	99	44	14	14	14	13	12	9	4	East	
	Southeast	42	82	100	100	83	53	22	14	14	13	12	9	4	Northeast	
	South	4	9	12	14	20	27	30	27	20	14	12	9	4	North	
	Southwest	4	9	12	13	14	14	14	53	83	100	100	82	42	Northwest	
	West	4	9	12	13	14	14	14	44	99	145	164	155	100	West	
	Northwest	4	9	12	13	14	14	14	16	46	89	123	131	93	Southwest	
	Horizontal	15	66	123	176	214	236	246	236	214	176	123	66	15	Horizontal	
AUG 24 & APR 20	North	6	8	11	13	13	14	14	14	13	13	11	8	6	South	FEB 20 & OCT 23
	Northeast	55	108	100	66	27	14	14	14	14	13	11	8	2	Southeast	
	East	66	147	165	148	102	46	14	14	13	13	11	8	2	East	
	Southeast	37	98	127	129	112	82	39	15	13	13	11	8	2	Northeast	
	South	2	8	13	27	47	58	63	58	47	27	13	8	2	North	
	Southwest	2	8	11	13	13	15	39	82	112	129	127	98	37	Northwest	
	West	2	8	11	13	13	14	14	46	102	148	165	147	66	West	
	Northwest	2	8	11	13	13	14	14	14	27	66	100	108	55	Southwest	
	Horizontal	6	47	107	161	200	225	235	225	200	161	107	47	6	Horizontal	
SEPT 22 & MAR 22	North	0	5	10	12	13	14	14	14	13	12	10	5	0	South	MAR 22 & SEPT 22
	Northeast	0	74	90	40	15	14	14	14	13	12	10	5	0	Southeast	
	East	0	124	158	144	103	48	14	14	13	12	10	5	0	East	
	Southeast	0	98	131	152	141	113	67	25	13	12	10	5	0	Northeast	
	South	0	9	18	60	82	98	105	98	82	60	18	9	0	North	
	Southwest	0	5	10	12	13	25	67	113	141	152	131	98	0	Northwest	
	West	0	5	10	12	13	14	14	48	103	144	158	124	0	West	
	Northwest	0	5	10	12	13	14	14	14	15	40	90	74	0	Southwest	
	Horizontal	0	25	81	135	179	202	212	202	179	135	81	25	0	Horizontal	
OCT 23 & FEB 20	North	0	3	11	12	13	14	14	13	12	11	8	3	0	South	APR 20 & AUG 24
	Northeast	0	33	39	18	12	13	14	13	12	11	8	3	0	Southeast	
	East	0	79	135	132	94	43	14	13	12	11	8	3	0	East	
	Southeast	0	73	142	163	159	136	92	47	15	11	8	3	0	Northeast	
	South	0	18	57	92	121	139	145	139	121	92	57	18	0	North	
	Southwest	0	3	8	11	15	47	92	136	159	163	142	73	0	Northwest	
	West	0	3	8	11	12	13	14	43	94	132	135	79	0	West	
	Northwest	0	3	8	11	12	13	14	13	12	18	39	33	0	Southwest	
	Horizontal	0	6	49	100	143	171	179	171	143	100	49	6	0	Horizontal	
NOV 21 & JAN 21	North	0	6	9	11	12	12	12	12	11	9	6	1	0	South	MAY 21 & JULY 23
	Northeast	0	8	16	9	11	12	12	12	11	9	6	1	0	Southeast	
	East	0	27	109	116	83	35	12	12	11	9	6	1	0	East	
	Southeast	0	28	127	161	162	143	104	64	23	9	6	1	0	Northeast	
	South	0	10	68	109	137	154	159	154	137	109	68	10	0	North	
	Southwest	0	1	6	9	23	64	104	143	162	161	127	28	0	Northwest	
	West	0	1	6	9	11	12	12	35	83	116	109	27	0	West	
	Northwest	0	1	6	9	11	12	12	12	11	9	16	8	0	Southwest	
	Horizontal	0	2	27	71	109	136	145	136	109	71	27	2	0	Horizontal	
DEC 22	North	0	0	4	9	11	12	12	12	11	9	4	0	0	South	JUNE 21
	Northeast	0	0	10	9	11	12	12	12	11	9	4	0	0	Southeast	
	East	0	0	92	105	80	32	12	12	11	9	4	0	0	East	
	Southeast	0	0	114	157	162	143	108	72	28	9	4	0	0	Northeast	
	South	0	0	64	113	142	159	163	159	142	113	64	0	0	North	
	Southwest	0	0	4	9	28	72	108	143	162	157	114	0	0	Northwest	
	West	0	0	4	9	11	12	12	32	80	105	92	0	0	West	
	Northwest	0	0	4	9	11	12	12	12	11	9	10	0	0	Southwest	
	Horizontal	0	0	19	60	97	122	131	122	97	60	19	0	0	Horizontal	

Solar Gain Correction	Steel Sash, or No Sash × 1/.85 or 1.17	Haze −15% (Max.)	Altitude +0.7% per 1000 Ft	Dewpoint Decrease From 67 F + 7% per 10 F	Dewpoint Increase From 67 F − 7% per 10 F	South Lat. Dec. or Jan. + 7%

Bold Face Values — Monthly Maximums   Boxed Values — Yearly maximums

TABLE 15—SOLAR HEAT GAIN THRU ORDINARY GLASS (Contd)

40°   Btu/(hr) (sq ft sash area)   **40°**

40° NORTH LATITUDE		AM						SUN TIME				PM			40° SOUTH LATITUDE	
Time of Year	Exposure	6	7	8	9	10	11	Noon	1	2	3	4	5	6	Exposure	Time of Year
JUNE 21	North	32	20	12	13	14	14	14	14	14	13	12	20	32	South	DEC 22
	Northeast	118	133	112	73	30	14	14	14	14	13	12	10	6	Southeast	
	East	126	161	162	142	95	44	14	14	14	13	12	10	6	East	
	Southeast	51	88	109	111	99	71	34	14	14	13	12	10	6	Northeast	
	South	6	10	12	19	35	44	54	44	35	19	12	10	6	North	
	Southwest	6	10	12	13	14	14	34	71	99	111	109	88	51	Northwest	
	West	6	10	12	13	14	14	14	44	95	142	162	161	126	West	
	Northwest	6	10	12	13	14	14	14	14	30	73	112	133	118	Southwest	
	Horizontal	3h	82	134	179	210	232	237	232	210	179	134	82	31	Horizontal	
JULY 23 & MAY 21	North	24	14	12	13	14	14	14	14	14	13	12	14	24	South	JAN 21 & NOV 21
	Northeast	106	127	105	66	26	14	14	14	14	13	12	10	5	Southeast	
	East	118	161	164	144	98	43	14	14	14	13	12	10	5	East	
	Southeast	54	96	119	125	110	82	42	15	14	13	12	10	5	Northeast	
	South	5	10	13	26	44	63	69	63	44	26	13	10	5	North	
	Southwest	5	10	12	13	14	15	42	82	110	125	119	96	54	Northwest	
	West	5	10	12	13	14	14	14	43	98	144	164	161	118	West	
	Northwest	5	10	12	13	14	14	14	14	26	66	105	127	106	Southwest	
	Horizontal	24	73	126	171	203	225	233	225	203	171	126	73	24	Horizontal	
AUG 24 & APR 20	North	7	8	11	13	14	14	14	14	14	13	11	8	7	South	FEB 20 & OCT 23
	Northeast	68	102	82	46	16	14	14	14	14	13	11	8	3	Southeast	
	East	84	147	162	145	101	45	14	14	14	13	11	8	3	East	
	Southeast	48	105	138	146	139	107	66	25	14	13	11	8	3	Northeast	
	South	3	8	24	51	89	97	102	97	89	51	24	8	3	North	
	Southwest	3	8	11	13	14	25	66	107	139	146	138	105	48	Northwest	
	West	3	8	11	13	14	14	14	45	101	145	162	147	84	West	
	Northwest	3	8	11	13	14	14	14	14	16	46	82	102	68	Southwest	
	Horizontal	9	47	100	150	185	205	214	205	185	150	100	47	9	Horizontal	
SEPT 22 & MAR 22	North	0	5	9	12	13	13	14	13	13	12	9	5	0	South	MAR 22 & SEPT 22
	Northeast	0	51	58	26	13	13	13	13	13	12	9	5	0	Southeast	
	East	0	116	149	139	99	45	14	13	13	12	9	5	0	East	
	Southeast	0	95	144	162	157	133	90	41	14	12	9	5	0	Northeast	
	South	0	12	44	81	110	122	140	122	110	81	44	12	0	North	
	Southwest	0	5	9	12	14	41	90	133	157	162	144	95	0	Northwest	
	West	0	5	9	12	13	13	14	45	99	139	149	116	0	West	
	Northwest	0	5	9	12	13	13	14	13	13	26	58	51	0	Southwest	
	Horizontal	0	21	67	124	153	176	183	176	153	124	67	21	0	Horizontal	
OCT 23 & FEB 20	North	0	2	6	10	11	12	12	12	11	10	6	2	0	South	APR 20 & AUG 24
	Northeast	0	35	33	12	11	12	12	12	11	10	6	2	0	Southeast	
	East	0	85	117	122	88	39	12	12	11	10	6	2	0	East	
	Southeast	0	81	132	161	163	144	107	63	20	10	6	2	0	Northeast	
	South	0	21	59	104	137	154	162	154	137	104	59	21	0	North	
	Southwest	0	2	6	10	20	63	107	144	163	161	132	81	0	Northwest	
	West	0	2	6	10	11	12	12	39	88	122	117	85	0	West	
	Northwest	0	2	6	10	11	12	12	12	11	12	33	35	0	Southwest	
	Horizontal	0	8	29	64	101	123	129	123	101	64	29	8	0	Horizontal	
NOV 21 & JAN 21	North	0	0	3	7	9	10	11	10	9	7	3	0	0	South	MAY 21 & JULY 23
	Northeast	0	0	12	7	9	10	11	11	9	7	3	0	0	Southeast	
	East	0	0	91	100	74	33	11	10	9	7	3	0	0	East	
	Southeast	0	0	109	144	156	144	116	70	27	7	3	0	0	Northeast	
	South	0	0	59	104	139	158	166	158	139	104	59	0	0	North	
	Southwest	0	0	3	7	27	70	116	144	156	144	109	0	0	Northwest	
	West	0	0	3	7	9	10	11	33	74	100	91	0	0	West	
	Northwest	0	0	3	7	9	10	11	10	9	7	12	0	0	Southwest	
	Horizontal	0	0	16	43	73	92	103	92	73	43	16	0	0	Horizontal	
DEC 22	North	0	0	2	6	9	10	10	10	9	6	2	0	0	South	JUNE 21
	Northeast	0	0	7	6	9	10	10	10	9	6	2	0	0	Southeast	
	East	0	0	72	86	68	31	10	10	9	6	2	0	0	East	
	Southeast	0	0	88	134	148	142	115	73	30	7	2	0	0	Northeast	
	South	0	0	51	99	134	158	165	158	134	99	51	0	0	North	
	Southwest	0	0	2	7	30	73	115	142	148	134	88	0	0	Northwest	
	West	0	0	2	6	9	10	10	31	68	86	72	0	0	West	
	Northwest	0	0	2	6	9	10	10	10	9	6	7	0	0	Southwest	
	Horizontal	0	0	8	32	55	76	85	76	55	32	8	0	0	Horizontal	

Solar Gain Correction	Steel Sash, or No Sash × 1/.85 or 1.17	Haze −15% (Max.)	Altitude +0.7% per 1000 Ft	Dewpoint Decrease From 67 F + 7% per 10 F	Dewpoint Increase From 67 F − 7% per 10 F	South Lat. Dec. or Jan. + 7%

Bold Face Values — Monthly Maximums   Boxed Values — Yearly maximums

TABLE 15—SOLAR HEAT GAIN THRU ORDINARY GLASS (Contd)

50°   Btu/(hr) (sq ft sash area)   **50°**

50° NORTH LATITUDE		AM						SUN TIME				PM			50° SOUTH LATITUDE	
Time of Year	Exposure	6	7	8	9	10	11	Noon	1	2	3	4	5	6	Exposure	Time of Year
JUNE 21	North	29	12	12	13	14	14	14	14	14	13	12	12	29	South	DEC 22
	Northeast	126	125	94	50	16	14	14	14	14	13	12	10	8	Southeast	
	East	139	164	162	136	94	41	14	14	14	13	12	10	8	East	
	Southeast	64	102	126	135	124	98	61	23	14	13	12	10	8	Northeast	
	South	8	10	16	39	68	87	93	87	68	39	16	10	8	North	
	Southwest	8	10	12	13	14	23	61	98	124	135	126	102	64	Northwest	
	West	8	10	12	13	14	14	14	41	94	136	162	164	139	West	
	Northwest	8	10	12	13	14	14	14	14	16	50	94	125	126	Southwest	
	Horizontal	44	86	133	173	197	214	220	214	197	173	133	86	44	Horizontal	
JULY 23 & MAY 21	North	21	11	12	13	14	14	14	14	14	13	12	11	21	South	JAN 21 & NOV 21
	Northeast	114	117	87	44	15	14	14	14	14	13	12	10	6	Southeast	
	East	131	161	163	141	96	43	14	14	14	13	12	10	6	East	
	Southeast	65	107	134	143	136	109	70	26	14	13	12	10	6	Northeast	
	South	6	10	21	50	80	98	106	98	80	50	21	10	6	North	
	Southwest	6	10	12	13	14	26	70	109	136	143	134	107	65	Northwest	
	West	6	10	12	13	14	14	14	43	96	141	163	161	131	West	
	Northwest	6	10	12	13	14	14	14	14	15	44	87	117	114	Southwest	
	Horizontal	33	75	119	159	188	205	211	205	188	159	119	75	33	Horizontal	
AUG 24 & APR 20	North	8	8	10	12	13	14	14	14	13	12	10	8	8	South	FEB 20 & OCT 23
	Northeast	76	94	70	31	13	14	14	14	13	12	10	8	4	Southeast	
	East	94	145	158	141	98	45	14	14	13	12	10	8	4	East	
	Southeast	53	111	144	157	153	132	89	40	13	12	10	8	4	Northeast	
	South	4	9	36	73	105	130	138	130	105	73	36	9	4	North	
	Southwest	4	8	10	12	13	40	89	132	153	157	144	111	53	Northwest	
	West	4	8	10	12	13	14	14	45	98	141	158	145	94	West	
	Northwest	4	8	10	12	13	14	14	14	13	31	70	94	76	Southwest	
	Horizontal	13	46	89	131	160	179	185	179	160	131	89	46	13	Horizontal	
SEPT 22 & MAR 22	North	0	4	8	10	12	12	12	12	12	10	8	4	0	South	MAR 22 & SEPT 22
	Northeast	0	58	46	16	12	12	12	12	12	10	8	4	0	Southeast	
	East	0	102	138	130	93	43	12	12	12	10	8	4	0	East	
	Southeast	0	86	139	162	163	145	105	56	17	10	8	4	0	Northeast	
	South	0	11	51	93	131	150	158	150	131	93	51	11	0	North	
	Southwest	0	4	8	10	17	56	105	145	163	162	139	86	0	Northwest	
	West	0	4	8	10	12	12	12	43	93	130	138	102	0	West	
	Northwest	0	4	8	10	12	12	12	12	12	16	46	58	0	Southwest	
	Horizontal	0	15	49	88	118	140	148	140	118	88	49	15	0	Horizontal	
OCT 23 & FEB 20	North	0	0	4	7	9	10	11	10	9	7	4	0	0	South	APR 20 & AUG 24
	Northeast	0	29	20	7	9	10	11	10	9	7	4	0	0	Southeast	
	East	0	73	99	105	79	35	11	10	9	7	4	0	0	East	
	Southeast	0	69	111	145	157	144	115	69	24	7	4	0	0	Northeast	
	South	0	17	53	99	137	157	167	157	137	99	53	17	0	North	
	Southwest	0	0	4	7	24	69	115	144	157	145	111	69	0	Northwest	
	West	0	0	4	7	9	10	11	35	79	105	99	73	0	West	
	Northwest	0	0	4	7	9	10	11	10	9	7	20	29	0	Southwest	
	Horizontal	0	2	19	45	72	86	94	86	72	45	19	2	0	Horizontal	
NOV 21 & JAN 21	North	0	0	1	4	6	8	9	8	6	4	1	0	0	South	MAY 21 & JULY 23
	Northeast	0	0	5	4	6	8	9	8	6	4	1	0	0	Southeast	
	East	0	0	51	64	57	28	9	8	6	4	1	0	0	East	
	Southeast	0	0	62	95	127	127	107	67	21	4	1	0	0	Northeast	
	South	0	0	34	70	116	143	153	143	116	70	34	0	0	North	
	Southwest	0	0	1	4	21	67	107	127	127	95	62	0	0	Northwest	
	West	0	0	1	4	6	8	9	28	57	64	51	0	0	West	
	Northwest	0	0	1	4	6	8	9	8	6	4	5	0	0	Southwest	
	Horizontal	0	0	4	13	30	47	53	47	30	13	4	0	0	Horizontal	
DEC 22	North	0	0	0	3	5	6	7	6	5	3	0	0	0	South	JUNE 21
	Northeast	0	0	0	3	5	6	7	6	5	3	0	0	0	Southeast	
	East	0	0	0	27	47	23	7	6	5	3	0	0	0	East	
	Southeast	0	0	0	41	107	116	100	62	25	3	0	0	0	Northeast	
	South	0	0	0	31	99	131	141	131	99	31	0	0	0	North	
	Southwest	0	0	0	3	25	62	100	116	107	41	0	0	0	Northwest	
	West	0	0	0	3	5	6	7	23	47	27	0	0	0	West	
	Northwest	0	0	0	3	5	6	7	6	5	3	0	0	0	Southwest	
	Horizontal	0	0	0	5	19	33	40	33	19	5	0	0	0	Horizontal	
Solar Gain Correction	Steel Sash, or No Sash × 1/.85 or 1.17	Haze −15% (Max.)			Altitude +0.7% per 1000 Ft			Dewpoint Decrease From 67 F + 7% per 10 F			Dewpoint Increase From 67 F − 7% per 10 F				South Lat. Dec. or Jan. + 7%	

Bold Face Values — Monthly Maximums   Boxed Values — Yearly maximums

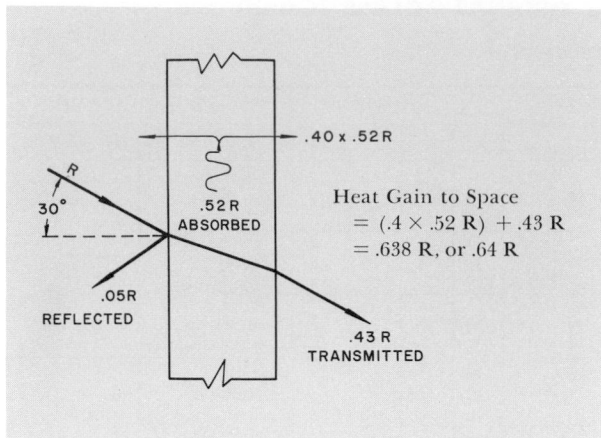

FIG. 15 — REACTION ON SOLAR HEAT (R), 52% HEAT
ABSORBING GLASS, 30° ANGLE OF INCIDENCE

ALL GLASS TYPES — WITH AND WITHOUT SHADING DEVICES

Glass, *other than ordinary glass*, absorbs more solar heat because it

1. May be thicker, or
2. May be specially treated to absorb solar heat (heat absorbing glass).

These special glass types reduce the transmitted solar heat but increase the amount of absorbed solar heat flowing into the space. Normally they reflect slightly less than ordinary glass because part of the reflection takes place on the inside surface. A portion of heat reflected from the inside surface is absorbed in passing back through the glass. The overall effect, however, is to reduce the solar heat gain to the conditioned space as shown in *Fig. 15*. (Refer to *Item 8, page 51*, for absorptivity, reflectivity and transmissibility of common types of glass at 30° angle of incidence.)

The solar heat gain factor through 52% heat absorbing glass as compared to ordinary glass is .64R/.88R = .728 or .73. This multiplier (.73) is used with *Table 15* to determine the solar heat gain thru 52% heat absorbing glass. Multipliers for various types of glass are listed in *Table 16*.

The effectiveness of a *shading device* depends on its ability to keep solar heat from the conditioned space. All shading devices reflect and absorb a major portion of the solar gain, leaving a small portion to be transmitted. The outdoor shading devices are much more effective than the inside devices because all of the reflected solar heat is kept out and the absorbed heat is dissipated to the outdoor air. Inside devices necessarily dissipate their absorbed heat within the conditioned space and

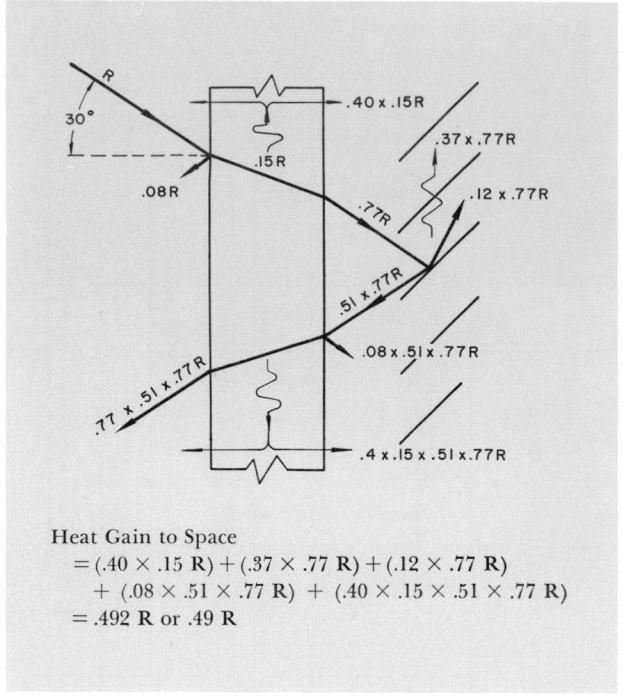

Heat Gain to Space
$$= (.40 \times .15 \text{ R}) + (.37 \times .77 \text{ R}) + (.12 \times .77 \text{ R})$$
$$+ (.08 \times .51 \times .77 \text{ R}) + (.40 \times .15 \times .51 \times .77 \text{ R})$$
$$= .492 \text{ R or } .49 \text{ R}$$

FIG. 16 — REACTION ON SOLAR HEAT (R), ¼-INCH
PLATE GLASS, WHITE VENETIAN BLIND, 30° ANGLE
OF INCIDENCE

must also reflect the solar heat back through the glass *(Fig. 16)* wherein some of it is absorbed. (Refer to *Item 8, page 51*, for absorptivity, reflectivity and transmissibility of common shading devices at 30° angle of incidence.)

The solar heat gain thru glass with an inside shading device may be expressed as follows:

$$Q = [.4a_g + t_g (a_{sd} + t_{sd} + r_g r_{sd} + .4a_g r_{sd})] \frac{R}{.88}$$

where:

Q = solar heat gain to space, Btu/(hr)(sq ft)
R = total solar intensity, Btu/(hr)(sq ft), (from *Table 15*)
a = solar absorptivity
t = solar transmissibility
r = solar reflectivity
g = glass
sd = shading device
$.88$ = conversion factor from *Fig. 12*

For drapes the above formula changes as follows, caused by the hot air space between glass and drapes:

$$Q = [.24a_g + t_g (.85a_{sd} + t_{sd} + r_g r_{sd} + .24a_g r_{sd})] \frac{R}{.88}$$

The transmission factor U for glass with 100% drape is 0.80 Btu/(hr) (sq ft) (F).

The solar heat gain factor thru the combination in *Fig. 16* as compared to ordinary glass is .49R/.88R = .557 or .56. (Refer to *Table 16* for ¼-inch regular plate glass with a white venetian blind.)

NOTE: Actually the reaction on the solar heat reflected back through the glass from the blind is not always identical to the first pass as assumed in this example. The first pass through the glass filters out most of solar radiation that is to be absorbed in the glass, and the second pass absorbs somewhat less. For simplicity, the reaction is assumed identical, since the quantities are normally small on the second pass.

Basis of Table 16
Over-all Factors for Solar Heat Gain thru Glass, With and Without Shading Devices

The factors in *Table 16* are based on:

1. An outdoor film coefficient of 2.8 Btu/(hr) (sq ft) (deg F) at 5 mph wind velocity.

2. An inside film coefficient of 1.8 Btu/(hr)(sq ft) (deg F), 100-200 fpm. This is not 1.47 as normally used, since the present practice in well designed systems is to sweep the window with a stream of air.

3. A 30° angle of incidence which is the angle at which most exposures peak. The 30° angle of incidence is approximately the balance point on reduction of solar heat coming through the atmosphere and the decreased transmissibility of glass. Above the 30° angle the transmissibility of glass decreases, and below the 30° angle the atmosphere absorbs or reflects more.

4. All shading devices fully drawn, except roller shades. Experience indicates that roller shades are *seldom fully drawn*, so the factors have been slightly increased.

5. Venetian blind slats horizontal at 45° and shading screen slats horizontal at 17°.

6. Outdoor canvas awnings ventilated at sides and top. (See *Table 16* footnote.)

7. Since *Table 15* is based on the net solar heat gain thru ordinary glass, all calculated solar heat factors are divided by .88 *(Fig. 12)*.

8. The average absorptivity, reflectivity and transmissability for common glass and shading devices at a 30° angle of incidence along with shading factors appear in the table below.

Use of Table 16
—Over-all Factors for Solar Heat Gain thru Glass, With and Without Shading Devices

The factors in *Table 16* are multiplied by the values in *Table 15* to determine the solar heat gain thru different combinations of glass and shading devices. The correction factors listed under *Table 15* are to be used if applicable. Transmission due to temperature difference between the inside and outdoor air must be added to the solar heat gain to determine total gain thru glass.

Example 3 — Partially Drawn Shades

Occasionally it is necessary to estimate the cooling load in a building where the blinds are not to be fully drawn. The procedure is illustrated in the following example:

Given:
West exposure, 40° North latitude
Thermopane window with white venetian blind on inside, ¾ drawn.

Find:
Peak solar heat gain.

Solution:
By inspection of *Table 15*, the boxed boldface values for peak solar heat gain, occurring at 4:00 p.m. on July 23

= 164 Btu/(hr)(sq ft)

TYPES OF GLASS OR SHADING DEVICES*	Absorptivity (a)	Reflectivity (r)	Transmissibility (t)	Solar Factor†
Ordinary Glass	.06	.08	.86	1.00
Regular Plate, ¼″	.15	.08	.77	.94
Glass, Heat Absorbing	by mfg.	.05	(1 − .05 − a)	—
Venetian Blind, Light Color	.37	.51	.12	.56‡
Medium Color	.58	.39	.03	.65‡
Dark Color	.72	.27	.01	.75‡
Fiberglass Cloth, Off White (5.72 - 61/58)	.05	.60	.35	.48‡
Cotton Cloth, Beige (6.18 - 91/36)	.26	.51	.23	.56‡
Fiberglass Cloth, Light Gray	.30	.47	.23	.59‡
Fiberglass Cloth, Tan (7.55 - 57/29)	.44	.42	.14	.64‡
Glass Cloth, White, Golden Stripes	.05	.41	.54	.65‡
Fiberglass Cloth, Dark Gray	.60	.29	.11	.75‡
Dacron Cloth, White (1.8 - 86/81)	.02	.28	.70	.76‡
Cotton Cloth, Dark Green, Vinyl Coated (similar to roller shade)	.85	.15	.00	.88‡
Cotton Cloth, Dark Green (6.06 - 91/36)	.02	.28	.70	.76‡

*Factors for various draperies are given for guidance only since the actual drapery material may be different in color and texture; figures in parentheses are ounces per sq yd, and yarn count warp/filling. Consult manufacturers for actual values.

†Compared to ordinary glass.

‡For a shading device in combination with ordinary glass.

Thermopane windows have no sash; therefore, sash area correction = 1/.85 (bottom *Table 15*).

In this example, ¾ of the window is covered with the venetian blind and ¼ is not; therefore, the solar heat gain factor equals ¾ of the overall factor + ¼ of the glass factor.

Factor for ¾ drawn = (¾ × .52) + (¼ × .80) (*Table 16*)

                     = .59

Solar heat gain   = $164 \times \dfrac{.59}{.85}$

                     = 114 Btu/(hr) (sq ft).

Example 4 — Peak Solar Heat Gain thru Solex "R" Glass

Given:

    West exposure, 40° North latitude

    ¼" Solex "R" glass in steel sash, double hung window

Find:

    Peak solar heat gain.

Solution:

    By inspection of *Table 15* the boxed boldface value for peak solar heat gain, occurring at 4:00 p.m. on July 23

        = 164 Btu/(hr) (sq ft).

    Steel sash window correction = 1/.85 (bottom *Table 15*).

    Solex "R" glass absorbs 50.9% of the solar heat (footnotes to *Table 16*) which places this glass in the 48% to 56% absorbing range.

    From *Table 16*, the factor = .73.

    Solar heat gain = $\dfrac{164 \times .73}{.85}$ = 141 Btu/(hr) (sq ft)

TABLE 16—OVER-ALL FACTORS FOR SOLAR HEAT GAIN THRU GLASS
WITH AND WITHOUT SHADING DEVICES*

Apply Factors to Table 15

Outdoor wind velocity, 5 mph — Angle of incidence, 30° — Shading devices fully covering window

TYPE OF GLASS	GLASS FACTOR NO SHADE	INSIDE VENETIAN BLIND* 45° horiz. or vertical or ROLLER SHADE			OUTSIDE VENETIAN BLIND 45° horiz. slats		OUTSIDE SHADING SCREEN† 17° horiz. slats		OUTSIDE AWNING‡ vent. sides & top	
		Light Color	Medium Color	Dark Color	Light Color	Light on Outside Dark on Inside	Medium** Color	Dark § Color	Light Color	Med. or Dark Color
ORDINARY GLASS	1.00	.56	.65	.75	.15	.13	.22	.15	.20	.25
REGULAR PLATE (¼ inch)	.94	.56	.65	.74	.14	.12	.21	.14	.19	.24
HEAT ABSORBING GLASS††										
40 to 48% Absorbing	.80	.56	.62	.72	.12	.11	.18	.12	.16	.20
48 to 56% Absorbing	.73	.53	.59	.62	.11	.10	.16	.11	.15	.18
56 to 70% Absorbing	.62	.51	.54	.56	.10	.10	.14	.10	.12	.16
DOUBLE PANE										
Ordinary Glass	.90	.54	.61	.67	.14	.12	.20	.14	.18	.22
Regular Plate	.80	.52	.59	.65	.12	.11	.18	.12	.16	.20
48 to 56% Absorbing outside; Ordinary Glass inside.	.52	.36	.39	.43	.10	.10	.11	.10	.10	.13
48 to 56% Absorbing outside; Regular Plate inside.	.50	.36	.39	.43	.10	.10	.11	.10	.10	.12
TRIPLE PANE										
Ordinary Glass	.83	.48	.56	.64	.12	.11	.18	.12	.16	.20
Regular Plate	.69	.47	.52	.57	.10	.10	.15	.10	.14	.17
PAINTED GLASS										
Light Color	.28									
Medium Color	.39									
Dark Color	.50									
STAINED GLASS‡‡										
Amber Color	.70									
Dark Red	.56									
Dark Blue	.60									
Dark Green	.32									
Greyed Green	.46									
Light Opalescent	.43									
Dark Opalescent	.37									

Footnotes for *Table 16* appear on next page.

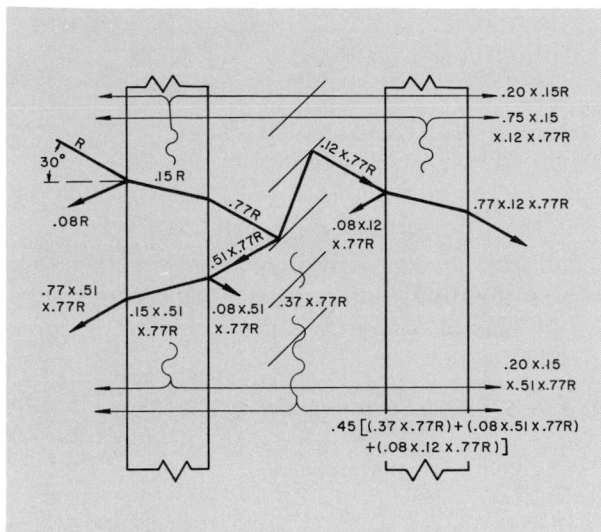

FIG. 17 — REACTION ON SOLAR HEAT (R), ¼-INCH
PLATE GLASS, WHITE VENETIAN BLIND, ¼-INCH
PLATE GLASS, 30° ANGLE OF INCIDENCE

APPROXIMATION OF FACTORS FOR COMBINATIONS NOT FOUND IN TABLE 16

Occasionally combinations of shading devices and types of glass may be encountered that are not covered in *Table 16*. These factors can be approximated (1) by using the solar heat gain flow diagrams in *Fig. 15 and 16*, (2) by applying the absorptivity, reflectivity and transmissibility of glass and shades listed in the table on *page 51*, or determined from manufacturer, and (3) by distributing heat absorbed within the dead air space and glass panes *(Fig. 17)*.

Example 5 — Approximation of Over-all Factor
Given:
A combination as in *Fig. 16* backed on the inside with another pane of ¼-inch regular plate glass.

Find:
The over-all factor.

Solution:
Figure 17 shows the distribution of solar heat. The heat absorbed between the glass panes (dead air space) is divided 45% and 55% respectively between the in and out flow. The heat absorbed within the glass panes is divided 20% in and 80% out for the outer pane, and 75% in and 25% out for the inner pane. These divisions are based on reasoning partially stated in the notes under *Fig. 13*, which assume the outdoor film coefficient of 2.8 Btu/ (hr) (sq ft) (deg F), the indoor film coefficient of 1.8 Btu/ (hr) (sq ft) (deg F), and the over-all thermal conductance of the air space of 1.37 Btu/ (hr) (sq ft) (deg F).

Heat gain to space *(Fig. 17)*
$$= (.75 \times .15 \times .12 \times .77R) + (.77 \times .12 \times .77R)$$
$$+ .45 [(.37 \times .77R) + (.08 \times .51 \times .77R)$$
$$+ (.08 \times .12 \times .77R)]$$
$$+ .20 [(.15R) + (.15 \times .51 \times .77R)]$$
$$= .2684R \text{ or } .27R$$

Solar heat gain factor as compared to ordinary glass
$$= .27R/.88R = .31$$

Equations: Solar Gain Without Shades = (Solar Data from Table 15) × (Glass Factor from table)
Solar Gain With Shades = (Solar Data from Table 15) × (Over-all Factor from table)
Solar Gain With Shades Partially Drawn = (Solar Data from Table 15) ×
[(Fraction Drawn × Over-all Factor) + (1 — Fraction Drawn) × (Glass Factor)]

Footnotes for Table 16:

*Shading devices fully drawn except roller shades. For fully drawn roller shades, multiply light colors by .73, medium colors by .95, and dark colors by 1.08.

†Factors for solar altitude angles of 40° or greater. At solar altitudes below 40°, some direct solar rays pass thru the slats. Use following multipliers:—

MULTIPLIERS FOR SOLAR ALTITUDES BELOW 40°

Approximate Sun Time, July 23			Solar Altitude Angle (deg)	Multiplier	
30° Lat.	40° Lat.	50° Lat.		Med. Color	Dark Color
6:00 a.m. 6:00 p.m.	5:45 a.m. 6:15 p.m.	5:30 a.m. 6:30 p.m.	10	2.09	3.46
6:45 a.m. 5:15 p.m.	6:40 a.m. 5:20 p.m.	6:30 a.m. 5:30 p.m.	20	1.59	2.66
7:30 a.m. 4:30 p.m.	7:30 a.m. 4:30 p.m.	7:30 a.m. 4:30 p.m.	30	1.09	1.67

‡With outside canvas awnings tight against building on sides and top, multiply over-all factor by 1.4.

§Commercial shade bronze. Metal slats 0.05 inches wide, 17 per inch.

**Commercial shade, aluminum. Metal slats 0.057 inches wide, 17.5 per inch.

††Most heat absorbing glass used in comfort air conditioning is in the 40% to 56% range; industrial applications normally use 56% to 70%. The following table presents the absorption qualities of the most common glass types:—

SOLAR RADIATION ABSORBED BY HEAT ABSORBING GLASS

Trade Name or Description	Manufacturer	Thickness (in.)	Color	Solar Radiation Absorbed (%)
Aklo	Blue Ridge Glass Corp.	⅛	Pale Blue-Green	56.6
Aklo	Blue Ridge Glass Corp.	¼	Pale Blue-Green	69.7
Coolite	Mississippi Glass Co.	⅛	Light Blue	58.4
Coolite	Mississippi Glass Co.	¼	Light Blue	70.4
L.O.F.	Libbey-Owens-Ford	¼	Pale Blue-Green	48.2
Solex R	Pittsburgh Plate Glass Co.	¼	Pale Green	50.9

‡‡With multicolor windows, use the predominant color.

GLASS BLOCK

Glass block differs from sheet glass in that there is an appreciable absorption of solar heat and a fairly long time lag before the heat reaches the inside (about 3 hours). This is primarily caused by the thermal storage capacity of the glass block itself. The high absorption of heat increases the inside surface temperature of the sunlit glass block which may require lower room temperatures to maintain comfort conditions as explained in *Chapter 2*.

Shading devices on the outdoor side of glass block are almost as effective as with any other kind of glass since they keep the heat away from the glass. Shading devices on the inside are not effective in reducing the heat gain because most of the heat reflected is absorbed in the glass block.

Basis of Table 17
— Solar Heat Gain Factors for Glass Block, With and Without Shading Devices

The factors in *Table 17* are the average of tests conducted by the ASHAE on several types of glass block.

Since glass block windows have no sash, the factors in *Table 17* have been increased to include the 1/.85 multiplier in *Table 15*.

Use of Table 17
— Solar Heat Gain Factors for Glass Block, With and Without Shading Devices

The factors in *Table 17* are used to determine the solar heat gain thru all types of glass block.

The transmission of heat caused by a difference between the inside and outdoor temperatures must also be figured, using the appropriate "U" value, *Chapter 5*.

Example 6 — Peak Solar Heat Gain, Glass Block

Given:
   West exposure, 40° North latitude
   Glass block window

Find:
   Peak solar heat gain

Solution:
   By inspection of *Table 15*, the peak solar heat gain occurs on July 23.
   Solar heat gain
      At 4:00 p.m. = $(.39 \times 164) + (.21 \times 43) = 73$
      At 5:00 p.m. = $(.39 \times 161) + (.21 \times 98) = 84$
      At 6:00 p.m. = $(.39 \times 118) + (.21 \times 144) = 76$
Peak solar heat gain occurs at 5:00 p.m. on July 23.

TABLE 17—SOLAR HEAT GAIN FACTORS FOR GLASS BLOCK

WITH AND WITHOUT SHADING DEVICES*

Apply Factors to Table 15

EXPOSURE IN NORTH LATITUDES		Instantaneous Transmission Factor (B_i)	Absorption Transmission		EXPOSURE IN SOUTH LATITUDES	
			Factor (B_a)	Time Lag Hours		
Northeast		.27	.24	3.0	Southeast	
East		.39	.21	3.0	East	
Southeast		.35	.22	3.0	Northeast	
South					North	
	Summer †	.27	.24	3.0		Summer †
	Winter †	.39	.22	3.0		Winter †
Southwest		.35	.22	3.0	Northwest	
West		.39	.21	3.0	West	
Northwest		.27	.24	3.0	Southwest	

*Factors include correction for no sash with glass block windows.

†Use the summer factors for all latitudes, North or South. Use the winter factor for intermediate seasons, 30° to 50° North or South latitude.

Equations:
   Solar heat gain *without shading devices*
      $= (B_i \times I_i) + (B_a \times I_a)$
   Solar heat gain *with outdoor shading devices*
      $= (B_i \times I_i + B_a \times I_a) \times .25$
   Solar heat gain *with inside shading devices*
      $= (B_i \times I_i + B_a \times I_a) \times .90$

Where:
   B_i = Instantaneous transmission factor from *Table 17*.
   B_a = Absorption transmission factor from *Table 17*.
   I_i = Solar heat gain value from *Table 15* for the desired time and wall facing.
   I_a = Solar heat gain value from *Table 15* for 3 hours earlier than I_i and same wall facing.

SHADING FROM REVEALS, OVERHANGS, FINS AND ADJACENT BUILDINGS

All windows are shaded to a greater or lesser degree by the projections close to it and by buildings around it. This shading reduces the solar heat gain through these windows by keeping the direct rays of the sun off part or all of the window. The shaded portion has only the diffuse component striking it. Shading of windows is significant in monumental type buildings where the reveal may be large, even at the time of peak solar heat gain. *Chart 1*, this chapter, is presented to simplify the determination of the shading of windows by these projections.

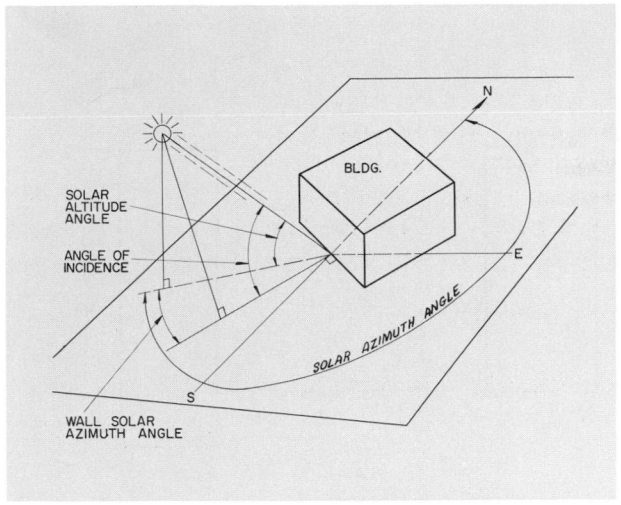

Fig. 18 — Solar Angles

Basis of Chart 1
— Shading from Reveals, Overhangs, Fins and Adjacent Buildings

The location of the sun is defined by the solar azimuth angle and the solar altitude angle as shown in *Fig. 18*. The solar azimuth angle is the angle in a horizontal plane between North and the vertical plane passing through the sun and the point on earth. The solar altitude angle is the angle in a vertical plane between the sun and a horizontal plane through a point on earth. The location of the sun with respect to the particular wall facing is defined by the wall solar azimuth angle and the solar altitude angle. The wall solar azimuth angle is the angle in the horizontal plane between the perpendicular to the wall and the vertical plane passing through the sun and the point on earth.

The shading of a window by a vertical projection alongside the window (see *Fig. 19*) is the tangent of the wall solar azimuth angle (B), times depth of the projection. The shading of a window by a horizontal projection above the window is the tangent of angle (X), a resultant of the combined effects of the altitude angle (A) and the wall solar azimuth angle (B), times the depth of the projection.

$$\text{Tan X} = \frac{\text{Tan A, solar altitude angle}}{\text{Cos B, wall solar azimuth angle}}$$

The upper part of *Chart 1* determines the tangent of the wall solar azimuth angle and the bottom part determines tan X.

Use of Chart 1
— Shading from Reveals, Overhangs, Fins and Adjacent Buildings

The procedure to determine the top and side shading from *Chart 1* is.

1. Determine the solar azimuth and altitude angles from *Table 18*.

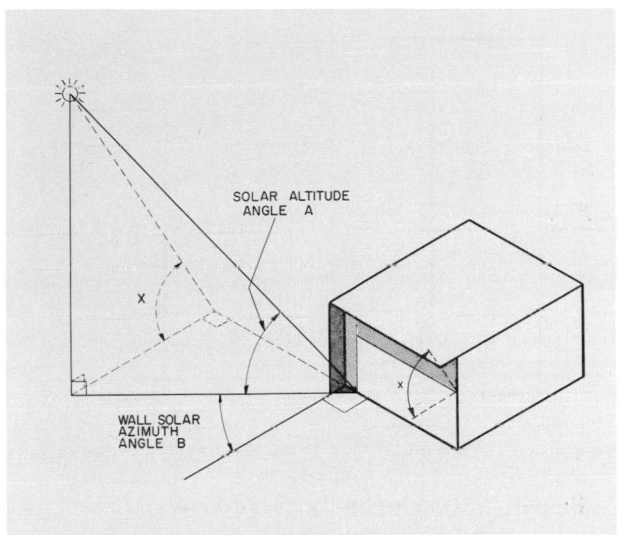

Fig. 19 — Shading by Wall Projections

2. Locate the solar azimuth angle on the scale in upper part of *Chart 1*.

3. Proceed horizontally to the exposure desired.

4. Drop vertically to "Shading from Side" scale.

5. Multiply the depth of the projection (plan view) by the "Shading from Side."

6. Locate the solar altitude angle on the scale in lower part of *Chart 1*.

7. Move horizontally until the "Shading from Side" value (45 deg. lines) determined in Step 4 is intersected.

8. Drop vertically to "Shading from Top" from intersection.

9. Multiply the depth of the projection (elevation view) by the "Shading from Top."

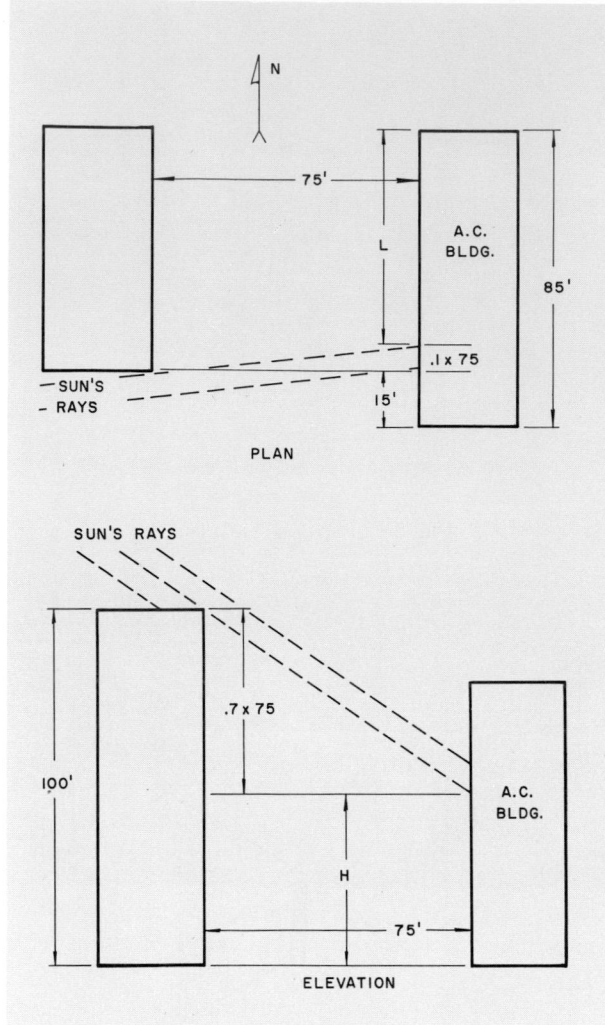

FIG. 20 — SHADING OF BUILDING BY ADJACENT
BUILDING

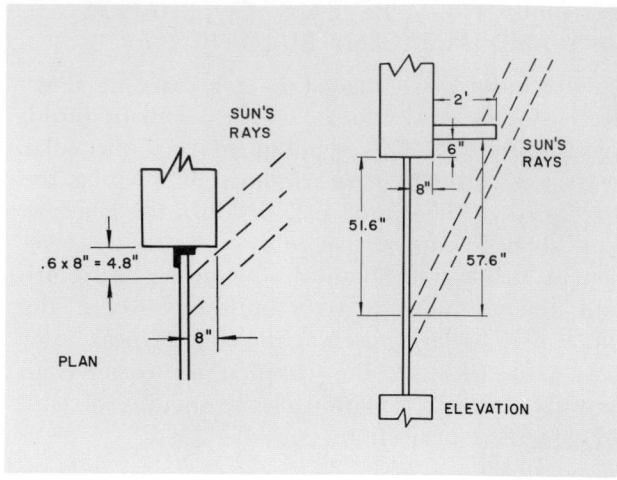

FIG. 21 — SHADING OF REVEAL AND OVERHANG

Length of building in shade, L
= 85 − 15 − (.1 × 75) = 62.5 ft

Height of building in shade, H = 100 − (75 × .7) = 47.5 ft

The air conditioned building is shaded to a height of 47.5 ft and 62.5 ft along the face at 4:00 p.m. on July 23.

Example 8 — Shading of Window by Reveals

Given:
A steel casement window on the west side with an 8-inch reveal.

Find:
Shading by the reveal at 2 p.m. on July 23, 40° North Latitude.

Solution:
From *Table 18*, solar azimuth angle = 242°
                 solar altitude angle = 57°
From *Chart 1*, shading from side reveal = .6 × 8 = 4.8 in.
                 shading from top reveal = 1.8 × 8 = 14.4 in.

Example 9 — Shading of Window by Overhang and Reveal

Given:
The same window as in *Example 8* with a 2 ft overhang 6 inches above the window.

Find:
Shading by reveal and overhang at 2 p.m. on July 23, 40° North Latitude.

Solution:
Refer to *Fig. 21*.

Shading from side reveal (same as *Example* 8) = 4.8 in.

Shading from overhang = 1.8 × (24 + 8) = 57.6 in.

Since the overhang is 6 inches above the window, the portion of window shaded = 57.6 − 6.0 = 51.6 in.

Example 7 — Shading of Building by Adjacent Building

Given:
Buildings located as shown in *Fig. 20*.

Find:
Shading at 4 p.m., July 23, of building to be air conditioned.

Solution:
It is recommended that the building plans and elevations be sketched to scale with approximate location of the sun, to enable the engineer to visualize the shading conditions.
From *Table 18*, solar azimuth angle = 267°
                 solar altitude angle = 35°
From *Chart 1*, shading from side = .1 ft/ft
                 shading from top = .7 ft/ft

CHART 1 — SHADING FROM REVEALS, OVERHANGS, FINS AND ADJACENT BUILDINGS

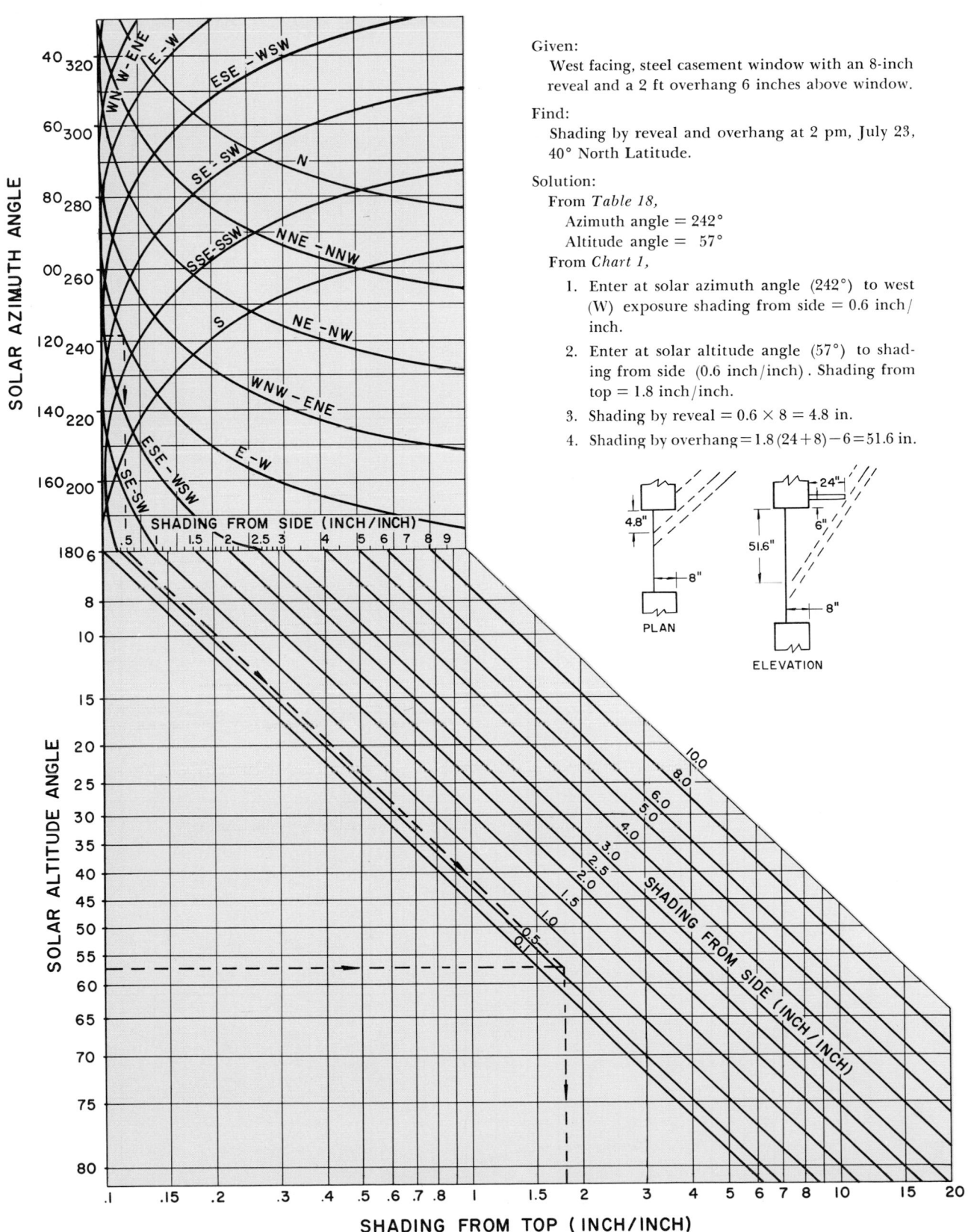

Given:

West facing, steel casement window with an 8-inch reveal and a 2 ft overhang 6 inches above window.

Find:

Shading by reveal and overhang at 2 pm, July 23, 40° North Latitude.

Solution:

From *Table 18,*

Azimuth angle = 242°

Altitude angle = 57°

From *Chart 1,*

1. Enter at solar azimuth angle (242°) to west (W) exposure shading from side = 0.6 inch/inch.

2. Enter at solar altitude angle (57°) to shading from side (0.6 inch/inch). Shading from top = 1.8 inch/inch.

3. Shading by reveal = 0.6 × 8 = 4.8 in.

4. Shading by overhang = 1.8 (24 + 8) − 6 = 51.6 in.

PLAN

ELEVATION

TABLE 18—SOLAR ALTITUDE AND AZIMUTH ANGLES

NORTH* LATITUDE	SUN TIME	Jan. 21 Alt	Az	Feb. 20 Alt	Az	Mar. 22 Alt	Az	Apr. 20 Alt	Az	May 21 Alt	Az	June 21 Alt	Az	July 23 Alt	Az	Aug. 24 Alt	Az	Sept. 22 Alt	Az	Oct. 23 Alt	Az	Nov. 21 Alt	Az	Dec. 22 Alt	Az	SUN TIME
LAT 0°	6 AM																									6 AM
	7	14	111	15	102	15	90	15	78	14	69	14	66	14	69	15	78	15	90	15	102	14	111	14	114	7
	8	28	113	30	103	30	89	30	77	28	67	27	63	28	67	30	77	30	89	30	103	28	113	27	117	8
	9	42	117	44	106	45	89	44	74	42	63	41	58	42	63	44	74	45	89	44	106	42	117	41	122	9
	10	54	126	58	112	60	89	58	68	54	54	53	49	54	54	58	68	60	89	58	112	54	126	53	131	10
	11	65	144	71	127	75	88	71	53	65	36	62	32	65	36	71	53	75	88	71	127	65	144	62	148	11
	12 N	70	180	79	180	90	0	79	0	70	0	67	0	70	0	79	0	90	0	79	180	70	180	67	180	12 N
	1 PM	65	216	71	233	75	272	71	307	65	324	62	328	65	324	71	307	75	272	71	233	65	216	62	212	1 PM
	2	54	234	58	248	60	271	58	292	54	306	53	311	54	306	58	292	60	271	58	248	54	234	53	229	2
	3	42	243	44	254	45	271	44	286	42	297	41	302	42	297	44	286	45	271	44	254	42	243	41	238	3
	4	28	247	30	257	30	271	30	283	28	293	27	297	28	293	30	283	30	271	30	257	28	247	27	243	4
	5	14	249	15	258	15	270	15	282	14	291	14	294	14	291	15	282	15	270	15	258	14	249	14	246	5
	6																									6
LAT 10°	6 AM					1	90	2	78	3	70	4	67	3	70	2	78	1	90							6 AM
	7	10	113	12	103	15	92	16	81	17	72	18	68	17	72	16	81	15	92	12	103	10	113	9	116	7
	8	24	117	27	108	30	95	31	83	32	72	32	68	32	72	31	83	30	95	27	108	24	117	23	121	8
	9	37	124	41	115	44	99	46	84	46	72	45	67	46	72	46	84	44	99	41	115	37	124	35	128	9
	10	48	136	54	125	59	106	61	84	60	67	58	61	60	67	61	84	59	106	54	125	48	136	46	139	10
	11	57	155	64	144	72	122	75	84	73	53	70	44	73	53	75	84	72	122	64	144	57	155	55	156	11
	12 N	60	180	69	180	80	180	89	0	80	0	77	0	80	0	89	0	80	180	69	180	60	180	57	180	12 N
	1 PM	57	205	64	216	72	238	75	276	73	307	70	316	73	307	75	276	72	238	64	216	57	205	53	204	1 PM
	2	48	224	54	235	59	254	61	276	60	293	58	299	60	293	61	276	59	254	54	235	48	224	46	221	2
	3	37	236	41	245	44	261	46	276	46	288	45	293	46	288	46	276	44	261	41	245	37	236	35	232	3
	4	24	243	27	252	30	265	31	277	32	288	32	292	32	288	31	277	30	265	27	252	24	243	23	239	4
	5	10	247	12	257	15	268	16	279	17	288	18	292	17	288	16	279	15	268	12	257	10	247	9	244	5
	6					1	270	2	282	3	290	4	293	3	290	2	282	1	270							6
LAT 20°	6 AM							4	79	7	71	8	68	7	71	4	79									6 AM
	7	6	114	10	106	14	95	18	84	20	75	21	72	20	75	18	84	14	95	10	106	6	114	5	117	7
	8	19	121	23	112	28	101	32	89	34	79	35	75	34	79	32	89	28	101	23	112	19	121	17	124	8
	9	30	130	36	121	42	108	46	94	48	82	48	77	48	82	46	94	42	108	36	121	30	130	28	133	9
	10	40	142	47	133	55	120	59	102	62	85	62	77	62	85	59	102	55	120	47	133	40	142	38	145	10
	11	47	158	55	152	66	143	72	117	75	88	76	74	75	88	72	117	66	143	55	152	47	158	44	163	11
	12 N	50	180	59	180	70	180	81	180	90	0	87	0	90	0	81	180	70	180	59	180	50	180	47	180	12 N
	1 PM	47	202	55	208	66	217	72	243	75	272	76	286	75	272	72	243	66	217	55	208	47	202	44	197	1 PM
	2	40	218	47	227	55	240	59	258	62	275	62	283	62	275	59	258	55	240	47	227	40	218	38	215	2
	3	30	230	36	239	42	252	46	266	48	278	48	283	48	278	46	266	42	252	36	239	30	230	28	227	3
	4	19	239	23	248	28	259	32	271	34	281	35	285	34	281	32	271	28	259	23	248	19	239	17	236	4
	5	6	246	10	254	14	265	18	276	20	285	21	288	20	285	18	276	14	265	10	254	6	246	5	243	5
	6							4	281	7	289	8	292	7	289	4	281									6
LAT 30°	6 AM							6	80	10	72	11	69	10	72	6	80									6 AM
	7	2	115	7	107	13	97	19	87	23	79	24	76	23	79	19	87	13	97	7	107	2	115			7
	8	14	124	19	116	26	106	31	95	35	86	37	82	35	86	31	95	26	106	19	116	14	124	11	126	8
	9	24	134	30	127	38	116	44	104	48	93	49	88	48	93	44	104	38	116	30	127	24	134	21	136	9
	10	32	146	40	141	49	130	56	117	61	103	62	96	61	103	56	117	49	130	40	141	32	146	29	149	10
	11	38	162	46	159	57	151	67	140	73	122	75	112	73	122	67	140	57	151	46	159	38	162	35	164	11
	12 N	40	180	49	180	60	180	71	180	80	180	83	180	80	180	71	180	60	180	49	180	40	180	37	180	12 N
	1 PM	38	198	46	201	57	209	67	220	73	238	75	248	73	238	67	220	57	209	46	201	38	198	35	196	1 PM
	2	32	214	40	219	49	230	56	243	61	257	62	264	61	257	56	243	49	230	40	219	32	214	29	211	2
	3	24	226	30	233	38	244	44	256	48	267	49	272	48	267	44	256	38	244	30	233	24	226	21	224	3
	4	14	236	19	244	26	254	31	265	35	274	37	278	35	274	31	265	26	254	19	244	14	236	11	234	4
	5	2	245	7	253	13	263	19	273	23	281	24	284	23	281	19	273	13	263	7	253	2	245			5
	6							6	280	10	288	11	291	10	288	6	280									6
LAT 40°	6 AM							7	81	13	74	15	72	13	74	7	81									6 AM
	7			5	110	12	99	19	91	24	83	26	80	24	83	19	91	12	99	5	110					7
	8	8	125	15	119	23	110	30	102	35	93	37	89	35	93	30	102	23	110	15	119	8	125	5	127	8
	9	17	136	24	131	33	122	41	113	47	104	49	100	47	104	41	113	33	122	24	131	17	136	14	138	9
	10	24	149	32	145	42	138	51	129	57	116	59	114	57	116	51	129	42	138	32	145	24	149	21	151	10
	11	28	164	37	162	48	157	58	151	66	143	69	138	66	143	58	151	48	157	37	162	28	164	25	165	11
	12 N	30	180	39	180	50	180	61	180	70	180	73	180	70	180	61	180	50	180	39	180	30	180	27	180	12 N
	1 PM	28	196	37	198	48	203	58	209	66	217	69	222	66	217	58	209	48	203	37	198	28	196	25	195	1 PM
	2	24	211	32	215	42	222	51	231	57	242	60	246	57	242	51	231	42	222	32	215	24	211	21	209	2
	3	17	224	24	229	33	238	41	247	47	256	49	260	47	256	41	247	33	238	24	229	17	224	14	222	3
	4	8	235	15	241	23	250	30	258	35	267	37	271	35	267	30	258	23	250	15	241	8	235	5	233	4
	5			5	250	12	261	19	269	24	277	26	280	24	277	19	269	12	261	5	250					5
	6							7	279	13	286	15	288	13	286	7	279									6
LAT 50°	6 AM							9	83	15	77	18	74	15	77	9	83									6 AM
	7					10	101	18	94	25	88	27	85	25	88	18	94	10	101							7
	8	3	125	10	121	19	114	28	106	34	100	37	97	34	100	28	106	19	114	10	121	3	125			8
	9	10	138	17	134	27	127	37	120	44	114	46	110	44	114	37	120	27	127	17	134	10	138	6	139	9
	10	15	151	23	148	34	143	44	137	52	131	55	128	52	131	44	137	34	143	23	148	15	151	12	152	10
	11	19	165	27	164	39	160	49	157	58	152	61	151	58	152	49	157	39	160	27	164	19	165	15	166	11
	12 N	20	180	29	180	40	180	51	180	60	180	63	180	60	180	51	180	40	180	29	180	20	180	17	180	12 N
	1 PM	19	195	27	196	39	200	49	203	58	208	61	209	58	208	49	203	39	200	27	196	19	195	15	194	1 PM
	2	15	209	23	212	34	217	44	223	52	229	55	232	52	229	44	223	34	217	23	212	15	209	12	208	2
	3	10	222	17	226	27	233	37	240	44	246	46	250	44	246	37	240	27	233	17	226	10	222	6	221	3
	4	3	235	10	239	19	246	28	254	34	260	37	263	34	260	28	254	19	246	10	239	3	235			4
	5					10	259	18	266	25	272	27	275	25	272	18	266	10	259							5
	6							9	277	15	283	18	286	15	283	9	277									6
SOUTH* LATITUDE	SUN TIME	July 23		Aug. 24		Sept. 22		Oct. 23		Nov. 21		Dec. 22		Jan. 21		Feb. 20		Mar. 22		Apr. 20		May 21		June 21		SUN TIME

*Use months indicated at top for North Latitudes; and use months at bottom for South Latitudes.

CHAPTER 5. HEAT AND WATER VAPOR FLOW THRU STRUCTURES

This chapter presents the methods and data for determining the sensible and latent heat gain or loss thru the outdoor structures of a building or thru a structure surrounding a space within the building. It also presents data for determining and preventing water vapor condensation on the enclosure surfaces or within the structure materials.

Heat flows from one point to another whenever a temperature difference exists between the two points; the direction of flow is always towards the lower temperature. Water vapor also flows from one point to another whenever a difference in vapor pressure exists between the two points; the direction of flow is towards the point of low vapor pressure. The rate at which the heat or water vapor will flow varies with the resistance to flow between the two points in the material. If the temperature and vapor pressure of the water vapor correspond to saturation conditions at any point, condensation occurs.

HEAT FLOW THRU BUILDING STRUCTURES

Heat gain thru the exterior construction (walls and roof) is normally calculated at the time of greatest heat flow. It is caused by solar heat being absorbed at the exterior surface and by the temperature difference between the outdoor and indoor air. Both heat sources are highly variable thruout any one day and, therefore, result in unsteady state heat flow thru the exterior construction. This unsteady state flow is difficult to evaluate for each individual situation; however, it can be handled best by means of an equivalent temperature difference across the structure.

The equivalent temperature difference is that temperature difference which results in the total heat flow thru the structure as caused by the variable solar radiation and outdoor temperature. The equivalent temperature difference across the structure must take into account the different types of construction and exposures, time of day, location of the building (latitude), and design conditions. The heat flow thru the structure may then be calculated, using the steady state heat flow equation with the equivalent temperature difference.

$$q = UA\Delta t_e$$

where q = heat flow, Btu/hr
U = transmission coefficient, Btu/(hr)(sq ft)(deg F temp diff)
A = area of surface, sq ft
Δt_e = equiv temp diff F

Heat loss thru the exterior construction (walls and roof) is normally calculated at the time of greatest heat flow. This occurs early in the morning after a few hours of very low outdoor temperatures. This approaches steady state heat flow conditions, and for all practical purposes may be assumed as such.

Heat flow thru the interior construction (floors, ceilings and partitions) is caused by a difference in temperature of the air on both sides of the structure. This temperature difference is essentially constant thruout the day and, therefore, the heat flow can be determined from the steady state heat flow equation, using the actual temperatures on either side.

EQUIVALENT TEMPERATURE DIFFERENCE — SUNLIT AND SHADED WALLS AND ROOFS

The process of transferring heat thru a wall under indicated unsteady state conditions may be visualized by picturing a 12-inch brick wall sliced into 12 one-inch sections. *Assume that temperatures in each slice are all equal at the beginning, and that the indoor and outdoor temperatures remain constant.*

When the sun shines on this wall, most of the solar heat is absorbed in the first slice, *Fig. 22*. This raises the temperature of the first slice above that of the outdoor air and the second slice, causing heat to flow to the outdoor air and also to the second slice, *Fig. 23*. The amount of heat flowing in either direction depends on the resistance to heat flow within the wall and thru the outdoor air film. The heat flow into the second slice, in turn, raises its temperature, causing heat to flow into the third slice, *Fig. 24*. This process of absorbing heat and passing some on to the next slice continues thru the wall to the last or 12th slice where the remaining heat is transferred to the inside by convection and radiation. For this particular wall, it takes approximately 7 hours for

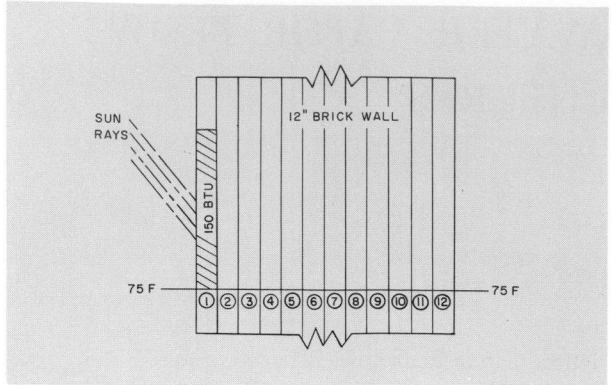

FIG. 22 — SOLAR HEAT ABSORBED IN FIRST SLICE

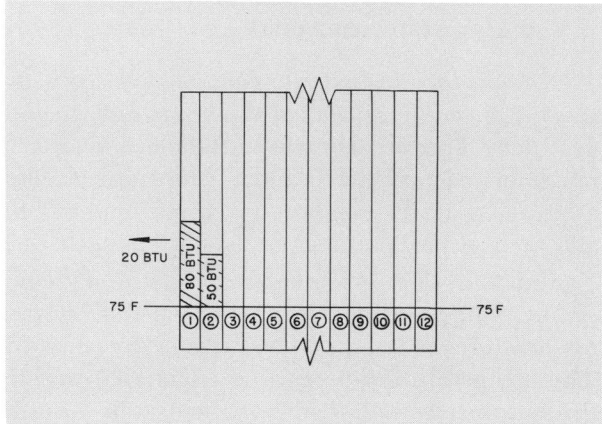

FIG. 23 — BEHAVIOR OF ABSORBED SOLAR HEAT
DURING SECOND TIME INTERVAL

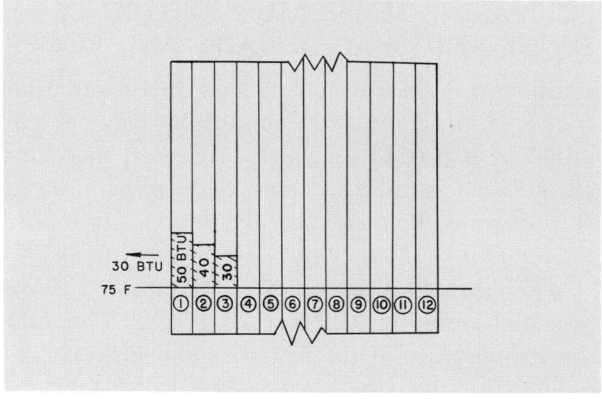

FIG. 24 — BEHAVIOR OF ABSORBED SOLAR HEAT
DURING THIRD TIME INTERVAL

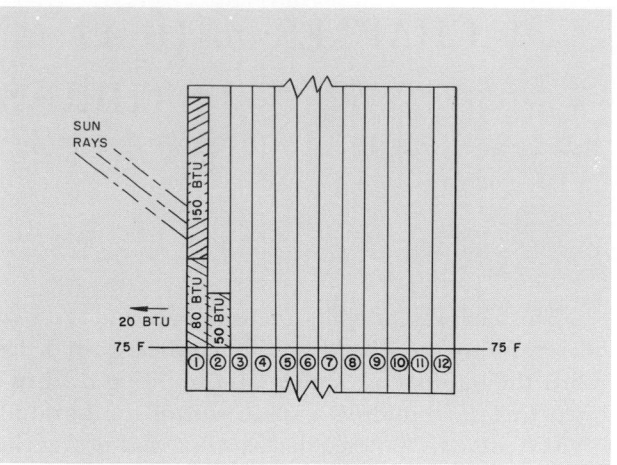

FIG. 25 — BEHAVIOR OF ABSORBED SOLAR HEAT DURING
SECOND TIME INTERVAL PLUS ADDITIONAL SOLAR
HEAT ABSORBED DURING THIS INTERVAL

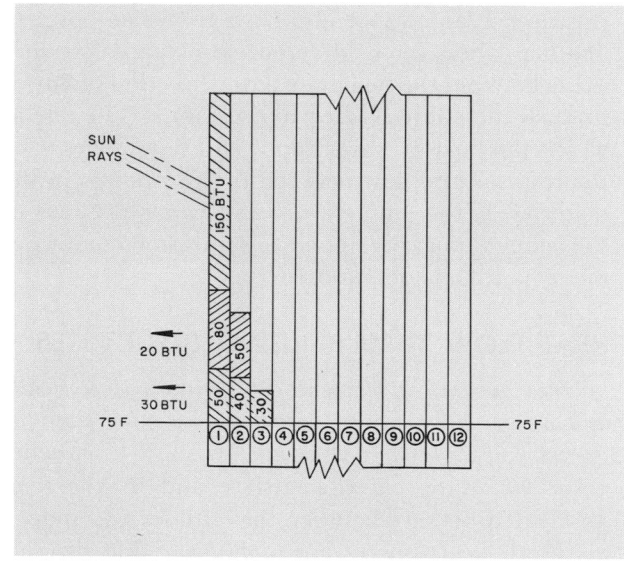

FIG. 26 — BEHAVIOR OF ABSORBED SOLAR HEAT DURING
THIRD TIME INTERVAL PLUS ADDITIONAL SOLAR HEAT
ABSORBED DURING THIS INTERVAL

solar heat to pass thru the wall into the room. Because each slice must absorb some heat before passing it on, the magnitude of heat released to inside space would be reduced to about 10% of that absorbed in the slice exposed to the sun.

These diagrams do not account for possible changes in solar intensity or outdoor temperature.

The solar heat absorbed at each time interval by the outdoor surface of the wall throughout the day goes thru this same process. *Figs. 25 and 26* show the total solar heat flow during the second and third time intervals.

A rise in outdoor temperature reduces the amount of absorbed heat going to the outdoors and more flows thru the wall.

This same process occurs with any type of wall construction to a greater or lesser degree, depending on the resistance to heat flow thru the wall and the thermal capacity of the wall.

NOTE: The thermal capacity of a wall or roof is the density of the material in the wall or roof, times the specific heat of the material, times the volume.

This progression of heat gain to the interior may occur over the full 24-hour period, and may result in a heat gain to the space during the night. If the equipment is operated less than 24 hours, i.e. either skipping the peak load requirement or as a routine procedure, the nighttime radiation to the sky and the lowering of the outdoor temperature may decrease the transmission gain and often may reverse it. Therefore, the heat gain estimate (sun and transmission thru the roof and outdoor walls), even with equipment operating less than 24 hours, may be evaluated by the use of the equivalent temperature data presented in *Tables 19 and 20*.

Basis of Tables 19 and 20
— Equivalent Temperature Difference for Sunlit and Shaded Walls and Roofs

Tables 19 and 20 are analogue computer calculations using Schmidt's method based on the following conditions:

1. Solar heat in July at 40° North latitude.
2. Outdoor daily range of dry-bulb temperatures, 20 deg F.
3. Maximum outdoor temperature of 95 F db and a design indoor temperature of 80 F db, i.e. a design difference of 15 deg F.
4. Dark color walls and roofs with absorptivity of 0.90. For light color, absorptivity is 0.50; for medium color, 0.70.
5. Sun time.

The specific heat of most construction materials is approximately 0.20 Btu/(lb)(deg F); the thermal capacity of typical walls or roofs is proportional to the weight per sq ft; this permits easy interpolation.

Use of Tables 19 and 20
— Equivalent Temperature Difference for Sunlit and Shaded Walls and Roofs

The equivalent temperature differences in *Tables 19 and 20* are multiplied by the transmission coefficients listed in *Tables 21 thru 33* to determine the heat gain thru walls and roofs per sq ft of area during the summer. The total weight per sq ft of walls and roofs is obtained by adding the weights per sq ft of each component of a given structure. These weights are shown in italics and parentheses in *Tables 21 thru 33*.

Example 1 — Equivalent Temperature Difference, Roof
Given:
A flat roof exposed to the sun, with built-up roofing, 1½ in. insulation, 3 in. wood deck and suspended acoustical tile ceiling.
Room design temperature = 80 F db
Outdoor design temperature = 95 F db
Daily range = 20 deg F

Find:
Equivalent temperature difference at 4 p.m. July.

Solution:
Wt/sq ft = 8 + 2 + 2 = 12 lb/sq ft (*Table 27, page 71*)

Equivalent temperature difference
= 43 deg F (*Table 20,* interpolated)

Example 2 — Daily Range and Design Temperature Difference Correction

At times the daily range may be more or less than 20 deg F; the difference between outdoor and room design temperatures may be more or less than 15 deg F. The corrections to be applied to the equivalent temperature difference for combinations of these two variables are listed in the notes following *Tables 19 and 20*.

Given:
The same roof as in *Example 1*
Room design temperature = 78 F db
Outdoor design temperature = 95 F db
Daily range = 26 deg F

Find:
Equivalent temperature difference under changed conditions

Solution:
Design temperature difference = 17 deg F
Daily range = 26 deg F
Correction to equivalent temperature difference
= −1 deg F (*Table 20A,* interpolated)
Equivalent temperature difference = 43 − 1 = 42 deg F

Example 3 — Other Months and Latitudes

Occasionally the heat gain thru a wall or roof must be known for months and latitudes other than those listed in *Note 3* following *Table 20*. This equivalent temperature difference is determined from the equation in *Note 3*. This equation adjusts the equivalent temperature difference for solar radiation only. Additional correction may have to be made for differences between outdoor and indoor design temperatures other than 15 deg F. Refer to *Tables 19 and 20, pages 62 and 63,* and to the correction *Table 20A*. Corrections for these differences must be made first; then the corrected equivalent temperature differences for both sun and shade must be applied in corrections for latitude.

Given:
12 in. common brick wall facing west, with no interior finish, located in New Orleans, 30° North latitude.

Find:
Equivalent temperature difference in November at 12 noon.

Solution:
The correction for design temperature difference is as follows:

Example 3, contd

Summer design dry-bulb for New Orleans
= 95 F db *(Table 1, page 11)*

Winter design dry-bulb for New Orleans
= 20 F db *(Table 1, page 11)*

Yearly range = 75 deg F

Correction in outdoor design temperature for November and a yearly range of 75 deg F
= −15F *(Table 3, page 19)*

Outdoor design dry-bulb temperature in November at 3 p.m.
= 95 − 15 = 80 F

With an 80 F db room design, the outdoor to indoor difference is 80 − 80 = 0 deg F

Average daily range in New Orleans
= 13 deg F *(Table 1, page 11)*

The design difference of 0 deg F and a 13 deg F daily range results in a −11.5 deg F addition to the equivalent temperature difference, by interpolation in *Table 20A.*

Equivalent temperature differences for 12 in. brick wall in New Orleans at 12 noon in November:

Δt_{em} for west wall in sun
= 7 *(Table 19)* − 11.5 = − 4.5 deg F

TABLE 19—EQUIVALENT TEMPERATURE DIFFERENCE (DEG F)

FOR DARK COLORED†, SUNLIT AND SHADED WALLS*

Based on Dark Colored Walls; 95 F db Outdoor Design Temp; Constant 80 F db Room Temp;
20 deg F Daily Range; 24-hour Operation; July and 40° N. Lat.†

EXPOSURE	WEIGHT OF WALL‡ (lb/sq ft)	AM 6	7	8	9	10	11	12	PM 1	2	3	4	5	6	7	8	9	10	11	12	AM 1	2	3	4	5
Northeast	20	5	15	22	23	24	19	14	13	12	13	14	14	14	12	10	8	6	4	2	0	−2	−3	−4	−2
	60	−1	−2	−2	5	24	22	20	15	10	11	12	13	14	13	12	11	10	8	6	4	2	1	0	−1
	100	4	3	4	4	4	10	16	15	14	12	10	11	12	12	12	11	10	9	8	7	6	6	5	5
	140	5	5	6	6	6	6	6	10	14	16	14	12	10	10	10	10	10	10	10	9	9	8	7	7
East	20	1	17	30	33	36	35	32	20	12	13	14	14	14	12	10	8	6	4	2	0	−1	−2	−3	−3
	60	−1	−1	0	21	30	31	31	19	14	13	12	13	14	13	12	11	10	8	5	4	3	1	1	0
	100	5	5	6	8	14	20	24	25	24	20	18	16	14	14	14	13	12	11	10	9	8	7	7	6
	140	11	10	10	9	8	9	10	15	18	19	18	17	16	14	12	13	14	14	14	13	13	12	12	12
Southeast	20	10	6	13	19	26	27	28	26	24	19	16	15	14	12	10	8	6	4	2	0	−1	−1	−2	−2
	60	1	1	0	13	20	24	28	26	25	21	18	15	14	13	12	11	10	8	6	5	4	3	3	2
	100	7	7	6	6	6	11	16	17	18	19	18	16	14	13	12	11	10	10	10	9	9	8	8	7
	140	9	8	8	8	8	7	6	11	14	15	16	18	16	15	14	13	12	12	12	11	11	10	10	9
South	20	−1	−2	−4	1	4	14	22	27	30	28	26	20	16	12	10	7	6	3	2	1	1	0	0	−1
	60	−1	−3	−4	−3	−2	7	12	20	24	25	26	23	20	15	12	10	8	6	4	2	1	1	0	−1
	100	4	4	2	2	2	3	4	8	12	15	16	18	18	15	14	11	10	9	8	7	6	6	6	5
	140	7	6	6	5	4	4	4	4	4	7	10	13	14	15	16	16	14	12	10	10	9	9	8	7
Southwest	20	−2	−4	−4	−2	0	4	6	19	26	34	40	41	42	30	24	12	6	4	2	1	1	0	−1	−1
	60	2	1	0	0	0	1	2	8	12	24	32	35	36	35	34	20	10	7	6	5	4	4	3	3
	100	7	5	6	5	4	5	6	7	8	12	14	19	22	23	24	23	22	15	10	10	9	9	8	7
	140	8	8	8	8	8	7	6	6	6	7	8	9	10	15	18	19	20	13	8	8	8	8	8	8
West	20	−2	−3	−4	−2	0	3	6	14	20	32	40	45	48	34	22	14	8	5	2	1	0	0	−1	−1
	60	2	1	0	0	0	2	4	7	10	19	26	34	40	41	36	28	16	10	6	5	4	3	3	2
	100	7	7	6	6	6	6	6	7	8	10	12	17	20	25	28	27	26	19	14	12	11	10	9	8
	140	12	11	10	9	8	8	8	9	10	10	10	11	12	14	16	21	22	23	22	20	18	16	15	13
Northwest	20	−3	−4	−4	−2	0	3	6	10	12	19	24	33	40	37	34	18	6	4	2	0	−1	−1	−2	−2
	60	−2	−3	−4	−3	−2	0	2	6	8	10	12	21	30	31	32	21	12	8	6	4	3	1	0	−1
	100	5	4	4	4	4	4	4	4	4	5	6	9	12	17	20	21	22	14	8	7	7	6	6	5
	140	8	7	6	6	6	6	6	6	6	6	6	7	8	9	10	14	18	19	20	16	13	11	10	9
North (Shade)	20	−3	−3	−4	−3	−2	1	4	8	10	12	14	13	12	10	8	6	4	2	0	0	−1	−1	−2	−2
	60	−3	−3	−4	−3	−2	−1	0	3	6	8	10	11	12	12	12	10	8	6	4	2	1	0	−1	−2
	100	1	1	0	0	0	0	0	1	2	3	4	5	5	5	8	7	6	5	4	3	3	2	2	1
	140	1	1	0	0	0	0	0	0	0	1	2	3	4	5	6	7	8	7	6	4	3	2	2	1
		6	7	8	9	10	11	12	1	2	3	4	5	6	7	8	9	10	11	12	1	2	3	4	5
		AM							PM												AM				

SUN TIME

Equation: Heat Gain Thru Walls, Btu/hr = (Area, sq ft) × (equivalent temp diff) × (transmission coefficient U, *Tables 21 thru 25*)

*All values are for both insulated and uninsulated walls.

†For other conditions, refer to corrections on *page 64*.

‡"Weight per sq ft" values for common types of construction are listed in *Tables 21 thru 25*.
For wall constructions less than 20 lb/sq ft, use listed values of 20 lb/sq ft.

Δt_{es} for west wall in shade
$= 0 \ (Table \ 19) - 11.5 = -11.5 \ \text{deg F}$

No correction is needed for the time of day; this is accounted for in *Table 19*.

The correction for different solar intensity is

$$\Delta t_e = \Delta t_{es} + \frac{R_s}{R_m} (\Delta t_{em} - \Delta t_{es}) = \frac{R_s}{R_m} \Delta t_{em} + \left(1 - \frac{R_s}{R_m}\right) \Delta t_{es}$$

Wt/sq ft of wall $= 120 \ \text{lb/sq ft} \ (Table \ 21)$
$\Delta t_{es} = -11.5 \ \text{deg F as corrected} \ (Tables \ 19 \ and \ 20A)$
$\Delta t_{em} = -4.5 \ \text{deg F as corrected} \ (Tables \ 19 \ and \ 20A)$
$R_s = 116 \ \text{Btu/hr} \ (Table \ 15, \ page \ 44)$
$R_m = 164 \ \text{Btu/hr} \ (Table \ 15, \ page \ 44)$
$\Delta t_e = -11.5 + \frac{116}{164}[-4.5 - (-11.5)]$
$\quad\quad = -6.5 \ \text{deg F (November, 12 Noon)}$

TABLE 20—EQUIVALENT TEMPERATURE DIFFERENCE (DEG F)

FOR DARK COLORED†, SUNLIT AND SHADED ROOFS*

Based on 95 F db Outdoor Design Temp; Constant 80 F db Room Temp; 20 deg F Daily Range;
24-hour Operation; July and 40° N. Lat.†

CONDI-TION	WEIGHT OF ROOF‡ (lb/sq ft)	SUN TIME																						
		AM							PM													AM			
		6	7	8	9	10	11	12	1	2	3	4	5	6	7	8	9	10	11	12	1	2	3	4	5
Exposed to Sun	10	−4	−6	−7	−5	−1	7	15	24	32	38	43	46	45	41	35	28	22	16	10	7	3	1	−1	−3
	20	0	−1	−2	−1	2	9	16	23	30	36	41	43	43	40	35	30	25	20	15	12	8	6	4	2
	40	4	3	2	3	6	10	16	23	28	33	38	40	41	39	35	32	28	24	20	17	13	11	9	6
	60	9	8	6	7	8	11	16	22	27	31	35	38	39	38	36	34	31	28	25	22	18	16	13	11
	80	13	12	11	11	12	13	16	22	26	28	32	35	37	37	35	34	34	32	30	27	23	20	18	14
Covered with Water	20	−5	−2	0	2	4	10	16	19	22	20	18	16	14	12	10	6	2	1	1	−1	−2	−3	−4	−5
	40	−3	−2	−1	−1	0	5	10	13	15	15	16	15	15	14	12	10	7	5	3	1	−1	−2	−3	−3
	60	−1	−2	−2	−2	−2	2	5	7	10	12	14	15	16	15	14	12	10	8	6	4	3	2	1	0
Sprayed	20	−4	−2	0	2	4	8	12	15	18	17	16	15	14	12	10	6	2	1	0	−1	−2	−2	−3	−3
	40	−2	−2	−1	−1	0	2	5	9	13	14	14	14	14	13	12	9	7	5	3	1	0	0	−1	−1
	60	−1	−2	−2	−2	−2	0	2	5	8	10	12	13	14	13	12	11	10	8	6	4	2	1	0	−1
Shaded	20	−5	−5	−4	−2	0	2	6	9	12	13	14	13	12	10	8	5	2	1	0	−1	−3	−4	−5	−5
	40	−5	−5	−4	−3	−2	0	2	5	8	10	12	13	12	11	10	8	6	4	2	0	−1	−3	−4	−5
	60	−3	−3	−2	−2	−2	−1	0	2	4	6	8	9	10	10	10	9	8	6	4	2	1	0	−1	−2
		6	7	8	9	10	11	12	1	2	3	4	5	6	7	8	9	10	11	12	1	2	3	4	5
		AM							PM													AM			
		SUN TIME																							

Equation: Heat Gain Thru Roofs, Btu/hr = (Area, sq ft) × (equivalent temp diff) × (transmission coefficient U, *Tables 27 or 28*)

*With attic ventilated and ceiling insulated roofs, reduce equivalent temp diff 25%.
 For peaked roofs, use the roof area projected on a horizontal plane.

†For other conditions, refer to corrections below and on *page 64*.

‡"Weight per sq ft" values for common types of construction are listed in *Tables 27 or 28*.

TABLE 20A—CORRECTIONS TO EQUIVALENT TEMPERATURES (DEG F)

OUTDOOR DESIGN FOR MONTH AT 3 P.M. MINUS ROOM TEMP (deg F)	DAILY RANGE (deg F)																
	8	10	12	14	16	18	20	22	24	26	28	30	32	34	36	38	40
−30	−39	−40	−41	−42	−43	−44	−45	−46	−47	−48	−49	−50	−51	−52	−53	−54	−55
−20	−29	−30	−31	−32	−33	−34	−35	−36	−37	−38	−39	−40	−41	−42	−43	−44	−45
−10	−19	−20	−21	−22	−23	−24	−25	−26	−27	−28	−29	−30	−31	−32	−33	−34	−35
0	−9	−10	−11	−12	−13	−14	−15	−16	−17	−18	−19	−20	−21	−22	−23	−24	−25
5	−4	−5	−6	−7	−8	−9	−10	−11	−12	−13	−14	−15	−16	−17	−18	−19	−20
10	1	0	−1	−2	−3	−4	−5	−6	−7	−8	−9	−10	−11	−12	−13	−14	−15
15	6	5	4	3	2	1	0	−1	−2	−3	−4	−5	−6	−7	−8	−9	−10
20	11	10	9	8	7	6	5	4	3	2	1	0	−1	−2	−3	−4	−5
25	16	15	14	13	12	11	10	9	8	7	6	5	4	3	2	1	0
30	21	20	19	18	17	16	15	14	13	12	11	10	9	8	7	6	5
35	26	25	24	23	22	21	20	19	18	17	16	15	14	13	12	11	10
40	31	30	29	28	27	26	25	24	23	22	21	20	19	18	17	16	15

Corrections to Equivalent Temperature Differences in Tables 19 & 20 for Conditions Other Than Basis of Table

1. Outdoor Design Temperature Minus Room Temperature *Greater or Less* Than 15 deg F db, and/or Daily Range *Greater or Less* Than 20 deg F db:

   Add the corrections listed in *Table 20A*, where the outdoor design temperature (*Table 1, page 10*) minus the room or indoor design temperature (*table 4, page 20*) is different from 15 deg F db, or the daily range is different from the 20 deg F db on which *Tables 19 and 20* are based.

   This correction is to be applied to both equivalent temperature difference values, exposed to sun and shaded walls or roof.

2. Shaded walls

   For shaded walls on any exposure, use the values of equivalent temperature difference listed for north (shade), corrected if necessary as shown in Correction 1.

3. Latitudes other than 40° North and for other months with different solar intensities. *Tables 19 and 20* values are approximately correct for the east or west wall in any latitude during the hottest weather. In lower latitudes when the maximum solar altitude is 80° to 90° (the maximum occurs at noon), the temperature difference for either south or north wall is approximately the same as a north or shade wall. See *Table 18* for solar altitude angles.

   The temperature differential Δt_e for any wall facing or roof and for any latitude for any month is approximated as follows:

   $$\Delta t_e = \Delta t_{es} + \frac{R_s}{R_m}(\Delta t_{em} - \Delta t_{es}) = \frac{R_s}{R_m}\Delta t_{em} + (1 - \frac{R_s}{R_m})\Delta t_{es}$$

   where

   Δt_e = equivalent temperature difference for month and time of day desired.

   Δt_{es} = equivalent temperature difference for same wall or roof in shade at desired time of day, corrected if necessary for design conditions.

   Δt_{em} = equivalent temperature difference for wall or roof exposed to the sun for desired time of day, corrected if necessary for design conditions.

   R_s = maximum solar heat gain in Btu/(hr)(sq ft) thru glass for wall facing or horizontal for roofs, for month and latitude desired, *Table 15, page 44*, or *Table 6, page 29*.

   R_m = maximum solar heat gain in Btu/(hr)(sq ft) thru glass for wall facing or horizontal for roofs, for July at 40° North latitude, *Table 15, page 44*, or *Table 6, page 29*.

   Example 3 illustrates the procedure.

4. Light or medium color wall or roof
   Light color wall or roof:

   $$\Delta t_e = \Delta t_{es} + \frac{.50}{.90}(\Delta t_{em} - \Delta t_{es}) = .55\,\Delta t_{em} + .45\,\Delta t_{es}$$

   Medium color wall or roof:

   $$\Delta t_e = \Delta t_{es} + \frac{.70}{.90}(\Delta t_{em} - \Delta t_{es}) = .78\,\Delta t_{em} + .22\,\Delta t_{es}$$

   where:

   Δt_e = equivalent temperature difference for color of wall or roof desired.

Δt_{es} = equivalent temperature difference for same wall or roof in shade at desired time of day, corrected if necessary for design conditions.

Δt_{em} = equivalent temperature difference for wall or roof exposed to the sun for the desired time of day, corrected if necessary for design conditions.

Note: Light color = white, cream, etc.
      Medium color = light green, light blue, gray, etc.
      Dark color = dark blue, dark red, dark brown, etc.

5. Other latitude, other month, light or medium color walls or roof.

   The combined formulae are:

   Light color walls or roof

   $$\Delta t_e = .55\frac{R_s}{R_m}\Delta t_{em} + (1 - .55\frac{R_s}{R_m})\Delta t_{es}$$

   Medium color walls or roof.

   $$\Delta t_e = .78\frac{R_s}{R_m}\Delta t_{em} + (1 - .78\frac{R_s}{R_m})\Delta t_{es}$$

6. For South latitudes, use the following exposure values from *Table 19*:

South Latitude	Use Exposure Value
Northeast	Southeast
East	East
Southeast	Northeast
South	North (shade)
Southwest	Northwest
West	West
Northwest	Southwest
North (shade)	South

TRANSMISSION COEFFICIENT U

Transmission coefficient or U value is the rate at which heat is transferred thru a building structure in Btu/(hr)(sq ft)(deg F temp diff). The rate times the temperature difference is the heat flow thru the structure. The reciprocal of the U value for any wall is the total resistance of this wall to the flow of heat. The total resistance of any wall to heat flow is the summation of the resistance in each component of the structure and the resistances of the outdoor and inside surface films. The transmission coefficients listed in *Tables 21 thru 33* have been calculated for the most common types of construction.

Basis of Tables 21 thru 33
— Transmission Coefficients U for Walls, Roofs, Partitions, Ceilings, Floors, Doors, and Windows

Tables 21 thru 33 contain calculated U values based on the resistance listed in *Table 34, page 78*. The resistance of the outdoor surface film coefficient for summer and winter conditions and the inside surface film is listed in *Table 34*.

Note: The difference between summer and winter transmission coefficients for a typical wall is negligible. For example, with a transmission coefficient of 0.3 Btu/(hr)(sq ft) (F) for winter conditions, the coefficient for summer conditions will be:

1. Thermal resistance R (winter) of wall

$$= \frac{1}{U} = \frac{1}{0.3} = 3.33$$

2. Outdoor film thermal resistance (winter)
   $= 0.17$ (Table 34)

3. Thermal resistance of wall without outdoor air film (winter) $= 3.33 - 0.17 = 3.16$

4. Outdoor film thermal resistance (summer)
   $= 0.25$ (Table 34)

5. Thermal resistance of wall with outdoor air film (summer) $= 3.16 + 0.25 = 3.41$

6. Transmission coefficient U of wall in summer

$$= \frac{1}{R} = \frac{1}{3.41} = 0.294$$

7. Difference between summer and winter transmission becomes greater with larger U values and less with smaller U values.

Use of Tables 21 thru 33
— Transmission Coefficients U for Walls, Roofs, Partitions, Ceilings, Floors, Doors, and Windows

The transmission coefficients may be used for calculating the heat flow for both summer and winter conditions for the average application.

Example 4 — Transmission Coefficients

Given:
   Masonry partition made of 8 in. hollow clay tile, both sides finished, metal lath plastered on furring with ¾ in. sand plaster.

Find:
   Transmission coefficient

Solution:
   Transmission coefficient U
      $= 0.18$ Btu/(hr)(sq ft)(deg F), *Table 26, page 70*

Example 5 — Transmission Coefficient, Addition of Insulation

The transmission coefficients listed in *Tables 21 thru 30* do not include insulation (except for flat roofs, *Table 27, page 71*).

Frequently, fibrous insulation or reflective insulation is included in the exterior building structure. The transmission coefficient for the typical constructions listed in *Tables 21 thru 30*, with insulation, may be determined from *Table 31, page 75*.

Given:
   Masonry wall consisting of 4 in. face brick, 8 in. concrete cinder block, metal lath plastered on furring with ¾ in. sand plaster and 3 in. of fibrous insulation in the stud space.

Find:
   Transmission coefficient.

Solution:
   Refer to *Tables 22 and 31*.
      U value for wall without insulation
         $= 0.24$ Btu/(hr)(sq ft)(deg F)
      U value for wall with insulation
         $= 0.07$ Btu/(hr)(sq ft)(deg F)

TABLE 21—TRANSMISSION COEFFICIENT U—MASONRY WALLS*

FOR SUMMER AND WINTER

Btu/(hr) (sq ft) (deg F temp diff)

All numbers in parentheses indicate weight per sq ft. Total weight per sq ft is sum of wall and finishes.

EXTERIOR FINISH		THICK-NESS (inches) and WEIGHT (lb per sq ft)	None	3/8" Gypsum Board (Plaster Board) (2)	5/8" Plaster on Wall		Metal Lath Plastered on Furring		3/8" Gypsum or Wood Lath Plastered on Furring		Insulating Board Plain or Plastered on Furring	
					Sand Agg (6)	Lt Wt Agg (3)	3/4" Sand Plaster (7)	3/4" Lt Wt Plaster (3)	1/2" Sand Plaster (7)	1/2" Lt Wt Plaster (2)	1/2" Board (2)	1" Board (4)
SOLID BRICK	Face & Common	8 (87)	.48	.41	.45	.41	.31	.28	.29	.27	.22	.16
		12 (123)	.35	.31	.33	.30	.25	.23	.23	.22	.19	.14
		16 (173)	.27	.25	.26	.25	.21	.19	.20	.19	.16	.13
	Common Only	8 (80)	.41	.36	.39	.35	.28	.26	.26	.25	.21	.15
		12 (120)	.31	.28	.30	.27	.23	.22	.22	.21	.18	.14
		16 (160)	.25	.23	.24	.23	.19	.18	.18	.18	.16	.12
STONE		8 (100)	.67	.55	.63	.53	.39	.34	.35	.32	.26	.18
		12 (150)	.55	.47	.52	.46	.34	.31	.31	.29	.24	.17
		16 (200)	.47	.41	.45	.40	.31	.28	.28	.27	.22	.16
		24 (300)	.36	.32	.35	.32	.26	.24	.24	.23	.19	.15
ADOBE-BLOCKS OR BRICK		8 (26)	.34	.30	.32	.30	.25	.23	.23	.22	18	.12
		12 (40)	.25	.23	.24	.23	.20	.18	.18	.18	.15	.14
POURED CONCRETE	140 lb/cu ft	6 (70)	.75	.55	.69	.58	.41	.36	.37	.34	.27	.18
		8 (93)	.67	.49	.63	.53	.39	.34	.35	.32	.26	.17
		10 (117)	.61	.44	.57	.49	.36	.32	.33	.31	.25	.17
		12 (140)	.55	.40	.52	.45	.34	.31	.31	.29	.24	.16
	80 lb/cu ft	6 (40)	.31	.28	.30	.27	.23	.21	.22	.21	.18	.14
		8 (53)	.25	.23	.24	.23	.19	.18	.18	.18	.16	.12
		10 (66)	.21	.19	.20	.19	.17	.16	.15	.14	.14	.11
		12 (80)	.18	.17	.17	.15	.15	.14	.14	.14	.12	.10
	30 lb/cu ft	6 (15)	.13	.13	.13	.13	.12	.11	.11	.11	.13	.09
		8 (20)	.10	.10	.10	.10	.09	.09	.09	.09	.10	.07
		10 (25)	.08	.08	.08	.08	.08	.07	.08	.07	.08	.06
		12 (30)	.07	.07	.07	.07	.07	.07	.06	.06	.07	.06
HOLLOW CONCRETE BLOCKS	Sand & Gravel Agg	8 (43)	.52	.44	.48	.43	.33	.29	.30	.28	.23	.17
		12 (63)	.47	.41	.45	.40	.31	.28	.28	.27	.22	.16
	Cinder Agg	8 (37)	.39	.35	.37	.34	.27	.25	.25	.24	.20	.15
		12 (53)	.36	.33	.35	.32	.26	.24	.23	.23	.19	.15
	Lt Wt Agg	8 (32)	.35	.32	.34	.31	.26	.23	.24	.22	.19	.15
		12 (43)	.32	.29	.31	.28	.24	.22	.22	.21	.18	.14
STUCCO ON HOLLOW CLAY TILE		8 (39)	.36	.32	.34	.32	.26	.24	.24	.23	.19	.15
		10 (44)	.32	.29	.31	.28	.23	.22	.22	.21	.18	.14
		12 (49)	.29	.27	.28	.26	.22	.20	.21	.20	.17	.13

1958 ASHAE Guide

Equations: Heat Gain, Btu/hr = (Area, sq ft) × (U value) × (equivalent temp diff, *Table 19*)

Heat Loss, Btu/hr = (Area, sq ft) × (U value) × (outdoor temp — inside temp)

*For addition of insulation and air spaces to above walls, refer to *Table 31, page 75.*

CHAPTER 5. HEAT AND WATER VAPOR FLOW THRU STRUCTURES

TABLE 22—TRANSMISSION COEFFICIENT U—MASONRY VENEER WALLS*

FOR SUMMER AND WINTER

Btu/(hr) (sq ft) (deg F temp diff)

All numbers in parentheses indicate weight per sq ft. Total weight per sq ft is sum of wall and finishes.

EXTERIOR FINISH	BACKING	THICKNESS (inches) and WEIGHT (lb per sq ft)	None	Gypsum Board (Plaster Board) (2)	5/8" Plaster on Wall		Metal Lath Plastered on Furring		3/8" Gypsum or Wood Lath Plastered on Furring		Insulating Board Plain or Plastered on Furring	
					Sand Agg (6)	Lt Wt Agg (3)	3/4" Sand Plaster(7)	3/4" Lt Wt Plaster(3)	1/2" Sand Plaster(7)	1/2" Lt Wt Plaster(2)	1/2" Board (2)	1" Board (4)
4" Face Brick (43) —or— 4" Stone (50) —or— Precast Concrete (Sand Agg) 4" & 6" (39) (58)	Concrete Block (Cinder Agg)	4 (20)	.41	.37	.39	.35	.28	.26	.26	.25	.21	.16
		8 (37)	.33	.30	.32	.29	.24	.22	.23	.21	.18	.14
		12 (53)	.31	.29	.30	.28	.23	.21	.22	.21	.18	.14
	(Lt Wt Agg)	4 (17)	.35	.32	.34	.31	.25	.23	.24	.22	.19	.15
		8 (32)	.30	.28	.29	.27	.23	.21	.21	.20	.17	.14
		12 (43)	.28	.26	.27	.25	.21	.20	.20	.19	.17	.13
	(Sand & Gravel Agg)	4 (23)	.49	.44	.46	.41	.32	.29	.29	.27	.22	.17
		8 (43)	.41	.37	.39	.35	.28	.26	.26	.25	.21	.16
		12 (63)	.38	.35	.37	.33	.27	.25	.25	.24	.20	.15
	Hollow Clay Tile	4 (16)	.41	.37	.39	.35	.28	.26	.26	.25	.21	.16
		8 (30)	.31	.29	.30	.28	.23	.22	.22	.21	.18	.14
		12 (40)	.26	.25	.25	.24	.20	.19	.19	.18	.16	.13
	Concrete (Lt Wt Agg) 80 lb/cu ft	4 (26)	.35	.31	.34	.31	.25	.23	.24	.22	.19	.15
		6 (40)	.27	.25	.27	.25	.21	.20	.20	.19	.16	.13
		8 (54)	.22	.21	.22	.21	.18	.17	.17	.16	.14	.12
	(Sand & Gravel Agg)	4 (47)	.60	.53	.56	.49	.36	.32	.33	.31	.25	.18
		6 (70)	.55	.49	.52	.45	.34	.31	.32	.29	.24	.17
		8 (95)	.51	.45	.48	.42	.32	.29	.30	.28	.23	.17
	Common Brick	4 (40)	.49	.42	.46	.41	.32	.29	.29	.27	.22	.16
		8 (80)	.35	.31	.34	.31	.25	.23	.24	.22	.19	.15
4" Common Brick (40) —or— Precast Concrete (Sand Agg) 8" & 10" (78) (98) —or— 4" Concrete Block (23) (Sand Agg) —or— 8" Stone (100)	Concrete Block (Cinder Agg)	4 (20)	.36	.33	.35	.32	.26	.24	.24	.23	.19	.15
		8 (37)	.29	.28	.29	.26	.22	.21	.21	.20	.17	.14
		12 (53)	.28	.26	.27	.25	.21	.20	.20	.19	.17	.13
	(Lt Wt Agg)	4 (17)	.32	.29	.30	.28	.23	.22	.22	.21	.18	.14
		8 (32)	.27	.26	.26	.25	.21	.20	.20	.19	.17	.13
		12 (43)	.25	.24	.25	.23	.20	.19	.19	.18	.16	.13
	(Sand & Gravel Agg)	4 (23)	.42	.38	.40	.36	.29	.26	.27	.25	.21	.16
		8 (43)	.36	.33	.35	.32	.26	.24	.24	.23	.19	.15
		12 (63)	.34	.32	.33	.30	.25	.23	.23	.22	.19	.15
	Hollow Clay Tile	4 (16)	.36	.33	.35	.32	.26	.24	.24	.23	.19	.15
		8 (30)	.28	.27	.28	.26	.22	.20	.20	.19	.17	.13
		12 (40)	.24	.23	.23	.22	.19	.18	.18	.17	.15	.12
	Concrete (Lt Wt Agg) 80 lb/cu ft	4 (26)	.32	.29	.30	.28	.23	.22	.22	.21	.18	.14
		6 (40)	.25	.23	.25	.23	.20	.18	.19	.18	.15	.13
		8 (54)	.21	.20	.20	.19	.17	.16	.16	.16	.14	.11
	(Sand & Gravel Agg)	4 (47)	.50	.45	.48	.42	.32	.29	.30	.28	.23	.17
		6 (70)	.47	.42	.44	.39	.31	.28	.29	.27	.22	.17
		8 (95)	.43	.40	.41	.37	.29	.27	.28	.26	.21	.16
	Common Brick	4 (40)	.42	.37	.40	.36	.29	.26	.27	.26	.21	.16
		8 (80)	.32	.29	.30	.28	.23	.22	.22	.21	.18	.14

1958 ASHAE Guide

Equations: Heat Gain, Btu/hr = (Area, sq ft) × (U value) × (equivalent temp diff, *Table 19*)

Heat Loss, Btu/hr = (Area, sq ft) × (U value) × (outdoor temp — inside temp)

*For addition of insulation and air spaces to walls, refer to *Table 31*, page 75.

TABLE 23—TRANSMISSION COEFFICIENT U—LIGHT CONSTRUCTION, INDUSTRIAL WALLS*†

FOR SUMMER AND WINTER

Btu/(hr) (sq ft) (deg F temp diff)

All numbers in parentheses indicate weight per sq ft. Total weight per sq ft is sum of wall and finishes.

EXTERIOR FINISH	SHEATHING	WEIGHT (lb per sq ft)	None	Flat Iron (1)	Insulating Board ½" (2)	Insulating Board 25⁄32" (3)	Wood ¾" (2)
⅜" Corrugated Transite	None	(1)	1.16	.55	.32	.26	.36
	½" Ins. Board	(2)	.34	.26	.19	.17	.21
	25⁄32" Ins. Board	(2)	.27	.21	.17	.15	.18
24 Gauge Corrugated Iron	None	(1)	1.40	.60	.33	.27	.38
	½" Ins. Board	(2)	.36	.27	.20	.17	.21
	25⁄32" Ins. Board	(2)	.28	.22	.17	.15	.18
	¾" Wood	(3)	.46	.33	.22	.19	.24
¾" Wood Siding	None	(2)	.58	.37	.25	.21	.27

1958 ASHAE Guide

Equations: Heat Gain, Btu/hr = (Area, sq ft) × (U value) × (equivalent temp diff, Table 19).

Heat Loss, Btu/hr = (Area, sq ft) × (U value) × (outdoor temp — inside temp).

*For addition of air spaces and insulation to walls, refer to Table 31, page 75.

†Values apply when sealed with calking compound between sheets, and at ground and roof lines. When sheets are not sealed, increase U factors by 10%. These values may be used for roofs, heat flow up-winter; for heat flow down-summer, multiply above factors by 0.8.

TABLE 24—TRANSMISSION COEFFICIENT U—LIGHTWEIGHT, PREFABRICATED CURTAIN TYPE WALLS*

FOR SUMMER AND WINTER

Btu/(hr) (sq ft) (deg F temp diff)

All numbers in parentheses indicate weight per sq ft. Total weight per sq ft is sum of wall and finishes.

INSULATING CORE MATERIAL	DENSITY† (lb/cu ft)	METAL FACING (3) Core Thickness (in.) 1	2	3	4	METAL FACING WITH ¼" AIR SPACE (3) Core Thickness (in.) 1	2	3	4
Glass, Wood, Cotton Fibers	3	.21	.12	.08	.06	.19	.11	.08	.06
Paper Honeycomb	5	.39	.23	.17	.13	.32	.20	.15	.12
Paper Honeycomb with Perlite Fill, Foamglas	9	.29	.17	.12	.09	.25	.15	.11	.09
Fiberboard	15	.36	.21	.15	.12	.29	.19	.14	.11
Wood Shredded (Cemented in Preformed Slabs)	22	.31	.18	.13	.10	.25	.16	.12	.09
Expanded Vermiculite	7	.34	.20	.14	.11	.28	.18	.13	.10
Vermiculite or Perlite Concrete	20	.44	.27	.19	.15	.35	.23	.18	.14
	30	.51	.32	.24	.19	.39	.27	.21	.17
	40	.58	.38	.29	.23	.43	.31	.25	.20
	60	.69	.49	.38	.31	.49	.38	.31	.26

Equations: Heat Gain, Btu/hr = (Area, sq ft) × (U value) × (equivalent temp diff, Table 19).

Heat Loss, Btu/hr = (Area, sq ft) × (U value) × (outdoor temp — inside temp).

*For addition of insulation and air spaces to walls, refer to Table 31, page 75.

†Total weight per sq ft = $\dfrac{\text{core density} \times \text{core thickness}}{12}$ + 3 lb/sq ft

TABLE 25—TRANSMISSION COEFFICIENT U—FRAME WALLS AND PARTITIONS*

FOR SUMMER AND WINTER

Btu/(hr) (sq ft) (deg F temp diff)

All numbers in parentheses indicate weight per sq ft. Total weight per sq ft is sum of component materials.

EXTERIOR FINISH	SHEATHING	None — 3/4" Wood Panel (2)	3/8" Gypsum Board (Plaster Board) (2)	Metal Lath Plastered — 3/4" Sand Plaster(7)	Metal Lath Plastered — 3/4" Lt Wt Plaster(3)	3/8" Gypsum or Wood Lath Plastered — 1/2" Sand Plaster(7)	3/8" Gypsum or Wood Lath Plastered — 1/2" Lt Wt Plaster(2)	Insulating Board Plain or Plastered — 1/2" Board (2)	Insulating Board Plain or Plastered — 1" Board (4)
1" Stucco (10) OR Asbestos Cement Siding (1) OR Asphalt Roll Siding (2)	None, Building Paper	.91	.33	.42	.45	.39	.40	.37	.29
	5/16" Plywood (1) or 1/2" Gyp (2)	.68	.30	.37	.40	.35	.36	.33	.26
	25/32" Wood & Bldg Paper (2)	.48	.25	.30	.31	.28	.29	.27	.22
	1/2" Insulating Board (2)	.42	.23	.27	.29	.26	.27	.25	.21
	25/32" Insulating Board (3)	.32	.20	.23	.24	.22	.22	.21	.18
4" Face Brick Veneer (43) OR 3/8" Plywood (1) OR Asphalt Siding (2)	None, Building Paper	.73	.30	.37	.40	.35	.36	.33	.26
	5/16" Plywood (1) or 1/2" Gyp (2)	.57	.28	.33	.36	.32	.32	.30	.24
	25/32" Wood & Bldg Paper (2)	.42	.23	.27	.29	.26	.27	.25	.21
	1/2" Insulating Board (2)	.38	.22	.25	.27	.25	.25	.24	.20
	25/32" Insulating Board (3)	.30	.19	.21	.22	.21	.21	.20	.17
Wood Siding (3) OR Wood Shingles (2) OR 3/4" Wood Panels (3)	None, Building Paper	.57	.27	.33	.35	.31	.32	.30	.24
	5/16" Plywood (1) or 1/2" Gyp (2)	.48	.25	.30	.31	.28	.29	.27	.22
	25/32" Wood & Bldg Paper	.36	.22	.25	.26	.24	.24	.23	.19
	1/2" Insulating Board (2)	.33	.20	.23	.24	.22	.23	.22	.18
	25/32" Insulating Board (3)	.27	.18	.20	.21	.19	.19	.19	.16
Wood Shingles Over 5/16" Insul Backer Board (3) OR Asphalt Insulated Siding (4)	None, Building Paper	.43	.24	.28	.29	.27	.27	.25	.21
	5/16" Plywood (1) or 1/2" Gyp (2)	.38	.22	.25	.27	.24	.25	.23	.19
	25/32" Wood & Bldg Paper	.30	.19	.22	.23	.21	.21	.20	.17
	1/2" Insulating Board (2)	.28	.18	.20	.21	.20	.20	.19	.16
	25/32" Insulating Board (3)	.23	.16	.18	.18	.17	.18	.17	.15

Last two columns for the four groups (1" Board (4)): .20 .19 .17 .16 .14 | .19 .18 .16 .15 .14 | .18 .17 .15 .14 .13 | .16 .15 .14 .13 .12

		3/4" Wood Panel (2)	3/8" Gypsum Board (2)	3/4" Sand Plaster(7)	3/4" Lt Wt Plaster(3)	1/2" Sand Plaster(7)	1/2" Lt Wt Plaster(2)	1/2" Board (2)	1" Board (4)
Single Partition (Finish on one side only)		.43	.60	.67	.55	.57	.50	.36	.23
Double Partition (Finish on both sides)		.24	.34	.39	.31	.32	.28	.19	.12

1958 ASHAE Guide

Equations: Walls—Heat Gain, Btu/hr = (Area, sq ft) × (U value) × (equivalent temp diff, *Table 19*).

　　　　　—Heat Loss, Btu/hr = (Area, sq ft) × (U value) × (outdoor temp—inside temp).

　　Partitions, unconditioned space adjacent—Heat Gain or Loss, Btu/hr = (Area sq ft) × (U value) × (outdoor temp—inside temp—5 F).

　　Partitions, kitchen or boiler room adjacent—Heat Gain, Btu/hr = (Area sq ft) × (U value)

　　　　　　　　　　　　　　　× (actual temp diff or outdoor temp—inside temp + 15 F to 25 F).

*For addition of insulation and air spaces to partitions, refer to *Table 31, page 75*.

TABLE 26—TRANSMISSION COEFFICIENT U—MASONRY PARTITIONS*

FOR SUMMER AND WINTER

Btu/(hr) (sq ft) (deg F temp diff)

All numbers in parentheses indicate weight per sq ft. Total weight per sq ft is sum of masonry unit and finish × 1 or 2 (finished one or both sides).

BACKING	THICK-NESS (inches) and WEIGHT (per sq ft)	Both Sides Un-finished	No. of Sides Finished	3/8" Gypsum Board (Plaster Board) (2)	5/8" Plaster on Wall Sand Agg (6)	5/8" Plaster on Wall Lt Wt Agg (3)	Metal Lath Plastered on Furring 3/4" Sand Plaster(7)	Metal Lath Plastered on Furring 3/4" Lt Wt Plaster(3)	3/8" Gypsum or Wood Lath Plastered on Furring 1/2" Sand Plaster(7)	3/8" Gypsum or Wood Lath Plastered on Furring 1/2" Lt Wt Plaster(2)	Insulating Board Plain or Plastered on Furring 1/2" Board(2)	Insulating Board Plain or Plastered on Furring 1" Board(4)
HOLLOW CONCRETE BLOCK — Cinder Agg	3 (17)	.45	One	.39	.43	.38	.30	.27	.28	.26	.21	.16
			Both	.35	.41	.33	.23	.20	.20	.18	.14	.10
	4 (20)	.40	One	.36	.39	.35	.28	.26	.26	.25	.20	.15
			Both	.32	.37	.31	.21	.19	.19	.18	.13	.11
	8 (37)	.32	One	.29	.31	.29	.24	.22	.22	.21	.18	.14
			Both	.27	.30	.26	.19	.17	.17	.16	.12	.09
	12 (53)	.31	One	.28	.30	.27	.23	.21	.22	.21	.17	.14
			Both	.26	.29	.25	.18	.16	.17	.15	.12	.09
Lt Wt Agg	3 (15)	.38	One	.34	.36	.33	.27	.25	.25	.24	.20	.15
			Both	.31	.35	.30	.21	.18	.19	.17	.13	.09
	4 (17)	.35	One	.31	.34	.31	.25	.23	.24	.22	.19	.15
			Both	.29	.32	.27	.20	.17	.17	.16	.13	.09
	8 (32)	.30	One	.27	.29	.27	.22	.21	.21	.20	.17	.14
			Both	.25	.28	.24	.18	.16	.16	.15	.12	.09
	12 (43)	.28	One	.25	.27	.25	.21	.20	.20	.19	.16	.13
			Both	.23	.26	.23	.17	.15	.16	.15	.12	.08
Sand & Gravel Agg	8 (43)	.40	One	.36	.39	.35	.28	.26	.26	.25	.20	.15
			Both	.32	.37	.31	.21	.19	.19	.18	.13	.11
	12 (63)	.38	One	.34	.36	.33	.27	.25	.25	.24	.19	.15
			Both	.30	.35	.29	.21	.18	.19	.17	.13	.09
HOLLOW CLAY TILE	3 (15)	.46	One	.40	.44	.39	.31	.28	.28	.27	.22	.16
			Both	.36	.42	.34	.23	.20	.20	.19	.14	.10
	4 (16)	.40	One	.36	.39	.35	.28	.26	.26	.25	.20	.15
			Both	.32	.37	.31	.21	.19	.19	.18	.13	.11
	6 (25)	.35	One	.31	.33	.31	.25	.23	.23	.22	.19	.15
			Both	.28	.32	.27	.20	.17	.18	.16	.13	.09
	8 (30)	.31	One	.28	.30	.28	.23	.22	.22	.21	.18	.14
			Both	.26	.29	.25	.18	.16	.17	.16	.12	.09
HOLLOW GYPSUM TILE	3 (9)	.37	One	.33	.35	.32	.26	.24	.24	.23	.19	.15
			Both	.30	.34	.29	.20	.18	.18	.13	.13	.09
	4 (13)	.33	One	.30	.32	.29	.24	.22	.23	.22	.18	.14
			Both	.27	.31	.26	.19	.17	.17	.16	.12	.09
SOLID GYPSUM PLASTER	1½						.61 (13)	.43 (6)				
	2						.58 (18)	.38 (8)				
	2½						.55 (22)	.34 (9)				

1958 ASHAE Guide

Equations: Partitions, unconditioned space adjacent: Heat Gain or Loss, Btu/hr = (Area, sq ft) × (U value) × (outdoor temp—inside temp—5 F).

Partitions, kitchen or boiler room adjacent: Heat Gain or Loss, Btu/hr = (Area, sq ft) × (U value)
× (actual temp diff or outdoor temp—inside temp + 15 F to 25 F).

*For addition of insulation and air spaces to partitions, refer to *Table 31, page 75.*

TABLE 27—TRANSMISSION COEFFICIENT U—FLAT ROOFS COVERED WITH BUILT-UP ROOFING*

FOR HEAT FLOW DOWN—SUMMER. FOR HEAT FLOW UP—WINTER (See Equation at Bottom of Page).

Btu/(hr) (sq ft) (deg F temp diff)

All numbers in parentheses indicate weight per sq ft. Total weight per sq ft is sum of roof, finish and insulation.

TYPE OF DECK	THICK-NESS OF DECK (inches) and WEIGHT (lb per sq ft)	CEILING †	INSULATION ON TOP OF DECK, INCHES						
			No Insulation	½ (1)	1 (1)	1½ (2)	2 (3)	2½ (3)	3 (4)
Flat Metal	1 (5)	None or Plaster (6)	.67	.35	.23	.18	.15	.12	.10
		Suspended Plaster (5)	.32	.22	.17	.14	.12	.10	.09
		Suspended Acou Tile (2)	.23	.18	.14	.12	.11	.09	.08
Preformed Slabs—Wood Fiber and Cement Binder	2 (4)	None or Plaster (6)	.20	.16	.13	.11	.10	.09	.08
		Suspended Plaster (5)	.15	.12	.11	.09	.08	.08	.07
		Suspended Acou Tile (2)	.13	.10	.09	.08	.08	.07	.06
	3 (7)	None or Plaster (6)	.14	.11	.10	.09	.08	.08	.07
		Suspended Plaster (5)	.12	.10	.09	.07	.07	.06	.05
		Suspended Acou Tile (2)	.10	.09	.08	.07	.07	.06	.05
Concrete (Sand & Gravel Agg)	4, 6, 8 (47),(70), (93)	None or Plaster (6)	.51	.30	.21	.16	.14	.12	.10
		Suspended Plaster (5)	.28	.20	.16	.13	.12	.10	.09
		Suspended Acou Tile(2)	.21	.16	.13	.11	.10	.09	.08
(Lt Wt Agg on Gypsum Board)	2 (9)	None or Plaster (6)	.27	.20	.15	.13	.11	.10	.08
		Suspended Plaster (5)	.18	.14	.12	.10	.09	.09	.08
		Suspended Acou Tile (2)	.15	.12	.11	.09	.08	.08	.07
	3 (13)	None or Plaster (6)	.21	.16	.13	.11	.10	.09	.08
		Suspended Plaster (5)	.15	.12	.11	.09	.08	.08	.07
		Suspended Acou Tile (2)	.13	.11	.10	.08	.08	.07	.06
	4 (16)	None or Plaster (6)	.17	.14	.11	.10	.09	.08	.07
		Suspended Plaster (5)	.13	.11	.10	.08	.08	.07	.06
		Suspended Acou Tile(2)	.12	.10	.09	.07	.07	.06	.05
Gypsum Slab on ½" Gypsum Board	2 (11)	None or Plaster (6)	.32	.22	.17	.14	.12	.10	.09
		Suspended Plaster (5)	.21	.17	.13	.11	.10	.09	.08
		Suspended Acou Tile (2)	.17	.13	.12	.10	.09	.08	.07
	3 (15)	None or Plaster (6)	.27	.19	.15	.13	.11	.10	.08
		Suspended Plaster (5)	.19	.15	.13	.11	.10	.09	.08
		Suspended Acou Tile (2)	.15	.12	.11	.09	.08	.08	.07
	4 (19)	None or Plaster (6)	.23	.17	.14	.12	.10	.09	.08
		Suspended Plaster (5)	.17	.13	.12	.10	.09	.08	.07
		Suspended Acou Tile (2)	.14	.12	.11	.09	.08	.08	.07
Wood	1 (3)	None or Plaster (6)	.40	.26	.19	.15	.13	.11	.09
		Suspended Plaster (5)	.24	.18	.14	.12	.11	.09	.08
		Suspended Acou Tile (2)	.19	.15	.13	.11	.10	.08	.07
	2 (5)	None or Plaster (6)	.28	.20	.16	.13	.11	.10	.08
		Suspended Plaster (5)	.19	.15	.13	.11	.10	.09	.07
		Suspended Acou Tile (2)	.16	.13	.11	.10	.09	.08	.07
	3 (8)	None or Plaster (6)	.21	.16	.13	.11	.10	.09	.08
		Suspended Plaster (5)	.16	.13	.11	.09	.09	.08	.07
		Suspended Acou Tile (2)	.13	.11	.10	.09	.08	.07	.06

1958 ASHAE Guide

Equations: Summer—(Heat Flow Down) Heat Gain, Btu/hr = (Area, sq ft) × (U value) × (equivalent temp diff, *Table 20*).

Winter—(Heat Flow Up) Heat Loss, Btu/hr = (Area, sq ft) × (U value × 1.1) × (outdoor temp—inside temp).

*For addition of air spaces or insulation to roofs, refer to *Table 31*, page 75.

†For suspended ½" insulation board, plain (.6) or with ½" sand aggregate plaster (5). use values of suspended acou tile.

TABLE 28—TRANSMISSION COEFFICIENT U—PITCHED ROOFS*

FOR HEAT FLOW DOWN—SUMMER. FOR HEAT FLOW UP—WINTER (See Equation at Bottom of Page)

Btu/(hr) (sq ft projected area) (deg F temp diff)

All numbers in parentheses indicate weight per sq ft. Total weight per sq ft is sum of component materials.

PITCHED ROOFS		CEILING										
		None	¾" Wood Panel (2)	⅜" Gypsum Board (Plaster Board) (2)	Metal Lath Plastered		⅜" Gypsum or Wood Lath Plastered		Insulating Board Plain or ½" Sand Agg Plastered		Acoustical Tile on Furring or ⅜" Gypsum	
EXTERIOR SURFACE	SHEATHING				¾" Sand Plaster (7)	¾" Lt Wt Plaster (3)	½" Sand Plaster (5)	½" Lt Wt Plaster (2)	½" Board (2)	1" Board (4)	½" Tile (2)	¾" Tile (3)
Asphalt Shingles, (2)	Bldg paper on ⁵⁄₁₆" plywood (2)	.51	.27	.30	.32	.29	.29	.28	.22	.17	.23	.21
	Bldg paper on ²⁵⁄₃₂" wood sheathing (3)	.30	.23	.26	.27	.25	.25	.24	.20	.16	.21	.19
Asbestos-Cement Shingles (3) or Asphalt Roll Roofing (1)	Bldg paper on ⁵⁄₁₆" plywood (2)	.59	.28	.34	.37	.33	.33	.31	.25	.18	.25	.22
	Bldg paper on ²⁵⁄₃₂" wood sheathing (3)	.45	.25	.29	.31	.28	.28	.27	.22	.17	.22	.20
Slates (8) Tile (10) or Sheet Metal (1)	Bldg paper on ⁵⁄₁₆" plywood (2)	.64	.29	.36	.38	.34	.35	.47	.26	.19	.26	.23
	Bldg paper on ²⁵⁄₃₂" wood sheathing (3)	.48	.25	.29	.31	.28	.28	.27	.22	.17	.23	.20
Wood Shingles (2)	Bldg paper on 1" x 4" strips (1)	.53	.26	.31	.33	.30	.30	.28	.23	.17	.24	.21
	Bldg paper on ⁵⁄₁₆" plywood (2)	.41	.23	.27	.29	.26	.27	.25	.21	.16	.21	.19
	Bldg paper on ²⁵⁄₃₂" wood sheathing (3)	.34	.21	.24	.25	.23	.23	.22	.19	.15	.19	.17

1958 ASHAE Guide

Equations: Summer (Heat Flow Down) Heat Gain, Btu/hr = (horizontal projected area, sq ft) × (U value) × (equivalent temp diff, *Table 20*).
Winter (Heat Flow Up) Heat Loss, Btu/hr = (horizontal projected area, sq ft) × (U value × 1.1) × (outdoor temp − inside temp).

*For addition of air spaces or insulation for above roofs, refer to *Table 31, page 75*.

TABLE 29—TRANSMISSION COEFFICIENT U—CEILING AND FLOOR, (Heat Flow Up)

Based on Still Air Both Sides, Btu/(hr) (sq ft) (deg F temp diff)

All numbers in parentheses indicate weight per sq ft. Total weight per sq ft is sum of ceiling and floor.

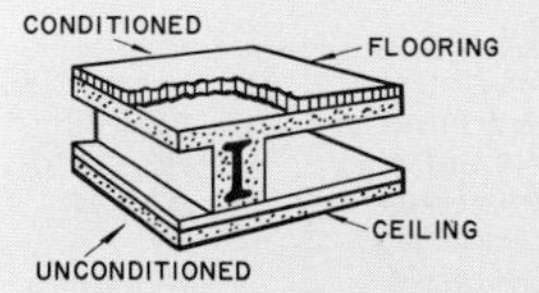

			MASONRY CEILING												
			Not Furred						Suspended or Furred						
		THICK-NESS (inches) and WEIGHT (lb per sq ft)	None or ½" Sand Plaster (5)	½" Lt Wt Plaster (3)	Acoustical Tile Glued		Metal Lath Plastered		⅜" Gypsum or Wood Lath Plastered		Insulating Board Plain or ½" Sand Agg Plastered		Acoustical Tile on Furring or ⅜" Gypsum	
					½" Tile (1)	¾" Tile (1)	¾" Sand Plaster (7)	¾" Lt Wt Plaster (3)	½" Sand Plaster (5)	½" Lt Wt Plaster (2)	½" Board (2)	1" Board (4)	½" Tile (1)	¾" Tile (1)
FLOOR	CONCRETE SUBFLOOR													
None or ⅛" Linoleum or Floor Tile	Sand Agg	2 (19)	.70	.53	.38	.31	.43	.38	.44	.41	.26	.19	.28	.24
		4 (39)	.63	.49	.36	.30	.41	.36	.41	.38	.25	.18	.26	.23
		6 (59)	.57	.45	.34	.28	.38	.34	.39	.36	.24	.18	.25	.22
		8 (79)	.52	.42	.32	.27	.36	.32	.37	.34	.23	.17	.24	.21
		10 (99)	.48	.39	.31	.26	.34	.31	.35	.32	.23	.17	.23	.21
	Lt Wt Agg 80 lb/ft³	2 (15)	.48	.39	.31	.26	.34	.31	.35	.32	.23	.17	.23	.21
		4 (28)	.35	.30	.25	.22	.27	.25	.27	.26	.19	.15	.20	.18
		6 (41)	.27	.24	.21	.18	.22	.21	.22	.21	.17	.13	.17	.15
13/16" Wood Block on Slab	Sand Agg	2 (20)	.47	.39	.30	.26	.33	.30	.33	.40	.22	.17	.23	.20
		4 (40)	.44	.36	.29	.25	.31	.28	.32	.38	.22	.16	.22	.20
		6 (60)	.41	.34	.28	.24	.30	.27	.30	.36	.21	.16	.22	.19
		8 (80)	.38	.33	.26	.23	.28	.26	.29	.34	.20	.15	.21	.19
		10 (100)	.36	.31	.25	.22	.27	.25	.27	.32	.19	.15	.20	.18
	Lt Wt Agg 80 lb/ft³	2 (16)	.36	.31	.25	.22	.27	.25	.27	.32	.19	.15	.20	.18
		4 (29)	.28	.25	.21	.19	.22	.21	.23	.26	.17	.13	.17	.16
		6 (42)	.23	.21	.18	.16	.19	.18	.19	.21	.15	.12	.15	.14
Floor Tile or ⅛" Linoleum on ⅝" Plywood on 2" x 2" Sleepers	Sand Agg	2 (22)	.32	.28	.23	.21	.31	.28	.32	.30	.18	.14	.18	.17
		4 (42)	.31	.27	.23	.20	.30	.27	.30	.28	.18	.14	.18	.17
		6 (62)	.29	.26	.22	.19	.20	.26	.29	.27	.17	.14	.18	.16
		8 (82)	.28	.25	.21	.19	.27	.25	.27	.26	.17	.13	.17	.16
		10 (102)	.27	.24	.20	.18	.26	.24	.26	.25	.16	.13	.17	.15
	Lt Wt Agg 80 lb/ft³	2 (19)	.27	.24	.20	.18	.26	.24	.26	.25	.16	.13	.17	.15
		4 (31)	.22	.20	.17	.16	.22	.20	.22	.21	.14	.12	.15	.14
		6 (44)	.19	.17	.15	.14	.18	.17	.19	.18	.13	.11	.13	.12
¾" Hardwood on 25/32" Subfloor on 2" x 2" Sleepers	Sand Agg	2 (24)	.26	.23	.20	.18	.25	.23	.25	.24	.16	.13	.16	.15
		4 (44)	.25	.22	.19	.17	.24	.22	.24	.23	.16	.13	.16	.15
		6 (64)	.24	.21	.19	.17	.23	.21	.23	.22	.15	.12	.16	.14
		8 (84)	.23	.21	.18	.16	.22	.21	.22	.21	.15	.12	.15	.14
		10 (104)	.22	.20	.17	.16	.21	.20	.22	.21	.14	.12	.15	.14
	Lt Wt Agg 80 lb/ft³	2 (20)	.22	.20	.17	.16	.21	.20	.22	.21	.14	.12	.15	.14
		4 (33)	.19	.17	.15	.14	.18	.17	.18	.18	.13	.11	.13	.12
		6 (46)	.16	.15	.14	.13	.16	.15	.16	.16	.12	.099	.12	.11

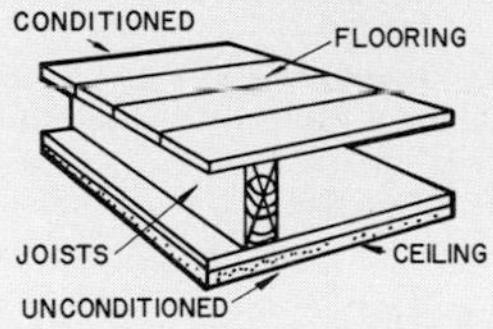

		FRAME CONSTRUCTION CEILING											
		Not Furred						Suspended or Furred					
		None	Acoustical Tile Glued		Metal Lath Plastered		⅜" Gypsum or Wood Lath Plastered		Insulating Board Plain or ½" Sand Agg Plastered		Acoustical Tile on Furring or ⅜" Gypsum	
			½" Tile (1)	¾" Tile (1)	¾" Sand Plaster (7)	¾" Lt Wt Plaster (3)	½" Sand Plaster (5)	½" Lt Wt Plaster (2)	½" Board (2)	1" Board (4)	½" Tile (1)	¾" Tile (1)
FLOOR	SUBFLOOR												
None	None				.74	.59	.61	.54	.37	.24	.39	.31
	25/32" Wood (2)	.45	.30	.26	.31	.28	.29	.27	.22	.17	.23	.20
	2" Wood (5)	.27	.20	.18	.22	.20	.20	.19	.17	.14	.17	.15
½" Ceramic Tile on 1½" Cement	25/32" Wood (21)	.38	.21	.19	.28	.26	.26	.24	.20	.16	.21	.19
	2" Wood (24)	.24	.19	.17	.20	.19	.19	.18	.16	.13	.16	.15
¾" Hardwood Floor or Linoleum on ⅝" Plywood	25/32" Wood (5)	.33	.24	.21	.25	.23	.23	.22	.18	.15	.19	.17
	2" Wood (7)	.22	.17	.16	.18	.17	.17	.17	.15	.12	.15	.14
⅛" Linoleum on ¼" Hardboard on ⅜" Insulating Board	25/32" Wood (5)	.28	.21	.19	.22	.20	.21	.20	.17	.14	.18	.16
	2" Wood (8)	.20	.16	.15	.17	.16	.16	.16	.14	.12	.14	.13

1958 ASHAE Guide

Equations: Heat flow up, Unconditioned space below: Heat Gain, Btu/hr = (Area, sq ft) × (U value) × (outdoor temp — inside temp — 5 F).

Kitchen or boiler room below: Heat Gain, Btu/hr = (Area, sq ft) × (U value)

× (actual temp diff, or outdoor temp — inside temp + 15 F to 25 F).

TABLE 30—TRANSMISSION COEFFICIENT U—CEILING AND FLOOR, (Heat Flow Down)

Based on Still Air Both Sides, Btu/(hr) (sq ft) (deg F temp diff)

All numbers in parentheses indicate weight per sq ft. Total weight per sq ft is sum of ceiling and floor.

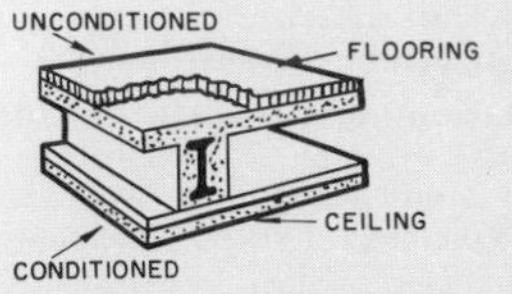

MASONRY CEILING

FLOOR	CONCRETE SUBFLOOR	THICKNESS (inches) and WEIGHT (lb per sq ft)	Not Furred: None or ½" Sand Plaster (5)	½" Lt Wt Plaster (3)	Acoustical Tile Glued: ½" Tile (1)	¾" Tile (1)	Metal Lath Plastered: ¾" Sand Plaster (7)	¾" Lt Wt Plaster (3)	3/8" Gypsum or Wood Lath Plastered: ½" Sand Plaster (5)	½" Lt Wt Plaster (2)	Insulating Board Plain or ½" Sand Agg Plastered: ½" Board (2)	1" Board (4)	Acoustical Tile on Furring or 3/8" Gypsum: ½" Tile (1)	¾" Tile (1)
None or 1/8" Linoleum or Floor Tile	Sand Agg	2 (19)	.48	.43	.31	.26	.32	.29	.30	.28	.23	.17	.23	.20
		4 (39)	.44	.40	.30	.25	.31	.28	.28	.27	.22	.17	.22	.20
		6 (59)	.41	.37	.28	.24	.29	.27	.27	.26	.21	.16	.22	.19
		8 (79)	.39	.35	.27	.23	.28	.26	.26	.25	.21	.16	.21	.19
		10 (99)	.36	.34	.26	.22	.27	.25	.25	.24	.20	.15	.20	.18
	Lt Wt Agg 80 lb/ft³	2 (15)	.36	.34	.26	.22	.27	.25	.25	.24	.20	.15	.20	.18
		4 (28)	.29	.26	.21	.19	.22	.21	.21	.20	.17	.14	.17	.16
		6 (41)	.23	.22	.18	.17	.19	.18	.18	.17	.15	.13	.15	.14
13/16" Wood Block on Slab	Sand Agg	2 (20)	.36	.33	.25	.22	.26	.24	.24	.23	.20	.15	.20	.18
		4 (40)	.33	.31	.24	.21	.25	.23	.23	.22	.19	.15	.19	.17
		6 (60)	.32	.29	.23	.21	.24	.22	.22	.21	.18	.15	.18	.17
		8 (80)	.30	.28	.23	.20	.23	.22	.22	.21	.18	.14	.18	.16
		10 (100)	.29	.27	.22	.19	.22	.21	.21	.20	.17	.14	.17	.16
	Lt Wt Agg 80 lb/ft³	2 (16)	.29	.27	.22	.19	.22	.21	.21	.20	.17	.14	.17	.16
		4 (29)	.23	.22	.19	.17	.19	.18	.18	.17	.15	.13	.15	.14
		6 (42)	.20	.19	.16	.15	.16	.16	.16	.15	.14	.11	.14	.13
Floor Tile or 1/8" Linoleum on 5/8" Plywood on 2" x 2" Sleepers	Sand Agg	2 (22)	.33	.31	.24	.21	.25	.23	.23	.22	.19	.15	.20	.17
		4 (42)	.32	.29	.23	.21	.24	.22	.22	.21	.18	.15	.19	.17
		6 (62)	.30	.28	.23	.20	.23	.21	.22	.21	.18	.14	.18	.16
		8 (82)	.29	.27	.22	.19	.22	.21	.21	.20	.17	.14	.18	.16
		10 (102)	.28	.26	.21	.19	.21	.20	.20	.19	.17	.13	.17	.15
	Lt Wt Agg 80 lb/ft³	2 (19)	.28	.26	.21	.19	.21	.20	.20	.19	.17	.13	.17	.15
		4 (31)	.22	.21	.18	.16	.18	.17	.17	.17	.15	.12	.15	.14
		6 (44)	.19	.18	.16	.14	.16	.15	.15	.15	.13	.11	.14	.13
3/4" Hardwood on 25/32" Subfloor on 2" x 2" Sleepers	Sand Agg	2 (24)	.26	.25	.20	.18	.20	.20	.20	.19	.16	.13	.17	.15
		4 (44)	.25	.24	.20	.18	.20	.19	.19	.18	.16	.13	.16	.15
		6 (64)	.24	.23	.19	.17	.19	.18	.19	.18	.15	.13	.16	.14
		8 (84)	.23	.22	.19	.17	.19	.18	.18	.17	.15	.12	.15	.14
		10 (104)	.22	.21	.18	.16	.18	.17	.17	.17	.14	.12	.15	.14
	Lt Wt Agg 80 lb/ft³	2 (20)	.22	.21	.18	.16	.18	.17	.17	.17	.14	.12	.15	.14
		4 (33)	.19	.18	.16	.14	.16	.15	.15	.15	.13	.11	.13	.12
		6 (46)	.16	.16	.14	.13	.14	.14	.14	.13	.12	.10	.12	.11

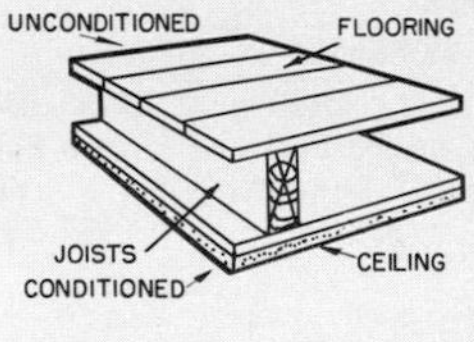

FRAME CONSTRUCTION CEILING

FLOOR	SUBFLOOR	Not Furred: None	Acoustical Tile Glued: ½" Tile (1)	¾" Tile (1)	Metal Lath Plastered: ¾" Sand Plaster (7)	¾" Lt Wt Plaster (3)	3/8" Gypsum or Wood Lath Plastered: ½" Sand Plaster (5)	½" Lt Wt Plaster (2)	Insulating Board Plain or ½" Sand Agg Plastered: ½" Board (2)	1" Board (4)	Acoustical Tile on Furring or 3/8" Gypsum: ½" Tile (1)	¾" Tile (1)
None	None				.51	.43	.44	.40	.31	.21	.31	.27
	25/32" Wood (2)	.35	.25	.22	.26	.24	.24	.23	.19	.15	.20	.17
	2" Wood (5)	.27	.18	.16	.19	.17	.18	.17	.15	.12	.15	.14
½" Ceramic Tile on 1½" Cement	25/32" Wood (21)	.38	.18	.17	.19	.18	.18	.17	.15	.12	.15	.14
	2" Wood (24)	.24	.14	.13	.15	.14	.14	.14	.12	.11	.12	.12
13/16" Hardwood Floor or Linoleum on 5/8" Plywood	25/32" Wood (5)	.33	.17	.16	.18	.17	.17	.16	.14	.12	.14	.13
	2" Wood (7)	.22	.14	.13	.14	.13	.13	.13	.12	.10	.12	.11
1/8" Linoleum on 1/4" Hardboard on 3/8" Insulating Board	25/32" Wood (5)	.29	.16	.15	.16	.15	.16	.15	.13	.11	.14	.13
	2" Wood (8)	.20	.13	.12	.13	.12	.13	.12	.11	.10	.11	.11

1958 ASHAE Guide

Equations: Heat flow down, unconditioned space above: Heat Gain, Btu/hr = (Area, sq ft) × (U value) × (outdoor temp — inside temp — 5 F).

Kitchen above: Heat Gain, Btu/hr = (Area, sq ft) × (U value) × (actual temp diff, or outdoor temp — inside temp + 15 F to 25 F).

TABLE 31—TRANSMISSION COEFFICIENT U—WITH INSULATION & AIR SPACES

SUMMER AND WINTER

Btu/(hr) (sq ft) (deg F temp diff)

| U Value Before Adding Insul. Wall, Ceiling, Roof Floor | Addition of Fibrous Insulation Thickness (Inches) | | | Add'n of Air Space 3/4" or more * | Addition of Reflective Sheets to Air Space (Aluminum Foil Average Emissivity = .05) Direction of Heat Flow | | | | | | | | |
| | | | | | Winter and Summer Horizontal | | | Summer Down | | | Winter Up | | |
	1	2	3		Added to one or both sides	One sheet in air space	Two sheets in air space	Added to one or both sides	One sheet in air space	Two sheets in air space	Added to one or both sides	One sheet in air space	Two sheets in air space
.60	.19	.11	.08	.38	.34	.18	.11	.12	.06	.05	.36	.20	.14
.58	.19	.11	.08	.37	.33	.18	.11	.12	.06	.05	.36	.20	.14
.56	.18	.11	.08	.36	.32	.18	.11	.11	.06	.05	.35	.20	.14
.54	.18	.11	.08	.36	.31	.17	.11	.11	.06	.05	.34	.19	.14
.52	.18	.11	.08	.35	.30	.17	.10	.11	.06	.05	.33	.19	.14
.50	.18	.11	.08	.34	.29	.17	.10	.11	.06	.05	.32	.19	.13
.48	.17	.11	.08	.33	.28	.16	.10	.11	.06	.04	.31	.18	.13
.46	.17	.10	.08	.32	.28	.16	.10	.11	.06	.04	.30	.18	.13
.44	.17	.10	.07	.31	.27	.16	.10	.11	.06	.04	.29	.18	.13
.42	.16	.10	.07	.30	.26	.15	.10	.11	.06	.04	.28	.17	.13
.40	.16	.10	.07	.29	.26	.15	.10	.10	.06	.04	.27	.17	.12
.38	.16	.10	.07	.28	.25	.15	.09	.10	.06	.04	.26	.17	.12
.36	.15	.10	.07	.27	.24	.14	.09	.10	.06	.04	.25	.16	.12
.34	.15	.10	.07	.26	.23	.14	.09	.10	.06	.04	.24	.16	.12
.32	.15	.10	.07	.25	.22	.13	.09	.10	.05	.04	.23	.15	.11
.30	.14	.09	.07	.23	.21	.13	.09	.10	.05	.04	.22	.15	.11
.28	.14	.09	.07	.22	.20	.13	.08	.09	.05	.04	.20	.14	.10
.26	.13	.09	.07	.21	.19	.12	.08	.09	.05	.04	.19	.13	.10
.24	.13	.09	.07	.20	.17	.12	.08	.09	.05	.04	.18	.13	.10
.22	.12	.08	.06	.18	.16	.11	.08	.08	.05	.04	.16	.12	.09
.20	.12	.08	.06	.17	.15	.10	.07	.08	.05	.04	.15	.11	.09
.18	.11	.08	.06	.15	.14	.10	.07	.08	.05	.04	.14	.11	.08
.16	.10	.07	.06	.14	.12	.09	.07	.07	.05	.04	.13	.10	.08
.14	.09	.07	.05	.12	.11	.08	.06	.07	.04	.04	.12	.09	.07
.12	.08	.06	.05	.11	.10	.08	.06	.06	.04	.03	.10	.08	.07
.10	.07	.06	.05	.09	.08	.07	.05	.06	.04	.03	.09	.07	.06

1958 ASHAE Guide

Insulation Added	Air Space Added	Reflective Sheets Added to One or Both Sides	Reflective Sheet in Air Space	Reflective Sheets in Air Space
AIR SPACES	AIR SPACE	AIR SPACE	AIR SPACES	AIR SPACES
INSULATION	DIVIDER	REFLECTIVE SHEETS	REFLECTIVE SHEETS	REFLECTIVE SHEETS

*Checked for summer conditions for up, down and horizontal heat flow. Error from above values is less than 1%.

TABLE 32—TRANSMISSION COEFFICIENT U—FLAT ROOFS WITH ROOF-DECK INSULATION

SUMMER AND WINTER

Btu/(hr) (sq ft) (deg F temp diff)

U VALUE OF ROOF BEFORE ADDING ROOF DECK INSULATION	Addition of Roof-Deck Insulation Thickness (in.)					
	½	1	1½	2	2½	3
.60	.33	.22	.17	.14	.12	.10
.50	.29	.21	.16	.14	.12	.10
.40	.26	.19	.15	.13	.11	.09
.35	.24	.18	.14	.12	.10	.09
.30	.21	.16	.13	.12	.10	.09
.25	.19	.15	.12	.11	.09	.08
.20	.16	.13	.11	.10	.09	.08
.15	.12	.11	.09	.08	.08	.07
.10	.09	.08	.07	.07	.06	.05

TABLE 33—TRANSMISSION COEFFICIENT U—WINDOWS, SKYLIGHTS, DOORS & GLASS BLOCK WALLS

Btu/(hr) (sq ft) (deg F temp diff)

GLASS											
	Vertical Glass							Horizontal Glass			
	Single	Double			Triple			Single		Double (¼")	
Air Space Thickness (in.)		¼	½	¾-4	¼	½	¾-4	Summer	Winter	Summer	Winter
Without Storm Windows	1.13	0.61	0.55	0.53	0.41	0.36	0.34	0.86	1.40	0.50	0.70
With Storm Windows	0.54							0.43	0.64		

DOORS		
Nominal Thickness of Wood (inches)	U Exposed Door	U With Storm Door
1	0.69	0.35
1¼	0.59	0.32
1½	0.52	0.30
1¾	0.51	0.30
2	0.46	0.28
2½	0.38	0.25
3	0.33	0.23
Glass (¾" Herculite)	1.05	0.43

HOLLOW GLASS BLOCK WALLS	
Description*	U
5¾x5¾x3⅞" Thick—Nominal Size 6x6x4 (14)	0.60
7¾x7¾x3⅞" Thick—Nominal Size 8x8x4 (14)	0.56
11¾x11¾x3⅞" Thick—Nominal Size 12x12x4 (16)	0.52
7¾x7¾x3⅞" Thick with glass fiber screen dividing the cavity (14)	0.48
11¾x11¾x3⅞" Thick with glass fiber screen dividing the cavity (16)	0.44

1958 ASHAE Guide

Equation: Heat Gain or Loss, Btu/hr = (Area, sq ft) × (U value) × (outdoor temp — inside temp)

*Italicized numbers in parentheses indicate weight in lb per sq ft.

CALCULATION OF TRANSMISSION COEFFICIENT U

For types of construction not listed in *Tables 21 thru 33,* calculate the *U* value as follows:

1. Determine the resistance of each component of a given structure and also the inside and outdoor air surface films from *Table 34.*

2. Add these resistances together,
$$R = r_1 + r_2 + r_3 + \ldots . r_n$$

3. Take the reciprocal, $U = \dfrac{1}{R}$

Basis of Table 34
— Thermal Resistance R, Building and Insulating Materials

Table 34 was extracted from the 1958 ASHAE Guide and the column "weight per sq ft" added.

Use of Table 34
— Thermal Resistance R, Building and Insulating Materials

The thermal resistances for building materials are listed in two columns. One column lists the thermal resistance per inch thickness, based on conductivity, while the other column lists the thermal resistance for a given thickness or construction, based on conductance.

Example 6 — Calculation of U Value

Given:
 A wall as per *Fig. 27*

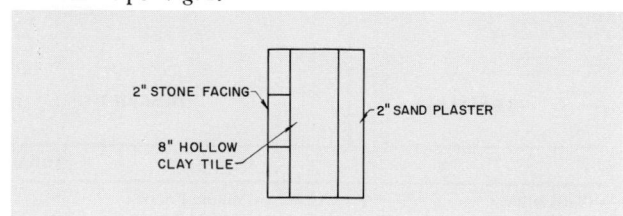

FIG. 27 — OUTDOOR WALL

Find:
 Transmission coefficient in summer.

Solution:
 Refer to *Table 34.*

	Resistance
Construction	R
1. Outdoor air surface (7½ mph wind)	0.25
2. Stone facing, 2 in. (2 × .08)	0.16
3. Hollow clay tile, 8"	1.85
4. Sand aggregate plaster, 2 in. (2 × .20)	0.40
5. Inside air surface (still air)	0.68
Total Resistance	3.34

$$U = \frac{1}{R} = \frac{1}{3.34} = 0.30 \text{ Btu/(hr)(sq ft)(deg F)}$$

TABLE 34—THERMAL RESISTANCES R—BUILDING AND INSULATING MATERIALS

(deg F per Btu) / (hr) (sq ft)

MATERIAL	DESCRIPTION	THICK-NESS (in.)	DENSITY (lb per cu ft)	WEIGHT (lb per sq ft)	RESISTANCE R Per Inch Thickness $\frac{1}{k}$	RESISTANCE R For Listed Thickness $\frac{1}{c}$
	BUILDING MATERIALS					
BUILDING BOARD Boards, Panels, Sheathing, etc	Asbestos-Cement Board		120	—	0.25	—
	Asbestos-Cement Board	⅛	120	1.25	—	0.03
	Gypsum or Plaster Board	⅜	50	1.58	—	0.32
	Gypsum or Plaster Board	½	50	2.08	—	0.45
	Plywood		34	—	1.25	—
	Plywood	¼	34	0.71	—	0.31
	Plywood	⅜	34	1.06	—	0.47
	Plywood	½	34	1.42	—	0.63
	Plywood or Wood Panels	¾	34	2.13	—	0.94
	Wood Fiber Board, Laminated or Homogeneous		26	—	2.38	—
			31	—	2.00	—
	Wood Fiber, Hardboard Type		65	—	0.72	—
	Wood Fiber, Hardboard Type	¼	65	1.35	—	0.18
	Wood, Fir or Pine Sheathing	²⁵⁄₃₂	32	2.08	—	0.98
	Wood, Fir or Pine	1⅝	32	4.34	—	2.03
BUILDING PAPER	Vapor Permeable Felt	—	—	—	—	0.06
	Vapor Seal, 2 Layers of Mopped 15 lb felt	—	—	—	—	0.12
	Vapor Seal, Plastic Film	—	—	—	—	Negl
WOODS	Maple, Oak, and Similar Hardwoods		45	—	0.91	—
	Fir, Pine, and Similar Softwoods		32	—	1.25	—
MASONRY UNITS	Brick, Common	4	120	40	—	.80
	Brick, Face	4	130	43	—	.44
	Clay Tile, Hollow:					
	1 Cell Deep	3	60	15	—	0.80
	1 Cell Deep	4	48	16	—	1.11
	2 Cells Deep	6	50	25	—	1.52
	2 Cells Deep	8	45	30	—	1.85
	2 Cells Deep	10	42	35	—	2.22
	3 Cells Deep	12	40	40	—	2.50
	Concrete Blocks, Three Oval Core Sand & Gravel Aggregate	3	76	19	—	0.40
		4	69	23	—	0.71
		6	64	32	—	0.91
		8	64	43	—	1.11
		12	63	63	—	1.28
	Cinder Aggregate	3	68	17	—	0.86
		4	60	20	—	1.11
		6	54	27	—	1.50
		8	56	37	—	1.72
		12	53	53	—	1.89
	Lightweight Aggregate (Expanded Shale, Clay, Slate or Slag; Pumice)	3	60	15	—	1.27
		4	52	17	—	1.50
		8	48	32	—	2.00
		12	43	43	—	2.27
	Gypsum Partition Tile:					
	3"x12"x30" solid	3	45	11	—	1.26
	3"x12"x30" 4-cell	3	35	9	—	1.35
	4"x12"x30" 3-cell	4	38	13	—	1.67
	Stone, Lime or Sand		150	—	0.08	—

TABLE 34—THERMAL RESISTANCES R—BUILDING AND INSULATING MATERIALS (Contd)

(deg F per Btu) / (hr) (sq ft)

MATERIAL	DESCRIPTION	THICK-NESS (in.)	DENSITY (lb per cu ft)	WEIGHT (lb per sq ft)	RESISTANCE R Per Inch Thickness $\frac{1}{k}$	RESISTANCE R For Listed Thickness $\frac{1}{c}$
	BUILDING MATERIALS, (CONT.)					
MASONRY MATERIALS Concretes	Cement Mortar		116	—	0.20	—
	Gypsum-Fiber Concrete 87½% gypsum, 12½% wood chips		51	—	0.60	—
	Lightweight Aggregates		120	—	0.19	—
	Including Expanded		100	—	0.28	—
	Shale, Clay or Slate		80	—	0.40	—
	Expanded Slag; Cinders		60	—	0.59	—
	Pumice; Perlite; Vermiculite		40	—	0.86	—
	Also, Cellular Concretes		30	—	1.11	—
			20	—	1.43	—
	Sand & Gravel or Stone Aggregate (Oven Dried)		140	—	0.11	—
	Sand & Gravel or Stone Aggregate (Not Dried)		140	—	0.08	—
	Stucco		116	—	0.20	—
PLASTERING MATERIALS	Cement Plaster, Sand Aggregate		116	—	0.20	—
	Sand Aggregate	½	116	4.8	—	0.10
	Sand Aggregate	¾	116	7.2	—	0.15
	Gypsum Plaster:					
	Lightweight Aggregate	½	45	1.88	—	0.32
	Lightweight Aggregate	⅝	45	2.34	—	0.39
	Lightweight Aggregate on Metal Lath	¾	45	2.80	—	0.47
	Perlite Aggregate		45	—	0.67	—
	Sand Aggregate		105	—	0.18	—
	Sand Aggregate	½	105	4.4	—	0.09
	Sand Aggregate	⅝	105	5.5	—	0.11
	Sand Aggregate on Metal Lath	¾	105	6.6	—	0.13
	Sand Aggregate on Wood Lath		105	—	—	0.40
	Vermiculite Aggregate		45	—	0.59	—
ROOFING	Asbestos-Cement Shingles		120	—	—	0.21
	Asphalt Roll Roofing		70	—	—	0.15
	Asphalt Shingles		70	—	—	0.44
	Built-up Roofing	⅜	70	2.2	—	0.33
	Slate	½	201	8.4	—	0.05
	Sheet Metal	—	—	—	Negl	—
	Wood Shingles		40	—	—	0.94
SIDING MATERIALS (On Flat Surface)	Shingles					
	Wood, 16", 7½" exposure		—	—	—	0.87
	Wood, Double, 16", 12" exposure		—	—	—	1.19
	Wood, Plus Insul Backer Board, ⁵⁄₁₆"		—	—	—	1.40
	Siding					
	Asbestos-Cement, ¼" lapped		—	—	—	0.21
	Asphalt Roll Siding		—	—	—	0.15
	Asphalt Insul Siding, ½" Board		—	—	—	1.45
	Wood, Drop, 1"x8"		—	—	—	0.79
	Wood, Bevel, ½"x8", lapped		—	—	—	0.81
	Wood, Bevel, ¾"x10", lapped		—	—	—	1.05
	Wood, Plywood, ⅜", lapped		—	—	—	0.59
	Structural Glass		—	—	—	0.10
FLOORING MATERIALS	Asphalt Tile	⅛	120	1.25	—	0.04
	Carpet and Fibrous Pad		—	—	—	2.08
	Carpet and Rubber Pad		—	—	—	1.23
	Ceramic Tile	1	—	—	—	0.08
	Cork Tile		25	—	2.22	—
	Cork Tile	⅛	25	0.26	—	0.28
	Felt, Flooring		—	—	—	0.06
	Floor Tile	⅛	—	—	—	0.05
	Linoleum	⅛	80	0.83	—	0.08
	Plywood Subfloor	⅝	34	1.77	—	0.78
	Rubber or Plastic Tile	⅛	110	1.15	—	0.02
	Terrazzo	1	140	11.7	—	0.08
	Wood Subfloor	²⁵⁄₃₂	32	2.08	—	0.98
	Wood, Hardwood Finish	¾	45	2.81	—	0.68

TABLE 34—THERMAL RESISTANCES R—BUILDING AND INSULATING MATERIALS (Contd)

(deg F per Btu) / (hr) (sq ft)

MATERIAL	DESCRIPTION	THICK-NESS (in.)	DENSITY (lb per cu ft)	WEIGHT (lb per sq ft)	RESISTANCE R	
					Per Inch Thickness $\frac{1}{k}$	For Listed Thickness $\frac{1}{c}$
INSULATING MATERIALS						
BLANKET AND BATT*	Cotton Fiber		0.8 - 2.0	—	3.85	—
	Mineral Wool, Fibrous Form Processed From Rock, Slag, or Glass		1.5 - 4.0	—	3.70	—
	Wood Fiber Wood Fiber, Multi-layer Stitched Expanded		3.2 - 3.6 1.5 - 2.0	— —	4.00 3.70	— —
BOARD AND SLABS	Glass Fiber		9.5	—	4.00	—
	Wood or Cane Fiber Acoustical Tile Acoustical Tile Interior Finish (Tile, Lath, Plank) Interior Finish (Tile, Lath, Plank)	½ ¾ ½	22.4 22.4 15.0 15.0	.93 1.4 — 0.62	— — 2.86 —	1.19 1.78 — 1.43
	Roof Deck Slab Sheathing (Impreg or Coated) Sheathing (Impreg or Coated) Sheathing (Impreg or Coated)	½ 25/32	20.0 20.0 20.0	— 0.83 1.31	2.63 — —	— 1.32 2.06
	Cellular Glass Cork Board (Without Added Binder) Hog Hair (With Asphalt Binder) Plastic (Foamed) Wood Shredded (Cemented in Preformed Slabs)		9.0 6.5 - 8.0 8.5 1.62 22.0	— — — — —	2.50 3.70 3.00 3.45 1.82	— — — — —
LOOSE FILL	Macerated Paper or Pulp Products Wood Fiber: Redwood, Hemlock, or Fir Mineral Wool (Glass, Slag, or Rock) Sawdust or Shavings Vermiculite (Expanded)		2.5 - 3.5 2.0 - 3.5 2.0 - 5.0 8.0 - 15.0 7.0	— — — — —	3.57 3.33 3.33 2.22 2.08	— — — — —
ROOF INSULATION	All Types Preformed, for use above deck Approximately Approximately Approximately Approximately Approximately Approximatley	½ 1 1½ 2 2½ 3	15.6 15.6 15.6 15.6 15.6 15.6	.7 1.3 1.9 2.6 3.2 3.9	— — — — — —	1.39 2.78 4.17 5.26 6.67 8.33
AIR						
AIR SPACES	POSITION HEAT FLOW Horizontal Up (Winter) Horizontal Up (Summer) Horizontal Down (Winter) Horizontal Down (Winter) Horizontal Down (Winter) Horizontal Down (Winter) Horizontal Down (Summer) Horizontal Down (Summer) Horizontal Down (Summer) Sloping 45° Up (Winter) Sloping 45° Down (Summer) Vertical Horiz. (Winter) Vertical Horiz. (Summer)	¾ - 4 ¾ - 4 ¾ 1½ 4 8 ¾ 1½ 4 ¾ - 4 ¾ - 4 ¾ - 4 ¾ - 4	— — — — — — — — — — — — —	— — — — — — — — — — — — —	— — — — — — — — — — — — —	0.85 0.78 1.02 1.15 1.23 1.25 0.85 0.93 0.99 0.90 0.89 0.97 0.86
AIR FILM	POSITION HEAT FLOW Horizontal Up Sloping 45° Up Vertical Horizontal Sloping 45° Down Horizontal Down		— — — — —	— — — — —	— — — — —	0.61 0.62 0.68 0.76 0.92
Still Air						
15 Mph Wind	Any Position (For Winter) Any Direction		—	—	—	0.17
7½ Mph Wind	Any Position (For Summer) Any Direction		—	—	—	0.25

*Includes paper backing and facing if any. In cases where the insulation forms a boundary (highly reflective) of an air space, refer to *Table 31, page 75*

HEAT LOSS THRU BASEMENT WALLS AND FLOORS BELOW THE GROUND LEVEL

The loss through the floor is normally small and relatively constant year round because the ground temperature under the floor varies only a little throughout the year. The ground is a very good heat sink and can absorb or lose a large amount of heat without an appreciable change in temperature at about the 8 ft level. Above the 8 ft level, the ground temperature varies with the outdoor temperature, with the greatest variation at the surface and a decreasing variation down to the 8 ft depth. The heat loss thru a basement wall may be appreciable and it is difficult to calculate because the ground temperature varies with depth. *Tables 35 thru 37* have been empirically calculated to simplify the evaluation of heat loss thru basement walls and floors.

The heat loss thru a slab floor is large around the perimeter and small in the center. This is because the ground temperature around the perimeter varies with the outdoor temperature, whereas the ground temperature in the middle remains relatively constant, as with basement floors.

Basis of Tables 35 thru 37
— Heat Loss thru Masonry Floors and Walls in Ground

Tables 35 thru 37 are based on empirical data. The perimeter factors listed in *Table 36* were developed by calculating the heat transmitted for each foot of wall to an 8 ft depth. The ground was assumed to decrease the transmission coefficient, thus adding resistance between the wall and the outdoor air. The transmission coefficients were then added to arrive at the perimeter factors.

Use of Tables 35 thru 37
— Heat Loss thru Masonry Floors and Walls in Ground

The transmission coefficients listed in *Table 35* may be used for any thickness of uninsulated masonry floors where there is good contact between the floor and the ground.

The perimeter factors listed in *Table 36* are used for estimating heat loss thru basement walls and the outside strip of basement floors. This factor can be used only when the space is heated continuously. If there is only occasional heating, calculate the heat loss using the wall or floor transmission coefficients as listed in *Tables 21 thru 33* and the temperature difference between the basement and outdoor air or ground as listed in *Table 37*.

The heat loss in a basement is determined by adding the heat transferred thru the floor, the walls and the outside strip of the floor and the portion of the wall above the ground level.

Example 7 — Heat Loss in a Basement

Given:

Basement — $100' \times 40' \times 9'$

Basement temp — 65 F db, heated continuously

Outdoor temp — 0° F db

Grade line — 6 ft above basement floor

Walls and floors — 12 in. concrete (80 lb/cu ft)

Find:

Heat loss from basement

Solution:

1. Heat loss above ground

$$H = UA_1 (t_b - t_{oa})$$
$$= 0.18 \times (200 + 80) \times 3 \times (65 - 0) = 9828 \text{ Btu/hr}$$

2. Heat loss thru walls and outside strip of floor below ground.

$$H = L_p Q (t_b - t_{oa})$$
$$= (200 + 80) \times 1.05 \times (65 - 0) \qquad = 19,100 \text{ Btu/hr}$$

3. Heat loss thru floor

$$H = UA_2 (t_b - t_g)$$
$$= 0.05 \times (100 \times 40) \times (65 - 55) \qquad = \underline{2000 \text{ Btu/hr}}$$

Total Heat Loss $= 30,928 \text{ Btu/hr}$

where U = Heat transmission coefficient of wall above ground (*Table 21*) and floor (*Table 35*) in Btu/(hr)(sq ft)(deg F)

A_1 = Area of wall above ground, sq ft

A_2 = Entire floor area, sq ft

L_p = Perimeter of wall, ft

Q = Perimeter factor (*Table 36*)

t_b = Basement dry-bulb temp, F

t_g = Ground temp, F, (*Table 37*)

t_{oa} = Outdoor design dry-bulb temp, F

TABLE 35—TRANSMISSION COEFFICIENT U— MASONRY FLOORS AND WALLS IN GROUND

(Use only in conjunction with *Table 36*)

Floor or Wall	Transmission Coefficient U Btu/(hr) (sq ft) (deg F)
*Basement Floor	.05
Portion of Wall exceeding 8 feet below ground level	.08

*Some additional floor loss is included in perimeter factor, see *Table 36*.

Equations:

Heat loss through floor, Btu/hr = (area of floor, sq ft) × (U value) × (basement — ground temp).

Heat loss through wall below 8 foot line, Btu/hr = (area of wall below 8 ft line, sq ft) × (U value) × (basement — ground temp).

NOTE: The factors in *Tables 35 and 36* may be used for any thickness of uninsulated masonry wall or floor, but there must be a good contact (no air space which may connect to the outdoors) between the ground and the floor or wall. Where the ground is dry and sandy, or where there is cinder fill along wall or where the wall has a low heat transmission coefficient, the perimeter factor may be reduced slightly.

TABLE 36—PERIMETER FACTORS

FOR ESTIMATING HEAT LOSS THROUGH BASEMENT WALLS AND OUTSIDE STRIP OF BASEMENT FLOOR

(Use only in conjunction with *Table 35*)

Distance of Floor From Ground Level	Perimeter Factor (q)
2 Feet above	.90
At ground level	.60
2 Feet below	.75
4 Feet below	.90
6 Feet below	1.05
8 Feet below	1.20

Equation:

Heat loss about perimeter, Btu/hr = (perimeter of wall, ft)
× (perimeter factor) × (basement − outdoor temp).

TABLE 37—GROUND TEMPERATURES

FOR ESTIMATING HEAT LOSS THROUGH BASEMENT FLOORS

Outdoor Design Temp (F)	−30	−20	−10	0	+10	+20
Ground Temp (F)	40	45	50	55	60	65

TRANSMISSION COEFFICIENTS — PIPES IN WATER OR BRINE

Heat transmission coefficients for copper and steel pipes are listed in *Tables 38 and 39*. These coefficients may be useful in applications such as cold water or brine storage systems and ice skating rinks.

Basis of Tables 38 and 39 — Transmission Coefficients, Pipes in Water or Brine

Table 38 is for ice coated pipes in water, based on a heat transfer film coefficient, inside the pipe, of 150 Btu/(hr)(sq ft internal pipe surface)(deg F).

Table 39 is for pipes in water or brine based on a heat transfer of 18 Btu/(hr)(sq ft external pipe surface) (deg F) in water, 14 Btu in brine. It is also based on a low rate of circulation on the outside of the pipe and 10 F to 15 F temperature difference between water or brine and refrigerant. High rates of circulation will increase the heat transfer rate. For special problems, consult heat transfer reference books.

TABLE 38—TRANSMISSION COEFFICIENT U—ICE COATED PIPES IN WATER

Btu/(hr) (lineal ft pipe) (deg F between 32 F db and refrig temp)

Inside film coefficient = 150 Btu/(hr) (sq ft) (deg F)

Copper Pipe Size (Inches O.D.)	Copper Pipe With Ice Thickness (Inches)				Steel Pipe Size Nominal (Inches)	Steel Pipe With Ice Thickness (Inches)				
	½	1	1½	2		½	1	1½	2	3
⅝	6.1	4.5	3.8	3.4	½	7.2	5.2	4.4	3.9	3.4
¾	7.1	5.1	4.2	3.8	¾	8.7	6.1	5.1	4.5	3.8
⅞	8.0	5.7	4.7	4.1	1	10.6	7.2	5.8	5.1	4.2
1⅛	9.8	6.7	5.4	4.7	1¼	13.0	8.6	6.8	5.9	4.8

TABLE 39—TRANSMISSION COEFFICIENT U—PIPES IMMERSED IN WATER OR BRINE

Btu/(hr) (lineal ft pipe) (deg F between 32 F db and refrig temp)

Outside water film coefficient = 18 Btu/(hr) (sq ft) (deg F)
Outside brine film coefficient = 14 Btu/(hr) (sq ft) (deg F)
Water refrigerant temp = 10 F to 15 F

Copper Pipe Size (Inches O.D.)	Pipes in Water	Steel Pipe Nominal Size (Inches)	Pipes in Water	Pipes in Brine
½	2.4	½	4.0	3.1
⅝	2.9	¾	5.0	3.9
¾	3.5	1	6.2	4.8
1⅛	5.3	1¼	7.8	6.1

WATER VAPOR FLOW THRU BUILDING STRUCTURES

Water vapor flows thru building structures, resulting in a latent load whenever a vapor pressure difference exists across a structure. The latent load from this source is usually insignificant in comfort applications and need be considered only in low or high dewpoint applications.

Water vapor flows from high to lower vapor pressure at a rate determined by the permeability of the structure. This process is quite similar to heat flow, except that there is transfer of mass with water vapor flow. As heat flow can be reduced by adding insulation, vapor flow can be reduced by vapor barriers. The vapor barrier may be paint (aluminum or asphalt), aluminum foil or galvanized iron. *It should always be placed on the side of a structure having the higher vapor pressure, to prevent the water vapor from flowing up to the barrier and condensing within the wall.*

Basis of Table 40
— Water Vapor Transmission thru Various Materials

The values for walls, floors, ceilings and partitions have been estimated from the source references listed in the bibliography. The resistance of a homogeneous material to water vapor transmission has been assumed to be directly proportional to the thickness, and it also has been assumed that there is no surface resistance to water vapor flow. The values for permeability of miscellaneous materials are based on test results.

NOTE: Some of the values for walls, roofs, etc., have been increased by a safety factor because conclusive data is not available.

Use of Table 40
— Water Vapor Transmission thru Various Materials

Table 40 is used to determine latent heat gain from water vapor transmission thru building structures in the high and low dewpoint applications where the air moisture content must be maintained.

Example 8 — Water Vapor Transmission

Given:

A 40 ft × 40 ft × 8 ft laboratory on second floor requiring inside design conditions of 40 F db, 50% rh, with the outdoor design conditions at 95 F db, 75 F wb. The outdoor wall is 12 inch brick with no windows. The partitions are metal lath and plaster on both sides of studs. Floor and ceiling are 4 inch concrete.

Find:

The latent heat gain from the water vapor transmission.

Solution:

Gr/lb at 95 F db, 75 F wb = 99 (psych chart)
Gr/lb at 40 F db, 50% rh = 18 (psych chart)
Moisture content difference = 81 gr/lb

Assume that the dewpoint in the areas surrounding the laboratory is uniform and equal to the outdoor dewpoint.

Latent heat gain:

$$\text{Outdoor wall} = \frac{40 \times 8}{100} \times 81 \times .04 \text{ (Table 40)}$$
$$= 10.4 \text{ Btu/hr}$$

$$\text{Floor and ceilings} = 2 \times \frac{40 \times 40}{100} \times 81 \times .10$$
$$= 259 \text{ Btu/hr}$$

$$\text{Partitions} = 3 \times \frac{40 \times 8}{100} \times 81 \times 1.0$$
$$= \underline{777 \text{ Btu/hr}}$$

$$\text{Total Latent Heat Gain} = 1046.4 \text{ Btu/hr}$$

TABLE 40—WATER VAPOR TRANSMISSION THRU VARIOUS MATERIALS

DESCRIPTION OF MATERIAL OR CONSTRUCTION	PERMEANCE Btu/(hr) (100 sq ft) (gr/lb diff) latent heat		
	No Vapor Seal Unless Noted Under Description	With 2 Coats Vapor-seal Paint on Smooth Inside Surface*	With Aluminum Foil Mounted on One Side of Paper Cemented to Wall†
WALLS			
Brick— 4 inches	.12	.075	.024
— 8 inches	.06	.046	.020
—12 inches	.04	.033	.017
—per inch of thickness	.49	— —	— —
Concrete— 6 inches	.067	.050	.021
—12 inches	.034	.029	.016
—per inch of thickness	.40	— —	— —
Frame—with plaster interior finish	.79	.16	.029
—same with asphalt coated insulating board lath	.42	.14	.028
Tile—hollow clay (face, glazed)—4 inches	.013	.012	.0091
—hollow clay (common)—4 inches	.24	.11	.025
—hollow clay, 4 inch face and 4 inch common	.012	.011	.0086
CEILINGS AND FLOORS			
Concrete—4 inches	.10	.067	.023
—8 inches	.051	.040	.019
Plaster on wood or metal lath on joist—no flooring	2.0	.18	.030
Plaster on wood or metal lath on joist—flooring	.50	.14	.028
Plaster on wood or metal lath on joists—double flooring	.40	.13	.028
PARTITIONS			
Insulating Board ½ inch on both sides of studding	4.0	.19	.030
Wood or metal lath and plaster on both sides of studding	1.0	.17	.029
ROOFS			
Concrete—2 inches, plus 3 layer felt roofing	.02	.018	.012
—6 inches, plus 3 layer felt roofing	.02	.018	.012
Shingles, sheathing, rafters—plus plaster on wood or metal lath	1.5	.18	.29
Wood—1 inch, plus 3 layer felt roofing	.02	.018	.012
—2 inches, plus 3 layer felt roofing	.02	.018	.012
MISCELLANEOUS			
Air Space, still air 3⅝ inch	3.6		
1 inch	13.0		
Building Materials			
Masonite—1 thickness, ⅛ inch	1.1	.17	.027
—5 thicknesses	.32		
Plaster on wood lath	1.1		
—plus 2 coats aluminum paint	— —	.12	
Plaster on gypsum lath	1.95	— —	
—ditto plus primer and 2 coats lead and oil paint	— —	.13	
Plywood—¼ inch Douglas fir (3 ply)	.63		
—ditto plus 2 coats asphalt paint	— —	.087	
—ditto plus 2 coats aluminum paint	— —	.13	
—½ inch Douglas fir (5 ply)	.27		
—ditto plus 2 coats asphalt paint	— —	.041	
—ditto plus 2 coats aluminum paint	— —	.12	
Wood—Pine .508 inch	.33		
—ditto plus 2 coats aluminum paint	— —	.046	
—spruce, .508 inch	.20		
Insulating Materials			
Corkboard, 1 inch thick	.63		
Interior finish insulating board, ½″	5.0 - 7.0		
—ditto plus 2 coats water emulsion paint	3.0 - 4.0		
—ditto plus 2 coats varnish base paint	.1 - 1.0		
—ditto plus 2 coats lead and oil paint	.17		
—ditto plus wall linoleum	.03 - .06		

TABLE 40—WATER VAPOR TRANSMISSION THRU VARIOUS MATERIALS (Contd)

DESCRIPTION OF MATERIAL OR CONSTRUCTION	PERMEANCE Btu/(hr) (100 sq ft) (gr/lb diff) latent heat		
	No Vapor Seal Unless Noted Under Description	With 2 Coats Vapor-seal Paint on Smooth Inside Surface*	With Aluminum Foil Mounted on One Side of Paper Cemented to Wall†
MISCELLANEOUS			
Insulating Materials, cont.			
Insulating board lath	4.6 - 8.2		
—ditto plus ½″ plaster	1.5		
—ditto plus ½″ plaster, sealer, and flat coat of paint	.16 - .31		
Insulating board sheathing, 25⁄32″	2.6 - 6.1		
—ditto plus asphalt coating both sides	.046 - 1.0		
Mineral wool (3⅝ inches thick), unprotected	3.5		
Packaging materials			
Cellophane, moisture proof	.01 - 0.25		
Glassine (1 ply waxed or 3 ply plain)	.0015 - .006		
Kraft paper soaked with parafin wax, 4.5 lbs per 100 sq ft	1.4 - 3.1		
Pliofilm	.01 - .025		
Paint Films			
2 coats aluminum paint, estimated	.05 - .2		
2 coats asphalt paint, estimated	.05 - .1		
2 coats lead and oil paint, estimated	.1 - .6		
2 coats water emulsion, estimated	5.0 - 8.0		
Papers			
Duplex or asphalt laminae (untreated) 30-30-30, 3.1 lb per 100 sq ft	.15 - .27		
—ditto 30-60-30, 4.2 lb per 100 sq ft	.051 - .091		
Kraft paper—1 sheet	8.1		
—2 sheets	5.1		
—aluminum foil on one side of sheet	.016		
—aluminum foil on both sides of sheet	.012		
Sheathing paper			
Asphalt impregnated and coated, 7 lb per 100 sq ft	.02 - .10		
Slaters felt, 6 lb per 100 sq ft, 50% saturated with tar	1.4		
Roofing Felt, saturated and coated with asphalt			
25 lb. per sq ft	.015		
50 lb. per sq ft	.011		
Tin sheet with 4 holes 1⁄16 diameter	.17		
Crack 12 inches long by 1⁄32 inches wide (approximated from above)	5.2		

*Painted surfaces: Two coats of a good vapor seal paint on a smooth surface give a fair vapor barrier. More surface treatment is required on a rough surface than on a smooth surface. Data indicates that either asphalt or aluminum paint are good for vapor seals.

†Aluminum Foil on Paper: This material should also be applied over a smooth surface and joints lapped and sealed with asphalt.
The vapor barrier should always be placed on the side of the wall having the higher vapor pressure if condensation of moisture in wall is possible.

Application: The heat gain due to water vapor transmission through walls may be neglected for the normal air conditioning or refrigeration job. This latent gain should be considered for air conditioning jobs where there is a great vapor pressure difference between the room and the outside, particularly when the dewpoint inside must be low. **Note that moisture gain due to infiltration usually is of much greater magnitude than moisture transmission through building structures.**

Conversion Factors: To convert above table values to: grain/(hr) (sq ft) (inch mercury vapor pressure difference), multiply by 9.8.
grain/(hr) (sq ft) (pounds per sq inch vapor pressure difference), multiply by 20.0.
To convert Btu latent heat to grains, multiply by 7000/1060 = 6.6.

CONDENSATION OF WATER VAPOR

Whenever there is a difference of temperature and pressure of water vapor across a structure, conditions may develop that lead to a condensation of moisture. This condensation occurs at the point of saturation temperature and pressure.

As water vapor flows thru the structure, its temperature decreases and, if at any point it reaches the dewpoint or saturation temperature, condensation begins. As condensation occurs, the vapor pressure decreases, thereby lowering the dewpoint or saturation temperature until it corresponds to the actual temperature. The rate at which condensation occurs is determined by the rate at which heat is removed from the point of condensation. As the vapor continues to condense, latent heat of condensation is released, causing the dry-bulb temperature of the material to rise.

To illustrate this, assume a frame wall with wood sheathing and shingles on the outside, plasterboard on the inside and fibrous insulation between the two. Also, assume that the inside conditions are 75 F db and 50% rh and the outdoor conditions are 0°F db and 80% rh. Refer to *Fig. 28*.

The temperature and vapor pressure gradient decreases approximately as shown by the solid and dashed lines until condensation begins (saturation point). At this point, the latent heat of condensation decreases the rate of temperature drop thru the insulation. This is approximately indicated by the dotted line.

Another cause of concealed condensation may be evaporation of water from the ground or damp locations. This water vapor may condense on the underside of the floor joints (usually near the edges where it is coldest) or may flow up thru the outdoor side of the walls because of stack effect and/or vapor pressure differences.

Concealed condensation may cause wood, iron and brickwork to deteriorate and insulation to lose its insulating value. These effects may be corrected by the following methods:

1. *Provide vapor barriers on the high vapor pressure side.*
2. In winter, ventilate the building to reduce the vapor pressure within. No great volume of air change is necessary, and normal infiltration alone is frequently all that is required.
3. In winter, ventilate the structure cavities to remove vapor that has entered. Outdoor air thru vents shielded from entrance of rain and insects may be used.

Condensation may also form on the surface of a building structure. Visible condensation occurs when the surface of any material is colder than the dewpoint temperature of the surrounding air. In winter, the condensation may collect on cold closet walls and attic roofs and is commonly observed as frost on window panes. *Fig. 29* illustrates the condensation on a window with inside winter design conditions of 70 F db and 40% rh. Point A represents the room conditions; point B, the dewpoint temperature of the thin film of water vapor adjacent to the window surface; and point C, the point at which frost or ice appears on the window.

Once the temperature drops below the dewpoint, the vapor pressure at the window surface is also reduced, thereby establishing a gradient of vapor pressure from the room air to the window surface. This gradient operates, in conjunction with the convec-

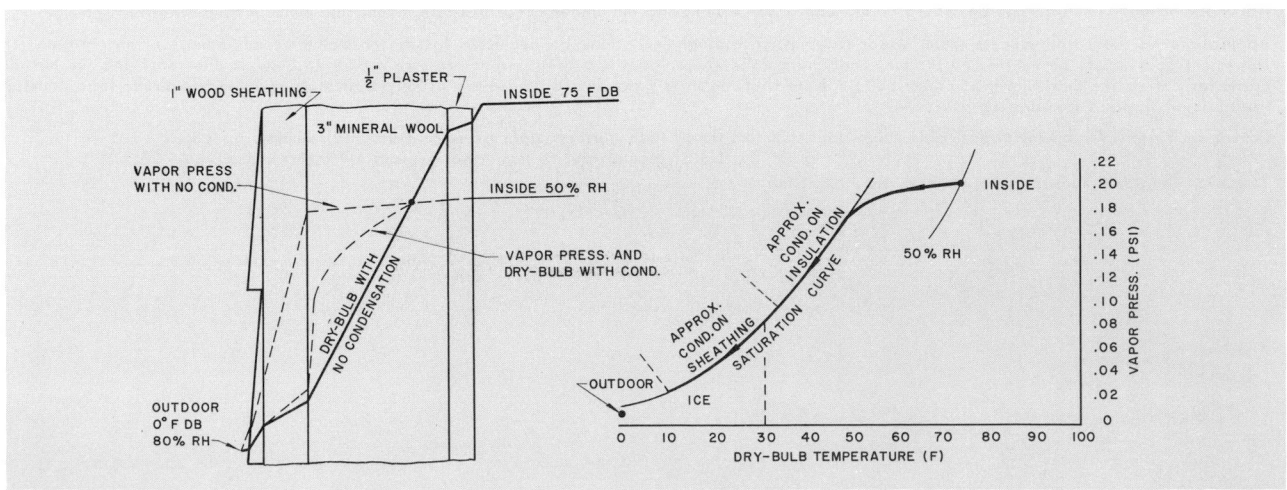

FIG. 28 — CONDENSATION WITHIN FRAME WALL

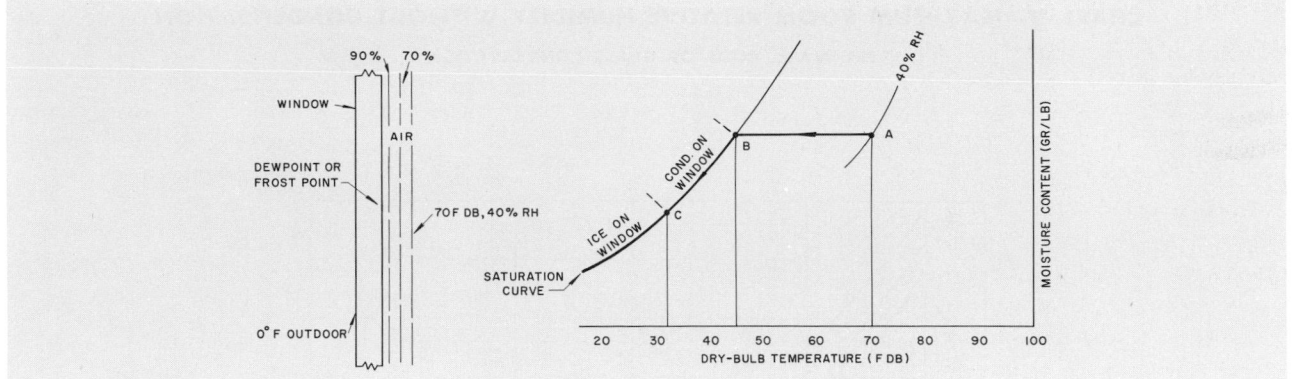

FIG. 29 — CONDENSATION ON WINDOW SURFACE

tive action within the room, to move water vapor continuously to the window surface to be condensed, as long as the concentration of the water vapor is maintained in a space.

Visible condensation is objectionable as it causes staining of surfaces, dripping on machinery and furnishings, and damage to materials in process of manufacture. Condensation of this type may be corrected by the following methods:

1. Increase the thermal resistance of walls, roofs and floors by adding insulation with vapor barriers to prevent condensation within the structures.

2. Increase the thermal resistance of glass by installing two or three panes with air space(s) between. In extreme cases, controlled heat, electric or other, may be applied between the glass of double glazed windows.

3. Maintain a room dewpoint lower than the lowest expected surface temperature in the room.

4. Decrease surface resistance by increasing the velocity of air passing over the surface. Decreasing the surface resistance increases the window surface temperature and brings it closer to the room dry-bulb temperature.

Basis of Chart 2
— Maximum Room RH; No Wall, Roof or Glass Condensation

Chart 2 has been calculated from the equation used to determine the maximum room dewpoint temperature that can exist with condensation.

$$t_{dp} = t_{rm} - \frac{U(t_{rm} - t_{oa})}{f_i}$$

where t_{dp} = dewpoint temp of room air, F db
t_{rm} = room temp, F
U = transmission coefficient,
Btu/(hr)(sq ft)(deg F)
t_{oa} = outdoor temp, F

f_i = inside air film or surface conductance,
Btu/(hr)(sq ft)(deg F)

Chart 2 is based upon a room dry-bulb temperature of 70 F db and an inside film conductance of 1.46 Btu/(hr)(sq ft)(deg F).

Use of Chart 2
— Maximum Room RH; No Wall, Roof or Glass Condensation

Chart 2 gives a rapid means of determining the maximum room relative humidity which can be maintained and yet avoid condensation with a 70 F db room.

Example 9 — Moisture Condensation

Given:
12 in. stone wall with ⅝ in. sand aggregate plaster
Room temp — 70 F db
Outdoor temp — 0° F db

Find:
Maximum room rh without wall condensation.

Solution:
Transmission coefficient U = 0.52 Btu/(hr)(sq ft)(deg F)
(Table 21, page 66)
Maximum room rh = 40.05%, *(Chart 2)*
Corrections in room relative humidity for room temperatures other than 70 F db are listed in the table under *Chart 2*. Values other than those listed may be interpolated.

Example 10 — Moisture Condensation

Given:
Same as *Example 9*, except room temp is 75 F db

Find:
Maximum room rh without wall condensation

Solution:
Transmission coefficient U = 0.52 Btu/(hr)(sq ft)(deg F)
(Example 9)
Maximum room rh for 70 F db room temp = 40.05%
(Example 9)
Rh correction for room temp of 75 F db with U factor of 0.52 = −1.57% (bottom *Chart 2*).
Maximum room rh = 40.05 − 1.57 = 38.48% or 38.5%

CHART 2—MAXIMUM ROOM RELATIVE HUMIDITY WITHOUT CONDENSATION

NO WALL, ROOF OR GLASS CONDENSATION

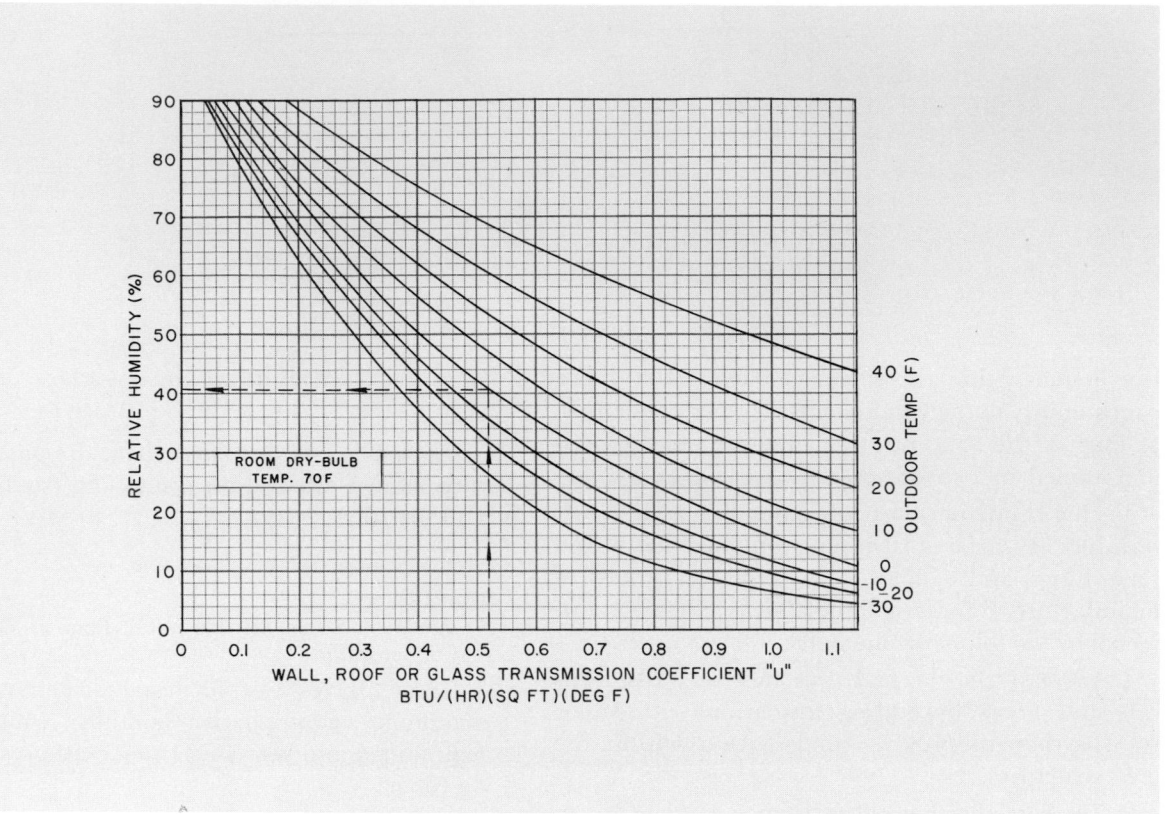

CORRECTION IN ROOM RH (%)

For Wall, Roof or Glass Transmission Coefficient U

Outdoor Temp (F db)	U = 1.1		U = .65		U = .35	
	Room Temp (F db)					
	60	80	60	80	60	80
−30	+1.0%	−1.0%	+1.5%	−2.0%	+2.5%	−2.0%
−20	+1.0	−1.5	+2.5	−2.5	+3.0	−2.0
−10	+2.0	−2.0	+3.5	−3.0	+3.0	−2.0
0	+3.5	−2.5	+4.0	−4.0	+3.5	−2.5
10	+5.0	−3.5	+5.0	−4.5	+4.0	−3.0
20	+7.0	−4.0	+6.5	−5.0	+4.5	−3.5
30	+9.0	−7.5	+8.5	−6.0	+5.0	−4.0
40	+12.0	−9.5	+9.5	−7.5	+6.0	−4.5

CHAPTER 6. INFILTRATION AND VENTILATION

The data in this chapter is based on ASHAE tests evaluating the infiltration and ventilation quantities of outdoor air. These outdoor air quantities normally have a different heat content than the air within the conditioned space and, therefore, impose a load on the air conditioning equipment.

In the case of infiltration, the load manifests itself directly within the conditioned space. The ventilation air, taken thru the conditioning apparatus, imposes a load both on the space thru apparatus bypass effect, and directly on the conditioning equipment.

The data in this chapter is based on ASHAE tests and years of practical experience.

INFILTRATION

Infiltration of air and particularly moisture into a conditioned space is frequently a source of sizable heat gain or loss. The quantity of infiltration air varies according to tightness of doors and windows, porosity of the building shell, height of the building, stairwells, elevators, direction and velocity of wind, and the amount of ventilation and exhaust air. Many of these cannot be accurately evaluated and must be based on the judgment of the estimator.

Generally, infiltration may be caused by wind velocity, or stack effort, or both:

1. *Wind Velocity* — The wind velocity builds up a pressure on the windward side of the building and a slight vacuum on the leeward side. The outdoor pressure build-up causes air to infiltrate thru crevices in the construction and cracks around the windows and doors. This, in turn, causes a slight build-up of pressure inside the building, resulting in an equal amount of exfiltration on the leeward side.

2. *Difference in Density or Stack Effect* — The variations in temperatures and humidities produce differences in density of air between inside and outside of the building. In tall buildings this density difference causes summer and winter infiltration and exfiltration as follows:

   Summer — Infiltration at the top and exfiltration at the bottom.

   Winter — Infiltration at the bottom and exfiltration at the top.

This opposite direction flow balances at some neutral point near the mid-height of the building. Air flow thru the building openings increases proportionately between the neutral point and the top and the neutral point and bottom of the building. The infiltration from stack effect is greatly influenced by the height of the building and the presence of open stairways and elevators.

The combined infiltration from wind velocity and stack effect is proportional to the square root of the sum of the heads acting on it.

The increased air infiltration flow caused by stack effect is evaluated by converting the stack effect force to an equivalent wind velocity, and then calculating the flow from the wind velocity data in the tables.

In buildings over 100 ft tall, the equivalent wind velocity may be calculated from the following formula, assuming a temperature difference of 70 F db (winter) and a neutral point at the mid-height of the building:

$$V_e = \sqrt{V^2 - 1.75a} \qquad \text{(for upper section of tall bldgs — winter)} \quad (1)$$

$$V_e = \sqrt{V^2 + 1.75b} \qquad \text{(for lower section of tall bldgs — winter)} \quad (2)$$

where V_e = equivalent wind velocity, mph
  V = wind velocity normally calculated for location, mph
  a = distance window is above mid-height, ft
  b = distance window is below mid-height, ft

NOTE: The total crackage is considered when calculating infiltration from stack effect.

INFILTRATION THRU WINDOWS AND DOORS, SUMMER

Infiltration during the summer is caused primarily by the wind velocity creating a pressure on the windward side. Stack effect is not normally a significant factor because the density difference is slight, (0.073 lb/cu ft at 75 F db, 50% rh and 0.070 lb/cu ft at 95 F db, 75 F wb). This small stack effect in tall buildings (over 100 ft) causes air to flow in the top and out the bottom. Therefore, the air infiltrating in the top of the building, because of the wind pressure, tends to flow down thru the building and out the doors on the street level, thereby offsetting some of the infiltration thru them.

In low buildings, air infiltrates thru open doors on the windward side unless sufficient outdoor air is introduced thru the air conditioning equipment to offset it; refer to *"Offsetting Infiltration with Outdoor Air."*

With *doors on opposite walls,* the infiltration can be considerable if the two are open at the same time.

Basis of Table 41
— Infiltration thru Windows and Doors, Summer

The data in *Tables 41a, b and c* is based on a wind velocity of 7.5 mph blowing directly at the window or door, and on observed crack widths around typical windows and doors. This data is derived from *Table 44* which lists infiltration thru cracks around windows and doors as established by ASHAE tests.

Table 41d shows values to be used for doors on opposite walls for various percentages of time that each door is open.

The data in *Table 41e* is based on actual tests of typical applications.

Use of Table 41
— Infiltration thru Windows and Doors, Summer

The data in *Table 41* is used to determine the infiltration thru windows and doors on the windward side with the wind blowing directly at them. When the wind direction is oblique to the windows or doors, multiply the values in *Tables 41a, b, c, d,* by 0.60 and apply to total areas. For specific locations, adjust the values in *Table 41* to the design wind velocity; refer to *Table 1, page 10.*

During the summer, infiltration is calculated for the windward side(s) only, because stack effect is small and, therefore, causes the infiltration air to flow in a downward direction in tall buildings (over 100 ft). Some of the air infiltrating thru the windows will exfiltrate thru the windows on the leeward side(s), while the remaining infiltration air flows out the doors, thus offsetting some of the infiltration thru the doors. To determine the net infiltration thru the doors, determine the infiltration thru the windows on the windward side, multiply this by .80, and subtract from the door infiltration. For low buildings the door infiltration on the windward side should be included in the estimate.

TABLE 41—INFILTRATION THRU WINDOWS AND DOORS—SUMMER*

7.5 mph Wind Velocity†

TABLE 41a—DOUBLE HUNG WINDOWS‡

DESCRIPTION	CFM PER SQ FT SASH AREA					
	Small—30" x 72"			Large—54" x 96"		
	No W-Strip	W-Strip	Storm Sash	No W-Strip	W-Strip	Storm Sash
Average Wood Sash	.43	.26	.22	.27	.17	.14
Poorly Fitted Wood Sash	1.20	.37	.60	.76	.24	.38
Metal Sash	.80	.35	.40	.51	.22	.25

TABLE 41b—CASEMENT TYPE WINDOWS‡

DESCRIPTION	CFM PER SQ FT SASH AREA									
	Percent Openable Area									
	0%	25%	33%	40%	45%	50%	60%	66%	75%	100%
Rolled Section—Steel Sash										
Industrial Pivoted	.33	.72	—	.99	—	—	—	1.45	—	2.6
Architectural Projected	—	.39	—	—	—	.55	.74	—	—	—
Residential	—	—	.28	—	—	.49	—	—	—	.63
Heavy Projected	—	—	—	—	.23	—	—	.32	.39	—
Hollow Metal—Vertically Pivoted	.27	.58	—	.82	—	—	—	1.2	—	2.2

ROLLED SECTION STEEL SASH WINDOWS

HOLLOW METAL WINDOW

   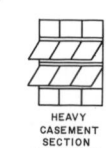 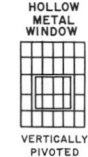

INDUSTRIAL PIVOTED   ARCHITECTURAL PROJECTED   RESIDENTIAL CASEMENT   HEAVY CASEMENT SECTION PROJECTED   VERTICALLY PIVOTED

REPRESENTATIVE TYPES OF WINDOWS
(VIEWED FROM OUTSIDE)

TABLE 41—INFILTRATION THRU WINDOWS AND DOORS—SUMMER* (Contd)

7.5 mph Wind Velocity†

TABLE 41c—DOORS ON ONE OR ADJACENT WALLS, FOR CORNER ENTRANCES

DESCRIPTION	CFM PER SQ FT AREA**		CFM	
			Standing Open	
	No Use	Average Use	No Vestibule	Vestibule
Revolving Doors—Normal Operation	.8	5.2	—	—
Panels Open	—	—	1,200	900
Glass Door—3⁄16″ Crack	4.5	10.0	700	500
Wood Door (3′ x 7′)	1.0	6.5	700	500
Small Factory Door	.75	6.5	—	—
Garage & Shipping Room Door	2.0	4.5	—	—
Ramp Garage Door	2.0	6.75	—	—

TABLE 41d—SWINGING DOORS ON OPPOSITE WALLS

% Time 2nd Door is Open	CFM PER PAIR OF DOORS					
	% Time 1st Door is Open					
	10	25	50	75	100	
10	100	250	500	750	1,000	
25	250	625	1250	1875	2,500	
50	500	1250	2500	3750	5,000	
75	750	1875	3750	5625	7,500	
100	1000	2500	5000	7500	10,000	

TABLE 41e—DOORS

APPLICATION	CFM PER PERSON IN ROOM PER DOOR		
	72″ Revolving Door	36″ Swinging Door	
		No Vestibule	Vestibule
Bank	6.5	8.0	6.0
Barber Shop	4.0	5.0	3.8
Candy and Soda	5.5	7.0	5.3
Cigar Store	20.0	30.0	22.5
Department Store (Small)	6.5	8.0	6.0
Dress Shop	2.0	2.5	1.9
Drug Store	5.5	7.0	5.3
Hospital Room	—	3.5	2.6
Lunch Room	4.0	5.0	3.8
Men's Shop	2.7	3.7	2.8
Restaurant	2.0	2.5	1.9
Shoe Store	2.7	3.5	2.6

*All values in *Table 41* are based on the wind blowing directly at the window or door. When the wind direction is oblique to the window or door, multiply the above values by 0.60 and use the total window and door area on the windward side(s).

†Based on a wind velocity of 7.5 mph. For design wind velocities different from the base, multiply the above values by the ratio of velocities.

‡Includes frame leakage where applicable.

**Vestibules may decrease the infiltration as much as 30% when the door usage is light. When door usage is heavy, the vestibule is of little value for reducing infiltration.

Example 1 — Infiltration in Tall Buildings, Summer

Given:

A 20-story building in New York City oriented true north. Building is 100 ft long and 100 ft wide with a floor-to-floor height of 12 ft. Wall area is 50% residential casement windows having 50% fixed sash. There are ten 7 ft x 3 ft swinging glass doors on the street level facing south.

Find:

Infiltration into the building thru doors and windows, disregarding outside air thru the equipment and the exhaust air quantity.

Solution:

The prevailing wind in New York City during the summer is south, 13 mph (*Table 1, page 10*).

Correction to *Table 1* values for wind velocity
= 13/7.5 = 1.73

Glass area on south side
= 20 × 12 × 100 × .50 = 12,000 sq ft

Infiltration thru windows
= 12,000 × .49 × 1.73 = 10,200 cfm (*Table 41b*)

Infiltration thru doors
= 10 × 7 × 3 × 10 × 1.73 = 3640 cfm (*Table 41c*)

Since this building is over 100 ft tall, net infiltration thru doors = 3640 − (10,200 × .80) = − 4520 cfm.

Therefore, there is no infiltration thru the doors on the street level *on design days,* only exfiltration.

OFFSETTING INFILTRATION WITH OUTDOOR AIR, SUMMER

Completely offsetting infiltration by the introduction of outdoor air thru the air conditioning apparatus is normally uneconomical except in buildings with few windows and doors. The outdoor air so introduced must develop a pressure equal to the wind velocity to offset infiltration. This pressure causes exfiltration thru the leeward walls at a rate equal to wind velocity. Therefore, in a four sided building with equal crack areas on each side and the wind blowing against one side, the amount of outdoor air introduced thru the apparatus must be a little more than three times the amount that infiltrates. Where the wind is blowing against two sides, the outdoor air must be a little more than equal to that which infiltrates.

Offsetting swinging door infiltration is not quite as difficult because air takes the path of least resistance, normally an open door. Most of the outdoor air introduced thru the apparatus flows out the door when it is opened. Also, in tall buildings the window infiltration tends to flow out the door.

The infiltration thru revolving doors is caused by displacement of the air in the door quadrants, is almost independent of wind velocity and, therefore, cannot be offset by outdoor air.

Basis of Table 42
— Offsetting Swinging Door Infiltration with Outdoor Air, Summer

Some of the outdoor air introduced thru the apparatus exfiltrates thru the cracks around the windows and in the construction on the leeward side. The outdoor air values have been increased by this amount for typical application as a result of experience.

Use of Table 42
— Offsetting Swinging Door Infiltration with Outdoor Air, Summer

Table 42 is used to determine the amount of outdoor air thru air conditioning apparatus required to offset infiltration thru swinging doors.

Example 2 — Offsetting Swinging Door Infiltration
Given:
A restaurant with 3000 cfm outdoor air being introduced thru the air conditioning apparatus. Exhaust fans in the kitchen remove 2000 cfm. Two 7 ft x 3 ft glass swinging doors face the prevailing wind direction. At peak load conditions, there are 300 people in the restaurant.

Find:
The net infiltration thru the outside doors.

Solution:
Infiltration thru doors = 300 × 2.5 = 750 cfm (*Table 41e*)
Net outdoor air = 3000 − 2000 = 1000 cfm
Only 975 cfm of outdoor air is required to offset 750 cfm of door infiltration (*Table 42*).
Therefore, there will be no net infiltration thru the outside doors unless there are windows on the leeward side. If there are windows in the building, calculate as outlined in *Example 1.*

TABLE 42—OFFSETTING SWINGING DOOR INFILTRATION WITH OUTDOOR AIR—SUMMER

Net Outdoor Air* (cfm)	Door Infiltration (cfm)	Net Outdoor Air* (cfm)	Door Infiltration (cfm)
140	100	1370	1100
270	200	1480	1200
410	300	1560	1300
530	400	1670	1400
660	500	1760	1500
790	600	1890	1600
920	700	2070	1800
1030	800	2250	2000
1150	900	2450	2200
1260	1000	2650	2400

*Net outdoor air is equal to the outdoor air quantity introduced thru the apparatus minus the exhaust air quantity.

INFILTRATION THRU WINDOWS AND DOORS, WINTER

Infiltration thru windows and doors during the winter is caused by the wind velocity and also stack effect. The temperature differences during the winter are considerably greater than in summer and, therefore, the density difference is greater; at 75 F db and 30% rh, density is .0738; at 0°F db, 40% rh, density is .0865. Stack effect causes air to flow in at the bottom and out at the top, and in many cases requires spot heating at the doors on the street level to maintain conditions. In applications where there is considerable infiltration on the street level, much of the infiltration thru the windows in the upper levels will be offset.

Basis of Table 43
— Infiltration thru Windows and Doors, Winter

The data in *Table 43* is based on a wind velocity of 15 mph blowing directly at the window or door and on observed crack widths around typical windows and doors. The infiltration thru these cracks is calculated from *Table 44* which is based on ASHAE tests.

Use of Table 43
— Infiltration thru Windows and Doors, Winter

Table 43 is used to determine the infiltration of air thru windows and doors on the windward side during the winter. The stack effect in tall buildings increases the infiltration thru the doors and windows on the lower levels and decreases it on the upper levels. Therefore, whenever the door infiltration is increased, the infiltration thru the upper levels must be decreased by 80% of the net increase in door infiltration. The infiltration from stack effect on the leeward sides of the building is determined by using the difference between the equivalent velocity (V_e) and the actual velocity (V) as outlined in *Example 3*. The data in *Table 43* is based on the wind blowing directly at the windows and doors. When the wind direction is oblique to the windows and doors, multiply the values by 0.60 and use the total window and door area on the windward sides.

Example 3 — Infiltration in Tall Buildings, Winter
Given:
   The building described in *Example 1*.
Find:
   The infiltration thru the doors and windows.
Solution:
   The prevailing wind in New York City during the winter is NW at 16.8 mph (*Table 1, page 10*)

Correction on *Table 43* for wind velocity is 16.8/15 = 1.12. Since the wind is coming from the Northwest, the crackage on the north and west sides will allow infiltration but the wind is only 60% effective. Correction for wind direction is .6.

Since this building is over 100 ft tall, stack effect causes infiltration on all sides at the lower levels and exfiltration at the upper levels. The total infiltration on the windward sides remains the same because the increase at the bottom is exactly equal to the decrease at the top. (For a floor-by-floor analysis, use equivalent wind velocity formulas.) Infiltration thru windows on the windward sides of the lower levels

$$= 12,000 \times 2 \times 1.12 \times .6 \times .98 = 15,810 \text{ cfm.}$$

The total infiltration thru the windows on the leeward sides of the building is equal to the difference between the equivalent velocity at the first floor and the design velocity at the midpoint of the building.

$$V_e = \sqrt{V^2 + 1.75b}$$

$$= \sqrt{(16.8)^2 + \left(1.75 \times \frac{240}{2}\right)} = 22.2 \text{ mph}$$

$$V_e - V = 22.2 - 16.8 = 5.4 \text{ mph}$$

Total infiltration thru windows in lower half of building (upper half is exfiltration) on leeward side
$$= 12,000 \times 2 \times \tfrac{1}{2} \times (5.4/15) \times \tfrac{1}{2} \times .98$$
$$= 2160 \text{ cfm } (Table\ 43)$$

NOTE: This is the total infiltration thru the windows on the leeward side. A floor-by-floor analysis should be made to balance the system to maintain proper conditions on each floor.

   The infiltration thru the doors on the street level (on leeward side)
   $$= 10 \times 7 \times 3 \times (5.4/15) \times 30$$
   $$= 2310 \text{ cfm } (Table\ 43c, \text{ average use, 1 and 2 story building}).$$

Example 4 — Offsetting Infiltration with Outdoor Air
Any outdoor air mechanically introduced into the building offsets some of the infiltration. In *Example 3* all of the outdoor air is effective in reducing the window infiltration. Infiltration is reduced on two windward sides, and the air introduced thru the apparatus exfiltrates thru the other two sides.

Given:
   The building described in *Example 1* with .25 cfm/sq ft supplied thru the apparatus and 40,000 cfm being exhausted from the building.

Find:
   The net infiltration into this building.

Solution:
   Net outdoor air = (.25 × 10,000 × 20) − 40,000 = 10,000 cfm
   Net infiltration thru windows
   $$= 15,800 + 2160 - 10,000 = 7970 \text{ cfm}$$
   Net infiltration thru doors = 2310 cfm (*Example 3*)
   Net infiltration into building = 7970 + 2310 = 10,280 cfm

TABLE 43—INFILTRATION THRU WINDOWS AND DOORS—WINTER*

15 mph Wind Velocity†

TABLE 43a—DOUBLE HUNG WINDOWS ON WINDWARD SIDE‡

DESCRIPTION	CFM PER SQ FT AREA					
	Small—30″ x 72″			Large—54″ x 96″		
	No W-Strip	W-Strip	Storm Sash	No W-Strip	W-Strip	Storm Sash
Average Wood Sash	.85	.52	.42	.53	.33	.26
Poorly Fitted Wood Sash	2.4	.74	1.2	1.52	.47	.74
Metal Sash	1.60	.69	.80	1.01	.44	.50

NOTE: W-Strip denotes weatherstrip.

TABLE 43b—CASEMENT TYPE WINDOWS ON WINDWARD SIDE‡

DESCRIPTION	CFM PER SQ FT AREA									
	Percent Ventilated Area									
	0%	25%	33%	40%	45%	50%	60%	66%	75%	100%
Rolled Section—Steel Sash										
Industrial Pivoted	.65	1.44	—	1.98	—	—	—	2.9	—	5.2
Architectural Projected	—	.78	—	—	—	1.1	1.48	—	—	—
Residential	—	—	.56	—	—	.98	—	—	—	1.26
Heavy Projected	—	—	—	—	.45	—	—	.63	.78	—
Hollow Metal—Vertically Pivoted	.54	1.19	—	1.64	—	—	—	2.4		4.3

TABLE 43c—DOORS ON ONE OR ADJACENT WINDWARD SIDES‡

DESCRIPTION	CFM PER SQ FT AREA**				
	Infrequent Use	Average Use			
		1 & 2 Story Bldg.	Tall Building (ft)		
			50	100	200
Revolving Door	1.6	10.5	12.6	14.2	17.3
Glass Door—(³⁄₁₆″ Crack)	9.0	30.0	36.0	40.5	49.5
Wood Door 3′ x 7′	2.0	13.0	15.5	17.5	21.5
Small Factory Door	1.5	13.0			
Garage & Shipping Room Door	4.0	9.0			
Ramp Garage Door	4.0	13.5			

*All values in *Table 43* are based on the wind blowing directly at the window or door. When the prevailing wind direction is oblique to the window or doors, multiply the above values by 0.60 and use the total window and door area on the windward side(s).

†Based on a wind velocity of 15 mph. For design wind velocities different from the base, multiply the table values by the ratio of velocities.

‡Stack effect in tall buildings may also cause infiltration on the leeward side. To evaluate this, determine the equivalent velocity (V_e) and subtract the design velocity (V). The equivalent velocity is:

$$V_e = \sqrt{V^2 - 1.75a} \text{ (upper section)}$$
$$V_e = \sqrt{V^2 + 1.75b} \text{ (lower section)}$$

Where a and b are the distances above and below the mid-height of the building, respectively, in ft.

Multiply the table values by the ratio (V_e − V)/15 for doors and one half of the windows on the leeward side of the building. (Use values under "1 and 2 Story Bldgs" for doors on leeward side of tall buildings.)

**Doors on opposite sides increase the above values 25%. Vestibules may decrease the infiltration as much as 30% when door usage is light. If door usage is heavy, the vestibule is of little value in reducing infiltration. Heat added to the vestibule will help maintain room temperature near the door.

INFILTRATION — CRACK METHOD (Summer or Winter)

The crack method of evaluating infiltration is more accurate than the area methods. It is difficult to establish the exact crack dimensions but, in certain close tolerance applications, it may be necessary to evaluate the load accurately. The crack method is applicable both summer and winter.

Basis of Table 44
— Infiltration thru Windows and Doors, Crack Method

The data on windows in *Table 44* are based on ASHAE tests. These test results have been reduced 20% because, as infiltration occurs on one side, a certain amount of pressure builds up in the building, thereby reducing the infiltration. The data on glass and factory doors has been calculated from observed typical crack widths.

Use of Table 44
— Infiltration thru Windows and Doors, Crack Method

Table 44 is used to determine the infiltration thru the doors and windows listed. This table does not take into account winter stack effect which must be evaluated separately, using the equivalent wind velocity formulas previously presented.

Example 5 — Infiltration thru Windows, Crack Method
Given:
  A 4 ft x 7 ft residential casement window facing south.

Find:
  The infiltration thru this window.

Solution:
  Assume the crack widths are measured as follows:
    Window frame — none, well sealed
    Window openable area — 1/32 in. crack; length, 20 ft

  Assume the wind velocity is 30 mph due south.
    Infiltration thru window = $20 \times 2.1 = 42$ cfm *(Table 44)*

TABLE 44—INFILTRATION THRU WINDOWS AND DOORS—CRACK METHOD—SUMMER—WINTER*

TABLE 44a—DOUBLE HUNG WINDOWS—UNLOCKED ON WINDWARD SIDE

TYPE OF DOUBLE HUNG WINDOW	CFM PER LINEAR FOOT OF CRACK											
	Wind Velocity—Mph											
	5		10		15		20		25		30	
	No W-Strip	W-Strip	No W-Strip	W-Strip	No W-Strip	W-Strip	No W-Strip	W-Strip	No W-Strip	W-Strip	No W-Strip	W-Strip
Wood Sash												
Average Window	.12	.07	.35	.22	.65	.40	.98	.60	1.33	.82	1.73	1.05
Poorly Fitted Window	.45	.10	1.15	.32	1.85	.57	2.60	.85	3.30	1.18	4.20	1.53
Poorly Fitted—with Storm Sash	.23	.05	.57	.16	.93	.29	1.30	.43	1.60	.59	2.10	.76
Metal Sash	.33	.10	.78	.32	1.23	.53	1.73	.77	2.3	1.00	2.8	1.27

TABLE 44b—CASEMENT TYPE WINDOWS ON WINDWARD SIDE

TYPE OF CASEMENT WINDOW AND TYPICAL CRACK SIZE		CFM PER LINEAR FOOT OF CRACK					
		Wind Velocity—Mph					
		5	10	15	20	25	30
Rolled Section—Steel Sash							
Industrial Pivoted	1/16″ crack	.87	1.80	2.9	4.1	5.1	6.2
Architectural Projected	1/32″ crack	.25	.60	1.03	1.43	1.86	2.3
Architectural Projected	3/64″ crack	.33	.87	1.47	1.93	2.5	3.0
Residential Casement	1/64″ crack	.10	.30	.55	.78	1.00	1.23
Residential Casement	1/32″ crack	.23	.53	.87	1.27	1.67	2.10
Heavy Casement Section Projected	1/64″ crack	.05	.17	.30	.43	.58	.80
Heavy Casement Section Projected	1/32″ crack	.13	.40	.63	.90	1.20	1.53
Hollow Metal—Vertically Pivoted		.50	1.46	2.40	3.10	3.70	4.00

*Infiltration caused by stack effect must be calculated separately during the winter.
†No allowance has been made for usage. See *Table 43* for infiltration due to usage.

TABLE 44—INFILTRATION THRU WINDOWS AND DOORS—CRACK METHOD—SUMMER—WINTER*
(Contd)

TABLE 44c—DOORS† ON WINDWARD SIDE

TYPE OF DOOR		CFM PER LINEAR FOOT OF CRACK					
		Wind Velocity— mph					
		5	10	15	20	25	30
Glass Door—Herculite							
Good Installation	⅛" crack	3.2	6.4	9.6	13.0	16.0	19.0
Average Installation	3⁄16" crack	4.8	10.0	14.0	20.0	24.0	29.0
Poor Installation	¼" crack	6.4	13.0	19.0	26.0	26.0	38.0
Ordinary Wood or Metal							
Well Fitted—W-Strip		.45	.60	.90	1.3	1.7	2.1
Well Fitted—No W-Strip		.90	1.2	1.8	2.6	3.3	4.2
Poorly Fitted—No W-Strip		.90	2.3	3.7	5.2	6.6	8.4
Factory Door	⅛" crack	3.2	6.4	9.6	13.0	16.0	19.0

VENTILATION

VENTILATION STANDARDS

The introduction of outdoor air for ventilation of conditioned spaces is necessary to dilute the odors given off by people, smoking and other internal air contaminants.

The amount of ventilation required varies primarily with the total number of people, the ceiling height and the number of people smoking. People give off body odors which require a minimum of 5 cfm per person for satisfactory dilution. Seven and one half cfm per person is recommended. This is based on a population density of 50 to 75 sq ft per person and a typical ceiling height of 8 ft. With greater population densities, the ventilation quantity should be increased. When people smoke, the additional odors given off by cigarettes or cigars require a minimum of 15 to 25 cfm per person. In special gathering rooms with heavy smoking, 30 to 50 cfm per person is recommended.

Basis of Table 45
—Ventilation Standards

The data in *Table 45* is based on test observation of the clean outdoor air required to maintain satisfactory odor levels with people smoking and not smoking. These test results were then extrapolated for typical concentrations of people, both smoking and not smoking, for the applications listed.

Use of Table 45
—Ventilation Standards

Table 45 is used to determine the minimum and recommended ventilation air quantity for the listed applications. In applications where the minimum values are used and the minimum cfm per person and cfm per sq ft of floor area are listed, use the larger minimum quantity. Where the crowd density is greater than normal or where better than satisfactory conditions are desired, use the recommended values.

SCHEDULED VENTILATION

In comfort applications, where local codes permit, it is possible to reduce the capacity requirements of the installed equipment by reducing the ventilation air quantity at the time of peak load. This quantity can be reduced at times of peak to, in effect, minimize the outdoor air load. At times other than peak load, the calculated outdoor air quantity is used. Scheduled ventilation is recommended *only for installations operating more than 12 hours or 3 hours longer than occupancy,* to allow some time for flushing out the building when no odors are being generated. It has been found, by tests, that few complaints of stuffiness are encountered when the outdoor air quantity is reduced for short periods of time, provided the flushing period is available. It is recommended that the outdoor air quantity be reduced to no less than 40% of the recommended quantity as listed in *Table 45*.

The procedure for estimating and controlling scheduled ventilation is as follows:

1. In estimating the cooling load, reduce the air quantity at design conditions to a minimum of 40% of the recommended air quantity.

2. Use a dry-bulb thermostat following the cooling and dehumidifying apparatus to control the leaving dewpoint such that:

a. With the dewpoint at design, the damper motor closes the outdoor air damper to 40% of the design ventilation air quantity.

b. As the dewpoint decreases below design, the outdoor air damper opens to the design setting.

Example 6 — Ventilation Air Quantity, Office Space

Given:
A 5000 sq ft office with a ceiling height of 8 ft and 50 people. Approximately 40% of the people smoke.

Find:
The ventilation air quantity.

Solution:
The population density is typical, 100 sq ft per person, but the number of smokers is considerable.

Recommended ventilation = 50 × 15 = 750 cfm *(Table 45)*

Minimum ventilation   = 50 × 10 = 500 cfm *(Table 45)*

500 cfm will more than likely not maintain satisfactory conditions within the space because the number of smokers is considerable. Therefore, 750 cfm should be used in this application.

NOTE: Many applications have exhaust fans. This means that the outdoor air quantity must at least equal the exhausted air; otherwise the infiltration rate will increase. *Tables 46 and 47* list the approximate capacities of typical exhaust fans. The data in these tables were obtained from published ratings of several manufacturers of exhaust fans.

TABLE 45—VENTILATION STANDARDS

APPLICATION		SMOKING	CFM PER PERSON		CFM PER SQ FT OF FLOOR Minimum*
			Recommended	Minimum*	
Apartment	Average	Some	20	15	—
	De Luxe	Some	30	25	.33
Banking Space		Occasional	10	7½	—
Barber Shops		Considerable	15	10	—
Beauty Parlors		Occasional	10	7½	—
Broker's Board Rooms		Very Heavy	50	30	—
Cocktail Bars		Heavy	30	25	—
Corridors (Supply or Exhaust)			—	—	.25
Department Stores		None	7½	5	.05
Directors Rooms		Extreme	50	30	—
Drug Stores †		Considerable	10	7½	—
Factories ‡§		None	10	7½	.10
Five and Ten Cent Stores		None	7½	5	—
Funeral Parlors		None	10	7½	—
Garage ‡		—	—	—	1.0
Hospitals	Operating Rooms ‡**	None	—	—	2.0
	Private Rooms	None	30	25	.33
	Wards	None	20	15	—
Hotel Rooms		Heavy	30	25	.33
Kitchen	Restaurant †	—	—	—	4.0
	Residence	—	—	—	2.0
Laboratories †		Some	20	15	—
Meeting Rooms		Very Heavy	50	30	1.25
Office	General	Some	15	10	—
	Private	None	25	15	.25
	Private	Considerable	30	25	.25
Restaurant	Cafeteria †	Considerable	12	10	—
	Dining Room †	Considerable	15	12	—
School Rooms ‡		None	—	—	—
Shop Retail		None	10	7½	—
Theater ‡		None	7½	5	—
Theater		Some	15	10	—
Toilets ‡ (Exhaust)		—	—	—	2.0

*When minimum is used, use the larger.

‡See local codes which may govern.

†May be governed by exhaust.

§Use these values unless governed by other sources of contamination or by local codes.

**All outdoor air is recommended to overcome explosion hazard of anesthetics.

TABLE 46—CENTRIFUGAL FAN CAPACITIES

Inlet Diameter (in.)	Capacity* (cfm)	Motor Horsepower Range	Outlet Velocity Range (fpm)
4	50- 250	1/70-1/20	800-2000
6	100- 550	1/20-1/6	500-2500
8	300-1000	1/20-½	850-2900
10	600-2800	1/5-2	950-4300
12†	800-1600	⅛-½	1000-2000
15†	1200-2500	¼-1	1000-2000
18†	1700-3600	¼-1¼	1000-2000
21†	2300-5000	⅓-1½	1000-2000

*These typical air capacities were obtained from published ratings of several manufacturers of nationally known exhaust fans, single width, single inlet. Range of static pressures ¼ to 1¼ inches. Fans with inlet diameter 10 inches and smaller are direct connected.

†The capacity of these fans has been arbitrarily taken at 1000 fpm minimum and 2000 fpm maximum outlet velocity. For these fans the usual selection probably is approximately 1500 fpm outlet velocity for ventilation.

TABLE 47—PROPELLER FAN CAPACITIES—FREE DELIVERY

Fan Diameter (in.)	Speed (rpm)	Capacity* (cfm)
8	1500	500
12	1140	825
12	1725	1100
16	855	1000
16	1140	1500
18	850	1800
18	1140	2350
20	850	2400
20	1140	2750
20	1620	3300

*The capacities of fans of various manufacturers may vary ±10% from the values given above.

CHAPTER 7. INTERNAL AND SYSTEM HEAT GAIN

INTERNAL HEAT GAIN

Internal heat gain is the sensible and latent heat released within the air conditioned space by the occupants, lights, appliances, machines, pipes, etc. This chapter outlines the procedures for determining the instantaneous *heat gain* from these sources. A portion of the heat gain from internal sources is radiant heat which is partially absorbed in the building structure, thereby reducing the instantaneous heat gain. *Chapter 3, "Heat Storage, Diversity and Stratification,"* contains the data and methods for estimating the actual cooling load from the heat sources referred to in the following text.

PEOPLE

Heat is generated within the human body by oxidation, commonly called metabolic rate. The metabolic rate varies with the individual and with his activity level. The normal body processes are performed most efficiently at a deep tissue temperature of about 98.6 F; this temperature may vary only thru a narrow range. However, the human body is capable of maintaining this temperature, thru a wide ambient temperature range, by conserving or dissipating the heat generated within itself.

This heat is carried to the surface of the body by the blood stream and is dissipated by:

1. Radiation from the body surface to the surrounding surfaces.
2. Convection from the body surface and the respiratory tract to the surrounding air.
3. Evaporation of moisture from the body surface and in the respiratory tract to the surrounding air.

The amount of heat dissipated by radiation and convection is determined by the difference in temperature between the body surface and its surroundings. The body surface temperature is regulated by the quantity of blood being pumped to the surface; the more blood, the higher the surface temperature up to a limit of about 96 F. The heat dissipated by evaporation is determined by the difference in vapor pressure between the body and the air.

Basis of Table 48
— Heat Gain from People

Table 48 is based on the metabolic rate of an average adult male, weighing 150 pounds, at different levels of activity, and generally for occupancies longer than 3 hours. These have been adjusted for typical compositions of mixed groups of males and females for the listed applications. The metabolic rate of women is about 85% of that for a male, and for children about 75%.

The heat gain for restaurant applications has been increased 30 Btu/hr sensible and 30 Btu/hr latent heat per person to include the food served.

The data in *Table 48* as noted are for continuous occupancy. The excess heat and moisture brought in by people, where short time occupancy is occurring (under 15 minutes), may increase the heat gain from people by as much as 10%.

Use of Table 48
— Heat Gain from People

To establish the proper heat gain, the room design temperature and the activity level of the occupants must be known.

Example 1 — Bowling Alley

Given:
A 10 lane bowling alley, 50 people, with a room design dry-bulb temperature of 75 F. Estimate one person per alley bowling, 20 of the remainder seated, and 20 standing.

Find:
Sensible and latent heat gain from people.

Solution:
Sensible heat gain = (10 × 525) + (20 × 240) + (20 × 280)
                                = 15,650 Btu/hr
Latent heat gain   = (10 × 925) + (20 × 160) + (20 × 270)
                                = 17,850 Btu/hr

LIGHTS

Lights generate sensible heat by the conversion of the electrical power input into light and heat. The heat is dissipated by radiation to the surrounding surfaces, by conduction into the adjacent materials and by convection to the surrounding air. The radiant portion of the light load is partially stored, and the convection portion may be stratified as described on *page 39*. Refer to *Table 12, page 35*, to determine the actual cooling load.

Incandescent lights convert approximately 10% of the power input into light with the rest being generated as heat within the bulb and dissipated by radiation, convection and conduction. About 80% of the power input is dissipated by radiation and only about 10% by convection and conduction, *Fig. 30*.

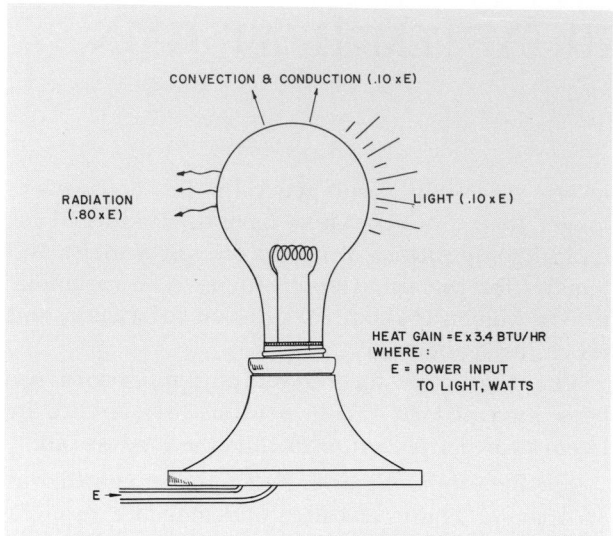

FIG. 30 – CONVERSION OF ELECTRIC POWER TO HEAT AND LIGHT WITH INCANDESCENT LIGHTS, APPROXIMATE

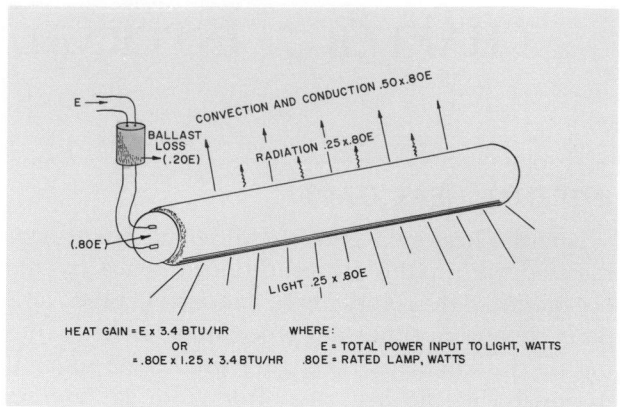

FIG. 31 – CONVERSION OF ELECTRIC POWER TO HEAT AND LIGHT WITH FLUORESCENT LIGHTS, APPROXIMATE

Fluorescent lights convert about 25% of the power input into light, with about 25% being dissipated by radiation to the surrounding surfaces. The other 50% is dissipated by conduction and convection. In addition to this, approximately 25% more heat is generated as heat in the ballast of the fluorescent lamp, *Fig. 31.*

Table 49 indicates the basis for arriving at the gross heat gain from fluorescent or incandescent lights.

TABLE 48—HEAT GAIN FROM PEOPLE

DEGREE OF ACTIVITY	TYPICAL APPLICATION	Metabolic Rate (Adult Male) Btu/hr	Average Adjusted Metabolic Rate* Btu/hr	ROOM DRY-BULB TEMPERATURE									
				82 F Btu/hr		80 F Btu/hr		78 F Btu/hr		75 F Btu/hr		70 F Btu/hr	
				Sensible	Latent	Sensible	Latent	Sensible	Latent	Sensible	Latent	Sensible	Latent
Seated at rest	Theater, Grade School	390	350	175	175	195	155	210	140	230	120	260	90
Seated, very light work	High School	450	400	180	220	195	205	215	185	240	160	275	125
Office worker	Offices, Hotels, Apts., College	475	450	180	270	200	250	215	235	245	205	285	165
Standing, walking slowly	Dept., Retail, or Variety Store	550											
Walking, seated	Drug Store	550	500	180	320	200	300	220	280	255	245	290	210
Standing, walking slowly	Bank	550											
Sedentary work	Restaurant†	500	550	190	360	220	330	240	310	280	270	320	230
Light bench work	Factory, light work	800	750	190	560	220	530	245	505	295	455	365	385
Moderate dancing	Dance Hall	900	850	220	630	245	605	275	575	325	525	400	450
Walking, 3 mph	Factory, fairly heavy work	1000	1000	270	730	300	700	330	670	380	620	460	540
Heavy work	Bowling Alley‡, Factory	1500	1450	450	1000	465	985	485	965	525	925	605	845

*Adjusted Metabolic Rate is the metabolic rate to be applied to a mixed group of people with a typical percent composition based on the following factors:
Metabolic rate, adult female = Metabolic rate, adult male × 0.85
Metabolic rate, children = Metabolic rate, adult male × 0.75

†Restaurant—Values for this application include 60 Btu per hr for food per Individual (30 Btu sensible and 30 Btu latent heat per hr).

‡Bowling—Assume one person per alley actually bowling and all others sitting, metabolic rate 400 Btu per hr; or standing, 550 Btu per hr.

TABLE 49—HEAT GAIN FROM LIGHTS

TYPE	HEAT GAIN* Btu/hr
Fluorescent	Total Light Watts × 1.25† × 3.4
Incandescent	Total Light Watts × 3.4

*Refer to *Tables 12 and 13, pages 35-37* to determine actual cooling load.
†Fluorescent light wattage is multiplied by 1.25 to include heat gain in ballast.

APPLIANCES

Most appliances contribute both sensible and latent heat to a space. Electric appliances contribute latent heat, only by virtue of the function they perform, that is, drying, cooking, etc., whereas gas burning appliances contribute additional moisture as a product of combustion. A properly designed hood with a positive exhaust system removes a considerable amount of the generated heat and moisture from most types of appliances.

Basis of Tables 50 thru 52 — Heat Gain from Restaurant Appliances and Miscellaneous Appliances

The data in these tables have been determined from manufacturers data, the American Gas Association data, Directory of Approved Gas Appliances and actual tests by Carrier Corporation.

TABLE 50—HEAT GAIN FROM RESTAURANT APPLIANCES

NOT HOODED*—ELECTRIC

APPLIANCE	OVERALL DIMENSIONS Less Legs and Handles (In.)	TYPE OF CONTROL	MISCELLANEOUS DATA	MFR MAX RATING Btu/hr	MAIN-TAIN-ING RATE Btu/hr	RECOM HEAT GAIN FOR AVG USE		
						Sensible Heat Btu/hr	Latent Heat Btu/hr	Total Heat Btu/hr
Coffee Brewer—½ gal		Man.		2240	306	900	220	1120
Warmer—½ gal		Man.		306	306	230	90	320
4 Coffee Brewing Units with 4½ gal Tank	20 x 30 x 26H	Auto.	Water heater—2000 watts Brewers—2960 watts	16900		4800	1200	6000
Coffee Urn—3 gal	15 Dia x 34H	Man.	Black finish	11900	3000	2600	1700	4300
—3 gal	12 x 23 oval x 21H	Auto.	Nickel plated	15300	2600	2200	1500	3700
—5 gal	18 Dia x 37H	Auto.	Nickel plated	17000	3600	3400	2300	5700
Doughnut Machine	22 x 22 x 57H	Auto.	Exhaust system to outdoors— ½ hp motor	16000		5000		5000
Egg Boiler	10 x 13 x 25H	Man.	Med. ht.—550 watts Low ht—275 watts	3740		1200	800	2000
Food Warmer with Plate Warmer, per sq ft top surface		Auto.	Insulated, separate heating unit for each pot. Plate warmer in base	1350	500	350	350	700
Food Warmer without Plate Warmer, per sq ft top surface		Auto.	Ditto, without plate warmer	1020	400	200	350	550
Fry Kettle—11½ lb fat	12 Dia x 14H	Auto.		8840	1100	1600	2400	4000
Fry Kettle—25 lb fat	16 x 18 x 12H	Auto.	Frying area 12" x 14"	23800	2000	3800	5700	9500
Griddle, Frying	18 x 18 x 8H	Auto.	Frying top 18" x 14"	8000	2800	3100	1700	4800
Grille, Meat	14 x 14 x 10H	Auto.	Cooking area 10" x 12"	10200	1900	3900	2100	6000
Grille, Sandwich	13 x 14 x 10H	Auto.	Grill area 12" x 12"	5600	1900	2700	700	3400
Roll Warmer	26 x 17 x 13H	Auto.	One drawer	1500	400	1100	100	1200
Toaster, Continuous	15 x 15 x 28H	Auto.	2 Slices wide— 360 slices/hr	7500	5000	5100	1300	6400
Toaster, Continuous	20 x 15 x 28H	Auto.	4 Slices wide— 720 slices/hr	10200	6000	6100	2600	8700
Toaster, Pop-Up	6 x 11 x 9H	Auto.	2 Slices	4150	1000	2450	450	2900
Waffle Iron	12 x 13 x 10H	Auto.	One waffle 7" dia	2480	600	1100	750	1850
Waffle Iron for Ice Cream Sandwich	14 x 13 x 10H	Auto.	12 Cakes, each 2½" x 3¾"	7500	1500	3100	2100	5200

*If properly designed positive exhaust hood is used, multiply recommended value by .50.

Use of Tables 50 thru 52
— Heat Gain from Restaurant Appliances and
Miscellaneous Appliances

The *Maintaining Rate* is the heat generated when
the appliance is being maintained at operating temperature but not being used.

The *Recommended for Average Use* values are
those which the appliance generates under normal
use. These appliances seldom operate at maximum
capacity during peak load since they are normally
warmed up prior to the peak.

The values in *Tables 50 thru 52* are for unhooded
appliances. If the appliance has a properly designed
positive exhaust hood, reduce the sensible and the
latent heat gains by 50%. A hood, to be effective,
should extend beyond the appliance approximately
4 inches per foot of height between the appliance
and the face of the hood. The lower edge should not
be higher than 4 feet above the appliance and the
average face velocity across the hood should not be
less than 70 fpm.

TABLE 51—HEAT GAIN FROM RESTAURANT APPLIANCES

NOT HOODED*—GAS BURNING AND STEAM HEATED

APPLIANCE	OVERALL DIMENSIONS Less Legs and Handles (In.)	TYPE OF CONTROL	MISCELLANEOUS DATA	MFR MAX RATING Btu/hr	MAINTAINING RATE Btu/hr	RECOM HEAT GAIN FOR AVG USE		
						Sensible Heat Btu/hr	Latent Heat Btu/hr	Total Heat Btu/hr
GAS BURNING								
Coffee Brewer—½ gal Warmer—½ gal		Man. Man.	Combination brewer and warmer	3400 500	500	1350 400	350 100	1700 500
Coffee Brewer Unit with Tank	19 x 30 x 26H		4 Brewers and 4½ gal tank			7200	1800	9000
Coffee Urn—3 gal	15" Dia x 34H	Auto.	Black finish	3200	3900	2900	2900	5800
Coffee Urn—3 gal	12 x 23 oval x 21H	Auto.	Nickel plated		3400	2500	2500	5000
Coffee Urn—5 gal	18 Dia x 37H	Auto.	Nickel plated		4700	3900	3900	7800
Food Warmer, Values per sq ft top surface		Man.	Water bath type	2000	900	850	450	1300
Fry Kettle—15 lb fat	12 x 20 x 18H	Auto.	Frying area 10 x 10	14250	3000	4200	2800	7000
Fry Kettle—28 lb fat	15 x 35 x 11H	Auto.	Frying area 11 x 16	24000	4500	7200	4800	12000
Grill—Broil-O-Grill Top Burner Bottom Burner	22 x 14 x 17H (1.4 sq ft grill surface)	Man.	Insulated 22,000 Btu/hr 15,000 Btu/hr	37000		14400	3600	18000
Stoves, Short Order—Open Top. Values per sq ft top surface		Man.	Ring type burners 12000 to 22000 Btu/ea	14000		4200	4200	8400
Stoves, Short Order—Closed Top. Values per sq ft top surface		Man.	Ring type burners 10000 to 12000 Btu/ea	11000		3300	3300	6600
Toaster, Continuous	15 x 15 x 28H	Auto.	2 Slices wide—360 slices/hr	12000	10000	7700	3300	11000
STEAM HEATED								
Coffee Urn—3 gal —3 gal —5 gal	15 Dia x 34H 12 x 23 oval x 21H 18 Dia x 37H	Auto. Auto. Auto.	Black finish Nickel plated Nickel plated			2900 2400 3400	1900 1600 2300	4800 4000 5700
Coffee Urn—3 gal —3 gal —5 gal	15 Dia x 34H 12 x 23 oval x 21H 18 Dia x 37H	Man. Man. Man.	Black finish Nickel plated Nickel plated			3100 2600 3700	3100 2600 3700	6200 5200 7400
Food Warmer, per sq ft top surface		Auto.				400	500	900
Food Warmer, per sq ft top surface		Man.				450	1150	1500

*If properly designed positive exhaust hood is used, multiply recommended value by .50.

TABLE 52—HEAT GAIN FROM MISCELLANEOUS APPLIANCES

NOT HOODED*

APPLIANCE	TYPE OF CONTROL	MISCELLANEOUS DATA	MFR MAX RATING Btu/hr	RECOM HEAT GAIN FOR AVG USE		
				Sensible Heat Btu/hr	Latent Heat Btu/hr	Total Heat Btu/hr
ELECTRIC						
Hair Dryer, Blower Type 15 amps, 115 volts AC	Man.	Fan 165 watts, (low 915 watts, high 1580 watts)	5,370	2,300	400	2,700
Hair Dryer, helmet type, 6.5 amps, 115 volts AC	Man.	Fan 80 watts, (low 300 watts, high 710 watts)	2,400	1,870	330	2,200
Permanent Wave Machine	Man.	60 heaters at 25 watts each, 36 in normal use	5,100	850	150	1,000
Pressurized Instrument Washer and Sterilizer		11" x 11" x 22"		12,000	23,460	35,460
Neon Sign, per linear ft tube		½" outside dia		30		30
		⅜" outside dia		60		60
Solution and/or Blanket Warmer		18" x 30" x 72"		1,200	3,000	4,200
		18" x 24" x 72"		1,050	2,400	3,450
Sterilizer Dressing	Auto.	16" x 24"		9,600	8,700	18,300
	Auto.	20" x 36"		23,300	24,000	47,300
Sterilizer, Rectangular Bulk	Auto.	24" x 24" x 36"		34,800	21,000	55,800
	Auto.	24" x 24" x 48"		41,700	27,000	68,700
	Auto.	24" x 36" x 48"		56,200	36,000	92,200
	Auto.	24" x 36" x 60"		68,500	45,000	113,500
	Auto.	36" x 42" x 84"		161,700	97,500	259,200
	Auto.	42" x 48" x 96"		184,000	140,000	324,000
	Auto.	48" x 54" x 96"		210,000	180,000	390,000
Sterilizer, Water	Auto.	10 gallon		4,100	16,500	20,600
	Auto.	15 gallon		6,100	24,600	30,700
Sterilizer, Instrument	Auto.	6" x 8" x 17"		2,700	2,400	5,100
	Auto.	9" x 10" x 20"		5,100	3,900	9,000
	Auto.	10" x 12" x 22"		8,100	5,900	14,000
	Auto.	10" x 12" x 36"		10,200	9,400	19,600
	Auto.	12" x 16" x 24"		9,200	8,600	17,800
Sterilizer, Utensil	Auto.	16" x 16" x 24"		10,600	20,400	31,000
	Auto.	20" x 20" x 24"		12,300	25,600	37,900
Sterilizer, Hot Air	Auto.	Model 120 Amer Sterilizer Co		2,000	4,200	6,200
	Auto.	Model 100 Amer Sterilizer Co		1,200	2,100	3,300
Water Still		5 gal/hour		1,700	2,700	4,400
X-ray Machines, for making pictures		Physicians and Dentists office		None	None	None
X-ray Machines, for therapy		Heat load may be appreciable—write mfg for data				
GAS BURNING						
Burners, Laboratory small bunsen	Man.	⁷⁄₁₆ dia barrel with manufactured gas	1,800	960	240	1,200
small bunsen	Man.	⁷⁄₁₆ dia with nat gas	3,000	1,680	420	2,100
fishtail burner	Man.	⁷⁄₁₆ dia with nat gas	3,500	1,960	490	2,450
fishtail burner	Man.	⁷⁄₁₆ dia bar with nat gas	5,500	3,080	770	3,850
large bunsen	Man.	1½ dia mouth, adj orifice	6,000	3,350	850	4,200
Cigar Lighter	Man.	Continuous flame type	2,500	900	100	1,000
Hair Dryer System 5 helmets	Auto.	Consists of heater & fan which blows hot air thru duct system to helmets	33,000	15,000	4,000	19,000
10 helmets	Auto.			21,000	6,000	27,000

*If properly designed positive exhaust hood is used, multiply recommended value by .50

Example 2 — Restaurant

Given:

A restaurant with the following electric appliances with a properly designed positive exhaust hood on each:
1. Two 5-gallon coffee urns, both used in the morning, only one used either in the afternoon or evening.
2. One 20 sq ft food warmer without plate warmer.
3. Two 24 x 20 x 10 inch frying griddles.
4. One 4-slice pop-up toaster, used only in the morning.
5. Two 25 lb deep fat, fry kettles.

Find:

Heat gain from these appliances during the afternoon and evening meal.

Solution:

Use *Table 50*.   Sensible Latent

1. Coffee Urn — only one in use:
   Sensible heat gain $= 3400 \times .50 =$ 1700
   Latent heat gain $= 2300 \times .50 =$ 1150

2. Food Warmer:
   Sensible heat gain $= 20 \times 200 \times .50 =$ 2000
   Latent heat gain $= 20 \times 350 \times .50 =$ 3500

3. Frying Griddles:
   Sensible heat gain $= 2 \times 5300 \times .50 =$ 5300
   Latent heat gain $= 2 \times 2900 \times .50 =$ 2900

4. Toaster — not in use

5. Fry Kettles:
   Sensible heat gain $= 2 \times 3800 \times .50 =$ 3800
   Latent heat gain $= 2 \times 5700 \times .50 =$ 5700

Total sensible heat gain $=$ 12,800

Total latent heat gain $=$ 13,250

ELECTRIC MOTORS

Electric motors contribute sensible heat to a space by converting the electrical power input to heat. Some of this power input is dissipated as heat in the motor frame and can be evaluated as

$$\text{input} \times (1 - \text{motor eff}).$$

The rest of the power input (brake horsepower or motor output) is dissipated by the driven machine and in the drive mechanism. The driven machine utilizes this motor output to do work which may or may not result in a heat gain to the space.

Motors driving fans and pumps: The power input increases the pressure and velocity of the fluid and the temperature of the fluid.

The increased energy level in the fluid is degenerated in pressure drop throughout the system and appears as a heat gain to the fluid at the point where pressure drop occurs. This heat gain does not appear as a temperature rise because, as the pressure reduces, the fluid expands. The fluid expansion is a cooling process which exactly offsets the heat generated by friction. The heat of compression required to increase the energy level is generated at the fan or pump *and* is a heat gain at this point.

If the fluid is conveyed outside of the air conditioned space, only the inefficiency of the motor driving fan or pump should be included in room sensible heat gain.

If the temperature of the fluid is maintained by a separate source, these heat gains to the fluid heat of compression are a load on this separate source only.

The heat gain or loss from the system should be calculated separately (*"System Heat Gain," p. 110*).

Motors driving process machinery (lathe, punch press, etc.): The total power input to the machine is dissipated as heat at the machine. If the product is removed from the conditioned space at a higher temperature than it came in, some of the heat input into the machine is removed and should not be considered a heat gain to the conditioned space. The heat added to a product is determined by multiplying the number of pounds of material handled per hour by the specific heat and temperature rise.

Basis of Table 53
— Heat Gain from Electric Motors

Table 53 is based on average efficiencies of squirrel cage induction open type integral horsepower and fractional horsepower motors. Power supply for fractional horsepower motors is 110 or 220 volts, 60 cycle, single phase; for integral horsepower motors, 208, 220, or 440 volts, 60 cycle, 2 or 3 phase general purpose and constant speed, 1160 or 1750 rpm. This table may also be applied with reasonable accuracy to 50 cycle, single phase a-c, 50 and 60 cycle enclosed and fractional horsepower polyphase motors.

Use of Table 53
— Heat Gain from Electric Motors

The data in *Table 53* includes the heat gain from electric motors and their driven machines when both the motor and the driven machine are in the conditioned space, or when only the driven machine is in the conditioned space, or when only the motor is in the conditioned space.

Caution: The power input to electric motors does not necessarily equal the rated horsepower divided by the motor efficiency. Frequently these motors may be operating under a continuous overload, or may be operating at less than rated capacity. It is always advisable to measure the power input wherever possible. This is especially important in estimates for industrial installations where the motor-machine load is normally a major portion of the cooling load.

When readings are obtained directly in watts and when both motors and driven machines are in the air conditioned space, the heat gain is equal to the number of watts times the factor 3.4 Btu/(watt)(hr).

When the machine is in the conditioned space and the motor outside, multiply the watts by the motor efficiency and by the factor 3.4 to determine heat gain to the space.

When the machine is outside the conditioned space, multiply the watts by one minus the motor efficiency and by the factor 3.4.

Although the results are less accurate, it may be expedient to obtain power input measurements using a clamp-on ammeter and voltmeter. These instruments permit instantaneous readings only. They afford means for determining the load factor but the usage factor must be obtained by a careful investigation of the operating conditions.

TABLE 53—HEAT GAIN FROM ELECTRIC MOTORS

CONTINUOUS OPERATION*

NAMEPLATE† OR BRAKE HORSEPOWER	FULL LOAD MOTOR EFFICIENCY PERCENT	LOCATION OF EQUIPMENT WITH RESPECT TO CONDITIONED SPACE OR AIR STREAM‡		
		Motor In - Driven Machine in $\dfrac{HP \times 2545}{\% \text{ Eff}}$	Motor Out - Driven Machine in $HP \times 2545$	Motor In - Driven Machine out $\dfrac{HP \times 2545 \, (1 - \% \text{ Eff})}{\% \text{ Eff}}$
		Btu per Hour		
1/20	40	320	130	190
1/12	49	430	210	220
1/8	55	580	320	260
1/6	60	710	430	280
1/4	64	1,000	640	360
1/3	66	1,290	850	440
1/2	70	1,820	1,280	540
3/4	72	2,680	1,930	750
1	79	3,220	2,540	680
1 1/2	80	4,770	3,820	950
2	80	6,380	5,100	1,280
3	81	9,450	7,650	1,800
5	82	15,600	12,800	2,800
7 1/2	85	22,500	19,100	3,400
10	85	30,000	25,500	4,500
15	86	44,500	38,200	6,300
20	87	58,500	51,000	7,500
25	88	72,400	63,600	8,800
30	89	85,800	76,400	9,400
40	89	115,000	102,000	13,000
50	89	143,000	127,000	16,000
60	89	172,000	153,000	19,000
75	90	212,000	191,000	21,000
100	90	284,000	255,000	29,000
125	90	354,000	318,000	36,000
150	91	420,000	382,000	38,000
200	91	560,000	510,000	50,000
250	91	700,000	636,000	64,000

*For intermittent operation, an appropriate usage factor should be used, preferably measured.

†If motors are overloaded and amount of overloading is unknown, multiply the above heat gain factors by the following maximum service factors:

Maximum Service Factors

Horsepower	1/20-1/8	1/6-1/3	1/2-3/4	1	1 1/2-2	3-250
AC Open Type	1.4	1.35	1.25	1.25	1.20	1.15
DC Open Type	—	—	—	1.15	1.15	1.15

No overload is allowable with enclosed motors.

‡For a *fan* or *pump* in air conditioned space, exhausting air and pumping fluid to outside of space, use values in last column.

The following is a conversion table which can be used to determine load factors from measurements:

TO FIND →	HP OUTPUT	KILOWATTS INPUT
Direct Current	$\dfrac{I \times E \times \text{eff}}{746}$	$\dfrac{I \times E}{1{,}000}$
1 Phase	$\dfrac{I \times E \times \text{pf} \times \text{eff}}{746}$	$\dfrac{I \times E \times \text{pf}}{1{,}000}$
3 or 4 Wire 3 Phase	$\dfrac{I \times E \times \text{pf} \times \text{eff} \times 1.73}{746}$	$\dfrac{I \times E \times \text{pf} \times 1.73}{1{,}000}$
4 Wire 2 Phase	$\dfrac{I \times E \times \text{pf} \times \text{eff} \times 2}{746}$	$\dfrac{I \times E \times 2 \times \text{pf}}{1{,}000}$

Where I = amperes   eff = efficiency
       E = volts   pf = power factor

NOTE: For 2 phase, 3 wire circuit, common conductor current is 1.41 times that in either of the other two conductors.

Example 3 — Electric Motor Heat Gain in a Factory (Motor Bhp Established by a Survey)

Given:

1. Forty-five 10 hp motors operated at 80% rated capacity, driving various types of machines located within air conditioned space (lathes, screw machines, etc.).
   Five 10 hp motors operated at 80% rated capacity, driving screw machines, each handling 5000 lbs of bronze per hr. Both the final product and the shavings from the screw machines are removed from the space on conveyor belts. Rise in bronze temperature is 30 F; sp ht is .01 Btu/(lb)(F).

2. Ten 5 hp motors (5 bhp) driving fans, exhausting air to the outdoors.

3. Three 20 hp motors (20 bhp) driving process water pumps, water discarded outdoors.

Find:
Total heat gain from motors.

Solution:
Use *Table 53*.

	Sensible Heat Gain Btu/hr
1. Machines — Heat gain to space $= 45 \times 30{,}000 \times .80 =$	1,080,000

Heat gain from screw machines
$= 5 \times 30{,}000 \times .80 = 120{,}000$ Btu/hr

Heat removed from space from screw machine work
$= 5000 \times 5 \times 30 \times .01 = 7{,}500$ Btu/hr

Net heat gain from screw machines to space
$= 120{,}000 - 7500 =$ 112,500

2. Fan exhausting air to the outdoors:
   Heat gain to space $= 10 \times 2800 =$ 28,000

3. Process water pumped to outside air conditioned space
   Heat gain to space $= 3 \times 7500 =$ 22,500

Total heat gain from motors on machines, fans, and pumps = ... 1,243,000

NOTE: If the process water were to be recirculated and cooled in the circuit from an outside source, the heat gain to the water
$$3 \times (58{,}500 - 7500) = 153{,}000 \text{ Btu/hr}$$
would become a load on this outside source.

PIPING, TANKS AND EVAPORATION OF WATER FROM A FREE SURFACE

Hot pipes and tanks add sensible heat to a space by convection and radiation. Conversely, cold pipes remove sensible heat. All open tanks containing hot water contribute not only sensible heat but also latent heat due to evaporation.

In industrial plants, furnaces or dryers are often encountered. These contribute sensible heat to the space by convection and radiation from the outside surfaces, and frequently dryers also contribute sensible and latent heat from the drying process.

Basis of Tables 54 thru 58
— Heat Gain from Piping, Tanks and Evaporation of Water

Table 54 is based on nominal flow in the pipe and a convection heat flow from a horizontal pipe of —

$$1.016 \times \left(\frac{1}{\text{Dia}}\right)^{.2} \times \left(\frac{1}{T_1}\right)^{.181}$$

$\times$ (temp diff between hot water or steam and room).

The radiation from horizontal pipes is expressed by —

$$17.23 \times 10^{-10} \times \text{emissivity} \times (T_1^{\,4} - T_2^{\,4})$$

where T_1 = room surface temp, deg R
       T_2 = pipe surface temp, deg R

Tables 55 and 56 are based on the same equation and an insulation resistance of approximately 2.5 per inch of thickness for 85% magnesia and 2.9 per inch of thickness with moulded type.

Caution: Tables 55 and 56 do not include an allowance for fittings. A safety factor of 10% should be added for pipe runs having numerous fittings.

Table 57 is based on an emissivity of 0.9 for painted metal and painted or bare wood and concrete. The emissivity of chrome, bright nickel plate, stainless steel, or galvanized iron is 0.4. The resistance (r) of wood is 0.833 per inch and of concrete 0.08 per inch. The metal surface temperature has been assumed equal to the water temperature.

NOTE: The heat gain from furnaces and ovens can be estimated from *Table 57*, using the outside temperature of furnace and oven.

Table 58 is based on the following formula for still air: Heat of evaporation = 95 (vapor pressure differential between water and air), where vapor pressure is expressed in inches of mercury, and the room conditions are 75 F db and 50% rh.

Use of Tables 54 thru 58
— Heat Gain from Piping, Tanks and Evaporation of Water

Example 4 — Heat Gain from Hot Water Pipe and Storage Tank

Given:
Room conditions — 75 F db, 50% rh
50 ft of 10-inch uninsulated hot water (125 F) pipe.
The hot water is stored in a 10 ft wide x 20 ft long x 10 ft high, painted metal tank with the top open to the atmosphere. The tank is supported on open steel framework.

Find:
Sensible and latent heat gain

Solution:

Use *Tables 54, 57 and 58*	Btu/hr
Piping — Sensible heat gain = 50 × 50 × 4.76 =	11,900
Tank — Sensible heat gain, sides = (20 × 10 × 2) + (10 × 10 × 2) × 50 × 1.8 =	54,000
— Sensible heat gain, bottom = (20 × 10) × 50 × 1.5 =	15,000
Total sensible heat gain =	80,900
Total latent heat gain, top = (20 × 10) × 330 =	66,000

STEAM

When steam is escaping into the conditioned space, the room sensible heat gain is only that heat represented by the difference in heat content of steam at the steam temperature and at the room dry-bulb temperature (lb/hr × temp diff × .45). The latent heat gain is equal to the pounds per hour escaping times 1050 Btu/lb.

MOISTURE ABSORPTION

When moisture (regain) is absorbed by hygroscopic materials, sensible heat is added to the space. The heat so gained is equal to the latent heat of vaporization which is approximately 1050 Btu/lb times the pounds of water absorbed. This sensible heat is an addition to room sensible heat, and a deduction from room latent heat if the hygroscopic material is removed from the conditioned space.

LATENT HEAT GAIN — CREDIT TO ROOM SENSIBLE HEAT

Some forms of latent heat gain reduce room sensible heat. Moisture evaporating at the room wet-bulb temperature (not heated or cooled from external source) utilizes room sensible heat for heat of evaporation. This form of latent heat gain should be deducted from room sensible heat and added to room latent heat. This does not change the total room heat gain, but may have considerable effect on the sensible heat factor.

When the evaporation of moisture derives its heat from another source such as steam or electric heating coils, only the latent heat gain to the room is figured; room sensible heat is not reduced. The power input to the steam or electric coils balances the heat of evaporation except for the initial warmup of the water.

TABLE 54—HEAT TRANSMISSION COEFFICIENTS FOR BARE STEEL PIPES

Btu/(hr) (linear ft) (deg F diff between pipe and surrounding air)

NOMINAL PIPE SIZE (in.)	HOT WATER				STEAM		
	120 F	150 F	180 F	210 F	5 psig 227 F	50 psig 300 F	100 psig 338 F
	TEMPERATURE DIFFERENCE*						
	50 F	80 F	110 F	140 F	157 F	230 F	268 F
½	0.46	0.50	0.55	0.58	0.61	0.71	0.76
¾	0.56	0.61	0.67	0.72	0.75	0.87	0.93
1	0.68	0.74	0.82	0.88	0.92	1.07	1.15
1¼	0.85	0.92	1.01	1.09	1.14	1.32	1.43
1½	0.96	1.04	1.15	1.23	1.29	1.49	1.63
2	1.18	1.28	1.41	1.51	1.58	1.84	1.99
2½	1.40	1.53	1.68	1.80	1.88	2.19	2.36
3	1.68	1.83	2.01	2.15	2.26	2.63	2.84
3½	1.90	2.06	2.22	2.43	2.55	2.97	3.22
4	2.12	2.30	2.53	2.72	2.85	3.32	3.59
5	2.58	2.80	3.08	3.30	3.47	4.05	4.39
6	3.04	3.29	3.63	3.89	4.07	4.77	5.16
8	3.88	4.22	4.64	4.96	5.21	6.10	6.61
10	4.76	5.18	5.68	6.09	6.41	7.49	8.12
12	5.59	6.07	6.67	7.15	7.50	8.80	9.53

*At 70 F db room temperature

TABLE 55—HEAT TRANSMISSION COEFFICIENTS FOR INSULATED PIPES*

Btu/(hr) (linear ft) (deg F diff between pipe and room)

IRON PIPE SIZE (in.)	85 PERCENT MAGNESIA INSULATION†		
	1 In. Thick	1½ In. Thick	2 In. Thick
½	0.16	0.14	0.12
¾	0.18	0.15	0.13
1	0.20	0.17	0.15
1¼	0.24	0.20	0.17
1½	0.26	0.21	0.18
2	0.30	0.24	0.21
2½	0.35	0.27	0.24
3	0.40	0.32	0.27
3½	0.45	0.35	0.30
4	0.49	0.38	0.32
5	0.59	0.45	0.38
6	0.68	0.52	0.43
8	0.85	0.65	0.53
10	1.04	0.78	0.64
12	1.22	0.90	0.73

*No allowance for fittings. This table applies only to straight runs of pipe. When numerous fittings exist, a suitable safety factor must be included. This added heat gain at the fittings may be as much as 10%. Generally this table can be used without adding this safety factor.

†Other insulation. If other types of insulation are used, multiply the above values by the factors shown in the following table:

MATERIAL	PIPE COVERING FACTORS
Corrugated Asbestos (Air Cell)	
4 Ply per inch	1.36
6 Ply per inch	1.23
8 Ply per inch	1.19
Laminated Asbestos (Sponge Felt)	0.98
Mineral Wool	1.00
Diatomaceous Silica (Super-X)	1.36
Brown Asbestos Fiber (Wool Felt)	0.88

TABLE 56—HEAT TRANSMISSION COEFFICIENTS FOR INSULATED COLD PIPES*

MOULDED TYPE†

Btu/(hr) (linear ft) (deg F diff between pipe and room)

IRON PIPE SIZE (in.)	ICE WATER		BRINE		HEAVY BRINE	
	Actual Thickness of Insulation (In.)	Coefficient	Actual Thickness of Insulation (In.)	Coefficient	Actual Thickness of Insulation (In.)	Coefficient
½	1.5	0.11	2.0	0.10	2.8	0.09
¾	1.6	0.12	2.0	0.11	2.9	0.09
1	1.6	0.14	2.0	0.12	3.0	0.10
1¼	1.6	0.16	2.4	0.13	3.1	0.11
1½	1.5	0.17	2.5	0.13	3.2	0.12
2	1.5	0.20	2.5	0.15	3.3	0.13
2½	1.5	0.23	2.6	0.17	3.3	0.15
3	1.5	0.27	2.7	0.19	3.4	0.16
3½	1.5	0.29	2.9	0.19	3.5	0.18
4	1.7	0.30	2.9	0.21	3.7	0.18
5	1.7	0.35	3.0	0.24	3.9	0.20
6	1.7	0.40	3.0	0.26	4.0	0.23
8	1.9	0.46	3.0	0.32	4.0	0.26
10	1.9	0.56	3.0	0.38	4.0	0.31
12	1.9	0.65	3.0	0.44	4.0	0.36

*No allowance for fittings. This table applies only to straight runs of pipe. When numerous fittings exist, a suitable safety factor must be included. This added heat gain at the fittings may be as much as 10%. Generally this table can be used without adding this safety factor.

†Insulation material. Values in this table are based on a material having a conductivity k=0.30. However, a 15% safety factor was added to this k value to compensate for seams and imperfect workmanship. The table applies to either cork covering (k=0.29), or mineral wool board (k=0.32). The thickness given above is for molded mineral wool board which is usually some 5 to 10% greater than molded cork board.

TABLE 57—HEAT TRANSMISSION COEFFICIENTS FOR UNINSULATED TANKS

SENSIBLE HEAT GAIN*

Btu/(hr) (sq ft) (deg F diff between liquid and room)

CONSTRUCTION	METAL								WOOD 2½ in. Thick				CONCRETE 6 in. Thick			
	Painted				Bright (Nickel)				Painted or Bare				Painted or Bare			
	Temp Diff				Temp Diff				Temp Diff				Temp Diff			
	50 F	100 F	150 F	200 F	50 F	100 F	150 F	200 F	50 F	100 F	150 F	200 F	50 F	100 F	150 F	200 F
Vertical(Sides)	1.8	2.0	2.3	2.6	1.3	1.7	1.6	1.7	.37	.37	.37	.37	.91	.93	.96	.97
Top	2.1	2.4	2.7	2.9	1.6	1.4	1.9	2.1	.38	.38	.38	.38	.99	1.0	1.0	1.1
Bottom	1.5	1.7	2.0	2.2	0.97	1.1	1.3	1.4	.35	.36	.36	.36	.83	.86	.88	.90

*To estimate latent heat load if water is being evaporated, see Table 58

TABLE 58—EVAPORATION FROM A FREE WATER SURFACE—LATENT HEAT GAIN

STILL AIR, ROOM AT 75 F db, 50% RH

WATER TEMP	75 F	100 F	125 F	150 F	175 F	200 F
Btu/(hr)(sq ft)	42	140	330	680	1260	2190

SYSTEM HEAT GAIN

The system heat gain is considered as the heat added to or lost by the system components, such as the ducts, piping, air conditioning fan, and pump, etc. This heat gain must be estimated and included in the load estimate but can be accurately evaluated only after the system has been designed.

SUPPLY AIR DUCT HEAT GAIN

The supply duct normally has 50 F db to 60 F db air flowing through it. The duct may pass through an unconditioned space having a temperature of, say, 90 F db and up. This results in a heat gain to the duct before it reaches the space to be conditioned. This, in effect, reduces the cooling capacity of the conditioned air. To compensate for it, the cooling capacity of the air quantity must be increased. It is recommended that long runs of ducts in unconditioned spaces be insulated to minimize heat gain.

Basis or Chart 3
— Percent Room Sensible Heat to be Added for Heat Gain to Supply Duct

Chart 3 is based on a difference of 30 F db between supply air and unconditioned space, a supply duct velocity of 1800 fpm in a square duct, still air on the outside of the duct and a supply air rise of 17 F db. Correction factors for different room temperatures, duct velocities and temperature differences are included below *Chart 3*. Values are plotted for use with uninsulated, furred and insulated ducts.

Use of Chart 3
— Percent Room Sensible Heat to be Added for Heat Gain to Supply Duct

To use this chart, evaluate the length of duct running thru the unconditioned space, the temperature of unconditioned space, the duct velocity, the supply air temperature, and room sensible heat subtotal.

Example 5 — Heat Gain to Supply Duct
Given:
    20 ft of uninsulated duct in unconditioned space at 100 F db
    Duct velocity — 2000 fpm
    Supply air temperature — 60 F db
    Room sensible heat gain — 100,000 Btu/hr

Find:
    Percent addition to room sensible heat

Solution:
    The supply air to unconditioned space temperature difference = 100 − 60 = 40 F db

    From *Chart 3*, percent addition = 4.5%
    Correction for 40 F db temperature difference and 2000 fpm duct velocity = 1.26

    Actual percent addition = 4.5 × 1.26 = 5.7%

CHART 3—HEAT GAIN TO SUPPLY DUCT

Percent of Room Sensible Heat

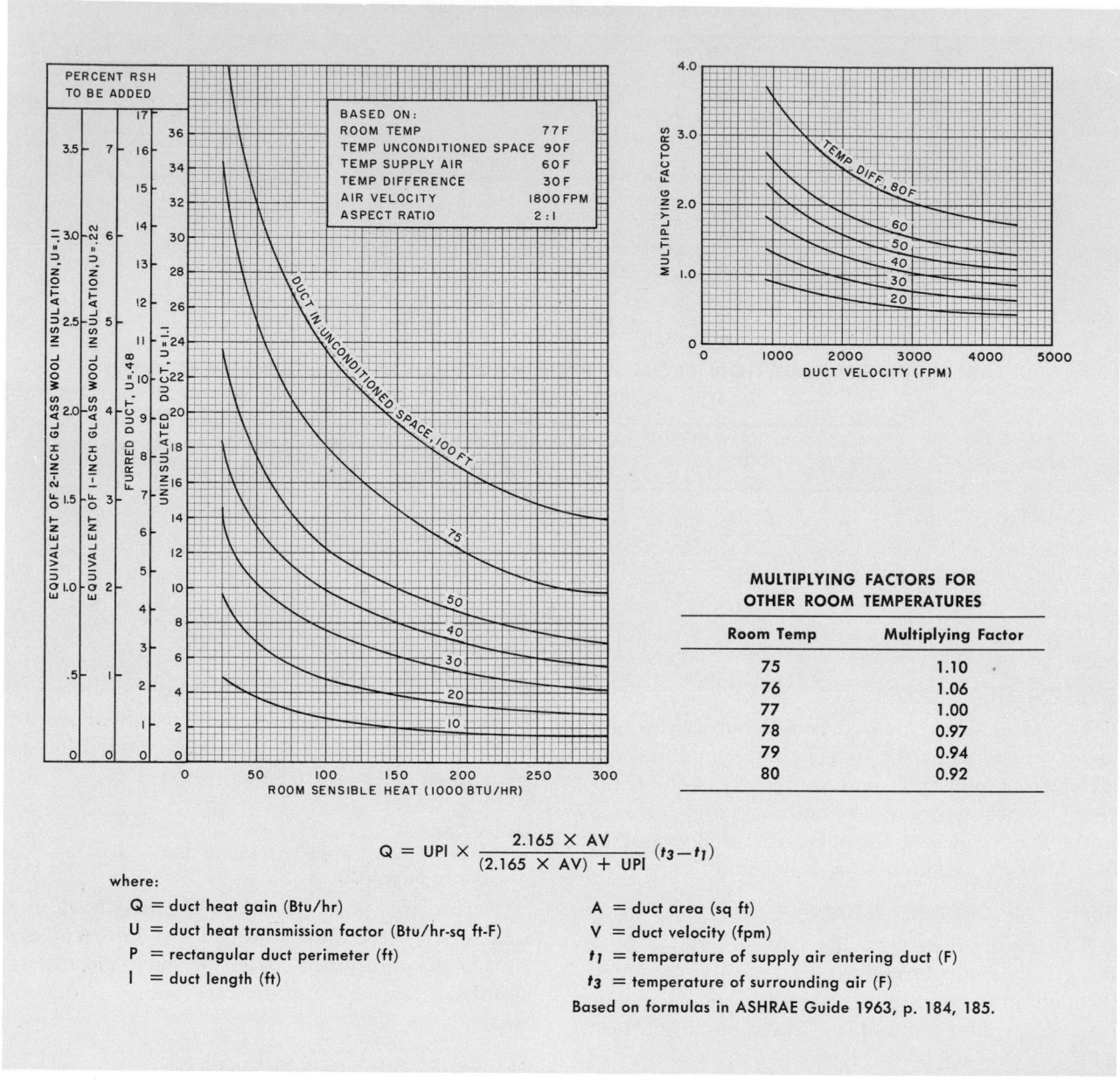

MULTIPLYING FACTORS FOR OTHER ROOM TEMPERATURES	
Room Temp	Multiplying Factor
75	1.10
76	1.06
77	1.00
78	0.97
79	0.94
80	0.92

$$Q = UPl \times \frac{2.165 \times AV}{(2.165 \times AV) + UPl} (t_3 - t_1)$$

where:

Q = duct heat gain (Btu/hr)
U = duct heat transmission factor (Btu/hr-sq ft-F)
P = rectangular duct perimeter (ft)
l = duct length (ft)

A = duct area (sq ft)
V = duct velocity (fpm)
t_1 = temperature of supply air entering duct (F)
t_3 = temperature of surrounding air (F)

Based on formulas in ASHRAE Guide 1963, p. 184, 185.

SUPPLY AIR DUCT LEAKAGE LOSS

Air leakage from the supply duct may be a serious loss of cooling effect, except when it leaks into the conditioned space. This loss of cooling effect must be added to the room sensible and latent heat load.

Experience indicates that the average air leakage from the entire length of supply ducts, whether large or small systems, averages around 10% of the supply air quantity. Smaller leakage per foot of length for larger perimeter ducts appears to be counterbalanced by the longer length of run. Individual work-

manship is the greatest variable, and duct leakages from 5% to 30% have been found. The following is a guide to the evaluation of duct leakages under various conditions:

1. Bare ducts within conditioned space — usually not necessary to figure leakage.

2. Furred or insulated ducts within conditioned space — a matter of judgment, depending on whether the leakage air actually gets into the room.

TABLE 59—HEAT GAIN FROM AIR CONDITIONING FAN HORSEPOWER, DRAW-THRU SYSTEM‡‡

	FAN TOTAL PRESSURE† (In. of Water)	CENTRAL STATION SYSTEMS‡					APPLIED OR UNITARY SYSTEM**				
		Temp Diff Room to Supply Air					Temp Diff Room to Supply Air				
		10 F	15 F	20 F	25 F	30 F	10 F	15 F	20 F	25 F	30 F
		PERCENT OF ROOM SENSIBLE HEAT *									
Fan Motor Not in Conditioned Space or Air Stream	0.50	1.2	0.8	0.6	0.5	0.4	2.2	1.5	1.1	0.9	0.7
	0.75	1.9	1.3	1.0	0.8	0.6	3.5	2.4	1.8	1.4	1.2
	1.00	2.7	1.8	1.4	1.1	0.9	4.8	3.2	2.4	1.9	1.6
	1.25	3.9	2.6	1.9	1.6	1.3	6.5	4.3	3.2	2.6	2.2
	1.50	4.6	3.1	2.3	1.9	1.6	7.8	5.2	3.9	3.1	2.6
	1.75	5.4	3.6	2.7	2.2	1.8	9.1	6.1	4.6	3.6	3.0
	2.00	6.2	4.1	3.1	2.5	2.1	10.4	6.9	5.2	4.2	3.5
	3.00	10.4	6.9	5.2	4.2	3.5	16.7	11.2	8.4	6.7	5.6
	4.00	15.3	10.2	7.7	6.1	5.1					
	5.00	19.2	12.8	9.6	7.7	6.4					
	6.00	24.4	16.3	12.2	9.9	8.2					
	8.00	38.0	25.4	19.0	15.2	12.7					
Fan Motor†† in Conditioned Space or Air Stream	0.50	1.6	1.1	0.8	0.6	0.5	2.7	1.8	1.4	1.1	0.9
	0.75	2.6	1.8	1.3	1.1	0.9	4.2	2.8	2.1	1.7	1.4
	1.00	3.6	2.4	1.8	1.5	1.2	5.8	3.8	2.9	2.3	1.9
	1.25	5.0	3.4	2.5	2.0	1.7	7.6	5.1	3.8	3.1	2.6
	1.50	6.0	4.0	3.0	2.4	2.0	9.2	6.1	4.6	3.7	3.1
	1.75	7.0	4.7	3.5	2.8	2.4	10.7	7.2	5.4	4.3	3.6
	2.00	8.0	5.4	4.0	3.2	2.7	12.2	8.2	6.1	4.9	4.1
	3.00	13.2	8.8	6.6	5.3	4.4	19.5	13.1	9.8	7.8	6.5
	4.00	19.0	12.7	9.5	7.6	6.4					
	5.00	23.8	15.9	11.9	9.5	8.0					
	6.00	30.0	20.0	15.0	12.0	10.0					
	8.00	45.5	30.3	22.8	18.2	15.2					

*Excludes from heat gain, typical values for bearing losses, etc. which are dissipated in apparatus room.

†Fan Total Pressure equals fan static pressure plus velocity pressure at fan discharge. Below 1200 fpm the fan total pressure is approximately equal to the fan static. Above 1200 fpm the total pressure should be figured.

‡70% fan efficiency assumed.

**50% fan efficiency assumed.

††80% motor and drive efficiency assumed.

‡‡For draw-thru systems, this heat is an addition to the supply air heat gain and is added to the room sensible heat. For blow-thru systems this fan heat is added to the grand total heat; use the RSH times the percent listed and add to the GTH.

3. All ducts outside the conditioned space — assume 10% leakage. This leakage is a total loss and the full amount must be included. When only part of the supply duct is outside the conditioned space, include that fraction of 10% as the leakage. (Fraction is ratio of length outside of conditioned space to total length of supply duct.)

HEAT GAIN FROM AIR CONDITIONING FAN HORSEPOWER

The inefficiency of the air conditioning equipment fan and the heat of compression adds heat to the system as described under *"Electric Motors."* In the case of draw-through systems, this heat is an addition to the supply air heat gain and should be added to the room sensible heat. With blow-through systems (fan blowing air through the coil, etc.) the fan heat added is a load on the dehumidifier and,

therefore, should be added to the grand total heat (see *"Percent Addition to Grand Total Heat"*).

Basis of Table 59
— Heat Gain from Air Conditioning Fan Horsepower

The air conditioning fan adds heat to the system in the following manner:

1. Immediate temperature rise in the air due to the inefficiency of the fan.

2. Energy gain in the air as a pressure and/or velocity rise.

3. With the motor and drive in the air stream or conditioned space, the heat generated by the inefficiency of the motor and drive is also an immediate heat gain.

The fan efficiencies are about 70% for central station type fans and about 50% for packaged equipment fans.

Use of Table 59
— Heat Gain from Air Conditioning Fan Horsepower

The approximate system pressure loss and dehumidified air rise (room minus supply air temperature) differential must be estimated from the system characteristics and type of application. These should be checked from the final system design.

The normal comfort application has a dehumidified air rise of between 15 F db and 25 F db and the fan total pressure depends on the amount of ductwork involved, the number of fittings (elbows, etc.) in the ductwork and the type of air distribution system used. Normally, the fan total pressure can be approximated as follows:

1. No ductwork (packaged equipment) — 0.5 to 1.00 inches of water.
2. Moderate amount of ductwork, low velocity systems — 0.75 to 1.50 inches of water.
3. Considerable ductwork, low velocity system — 1.25 to 2.00 inches of water.
4. Moderate amount of ductwork, high pressure system — 2.00 to 4.00 inches of water.
5. Considerable ductwork, high pressure system — 3.00 to 6.00 inches of water.

Example 6 — Heat Gain from Air Conditioning Fan Horsepower

Given:
　　Same data as *Example 5*
　　80 ft of supply duct in conditioned space

Find:
　　Percent addition to room sensible heat.

Solution:
　　Assume 1.50 inches of water, fan total pressure, and 20 F db dehumidifier rise. Refer to *Table 59*.
　　Heat gain from fan horsepower = 2.3%

SAFETY FACTOR AND PERCENT ADDITIONS TO ROOM SENSIBLE AND LATENT HEAT

A safety factor to be added to the room sensible heat sub-total should be considered as strictly a factor of probable error in the survey or estimate, and should usually be between 0% and 5%.

The total room sensible heat is the sub-total plus percentage additions to allow for (1) supply duct heat gain, (2) supply duct leakage losses, (3) fan horsepower and (4) safety factor, as explained in the preceding paragraph.

Example 7 — Percent Addition to Room Sensible Heat
Given:
　　Same data as *Examples 5 and 6*

Find:
　　Percent addition to room sensible heat gain sub-total

Solution:

Supply duct heat gain	= 5.7%
Supply duct leakage (20 ft duct of total 100 ft)	= 2.0%
Fan horsepower	= 2.3%
Safety factor	= 0.0%
Total percent addition to RSH	= 10.0%

The percent additions to room latent heat for supply duct leakage loss and safety factor should be the same as the corresponding percent additions to room sensible heat.

RETURN AIR DUCT HEAT AND LEAKAGE GAIN

The evaluation of heat and leakage effects on return air ducts is made in the same manner as for supply air ducts, except that the process is reversed; there is inward gain of hot moist air instead of loss of cooling effect.

Chart 3 can be used to approximate heat gain to the return duct system in terms of percent of RSH, using the following procedure:

1. Using RSH and the length of return air duct, use *Chart 3* to establish the percent heat gain.
2. Use the multiplying factor from table below *Chart 3* to adjust the percent heat gain for actual temperature difference between the air surrounding the return air duct and the air inside the duct, and also for the actual velocity.
3. Multiply the resulting percentage of heat gain by the ratio of RSH to GTH.
4. Apply the resulting heat gain percentage to GTH.

To determine the return air duct leakage, apply the following reasoning:

1. Bare duct within conditioned space — no inleakage.
2. Furred duct within conditioned space or furred space used for return air — a matter of judgment, depending on whether the furred space may connect to unconditioned space.
3. Ducts outside conditioned space — assume up to 3% inleakage, depending on the length of duct. If there is only a short connection between conditioned space and apparatus, inleakage may be disregarded. If there is a long run of duct, then apply judgment as to the amount of inleakage.

HEAT GAIN FROM DEHUMIDIFIER PUMP HORSEPOWER

With dehumidifier systems, the horsepower required to pump the water adds heat to the system as outlined under *"Electric Motors"*. This heat will be an addition to the grand total heat.

TABLE 60—HEAT GAIN FROM DEHUMIDIFIER PUMP HORSEPOWER

PUMP HEAD (ft)	SMALL PUMPS* 0-100 GPM					LARGE PUMPS† 100 GPM AND LARGER				
	CHILLED WATER TEMP RISE					CHILLED WATER TEMP RISE				
	5 F	7 F	10 F	12 F	15 F	5 F	7 F	10 F	12 F	15 F
	PERCENT OF GRAND TOTAL HEAT									
35	2.0	1.5	1.0	1.0	0.5	1.5	1.0	0.5	0.5	0.5
70	3.5	2.5	2.0	1.5	1.0	2.5	2.0	1.5	1.0	1.0
100	5.0	4.0	2.5	2.0	1.5	4.0	3.0	2.0	1.5	1.0

*Efficiency 50% †Efficiency 70%

Basis of Table 60
— Heat Gain from Dehumidifier Pump Horsepower

Table 60 is based on pump efficiencies of 50% for small pumps and 70% for large pumps. Small pumps are considered to have a capacity of less than 100 gallons; large pumps, more than 100 gallons.

Use of Table 60
— Heat Gain from Dehumidifier Pump Horsepower

The chilled water temperature rise in the dehumidifier and the pump head must be approximated to use *Table 60*.

1. Large systems with considerable piping and fittings may require up to 100 ft pump head; normally, 70 ft head is the average.

2. The normal water temperature rise in the dehumidifier is between 7 F and 12 F. Applications using large amounts of water have a lower rise; those using small amounts of water have a higher rise.

PERCENT ADDITION TO GRAND TOTAL HEAT

The percent additions to the grand total heat to compensate for various external losses consist of heat and leakage gain to return air ducts, heat gain from the dehumidifier pump horsepower, and the heat gain to the dehumidifier and piping system.

These heat gains can be estimated as follows:

1. Heat and leakage gain to return air ducts, see above.

2. Heat gain from dehumidifier pump horsepower, *Table 60*.

3. Dehumidifier and piping losses:
   a. Very little external piping — 1% of GTH.
   b. Average external piping — 2% of GTH.
   c. Extensive external piping — 4% of GTH.

4. Blow-through fan system — add percent room sensible heat from *Table 59* to GTH.

5. Dehumidifier in conditioned apparatus room — reduce the above percentages by one half.

CHAPTER 8. APPLIED PSYCHROMETRICS

The preceding chapters contain the practical data to properly evaluate the heating and cooling loads. They also recommend outdoor air quantities for ventilation purposes in areas where state, city or local codes do not exist.

This chapter describes practical psychrometrics as applied to apparatus selection. It is divided into three parts:

1. *Description of terms, processes and factors* — as encountered in normal air conditioning applications.

2. *Air conditioning apparatus* — factors affecting common processes and the effect of these factors on selection of air conditioning equipment.

3. *Psychrometrics of partial load control* — the effect of partial load on equipment selection and on the common processes.

To help recognize terms, factors and processes described in this chapter, a brief definition of psychrometrics is offered at this point, along with an illustration and definition of terms appearing on a standard psychrometric chart (*Fig. 32*).

Dry-bulb Temperature — The temperature of air as registered by an ordinary thermometer.

Wet-bulb Temperature — The temperature registered by a thermometer whose bulb is covered by a wetted wick and exposed to a current of rapidly moving air.

Dewpoint Temperature — The temperature at which condensation of moisture begins when the air is cooled.

Relative Humidity — Ratio of the actual water vapor pressure of the air to the saturated water vapor pressure of the air at the same temperature.

Specific Humidity or Moisture Content — The weight of water vapor in grains or pounds of moisture per pound of dry air.

Enthalpy — A thermal property indicating the quantity of heat in the air above an arbitrary datum, in Btu per pound of dry air. The datum for dry air is 0°F and, for the moisture content, 32 F water.

Enthalpy Deviation — Enthalpy indicated above, for any given condition, is the enthalpy of saturation. It should be corrected by the enthalpy deviation due to the air not being in the saturated state. Enthalpy deviation is in Btu per pound of dry air. Enthalpy deviation is applied where extreme accuracy is required; however, on normal air conditioning estimates it is omitted.

Specific Volume — The cubic feet of the mixture per pound of dry air.

Sensible Heat Factor — The ratio of sensible to total heat.

Alignment Circle — Located at 80 F db and 50% rh and used in conjunction with the sensible heat factor to plot the various air conditioning process lines.

Pounds of Dry Air — The basis for all psychrometric calculations. Remains constant during all psychrometric processes.

The dry-bulb, wet-bulb, and dewpoint temperatures and the relative humidity are so related that, if two properties are known, all other properties shown may then be determined. When air is saturated, dry-bulb, wet-bulb, and dewpoint temperatures are all equal.

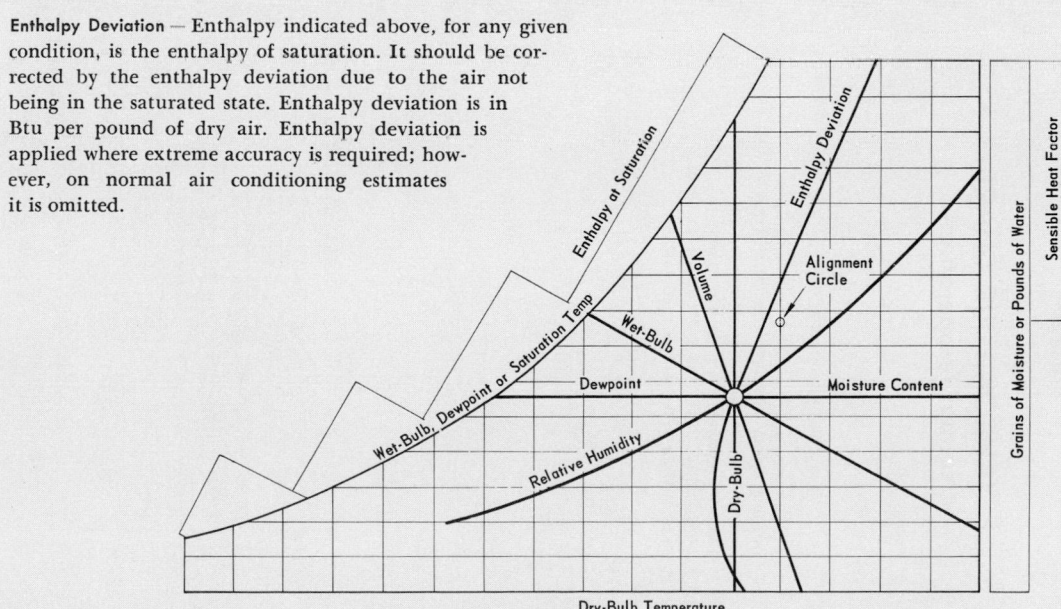

FIG. 32 — SKELETON PSYCHROMETRIC CHART

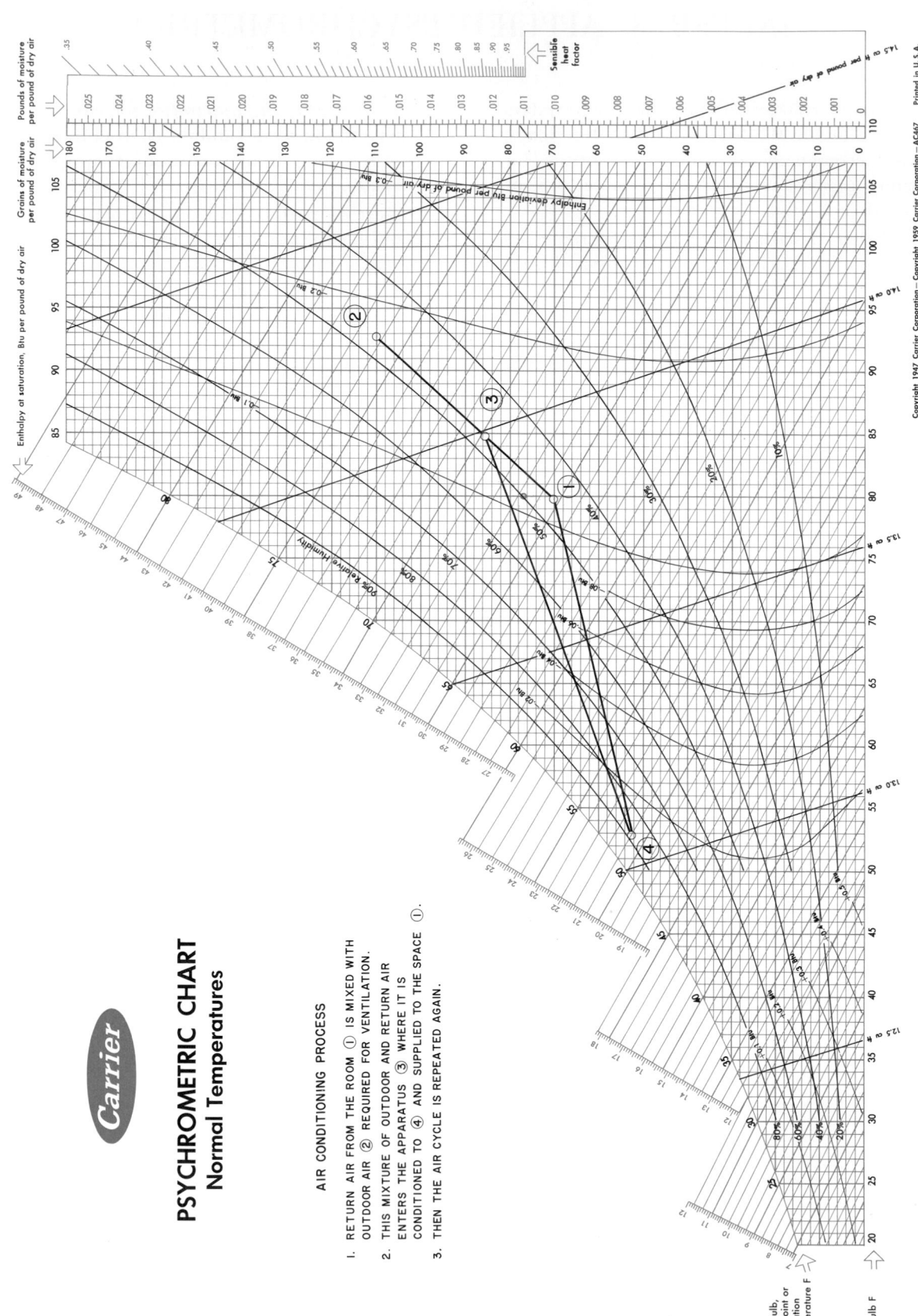

PSYCHROMETRIC CHART
Normal Temperatures

AIR CONDITIONING PROCESS

1. RETURN AIR FROM THE ROOM ① IS MIXED WITH
   OUTDOOR AIR ② REQUIRED FOR VENTILATION.

2. THIS MIXTURE OF OUTDOOR AND RETURN AIR
   ENTERS THE APPARATUS ③ WHERE IT IS
   CONDITIONED TO ④ AND SUPPLIED TO THE SPACE ①.

3. THEN THE AIR CYCLE IS REPEATED AGAIN.

FIG. 33 — TYPICAL AIR CONDITIONING PROCESS TRACED ON A STANDARD PSYCHROMETRIC CHART

DEFINITION

Psychrometrics is the science involving thermodynamic properties of moist air and the effect of atmospheric moisture on materials and human comfort. As it applies to this chapter, the definition must be broadened to include the method of controlling the thermal properties of moist air.

AIR CONDITIONING PROCESSES

Fig. 33 shows a typical air conditioning process traced on a psychrometric chart. Outdoor air *(2)** is mixed with return air from the room *(1)* and enters the apparatus *(3)*. Air flows through the conditioning apparatus *(3 - 4)* and is supplied to the space *(4)*. The air supplied to the space moves along line *(4 - 1)* as it picks up the room loads, and the cycle is re-

peated. Normally most of the air supplied to the space by the air conditioning system is returned to the conditioning apparatus. There, it is mixed with outdoor air required for ventilation. The mixture then passes thru the apparatus where heat and moisture are added or removed, as required, to maintain the desired conditions.

The selection of proper equipment to accomplish this conditioning and to control the thermodynamic properties of the air depends upon a variety of elements. However, only those which affect the psychrometric properties of air will be discussed in this chapter. These elements are: room sensible heat factor (RSHF)†, grand sensible heat factor (GSHF), effective surface temperature (t_{es}), bypass factor (BF), and effective sensible heat factor (ESHF).

DESCRIPTION OF TERMS, PROCESSES AND FACTORS

SENSIBLE HEAT FACTOR

The thermal properties of air can be separated into latent and sensible heat. The term *sensible heat factor* is the ratio of sensible to total heat, where total heat is the sum of sensible and latent heat. This ratio may be expressed as:

$$SHF = \frac{SH}{SH + LH} = \frac{SH}{TH}$$

where: SHF = sensible heat factor
SH = sensible heat
LH = latent heat
TH = total heat

ROOM SENSIBLE HEAT FACTOR (RSHF)

The *room sensible heat factor* is the ratio of room sensible heat to the summation of room sensible and room latent heat. This ratio is expressed in the following formula:

$$RSHF = \frac{RSH}{RSH + RLH} = \frac{RSH}{RTH}$$

The supply air to a conditioned space must have the capacity to offset simultaneously both the room sensible and room latent heat loads. The room and the supply air conditions to the space may be plotted on the standard psychrometric chart and these points connected with a straight line *(1 - 2)*,

Fig. 34. This line represents the psychrometric process of the supply air within the conditioned space and is called the room sensible heat factor line.

The slope of the RSHF line illustrates the ratio of sensible to latent loads within the space and is illustrated in *Fig. 34* by Δh_s (sensible heat) and Δh_l (latent heat). Thus, if adequate air is supplied to offset these room loads, the room requirements will

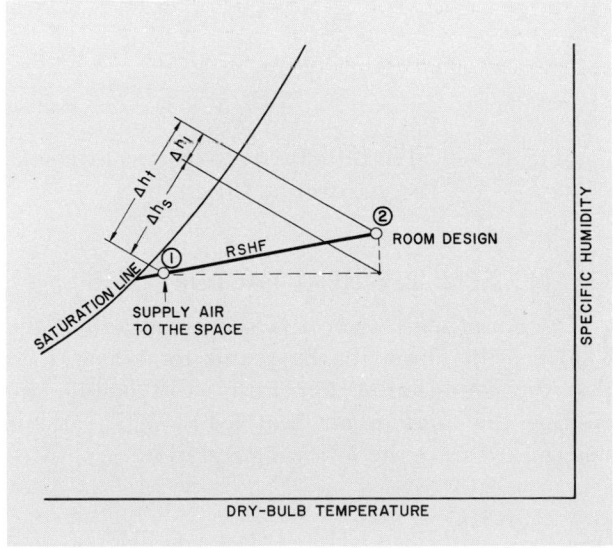

Fig. 34 — RSHF Line Plotted Between Room and Supply Air Conditions

---

*One italic number in parentheses represents a point, and two italic numbers in parentheses represent a line, plotted on the accompanying psychrometric chart examples.

†Refer to *page 149* for a description of all abbreviations and symbols used in this chapter.

be satisfied, provided both the dry- and wet-bulb temperatures of the supply air fall on this line.

The room sensible heat factor line can also be drawn on the psychrometric chart without knowing the condition of supply air. The following procedure illustrates how to plot this line, using the calculated RSHF, the room design conditions, the sensible heat factor scale in the upper right hand corner of the psychrometric chart, and the alignment circle at 80 F dry-bulb and 50% relative humidity:

1. Draw a base line thru the alignment circle and the calculated RSHF shown on the sensible heat factor scale in the upper right corner of psychrometric chart (1 - 2), Fig. 35.

2. Draw the actual room sensible heat factor line thru the room design conditions parallel to the base line in Step 1 (3 - 4), Fig. 35. As shown, this line may be drawn to the saturation line on the psychrometric chart.

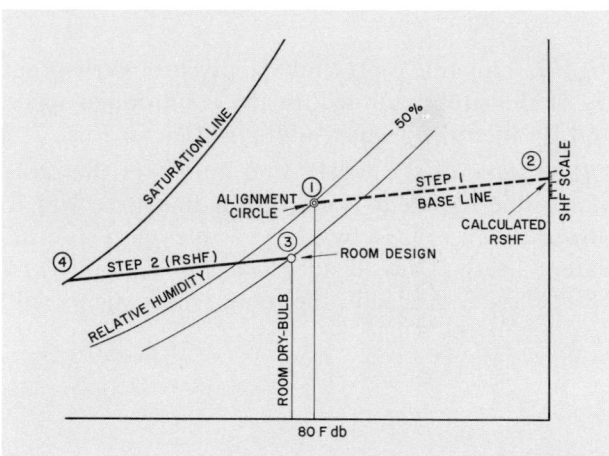

FIG. 35 — RSHF LINE PLOTTED ON SKELETON
PSYCHROMETRIC CHART

GRAND SENSIBLE HEAT FACTOR (GSHF)

The *grand sensible heat factor* is the ratio of the total sensible heat to the grand total heat load that the conditioning apparatus must handle, including the outdoor air heat loads. This ratio is determined from the following equation:

$$GSHF = \frac{TSH}{TLH + TSH} = \frac{TSH}{GTH}$$

Air passing thru the conditioning apparatus increases or decreases in temperature and/or moisture content. The amount of rise or fall is determined by the total sensible and latent heat loads that the conditioning apparatus must handle. The condition

of the air entering the apparatus (mixture condition of outdoor and return room air) and the condition of the air leaving the apparatus may be plotted on the psychrometric chart and connected by a straight line (1 - 2), Fig. 36. This line represents the psychrometric process of the air as it passes through the conditioning apparatus, and is referred to as the grand sensible heat factor line.

The slope of the GSHF line represents the ratio of sensible and latent heat that the apparatus must handle. This is illustrated in Fig. 36 by Δh_s (sensible heat) and Δh_l (latent heat).

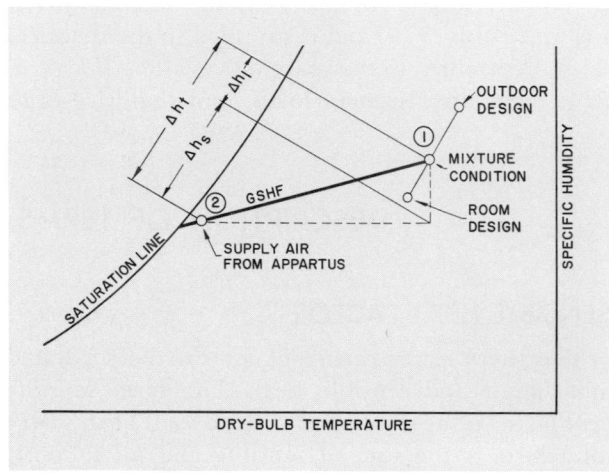

FIG. 36 — GSHF LINE PLOTTED BETWEEN MIXTURE
CONDITIONS TO APPARATUS AND LEAVING
CONDITION FROM APPARATUS

The grand sensible heat factor line can be plotted on the psychrometric chart without knowing the condition of supply air, in much the same manner as the RSHF line. Fig. 37, Step 1 (1 - 2) and Step 2 (3 - 4) show the procedure, using the calculated GSHF, the mixture condition of air to the apparatus, the sensible heat factor scale, and the alignment circle on the psychrometric chart. The resulting GSHF line is plotted thru the mixture conditions of the air to the apparatus.

REQUIRED AIR QUANTITY

The air quantity required to offset simultaneously the room sensible and latent loads and the air quantity required thru the apparatus to handle the total sensible and latent loads may be calculated, using the conditions on their respective RSHF and GSHF lines. For a particular application, when both the RSHF and GSHF ratio lines are plotted on the psychrometric chart, the intersection of the two lines (1) Fig. 38, represents the condition of the supply air to

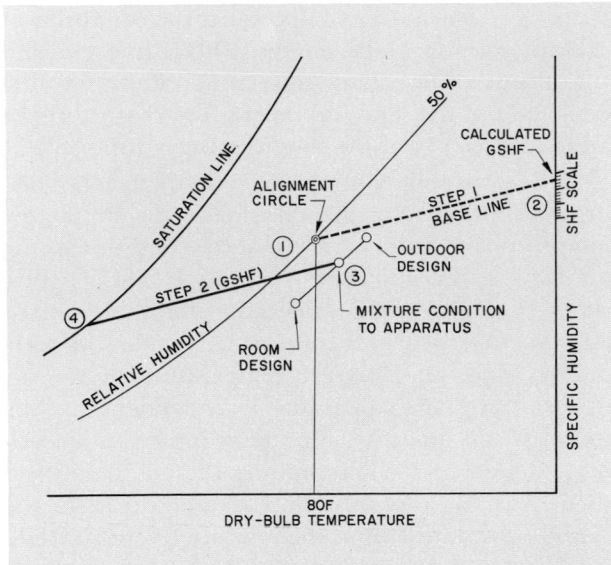

FIG. 37 — GSHF LINE PLOTTED ON SKELETON
PSYCHROMETRIC CHART

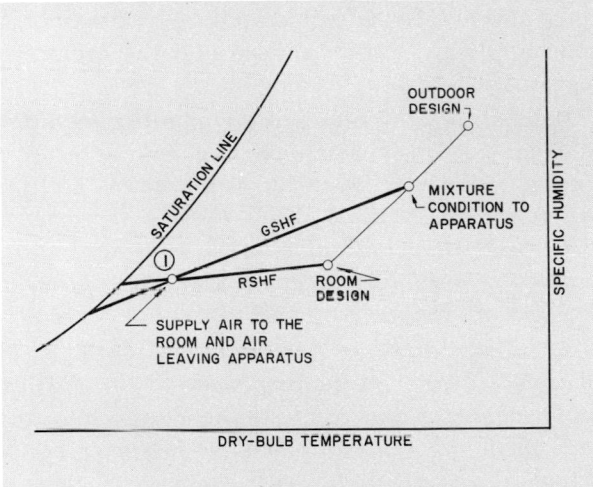

FIG. 38 — RSHF AND GSHF LINES PLOTTED ON
SKELETON PSYCHROMETRIC CHART

the space. It is also the condition of the air leaving the apparatus.

This neglects fan and duct heat gain, duct leakage losses, etc. In actual practice, these heat gains and losses are taken into account in estimating the cooling load. *Chapter 7* gives the necessary data for evaluating these supplementary loads. Therefore, the temperature of the air leaving the apparatus is not necessarily equal to the temperature of the air supplied to the space as indicated in *Fig. 38.*

Fig. 39 illustrates what actually happens when

these supplementary loads are considered in plotting the RSHF and GSHF lines.

Point *(1)* is the condition of air leaving the apparatus and point *(2)* is the condition of supply air to the space. Line *(1 - 2)* represents the temperature rise of the air stream resulting from fan horsepower and heat gain to the duct.

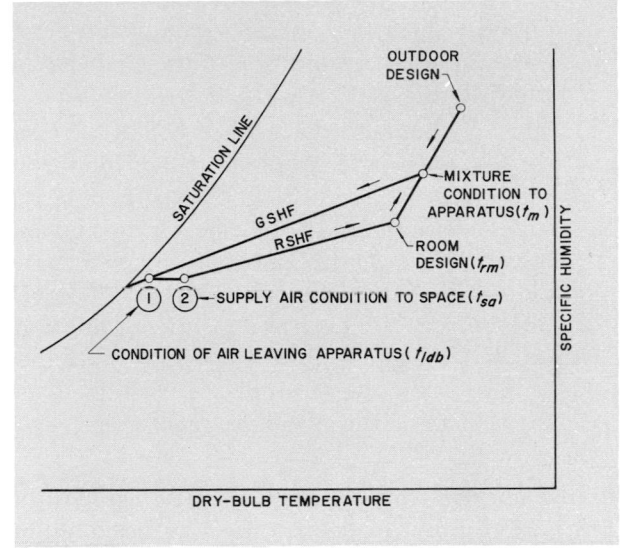

FIG. 39 — RSHF AND GSHF LINES PLOTTED
WITH SUPPLEMENTARY LOAD LINE

The air quantity required to satisfy the room load may be calculated from the following equation:

$$cfm_{sa} = \frac{\text{RSH}}{1.08 \ (t_{rm} - t_{sa})}$$

The air quantity required thru the conditioning apparatus to satisfy the total air conditioning load (including the supplementary loads) is calculated from the following equation:

$$cfm_{da} = \frac{\text{TSH}}{1.08 \ (t_m - t_{ldb})}$$

The required air quantity supplied to the space is equal to the air quantity required thru the apparatus, neglecting leakage losses. The above equation contains the term t_m which is the mixture condition of air entering the apparatus. With the exception of an all outdoor air application, the term t_m can only be determined by trial and error.

One possible procedure to determine the mixture temperature and the air quantities is outlined below. This procedure illustrates one method of apparatus selection and is presented to show how cumbersome and time consuming it may be.

1. Assume a rise $(t_{rm} - t_{sa})$ in the supply air to the space, and calculate the supply air quantity (cfm_{sa}) to the space.

2. Use this air quantity to calculate the mixture condition of the air (t_m) to the space, (*Equation 1, page 150*).

3. Substitute this supply air quantity and mixture condition of the air in the formula for air quantity thru the apparatus (cfm_{da}) and determine the leaving condition of the air from the conditioning apparatus (t_{ldb}).

4. The rise between the leaving condition from the apparatus and supply air condition to the space $(t_{sa} - t_{ldb})$ must be able to handle the supplementary loads (duct heat gain and fan heat). These temperatures (t_{ldb}, t_{sa}) may be plotted on their respective GSHF and RSHF lines (*Fig. 39*) to determine if these conditions can handle the supplementary loads. If they cannot, a new rise in supply air is assumed and the trial-and-error procedure repeated.

In a normal, well designed, tight system this difference in supply air temperature and the condition of the air leaving the apparatus $(t_{sa} - t_{ldb})$ is usually not more than a few degrees. To simplify the discussion on the interrelationship of RSHF and GSHF, the supplementary loads have been neglected in the various discussions, formulas and problems in the remainder of this chapter. It can not be overemphasized, however, that these supplementary loads must be recognized when estimating the cooling and heating loads. These loads are taken into account on the air conditioning load estimate in *Chapter 1*, and are evaluated in *Chapter 7*.

The RSHF ratio will be constant (at full load) under a specified set of conditions; however, the GSHF ratio may increase or decrease as the outdoor air quantity and mixture conditions are varied for design purposes. As the GSHF ratio changes, the supply air condition to the space varies along the RSHF line (*Fig. 38*).

The difference in temperature between the room and the air supply to the room determines the air quantity required to satisfy the room sensible and room latent loads. As this temperature difference increases (supplying colder air, since the room conditions are fixed), the required air quantity to the space decreases. This temperature difference can increase up to a limit where the RSHF line crosses the saturation line on the psychrometric chart, *Fig. 38*; assuming, of course, that the available conditioning equipment is able to take the air to 100%

saturation. Since this is impossible, the condition of the air normally falls on the RSHF line close to the saturation line. How close to the saturation line depends on the physical operating characteristics and the efficiency of the conditioning equipment.

In determining the required air quantity, when neglecting the supplementary loads, the supply air temperature is assumed to equal the condition of the air leaving the apparatus $(t_{sa} - t_{ldb})$. This is illustrated in *Fig. 38*. The calculation for the required air quantity still remains a trial-and-error procedure, since the mixture temperature of the air (t_m) entering the apparatus is dependent on the required air quantity. The same procedure previously described for determining the air quantity is used. Assume a supply air rise and calculate the supply air quantity and the mixture temperature to the conditioning apparatus. Substitute the supply air quantity and mixture temperature in the equation for determining the air quantity thru the apparatus, and calculate the leaving condition of the air from the apparatus. This temperature must equal the supply air temperature; if it does not, a new supply air rise is assumed and the procedure repeated.

Determining the required air quantity by either method previously described is a tedious process, since it involves a trial-and-error procedure, plotting the RSHF and GSHF ratios on a psychrometric chart, and in actual practice accounting for the supplementary loads in determining the supply air, mixture and leaving air temperatures.

This procedure has been simplified, however, by relating all the conditioning loads to the physical performance of the conditioning equipment, and then including this equipment performance in the actual calculation of the load.

This relationship is generally recognized as a psychrometric correlation of loads to equipment performance. The correlation is accomplished by calculating the "effective surface temperature," "bypass factor" and "effective sensible heat factor." These alone will permit the simplified calculation of supply air quantity.

EFFECTIVE SURFACE TEMPERATURE (t_{es})

The surface temperature of the conditioning equipment varies throughout the surface of the apparatus as the air comes in contact with it. However, the effective surface temperature can be considered to be the uniform surface temperature which would produce the same leaving air conditions as the nonuniform surface temperature that actually occurs when the apparatus is in operation. This is more clearly understood by illustrating the heat transfer effect between the air and the cooling (or heating) medium. *Fig. 40* illustrates this process and is applicable to a chilled water cooling medium with the supply air counterflow in relation to the chilled water.

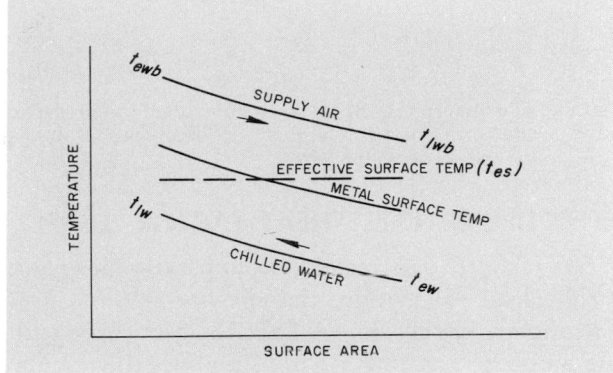

FIG. 40 — RELATIONSHIP OF EFFECTIVE SURFACE TEMP
TO SUPPLY AIR AND CHILLED WATER

The relationship shown in *Fig. 40* may also be illustrated for heating, direct expansion cooling and for air flowing parallel to the cooling or heating medium. The direction, slope and position of the lines change, but the theory is identical.

Since conditioning the air thru the apparatus reduces to the basic principle of heat transfer between the heating or cooling media of the conditioning apparatus and the air thru that apparatus, there must be a common reference point. This point is the effective surface temperature of the apparatus. The two heat transfers are relatively independent of each other, but are quantitatively equal when referred to the effective surface temperature.

Therefore, to obtain the most economical apparatus selection, the effective surface temperature is used in calculating the required air quantity and in selecting the apparatus.

For applications involving cooling and dehumidification, the effective surface temperature is at the point where the GSHF line crosses the saturation line on the psychrometric chart *(Fig. 36)*. As such, this effective surface temperature is considered to be the dewpoint of the apparatus, and hence the term apparatus dewpoint (adp) has come into common usage for cooling and dehumidifying processes.

Since cooling and dehumidification is one of the most common applications for central station apparatus, the "Air Conditioning Load Estimate" form, *Fig. 44*, is designed around the term apparatus

dewpoint (adp). The term is used exclusively in this chapter when referring to cooling and dehumidifying applications. The psychrometrics of air can be applied equally well to other types of heat transfer applications such as sensible heating, evaporative cooling, sensible cooling, etc., but for these applications the effective surface temperature will not necessarily fall on the saturation line.

BYPASS FACTOR (BF)

Bypass factor is a function of the physical and operating characteristics of the conditioning apparatus and, as such, represents that portion of the air which is considered to pass through the conditioning apparatus completely unaltered.

The physical and operating characteristics affecting the bypass factor are as follows:

1. A decreasing amount of available apparatus heat transfer surface results in an increase in bypass factor, i.e. less rows of coil, less coil surface area, wider spacing of coil tubes.

2. A decrease in the velocity of air through the conditioning apparatus results in a decrease in bypass factor, i.e. more time for the air to contact the heat transfer surface.

Decreasing or increasing the amount of heat transfer surface has a greater effect on bypass factor than varying the velocity of air through the apparatus.

There is a psychrometric relationship of bypass factor to GSHF and RSHF. Under specified room, outdoor design conditions and quantity of outdoor air, RSHF and GSHF are fixed. The position of RSHF is also fixed, but the relative position of GSHF may vary as the supply air quantity and supply air condition change.

To properly maintain room design conditions, the air must be supplied to the space at some point along the RSHF line. Therefore, as the bypass factor varies, the relative position of GSHF to RSHF changes, as shown by the dotted lines in *Fig. 41*. As the position of GSHF changes, the entering and leaving air conditions at the apparatus, the required air quantity, bypass factor and the apparatus dewpoint also change.

The effect of varying the bypass factor on the conditioning equipment is as follows:

1. Smaller bypass factor —
   a. Higher adp — DX equipment selected for higher refrigerant temperature and chilled water equipment would be selected for less or higher temperature chilled water. Possibly smaller refrigeration machine.

b. Less air — smaller fan and fan motor.

c. More heat transfer surface — more rows of coil or more coil surface available.

d. Smaller piping if less chilled water is used.

2. Larger bypass factor —

a. Lower adp — Lower refrigerant temperature to select DX equipment, and more water or lower temperature for chilled water equipment. Possibly larger refrigeration machine.

b. More air —larger fan and fan motor.

c. Less heat transfer surface — less rows of coil or less coil surface available.

d. Larger piping if more chilled water is used.

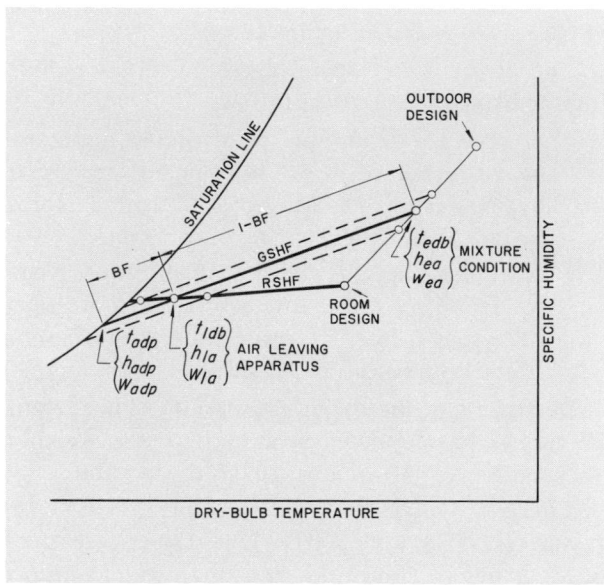

FIG. 41 — RSHF AND GSHF LINES PLOTTED ON SKELETON PSYCHROMETRIC CHART

It is, therefore, an economic balance of first cost and operating cost in selecting the proper bypass factor for a particular application. *Table 62, page 127,* lists suggested bypass factors for various applications and is a guide for the engineer to proper bypass factor selection for use in load calculations.

Tables have also been prepared to illustrate the various configurations of heat transfer surfaces and the resulting bypass factor for different air velocities. *Table 61, page 127,* lists bypass factors for various coil surfaces. Spray washer equipment is normally rated in terms of saturation efficiency which is the complement of bypass factor (1 − BF). *Table 63, page 136,* is a guide to representative saturation efficiencies for various spray arrangements.

As previously indicated, the entering and leaving air conditions at the conditioning apparatus and the apparatus dewpoint are related psychromet-

rically to the bypass factor. Although it is recognized that bypass factor is not a true straight line function, it can be accurately evaluated mathematically from the following equations:

$$BF = \frac{t_{ldb} - t_{adp}}{t_{edb} - t_{adp}} = \frac{h_{la} - h_{adp}}{h_{ea} - h_{adp}} = \frac{W_{la} - W_{adp}}{W_{ea} - W_{adp}}$$

and

$$1-BF = \frac{t_{edb} - t_{ldb}}{t_{edb} - t_{adp}} = \frac{h_{ea} - h_{la}}{h_{ea} - h_{adp}} = \frac{W_{ea} - W_{la}}{W_{ea} - W_{adp}}$$

NOTE: The quantity (1−BF) is frequently called contact factor and is considered to be that portion of the air leaving the apparatus at the adp.

EFFECTIVE SENSIBLE HEAT FACTOR (ESHF)

To relate bypass factor and apparatus dewpoint to the load calculation, the *effective sensible heat factor* term was developed. ESHF is interwoven with BF and adp, and thus greatly simplifies the calculation of air quantity and apparatus selection.

The effective sensible heat factor is the ratio of effective room sensible heat to the effective room sensible and latent heats. Effective room sensible heat is composed of room sensible heat (see RSHF) plus that portion of the outdoor air sensible load which is considered as being bypassed, unaltered, thru the conditioning apparatus. The effective room latent heat is composed of the room latent heat (see RSHF) plus that portion of the outdoor air latent heat load which is considered as being bypassed, unaltered, thru the conditioning apparatus. This ratio is expressed in the following formula:

$$ESHF = \frac{ERSH}{ERSH + ERLH} = \frac{ERSH}{ERTH}$$

The bypassed outdoor air loads that are included in the calculation of ESHF are, in effect, loads imposed on the conditioned space in exactly the same manner as the infiltration load. The infiltration load comes thru the doors and windows; the bypassed outdoor air load is supplied to the space thru the air distribution system.

Plotting RSHF and GSHF on the psychrometric chart defines the adp and BF as explained previously. Drawing a straight line between the adp and room design conditions *(1 - 2), Fig. 42* represents the ESHF ratio. The interrelationship of RSHF and GSHF to BF, adp and ESHF is graphically illustrated in *Fig. 42.*

The effective sensible heat factor line may also be drawn on the psychrometric chart without initially knowing the adp. The procedure is identical to the one described for RSHF on *page 118.* The calculated ESHF, however, is plotted thru the room design conditions to the saturation line *(1 - 2), Fig. 43,* thus indicating the adp.

Tables have been prepared to simplify the method of determining adp from ESHF. Adp can be obtained by entering *Table 65* at room design conditions and at the calculated ESHF. It is not necessary to plot ESHF on a psychrometric chart.

AIR QUANTITY USING ESHF, ADP AND BF

A simplified approach for determining the required air quantity is to use the psychrometric corre-

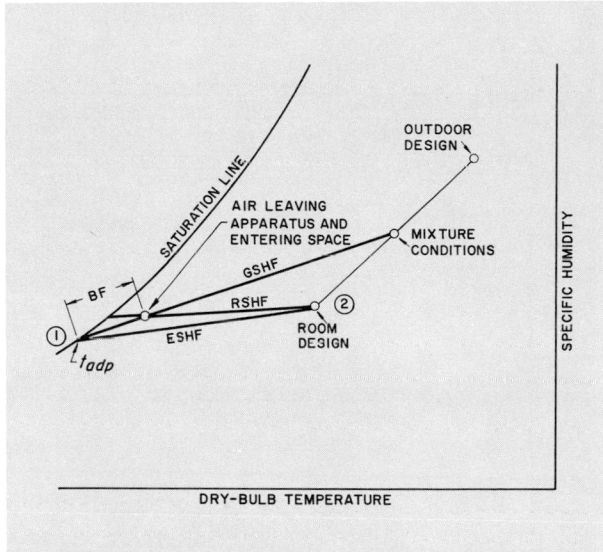

FIG. 42 — RSHF, GSHF AND ESHF LINES PLOTTED ON SKELETON PSYCHROMETRIC CHART

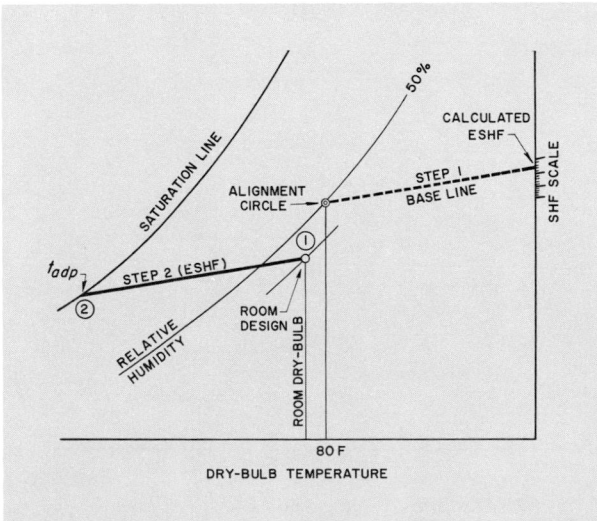

FIG. 43 — ESHF LINE PLOTTED ON SKELETON PSYCHROMETRIC CHART

lation of effective sensible heat factor, apparatus dewpoint and bypass factor. Previously in this chapter, the interrelationship of ESHF, BF and adp was shown with GSHF and RSHF. These two factors need not be calculated to determine the required air quantity, since the use of ESHF, BF and adp results in the same air quantity.

The formula for calculating air quantity, using BF and t_{adp}, is:

$$cfm_{da} = \frac{ERSH}{1.08 \ (t_{rm} - t_{adp}) \ (1 - BF)}$$

(ESHF is used to determine t_{adp}.)

This air quantity simultaneously offsets the room sensible and room latent loads, and also handles the total sensible and latent loads for which the conditioning apparatus is designed, including the outdoor air loads and the supplementary loads.

AIR CONDITIONING LOAD ESTIMATE FORM

The "Air Conditioning Load Estimate" form is designed for cooling and dehumidifying applications, and may be used for psychrometric calculations. Normally, only ESHF, BF and adp are required to determine air quantity and to select the apparatus. But for those instances when it is desirable to know RSHF and GSHF, this form is designed so that these factors may also be calculated. *Fig. 44,* in conjunction with the following items, explains how each factor is calculated. (The circled numbers correspond to numbers in *Fig. 44.*)

1. $RSHF = \dfrac{RSH}{RSH + RLH} = \dfrac{①}{① + ②}$

2. $GSHF = \dfrac{TSH}{GTH} = \dfrac{③ + ④}{⑤}$

3. $ESHF = \dfrac{ERSH}{ERSH + ERLH} = \dfrac{ERSH}{ERTH}$

   $⑧ = \dfrac{③}{③ + ⑥} = \dfrac{③}{⑦}$

4. Adp located where ESHF crosses the saturation line, or from *Table 65.* ESHF ⑧ and room conditions ⑨ give adp ⑩.

5. BF ⑪ used in the outdoor air calculations is obtained from the equipment performance table or charts. Typical bypass factors for different surfaces and for various applications are given on *page 127.* These are to guide the engineer and may be used in the outdoor air calculation when the actual equipment performance tables are not readily available.

SHEET_____ $\widehat{Carrier}$ DATE_____

PREPARED BY_____OFFICE_____   PROP NO._____JOB NO._____

NAME OF JOB_____

LOCATION_____   APPROVED_____

| SPACE USED FOR_____ | | | | | |
| SIZE ___ × ___ = SQ FT × ___ = CU FT | | | | | |

ITEM	AREA OR QUANTITY	SUN GAIN OR TEMP. DIFF.	FACTOR		BTU/HOUR
SOLAR GAIN—GLASS					
GLASS	SQ FT ×	×			
GLASS	SQ FT ×	×			
GLASS	SQ FT ×	×			
GLASS	SQ FT ×	×			
SKYLIGHT	SQ FT ×	×			
SOLAR & TRANS. GAIN—WALLS & ROOF					
WALL	SQ FT ×	×			
WALL	SQ FT ×	×			
WALL	SQ FT ×	×			
WALL	SQ FT ×	×			
ROOF—SUN	SQ FT ×	×			
ROOF—SHADED	SQ FT ×	×			
TRANS. GAIN—EXCEPT WALLS & ROOF					
ALL GLASS	SQ FT ×	×			
PARTITION	SQ FT ×	×			
CEILING	SQ FT ×	×			
FLOOR	SQ FT ×	×			
INFILTRATION	CFM ×	×	1.08		
INTERNAL HEAT					
PEOPLE		PEOPLE ×			
POWER		HP OR KW ×			
LIGHTS		WATTS × 3.4 ×			
APPLIANCES, ETC.		×			
ADDITIONAL HEAT GAINS		×			
			SUB TOTAL		
STORAGE	SQ FT ×	× (—)			
			SUB TOTAL		
SAFETY FACTOR	%				
ROOM SENSIBLE HEAT ■					①

SUPPLY DUCT HEAT GAIN ___% +	SUPPLY DUCT LEAK. LOSS ___% +	FAN H.P. ___%			
OUTDOOR AIR	CFM ×	F ×	⑪ BF × 1.08		③
EFFECTIVE ROOM SENSIBLE HEAT ■					

LATENT HEAT					
INFILTRATION	CFM ×		GR/LB × 0.68		
PEOPLE		PEOPLE ×			
STEAM		LB/HR × 1050			
APPLIANCES, ETC.					
ADDITIONAL HEAT GAINS					
VAPOR TRANS.	SQ FT × 1/100 ×	GR/LB ×			
			SUB TOTAL		
SAFETY FACTOR	%				②
ROOM LATENT HEAT					
SUPPLY DUCT LEAKAGE LOSS			%		
OUTDOOR AIR	CFM ×	GR/LB ×	⑪ BF × 0.68		⑥ ⑦
EFFECTIVE ROOM LATENT HEAT					
EFFECTIVE ROOM TOTAL HEAT ■					④

OUTDOOR AIR HEAT					
SENSIBLE:	CFM ×	F × (1 —	⑪ BF) × 1.08		
LATENT:	CFM ×	GR/LB × (1 —	⑪ BF) × 0.68		
RETURN DUCT HEAT GAIN ___% +	RETURN DUCT LEAK. GAIN ___% +	HP PUMP ___% +	SUB TOTAL DEHUM. & PIPE LOSS ___%		⑤
GRAND TOTAL HEAT ■					

(Arrows labeled: ⑯ ⑨, ⑮, ⑧, ⑩, ⑫, ⑬, ⑭, ⑰, ⑱)

ESTIMATE FOR	LOCAL TIME SUN TIME	PEAK LOAD			LOCAL TIME SUN TIME
HOURS OF OPERATION					
CONDITIONS	DB	WB	% RH	DP	GR/LB
OUTDOOR (OA)					
ROOM (RM)					
DIFFERENCE		X X X	X X X	X X X	

OUTDOOR AIR

VENTILATION _____ PEOPLE × _____ CFM/PERSON = _____
_____ SQ FT × _____ CFM/SQ FT = _____
CFM VENTILATION ■

SWINGING REVOLVING DOORS _____ PEOPLE × _____ CFM/PERSON = _____
OPEN DOORS _____ DOORS × _____ CFM/DOOR = _____
INFILTRATION   EXHAUST FAN _____
CRACK _____ FEET × _____ CFM/FT = _____
CFM INFILTRATION ■

CFM OUTDOOR AIR THRU APPARATUS ■ _____ CFM$_{OA}$

APPARATUS DEWPOINT

ESHF   EFFECTIVE SENS HEAT FACTOR = $\dfrac{③ \text{ EFFECTIVE ROOM SENS. HEAT}}{⑦ \text{ EFFECTIVE ROOM TOTAL HEAT}}$ = _____

ADP   INDICATED ADP = _____ F   SELECTED ADP = _____ F

DEHUMIDIFIED AIR QUANTITY

TEMP. RISE $(1 - \dfrac{⑪}{} BF) \times (T_{RM} \; ⑨ \; F - T_{ADP} \; ⑩ \; F)$ = _____ F

DEHUM. CFM $\dfrac{③ \text{ EFFECTIVE ROOM SENS. HEAT}}{1.08 \times ⑫ \text{ F TEMP. RISE}}$ = _____ CFM$_{DA}$

OUTLET TEMP. DIFF. $\dfrac{① \text{ ROOM SENS. HEAT}}{1.08 \times ⑬ \text{ CFM}_{DA}}$ = _____ F (RM—OUTLET AIR)*

SUPPLY AIR QUANTITY

SUPPLY CFM $\dfrac{① \text{ ROOM SENS. HEAT}}{1.08 \times \text{ F DESIRED DIFF}}$ = _____ CFM$_{SA}$

BYPASS CFM $⑭ \; \text{CFM}_{SA} - ⑬ \; \text{CFM}_{DA}$ = _____ CFM$_{BA}$

RESULTING ENT & LVG CONDITIONS AT APPARATUS

EDB $T_{RM} \; ⑨ \; F + \dfrac{⑮ \; \text{CFM}_{OA}}{⑬ \text{ or } ⑭ \; \text{CFM†}} \times (T_{OA} \; ⑯ \; F - T_{RM} \; ⑨ \; F) = T_{EDB}$ ___ F

LDB $T_{ADP} \; ⑩ \; F + \dfrac{⑪}{} BF \times (T_{EDB} \; ⑰ \; F - T_{ADP} \; ⑩ \; F) = T_{LDB}$ ___ F

FROM PSYCH. CHART: T_{EWB} _____ F, T_{LWB} _____ F

NOTES

*IF THIS ΔT IS TOO HIGH, DETERMINE SUPPLY CFM FOR DESIRED DIFFERENCE BY SUPPLY AIR QUANTITY FORMULA.
†WHEN BYPASSING A MIXTURE OF OUTDOOR AND RETURN AIR, USE SUPPLY CFM. WHEN BYPASSING RETURN AIR ONLY, USE DEHUMIDIFIED CFM.

Form E 20

NOTE: The circled numbers are explained on the previous page under "Air Conditioning Load Estimate" form.

FIG. 44 — AIR CONDITIONING LOAD ESTIMATE

6. $$cfm_{da} = \frac{ERSH}{1.08\ (t_{rm} - t_{adp})\ (1 - BF)}$$

$$⑬ = \frac{③}{1.08\ (⑨ - ⑩)\ (1 - ⑪)}$$

Once the dehumidified air quantity is calculated, the conditioning apparatus may be selected. The usual procedure is to use the grand total heat ⑤, dehumidified air quantity ⑬, and the apparatus dewpoint ⑩, to select the apparatus.

Since guides are available, the bypass factor of the apparatus selected is usually in close agreement with the originally assumed bypass factor. If, because of some peculiarity in loading in a particular application, there is a wide divergence in bypass factor, that portion of the load estimate form involving bypass factor should be adjusted accordingly.

7. Outlet temperature difference — *Fig. 44* shows a calculation for determining the temperature difference between room design dry-bulb and the supply air dry-bulb to the room. Frequently a maximum temperature difference is established for the application involved. If the outlet temperature difference calculation is larger than desired, the total air quantity in the system is increased by bypassing air around the conditioning apparatus. This temperature difference calculation is:

$$\text{Outlet temp diff} = \frac{RSH}{1.08 \times cfm_{da}}$$

$$= \frac{①}{1.08 \times ⑬}$$

8. Total air quantity when outlet temperature difference is greater than desired — The calculation for the total supply air quantity for a desired temperature difference (between room and outlet) is:

$$cfm_{sa} = \frac{RSH}{1.08 \times \Delta t} = \frac{①}{1.08 \times \Delta t}$$

The amount of air that must be bypassed around the conditioning apparatus to maintain this desired temperature difference (Δt) is the difference between cfm_{sa} and cfm_{da}.

9. Entering and leaving conditions at the apparatus — Often it is desired to specify the selected conditioning apparatus in terms of entering and leaving air conditions at the apparatus. Once the apparatus has been selected from ESHF, adp, BF and GTH, the entering and

leaving air conditions are easily determined. The calculations for the entering and leaving dry-bulb temperatures at the apparatus are illustrated in *Fig. 44*.

The entering dry-bulb calculation contains the term "cfm†"*. This air quantity "cfm†" depends on whether a mixture of outdoor and return air or return air only is bypassed around the conditioning apparatus.

The total supply air quantity cfm_{sa} ⑭ is used for "cfm†" when bypassing a mixture of outdoor and return air. *Fig. 45* is a schematic sketch of a system bypassing a mixture of outdoor and return air.

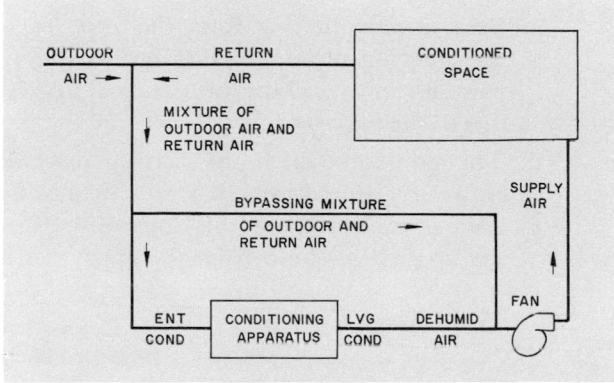

FIG. 45 — BYPASSING MIXTURE OF OUTDOOR AND RETURN AIR

When bypassing a mixture of return air only or when there is no need for a bypass around the apparatus, use the cfm_{da} ⑬ for the value of "cfm†". *Fig. 46* is a schematic sketch of a system bypassing room return air only.

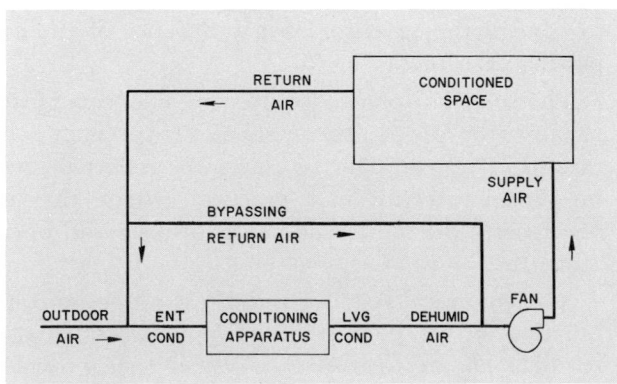

FIG. 46 — BYPASSING RETURN AIR ONLY OR NO FIXED BYPASS

*"cfm†" is a symbol appearing in the equation next to ⑰ in *Fig. 44*.

The entering and leaving wet-bulb temperatures at the apparatus are determined on the standard psychrometric chart, once the entering and leaving dry-bulb temperatures are calculated. The procedure for determining the wet-bulb temperatures at the apparatus is illustrated in *Fig. 47* and described in the following items:

a. Draw a straight line connecting room design conditions and outdoor design conditions.

b. The point at which entering dry-bulb crosses the line plotted in *Step a* defines the entering conditions to the apparatus. The entering wet-bulb is read on the psychrometric chart.

c. Draw a straight line from the adp ⑩ to the entering mixture conditions at the apparatus (*Step b.*) (This line defines the GSHF line of the apparatus.)

d. The point at which the leaving dry-bulb crosses the line drawn in *Step c* defines the leaving conditions of the apparatus. Read the leaving wet-bulb from the apparatus at

this point. (This point defines the intersection of the RSHF and GSHF as described previously.)

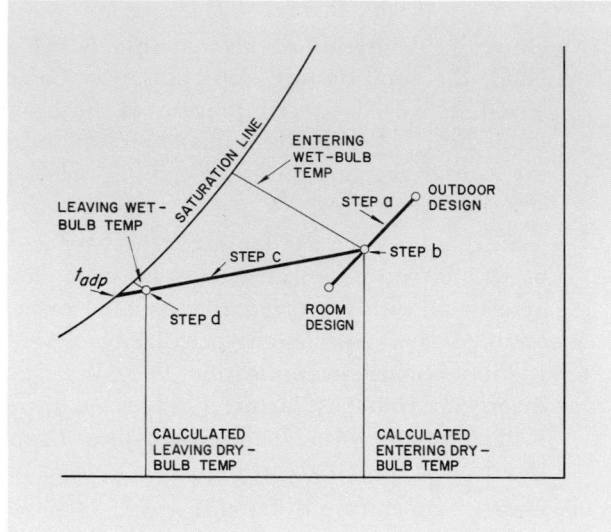

FIG. 47 — ENTERING AND LEAVING CONDITIONS
AT APPARATUS

AIR CONDITIONING APPARATUS

The following section describes the characteristic psychrometric performance of air conditioning equipment.

Coils, sprays and sorbent dehumidifiers are the three basic types of heat transfer equipment required for air conditioning applications. These components may be used singly or in combination to control the psychrometric properties of the air passing thru them.

The selection of this equipment is normally determined by the requirements of the specific application. The components must be selected and integrated to result in a practical system; that is, one having the most economical owning and operating cost.

An economical system requires the optimum combination of air conditioning components. It also requires an air distribution system that provides good air distribution within the conditioned space, using a practical rise between supply air and room air temperatures.

Since the only known items are the load in the space and the conditions to be maintained within

the space, the selection of the various components is based on these items. Normally, performance requirements are established and then equipment is selected to meet the requirements.

COIL CHARACTERISTICS

In the operation of coils, air is drawn or forced over a series of tubes thru which chilled water, brine, volatile refrigerant, hot water or steam is flowing. As the air passes over the surface of the coil, it is cooled, cooled and dehumidified, or heated, depending upon the temperature of the media flowing thru the tubes. The media in turn is heated or cooled in the process.

The amount of coil surface not only affects the heat transfer but also the bypass factor of the coil. The bypass factor, as previously explained, is the measure of air side performance. Consequently, it is a function of the type and amount of coil surface and the time available for contact as the air passes thru the coil. *Table 61* gives approximate bypass factors for various finned coil surfaces and air velocities.

TABLE 61—TYPICAL BYPASS FACTORS

(For Finned Coils)

DEPTH OF COILS (rows)	WITHOUT SPRAYS		WITH SPRAYS*	
	8 fins/in.	14 fins/in.	8 fins/in.	14 fins/in.
	Velocity (fpm)			
	300 - 700	300 - 700	300 - 700	300 - 700
2	.42 - .55	.22 - .38		
3	.27 - .40	.10 - .23		
4	.19 - .30	.05 - .14	.12 - .22	.03 - .10
5	.12 - .23	.02 - .09	.08 - .14	.01 - .08
6	.08 - .18	.01 - .06	.06 - .11	.01 - .05
8	.03 - .08		.02 - .05	

*The bypass factor with spray-coils is decreased because the spray provides more surface for contacting the air.

These bypass factors apply to coils with ⅝ in. O.D. tubes and spaced on approximately 1¼ in. centers. The values are approximate. Bypass factors for coils with plate fins, or for combinations other than those shown, should be obtained from the coil manufacturer.

Table 61 contains bypass factors for a wide range of coils. This range is offered to provide sufficient latitude in selecting coils for the most economical system. *Table 62* lists some of the more common applications with representative coil bypass factors. This table is intended only as a guide for the design engineer.

TABLE 62—TYPICAL BYPASS FACTORS

(For Various Applications)

COIL BYPASS FACTOR	TYPE OF APPLICATION	EXAMPLE
0.30 to 0.50	A *small* total load or a load that is somewhat larger with a low sensible heat factor (high latent load).	Residence
0.20 to 0.30	Typical comfort application with a *relatively small* total load or a low sensible heat factor with a somewhat larger load.	Residence, Small Retail Shop, Factory
0.10 to 0.20	Typical comfort application.	Dept. Store, Bank, Factory
0.05 to 0.10	Applications with high internal sensible loads or requiring a large amount of outdoor air for ventilation.	Dept. Store, Restaurant, Factory
0 to 0.10	All outdoor air applications.	Hospital Operating Room, Factory

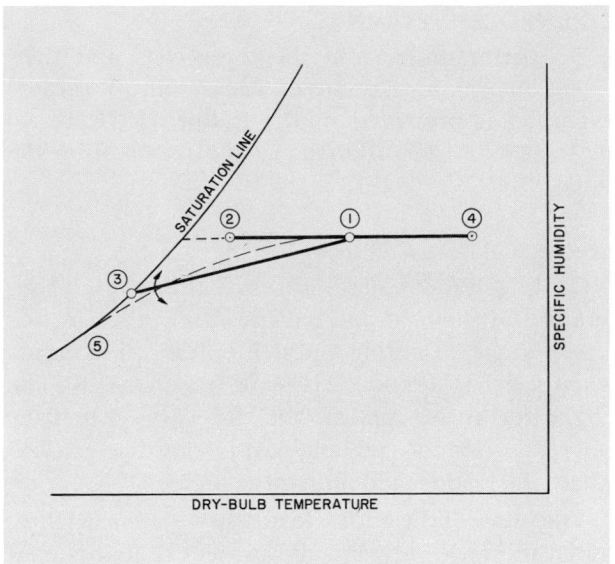

Fig. 48 — Coil Processes

COIL PROCESSES

Coils are capable of heating or cooling air at a constant moisture content, or simultaneously cooling and dehumidifying the air. They are used to control dry-bulb temperature and maximum relative humidity at peak load conditions. Since coils alone cannot *raise* the moisture content of the air, a water spray on the coil surface must be added if humidification is required. If this spray water is recirculated, it will not materially affect the psychrometric process when the air is being cooled and dehumidified.

Fig. 48 illustrates the various processes that can be accomplished by using coils.

Sensible Cooling

The first process, illustrated by line *(1 - 2)*, represents a sensible cooling application in which the heat is removed from the air at a constant moisture content.

Cooling and Dehumidification

Line *(1 - 3)* represents a cooling and dehumidification process in which there is a simultaneous removal of heat and moisture from the air.

For practical considerations, line *(1 - 3)* has been plotted as a straight line. It is, in effect, a line that starts at point *(1)* and curves toward the saturation line below point *(3)*. This is indicated by line *(1 - 5)*.

Sensible Heating

Sensible heating is illustrated by line *(1 - 4)*; heat is added to the air at constant moisture content.

COIL PROCESS EXAMPLES

To better understand these processes and their variations, a description of each with illustrated examples is presented in the following: (Refer to *page 149* for definition of symbols and abbreviations.)

Cooling and Dehumidification

Cooling and dehumidification is the simultaneous removal of the heat and moisture from the air, line *(1 - 3)*, *Fig. 48*. Cooling and dehumidification occurs when the ESHF and GSHF are less than 1.0. The ESHF for these applications can vary from 0.95, where the load is predominantly sensible, to 0.45 where the load is predominantly latent.

The air conditioning load estimate form illustrated in *Fig. 44* presents the procedure that is used to determine the ESHF, dehumidified air quantity, and entering and leaving air conditions at the apparatus. *Example 1* illustrates the psychrometrics involved in establishing these values.

Example 1 — Cooling and Dehumidification

Given:

    Application — 5¢ & 10¢ Store
    Location — Bloomfield, N. J.
    Summer design — 95 F db, 75 F wb
    Inside design — 75 F db, 50% rh

    RSH — 200,000 Btu/hr
    RLH — 50,000 Btu/hr
    Ventilation — 2,000 cfm_{oa}

Find:

1. Outdoor air load (OATH)
2. Grand total heat (GTH)
3. Effective sensible heat factor (ESHF)
4. Apparatus dewpoint temperature (t_{adp})
5. Dehumidified air quantity (cfm_{da})
6. Entering and leaving conditions at the apparatus $(t_{edb}, t_{ewb}, t_{ldb}, t_{lwb})$

Solution:

1. $\text{OASH} = 1.08 \times 2000 \times (95 - 75) = 43{,}200 \text{ Btu/hr}$ (14)
    $\text{OALH} = .68 \times 2000 \times (99 - 65) = 46{,}200 \text{ Btu/hr}$ (15)
    $\text{OATH} = 43{,}200 + 46{,}200 = 89{,}400 \text{ Btu/hr}$ (17)

2. $\text{TSH} = 200{,}000 + 43{,}200 = 243{,}200 \text{ Btu/hr}$ (7)
    $\text{TLH} = 50{,}000 + 46{,}200 = 96{,}200 \text{ Btu/hr}$ (8)
    $\text{GTH} = 243{,}200 + 96{,}200 = 339{,}400 \text{ Btu/hr}$ (9)

3. Assume a bypass factor of 0.15 from *Table 62*.

$$\text{ESHF} = \frac{200{,}000 + (.15)\,(43{,}200)}{200{,}000 + (.15)\,(43{,}200) + 50{,}000 + (.15)\,(46{,}200)}$$

$$= .785 \qquad (26)$$

4. Determine the apparatus dewpoint from the room design conditions and the ESHF, by either plotting on the psychrometric chart or using *Table 65*. *Fig. 49* illustrates the ESHF plotted on the psychrometric chart.

$$t_{adp} = 50 \text{ F}$$

NOTE: Numbers in parentheses at right edge of column refer to equations beginning on *page 150*.

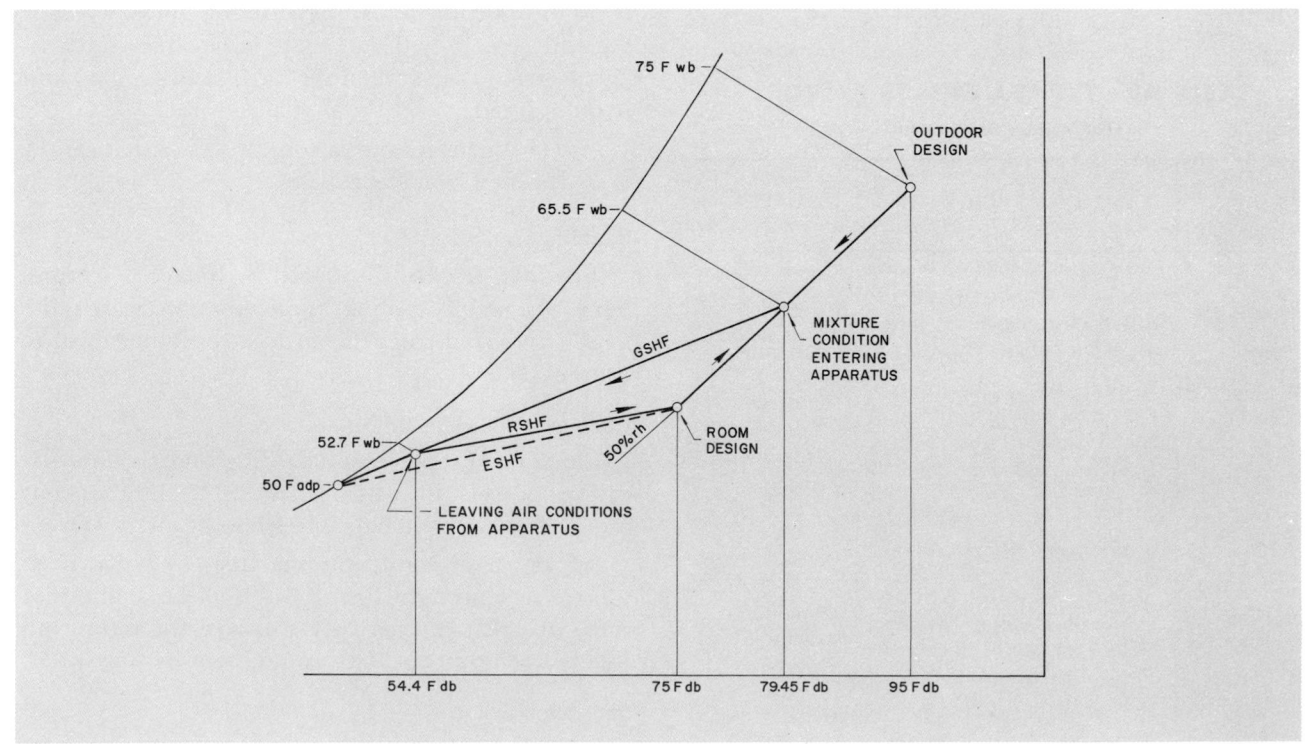

FIG. 49 -- COOLING AND DEHUMIDIFICATION

6. Assume for this example that the apparatus selected for 9,000 cfm, 50 F adp, and GTH = 339,400, has a bypass factor that is equal, or nearly equal, to the assumed BF = 0.15. Also, assume that it is not necessary to physically bypass air around the apparatus.

$$t_{edb} = \frac{(2000 \times 95) + (7000 \times 75)}{9000} = 79.45 \text{ F db} \qquad (31)$$

Read t_{ewb} where the t_{edb} crosses the straight line plotted between the outdoor and room design conditions on the psychrometric chart, *Fig. 49*.

$$t_{ewb} = 65.5 \text{ F wb}$$

$$t_{ldb} = 50 + .15 (79.45 - 50) = 54.4 \text{ F db} \qquad (32)$$

Determine the t_{lwb} by drawing a straight line between the adp and the entering conditions at the apparatus. (This is the GSHF line.) Where t_{ldb} intersects this line, read t_{lwb}.

$$t_{lwb} = 52.7 \text{ F wb}$$

Cooling and Dehumidification — High Latent Load Application

On some applications a special situation exists if the ESHF and GSHF lines do not intersect the saturation line when plotted on the psychrometric chart or if they do the adp is absurdly low. This may occur where the latent load is high with respect to the total loads (dance halls, etc.). In such applications, an appropriate apparatus dewpoint is selected and the air is reheated to the RSHF line. Occasionally, altering the room design conditions eliminates the need for reheat, or reduces the quantity of reheat required. Similarly, the utilization of a large air side surface (low bypass factor) coil may eliminate the need for reheat or reduce the required reheat.

Once the ventilation air requirement is determined, and if the supply air quantity is not fixed, the best approach to determining the apparatus dewpoint is to assume a maximum allowable temperature difference between the supply air and the room. Then, calculate the supply air conditions to the space. The supply air conditions to the space must fall on the RSHF line to properly offset the sensible and latent loads in the space.

There are four criteria which should be examined, to aid in establishing the supply air requirements to the space. These are:

1. Air movement in the space.

2. Maximum temperature difference between the supply air and the room.

3. The selected adp should provide an economical refrigeration machine selection.

4. In some cases, the ventilation air quantity required may result in an all outdoor air application.

Example 2 is a laboratory application with a high latent load. In this example the ESHF intersects the saturation line, but the resulting adp is too low.

Example 2 — Cooling and Dehumidification — High Latent Load

Given:

Application — Laboratory
Location — Bangor, Maine
Summer design — 90 F db, 73 F wb
Inside design — 75 F db, 50% rh
RSH — 120,000 Btu/hr
RLH — 65,000 Btu/hr
Ventilation — 2,500 cfm_{oa}
Temp. diff. between room and supply air, 20 F maximum

Find:

1. Outdoor air load (OATH)
2. Effective sensible heat factor (ESHF)
3. Apparatus dewpoint (t_{adp})
4. Reheat required
5. Supply air quantity (cfm_{sa})
6. Entering conditions to coil $(t_{edb}, t_{ewb}, W_{ea})$
7. Leaving conditions from coil (t_{ldb}, t_{lwb})
8. Supply air condition to the space (t_{sa}, W_{sa})
9. Grand total heat (GTH)

Solution:

1. OASH $= 1.08 \times 2500 \times (90\text{-}75) = 40,500 \text{ Btu/hr}$ (14)
   OALH $= .68 \times 2500 \times (95\text{-}65) = 51,000 \text{ Btu/hr}$ (15)
   OATH $= 40,500 + 51,000 = 91,500 \text{ Btu/hr}$ (17)

2. Assume a bypass factor of 0.05 because of high latent load.

$$\text{ESHF} = \frac{120,000 + (.05)(40,500)}{120,000 + (.05)(40,500) + 65,000 + (.05)(51,000)}$$
$$= .645 \qquad (26)$$

When plotted on the psychrometric chart, this ESHF (.645 intersects the saturation curve at 35 F. With such a low adp an appropriate apparatus dewpoint should be selected and the air reheated to the RSHF line.

3. Refer to Table 65. For inside design conditions of 75 F db, 50% rh, an ESHF of .74 results in an adp of 48 F which is a reasonable minimum figure.

4. Determine amount of reheat (Btu/hr) required to produce an ESHF of .74.

$$\text{ESHF (74)} =$$
$$\frac{120,000 + .05 (40,500) + \text{reheat}}{120,000 + .05 (40,500) + \text{reheat} + 65,000 + (.05) 51,000}$$
$$.74 = \frac{122,025 + \text{reheat}}{189,575 + \text{reheat}} \qquad (25)$$
$$\text{reheat} = 70,230 \text{ Btu/hr}$$

5. Determine dehumidifier air quantity (cfm_{da})

$$cfm_{da} = \frac{\text{ERSH}}{1.08 \times (1 - \text{BF})(t_{rm} - t_{adp})} \qquad (36)$$
$$= \frac{122,025 + 70,230}{1.08 (1 - .05)(75\text{-}48)} = 6940 \text{ cfm}$$

cfm_{da} is also cfm_{sa} when no air is to be physically bypassed around the cooling coil.

6. $t_{edb} = \dfrac{(2500 \times 90) + (4440 \times 75)}{6940}$

$$= 80.4 \qquad (31)$$

NOTE: Numbers in parentheses at right edge of column refer to equations beginning on *page 150*.

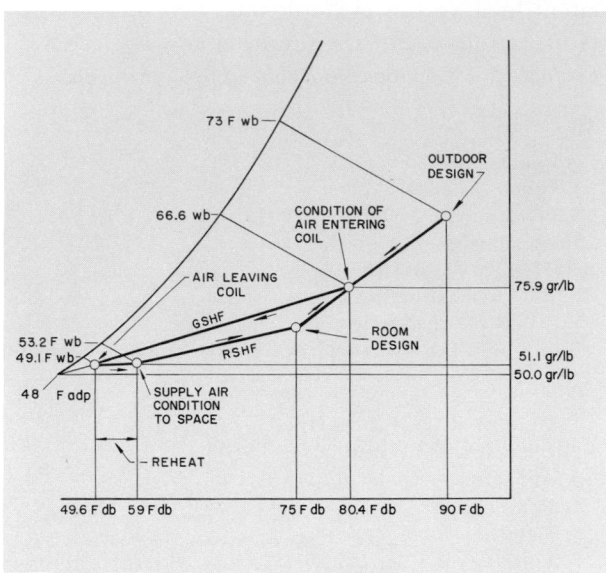

Fig. 50 — Cooling and Dehumidification
with High Latent Load

Read t_{ewb} where the t_{edb} crosses the straight line plotted between the outdoor air and room design conditions on the psychrometric chart, *Fig. 50*.

$t_{ewb} = 66.6$ F

The moisture content at the entering conditions to the coil is read from the psychrometric chart.

$W_{ea} = 75.9$ gr/lb

7. Determine leaving conditions of air from cooling coil.

$$t_{ldb} = t_{adp} + \text{BF} \, (t_{edb} - t_{adp}) \qquad (32)$$
$$= 48 + .05 \, (80.4 - 48)$$
$$= 49.6$$

$$h_{sa} = h_{adp} + \text{BF} \, (h_{ea} - h_{adp}) \qquad (34)$$
$$= 19.21 + .05 \, (31.3 - 19.21)$$
$$= 19.82$$

$$t_{lwb} = 49.1 \text{ F}$$

8. Determine supply air temperature to space

$$t_{sa} = t_{rm} - \frac{\text{RSH}}{1.08 \, (cfm_{sa})} \qquad (35)$$
$$= 75 - \frac{(120,000)}{1.08 \, (6940)}$$
$$= 59 \text{ F}$$

t_{sa} should also equal $t_{ldb} + \dfrac{\text{reheat}}{1.08 \, (cfm_{sa})}$

$$= 49.6 + \frac{70230}{1.08 \, (6940)}$$
$$= 59 \text{ F}$$

$W_{sa} = 51.1$ gr/lb

Temp. diff between room and supply air

$$= t_{rm} - t_{sa} = 75 - 59 = 16 \text{ F}$$
$$= < 20 \text{ F}$$

9. GTH $= 4.45 \times 6940 \, (31.3 - 19.82) = 354,500$ Btu/hr (24)

Cooling and Dehumidification — Using All Outdoor Air

In some applications it may be necessary to supply all outdoor air; for example, a hospital operating room, or an area that requires large quantities of ventilation air. For such applications, the ventilation or code requirements may be equal to, or more than, the air quantity required to handle the room loads.

Items 1 thru 5 explain the procedure for determining the dehumidified air requirements using the "Air Conditioning Load Estimate" form when all outdoor air is required.

1. Calculate the various loads and determine the apparatus dewpoint and dehumidified air quantity.

2. If the dehumidified air quantity is *equal* to the outdoor air requirements, the solution is self-evident.

3. If the dehumidified air quantity is *less* than the outdoor air requirements, a coil with a larger bypass factor should be investigated when the difference in air quantities is small. If a large difference exists, however, reheat is required. This situation sometimes occurs when the application requires large exhaust air quantities.

4. If the dehumidified air quantity is *greater* than the outdoor air requirements, substitute cfm_{da} for cfm_{oa} in the outdoor air load calculations.

5. Use the recalculated outdoor air loads to determine a new apparatus dewpoint and dehumidified air quantity. This new dehumidified air quantity should check reasonably close to the cfm_{da} in *Item 1*.

A special situation may arise when the condition explained in *Item 4* occurs. This happens when the ESHF, as plotted on the psychrometric chart, does not intersect the saturation line. This situation is handled in a manner similar to that previously described under *"Cooling and Dehumidification — High Latent Load Application."*

Example 3 illustrates an application where codes specify that all outdoor air be supplied to the space.

Example 3 — Cooling and Dehumidification — All Outdoor Air

Given:
    Application — Laboratory
    Location — Wheeling, West Virginia
    Summer design — 95 F db, 75 F wb
    Inside design — 75 F db, 55% rh
    RSH — 50,000 Btu/hr
    RLH — 11,000 Btu/hr
    Ventilation — 1600 cfm_{oa}
    All outdoor air to be supplied to space.

Find:

1. Outdoor air load (OATH)
2. Effective sensible heat factor (ESHF)
3. Apparatus dewpoint (t_{adp})
4. Dehumidified air quantity (cfm_{da})
5. Recalculated outdoor air load (OATH)
6. Recalculated effective sensible heat factor (ESHF)
7. Final apparatus dewpoint temperature (t_{adp})
8. Recalculated dehumidified air quantity (cfm_{da})

Solution:

1. OASH = $1.08 \times 1600 \times (95 - 75)$ = 34,600 Btu/hr   (14)
   OALH = $.68 \times 1600 \times (98.5 - 71)$ = 30,000 Btu/hr   (15)
   OATH = $34,600 + 30,000$ = 64,600 Btu/hr   (17)

2. Assume a bypass factor of 0.05 from *Tables 61 and 62.*

$$\text{ESHF} = \frac{50,000 + (.05)(34,600)}{50,000 + (.05)(34,600) + 11,000 + (.05)(30,000)}$$

$$= .81 \qquad (26)$$

3. *Table 65* shows that, at the given room design conditions and effective sensible heat factor, t_{adp} = 54.5 F.

4. $cfm_{da} = \dfrac{50,000 + (.05)(34,600)}{1.08(1 - .05)(75 - 54.5)} = 2450 \text{ cfm}$ (36)

Since 2450 cfm is larger than the ventilation requirements, and by code all OA is required, the OA loads, the adp, and the dehumidified air quantity must be recalculated using 2450 cfm as the OA requirements.

5. Recalculating outdoor air load

OASH = $1.08 \times 2450 \times (95 - 75)$ = 53,000 Btu/hr   (14)
OALH = $.68 \times 2450 \times (98.5 - 71)$ = 46,000 Btu/hr   (15)
OATH = $53,000 + 46,000$ = 99,000 Btu/hr   (17)

6. $\text{ESHF} = \dfrac{50,000 + (.05)(53,000)}{(50,000) + (.05)(53,000) + 11,000 + (.05)(46,000)}$

$$= .80 \qquad (26)$$

7. t_{adp} = 54 F

8. $cfm_{da} = \dfrac{50,000 + (.05)(53,000)}{1.08(1 - .05)(75 - 54)} = 2500 \text{ cfm}$ (36)

This checks reasonably close to the value in *Step 4*, and recalculation is not necessary.

Cooling With Humidification

Cooling with humidification may be required at partial load operation to make up a deficiency in the room latent load. It may also be used at design conditions for industrial applications having relatively high sensible loads and high room relative humidity requirements. Without humidification, excessively high supply air quantities may be required. This not only creates air distribution problems but also is often economically unsound. Excessive supply air quantity requirements can be avoided by introducing moisture into the space to convert sensible heat to latent heat. This is some-

times referred to as a "split system." The moisture is introduced into the space by using steam or electric humidifiers or auxiliary sprays.

When humidification is performed in the space, the room sensible load is decreased by an amount equal to the latent heat added, since the process is merely an interchange of heat. The humidifier motor adds sensible heat to the room but the amount is negligible and is usually ignored.

Where humidification is required at design to reduce the air quantity, then a credit to the room sensible heat should be taken in the amount of the latent heat from the added moisture. No credit to the room sensible load is taken when humidification is used to make up a deficiency in the room latent load during partial load operation.

When the humidifiers and sprays are used to reduce the required air quantity, the latent load introduced into the space is added to the room latent load.

When the humidifier or sprays are operated only to make up the room deficiency, the latent load introduced into the room by the humidifier or auxiliary sprays in the space is not added to the room latent load.

The introduction of this moisture into the space to reduce the required air quantity decreases the RSHF, ESHF and the apparatus dewpoint. This method of reducing the required air quantity is normally advantageous when designing for high room relative humidities.

The method of determining the amount of moisture necessary to reduce the required air quantity results in a trial-and-error procedure. The method is outlined in the following steps:

1. Assume an amount of moisture to be added and determine the latent heat available from this moisture. *Table 64* gives the maximum moisture that may be added to a space without causing condensation on supply air ducts and equipment.

2. Deduct this assumed latent heat from the original effective room sensible heat and use the difference in the following equation for ERSH to determine t_{adp}.

$$t_{adp} = t_{rm} - \frac{\text{ERSH}}{1.08 \times (1 - \text{BF}) \, cfm_{da}}$$

Cfm_{da} is the reduced air quantity permissible in the air distribution system.

3. The ESHF is obtained from a psychrometric chart or *Table 65,* using the apparatus dewpoint (from *Step 2*) and room design conditions.

4. The new effective room latent load is determined from the following equation:

$$ERLH = ERSH \times \frac{1 - ESHF}{ESHF}$$

The ERSH is from *Step 2* and ESHF is from *Step 3*.

5. Deduct the original ERLH (before adding sprays or humidifier in the space) from the new effective room latent heat in *Step 4*. The result is equal to the latent heat from the added moisture, and must check with the value assumed in *Step 1*. If it does not check, assume another value and repeat the procedure.

Example 4 illustrates the procedure for investigating an application where humidification is accomplished within the space to reduce the air quantity.

Example 4 — Cooling With Humidification in the Space
Given:
Application — A high humidity chamber
Location — St. Louis, Missouri
Summer design — 95 F db, 78 F wb
Inside design — 70 F db, 70% rh
RSH — 160,000 Btu/hr
RLH — 10,000 Btu/hr
RSHF — .94
Ventilation — 4000 cfm_{oa}

Find:
A. When space humidification is not used:
1. Outdoor air load (OATH)
2. Grand total heat (GTH)
3. Effective sensible heat factor (ESHF)
4. Apparatus dewpoint (t_{adp})
5. Dehumidified air quantity (cfm_{da})
6. Entering and leaving conditions at the apparatus (t_{edb}, t_{ewb}, t_{ldb}, t_{lwb})

B. When humidification is used in the space:
1. Determine maximum air quantity and assume an amount of moisture added to the space and latent heat from this moisture.
2. New effective room sensible heat (ERSH)
3. New apparatus dewpoint (t_{adp})
4. New effective sensible heat factor (ESHF)
5. New effective room latent heat (ERLH)
6. Check calculated latent heat from the moisture added with amount assumed in *Item 1*.
7. Theoretical conditions of the air entering the evaporative humidifier before humidification.
8. Entering and leaving conditions at the apparatus (t_{edb}, t_{ewb}, t_{ldb}, t_{lwb})

Solution:
A. When space humidification is not used:
1. OASH = 1.08 × 4000 × (95--70) = 108,000 Btu/hr (14)
   OALH = .68 × 4000 × (117−77) = 109,000 Btu/hr (15)
   OATH= 108,000 + 109,000   = 217,000 Btu/hr (17)

NOTE: Numbers in parentheses at right edge of column refer to equations beginning on *page 150*.

2. GTH = 160,000 + 10,000 + 108,000 + 109,000 (9)
      = 387,000 Btu/hr

3. Assume a bypass factor of 0.05 from *Tables 61 and 62*.

$$ESHF = \frac{160,000+(.05)(108,000)}{160,000+10,000+(.05)(108,000)+(.05)(109,000)}$$
$$= .92 \quad (26)$$

4. Plot the ESHF on a psychrometric chart and read the adp (dotted line in *Fig. 51*).
$$t_{adp} = 59.5 \text{ F}$$

5. $cfm_{da} = \frac{160,000 + (.05)(108,000)}{1.08(1-.05)(70-59.5)} = 15,400$ cfm (36)

6. $t_{edb} = \frac{(4000 \times 95) + (11,400 \times 70)}{15,400} = 76.7$ F db (31)

Read t_{ewb} where the t_{edb} crosses the straight line plotted between the outdoor and room design conditions on the psychrometric chart (*Fig. 51*).
$$t_{ewb} = 67.9 \text{ F wb}$$
$$t_{ldb} = 59.5 + .05(76.7 - 59.5) = 60.4 \text{ F db} \quad (32)$$

Determine the t_{lwb} by drawing a straight line between the adp and the entering conditions to the apparatus (the GSHF line). Where t_{ldb} intersects this line, read the t_{lwb} (*Fig. 51*).
$$t_{lwb} = 60 \text{ F wb}$$

B. When humidification is used in the space:
1. Assume, for the purpose of illustration in this problem, that the maximum air quantity permitted in the air distribution system is 10,000 cfm. Assume 5 grains of moisture per pound of dry air is to be added to convert sensible to latent heat. The latent heat is calculated by multiplying the air quantity times the moisture added times the factor .68.
5 × 10,000 × .68 = 34,000 Btu/hr

2. New ERSH = Original ERSH − latent heat of added moisture
   = [160,000 + (.05 × 108,000)] − 34,000
   = 131,400 Btu/hr

3. $t_{adp} = 70 - \frac{131,400}{1.08(1-.05)(10,000)} = 57.2$ F (36)

4. ESHF is read from the psychrometric chart as .73 (dotted line in *Fig. 52*).

5. New ERLH = New ERSH × $\frac{1 - ESHF}{ESHF}$
   = 131,400 × $\frac{1 - .73}{.73}$
   = 48,600 Btu/hr

6. Check for latent heat of added moisture.
   Latent heat of added moisture
   = New ERLH − Original ERLH
   = 48,600 − [10,000 + (.05 × 109,000)]
   = 33,200 Btu/hr

This checks reasonably close with the assumed value in *Step 1* (34,000 Btu/hr).

7. Psychrometrically, it can be assumed that the atomized water from the spray heads in the space absorbs part of the room sensible heat and turns into water vapor at the final room wet-bulb temperature. The theoretical dry-bulb of the air entering the sprays is

at the intersection of the room design wet-bulb line and the moisture content of the air entering the sprays. This moisture content is determined by subtracting the moisture added by the room sprays from the room design moisture content.

Moisture content of air entering humidifier
= 77 − 5 = 72 gr/lb.

The theoretical dry-bulb is determined from the psychrometric chart as 73.3 db, illustrated on *Fig. 52*.

8. $t_{edb} = \dfrac{(4000 \times 95) + (6000 \times 70)}{10,000} = 80$ F db (31)

Read t_{ewb} where the t_{edb} crosses the straight line plotted between the outdoor and room design conditions on the psychrometric chart *(Fig. 52)*.

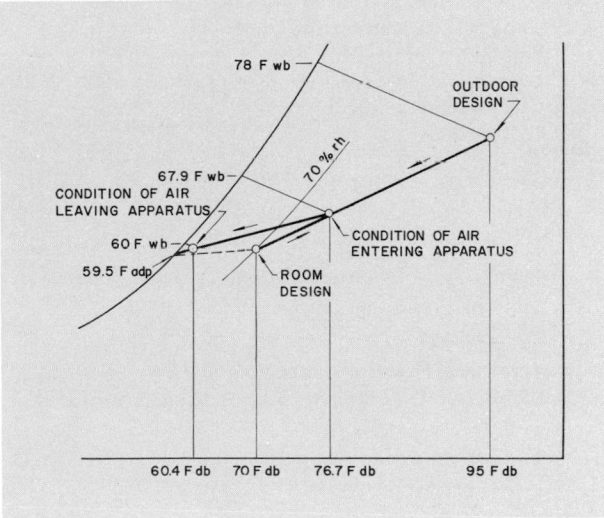

FIG. 51 — COOLING AND DEHUMIDIFICATION
ADDING NO MOISTURE TO THE SPACE

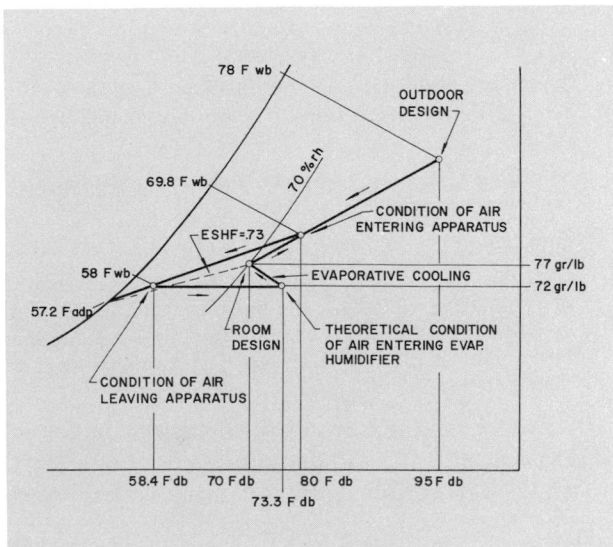

FIG. 52 — COOLING AND DEHUMIDIFICATION
ADDING MOISTURE INTO THE SPACE

$t_{ewb} = 69.8$ F wb

$t_{ldb} = 57.2 + (.05)(80 - 57.2) = 58.4$ F db (32)

Determine t_{lwb} by drawing a straight line between the adp and the entering conditions to the apparatus (GSHF line). Where t_{ldb} intersects this line, read the t_{lwb} *(Fig. 52)*.

$t_{lwb} = 58$ F wb

The straight line connecting the leaving conditions at the apparatus with the theoretical condition of the air entering the evaporative humidifier represents the theoretical process line of the air. This theoretical condition of the air entering the humidifier represents what the room conditions are if the humidifier is not operating. The slope of this theoretical process line is the same as RSHF (.94).

The heavy lines on *Fig. 52* illustrate the theoretical air cycle as air passes through the conditioning apparatus to the evaporative humidifier, then to the room, and finally back to the apparatus where the return air is mixed with the ventilation air. Actually, if a straight line were drawn from the leaving conditions of the apparatus (58.4 F db, 58 F wb) to the room design conditions, this line would be the RSHF line and would be the process line for the supply air as it picks up the sensible and latent loads in the space (including the latent heat added by the sprays).

The following two methods of laying out the system are recommended when the humidifier is to be used for both partial load control and reducing the air quantity.

1. Use two humidifiers; one to operate continuously, adding the moisture to reduce the air quantity, and the other to operate intermittently to control the humidity. The humidifier used for partial load is sized for the effective room latent load, not including that produced by the other humidifier. If the winter requirements for moisture addition are larger than summer requirements, then the humidifier is selected for these conditions. This method of using two humidifiers gives the best control.

2. Use one humidifier of sufficient capacity to handle the effective room latent heat plus the calculated amount of latent heat from the added moisture required to reduce the air quantity. In *Part B, Step 5*, the humidifier would be sized for a latent load of 48,600 Btu/hr.

Sensible Cooling

A sensible cooling process is one that removes heat from the air at a constant moisture content, line *(1 - 2), Fig. 48.* Sensible cooling occurs when either of the following conditions exist:

1. The ESHF calculated on the air conditioning load estimate form is equal to 1.0.

or

2. The entering and leaving conditions at the apparatus, as checked or plotted on the psychrometric chart, indicate a GSHF equal to 1.0.

In a sensible cooling application the ESHF, GSHF and RSHF all equal 1.0. When the RSHF equals 1.0, however, it does not necessarily indicate a sensible cooling process because latent load, introduced by outdoor air, can give a GSHF less than 1.

The apparatus dewpoint is referred to as the effective surface temperature (t_{es}) in sensible cooling applications. The effective surface temperature must be equal to, or higher than, the dewpoint temperature. In most instances, the t_{es} does not lie on the saturation line and, therefore, will not be the dewpoint of the apparatus. Whether or not t_{es} lies on the saturation line depends entirely on the bypass factor of the coil selected for the application. However, the calculations for ESHF, adp and cfm_{da} may still be performed on the air conditioning load estimate form by substituting the term t_{es} for t_{adp}. The use of the term cfm_{da} in a sensible cooling application should not be construed to indicate that dehumidification is occurring. It is used in the "Air Conditioning Load Estimate" form and in *Example 5* to determine the air quantity required thru the apparatus to offset the conditioning loads.

The leaving air conditions from the coil are dictated by the room design conditions, the load and the required air quantity. The entering and leaving dry-bulb temperatures at the apparatus should always be used to determine the effective surface temperature of the coil. Since this is strictly a sensible heat process, it is straight line function occurring at constant moisture content. Introducing wet-bulb into the calculation results in an erroneously low t_{es}. If this erroneous t_{es} is used to select the conditioning equipment, then —

1. Direct expansion equipment would be selected at a lower refrigerant temperature than actually required.

2. Chilled water equipment would be selected for colder or more chilled water than actually required.

Example 5 illustrates the method of determining the effective surface temperature for a sensible cooling application.

Example 5 — Sensible Cooling

Given:

Location — Bakersfield, California
Summer design — 105 F db, 70 F wb
Inside design — 75 F db, 50% maximum rh
RSH — 200,000 Btu/hr
RLH — none
Ventilation — 13,000 cfm_{oa}

Find:

1. Outdoor air load (OATH)
2. Effective sensible heat factor (ESHF)
3. Effective surface temperature (t_{es})*
4. Dehumidified air quantity (cfm_{da})
5. Effective surface temperature (t_{es})*
6. Supply air temperature (t_{sa})

Solution:

1. OASH $= 1.08 \times 13,000 \, (105 - 75) = 420,000$ Btu/hr   (14)
   OALH $= .68 \times 13,000 \, (54 - 54) = 0$ (15)
   OATH $= 420,000$ Btu/hr   (17)

2. Assume a bypass factor of 0.05 from *Tables 61 and 62.*

$$\text{ESHF} = \frac{200,000 + (.05)\,(420,000)}{200,000 + (.05)\,(420,000)} = 1.0 \qquad (26)$$

3. Plot the ESHF to the saturation line on the psychrometric chart. The effective surface temperature is read as $t_{es} = 50.3$ F, *Fig. 53.*

4. $cfm_{da} = \dfrac{200,000 + (.05)\,(420,000)}{1.08 \times (1 - .05) \times (75 - 50.3)} = 8650$ cfm   (36)

5. Since the dehumidified air quantity is less than the outdoor ventilation requirements, substitute the cfm_{oa} for cfm_{da} in *Step 4.* This results in a new effective surface temperature which does not lie on the saturation line. This is illustrated in the following equation:

$$t_{es} = 75 - \frac{200,000 + (.05)\,(420,000)}{1.08 \times (1 - .05) \times 13,000} = 58.4 \text{ F} \qquad (36)$$

This temperature falls on the ESHF line, which is also the GSHF and RSHF lines in a sensible cooling application.

6. Substitute t_{es} for t_{adp} in equation (28), and calculate the supply air condition t_{sa} as follows:

$$t_{sa} = 105 - (1 - .05)\,(105 - 58.4) = 60.7 \text{ F db} \qquad (28)$$

t_{sa} is the same as the t_{ldb} which is the leaving air condition from the coil. The wet-bulb temperature of the air supplied to the space is 54 F wb. This is read at the intersection of the supply air dry-bulb temperature and the ESHF line, *Fig. 53.*

In *Example 5*, the assumed .05 bypass factor is used to determine t_{es} and dehumidified air quantity. Since the dehumidified air quantity is less than

*The terms t_{es} is substituted for t_{adp} on the air conditioning load estimate form for sensible cooling application.

NOTE: Numbers in parentheses at right edge of column refer to equations beginning on *page 150.*

ventilation air requirement, the .05 bypass factor is used again to determine a new t_{es}, substituting the ventilation air requirement for the dehumidified air quantity. The new t_{es} is 58.4 F.

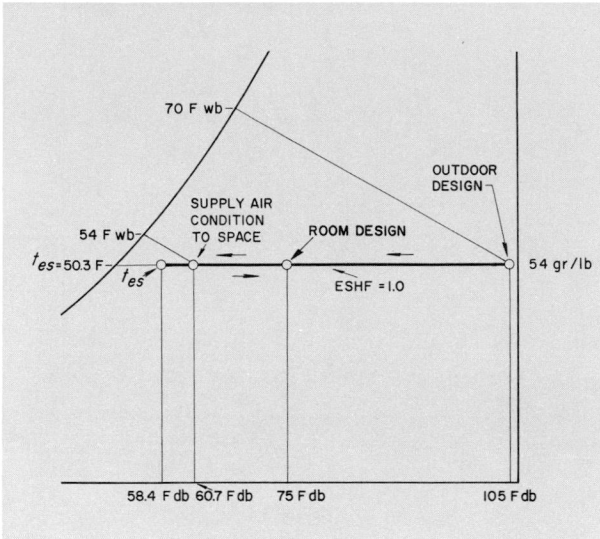

FIG. 53 — SENSIBLE COOLING

If a coil with a higher bypass factor is substituted in *Example 5*, a lower t_{es} results. Under these conditions, it becomes a question of economic balance when determining which coil selection and which refrigerant temperature is the best for the application. For instance, the maximum possible coil bypass factor that can be used is .19. This still results in a t_{es} above 50.3, and at the same time maintains a dehumidified air cfm of 13,000 which equals the ventilation requirements.

SPRAY CHARACTERISTICS

In the operation of spray type equipment, air is drawn or forced thru a chamber where water is sprayed thru nozzles into the air stream. The spray nozzles may be arranged within the chamber to spray the water counter to air flow, parallel to air flow, or in a pattern that is a combination of these two. Generally, the counter-flow sprays are the most efficient; parallel flow sprays are the least efficient; and when both are employed, the efficiency falls somewhere in between these extremes.

SATURATION EFFICIENCY

In a spray chamber, air is brought into contact with a dense spray of water. The air approaches the state of complete saturation. The degree of saturation is termed saturation efficiency (sometimes called contact or performance factor). Saturation efficiency is, therefore, a measure of the spray chamber effi-

ciency. It can be considered to represent that portion of the air passing thru the spray chamber which contacts the spray water surface. This contacted air is considered to be leaving the spray chamber at the effective surface temperature of the spray water. This effective surface temperature is the temperature at complete saturation of the air.

Though not a straight line function, the effect of saturation efficiency on the leaving air conditions from a spray chamber may be determined with a sufficient degree of accuracy from the following equation:

$$\text{Sat Eff} = \frac{t_{edb} - t_{ldb}}{t_{edb} - t_{es}} = \frac{W_{ea} - W_{la}}{W_{ea} - W_{es}} = \frac{h_{ea} - h_{la}}{h_{ea} - h_{es}}$$

The saturation efficiency is the complement of bypass factor, and with spray equipment the bypass factor is used in the calculation of the cooling load. Bypass factor, therefore, represents that portion of the air passing thru the spray equipment which is considered to be leaving the spray chamber completely unaltered from its entering condition.

This efficiency of the sprays in the spray chamber is dependent on the spray surface available and on the time available for the air to contact the spray water surface. The available surface is determined by the water particle size in the spray mist (pressure at the spray nozzle and the nozzle size), the quantity of water sprayed, number of banks of nozzles, and the number of nozzles in each bank. The time available for contact depends on the velocity of the air thru the chamber, the length of the effective spray chamber, and the direction of the sprays relative to the air flow. As the available surface decreases or as the time available for contact decreases, the saturation efficiency of the spray chamber decreases. *Table 63* illustrates the relative efficiency of different spray chamber arrangements.

The relationship of the spray water temperatures to the air temperatures is essential in understanding the psychrometrics of the various spray processes. It can be assumed that the leaving water temperature from a spray chamber, after it has contacted the air, is equal to the leaving air wet-bulb temperature. The leaving water temperature will not usually vary more than a degree from the leaving air wet-bulb temperature. Then the entering water temperature is, therefore, dependent on the water quantity and the heat required to be added or removed from the air.

Table 63 illustrates the relative efficiency of different spray chamber arrangements.

TABLE 63—TYPICAL SATURATION EFFICIENCY*

For Spray Chambers

NO. OF BANKS	DIRECTION OF WATER SPRAY	1/4" NOZZLE (25 psig Nozzle Pressure 3 gpm/sq ft†)		1/8" NOZZLE (30 psig Nozzle Pressure 2.5 gpm/sq ft†)	
		Velocity‡ (fpm)			
		300	700	300	700
1	Parallel	70%	50%	80%	60%
	Counter	75%	65%	82%	70%
2	Parallel	90%	85%	92%	87%
	Opposing	98%	92%	98%	93%
	Counter	99%	93%	99%	94%

*Saturation efficiency = 1 − BF

†Gpm/sq ft of chamber face area

‡Velocities above 700 fpm and below 300 fpm normally do not permit eliminators to adequately remove moisture from the air. Reference to manufacturers' data is suggested for limiting velocity and performance.

SPRAY PROCESSES

Sprays are capable of cooling and dehumidifying, sensible cooling, cooling and humidifying, and heating and humidifying. Sensible cooling may be accomplished only when the entering air dewpoint is the same as the effective surface temperature of the spray water.

The various spray processes are represented on the psychrometric chart in *Fig. 54*. All process lines must go toward the saturation line, in order to be at or near saturation.

Adiabatic Saturation or Evaporative Cooling

Line *(1 - 2)* represents the evaporative cooling process. This process occurs when air passes thru a spray chamber where heat has not been added to or removed from the spray water. (This does not include heat gain from the water pump and thru the apparatus casing.) When plotted on the psychrometric chart, this line approximately follows up the line of the wet-bulb temperature of the air entering the spray chamber. The spray water temperature remains essentially constant at this wet-bulb temperature.

Cooling and Humidification — With Chilled Spray Water

If the spray water receives limited cooling before it is sprayed into the air stream, the slope of the process line will move down from the evaporative cooling line. This process is represented by line *(1 - 3)*. Limited cooling causes the leaving air to be lower in dry- and wet-bulb temperatures, but higher in moisture content, than the air entering the spray chamber.

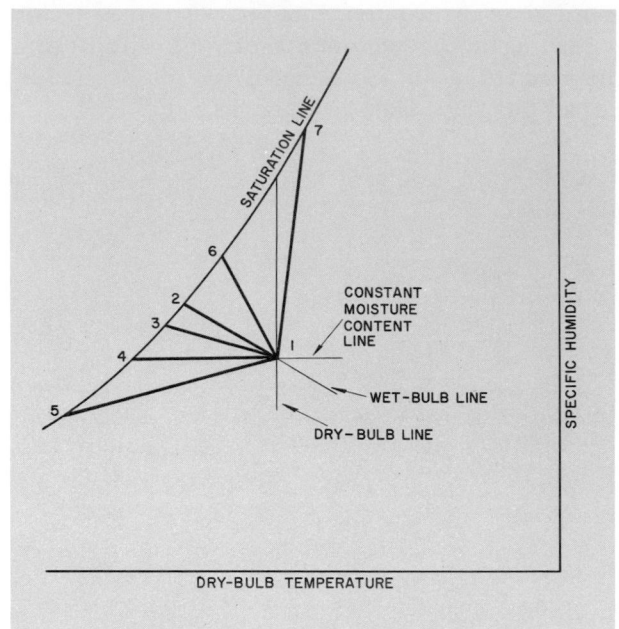

FIG. 54 — SPRAY PROCESSES

Sensible Cooling

If the spray water is cooled further, sensible cooling occurs. This process is represented by line *(1 - 4)*. Sensible cooling occurs only when the entering air dewpoint is equal to the effective surface temperature of the spray water; this condition is rare. In a sensible cooling process, the air leaving the spray chamber is lower in dry- and wet-bulb temperatures but equal in moisture content to the entering air.

Cooling and Dehumidification

If the spray water is cooled still further, cooling and dehumidification takes place. This is illustrated by line *(1 - 5)*. The leaving air is lower in dry- and wet-bulb temperatures and in moisture content than the air entering the spray chamber.

Cooling and Humidification — With Heated Spray Water

When the spray water is heated to a limited degree before it is sprayed into the air stream, the slope of the process line rises to a point above the evaporative cooling line. This is illustrated by line *(1 - 6)*. Note that the leaving air is lower in dry-bulb temperature, but higher in wet-bulb temperature and moisture content, than the air entering the spray chamber.

Heating and Humidification

If the spray water is sufficiently heated, a heating and humidification process results. This is represented by line *(1 - 7)*. In this process the dry-bulb temperature, wet-bulb temperature, and moisture content of the leaving air is greater than that of the entering air.

SPRAY PROCESS EXAMPLES

The following descriptions and examples provide a better understanding of the various psychrometric processes involved in spray washer equipment.

Cooling and Dehumidification

When a spray chamber is to be used for cooling and dehumidification, the procedure for estimating the load and selecting the equipment is identical to the procedure described on *page 128* for coils. The "Air Conditioning Load Estimate" form is used to evaluate the load; bypass factor is determined by subtracting the selected saturation efficiency from one. Spray chamber dehumidifiers may not be rated in terms of apparatus dewpoint but in terms of entering and leaving wet-bulb temperatures at the apparatus. The apparatus dewpoint must still be determined, however, to evaluate properly the entering and leaving wet-bulb temperatures and the dehumidified air quantity.

Although originally prepared to exemplify the operation of a coil, *Example 1, page 128,* is also typical of the cooling and dehumidifying process using sprays.

Cooling and Dehumidification — Using All Outdoor Air

When a spray chamber is to be used for cooling and dehumidifying with all outdoor air, the procedure for determining adp, entering and leaving conditions at the chamber, ESHF and cfm_{da} is identical to the procedure for determining these items for coils using all outdoor air. Therefore, the description on *page 130* and *Example 3* may be used to analyze this type of application.

Evaporative Cooling

An evaporative cooling application is the simultaneous removal of sensible heat and the addition of moisture to the air, line *(1 - 2), Fig. 54.* The spray water temperature remains essentially constant at the wet-bulb temperature of the air. This is a process in which heat is not added to or removed from the spray water. (Heat gain from the water pump and heat gain thru the apparatus casing are not included.)

Evaporative cooling is commonly used for those applications where the relative humidity is to be controlled but where no control is required for the room dry-bulb temperature, except to hold it above a predetermined minimum. When the dry-bulb

temperature is to be maintained during the winter or intermediate season, heat must be available to the system. This is usually accomplished by adding a reheat coil. When relative humidity is to be maintained in addition to room dry-bulb during the winter or intermediate season, a combination of preheat and reheat coils, or a reheat coil and spray water heating, is required. The latter method changes the process from evaporative cooling to one of the humidification processes illustrated by lines *(1 - 6)* or *(1 - 7)* in *Fig. 54.*

Evaporative cooling may be used in industrial applications where the humidity alone is critical, and also in dry climates where evaporative cooling gives some measure of relief by removing sensible heat.

Example 6 illustrates an industrial application designed to maintain the space relative humidity only.

Example 6 — Evaporative Cooling

Given:
    An industrial application
    Location — Columbia, South Carolina
    Summer design — 95 F db, 75 F wb
    Inside design — 55% rh
    RSH — 2,100,000 Btu/hr
    RSHF — 1.0
    Use all outdoor air at design load conditions

Find:
    1. Room dry-bulb temperature at design (t_{rm})
    2. Supply air quantity (cfm_{sa})

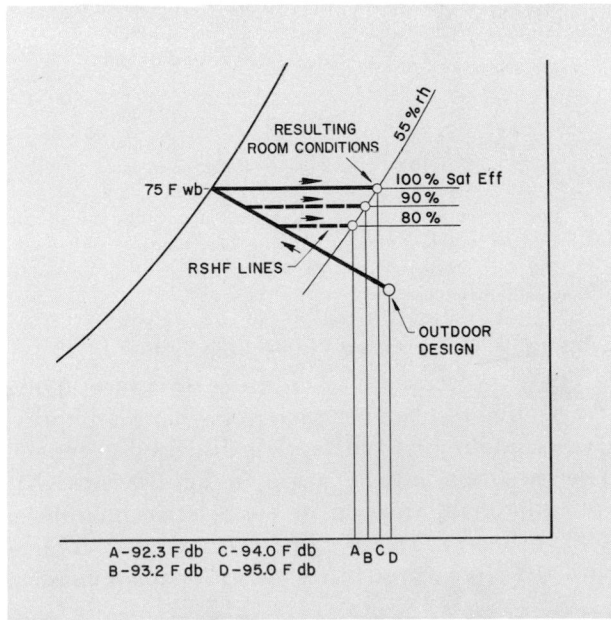

Fig. 55 — Evaporative Cooling, With Varying Saturation Efficiency

Solution:

1. Determine the room dry-bulb temperature by compromising between the spray saturation efficiency, the acceptable room dry-bulb temperature, and the supply air quantity. To evaluate these items, use the following equation to determine the leaving conditions from the spray for various saturation efficiencies:

$$t_{ldb} = t_{edb} - (\text{Sat Eff})(t_{edb} - t_{ewb})^*$$

The room dry-bulb temperature in the following table results from various spray saturation efficiencies and is determined by plotting the RSHF thru the various leaving conditions, to the design relative humidity, *Fig. 55*. Note that the supply air temperature rise decreases more rapidly than the room dry-bulb temperature. Correspondingly, as the supply air temperature rise decreases, the supply air quantity increases in the same proportion.

SAT EFF (%)	DRY-BULB TEMP LEAVING SPRAYS (t_{ldb})	SUPPLY AIR TEMP RISE (Δt)	ROOM DRY-BULB TEMP AT 55% RH (t_{rm})
100	75	19	94
95	76	17.6	93.6
90	77	16.2	93.2
85	78	14.7	92.7
80	79	13.3	92.3

2. Calculate the supply air quantity for the various temperature rises from the following equation:

$$cfm_{sa} = \frac{RSH}{1.08(t_{rm} - t_{ldb})}$$

SUPPLY AIR TEMP RISE ($t_{rm} - t_{ldb}$)	SUPPLY AIR QUANTITY (cfm_{sa})
19	102,400
17.6	110,600
16.2	120,000
14.7	132,300
13.3	146,200

The spray chamber and supply air quantity should then be selected to result in the best owning and operating costs. The selection is based primarily on economic considerations.

Evaporative Cooling Used With A Split System

There are occasions when using straight evaporative cooling results in excessive air quantity requirements and an unsatisfactory air distribution system. This situation usually arises in applications that are to be maintained at higher relative humidities (70% or more). To use straight evaporative cooling with the large air quantity, or to use a split system

*This equation is applicable only to evaporative cooling applications where the entering air wet-bulb temperature, the leaving air wet-bulb temperature, and the entering and leaving water temperature to the sprays are all equal.

with the auxiliary sprays in the space, becomes a problem of economics which should be analyzed for each particular application.

When a split system is used, supplemental spray heads are usually added to the straight evaporative cooling system. These spray heads atomize water and add supplementary moisture directly to the room. This added moisture is evaporated at the final room wet-bulb temperature, and the room sensible heat is reduced by the amount of heat required to evaporate the sprayed water.

Table 64 gives the recommended maximum moisture to be added, based on a 65 F db room temperature or over, without causing condensation on the ductwork.

TABLE 64—MAXIMUM RECOMMENDED MOISTURE ADDED TO SUPPLY AIR

Without Causing Condensation on Ducts†

ROOM DESIGN RH	MOISTURE Gr/Cu Ft Dry Air	ROOM DESIGN RH	MOISTURE Gr/Cu Ft Dry Air
85	1.25	65	1.50
80	1.30	60	1.60
75	1.35	55	1.70
70	1.40	50	1.80

†These are arbitrary limits which have been established by a combination of theory and field experience. These limits apply where the room dry-bulb temperature is 65 F db or over.

As a rule of thumb, the air is reduced in temperature approximately 8.3 F for every grain of moisture per cubic foot added. This value is often used as a check on the final room temperature as read from the psychrometric chart.

Example 7 illustrates an evaporative cooling application with supplemental spray heads used in the space.

Example 7—Evaporative Cooling—With Auxiliary Sprays

Given:

An industrial application
Location — Columbia, South Carolina
Summer design — 95 F db, 75 F wb
Inside design — 70% rh
RSH — 2,100,000 Btu/hr
RSHF — 1.0
Moisture added by auxiliary spray heads — 19 gr/lb (13.9 cu ft/lb × 1.4 gr/cu ft)
Use all outdoor air thru a spray chamber with 90% saturation efficiency.

Find:

1. Leaving conditions from spray chamber (t_{ldb}, t_{lwb})
2. Room dry-bulb temperature (t_{rm})
3. Supply air quantity (cfm_{sa}) with auxiliary sprays
4. Supply air quantity (cfm_{sa}) without auxiliary sprays

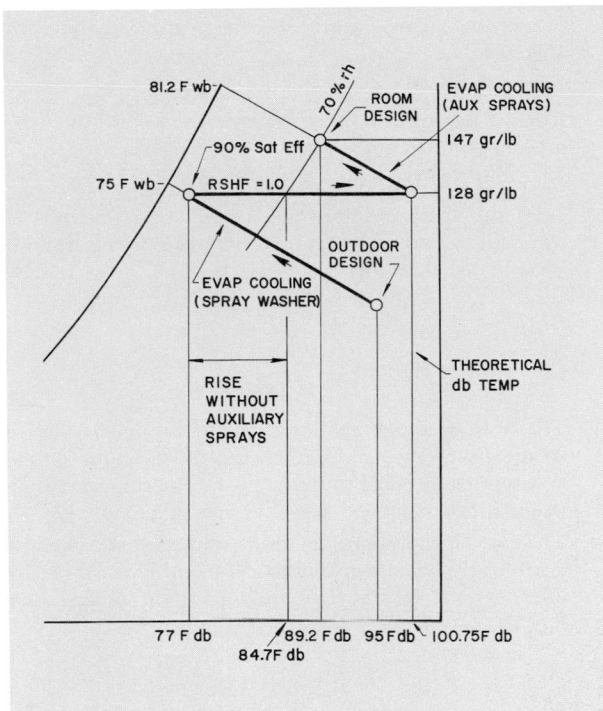

FIG. 56 — EVAPORATIVE COOLING, WITH AUXILIARY
SPRAYS WITHIN THE SPACE

Solution:

1. $t_{ldb} = t_{edb} - (\text{Sat Eff})(t_{edb} - t_{ewb})$
   $= 95 - .90(95 - 75) = 77 \text{ F db}$

t_{lwb} is the same as the t_{ewb} in an evaporative cooling process, *Fig. 56*.

2. Room dry-bulb temperature is evaluated by determining the moisture content of the space.

$$W_{rm} = W_{sa} + 19 = 128 + 19 = 147 \text{ gr/lb}$$

The 19 gr/lb is the moisture added to the space by the auxiliary spray heads.
The t_{rm} is the point on the psychrometric chart where the W_{rm} intersects the 70% design relative humidity line, *Fig. 56*.

$$t_{rm} = 89.2 \text{ F db}$$

3. Psychrometrically, it can be assumed that the atomized water from the spray heads absorbs part of the room sensible heat and turns into water vapor at the final room wet-bulb temperature. The intersection of this wet-bulb temperature with the moisture content of the air leaving the evaporative cooler is the theoretical dry-bulb equivalent temperature if the auxiliary sprays were not operating. The difference between this theoretical dry-bulb equivalent temperature and the temperature of the spray chamber, t_{ldb}, is used to determine the supply air quantity.

t_{ldb} (from spray chamber) = 77 F.

The theoretical dry-bulb temp is 100.75 F, *Fig. 56*.

Temp rise = 23.75 F db

$$cfm_{sa} = \frac{RSH}{1.08 \times \text{temp rise}} = \frac{2,100,000}{1.08 \times 23.75} = 82,000 \text{ cfm}$$

4. If no auxiliary sprays were to be used, the room design dry-bulb would be where the RSHF line intersects the room design relative humidity. From *Fig. 56*, the room dry-bulb is read

$$t_{rm} = 84.7 \text{ F db}$$

The supply air quantity required to maintain the room design relative humidity is determined from the following equation:

$$cfm_{sa} = \frac{RSH}{1.08(t_{rm} - t_{ldb})} = \frac{2,100,000}{1.08(84.7 - 77)}$$
$$= 253,000 \text{ cfm}$$

This air quantity is over three times the air quantity required when auxiliary sprays are used in the space. However, it should be noted that, by reducing the air quantity, the room dry-bulb temperature increased from 84.7 F to 89.2 F.

Heating and Humidification — With Sprays

A heating and humidifying application is one in which heat and moisture are simultaneously added to the air, line *(1 - 7)*, *Fig. 54*. This may be required during the intermediate and winter seasons or during partial loads where both the dry-bulb temperature and relative humidity are to be maintained.

Heating and humidification may be accomplished by either of the following methods:

1. Add heat to the spray water before it is sprayed into the air stream.
2. Preheat the air with a steam or hot water coil and then evaporatively cool it in the spray chamber.

Spray water is heated, by a steam to water interchanger or by direct injection of steam into the water system. Since the supply air quantity and the spray water quantity have been determined from the summer design conditions, the only other requirement is to determine the amount of heat to be added to the spray water or to the preheater.

For applications requiring humidification, the room latent load is usually not calculated and the room sensible heat factor is assumed to be 1.0.

Example 8 illustrates the psychrometric calculations for a heating and humidifying application when the spray water is heated. It should be noted that this type of application occurs only when the quantity of outdoor air required is large in relation to the total air quantity.

Example 8 — Heating and Humidification — With Heated Spray Water

Given:
  An industrial application
  Location — Richmond, Virginia

Winter design — 15 F db
Inside design — 72 F db, 35% rh
Ventilation — 50,000 cfm_{oa} (see explanation above)
Supply air — 85,000 cfm_{sa}
Design room heat loss — 2,500,000 Btu/hr
Spray saturation efficiency — 95%
RSHF (winter conditions) — 1.0
Make-up water — 65 F

Find:

1. Supply air conditions to the space (t_{sa})
2. Entering and leaving spray water temperature (t_{ew}, t_{lw})
3. Heat added to spray water to select water heater.

Solution:

1. $t_{sa} = \dfrac{\text{design room heat loss}}{1.08 \times cfm_{sa}} + t_{rm}$

       $= \dfrac{2,500,000}{1.08 \times 85,000} + 72 = 99.2 \text{ F db}$

To determine the wet-bulb temperature, plot the RSHF line on the psychrometric chart and read the wet-bulb at the point where t_{sa} crosses this line (*Fig. 57*). Supply air wet-bulb to the space = 65.8 F wb.

2. To determine the entering and leaving spray water temperature, calculate the entering and leaving air conditions at the spray chamber:

$$t_{edb} = \frac{(15 \times 50,000) + (72 \times 35,000)}{85,000} = 38.5 \text{ F db} \qquad (31)$$

To determine wet-bulb temperature of the air entering the spray chamber, plot the mixture line of outdoor and return room air on the psychrometric chart, and read

the wet-bulb temperature where t_{edb} crosses the mixture line, *Fig. 54*.

$$t_{ewb} = 32.4 \text{ F wb}$$

The air leaving the spray chamber must have the same moisture content as the air in the room.

$$W_{rm} = W_{la} = 41 \text{ gr/lb}$$

Since the spray chamber has a saturation efficiency of 95%, the moisture content of completely saturated air is calculated as follows:

$$W_{sat} = \frac{W_{la} - W_{ea}}{\text{Sat Eff}} + W_{ea}$$

$$= \frac{41 - 17}{.95} + 17 = 42.3 \text{ gr/lb}$$

The heating and humidification process line is plotted on the psychrometric chart between the moisture content of saturated air (42.3 gr/lb) and the entering conditions to the spray chamber (38.5 F db and 32.4 F wb), *Fig. 57*.

The leaving conditions are read from the psychrometric chart where the room moisture content line (41 gr/lb) intersects the heating and humidification process line, *Fig. 57*.

$$t_{ldb} = 43.6 \text{ F db}$$

$$t_{lwb} = 43.4 \text{ F wb}$$

The temperature of the leaving spray water is approximately equal to the wet-bulb temperature of the air leaving the spray chamber.

$$t_{lw} = 43.4 \text{ F}$$

NOTE: Numbers in parentheses at right edge of column refer to equations beginning on *page 150*.

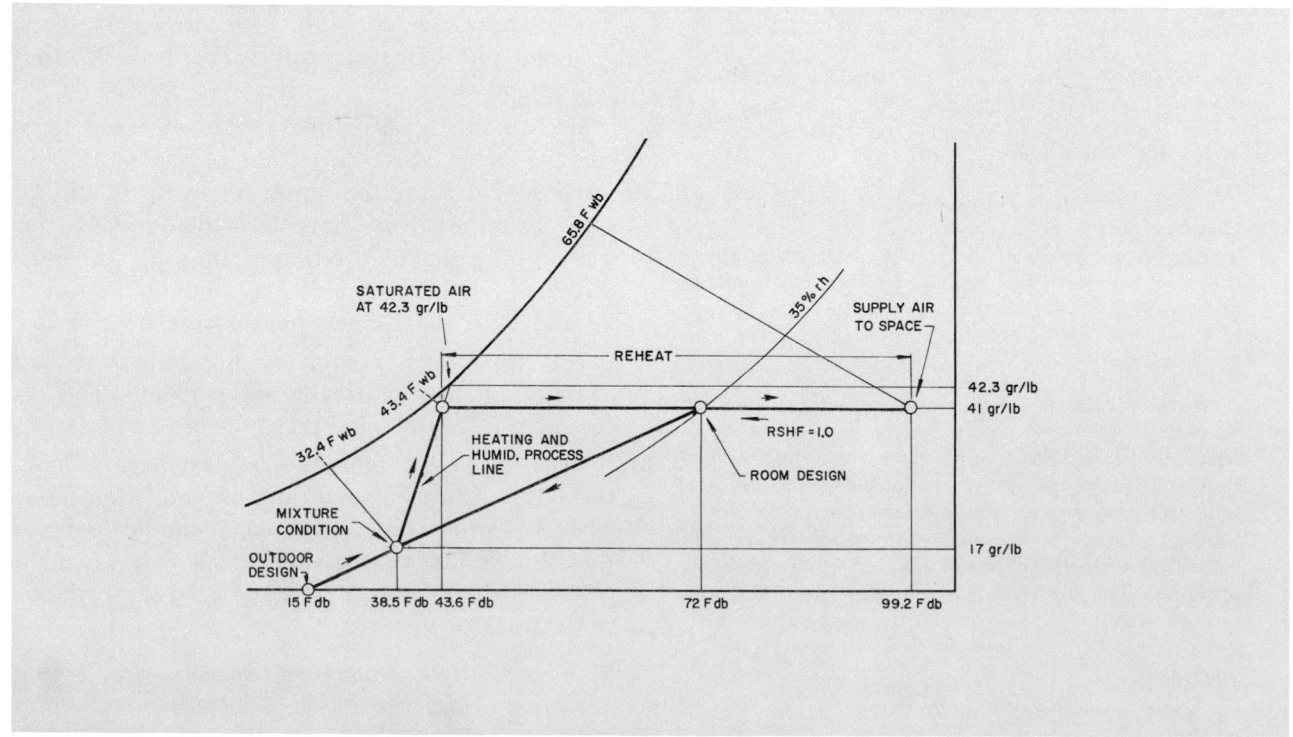

FIG. 57 — HEATING AND HUMIDIFICATION, WITH HEATING SPRAY WATER

The temperature of the entering spray water is dependent on the water quantity and the heat to be added or removed from the air. In this type of application, the water quantity is usually dictated by the cooling load design requirements. Assume, for illustration purposes, that this spray washer is selected for 110 gpm for cooling.

The heat added to the air as it passes through the washer

$$= cfm_{sa} \times 4.45 \times (h_{la} - h_{ea})$$
$$= 85,000 \times 4.45 \times (16.85 - 12)$$
$$= 1,830,000 \text{ Btu/hr}$$

The entering water temperature is determined from the following equation:

$$t_{ew} = t_{lw} + \frac{\text{heat added to air}}{500 \times \text{gpm}}$$
$$= 43.4 + \frac{1,830,000}{500 \times 110}$$
$$= 76.8 \text{ F}$$

3. The heat added to the spray water (for selecting spray water heater) is equal to the heat added to the air plus the heat added to the make-up water. The amount of make-up water is equal to the amount of moisture evaporated into the air and is determined from the following equation:

$$\text{Make-up water} = \frac{cfm_{sa} (W_{la} - W_{ea})}{7000 \times 12.7 \times 8.34}$$

where:

W_{ea}, W_{la} = moisture content of the air entering and leaving the spray washer in grains per pound of dry air

7000 = grains of moisture per pound of dry air

12.7 = volume of the mixture in cubic feet per pound of dry air, determined from psychrometric chart

8.34 = water in pounds per gallon

$$\text{Make-up water} = \frac{85,000 (41 - 17)}{7000 \times 12.7 \times 8.34} = 2.8 \text{ gpm}$$

The heat added to the make-up spray water is determined from the following equation:

Heat added to make-up water

$$= \text{gpm} \times 500 (t_{ew} - \text{make-up water temp})$$
$$= 2.8 \times 500 (76.8 - 65)$$
$$= 16,200 \text{ Btu/hr}$$

To select a water heater, the total amount of heat added to the spray water is determined by totaling the heat added to the air and the heat added to the make-up spray water.

Heat added to spray water

$$= 1,830,000 + 16,200$$
$$= 1,846,200 \text{ Btu/hr}$$

If the make-up water was at a higher temperature than the required entering water temperature to the sprays, then a credit to the heat added to the spray water may be taken.

In this example a reheat coil is required to heat the air leaving the spray chamber, at 43.6 F db and at a constant moisture content of 41 gr/lb, to the required supply air temperature of 99.2 F db.

The requirements of the application illustrated in *Example 8* can also be met by preheating the

outdoor air and mixing it with the return air from the space. This mixture must then be evaporatively cooled to the room dewpoint (or room moisture content). And finally, the air leaving the spray chamber must be reheated to the required supply air temperature.

SORBENT DEHUMIDIFIERS

Sorbent dehumidifiers contain liquid absorbent or solid adsorbent which are either sprayed directly into, or located in, the path of the air stream. The liquid absorbent changes either physically or chemically, or both, during the sorption process. The solid adsorbent does not change during the sorption process.

As moist air comes in contact with either the liquid absorbent or solid adsorbent, moisture is removed from the air by the difference in vapor pressure between the air stream and the sorbent. As

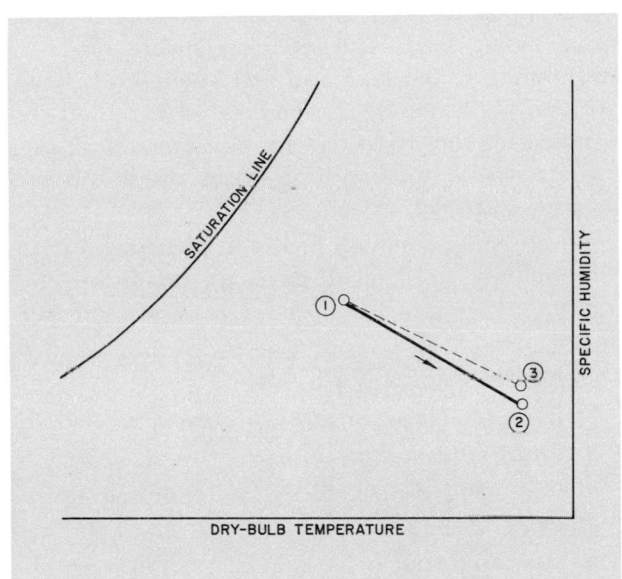

Fig. 58 — Sorbent Dehumidification Processes

this moisture condenses, latent heat of condensation is liberated, causing a rise in the temperature of the air stream and the sorbent material. This process occurs at a wet-bulb temperature that is approximately constant. However, instead of adding moisture to the air as in an evaporative cooling process, the reverse occurs. Heat is added to the air and moisture is removed from the air stream; thus it is a dehumidification and heating process as illustrated in *Fig. 58.* Line *(1 - 2)* is the theoretical process and the dotted line *(1 - 3)* approximates what actually happens. Line *(1 - 3)* can vary, depending on the type of sorbent used.

PSYCHROMETRICS OF PARTIAL LOAD CONTROL

The apparatus required to maintain proper space conditions is normally selected for peak load operation. Actually, peak load occurs but a few times each year and operation is predominantly at partial load conditions. Partial load may be caused by a reduction in sensible or latent loads in the space, or in the outdoor air load. It may also be caused by a reduction in these loads in any combination.

PARTIAL LOAD ANALYSIS

Since the system operates at partial load most of the time and must maintain conditions commensurate with job requirements, partial load analysis is at least as important as the selection of equipment. Partial load analysis should include a study of resultant room conditions at minimum total load. Usually this will be sufficient. Certain applications, however, should be evaluated at minimum latent load with design sensible load, or minimum sensible load and full latent load. Realistic minimum and maximum loads should be assumed for the particular application so that, psychrometrically, the resulting room conditions are properly analyzed.

The six most common methods, used singly or in combination, of controlling space conditions for cooling applications at partial load are the following:

1. Reheat the supply air.
2. Bypass the heat transfer equipment.
3. Control the volume of the supply air.
4. Use on-off control of the air handling equipment.
5. Use on-off control of the refrigeration machine.
6. Control the refrigeration capacity.

The type of control selected for a specific application depends on the nature of the loads, the conditions to be maintained within the space, and available plant facilities.

REHEAT CONTROL

Reheat control maintains the dry-bulb temperature within the space by replacing any decrease in the sensible loads by an artificial load. As the internal latent load and/or the outdoor latent load decreases, the space relative humidity decreases. If humidity is to be maintained, rehumidifying is required in addition to reheat. This was described previously under *"Spray Process, Heating and Humidifying."*

Figure 59 illustrates the psychrometrics of reheat control. The solid lines represent the process at design load, and the broken lines indicate the resulting process at partial load. The RSHF value, plotted from room design conditions to point *(2)*, must be calculated for the minimum practical room sensible load. The room thermostat then controls the temperature of the air leaving the reheat coil along line *(1 - 2)*. This type of control is applicable for any RSHF ratio that intersects line *(1 - 2)*.

If the internal latent loads decrease, the resulting room conditions are at point *(3)*, and the new RSHF process line is along line *(2 - 3)*. However, if humidity is to be maintained within the space, the reduced latent load is compensated by humidifying, thus returning to the design room conditions.

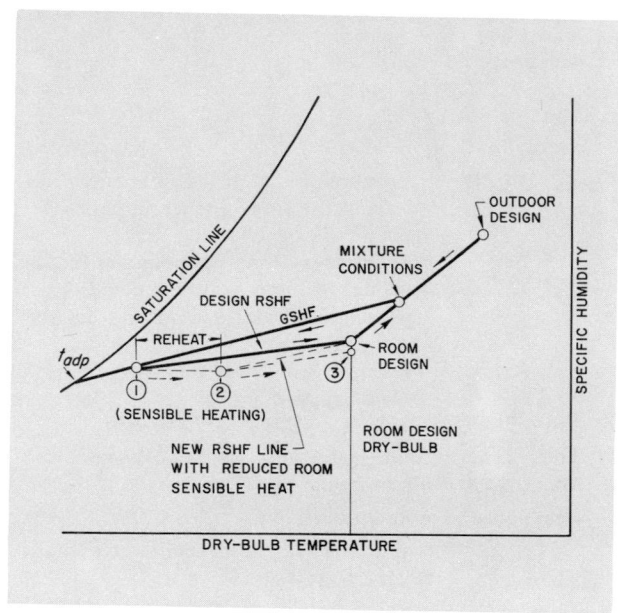

FIG. 59 — PSYCHROMETRICS OF REHEAT CONTROL

BYPASS CONTROL

Bypass control maintains the dry-bulb temperature within the space by modulating the amount of air to be cooled, thus varying the supply air temperature to the space. *Fig. 60* illustrates one method of bypass control when bypassing return air only.

Bypass control may also be accomplished by bypassing a mixture of outdoor and return air around the heat transfer equipment. This method of control is inferior to bypassing return air only since it introduces raw unconditioned air into the space, thus allowing an increase in room relative humidity.

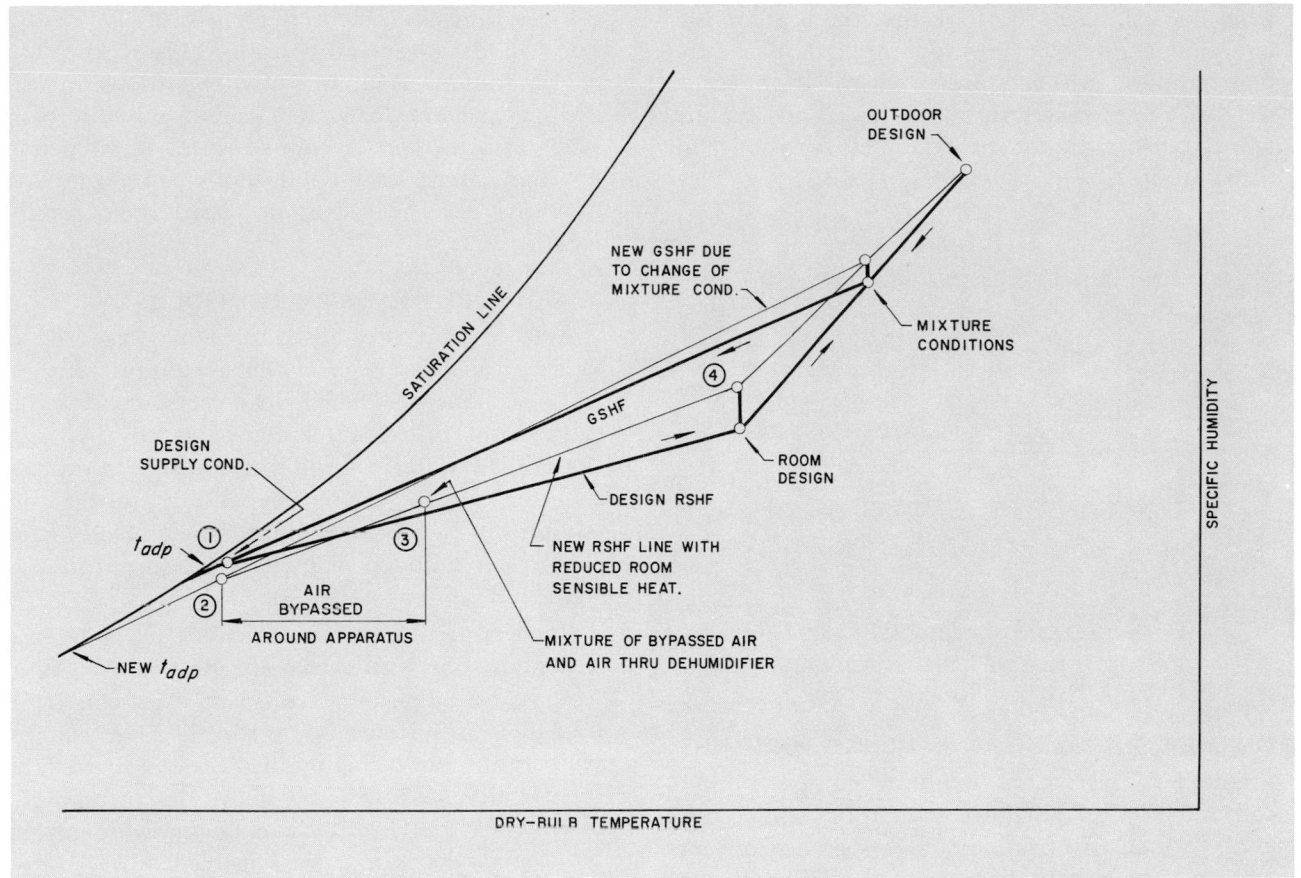

FIG. 60 — PSYCHROMETRICS OF BYPASS CONTROL WITH RETURN AIR ONLY

A reduction in room sensible load causes the bypass control to reduce the amount of air thru the dehumidifier. This reduced air quantity results in equipment operation at a lower apparatus dewpoint. Also, the air leaves the dehumidifier at a lower temperature so that there is a tendency to adjust for a decrease in sensible load that is proportionately greater than the decrease in latent load.

Bypass control maintains the room dry-bulb temperature but does not prevent the relative humidity from rising above design. With bypass control, therefore, increased relative humidity occurs under conditions of decreasing room sensible load and relatively constant room latent load and outdoor air load.

The heavy lines in *Fig. 60* represent the cycle for design conditions. The light lines illustrate the initial cycle of the air when bypass control first begins to function. The new room conditions, mixture conditions and apparatus dewpoint continue to change until the equilibrium point is reached.

Point *(2)* on *Figs. 60 and 61* is the condition of air leaving the dehumidifier. This is a result of a smaller bypass factor and lower apparatus dewpoint caused by less air thru the cooling equipment and a smaller load on the equipment. Line *(2 - 3 - 4)* represents the new RSHF line caused by the reduced room sensible load. Point *(3)* falls on the new RSHF line when bypassing return air only.

Bypassing a mixture of outdoor and return air causes the mixture point *(3)* to fall on the GSHF line, *Fig. 60*. The air is then supplied to the space along the new RSHF line (not shown in *Fig. 60*) at a higher moisture content than the air supplied when bypassing return air only. Thus it can be readily observed that humidity control is further hindered with the introduction of unconditioned outdoor air into the space.

VOLUME CONTROL
Volume control of the supply air quantity provides essentially the same type of control that results

from bypassing return air around the heat transfer equipment, *Fig. 60*. However, this type of control may produce problems in air distribution within the space and, therefore, the required air quantity at partial load should be evaluated for proper air distribution.

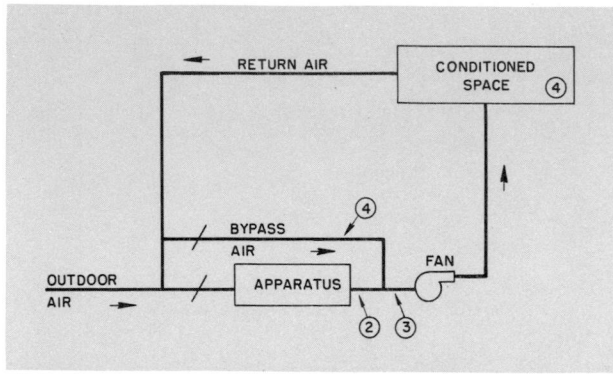

FIG. 61 — SCHEMATIC SKETCH OF BYPASS CONTROL WITH BYPASS OF RETURN AIR ONLY

ON-OFF CONTROL OF AIR HANDLING EQUIPMENT

On-off control of air handling equipment (fan-coil units) results in a fluctuating room temperature and space relative humidity. During the "off" operation the ventilation air supply is shut off, but chilled water continues to flow thru the coils. This method of control is not recommended for high latent load applications, as control of humidity may be lost at reduced room sensible loads.

ON-OFF CONTROL OF REFRIGERATION EQUIPMENT

On-off control of refrigeration equipment (large packaged equipment) results in a fluctuating room temperature and space relative humidity. During the "off" operation air is available for ventilation purposes but the coil does not provide cooling. Thus, any outdoor air in the system is introduced into the space unconditioned. Also the condensed moisture that remains on the cooling coil, when the refrigeration equipment is turned off, is re-evaporated in the warm air stream. This is known as re-evaporation. Both of these conditions increase the space latent load, and excessive humidity results. This method of control is not recommended for high latent load applications since control of humidity may be lost at decreased room sensible loads.

REFRIGERATION CAPACITY CONTROL

Refrigeration capacity control may be used on either chilled water or direct expansion refrigeration equipment. Partial load control is accomplished on chilled water equipment by bypassing the chilled water around the air side equipment (fan-coil units). Direct expansion refrigeration equipment is controlled either by unloading the compressor cylinders or by back pressure regulation in the refrigerant suction line.

Refrigeration capacity control is normally used in combination with bypass or reheat control. When used in combination, results are excellent. When used alone, results are not as effective. For example, temperature can be maintained reasonably well, but relative humidity will rise above design at partial load conditions, because the latent load may not reduce in proportion to the sensible load.

PARTIAL LOAD CONTROL

Generally, reheat control is more expensive but provides the best control of conditions in the space. Bypass control, volume control and refrigeration capacity control provide reasonably good humidity control in average or high sensible heat factor applications, and poor humidity control in low sensible heat factor applications. On-off control usually results in the least desirable method of maintaining space conditions. However, this type of control is frequently used for high sensible heat factor applications with reasonably satisfactory results.

TABLE 65—APPARATUS DEWPOINTS

90 - 80 F DB

EFFECTIVE SENSIBLE HEAT FACTOR AND APPARATUS DEWPOINT*

DB (F)	RH (%)	WB (F)	W (gr/lb)										
90	20	62.7	42.0	ESHF	1.00	.96	.92	.90	.88	.86	.84	.82	.81
				ADP	43.5	41	39	37	35	32	29	24	22
	25	65.1	52.7	ESHF	1.00	.96	.92	.88	.84	.82	.80	.78	.75
				ADP	49.6	48	46	44	41	39	36	32	22
	30	67.3	63.6	ESHF	1.00	.92	.87	.83	.80	.76	.74	.72	.70
				ADP	54.5	52	50	48	46	42	38	34	24
	35	69.3	74.2	ESHF	1.00	.92	.85	.81	.76	.73	.71	.69	.66
				ADP	58.8	57	55	53	50	48	45	42	33
	40	71.2	84.8	ESHF	1.00	.92	.83	.78	.74	.69	.66	.63	.62
				ADP	62.4	61	59	57	55	52	48	44	40
	45	73.0	95.5	ESHF	1.00	.92	.82	.76	.70	.66	.62	.60	.58
				ADP	65.8	65	63	61	59	56	52	49	43
	50	74.9	106.4	ESHF	1.00	.92	.78	.68	.64	.60	.58	.56	.54
				ADP	68.9	68	66	63	61	58	56	53	47
	55	76.7	117.5	ESHF	1.00	.92	.76	.68	.64	.57	.54	.52	.50
				ADP	71.6	71	69	67	66	62	59	57	50
	60	78.4	128.4	ESHF	1.00	.86	.68	.60	.56	.52	.50	.48	.46
				ADP	74.2	73	71	69	67	64	62	59	50
	65	80.0	139.6	ESHF	1.00	.75	.68	.62	.55	.50	.47	.45	.43
				ADP	76.8	75	74	73	71	69	66	64	59
	70	81.6	151.0	ESHF	1.00	.78	.66	.60	.52	.47	.43	.41	.39
				ADP	79.0	78	77	76	74	72	69	66	58

DB (F)	RH (%)	WB (F)	W (gr/lb)										
85	20	59.6	35.8	ESHF	1.00	.98	.95	.92	.90	.88	.87	.86	.84
				ADP	39.4	38	36	34	32	30	28	26	22
	25	61.7	44.8	ESHF	1.00	.98	.93	.90	.86	.84	.82	.80	.78
				ADP	45.2	44	42	40	37	35	32	28	20
	30	63.7	54.1	ESHF	1.00	.94	.89	.85	.81	.79	.77	.75	.73
				ADP	50.2	48	46	44	40	38	35	31	22
	35	65.5	62.9	ESHF	1.00	.92	.86	.82	.78	.74	.72	.70	.69
				ADP	54.1	52	50	48	45	41	38	32	27
	40	67.4	71.7	ESHF	1.00	.92	.84	.79	.76	.73	.69	.67	.65
				ADP	57.9	56	54	52	50	48	44	40	32
	45	69.1	81.1	ESHF	1.00	.92	.83	.77	.72	.68	.64	.62	.61
				ADP	61.2	60	58	56	54	51	46	41	36
	50	70.8	90.1	ESHF	1.00	.92	.80	.73	.68	.64	.61	.59	.57
				ADP	64.2	63	61	59	57	54	51	48	39
	55	72.3	99.4	ESHF	1.00	.92	.83	.73	.67	.60	.57	.56	.54
				ADP	66.9	66	65	63	61	57	54	52	47
	60	73.9	108.8	ESHF	1.00	.92	.76	.67	.61	.56	.54	.52	.50
				ADP	69.5	69	67	65	63	60	58	55	49
	65	75.5	118.2	ESHF	1.00	.88	.69	.61	.56	.53	.50	.48	.47
				ADP	71.9	71	69	67	65	63	61	58	54
	70	77.0	127.6	ESHF	1.00	.81	.63	.55	.51	.49	.47	.45	.43
				ADP	74.0	73	71	69	67	66	64	62	55

DB (F)	RH (%)	WB (F)	W (gr/lb)										
82	35	63.3	57.0	ESHF	1.00	.92	.88	.84	.80	.76	.74	.72	.71
				ADP	51.6	49	48	46	43	39	36	31	27
	40	65.0	65.1	ESHF	1.00	.90	.87	.82	.78	.74	.71	.69	.67
				ADP	55.2	53	52	50	48	45	41	38	31

DB (F)	RH (%)	WB (F)	W (gr/lb)										
82	45	66.7	73.5	ESHF	1.00	.91	.87	.80	.75	.72	.68	.65	.63
				ADP	58.5	57	56	54	52	50	46	41	53
	50	68.3	81.9	ESHF	1.00	.90	.80	.74	.70	.64	.62	.60	.59
				ADP	61.5	60	58	56	54	50	47	42	37
	55	69.8	90.2	ESHF	1.00	.90	.83	.74	.68	.64	.61	.58	.56
				ADP	64.2	63	62	60	58	56	54	50	44
	60	71.3	98.5	ESHF	1.00	.92	.76	.68	.63	.59	.56	.54	.52
				ADP	66.7	66	64	62	60	58	55	52	44
	65	72.8	107.0	ESHF	1.00	.86	.71	.63	.58	.54	.52	.51	.49
				ADP	69.1	68	66	64	62	60	58	56	51
	70	74.2	115.5	ESHF	1.00	.80	.71	.65	.60	.54	.51	.48	.46
				ADP	71.2	70	69	68	67	65	63	60	56

DB (F)	RH (%)	WB (F)	W (gr/lb)										
81	35	62.5	55.2	ESHF	1.00	.94	.89	.84	.81	.77	.75	.73	.71
				ADP	50.8	49	47	45	43	39	36	32	21
	40	64.2	63.2	ESHF	1.00	.94	.87	.82	.78	.75	.72	.69	.67
				ADP	54.4	53	51	49	47	45	41	36	23
	45	65.9	71.2	ESHF	1.00	.96	.91	.83	.78	.74	.70	.67	.64
				ADP	57.6	57	56	54	52	50	47	43	36
	50	67.5	79.0	ESHF	1.00	.90	.84	.80	.74	.70	.66	.62	.60
				ADP	60.5	59	58	57	55	53	50	45	38
	55	69.0	87.4	ESHF	1.00	.90	.77	.71	.66	.62	.60	.58	.56
				ADP	63.2	62	60	58	56	53	51	47	35
	60	70.5	95.4	ESHF	1.00	.92	.77	.68	.63	.59	.56	.54	.53
				ADP	65.8	65	63	61	59	56	53	50	46
	65	71.9	103.7	ESHF	1.00	.85	.76	.71	.66	.60	.56	.52	.50
				ADP	68.2	67	66	65	64	62	60	56	52
	70	73.3	111.9	ESHF	1.00	.80	.71	.61	.55	.52	.48	.47	.46
				ADP	70.3	69	68	66	64	62	58	56	52

DB (F)	RH (%)	WB (F)	W (gr/lb)										
80	20	56.4	30.4	ESHF	1.00	.98	.95	.93	.91	.89	.88	.87	.86
				ADP	35.4	34	32	30	28	26	24	22	20
	25	58.3	38.0	ESHF	1.00	.96	.93	.90	.88	.86	.84	.82	.81
				ADP	40.9	39	37	35	33	31	28	24	21
	30	60.0	45.6	ESHF	1.00	.96	.91	.88	.85	.83	.80	.78	.76
				ADP	45.7	44	42	40	38	36	32	28	21
	35	61.8	53.5	ESHF	1.00	.94	.88	.85	.82	.79	.77	.73	.72
				ADP	49.8	48	46	44	42	40	37	29	24
	40	63.5	61.2	ESHF	1.00	.94	.90	.84	.80	.76	.73	.70	.68
				ADP	53.5	52	51	49	47	44	41	36	28
	45	65.1	68.9	ESHF	1.00	.96	.87	.81	.76	.73	.70	.67	.64
				ADP	56.8	56	54	52	50	48	45	41	31
	50	66.7	76.7	ESHF	1.00	.89	.80	.74	.70	.66	.64	.62	.61
				ADP	59.7	58	56	54	52	49	46	42	39
	55	68.2	84.6	ESHF	1.00	.89	.82	.74	.69	.65	.61	.59	.57
				ADP	62.3	61	60	58	56	54	50	47	40
	60	69.6	92.3	ESHF	1.00	.91	.83	.72	.66	.62	.59	.57	.54
				ADP	64.8	64	63	61	59	57	55	53	47
	65	71.1	100.4	ESHF	1.00	.85	.76	.71	.63	.58	.55	.52	.50
				ADP	67.2	66	65	64	62	60	58	54	47
	70	72.4	108.3	ESHF	1.00	.78	.71	.65	.61	.55	.52	.49	.47
				ADP	69.4	68	67	66	65	63	61	58	53

See page 147 for notes.

TABLE 65—APPARATUS DEWPOINTS (Continued)

79 - 72 F DB

DB 79

DB (F)	RH (%)	WB (F)	W (gr/lb)										
79	35	61.0	51.5	ESHF	1.00	.96	.91	.89	.85	.82	.78	.75	.73
				ADP	48.9	48	46	45	43	41	37	32	26
	40	62.7	59.2	ESHF	1.00	.97	.90	.84	.80	.76	.74	.71	.69
				ADP	52.7	52	50	48	46	43	41	36	29
	45	64.3	66.7	ESHF	1.00	.91	.83	.78	.75	.72	.70	.67	.65
				ADP	55.9	54	52	50	48	46	44	39	32
	50	65.9	74.2	ESHF	1.00	.89	.80	.75	.71	.68	.66	.63	.61
				ADP	58.9	57	55	53	51	49	47	42	33
	55	67.4	81.9	ESHF	1.00	.96	.82	.74	.69	.66	.63	.60	.58
				ADP	61.4	61	59	57	55	53	51	47	41
	60	68.8	89.3	ESHF	1.00	.90	.76	.69	.64	.61	.57	.55	.54
				ADP	63.9	63	61	59	57	55	51	47	41
	65	70.2	97.0	ESHF	1.00	.84	.71	.64	.59	.56	.54	.52	.51
				ADP	66.3	65	63	61	59	57	55	51	48
	70	71.6	104.8	ESHF	1.00	.81	.71	.65	.58	.54	.52	.50	.48
				ADP	68.5	67	66	65	63	61	59	57	53

DB 78

DB (F)	RH (%)	WB (F)	W (gr/lb)										
78	35	60.3	50.0	ESHF	1.00	.96	.91	.87	.83	.79	.77	.75	.73
				ADP	48.2	47	45	43	41	37	35	31	22
	40	61.9	57.3	ESHF	1.00	.93	.87	.82	.79	.77	.73	.71	.69
				ADP	51.7	50	48	46	44	42	38	34	25
	45	63.5	64.6	ESHF	1.00	.95	.86	.81	.76	.74	.70	.68	.66
				ADP	55.0	54	52	50	48	46	42	39	34
	50	65.0	71.9	ESHF	1.00	.94	.83	.76	.73	.70	.67	.64	.62
				ADP	57.9	57	55	53	51	49	47	42	36
	55	66.6	79.2	ESHF	1.00	.96	.83	.75	.70	.65	.62	.60	.59
				ADP	60.5	60	58	56	54	51	48	44	41
	60	67.9	86.4	ESHF	1.00	.90	.82	.76	.69	.64	.60	.57	.55
				ADP	63.0	62	61	60	58	56	53	49	42
	65	69.3	93.8	ESHF	1.00	.85	.77	.71	.67	.62	.58	.54	.52
				ADP	65.2	64	63	62	61	59	57	53	48
	70	70.6	101.2	ESHF	1.00	.71	.66	.62	.59	.55	.52	.50	.48
				ADP	67.5	65	64	63	62	60	58	55	48

DB 77

DB (F)	RH (%)	WB (F)	W (gr/lb)										
77	35	59.6	48.3	ESHF	1.00	.96	.91	.87	.83	.79	.77	.75	.74
				ADP	47.3	46	44	42	40	36	33	28	24
	40	61.2	55.5	ESHF	1.00	.96	.89	.84	.81	.78	.76	.73	.70
				ADP	50.9	50	48	46	44	42	40	36	27
	45	62.7	62.4	ESHF	1.00	.94	.86	.81	.77	.74	.72	.69	.66
				ADP	54.1	53	51	49	47	45	43	39	29
	50	64.2	69.7	ESHF	1.00	.94	.84	.77	.73	.70	.68	.65	.63
				ADP	57.0	56	54	52	50	48	46	42	37
	55	65.6	76.6	ESHF	1.00	.95	.83	.75	.70	.67	.63	.61	.59
				ADP	59.6	59	57	55	53	51	48	44	37
	60	67.1	83.6	ESHF	1.00	.89	.82	.77	.73	.67	.62	.58	.56
				ADP	62.0	61	60	59	58	56	53	48	43
	65	68.5	90.8	ESHF	1.00	.84	.72	.64	.60	.57	.55	.54	.53
				ADP	64.4	63	61	59	57	55	53	51	48
	70	69.8	97.9	ESHF	1.00	.79	.66	.60	.55	.53	.51	.50	.49
				ADP	66.5	65	63	61	59	57	55	53	49

DB 76

DB (F)	RH (%)	WB (F)	W (gr/lb)										
76	35	58.9	46.7	ESHF	1.00	.96	.91	.87	.84	.81	.79	.77	.74
				ADP	46.3	45	43	41	39	37	34	31	21
	40	60.4	53.7	ESHF	1.00	.96	.89	.84	.81	.78	.76	.72	.70
				ADP	49.9	49	47	45	43	41	39	32	22
	45	61.9	60.4	ESHF	1.00	.94	.86	.81	.77	.74	.71	.69	.67
				ADP	53.2	52	50	48	46	44	40	37	31
	50	63.4	67.4	ESHF	1.00	.93	.83	.77	.73	.69	.67	.65	.63
				ADP	56.2	55	53	51	49	46	43	40	32
	55	64.9	74.0	ESHF	1.00	.94	.82	.75	.70	.67	.65	.62	.60
				ADP	58.7	58	56	54	52	50	48	44	38
	60	66.2	80.9	ESHF	1.00	.90	.77	.70	.66	.62	.60	.58	.57
				ADP	61.1	60	58	56	54	52	49	46	43
	65	67.6	87.6	ESHF	1.00	.84	.72	.65	.61	.58	.56	.54	.53
				ADP	63.4	62	60	58	56	54	52	48	43
	70	68.9	94.6	ESHF	1.00	.80	.67	.60	.56	.54	.52	.51	.50
				ADP	65.5	64	62	60	58	56	54	52	49

DB 75

DB (F)	RH (%)	WB (F)	W (gr/lb)										
75	20	53.2	25.7	ESHF	1.00	.98	.96	.94	.92	.90	.89		
				ADP	31.5	30	28	26	24	22	20		
	25	54.8	32.1	ESHF	1.00	.95	.92	.90	.88	.86	.84		
				ADP	36.9	34	32	30	28	25	21		
	30	56.5	38.5	ESHF	1.00	.97	.93	.90	.87	.85	.82	.80	.79
				ADP	41.4	40	38	36	34	32	28	24	20
	35	58.1	45.2	ESHF	1.00	.96	.91	.87	.84	.80	.78	.76	.75
				ADP	45.5	44	42	40	38	34	31	27	22
	40	59.6	51.8	ESHF	1.00	.96	.89	.84	.81	.79	.76	.73	.71
				ADP	49.1	48	46	44	42	40	37	32	24
	45	61.1	58.2	ESHF	1.00	.94	.87	.81	.77	.75	.72	.69	.67
				ADP	52.2	51	49	47	45	43	40	35	21
	50	62.6	65.0	ESHF	1.00	.92	.84	.78	.74	.71	.69	.66	.64
				ADP	55.2	54	52	50	48	46	44	40	34
	55	64.0	71.5	ESHF	1.00	.94	.87	.78	.73	.69	.65	.63	.61
				ADP	57.8	57	56	54	52	50	47	44	39
	60	65.3	77.9	ESHF	1.00	.90	.77	.71	.66	.63	.61	.59	.58
				ADP	60.1	59	57	55	53	51	49	46	43
	65	66.7	84.8	ESHF	1.00	.84	.72	.65	.61	.59	.57	.55	.54
				ADP	62.4	61	59	57	55	53	51	48	44
	70	68.0	91.2	ESHF	1.00	.80	.73	.68	.61	.57	.54	.52	.51
				ADP	64.5	63	62	61	59	57	55	52	49

DB 72

DB (F)	RH (%)	WB (F)	W (gr/lb)										
72	35	55.9	40.8	ESHF	1.00	.98	.93	.89	.86	.83	.81	.79	.77
				ADP	42.8	42	40	38	36	34	31	28	22
	40	57.3	46.7	ESHF	1.00	.95	.92	.87	.84	.81	.77	.75	.73
				ADP	46.3	45	44	42	40	38	34	30	23
	45	58.7	52.7	ESHF	1.00	.94	.87	.82	.79	.76	.74	.71	.69
				ADP	49.5	48	46	44	42	40	38	32	22
	50	60.1	58.8	ESHF	1.00	.92	.88	.81	.77	.73	.70	.68	.66
				ADP	52.4	51	50	48	46	43	40	37	30
	55	61.4	64.4	ESHF	1.00	.93	.83	.77	.72	.68	.66	.64	.63
				ADP	54.9	54	52	50	48	45	42	39	36
	60	62.7	70.2	ESHF	1.00	.89	.79	.72	.68	.65	.63	.61	.60
				ADP	57.3	56	54	52	50	48	46	42	39

The column headers for each group are: **ROOM CONDITIONS** (DB (F), RH (%), WB (F), W (gr/lb)) and **EFFECTIVE SENSIBLE HEAT FACTOR AND APPARATUS DEWPOINT***

See page 147 for notes.

TABLE 65—APPARATUS DEWPOINTS (Continued)

72 - 55 F DB

DB (F)	RH (%)	WB (F)	W (gr/lb)										
72	65	64.0	76.3	ESHF	1.00	.84	.73	.67	.63	.61	.59	.58	
				ADP	59.5	58	56	54	52	50	48	47	
	70	65.2	82.3	ESHF	1.00	.80	.69	.62	.59	.56	.54	.53	
				ADP	61.6	60	58	56	54	51	48	44	

DB (F)	RH (%)	WB (F)	W (gr/lb)										
70	20	49.9	21.6	ESHF	1.00	.98	.96	.94	.93				
				ADP	27.6	26	24	22	21				
	25	51.5	27.0	ESHF	1.00	.97	.94	.92	.90	.88			
				ADP	33.7	31	29	27	25	22			
	30	53.0	32.8	ESHF	1.00	.98	.94	.91	.88	.86	.84	.82	
				ADP	37.1	36	34	32	30	27	25	20	
	35	54.4	38.0	ESHF	1.00	.97	.93	.89	.86	.84	.82	.80	.78
				ADP	41.1	40	38	36	34	32	30	27	22
	40	55.8	43.5	ESHF	1.00	.95	.90	.86	.83	.80	.78	.76	.74
				ADP	44.5	43	41	39	37	35	32	29	22
	45	57.1	49.1	ESHF	1.00	.93	.87	.82	.79	.77	.75	.73	.71
				ADP	47.7	46	44	42	40	38	36	33	27
	50	58.5	54.8	ESHF	1.00	.92	.84	.80	.76	.74	.71	.69	.67
				ADP	50.5	49	47	45	43	41	38	35	25
	55	59.7	60.1	ESHF	1.00	.93	.83	.77	.73	.71	.68	.66	.64
				ADP	53.1	52	50	48	46	44	42	38	32
	60	60.9	65.5	ESHF	1.00	.89	.79	.73	.69	.66	.64	.62	.61
				ADP	55.4	54	52	50	48	46	43	40	36
	65	62.2	71.1	ESHF	1.00	.93	.78	.71	.66	.63	.61	.59	.58
				ADP	57.7	57	55	53	51	49	47	44	40
	70	63.4	76.9	ESHF	1.00	.90	.74	.66	.61	.59	.57	.56	.55
				ADP	59.8	59	57	55	53	51	49	47	45
	75	64.5	82.5	ESHF	1.00	.88	.70	.62	.57	.55	.53	.52	.51
				ADP	61.7	61	59	57	55	53	51	49	44
	80	65.7	88.0	ESHF	1.00	.87	.73	.65	.60	.54	.51	.49	.48
				ADP	63.5	63	62	61	60	58	56	53	49
	85	66.8	93.7	ESHF	1.00	.71	.56	.52	.50	.48	.47	.46	.45
				ADP	65.3	64	62	61	60	59	58	57	54
	90	67.9	99.3	ESHF	1.00	.66	.56	.50	.47	.45	.43	.42	.41
				ADP	66.9	66	65	64	63	62	61	60	56
	95	69.0	105.0	ESHF	1.00	.60	.47	.42	.39	.38	.37		
				ADP	68.5	68	67	66	65	64	62		

DB (F)	RH (%)	WB (F)	W (gr/lb)										
65	60	56.6	55.0	ESHF	1.00	.95	.84	.77	.73	.70	.68	.66	.65
				ADP	50.6	50	48	46	44	42	39	36	34
	65	57.7	59.7	ESHF	1.00	.92	.85	.80	.73	.69	.66	.64	.62
				ADP	52.9	52	51	50	48	46	44	41	37
	70	58.9	64.5	ESHF	1.00	.89	.80	.76	.69	.65	.62	.60	.58
				ADP	55.0	54	53	52	50	48	46	43	37
	75	59.9	69.2	ESHF	1.00	.88	.78	.72	.65	.61	.58	.56	.55
				ADP	56.9	56	55	54	52	50	48	45	41
	80	51.0	73.8	ESHF	1.00	.75	.68	.63	.60	.58	.55	.53	.52
				ADP	58.7	57	56	55	54	53	51	48	46
	85	62.0	78.6	ESHF	1.00	.71	.63	.58	.55	.52	.50	.49	
				ADP	60.3	59	58	57	56	54	52	50	
	90	63.0	83.2	ESHF	1.00	.70	.58	.53	.50	.48	.46	.45	
				ADP	61.9	61	60	59	58	57	55	53	
	95	64.0	88.0	ESHF	1.00	.69	.51	.46	.43	.42	.41		
				ADP	63.5	63	62	61	60	59	58		

DB (F)	RH (%)	WB (F)	W (gr/lb)										
60	60	52.3	46.2	ESHF	1.00	.94	.89	.81	.77	.74	.72	.70	.68
				ADP	46.0	45	44	42	40	38	36	34	28
	65	53.3	50.0	ESHF	1.00	.91	.86	.78	.74	.70	.69	.67	.65
				ADP	48.1	47	46	44	42	40	39	36	31
	70	54.3	53.9	ESHF	1.00	.89	.83	.74	.70	.67	.65	.63	.62
				ADP	50.1	49	48	46	44	42	40	37	34
	75	55.3	57.8	ESHF	1.00	.79	.74	.71	.68	.64	.62	.60	.59
				ADP	52.0	50	49	48	47	45	43	40	37
	80	56.3	61.7	ESHF	1.00	.85	.76	.70	.66	.61	.59	.57	.56
				ADP	53.8	53	52	51	50	48	46	44	41
	85	57.2	65.5	ESHF	1.00	.75	.67	.63	.57	.56	.54	.53	
				ADP	55.4	54	53	52	50	49	47	45	
	90	58.2	69.4	ESHF	1.00	.72	.62	.57	.54	.52	.50	.49	
				ADP	57.0	56	55	54	53	52	50	47	
	95	59.1	73.5	ESHF	1.00	.69	.55	.49	.47	.46	.45		
				ADP	58.5	58	57	56	55	54	52		

DB (F)	RH (%)	WB (F)	W (gr/lb)										
55	60	47.9	38.4	ESHF	1.00	.93	.89	.85	.80	.77	.75	.73	.71
				ADP	41.3	40	39	38	36	34	32	29	23
	65	48.8	41.4	ESHF	1.00	.91	.86	.83	.78	.74	.72	.70	.68
				ADP	43.3	42	41	40	38	36	34	31	24
	70	49.7	44.6	ESHF	1.00	.90	.84	.80	.74	.71	.69	.67	.66
				ADP	45.2	44	43	42	40	38	36	33	31
	75	50.6	48.0	ESHF	1.00	.89	.82	.74	.69	.66	.65	.64	.63
				ADP	47.1	46	45	43	41	39	37	36	34
	80	51.5	51.2	ESHF	1.00	.88	.79	.74	.67	.64	.62	.61	.60
				ADP	48.8	48	47	46	44	42	40	39	37
	85	52.4	54.5	ESHF	1.00	.77	.70	.66	.63	.60	.58	.57	
				ADP	50.4	49	48	47	46	44	42	40	
	90	53.2	57.7	ESHF	1.00	.76	.67	.61	.58	.55	.54	.53	
				ADP	52.0	51	50	49	48	46	44	41	
	95	54.2	61.2	ESHF	1.00	.69	.58	.54	.51	.49			
				ADP	53.6	53	52	51	50	48			

*The values shown in the gray areas indicate the lowest effective sensible heat factor possible without the use of reheat. This limiting condition is the lowest effective sensible heat factor line that intersects the saturation curve. Note that the room dewpoint is equal to the required apparatus dewpoint for an effective sensible heat factor of 1.0.

NOTES FOR TABLE 65:

1. **For Room Conditions Not Given;** The apparatus dewpoint may be determined from the scale on the chart, or may be calculated as shown in the following equation:

$$ESHF = \frac{1}{1 + .628 \frac{(W_{rm} - W_{adp})}{(t_{rm} - t_{adp})}}$$

This equation in more familiar form is:

$$ESHF = \frac{0.244 \, (t_{rm} - t_{adp})}{0.244 \, (t_{rm} - t_{adp}) + \frac{1076}{7000} \, (W_{rm} - W_{adp})}$$

(Cont.)

where W_{rm} = room moisture content, gr/lb of dry air

W_{adp} = moisture content at apparatus dewpoint, gr/lb of dry air

t_{rm} = room dry-bulb temperature

t_{adp} = apparatus dewpoint temperature

0.244 = specific heat of moist air at 55 F dewpoint, Btu per deg F per lb of dry air

1076 = average heat removal required to condense one pound of water vapor from the room air

7000 = grains per pound.

2. **For High Elevations.** For effective sensible heat factors at high elevations, see *Table 66*.

3. **For Apparatus Dewpoint Below Freezing.** The latent heat of fusion of the moisture removed is not included in the calculation of apparatus dewpoint below freezing or in the calculation of room load, in order to simplify estimating procedures. Use the same equation as in Note 1. The selection of equipment on a basis of 16 to 18 hour operating time provides a safety factor large enough to cover the omission of this latent heat of fusion, which is a small part of the total load.

TABLE 66—EQUIVALENT EFFECTIVE SENSIBLE HEAT FACTORS FOR VARIOUS ELEVATIONS*

For use with sea level psychrometric chart or tables

Effective Sensible Heat Factor from Air Conditioning Load Estimate	Elevation (Feet) and Barometric Pressure (Inches of Hg) at Installation									
	1000 (28.86)	2000 (27.82)	3000 (26.82)	4000 (25.84)	5000 (24.89)	6000 (23.98)	7000 (23.09)	8000 (22.12)	9000 (21.39)	10000 (20.57)
	Equivalent Effective Sensible Heat Factor Referred to a Sea Level Psychrometric Chart or Tables									
.95	.95	.95	.95	.96	.96	.96	.96	.96	.96	.96
.90	.90	.91	.91	.91	.92	.92	.92	.92	.93	.93
.85	.85	.86	.86	.87	.87	.88	.88	.88	.89	.89
.80	.81	.81	.82	.82	.83	.83	.84	.84	.85	.85
.75	.76	.76	.77	.78	.78	.79	.80	.80	.81	.81
.70	.71	.72	.72	.73	.74	.75	.75	.76	.77	.77
.65	.66	.67	.68	.68	.69	.70	.71	.71	.72	.73
.60	.61	.62	.63	.64	.64	.65	.66	.67	.68	.69
.55	.56	.57	.58	.59	.60	.61	.61	.62	.63	.64
.50	.51	.52	.53	.54	.55	.56	.57	.57	.58	.59

*Values obtained by use of equation

$$ESHF_e = \frac{1}{\dfrac{(p_1)\,(1-ESHF)}{(p_0)\,(ESHF)} + 1}$$

Where p_0 = barometric pressure at sea level

p_1 = barometric pressure at high elevation

ESHF = ESHF obtained from air conditioning load estimate

$ESHF_e$ = equivalent ESHF referred to a sea level psychrometric chart or *Table 66*

NOTES FOR TABLE 66:

1. The required apparatus dewpoint for the high elevation is determined from the sea level chart or *Table 65* by use of the equivalent effective sensible heat factor. The relative humidity and dry-bulb temperature must be used to define the room condition when using this table because the above equation was derived on this basis. The room wet-bulb temperature must not be used because the wet-bulb temperature corresponding to any particular condition, for example, 75 F db, 40% rh, at a high elevation is lower (except for saturation) than that corresponding to the same condition (75 F db, 40% rh) at sea level. For the same value of room relative humidity and dry-bulb temperature, and the same apparatus dewpoint, there is a greater difference in moisture content between the two conditions at high elevation than at sea level. Therefore, a higher apparatus dewpoint is required at high elevation for a given effective sensible heat factor.

2. Air conditioning load estimate (See *Fig. 44*). The factors 1.08 and .68 on the air conditioning load estimate should be multiplied by the direct ratio of the barometric pressures $\frac{(p_1)}{(p_0)}$. Using this method, it is assumed that the air quantity (cfm) is measured at actual conditions rather than at standard air conditions. The outdoor and room moisture contents, grains per pound, must also be corrected for high elevations.

3. Reheat—Where the equivalent effective sensible heat factor is lower than the shaded values in *Table 65*, reheat is required.

ABBREVIATIONS

adp	apparatus dewpoint
BF	bypass factor
(BF) (OALH)	bypassed outdoor air latent heat
(BF) (OASH)	bypassed outdoor air sensible heat
(BF) (OATH)	bypassed outdoor air total heat
Btu/hr	British thermal units per hour
cfm	cubic feet per minute
db	dry-bulb
dp	dewpoint
ERLH	effective room latent heat
ERSH	effective room sensible heat
ERTH	effective room total heat
ESHF	effective sensible heat factor
F	Fahrenheit degrees
fpm	feet per minute
gpm	gallons per minute
gr/lb	grains per pound
GSHF	grand sensible heat factor
GTH	grand total heat
GTHS	grand total heat supplement
OALH	outdoor air latent heat
OASH	outdoor air sensible heat
OATH	outdoor air total heat
rh	relative humidity
RLH	room latent heat
RLHS	room latent heat supplement
RSH	room sensible heat
RSHF	room sensible heat factor
RSHS	room sensible heat supplement
RTH	room total heat
Sat Eff	saturation efficiency of sprays
SHF	sensible heat factor
TLH	total latent heat
TSH	total sensible heat
wb	wet-bulb

SYMBOLS

cfm_{ba}	bypassed air quantity around apparatus
cfm_{da}	dehumidified air quantity
cfm_{oa}	outdoor air quantity
cfm_{ra}	return air quantity
cfm_{sa}	supply air quantity
h	specific enthalpy
h_{adp}	apparatus dewpoint enthalpy
h_{es}	effective surface temperature enthalpy
h_{ea}	entering air enthalpy
h_{la}	leaving air enthalpy
h_m	mixture of outdoor and return air enthalpy
h_{oa}	outdoor air enthalpy
h_{rm}	room air enthalpy
h_{sa}	supply air enthalpy
t	temperature
t_{adp}	apparatus dewpoint temperature
t_{edb}	entering dry-bulb temperature
t_{es}	effective surface temperature
t_{ew}	entering water temperature
t_{ewb}	entering wet-bulb temperature
t_{ldb}	leaving dry-bulb temperature
t_{lw}	leaving water temperature
t_{lwb}	leaving wet-bulb temperature
t_m	mixture of outdoor and return air dry-bulb temperature
t_{oa}	outdoor air dry-bulb temperature
t_{rm}	room dry-bulb temperature
t_{sa}	supply air dry-bulb temperature
W	moisture content or specific humidity
W_{adp}	apparatus dewpoint moisture content
W_{ea}	entering air moisture content
W_{es}	effective surface temperature moisture content
W_{la}	leaving air moisture content
W_m	mixture of outdoor and return air moisture content
W_{oa}	outdoor air moisture content
W_{rm}	room moisture content
W_{sa}	supply air moisture content

PSYCHROMETRIC FORMULAS

A. AIR MIXING EQUATIONS (Outdoor and Return Air)

$$t_m = \frac{(cfm_{oa} \times t_{oa}) + (cfm_{ra} \times t_{rm})}{cfm_{sa}} \quad (1)$$

$$h_m = \frac{(cfm_{oa} \times h_{oa}) + (cfm_{ra} \times h_{rm})}{cfm_{sa}} \quad (2)$$

$$W_m = \frac{(cfm_{oa} \times W_{oa}) + (cfm_{ra} \times W_{rm})}{cfm_{sa}} \quad (3)$$

B. COOLING LOAD EQUATIONS

$$\text{ERSH} = \text{RSH} + (\text{BF})(\text{OASH}) + \text{RSHS*} \quad (4)$$

$$\text{ERLH} = \text{RLH} + (\text{BF})(\text{OALH}) + \text{RLHS*} \quad (5)$$

$$\text{ERTH} = \text{ERLH} + \text{ERSH} \quad (6)$$

$$\text{TSH} = \text{RSH} + \text{OASH} + \text{RSHS*} \quad (7)$$

$$\text{TLH} = \text{RLH} + \text{OALH} + \text{RLHS*} \quad (8)$$

$$\text{GTH} = \text{TSH} + \text{TLH} + \text{GTHS*} \quad (9)$$

$$\text{RSH} = 1.08\dagger \times cfm_{sa} \times (t_{rm} - t_{sa}) \quad (10)$$

$$\text{RLH} = .68\dagger \times cfm_{sa} \times (W_{rm} - W_{sa}) \quad (11)$$

$$\text{RTH} = 4.45\dagger \times cfm_{sa} \times (h_{rm} - h_{sa}) \quad (12)$$

$$\text{RTH} = \text{RSH} + \text{RLH} \quad (13)$$

$$\text{OASH} = 1.08 \times cfm_{oa}(t_{oa} - t_{rm}) \quad (14)$$

$$\text{OALH} = .68 \times cfm_{oa}(W_{oa} - W_{rm}) \quad (15)$$

$$\text{OATH} = 4.45 \times cfm_{oa}(h_{oa} - h_{rm}) \quad (16)$$

$$\text{OATH} = \text{OASH} + \text{OALH} \quad (17)$$

$$(\text{BF})(\text{OATH}) = (\text{BF})(\text{OASH}) + (\text{BF})(\text{OALH}) \quad (18)$$

$$\text{ERSH} = 1.08 \times cfm_{da}\ddagger \times (t_{rm} - t_{adp})(1 - \text{BF}) \quad (19)$$

$$\text{ERLH} = .68 \times cfm_{da}\ddagger \times (W_{rm} - W_{adp})(1 - \text{BF}) \quad (20)$$

$$\text{ERTH} = 4.45 \times cfm_{da}\ddagger \times (h_{rm} - h_{adp})(1 - \text{BF}) \quad (21)$$

$$\text{TSH} = 1.08 \times cfm_{da}\ddagger \times (t_{edb} - t_{ldb})** \quad (22)$$

$$\text{TLH} = .68 \times cfm_{da}\ddagger \times (W_{ea} - W_{la})** \quad (23)$$

$$\text{GTH} = 4.45 \times cfm_{da}\ddagger \times (h_{ea} - h_{la})** \quad (24)$$

C. SENSIBLE HEAT FACTOR EQUATIONS

$$\text{RSHF} = \frac{\text{RSH}}{\text{RSH} + \text{RLH}} = \frac{\text{RSH}}{\text{RTH}} \quad (25)$$

$$\text{ESHF} = \frac{\text{ERSH}}{\text{ERSH} + \text{ERLH}} = \frac{\text{ERSH}}{\text{ERTH}} \quad (26)$$

$$\text{GSHF} = \frac{\text{TSH}}{\text{TSH} + \text{TLH}} = \frac{\text{TSH}}{\text{GTH}} \quad (27)$$

D. BYPASS FACTOR EQUATIONS

$$\text{BF} = \frac{t_{ldb} - t_{adp}}{t_{edb} - t_{adp}}; \ (1 - \text{BF}) = \frac{t_{edb} - t_{ldb}}{t_{edb} - t_{adp}} \quad (28)$$

$$\text{BF} = \frac{W_{la} - W_{adp}}{W_{ea} - W_{adp}}; \ (1 - \text{BF}) = \frac{W_{ea} - W_{la}}{W_{ea} - W_{adp}} \quad (29)$$

$$\text{BF} = \frac{h_{la} - h_{adp}}{h_{ea} - h_{adp}}; \ (1 - \text{BF}) = \frac{h_{ea} - h_{la}}{h_{ea} - h_{adp}} \quad (30)$$

E. TEMPERATURE EQUATIONS AT APPARATUS

$$t_{edb}** = \frac{(cfm_{oa} \times t_{oa}) + (cfm_{ra} \times t_{rm})}{cfm_{sa}\ddagger} \quad (31)$$

$$t_{ldb} = t_{adp} + \text{BF}\,(t_{edb} - t_{adp}) \quad (32)$$

t_{ewb} and t_{lwb} correspond to the calculated values of h_{ea} and h_{la} on the psychrometric chart.

$$h_{ea}** = \frac{(cfm_{oa} \times h_{oa}) + (cfm_{ra} \times h_{rm})}{cfm_{sa}\ddagger} \quad (33)$$

$$h_{la} = h_{adp} + \text{BF}\,(h_{ea} - h_{adp}) \quad (34)$$

F. TEMPERATURE EQUATIONS FOR SUPPLY AIR

$$t_{sa} = t_{rm} - \frac{\text{RSH}}{1.08\,(cfm_{sa}\ddagger)} \quad (35)$$

---

*RSHS, RLHS and GTHS are supplementary loads due to duct heat gain, duct leakage loss, fan and pump horsepower gains, etc. To simplify the various examples, these supplementary loads have *not* been used in the calculations. However, in actual practice, these supplementary loads should be used where appropriate. *Chapter 7* gives the values for the various supplementary loads. *Fig. 1, Chapter 1*, illustrates the method of accounting for these supplementary loads on the air conditioning load estimate.

†*Item H, page 151*, gives the derivation of these air constants.

‡When no air is to be physically bypassed around the conditioning apparatus, $cfm_{da} = cfm_{sa}$.

**When t_m, W_m and h_m are equal to the entering conditions at the cooling apparatus, they may be substituted for t_{edb}, W_{ea} and h_{ea} respectively.

G. AIR QUANTITY EQUATIONS

$$cfm_{da} = \frac{ERSH}{1.08 \times (1 - BF)(t_{rm} - t_{adp})} \tag{36}$$

$$cfm_{da} = \frac{ERLH}{.68 \times (1 - BF)(W_{rm} - W_{adp})} \tag{37}$$

$$cfm_{da} = \frac{ERTH}{4.45 \times (1 - BF)(h_{rm} - h_{adp})} \tag{38}$$

$$cfm_{da\ddagger} = \frac{TSH}{1.08(t_{edb} - t_{ldb})} \tag{39}$$

$$cfm_{da\ddagger} = \frac{TLH}{.68(W_{ea} - W_{la})} \tag{40}$$

$$cfm_{da\ddagger} = \frac{GTH}{4.45(h_{ea} - h_{la})} \tag{41}$$

$$cfm_{sa} = \frac{RSH}{1.08 \times (t_{rm} - t_{sa})} \tag{42}$$

$$cfm_{sa} = \frac{RLH}{.68 \times (W_{rm} - W_{sa})} \tag{43}$$

$$cfm_{sa} = \frac{RTH}{4.45 \times (h_{rm} - h_{sa})} \tag{44}$$

$$cfm_{ba} = cfm_{sa} - cfm_{da} \tag{45}$$

Note: cfm_{da} will be less than cfm_{sa} only when air is physically bypassed around the conditioning apparatus.

$$cfm_{sa} = cfm_{oa} + cfm_{ra} \tag{46}$$

H. DERIVATION OF AIR CONSTANTS

$$1.08 = .244 \times \frac{60}{13.5}$$

where .244 = specific heat of moist air at 70 F db and 50% rh, Btu/(deg F) (lb dry air)

60 = min/hr

13.5 = specific volume of moist air at 70 F db and 50% rh

$$.68 = \frac{60}{13.5} \times \frac{1076}{7000}$$

where 60 = min/hr

13.5 = specific volume of moist air at 70 F db and 50% rh

1076 = average heat removal required to condense one pound of water vapor from the room air

7000 = grains per pound

$$4.45 = \frac{60}{13.5}$$

where 60 = min/hr

13.5 = specific volume of moist air at 70 F db and 50% rh

BIBLIOGRAPHY

CHAPTER 2

TABLE 1

1. Air Conditioning and Refrigeration Institute, Application Engineering Standard 530-56 Air Conditioning. 1956.
2. *Heating, Ventilating and Air Conditioning Guide,* Chapters 12 and 13, 1956.
3. Summer Weather Data — Marley Company.

TABLE 4

1. Conditions for Comfort, by C. S. Leopold; *Heating, Piping and Air Conditioning,* June 1947, p. 117.
2. The Mechanism of Heat Loss and Temperature Regulation, by E. F. DuBois; Transactions of the Association of American Physicians, Vol. 51, 1936, p. 252.
3. The Relative Influence of Radiation and Convection upon the Temperature Regulation of the Clothed Body, by C. E. A. Winslow, L. P. Herrington, and A. P. Gagge; *American Journal of Physiology,* Vol. 124, October 1938, p. 51.
4. Reactions of Office Workers to Air Conditioning in South Texas, by A. J. Rummel, F. E. Giesecke, W. H. Badgett, and A. T. Moses; *ASHVE Trans.,* Vol. 45, 1939, p. 459.
5. Shock Experiences of 275 Workers After Entering and Leaving Cooled and Air Conditioned Offices, by A. B. Newton, F. C. Houghten, C. Gutherlet, R. W. Qualley, and M. C. W. Tomlinson; *ASHVE Trans.,* Vol. 44, 1938, p. 571.
6. How to Use the Effective Temperature Index and Comfort Charts, by C. P. Yaglou, W. H. Carrier, Dr. E. V. Hill, F. C. Houghten, and J. H. Walker; *ASHVE Trans.,* 1932, p. 411.
7. Heat and Moisture Losses from the Human Body and Their Relation to Air Conditioning Problems, by F. C. Houghten, W. W. Teague, W. E. Miller and W. P. Yant; *ASHVE Trans.,* 1929, p. 245.
8. Thermal Exchanges Between the Human Body and Its Atmospheric Environment, by F. C. Houghten, W. W. Teague, W. E. Miller and W. P. Yant; *American Journal of Physiology,* Vol. 88, 1929, p. 386.

TABLE 5

1. *Heating, Ventilating and Air Conditioning Engineers Guide,* Chapter 45, 1956.
2. *Air Conditioning and Refrigerating Data Book,* 1955.

CHAPTER 3

TABLES 6 THRU 12

1. *Heat Transfer,* by Max Jakob, Vol. 1, John Wiley & Sons, Inc., New York, N. Y., 1949.
2. The Solution of Transient Heat Conduction Problems by Finite Differences, by G. A. Hawkins and J. T. Agnew. Purdue Univ., Eng. Exp. Sta. Bulletin No. 98, 1946.
3. Hydraulic Analogue for the Solution of Problems of Thermal Storage, Radiation, Convection and Conduction, by C. S. Leopold; *ASHAE Transactions,* Vol. 54, 1948, p. 389.
4. Circuit Analysis Applied to Load Estimating, by H. B. Nottage and G. V. Parmelee; *ASHAE Transactions,* Vol. 60, 1954, pp. 59-102.

5. Circuit Analysis Applied to Load Estimating, Part II—Influence of Transmitted Solar Radiation, by H. B. Nottage and G. V. Parmelee; *ASHAE Transactions,* Vol. 61, 1955, pp. 128-139.
6. Thermal Circuit Analysis for Developing Application Engineering Information, by Stanley F. Gilman and O. William Clausen; *Heating, Piping and Air Conditioning,* June 1957, pp. 153-160.
7. Temperature Changes in Refrigerated Rooms During Pulldown Period, by J. L. Threlkeld and T. Kusada; *Refrigerating Engineering,* July 1956, p. 35.
8. Heat Transmission as influenced by Heat Capacity and Solar Radiation by F. C. Houghton, J. L. Blackshaw, E. M. Pugh, and P. McDemott; *ASHAE Transactions,* Vol. 38, 1932, p. 263.
9. Cooling Load from Sunlit Glass, by C. O. Mackey and N. R. Gay; *ASHAE Transactions,* Vol. 58, 1952, p. 321.
10. Analysis of an Air Conditioning Thermal Circuit by an Electronic Differential Analyzer, by G. V. Parmelee, P. Vance, and A. N. Cerny; *Heating, Piping and Air Conditioning,* Sept. 1956, p. 117.

TABLE 15

1. The Transmission of Solar Radiation Through Flat Glass Under Summer Conditions, by G. V. Parmelee; *Heating, Piping and Air Conditioning,* October-November 1945. Also *ASHVE Trans.,* 1945, Vol. 51, p. 317.
2. Measurements of Solar Radiation Intensity and Determinations of its Depletion by the Atmosphere, by H. H. Kimball; *Monthly Weather Review,* February 1930, Vol. 58, p. 52.
3. Review of United States Weather Bureau Solar Radiation Investigations, by I. F. Hand; *Monthly Weather Review,* December 1937, Vol. 65, p. 430.
4. *Smithsonian Meteorological Tables,* 5th Revised edition, p. LXXXIV.
5. Pyrheliometers and Pyrheliometric Measurements, by I. F. Hand; *U. S. Weather Bureau,* November 1946.
6. Proposed Standard Solar Radiation Curves for Engineering Use, by Parry Moon; *Journal of the Franklin Institute,* November 1940, Vol. 230, No. 5, pp. 586-617.
7. Performance of Flat Plate Solar Heat Collectors, by H. C. Hottel and B. B. Woertz; *ASME Trans.,* February 1942, Vol. 64, pp. 91-104.
8. Where is the Sun?, by M. J. Wilson and J. M. Van Swaay; *Heating and Ventilating,* May and June 1942.
9. Summer Weather Data and Sol-Air Temperature — Study of Data for New York City, by C. O. Mackey and E. B. Watson; *Heating, Piping and Air Conditioning,* Nov. 1944, p. 651. Also *ASHVE Trans.,* 1945, Vol. 51, p. 75.
10. Summer Weather Data and Sol-Air Temperature — Study of Data for Lincoln, Nebraska, by C. O. Mackey and E. B. Watson; *Heating, Piping and Air Conditioning,* January 1945, p. 42. Also *ASHVE Trans.,* 1945, Vol. 51, p. 93.

TABLE 16

1. An Experimental Study of Flat-type Sun Shades, by G. V. Parmalee, W. W. Aubele and D. J. Vild; *Heating, Piping and Air Conditioning,* January 1953.

2. Design Data for Slat-type Sun Shades for Use in Load Estimating, by G. V. Parmalee and D. J. Vild; *Heating, Piping and Air Conditioning*, September 1953.

3. The Transmission of Solar Radiation Through Flat Glass under Summer Conditions, by G. V. Parmalee; *Heating, Piping and Air Conditioning*, October, November 1945. Also *ASHVE Trans.*, 1945, Vol. 51, p. 317.

4. Heat Gain Through Western Windows With and Without Shading, by F. C. Houghten and David Shore; *Heating, Piping and Air Conditioning*, April 1941, p. 256. Also *ASHVE Trans.*, 1941, pp. 251-274.

5. Studies of Solar Radiation Through Bare and Shaded Windows, by F. C. Houghten, C. Gutberlet, and J. Blackshaw; *ASHVE Trans.*, 1934, Vol. 40, pp. 101-116.

6. Solar Heat Gain Factors For Windows With Drapes, by N. Ozisik and L. F. Schutrum; paper presented at ASHRAE meeting, Dallas, Texas, February, 1-4, 1960.

TABLE 17

1. Heat Gain Through Glass Blocks by Solar Radiation and Transmittance, by F. C. Houghten, David Shore, H. J. Olson and Burt Gunst; *ASHVE Trans.*, 1940, pp. 83-107.

CHART 1 AND TABLE 18

1. Tables of Computed Altitude and Azimuth, Volume I to V inclusive (0° to 50° Latitude, 10° per volume); U. S. Navy Department — Hydrographic Office, No. 214.

2. Where is the Sun?, by M. J. Wilson and J. M. Van Swaay; *Heating and Ventilating*, May, June 1942.

CHAPTER 5

TABLES 19 AND 20

1. Periodic Heat Flow — Homogeneous Walls or Roofs, by C. O. Mackey and L. T. Wright, Jr.; *Heating, Piping and Air Conditioning*, September 1944, p. 546. Also *ASHVE Trans.*, 1944, Vol. 50, p. 293.

2. Periodic Heat Flow — Composite Walls or Roofs, by C. O. Mackey and L. T. Wright, Jr.; *Heating, Piping and Air Conditioning*, June 1946, p. 107.

3. Summer Cooling Load as Affected by Heat Gain Through Dry, Sprinkled, and Water Covered Roofs, by F. C. Houghten, H. T. Olson, and Carl Gutberlet; *Heating, Piping and Air Conditioning*, July 1940. Also *ASHVE Trans.*, 1940, p. 231.

4. Summer Weather Data and Sol-Air Temperature — Study of Data for New York City, by C. O. Mackey and E. B. Watson; *Heating, Piping and Air Conditioning*, November 1944, p. 651. Also *ASHVE Trans.*, 1945, Vol. 51, p. 75.

5. Summer Weather Data and Sol-Air Temperature — Study of Data for Lincoln, Nebraska, by C. O. Mackey and E. B. Watson; *Heating, Piping and Air Conditioning*, January 1945, p. 42. Also *ASHVE Trans.*, 1945, Vol. 51, p. 93.

6. Estimating Heat Flow Through Sunlit Walls, by C. O. Mackey and L. T. Wright; *Heating and Ventilating*, March, April, May 1940.

7. Heat Transmission As Influenced by Heat Capacity and Solar Radiation, by F. C. Houghten and others; *ASHVE Trans.*, 1932, pp. 231-284.

8. Heat Gain Through Walls and Roofs as Affected by Solar Radiation, by F. C. Houghten and others; *ASHVE Trans.*, 1942, Vol. 48, pp. 21-105.

9. Solar Heat Gain Through Walls and Roofs for Cooling Load Calculations, by James P. Stewart; *Heating, Piping and Air Conditioning*, August 1948.

TABLES 35 AND 36

1. Heat Loss Through Basement Walls and Floors, by F. C. Houghten, S. I. Taimuty, C. Gutberlet and C. J. Brown; *ASHVE Trans.*, Vol. 48, 1942, p. 369.

2. Measurements of Heat Losses From Slab Floors, by R. S. Dill, W. C. Robinson and H. E. Robinson; U. S. National Bureau of Standards, Report BMS103, 1945.

3. Heat Losses Through Floors of Basementless Building; *Heating and Ventilating*, Vol. 42, September 1945, p. 89.

TABLE 38

1. Ice Formation on Pipe Surfaces by S. Lewis Elmer, Jr.; *Refrigerating Engineering*, July 1932, p. 17.

2. Notes on the Formation of Ice on Pipe Surfaces by F. Raseri; *Refrigerating Engineering*, January 1933, p. 21.

TABLE 40

1. Comparative Resistance to Vapor Transmission of Various Building Materials, by L. V. Teesdale; *ASHVE Trans.*, Vol. 49, 1943, p. 124.

2. The Relation of Wall Construction to Moisture Accumulation in Fill Type Insulation, by Henry J. Barre; Research Bulletin 271, Agricultural Experiment Station, Iowa State College; Ames, Iowa, April 1940.

3. Vapor Transmission Analysis of Structural Insulating Board, by F. P. Rowley and C. E. Lund; Bulletin No. 22, University of Minnesota Engineering Experiment Station, October 1944.

4. The Diffusion of Water Vapor Through Various Building Materials by J. D. Babbitt; *Canadian Journal of Research*, Vol. 17, Sec. A, pp. 15-32, February 1939. Also Permeability of Building Paper to Water Vapor, by J. D. Babbitt; *Canadian Journal of Research*, Vol. 18, Sect. A, May 1940, pp. 90-97.

5. Comparison of Methods for the Determination of Water Vapor Permeability, by Sears, Schlagenhauf, Givens and Yett; *Paper Trade Journal*, Vol. 118, TAPPI Section, pp. 27-28, January 20, 1944.

6. Comparison of Methods for the Determination of Water Vapor Permeability, by C. J. Weber; *Paper Trade Journal*, Vol. 118, TAPPI Section, pp. 24-26, January 20, 1944.

7. International Critical Tables.

8. Vapor Barriers with Annotated Bibliography, by J. Louis York, University of Michigan, for Office of Production, Research and Development, War Production Board, Washington, D. C., February 1, 1945.

9. How to Overcome Condensation in Building Walls and Attics, by L. V. Teesdale; *Heating and Ventilating*, April 1939.

10. Moisture Condensation in Building Walls, by H. W. Wooley; National Bureau of Standards, Report BMS63, 1940.

CHART 2

1. Preventing Condensation on Interior Building Surfaces, by P. D. Close; *ASHVE Trans.*, Vol. 36, 1930.

2. Permissible Relative Humidities in Humidified Buildings, by P. D. Close; *Heating, Piping and Air Conditioning*, December 1939, p. 766.

3. Methods of Moisture Control and Their Application to Building Construction, by F. B. Rowley, A. B. Algren, and C. E. Lund; University of Minnesota Engineering Experiment Station, Bulletin No. 17.

4. Condensation Within Walls, by F. B. Rowley, A. B. Algren, and C. E. Lund; *ASHVE Trans.*, Vol. 44, 1938.

CHAPTER 6

TABLES 41, 43 AND 44

1. The Infiltration Problem of Multiple Entrances, by A. M. Simpson and K. B. Atkinson; *Heating, Piping and Air Conditioning,* June 1936, p. 345.

2. Air Leakage Studies on Metal Windows in a Modern Office Building, by F. C. Houghten and M. E. O'Connell; *ASHVE Trans.,* Vol. 34, 1928, p. 321.

3. The Weathertightness of Rolled Section Steel Windows, by J. E. Emswiler and W. C. Randall; *ASHVE Trans.,* Vol. 34, 1928, p. 527.

4. Effect of Frame Calking and Storm Windows on Infiltration Around and Through Windows, by W. M. Richtmann and C. Braatz; *ASHVE Trans.,* Vol. 34, 1928, p. 547.

5. Air Infiltration Through Double-Hung Wood Windows, by G. L. Larson, D. W. Nelson and R. W. Kubasta; *ASHVE Trans.,* Vol. 37, 1931, p. 571.

6. Pressure Differences Across Windows in Relation to Wind Velocity, by J. E. Emswiler and W. C. Randall; *ASHVE Trans.,* Vol. 36, 1930, p. 83.

7. Air Infiltration Through Steel Frame Windows, by D. O. Rusk, V. H. Cherry and L. Boelter; *ASHVE Trans.,* Vol. 39, 1933, p. 169.

TABLE 45

1. Code of Minimum Requirements for Comfort Air Conditioning; *ASHVE Trans.,* Vol. 44, 1938, p. 27.

2. Ventilation Requirements, by C. P. Yaglou, E. C. Riley and D. J. Coggins; *ASHVE Trans.,* Vol. 42, 1936, p. 133.

3. Control of Physical Hazards of Anesthesia, by R. M. Tovell and A. W. Friend; *Canadian Medical Association Journal,* 46:560, 1942.

4. Air Conditioning Requirements of An Operating Room and Recovery Ward, by F. C. Houghten and W. Leigh Cook Jr.; *ASHVE Trans.,* Vol. 45, 1939, p. 161.

5. Code of Minimum Requirements for Heating and Ventilating Garages; *ASHVE Trans.,* Vol. 41, 1935, p. 30.

CHAPTER 7

TABLE 48

1. Heat and Moisture Losses From the Human Body and Their Relation to Air Conditioning Problems, by F. C. Houghten, W. W. Teague, W. E. Miller and W. P. Yant; *ASHVE Trans.,* Vol. 35, 1929, p. 245.

2. Thermal Exchanges Between the Human Body and Its Atmospheric Environment, by F. C. Houghten, W. W. Teague, W. E. Miller and W. P. Yant; *American Journal of Physiology,* Vol. 88, 1929, p. 386.

3. Heat and Moisture Losses From Men at Work and Application to Air Conditioning Problems, by F. C. Houghten, W. W. Teague, W. E. Miller and W. P. Yant; *ASHVE Trans.,* Vol. 37, 1931, p. 54.

4. Air Conditioning in Industry, by W. L. Fleischer, A. E. Stacey, Jr., F. C. Houghten and M. B. Ferberber; *ASHVE Trans.,* Vol. 45, 1939, p. 59.

5. *Physiological Basis of Medical Practice,* Best and Taylor.

6. Tables, Factors and Formulas for Computing Respiratory Exchange and Biological Transformations of Energy, by Thorne M. Carpenter; Carnegie Institution, Washington, D. C.

TABLE 49

1. Westinghouse data (1952) for 110v to 125v, 60 cycle, a.c. current.

TABLE 50

1. *Exhaust Hoods,* by J. M. Dalla Valle.

2. Reducing Heat Loads in Industrial Air Conditioning, by L. R. St. Onge; *Refrigerating Engineering,* January 1946, p. 35.

TABLE 51

1. *Helpful Hints to Fried Food Fame,* by the Edison General Electric Appliance Company.

TABLE 53

1. Motors and Generators, National Electric Manufacturers Association Standards Publication, No. MG-1 — 1955, Part 4, p. 10.

TABLE 54

1. Heat Loss from Copper Piping, by R. H. Heilman; *Heating, Piping and Air Conditioning,* September 1935, p. 458.

CHAPTER 8

Rational Psychrometric Formulae, by Willis H. Carrier; *ASME Trans.,* Vol. 23, 1911, p. 1005.

Psychrometric Factors in the Air Conditioning Estimate, by C. M. Ashley; *ASHVE Trans.,* Vol. 55, 1949.

Part 2
AIR DISTRIBUTION

CHAPTER 1. AIR HANDLING APPARATUS

This chapter describes the location and layout of air handling apparatus from the outdoor air intake thru the fan discharge on a standard air conditioning system. Construction details are also included for convenience.

Air handling apparatus can be of three types: (1) built-up apparatus where the casing for the conditioning equipment is fabricated and installed at or near the job site; (2) fan coil equipment that is manufactured and shipped to the job site, either completely or partially assembled; and (3) self-contained equipment which is shipped to the job site completely assembled.

This chapter is primarily concerned with built-up apparatus; fan coil and self-contained equipment are discussed in *Part 6*. In addition to the built-up apparatus, items such as outdoor air louvers, dampers, and fan discharge connections are also discussed in this chapter. These items are applied to all types of apparatus.

Equipment location and equipment layout must be carefully studied when designing air handling apparatus. These two items are discussed in detail in the following pages.

LOCATION

The location of the air handling apparatus directly influences the economic and sound level aspects of any system.

ECONOMIC CONSIDERATION

The air handling apparatus should be centrally located to obtain a minimum-first-cost system. In a few instances, however, it may be necessary to locate the apparatus, refrigeration machine, and cooling tower in one area, to achieve optimum system cost. When the three components are grouped in one location, the cost of extra ductwork is offset by the reduced piping cost. In addition, when the complete system becomes large enough to require more than one refrigeration machine, grouping the mechanical equipment on more than one floor becomes practical. This design is often used in large buildings. The upper floor equipment handles approximately the top 20 to 30 floors, and the lower floor equipment is used for the lower 20 to 30 floors.

Occasionally a system is designed requiring a grouping of several units in one location, and the use of a single unit in a remote location. This condition should be carefully studied to obtain the optimum coil selection-versus-piping cost for the remotely located unit. Often, the cost of extra coil surface is more than offset by the lower pipe cost for the smaller water quantity resulting when the extra surface coil is used.

SOUND LEVEL CONSIDERATIONS

It is extremely important to locate the air handling apparatus in areas where reasonable sound levels can be tolerated. Locating apparatus adjacent to conference rooms, sleeping quarters and broadcasting studios is not recommended. The following items point up the conditions that are usually created by improper location; these conditions can be eliminated by careful planning when making the initial placement of equipment:

1. The cost of correcting a sound or vibration problem after installation is much more than than the original cost of preventing it.
2. It may be impossible to completely correct the sound level, once the job is installed.
3. The owner may not be convinced even after the trouble has been corrected.

The following practices are recommended to help avoid sound problems for equipment rooms located on upper floors.

1. In new construction, locate the steel floor framing to match equipment supports designed for weights, reactions and speeds to be used. This transfers the loads to the building columns.
2. In existing buildings, use of existing floor slabs should be avoided. Floor deflection can, at times, magnify vibration in the building structure. Supplemental steel framing is often necessary to avoid this problem.
3. Equipment rooms adjacent to occupied spaces should be acoustically treated.
4. In apartments, hotels, hospitals and similar buildings, non-bearing partition walls should be separated at the floor and ceilings adjoining occupied spaces by resilient materials to avoid transmission of noise vibration.

5. Bearing walls adjacent to equipment rooms should be acoustically treated on the occupied side of the wall.

LAYOUT

Package equipment is usually factory shipped with all of the major equipment elements in one unit. With this arrangement, the installation can be completed by merely connecting the ductwork and assembling and installing the accessories.

In a central station system, however, a complete, workable and pleasing layout must be made of all major components. This involves considerations usually not present in the unitary equipment installation.

The shape and cross-sectional area of the air handling equipment are the factors that determine the dimensions of the layout. The dehumidifier assembly or the air cleaning equipment usually dictates the overall shape and dimensions. A superior air handling system design has a regular shape. A typical apparatus is shown in *Fig. 1*. The shape shown allows for a saving in sheet metal fabrication time and, therefore, is considered to be better industrial design. Its clean lines give a more workmanlike appearance. From a functional standpoint, an irregular shaped casing tends to cause air stratification and irregular flow patterns.

The most important rule in locating the equipment for the air handling apparatus is to arrange the equipment along a center line for the best air flow conditions. This arrangement keeps plenum pressure losses to a minimum, and is illustrated in *Fig. 1*.

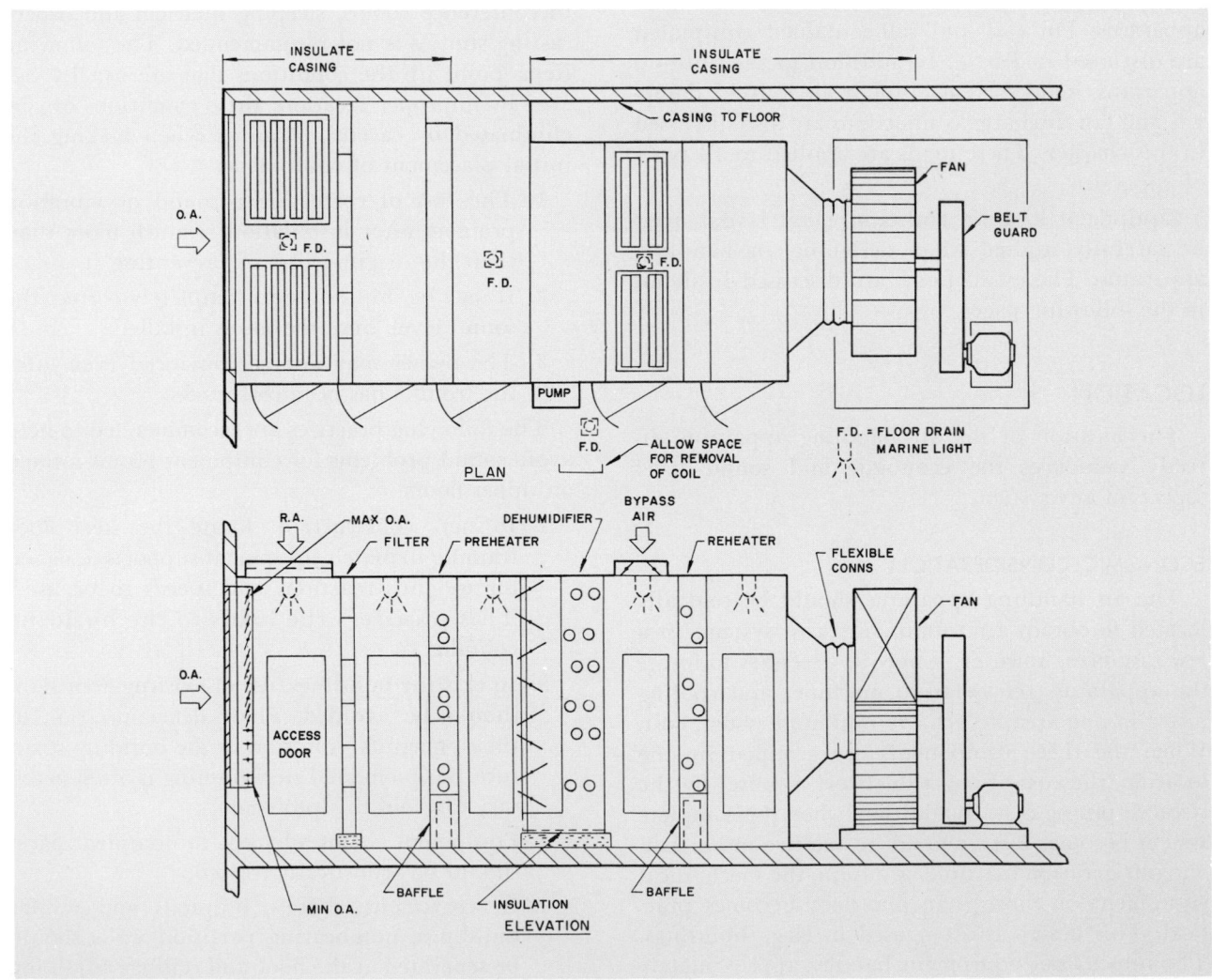

FIG. 1 — TYPICAL CENTRAL STATION EQUIPMENT

EQUIPMENT

This section describes available central station apparatus equipment and recommends suitable application of the various components.

OUTDOOR AIR LOUVERS AND SCREEN

Fig. 2 illustrates outdoor air louvers that minimize the entry of snow and water into the equipment. It is impossible to completely eliminate all moisture with vertical louvers, and this is usually not necessary. The screen is added to arrest most foreign materials such as paper, trash and birds. Often the type of screen required and the mesh are specified by codes.

The screen and louver is located sufficiently above the roof to minimize the pickup of roof dust and the probability of snow piling up and subsequently entering the louver during winter operation. This height is determined by the annual snowfall. However, a minimum of 2.5 feet is recommended for most areas. In some locations, doors are added outside the louver for closure during extreme in-

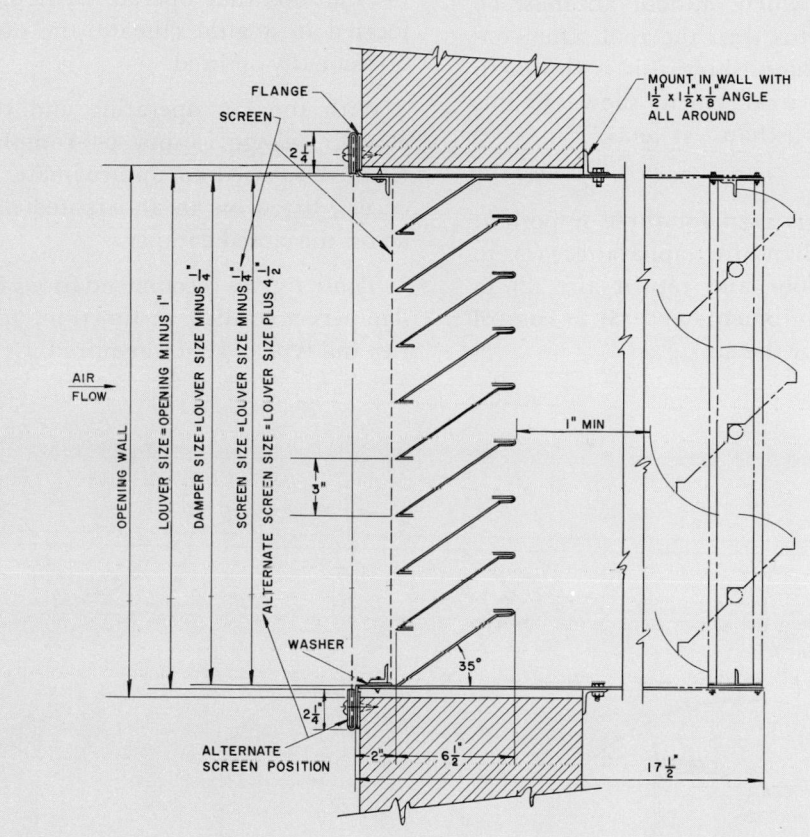

MATERIAL SPECIFICATIONS

Maximum Over-all Height	91½″
Maximum Over-all Width	95″
Blades	22 U.S. gage steel*
Frame	18 U.S. gage steel*
Screen	½″ #16 wire mesh
Screen Frame	1″ x 1″ x ⅛″ angle
Braces	1″ x ⅛″ band iron

*Equivalent strength aluminum may be substituted.

SCREEN AND BRACES

LOUVER WIDTH (in.)	NUMBER OF SCREENS†	NUMBER OF BRACES‡
0 - 30	1	0
31 - 47	1	1
48 - 60	2	1
61 - 95	2	2
Over 95	2 equal length louvers	

†Screens over 60″ high have center horizontal stiffening braces of 1″ x 1″ x ⅛″ angle.

‡Braces spaced evenly on front and back of louver and tack welded to blade edges.

FIG. 2 — OUTDOOR AIR LOUVER AND SCREEN

clement weather such as hurricanes and blizzards.

It is best to locate the outdoor air louver in such a manner that cross contamination from exhaust fan to louver does not occur, specifically toilet and kitchen exhaust. In addition, the outdoor air intake is located to minimize pulling air over a long stretch of roof since this increases the outdoor air load during summer operation.

Chart 1 is used to estimate the air pressure loss at various face velocities when the outdoor louvers are constructed, as shown in *Fig. 2*.

There are occasions when outdoor air must be drawn into the apparatus thru the roof. One convenient method of accomplishing this is shown in *Fig. 3*. The gooseneck arrangement shown in this figure is also useful for exhaust systems.

LOUVER DAMPERS

The louver damper is used for three important functions in the air handling apparatus: (1) to control and mix outdoor and return air; (2) to bypass heat transfer equipment; and (3) to control air quantities handled by the fan.

Fig. 4 shows two damper blade arrangements. The single action damper is used in locations where the damper is either fully open or fully closed. A double-acting damper is used where control of air flow is required. This arrangement is superior since the air flow is throttled more or less in proportion to the blade position, whereas the single action type damper tends to divert the air and does little or no throttling until the blades are nearly closed.

Outdoor and return air dampers are located so that good mixing of the two air streams occurs. On installations that operate 24 hours a day and are located in a mild climate, the outdoor damper is occasionally omitted.

With the fan operating and the damper fully closed, leakage cannot be completely eliminated. *Chart 2* is used to approximate the leakage that occurs, based on an anticipated pressure difference across the closed damper.

Table 1 gives recommendations for various louver dampers according to function, application, velocities and type of action required.

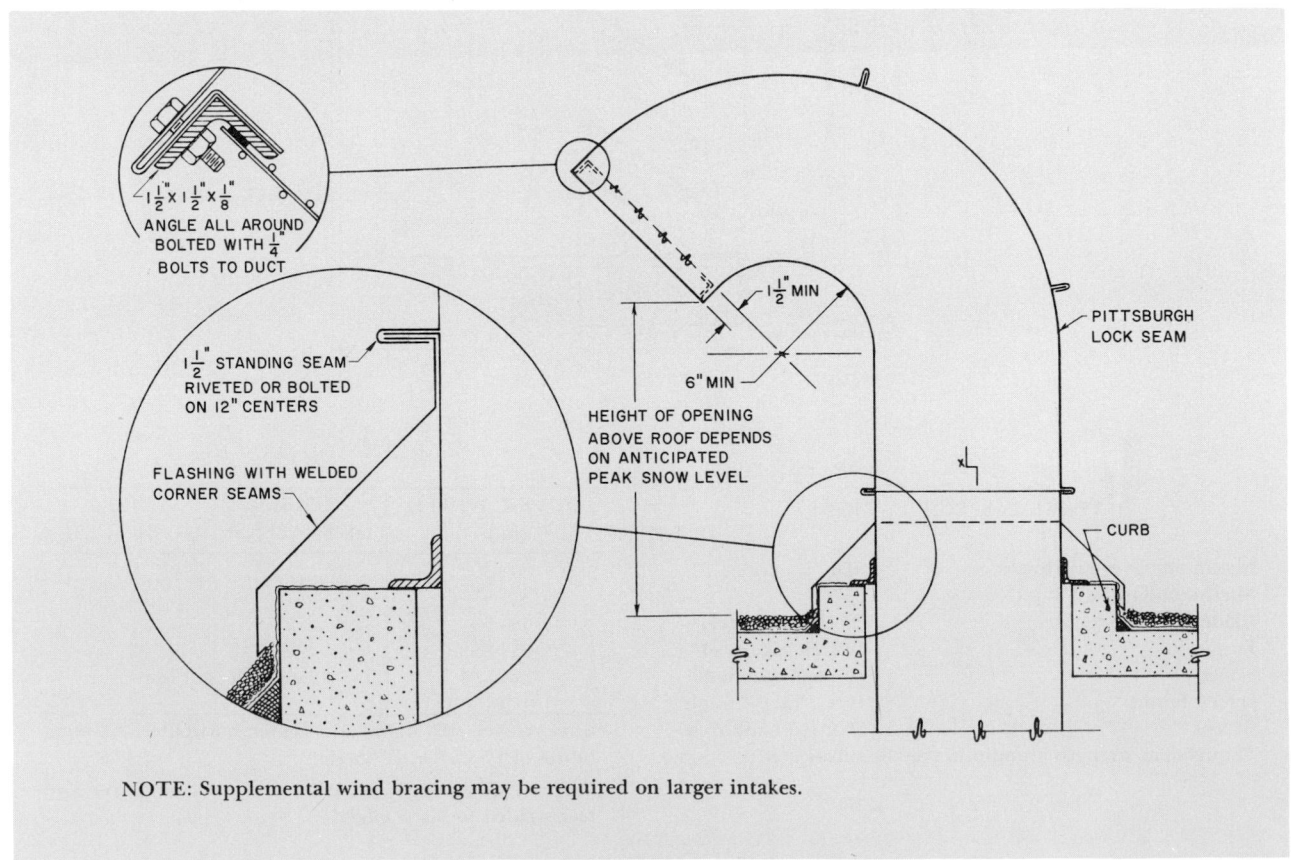

NOTE: Supplemental wind bracing may be required on larger intakes.

FIG. 3 — GOOSENECK OUTSIDE AIR INTAKE

CHART 1—LOUVER PRESSURE DROP

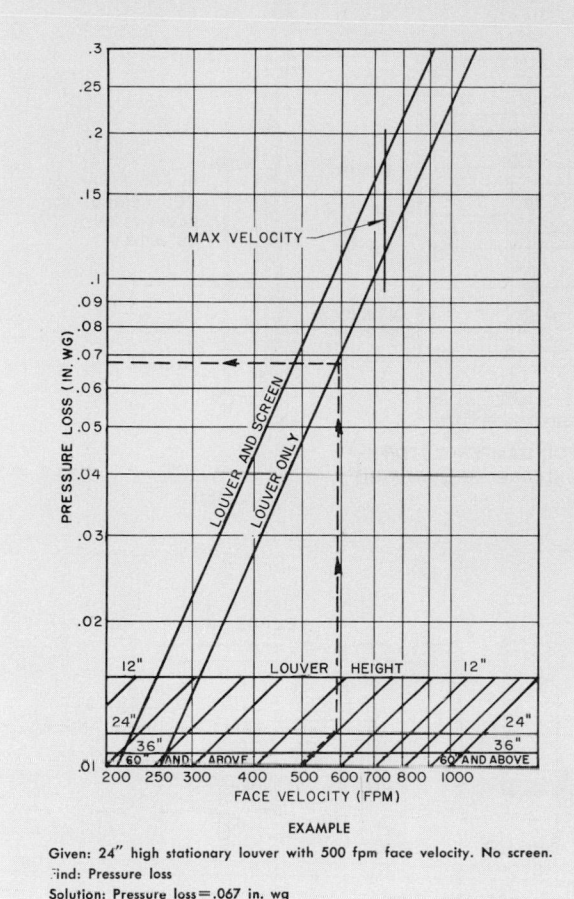

EXAMPLE

Given: 24″ high stationary louver with 500 fpm face velocity. No screen.
Find: Pressure loss
Solution: Pressure loss = .067 in. wg

CHART 2—LOUVER DAMPER LEAKAGE

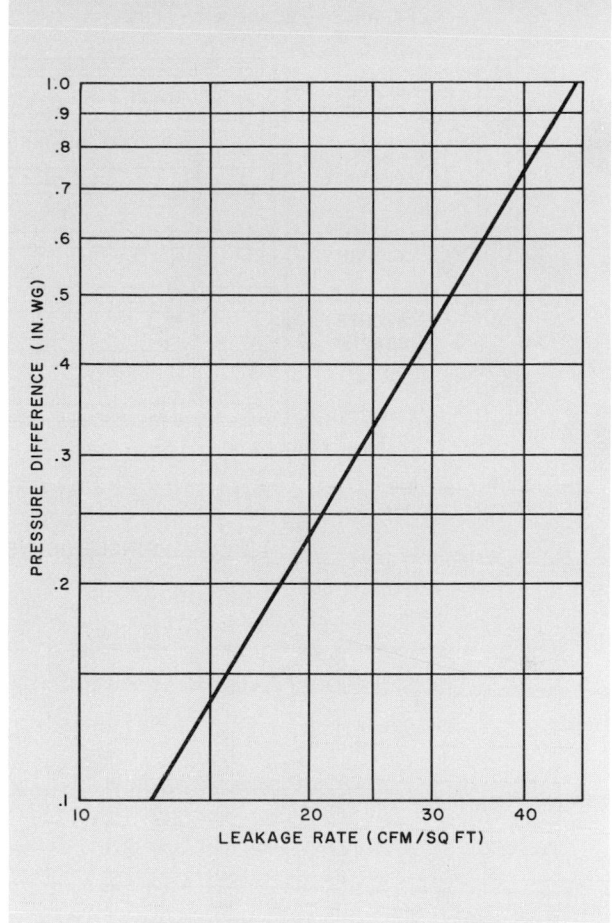

TABLE 1—LOUVER DAMPERS

FUNCTION OR LOCATION	APPLICATION	VELOCITY* (fpm)	REMARKS
Minimum Outdoor Air	Ventilation	500-800	The higher limit may be used with short outdoor air duct connection and long return air duct. May be single acting damper.
Maximum Outdoor Air	Permissible system resistance and balance	500-800	Should be double acting when used for throttling.
All Outdoor Air	Permissible system resistance and balance	500-800	Single acting damper may be used.
Return Air	Permissible system resistance and balance	800-1200	May be higher velocity with short return duct and long outdoor air duct. Should be double acting damper.
Dehumidifier Face	Control space conditions	400-800	Should equal cross-sectional area of dehumidifier. Should be a double acting damper.
Dehumidifier Bypass	System balance	1500-2500	Should balance resistance of dehumidifier plus humidifier face damper. Should be double acting.
Heater Bypass	Balance	1000-1500	Should balance resistance at heater. Should be double acting.
Fan Suction or Discharge or Located in Duct	Available duct area	same as duct	Use double acting damper.

* Recommend velocity through a fully open damper.

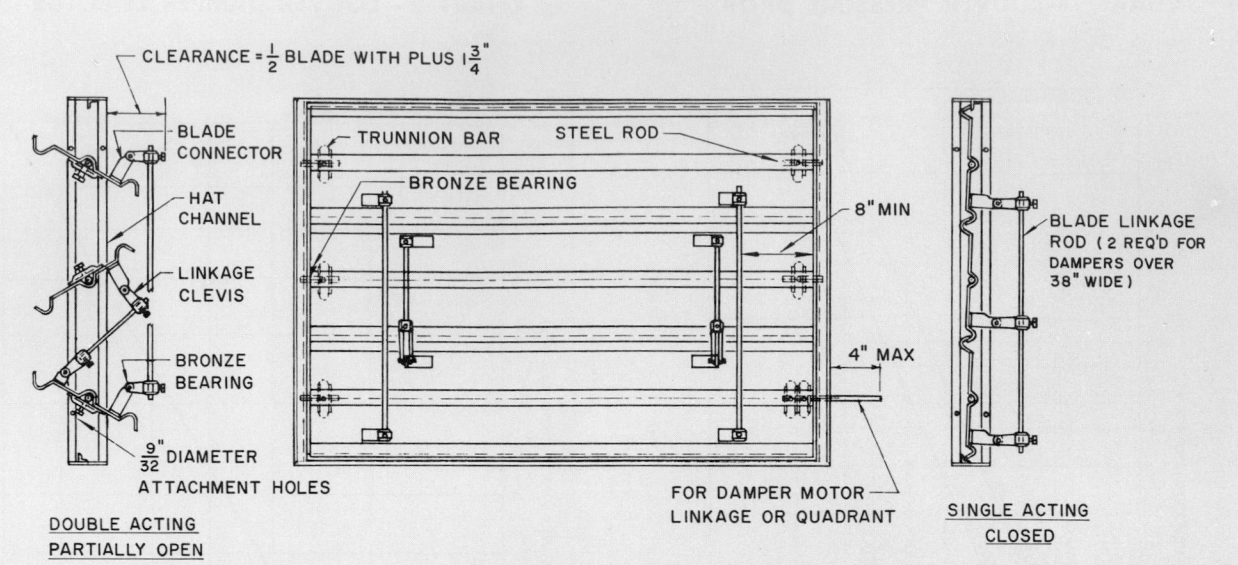

SINGLE LOUVER DAMPER

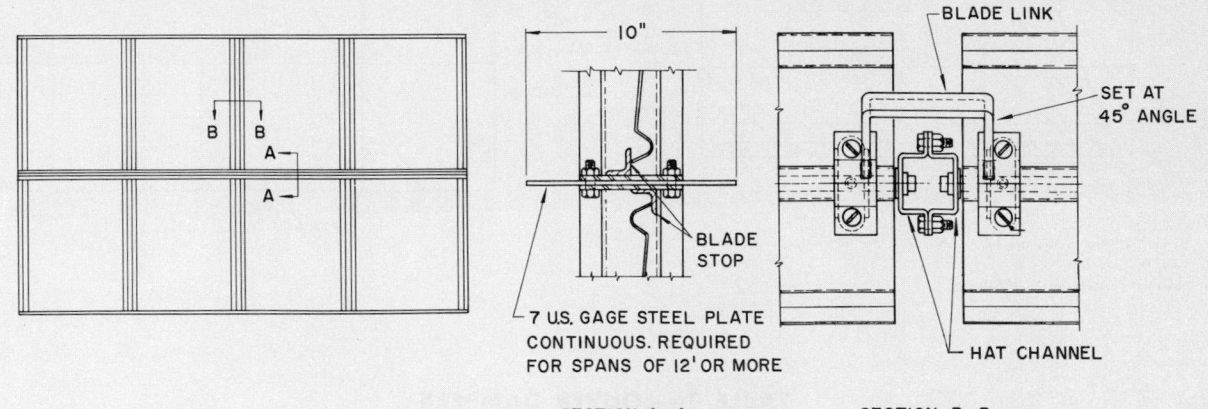

SECTION A–A

SECTION B–B

MULTIPLE LOUVER DAMPER ASSEMBLY
(FOR ASSEMBLY EXCEEDING MAXIMUM DIMENSIONS)

MATERIAL SPECIFICATIONS

Maximum Over-all Height	91½"
Maximum Over-all Width	50"
Maximum Blade Width	12"
Frame — Top and Bottom	3" x ⅛" flat bar
— Sides	3" x ⅞" x ⅛" hat channel
Blades	16 U.S. gage steel
Bearing	Oil-retaining porous bronze
Blade Linkage Rod	5/16" dia. CRS
Trunnion	Die-formed steel
Blade Link (Multi-section)	Stainless steel bar

BLADES

DAMPER HEIGHT (in.)	NUMBER OF BLADES
To and incl. 12-11/16	1
12¾ thru 21½	2
21 9/16 thru 31½	3
31 9/16 thru 41½	4
41 9/16 thru 51½	5
51 9/16 thru 61½	6
61 9/16 thru 71½	7
71 9/16 thru 81½	8
81 9/16 thru 91½	9

FIG. 4 — LOUVER DAMPER ARRANGEMENTS

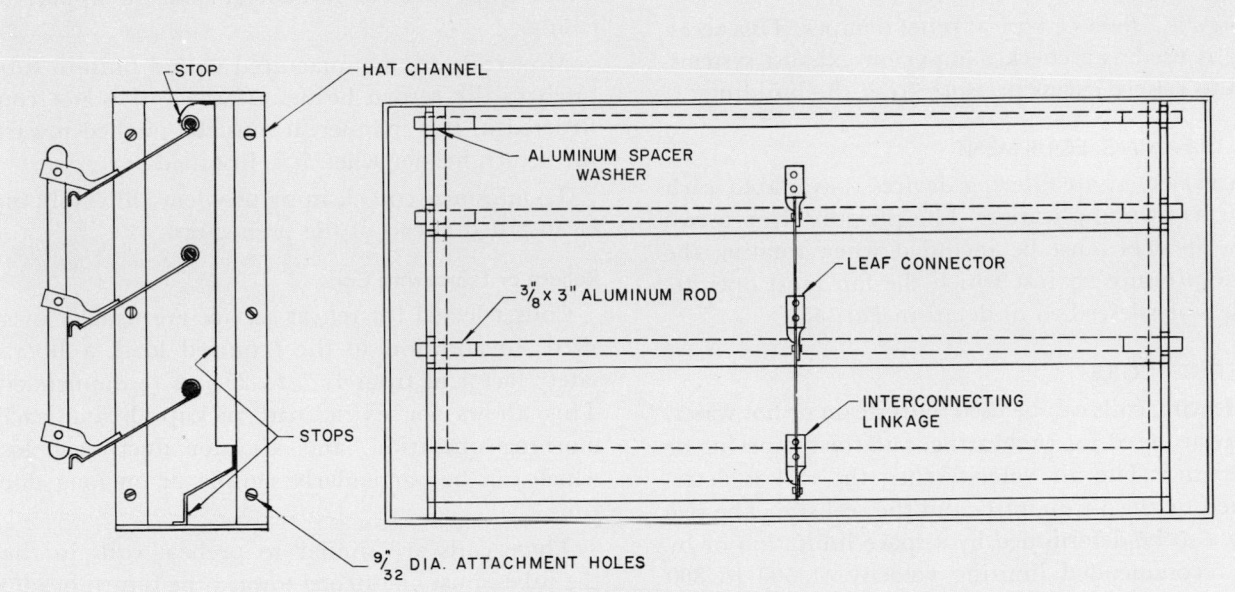

STOP   HAT CHANNEL

ALUMINUM SPACER WASHER

$\frac{3}{8}"$ x 3" ALUMINUM ROD

LEAF CONNECTOR

INTERCONNECTING LINKAGE

STOPS

$\frac{9}{32}"$ DIA. ATTACHMENT HOLES

SINGLE RELIEF DAMPER

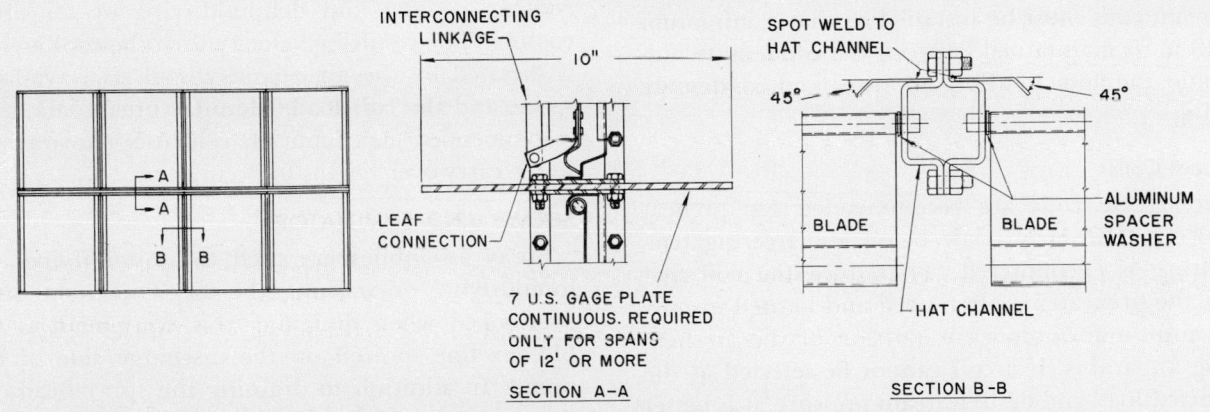

INTERCONNECTING LINKAGE

10"

LEAF CONNECTION

7 U.S. GAGE PLATE CONTINUOUS. REQUIRED ONLY FOR SPANS OF 12' OR MORE

SECTION A-A

SPOT WELD TO HAT CHANNEL

45°   45°

BLADE   BLADE

ALUMINUM SPACER WASHER

HAT CHANNEL

SECTION B-B

MULTIPLE RELIEF DAMPER ASSEMBLY
(FOR ASSEMBLY EXCEEDING MAXIMUM DIMENSION)

MATERIAL SPECIFICATIONS

Maximum Over-all Height	91½"
Maximum Over-all Width	40"
Maximum Blade Width	3½"
Frame — Top and Bottom	3" wide, 11 gage black iron
— Sides	3" x ⅞" x ⅛" hat channel
Blades	22 B & S gage aluminum
Blade Linkage Rod	½" wide, 0.050" aluminum
Spacer Washer	⅜" ID x ½" OD aluminum

PRESSURE DROP

FACE VELOCITY (fpm)	PRESSURE DROP (in. wg)
400	.067
500	.084
600	.120
700	.160
800	.200
900	.256

FIG. 5 — RELIEF DAMPER

RELIEF DAMPERS

Figure 5 shows a typical relief damper. This accessory is used as a check damper on exhaust systems, and to relieve excess pressure from the building.

AIR CLEANING EQUIPMENT

A variety of air filtering devices is available, each with its own application. The pressure drop across these devices must be included when totaling the static pressure against which the fan must operate. Filters are described in detail in *Part 6*.

HEATING COILS

Heating coils can be used with steam or hot water. They are used for preheating, and for tempering or reheating. The air velocity thru the coil is determined by the air quantity and the coil size. The size may also be determined by a space limitation or by the recommended limiting velocity of 500 to 800 fpm. The number of rows and fin spacing is determined by the required temperature rise. Manufacturer's data lists pressure drop and capacity for easy selection.

Steam coils must be installed so that a minimum of 18 in. is maintained between the condensate outlet and the floor to allow for traps and condensate piping.

Preheat Coils

Non-freeze coils are recommended for preheat service, particularly if air below the freezing temperature is encountered. To reduce the coil first cost, the preheater is often sized and located in only the minimum outdoor air portion of the air handling apparatus. If a coil cannot be selected at the required load and desired steam pressure, it is better to make a selection that is slightly undersize than one that is oversize. An undersized coil aids in preventing coil freeze-up.

The use of two coils for preheating also minimizes the possibility of freeze-up. The first coil is deliberately selected to operate with full steam pressure at all times during winter operation. In this instance, the air is heated from outdoor design to above the freezing temperature. The second coil is selected to heat the air from the freezing temperature to the desired leaving temperature. The temperature of the air leaving the second coil is automatically controlled. Refer to *Part 3, "Freeze-up Protection."*

In addition to the normal steam trap required to drain the coil return header, a steam supply trap immediately ahead of the coil is recommended.

These traps must be located outside the apparatus casing.

Most coils are manufactured with a built-in tube pitch to the return header. If the coil is not constructed in this manner, it must be pitched toward the return header when it is installed.

To minimize coil cleaning problems, filters should be installed ahead of the preheaters.

Reheat or Tempering Coils

Coils selected for reheat service are usually oversized. In addition to the required load, a liberal safety factor of from 15% to 25% is recommended. This allows for extra load pickup during early morning operation, and also for duct heat loss which can be particularly significant on long duct runs.

These coils are similar to preheat coils in that the tubes must be pitched toward the return header.

COOLING COILS

Cooling coils are used with chilled water, well water or direct expansion for the purpose of precooling, cooling and dehumidifying or for aftercooling. The resulting velocity thru the cooling coil is dictated by the air quantity, coil size, available space, and the coil load. Manufacturer's data gives recommended maximum air velocities above which water carry-over begins to occur.

SPRAYS AND ELIMINATORS

Spray assemblies are used for humidifying, dehumidifying or washing the air. One item often overlooked when designing this equipment is the bleeder line located on the discharge side of the pump. In addition to draining the spray heads on shutdown, this line controls the water concentrates in the spray pan. See *Part 5, Water Conditioning*. Eliminators are used after spray chambers to prevent entrained water from entering the duct system.

AIR BYPASS

An air bypass is used for two purposes: (1) to increase room air circulation and (2) to control leaving air temperature.

The fixed bypass is used when increased air circulation is required in a given space. It permits return air from the room to flow thru the fan without first passing thru a heat exchange device. This arrangement prevents stagnation in the space and maintains a reasonable room circulation factor.

The total airway resistance for this type system is the sum of the total resistance thru the ductwork and air handling apparatus. Therefore, the resistance thru the bypass is normally designed to balance the resistance of the components bypassed. This can be accomplished by using a balancing damper and by varying the size of the bypass opening.

The following formula is suggested for use in sizing the bypass opening :

$$A = \frac{cfm}{581 \sqrt{\dfrac{h}{.0707}}}$$

where: A = damper opening (sq ft)
cfm = maximum required air quantity thru bypass
h = design pressure drop (in. wg) thru bypassed equipment

Temperature control with bypassed air is accomplished with either a face and bypass damper or a controlled bypass damper alone. However, the face and bypass damper arrangement is recommended, since the bypass area becomes very large, and it is difficult to accommodate the required air flow thru the bypass at small partial loads. Even where a controlled face and bypass damper is used, leakage approaching 5% of design air quantity passes thru the face damper when the face damper is closed. This 5% air quantity normally is included when the fan is selected.

See *Part 6* for systems having a variable air flow to determine fan selection and brake horsepower requirements.

FANS

Properly designed approaches and discharges from fans are required for rated fan performance in addition to minimizing noise generation. *Figures 6 and 7* indicate several possible layouts for varying degrees of fan performance. In addition, these figures indicate recommended location of double width fans in plenums.

Fans in basement locations require vibration isolation based on the blade frequency. Usually cork or rubber isolators are satisfactory for this service. On upper floor locations, however, spring mounted concrete bases designed to absorb the lowest natural frequency are recommended.

The importance of controlling sound and vibration cannot be overstressed, particularly on upper floors. The number of fans involved in one location and the horsepower required for these fans directly determines the quality of sound and vibration control needed.

Small direct connected fans, due to higher operating speed, are generally satisfactorily isolated by rubber or cork.

In addition, all types of fans must have flexible connections to the discharge ductwork and, where required, must have flexible connections to the intake ductwork. Details of a recommended flexible connection are shown in *Fig. 8*.

Unitary equipment should be located near columns or over main beams to limit the floor deflection. Rubber or cork properly loaded usually gives the required deflection for efficient operation.

FAN MOTOR AND DRIVE

A proper motor and drive selection aids in long life and minimum service requirements. Direct drive fans are normally used on applications where exact air quantities are not required, because ample energy (steam or hot water, etc.) is available at more than enough temperature difference to compensate for any lack of air quantity that exists. This applies, for example, to a unit heater application. Direct drive fans are also used on applications where system resistance can be accurately determined. However, most air conditioning applications use belt drives.

V-belts must be applied in matched sets and used on balanced sheaves to minimize vibration problems and to assure long life. They are particularly useful on applications where adjustments may be required to obtain more exact air quantities. These adjustments can be accomplished by varying the pitch diameter on adjustable sheaves, or by changing one or both sheaves on a fixed sheave drive.

Belt guards are required for safety on all V-belt drives, and coupling guards are required for direct drive equipment. *Figure 9* illustrates a two-piece belt guard.

The fan motor must be selected for the maximum anticipated brake horsepower requirements of the fan. The motor must be large enough to operate within its rated horsepower capacity. Since the fan motor runs continuously, the normal 15% overload allowed by NEMA should be reserved for drive losses and reductions in line voltages. Normal torque motors are used for fan duty.

APPARATUS CASING

The apparatus casing on central station equipment must be designed to avoid restrictions in air flow. In addition, it must have adequate strength to prevent collapse or bowing under maximum operating conditions.

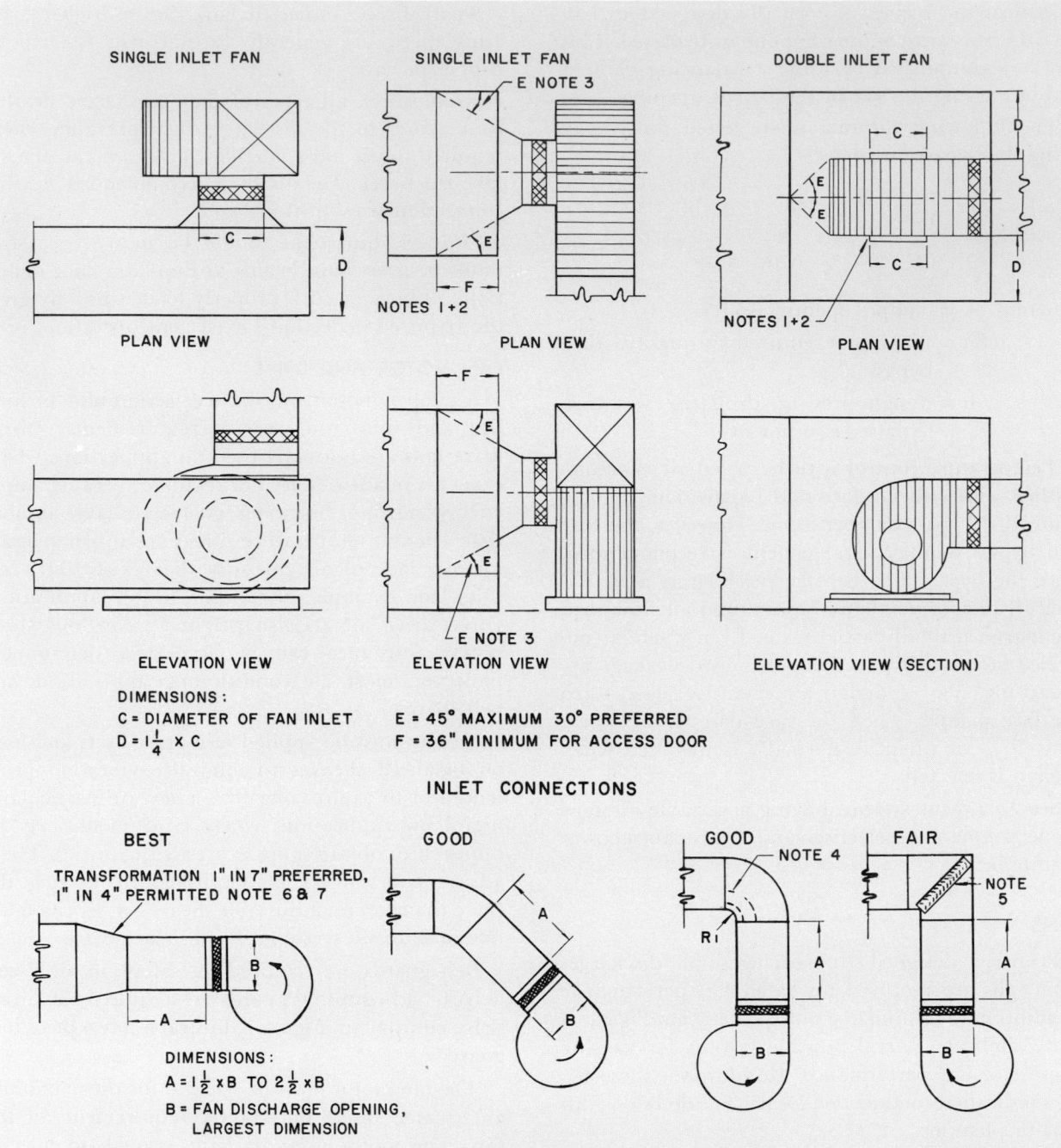

FIG. 6 — SINGLE FAN INLET AND DISCHARGE CONNECTIONS

NOTES:
1. Fan should be centered in casing to provide good flow conditions.
2. All equipment should be centered for best performance.
3. Angle "E" is used to determine "F" distance between equipment and fan.
4. R_1 = 6" minimum. Vane spacing determined from *Chart 6*.
5. Use square vaned elbow for best results, with take-off in opposite direction to fan rotation.
6. Slope of 1" in 4" recommended for low velocity.
7. Slope of 1" in 7" recommended for high velocity.

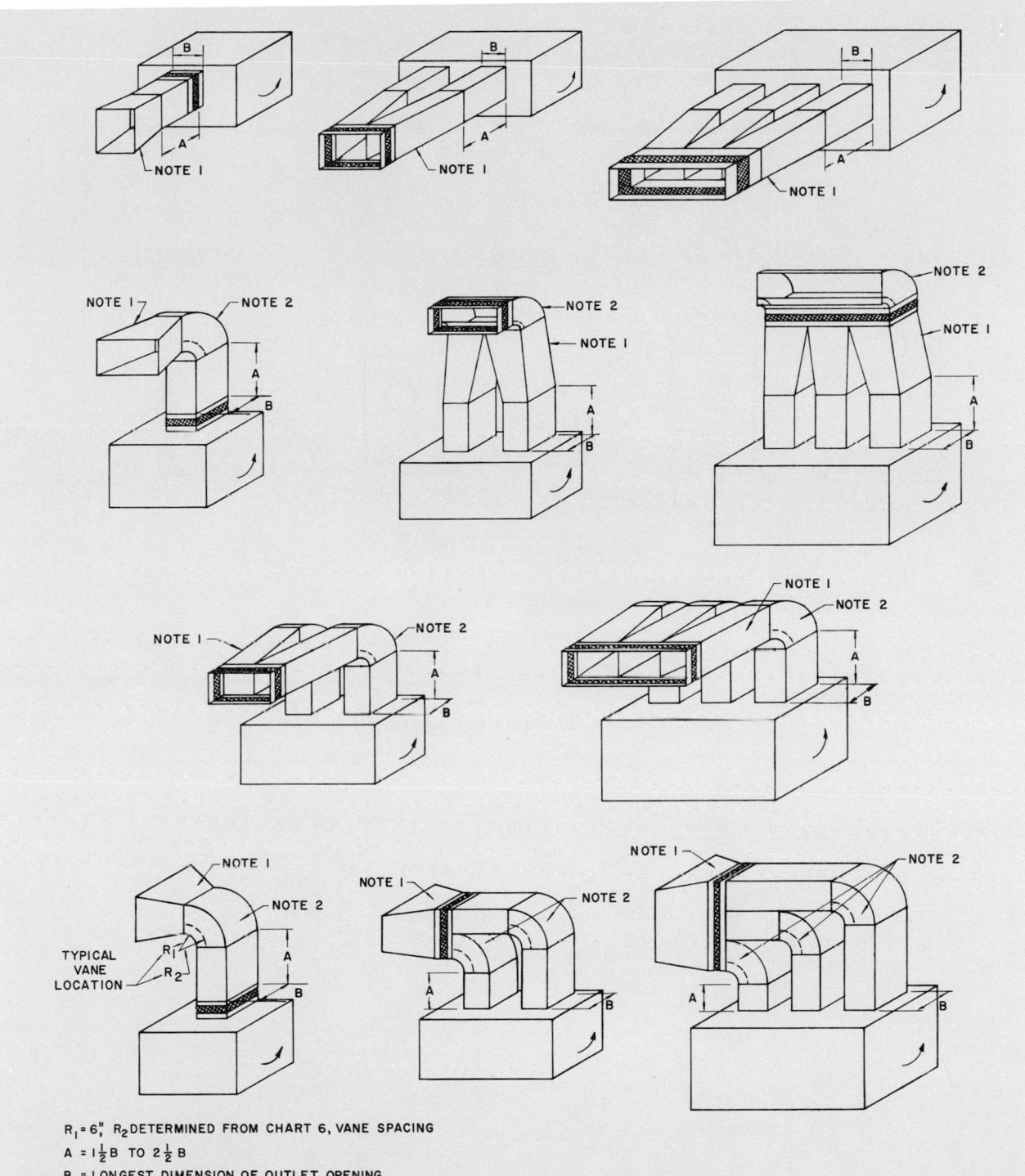

R₁ = 6", R₂ DETERMINED FROM CHART 6, VANE SPACING
$A = 1\frac{1}{2}B$ TO $2\frac{1}{2}B$
B = LONGEST DIMENSION OF OUTLET OPENING

NOTES:

1. Transformations to supply duct have maximum slope of 1" in 7".

2. Square elbows with double thickness vanes may be substituted.

3. Do not install ducts so that the air flow is counter to fan rotation. If necessary, turn fan section.

4. Transformations and units shall be adequately supported so no weight is on the flexible fan connection.

FIG. 7 — MULTIPLE FAN UNIT DISCHARGE CONNECTIONS

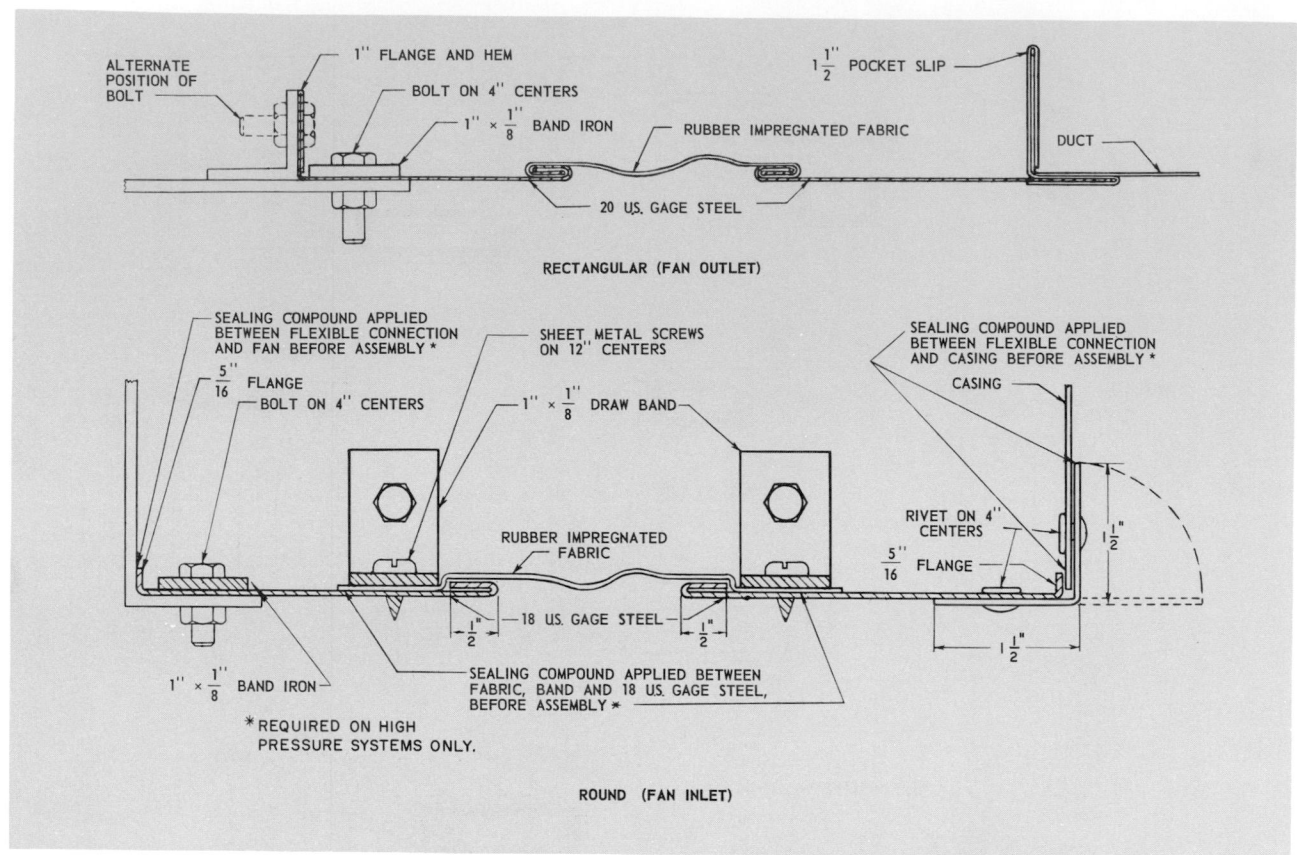

FIG. 8 — FLEXIBLE CONNECTION

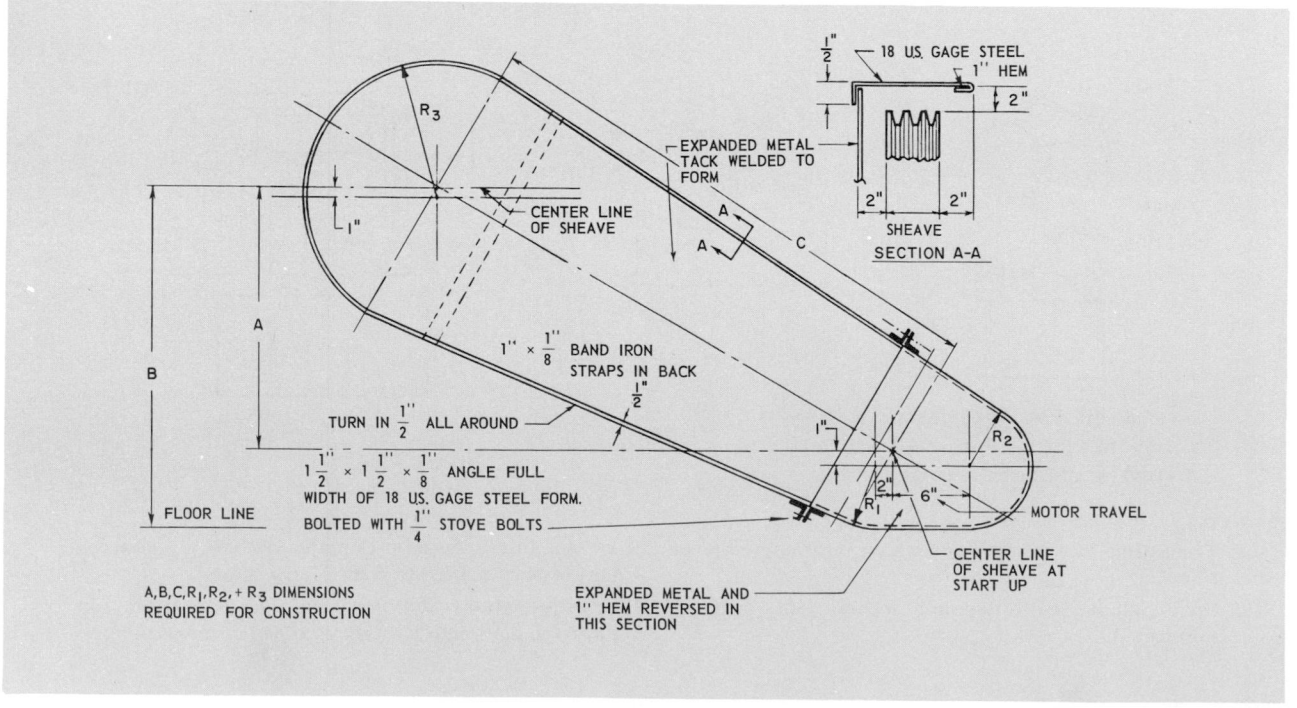

FIG. 9 — TWO-PIECE BELT GUARD

Each sheet of material should be fabricated as a panel and joined together, as illustrated in *Fig 10,* by standing seams, bolted or riveted on 12 in. centers. Normally, seams perpendicular to air flow are placed outside of the casing. Side walls over 6 ft high and roof spans over 6 ft wide require supplemental reinforcing as shown in *Table 2.* Diagonal angle braces as illustrated in *Fig. 11* may also be required.

The recommended construction of apparatus casings and connections between equipment components (except when mounted in the ducts) is 18 U.S. gage steel or 16 B & S gage aluminum. Aluminum in contact with galvanized steel at connections to spray type equipment requires that the inside of the casing be coated with an isolating material for a distance of 6 in. from the point of contact.

CONNECTIONS TO MASONRY

A concrete curb is recommended to protect insulation from deteriorating where the apparatus casing joins the floor. It also provides a uniform surface for attaching the casing; this conserves fabrication time. *Figure 12* illustrates the recommended method of attaching a casing to the curb.

When an equipment room wall is used as one side of the apparatus, the casing is attached as shown in *Fig. 13.*

The degree of tightness required for an apparatus casing depends on the air conditioning application.

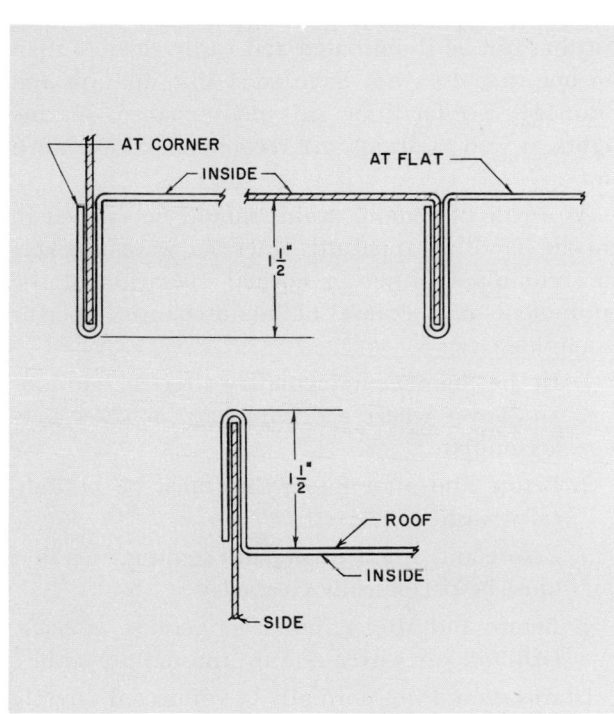

Fig. 10 — Apparatus Casing Seams

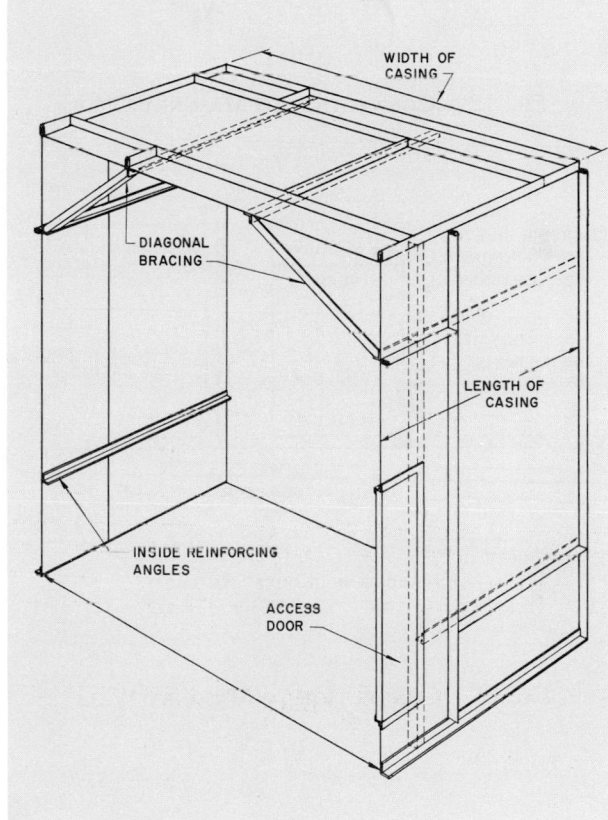

Fig. 11 — Apparatus Casing

TABLE 2—SUPPLEMENTAL REINFORCING FOR APPARATUS CASING

SIDE WALL HEIGHT OR ROOF WIDTH (ft)	NUMBER OF ANGLES*	ANGLE SPACING	CASTING LENGTHS	DIAGONAL ANGLE BRACES* (pairs)
6 to 8	1	middle	—	—
8 to 12	2	⅓ points	—	—
over 12	variable	4 ft centers	3 & 4 panels	1
			5 & 6 panels	2
			7 & 8 panels	3

*For lengths up to 12 ft., use 1½ x 1½ x ⅛ in. angle.   For lengths over 12 ft., use 1¾ x 1¾ x 3/16 in. angle.

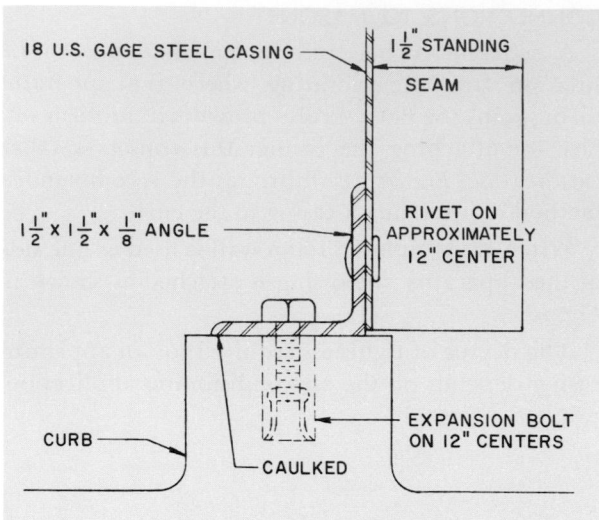

Fig. 12 — Connection to Masonry Curb

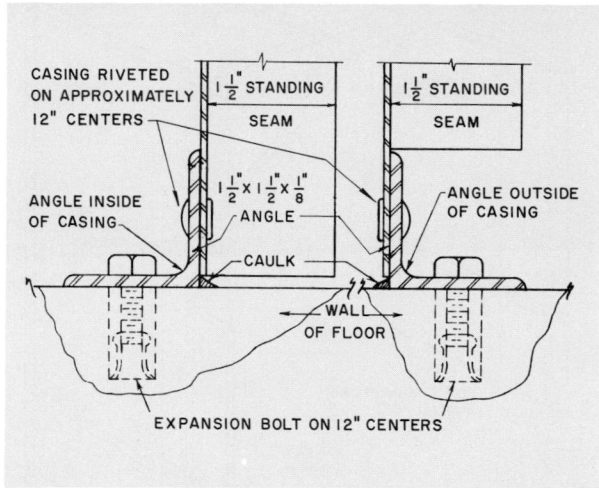

Fig. 13 — Connection to Masonry Wall

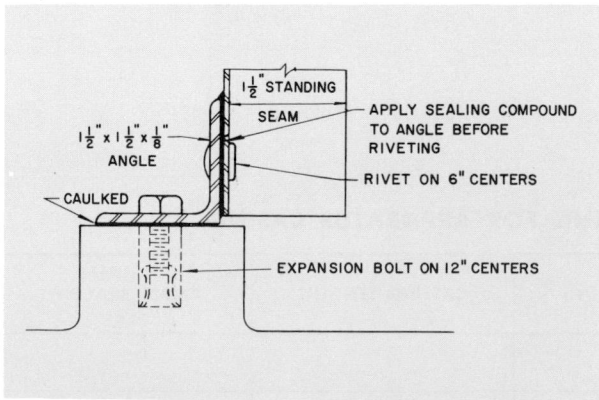

Fig. 14 — Low Dewpoint Masonry Curb
Connections

For instance, on a pull-thru system, leakage between the dehumidifier and the fan cannot be tolerated if the apparatus is in a humid non-conditioned space. Also, as the negative pressure at the fan intake increases, the less the leakage that can be tolerated. If the apparatus is located in a return air plenum, normal construction as shown in *Figs. 12 and 13* can be used. Corresponding construction practice for equipment requiring extreme care is shown in *Figs. 14, 15 and 16*.

In addition to the construction required for leakage at seams, pipes passing thru the casing at cooling coil connections must be sealed as shown in *Fig. 17*. This applies in applications where the temperature difference between the room and supply air temperatures is 20 F and greater.

DRAINS AND MARINE LIGHTS

Upkeep and maintenance is better on an apparatus that can be illuminated and easily cleaned than on one that does not have good illumination and drainage. To facilitate this maintenance, marine lights, as well as drains, are recommended as shown in *Fig. 1*.

As a rule of thumb, drains should be located in the air handling apparatus wherever water is likely to accumulate, either in normal operation of the equipment or because of maintenance. Specific examples are:

1. In the chamber immediately after the outdoor air louver where a driving rain or snow may accumulate.
2. Before and after filters that must be periodically washed.
3. Before and after heating and cooling coils that must be periodically cleaned.
4. Before and after eliminators because of backlash and carry-over due to unusual air eddies.

Drains should not normally be connected directly to sewers. Instead, an open site drain should be used as illustrated in *Part 3*.

INSULATION

Insulation is required ahead of the preheater and vapor sealed for condensation during winter operation. Normally, the section of the casing from the preheater to the dehumidifier is not insulated. The dehumidifier, the fan and connecting casing must be insulated and vapor sealed; fan access doors are not insulated, however. The bottoms and sides of the dehumidifier condensate pan must also be insulated, and all parts of the building surfaces that are used to form part of the apparatus casing must be insulated and vapor sealed.

SERVICE

Equipment service is essential and space must be provided to accomplish this service. It is recommended that minimum clearances be maintained so that access to all equipment is available. In addition, provision should be made so that equipment can be removed without dismantling the complete apparatus. Access must be provided for heating and cooling coils, steam traps, damper motors and linkages, control valves, bearings, fan motors, fans and similar components.

Service access doors as illustrated in *Fig. 18* are recommended, and are located in casing sections as shown in *Fig. 1*.

To conserve floor space, the entrance to the equipment room is often located so that coils can be removed directly thru the equipment room doors. This arrangement requires less space than otherwise possible.

If the equipment room is not arranged as described, space must be allowed to clean the coil tubes mechanically. This applies to installations that have removable water headers.

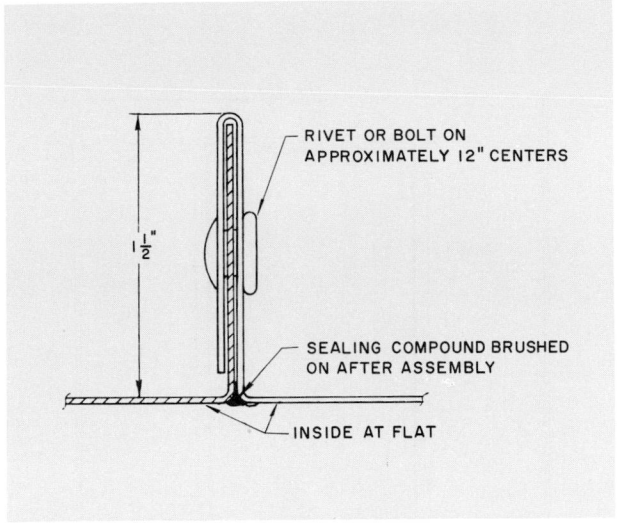

Fig. 16 — Sealing Standing Seams

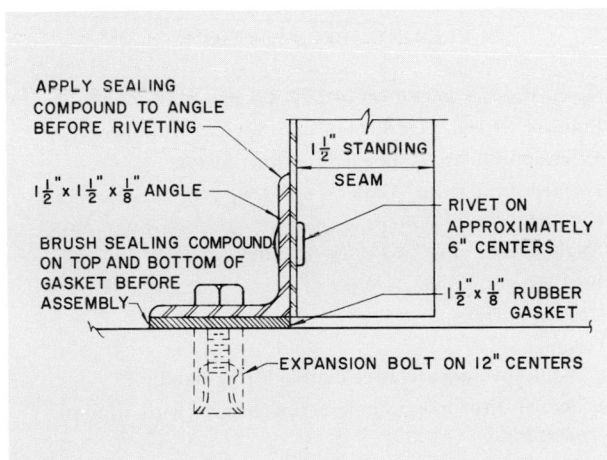

Fig. 15 — Low Dewpoint Masonry Wall Connections

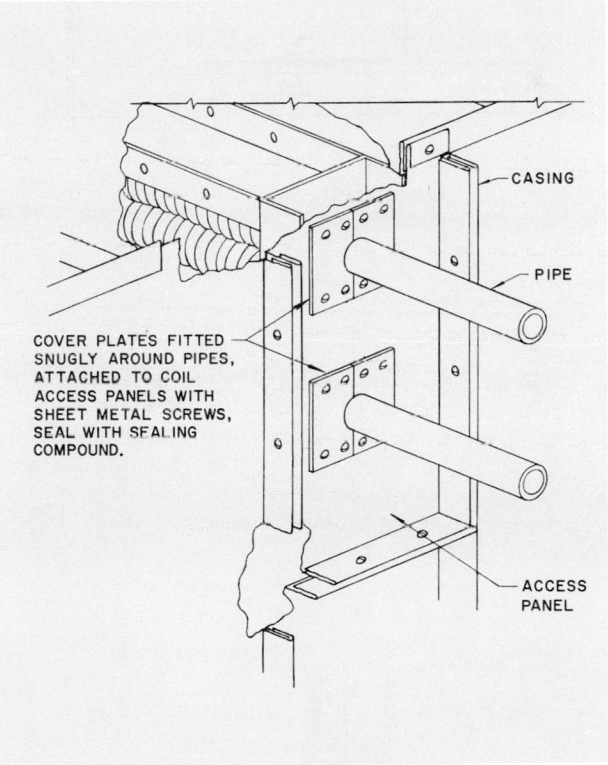

Fig. 17 — Sealing Pipe Connections

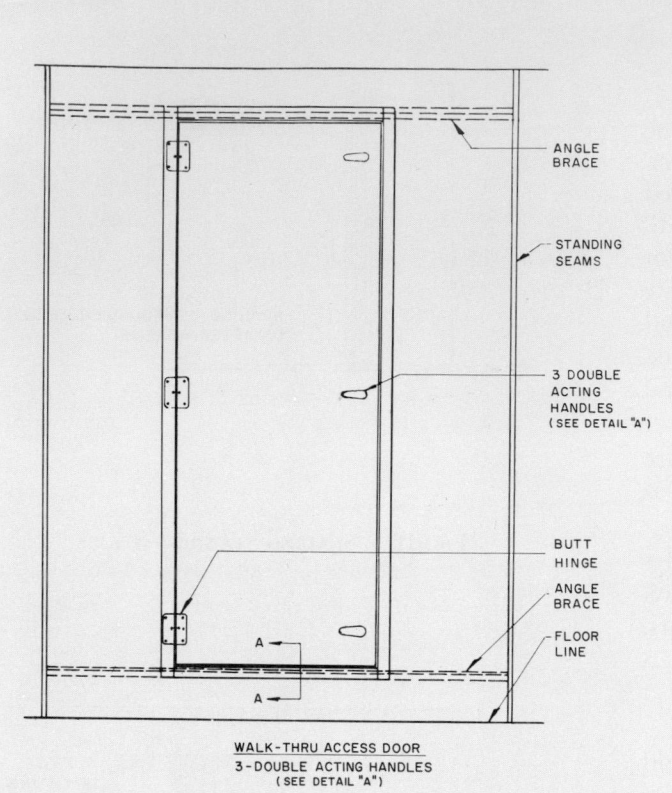

WALK-THRU ACCESS DOOR
3-DOUBLE ACTING HANDLES
(SEE DETAIL "A")

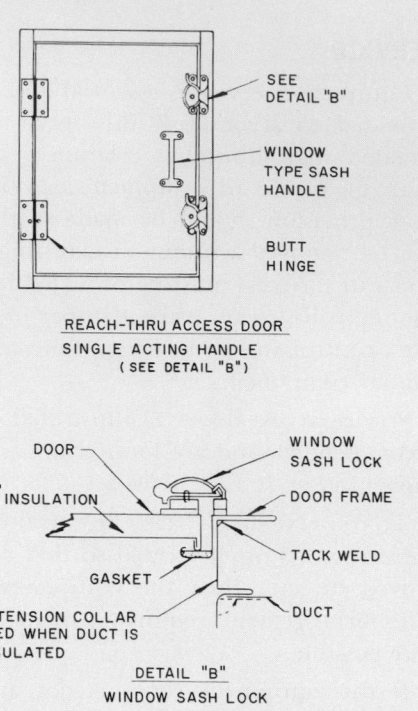

REACH-THRU ACCESS DOOR
SINGLE ACTING HANDLE
(SEE DETAIL "B")

DETAIL "B"
WINDOW SASH LOCK

EXTENSION COLLAR LENGTH (SEE DETAIL B) IS DETERMINED
BY INSULATION THICKNESS USED ON DUCT OR CASING.

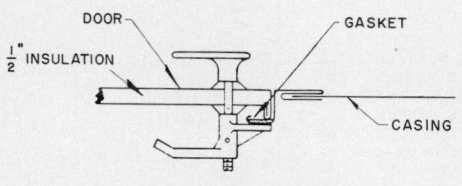

DETAIL "A"
DOUBLE ACTING DOOR HANDLES

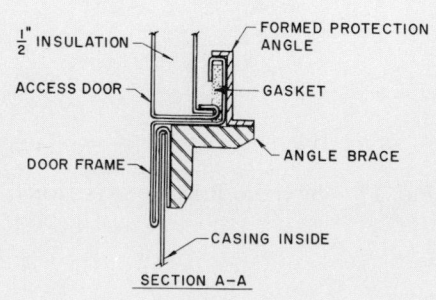

SECTION A–A

MATERIAL SPECIFICATION

1. Door — 24 U.S. gage steel or 22 B & S gage aluminum.
2. Frame — 24 U.S. gage steel or 22 B & S gage aluminum.
3. Extension Collar — Same gage as duct metal.
4. Formed Protection Angle — 18 U.S. gage steel or 16 B & S gage aluminum.
5. Angle Brace — $1\frac{3}{4}$" x $1\frac{3}{4}$" x $\frac{1}{8}$" angle.
6. Butt Hinges — Steel.
7. Gaskets — Felt.
8. Fastener —
   a. Walk-thru door: Three double acting handles.
   b. Reach-thru door: Single acting handle with window sash lock.

Walk-thru Access Doors
    Normal size — 22" W x 58" H

Reach-thru Access Doors
    Normal sizes

W	H
10"	12"
12"	16"
16"	24"

FIG. 18 — ACCESS DOORS

CHAPTER 2. AIR DUCT DESIGN

The function of a duct system is to transmit air from the air handling apparatus to the space to be conditioned.

To fulfill this function in a practical manner, the system must be designed within the prescribed limits of available space, friction loss, velocity, sound level, heat and leakage losses and gains.

This chapter discusses these practical design criteria and also considers economic balance between first cost and operating cost. In addition, it offers recommended construction for various types of duct systems.

GENERAL SYSTEM DESIGN

CLASSIFICATION

Supply and return duct systems are classified with respect to the velocity and pressure of the air within the duct.

Velocity

There are two types of air transmission systems used for air conditioning applications. They are called conventional or low velocity and high velocity systems. The dividing line between these systems is rather nebulous but, for the purpose of this chapter, the following initial supply air velocities are offered as a guide:

1. Commercial comfort air conditioning
   a. Low velocity — up to 2500 fpm. Normally between 1200 and 2200 fpm.
   b. High velocity — above 2500 fpm.
2. Factory comfort air conditioning
   a. Low velocity — up to 2500 fpm. Normally between 2200 and 2500 fpm.
   b. High velocity — 2500 to 5000 fpm.

Normally, return air systems for both low and high velocity supply air systems are designed as low velocity systems. The velocity range for commercial and factory comfort application is as follows:

1. Commercial comfort air conditioning — low velocity up to 2000 fpm. Normally between 1500 and 1800 fpm.
2. Factory comfort air conditioning — low velocity up to 2500 fpm. Normally between 1800 and 2200 fpm.

Pressure

Air distribution systems are divided into three pressure categories; low, medium and high. These divisions have the same pressure ranges as Class I, II and III fans as indicated:

1. Low pressure — up to 3¾ in. wg — Class I fan
2. Medium pressure — 3¾ to 6¾ in. wg — Class II fan
3. High pressure — 6¾ to 12¼ in. wg — Class III fan

These pressure ranges are total pressure, including the losses thru the air handling apparatus, ductwork and the air terminal in the space.

AVAILABLE SPACE AND ARCHITECTURAL APPEARANCE

The space allotted for the supply and return air conditioning ducts, and the appearance of these ducts, often dictates system layout and, in some instances, the type of system. In hotels and office buildings where space is at a premium, a high velocity system with induction units using small round ducts is often the most practical.

Some applications require the ductwork to be exposed and attached to the ceiling, such as in an existing department store or existing office building. For this type of application, streamline rectangular ductwork is ideal. Streamline ductwork is constructed to give the appearance of a beam on the ceiling. It has a smooth exterior surface with the duct joints fabricated inside the duct. This ductwork is laid out with a minimum number of reductions in size to maintain the beam appearance.

Duct appearance and space allocation in factory air conditioning is usually of secondary importance. A conventional system using rectangular ductwork is probably the most economical design for this application.

ECONOMIC FACTORS INFLUENCING DUCT LAYOUT

The balance between first cost and operating cost must be considered in conjunction with the available space for the ductwork to determine the best air distribution system. Each application is different and must be analyzed separately; only general rules or principles can be given to guide the engineer in selecting the proper system. The

CHART 3—DUCT HEAT GAIN VS ASPECT RATIO

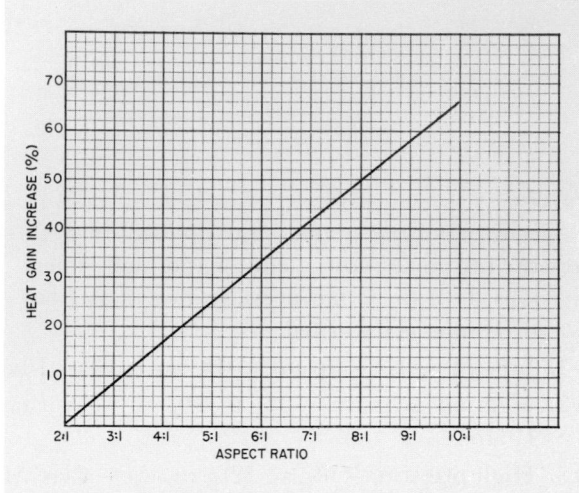

following items directly influence the first and operating cost:

1. Heat gain or loss from the duct
2. Aspect ratio of the duct
3. Duct friction rate
4. Type of fittings

Heat Gain or Loss

The heat gains or losses in the supply and return duct system can be considerable. This occurs not only if the duct passes thru an unconditioned space but also on long duct runs within the conditioned space. The transfer of heat takes place from the space to the air in the duct when cooling, and from the air to the space when heating.

An allowance must be made for duct heat gain for that portion of the duct in the unconditioned space when estimating the air conditioning load. The method of making this allowance is presented in *Part I, Load Estimating*. This allowance for duct heat gain increases the cooling capacity of the air. This increase then requires a larger air quantity or lower supply air temperature or both.

To compensate for the cooling or heating effect of the duct surface, a redistribution of the air to the supply outlets is sometimes required in the initial design of the duct system.

The following general guides are offered to help the engineer understand the various factors influencing duct design:

1. Larger duct aspect ratios have more heat gain than ducts with small aspect ratios, with each carrying the same air quantity. *Chart 3* illustrates this relationship.

2. Ducts carrying small air quantities at a low velocity have the greatest heat gain.

3. The addition of insulation to the duct decreases duct heat gain; for example, insulating the duct with a material that has a *U* value of .12 decreases heat gain 90%.

It is, therefore, good practice to design the duct system for low aspect ratios and higher velocities to minimize heat gain to the duct. If the duct is to run thru an unconditioned area, it should be insulated.

Aspect Ratio

The aspect ratio is the ratio of the long side to the short side of a duct. This ratio is an important factor to be considered in the initial design. Increasing the aspect ratio increases both the installed cost and the operating cost of the system.

The installed or first cost of the ductwork depends on the amount of material used and the difficulty experienced in fabricating the ducts. *Table 6* reflects these factors. This table also contains duct class, cross-section area for various round duct sizes and the equivalent diameter of round duct for rectangular ducts. The large numbers in the table are the duct class.

The duct construction class varies from 1 to 6 and depends on the maximum duct side and the semi-perimeter of the ductwork. This is illustrated as follows:

DUCT CLASS	MAX. SIDE (in.)	SEMI-PERIMETER (in.)
1	6 - 17½	10 - 23
2	12 - 24	24 - 46
3	26 - 40	32 - 46
4	24 - 88	48 - 94
5	48 - 90	96 - 176
6	90 - 144	96 - 238

Duct class is a numerical representation of relative first costs of the ductwork. The larger the duct class, the more expensive the duct. If the duct class is increased but the duct area and capacity remain the same, the following items may be increased:

1. Semi-perimeter and duct surface
2. Weight of material
3. Gage of metal
4. Amount of insulation required

Therefore, for best economics the duct system should be designed for the lowest duct class at the smallest aspect ratio possible. *Example 1* illustrates the effect on first cost of varying the aspect ratio for a specified air quantity and static pressure requirement.

Example 1 — Effect of Aspect Ratio on First Cost of the Ductwork

Given:

    Duct cross-section area — 5.86 sq ft
    Space available — unlimited
    Low velocity duct system

Find:

    The duct dimensions, class, surface area, weight and gage of metal required.

Solution:

  1. Enter *Table 6* at 5.86 sq ft and determine the rectangular duct dimensions and duct class (see tabulation).
  2. Determine recommended metal gages from *Tables 14 and 15* (see tabulation).
  3. Determine weight of metal from *Table 18* (see tabulation).

DIMENSION (in.)	AREA (sq ft)	ASPECT RATIO	DUCT CONSTR. CLASS
94 x 12	5.86	7.8:1	6
84 x 13	5.86	6.5:1	5
76 x 14	5.86	5.4:1	4
42 x 22	5.86	1.9:1	4
30 x 30	5.86	1:1	4
32.8 (round)	5.86	—	—

DIMENSION (in.)	GAGE (U.S.)	SURFACE AREA (sq ft/ft)	WEIGHT (lb/ft)
94 x 12	18	17.7	38.3
84 x 13	20	16.2	26.8
76 x 14	20	15.0	24.8
42 x 22	22	10.7	15.1
30 x 30	24	10.0	11.6
32.8 (round)	20	8.6	14.3

When the aspect ratio increases from 1:1 to 8:1, the surface area and insulation requirements increase 70% and the weight of metal increases over three and one-half times. This example also points out that it is possible to construct Class 4 duct, for the given area, with three different sheet metal gages. Therefore, for lowest first cost, ductwork should be designed for the lowest class, smallest aspect ratio and for the lightest gage metal recommended.

Chart 4 illustrates the percent increase in installed cost for changing the aspect ratio of rectangular duct. The installed cost of round duct is also included in this chart. The curve is based on installed cost of 100 ft of round and rectangular duct with various aspect ratios of equal air handling capacities. The installed cost of rectangular duct with an aspect ratio of 1:1 is used as the 100% cost.

Friction Rate

The operating costs of an air distribution system can be adversely influenced when the rectangular

CHART 4—INSTALLED COST VS ASPECT RATIO

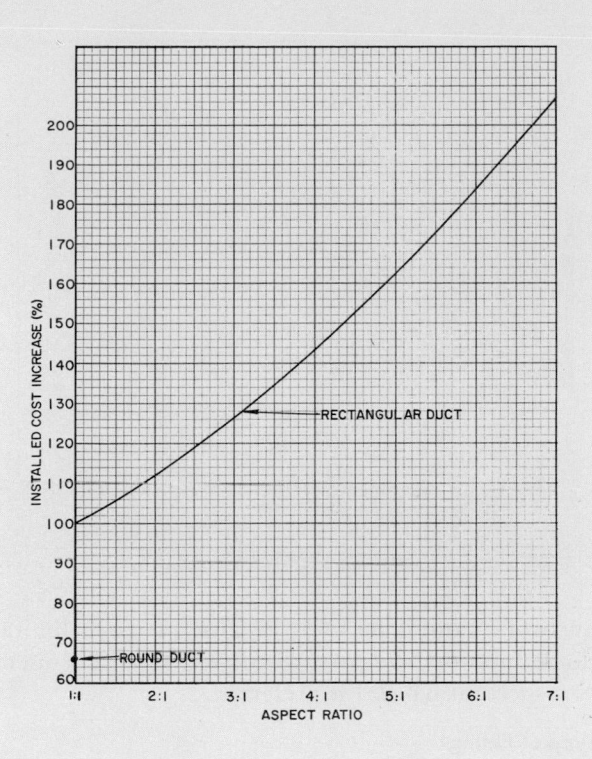

duct sizes are not determined from the table of circular equivalents *(Table 6)*. This table is used to obtain rectangular duct sizes that have the same friction rate and capacity as the equivalent round duct. For example, assume that the required duct area for a system is 480 sq in. and the rectangular duct dimensions are determined directly from this area. The following tabulation shows the resulting equivalent duct diameters and friction rate when 4000 cfm of air is handled in the selected ducts:

DUCT DIM. (in.)	EQUIV ROUND DUCT DIAM (in.)	FRICTION RATE (in. wg/100 ft)	ASPECT RATIO
24 x 20	23.9	.090	1.2:1
30 x 16	23.7	.095	1.9:1
48 x 10	22.3	.125	4.8:1
80 x 6	20.1	.210	13.3:1

If a total static pressure requirement of 1 in., based on 100 ft of duct and other equipment is assumed for the above system, the operating cost increases as the aspect ratio increases. This is shown in *Chart 5*.

Therefore, the lowest owning and operating cost occurs where round or *Spira-Pipe* is used. If round

CHART 5—OPERATING COST VS ASPECT RATIO

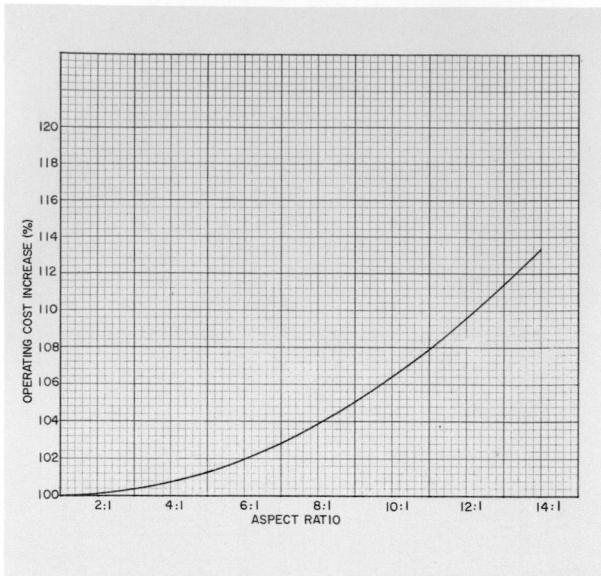

ductwork cannot be used because of space limitations, ductwork as square as possible should be used. An aspect ratio of 1:1 is preferred.

Type of Fittings

In general, fittings can be divided into Class A and Class B as shown in *Table 3*. For the lowest first cost it is desirable to use those fittings shown as Class A since fabrication time for a Class B fitting is approximately 2.5 times that of a Class A fitting.

TABLE 3—DUCT FITTING CLASSES

CLASS A—NO VANED FITTINGS	
Any fitting with constant cross-section dimensions.	
Any fitting with changing radius and constant width.	
Fittings with straight sides and seams.	

CLASS B—ALL VANED FITTINGS	
Any fitting with concentric radii, and changing width.	
Any fitting with eccentric radii and changing width.	

DUCT LAYOUT CONSIDERATIONS

There are several items in duct layout that should be considered before sizing the ductwork. These include duct transformations, elbows, fittings, take-offs, duct condensation and air control.

Transformations

Duct transformations are used to change the shape of a duct or to increase or decrease the duct area. When the shape of a rectangular duct is changed but the cross-sectional area remains the same, a slope of 1 in. in 7 in. is recommended for the sides of the transformation piece as shown in *Fig. 19*. If this slope cannot be maintained, a maximum slope of 1 in. in 4 in. should not be exceeded.

Often ducts must be reduced in size to avoid obstructions. It is good practice not to reduce the duct more than 20% of the original area. The recommended slope of the transformation is 1 in. in 7 in. when reducing the duct area. Where it is impossible to maintain this slope, it may be increased to a maximum of 1 in. in 4 in. When the duct area is increased, the slope of the transformation is not to exceed 1 in. in 7 in. *Fig. 20* illustrates a rectangular duct transformation to avoid an obstruction, and *Fig. 21* shows a round-to-rectangular transformation to avoid an obstruction.

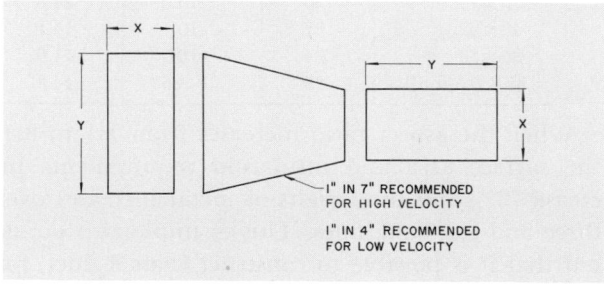

FIG. 19 — DUCT TRANSFORMATION

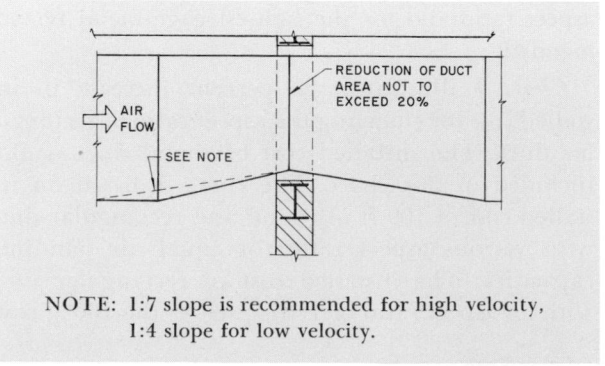

NOTE: 1:7 slope is recommended for high velocity, 1:4 slope for low velocity.

FIG. 20 — RECTANGULAR DUCT TRANSFORMATION TO AVOID OBSTRUCTION

In some air distribution systems, equipment such as heating coils is installed in the ductwork. Normally the equipment is larger than the ductwork and the duct area must be increased. The slope of the transformation piece on the upstream side of the equipment is limited to 30° as shown in *Fig. 22.* On the leaving side the slope should be not more than 45°.

Duct Reduction Increments

Accepted methods of duct design usually indicate a reduction in duct area after each terminal and branch take-off. Unless a reduction of at least 2 in. can be made, however, it is recommended that the original duct size be maintained. Savings in installed cost of as much as 25% can be realized by running the duct at the same size for several terminals.

All duct sizes should be even dimensions and all reductions should be in 2 in. increments, preferably in one dimension only. The recommended minimum duct size is 8 in. x 10 in. for fabricated shop ductwork.

Obstructions

Locating pipes, electrical conduit, structural members and other items inside the ductwork should always be avoided, especially in elbows and tees. Obstruction of any kind must be avoided inside

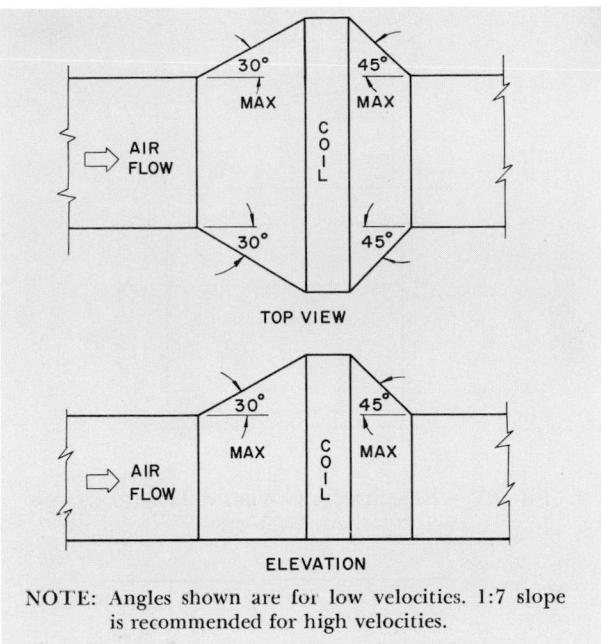

NOTE: Angles shown are for low velocities. 1:7 slope is recommended for high velocities.

FIG. 22 — DUCT TRANSFORMATION WITH EQUIPMENT IN THE DUCT

high velocity ducts. Obstructions cause unnecessary pressure loss and, in a high velocity system, may also be a source of noise in the air stream.

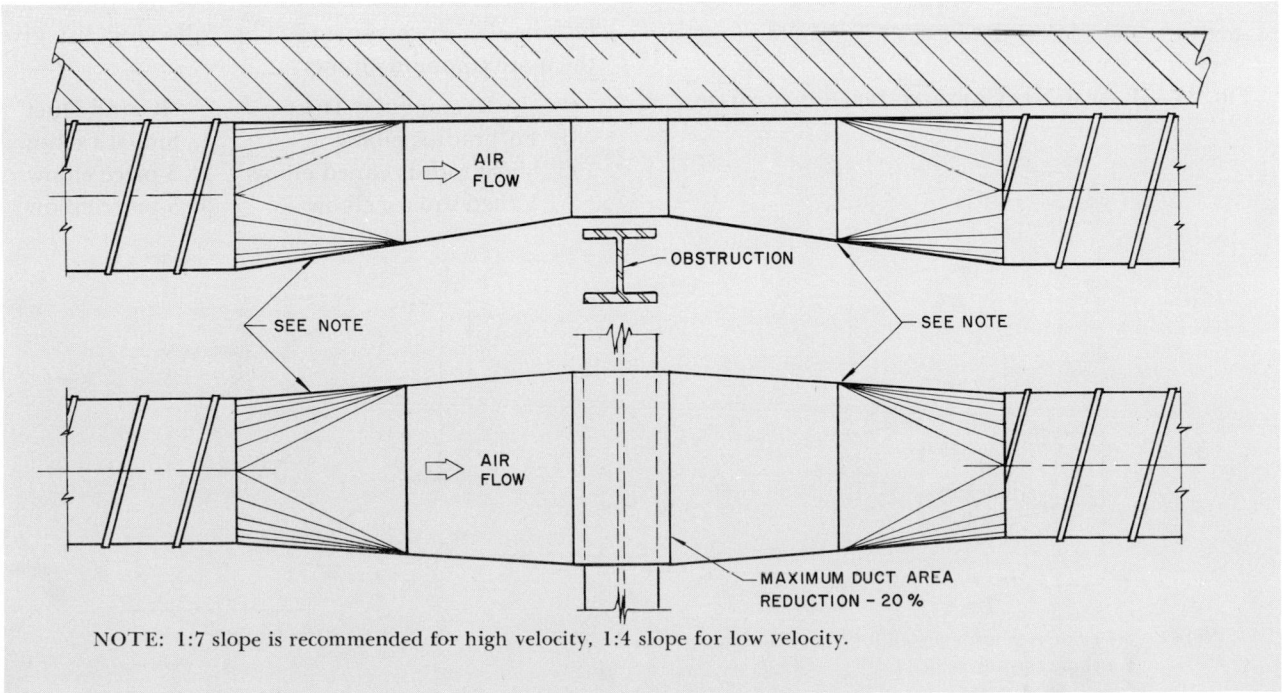

NOTE: 1:7 slope is recommended for high velocity, 1:4 slope for low velocity.

FIG. 21 — ROUND DUCT TRANSFORMATION TO AVOID OBSTRUCTION

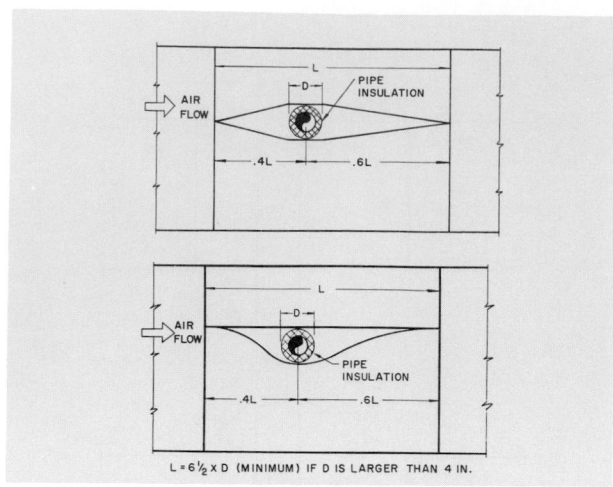

FIG. 23 — EASEMENTS COVERING OBSTRUCTIONS

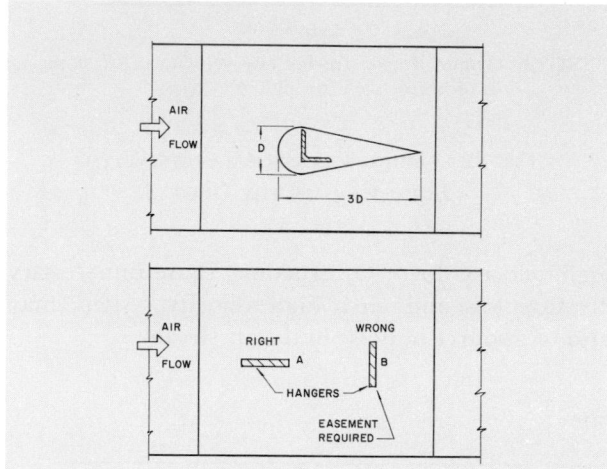

FIG. 24 — EASEMENTS COVERING IRREGULAR SHAPES

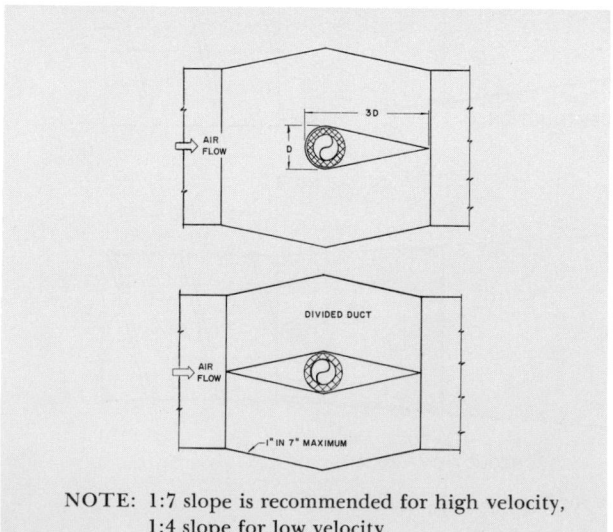

NOTE: 1:7 slope is recommended for high velocity,
1:4 slope for low velocity.

FIG. 25 — DUCT TRANSFORMED FOR EASEMENTS

In those few instances in which obstructions must pass thru the duct, use the following recommendations:

1. Cover all pipes and circular obstructions over 4 in. in diameter with an easement. Two typical easements are illustrated in *Fig. 23*.

2. Cover any flat or irregular shapes having a width exceeding 3 in. with an easement. Hangers or stays in the duct should be parallel to the air flow. If this is not possible, they should be covered with an easement. *Fig. 24* shows a tear drop-shaped easement covering an angle. Hanger *"B"* also requires an easement.

3. If the easement exceeds 20% of the duct area, the duct is transformed or split into two ducts. When the duct is split or transformed, the original area should be maintained. *Fig. 25* illustrates a duct transformed and a duct split to accommodate the easement. In the second case, the split duct acts as the easement. When the duct is split or transformed, slope recommendations for transformations should be followed.

4. If an obstruction restricts only the corner of the duct, that part of the duct is transformed to avoid the obstruction. The reduction in duct area must not exceed 20% of the original area.

Elbows

A variety of elbows is available for round and rectangular duct systems. The following list gives the more common elbows:

Rectangular Duct	Round Duct
1. Full radius elbow	1. Smooth elbow
2. Short radius vaned elbow	2. 3-piece elbow
3. Vaned square elbow	3. 5-piece elbow

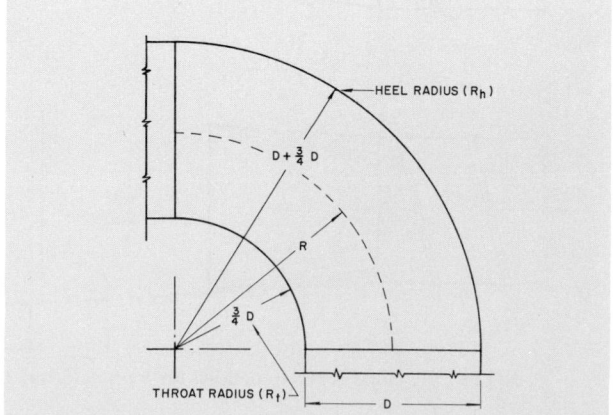

FIG. 26 — FULL RADIUS RECTANGULAR ELBOW

The elbows are listed in order of minimum cost. This sequence does not necessarily indicate the minimum pressure drop thru the elbow. *Tables 9 thru 12* show the losses for the various rectangular and round elbows.

Full radius elbows *(Fig. 26)* are constructed with a throat radius equal to ¾ of the duct dimension in the direction of the turn. An elbow having this throat radius has an R/D ratio of 1.25. This is considered to be an optimum ratio.

The short radius vaned elbow is shown in *Fig. 27*. This elbow can have one, two or three turning vanes. The vanes extend the full curvature of the elbow and their location is determined from *Chart 6. Example 2* illustrates the use of *Chart 6* in determining the location of the vanes in the elbow in *Fig. 28*.

Example 2 — Locating Vanes in a Rectangular Elbow

Given:
Rectangular elbow shown in *Fig. 28*.
Throat radius (R_t) — 3 in.
Duct dimension in direction of turn — 20 in.
Heel radius (R_h) — 23 in.

Find:
1. Spacing for two vanes.
2. R/D ratio of elbow.

Solution:
1. Enter *Chart 6* at $R_t = 3$ in. and $R_h = 23$ in. Read vane spacing for R_1 and R_2 (dotted line on chart).

$$R_1 = 6 \text{ in.} \quad R_2 = 12 \text{ in.}$$

2. The centerline radius R of the elbow equals 13 in. Therefore $R/D = 13/20 = .65$.

Although a throat radius is recommended, there may be instances in which a square corner is imperative. *Chart 6* can still be used to locate the vanes. A throat radius is assumed to equal one-tenth of the heel radius. *Example 3* illustrates vane location in an elbow with a square throat.

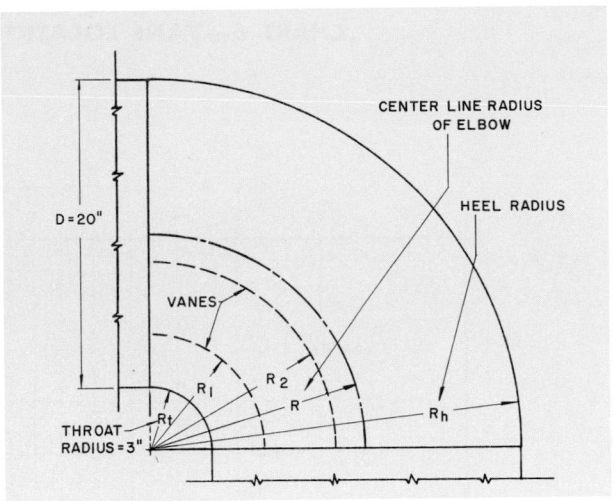

FIG. 28 — RECTANGULAR ELBOW VANE LOCATION

Example 3 — Locating Vanes in a Rectangular Elbow with a Square Throat

Given:
Elbow shown in *Fig. 29*.
Throat radius — none
Heel radius — 20 in.
Duct dimension in direction of turn — 20 in.

Find:
Vane spacing

Solution:
Assume a throat radius equal to .1 of the heel radius:
.1 x 20 = 2 in.
Enter *Chart 6* at $R_t = 2$, and $R_h = 20$ in. Read vane spacing for R_1 and R_2.

$$R_1 = 4.5 \text{ in.} \quad R_2 = 9.5 \text{ in.}$$

In addition a third vane is located at 2 in. which is the assumed throat radius.

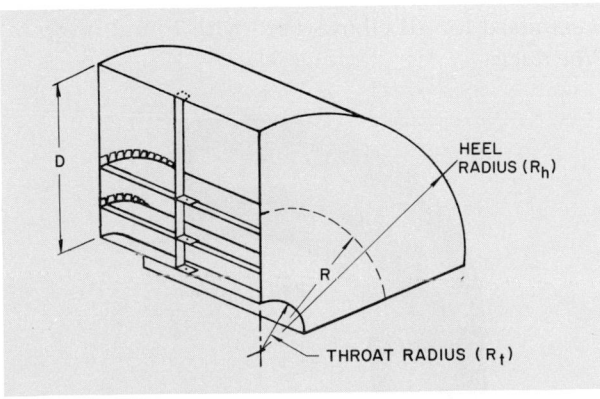

FIG. 27 — SHORT RADIUS VANED ELBOW

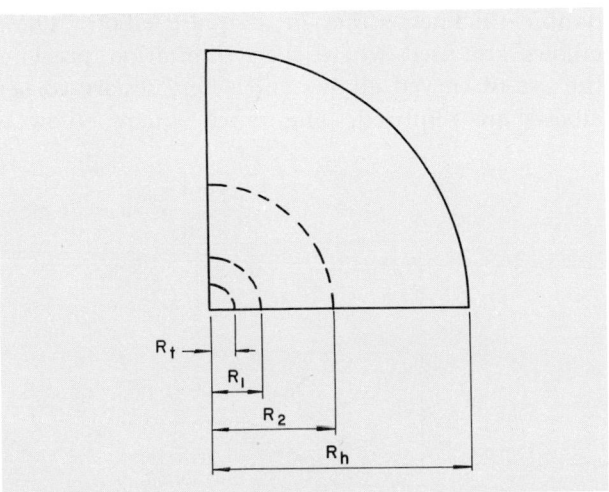

FIG. 29 — RECTANGULAR ELBOW WITH NO THROAT RADIUS

CHART 6—VANE LOCATION FOR RECTANGULAR ELBOWS

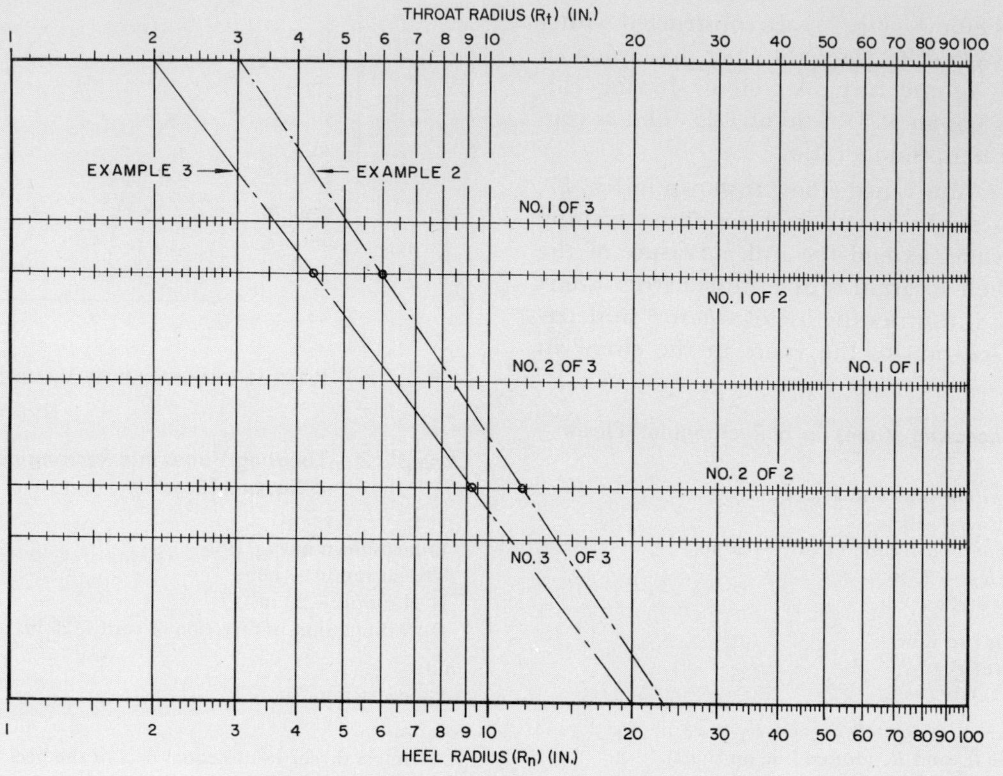

THROAT RADIUS (R$_t$) (IN.)

EXAMPLE 3 →   ← EXAMPLE 2

NO. I OF 3

NO. I OF 2

NO. 2 OF 3   NO. I OF I

NO. 2 OF 2

NO. 3 OF 3

HEEL RADIUS (R$_n$) (IN.)

From *Fan Engineering*, Buffalo Forge Co.

A vaned square elbow has either double or single thickness closely spaced vanes. *Fig. 30* illustrates double thickness vanes in a square elbow. These elbows are used where space limitation prevents the use of curved elbows and where square corner elbows are required. The vaned square elbow is expensive to construct and usually has a higher pressure drop than the vaned short radius elbow and the standard elbow (R/D = 1.25).

Smooth elbows are recommended for round or *Spira-Pipe* systems. *Fig. 31* illustrates a 90° smooth elbow with a R/D ratio of 1.5. This R/D ratio is standard for all elbows used with round or *Spira-Pipe* duct.

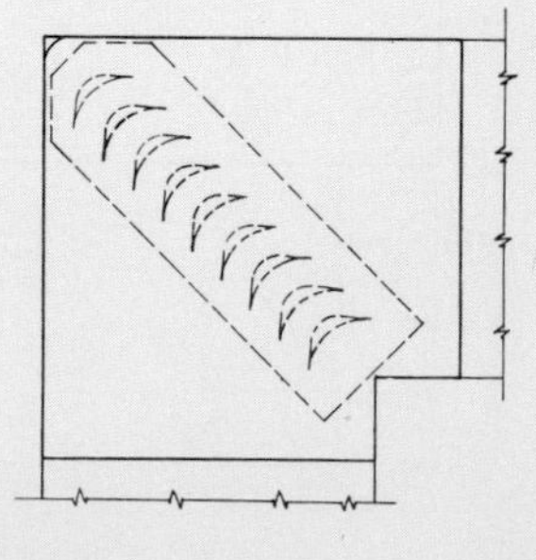

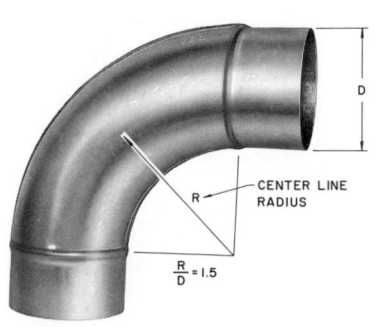

FIG. 30 — VANED SQUARE ELBOW

FIG. 31 — 90° SMOOTH ELBOW

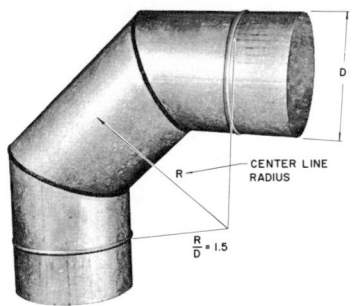

FIG. 32 — 90° 3-PIECE ELBOW

A 3-piece elbow *(Fig. 32)* has the same R/D ratio as a smooth elbow but has the highest pressure drop of either the smooth or 5-piece elbow *(Fig. 33)*. This elbow is second in construction costs and should be used when smooth elbows are unavailable.

The 5-piece elbow *(Fig. 33)* has the highest first cost of all three types. It is used only when it is necessary to reduce the pressure drop below that of the 3-piece elbow, and when smooth elbows are not available.

A 45° elbow is either smooth *(Fig. 34)* or 3-piece *(Fig. 35)*. A smooth 45° elbow is lower in first cost and pressure drop than the 3-piece 45° elbow. A 3-piece elbow is used when smooth elbows are not available.

Take-offs

There are several types of take-offs commonly used in rectangular duct systems. The recommendations given for rectangular elbows apply to take-offs. *Fig. 36* illustrates the more common take-offs. *Fig. 36A* is a take-off using a full radius elbow. In *Figs. 36A and 36B* the heel and throat radii originate from two different points since D is larger than D_1. The principal difference in *Figs. 36A and 36B* is that the take-off extends into the duct in *Fig. 36B* and there is no reduction in the main duct.

Figure 36C illustrates a tap-in take-off with no part of the take-off extending into the duct. This

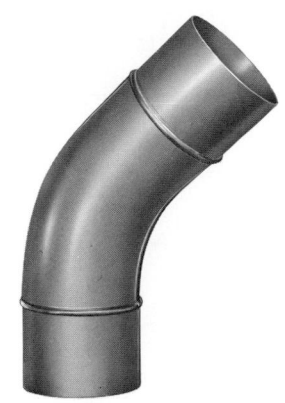

FIG. 34 — 45° SMOOTH ELBOW

type is often used when the quantity of air to be taken into the branch is small. The square elbow take-off *(Fig. 36D)* is the least desirable from a cost and pressure drop standpoint. It is limited in application to the condition in which space limitation prevents the use of a full radius elbow take-off.

A straight take-off *(Fig. 37)* is seldom used for duct branches. Its use is quite common, however, when a branch has only one outlet. In this instance it is called a collar. A splitter damper can be added for better control of the air to the take-off.

There are two varieties of take-offs for round and *Spira-Pipe* duct systems: the 90° tee *(Fig. 38)* and the 90° conical tee *(Fig. 39)*. A 90° conical tee is used when the air velocity in the branch exceeds 4000 fpm or when a smaller pressure drop than the straight take-off is required. Crosses with the take-offs located at 180°, 135° and 90° to each other are shown in *Fig. 40*.

When the duct system is designed, it may be necessary to reduce the duct size at certain take-offs. The reduction may be accomplished at the take-off *(Fig. 41)* or immediately after the take-off *(Fig. 42)*. Reduction at the take-off is recommended since it eliminates one fitting.

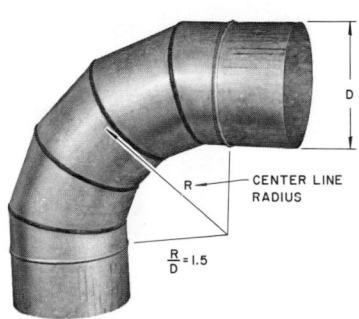

FIG. 33 — 90° 5-PIECE ELBOW

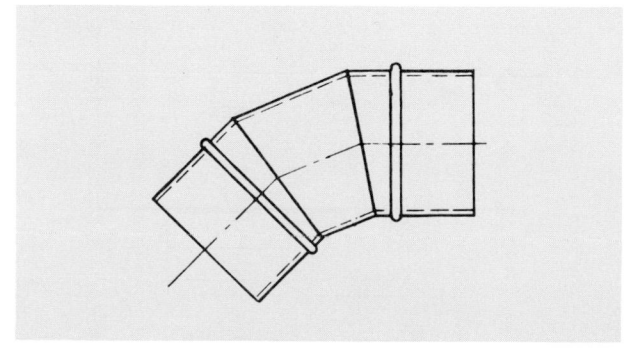

FIG. 35 — 45° 3-PIECE ELBOW

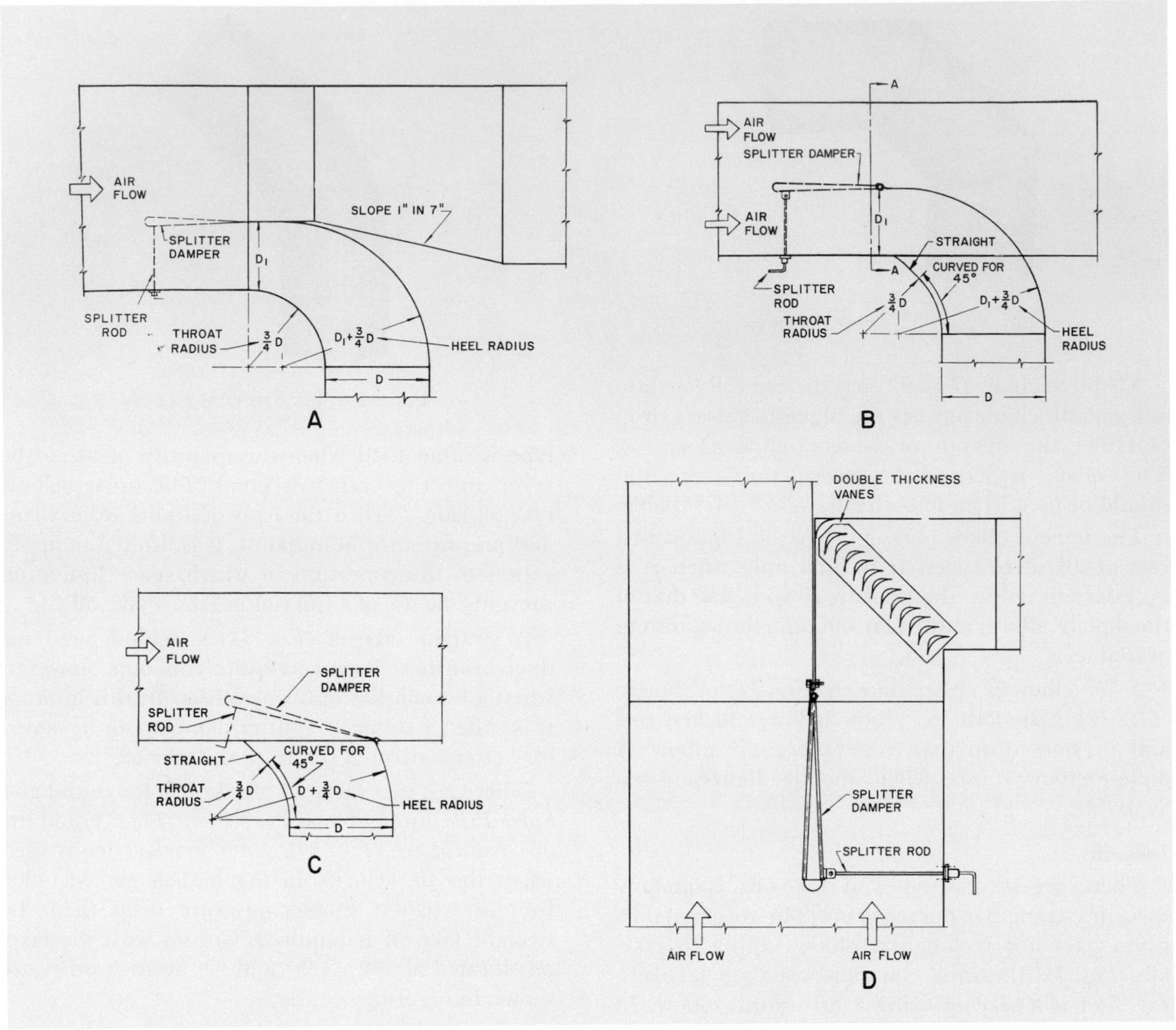

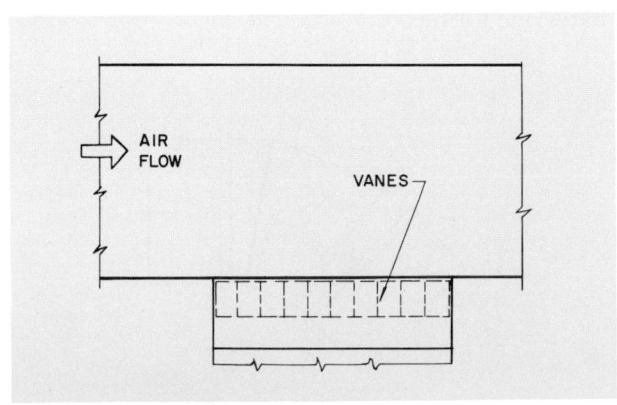

AIR FLOW

SPLITTER DAMPER

SLOPE 1" IN 7"

D_1

SPLITTER ROD

THROAT RADIUS

$\frac{3}{4}$ D

$D_1 + \frac{3}{4}$ D

HEEL RADIUS

D

A

A

AIR FLOW

SPLITTER DAMPER

D_1

STRAIGHT

CURVED FOR 45°

A

SPLITTER ROD

$\frac{3}{4}$ D

THROAT RADIUS

$D_1 + \frac{3}{4}$ D

HEEL RADIUS

D

B

AIR FLOW

SPLITTER DAMPER

SPLITTER ROD

STRAIGHT

CURVED FOR 45°

THROAT RADIUS

$\frac{3}{4}$ D

D + $\frac{3}{4}$ D

HEEL RADIUS

D

C

DOUBLE THICKNESS VANES

SPLITTER DAMPER

SPLITTER ROD

AIR FLOW   AIR FLOW

D

FIG. 36 — TYPICAL TAKE-OFFS

AIR FLOW

VANES

FIG. 37 — OUTLET COLLAR

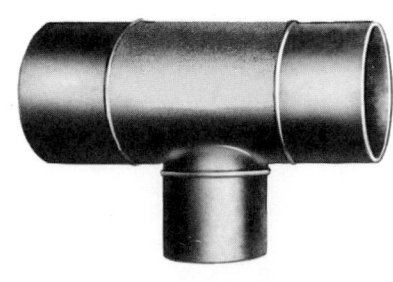

FIG. 38 — 90° TEE

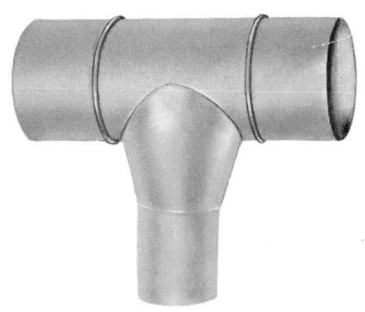

FIG. 39 — 90° CONICAL TEE

 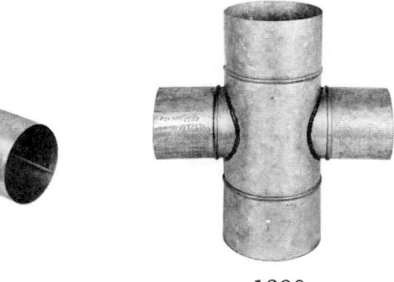

90°   180°

FIG. 40 — CROSSES

TABLE 4—MAXIMUM DIFFERENCE BETWEEN SUPPLY AIR TEMPERATURE AND ROOM DEWPOINT WITHOUT CONDENSING MOISTURE ON DUCTS (F)

AIR CONDITIONS SURROUNDING DUCT		AIR VELOCITY IN STRAIGHT RUN OF DUCT (FPM)*											
		Painted	Bright Metal	Painted	Bright Metal	Painted	Bright Metal	Painted	Bright Metal	Painted	Bright Metal	Painted	Bright Metal
DB (F)	RH (%)	400		800		1200		1600		2000		3000	
	45	20	15	15	9	11	8	8	5	7	4	5	3
	50	18	13	13	8	10	7	7	5	6	4	4	3
	55	15	11	11	7	8	6	6	4	5	3	4	2
74 - 100	60	13	10	10	6	7	5	5	3	4	3	3	2
	70	9	7	7	4	5	4	4	2	3	2	2	2
	80	6	4	4	3	3	2	2	2	2	1	2	1
	85	4	3	3	2	2	2	2	1	2	1	1	1
VALUE OF $\frac{f_2}{U} - 1$		.90	.66	.66	.42	.49	.31	.37	.24	.31	.20	.23	.15

*For elbows and other fittings, see Notes 4 and 7.

EQUATION: $t_{dp} - t_{sa} = (t_{rm} - t_{dp}) \left(\dfrac{f_2}{U} - 1 \right)$

where: t_{dp} = duct surface temp, assumed equal to room dewpoint.
   t_{sa} = supply air dry-bulb temp in duct.
   t_{rm} = room dry-bulb temp.

U = overall heat transmission coefficient of duct Btu/(hr) (sq ft) (deg F)

f_2 = film heat transmission coefficient on outside of duct, Btu/(hr) (sq ft) (deg F) = 1.65 for painted ducts and 1.05 for bright metal ducts.

NOTES:

1. Exceptional Cases: Condensation will occur at a lower relative humidity than indicated in the table when f_2 falls below the average value of 1.65 for painted ducts and 1.05 for bright metal ducts. The radiation component of f_2 will decrease when the duct is exposed to surfaces colder than the room air, as near a cold wall. The convection component will decrease for the top of ducts, and also where the air flow is obstructed, such as a duct running very close to a partition. If either condition exists, use value given for a relative humidity 5% less than the relative humidity in the room. If both conditions exist, use value given for 10% lower relative humidity.

2. Source: Calculated using film heat transmission coefficient on inside of duct ranging from 1.5 to 7.2 Btu/(hr) (sq ft) (deg F). The above equation is based on the principle that the temperature drop through any layer is directly proportional to its thermal resistance. It is assumed that the air movement surrounding the outside of the duct does not exceed approximately 50 fpm.

3. For Room Conditions Not Given: Use the above equation and the values of $f_2/U - 1$ shown at the bottom of the table.

4. Application: Use for bare ducts, not furred or insulated. Use the values for bright metal ducts for both unpainted aluminum and unpainted galvanized ducts. Condensation at elbows, transformations and other fittings will occur at a higher supply air temperature because of the higher inside film heat transmission coefficient due to the air impinging against the elbow or fitting. For low velocity fittings, assume an equivalent velocity of two times the straight run velocity and use the above table. For higher velocity fittings where straight run velocity is 1500 fpm and above, keep the supply air temperature no more than one degree lower than the room dewpoint. Transformations having a slope less than one in six may be considered as a straight run.

5. Bypass Factor and Fan Heat: The air leaving the dehumidifier will be higher than the apparatus dewpoint temperature when the bypass factor is greater than zero. Treat this as a mixture problem. Whenever the fan is on the leaving side of the dehumidifier, the supply air temperature is usually at least one to four degrees higher than the air leaving the dehumidifier, and can be calculated using the fan brake horsepower.

6. Dripping: Condensation will generally not be severe enough to cause dripping unless the surface temperature is two to three degrees below the room dewpoint. Note that the table is based on the duct surface temperature equal to the room dewpoint in estimating the possibility of dripping. It is recommended that the surface temperature be kept above the room dewpoint.

7. Elimination of Condensation: The supply air temperature must be high enough to prevent condensation at elbows and fittings. Occasionally, it might be desirable to insulate only the elbows or fittings. If moisture is expected to condense only at the fittings, apply insulation (½″ thick usually sufficient) either to the inside or outside of duct at the fitting and for a distance downstream equal to 1.5 times the duct perimeter. If condensation occurs on a straight run, the thickness of insulation required can be found by solving the above equation for U.

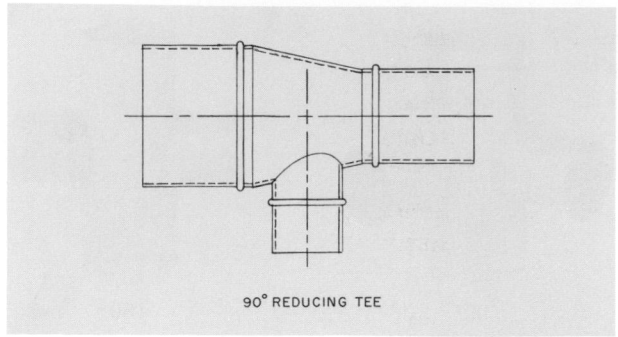

Fig. 41 — Reducing Duct Size At Take-off

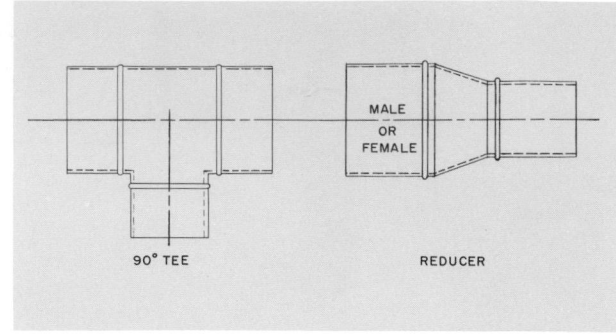

Fig. 42 — Reducing Duct Size After Take-off

Air Control

In low velocity air distribution systems the flow of air to the branch take-off is regulated by a splitter damper. The position of the splitter damper is adjusted by use of the splitter rod. Splitter dampers for rectangular duct systems are illustrated in *Fig. 36*. Pivot type dampers are sometimes installed in the branch line to control flow. When these are used, splitter dampers are omitted. Splitter dampers are preferred in low velocity systems, and pivot type or volume dampers are used in high velocity systems.

In high velocity systems, balancing or volume dampers are required at the air conditioning terminals to regulate the air quantity.

Duct Condensation

Ducts may "sweat" when the surface temperature of the duct is below the dewpoint of the surrounding air. *Table 4* lists the maximum difference between supply air temperature and room dewpoint without condensing moisture on the duct for various duct velocities. See the notes below the table for application of the data contained in the table. *Table 5* lists various U factors for common insulating material. It can be used in conjunction with *Table 4* to determine required insulation to eliminate condensation.

TABLE 5—DUCT HEAT TRANSMISSION COEFFICIENTS

TYPE DUCT INSULATION	FINISH	TOTAL THICKNESS (in.)	WEIGHT (lb/sq ft)	K*	U†
Uninsulated Sheet Metal	None	—	—	—	1.14‡
	Metal lath and plaster—¾″	—	—	—	.99
	Wood lath and plaster—¾″	—	—	—	.79
Corkboard	None	1	0.7	.28	.22
	None	2	1.4	—	.12
	Plaster—⅜″	1	2.2	.28	.22
	Plaster—⅜″	2	2.9	—	.12
Corrugated Asbestos Paper (air cell)	None	1	0.73	.50	.34
	None	2	1.46	—	.20
Rock Cork	None	1	1.35	.35	.23
	None	2	2.7	—	.13
	Plaster—⅜″	1	2.9	.35	.23
	Plaster—⅜″	2	4.2	—	.13
Mineral Wool Blanket	None	1	1.17	.28	.22
	None	2	2.35	—	.12
Glass Fiber	None	1	.08	.27	.21
	None	2	.17	—	.10
85% Magnesia	None	1	1.0	.39	.26

*Conductivity of insulating material only (per in.)
†Overall U for still air outside duct and 1200 fpm inside duct.
‡Uninsulated Bare Duct.

Air Velocity (fpm)	400	800	1200	1600	2000
Overall U	.98	1.08	1.14	1.19	1.22

DUCT SYSTEM ACCESSORIES

Fire dampers, access doors and sound absorbers are accessories which may be required in a duct system but do not materially affect the design, unless there are several dampers in series. For this arrangement, the additional resistance to air flow must be recognized when selecting the fan.

Fire Dampers

Normally local or state codes dictate the use, location and construction of fire dampers for an air distribution system. The National Board of Fire Underwriters describes the general construction and installation practices in pamphlet NBFU 90A.

There are two principal types of fire dampers used in rectangular ductwork:

1. The rectangular pivot damper *(Fig. 43)* which may be used in either the vertical or horizontal position.

2. The rectangular louver fire damper which may be used only in the horizontal position *(Fig. 44)*.

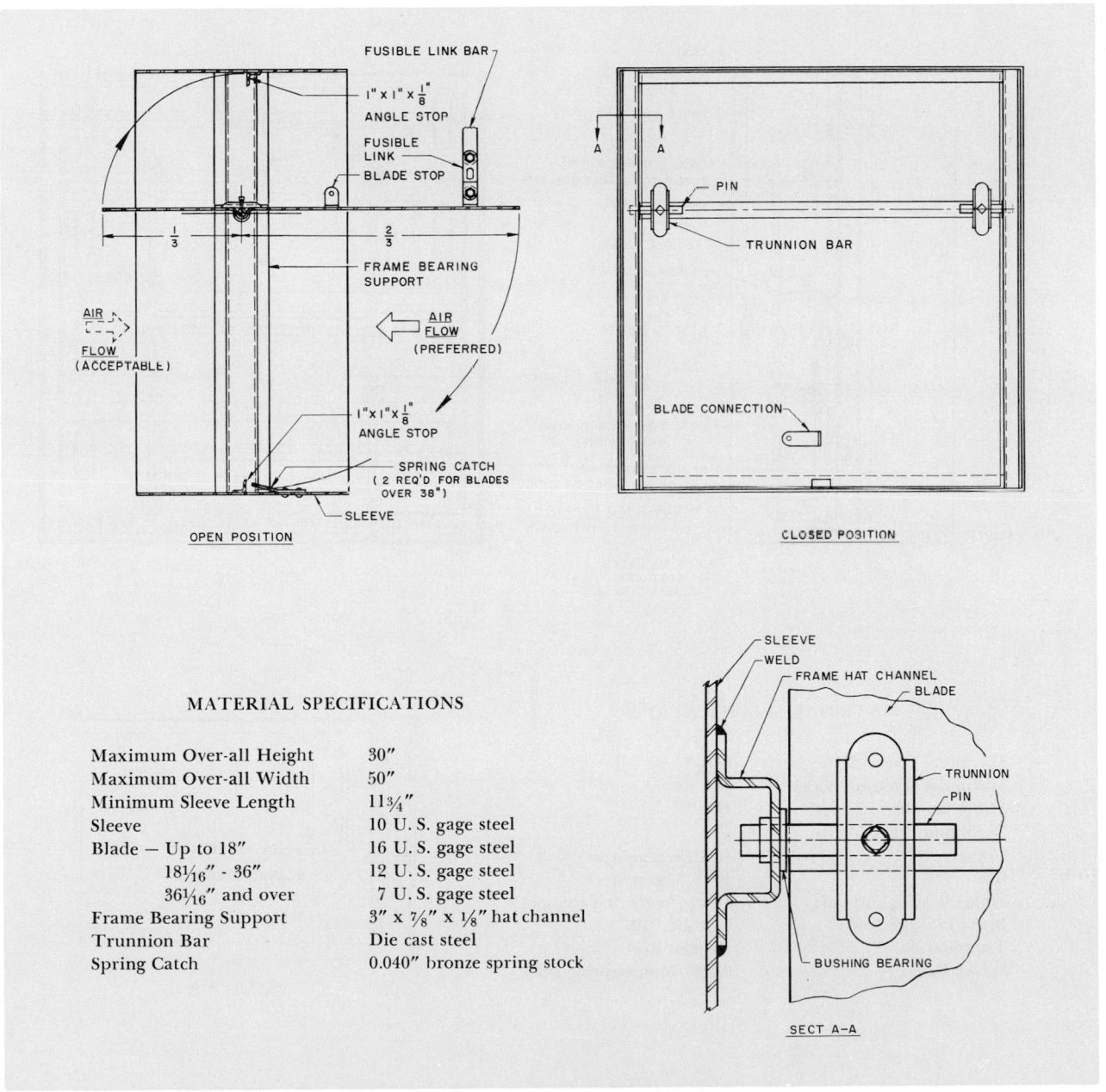

MATERIAL SPECIFICATIONS

Maximum Over-all Height	30"
Maximum Over-all Width	50"
Minimum Sleeve Length	11¾"
Sleeve	10 U.S. gage steel
Blade — Up to 18"	16 U.S. gage steel
18¹⁄₁₆" - 36"	12 U.S. gage steel
36¹⁄₁₆" and over	7 U.S. gage steel
Frame Bearing Support	3" x ⅞" x ⅛" hat channel
Trunnion Bar	Die cast steel
Spring Catch	0.040" bronze spring stock

FIG. 43 — RECTANGULAR PIVOT FIRE DAMPER

Figure 45 illustrates a pivot fire damper for round duct systems. This damper may be used in either the horizontal or vertical position.

Access Doors

Access doors or access panels are required in duct systems before and after equipment installed in ducts. Access panels are also required for access to fusible links in fire dampers.

DUCT DESIGN

This section presents the necessary data for designing low and high velocity duct systems. This data includes the standard air friction charts, recommended design velocities, losses thru elbows and fittings, and the common methods of designing the air distribution systems. Information is given also for evaluating the effects of duct heat gain and altitude on system design.

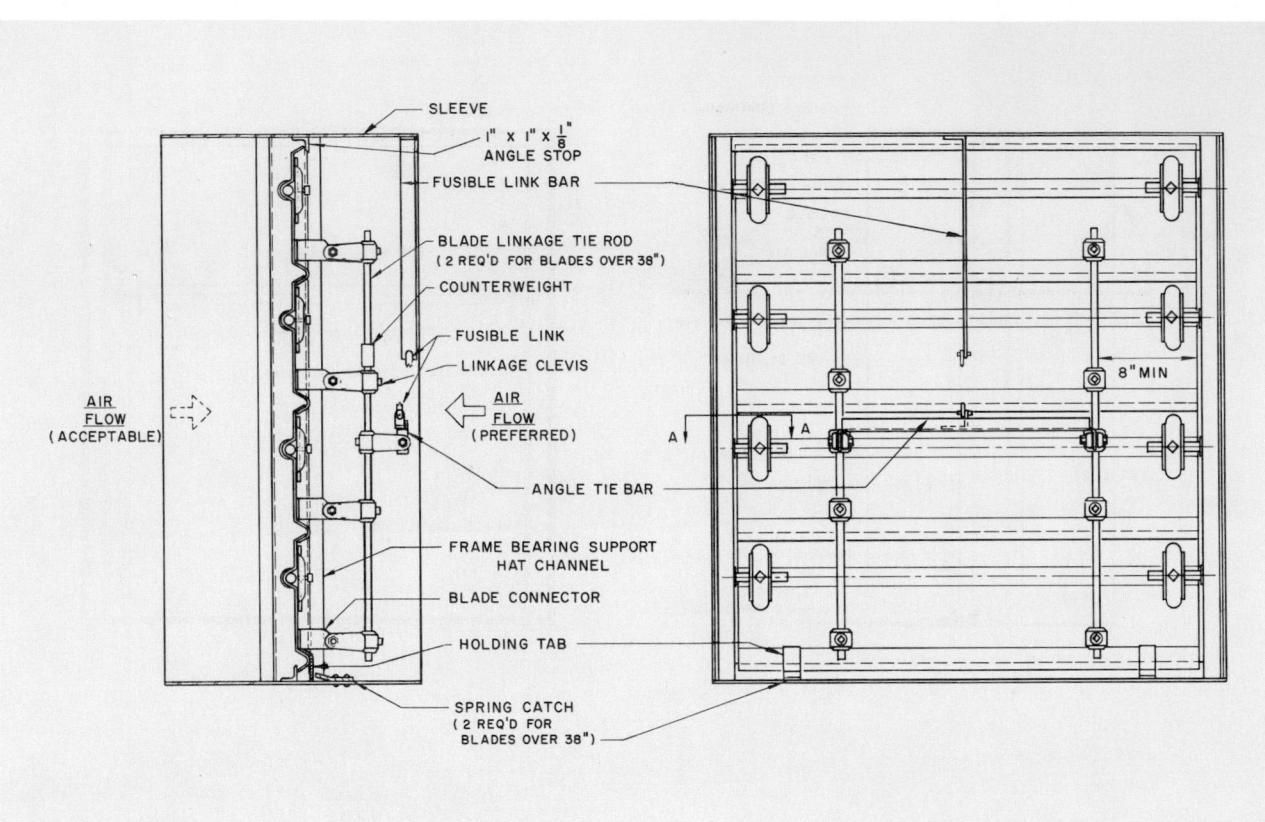

MATERIAL SPECIFICATIONS

Maximum Over-all Height	91½″
Maximum Over-all Width	50″
Minimum Sleeve Length	11¾″
Maximum Blade Width	6″
Sleeve	10 U. S. gage steel
Blade	16 U. S. gage steel
Frame Bearing Support	3″ x ⅞″ x ⅛″ hat channel
Blade Linkage Rod	5⁄16″ die CRS
Trunnion Bar	Die cast steel
Spring Catch	0.040″ bronze spring stock

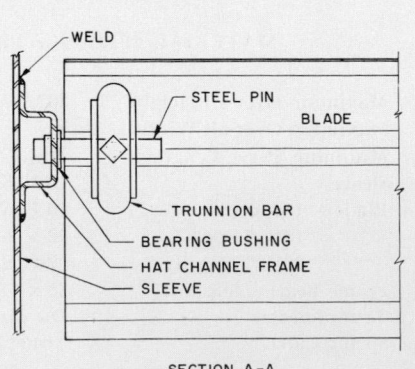

SECTION A-A

FIG. 44 — RECTANGULAR LOUVER FIRE DAMPER

Friction Chart

In any duct section thru which air is flowing, there is a continuous loss of pressure. This loss is called duct friction loss and depends on the following:

1. Air velocity
2. Duct size
3. Interior surface roughness
4. Duct length

Varying any one of these four factors influences the friction loss in the ductwork. The relationship of these factors is illustrated in the following equation:

$$\Delta P = 0.03\, f \left(\frac{L}{d^{1.22}} \right) \left(\frac{V}{1000} \right)^{1.82}$$

where: ΔP = friction loss (in. wg)

f = interior surface roughness (0.9 for galvanized duct)

L = length of duct (ft)

d = duct diameter (in.), equivalent diam. for rectangular ductwork

V = air velocity (fpm)

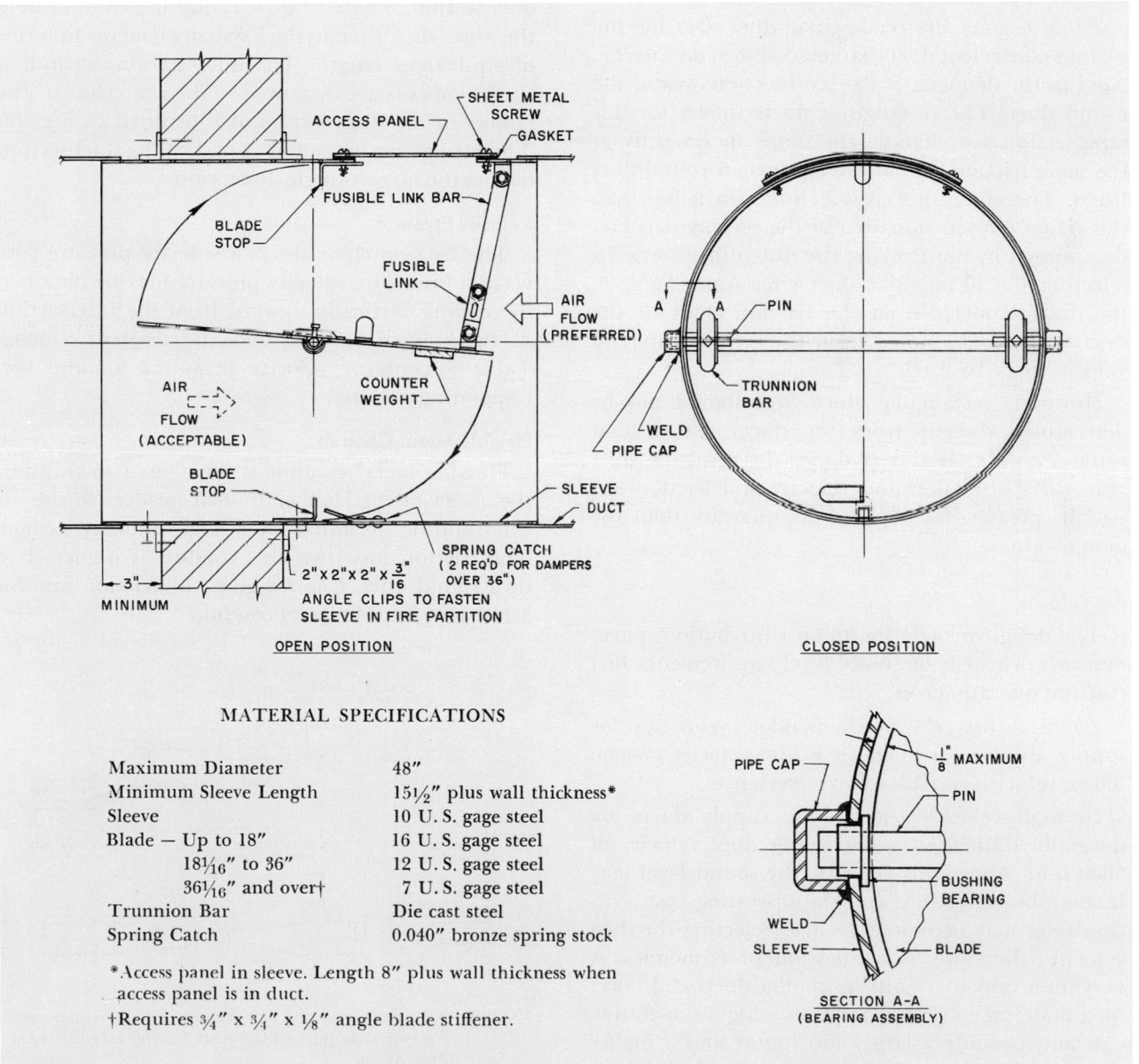

OPEN POSITION   CLOSED POSITION

MATERIAL SPECIFICATIONS

Maximum Diameter	48″
Minimum Sleeve Length	15½″ plus wall thickness*
Sleeve	10 U. S. gage steel
Blade — Up to 18″	16 U. S. gage steel
18¹⁄₁₆″ to 36″	12 U. S. gage steel
36¹⁄₁₆″ and over†	7 U. S. gage steel
Trunnion Bar	Die cast steel
Spring Catch	0.040″ bronze spring stock

*Access panel in sleeve. Length 8″ plus wall thickness when access panel is in duct.

†Requires ¾″ x ¾″ x ⅛″ angle blade stiffener.

SECTION A-A
(BEARING ASSEMBLY)

FIG. 45 — ROUND PIVOT FIRE DAMPER

This equation is used to construct the standard friction chart *(Chart 7)* based on galvanized duct and air at 70 F and 29.92 in. Hg. This chart may be used for systems handling air from 30 to 120 F and for altitudes up to 2000 ft without correcting the air density. *Page 59* contains the data for designing high altitude air distribution systems.

Air Quantity

The total supply air quantity and the quantity required for each space is determined from the air conditioning load estimate in *Part I.*

Duct Diameter

Table 6 gives the rectangular duct sizes for the various equivalent duct diameters shown on *Chart 7*. Next to the diameter is the cross-section area of the round duct. The rectangular ducts shown for this cross-section area handle the same air quantity at the same friction rate as the equivalent round duct listed. Therefore, this cross-section area is less than the actual cross-section area of the rectangular duct determined by multiplying the duct dimensions. In selecting the rectangular duct sizes from *Table 6,* the duct diameter from the friction chart or the duct area as determined from the air quantity and velocity may be used.

However, rectangular duct sizes should not be determined directly from the duct area without using *Table 6*. If this is done, the resulting duct sizes will be smaller, and velocity and friction loss will be greater, for a given air quantity than the design values.

Air Velocity

The design velocity for an air distribution system depends primarily on sound level requirements, first cost and operating cost.

Table 7 lists the recommended velocities for supply and return ducts in a low velocity system. These velocities are based on experience.

In high velocity systems, the supply ducts are normally limited to a maximum duct velocity of 5000 fpm. Above this velocity, the sound level may become objectionable and the operating cost (friction rate) may become excessive. Selecting the duct velocity, therefore, is a question of economics. A very high velocity results in smaller ducts and lower duct material cost but it requires a higher operating cost and possibly a larger fan motor and a higher class fan. If a lower duct velocity is used, the ducts must be larger but the operating cost decreases and the fan motor and fan class may be smaller.

The return ducts for a high velocity supply system have the same design velocity recommendations as listed in *Table 7* for a low velocity system, unless extensive sound treatment is provided to use higher velocities.

Friction Rate

The friction rate on the friction chart is given in terms of inches of water per 100 ft of equivalent length of duct. To determine the loss in any section of ductwork, the total equivalent length in that section is multiplied by the friction rate which gives the friction loss. The total equivalent length of duct includes all elbows and fittings that may be in the duct section. *Tables 9 thru 12* are used to evaluate the losses thru various duct system elements in terms of equivalent length. The duct sections including these elements are measured to the centerline of the elbows in the duct section as illustrated in *Fig. 46.* The fittings are measured as part of the duct section having the largest single dimension.

Velocity Pressure

The friction chart shows a velocity pressure conversion line. The velocity pressure may be obtained by reading vertically upward from the intersection of the conversion line and the desired velocity. *Table 8* contains velocity pressures for the corresponding velocity.

Flexible Metal Conduit

Flexible metal conduit is often used to transmit the air from the riser or branch headers to the air conditioning terminal in a high velocity system. The friction loss thru this conduit is higher than thru round duct. *Chart 8* gives the friction rate for 3 and 4 in. flexible metal conduit.

(Continued on page 38)

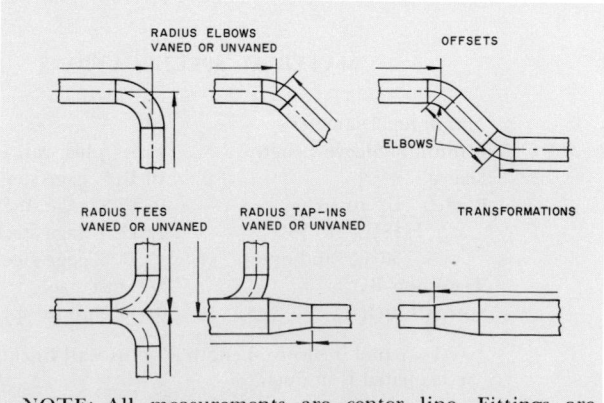

NOTE: All measurements are center line. Fittings are measured as part of the duct having largest single dimension.

FIG. 46 — GUIDE FOR MEASURING DUCT LENGTHS

CHART 7—FRICTION LOSS FOR ROUND DUCT

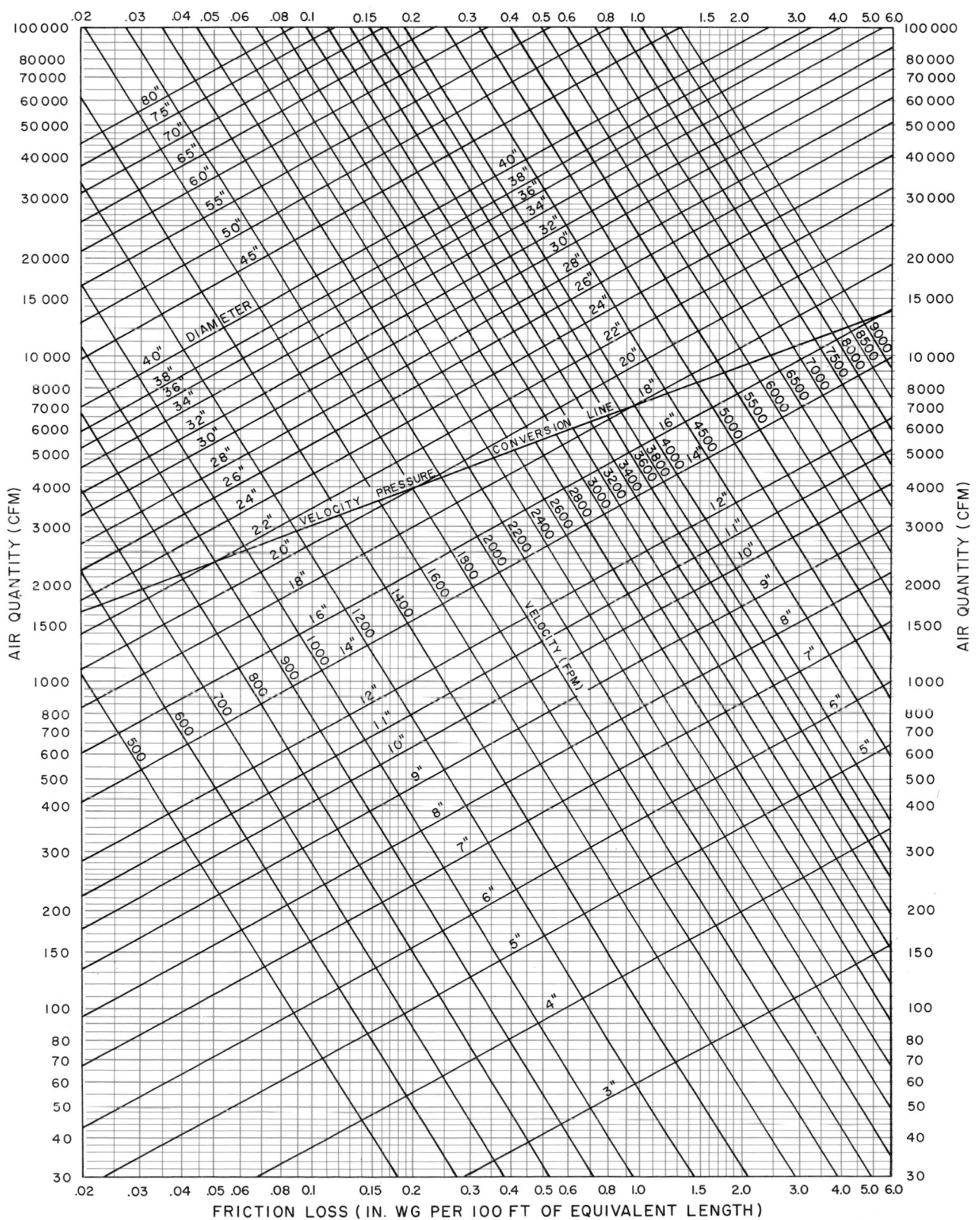

TABLE 6—DUCT DIMENSIONS, SECTION AREA, CIRCULAR EQUIVALENT DIAMETER,* AND DUCT CLASS†

SIDE	6 Area sq ft	6 Diam in.	8 Area sq ft	8 Diam in.	10 Area sq ft	10 Diam in.	12 Area sq ft	12 Diam in.	14 Area sq ft	14 Diam in.	16 Area sq ft	16 Diam in.	18 Area sq ft	18 Diam in.	20 Area sq ft	20 Diam in.	22 Area sq ft	22 Diam in.
10	.39	8.4	.52	9.8	.65	10.9												
12	.45	9.1	.62	10.7	.77	11.9	.94	13.1										
14	.52	9.8	.72	11.5	.91	12.9	1.09	14.2	1.28	15.3								
16	.59	10.4	.81	12.2	1.02	13.7	1.24	15.1	1.45	16.3	1.67	17.5						
18	.66	11.0	.91	12.9	1.15	14.5	1.40	16.0	1.63	17.3	1.87	18.5	2.12	19.7				
20	.72	11.5	.99	13.5	1.26	15.2	1.54	16.8	1.81	18.2	2.07	19.5	2.34	20.7	2.61	21.9		
22	.78	12.0	1.08	14.1	1.38	15.9	1.69	17.6	1.99	19.1	2.27	20.4	2.57	21.7	2.86	22.9	3.17	24.1
24	.84	12.4	1.16	14.6	1.50	16.6	1.83	18.3	2.14	19.8	2.47	21.3	2.78	22.6	3.11	23.9	3.43	25.1
26	.89	12.8	1.26	15.2	1.61	17.2	1.97	19.0	2.31	20.6	2.66	22.1	3.01	23.5	3.35	24.8	3.71	26.1
28	.95	13.2	1.33	15.6	1.71	17.7	2.09	19.6	2.47	21.3	2.86	22.9	3.25	24.4	3.60	25.7	4.00	27.1
30	1.01	13.6	1.41	16.1	1.82	18.3	2.22	20.2	2.64	22.0	3.06	23.7	3.46	25.2	3.89	26.7	4.27	28.0
32	1.07	14.0	1.48	16.5	1.93	18.8	2.36	20.8	2.81	22.7	3.25	24.4	3.68	26.0	4.12	27.5	4.55	28.9
34	1.13	14.4	1.58	17.0	2.03	19.3	2.49	21.4	2.96	23.3	3.43	25.1	3.89	26.7	4.37	28.3	4.81	29.7
36	1.18	14.7	1.65	17.4	2.14	19.8	2.61	21.9	3.11	23.9	3.63	25.8	4.09	27.4	4.58	29.0	5.07	30.5
38	1.23	15.0	1.73	17.8	2.25	20.3	2.76	22.5	3.27	24.5	3.80	26.4	4.30	28.1	4.84	29.8	5.37	31.4
40	1.28	15.3	1.81	18.2	2.33	20.7	2.88	23.0	3.43	25.1	3.97	27.0	4.52	28.8	5.07	30.5	5.62	32.1
42	1.33	15.6	1.86	18.5	2.43	21.1	2.98	23.4	3.57	25.6	4.15	27.6	4.71	29.4	5.31	31.2	5.86	32.8
44	1.38	15.9	1.95	18.9	2.52	21.5	3.11	23.9	3.71	26.1	4.33	28.2	4.90	30.0	5.55	31.9	6.12	33.5
46	1.43	16.2	2.01	19.2	2.61	21.9	3.22	24.3	3.88	26.7	4.49	28.7	5.10	30.6	5.76	32.5	6.37	34.2
48	1.48	16.5	2.09	19.6	2.71	22.3	3.35	24.8	4.03	27.2	4.65	29.2	5.30	31.2	5.97	33.1	6.64	34.9
50			2.16	19.9	2.81	22.7	3.46	25.2	4.15	27.6	4.84	29.8	5.51	31.8	6.19	33.7	6.87	35.5
52			2.22	20.2	2.91	23.1	3.57	25.6	4.30	28.1	5.00	30.3	5.72	32.4	6.41	34.3	7.14	36.0
54			2.29	20.5	2.98	23.4	3.71	26.1	4.43	28.5	5.17	30.8	5.90	32.9	6.64	34.9	7.38	36.8
56			2.38	20.9	3.09	23.8	3.83	26.5	4.55	28.9	5.31	31.2	6.08	33.4	6.87	35.5	7.62	37.4
58			2.43	21.1	3.19	24.2	3.94	26.9	4.68	29.3	5.48	31.7	6.26	33.9	7.06	36.0	7.87	38.0
60			2.50	21.4	3.27	24.5	4.06	27.3	4.84	29.8	5.65	32.2	6.50	34.5	7.26	36.5	8.12	38.6
64			2.64	22.0	3.46	25.2	4.24	27.9	5.10	30.6	5.91	33.1	6.87	35.5	7.71	37.6	8.59	39.7
68					3.63	25.8	4.49	28.7	5.37	31.4	6.26	33.9	7.18	36.3	8.12	38.6	9.03	40.7
72					3.83	26.5	4.71	29.4	5.69	32.3	6.60	34.8	7.54	37.2	8.50	39.5	9.52	41.8
76					4.09	27.4	4.91	30.0	5.86	32.8	6.83	35.4	7.95	38.2	8.90	40.4	9.98	42.8
80					4.15	27.6	5.17	30.8	6.15	33.6	7.22	36.4	8.29	39.0	9.21	41.1	10.4	43.8
84							5.41	31.5	6.41	34.5	7.54	37.2	8.55	39.6	9.75	42.3	10.8	44.6
88							5.58	32.0	6.64	34.9	7.87	38.0	8.94	40.5	10.1	43.1	11.2	45.4
92							5.79	32.6	6.91	35.6	8.12	38.6	9.39	41.5	10.4	43.8	11.7	46.3
96							5.90	33.0	7.14	36.2	8.40	39.2	9.70	42.1	10.8	44.5	12.1	47.2
100									7.40	36.9	8.50	39.5	9.80	42.5	11.3	45.5	12.3	47.6
104									7.60	37.4	8.90	40.5	10.3	43.5	11.6	46.2	13.0	48.8
108									7.90	38.0	9.20	41.2	10.6	44.0	12.0	47.0	13.4	49.6
112									8.10	38.6	9.50	41.8	10.9	44.7	12.3	47.5	13.8	50.3
116											9.80	42.4	11.3	45.5	12.6	48.1	14.3	51.3
120											10.0	42.8	11.5	46.0	13.1	49.1	14.4	51.5
124											10.3	43.5	11.9	46.7	13.4	49.6	15.0	52.4
128											10.6	44.1	12.1	47.1	13.8	50.4	15.5	53.3
132													12.5	47.9	14.1	50.9	15.8	53.9
136													12.8	48.5	14.5	51.6	16.2	54.5
140													13.0	48.8	14.7	52.0	16.5	55.0
144													13.3	49.4	15.2	52.9	16.8	55.6

*Circular equivalent diameter (d_e). Calculated from $d_e = 1.3 \dfrac{(ab)^{.625}}{(a + b)^{.25}}$ †Large numbers in table are duct class.

TABLE 6. DUCT DIMENSIONS, SECTION AREA, CIRCULAR EQUIVALENT DIAMETER,* AND DUCT CLASS† (Cont.)

SIDE	24 Area sq ft	24 Diam in.	26 Area sq ft	26 Diam in.	28 Area sq ft	28 Diam in.	30 Area sq ft	30 Diam in.	32 Area sq ft	32 Diam in.	34 Area sq ft	34 Diam in.	36 Area sq ft	36 Diam in.	38 Area sq ft	38 Diam in.	40 Area sq ft	40 Diam in.
10																		
12																		
14																		
16																		
18																		
20																		
22																		
24	3.74	26.2																
26	4.03	27.2	4.40	28.4														
28	4.33	28.2	4.74	29.5	5.10	30.6												
30	4.68	29.3	5.07	30.5	5.44	31.6	5.86	32.8										
32	4.94	30.1	5.37	31.4	5.79	32.6	6.23	33.8	6.68	35.0								
34	5.24	31.0	5.69	32.3	6.15	33.6	6.60	34.8	7.06	36.0	7.54	37.2						
36	5.58	32.0	5.94	33.0	6.52	34.6	6.99	35.8	7.46	37.0	7.95	38.2	8.46	39.4				
38	5.86	32.8	6.38	34.2	6.87	35.5	7.34	36.7	7.87	38.0	8.37	39.2	8.89	40.4	9.43	41.6		
40	6.15	33.6	6.71	35.1	7.22	36.4	7.71	37.6	8.29	39.0	8.81	40.2	9.34	41.4	9.89	42.6	10.5	43.8
42	6.45	34.4	7.03	35.9	7.58	37.3	8.12	38.6	8.68	39.9	9.21	41.1	9.80	42.4	10.4	43.6	11.0	44.8
44	6.75	35.2	7.34	36.7	7.91	38.1	8.50	39.5	9.07	40.8	9.61	42.0	10.3	43.4	10.8	44.6	11.4	45.8
46	7.03	35.9	7.63	37.4	8.25	38.9	8.85	40.3	9.48	41.7	10.1	43.0	10.7	44.3	11.3	45.6	11.9	46.8
48	7.30	36.6	7.95	38.2	8.59	39.7	9.25	41.2	9.89	42.6	10.5	43.9	11.1	45.2	11.8	46.5	12.4	47.8
50	7.58	37.3	0.25	38.9	8.90	40.4	9.61	42.0	10.3	43.5	10.9	44.8	11.6	46.1	12.2	47.4	13.0	48.8
52	7.87	38.0	8.55	39.6	9.25	41.2	9.98	42.8	10.7	44.3	11.4	45.7	12.1	47.1	12.7	48.3	13.5	49.7
54	8.16	38.7	8.85	40.3	9.61	42.0	10.4	43.6	11.0	45.0	11.8	46.5	12.6	48.0	13.2	49.2	14.0	50.6
56	8.42	39.3	9.16	41.0	9.94	42.7	10.7	44.3	11.4	45.8	12.2	47.3	13.0	48.8	13.7	50.1	14.5	51.5
58	8.63	39.8	9.48	41.7	10.3	43.4	11.0	45.0	11.8	46.6	12.6	48.1	13.4	49.6	14.2	51.0	15.0	52.4
60	8.89	40.4	9.75	42.3	10.5	44.0	11.4	45.8	12.2	47.3	13.0	48.9	13.8	50.4	14.6	51.8	15.5	53.3
64	9.43	41.6	10.3	43.5	11.2	45.4	12.1	47.2	12.9	48.7	13.8	50.4	14.7	52.0	15.5	53.4	16.5	55.0
68	9.98	42.8	10.9	44.7	11.8	46.6	12.8	48.4	13.7	50.2	14.6	51.8	15.6	53.5	16.5	55.0	17.5	56.6
72	10.4	43.8	11.5	45.9	12.4	47.8	13.5	49.7	14.4	51.5	15.4	53.2	16.4	54.9	17.4	56.5	18.3	58.0
76	10.8	44.9	12.0	47.0	13.1	49.0	14.1	50.8	15.1	52.7	16.2	54.6	17.3	56.3	18.3	57.9	19.3	59.5
80	11.5	46.0	12.6	48.0	13.7	50.1	14.7	52.0	15.8	53.9	17.0	55.8	18.1	57.6	19.2	59.3	20.3	61.0
84	12.0	46.9	13.2	49.2	14.2	51.1	15.4	53.2	16.5	55.0	17.7	57.0	18.9	58.9	20.1	60.7	21.2	62.4
88	12.5	47.9	13.7	50.1	14.8	52.2	16.1	54.3	17.3	56.3	18.5	58.2	19.7	60.1	20.9	62.0	22.1	63.7
92	12.9	48.7	14.2	51.1	15.5	53.4	16.7	55.4	18.0	57.4	19.2	59.4	20.5	61.3	21.8	63.2	23.0	65.0
96	13.3	49.5	14.8	52.2	15.9	54.0	17.2	56.2	18.6	58.5	19.7	60.2	21.1	62.2	22.7	64.5	24.0	66.3
100	13.9	50.6	15.0	52.5	16.7	55.3	17.9	57.3	19.2	59.4	20.6	61.5	21.6	63.0	23.4	65.5	24.8	67.5
104	14.6	51.8	15.8	53.9	17.1	56.0	18.6	58.5	19.9	60.5	21.4	62.6	22.7	64.5	24.1	66.5	25.6	68.5
108	14.8	52.1	16.2	54.6	17.6	56.8	19.2	59.4	20.5	61.4	22.0	63.5	23.5	65.7	24.8	67.5	26.5	69.7
112	15.1	52.7	16.8	55.5	18.3	58.0	19.7	60.1	21.1	62.3	22.5	64.3	24.5	67.0	25.7	68.7	27.1	70.5
116	15.8	53.9	17.3	56.4	18.9	58.9	20.3	61.1	22.0	63.6	23.5	65.7	24.8	67.5	26.2	69.4	28.2	71.9
120	16.2	54.6	17.8	57.1	19.4	59.6	20.9	62.0	22.7	64.5	24.2	66.7	26.1	69.2	27.2	70.6	29.0	73.0
124	16.6	55.2	18.4	58.1	19.8	60.3	21.6	63.0	23.2	65.4	25.2	68.0	26.5	69.8	28.2	71.9	29.8	74.0
128	17.1	56.0	18.8	58.8	20.3	61.1	22.3	64.0	23.7	66.0	25.6	68.6	27.3	70.8	28.7	72.6	30.2	74.5
132	17.4	56.5	19.3	59.5	20.8	61.8	22.6	64.4	24.5	67.0	26.3	69.5	28.2	72.0	29.8	74.0	32.0	76.6
136	17.9	57.3	19.7	60.2	21.4	62.7	23.0	65.0	25.1	67.9	26.9	70.3	28.7	72.6	30.5	74.8	32.6	77.3
140	18.5	58.2	20.3	61.0	22.3	64.0	24.1	66.5	25.9	69.0	27.5	71.1	29.4	73.5	31.5	76.0	33.4	78.3
144	18.8	58.7	20.6	61.5	22.7	64.5	24.8	67.5	26.3	69.5	28.2	72.0	29.9	74.1	32.0	76.6	34.0	79.0

*Circular equivalent diameter (d$_e$). Calculated from $d_e = 1.3 \dfrac{(ab)^{.625}}{(a + b)^{.25}}$ †Large numbers in table are duct class.

TABLE 6. DUCT DIMENSIONS, SECTION AREA, CIRCULAR EQUIVALENT DIAMETER,* AND DUCT CLASS† (Cont.)

SIDE	42 Area sq ft	42 Diam in.	44 Area sq ft	44 Diam in.	46 Area sq ft	46 Diam in.	48 Area sq ft	48 Diam in.	50 Area sq ft	50 Diam in.	52 Area sq ft	52 Diam in.	54 Area sq ft	54 Diam in.	56 Area sq ft	56 Diam in.	58 Area sq ft	58 Diam in.
42	11.5	45.9																
44	12.0	46.9	12.6	48.1														
46	12.5	47.9	13.1	49.1	13.8	50.3												
48	13.0	48.9	13.7	50.2	14.3	51.3	15.1	52.6										
50	13.5	49.8	14.3	51.2	14.9	52.3	15.7	53.6	16.3	54.7								
52	14.1	50.8	14.8	52.2	15.5	53.3	16.2	54.6	17.0	55.8	17.6	56.9						
54	14.6	51.8	15.4	53.2	16.1	54.3	16.8	55.6	17.6	56.8	18.3	57.9	19.2	59.4				
56	15.1	52.7	15.9	54.1	16.7	55.3	17.4	56.5	18.2	57.8	18.9	58.9	19.6	60.0	20.5	61.3		
58	15.7	53.7	16.5	55.0	17.2	56.2	18.0	57.5	18.8	58.8	19.6	60.0	20.4	61.2	21.1	62.3	22.0	63.5
60	16.2	54.6	17.0	55.9	17.8	57.1	18.6	58.5	19.5	59.8	20.3	61.0	21.1	62.2	21.8	63.3	22.5	64.3
64	17.3	56.4	18.1	57.7	19.0	59.0	19.8	60.3	20.7	61.6	21.6	62.9	22.4	64.1	23.2	65.3	24.4	66.9
68	18.3	58.0	19.3	59.5	20.1	60.8	21.1	62.1	21.9	63.4	22.9	64.8	23.8	66.1	24.7	67.3	25.5	68.4
72	19.4	59.6	20.3	61.1	21.4	62.6	22.2	63.9	23.1	65.2	24.2	66.6	25.1	67.9	26.1	69.2	27.1	70.5
76	20.4	61.2	21.4	62.7	22.4	64.1	23.4	65.6	24.5	67.0	25.5	68.4	26.4	69.6	27.5	71.0	28.9	72.8
80	21.4	62.7	22.4	64.1	23.5	65.7	24.6	67.2	25.7	68.7	26.8	70.1	28.1	71.8	28.8	72.7	30.1	74.3
84	22.4	64.1	23.5	65.7	24.7	67.3	25.8	68.8	26.9	70.3	28.1	71.8	29.1	73.1	30.2	74.5	31.5	76.0
88	23.3	65.4	24.5	67.0	25.7	68.7	26.9	70.3	28.1	71.8	29.4	73.4	30.6	74.9	31.7	76.3	32.7	77.5
92	24.3	66.8	25.6	68.5	26.8	70.1	28.1	71.8	29.3	73.3	30.6	74.9	31.9	76.5	33.1	77.9	34.2	79.2
96	25.2	68.0	26.7	70.0	27.6	71.1	29.4	73.5	30.2	74.5	31.8	76.4	33.2	78.0	33.9	78.9	35.7	80.9
100	26.0	69.1	27.1	70.5	29.0	72.9	30.2	74.5	31.6	76.1	32.7	77.5	33.8	78.7	35.5	80.7	36.6	82.0
104	27.1	70.5	28.4	72.2	29.4	74.0	31.1	75.5	32.7	77.5	34.0	79.0	35.8	81.0	37.1	82.5	38.5	84.1
108	28.0	71.7	29.5	73.6	30.6	74.9	32.3	77.0	33.3	78.2	35.3	80.5	36.6	82.0	38.5	84.0	39.8	85.5
112	29.2	73.2	30.2	74.5	31.9	76.5	33.1	78.0	34.9	80.0	36.6	82.0	38.0	83.5	39.8	85.5	40.8	86.5
116	30.0	74.2	32.0	76.6	32.7	77.5	34.0	79.0	35.9	81.2	38.0	83.5	39.8	85.5	41.0	86.7	42.4	88.2
120	30.7	75.0	32.7	77.5	33.6	78.5	35.8	81.0	37.4	82.9	39.4	85.0	40.9	86.6	41.9	87.7	43.6	89.4
124	31.5	76.0	33.6	78.5	34.4	79.5	36.5	81.8	38.5	84.1	40.7	86.1	41.5	87.3	43.3	89.1	44.6	90.5
128	32.1	76.8	34.0	79.0	36.2	81.5	27.5	83.0	39.2	84.8	41.4	87.2	42.9	88.7	44.6	90.5	46.6	92.5
132	33.2	78.0	34.9	80.0	36.9	82.3	38.8	84.4	40.7	86.4	42.7	88.5	44.1	90.0	45.9	91.9	48.0	93.9
136	34.0	79.0	35.6	80.8	38.0	83.5	39.7	85.4	41.7	87.5	43.8	89.7	44.8	90.7	47.4	93.3	49.7	95.5
140	35.3	80.5	37.0	82.4	38.8	84.4	40.5	86.2	42.4	88.2	44.9	90.8	46.5	92.4	48.6	94.4	50.3	96.1
144	35.8	81.1	37.8	83.3	40.0	85.7	41.4	87.2	44.1	90.0	45.6	91.5	47.8	93.7	49.7	95.5	51.5	97.2

SIDE	60 Area sq ft	60 Diam in.	64 Area sq ft	64 Diam in.	68 Area sq ft	68 Diam in.	72 Area sq ft	72 Diam in.	76 Area sq ft	76 Diam in.	80 Area sq ft	80 Diam in.	84 Area sq ft	84 Diam in.	88 Area sq ft	88 Diam in.	92 Area sq ft	92 Diam in.
42																		
44																		
46																		
48																		
50																		
52																		
54																		
56																		
58																		
60	23.5	65.7																
64	25.0	67.7	26.7	70.0														
68	26.5	69.7	28.3	72.1	30.2	74.4												
72	28.0	71.7	29.9	74.1	31.8	76.4	33.8	78.8										
76	29.5	73.6	31.6	76.1	33.5	78.4	35.7	80.9	37.7	83.2								
80	31.0	75.4	33.2	78.1	35.2	80.4	37.4	82.8	39.6	85.3	41.7	87.5						
84	32.5	77.2	34.8	79.9	37.0	82.4	39.2	84.8	41.4	87.2	43.7	89.6	46.0	91.9				
88	34.0	79.0	36.3	81.6	38.6	84.2	41.1	87.0	43.4	89.2	45.7	91.6	48.0	93.9	50.5	96.3		
92	35.6	80.8	37.9	83.4	40.3	86.0	42.9	88.7	45.3	91.2	47.7	93.6	50.1	95.9	52.7	98.3	55.1	100.5
96	37.0	82.4	39.8	85.5	42.1	87.9	44.6	90.5	47.5	93.4	49.8	95.6	51.9	97.6	55.2	100.6	57.8	103.0
100	38.4	83.9	41.2	87.0	44.3	90.2	47.5	93.4	50.2	96.0	51.9	97.6	53.3	98.9	56.7	102.0	60.1	105.0
104	40.3	86.0	42.8	88.6	46.1	92.0	48.2	94.0	51.5	97.2	53.6	99.2	57.3	102.5	59.5	104.5	62.4	107.0
108	41.7	87.5	44.1	90.0	46.9	92.8	50.1	95.9	53.0	98.6	55.6	101.0	58.5	103.6	61.0	105.8	64.7	109.0
112	42.3	88.1	45.3	91.2	48.9	94.7	51.7	97.4	54.3	99.8	57.4	102.6	58.9	104.0	63.8	108.2	67.1	111.0
116	44.1	90.0	47.6	93.5	51.1	96.8	53.7	99.3	57.0	102.3	60.1	105.0	63.3	107.8	66.2	110.2	69.3	112.8
120	45.5	91.4	49.7	95.5	51.8	97.5	55.8	101.2	58.9	104.0	62.4	107.0	65.5	109.6	69.0	112.5	72.1	115.0
124	47.1	93.0	49.8	95.6	53.8	99.4	56.7	102.0	60.1	105.0	63.6	108.0	66.2	110.2	69.3	112.8	73.3	116.0
128	47.6	93.5	51.3	97.0	55.4	100.8	58.7	103.8	61.8	106.5	65.5	109.6	68.1	111.8	72.3	115.2	76.3	118.3
132	49.7	95.5	53.0	98.6	56.3	101.6	60.1	105.0	64.2	108.5	68.4	112.0	71.8	114.8	74.6	117.0	78.5	120.0
136	50.3	96.1	54.9	100.4	58.9	104.0	62.2	106.8	64.7	109.0	69.6	113.0	72.8	115.6	76.7	118.6	81.5	122.3
140	52.4	98.1	55.6	101.0	60.4	105.3	63.8	108.2	67.8	111.5	71.4	114.5	75.6	117.8	79.1	120.5	82.7	123.2
144	54.1	99.6	57.8	103.0	61.2	106.0	64.7	109.0	69.1	112.6	73.3	116.0	78.0	119.6	81.1	122.0	85.2	125.0

*Circular equivalent diameter (d_e). Calculated from $d_e = 1.3 \dfrac{(ab)^{.625}}{(a + b)^{.25}}$ †Large numbers in table are duct class.

TABLE 7—RECOMMENDED MAXIMUM DUCT VELOCITIES FOR LOW VELOCITY SYSTEMS (FPM)

APPLICATION	CONTROLLING FACTOR NOISE GENERATION Main Ducts	CONTROLLING FACTOR—DUCT FRICTION			
		Main Ducts		Branch Ducts	
		Supply	Return	Supply	Return
Residences	600	1000	800	600	600
Apartments Hotel Bedrooms Hospital Bedrooms	1000	1500	1300	1200	1000
Private Offices Directors Rooms Libraries	1200	2000	1500	1600	1200
Theatres Auditoriums	800	1300	1100	1000	800
General Offices High Class Restaurants High Class Stores Banks	1500	2000	1500	1600	1200
Average Stores Cafeterias	1800	2000	1500	1600	1200
Industrial	2500	3000	1800	2200	1500

TABLE 8—VELOCITY PRESSURES

VELOCITY PRESSURE (in. wg)	VELOCITY (Ft/Min)	VELOCITY PRESSURE (in. wg)	VELOCITY (Ft/Min)	VELOCITY PRESSURE (in. wg)	VELOCITY (Ft/Min)	VELOCITY PRESSURE (in. wg.)	VELOCITY (Ft/Min)
.01	400	.29	2150	.58	3050	1.28	4530
.02	565	.30	2190	.60	3100	1.32	4600
.03	695	.31	2230	.62	3150	1.36	4670
.04	800	.32	2260	.64	3200	1.40	4730
.05	895	.33	2300	.66	3250	1.44	4800
.06	980	.34	2330	.68	3300	1.48	4870
.07	1060	.35	2370	.70	3350	1.52	4930
.08	1130	.36	2400	.72	3390	1.56	5000
.09	1200	.37	2440	.74	3440	1.60	5060
.10	1270	.38	2470	.76	3490	1.64	5120
.11	1330	.39	2500	.78	3530	1.68	5190
.12	1390	.40	2530	.80	3580	1.72	5250
.13	1440	.41	2560	.82	3620	1.76	5310
.14	1500	.42	2590	.84	3670	1.80	5370
.15	1550	.43	2620	.86	3710	1.84	5430
.16	1600	.44	2650	.88	3750	1.88	5490
.17	1650	.45	2680	.90	3790	1.92	5550
.18	1700	.46	2710	.92	3840	1.96	5600
.19	1740	.47	2740	.94	3880	2.00	5660
.20	1790	.48	2770	.96	3920	2.04	5710
.21	1830	.49	2800	.98	3960	2.08	5770
.22	1880	.50	2830	1.00	4000	2.12	5830
.23	1920	.51	2860	1.04	4080	2.16	5880
.24	1960	.52	2880	1.08	4160	2.20	5940
.25	2000	.53	2910	1.12	4230	2.24	5990
.26	2040	.54	2940	1.16	4310	2.28	6040
.27	2080	.55	2970	1.20	4380		
.28	2120	.56	2990	1.24	4460		

NOTES: 1. Data for standard air (29.92 in. Hg and 70 F)

2. Data derived from the following equation: $h_v = \left(\dfrac{V}{4005}\right)^2$ where: V = velocity in fpm.

h_v = pressure difference termed "velocity head" (in. wg).

CHART 8—PRESSURE DROP THRU FLEXIBLE CONDUIT

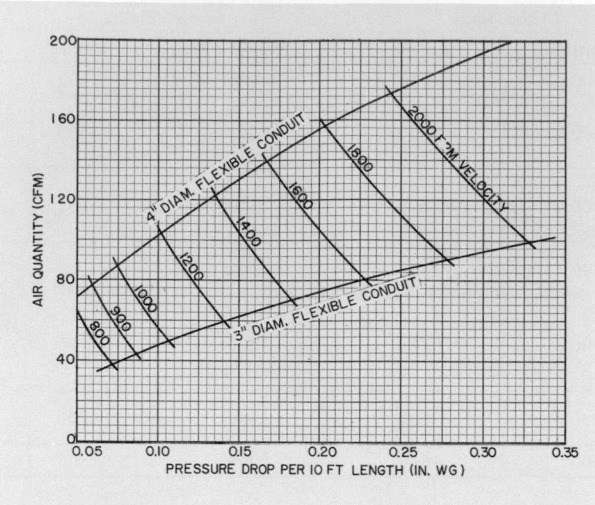

FAN CONVERSION LOSS OR GAIN

In addition to the calculations shown for determining the required static pressure at the fan discharge in *Example 4*, a fan conversion loss or gain must be included. This conversion quantity can be a significant amount, particularly on a high velocity system. It is determined by the following equations.

If the velocity in the duct is higher than the fan outlet velocity, use the following formula for the additional static pressure required:

$$\text{Loss} = 1.1 \left[\left(\frac{V_d}{4000} \right)^2 - \left(\frac{V_f}{4000} \right)^2 \right]$$

where V_d = duct velocity
V_f = fan outlet velocity
Loss = in. wg

If the fan discharge velocity is higher than the duct velocity, use the following formula for the credit taken to the static pressure required:

$$\text{Gain} = .75 \left[\left(\frac{V_f}{4000} \right)^2 - \left(\frac{V_d}{4000} \right)^2 \right]$$

DUCT SYSTEM ELEMENT FRICTION LOSS

Friction loss thru any fitting is expressed in terms of equivalent length of duct. This method provides units that can be used with the friction chart to determine the loss in a section of duct containing elbows and fittings. *Table 12* gives the friction losses for rectangular elbows, and *Table 11* gives the losses for standard round elbows. The friction losses in *Tables 11 and 12* are given in terms of additional equivalent length of straight duct. This loss for the elbow is added to the straight run of duct to obtain the total equivalent length of duct. The straight run

of duct is measured to the intersection of the center line of the fittings. *Fig. 46* gives the guides for measuring duct lengths.

Tables 9 and 10 list the friction losses for other size elbows or other R/D ratios. *Table 10* presents the friction losses of rectangular elbows and elbow combinations in terms of L/D. *Table 10* also includes the losses and regains for various duct shapes, entrances and exits, and items located in the air stream of the duct. This loss or regain is expressed in the number of velocity heads and is represented by "*n*". This loss or regain may be converted into equivalent length of duct by the equation at the end of the table and added or subtracted from the actual duct length.

Table 9 gives the loss of round elbows in terms of L/D, the additional equivalent length to the diameter of the elbow. The loss for round tees and crosses are in terms of the number of velocity heads ("*n*"). The equation for converting the loss in velocity head to additional equivalent length of duct is located at the bottom of the table.

In high velocity systems it is often desirable to have the pressure drop in round elbows, tees, and crosses in inches of water. These losses may be obtained from *Chart 9* for standard round fittings.

DESIGN METHODS

The general procedure for designing any duct system is to keep the layout as simple as possible and make the duct runs symmetrical. Supply terminals are located to provide proper room air distribution (*Chapter 3*), and ducts are laid out to connect these outlets. The ductwork should be located to avoid structural members and equipment.

The design of a low velocity supply air system may be accomplished by any one of the three following methods:

1. Velocity reduction
2. Equal friction
3. Static regain

The three methods result in different levels of accuracy, economy and use.

The equal friction method is recommended for return and exhaust air systems.

LOW VELOCITY DUCT SYSTEMS

Velocity Reduction Method

The procedure for designing the duct system by this method is to select a starting velocity at the fan discharge and make arbitrary reductions in velocity down the duct run. The starting velocity selected should not exceed those in *Table 7*. Equiva-

(Continued on page 46)

TABLE 9—FRICTION OF ROUND DUCT SYSTEM ELEMENTS

ELEMENT	CONDITION	L/D RATIO*
90° Smooth Elbow	R/D = 1.5	9
90° 3-Piece Elbow	R/D = 1.5	24
90° 5-Piece Elbow	R/D = 1.5	12
45° 3-Piece Elbow	R/D = 1.5	6
45° Smooth Elbow	R/D = 1.5	4.5
90° Miter Elbow	Vaned Not Vaned	22 65

ELEMENT	CONDITION	VALUE OF n†
90° Tee‡ and 90°, 135° & 180° Cross‡ Pressure Loss Thru Branch = nhv_2	$\dfrac{V_2}{V_1} = \begin{cases} 0.2 \\ 0.5 \\ 1.0 \\ 5.0 \end{cases}$	4.0 2.0 1.75 1.6
45° Tee‡ Pressure Loss Thru Branch = nhv_2	$\dfrac{V_2}{V_1} = \begin{cases} 0.8 \\ 1.0 \\ 2.0 \\ 3.0 \end{cases}$	.10 .44 1.21 1.47
90° Conical Tee and 180° Conical Cross Pressure Loss Thru Branch = nhv_2	$\dfrac{V_2}{V_1} = \begin{cases} 0.5 \\ 1.0 \\ 2.0 \\ 5.0 \end{cases}$	0.2 0.5 1.0 1.2

Notes on page 42.

TABLE 10—FRICTION OF RECTANGULAR DUCT SYSTEM ELEMENTS

ELEMENT	CONDITIONS		L/D RATIO †

Rectangular Radius Elbow

W/D	R/D				
	.5	.75	1.00	1.25*	1.50
	L/D Ratio				
.5	33	14	9	5	4
1	45	18	11	7	4
3	80	30	14	8	5
6	125	40	18	12	7

Rectangular Vaned Radius Elbow

Number of Vanes	R/D			
	.50	.75	1.00	1.50
	L/D Ratio			
1	18	10	8	7
2	12	8	7	7
3	10	7	7	6

X° Elbow — Vaned or Unvaned Radius Elbow — X/90 times value for similar 90° elbow

Rectangular Square Elbow

Condition	Value
No Vanes	60
Single Thickness Turning Vanes	15
Double Thickness Turning Vanes	10

Double Elbow — W/D = 1, R/D = 1.25*

Condition	Value
S = O	15
S = D	10

Double Elbow — W/D = 1, R/D = 1.25*

Condition	Value
S = O	20
S = D	22

Double Elbow — W/D = 1, R/D = 1.25* For Both

Condition	Value
S = O	15
S = D	16

Double Elbow — W/D = 2, R₁/D = 1.25*, R₂/D = .5

Condition	Value
Direction of Arrow	45
Reverse Direction	40

Double Elbow — W/D = 4, R/D = 1.25* for both elbows

Condition	Value
Direction of Arrow	17
Reverse Direction	18

TABLE 10—FRICTION OF RECTANGULAR DUCT SYSTEM ELEMENTS (Contd)

ELEMENT	CONDITIONS	VALUE OF n‡

Transformer

$V_2 = V_1$
S.P. Loss $= nhv_1$

.15

Expansion

"n"

		Angle "a"				
v_2/v_1	5°	10°	15°	20°	30°	40°
.20	.83	.74	.68	.62	.52	.45
.40	.89	.83	.78	.74	.68	.64
.60	.93	.87	.84	.82	.79	.77

S.P. Regain $= n(hv_1 - hv_2)$

Contraction

a	30°	45°	60°
n	1.02††	1.04	1.07

S.P. Loss $= n(hv_2 - hv_1)$ ††Slope 1″ in 4″

Abrupt Entrance

.35

Bellmouth Entrance

S.P. Loss $= nhv_1$

.03

Abrupt Exit

Bellmouth Exit

S.P. Loss or Regain Considered Zero

Re-Entrant Entrance

S.P. Loss $- nhv_1$

.85

Sharp Edge Round Orifice

A_2/A_1	0	.25	.50	.75	1.00
n	2.5	2.3	1.9	1.1	0

S.P. Loss $= nhv_2$

Abrupt Contraction

V_1/V_2	0	.25	.50	.75
n	1.34	1.24	.96	.52

S.P. Loss $= nhv_2$

Abrupt Expansion

V_2/V_1	.20	.40	.60	.80
n	.32	.48	.48	.32

S.P. Regain $= nhv_1$

Pipe Running Thru Duct

E/D	.10	.25	.50
n	.20	.55	2.00

S.P. Loss $= nhv_1$

Bar Running Thru Duct

E/D	.10	.25	.50
n	.7	1.4	4.00

S.P. Loss $= nhv_1$

Easement Over Obstruction

E/D	.10	.25	.50
n	.07	.23	.90

S.P. Loss $= nhv_1$

Notes on *page 42.*

NOTES FOR TABLE 9

*L and D are in feet. D is the elbow diameter. L is the additional equivalent length of duct added to the measured length. The equivalent length L equals D in feet times the ratio listed.

†The value of n is the loss in velocity heads and may be converted to additional equivalent length of duct by the following equation.

$$L = n \times \frac{h_v \times 100}{h_f}$$

where: L = additional equivalent length, ft

h_v = velocity pressure at V_2, in. wg (conversion line on *Chart 7* or *Table 8*).

h_f = friction loss/100 ft, duct diameter at V_2, in. wg (*Chart 7*).

n = value for tee or cross

‡Tee or cross may be either reduced or the same size in the straight thru portion

NOTES FOR TABLE 10:

*1.25 is standard for an unvaned full radius elbow.

†L and D are in feet. D is the duct dimension illustrated in the drawing. L is the additional equivalent length of duct added to the measured duct. The equivalent length L equals D in feet times the ratio listed.

‡The value of n is the number of velocity heads or differences in velocity heads lost or gained at a fitting, and may be converted to additional equivalent length of duct by the following equation:

$$L = n \times \frac{h_v \times 100}{h_f}$$

where: L = additional equivalent length, ft.

h_v = velocity pressure for V_1 or V_2, in. wg (conversion line on *Chart 7* or *Table 8*).

h_f = friction loss/100 ft, duct cross section at h_v, in. wg (*Chart 7*).

n = value for particular fitting.

TABLE 11—FRICTION OF ROUND ELBOWS

ELBOW DIAMETER (in.)	90° SMOOTH R/D = 1.5	90° 5-PIECE R/D = 1.5	90° 3-PIECE R/D = 1.5	45° 3-PIECE R/D = 1.5	45° SMOOTH R/D = 1.5
	ADDITIONAL EQUIVALENT LENGTH OF STRAIGHT DUCT (FT)				
3	2.3	3	6	1.5	1.1
4	3	4	8	2	1.5
5	3.8	5	10	2.5	1.9
6	4.5	6	12	3	2.3
7	5.3	7	14	3.5	2.6
8	6	8	16	4	3
9	—	9	18	4.5	—
10	—	10	20	5	—
11	—	11	22	5.5	—
12	—	12	24	6	—
14	—	14	28	7	—
16	—	16	32	8	—
18	—	18	36	9	—
20	—	20	40	10	—
22	—	22	44	11	—
24	—	24	48	12	—

TABLE 12—FRICTION OF RECTANGULAR ELBOWS

DUCT DIMENSIONS (in.)		RADIUS ELBOW NO VANES	RADIUS ELBOW—WITH VANES‡				SQUARE ELBOWS‡	
W	D	Radius Ratio† R/D = 1.25	Rₜ = 6″ (Recommended)	Vanes	Rₜ = 3″ (Acceptable)	Vanes	Double Thickness Turning Vanes	Single Thickness Turning Vanes
		ADDITIONAL EQUIVALENT LENGTH OF STRAIGHT DUCT (FT)						
96	48	31	45	2	43	3	40	60
	36	25	36	2	31	3	30	45
	30	22	31	2	38	2	25	37
	24	19	33	1	29	2	20	30
	20	16	28	1	25	2	17	25
72	48	28	44	2	41	3	35	60
	36	23	33	2	29	3	29	45
	30	21	28	2	33	2	25	37
	24	17	29	1	25	2	21	30
	20	15	23	1	19	2	18	25
	16	13	18	1	16	2	15	20
	12	12			15	1	11	15
60	48	27	41	2	39	3	33	60
	36	22	31	2	27	3	27	45
	30	19	25	2	31	2	23	37
	24	16	27	1	26	2	20	30
	20	14	22	1	21	2	17	25
	16	12	16	1	15	2	13	20
	12	10			14	1	10	15
48	96*	45	35	3			29	60
	48	26	35	2	34	3	29	60
	36	20	26	2	22	3	23	45
	30	18	23	2	28	2	21	37
	24	15	24	1	21	2	18	30
	20	14	19	1	17	2	15	25
	16	11	15	1	14	2	12	20
	12	9			13	1	10	15
	10	8			11	1	8	12
	8	8			9	1	7	10
42	42	23	28	2	26	3	24	53
	36	20	24	2	21	3	22	45
	30	17	21	2	26	2	20	37
	24	15	21	1	19	2	16	30
	20	13	18	1	16	2	14	25
	16	11	14	1	13	2	12	20
	12	9			13	1	9	15
	10	8			10	1	8	12
	8	7			8	1	6	10
36	72*	34	27	3			20	45
	36	19	22	2	19	3	20	45
	30	16	19	2	22	2	18	37
	24	14	20	1	22	2	15	30
	20	12	17	1	15	2	13	25
	16	10	13	1	12	2	11	20
	12	9			12	1	9	15
	10	8			9	1	8	12
	8	7			8	1	6	10
32	32	17	19	2	16	3	17	40
	30	16	18	2	21	2	17	37
	24	14	19	1	17	2	15	30
	20	12	16	1	14	2	12	25
	16	10	12	1	12	2	11	20
	12	8			12	1	8	15
	10	7			9	1	7	12
	8	6			8	1	6	10

TABLE 12—FRICTION OF RECTANGULAR ELBOWS (CONT.)

DUCT DIMENSIONS (in.)		RADIUS ELBOW NO VANES	RADIUS ELBOW—WITH VANES‡				SQUARE ELBOWS‡	
W	D	Radius Ratio† R/D = 1.25	Rₜ = 6" (Recommended)	Vanes	Rₜ = 3" (Acceptable)	Vanes	Double Thickness Turning Vanes	Single Thickness Turning Vanes
		ADDITIONAL EQUIVALENT LENGTH OF STRAIGHT DUCT (FT)						
28	28	15	14	2	17	2	14	34
	24	13	17	1	15	2	13	30
	20	12	15	1	13	2	12	25
	16	10	11	1	11	2	10	20
	12	8			11	1	8	15
	10	7			9	1	7	12
	8	6			8	1	6	10
24	96*	38	19	3			23	80
	72*	32	17	3			21	72
	48*	22	20	2	20	3	18	62
	24	13	16	1	14	2	12	30
	20	11	13	1	12	2	10	25
	16	10	11	1	10	2	9	20
	12	8			10	1	8	15
	10	7			8	1	7	12
	8	6			7	1	6	10
	6	5					4	8
20	80*	32	16	3			19	66
	60*	26	19	2			17	58
	40*	22	15	2	14	3	14	49
	20	11	12	1	10	2	10	25
	16	9	9	1	9	2	8	20
	12	7			9	1	7	15
	10	6			8	1	6	12
	8	5			7	1	5	10
	6	4					4	8
16	64*	26	9	3			14	48
	48*	21	12	2	12	3	12	43
	32*	15	11	2	9	3	11	38
	16	9	8	1	8	2	7	20
	12	7			8	1	6	15
	10	6			6	1	5	12
	8	5			6	1	5	10
	6	4					4	8
12	48*	19	8	2	8	3	10	33
	36*	16	7	2	7	3	9	30
	24*	11	8	1	8	2	8	26
	12	7			7	1	5	15
	10	6			5	1	5	12
	8	5			5	1	4	10
	6	4					3	8
10	40*	19	6	2	6	3	8	27
	30*	13	6	2	8	2	7	24
	20*	9	7	1	6	2	6	21
	10	5			5	1	4	12
	8	4			5	1	4	10
	6	4					3	8
8	32*	13	5	2	4	3	6	21
	24*	11	6	1	5	2	6	19
	16*	8	4	1	5	2	5	16
	8	4			4	1	3	10
	6	3					3	8
6	24*	10	4	1	4	2	4	15
	18*	8	3	1	4	2	4	13
	12*	6			4	1	3	11
	6	3					3	8

*Denotes Hard Bends as shown

Hard Bend   Easy Bend

†For other radius ratios, see *Table 10*.

‡For other sizes, see *Table 10*.

Vanes must be located as illustrated in *Chart 6, page 24*, to have these minimum losses.

CHART 9—LOSSES FOR ROUND FITTINGS
Elbows, Tees and Crosses

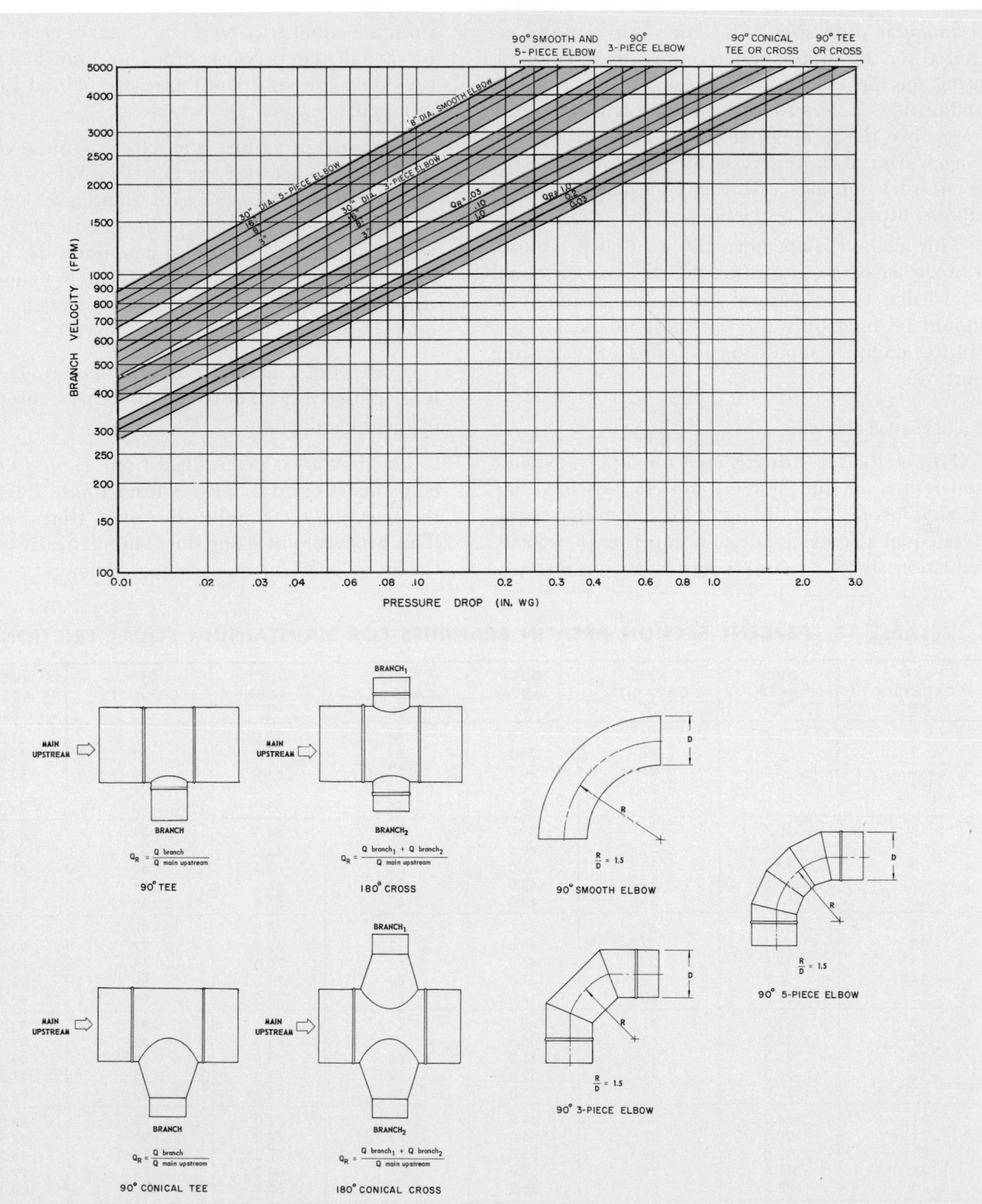

NOTES: 1. Loss for tee or cross is a function of the velocity in the branch. This represents the loss in static pressure from the main upstream to the branch. Q_R is the ratio of air quantity of the branch to the main upstream.
2. Loss for 45° smooth elbow is equal to one-half the loss for a 90° smooth elbow.
3. Loss for 45° 3-piece elbow is equal to one-half the loss for a 90° 5-piece elbow.

lent round diameters may be obtained from *Chart 7* using air velocity and air quantity. *Table 6* is used with the equivalent round diameter to select the rectangular duct sizes. The fan static pressure required for the supply is determined by calculation, using the longest run of duct including all elbows and fittings. *Tables 10 and 12* are used to obtain the losses thru the rectangular elbows and fittings. The longest run is not necessarily the run with the greatest friction loss, as shorter runs may have more elbows, fittings and restrictions.

This method is not normally used, as it requires a broad background of duct design experience and knowledge to be within reasonable accuracy. It should be used only for the most simple layouts. Splitter dampers should be included for balancing purposes.

Equal Friction Method

This method of sizing is used for supply, exhaust and return air duct systems and employs the same friction loss per foot of length for the entire system. The equal friction method is superior to velocity reduction since it requires less balancing for sym-metrical layouts. If a design has a mixture of short and long runs, the shortest run requires considerable dampering. Such a system is difficult to balance since the equal friction method makes no provision for equalizing pressure drops in branches or for providing the same static pressure behind each air terminal.

The usual procedure is to select an initial velocity in the main duct near the fan. This velocity should be selected from *Table 7* with sound level being the limiting factor. *Chart 7* is used with this initial velocity and air quantity to determine the friction rate. This same friction loss is then maintained throughout the system and the equivalent round duct diameter is selected from *Chart 7*.

To expedite equal friction calculations, *Table 13* is often used instead of the friction chart; this results in the same duct sizes.

The duct areas determined from *Table 13* or the equivalent round diameters from *Chart 7* are used to select the rectangular duct sizes from *Table 6*. This procedure of sizing duct automatically reduces the air velocity in the direction of flow.

TABLE 13—PERCENT SECTION AREA IN BRANCHES FOR MAINTAINING EQUAL FRICTION

CFM CAPACITY %	DUCT AREA %	CFM CAPACITY %	DUCT AREA %	CFM CAPACITY %	DUCT AREA %	CFM CAPACITY %	DUCT AREA %
1	2.0	26	33.5	51	59.0	76	81.0
2	3.5	27	34.5	52	60.0	77	82.0
3	5.5	28	35.5	53	61.0	78	83.0
4	7.0	29	36.5	54	62.0	79	84.0
5	9.0	30	37.5	55	63.0	80	84.5
6	10.5	31	39.0	56	64.0	81	85.5
7	11.5	32	40.0	57	65.0	82	86.0
8	13.0	33	41.0	58	65.5	83	87.0
9	14.5	34	42.0	59	66.5	84	87.5
10	16.5	35	43.0	60	67.5	85	88.5
11	17.5	36	44.0	61	68.0	86	89.5
12	18.5	37	45.0	62	69.0	87	90.0
13	19.5	38	46.0	63	70.0	88	90.5
14	20.5	39	47.0	64	71.0	89	91.5
15	21.5	40	48.0	65	71.5	90	92.0
16	23.0	41	49.0	66	72.5	91	93.0
17	24.0	42	50.0	67	73.5	92	94.0
18	25.0	43	51.0	68	74.5	93	94.5
19	26.0	44	52.0	69	75.5	94	95.0
20	27.0	45	53.0	70	76.5	95	96.0
21	28.0	46	54.0	71	77.0	96	96.5
22	29.5	47	55.0	72	78.0	97	97.5
23	30.5	48	56.0	73	79.0	98	98.0
24	31.5	59	57.0	74	80.0	99	99.0
25	32.5	50	58.0	75	80.5	100	100.0

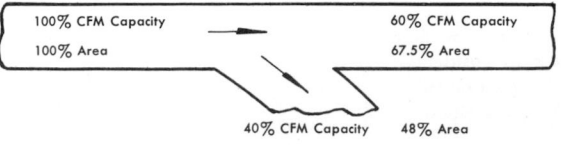

To determine the total friction loss in the duct system that the fan must overcome, it is necessary to calculate the loss in the duct run having the highest resistance. The friction loss thru all elbows and fittings in the section must be included.

Example 4 — Equal Friction Method of Designing Ducts

Given:

Duct systems for general office (*Fig. 47*).
Total air quantity — 5400 cfm
18 air terminals — 300 cfm each
Operating pressure for all terminals — 0.15 in. wg
Radius elbows, R/D = 1.25

Find:

1. Initial duct velocity, area, size and friction rate in the duct section from the fan to the first branch.
2. Size of remaining duct runs.
3. Total equivalent length of duct run with highest resistance.
4. Total static pressure required at fan discharge.

Solution:

1. From *Table 7* select an initial velocity of 1700 fpm.

$$\text{Duct area} = \frac{5400 \text{ cfm}}{1700 \text{ fpm}} = 3.18 \text{ sq ft}$$

From *Table 6,* select a duct size — 22 in. x 22 in.

Initial friction rate is determined from *Chart 7* using the air quantity (5400), and the equivalent round duct diameter from *Table 6*. Equivalent round duct diameter = 24.1 in.

Friction rate = .145 in. wg per 100 ft of equivalent length.

2. The duct areas are calculated using *Table 13* and duct sizes are determined from *Table 6*. The following tabulates the design information:

DUCT SECTION	AIR QUANTITY (cfm)	CFM* CAPACITY (%)
To A	5400	100
A - B	3600	67
B - 13	1800	33
13 - 14	1500	28
14 - 15	1200	22
15 - 16	900	17
16 - 17	600	11
17 - 18	300	6

DUCT SECTION	DUCT AREA (%)	AREA† (sq ft)	DUCT SIZE‡ (in.)
To A	100.0	3.18	22 x 22
A - B	73.5	2.43	22 x 16
B - 13	41.0	1.3	22 x 10
13 - 14	35.5	1.12	18 x 10
14 - 15	29.5	.94	14 x 10
15 - 16	24.0	.76	12 x 10
16 - 17	17.5	.56	8 x 10
17 - 18	10.5	.33	8 x 10

*Percent of cfm = $\dfrac{\text{air quantity in duct section}}{\text{total air quantity}}$

†Duct area = percent of area times initial duct area (fan to A)

‡Refer to *page 21* for reducing duct sizes.

Duct sections *B thru 12* and *A thru 6* have the same dimension as the corresponding duct sections in *B thru 18*.

3. It appears that the duct run from the fan to terminal *18* has the highest resistance. *Tables 10 and 12* are used to determine the losses thru the fittings. The following list is a tabulation of the total equivalent length in this duct run:

DUCT SECTION	ITEM	LENGTH (ft)	ADD. EQUIV. LENGTH (ft)
To A	Duct	60	
	Elbow		12
A - B	Duct	20	
B - 13	Duct	30	
—	Elbow		7
13 - 14	Duct	20	
14 - 15	Duct	20	
15 - 16	Duct	20	
16 - 17	Duct	20	
17 - 18	Duct	20	
Total		210	19

4. The total friction loss in the ductwork from the fan to last terminal *18* is shown in the following:

Loss = total equiv length × friction rate

$$= 229 \text{ ft} \times \frac{.145 \text{ in. wg}}{100 \text{ ft}} = .332 \text{ or } .33 \text{ in. wg}$$

Total static pressure required at fan discharge is the sum of the terminal operating pressure and the loss in the ductwork. Credit can be taken for the velocity regain between the first and last sections of duct:

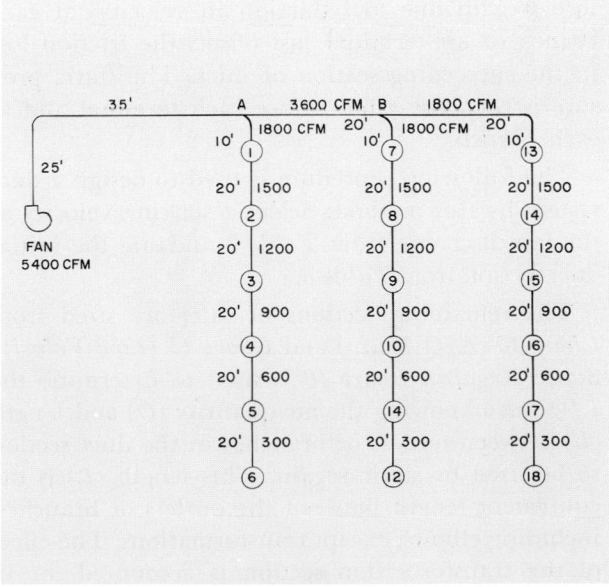

FIG. 47 — DUCT LAYOUT FOR LOW VELOCITY SYSTEM
(EXAMPLES 3 AND 4)

Velocity in initial section = 1700 fpm
Velocity in last section = 590 fpm
Using a 75% regain coefficient,

$$\text{Regain} = .75\left[\left(\frac{1700}{4000}\right)^2 - \left(\frac{590}{4000}\right)^2\right]$$
$$= .75\,(.18 - .02) = .12 \text{ in. wg}$$

Therefore, the total static pressure at fan discharge:
$$= \text{duct friction} + \text{terminal pressure} - \text{regain}$$
$$= .33 + .15 - .12$$
$$= .36 \text{ in. wg}$$

The equal friction method does not satisfy the design criteria of uniform static pressure at all branches and air terminals. To obtain the proper air quantity at the beginning of each branch, it is necessary to include a splitter damper to regulate the flow to the branch. It may also be necessary to have a control device (vanes, volume damper, or adjustable terminal volume control) to regulate the flow at each terminal for proper air distribution.

In *Example 4*, if the fan selected has a discharge velocity of 2000 fpm, the net credit to the total static pressure required is determined as described under *"Fan Conversion Loss or Gain."*

$$\text{Gain} = .75\left[\left(\frac{2000}{4000}\right)^2 - \left(\frac{1700}{4000}\right)^2\right]$$
$$= .75\,(.25 - .18) = .05 \text{ in. wg}$$

Static Regain Method

The basic principle of the static regain method is to size a duct run so that the increase in static pressure (regain due to reduction in velocity) at each branch or air terminal just offsets the friction loss in the succeeding section of duct. The static pressure is then the same before each terminal and at each branch.

The following procedure is used to design a duct system by this method: Select a starting velocity at the fan discharge from *Table 7* and size the initial duct section from *Table 6*.

The remaining sections of duct are sized from *Chart 10 (L/Q Ratio)* and *Chart 11 (Low Velocity Static Regain)*. *Chart 10* is used to determine the L/Q ratio knowing the air quantity (Q) and length (L) between outlets or branches in the duct section to be sized by static regain. This length (L) is the equivalent length between the outlets or branches, including elbows, except transformations. The effect of the transformation section is accounted for in *"Chart 11 — Static Regain."* This assumes that the transformation section is laid out according to the recommendation presented in this chapter.

Chart 11 is used to determine the velocity in the duct section that is being sized. The values of the L/Q ratio (*Chart 10*) and the velocity (V_1) in the duct section immediately before the one being sized are used in *Chart 11*. The velocity (V_2) determined from *Chart 11* is used with the air quantity to arrive at the duct area. This duct area is used in *Table 6* to size the rectangular duct and to obtain the equivalent round duct size. By using this duct size, the friction loss thru the length of duct equals the increase in static pressure due to the velocity change after each branch take-off and outlet. However, there are instances when the reduction in area is too small to warrant a change in duct size after the outlet, or possibly when the duct area is reduced more than is called for. This gives a gain or loss for the particular duct section that the fan must handle. Normally, this loss or gain is small and, in most instances, can be neglected.

Instead of designing a duct system for zero gain or loss, it is possible to design for a constant loss or gain thru all or part of the system. Designing for a constant loss increases operating cost and balancing time and may increase the fan motor size. Although not normally recommended, sizing for a constant loss reduces the duct size.

Example 5 — Static Regain Method of Designing Ducts

Given:
    Duct layout *(Example 4* and *Fig. 47)*
    Total air quantity — 5400 cfm
    Velocity in initial duct section — 1700 fpm *(Example 4)*
    Unvaned radius elbow, R/D = 1.25
    18 air terminals — 300 cfm each
    Operating pressure for all terminals — 0.15 in. wg

Find:
    1. Duct sizes.
    2. Total static pressure required at fan discharge.

Solution:
    1. Using an initial velocity of 1700 fpm and knowing the air quantity (5400 cfm), the initial duct area after the fan discharge equals 3.18 sq ft. From *Table 6*, a duct size of 22" x 22" is selected. The equivalent round duct size from *Table 6* is 24.1 in. and the friction rate from *Chart 7* is 0.145 in. wg per 100 ft of equivalent length.
    The equivalent length of duct from the fan discharge to the first branch:
        = duct length + additional length due to fittings
        = 60 + 12 = 72 ft
    The friction loss in the duct section up to the first branch:
        = equiv length of duct × friction rate
        $$= 72 \times \frac{0.145}{100} = .104 \text{ in. wg}$$
    The remaining duct sections are now sized.
    The longest duct run (*A* to outlet *18*, *Fig. 47*) should be sized first. In this example, it is desirable to have the

CHART 10—L/Q RATIO

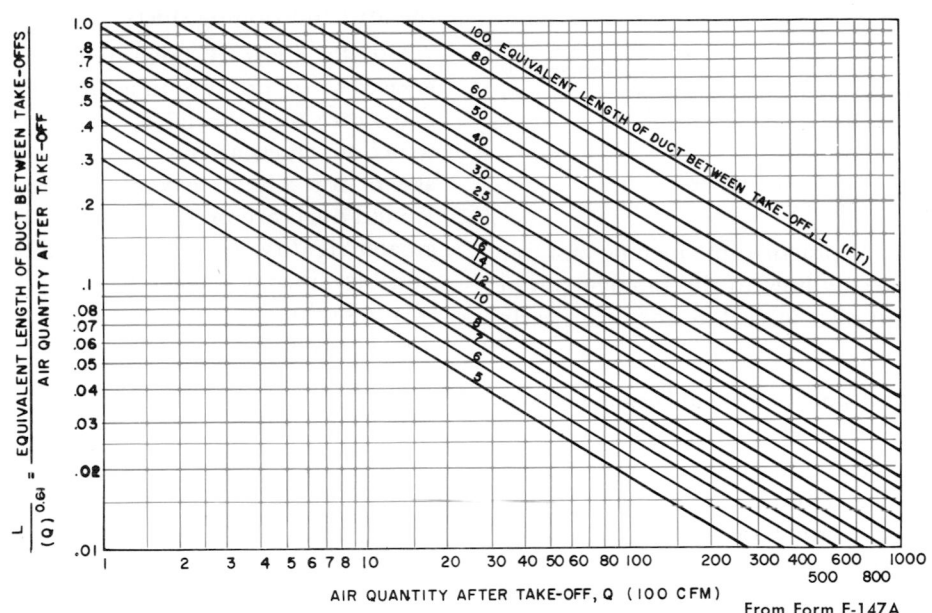

From Form E-147A

CHART 11—LOW VELOCITY STATIC REGAIN

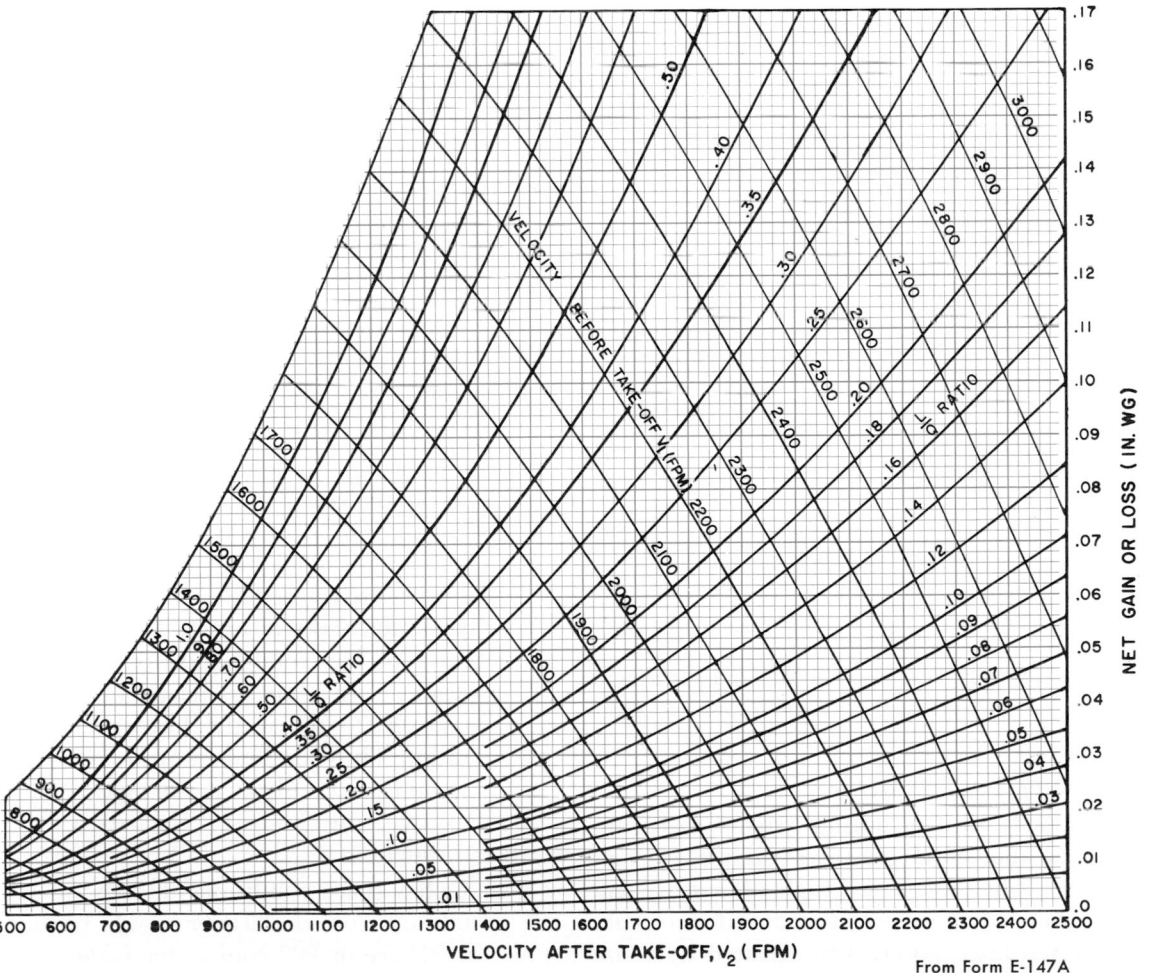

From Form E-147A

static pressure in the duct immediately before outlets *1* and *7* equal to the static pressure before outlet *13*.

Figure 48 tabulates the duct sizes.

2. The total pressure required at the fan discharge is equal to the sum of the friction loss in the initial duct section plus the terminal operating pressure.

Fan discharge pressure:
= friction loss + terminal pressure
= .104 + .15 = .25 in. wg

It is good design practice to include splitter dampers to regulate the flow to the branches, even though the static pressure at each terminal is nearly equal.

Comparison of Static Regain and Equal Friction Methods

Examples 4 and 5 show that the header duct sizes determined by the equal friction or static regain method are the same. However, the branch ducts, sized by static regain, are larger than the branch ducts sized by equal friction.

Figure 49 shows a comparison of duct sizes and weights established by the two methods.

The weight of sheet metal required for the system designed by static regain is approximately 13% more than the system designed by equal friction. However, this increase in first cost is offset by reduced balancing time and operating cost.

If it is assumed that a low velocity air handling system is used in *Examples 3 and 4* and that a design air flow of 5400 cfm requires a static pressure of 1.5 in. wg, the increased horsepower required for other equal friction design is determined in the following manner.

	STATIC REGAIN METHOD S.P. (in. wg)	EQUAL FRICTION METHOD S.P. (in. wg)
Air handling equipment	1.5	1.5
Duct friction	.104	.33
Terminal pressure	.15	.15
Static regain credit	—	—.12
Total	1.75	1.86

$$\text{Additional hp} = \frac{1.86 - 1.75}{1.75} = 6.3\% \text{ approx.}$$

A 6% increase in horsepower often indicates a larger fan motor and subsequent increased electrical transmission costs.

HIGH VELOCITY DUCT SYSTEMS

A high velocity air distribution system uses higher air velocities and static pressures than a conventional system. The design of a high velocity system involves a compromise between reduced duct sizes and higher fan horsepower. The reduced duct size

is a savings in building space normally allotted to the air conditioning ducts.

Usually Class II fans are required for the increased static pressure in a high velocity system and extra care must be taken in duct layout and construction. Ducts are normally sealed to prevent leakage of air which may cause objectionable noise. Round ducts are preferred to rectangular because of greater rigidity. *Spira-Pipe* should be used whenever possible, since it is made of lighter gage metal than corresponding round and rectangular ducts, and does not require bracing.

Symmetry is a very important consideration when designing a duct system. Maintaining as symmetrical a system as possible reduces balancing time, design time and layout. Using the maximum amount of symmetrical duct runs also reduces construction and installation costs.

Particular care must be given to the selection and location of fittings to avoid excessive pressure drops and possible noise problems. *Figure 50* illustrates the minimum distance of six duct diameters between elbows and 90° tees. If a 90° conical tee is used, the next fitting in the direction of air flow may be located a minimum of one-half duct diameter away *(Fig. 51)*. The use of a conical tee is limited to header ductwork and then only for increased initial velocities in the riser.

When laying out the header ductwork for a high velocity system, there are certain factors that must be considered:

1. The design friction losses from the fan discharge to a point immediately upstream of the first riser take-off from each branch header should be as nearly equal as possible. These points of the same friction loss are shown in *Fig. 52*.

2. To satisfy the above principle when applied to multiple headers leaving the fan, and to take maximum advantage of allowable high velocity, adhere to the following basic rule wherever possible: Make as nearly equal as possible the ratio of the total equivalent length of each header run (fan discharge to the first riser take-off) to the initial header diameter (L/D ratio). Thus the longest *Spira-Pipe* header run should preferably have the highest air quantity so that the highest velocities can be used throughout.

3. Unless space conditions dictate otherwise, the take-off from the header should be made using a 90° tee or 90° conical tee rather than a 45°

Total S.P. Loss for supply duct system = S.P. for critical duct ___ in. wg plus air outlet S.P. loss ___ in. wg = ___ in. wg.

1	2	3	4	5		6		7	8	9
SECTION NO.	AIR QUAN-TITY Q (cfm)	EQUIV. LENGTH L (ft)	$\frac{L}{Q}$ RATIO	VELOCITY V (fpm)		AREA (sq ft)		DUCT DIAM. OR RECT. SIZE† (in.)	FRICTION LOSS OR TAKE-OFF TO TAKE-OFF S.P. CHANGE (in. wg)	TOTAL S.P. LOSS IN DUCT (in. wg)
				Indicated	Selected	Indicated	Selected			
Fan to A	5400	72		1700		3.18		22 x 22	0.104	0.104
A - B	3600	20	.135	1510		2.38		22 x 16		
B - 13	1800	37*	.39	1170		1.54		22 x 10		
13 - 14	1500	20	.23	1000		1.50		22 x 10		
14 - 15	1200	20	.26	850		1.41		22 x 10		
15 - 16	900	20	.32	720		1.25		20 x 10		
16 - 17	600	20	.41	590		1.01		16 x 10		
17 - 18	300	20	.63	480		.63		10 x 10		
B - 7	1800	17*						22 x 10		
7 - 8	1500	20						22 x 10		
8 - 9	1200	20						22 x 10		
9 - 10	900	20						20 x 10		
10 - 11	600	20						16 x 10		
11 - 12	300	20						10 x 10		
A - 1	1800	17*						22 x 10		
1 - 2	1500	20						22 x 10		
2 - 3	1200	20						22 x 10		
3 - 4	900	20						20 x 10		
4 - 5	600	20						16 x 10		
5 - 6	300	20						10 x 10		

From Form E-147

*Duct size is assumed to determine loss thru elbow.
†Duct sizes from *Table 6.* Longest duct run is sized first. Remaining duct sections are the same size, as they are symmetrical to branch *B thru 18.* If other branches are not symmetrical and handle different air quantities, an initial velocity is assumed at the beginning of the branch. This velocity is somewhat less than the velocity in the header before take-off.

FIG. 48 — DUCT SIZING CALCULATION FORM

DUCT SECTION	EQUAL FRICTION METHOD		STATIC REGAIN METHOD	
	Duct Dimensions (in.)	Duct Weight (lb)	Duct Dimensions (in.)	Duct Weight (lb)
To A	22 x 22	592	22 x 22	592
A to B	22 x 16	179	22 x 16	179
A-1, B-7, B-13	22 x 10	394	22 x 10	394
1-2, 7-8, 13-14	18 x 10	411	22 x 10	438
2-3, 8-9, 14-15	14 x 10	360	22 x 10	438
3-4, 9-10, 15-16	12 x 10	321	20 x 10	435
4-5, 10-11, 16-17	8 x 10	270	16 x 10	384
5-6, 11-12, 17-18	8 x 10	270	10 x 10	297
Total weight of duct*		2797		3157
Allow 15% for scrap		420		475
Total wt of sheet metal		3217		3632

*Total weight includes transformation and elbows.

FIG. 49 — COMPARISON OF DUCT SIZING METHODS

tee. By using 90° fittings, the pressure drop to the branch throughout the system is more uniform. In addition, two fittings are normally required when a 45° tee is used and only one when a 90° fitting is used, resulting in lower first cost.

The design of a high velocity system is basically the same as a low velocity duct system designed for static regain. The air velocity is reduced at each take-off to the riser and air terminals. This reduction in velocity results in a recovery of static pressure (velocity regain) which offsets the friction loss in the succeeding duct section.

The initial starting velocity in the supply header depends on the number of hours of operation. To achieve an economic balance between first cost and

operating cost, lower air velocities in the header are recommended for 24-hour operation where space permits. When a 90° conical tee is used instead of a 90° tee for the header to branch take-off, a higher initial starting velocity in the branch is recommended. The following table suggests initial velocities for header and branch duct sizing:

RECOMMENDED INITIAL VELOCITIES USED WITH CHARTS 12 AND 13 (fpm)	
HEADER	
12 hr. operation	3000 - 4000
24 hr. operation	2000 - 3500
BRANCH*	
90° conical tee	4000 - 5000
90° tee	3500 - 4000
TAKE-OFFS TO TERMINALS	2000 maximum

*Branches are defined as a branch header or riser having 4 to 5 or more take-offs to terminals.

Static regain charts are presented for the design of high velocity systems. *Chart 12* is used for designing branches and *Chart 13* is used for header design. The basic difference in the two charts is the air quantity for the duct sections.

Chart 12 is used for sizing risers and branch headers handling 6000 cfm or less. The chart is based on 12 ft increments between take-offs to the air terminals in the branches or take-offs to the risers in branch headers. A scale is provided to correct for spacings more or less than 12 ft.

Chart 13 is used to size headers, and has an air quantity range of 1000 to 40,000 cfm. The chart is based on 20 ft increments between branches. A correction scale at the top of the chart is used when take-off to branch is more or less than 20 ft.

Examples 6 and 7 are presented to illustrate the use of these two charts. *Example 6* is a branch sizing problem for the duct layout in *Fig. 53*, and *Example 7* is a header layout (*Fig. 55*).

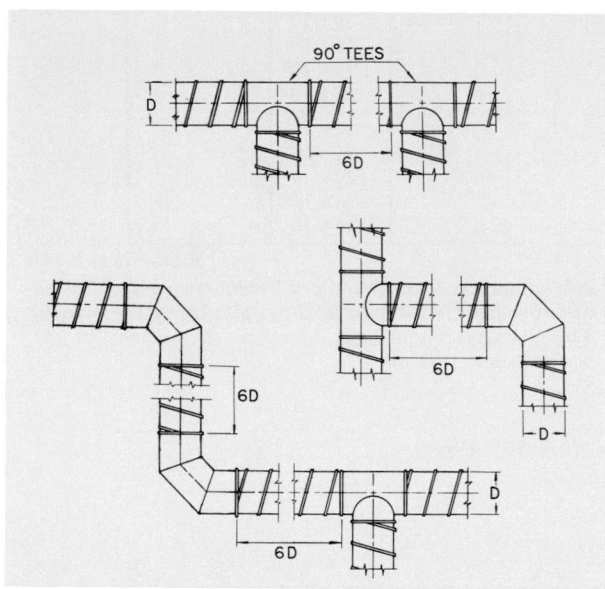

FIG. 50 — SPACING OF FITTINGS IN DUCT RUN

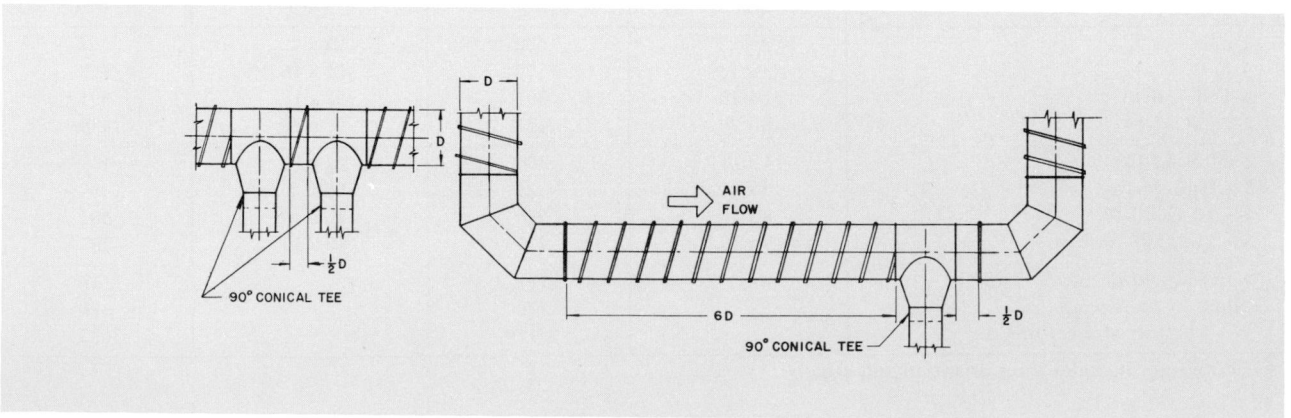

FIG. 51 — SPACING OF FITTINGS WHEN USING 90° CONICAL TEE

CHART 12—BRANCH HIGH VELOCITY STATIC REGAIN

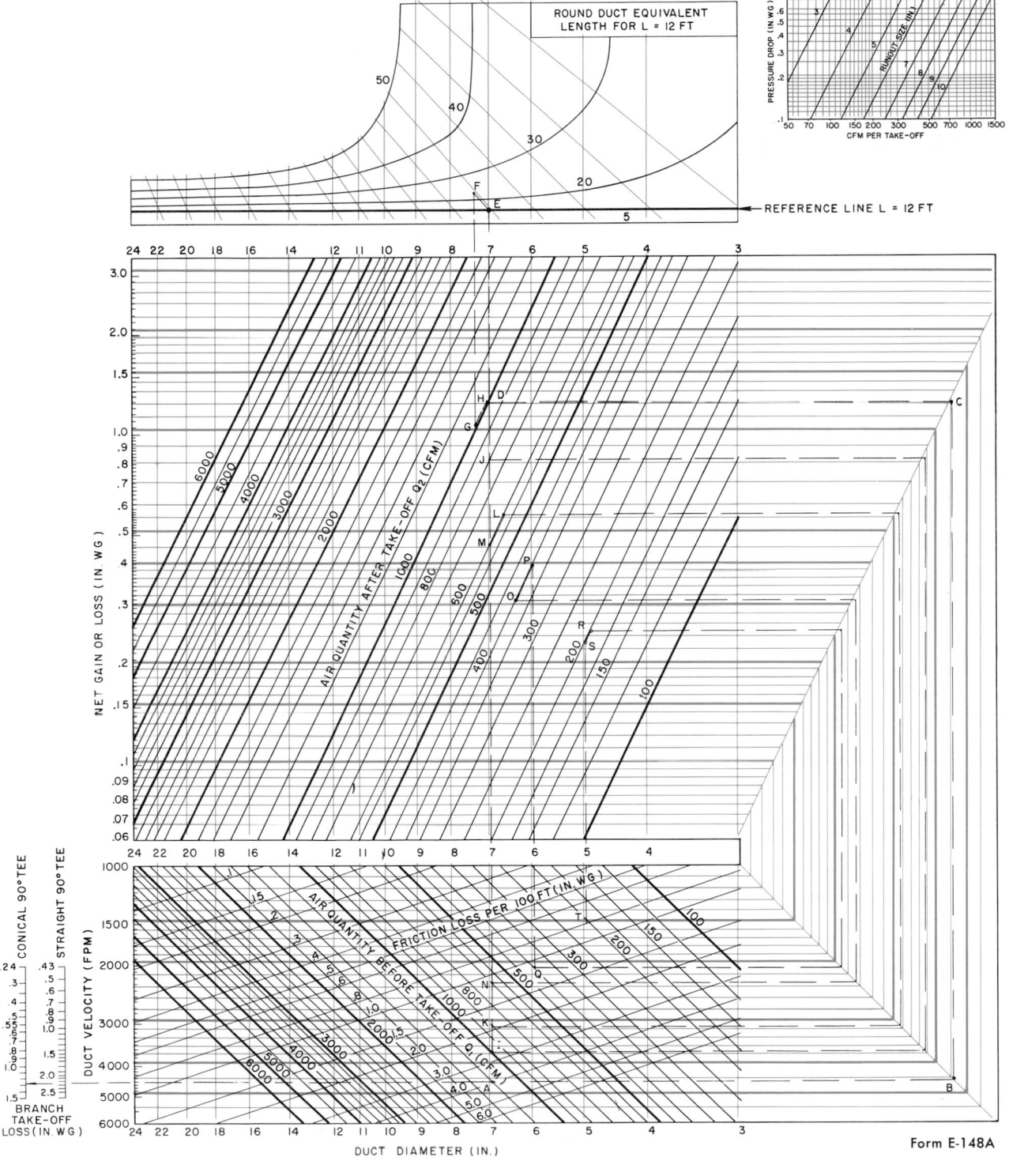

Form E-148A

CHART 13—HEADER HIGH VELOCITY STATIC REGAIN

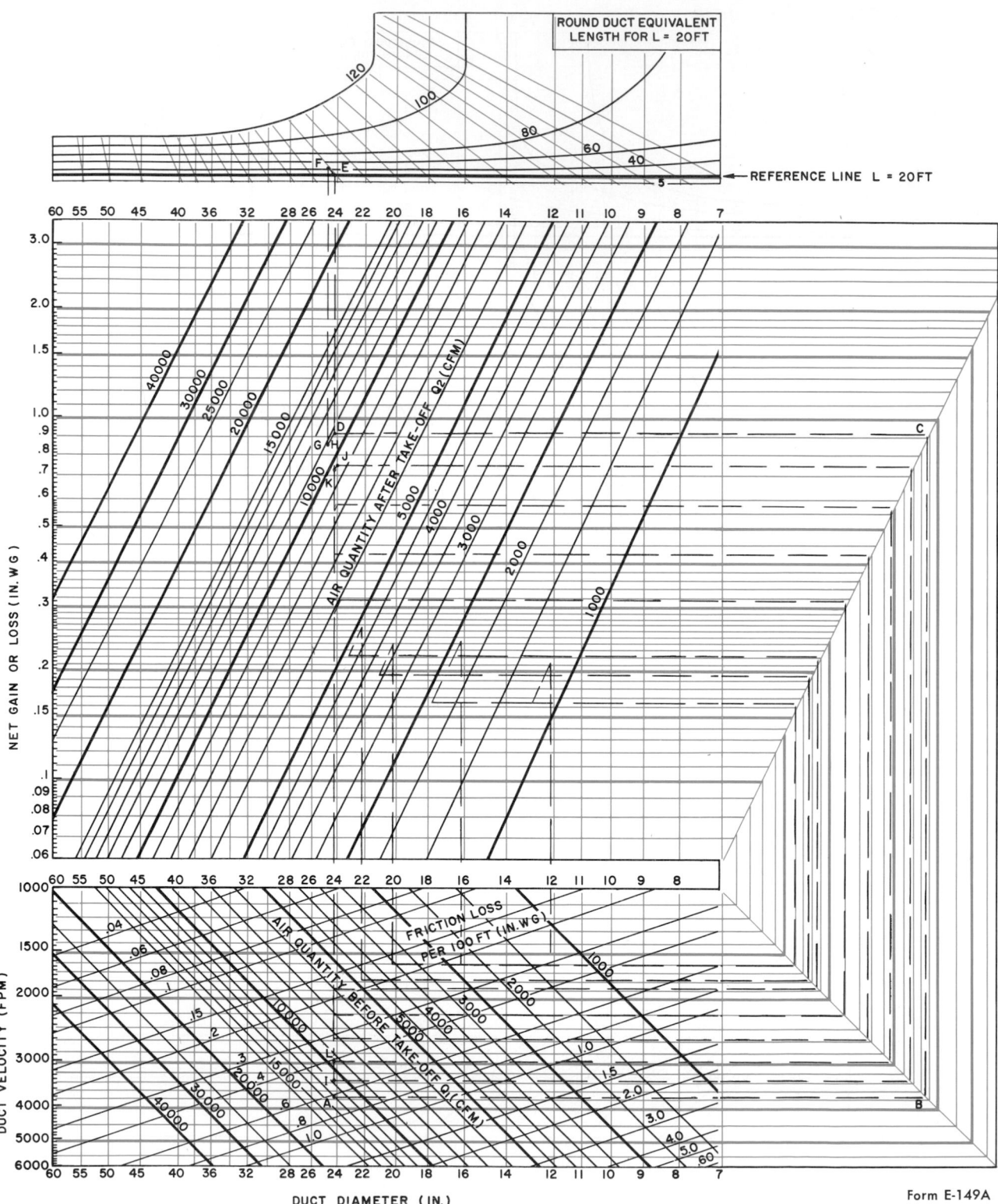

Form E-149A

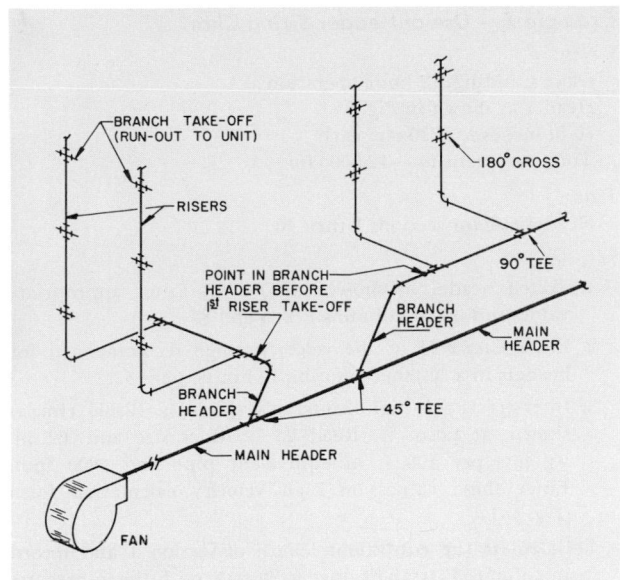

FIG. 52 — HIGH VELOCITY HEADERS AND BRANCHES

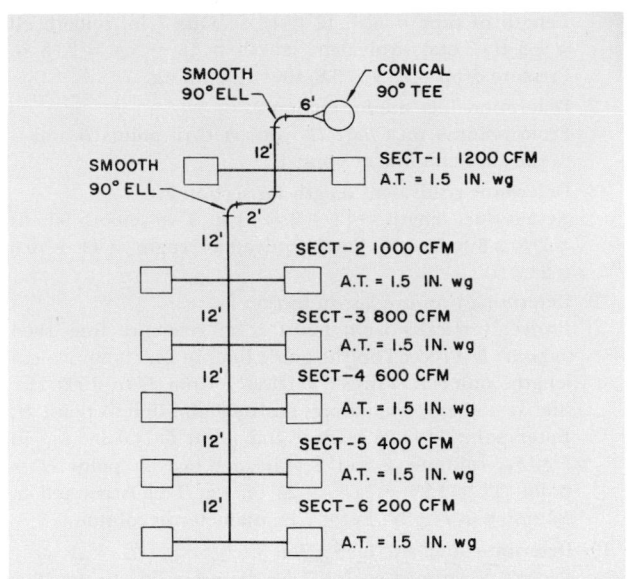

FIG. 53 — BRANCH DUCT FOR EXAMPLE 6

Example 6 — Use of Branch Duct Sizing Chart

Given:
    Office building riser as shown in *Fig. 53*
    12 air terminals — 100 cfm each
    Total air quantity — 1200 cfm
    Air terminal static pressure — 1.5 in. wg

Find:
    Duct sizes for Sections 1 thru 6, *Fig. 53.*

Solution:
    1. Sketch branch as shown in *Fig. 53.* Enter appropriate
       values in columns 2, 3 and 8, *Fig. 54.*

2. Enter *Chart 12* at the velocity range recommended for branch risers with a 90° conical tee, *page 52.*

3. Intersect the initial branch air quantity, 1200 cfm, as shown at point *A.* Read 7 in. duct size and 3.8 in. wg loss per 100 ft of equivalent pipe and 4500 fpm velocity. Enter these values on the high velocity calculations form *(Fig. 54).*

4. From point *A,* determine header take-off loss by projecting horizontally to the left of point *A* and read 1.25 in. wg.

5. Enter 1.25 in. wg in *Fig. 54* for section 1.

6. Determine equivalent length from the header to the first air terminal take-off:

INITIAL CONDITIONS: Cfm 1200; Duct Size 7 in.; Velocity 4500 fpm.									
1	2	3	4	5	6	7	8	9	10
BRANCH SECT. NO.	AIR QUANTITY Q	EQUIV. DUCT LENGTH L	PRESSURE READING (in. wg)		TAKE-OFF TO TAKE-OFF S.P. CHANGE (4 minus 5)	S.P. AHEAD OF TAKE-OFF	AIR TERMINAL PRESS.	DUCT SIZE	VELOCITY V
			Initial	Selected					
	(cfm)	(ft)			(in. wg)	(in. wg)	(in. wg)	(in.)	(fpm)
1	1200	23.3	Branch Take-Off F. L. = 1.25			2.14	1.5	7	4500
			Duct Friction Loss = .89						
2	1000	24.6	1.0	1.25	− 0.25	2.39	1.5	7	3700
3	800	12	0.84	0.84	0.0	2.39	1.5	7	3050
4	600	12	0.57	0.47	+ 0.1	2.29	1.5	7	2300
5	400	12	0.32	0.40	− 0.085	2.37	1.5	6	2050
6	200	12	0.26	0.24	+ 0.02	2.35	1.5	5	1475
			Maximum S. P. is at Section 2:		2.39	+	1.5	+ .19 = 4.08	

From Form E-148

FIG. 54 — HIGH VELOCITY BRANCH SIZING CALCULATIONS

Length of pipe = 6 + 12 = 18 ft. One 7 in. smooth ell = 5.3 ft. Total equivalent length = 18 + 5.3 = 23.3 ft. Pressure drop = 23.3 × 3.8/100 = .89 in. wg.

7. Determine duct size for section 2:
   From point *A* on *Chart 12*, project thru points *B* and *C* to the 1000 cfm line at point *D*.

8. Determine equivalent length for section 2:
   Actual duct length = 12 + 2 = 14 ft. Two smooth 90 ells = 2 × 5.3 = 10.6 ft. Total equivalent length = 14 + 10.6 = 24.6 ft.

9. Determine pressure loss in section 2:
   Project vertically from point *D* to reference line, then to point *E*. Proceed on the guide lines to 24.6 ft equivalent length, point *F*. Project vertically from *F* to 1000 cfm line at point *G*, then along the 1000 cfm line to point *H*. Enter point *H* (1.25 in. wg) and point *G* (1.0 in. wg) in *Fig. 54*, columns 4 and 5. The net loss is "point *H* — point *G*" = 1.25 − 1.00 = .25 in. wg. This is entered in column 6 of *Fig. 54*. Enter 7 in. diameter in column 9.

10. Determine duct size for section 3:
    Project downward on the 7 in. diameter line to the 1000 cfm line, points *H* to *I*.

11. Project along guide lines at the right side of the chart from *I* to the 800 cfm line at point *J*. Duct size is 7 in. Enter appropriate values from the chart in columns 4, 5, 6 and 9 of *Fig. 54*.

12. Determine duct size for section 4:
    Project downward on the 7 in. diameter line to the 800 cfm line, points *J* to *K*.

13. Project along guide lines at the right side of the chart from point *K* to the 600 cfm line at point *L*. Project along the 600 cfm line to the 7 in. diameter line, point *L* to *M*. This results in a static regain of .57 − .47 = .10 in. wg. Duct size for section 4 is 7 in. Enter appropriate values in *Fig. 54*, columns 4, 5, 6, 7 and 9.

    NOTE: If the 600 cfm line is projected from point *L* to the 6 in. diameter line, a net loss of .88 − .45 = .43 in. wg results. This friction loss unnecessarily penalizes the system. Therefore, the projection from *L* is made to the 7 in. diameter line.

14. Determine duct size for section 5: Project downward from *M* to 600 cfm line, point *N*. Project along guide lines to 400 cfm line, point *O*. Continue along the 400 cfm line to the 6 in. diameter line, point *O* to *P*. This results in a static pressure loss of .40 − .315 = .085 in. wg. Duct size is 6 in. Enter the appropriate values in *Fig. 54*, columns 4, 5, 6, 7 and 9.

    NOTE: If the 400 cfm is projected from point *O* to the 7 in. diameter line, a net regain of .315 − .20 = .115 in. wg results. Therefore, the 6 in. size is used to save on first cost since the net loss using the 6 in. size is insignificant.

15. Determine duct size for section 6:
    Duct size is 5 in. as determined from point *S*.

16. Determine velocities for duct sections 1–6 from points *A*, *I*, *K*, *N*, *Q* and *T* respectively;, enter in column 10.

17. Determine take-off and runout pressure drop by entering upper right hand portion of Chart 12 at 100 cfm and read a pressure drop of .19 in. wg for a 4 in. runout size.

18. Add 2.39 in. wg (maximum from column 7) plus 1.5 in. wg (column 8) plus .19 (take-off and runout drop) to find 4.08 in. wg (total branch S.P.).

Example 7 — Use of Header Sizing Chart

Given:
   Office building, 12-hour operation
   Header as shown in *Fig. 55*
   10 branches — 1200 cfm each
   Total air quantity — 12,000 cfm

Find:
   Header size for sections 1 thru 10

Solution:

1. Sketch header as shown in *Fig. 55*. Enter appropriate values in *Fig. 56*, columns 1, 2, 3 and 8.

2. Enter *Chart 13* at the velocity range recommended for headers in a system operating 12 hours, *page 52*.

3. Intersect the initial header air quantity 12,000 cfm as shown, at point *A*. Read 24 in. duct size and .62 in. wg loss per 100 ft of equivalent pipe and 3800 fpm. Enter these values on high velocity calculation form (*Fig. 56*).

4. Calculate the equivalent length of section 1 and record in column 3; straight duct = 20 feet, no fittings; pressure drop = 20 × .62 = .124 in. wg.

5. Size duct section 2: From point *A* on chart, project thru points *B* and *C* to the 10,800 cfm line at point *D*.

6. Determine equivalent length for section 2: Actual length = 20 ft. One 5-piece 90° ell = 24 ft. Total equivalent length = 20 + 24 = 44 ft.

7. Determine pressure loss in section 2: Project vertically from point *D* to reference line, point *E*. Proceed on the guide lines to 44 ft equivalent length, point *F*. Project vertically from *F* to 10,800 cfm line at point *G*, then along the 10,800 cfm line to point *H*. Enter the net loss read from point *G* (.84) and from point *H* (.90) in columns 4 and 5, *Fig. 56*. The net loss is "point *H* — point *G*" = .90 − .84 = .06 in. wg. This is entered in column 6, *Fig. 56*. Enter 24 in. diameter in column 9.

8. Determine duct size for section 3: Project downward on the 24 in. line to the 10,800 cfm line, point *H* to *I*. Project along the guide lines at the right side of the chart from *I* to the 9600 cfm line at point *J*. Enter appropriate values from the chart in columns 4, 5, 6 and 9, *Fig. 56*.

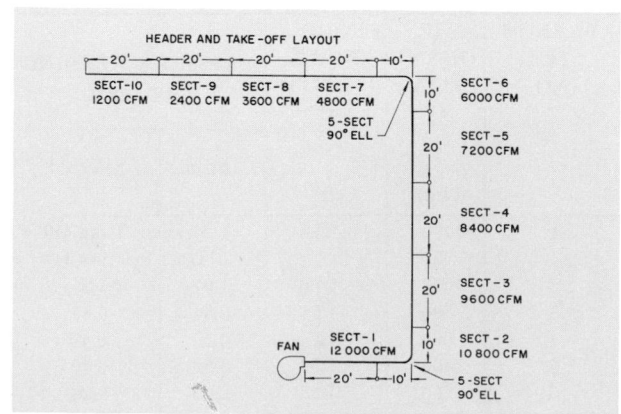

Fig. 55 — High Velocity Duct System — Header Static Regain Method Sizing

INITIAL CONDITIONS: Cfm 12,000; Duct Size 24 in.; Velocity 3800 fpm.

1	2	3	4	5	6	7	8	9	10
HEADER SECT. NO.	AIR QUAN-TITY Q	EQUIV. DUCT LENGTH L	PRESSURE READING (in. wg)		TAKE-OFF TO TAKE-OFF S.P. CHANGE (4 minus 5)	S.P. AHEAD OF TAKE-OFF	BRANCH S.P.	DUCT SIZE	VELOC-ITY V
	(cfm)	(ft)	Initial	Selected	(in. wg)	(in. wg)	(in. wg)	(in.)	(fpm)
1	12000	20	Duct Friction = 0.124			0.124	4.08	24	3800
2	10800	44	0.84	0.90	−0.06	0.184	4.08	24	3400
3	9600	20	0.74	0.70	+0.04	0.144	4.08	24	3000
4	8400	20	0.57	0.55	+0.02	0.124	4.08	24	2600
5	7200	20	0.42	0.42	0.0	0.124	4.08	24	2250
6	6000	44	0.31	0.30	+0.01	0.114	4.08	24	1900
7	4800	20	0.22	0.26	−0.04	0.154	4.08	22	1800
8	3600	20	0.195	0.23	−0.035	0.189	4.08	20	1650
9	2400	20	0.165	0.24	−0.075	0.264	4.08	16	1650
10	1200	20	0.165	0.21	−0.045	0.309	4.08	12	1500

Maximum S.P. at Section 10 = 0.31   +   4.08 = 4.39

From Form E-149

FIG. 56 — HIGH VELOCITY HEADER SIZING CALCULATIONS

9. Determine duct sizes for sections 4 thru 10 in a manner similar to Step 8, using the listed air quantities and equivalent lengths. One exception is duct section 6. Since its equivalent length is 44 feet, use the method outlined in Steps 5, 6 and 7 to determine the pressure drop. In addition, see *Example 5*, Steps 13 and 14, for explanation when the chart indicates a duct diameter other than those listed, for instance 23 inches.

DUCT HEAT GAIN AND AIR LEAKAGE

Whenever the air inside the duct system is at a temperature different than the air surrounding the duct, heat flows in or out of the duct. As the load is calculated, an allowance is made for this heat gain or loss. In addition, air leakage is also included in the calculated load. The load allowance required and guides to conditions under which an allowance should be made for both heat gain or loss and duct leakage are included in *Part I, System Heat Gain*.

Chart 14 is used to determine the temperature rise or drop for bare duct that has an aspect ratio of 2:1. In addition, correction factors for other aspect ratios and insulated duct are given in the notes to the chart.

Example 8 — Calculations for Supply Duct

Given:
Supply air quantity from load estimate form — 1650 cfm
Supply duct heat gain from load estimate form — 5%
Supply duct leakage from load estimate form — 5%
Unconditioned space temp — 95 F
Room air temperature — 78 F
Duct insulation *U* value — .24
Duct shown in *Fig. 57*.

Find:
Air quantities at each outlet

Solution:
1. Room air quantity required at 60 F
$$= \frac{1650}{1 + .05 + .05} = 1500 \text{ cfm}$$

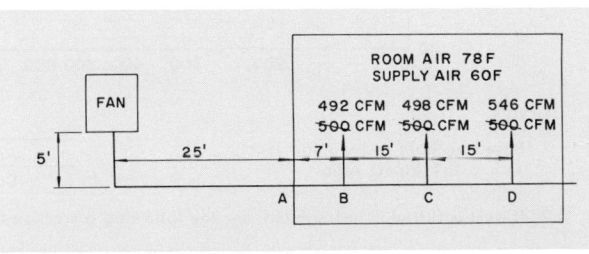

FIG. 57 — DUCT HEAT GAIN AND AIR LEAKAGE

Normally a 10% leakage allowance is used if the complete duct is outside the room. Since a large portion of the duct is within the room, 5% is used in this example.

2. Determine the temperature rise from A to B: Select an initial starting velocity from *Table 7* (assume 1400 fpm). Calculate the temperature rise from the fan to the room. Enter *Chart 14* at 1500 cfm; project vertically to 1400 fpm and read .27 degrees temperature change per 100 ft per degree F difference. Using aspect ratio of 2:1, temperature rise

$$= \frac{30 \text{ ft}}{100 \text{ ft}} \times .27 \text{ F change} \times .185 \times (95 - 60) = .52 \text{ F}$$

Air temperature entering room = 60.52 F
Actual air quantity entering room

$$= \frac{78 - 60}{78 - 60.52} \times 1500 = 1540 \text{ cfm}$$

Air temperature rise from A to B

$$= \frac{7}{100} \times 17.48 \times .27 = .33 \text{ F}$$

Supply air temperature diff to outlet B

$$= 78 - (60.52 + .33) = 17.15 \text{ F}$$

Required air quantity to outlet B

$$= 500 \times \frac{18}{17.2} = 522 \text{ cfm}$$

with no allowance for cooling from the duct.

Outlet B cfm with allowance for duct cooling

$$= 522 - \left(1540 \times \frac{.33}{17.2}\right) = 492 \text{ cfm}$$

3. Determine cfm for outlet C: Use equal friction method to determine velocity in second section of duct, with $1540 - 492 = 1048$ cfm; velocity = 1280 fpm.
Determine temperature rise at outlet: From *Chart 14*, read .32 for 1280 fpm and 1040 cfm. Temperature rise

$$= .32 \times 17.2 \times \frac{15}{100} = .83 \text{ F}$$

Supply air temperature diff = 17.2 − .8 = 16.4 F
Outlet cfm adjusted for temperature rise

$$= 500 \times \frac{18}{16.4} = 550 \text{ cfm}$$

Allowance for duct cooling

$$= 550 - \left(1048 \times \frac{.8}{16.4}\right) = 498 \text{ cfm}$$

4. Determine cfm for outlet D:
Use equal friction method to determine velocity in third section of duct with $1048 - 498 = 550$ cfm; velocity = 1180 fpm.
Determine temperature rise at outlet:
From *Chart 14*, read .43 F for 1180 fpm and 550 cfm. Temperature rise

$$= .43 \times 16.4 \times \frac{15}{100} = 1.06 \text{ F}$$

Supply air temperature diff = 16.4 − 1.1 = 15.3 F

CHART 14—DUCT HEAT GAIN OR LOSS

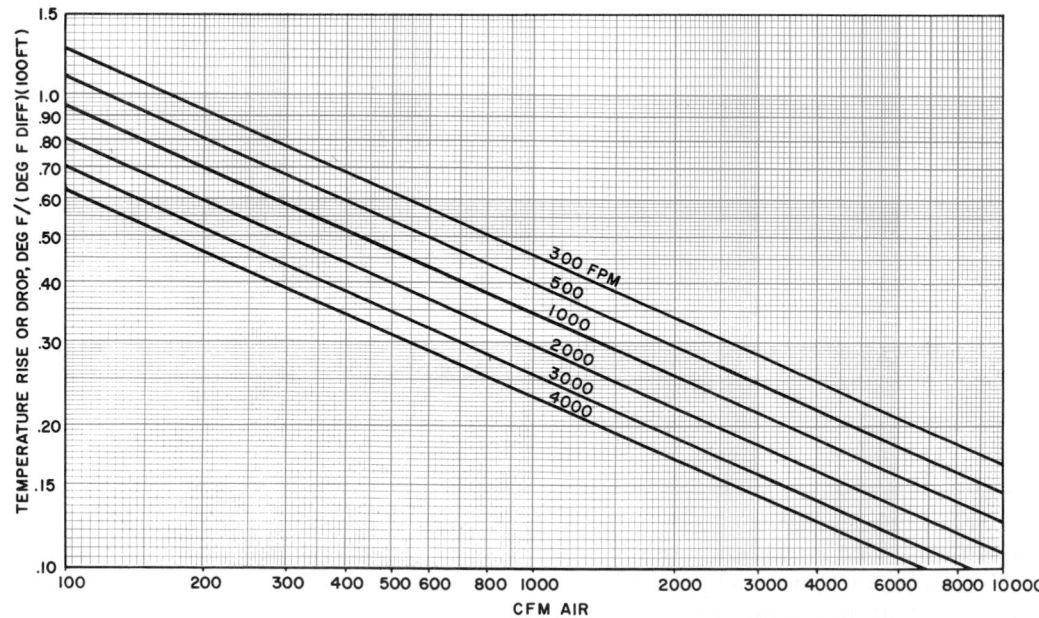

NOTES:

1. Based on bare rectangular duct with a 2:1 aspect ratio.

Aspect Ratio	Round	1:1	3:1	4:1	5:1	6:1	7:1	8:1	9:1	10:1
Correction	.83	.92	1.1	1.18	1.26	1.35	1.43	1.5	1.58	1.65

Aspect Ratio Correction

2. If duct is furred-in or insulated, use the following correction factors: Furred-in duct — .45
                                                                 Insulated (U = .27) — .185
                                                                 Insulated (U = .13) — .10

3. For air quantities greater than 10,000 cfm, divide air quantity by 100 and multiply degree change by 0.1.

Outlet cfm adjusted for temperature rise

$$= 500 \times \frac{18}{15.3} = 588 \text{ cfm}$$

Allowance for duct cooling

$$= 588 - \left(588 \times \frac{1.1}{15.3} \right) = 546 \text{ cfm}$$

5. Check for total cfm:

$$492 + 498 + 546 = 1536 \text{ cfm}$$

This compares favorably with the 1540 cfm entering room. *Fig. 57* shows original and corrected outlet air quantities.

HIGH ALTITUDE DUCT DESIGN

When an air distribution system is designed to operate above 2000 feet altitude, below 30 F, or above 120 F temperature, the friction factor obtained from *Chart 7, page 33* using the actual air quantity at final conditions must be corrected for air density. *Chart 15* presents correction factors for temperature and altitude. The factors are multiplied together when a system is at high altitude and also operates outside the temperature range.

CHART 15—AIR DENSITY CORRECTION FACTORS

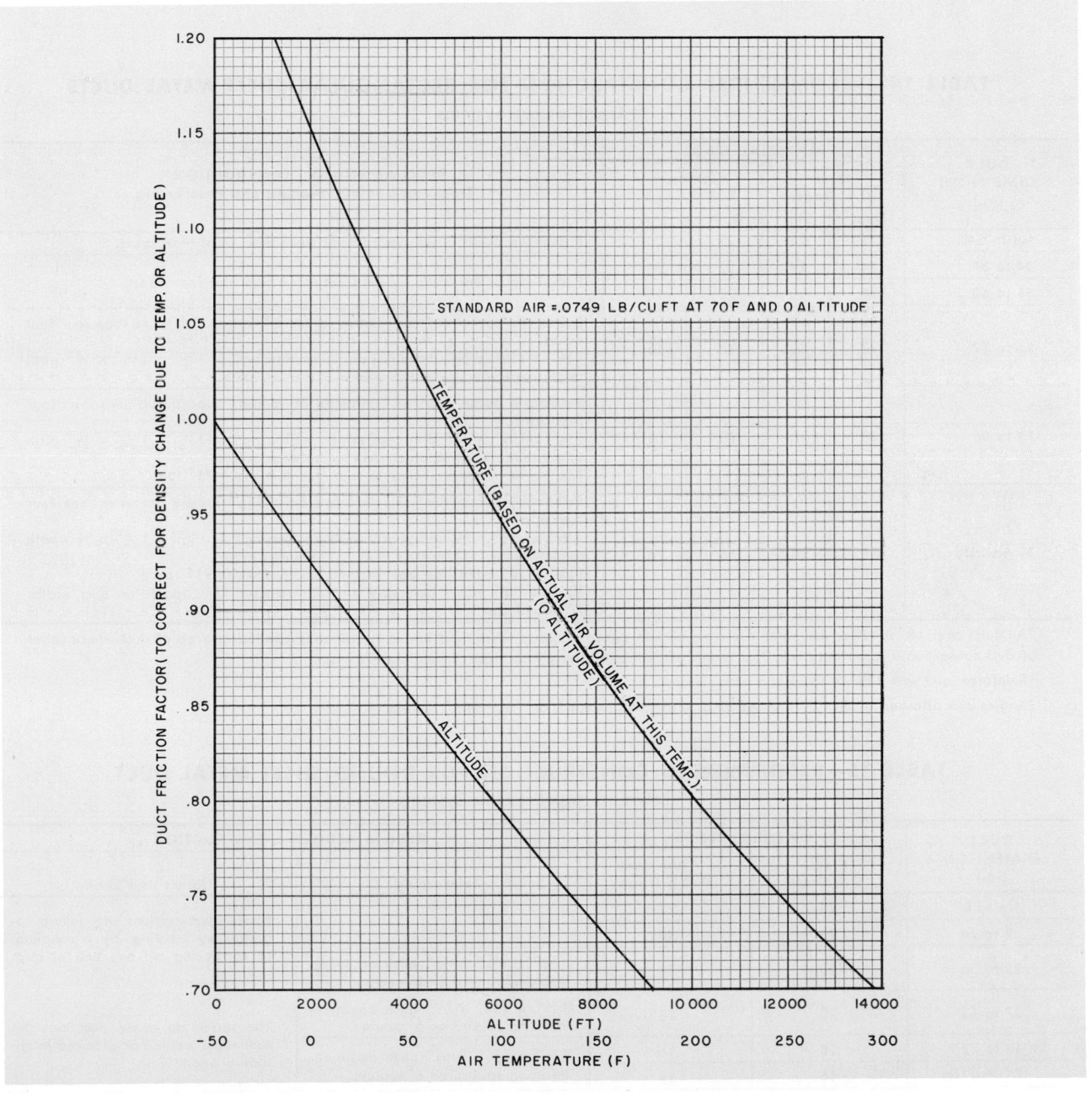

DUCT CONSTRUCTION

The sheet metal gage used in the ducts and the reinforcing required depends on the pressure conditions of the system. There is also a wide variety of joints and seams used to form the ducts which also depend on pressure conditions in the duct system.

Low Pressure Systems

Table 14 lists the recommended construction for rectangular ducts made of aluminum or steel. The method of bracing and reinforcing and types of joints and seams are included in the table. Round duct and *Spira-Pipe* construction are included in

Tables 15 and 16 which apply for low and high pressure systems. *Fig. 58* illustrates the more common seams and joints used in low pressure systems.

TABLE 16—MATERIAL GAGE FOR SPIRA-PIPE DUCT

Low and High Pressure Systems

DUCT DIMENSION (in.)	DUCT MATERIAL GAGE	
	Steel U.S. Gage	Aluminum B & S Gage
Up to 8	26	22
9 to 24	24	20
26 to 32	22	18

TABLE 14—RECOMMENDED CONSTRUCTION FOR RECTANGULAR SHEET METAL DUCTS

Low Pressure Systems

DUCT DIMENSION (in.)	MATERIAL GAGE				RECOMMENDED CONSTRUCTION* Transverse Joints, Bracing and Reinforcing
	Steel U.S. Gage		Aluminum B & S Gage		
	Duct	Slip	Duct	Slip	
Up to 24	24	24	22	20	Pocket slip or Bar-S slip, spaced not more than eight feet apart.
24 to 30	24	24	22	20	Pocket slip or Bar-S slip, spaced not more than four feet apart.
31 to 60	22	22	20	18	
61 to 72	20	20	18	16	Reinforced pocket slip† or reinforced Bar-S†, spaced not more than four feet apart. 1½" x 1½" x ⅛" diagonal angle reinforcing‡ or 1½" x 1½" x ⅛" girth angle reinforcing‡ located midway between joints.
73 to 90	20	20	18	16	Reinforced pocket slip† or reinforced Bar-S slip† spaced not more than four feet apart. 1½" x 1½" x ⅛" diagonal angle reinforcing† or 1½" x 1½" x ⅛" girth angle reinforcing‡ located midway between joints. 1¼" x ⅛" band iron stay bracing for duct width 73" to 90".
91 and Up	18	20	16	16	Reinforced pocket slip† or reinforced Bar-S slip† spaced not more than four feet apart. 1½" x 1½" x ⅛" diagonal angle reinforcing‡ or 1½" x 1½" x ⅛" girth angle reinforcing‡ located midway between joints. 1¼" x ⅛" band iron stay bracing for duct width 91" to 120". 1¼" x ⅛" band iron stay bracing spaced 48" apart for duct widths 121" and up.

*All ducts over 18″ in either dimension are cross-broken, except those to which rigid board insulation is applied or area of duct where outlet or duct connection is to be installed. Duct seams are either Pittsburg lock seam or longitudinal seam.

†Reinforce joint with 1¼" x ⅛" band iron.

‡Angles are attached to duct by tack welding, sheet metal screws, or rivets on 6″ centers.

TABLE 15—RECOMMENDED CONSTRUCTION FOR ROUND SHEET METAL DUCT

Low and High Pressure Systems

DUCT DIMENSION (in.)	MATERIAL GAGE		RECOMMENDED CONSTRUCTION	
	Steel U.S. Gage	Aluminum B & S Gage	Reinforcing	Joints and Seams
Up to 8	24	22		Round duct sections are joined together by welding, by a coupling, or by belling out one end of duct.
9 to 24	22	20		
25 to 36	20	18	1¼" x 1¼" x ⅛" girth angle reinforcing spaced on 8' centers.	
37 to 48	20	18	1¼" x 1¼" x ⅛" girth angle reinforcing spaced on 6' centers.	The seams on round duct may be continuous welded or grooved longitudinal seam.
49 to 72	18	16	1½" x 1½" x ⅛" girth angle reinforcing spaced on 4' centers.	
73 and Up	16	14		

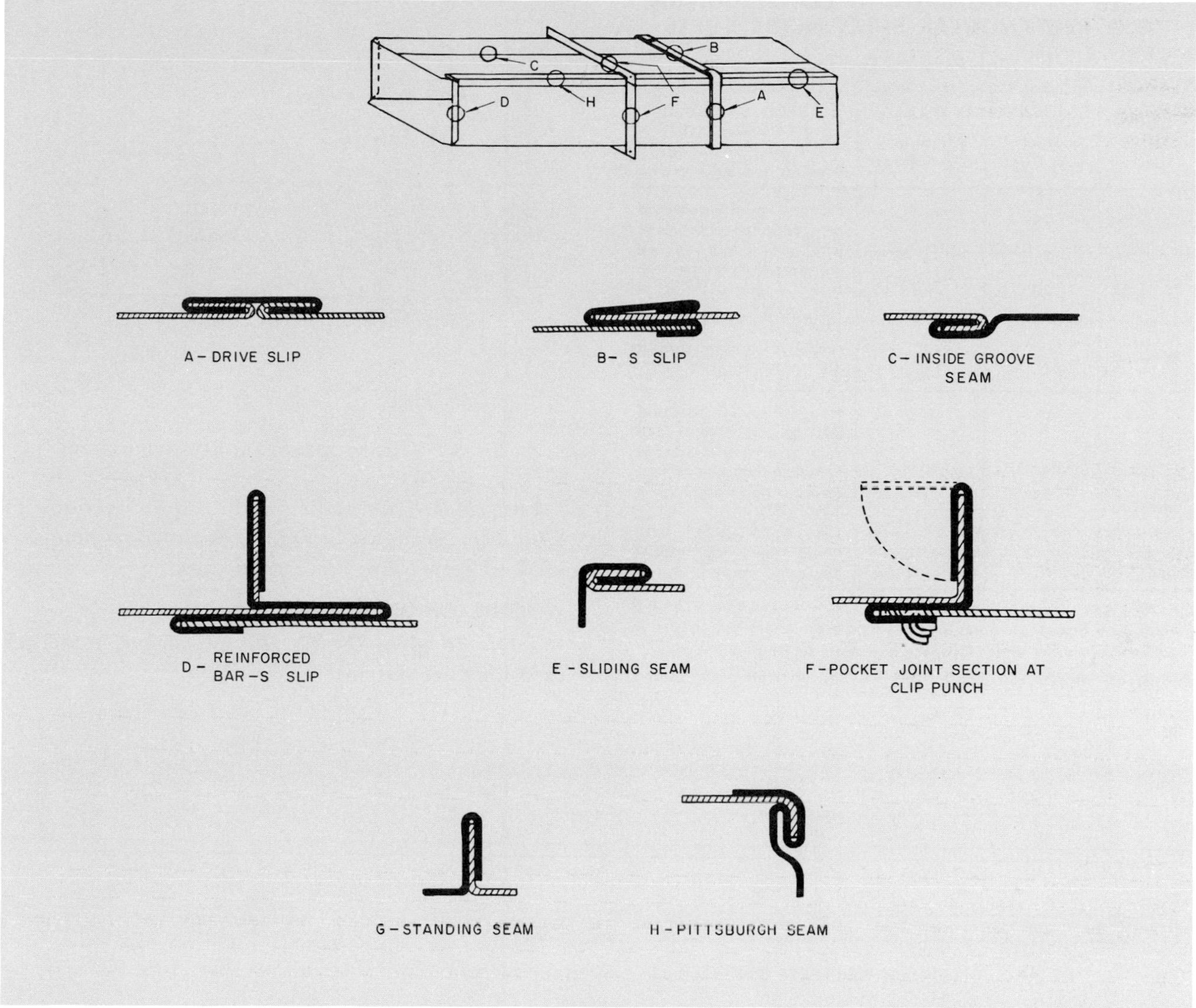

A – DRIVE SLIP

B – S SLIP

C – INSIDE GROOVE SEAM

D – REINFORCED BAR–S SLIP

E – SLIDING SEAM

F – POCKET JOINT SECTION AT CLIP PUNCH

G – STANDING SEAM

H – PITTSBURGH SEAM

FIG. 58 — JOINTS AND SEAMS FOR LOW PRESSURE SYSTEM

High Pressure Systems

Table 17 contains the construction recommendations for rectangular duct made of aluminum or steel. The table includes the required reinforcing and bracing and types of joints and seams used in high pressure duct systems.

Fig. 59 shows the common joint used for rectangular ducts in high pressure systems. The ducts are constructed with a Pittsburg lock or grooved longitudinal seams *(Fig. 58)*.

Table 15 shows the recommended duct construction for round ducts. The data applies for either high or low pressure systems. *Fig. 60* illustrates the seams and joints used in round duct systems. The duct materials for *Spira-Pipe* are given in *Table 16*.

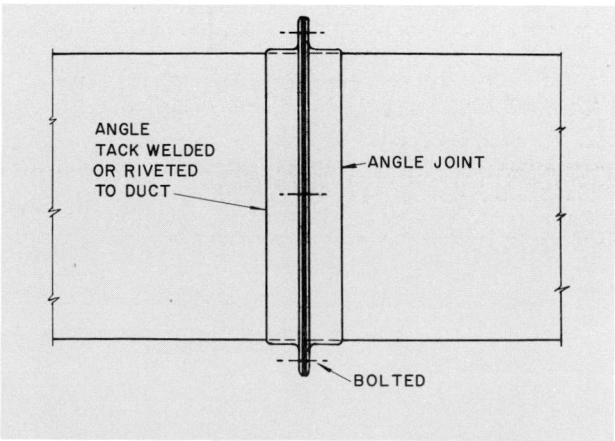

ANGLE TACK WELDED OR RIVETED TO DUCT

ANGLE JOINT

BOLTED

FIG. 59 — JOINT FOR HIGH PRESSURE SYSTEM

TABLE 17—RECOMMENDED CONSTRUCTION FOR RECTANGULAR SHEET METAL DUCTS

High Pressure Systems

DUCT DIM (in.)	MATERIAL GAGE		RECOMMENDED CONSTRUCTION* Transverse Joints Bracing and Reinforcing
	Steel U.S. Gage	Aluminum B & S Gage	
Up to 24	22	20	Flanged angle gasketed joint or butt welded joint with girth angle, spaced not more than twelve feet apart. Angles are 1½" x 1½" x ⅛"†. 1½" x 1½" x ⅛" girth angle reinforcing spaced 38" to 40" apart†.
25 to 48	20	18	
49 to 60	18	16	
61 and Up	18	16	Flanged angle gasketed joint or butt welded joint with girth angle, spaced not more than twelve feet apart. Angles are 1½" x 1½" x 3⁄16"†. 1½" x 1½" x 3⁄16" girth angle reinforcing spaced 38" x 40" apart†.

*All ducts over 18" in either dimension are cross-broken except those to which rigid board insulation is applied or area where outlets are installed. Seams are either Pittsburg lock seam or longitudinal seam.

†Angle are attached to duct by tack welding or rivets on 6" centers.

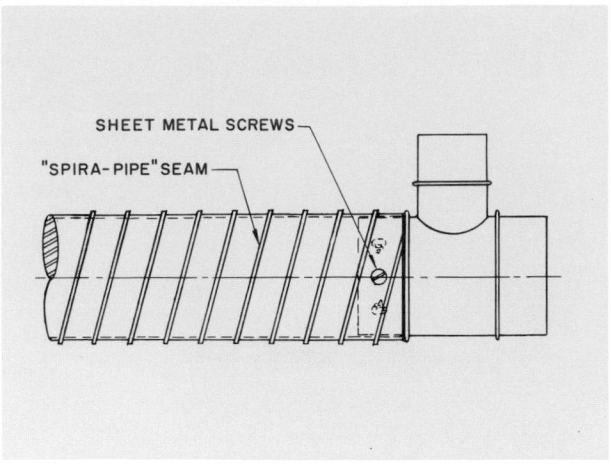

FIG. 61 — JOINT AND SEAM FOR SPIRA-PIPE

Fittings are normally used to join sections of *Spira-Pipe* as shown in *Fig. 61*. Sealing compound is used to join *Spira-Pipe* to fittings.

WEIGHTS OF DUCT MATERIALS

Table 18 gives the weights of various materials used for duct systems.

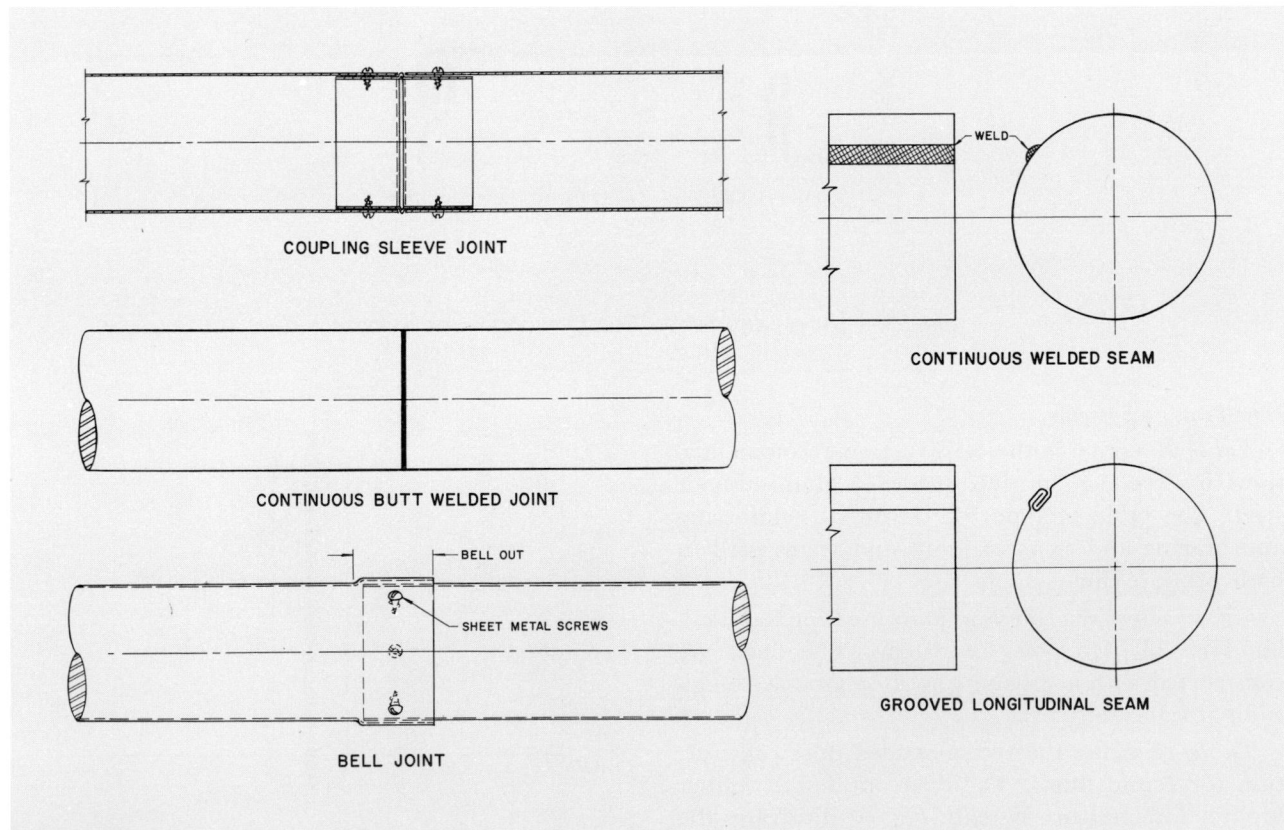

FIG. 60 — ROUND DUCT JOINTS AND SEAMS

TABLE 18—WEIGHTS OF DUCT MATERIAL

WEIGHT (lb/sq ft)	GAGE (THICKNESS) (in.)	WEIGHT PER SHEET (lb)		
		36 x 96	48 x 96	48 x 120
GALVANIZED STEEL, U.S. GAGE				
.906	26 ga. (.022)	21.8	29.0	36.2
1.156	24 ga. (.028)	27.7	37.0	46.2
1.406	22 ga. (.034)	33.8	45.0	56.2
1.656	20 ga. (.040)	39.7	53.0	66.2
2.156	18 ga. (.052)	51.6	70.0	86.2
2.656	16 ga. (.064)	63.6	85.0	102.2
3.281	14 ga. (.080)	78.8	105.0	131.2
HOT ROLLED STEEL, U.S. GAGE				
.750	26 ga. (.0179)	18.0	24.0	30.0
1.000	24 ga. (.0239)	24.0	32.0	40.0
1.250	22 ga. (.0299)	30.0	40.0	50.0
1.500	20 ga. (.0359)	36.0	48.0	60.0
2.000	18 ga. (.0478)	48.0	64.0	80.0
2.500	16 ga. (.0596)	60.0	80.0	100.0
3.125	14 ga. (.0747)	78.0	104.0	130.0
5.625	10 ga. (.1345)	135.0	180.0	225.0
ALUMINUM, B & S GAGE (3S)				
.288	24 ga. (.020)	6.9	9.2	11.5
.355	22 ga. (.025)	8.6	11.3	14.2
.456	20 ga. (.032)	11.0	14.6	18.2
.575	18 ga. (.040)	13.8	18.4	23.0
.724	16 ga. (.051)	17.4	23.2	29.0
.914	14 ga. (.064)	22.0	29.2	36.6
1.03	12 ga. (.071)	24.7	33.0	41.3
STAINLESS STEEL, U.S. GAGE (302)				
.66	28 ga. (.016)	15.8	21.1	26.4
.79	26 ga. (.019)	18.9	25.2	31.6
1.05	24 ga. (.025)	25.2	33.6	42.0
1.31	22 ga. (.031)	31.5	42.0	52.5
1.58	20 ga. (.038)	37.8	50.4	63.0
2.10	18 ga. (.050)	50.4	61.2	84.0
2.63	16 ga. (.063)	63.0	84.0	105.0
3.28	14 ga. (.078)	78.7	104.9	131.2
COPPER, OZ/SQ FT				
1.00	16 oz. (.0216)	24.0	32.0	40.0
1.25	20 oz. (.027)	30.0	40.0	50.0
1.50	24 oz. (.0323)	36.0	48.0	64.0
2.00	32 oz. (.0432)	48.0	64.0	80.0
2.25	36 oz. (.0486)	54.0	72.0	90.0
2.50	40 oz. (.0540)	60.0	80.0	100.0

CHAPTER 3. ROOM AIR DISTRIBUTION

This chapter discusses the distribution of conditioned air after it has been transmitted to the room. The discussion includes proper room air distribution, principles of air distribution, and types and location of outlets.

REQUIREMENTS NECESSARY FOR GOOD AIR DISTRIBUTION

TEMPERATURE

Recommended standards for room design conditions are listed in *Part I, Chapter 2*. The air distributing system must be designed to hold the temperature within tolerable limits of the above recommendations. In a single space a variation of 2 F at different locations in the occupied zone is about the maximum that is tolerated without complaints. For a group of rooms located within a space, a maximum of 3 F between rooms is not unusual. Temperature variations are generally more objectionable in the heating season than in the cooling season.

Temperature fluctuations are more noticeable than temperature variations. These fluctuations are usually a function of the temperature control system. When they are accompanied by air movements on the high end of the recommended velocities, they may result in complaints of drafts.

AIR VELOCITY

Table 19 shows room air velocities. It also illustrates occupant reaction to various room air velocities in the occupied zone.

AIR DIRECTION

Table 19 shows that air motion is desirable and actually necessary. *Fig. 62* is a guide to the most desirable air direction for a seated person.

PRINCIPLES OF AIR DISTRIBUTION

The following section describes the principles of air distribution.

BLOW

Blow is the horizontal distance that an air stream travels on leaving an outlet. This distance is measured from the outlet to a point at which the velocity of the air stream has reached a definite minimum value. This velocity is 50 fpm and is measured at 6.5 ft. above the floor.

TABLE 19—OCCUPIED ZONE ROOM AIR VELOCITIES

ROOM AIR VELOCITY (fpm)	REACTION	RECOMMENDED APPLICATION
0-16	Complaints about stagnant air	none
25	Ideal design—favorable	all commercial applications
25-50	Probably favorable but 50 fpm is approaching maximum tolerable velocity for seated persons	all commercial applications
65	Unfavorable—light papers are blown off a desk	
75	Upper limit for people moving about slowly—favorable	retail and dept. store
75-300	Some factory air conditioning installations—favorable	factory air conditioning higher velocities for spot cooling

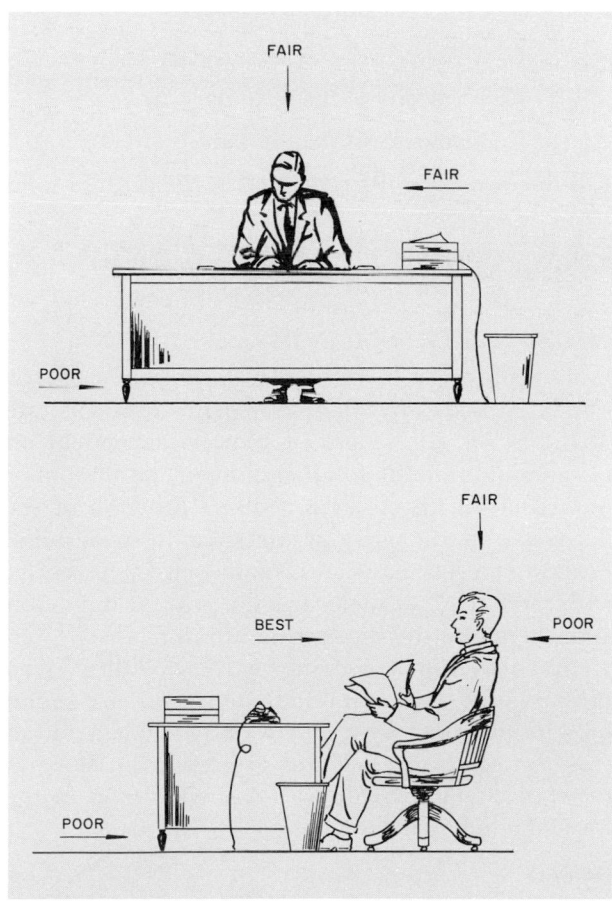

FIG. 62 – DESIRABLE AIR DIRECTION

Blow is proportional to the velocity of the primary air as it leaves the outlet, and is independent of the temperature difference between the supply air and the room air.

DROP

Drop, or rise, is the vertical distance the air moves between the time it leaves the outlet and the time it reaches the end of its blow.

INDUCTION

Induction is the entrainment of room air by the air ejected from the outlet and is a result of the velocity of the outlet air. The air coming directly from the outlet is called primary air. The room air which is picked up and carried along by the primary air is called secondary air. The entire stream, composed of a mixture of primary and secondary air, is called total air.

Induction is expressed by the momentum equation:

$$M_1 V_1 + M_2 V_2 = (M_1 + M_2) \times V_3$$

where M_1 = mass of the primary air

M_2 = mass of the secondary air

V_1 = velocity of the primary air

V_2 = velocity of the secondary air

V_3 = velocity of the total air

Induction ratio (R) is defined as the ratio of total air to primary air;

$$R = \frac{\text{total air}}{\text{primary air}} = \frac{\text{primary + secondary air}}{\text{primary air}}$$

IMPORTANCE OF INDUCTION

Since blow is a function of velocity and since the rate of decrease of velocity is dependent on the rate of induction, the length of blow is dependent on the amount of induction that occurs. The amount of induction for an outlet is a direct function of the perimeter of the primary air stream cross-section. For two outlets having the same area, the outlet with the larger perimeter has the greatest induction and, therefore, the shortest blow. Thus, for a given air quantity discharged into a room with a given pressure, the minimum induction and maximum blow is obtained by a single outlet with a round cross-section. Conversely, the greatest induction and the shortest blow occur with a single outlet in the form of a long narrow slot.

SPREAD

Spread is the angle of divergence of the air stream after it leaves the outlet. Horizontal spread is di-

vergence in the horizontal plane and vertical spread is divergence in the vertical plane. Spread is the included angle measured in degrees.

Spread is the result of the momentum law. *Fig. 63* is an illustration of the effect of induction on stream area and air velocity.

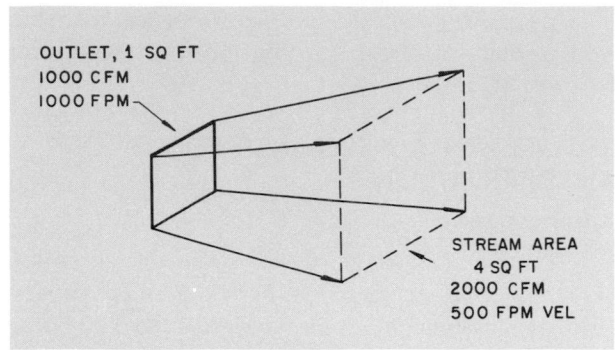

FIG. 63 — EFFECT OF INDUCTION

Example 1 — Effect of Induction

Given:

   1000 cfm primary air

   1000 cfm secondary air

   1000 fpm primary air velocity

     0 fpm secondary air velocity

Find:

   The velocity and area of the total air stream when 1000 cfm of primary and 1000 cfm of secondary air are mixed.

Solution:

   Area of the initial primary air stream before induction

$$= \frac{M_1}{V_1} = \frac{1000}{1000} = 1 \text{ sq ft}$$

   Substituting in the momentum equation

$$(1000 \times 1000) + (1000 \times 0) = (1000 + 1000)\, V_3$$

$$V_3 = 500$$

   Area of the total air stream

$$= \frac{M_1 + M_2}{V_3} = \frac{1000 + 1000}{500} = 4 \text{ sq ft}$$

An outlet discharging air uniformly forward, no diverging or converging vane setting, results in a spread of about an 18° to 20° included angle in both planes. This is equal to a spread of about one foot in every six feet of blow. Type and shape of outlet has an influence on this included angle, but for nearly all outlets it holds to somewhere between 15° and 23°.

INFLUENCE OF VANES ON OUTLET PERFORMANCE

Straight Vanes

Outlets with vanes set at a straight angle result in a spread of approximately 19° in both the horizontal and vertical plane (*Fig. 64*).

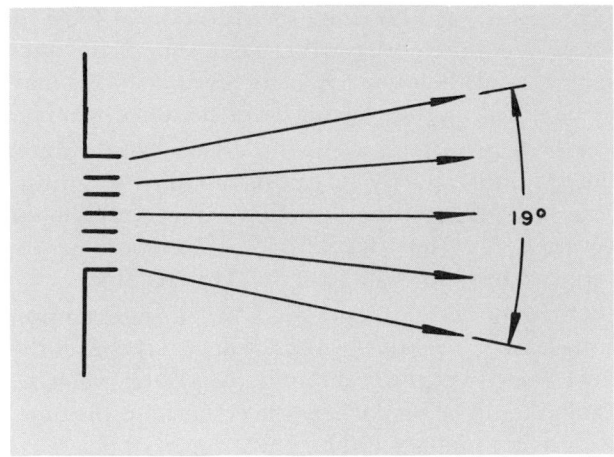

FIG. 64 — SPREAD WITH STRAIGHT VANES

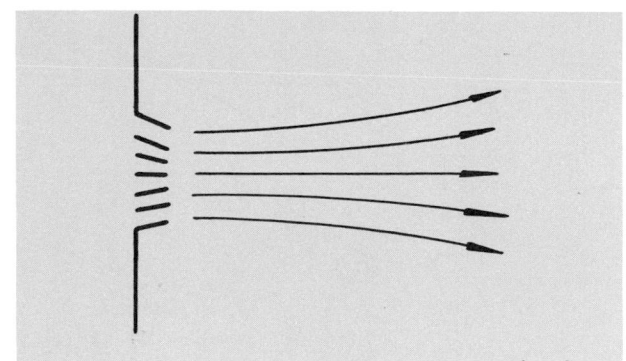

FIG. 65 — SPREAD WITH CONVERGING VANES

Converging Vanes

Outlets with vanes set to direct the discharge air *(Fig. 65)* result in approximately the same spread (19°) as when the vanes are set straight. However, the resulting blow is approximately 15% longer than the straight vane setting.

Diverging Vanes

Outlets with vanes set to give an angular spread to the discharge air have a marked effect on direction and distance of travel. Vertical vanes with the end vanes set at a 45° angle, and all other vanes set at intermediate angles to give a fanning effect, produce an air stream with a horizontal included angle of approximately 60° *(Fig. 66)*. Under this condition the blow is reduced about 50%. Outlets with end vanes set at angles less than 45°, and all other vanes set at intermediate angles to give a fanning effect, have a blow correspondingly larger than the 45° vane setting, but less than a straight vane setting.

Where diverging vanes are used, the free outlet area is reduced; therefore, the air quantity is less than for straight vanes unless the pressure is increased. To miss an obstruction or to direct the air in a particular direction, all vanes can be set for a specific angle as illustrated in *Fig. 67*. Notice that the spread angle is still approximately 19°.

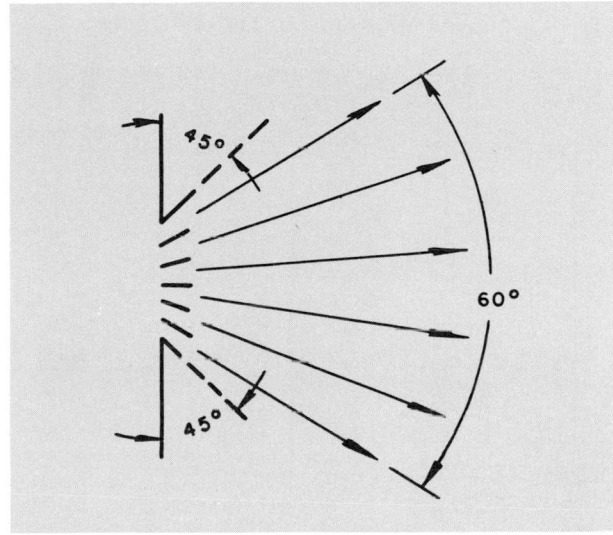

FIG. 66 — SPREAD WITH DIVERGING VANES

INFLUENCE OF DUCT VELOCITY ON OUTLET PERFORMANCE

An outlet is designed to distribute air that has been supplied to it with velocity, pressure and direction, within limits that enable it to completely perform its function. However, an outlet is not designed to correct unreasonable conditions of flow in the air supplied to it.

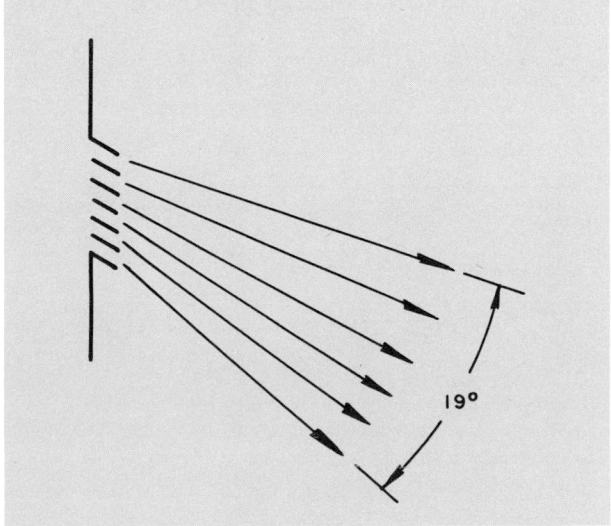

FIG. 67 — SPREAD WITH STRAIGHT VANES SET AT AN ANGLE

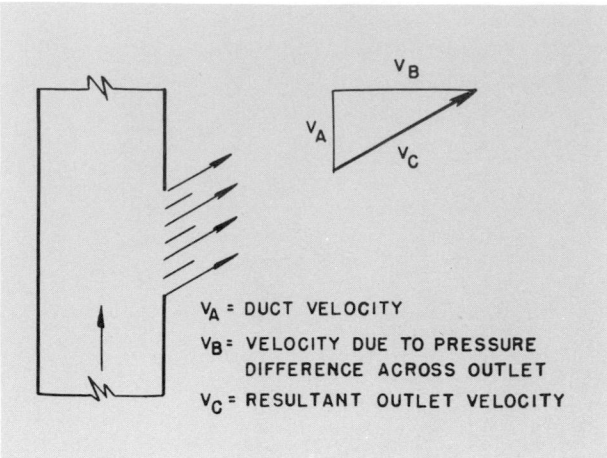

FIG. 68 — OUTLET LOCATED IN DUCT

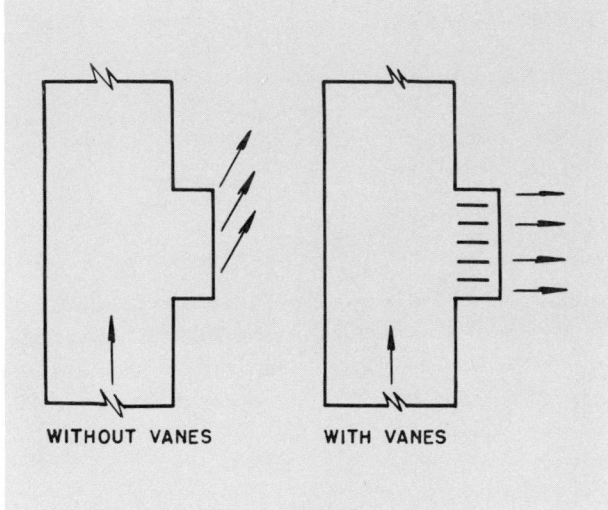

FIG. 69 — COLLAR FOR OUTLETS

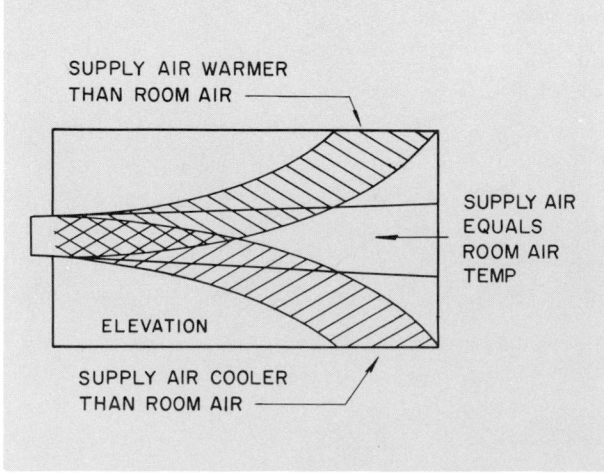

FIG. 70 — AIR STREAM PATTERNS FOR VARIOUS
TEMPERATURE DIFFERENTIALS

Where an outlet without vanes is located directly against the side of a duct, the direction of blow of the air from the outlet is the vector sum of the duct velocity and the outlet velocity *(Fig. 68)*. This may be modified by the peculiarity of the duct opening.

Where an outlet is applied to the face of the duct, the resultant velocity V_C can be modified by adjustable vanes behind the outlet. Whether they should be applied or not depends on the amount of divergence from straight blow that is acceptable.

Often outlets are mounted on short extension collars away from the face of the duct. Whenever the duct velocity exceeds the outlet discharge velocity, vanes should be used where the collar joins the duct. Results are indicated in *Fig. 69*.

IMPORTANCE OF CORRECT BLOW

Normally it is not necessary to blow the entire length or width of a room. A good rule of thumb to follow is to blow ¾ of the distance to the opposite wall. Exceptions occur, however, when there are local sources of heat at the end of the room opposite the outlet. These sources can be equipment heat and open doors. Under these circumstances, overblow may be required and caution must be exercised to prevent draft conditions.

SUPPLY TEMPERATURE DIFFERENTIAL

The allowable supply temperature difference that can be tolerated between the room and the supply air depends to a great extent on (1) outlet induction ratio, (2) obstructions in the path of the primary air, and (3) the ceiling height. *Fig. 70* indicates the effect of changing the supply air temperature from warm to cold.

Since induction depends on the outlet velocity, there is a supply temperature differential which must be specified to give satisfactory results.

TOTAL ROOM AIR MOVEMENT

The object of room air distribution is to provide satisfactory room air motion within the occupied zone, and is accomplished by relating the outlet characteristics and performance to the room air motion as follows:

1. Total air in circulation

    = outlet cfm × induction ratio.

2. Average room velocity

$$= \frac{1.4 \times \text{total cfm in circulation}}{\text{area of wall opposite outlet(s)}}$$

3. $K = \dfrac{\text{average room velocity}}{1.4 \times \text{induction ratio}}$

$= \dfrac{\text{outlet cfm}}{\text{clear area of wall opposite outlet(s)}}$

where K is the room circulation factor expressed in primary air cfm/sq ft of wall opposite the outlet.

The multiplier 1.4 allows for the blocking caused by the air stream. Note that the clear wall area is indicated in the equation and all obstruction must be deducted. See *Note 8, Table 21*.

Table 19 indicates that the average room air movement should be kept between 15 and 50 fpm for most applications. Tests have been performed on outlets at various outlet velocities to determine performance characteristics. The results of such tests on a specific series of wall outlets (*Fig. 93*) are shown in the rating tables at the end of this chapter. This rating data can be successfully used for outlets having the nominal dimensions and free area indicated in *Table 21*. An example illustrating outlet selection accompanies the table. The K factor as indicated in *Item 3* is shown at the bottom of the rating table as maximum and minimum cfm/sq ft of outlet wall area.

TYPES OF OUTLETS

PERFORATED GRILLE

This grille has a small vane ratio (usually from 0.05 to 0.20) and, therefore, has little directional effect. Consequently, it is used principally as an exhaust or return grille but seldom as a supply grille. When a manual shut-off damper backs up this grille, it becomes a register.

FIXED BAR GRILLE

The fixed bar grille is used satisfactorily in locations where flow direction is not critical or can be predetermined. A vane ratio of one or more is desirable. To obstruct the line of sight into the duct interior, closely spaced vanes are preferred.

ADJUSTABLE BAR GRILLE

This grille is the most desirable for side wall location. Since it is available with both horizontal and vertical adjustable bars, minor air motion problems can be quickly corrected by adjusting the vanes.

SLOTTED OUTLET

This outlet may have multiple slots widely spaced, resulting in about 10% free area. Performance is about the same as for a bar grille of the same cfm

and static pressure, but the blow is shorter because of greater induction at the outlet face.

Another design to effect early completion of induction is the long single, or double, horizontal slot. It is particularly advantageous where low ceiling heights exist and outlet height is limited, or where objections to the appearance of grilles are raised.

EJECTOR OUTLET

The ejector outlet operates at a high pressure to obtain a high induction ratio and is primarily used for industrial work and spot cooling. When applied to spot cooling, a high degree of ejector flexibility is desired.

INTERNAL INDUCTION OUTLET

Where a sufficiently high air pressure is used, room air is induced thru auxiliary openings into the outlet. Here it is mixed with primary air, and discharged into the room at a lower temperature differential than the primary stream. Induction progresses in two steps, one in the outlet casing and the other after the air leaves the outlet.

CEILING OUTLETS
Pan Outlet

This simple design of ceiling distribution makes use of a duct collar with a pan under it. Air passes from the plenum thru the duct collar and splashes against the pan. The pan should be of sufficient diameter to hide the duct opening from sight, and also should be adjustable in distance from the ceiling. Pans may be perforated to permit part of the air to diffuse downward. Advantages of the pan outlet are low cost and ability to hide the air opening. Disadvantages are lack of uniform air direction because of poor approach conditions and the tendency to streak ceilings.

Ceiling Diffuser

These outlets are improvements over the pan type. They hasten induction somewhat by supplying air in multiple layers. Approach conditions must be good to secure even distribution. Frequently they are combined with lighting fixtures, and are available with an internal induction feature. See *Fig. 71*.

Perforated Ceilings and Panels

Various types of perforated ceilings for the introduction of conditioned air for comfort and industrial systems are available. The principal feature of this method of handling air is that a greater volume of air per square foot of floor area can be introduced at a lower temperature, with a minimum of movement in the occupied zone and with less

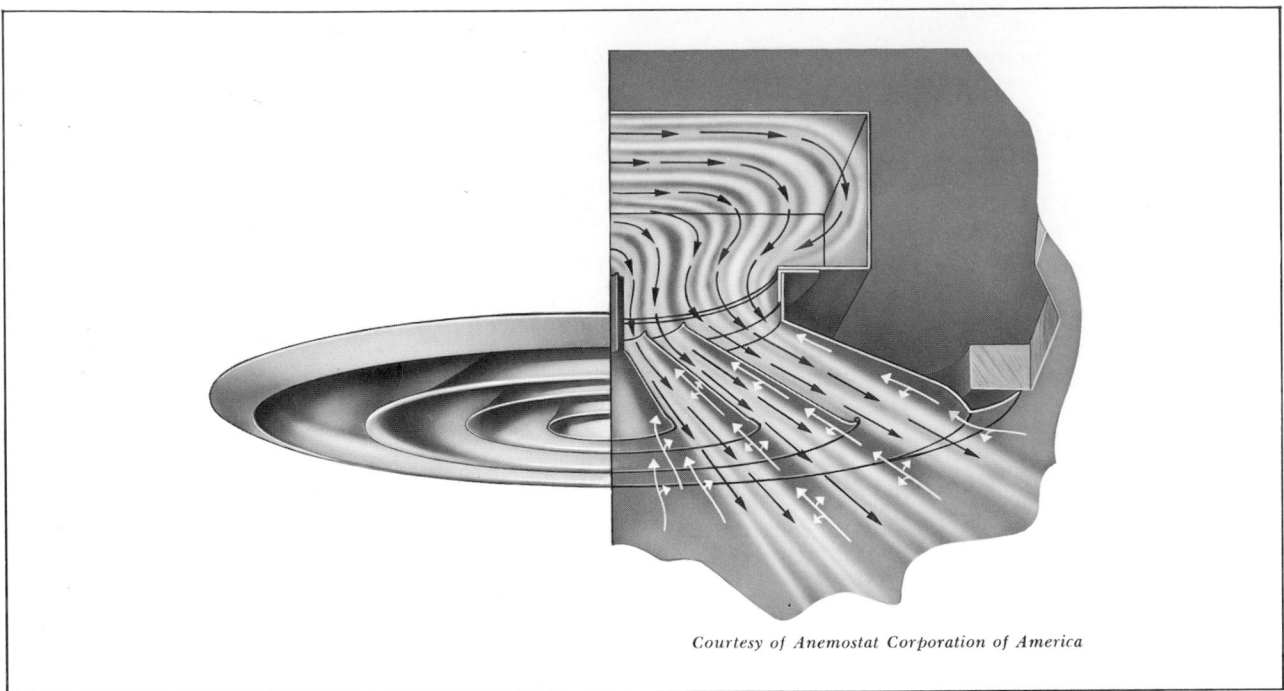

Courtesy of Anemostat Corporation of America

FIG. 71 — INTERNAL INDUCTION CEILING DIFFUSER

danger of draft. Since discharge velocity is low, induction is low. Therefore, care must be taken to provide adequate room air motion in excess of 15 fpm.

Duct designed for a perforated ceiling is the same as duct designed for a standard ceiling. To obtain adequate supply to all areas, the same care necessary for conventional systems must be taken in laying out ducts for the perforated ceiling. The ceiling panels should not be depended upon to obtain proper air distribution, since they cannot convey air to areas not otherwise properly supplied. Perforated panels do assist in "spreading out" the air supply and, therefore, comparatively large temperature differentials may be used, even with low ceiling heights.

APPLICATION OF CEILING DIFFUSERS

Installations using ceiling diffusers normally result in fewer complaints of drafts than those using side wall terminals. To eliminate or minimize these complaints, the following recommendations should be considered when applying ceiling diffusers.

BLOW

Select ceiling diffusers for a conservative blow, generally not over 75% of the tabulated value. Over-blow may cause problems on many installations; under-blow seldom does.

PRESSURE DROPS

Most rating tables express the pressure drop thru the outlet only and do not include the pressure drop necessary to force the air out of the duct thru the collar and outlet and into the room. Therefore, it is recommended that rated pressure drops be carefully investigated and the proper safety factor applied when necessary.

DIFFUSER APPROACH

An important criterion for good diffuser performance is the proper approach condition. This means either a collar of at least 4 times the duct diameter, or good turning vanes. If vanes are used, they must be placed perpendicular to the air flow at the upper end of the collar and spaced approximately 2 in. apart.

OBSTRUCTIONS

Where obstructions to the flow of air from the diffuser occur, blank off a small portion of the diffuser at the point at which the obstruction is located. Clip-on baffles are usually provided for this purpose.

OUTLET NOISE LIMITATIONS

One important criterion affecting the choice of an outlet is its sound level. *Table 20* shows recommended outlet velocities that result in acceptable sound levels for various types of applications.

TABLE 20—RECOMMENDED OUTLET VELOCITIES

APPLICATION	TERMINAL VELOCITY (FPM)
Broadcast studios	300-500
Residences	500-750
Apartments	500-750
Churches	500-750
Hotel bedrooms	500-750
Legitimate theaters	500-750
Private offices, acoustically treated	500-750
Private offices, not treated	500-800
Motion picture theaters	1000
General offices	1000-1250
Dept. stores, upper floors	1500
Dept. stores, main floor	2000

OUTLET LOCATIONS

Interior architecture, building construction and dirt streaking possibilities necessarily influence the layout and location of the outlet. However desirable it may be to locate an outlet in a given spot, these items may prevent such location.

After all the foregoing limitations have been successfully dealt with, the air distribution principles which relate to flow, drop, capacity and room air circulation create further limitations in designing an acceptable air distribution system. These are tabulated in the rating tables at the end of the chapter.

Local loads due to people concentration, equipment heat, outside walls and window locations frequently modify the choice of outlet location. The downdraft from a cold wall or a glass window (*Fig. 72*) can reach velocities of over 200 fpm, causing discomfort to occupants. Unless this downdraft is overcome, complaints of cold feet result. In northern climates this is accomplished by supplementary radiation, or by an outlet located under a window as illustrated in *Fig. 73.*

FIG. 72 — DOWNDRAFT FROM COLD WINDOW

FIG. 73 — DISCHARGE AIR OFFSETTING WINDOW DOWNDRAFT

Another item to consider when choosing an outlet location is the radiant effect from cold or warm surfaces. During the heating season an outlet discharging warm air under a cold window raises the surface temperature and reduces the feeling of discomfort.

The following describe four typical applications of specific outlet types.

CEILING DIFFUSERS

Ceiling diffusers may be applied to exposed duct, furred duct, or duct concealed in a ceiling. Although wall outlets are installed on exposed and furred duct, they are seldom applied to blow directly downward unless complete mixing is accomplished before the air reaches the occupied zone.

WALL OUTLETS

A high location for wall outlets is preferred where a ceiling is free from obstructions. Where beams are encountered, move the outlet down so that the air stream is horizontal and free from obstruction. If this is not done and if vanes are used to direct the air stream downward, the air enters the occupied zone at an angle and strikes the occupants too quickly. This is shown in *Fig. 74*.

Wall outlets located near the floor *(Fig. 75)* are suitable for heating but not for cooling, unless the air is directed upward at a steep angle. The angle must be such that either the air does not strike occupants directly or the secondary induced stream does not cause an objectionable draft.

WINDOW OUTLETS

Where single glass is used, window outlets are preferred to either wall or ceiling distribution to offset the pronounced downdraft during the winter. The air should be directed with vanes at an angle of 15° or 20° from the vertical into the room.

FLOOR OUTLETS

Where people are seated as in a theater, floor outlet distribution is not permissible. Where people are walking about, it is possible to introduce air at the floor level; for example, in stores where air is directed horizontally thru a slot under a counter. In this application, however, a very low temperature differential of not more than 5 or 6 degrees must be used. Maintaining this maximum is usually uneconomical because of the large air volume required. However, if air is directed upward behind the counter and diffused at an elevation above the occupied zone, the temperature differential may be increased approximately 5 times. Another disadvantage is that floor outlets become dirt collectors.

SPECIFIC APPLICATIONS

If the principles described in the previous paragraphs are properly applied, problems after installation will be at a minimum. Basically, the higher the ceiling the fewer the number of problems encountered, and consequently liberties may be taken at little or no risk when designing the system. However, with ceiling heights of approximately 12 feet or less, greater care must be exercised to minimize problems.

Experience has shown that ceiling diffusers are easier to apply than side wall outlets, and are preferred when air quantities approach 2 cfm/sq ft of floor area.

The following general remarks about specific ap-

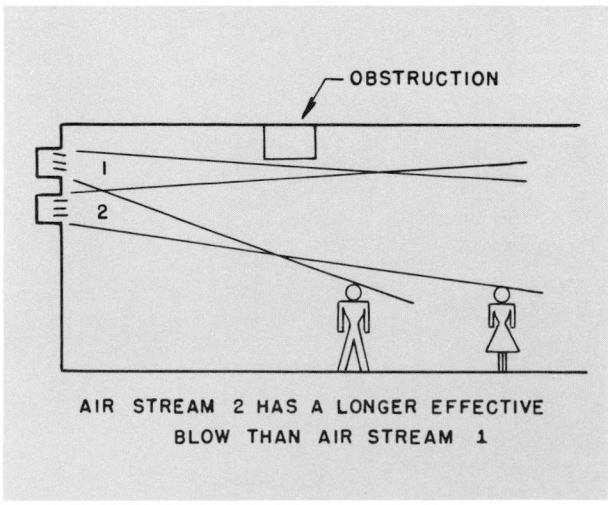

FIG. 74 — WALL OUTLET IN ROOM WITH CEILING OBSTRUCTION

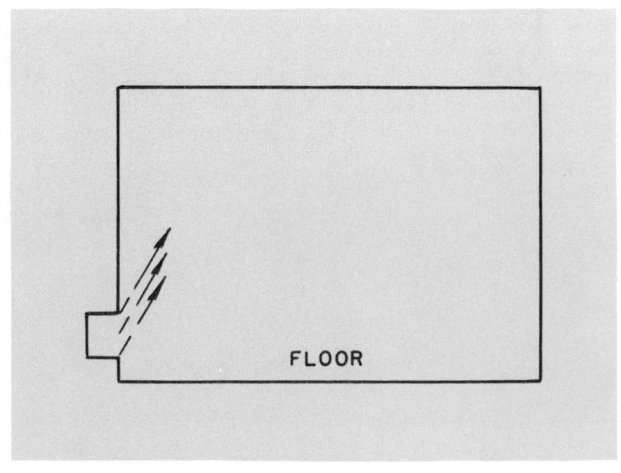

FIG. 75 — WALL OUTLET NEAR FLOOR

plications are the result of thousands of installations and are offered as a guide for better air distribution. Apartments, hotels and office buildings are discussed in relation to specific location of sources of air supply common to these types of buildings. Banks, restaurants, and department and specialty stores are discussed in more general terms, although the common sources of locations of outlets previously discussed can be applied.

APARTMENTS, HOTELS AND OFFICE BUILDINGS

1. Corridor Supply — No direct radiation *(Fig. 76)*:
   Advantage — Low cost.
   Disadvantage — Very poor in winter. Downdraft under the window accentuated by the outlet blow.
   Precaution — Blow must be not more than 75% of the room length.

2. Corridor Supply — Direct radiation under windows *(Fig. 77)*:
   Advantage — Offset of downdraft under window in winter when the heat is on.
   Disadvantage — Slight downdraft still occurs during intermediate season or whenever radiation is shut off during cool weather.
   Precaution — Do not blow more than 75% of room length.

3. Duct above window blowing toward corridor *(Fig. 78)*:
   Advantage — Somewhat better than corridor distribution but does not prevent winter downdraft unless supplemented by direct radiation.
   Disadvantage — Nearly as expensive (considering building alterations) as window outlet which results in better air distribution.

4. Window outlet *(Fig. 79)*:
   Advantage — Eliminates winter downdraft — by far the best method of distribution.
   Disadvantage — May not be economical for multiple windows.

5. Return grille:
   Where return air thru the corridor is permissible and return ducts are not used, it is necessary to use relief grilles or to undercut office doors.

In apartments and hotels, codes must be checked before using the corridor, as a return plenum. Even if codes permit, it is not good engineering practice to use the corridor as a return plenum.

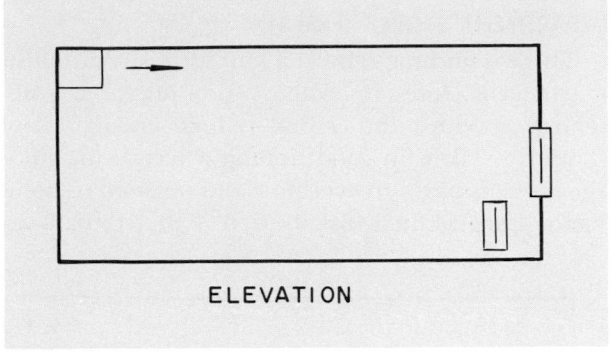

Fig. 77 — Corridor Air Supply with Direct Radiation

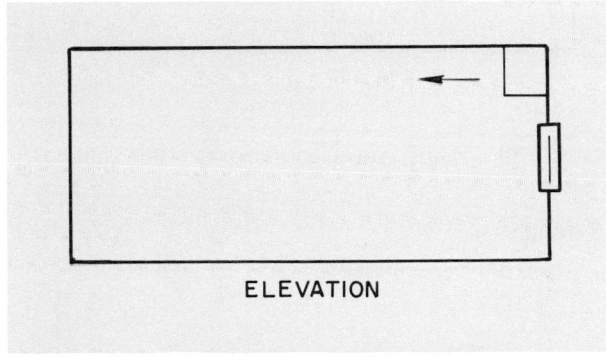

Fig. 78 — Duct Above Window, Blowing Toward Corridor

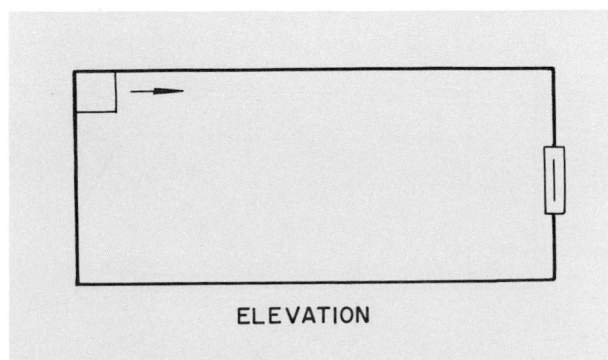

Fig. 76 — Corridor Air Supply

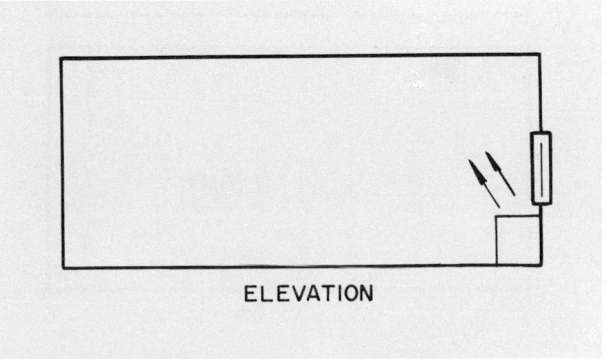

Fig. 79 — Window Outlet

BANKS (FIG. 80)

Often the center bank space has a high ceiling with an electrical load. In this case, use of side wall outlets part of the way up the wall may result in segregating some of the ceiling load and keeping it out of the occupied zone, thus permitting some reduction in cooling load. This location of outlets part way up the wall is used with ceiling heights in excess of 20 ft.

DEPARTMENT STORES (FIG. 81)

There is nothing critical about air distribution in department stores if ordinary precautions are observed, provided the ceiling is high enough. Care should be taken in conditioning a mezzanine since the outlet is likely to overblow and not cool its occupants. Longitudinal distribution is preferred. Base-

ments may give trouble due to low ceilings and pipe obstructions. Main floors usually require more air near doors.

RESTAURANTS (FIG. 82)

Great care must be taken in locating outlets with respect to exhaust hoods or kitchen pass-thru windows. Usually the air velocities over such openings are low and excessive disturbance due to direct blow or induction from outlets may pull air out of them into the conditioned space.

STORES

1. Outlets at rear blowing toward door *(Fig. 83)*:
   Requirements — Unobstructed ceiling.
   Disadvantage — May result in high room circulation factor K.
   Precaution — Blow must be sized for the entire

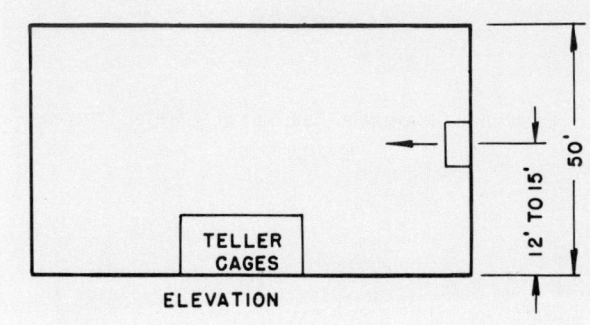

FIG. 80 — AIR DISTRIBUTION WITH HIGH CEILING

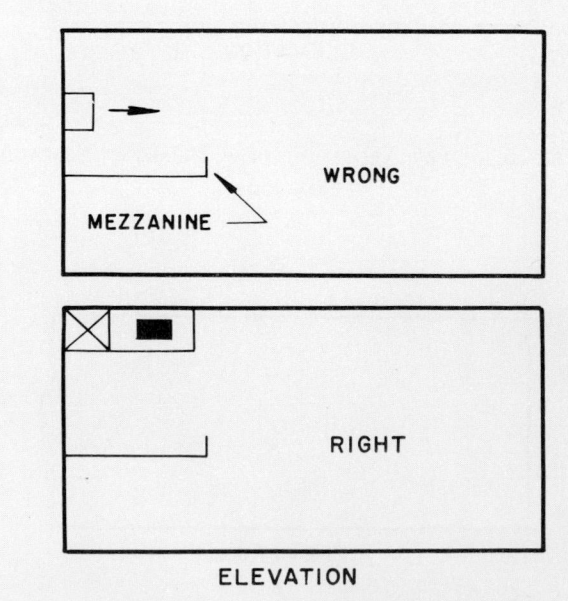

FIG. 81 — MEZZANINE AIR DISTRIBUTION

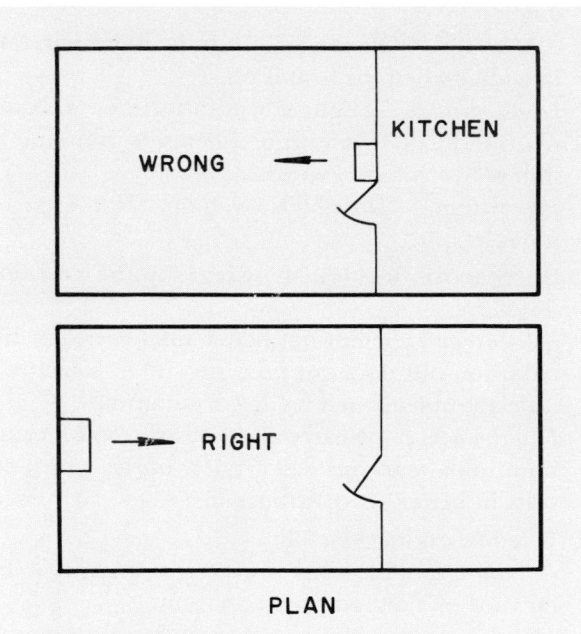

FIG. 82 — RESTAURANT AIR DISTRIBUTION

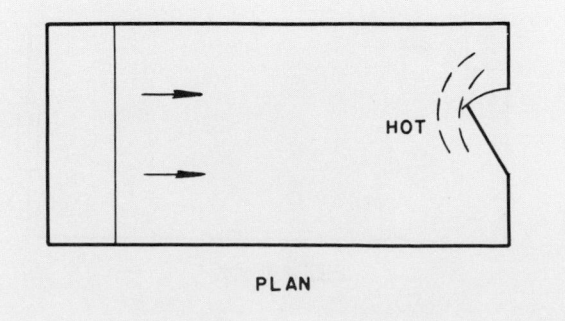

FIG. 83 — AIR DISTRIBUTION FROM REAR OF STORE

length of the room; otherwise a hot zone may occur due to infiltration at the doorway. Care must be taken to avoid downdrafts near walls.

2. Outlets over door blowing toward rear *(Fig. 84)*:
   Requirements — Unobstructed ceiling.
   Disadvantage — May result in high room circulation.
   Precaution — Excessive infiltration may occur due to induction from doorway.

3. Outlets blowing from each end toward center *(Fig. 85)*:
   Advantage — Moderate room circulation.
   Precaution — There may be a downdraft in the center. Outlets should be sized for blow not greater than 40 percent of the total length of the room.

4. Center outlets blowing toward each end *(Fig. 86)*:
   Advantage — Moderate room circulation.

5. Duct along side wall blowing across the store *(Fig. 87)*:
   Advantage — Moderate room circulation.
   Precaution — Overblow may cause downdrafts on the opposite wall.

6. Ceiling diffusers *(Fig. 88)*:
   Requirements — Necessary where ceiling is badly cut up.

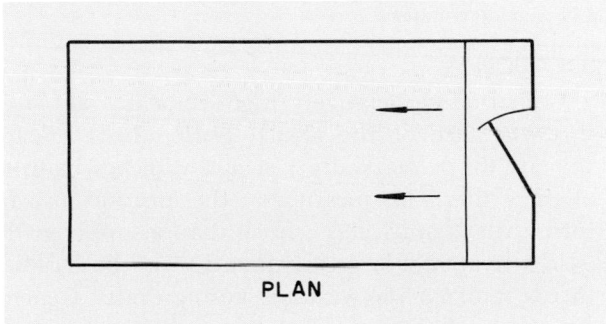

FIG. 84 — AIR DISTRIBUTION FROM OVER THE DOOR

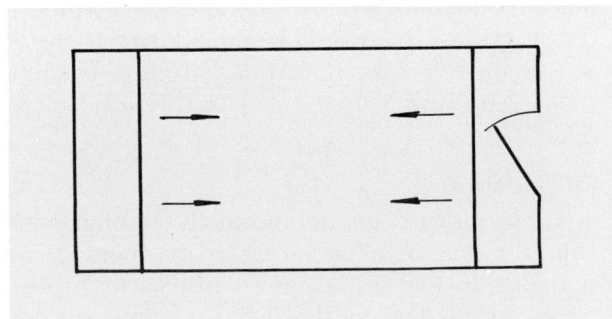

FIG. 85 — AIR DISTRIBUTION FROM EACH END OF STORE

Advantage — Best air distribution.
Disadvantage — High cost.

THEATERS

1. Ejector system for small theaters, no balcony *(Fig. 89)*:
   Requirements — Unobstructed ceiling and ability to locate outlets in the rear wall.
   Advantage — Low cost.
   Precaution — Possibility of dead spots at front and back of theater. Use mushrooms for return air under seats if excavated. In northern climates direct radiation may be advisable along the sides.

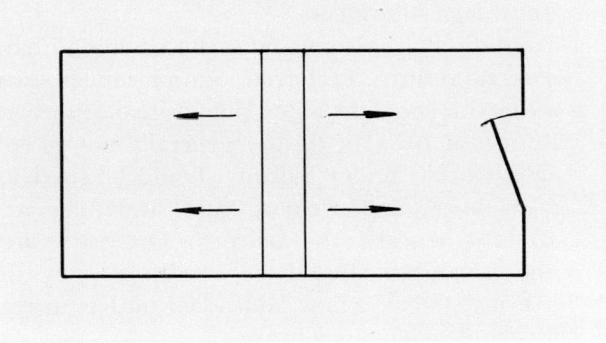

FIG. 86 — AIR DISTRIBUTION FROM CENTER OF STORE

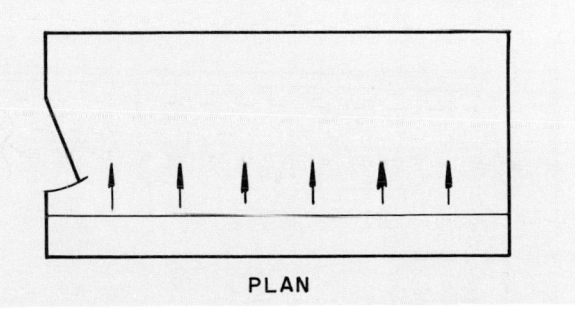

FIG. 87 — AIR DISTRIBUTION FROM SIDEWALL OUTLETS

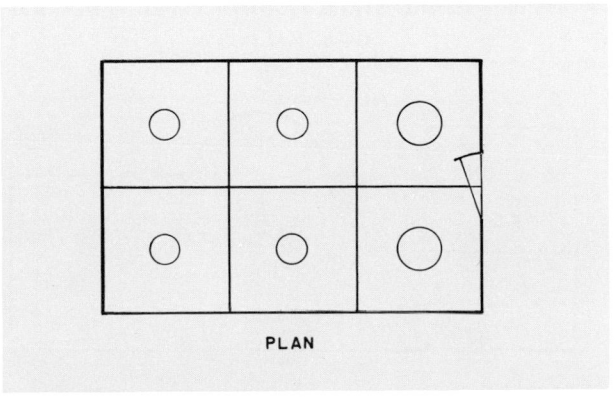

FIG. 88 — AIR DISTRIBUTION FROM CEILING DIFFUSERS

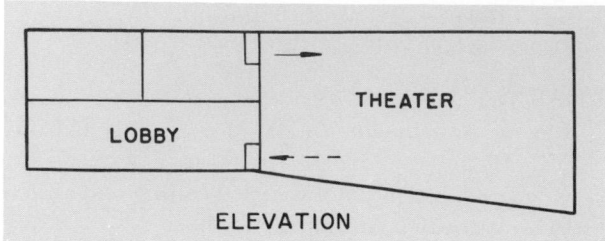

FIG. 89 — AIR DISTRIBUTION FOR SMALL THEATRES

2. Ejector system for large theaters with balcony (*Fig. 90*):
   Requirements — Unobstructed ceiling.
   Advantage — Low cost.
   Precaution — Balcony and orchestra should have separate returns. Preferred location, under seats; acceptable location, along sides or rear of theater. Return at front of theater generally not acceptable. Outlets under balcony should be sized for distribution and blow to cover only the area directly beneath the balcony. Orchestra area under balcony should be conditioned by the balcony system. Allow additional outlets in rear for standees when necessary.
3. Overhead system (*Fig. 91*):
   Requirements — Necessary when ceiling is obstructed.

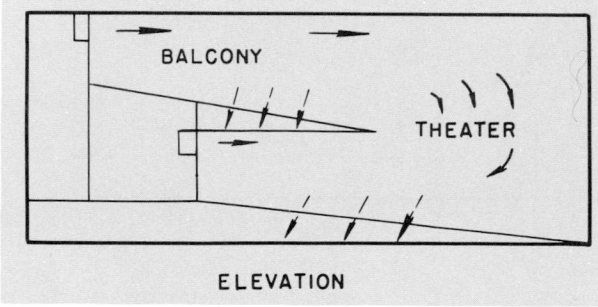

FIG. 90 — AIR DISTRIBUTION FOR LARGE THEATRES
WITH BALCONY

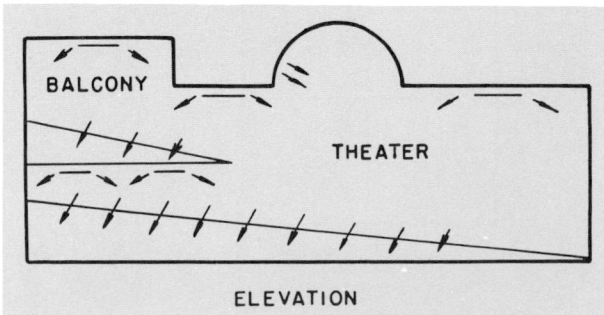

FIG. 91 — OVERHEAD AIR DISTRIBUTION

Advantage — Complete coverage, no dead spots.
Disadvantage — Higher first cost.
Precaution — Air must not strike obstructions with a velocity force that causes deflection and drafts in the occupied zone. Temperature differentials must be limited in regions of low ceiling heights. Use low outlet velocities.

RETURN GRILLES

Velocities thru return grilles depend on (1) the static pressure loss allowed and (2) the effect on occupants or materials in the room.

In determining the pressure loss, computations should be based on the free velocity thru the grille, not on the face velocity, since the orifice coefficient may approach 0.7.

In general the following velocities may be used:

GRILLE LOCATION	FPM OVER GROSS AREA
Commercial	
Above occupied zone	800 and above
Within occupied zone not near seats	600-800
Within occupied zone near seats	400-600
Door or wall louvers	500-1000
Undercutting of doors	600*
Industrial	800 and above
Residential	400

*Thru undercut area

LOCATION

Even though relatively high velocities are used thru the face of the return grille, the approach velocity drops markedly just a few inches in front of the grille. This means that the location of a return grille is much less critical than a supply grille. Also a relatively large air quantity can be handled thru a return grille without causing drafts. General drift toward the return grille must be within acceptable limits of less than 50 fpm; otherwise complaints resulting from drafts may result. *Fig. 92* indicates the fall-off in velocity as distance from the return grille is increased. It also illustrates the approximate velocities at various distances from the grille, returning 500 cfm at a face velocity of 500 fpm.

Ceiling Return

These returns are not normally recommended. Difficulty may be expected when the room circulation due to low induction is insufficient to cause warm air to flow to the floor in winter. Also, a poorly located ceiling return is likely to bypass the cold air in summer or warm air in winter before it has time to accomplish its work.

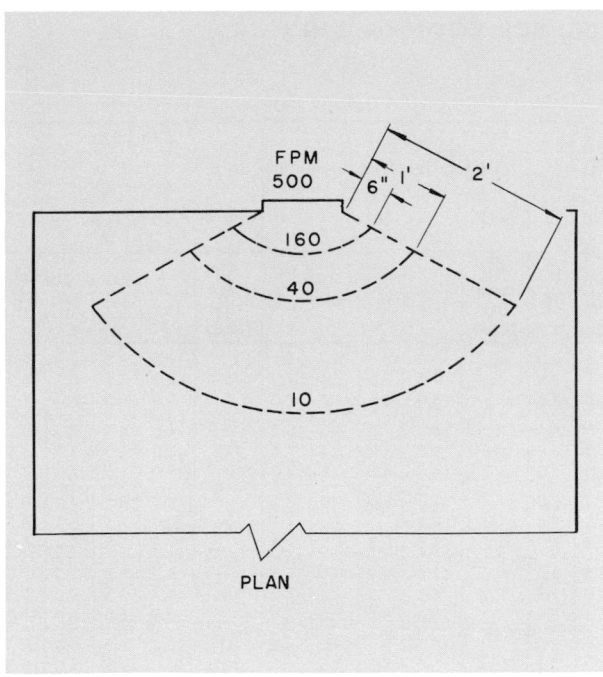

FIG. 92—VELOCITY FALL-OFF PER DISTANCE
FROM GRILLE

Wall Return

A wall return near the floor is the best location. Wall returns near the ceiling are almost as undesirable as ceiling returns. Differences due to poor mixing in the winter are counteracted by a low return since the cool floor air is withdrawn first and is replaced by the warmer upper air strata.

Floor Return

These should be avoided wherever possible because they are a catch-all for dirt and impose a severe strain both on the filter and cooling coils. Whenever floor returns are used, a low velocity settling chamber should be incorporated.

OUTLET SELECTION

The following example describes a method of selecting a wall outlet using the rating tables on *page 78*.

Example 2 — Wall Outlet Selection

Given:
   Small store
   Dimensions — 32 ft x 23 ft x 16 ft
   Ceiling — flat
   Load — equally distributed
   Air quantity — 2000 cfm
   Temp diff — 25 F

Find:
   Number of outlets
   Size of outlets
   Location

Solution:
   First, find the required blow in feet and the wall outlet area (air movement K factor). The minimum blow is 75% of the room width for the given condition of equally distributed heat load. Therefore, the minimum blow necessary is $\frac{3}{4} \times 23 = 17.3$ ft. The maximum blow is the width of the room. The outlet wall area K factor is equal to the cfm supplied divided by the outlet wall area:

$$\frac{2000}{32 \times 16} = 3.9 \text{ primary air cfm/sq ft wall area}$$

Enter *Table 21* and select one or more outlets to give a blow of at least 17.3 ft. Air movement must be such that the value of K equals 3.9 primary air cfm/sq ft and that this value falls within the maximum and minimum values which are shown at the bottom of the rating tables. The tables indicate that, to best satisfy conditions, four outlets, nominal size 24 in. x 6 in., are to be used. By interpolating, it is found that the four 24 in. x 6 in. outlets at 500 cfm have a range in blow of 17.5 to 34 ft. By adjusting the vanes the proper blow can be obtained. Also the velocity of the outlet is found to be about 775 fpm. This is well within the recommended maximum velocity of 1500 fpm, *Table 20*. The minimum ceiling height from the table is just over 9 ft. This is less than the height of the room; therefore, the outlet selection is satisfactory. The top of the outlets should be installed at least 12 in. from the ceiling, *(Note 8, Table 21)*.

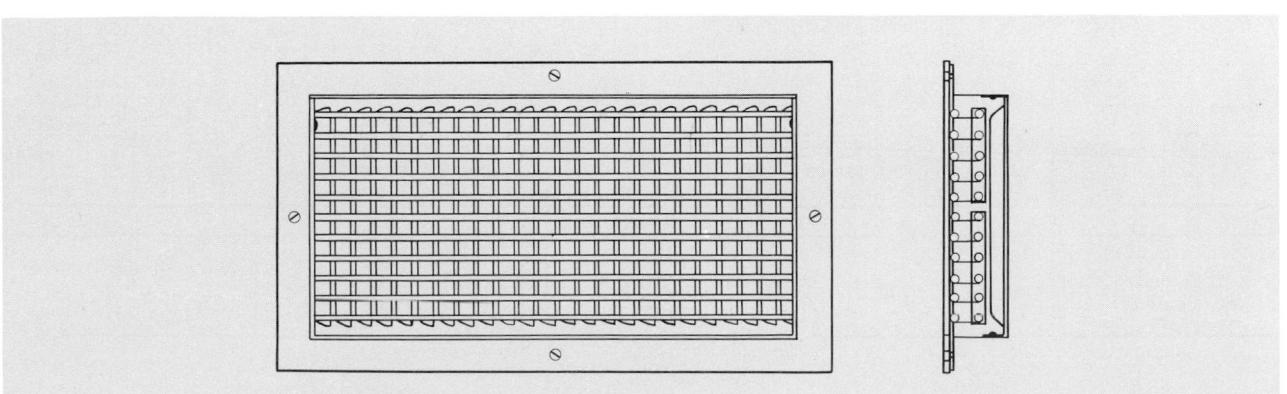

FIG. 93—WALL OUTLET ON WHICH RATINGS ARE BASED

TABLE 21—WALL OUTLET RATINGS, FOR COOLING ONLY

For Flat Ceilings

OUTLET VELOCITY		250 FPM					375 FPM					500 FPM					750 FPM				
STATIC PRESSURE STANDARD OUTLET		Str B = .01, 22½° = .01 45° = .01					Str B = .013, 22½° = .015 45° = .019					Str B = .024, 22½° = .028 45° = .035					Str B = .051, 22½° = .061 45° = .08				
STATIC PRESSURE WITH METERING PLATE		Str B = .01, 22½° = .015 45° = .028					Str B = .024, 22½° = .043 45° = .065					Str B = .061, 22½° = .082 45° = .118					Str B = .175, 22½° = .19 45° = .27				
Nom. Size of Outlet (and Free Area)	Vane Setting	Air Quantity (cfm)	Blow (ft)	Temp Diff (F) 15	20	25 Min Clg Ht	Air Quantity (cfm)	Blow (ft)	Temp Diff (F) 15	20	25 Min Clg Ht	Air Quantity (cfm)	Blow (ft)	Temp Diff (F) 15	20	25 Min Clg Ht	Air Quantity (cfm)	Blow (ft)	Temp Diff (F) 15	20	25 Min Clg Ht
8 x 4 (16.9)	Straight	30	3.5	6.5	7.0	7.0	44	7.0	7.5	7.5	8.0	59	10.0	7.5	8.0	8.5	89	17.0	8.5	9.0	9.0
	22½°		2.5	6.5	6.5	6.5		5.1	6.5	7.0	7.0		7.5	7.0	7.5	7.5		13.0	7.5	7.5	8.0
	45°		1.8	6.0	6.0	6.5		3.5	6.5	6.5	7.0		5.0	6.5	6.5	7.0		9.0	6.5	7.0	7.0
10 x 4 (21.7)	Straight	37	3.5	6.5	7.0	7.5	57	7.4	7.5	7.5	8.0	75	10.5	7.5	8.0	8.5	112	18.0	8.5	9.0	9.0
	22½°		2.5	6.5	6.5	7.0		5.5	6.5	7.0	7.0		8.0	7.0	7.5	7.5		13.0	7.5	8.0	8.0
	45°		1.8	6.0	6.0	6.5		3.7	6.5	6.5	7.0		5.4	6.5	6.5	7.0		9.0	6.5	7.0	7.0
12 x 4 (24.6)	Straight	44	3.5	6.5	7.0	7.5	68	7.5	7.5	7.5	8.0	91	11.0	8.0	8.0	8.5	136	18.0	8.5	9.0	9.5
	22½°		2.5	6.5	6.5	7.0		5.5	7.0	7.0	7.5		8.1	7.0	7.5	7.5		13.0	7.5	8.0	8.5
	45°		1.8	6.0	6.5	6.5		3.9	6.5	6.5	7.0		5.5	6.5	7.0	7.0		9.0	6.5	7.0	7.0
16 x 4 (35.9)	Straight	61	3.7	7.0	7.0	7.5	92	7.9	7.5	7.5	8.0	122	11.0	8.0	8.0	8.5	183	19.0	8.5	9.0	9.5
	22½°		2.7	6.5	6.5	7.0		6.0	7.0	7.0	7.5		8.1	7.0	7.5	7.5		14.0	7.5	8.0	8.5
	45°		2.0	6.0	6.5	6.5		4.0	6.5	6.5	7.0		5.5	6.5	7.0	7.0		10.0	6.5	7.0	7.5
20 x 4 (45.5)	Straight	77	4.0	7.0	7.0	7.5	115	8.0	7.5	8.0	8.0	154	11.5	8.0	8.0	8.5	231	20.0	8.5	9.0	9.5
	22½°		3.0	6.5	6.5	7.0		6.0	7.0	7.0	7.5		8.5	7.5	7.5	8.0		15.0	7.5	8.0	8.5
	45°		2.0	6.0	6.5	6.5		4.0	6.5	6.5	7.0		6.0	6.5	7.0	7.0		10.0	6.5	7.0	7.5
24 x 4 (55.0)	Straight	93	4.1	7.0	7.0	7.5	139	8.0	7.5	8.0	8.0	185	11.5	8.0	8.0	8.5	278	20.0	8.5	9.0	10.0
	22½°		3.1	6.5	7.0	7.5		6.0	7.0	7.0	7.5		8.5	7.5	7.5	8.0		15.0	7.5	8.0	8.5
	45°		2.0	6.0	6.5	6.5		4.0	6.5	6.5	7.0		6.0	6.5	7.0	7.0		10.6	6.5	7.0	7.5
30 x 4 (68.3)	Straight	116	4.2	7.0	7.0	7.5	175	8.0	7.5	8.0	8.0	233	12.0	8.0	8.0	8.5	349	21.0	8.5	9.5	10.0
	22½°		3.1	6.5	7.0	7.0		6.0	7.0	7.5	7.5		9.0	7.5	7.5	8.0		16.0	7.5	8.0	8.5
	45°		2.1	6.0	6.5	6.5		4.0	6.5	6.5	7.0		6.0	6.5	7.0	7.0		11.0	7.0	7.0	7.5
36 x 4 (83.5)	Straight	140	4.4	7.0	7.5	7.5	210	8.0	7.5	8.0	8.0	279	12.0	8.0	8.5	9.0	420	21.0	9.0	9.5	10.0
	22½°		3.3	6.5	7.0	7.0		6.0	7.0	7.5	7.5		9.0	7.5	7.5	8.0		16.0	7.5	8.0	8.5
	45°		2.2	6.0	6.5	6.5		4.0	6.5	6.5	7.0		6.0	6.5	7.0	7.0		11.0	7.0	7.0	7.5
8 x 6 (26.5)	Straight	52	5.0	7.5	7.5	8.0	77	9.5	8.0	8.0	8.5	103	13.0	8.5	9.0	9.0	155	24.0	8.5	10.0	10.5
	22½°		3.8	7.0	7.0	7.5		7.0	7.0	7.5	8.0		10.0	7.5	8.0	8.5		18.0	8.0	8.5	9.5
	45°		2.5	6.0	6.5	6.5		4.8	6.5	7.0	7.0		6.0	7.0	7.0	7.5		12.0	7.0	7.5·	8.0
10 x 6 (34.0)	Straight	66	5.5	7.5	8.0	8.0	98	10.0	8.0	8.5	9.0	131	15.0	9.0	9.5	10.0	196	27.0	10.0	10.5	11.5
	22½°		4.1	7.0	7.5	7.5		7.5	7.5	8.0	8.5		11.0	8.0	8.5	9.0		20.0	8.5	9.0	10.0
	45°		2.8	6.5	7.0	7.0		5.0	7.0	7.0	7.5		7.0	7.0	7.5	7.5		14.0	7.5	7.5	8.0
12 x 6 (41.6)	Straight	80	6.0	7.5	8.0	8.5	119	11.0	8.0	9.0	9.5	159	15.0	9.0	9.5	10.0	238	28.0	10.0	11.0	11.5
	22½°		4.5	7.0	7.5	7.5		8.1	7.5	8.0	8.5		11.0	8.0	8.5	9.0		21.0	9.0	9.5	10.0
	45°		3.0	6.5	7.0	7.0		5.5	7.0	7.0	7.5		7.0	7.0	7.5	7.5		14.0	7.5	8.0	8.0
16 x 6 (56.6)	Straight	107	6.2	8.0	8.0	8.5	161	12.0	8.5	9.0	9.5	214	16.0	9.5	10.0	10.5	321	30.0	11.0	11.5	12.5
	22½°		4.7	7.0	7.5	7.5		9.0	8.0	8.0	8.5		12.0	8.5	9.0	9.5		22.0	9.5	10.0	10.5
	45°		3.2	6.5	7.0	7.0		6.0	7.0	7.0	7.5		8.0	7.5	7.5	8.0		15.0	8.0	8.5	8.5
20 x 6 (71.5)	Straight	135	6.6	8.0	8.5	9.0	202	12.0	9.0	9.5	10.0	269	17.0	9.5	10.0	11.0	403	32.0	11.5	12.0	13.0
	22½°		5.0	7.5	7.5	8.0		9.0	8.0	8.5	9.0		13.0	8.5	9.0	9.5		24.0	9.5	10.0	11.0
	45°		3.2	7.0	7.0	7.5		6.6	7.0	7.5	7.5		9.0	7.5	8.0	8.0		16.0	8.0	8.5	9.0
24 x 6 (86.5)	Straight	162	7.0	8.0	8.5	9.0	243	13.0	9.0	9.5	10.0	324	18.0	10.0	10.5	11.0	486	33.0	12.0	12.5	13.0
	22½°		5.1	7.5	8.0	8.0		10.0	8.0	8.5	9.0		13.0	8.5	9.0	10.0		25.0	10.0	10.5	11.0
	45°		3.5	7.0	7.0	7.5		6.5	7.0	7.5	8.0		9.0	7.5	8.0	8.5		17.0	8.0	8.5	9.1
30 x 6 (109.0)	Straight	203	7.0	8.5	8.5	9.5	304	13.0	9.0	10.0	10.5	406	19.0	10.0	11.0	11.5	609	34.0	12.0	12.5	13.5
	22½°		5.4	7.5	8.0	8.0		10.0	8.0	9.0	9.0		14.0	9.0	9.5	10.0		25.0	10.0	10.5	11.5
	45°		3.5	7.0	7.0	7.5		6.5	7.5	7.5	8.0		10.0	7.5	8.0	8.5		17.0	8.0	9.0	9.0
36 x 6 (131.3)	Straight	245	7.1	8.5	9.0	9.5	368	13.0	9.5	10.0	10.5	490	19.0	10.0	11.0	12.0	735	35.0	12.0	13.0	14.0
	22½°		5.5	7.5	8.0	8.5		10.0	8.5	9.0	9.5		14.0	9.0	9.5	10.0		26.0	10.0	10.5	11.5
	45°		3.5	7.0	7.5	7.5		6.5	7.5	8.0	8.0		10.0	8.0	8.0	8.5		18.0	8.5	9.0	9.5

K FACTOR																					
Max Cfm/Sq Ft Outlet Wall Area		29.0					19.0					14.0					9.6				
Min Cfm/Sq Ft Outlet Wall Area		8.7					5.7					4.2					2.9				

TABLE 21—WALL OUTLET RATINGS, FOR COOLING ONLY (Cont.)

For Flat Ceilings

OUTLET VELOCITY		1000 FPM					1500 FPM					2000 FPM				
STATIC PRESSURE STANDARD OUTLET		Str B = .093, 22½° = .11, 45° = .14					Str B = .211, 22½° = .24, 45° = .32					Str B = .375, 22½° = .42, 45° = .565				
STATIC PRESSURE WITH METERING PLATE		Str B = .33, 22½° = .33, 45° = .475					Str B = .715, 22½° = .74, 45° = 1.15					Str B = 1.36				
Nom. Size of Outlet (and Free Area)	Vane Setting	Air Quantity (cfm)	Blow (ft)	Temp Diff (F) 15 Min Clg Ht	20	25	Air Quantity (cfm)	Blow (ft)	Temp Diff (F) 15 Min Clg Ht	20	25	Air Quantity (cfm)	Blow (ft)	Temp Diff (F) 15 Min Clg Ht	20	25
8 x 4 (16.9)	Straight	118	24	9.0	9.5	10.0	177	40	10.0	10.5	11.0	237	58	10.5	11.0	12.0
	22½°		18	7.5	8.0	8.5		30	8.5	9.0	9.5		44	8.5	9.0	9.5
	45°		12	6.5	7.0	7.5		20	7.0	7.0	7.5		29	7.0	7.0	7.5
10 x 4 (21.7)	Straight	150	26	9.0	9.5	10.0	224	42	10.0	10.5	11.5	299	60	10.5	11.5	12.0
	22½°		19	7.5	8.0	8.5		32	8.5	9.0	9.5		45	9.0	9.5	10.0
	45°		13	7.0	7.0	7.5		21	7.0	7.5	7.5		30	7.0	7.5	7.5
12 x 4 (24.6)	Straight	181	27	9.0	9.5	10.0	272	44	10.0	11.0	11.5	362	62	10.5	11.5	12.5
	22½°		20	8.0	8.5	9.0		33	8.5	9.0	9.5		47	9.0	9.5	10.0
	45°		14	7.0	7.0	7.5		22	7.0	7.5	7.5		31	7.5	7.5	8.0
16 x 4 (35.9)	Straight	244	28	9.0	10.0	10.5	366	46	10.0	11.0	12.0	488	65	11.0	12.0	12.5
	22½°		21	8.0	8.5	9.5		35	9.0	9.5	10.0		49	9.0	10.0	10.5
	45°		14	7.0	7.5	7.5		23	7.0	7.5	8.0		33	7.5	7.5	8.0
20 x 4 (45.5)	Straight	308	29	9.5	10.0	10.5	462	48	10.5	11.0	12.0	616	67	11.0	12.0	13.0
	22½°		22	8.0	8.5	9.5		36	9.0	9.5	10.0		50	9.5	10.0	10.5
	45°		15	7.0	7.5	7.5		24	7.5	7.5	8.0		34	7.5	8.0	8.0
24 x 4 (55.0)	Straight	370	30	9.5	10.0	10.5	556	49	10.5	11.5	12.0	740	68	11.5	12.0	13.0
	22½°		22	8.5	9.0	9.5		37	9.0	9.5	10.0		51	9.5	10.0	10.5
	45°		15	7.5	7.5	7.5		25	7.5	8.0	8.0		34	7.5	8.0	8.5
30 x 4 (68.3)	Straight	466	30	9.5	10.0	10.5	698	50	10.5	11.5	12.5	932	70	11.5	12.5	13.5
	22½°		22	8.5	9.0	9.5		37	9.0	9.5	10.0		53	9.5	10.0	11.0
	45°		15	7.0	7.5	7.5		25	7.5	8.0	8.0		35	7.5	8.0	8.5
36 x 4 (83.5)	Straight	558	31	9.5	10.0	11.0	840	51	11.0	11.5	12.5	1116	71	11.5	12.5	13.5
	22½°		23	8.5	9.0	9.5		38	9.0	9.5	10.0		53	9.5	10.5	11.0
	45°		16	7.0	7.5	8.0		26	7.5	8.0	8.0		36	7.5	8.0	8.5
8 x 6 (26.5)	Straight	206	36	9.5	11.0	12.0	310	59	12.0	12.5	13.5	412	82	12.5	14.0	15.0
	22½°		27	9.0	9.5	10.0		44	10.0	10.0	11.0		62	10.5	11.5	12.0
	45°		18	7.5	8.0	8.0		30	8.0	8.5	9.0		41	8.5	9.0	9.0
10 x 6 (34.0)	Straight	262	40	11.0	12.0	13.0	392	66	12.5	14.0	15.0	524	92	14.0	15.5	16.5
	22½°		30	9.5	10.0	11.0		50	10.5	10.5	12.0		69	11.5	12.5	13.0
	45°		20	8.0	8.5	9.0		33	8.5	9.0	9.5		46	9.0	9.5	10.0
12 x 6 (41.6)	Straight	318	41	11.5	12.5	13.5	476	67	13.0	14.0	15.5	636	94	14.5	15.5	17.0
	22½°		31	10.0	10.5	11.0		50	11.0	11.5	12.5		70	11.5	12.5	13.5
	45°		21	8.0	8.5	9.0		34	8.5	9.0	9.5		47	9.0	9.5	10.5
16 x 6 (56.6)	Straight	428	44	12.0	13.0	14.0	642	72	13.5	15.0	16.5	856	102	15.5	17.0	18.0
	22½°		33	10.0	11.0	11.5		54	11.5	12.0	13.0		77	12.5	13.5	14.5
	45°		22	8.0	9.0	9.0		36	9.0	9.5	10.0		51	9.5	10.0	11.0
20 x 6 (71.5)	Straight	538	47	12.5	13.5	14.5	806	77	14.5	16.0	17.0	1076	108	16.0	17.5	19.0
	22½°		35	10.5	11.5	12.0		58	12.0	12.5	14.0		81	13.0	14.0	15.0
	45°		24	8.5	9.0	9.5		39	9.5	10.0	10.5		54	10.0	10.5	11.0
24 x 6 (86.5)	Straight	648	48	13.0	14.0	15.5	972	79	15.0	16.5	17.5	1296	111	17.0	18.5	19.5
	22½°		36	10.5	11.5	12.5		59	12.0	13.0	14.5		83	13.5	14.5	15.5
	45°		24	8.5	9.5	9.5		40	9.5	10.0	10.5		56	10.0	10.5	11.5
30 x 6 (109.0)	Straight	812	50	13.0	14.5	15.5	1218	82	15.5	17.0	18.0	1624	115	17.5	19.0	20.5
	22½°		38	11.0	12.0	12.5		62	12.5	13.0	15.0		86	13.5	15.0	16.0
	45°		25	9.0	9.5	10.0		41	9.5	10.0	11.0		58	10.0	11.0	12.0
36 x 6 (131.3)	Straight	980	51	13.5	15.0	16.0	1470	84	16.0	17.5	19.0	1960	119	18.0	19.5	21.0
	22½°		38	11.0	12.0	13.0		63	13.0	13.5	15.0		89	14.0	15.0	16.5
	45°		26	9.0	9.5	10.0		42	10.0	10.5	11.0		60	10.5	11.0	12.0
K FACTOR																
Max Cfm/Sq Ft Outlet Wall Area		7.2					4.8					3.6				
Min Cfm/Sq Ft Outlet Wall Area		2.2					1.4					1.1				

NOTES:

1. When employing ratings for flat ceilings, it is understood that the front louvres are set to deflect the air upward toward the ceiling.

2. Blow indicates distance from outlet to the point where the air stream is substantially dissipated.

3. Underblow. It is not always necessary to blow the entire length of the room unless there are heat load sources at that end, equipment load, open doors, sun-glass, etc. Considering the concentration of room heat load on the basis of Btu/(hr) (sq ft), the outlet blow should cover 75% of the heat load.

4. Divergent Blow has vertical louvres straight forward in the center, with uniformly increasing angular deflection to a maximum at each end. The 45° divergence signifies an angular deflection at each end of the outlet of 45°, and similarly for 22½° divergence.

5. Velocity is based on effective face area.

6. Static Pressure is that pressure required to produce the indicated velocities and is measured in inches of water.

7. Measure ceiling height in the CLEAR only. This is the distance from the floor to the lowest ceiling beam or obstruction.

8. The Minimum Ceiling Height (table) is the minimum ceiling height which will give proper operation of the outlet for the given outlet velocity, vane setting, temperature difference, blow, and cfm. The actual measured ceiling height must be equal to or greater than the minimum ceiling height for the selection made. Preferably the top of an outlet should be not less than twice the outlet's height below the minimum ceiling height.

9. Cfm/Sq Ft Outlet Wall Area is the standard for judging total room air movement. The maximum value shown results in an air movement in the zone of occupancy of about 50 fpm. It is assumed that furniture, people, etc., obstruct 10% of the room cross-section. If the obstructions vary widely from 10%, the values of the cfm/sq ft outlet wall area should be tempered accordingly.

10. For applications requiring a limiting sound level—the outlet velocity is limited by the sound generated by the outlet.

TABLE 21—WALL OUTLET RATINGS, FOR COOLING ONLY (Cont.)

For Flat Ceilings

Outlet Velocity		250 FPM					375 FPM					500 FPM					750 FPM				
Static Pressure Standard Outlet		Str B = .01, 22½° = .01, 45° = .01					Str B = .013, 22½° = .015, 45° = .019					Str B = .024, 22½° = .028, 45° = .035					Str B = .051, 22½° = .061, 45° = .08				
Static Pressure With Metering Plate		Str B = .01, 22½° = .015, 45° = .028					Str B = .024, 22½° = .043, 45° = .065					Str B = .061, 22½° = .082, 45° = .118					Str B = .175, 22½° = .19, 45° = .27				
Nom. Size of Outlet (and Free Area)	Vane Setting	Air Quantity (cfm)	Blow (ft)	15	20	25	Air Quantity (cfm)	Blow (ft)	15	20	25	Air Quantity (cfm)	Blow (ft)	15	20	25	Air Quantity (cfm)	Blow (ft)	15	20	25
					Min Clg Ht					Min Clg Ht					Min Clg Ht					Min Clg Ht	
12 x 8 (56.7)	Straight	113	7.4	8.5	9.0	9.5	170	14.0	9.5	10.0	10.5	226	20.0	10.0	11.0	12.0	339	36	12.0	13.5	14.0
	22½°		5.5	7.5	8.0	8.5		10.0	8.5	9.0	9.5		15.0	9.0	10.0	10.0		27	10.0	11.0	12.0
	45°		3.7	7.0	7.5	7.5		7.0	7.5	8.0	8.0		11.0	8.0	8.5	8.5		18	8.5	9.0	9.5
16 x 8 (77.1)	Straight	155	8.0	9.0	9.5	10.0	231	15.0	10.0	10.5	11.0	308	22.0	11.0	12.0	12.5	463	40	13.5	14.5	15.5
	22½°		6.0	8.0	8.5	9.0		11.0	9.0	9.5	10.0		16.0	9.5	10.5	11.0		30	11.0	12.0	12.5
	45°		4.0	7.5	7.5	8.0		7.5	7.5	8.5	8.5		11.5	8.0	9.0	9.0		20	9.0	9.5	10.0
20 x 8 (97.6)	Straight	192	8.5	9.5	10.0	10.5	287	16.0	10.5	11.0	12.0	385	24.0	11.5	12.5	13.5	575	43	14.0	15.5	16.5
	22½°		6.5	8.5	9.0	9.5		12.0	9.0	10.0	10.5		18.0	10.0	10.5	11.5		32	11.5	12.5	13.5
	45°		4.3	7.5	8.0	8.0		8.0	8.0	8.5	9.0		12.5	8.5	9.0	10.0		22	9.5	10.0	10.5
24 x 8 (118.0)	Straight	231	9.0	9.5	10.0	10.5	346	17.0	10.5	11.5	12.5	460	25.0	12.0	13.0	14.0	692	45	14.5	16.0	17.0
	22½°		6.9	8.5	9.0	9.5		13.0	9.5	10.0	10.5		19.0	10.5	11.0	12.0		34	12.0	13.0	14.0
	45°		4.5	7.5	8.0	8.5		8.5	8.0	8.5	9.0		13.0	9.0	9.0	10.0		23	9.5	10.0	10.5
30 x 8 (149.0)	Straight	289	9.5	10.0	10.5	11.0	435	18.0	11.0	12.0	13.0	580	26.0	12.5	13.5	15.0	868	46	15.5	17.0	18.0
	22½°		7.0	9.0	9.5	10.0		13.0	9.5	10.0	10.5		19.0	10.5	11.5	12.5		35	12.5	13.5	15.0
	45°		4.7	8.0	8.0	8.5		9.0	8.5	9.0	9.0		13.5	9.0	9.5	10.0		23	10.0	10.5	11.0
36 x 8 (179.0)	Straight	350	9.9	10.0	11.0	11.5	525	18.0	11.5	12.0	13.0	702	27.0	13.0	14.0	15.5	1048	48	16.0	18.0	19.0
	22½°		7.5	9.0	9.5	10.0		13.0	9.5	10.5	11.0		20.0	11.0	12.0	12.5		36	13.0	14.0	15.0
	45°		5.0	8.0	8.5	9.0		9.0	8.5	9.0	9.5		14.0	9.0	9.5	10.0		24	10.0	10.5	11.5
16 x 10 (97.7)	Straight	198	9.6	9.5	10.0	11.5	297	18.0	11.5	12.5	13.0	396	27.0	13.0	14.0	15.5	595	48	15.5	18.0	19.0
	22½°		7.1	9.0	9.5	10.0		13.0	10.0	10.5	11.0		20.0	11.0	12.0	12.5		36	13.0	14.5	15.0
	45°		5.0	8.0	8.5	9.0		9.0	8.5	9.0	9.5		14.5	9.0	9.5	10.0		24	10.0	10.5	11.5
20 x 10 (124.0)	Straight	249	10.5	10.5	11.0	12.0	374	19.0	12.0	13.0	14.0	497	29.0	13.5	15.0	16.0	746	51	17.0	18.5	20.0
	22½°		8.0	9.5	10.0	10.5		14.0	10.5	11.0	12.0		22.0	12.0	12.5	13.5		38	13.5	15.0	16.5
	45°		5.2	8.0	8.5	9.0		9.5	9.0	9.5	9.5		15.0	9.5	10.0	10.5		26	10.5	11.0	12.0
24 x 10 (150.0)	Straight	300	11.0	11.0	12.0	12.5	450	21.0	12.5	13.5	15.0	600	30.0	14.5	16.0	17.0	899	55	18.5	20.0	21.0
	22½°		8.4	10.0	10.5	11.0		16.0	11.0	11.5	12.5		22.0	12.0	13.0	14.0		41	14.5	15.5	17.0
	45°		5.5	9.0	9.0	9.5		10.5	9.0	9.5	10.0		15.5	10.0	10.5	11.0		28	11.0	12.0	12.5
30 x 10 (195.0)	Straight	364	12.0	12.0	12.5	13.5	564	22.0	13.5	14.5	16.0	751	32.0	15.0	17.0	18.5	1126	58	19.5	21.5	23.0
	22½°		9.0	10.5	11.0	12.0		16.0	11.5	12.0	13.0		24.0	13.0	14.0	15.0		44	15.5	17.0	18.5
	45°		6.0	9.5	9.0	9.5		11.0	9.5	10.0	10.5		16.5	10.0	10.5	11.5		29	11.5	12.5	13.5
36 x 10 (227.0)	Straight	453	12.4	12.0	13.0	14.0	680	22.0	14.0	15.0	16.0	904	33.0	15.0	17.5	19.0	1355	60	20.0	22.0	23.5
	22½°		9.1	10.5	11.5	12.0		16.0	12.0	12.5	13.5		25.0	13.0	14.0	15.5		45	16.0	17.5	19.0
	45°		6.1	9.5	9.0	9.5		11.0	9.5	10.0	10.5		17.0	11.0	11.0	12.0		30	12.0	12.5	13.5
16 x 12 (118.0)	Straight	244	11.0	11.0	12.0	12.5	367	21.0	12.5	13.5	15.0	488	31.0	14.5	16.0	17.0	733	55	18.5	20.0	21.0
	22½°		8.1	10.0	10.5	11.0		16.0	11.0	11.5	12.5		23.0	12.0	13.0	14.0		41	14.5	16.0	17.0
	45°		5.5	8.5	9.0	9.5		11.0	9.0	9.5	10.0		16.0	10.0	10.5	11.0		28	11.0	12.0	12.5
20 x 12 (150.0)	Straight	307	12.1	12.0	13.0	14.0	460	22.0	14.0	15.0	16.0	613	33.0	16.0	17.5	19.0	918	60	20.0	22.0	23.5
	22½°		9.1	10.5	11.0	12.0		16.0	12.0	12.5	13.5		25.0	13.0	14.0	15.5		45	16.0	17.5	19.0
	45°		6.0	9.0	9.0	9.5		11.0	9.5	9.5	10.5		17.0	11.0	11.0	12.0		30	12.0	12.5	13.5
24 x 12 (181.0)	Straight	370	13.0	12.5	13.5	14.7	555	24.0	14.5	16.0	18.0	740	35.0	17.0	18.5	20.0	1110	64	21.5	24.0	25.0
	22½°		10.0	11.0	12.0	12.5		18.0	12.0	13.0	14.0		26.0	14.0	15.0	16.0		48	17.0	18.5	20.0
	45°		6.5	9.0	9.5	10.0		12.0	10.0	10.0	11.0		18.0	11.0	11.0	12.5		32	12.0	13.5	14.5
30 x 12 (228.0)	Straight	462	13.9	13.5	14.5	15.5	695	25.0	15.5	17.0	18.0	925	37.0	18.0	20.0	21.5	1388	68	23.0	26.0	27.5
	22½°		10.0	11.5	12.5	13.0		19.0	13.0	14.0	15.0		28.0	14.5	16.0	17.0		51	18.0	20.0	21.5
	45°		7.0	9.5	10.0	10.5		12.0	10.0	10.5	11.0		19.0	11.0	12.0	13.0		34	13.0	14.0	15.0
36 x 12 (275.0)	Straight	560	14.5	14.0	15.5	16.5	836	27.0	16.0	18.0	19.0	1115	39.0	19.5	21.0	22.5	1673	71	24.5	27.5	29.0
	22½°		11.0	12.0	13.0	14.0		20.0	13.5	14.5	15.5		29.0	15.5	16.5	18.0		53	19.0	21.0	22.5
	45°		8.0	9.5	10.0	10.5		13.0	10.5	11.0	11.5		20.0	11.5	12.0	13.0		36	13.0	14.5	15.5
K FACTOR																					
Max Cfm/Sq Ft Outlet Wall Area		29.0					19.0					14.0					9.6				
Min Cfm/Sq Ft Outlet Wall Area		8.7					5.7					4.2					2.9				

TABLE 21—WALL OUTLET RATINGS, FOR COOLING ONLY (Cont.)

For Flat Ceilings

OUTLET VELOCITY		1000 FPM					1500 FPM					2000 FPM				
STATIC PRESSURE STANDARD OUTLET		Str B = .093, 22½° = .11 45° = .14					Str B = .211, 22½° = .24 45° = .32					Str B = .375, 22½° = .42 45° = .565				
STATIC PRESSURE WITH METERING PLATE		Str B = .33, 22½° = .33 45° = .475					Str B = .715, 22½° = .74 45° = 1.15					Str B = 1.36				
Nom. Size of Outlet (and Free Area)	Vane Setting	Air Quantity (cfm)	Blow (ft)	Temp Diff (F) 15 Min Clg Ht	20	25	Air Quantity (cfm)	Blow (ft)	Temp Diff (F) 15 Min Clg Ht	20	25	Air Quantity (cfm)	Blow (ft)	Temp Diff (F) 15 Min Clg Ht	20	25
12 x 8 (56.7)	Straight	452	52	14.0	15.0	16.5	678	86	16.0	18.0	19.0	904	121	18.0	20.0	21.5
	22½°		39	11.5	12.5	13.5		65	13.0	14.0	15.5		91	14.0	15.5	17.0
	45°		26	9.0	10.0	10.0		43	10.0	10.5	11.5		62	10.5	11.5	12.0
16 x 8 (77.1)	Straight	616	57	15.0	16.5	18.0	926	95	18.0	19.5	21.0	1232	134	20.0	22.0	24.0
	22½°		43	12.5	13.5	14.5		71	14.0	14.0	17.0		100	15.5	17.0	18.5
	45°		29	9.5	10.0	11.0		48	10.5	11.5	12.0		67	11.0	12.0	13.0
20 x 8 (97.6)	Straight	770	62	16.0	18.0	19.5	1150	102	19.0	21.0	23.0	1540	144	21.0	23.5	26.0
	22½°		47	13.0	14.5	15.5		77	15.0	15.5	18.0		108	16.5	18.0	20.0
	45°		31	10.0	10.5	11.5		51	11.0	12.0	13.0		72	12.0	13.0	14.0
24 x 8 (118.0)	Straight	920	65	17.0	19.0	20.0	1384	107	20.0	22.5	24.5	1840	151	22.0	25.0	27.5
	22½°		49	13.5	15.0	16.0		80	15.5	16.5	18.5		113	17.5	19.0	20.5
	45°		33	10.5	11.0	12.0		54	11.5	12.5	13.0		76	12.0	13.5	14.5
30 x 8 (149.0)	Straight	1160	68	17.5	19.5	21.0	1736	111	21.0	23.5	25.5	2320	157	23.5	26.0	29.5
	22½°		51	14.0	15.5	17.0		83	16.5	17.5	19.5		118	18.0	20.0	21.5
	45°		34	10.5	11.5	12.5		56	11.5	13.0	13.5		79	13.0	14.0	15.0
36 x 8 (179.0)	Straight	1404	71	18.5	20.5	22.0	2096	116	21.5	25.0	26.5	2808	164	24.5	27.5	31.0
	22½°		53	14.5	16.0	17.5		87	17.0	18.0	20.0		123	19.0	21.0	22.5
	45°		36	11.0	12.0	12.5		58	12.0	13.0	14.0		82	13.0	14.5	15.5
16 x 10 (97.7)	Straight	792	71	18.0	20.5	22.0	1190	116	21.0	25.0	26.5	1584	164	24.0	27.5	31.0
	22½°		53	14.5	16.0	17.5		87	17.0	19.0	20.0		123	19.0	21.0	22.5
	45°		36	11.0	12.0	12.5		58	12.0	13.0	14.0		82	13.0	14.5	15.5
20 x 10 (124.0)	Straight	994	75	19.5	22.0	24.0	1492	122	23.0	26.5	29.0	1988	174	26.0	29.5	34.0
	22½°		56	15.5	17.0	18.5		92	18.0	19.0	21.5		130	20.0	22.5	24.0
	45°		38	11.5	12.5	13.5		61	13.0	14.0	15.0		87	14.0	15.5	16.5
24 x 10 (150.0)	Straight	1200	80	21.0	24.0	25.5	1798	131	24.5	28.5	31.0	2400	185	28.0	32.0	36.5
	22½°		60	16.5	18.5	19.5		98	19.0	20.0	22.5		139	21.5	24.0	25.5
	45°		40	12.0	13.0	14.0		66	13.5	14.5	15.5		93	14.5	16.5	17.0
30 x 10 (195.0)	Straight	1502	86	22.5	25.5	27.5	2252	139	26.5	30.5	34.0	3004	196	30.5	34.0	39.5
	22½°		65	18.0	19.5	21.0		104	20.5	21.5	24.5		147	23.0	26.0	28.0
	45°		43	12.5	14.0	15.0		70	14.0	15.5	16.5		98	15.5	17.0	18.5
36 x 10 (227.0)	Straight	1808	87	23.5	26.5	28.5	2710	142	27.0	31.5	35.0	3616	200	31.5	36.0	40.5
	22½°		65	18.5	20.0	21.5		106	21.0	23.0	25.0		150	23.5	27.0	29.0
	45°		44	13.0	14.5	15.0		71	14.5	16.0	17.0		100	15.5	17.5	19.0
16 x 12 (118.0)	Straight	976	81	21.0	24.0	25.5	1466	131	24.5	28.0	31.0	1952	158	28.0	32.0	36.5
	22½°		61	16.5	18.5	19.5		98	19.0	21.5	22.5		138	21.5	24.0	25.5
	45°		41	12.0	13.0	14.0		66	13.5	14.5	15.5		93	14.5	16.5	17.0
20 x 12 (150.0)	Straight	1226	87	23.5	26.5	28.5	1836	142	27.0	31.5	35.0	2452	200	31.5	36.0	40.5
	22½°		65	18.5	20.0	21.5		106	21.0	23.0	25.0		150	23.5	27.0	29.0
	45°		44	13.0	14.5	15.0		71	14.5	16.0	17.0		100	15.5	17.5	19.0
24 x 12 (181.0)	Straight	1480	93	25.0	28.5	31.0	2220	153	29.0	34.0	37.5	2960	213	34.5	38.5	42.5
	22½°		70	19.5	21.5	23.0		115	22.0	25.0	26.5		160	25.0	29.0	31.0
	45°		47	13.5	15.0	16.0		77	15.0	17.0	18.0		107	16.5	18.5	20.0
30 x 12 (228.0)	Straight	1850	98	27.0	31.0	34.0	2776	163	31.5	36.5	40.5	3700	226	38.0	43.0	47.0
	22½°		74	21.0	23.0	25.0		122	24.0	27.0	28.5		169	27.0	31.0	33.5
	45°		49	14.5	16.0	17.0		82	16.0	18.0	19.0		113	17.5	19.5	21.0
36 x 12 (275.0)	Straight	2230	103	29.0	33.0	36.0	3346	172	33.5	38.0	42.5	4460	238	40.0	45.0	50.0
	22½°		77	22.0	24.5	26.5		129	25.0	28.0	30.0		178	28.5	32.5	35.5
	45°		52	15.0	16.0	17.5		86	16.5	18.5	20.0		119	18.0	20.5	22.0
K FACTOR																
Max Cfm/Sq Ft Outlet Wall Area		7.2					4.8					3.6				
Min Cfm/Sq Ft Outlet Wall Area		2.2					1.4					1.1				

NOTES:

1. When employing ratings for flat ceilings, it is understood that the front louvres are set to deflect the air upward toward the ceiling.

2. **Blow** indicates distance from outlet to the point where the air stream is substantially dissipated.

3. **Underblow.** It is not always necessary to blow the entire length of the room unless there are heat load sources at that end, equipment load, open doors, sun-glass, etc. Considering the concentration of room heat load on the basis of Btu/(hr) (sq ft), the outlet blow should cover 75% of the heat load.

4. **Divergent Blow** has vertical louvres straight forward in the center, with uniformly increasing angular deflection to a maximum at each end. The 45° divergence signifies an angular deflection at each end of the outlet of 45°, and similarly for 22½° divergence.

5. **Velocity** is based on effective face area.

6. **Static Pressure** is that pressure required to produce the indicated velocities and is measured in inches of water.

7. Measure ceiling height in the CLEAR only. This is the distance from the floor to the lowest ceiling beam or obstruction.

8. **The Minimum Ceiling Height** (table) is the minimum ceiling height which will give proper operation of the outlet for the given outlet velocity, vane setting, temperature difference, blow, and cfm. The actual measured ceiling height must be equal to or greater than the minimum ceiling height for the selection made. Preferably the top of an outlet should be not less than twice the outlet's height below the minimum ceiling height.

9. **Cfm/Sq Ft Outlet Wall Area** is the standard for judging total room air movement. The maximum value shown results in an air movement in the zone of occupancy of about 50 fpm. It is assumed that furniture, people, etc., obstruct 10% of the room cross-section. If the obstructions vary widely from 10%, the values of the cfm/sq ft outlet wall area should be tempered accordingly.

10. **For applications requiring a limiting sound level**—the outlet velocity is limited by the sound generated by the outlet.

TABLE 21—WALL OUTLET RATINGS, FOR COOLING ONLY (Cont.)

For Beamed Ceilings

	250 FPM	375 FPM	500 FPM	750 FPM
OUTLET VELOCITY				
STATIC PRESSURE STANDARD OUTLET	Str B = .01, 22½° = .01, 45° = .01	Str B = .013, 22½° = .015, 45° = .019	Str B = .024, 22½° = .028, 45° = .035	Str B = .051, 22½° = .061, 45° = .08
STATIC PRESSURE WITH METERING PLATE	Str B = .01, 22½° = .015, 45° = .028	Str B = .024, 22½° = .043, 45° = .065	Str B = .061, 22½° = .082, 45° = .118	Str B = .175, 22½° = .19, 45° = .27

Nom. Size of Outlet (and Free Area)	Vane Setting	Air Quantity (cfm)	Blow (ft)	15	20	25	Air Quantity (cfm)	Blow (ft)	15	20	25	Air Quantity (cfm)	Blow (ft)	15	20	25	Air Quantity (cfm)	Blow (ft)	15	20	25
				Min Clg Ht					Min Clg Ht					Min Clg Ht					Min Clg Ht		
8 x 4 (16.9)	Straight	30	3.5	7.3	7.7	8.1	44	7.0	8.2	8.7	9.1	59	10.0	8.7	9.3	9.8	89	17.0	9.7	10.4	11.1
	22½°		2.5	6.9	7.2	7.5		5.1	7.5	7.9	8.2		7.5	8.0	8.4	8.8		13.0	8.5	9.1	9.6
	45°		1.8	6.5	6.8	7.0		3.5	6.5	6.5	7.5		5.0	7.2	7.5	7.8		9.0	7.4	7.8	8.1
10 x 4 (21.7)	Straight	37	3.5	7.4	7.7	8.2	57	7.4	8.2	8.8	9.2	75	10.5	8.8	9.4	9.9	112	18.0	9.8	10.5	11.2
	22½°		2.5	7.0	7.3	7.6		5.5	7.5	8.0	8.2		8.0	8.0	8.5	8.8		13.0	8.6	9.2	9.6
	45°		1.8	6.5	6.8	7.0		3.7	6.9	7.2	7.5		5.4	7.2	7.6	7.8		9.0	7.4	7.9	8.2
12 x 4 (24.6)	Straight	44	3.5	7.4	7.8	8.2	68	7.5	8.3	8.8	9.2	91	11.0	8.9	9.4	9.9	136	18.0	9.9	10.6	11.3
	22½°		2.5	7.1	7.3	7.6		5.5	7.6	8.0	8.3		8.1	8.1	8.5	8.9		13.0	8.6	9.2	9.8
	45°		1.8	6.5	6.9	7.0		3.9	7.0	7.2	7.5		5.5	7.2	7.6	7.9		9.0	7.5	7.9	8.2
16 x 4 (35.9)	Straight	61	3.7	7.5	7.9	8.3	92	7.9	8.4	8.9	9.4	122	11.0	9.0	9.6	10.0	183	19.0	10.0	10.8	11.5
	22½°		2.7	7.2	7.4	7.7		6.0	7.7	8.1	8.4		8.1	8.2	8.6	9.0		14.0	8.8	9.4	9.9
	45°		2.0	6.5	7.0	7.0		4.0	7.0	7.3	7.6		5.5	7.3	7.7	7.9		10.0	7.6	8.0	8.3
20 x 4 (45.5)	Straight	77	4.0	7.6	8.0	8.4	115	8.0	8.4	9.0	9.4	154	11.5	9.1	9.6	10.1	231	20.0	10.2	11.0	11.7
	22½°		3.0	7.2	7.5	7.8		6.0	7.8	8.2	8.5		8.5	8.3	8.7	9.1		15.0	8.9	9.5	10.0
	45°		2.0	6.6	7.0	7.1		4.0	7.0	7.4	7.6		6.0	7.4	7.8	8.0		10.0	7.6	8.1	8.4
24 x 4 (55.0)	Straight	93	4.1	7.6	8.0	8.4	139	8.0	8.5	9.0	9.5	185	11.5	9.2	9.6	10.2	278	20.0	10.2	11.0	11.8
	22½°		3.1	7.2	7.5	7.8		6.0	7.8	8.2	8.6		8.5	8.3	8.8	9.1		15.0	8.9	9.5	10.1
	45°		2.0	6.6	7.0	7.1		4.0	7.1	7.4	7.6		6.0	7.4	7.8	8.0		10.0	7.6	8.1	8.4
30 x 4 (68.3)	Straight	116	4.2	7.7	8.1	8.5	175	8.0	8.5	9.0	9.6	233	12.0	9.3	9.7	10.3	349	21.0	10.3	11.1	11.9
	22½°		3.1	7.3	7.6	7.9		6.0	7.9	8.3	8.6		9.0	8.4	8.8	9.2		16.0	9.0	9.6	10.2
	45°		2.1	6.6	7.0	7.2		4.0	7.1	7.5	7.7		4.6	7.4	7.8	8.0		11.0	7.7	8.2	8.5
36 x 4 (83.5)	Straight	140	4.4	7.7	8.2	8.5	210	8.0	8.5	9.1	9.6	279	12.0	9.3	9.8	10.4	420	21.0	10.4	11.2	12.0
	22½°		3.3	7.3	7.6	7.9		6.0	7.9	8.3	8.6		9.0	8.4	8.9	9.2		16.0	9.0	9.7	10.3
	45°		2.2	6.6	7.0	7.2		4.0	7.1	7.5	7.7		6.0	7.4	7.9	8.1		11.0	7.8	8.2	8.6
8 x 6 (26.5)	Straight	52	5.0	8.2	8.6	9.0	77	9.5	9.0	9.6	10.2	103	13.0	9.9	10.4	11.1	155	24.0	11.3	12.2	13.1
	22½°		3.8	7.6	8.0	8.3		7.0	8.3	8.8	9.2		10.0	8.8	9.4	9.8		18.0	9.7	10.4	11.2
	45°		2.5	6.8	7.3	7.4		4.8	7.4	7.8	8.0		6.0	7.7	8.2	8.5		12.0	8.2	8.6	9.2
10 x 6 (34.0)	Straight	66	5.5	8.6	9.2	9.6	98	10.0	9.6	10.2	10.9	131	15.0	10.5	11.2	11.9	196	27.0	12.3	13.3	14.2
	22½°		4.1	7.9	8.4	8.7		7.5	8.7	9.3	9.8		11.0	9.3	9.9	10.5		20.0	10.4	11.2	12.1
	45°		2.8	7.1	7.6	7.8		5.0	7.6	8.0	8.4		7.0	8.1	8.6	8.9		14.0	8.6	9.1	9.7
12 x 6 (41.6)	Straight	80	6.0	8.7	9.2	9.7	119	11.0	9.7	10.4	11.1	159	15.0	10.6	11.4	12.2	238	28.0	12.5	13.6	14.5
	22½°		4.5	8.0	8.4	8.8		8.1	8.8	9.4	9.8		11.0	9.4	10.1	10.7		21.0	10.6	11.4	12.2
	45°		3.0	7.2	7.6	7.8		5.5	7.7	8.1	8.5		7.0	8.1	8.6	9.0		14.0	8.7	9.2	9.8
16 x 6 (56.6)	Straight	107	6.2	9.1	9.6	10.2	161	12.0	10.1	10.8	11.6	214	16.0	11.2	12.0	12.8	321	30.0	13.3	14.5	15.5
	22½°		4.7	8.2	8.7	9.0		9.0	9.1	9.7	10.2		12.0	9.8	10.5	11.2		22.0	11.2	12.0	12.9
	45°		3.2	7.4	7.8	8.1		6.0	7.9	8.4	8.8		8.0	8.4	9.0	9.3		15.0	9.0	9.7	10.3
20 x 6 (71.5)	Straight	135	6.6	9.4	9.9	10.5	202	12.0	10.5	11.2	12.1	269	17.0	11.6	12.5	13.4	403	32.0	14.0	15.2	16.3
	22½°		5.0	8.4	9.0	9.3		9.0	9.3	10.0	10.5		13.0	10.1	10.9	11.6		24.0	11.6	12.5	13.5
	45°		3.2	7.6	8.0	8.3		6.6	8.1	8.6	9.0		9.0	8.6	9.2	9.6		16.0	9.3	10.0	10.6
24 x 6 (86.5)	Straight	162	7.0	9.6	10.1	10.7	243	13.0	10.7	11.5	12.3	324	18.0	11.9	12.8	13.7	486	33.0	14.4	15.7	16.9
	22½°		5.1	8.8	9.1	9.5		10.0	9.5	10.2	10.7		13.0	10.3	11.1	11.9		25.0	11.9	12.9	13.9
	45°		3.5	7.7	8.1	8.5		6.5	8.2	8.8	9.2		9.0	8.7	9.4	9.8		17.0	9.5	10.2	10.8
30 x 6 (109.0)	Straight	203	7.0	9.8	10.3	11.0	304	13.0	11.0	11.8	12.6	406	19.0	12.2	13.3	14.1	609	34.0	14.8	16.1	17.4
	22½°		5.4	8.7	9.3	9.7		10.0	9.6	10.4	10.9		14.0	10.6	11.4	12.2		25.0	12.1	13.3	14.2
	45°		3.5	7.8	8.2	8.6		6.5	8.4	8.9	9.4		10.0	8.9	9.6	10.0		17.0	9.7	10.4	11.1
36 x 6 (131.3)	Straight	245	7.1	9.9	10.5	11.1	368	13.0	11.2	12.0	12.4	490	19.0	12.5	13.6	14.7	735	35.0	15.2	16.6	18.0
	22½°		5.5	8.8	9.6	9.8		10.0	9.8	10.6	11.1		14.0	10.7	11.6	12.5		26.0	12.4	13.5	14.6
	45°		3.5	7.9	8.4	8.7		6.5	8.5	9.1	9.5		10.0	9.0	9.7	10.1		18.0	9.9	10.6	11.3

K FACTOR

	250 FPM	375 FPM	500 FPM	750 FPM
Max Cfm/Sq Ft Outlet Wall Area	29.0	19.0	14.0	9.6
Min Cfm/Sq Ft Outlet Wall Area	8.7	5.7	4.2	2.9

TABLE 21—WALL OUTLET RATINGS, FOR COOLING ONLY (Cont.)

For Beamed Ceilings

OUTLET VELOCITY	1000 FPM	1500 FPM	2000 FPM
STATIC PRESSURE STANDARD OUTLET	Str B = .093, 22½° = .11, 45° = .14	Str B = .211, 22½° = .24, 45° = .32	Str B = .375, 22½° = .42, 45° = .565
STATIC PRESSURE WITH METERING PLATE	Str B = .33, 22½° = .33, 45° = .475	Str B = .715, 22½° = .74, 45° = 1.15	Str B = 1.36

Nom. Size of Outlet (and Free Area)	Vane Setting	Air Quantity (cfm)	Blow (ft)	Temp Diff (F) 15 Min Clg Ht	20	25	Air Quantity (cfm)	Blow (ft)	Temp Diff (F) 15 Min Clg Ht	20	25	Air Quantity (cfm)	Blow (ft)	Temp Diff (F) 15 Min Clg Ht	20	25
8 x 4 (16.9)	Straight	118	24	10.4	11.2	11.9	177	40	11.9	12.9	13.8	237	58	12.7	13.9	15.0
	22½°		18	9.0	9.6	10.2		30	9.8	10.5	11.2		44	10.2	11.0	11.7
	45°		12	7.6	8.0	8.4		20	7.7	8.1	8.5		29	7.8	8.2	8.6
10 x 4 (21.7)	Straight	150	26	10.6	11.4	12.1	224	42	12.1	13.1	14.1	299	60	13.0	14.2	15.4
	22½°		19	9.1	9.7	10.4		32	10.0	10.7	11.4		45	10.4	11.3	12.0
	45°		13	7.7	8.2	8.5		21	7.8	8.4	8.7		30	8.0	8.5	8.9
12 x 4 (24.6)	Straight	181	27	10.7	11.6	12.3	272	44	12.3	13.4	14.4	362	62	13.3	14.5	15.7
	22½°		20	9.3	9.9	10.5		33	10.2	10.9	11.6		47	10.6	11.6	12.3
	45°		14	7.8	8.2	8.6		22	8.0	8.5	8.9		31	8.2	8.7	9.2
16 x 4 (35.9)	Straight	244	28	11.0	11.9	12.7	366	46	12.6	13.8	14.8	488	65	13.7	15.0	16.2
	22½°		21	9.5	10.1	10.7		35	10.4	11.3	12.0		49	11.0	12.0	12.8
	45°		14	7.8	8.4	8.6		23	8.2	8.7	9.2		33	8.5	9.0	9.5
20 x 4 (45.5)	Straight	308	29	11.2	12.1	12.9	462	48	12.9	14.1	15.1	616	67	14.0	15.3	16.6
	22½°		22	9.7	10.3	10.9		36	10.6	11.5	12.2		50	11.3	12.3	13.1
	45°		15	7.9	8.5	8.8		24	8.3	8.9	9.4		34	8.6	9.2	9.7
24 x 4 (55.0)	Straight	370	30	11.3	12.3	13.1	556	49	13.0	14.2	15.3	740	68	14.2	15.5	16.8
	22½°		22	9.8	10.4	11.0		37	10.8	11.6	12.4		51	11.4	12.4	13.3
	45°		15	8.0	8.5	8.8		25	8.4	9.0	9.4		34	8.7	9.3	9.9
30 x 4 (68.3)	Straight	466	30	11.4	12.4	13.3	698	50	13.2	14.4	15.5	932	70	14.4	15.7	17.0
	22½°		22	9.9	10.5	11.2		37	10.9	11.8	12.6		53	11.6	12.6	13.5
	45°		15	8.0	8.6	8.9		25	8.5	9.1	9.6		35	8.9	9.5	10.0
36 x 4 (83.5)	Straight	558	31	11.5	12.5	13.4	840	51	13.3	14.5	15.7	1116	71	14.5	15.9	17.2
	22½°		23	10.0	10.6	11.2		38	11.0	11.9	12.6		53	11.7	12.8	13.6
	45°		16	8.1	8.6	9.0		26	8.5	9.2	9.6		34	9.0	9.6	10.1
8 x 6 (26.5)	Straight	206	36	11.6	13.8	14.9	310	59	14.7	16.2	17.5	412	82	16.2	17.9	19.3
	22½°		27	10.8	11.5	9.2		44	12.0	13.1	14.0		62	13.0	14.2	15.3
	45°		18	8.6	9.2	9.6		30	9.3	10.0	10.6		41	9.9	10.6	11.3
10 x 6 (34.0)	Straight	262	40	13.8	15.2	16.5	392	66	16.1	17.9	19.5	524	92	18.1	20.0	21.7
	22½°		30	11.6	12.6	13.5		50	13.2	14.4	15.5		69	14.3	15.7	17.0
	45°		20	9.2	9.9	10.4		33	10.1	10.8	11.6		46	10.8	11.6	12.5
12 x 6 (41.6)	Straight	318	41	14.1	15.5	16.9	476	67	16.4	18.3	19.9	636	94	18.5	20.4	22.1
	22½°		31	11.8	12.8	13.7		50	13.4	14.7	15.7		70	14.6	16.0	17.3
	45°		21	9.3	10.0	10.5		34	10.2	11.0	11.8		47	11.0	11.8	12.7
16 x 6 (56.6)	Straight	428	44	15.0	16.4	18.1	642	72	17.6	19.7	21.3	856	102	20.0	22.1	23.9
	22½°		33	12.5	13.6	14.6		54	14.3	15.6	16.9		77	15.6	17.1	18.7
	45°		22	9.7	10.5	11.1		36	10.8	11.6	12.5		51	11.6	12.5	13.5
20 x 6 (71.5)	Straight	538	47	15.7	17.5	19.0	806	77	18.6	20.8	22.4	1076	108	21.1	23.3	25.2
	22½°		35	13.1	14.2	15.3		58	14.9	16.4	17.8		81	16.4	17.9	19.7
	45°		24	10.1	10.8	11.5		39	11.2	12.1	13.0		54	12.1	13.0	14.1
24 x 6 (86.5)	Straight	648	48	16.3	18.0	19.7	972	79	19.3	21.6	23.2	1296	111	21.9	24.2	26.1
	22½°		36	13.4	14.6	15.7		59	15.4	16.9	18.4		83	17.0	18.5	20.4
	45°		24	10.3	11.1	11.8		40	11.6	12.4	13.3		56	12.4	13.4	14.5
30 x 6 (109.0)	Straight	812	50	16.9	18.7	20.4	1218	82	20.0	22.4	24.0	1624	115	22.7	25.0	27.1
	22½°		38	13.8	15.0	16.2		62	15.9	17.4	19.0		86	17.5	19.1	21.1
	45°		25	10.6	11.4	12.1		41	11.7	12.6	13.7		58	12.7	13.7	14.9
36 x 6 (131.3)	Straight	980	51	17.3	19.2	21.0	1470	84	20.6	23.0	24.8	1960	119	23.4	25.8	28.0
	22½°		38	14.1	15.4	16.7		63	16.3	17.8	19.6		89	18.0	19.6	21.7
	45°		26	10.8	11.6	12.4		42	11.9	12.9	14.0		60	13.0	14.0	15.3

K FACTOR	1000 FPM	1500 FPM	2000 FPM
Max Cfm/Sq Ft Outlet Wall Area	7.2	4.8	3.6
Min Cfm/Sq Ft Outlet Wall Area	2.2	1.4	1.1

NOTES:

1. **Divergent Blow** has vertical louvres straight forward in the center, with uniformly increasing angular deflection to a maximum at each end. The 45° divergence signifies an angular deflection at each end of the outlet of 45°, and similarly for 22½° divergence.

2. **Blow** indicates distance from outlet to the point where the air stream is substantially dissipated.

3. **Underblow**. It is not always necessary to blow the entire length of the room unless there are heat load sources at that end, equipment load, open doors, sun-glass, etc. Considering the concentration of room heat load on the basis of Btu/(hr) (sq ft), the outlet blow should cover 75% of the heat load.

4. **Velocity** is based on effective face area.

5. **Static Pressure** is that pressure required to produce the indicated velocities and is measured in inches of water.

6. **Measure ceiling height in the CLEAR only**. This is the distance from the floor to the lowest ceiling beam or obstruction.

7. **The Minimum Ceiling Height** (table) is the minimum ceiling height which will give proper operation of the outlet for the given outlet velocity, vane setting, temperature difference, blow, and cfm. The actual measured ceiling height must be equal to or greater than the minimum ceiling height for the selection made. Preferably the top of an outlet should be not less than twice the outlet's height below the minimum ceiling height.

8. **Cfm/Sq Ft Outlet Wall Area** is the standard for judging total room air movement. The maximum value shown results in an air movement in the zone of occupancy of about 50 fpm. It is assumed that furniture, people, etc., obstruct 10% of the room cross-section. If the obstructions vary widely from 10%, the values of the cfm/sq ft outlet wall area should be tempered accordingly.

9. **For applications requiring a limiting sound level**—the outlet velocity is limited by the sound generated by the outlet.

TABLE 21—WALL OUTLET RATINGS, FOR COOLING ONLY (Cont.)

For Beamed Ceilings

Outlet Velocity		250 FPM					375 FPM					500 FPM					750 FPM				
Static Pressure Standard Outlet		Str B = .01, 22½° = .01 45° = .01					Str B = .013, 22½° = .015 45° = .019					Str B = .024, 22½° = .028 45° = .035					Str B = .051, 22½° = .061 45° = .08				
Static Pressure with Metering Plate		Str B = .01, 22½° = .015 45° = .028					Str B = .024, 22½° = .043 45° = .065					Str B = .061, 22½° = .082 45° = .118					Str B = .175, 22½° = .19 45° = .27				
Nom. Size of Outlet (and Free Area)	Vane Setting	Air Quantity (cfm)	Blow (ft)	Temp Diff (F) 15	20	25	Air Quantity (cfm)	Blow (ft)	Temp Diff (F) 15	20	25	Air Quantity (cfm)	Blow (ft)	Temp Diff (F) 15	20	25	Air Quantity (cfm)	Blow (ft)	Temp Diff (F) 15	20	25
				Min Clg Ht					Min Clg Ht					Min Clg Ht					Min Clg Ht		
12 x 8 (56.7)	Straight	113	7.4	10.0	10.6	11.3	170	14.0	11.3	12.1	13.0	226	20.0	12.6	13.8	14.7	339	36	15.4	17.0	18.3
	22½°		5.5	8.9	9.5	10.0		10.0	9.9	10.7	11.2		15.0	10.9	11.8	12.7		27	12.6	13.7	14.8
	45°		3.7	8.0	8.4	8.8		7.0	8.6	9.2	9.8		11.0	9.1	9.8	10.3		18	10.0	10.8	11.4
16 x 8 (77.1)	Straight	155	8.0	10.6	11.4	12.1	231	15.0	12.1	13.1	14.0	308	22.0	13.7	15.0	16.1	463	40	16.9	18.6	20.3
	22½°		6.0	9.4	10.1	10.7		11.0	10.4	11.4	11.9		16.0	11.7	12.7	13.7		30	13.7	15.0	16.2
	45°		4.0	8.4	8.9	9.3		7.5	9.0	9.7	10.1		11.5	9.6	10.4	10.9		20	10.6	11.5	12.2
20 x 8 (97.6)	Straight	192	8.5	11.1	12.0	12.7	287	16.0	12.7	13.8	14.9	385	24.0	14.6	15.9	17.2	575	43	18.1	20.0	21.7
	22½°		6.5	9.8	10.6	11.3		12.0	10.9	11.9	12.6		18.0	12.4	13.4	14.5		32	14.5	16.0	17.3
	45°		4.3	8.7	9.2	9.7		8.0	9.4	10.0	10.6		12.5	10.0	10.8	11.5		22	11.3	12.1	13.0
24 x 8 (118.0)	Straight	231	9.0	11.5	12.4	13.2	346	17.0	13.3	14.3	15.6	460	25.0	15.2	16.6	18.1	692	45	19.0	21.0	22.7
	22½°		6.9	10.2	10.9	11.7		13.0	11.4	12.3	13.2		19.0	12.8	13.9	15.1		34	15.2	16.8	18.2
	45°		4.5	8.9	9.4	10.0		8.5	9.6	10.2	10.8		13.0	10.4	11.0	11.8		23	11.6	12.5	13.4
30 x 8 (149.0)	Straight	289	9.5	12.0	12.8	13.8	435	18.0	13.8	14.9	16.3	580	26.0	15.9	17.3	19.0	868	46	20.0	22.0	23.9
	22½°		7.0	10.6	11.3	12.1		13.0	11.8	12.7	13.7		19.0	13.4	14.5	15.7		35	16.0	17.5	19.1
	45°		4.7	9.1	9.6	10.2		9.0	9.9	10.5	11.1		13.5	10.7	11.4	12.2		23	12.0	13.0	14.0
36 x 8 (179.0)	Straight	350	9.9	12.4	13.3	14.2	525	18.0	14.3	15.5	16.8	702	27.0	16.5	18.0	19.7	1048	48	20.8	22.9	24.8
	22½°		7.5	10.8	11.7	12.5		13.0	12.1	13.1	14.1		20.0	13.7	15.0	16.2		36	16.4	18.2	19.7
	45°		5.0	9.3	9.8	10.4		9.0	10.1	10.7	11.4		14.0	10.9	11.7	12.6		24	12.4	13.3	14.4
16 x 10 (97.7)	Straight	198	9.6	12.2	13.3	14.2	297	18.0	14.1	15.5	16.8	396	27.0	16.3	18.0	19.7	595	48	20.5	22.9	24.8
	22½°		7.1	10.8	11.7	12.5		13.0	12.1	13.1	14.1		20.0	13.7	15.0	16.2		36	16.4	18.2	19.7
	45°		5.0	9.3	9.8	10.4		9.0	10.1	10.7	11.4		14.5	10.9	11.7	12.6		24	12.4	13.3	14.4
20 x 10 (124.0)	Straight	249	10.5	13.1	14.1	15.2	374	19.0	15.2	16.5	18.0	497	29.0	17.5	19.3	21.2	746	51	22.4	24.6	26.6
	22½°		8.0	11.4	12.3	13.2		14.0	12.9	13.8	15.0		22.0	14.7	15.9	17.3		38	17.7	19.4	21.3
	45°		5.2	9.7	10.2	10.8		9.5	10.5	11.1	11.9		15.0	11.5	12.2	13.3		26	13.1	14.1	15.3
24 x 10 (150.0)	Straight	300	11.0	13.9	15.0	16.1	450	21.0	16.1	17.6	19.1	600	30.0	18.7	20.6	22.5	899	55	24.1	26.5	28.4
	22½°		8.4	12.0	13.0	13.9		16.0	13.6	14.6	15.9		22.0	15.4	16.8	18.3		41	18.8	20.7	22.4
	45°		5.5	10.0	10.4	11.2		10.5	10.8	11.5	12.2		15.5	11.9	12.6	13.8		28	13.7	14.8	16.1
30 x 10 (195.0)	Straight	364	12.0	14.8	16.0	17.3	564	22.0	17.3	18.9	20.5	751	32.0	20.0	22.1	24.4	1126	58	26.0	28.7	30.9
	22½°		9.0	12.7	13.8	14.8		16.0	14.5	15.5	16.9		24.0	16.5	18.0	19.6		44	20.3	22.3	24.4
	45°		6.0	10.4	10.9	11.7		11.0	11.3	12.0	12.8		16.5	12.6	13.3	14.6		29	14.5	15.7	17.1
36 x 10 (227.0)	Straight	453	12.4	15.2	16.4	17.7	680	22.0	17.7	19.5	21.1	904	33.0	20.5	22.7	25.0	1355	60	26.7	29.5	31.7
	22½°		9.1	13.0	14.0	15.1		16.0	14.8	15.8	17.3		25.0	16.9	18.3	20.0		45	20.8	23.0	25.1
	45°		6.1	10.5	11.0	11.8		11.0	11.4	12.1	13.0		17.0	13.7	13.5	14.9		30	14.8	16.1	17.4
16 x 12 (118.0)	Straight	244	11.0	13.9	15.0	16.1	367	21.0	16.1	17.6	19.1	488	31.0	18.7	20.6	22.5	733	55	24.1	26.5	28.4
	22½°		8.1	12.0	13.0	13.9		16.0	13.6	14.6	15.9		23.0	15.4	16.8	18.3		41	18.8	20.7	22.4
	45°		5.5	10.0	10.4	11.2		11.0	10.8	11.5	12.2		16.0	11.9	12.6	13.8		28	13.7	14.8	16.1
20 x 12 (150.0)	Straight	307	12.1	15.2	16.4	17.7	460	22.0	17.7	19.5	21.1	613	33.0	20.5	22.7	25.0	918	60	26.7	29.5	31.7
	22½°		9.1	13.0	14.0	15.1		16.0	14.8	15.8	17.3		25.0	16.9	18.3	20.0		45	20.8	23.0	25.1
	45°		6.0	10.5	11.0	11.8		11.0	11.4	12.1	13.0		17.0	13.7	13.5	14.9		30	14.8	16.1	17.4
24 x 12 (181.0)	Straight	370	13.0	16.1	17.4	18.8	555	24.0	18.7	20.7	23.4	740	35.0	21.8	24.5	26.7	1110	64	28.7	32.0	34.0
	22½°		10.0	13.6	14.8	16.0		18.0	15.6	16.8	18.3		26.0	17.8	19.5	21.2		48	22.1	24.5	26.7
	45°		6.5	10.8	11.4	12.2		12.0	11.8	12.6	13.5		18.0	13.3	14.1	15.5		32	15.5	17.0	18.4
30 x 12 (228.0)	Straight	462	13.9	17.1	18.7	20.2	695	25.0	20.0	22.2	24.0	925	37.0	23.3	26.4	28.8	1388	68	31.1	35.0	37.2
	22½°		10.0	14.4	15.6	16.9		19.0	16.4	17.7	19.4		28.0	19.0	20.7	22.6		51	23.8	26.3	28.8
	45°		7.0	11.2	11.9	12.7		12.0	12.3	13.2	14.1		19.0	13.9	14.8	16.3		34	16.3	18.0	19.3
36 x 12 (275.0)	Straight	560	14.5	18.0	19.8	21.3	836	27.0	21.0	23.6	25.3	1115	39.0	25.7	27.7	30.3	1673	71	33.0	37.3	39.7
	22½°		11.0	15.0	16.3	17.6		20.0	17.1	18.5	20.3		29.0	19.8	21.7	23.7		53	25.0	27.7	30.4
	45°		8.0	11.5	12.2	13.1		13.0	12.7	13.6	14.6		20.0	14.4	15.4	17.0		36	16.9	18.7	20.1
K FACTOR																					
Max Cfm/Sq Ft Outlet Wall Area		29.0					19.0					14.0					9.6				
Min Cfm/Sq Ft Outlet Wall Area		8.7					5.7					4.2					2.9				

TABLE 21—WALL OUTLET RATINGS, FOR COOLING ONLY (Cont.)

For Beamed Ceilings

OUTLET VELOCITY		1000 FPM					1500 FPM					2000 FPM				
STATIC PRESSURE STANDARD OUTLET		Str B = .093, 22½° = .11, 45° = .14					Str B = .211, 22½° = .24, 45° = .32					Str B = .375, 22½° = .42, 45° = .565				
STATIC PRESSURE WITH METERING PLATE		Str B = .33, 22½° = .33, 45° = .475					Str B = .715, 22½° = .74, 45° = 1.15					Str B = .375				
Nom. Size of Outlet (and Free Area)	Vane Setting	Air Quantity (cfm)	Blow (ft)	Temp Diff (F) 15 Min Clg Ht	20	25	Air Quantity (cfm)	Blow (ft)	Temp Diff (F) 15 Min Clg Ht	20	25	Air Quantity (cfm)	Blow (ft)	Temp Diff (F) 15 Min Clg Ht	20	25
12 x 8 (56.7)	Straight	452	52	17.7	19.5	21.4	678	86	21.0	23.5	25.3	904	121	23.9	26.4	28.6
	22½°		39	14.4	15.6	17.0		65	16.6	18.1	20.0		91	18.3	20.0	22.2
	45°		26	10.9	11.8	12.6		43	12.1	13.1	14.2		62	13.2	14.2	15.5
16 x 8 (77.1)	Straight	616	57	19.6	21.7	23.8	926	95	23.4	26.2	28.5	1232	134	26.6	29.4	32.2
	22½°		43	15.7	17.3	18.8		71	18.3	20.2	22.0		100	20.2	22.4	24.6
	45°		29	11.6	12.7	13.5		48	12.9	14.2	15.3		67	14.1	15.6	16.6
20 x 8 (97.6)	Straight	770	62	21.1	23.4	25.6	1150	102	25.1	28.2	31.1	1540	144	28.5	31.8	35.0
	22½°		47	16.8	18.5	20.1		77	19.6	21.7	23.5		108	21.7	24.2	26.4
	45°		31	12.4	13.4	14.5		51	13.8	15.0	16.3		72	15.0	16.5	17.9
24 x 8 (118.0)	Straight	920	65	22.1	24.7	27.0	1384	107	26.4	30.0	32.8	1840	151	29.9	33.6	37.5
	22½°		49	17.6	19.4	21.1		80	20.5	22.8	24.6		113	22.8	25.5	27.7
	45°		33	12.9	14.0	15.1		54	14.3	15.7	16.9		76	15.4	17.3	18.7
30 x 8 (149.0)	Straight	1160	68	23.3	26.0	28.4	1736	111	27.8	31.8	34.8	2320	157	31.4	35.5	40.4
	22½°		51	18.4	20.3	22.1		83	21.5	24.0	25.8		118	24.0	27.0	29.2
	45°		34	13.3	14.5	15.7		56	14.8	16.4	17.6		79	16.3	18.2	19.6
36 x 8 (179.0)	Straight	1404	71	24.3	27.2	29.5	2096	116	29.0	33.4	36.2	2808	164	32.8	37.3	42.2
	22½°		53	19.0	21.2	22.8		87	22.2	24.9	26.6		123	24.8	28.1	30.2
	45°		36	13.8	15.0	16.2		58	15.3	16.9	18.2		82	16.8	18.8	20.2
16 x 10 (97.7)	Straight	792	71	23.9	27.2	29.5	1190	116	28.5	33.4	36.2	1584	164	32.3	37.3	42.2
	22½°		53	19.0	21.2	22.8		87	22.2	24.9	26.6		123	24.8	28.1	30.2
	45°		36	13.8	15.0	16.2		58	15.3	16.9	18.2		82	16.8	18.8	20.2
20 x 10 (124.0)	Straight	994	75	26.0	29.6	32.0	1492	122	30.9	36.0	39.6	1988	174	35.3	40.2	46.6
	22½°		56	20.6	22.7	24.6		92	24.0	26.7	28.7		130	27.0	30.3	32.7
	45°		38	14.6	16.0	17.3		61	16.3	18.1	19.4		87	17.9	20.3	21.7
24 x 10 (150.0)	Straight	1200	80	28.0	32.0	34.2	1798	131	33.0	38.6	42.5	2400	185	38.1	43.4	50.0
	22½°		60	21.8	24.2	26.1		98	25.4	28.6	30.4		139	28.6	32.5	34.8
	45°		40	15.3	16.9	18.1		66	17.1	19.1	20.4		93	18.8	21.5	22.8
30 x 10 (195.0)	Straight	1502	86	30.2	34.6	37.4	2252	139	35.7	41.7	46.5	3004	196	41.7	47.0	54.3
	22½°		65	23.6	26.3	28.2		104	27.4	30.9	32.9		147	31.1	35.4	38.0
	45°		43	16.3	18.2	19.3		70	18.2	20.5	21.8		98	20.1	22.8	24.5
36 x 10 (227.0)	Straight	1808	87	31.3	36.0	38.5	2710	142	36.6	42.9	48.0	3616	200	43.0	49.1	55.8
	22½°		65	24.2	27.0	29.0		106	28.2	31.8	34.0		150	32.0	36.4	39.2
	45°		44	16.6	18.5	19.7		71	18.5	20.8	22.3		100	20.5	23.3	25.0
16 x 12 (118.0)	Straight	976	81	28.0	32.0	34.2	1466	131	33.0	38.6	42.5	1952	158	38.1	43.4	50.0
	22½°		61	21.8	24.2	26.1		98	25.4	28.6	30.4		138	28.6	32.5	34.8
	45°		41	15.3	16.0	18.1		66	17.1	19.1	20.4		93	18.8	21.5	22.8
20 x 12 (150.0)	Straight	1226	87	31.3	36.0	38.5	1836	142	36.6	42.9	48.0	2452	200	43.0	49.1	55.8
	22½°		65	24.2	27.0	29.0		106	28.2	31.8	34.0		150	32.0	36.4	39.2
	45°		44	16.6	18.5	19.7		71	18.5	20.8	22.3		100	20.5	23.3	25.0
24 x 12 (181.0)	Straight	1480	93	33.6	38.6	42.0	2220	153	39.4	46.3	51.5	2960	213	47.0	53.2	59.2
	22½°		70	25.8	28.9	31.0		115	29.9	34.0	36.2		160	34.2	39.2	42.1
	45°		47	17.4	19.5	20.8		77	19.5	22.0	23.6		107	21.7	24.6	26.5
30 x 12 (228.0)	Straight	1850	98	36.7	42.4	46.2	2776	163	42.8	50.0	55.8	3700	226	51.7	59.0	65.0
	22½°		74	27.8	31.1	33.6		122	32.2	36.5	39.1		169	37.0	42.3	45.8
	45°		49	18.4	20.6	22.1		82	20.7	23.4	25.0		113	22.9	26.0	28.1
36 x 12 (275.0)	Straight	2230	103	39.4	45.1	49.8	3346	172	45.7	52.2	58.7	4460	238	55.6	62.0	69.0
	22½°		77	29.4	33.0	35.7		129	33.8	38.4	41.4		178	39.0	44.7	48.7
	45°		52	19.1	21.0	23.0		86	21.5	24.3	26.2		119	23.9	27.2	29.5
K FACTOR																
Max Cfm/Sq Ft Outlet Wall Area		7.2					4.8					3.6				
Min Cfm/Sq Ft Outlet Wall Area		2.2					1.4					1.1				

NOTES:

1. **Divergent Blow** has vertical louvres straight forward in the center, with uniformly increasing angular deflection to a maximum at each end. The 45° divergence signifies an angular deflection at each end of the outlet of 45°, and similarly for 22½° divergence.

2. **Blow** indicates distance from outlet to the point where the air stream is substantially dissipated.

3. **Underblow.** It is not always necessary to blow the entire length of the room unless there are heat load sources at that end, equipment load, open doors, sun-glass, etc. Considering the concentration of room heat load on the basis of Btu/(hr) (sq ft), the outlet blow should cover 75% of the heat load.

4. **Velocity** is based on effective face area.

5. **Static Pressure** is that pressure required to produce the indicated velocities and is measured in inches of water.

6. **Measure ceiling height in the CLEAR only.** This is the distance from the floor to the lowest ceiling beam or obstruction.

7. **The Minimum Ceiling Height** (table) is the minimum ceiling height which will give proper operation of the outlet for the given outlet velocity, vane setting, temperature difference, blow, and cfm. The actual measured ceiling height must be equal to or greater than the minimum ceiling height for the selection made. Preferably the top of an outlet should be not less than twice the outlet's height below the minimum ceiling height.

8. **Cfm/Sq Ft Outlet Wall Area** is the standard for judging total room air movement. The maximum value shown results in an air movement in the zone of occupancy of about 50 fpm. It is assumed that furniture, people, etc., obstruct 10% of the room cross-section. If the obstructions vary widely from 10%, the values of the cfm/sq ft outlet wall area should be tempered accordingly.

9. **For applications requiring a limiting sound level**—the outlet velocity is limited by the sound generated by the outlet.

Part 3
PIPING DESIGN

CHAPTER 1. PIPING DESIGN—GENERAL

Piping characteristics that are common to normal air conditioning, heating and refrigeration systems are presented in this chapter. The areas discussed include piping material, service limitations, expansion, vibration, fittings, valves, and pressure losses. These areas are of prime consideration to the design engineer since they influence the piping life, maintenance cost and first cost.

The basic concepts of fluid flow and design information on the more specialized fields such as high temperature water or low temperature refrigeration systems are not included; this information is available in other authoritative sources.

GENERAL SYSTEM DESIGN

MATERIALS

The materials most commonly used in piping systems are the following:

1. Steel — black and galvanized
2. Wrought iron — black and galvanized
3. Copper — soft and hard

Table 1 illustrates the recommended materials for the various services. Minimum standards, as shown, should be maintained. *Table 2* contains the physical properties of steel pipe and *Table 3* lists the physical properties of copper tubing.

TABLE 1—RECOMMENDED PIPE AND FITTING MATERIALS FOR VARIOUS SERVICES

SERVICE		PIPE	FITTINGS
REFRIGERANTS 12, 22, AND 500	Suction Line	Hard copper tubing, Type L*	Wrought copper, wrought brass or tinned cast brass
		Steel pipe, standard wall Lap welded or seamless for sizes larger than 2 in. IPS	150 lb welding or threaded malleable iron
	Liquid Line	Hard copper tubing, Type L*	Wrought copper, wrought brass or tinned cast brass
		Steel pipe Extra strong wall for sizes 1½ in. IPS and smaller Standard wall for sizes larger than 1½ in. IPS Lap welded or seamless for sizes larger than 2 in. IPS	300 lb welding or threaded malleable iron
	Hot Gas Line	Hard copper tubing, Type L*	Wrought copper, wrought brass or tinned cast brass
		Steel pipe, standard wall Lap welded or seamless for sizes larger than 2 in. IPS	300 lb welding or threaded malleable iron
CHILLED WATER		Black or galvanized steel pipe†	Welding, galvanized; cast, malleable or black iron‡
		Hard copper tubing†	Cast brass, wrought copper or wrought brass
CONDENSER OR MAKE-UP WATER		Galvanized steel pipe†	Welding, galvanized; cast or malleable iron‡
		Hard copper tubing†	Cast brass, wrought copper or wrought brass
DRAIN OR CONDENSATE LINES		Galvanized steel pipe†	Galvanized, drainage; cast or malleable iron‡
		Hard copper tubing†	Cast brass, wrought copper or wrought brass
STEAM OR CONDENSATE		Black steel pipe†	Welding or cast iron‡
		Hard copper tubing†	Cast brass, wrought copper or wrought brass
HOT WATER		Black steel pipe	Welding or cast iron‡
		Hard copper tubing†	Cast brass, wrought copper or wrought brass

*Except for sizes ¼" and ⅜" OD where wall thicknesses of 0.30 and 0.32 in. are required. Soft copper refrigeration tubing may be used for sizes 1⅜" OD and smaller. Mechanical joints must not be used with soft copper tubing in sizes larger than ⅞" OD.

†Normally standard wall steel pipe or Type M hard copper tubing is satisfactory for air conditioning applications. However, the piping material selected should be checked for the design temperature-pressure ratings.

‡Normally 125 lb cast iron and 150 lb malleable iron fittings are satisfactory for the usual air conditioning application. However, the fitting material selected should be checked for the design temperature-pressure ratings.

TABLE 2—PHYSICAL PROPERTIES OF STEEL PIPE

NOM. PIPE SIZE (in.)	SCHEDULE NO.†	OUTSIDE DIAM (in.)	INSIDE DIAM (in.)	WALL THICK-NESS (in.)	WEIGHT OF PIPE (lb/ft)	WT OF WATER IN PIPE* (lb/ft)	OUTSIDE SURFACE (sq ft/ft)	INSIDE SURFACE (sq ft/ft)	TRANS-VERSE AREA (sq in.)
⅛	40(S)	.405	.269	.068	.244	.0246	.106	.0705	.0568
	80(X)	.405	.215	.095	.314	.0157	.106	.0563	.0364
¼	40(S)	.540	.364	.088	.424	.0451	.141	.0955	.1041
	80(X)	.540	.302	.119	.535	.0310	.141	.0794	.0716
⅜	40(S)	.675	.493	.091	.567	.0827	.177	.1295	.1910
	80(X)	.675	.423	.126	.738	.0609	.177	.1106	.1405
½	40(S)	.840	.622	.109	.850	.1316	.220	.1637	.3040
	80(X)	.840	.546	.147	1.087	.1013	.220	.1433	.2340
¾	40(S)	1.050	.824	.113	1.130	.2301	.275	.2168	.5330
	80(X)	1.050	.742	.154	1.473	.1875	.275	.1948	.4330
1	40(S)	1.315	1.049	.133	1.678	.3740	.344	.2740	.8640
	80(X)	1.315	.957	.179	2.171	.3112	.344	.2520	.7190
1¼	40(S)	1.660	1.380	.140	2.272	.6471	.434	.3620	1.495
	80(X)	1.660	1.278	.191	2.996	.5553	.434	.3356	1.283
1½	40(S)	1.900	1.610	.145	2.717	.8820	.497	.4213	2.036
	80(X)	1.900	1.500	.200	3.631	.7648	.497	.3927	1.767
2	40(S)	2.375	2.067	.154	3.652	1.452	.622	.5401	3.355
	80(X)	2.375	1.939	.218	5.022	1.279	.622	.5074	2.953
2½	40(S)	2.875	2.469	.203	5.79	2.072	.753	.6462	4.788
	80(X)	2.875	2.323	.276	7.66	1.834	.753	.6095	4.238
3	40(S)	3.500	3.068	.216	7.57	3.20	.916	.802	7.393
	80(X)	3.500	2.900	.300	10.25	2.86	.916	.761	6.605
3½	40(S)	4.000	3.548	.226	9.11	4.28	1.047	.929	9.89
	80(X)	4.000	3.364	.318	12.51	3.85	1.047	.880	8.89
4	40(S)	4.500	4.026	.237	10.79	5.51	1.178	1.055	12.73
	80(X)	4.500	3.826	.337	14.98	4.98	1.178	1.002	11.50
5	40(S)	5.563	5.047	.258	14.62	8.66	1.456	1.321	20.01
	80(X)	5.563	4.813	.375	20.78	7.87	1.456	1.260	18.19
6	40(S)	6.625	6.065	.280	18.97	12.51	1.735	1.587	28.99
	80(X)	6.625	5.761	.432	28.57	11.29	1.735	1.510	26.07
8	40(S)	8.625	7.981	.322	28.55	21.6	2.26	2.090	50.0
	80(X)	8.625	7.625	.500	43.39	19.8	2.26	2.006	45.6
10	40(S)	10.750	10.020	.365	40.48	34.1	2.81	2.62	78.9
	60(X)	10.750	9.750	.500	54.70	32.4	2.81	2.55	74.7
	80	10.750	9.564	.593	64.33	31.1	2.81	2.50	71.8
12	30(S)	12.750	12.090	.330	43.80	49.6	3.34	3.17	115.0
	40	12.750	11.938	.406	53.53	48.5	3.34	3.13	111.9
	(X)	12.750	11.750	.500	65.40	46.9	3.34	3.08	108.0
	80	12.750	11.376	.687	88.51	44.0	3.34	2.98	101.6
14	30(S)	14.000	13.250	.375	54.60	59.8	3.67	3.46	138.0
	40	14.000	13.125	.438	63.37	58.5	3.67	3.44	135.3
	(X)	14.000	13.000	.500	72.10	55.8	3.67	3.40	133.0
	80	14.000	12.500	.750	106.31	51.2	3.67	3.27	122.7
16	30(S)	16.000	15.250	.375	62.40	79.1	4.18	3.99	183.0
	40(X)	16.000	15.000	.500	82.77	76.5	4.18	3.93	176.7
	80	16.000	14.314	.843	136.46	69.7	4.18	3.75	160.9
18	(S)	18.000	17.250	.375	70.60	100.8	4.71	4.52	234.0
	(X)	18.000	17.000	.500	93.50	98.3	4.71	4.45	227.0
	40	18.000	16.874	.562	104.75	97.2	4.71	4.42	224.0
	80	18.000	16.126	.937	170.75	88.5	4.71	4.22	204.2
20	20(S)	20.000	19.250	.375	78.60	126.7	5.24	5.04	291.0
	30(X)	20.000	19.000	.500	104.20	122.5	5.24	4.97	284.0
	40	20.000	18.814	.593	122.91	120.4	5.24	4.93	278.0
	80	20.000	17.938	1.031	208.87	109.4	5.24	4.70	252.7
24	20(S)	24.000	23.250	.375	94.60	184.6	6.28	6.08	426.0
	(X)	24.000	23.000	.500	125.50	179.0	6.28	6.03	415.0
	40	24.000	22.626	.687	171.17	174.2	6.28	5.92	402.1
	80	24.000	21.564	1.218	296.36	158.2	6.28	5.65	365.2

*To change "Wt of Water in Pipe (lb/ft)" to "Gallons of Water in Pipe (gal/ft)," divide values in table by 8.34.

†S is designation of standard wall pipe.

X is designation of extra strong wall pipe.

TABLE 3—PHYSICAL PROPERTIES OF COPPER TUBING

CLASSIFICATION	NOM. TUBE SIZE (in.)	OUTSIDE DIAM (in.)	STUBBS GAGE	WALL THICK-NESS (in.)	INSIDE DIAM (in.)	TRANS-VERSE AREA (sq in.)	MINIMUM TEST PRESSURE (psi)	WEIGHT OF TUBE (lb/ft)	WT OF WATER IN TUBE* (lb/ft)	OUTSIDE SURFACE (sq ft/ft)
HARD	¼	⅜	23	.025	.325	.083	1000	.106	.036	.098
	⅜	½	23	.025	.450	.159	1000	.144	.069	.131
	½	⅝	22	.028	.569	.254	890	.203	.110	.164
	¾	⅞	21	.032	.811	.516	710	.328	.224	.229
	1	1⅛	20	.035	1.055	.874	600	.464	.379	.295
	1¼	1⅜	19	.042	1.291	1.309	590	.681	.566	.360
Govt. Type "M" 250 Lb Working Pressure	1½	1⅝	18	.049	1.527	1.831	580	.94	.793	.425
	2	2⅛	17	.058	2.009	3.17	520	1.46	1.372	.556
	2½	2⅝	16	.065	2.495	4.89	470	2.03	2.120	.687
	3	3⅛	15	.072	2.981	6.98	440	2.68	3.020	.818
	3½	3⅝	14	.083	3.459	9.40	430	3.58	4.060	.949
	4	4⅛	13	.095	3.935	12.16	430	4.66	5.262	1.08
	5	5⅛	12	.109	4.907	18.91	400	6.66	8.180	1.34
	6	6⅛		.122	5.881	27.16	375	8.91	11.750	1.60
	8	8⅛		.170	7.785	47.6	375	16.46	20.60	2.13
HARD	⅜	½	19	.035	.430	.146	1000	.198	.063	.131
	½	⅝		.040	.545	.233	1000	.284	.101	.164
	¾	⅞		.045	.785	.484	1000	.454	.209	.229
	1	1⅛		.050	1.025	.825	880	.653	.358	.295
	1¼	1⅜		.055	1.265	1.256	780	.882	.554	.360
Govt. Type "L" 250 Lb Working Pressure	1½	1⅝		.060	1.505	1.78	720	1.14	.770	.425
	2	2⅛		.070	1.985	3.094	640	1.75	1.338	.556
	2½	2⅝		.080	2.465	4.77	580	2.48	2.070	.687
	3	3⅛		.090	2.945	6.812	550	3.33	2.975	.818
	3½	3⅝		.100	3.425	9.213	530	4.29	4.000	.949
	4	4⅛		.110	3.905	11.97	510	5.38	5.180	1.08
	5	5⅛		.125	4.875	18.67	460	7.61	8.090	1.34
	6	6⅛		.140	5.845	26.83	430	10.20	11.610	1.60
HARD	¼	⅜	21	.032	.311	.076	1000	.133	.033	.098
	⅜	½	18	.049	.402	.127	1000	.269	.055	.131
	½	⅝	18	.049	.527	.218	1000	.344	.094	.164
	¾	⅞	16	.065	.745	.436	1000	.641	.189	.229
	1	1⅛	16	.065	.995	.778	780	.839	.336	.295
	1¼	1⅜	16	.065	1.245	1.217	630	1.04	.526	.360
Govt. Type "K" 400 Lb Working Pressure	1½	1⅝	15	.072	1.481	1.722	580	1.36	.745	.425
	2	2⅛	14	.083	1.959	3.014	510	2.06	1.300	.556
	2½	2⅝	13	.095	2.435	4.656	470	2.92	2.015	.687
	3	3⅛	12	.109	2.907	6.637	450	4.00	2.870	8.18
	3½	3⅝	11	.120	3.385	8.999	430	5.12	3.890	.949
	4	4⅛	10	.134	3.857	11.68	420	6.51	5.05	1.08
	5	5⅛		.160	4.805	18.13	400	9.67	7.80	1.34
	6	6⅛		.192	5.741	25.88	400	13.87	11.20	1.60
SOFT	¼	⅜	21	.032	.311	.076	1000	.133	.033	.098
	⅜	½	18	.049	.402	.127	1000	.269	.055	.131
	½	⅝	18	.049	.527	.218	1000	.344	.094	.164
	¾	⅞	16	.065	.745	.436	1000	.641	.189	.229
	1	1⅛	16	.065	.995	.778	780	.839	.336	.295
	1¼	1⅜	16	.065	1.245	1.217	630	1.04	.526	.360
Govt. Type "K" 250 Lb Working Pressure	1½	1⅝	15	.072	1.481	1.722	580	1.36	.745	.425
	2	2⅛	14	.083	1.959	3.014	510	2.06	1.300	.556
	2½	2⅝	13	.095	2.435	4.656	470	2.92	2.015	.687
	3	3⅛	12	.109	2.907	6.637	450	4.00	2.870	.818
	3½	3⅝	11	.120	3.385	8.999	430	5.12	3.89	.949
	4	4⅛	10	.134	3.857	11.68	420	6.51	5.05	1.08
	5	5⅛		.160	4.805	18.13	400	9.67	7.80	1.34
	6	6⅛		.192	5.741	25.88	400	13.87	11.2	1.60

*To change "Wt of Water in Tube (lb/ft)" to "Gallons of Water in Tube (gal/ft)," divide values in table by 8.34.

SERVICE LIMITATIONS

The safe working pressure and temperature for steel pipe and copper tubing, including fittings, are limited by the ASA codes. Check these codes when there is doubt about the ability of pipe, tubing, or fittings to withstand pressures and temperatures in a given installation. In many instances cost can be reduced and over-design eliminated. For example, if the working pressure is to be 175 psi at 250 F, a 150 psi, class A, carbon steel flange can be safely used since it can withstand a pressure of 225 psi at 250 F. If the code is not checked, a 300 psi flange must be specified because the 175 psi working pressure exceeds the 150 psi rating of the 150 psi flange.

The safe working pressure and temperature for copper tubing is dependent on the strength of the fittings and tube, the composition of the solder used for making a joint, and on the temperature of the fluid conveyed. *Table 4* indicates recommended service limits for copper tubing.

EXPANSION OF PIPING

All pipe lines which are subject to changes in temperature expand and contract. Where temperature changes are anticipated, piping members capable of absorbing the resultant movement must be included in the design. *Table 5* gives the thermal linear expansion of copper tubing and steel pipe.

There are three methods commonly used to absorb pipe expansion and contraction:

1. *Expansion loops and offsets* — *Table 6, page 6,* shows the copper expansion loop and offset sizes required for expansion travels up to six inches. *Chart 1* shows the sizes of expansion loops made of steel pipe and welding ells for expansion travels up to 10 inches.

TABLE 5—THERMAL LINEAR EXPANSION OF COPPER TUBING AND STEEL PIPE

(Inches per 100 feet)

TEMP RANGE (F)	COPPER TUBING	STEEL PIPE
0	0	0
50	.56	.37
100	1.12	.76
150	1.69	1.15
200	2.27	1.55
250	2.85	1.96
300	3.45	2.38
350	4.05	2.81
400	4.65	3.25
450	5.27	3.70
500	5.89	4.15

NOTE: Above data are based on expansion from 0°F but are sufficiently accurate for all other temperature ranges.

Chart 2 gives the sizes of offsets in steel piping for travels up to 3 inches. Expansion loop sizes may be reduced by cold-springing them into place. The pipe lines are cut short at about 50% of the expansion travel and the expansion loop is then sprung into place. Thus, the loops are subject to only one-half the stress when expanded or contracted.

2. *Expansion joints* — There are two types available, the slip type and the bellows type. The slip type expansion joint has several disadvantages: (a) It requires packing and lubrication, which dictates that it be placed in an accessible location; (b) Guides must be installed in the lines to prevent the pipes from bending and binding in the joint.

Bellows type expansion joints are very satisfactory for short travels, but must be guided or

TABLE 4—RECOMMENDED MAXIMUM SERVICE PRESSURE FOR VARIOUS SOLDER JOINTS

SOLDER USED IN JOINTS	SERVICE TEMP (F)	MAXIMUM SERVICE PRESSURE (PSI)			Steam
		Water			
		¼" to 1⅛" Incl.	1⅜" to 2⅛" Incl.	2⅝" to 4⅛" Incl.	All
50-50 Tin-Lead	100	200	175	150	—
	150	150	125	100	—
	200	100	90	75	—
	250	85	75	50	15
95-5 Tin-Antimony or 95-5 Tin-Lead	100	500	400	300	—
	150	400	350	275	—
	200	300	250	200	—
	250	200	175	150	15
Solders Melting At or Above 1100 F	350	270	190	155	120

Extracted from *American Standard Wrought-Copper and Wrought-Bronze Solder-Joint Fittings,* (ASA B16.22-1951), with the permission of the publisher, The American Society of Mechanical Engineers, 29 West 39th Street, New York 18, New York.

CHART 1—STEEL EXPANSION LOOPS

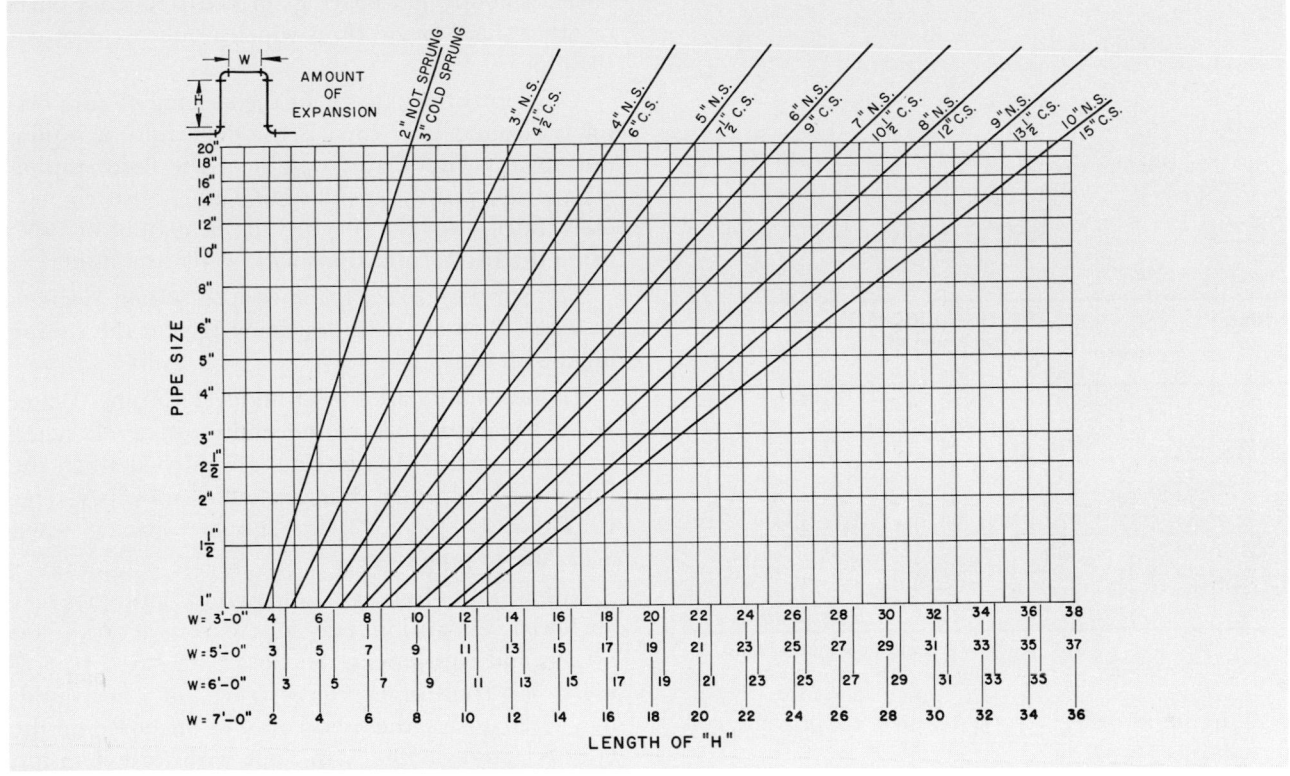

Data from Ric-Wil Co.

CHART 2—STEEL EXPANSION OFFSETS

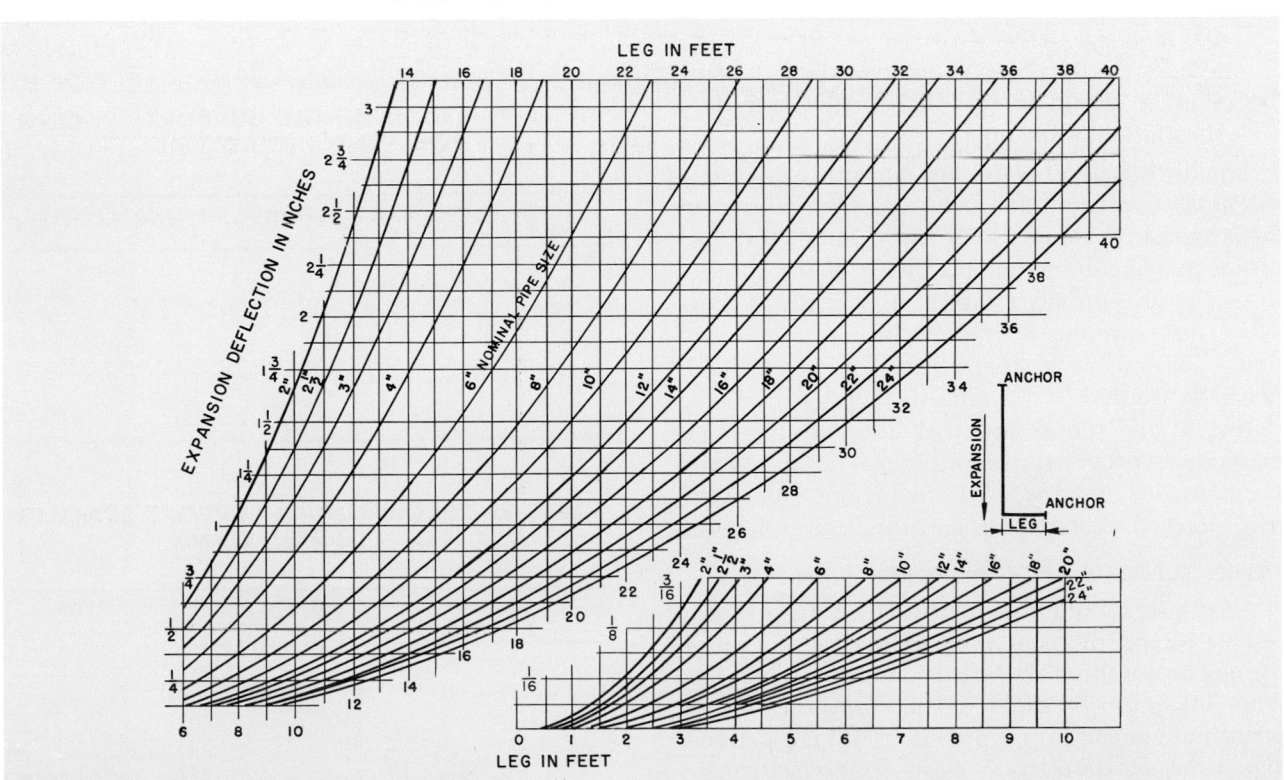

Data from Pittsburgh Pipe Coil & Bending Co.

TABLE 6—COPPER EXPANSION LOOPS AND OFFSETS

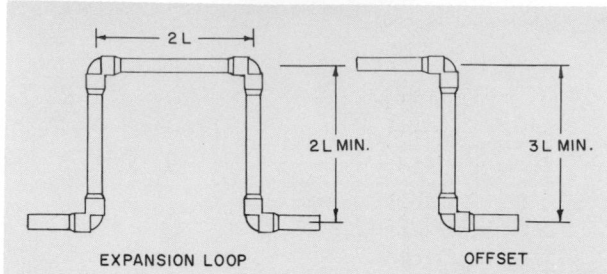

EXPANSION LOOP   OFFSET

TUBE OD (in.)	LENGTH—L (INCHES) For Travel of								
	½"	1"	1½"	2"	2½"	3"	4"	5"	6"
⅞	10	15	19	22	25	27	30	34	38
1⅛	11	16	20	24	27	29	33	38	42
1⅜	11	17	21	26	29	32	36	42	47
1⅝	12	18	23	28	31	35	39	46	51
2⅛	14	20	25	31	34	38	44	51	57
2⅝	16	22	27	32	37	42	47	56	62
3⅛	18	24	30	34	39	45	53	60	67
4⅛	20	28	34	39	44	48	58	66	75
5⅛	22	31	39	44	49	54	62	70	78
6⅛	24	34	42	48	54	59	68	76	83

Data from Mueller Brass Co.

in some other way restrained to prevent collapse.

3. *Flexible metal and rubber hose* — Flexible hose, to absorb expansion, is recommended for smaller size pipe or tubing only. It is not recommended for larger size pipe since an excessive length is required.

Where flexible hose is used to absorb expansion, it should be installed at right angles to the motion of the pipe.

The devices listed above are not always necessary to absorb expansion and contraction of piping. In fact they can be omitted in the great majority of piping systems by taking advantage of the changes in direction normally required in the layout. Consider, for example, a heat exchanger unit and a pump located 50 ft. apart. Sufficient flexibility is normally assured by running the piping up to the ceiling at the pump and back down at the heat exchanger, provided the piping is merely hung from hangers and anchored only at the ends where it is attached to the pump and the heat exchanger.

PIPING SUPPORTS AND ANCHORS

All piping should be supported with hangers that can withstand the combined weight of pipe, pipe fittings, valves, fluid in the pipe, and the insulation. They must also be capable of keeping the pipe in proper alignment when necessary. Where extreme expansion or contraction exists, roll-type hangers

or saddles should be used. The pipe supports must have a smooth, flat bearing surface, free from burrs or other sharp projections which would wear or cut the pipe.

The controlling factor in the spacing of supports for horizontal pipe lines is the deflection of piping due to its own weight, weight of the fluid, piping accessories, and the insulation. *Table 7* lists the recommended support spacing for Schedule 40 pipe, using the listed conditions with water as a fluid.

The support spacing for copper tubing is given in *Table 8* which includes the weight of the tubing filled with water.

Tables 7 and 8 are for "dead level" piping. Water and refrigerant lines are normally run level; steam lines are pitched. Water lines are pitched when the line must be drained. For pitched steam pipes, refer to *Table 25, page 82,* for support spacing when Schedule 40 pipe is used.

Unless pipe lines are adequately and properly anchored, expansion may put excessive strain on fittings and equipment. Anchors are located according to job conditions. For instance, on a tall building, i.e. 20 stories, the risers could be anchored on the 5th floor and on the 15th floor with an expansion device located at the 10th floor. This arrangement allows the riser to expand in both directions from the 5th and 15th floor, resulting in less pipe travel at headers, whether they are located at the top or bottom of the building or in both locations.

TABLE 7—RECOMMENDED SUPPORT SPACING FOR SCHEDULE 40 PIPE

NOMINAL PIPE SIZE (in.)	DISTANCE BETWEEN SUPPORTS (ft)
¾ - 1¼	8
1½ - 2½	10
3 - 3½	12
4 - 6	14
8 - 12	16
14 - 24	20

TABLE 8—RECOMMENDED SUPPORT SPACING FOR COPPER TUBING

TUBE OD (in.)	DISTANCE BETWEEN SUPPORTS (ft)
⅝	6
⅞ - 1⅛	8
1⅜ - 2⅛	10
2⅝ - 5⅛	12
6⅛ - 8⅛	14

On smaller buildings, i.e. 5 stories, risers are anchored but once. Usually this is done near the header, allowing the riser to grow in one direction only, either up or down depending on the header location.

The main point to consider when applying pipe support anchors and expansion joints is that expansion takes place on a temperature change. The greater the temperature change, the greater the expansion. The supports, anchors and guides are applied to restrain the expansion in a desired direction so that trouble does not develop because of negligent design or installation. For example, if a take-off connection from risers or headers is located close to floors, beams or columns as shown in *Fig. 1,* a change in temperature may cause a break in the take-off with subsequent loss of fluid and flooding damage. In this figure trouble develops when the riser expands greater than dimension "X." Proper consideration of these items is a must when designing piping systems.

VIBRATION ISOLATION OF PIPING SYSTEMS

The undesirable effects caused by vibration of the piping are:

1. Physical damage to the piping, which results in the rupture of joints. For refrigerant piping, loss of refrigerant charge results.
2. Transmission of noise thru the piping itself or thru the building construction where piping comes into direct contact.

It is always difficult to anticipate trouble resulting from vibration of the piping system. For this reason, recommendations made toward minimizing the effects of vibration are divided into two categories:

1. *Design consideration* — These involve design precautions that can prevent trouble effectively.

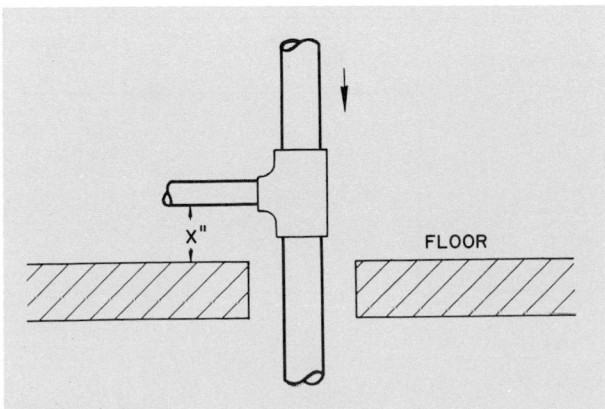

Fig. 1 — Take-off Too Close To Floor

2. *Remedies or repairs* — These are necessary where precautions are not taken initially or, in a minority of cases, where the precautions prove to be insufficient.

Design Considerations for Vibration Isolation

1. In all piping systems vibration has an originating source. This source is usually a moving component such as a water pump or a compressor. When designing to eliminate vibration, the method of supporting these moving components is the prime consideration. For example:

    a. The weight of the mass supporting the components should be heavy enough to minimize the intensity of the vibrations transmitted to the piping and to the surrounding structure. The heavier the stabilizing mass, the smaller the intensity of the vibration.

    b. Vibration isolators can also be used to minimize the intensity of vibration.

    c. A combination of both methods may be required.

2. Piping must be laid out so that none of the lines are subject to the push-pull action resulting from vibration. Push-pull vibration is best dampened in torsion or bending action.

3. The piping must be supported securely at the proper places. The supports should have a relatively wide bearing surface to avoid a swivel action and to prevent cutting action on the pipe.

    The support closest to the source of vibration should be an isolation hanger and the succeeding hangers should have isolation sheaths as illustrated in *Fig. 2, page 8.* Non-isolated hangers (straps or rods attached directly to the pipe) should not be used on piping systems with machinery having moving parts.

4. The piping must not touch any part of the building when passing thru walls, floors, or furring. Sleeves which contain isolating material should be used wherever this is anticipated. Isolation hangers should be used to suspend the piping from walls and ceilings to prevent transmission of vibration to the building.

    Isolation hangers are also used where access to piping is difficult after installation.

5. Flexible hose is often of value in absorbing vibration on smaller sizes of pipe. To be effective, these flexible connectors are installed at right angles to the direction of the vibration.

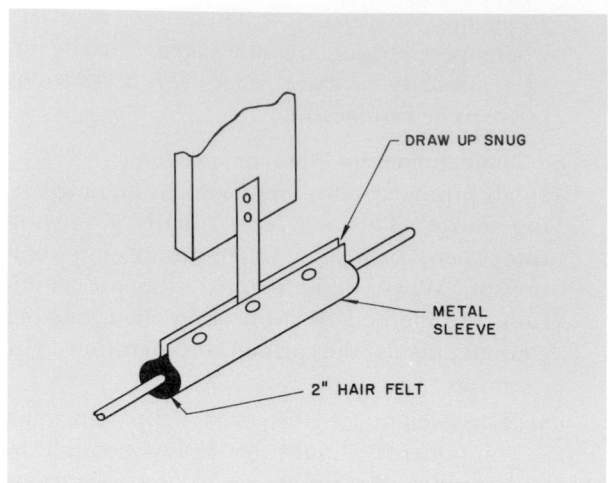

FIG. 2 — ISOLATED SHEATH PIPE HANGER

Where the vibration is not limited to one plane or direction, two flexible connectors are used and installed at right angles to each other. The flexible hose must not restrain the vibrating machine to which it is attached. At the opposite end of the hose or pair of hoses, a rigid but isolated anchor is secured to the connecting pipe to minimize vibration.

Generally, flexible hose is not recommended in systems subject to pressure conditions. Under pressure they become stiff and transmit vibration in the same manner as a straight length of pipe.

Flexible hose is not particularly efficient in absorbing vibration on larger sizes of pipe. Efficiency is impaired since the length-to-diameter ratio must be relatively large to obtain full flexibility from the hose. In practice the length which can be used is often limited, thus reducing its flexibility.

Remedies or Repairs for Vibration Isolation After Installation

1. Relocation of the piping supports by trial and error tends to dampen extraordinary pipe vibration. This relocation allows the piping to take up the vibration in bending and helps to correct any vibrations which cause mechanical resonance.

2. If relocation of the pipe supports does not eliminate the noise problem caused by vibration, there are several possible recommendations:

    a. The pipe may be isolated from the support by means of cork, hair felt, or pipe insulation as shown in *Fig. 2*.

    b. A weight may be added to the pipe before the first fixed support as illustrated in *Fig. 3*. This weight adds mass to the pipe, reducing vibration.

    c. Opposing isolation hangers may be added.

FITTINGS

Elbows are responsible for a large percentage of the pressure drop in the piping system. With equal velocities the magnitude of this pressure drop depends upon the sharpness of the turn. Long radius rather than short radius elbows are recommended wherever possible.

When laying out offsets, 45° ells are recommended over 90° ells wherever possible. See *Fig. 4*.

Tees should be installed to prevent "bullheading" as illustrated in *Fig. 5*. "Bullheading" causes turbulence which adds greatly to the pressure drop and may also introduce hammering in the line. If more than one tee is installed in the line, a straight length of 10 pipe diameters between tees is recommended. This is done to reduce unnecessary turbulence.

To facilitate erection and servicing, unions and flanges are included in the piping system. They are installed where equipment and piping accessories must be removed for servicing.

The various methods of joining fittings to the piping are described on *page 12*.

GENERAL PURPOSE VALVES

An important consideration in the design of the piping system is the selection of valves that give proper performance, long life and low maintenance.

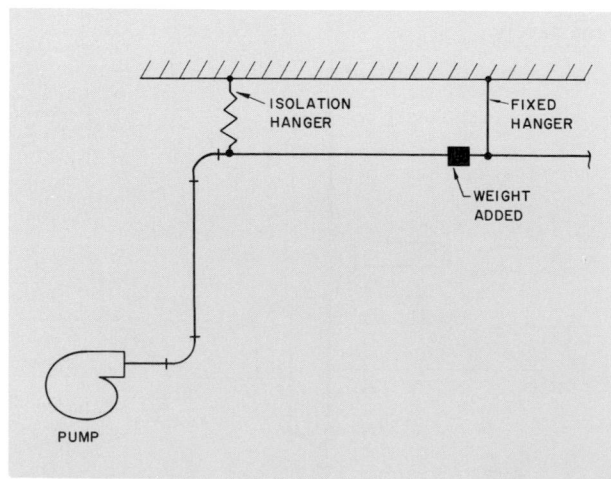

FIG. 3 — WEIGHT ADDED TO DAMPEN VIBRATION

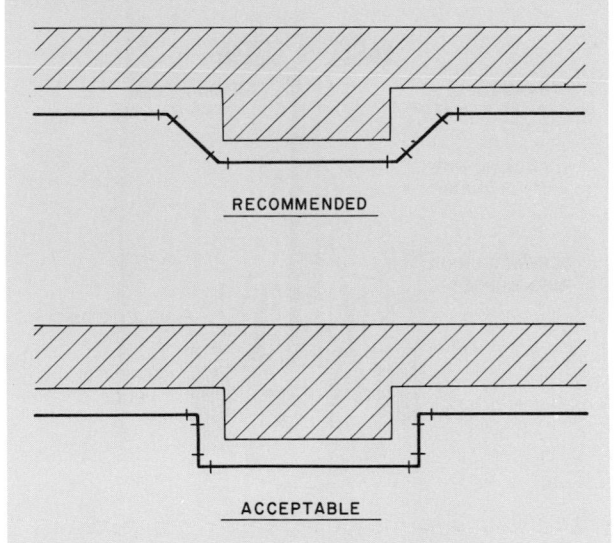

FIG. 4 — OFFSETS TO AVOID OBSTRUCTIONS

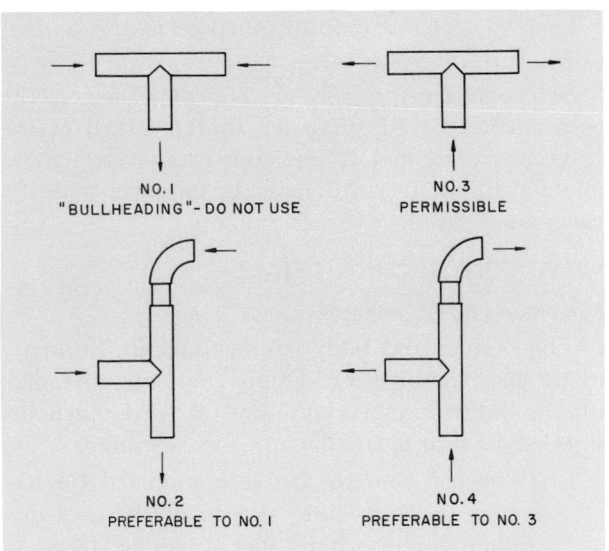

FIG. 5 — TEES

The design, construction and material of the valve determines whether or not it is suited for the particular application. *Table 9* is for quick reference in selecting a valve for a particular application. There

are six basic valves which are commonly used in piping systems. These are gate, globe, check, angle, "Y" and plug cock.

TABLE 9—GENERAL PURPOSE VALVES

	WATER	STEAM	REFRIGERANT*
VALVE CONSTRUCTION			
A. Bonnet and body connections			
1. Threaded	Satisfactory	Satisfactory (Low Press)	Not Recommended
2. Union	Satisfactory	Satisfactory	Not Recommended
3. Bolted	Satisfactory	Satisfactory	Satisfactory
4. Welded	Satisfactory	Satisfactory (High Press)	Satisfactory
5. Pressure-seal	Satisfactory	Satisfactory	Satisfactory
B. Valve stem, operation			
1. Rising stem, outside screw	Satisfactory	Satisfactory	Satisfactory
2. Rising stem, inside screw	Satisfactory	Satisfactory	Satisfactory
3. Non-rising stem, inside screw	Satisfactory (non-corrosive brines)	Not Recommended	Not Recommended
4. Sliding stem	Satisfactory	Satisfactory	Not Recommended
C. Valve connections to pipe			
1. Screwed	Satisfactory	Satisfactory	Not Recommended
2. Welded	Satisfactory	Satisfactory	Recommended
3. Brazed	Satisfactory	Satisfactory (Low Temp)	Recommended
4. Soldered	Satisfactory	Satisfactory (Low Temp)	Satisfactory
5. Flared	Satisfactory	Satisfactory	Satisfactory
6. Flanged	Satisfactory (non-corrosive brines)	Satisfactory	Satisfactory
DISC CONSTRUCTION			
Gate Valve			
1. Solid wedge	Satisfactory	Satisfactory	Not Recommended
2. Split wedge	Satisfactory	Satisfactory	Not Recommended
3. Flexible wedge	Satisfactory	Recommended	Not Recommended
4. Double disc, parallel seat	Satisfactory	Not Recommended	Not Recommended
Globe, Angle, "Y" Valve			
1. Plug disc	Satisfactory	Satisfactory	Satisfactory
2. Conventional (Narrow-seat)	Satisfactory	Not Recommended	Satisfactory
3. Needle valve	Satisfactory	Satisfactory	Satisfactory
4. Composition disc	Satisfactory	Satisfactory (Low Press)	Satisfactory
Plug Cock Valve	Satisfactory	Satisfactory	Not Recommended

*For Refrigerants 12, 22 and 500 only.

Each valve has a definite purpose in the control of fluids in the system.

Before discussing the various valve types, the construction details that are similar in all of the valves are presented. These construction details are made available to familiarize the engineer with the various aspects of valve selection.

VALVE CONSTRUCTION DETAILS

Bonnet and Body Connections

The bonnet and body connections are normally made in five different designs, namely threaded, union, bolted, welded and pressure-seal. Each design has its own particular use and advantage.

1. *Threaded bonnets* are recommended for low pressure service. They should not be used in a piping system where there may be frequent dismantling and reassembly of the valve, or where vibration, shock, and other severe conditions may strain and distort the valve body. Threaded bonnets are economical and very compact. *Fig. 6* illustrates a threaded or screwed-in bonnet and body connection in an angle valve.

2. *Union bonnet* and body construction is illustrated in *Fig. 7*. This type of bonnet is normally not made in sizes above 2 in. because it would require an extremely large wrench to dismantle. A union bonnet connection makes a sturdy, tight joint and is easily dismantled and reassembled.

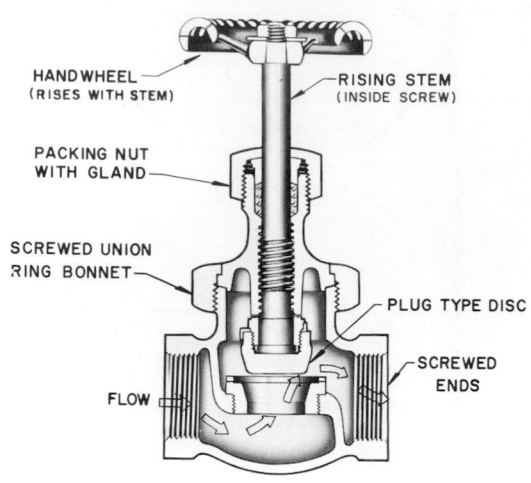

Fig. 7 — Globe Valve

3. *Bolted bonnets* are used on practically all large size valves; they are also available for small sizes. This type of joint is readily taken apart or reassembled. The bolted bonnet is practical for high working pressure and is of rugged, sturdy construction. *Fig. 8* is a gate valve illustrating a typical bolted bonnet and body valve construction.

4. *Welded bonnets* are used on small size steel valves only, and then usually for high pressure, high temperature steam service *(Fig. 9)*. Welded bonnet construction is difficult to dismantle

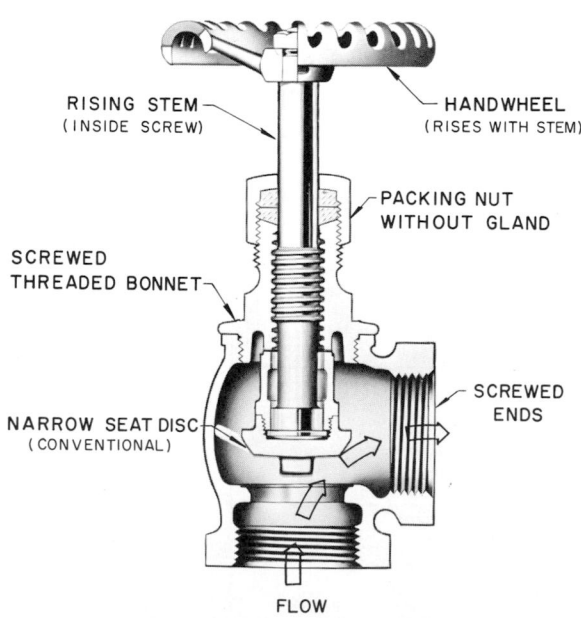

Fig. 6 — Angle Valve

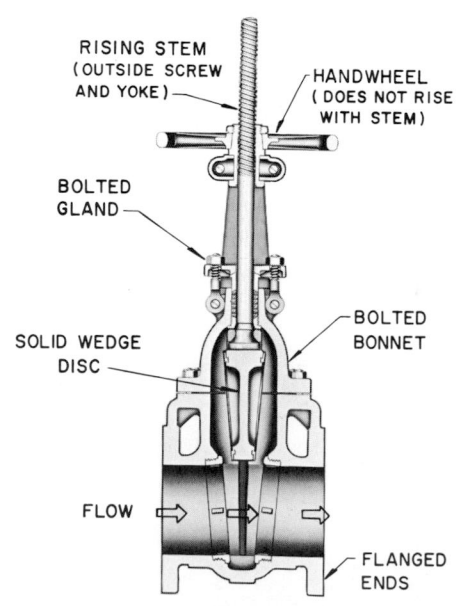

Fig. 8 — Gate Valve (Rising Stem)

Figures 6-10, courtesy of Crane Co.

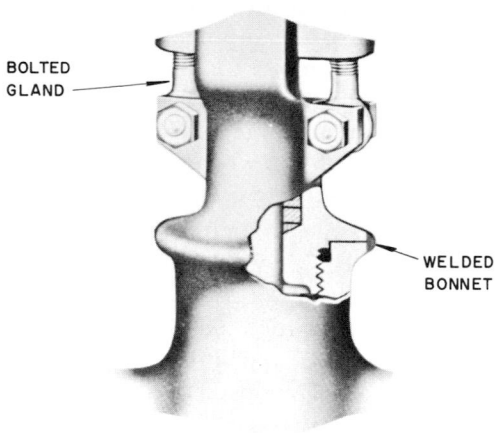

FIG. 9 — WELDED BONNET CONSTRUCTION

and reassemble. For this reason these valves are not available in larger sizes.

5. *Pressure-seal bonnets* are for high temperature steam. *Fig. 10* illustrates a pressure-seal bonnet and body construction used on a gate valve. Internal pressure keeps the bonnet joint tight. This type of bonnet construction simplifies "making" and "breaking" the bonnet joint in large high pressure valves.

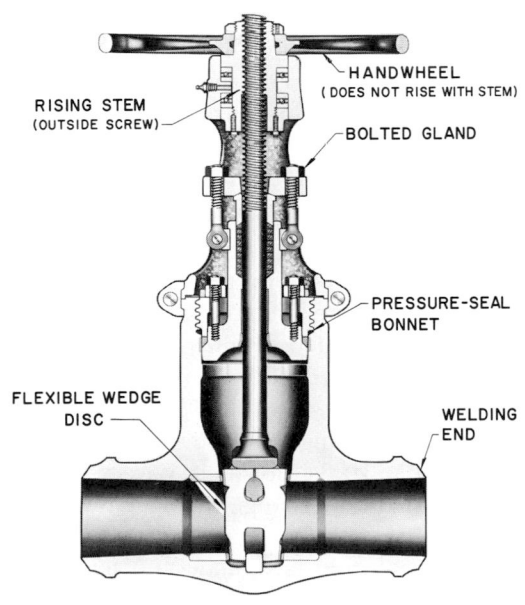

FIG. 10 — FLEXIBLE WEDGE DISC (GATE VALVE)

Valve Stem Operation

In most applications the type of stem operation does not affect fluid control. However, stem construction may be important where the need for indication of valve position is required or where head room is critical. There are four types of stem construction: rising stem with outside screw; rising stem with inside screw; non-rising stem with inside screw; and sliding stem (quick opening).

1. *Rising stem with outside screw* is shown in *Fig. 8.* The gate valve illustrated in this figure has the stem threads outside of the valve body in both the open and closed position. Stem threads are, therefore, not subject to corrosion, erosion, sediment, and galling from extreme temperature changes caused by elements in the line fluid flow. However, since the valve stem is outside the valve body, it is subject to damage when the valve is open. This type of stem is well suited to steam and high temperature, high pressure water service. A rising stem requires more headroom than a non-rising stem. The position of the stem indicates the position of the valve disc. The stem can be easily lubricated since it is outside the valve body.

2. *Rising stem with inside screw* is probably the most common type found in the smaller size valves. This type of stem is illustrated in an angle valve in *Fig. 6,* and in a globe valve in *Fig. 7.* The stem turns and rises on threads inside the valve body. The position of the stem also indicates position of the valve disc. The stem extends beyond the bonnet when the valve is open and, therefore, requires more headroom. In addition it is subject to damage.

3. *Non-rising stem with inside screw* is generally used on gate valves. It is undesirable for use with fluids that may corrode or erode the threads since the stem is in the path of flow. *Fig. 11* shows a gate valve that has a non-rising stem with the threads inside the valve body. The non-rising stem feature makes the valve ideally suited to applications where headroom is limited. Also, the stem cannot be easily damaged. The valve disc position is not indicated with this stem.

4. *Sliding stem (quick opening)* is useful where quick opening and closing is desirable. A lever and sliding stem is used which is suitable for both manual or power operation as illustrated in *Fig. 12.* The handwheel and stem threads are eliminated.

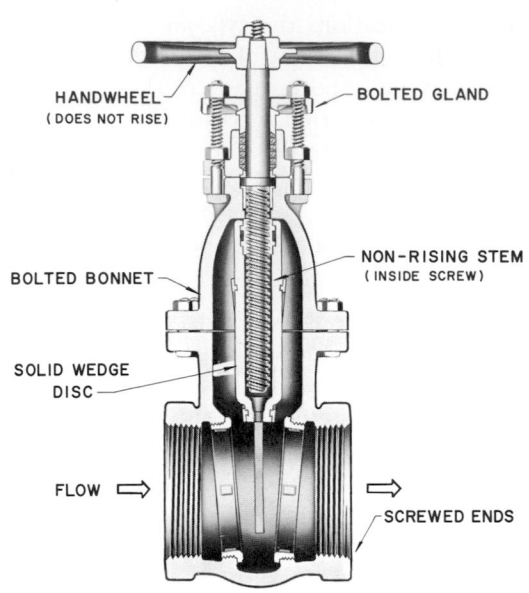

HANDWHEEL
(DOES NOT RISE)

BOLTED GLAND

BOLTED BONNET

NON-RISING STEM
(INSIDE SCREW)

SOLID WEDGE
DISC

FLOW ⇨

⇨

SCREWED ENDS

FIG. 11 — GATE VALVE (NON-RISING STEM)

Pipe Ends and Valve Connections

It is important to specify the proper end connection for valves and fittings. There are six standard methods of joints available. These are screwed, welded, brazed, soldered, flared, and flanged ends, and are described in the following:

1. *Screwed ends* are widely used and are suited for all pressures. To remove screwed end valves

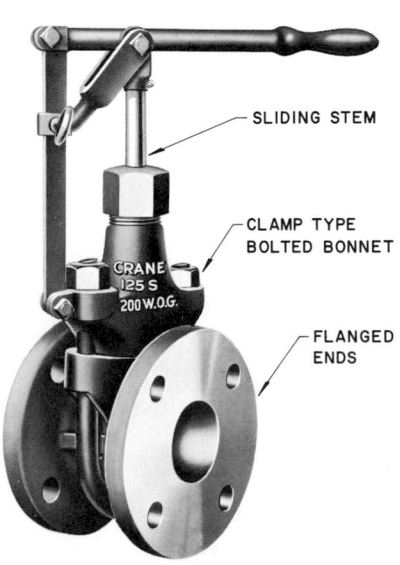

SLIDING STEM

CLAMP TYPE
BOLTED BONNET

CRANE
125 S
200 W.O.G.

FLANGED
ENDS

FIG. 12 — SLIDING STEM GATE VALVE

and fittings from the line, extra fittings (unions) are required to avoid dismantling a considerable portion of the piping. Screwed end connections are normally confined to smaller pipe sizes since it is more difficult to make up the screwed joint on large pipe sizes. *Fig. 7* is a globe valve with screwed ends that connect to pipe or other fittings.

2. *Welded ends* are available for steel pipe, fittings, and valves. They are used widely for all fitting connections, but for valves they are used mainly for high temperature and high pressure services. They are also used where a tight, leak-proof connection is required over a long period of time. The welded ends are available in two designs, butt weld or socket weld. Butt weld valves and fittings come in all sizes; socket weld ends are usually limited to the smaller size valves and fittings. *Fig. 10* illustrates a gate valve with ends suitable for welding.

3. *Brazed ends* are designed for brazing alloys. This type of joint is similar to the solder joint but can withstand higher temperature service because of the higher melting point of the brazing alloy. Brazing joints are used principally on brass valves and fittings.

4. *Soldered ends* for valve and fitting are restricted to copper pipe and also for low pressure service. The use of this type of joint for high temperature service is limited because of the low melting point of the solder.

5. *Flared end* connections for valves and fittings are commonly used on metal and plastic tubing. This type of connection is limited to pipe sizes up to 2 in. Flared connections have the advantage of being easily removed from the piping system at any time.

6. *Flanged ends* are higher in first cost than any of the other end connections. The installation cost is also greater because companion flanges, gaskets, nuts and bolts must be provided and installed. Flanged end connections, although made in small sizes, are generally used in larger size piping because they are easy to assemble and take down. It is very important to have matching flange facing for valves and fittings. Some of the common flange facings are plain face, raised face, male and female joint, tongue and groove joint, and a ring joint. Flange facings should never be mixed in making up a joint. *Fig. 8* illustrates a gate valve with a flanged end.

GATE VALVES

A gate valve is intended for use as a stop valve. It gives the best service when used in the fully open or closed position. *Figures 8 and 10 thru 14* are typical gate valves commonly used in piping practice.

An important feature of the gate valve is that there is less obstruction and turbulence within the valve and, therefore, a correspondingly lower pressure drop than other valves. With the valve wide open, the wedge or disc is lifted entirely out of the fluid stream, thus providing a straight flow area thru the valve.

Disc Construction

Gate valves should not be used for throttling flow except in an emergency. They are not designed for this type of service and consequently it is difficult to control flow with any degree of accuracy. Vibration and chattering of the disc occurs when the valve is used for throttling. This results in damage to the seating surface. Also, the lower edge of the disc may be subject to severe wire drawing effects. The wedge or disc in the gate valve is available in several forms: solid wedge, split wedge, flexible wedge, and double disc parallel seat. These are described in the following:

1. *Solid wedge disc* is the most common type. It has a strong, simple design and only one part. This type of disc is shown in *Figs. 8 and 11*. It can be installed in any position without danger of jamming or misalignment of parts. It is satisfactory for all types of service except where the possibility of extreme temperature changes exist. Under this condition it is subject to sticking.

2. *Split wedge disc* is designed to prevent sticking, but it is subject to undesirable vibration intensity. *Fig. 13* is a typical illustration of this type of disc.

3. *Flexible wedge disc* construction is illustrated in *Fig. 10*. This type of disc is primarily used for high temperature, high pressure applications and where extreme temperature changes are likely to occur. It is solid in the center portion and flexible around the outer edge. This design helps to eliminate sticking and permits the disc to open easily under all conditions.

4. *Double disc parallel seat (Fig. 14)* has an internal wedge between parallel discs. Wedge action damage at the seats is minimized and transferred to the internal wedge where reasonable wear does not prevent tight closure.

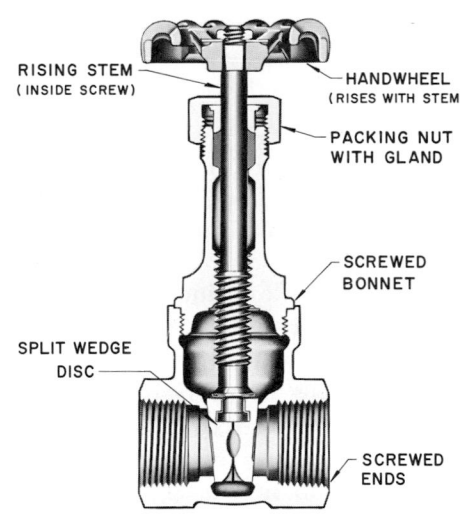

FIG. 13 — SPLIT WEDGE DISC (GATE VALVE)

The parallel sliding motion of the discs tends to clean the seating surfaces and prevents foreign material from being wedged between disc and seat.

Since the discs are loosely supported except when wedged closed, this design is subject to vibration of the disc assembly parts when partially open.

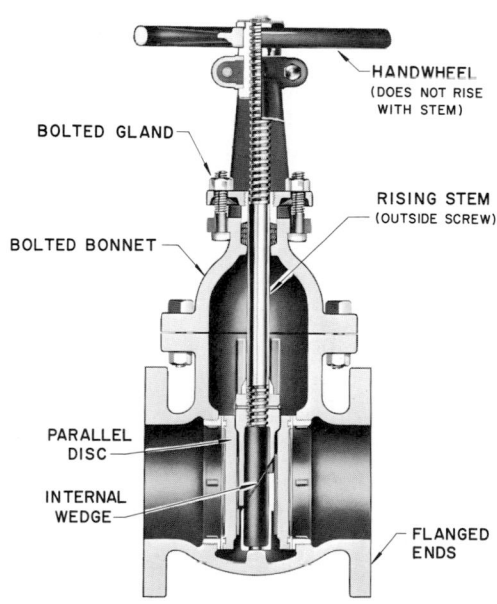

FIG. 14 — DOUBLE DISC PARALLEL SEAT
(GATE VALVE)

Figures 11-14, courtesy of Crane Co.

When used in steam service, the closed valve may trap steam between the discs where it condenses and creates a vacuum. This may result in leakage at the valve seats.

GLOBE, ANGLE AND "Y" VALVES

These three valves are of the same basic design, use and construction. They are primarily intended for throttling service and give close regulation of flow. The method of valve seating reduces wire drawing and seat erosion which is prevalent in gate valves when used for throttling service.

The angle or "Y" valve pattern is recommended for full flow service since it has a substantially lower pressure drop at this condition than the globe valve. Another advantage of the angle valve is that it can be located to replace an elbow, thus eliminating one fitting.

Fig. 7, page 10, is a typical illustration of a globe valve, and *Fig. 6, page 10,* shows an angle valve. The "Y" valve is illustrated in *Fig. 15.*

Globe, angle and "Y" valves can be opened or closed substantially faster than a gate valve because of the shorter lift of the disc. When valves are to be operated frequently or continuously, the globe valve provides the more convenient operation. The seating surfaces of the globe, angle or "Y" valve are less subject to wear and the discs and seats are more easily replaced than on the gate valve.

Disc Construction

There are several different disc and seating arrangements for globe, angle and "Y" valves, each of which has its own use and advantage. The different types are plug disc, narrow seat (or conventional disc), needle valve, and composition disc.

1. The *plug disc* has a wide bearing surface on a long, tapered disc and matching seat. This type of construction offers the greatest resistance to the cutting effects of dirt, scale and other foreign matter. The plug type disc is ideal for the toughest flow control service such as throttling, drip and drain lines, blow-off, and boiler feed lines. It is available in a wide variety of pressure temperature ranges. *Fig. 7, page 10,* shows a plug disc seating arrangement in a globe valve.

2. *Narrow seat* (or conventional disc) is illustrated in an angle valve in *Fig. 6.* This type of disc does not resist wire drawing or erosion in closely throttled high velocity flow. It is also subject to erosion from hard particles. The narrow seat disc design is not applicable for close throttling.

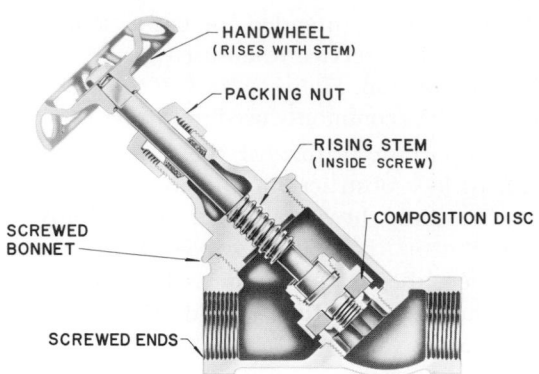

Courtesy of Jenkins Bros.

FIG. 15 — "Y" VALVE

3. *Needle valves,* sometimes referred to as expansion valves, are designed to give fine control of flow in small diameter piping. The disc is normally an integral part of the stem and has a sharp point which fits into the reduced area seat opening. *Fig. 16* is an angle valve with a needle type seating arrangement.

4. *Composition disc* is adaptable to many services by simply changing the material of the disc. It has the advantage of being able to seat tight with less power than the metal type discs. It is also less likely to be damaged by dirt or foreign material than the metal disc. A composition disc is suitable to all moderate pressure services but not for close regulating and throttling. *Fig. 15* shows a composition disc in a "Y" valve. This type of seating design is also illustrated in *Fig. 19* in a swing check valve and in *Fig. 20* in a lift check valve.

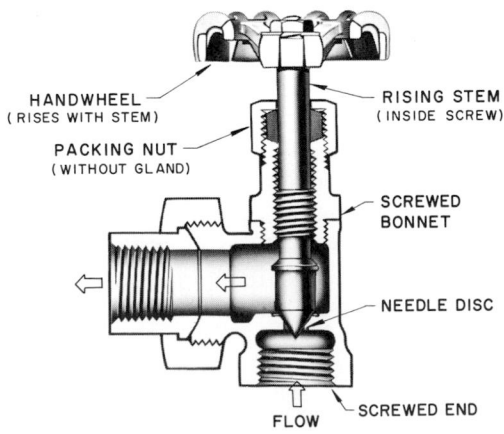

FIG. 16 — ANGLE VALVE WITH NEEDLE DISC

PLUG COCKS

Plug cocks are primarily used for balancing in a piping system not subject to frequent changes in flow. They are normally less expensive than globe type valves and the setting cannot be tampered with as easily as a globe valve.

Plug cocks have approximately the same line loss as a gate valve when in the fully open position. When partially closed for balancing, this line loss increases substantially. *Fig. 17* is a lubricated type plug valve.

REFRIGERANT VALVES

Refrigerant valves are back-seating globe valves of either the packed or diaphragm packless type. The packed valves are available with either a hand wheel or a wing type seal cap. The wing type seal cap is preferable since it provides the safety of an additional seal.

Where frequent operation of the valve is required, the diaphragm packless type is used. The diaphragm acts as a seal and is illustrated in the "Y" valve construction in *Fig. 18*. The refrigerant valve is available in sizes up to 1⅝ in. OD. For larger sizes the seal cap type packed valve is used.

CHECK VALVES

There are two basic designs of check valves, the swing check and the lift check.

The swing check valve may be used in a horizontal or a vertical line for upward flow. A typical swing check valve is illustrated in *Fig. 19*. The flow thru the swing check is in a straight line and without restriction at the seat. Swing checks are generally used in combination with gate valves.

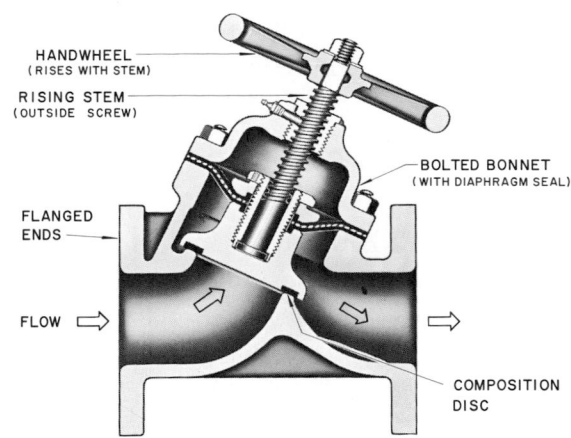

FIG. 18 – "Y" VALVE (DIAPHRAGM TYPE)

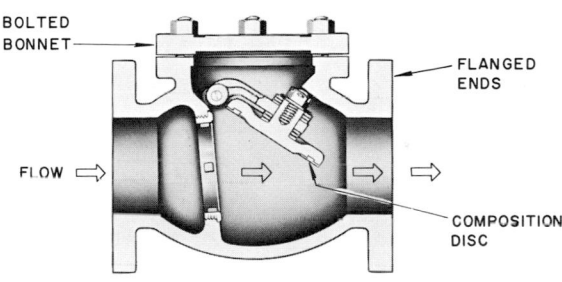

FIG. 19 – SWING CHECK VALVE

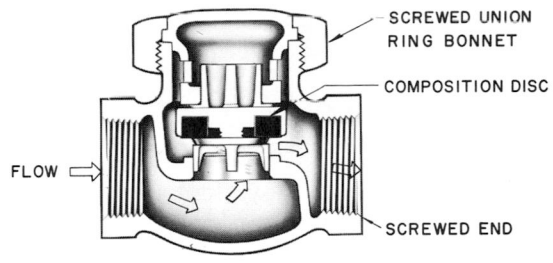

FIG. 20 – LIFT CHECK VALVE

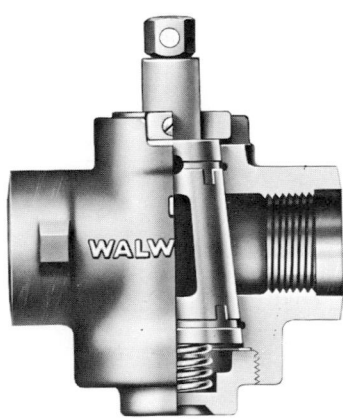

Courtesy of Walworth Co.

FIG. 17 – PLUG COCK

The lift check operates in a manner similar to that of a globe valve and, like the globe valve, its flow is restricted as illustrated in *Fig. 20*. The disc is seated by backflow or by gravity when there is no flow, and is free to rise and fall, depending on the pressure under it. The lift check should only be installed in horizontal pipe lines and usually in combination with globe, angle and "Y" valves.

Figures 16, 18-20, courtesy of Crane Co.

SPECIAL SERVICE VALVES

There are several types of valves commonly used in different piping applications that do not necessarily fall into the classification of general purpose valves. Expansion, relief, and solenoid valves are some of the more common special purpose valves.

A relief valve is held closed by a spring or some other means and is designed to automatically relieve the line or container pressure in excess of its setting. In general a relief valve should be installed wherever there is any danger of the fluid pressure rising above the design working pressure of the pipe fittings or pressure vessels.

VALVE AND FITTING PRESSURE LOSSES

To properly design any type of piping system conveying a fluid, the losses thru the valves and fittings in the system must be realistically evaluated. Tables have been prepared for determining these losses in terms of equivalent length of pipe. These values are then used with the correct friction chart for the particular fluid flowing thru the system.

Table 10 gives valve losses with screwed, flanged, flared, welded, soldered, or brazed connections.

Table 11 gives fitting losses with screwed, flanged, flared, welded, soldered, or brazed connections.

Table 12 lists the losses for special types of fittings sometimes encountered in piping applications.

TABLE 10—VALVE LOSSES IN EQUIVALENT FEET OF PIPE*
Screwed, Welded, Flanged, and Flared Connections

NOMINAL PIPE OR TUBE SIZE (in.)	GLOBE†	60° - Y	45° - Y	ANGLE†	GATE††	SWING CHECK‡	LIFT CHECK
⅜	17	8	6	6	0.6	5	
½	18	9	7	7	0.7	6	
¾	22	11	9	9	0.9	8	
1	29	15	12	12	1.0	10	Globe & Vertical Lift Same as Globe Valve**
1¼	38	20	15	15	1.5	14	
1½	43	24	18	18	1.8	16	
2	55	30	24	24	2.3	20	
2½	69	35	29	29	2.8	25	
3	84	43	35	35	3.2	30	
3½	100	50	41	41	4.0	35	
4	120	58	47	47	4.5	40	
5	140	71	58	58	6	50	
6	170	88	70	70	7	60	
8	220	115	85	85	9	80	
10	280	145	105	105	12	100	
12	320	165	130	130	13	120	Angle Lift Same as Angle Valve
14	360	185	155	155	15	135	
16	410	210	180	180	17	150	
18	460	240	200	200	19	165	
20	520	275	235	235	22	200	
24	610	320	265	265	25	240	

*Losses are for all valves in fully open position.

†These losses do not apply to valves with needle point type seats.

‡Losses also apply to the in-line, ball type check valve.

**For "Y" pattern globe lift check valve with seat approximately equal to the nominal pipe diameter, use values of 60° "Y" valve for loss.

††Regular and short pattern plug cock valves, when fully open, have same loss as gate valve. For valve losses of short pattern plug cocks above 6 ins. check manufacturer.

TABLE 11—FITTING LOSSES IN EQUIVALENT FEET OF PIPE
Screwed, Welded, Flanged, Flared, and Brazed Connections

NOMINAL PIPE OR TUBE SIZE (in.)	SMOOTH BEND ELBOWS						SMOOTH BEND TEES			
	90° Std*	90° Long Rad.†	90° Street*	45° Std*	45° Street*	180° Std*	Flow-Thru Branch	Straight-Thru Flow		
								No Reduction	Reduced ¼	Reduced ½
⅜	1.4	0.9	2.3	0.7	1.1	2.3	2.7	0.9	1.2	1.4
½	1.6	1.0	2.5	0.8	1.3	2.5	3.0	1.0	1.4	1.6
¾	2.0	1.4	3.2	0.9	1.6	3.2	4.0	1.4	1.9	2.0
1	2.6	1.7	4.1	1.3	2.1	4.1	5.0	1.7	2.3	2.6
1¼	3.3	2.3	5.6	1.7	3.0	5.6	7.0	2.3	3.1	3.3
1½	4.0	2.6	6.3	2.1	3.4	6.3	8.0	2.6	3.7	4.0
2	5.0	3.3	8.2	2.6	4.5	8.2	10	3.3	4.7	5.0
2½	6.0	4.1	10	3.2	5.2	10	12	4.1	5.6	6.0
3	7.5	5.0	12	4.0	6.4	12	15	5.0	7.0	7.5
3½	9.0	5.9	15	4.7	7.3	15	18	5.9	8.0	9.0
4	10	6.7	17	5.2	8.5	17	21	6.7	9.0	10
5	13	8.2	21	6.5	11	21	25	8.2	12	13
6	16	10	25	7.9	13	25	30	10	14	16
8	20	13	—	10	—	33	40	13	18	20
10	25	16	—	13	—	42	50	16	23	25
12	30	19	—	16	—	50	60	19	26	30
14	34	23	—	18	—	55	68	23	30	34
16	38	26	—	20	—	62	78	26	35	38
18	42	29	—	23	—	70	85	29	40	42
20	50	33	—	26	—	81	100	33	44	50
24	60	40	—	30	—	94	115	40	50	60

NOMINAL PIPE OR TUBE SIZE (in.)	MITRE ELBOWS			
	90° Ell	60° Ell	45° Ell	30° Ell
⅜	2.7	1.1	0.6	0.3
½	3.0	1.3	0.7	0.4
¾	4.0	1.6	0.9	0.5
1	5.0	2.1	1.0	0.7
1¼	7.0	3.0	1.5	0.9
1½	8.0	3.4	1.8	1.1
2	10	4.5	2.3	1.3
2½	12	5.2	2.8	1.7
3	15	6.4	3.2	2.0
3½	18	7.3	4.0	2.4
4	21	8.5	4.5	2.7
5	25	11	6.0	3.2
6	30	13	7.0	4.0
8	40	17	9.0	5.1
10	50	21	12	7.2
12	60	25	13	8.0
14	68	29	15	9.0
16	78	31	17	10
18	85	37	19	11
20	100	41	22	13
24	115	49	25	16

*R/D approximately equal to 1. †R/D approximately equal to 1.5.

TABLE 12—SPECIAL FITTING LOSSES IN EQUIVALENT FEET OF PIPE

NOM. PIPE OR TUBE SIZE (in.)	SUDDEN ENLARGEMENT* d/D			SUDDEN CONTRACTION* d/D			SHARP EDGE*		PIPE PROJECTION*	
	¼	½	¾	¼	½	¾	Entrance	Exit	Entrance	Exit
⅜	1.4	0.8	0.3	0.7	0.5	0.3	1.5	.8	1.5	1.1
½	1.8	1.1	0.4	0.9	0.7	0.4	1.8	1.0	1.8	1.5
¾	2.5	1.5	0.5	1.2	1.0	0.5	2.8	1.4	2.8	2.2
1	3.2	2.0	0.7	1.6	1.2	0.7	3.7	1.8	3.7	2.7
1¼	4.7	3.0	1.0	2.3	1.8	1.0	5.3	2.6	5.3	4.2
1½	5.8	3.6	1.2	2.9	2.2	1.2	6.6	3.3	6.6	5.0
2	8.0	4.8	1.6	4.0	3.0	1.6	9.0	4.4	9.0	6.8
2½	10	6.1	2.0	5.0	3.8	2.0	12	5.6	12	8.7
3	13	8.0	2.6	6.5	4.9	2.6	14	7.2	14	11
3½	15	9.2	3.0	7.7	6.0	3.0	17	8.5	17	13
4	17	11	3.8	9.0	6.8	3.8	20	10	20	16
5	24	15	5.0	12	9.0	5.0	27	14	27	20
6	29	22	6.0	15	11	6.0	33	19	33	25
8	—	25	8.5	—	15	8.5	47	24	47	35
10	—	32	11	—	20	11	60	29	60	46
12	—	41	13	—	25	13	73	37	73	57
14	—	—	16	—	—	16	86	45	86	66
16	—	—	18	—	—	18	96	50	96	77
18	—	—	20	—	—	20	115	58	115	90
20	—	—	—	—	—	—	142	70	142	108
24	—	—	—	—	—	—	163	83	163	130

*Enter table for losses at smallest diameter "d."

CHAPTER 2. WATER PIPING

This chapter presents the principles and currently accepted design techniques for water piping systems used in air conditioning applications. It also includes the various piping arrangements for air conditioning equipment and the standard accessories found in most water piping systems.

The principles and techniques described are applicable to chilled water and hot water heating systems. General piping principles and techniques are described in *Chapter 1*.

WATER PIPING SYSTEMS

Once-Thru and Recirculating

The water piping systems discussed here are divided into once-thru and recirculating types. In a once-thru system water passes thru the equipment only once and is discharged. In a recirculating system water is not discharged, but flows in a repeating circuit from the heat exchanger to the refrigeration equipment and back to the heat exchanger.

Open and Closed

Both types are further classified as open or closed systems. An open system is one in which the water flows into a reservoir open to the atmosphere; cooling towers and air washers are examples of reservoirs open to the atmosphere. A closed system is one in which the flow of water is not exposed to the atmosphere at any point. This system usually contains an expansion tank that is open to the atmosphere but the water area exposed is insignificant.

Water Return Arrangements

The recirculating system is further classified according to water return arrangements. When two or more units are piped together, one of the following piping arrangements may be used:

1. Reverse return piping.
2. Reverse return header with direct return risers.
3. Direct return piping.

If the units have the same or nearly the same pressure drop thru them, one of the reverse return methods of piping is recommended. However, if the units have different pressure drops or require balancing valves, then it is usually more economical to use a direct return.

Reverse return piping is recommended for most closed piping applications; it cannot be used on open systems. It is often the most economical design

on new construction. The length of the water circuit thru the supply and return piping is the same for all units. Since the water circuits are equal for each unit, the major advantage of a reverse return system is that it seldom requires balancing. *Fig. 21* is a schematic sketch of this system with units piped horizontally and vertically.

There are installations where it is both inconvenient and economically unsound to use a complete reverse return water piping system. This sometimes exists in a building where the first floor has previously been air conditioned. To avoid disturbing the first floor occupants, reverse return headers are located at the top of the building and direct return risers to the units are used. *Fig. 22* illustrates a reverse return header and direct return riser piping system.

In this system the flow rate is not equal for all units on a direct return riser. The difference in flow rate depends on the design pressure drop of the supply and return riser. This difference can be reduced to practical limits. The pressure drop across the riser includes the following: (1) the loss thru the supply and return runouts from the riser to the unit, (2) the loss thru the unit itself, and (3) the loss thru

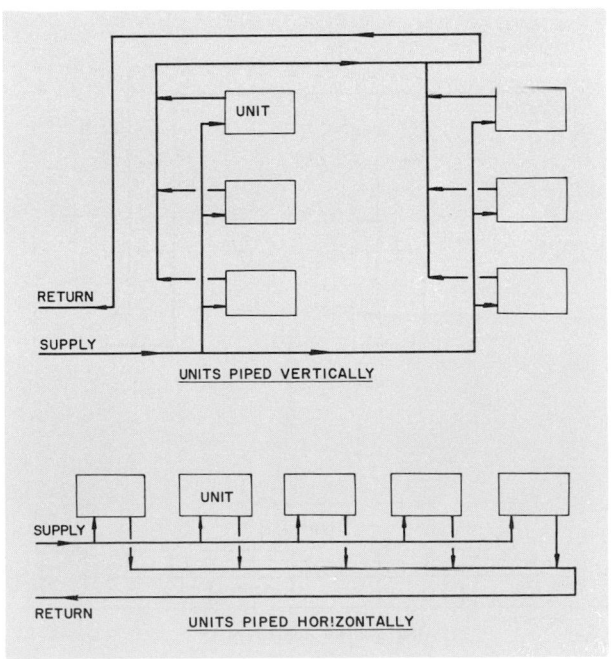

Fig. 21 — Reverse Return Piping

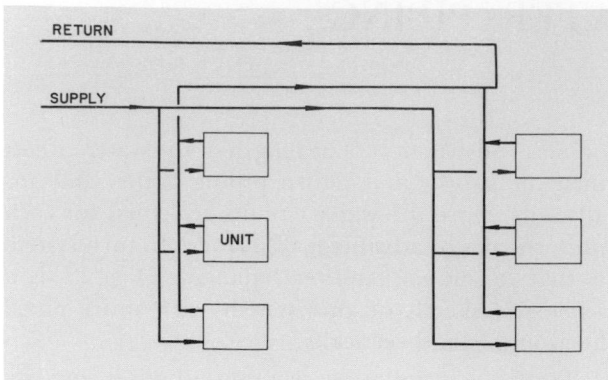

FIG. 22 — REVERSE RETURN HEADERS WITH DIRECT
RETURN RISERS

the fittings and valves. Excessive unbalance in the direct supply and return portion of the piping system may dictate the need for balancing valves or orifices.

To eliminate balancing valves, design the supply and return pressure drop equal to one-fourth the sum of the pressure drops of the preceding *Items 1, 2 and 3.*

Direct return piping is necessary for open piping systems and is recommended for some closed piping systems. A reverse return arrangement on an open system requires piping that is normally unnecessary, since the same atmospheric conditions exist at all open points of the system. A direct return is recom-

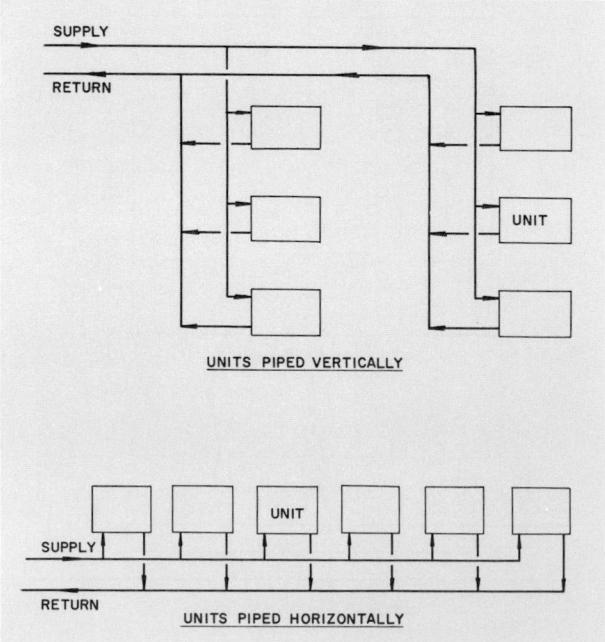

UNITS PIPED VERTICALLY

UNIT

SUPPLY

RETURN

UNITS PIPED HORIZONTALLY

FIG. 23 — DIRECT RETURN WATER PIPING SYSTEM

mended for a closed recirculating system where all the units require balancing valves and have different pressure drops. Several fan-coil units piped together and requiring different water flow rates, capacities and pressure drops is an example of this type of system.

The direct return piping system is inherently unbalanced and requires balancing valves or orifices, and provisions to measure the pressure drop in order to meter the water flow. Although material costs are lower in this system than in the two reverse return systems, engineering cost and balancing time often offset this advantage. *Fig. 23* illustrates units piped vertically and horizontally to a direct return.

CODES AND REGULATIONS

All applicable codes and regulations should be checked to determine acceptable piping practice for the particular application. Sometimes these codes and regulations dictate piping design, limit the pressure, or qualify the selection of materials and equipment.

WATER CONDITIONING

Normally all water piping systems must have adequate treatment to protect the various components against corrosion, scale, slime and algae. Water treatment should always be under the supervision of a water conditioning specialist. Periodic inspection of the water is required to maintain suitable quality. *Part 5* of this manual contains a discussion of the various aspects of water conditioning including cause, effect and remedies for corrosion, scale, slime and algae.

WATER PIPING DESIGN

There is a friction loss in any pipe thru which water is flowing. This loss depends on the following factors:

1. Water velocity
2. Pipe diameter
3. Interior surface roughness
4. Pipe length

System pressure has no effect on the head loss of the equipment in the system. However, higher than normal system pressures may dictate the use of heavier pipe, fittings and valves along with specially designed equipment.

To properly design a water piping system, the engineer must evaluate not only the pipe friction loss but the loss thru valves, fittings and other equipment. In addition to these friction losses, the use of diversity in reducing the water quantity and pipe size is to be considered in designing the water piping system.

PIPE FRICTION LOSS

The pipe friction loss in a system depends on water velocity, pipe diameter, interior surface roughness and pipe length. Varying any one of these factors influences the total friction loss in the pipe.

Most air conditioning applications use either steel pipe or copper tubing in the piping system. To evaluate the friction loss in steel pipe or copper tubing, refer to *Charts 3 thru 5* in this chapter.

Charts 3 and 4 are for Schedule 40 pipe up to 24 in. in diameter. *Chart 3* shows the friction losses for closed recirculation piping systems. The friction losses in *Chart 4* are for open once-thru and for open recirculation piping systems.

Chart 5 shows friction losses for Types K, L and M copper tubing when used in either open or closed water systems.

These charts show water velocity, pipe or tube diameter, and water quantity, in addition to the friction rate per 100 ft of equivalent pipe length. Knowing any two of these factors, the other two can be easily determined from the chart. The effect of inside roughness of the pipe or tube is considered in all these values.

The water quantity is determined from the air conditioning load and the water velocity by predetermined recommendations. These two factors are used to establish pipe size and friction rate.

Water Velocity

The velocities recommended for water piping depend on two conditions:

1. The service for which the pipe is to be used.
2. The effects of erosion.

Table 13 lists recommended velocity ranges for different services. The design of the water piping system is limited by the maximum permissible flow velocity. The maximum values listed in *Table 13* are based on established permissible sound levels of moving water and entrained air, and on the effects of erosion.

Erosion in water piping systems is the impingement on the inside surface of tube or pipe of rapidly moving water containing air bubbles, sand or other solid matter. In some cases this may mean complete deterioration of the tube or pipe walls, particularly on the bottom surface and at the elbows.

Since erosion is a function of time, water velocity, and suspended materials in the water, the selection of a design water velocity is a matter of judgment.

TABLE 13—RECOMMENDED WATER VELOCITY

SERVICE	VELOCITY RANGE (fps)
Pump discharge	8 - 12
Pump suction	4 - 7
Drain line	4 - 7
Header	4 - 15
Riser	3 - 10
General service	5 - 10
City water	3 - 7

The maximum water velocities presented in *Table 14* are based on many years of experience and they insure the attainment of optimum equipment life under normal conditions.

Friction Rate

The design of a water piping system is limited by the friction loss. Systems using city water must have the piping sized so as to provide the required flow rate at a pressure loss within the pressure available at the city main. This pressure or friction loss is to include all losses in the system, as condenser pressure drop, pipe and fitting losses, static head, and water meter drop. The total system pressure drop must be less than the city main pressure to have design water flow.

A recirculating system is sized to provide a reasonable balance between increased pumping horsepower due to high friction loss and increased piping first cost due to large pipe sizes. In large air conditioning applications this balance point is often taken as a maximum friction rate of 10 ft of water per 100 ft of equivalent pipe length.

In the average air conditioning application the installed cost of the water piping exceeds the cost of the water pumps and motors. The cost of increasing the pipe size of small pipe to reduce the friction rate is normally not too great, whereas the installed cost increases rapidly when the size of large pipe (approximately 4 in. and larger) is increased. Smaller pipes can be economically sized at lower friction rates (increasing the pipe size) than the larger pipes. In most applications economic considerations dictate that larger pipe be sized for higher flow rates

TABLE 14—MAXIMUM WATER VELOCITY TO MINIMIZE EROSION

NORMAL OPERATION (hr/yr)	WATER VELOCITY (fps)
1500	12
2000	11.5
3000	11
4000	10
6000	9
8000	8

CHART 3—FRICTION LOSS FOR CLOSED PIPING SYSTEMS
Schedule 40 Pipe

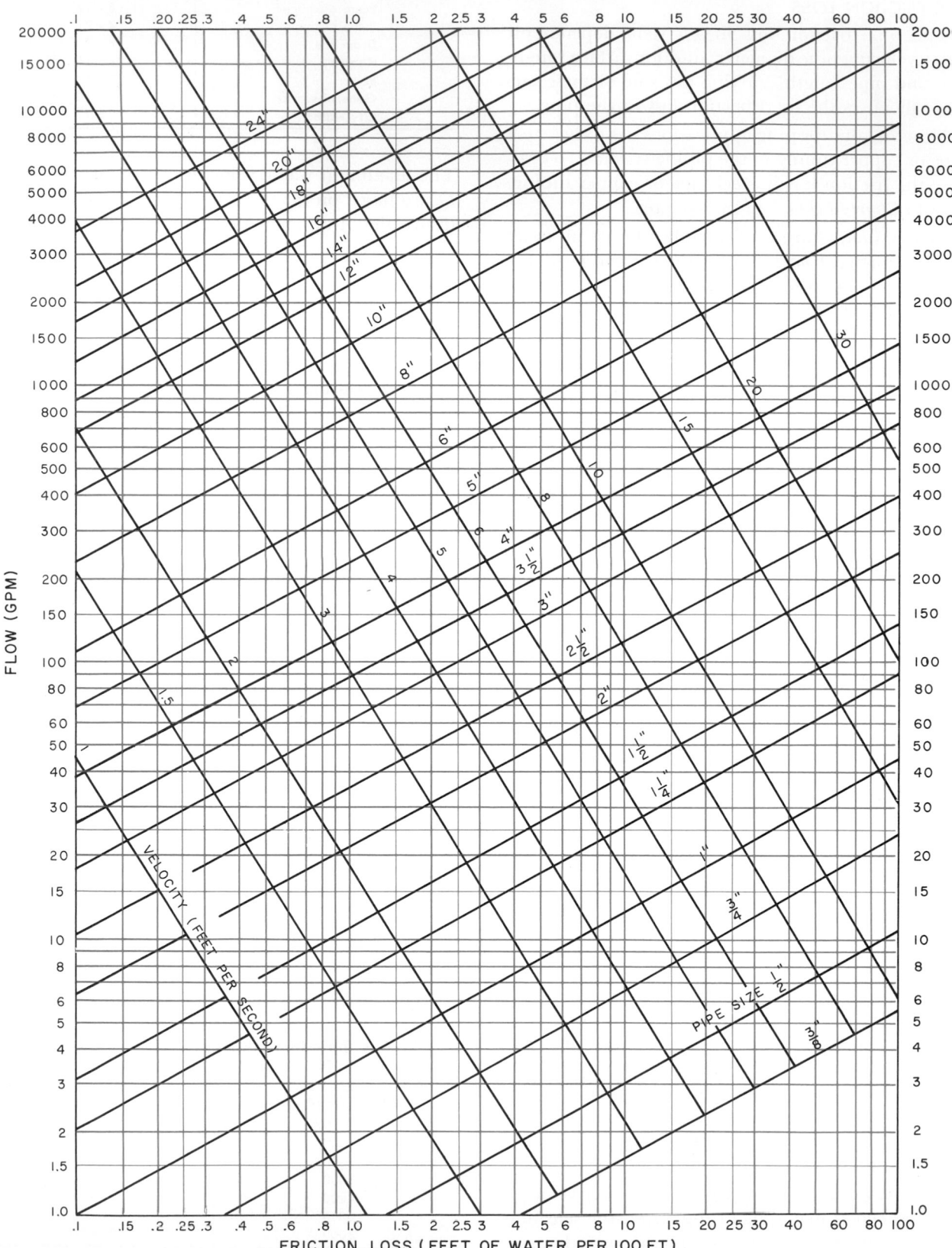

FLOW (GPM)

FRICTION LOSS (FEET OF WATER PER 100 FT)

CHART 4—FRICTION LOSS FOR OPEN PIPING SYSTEMS
Schedule 40 Pipe

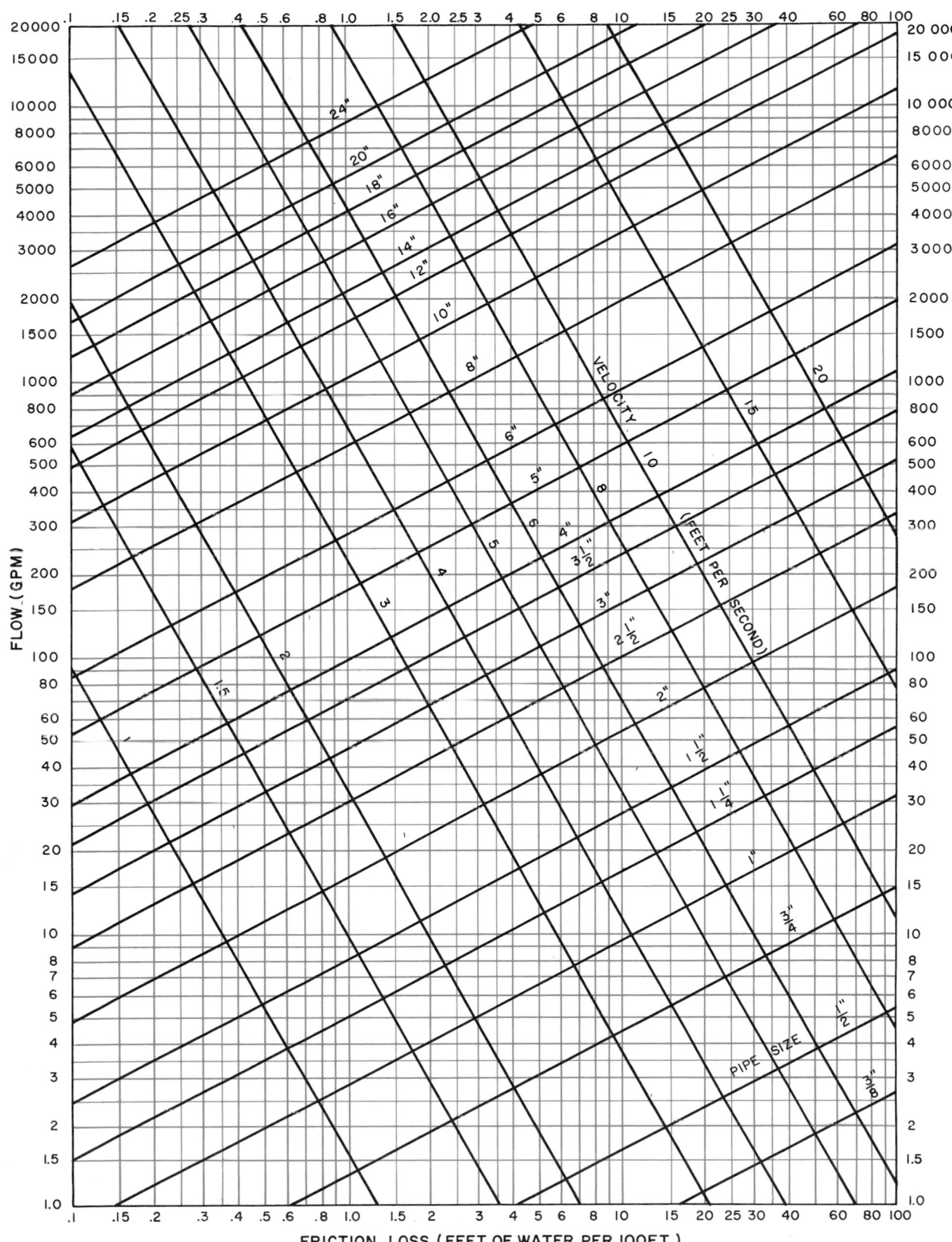

FRICTION LOSS (FEET OF WATER PER 100 FT.)

CHART 5—FRICTION LOSS FOR CLOSED AND OPEN PIPING SYSTEMS

Copper Tubing

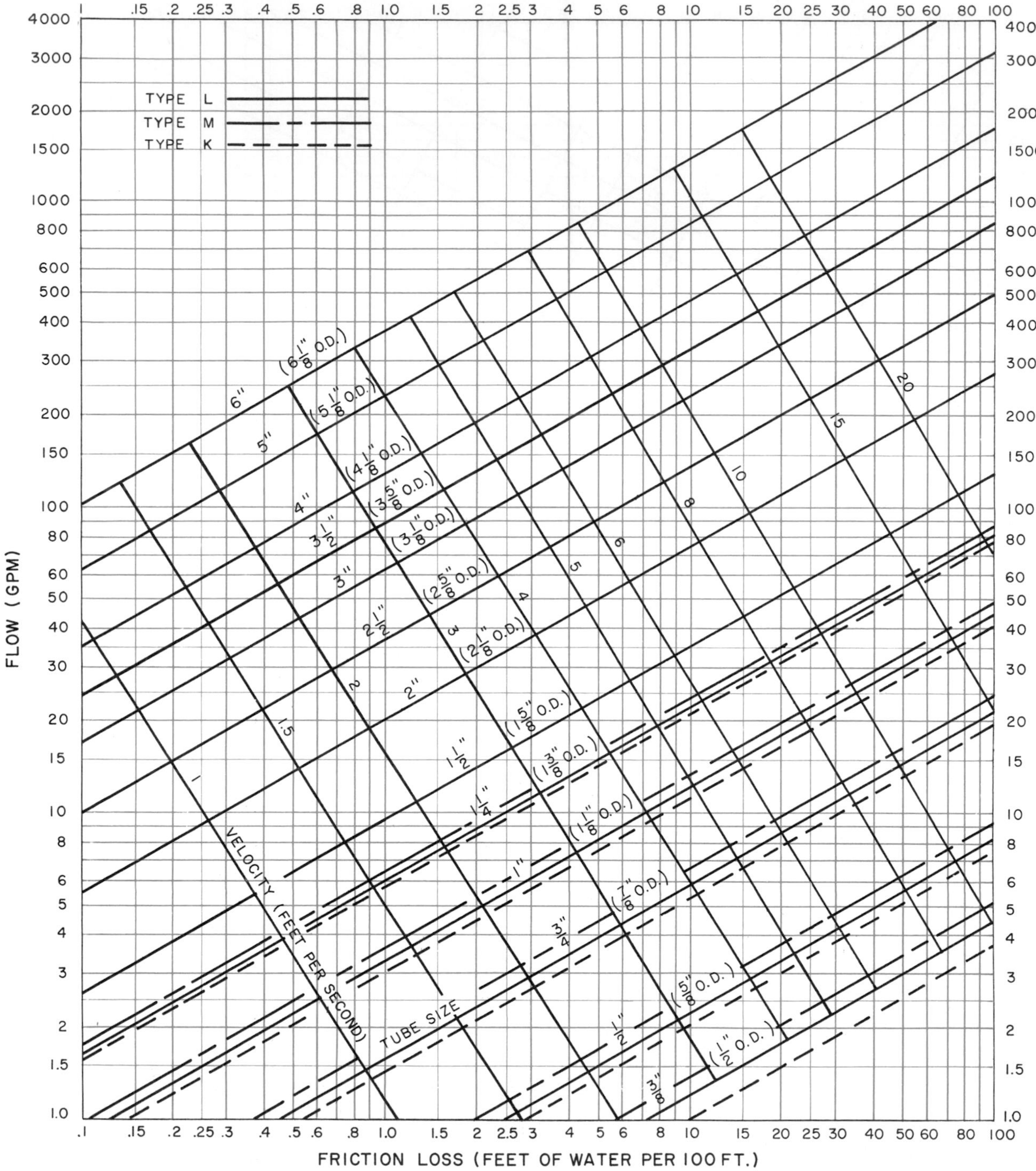

and pressure drops than smaller pipe which is sized for lower pressure drops and flow rates.

Exceptions to this general guide often occur. For example, appearance or physical limitations may dictate the use of small pipes. This is often done for short runs where the total pressure drop is not greatly influenced.

Each system should be analyzed separately to determine the economic balance between first cost (pipe size, pump and motor) and operating cost (pressure drop, pump and motor).

Pipe Length

To determine the friction loss in a water piping system, the engineer must calculate the straight lengths of pipe and evaluate the additional equivalent lengths of pipe due to fittings, valves and other elements in the piping system. *Tables 10, 11 and 12* give the additional equivalent lengths of pipe for these various components. The straight length of pipe is measured to the centerline of all fittings and valves (*Fig. 24*). The equivalent length of the components must be added to this straight length of pipe.

WATER PIPING DIVERSITY

When the air conditioning load is determined for each exposure of a building, it is assumed that the exposure is at peak load. Since the sun load is at a maximum on one exposure at a time, not all of the units on all the exposures require maximum water flow at the same time to handle the cooling load. Units on the same exposure normally require maximum flow at the same time; units on the adjoining or opposite exposures do not. Therefore, if the individual units are *automatically controlled* to vary the water quantity, the system water quantity actually required during normal operation is less than the total water quantity required for the peak design conditions for all the exposures. Good engineering design dictates that the water piping and the pump be sized for this reduced water quantity.

The principle of diversity allows the engineer to evaluate and calculate the reduced water quantity. In all water piping systems two conditions must be satisfied before diversity can be applied:

1. The water flow to the units must be automatically controlled to compensate for varying loads.

2. Diversity may only be applied to piping that supplies units on more than one exposure.

Figure 25 is a typical illustration of a header layout to which diversity may be applied. In this il-

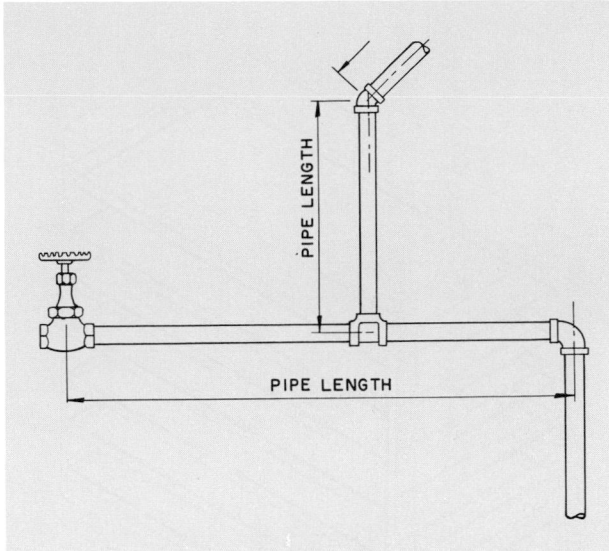

FIG. 24 — PIPE LENGTH MEASUREMENT

lustration the header piping supplies all four exposures. Assuming that the units supplied are automatically controlled, diversity is applied to the west, south and east exposures only. The last leg or exposure is never reduced in water quantity or pipe size since it requires full flow at some time during operation to meet design conditions.

Figure 26 illustrates another layout where diversity may be used to reduce pipe size and pump capacity. In this illustration diversity may be applied to the vertical supply and return headers and also to the supply and return branch headers at each floor. Diversity is not applied to pipe section 7-8 of both the supply and return vertical headers. In addition

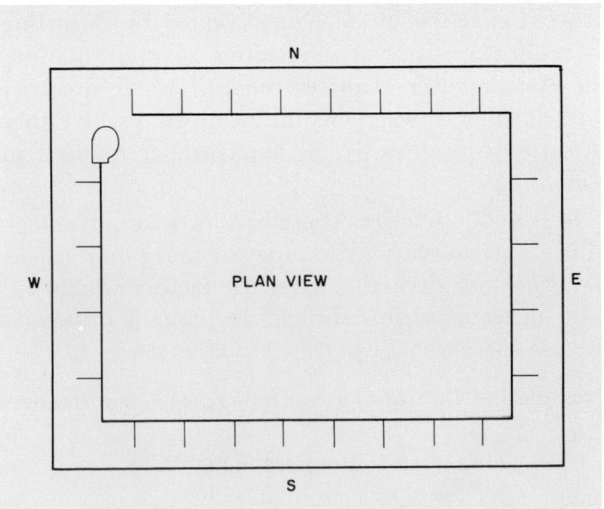

FIG. 25 — HEADER PIPING

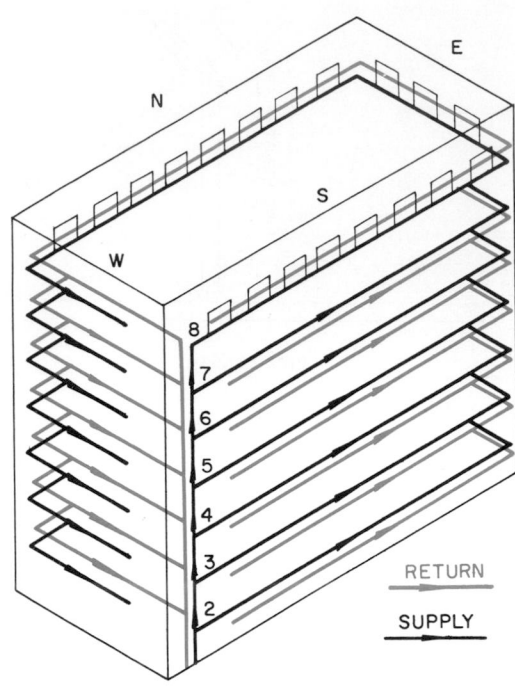

FIG. 26 — HORIZONTAL WATER PIPING LAYOUT

the south leg of the return piping and the west leg of the supply piping on each floor must be full size.

In any water piping system with automatically controlled units, the water requirements and pump head pressure varies. This is true whether or not diversity is applied. However the water requirements and pump head vary considerably more in a system in which diversity is not considered.

In a system in which diversity is not applied, greater emphasis is required for pump controls to prevent excessive noise being created by throttling valves or excessive water velocities. In addition, since the system never requires the full water quantity for which it is designed, the pump must be either throttled continuously, or bypassed, or reduced in size.

It is good practice, therefore, to take advantage of diversity to reduce the pipe size and pump capacity. *Chart 6* gives the diversity factors which are used in water piping design. *Example 1* illustrates the use of *Chart 6*.

Example 1 — Diversity Factors for Water Piping Headers

Given:
Water piping layout as illustrated in *Fig. 27*.

Find:
1. Diversity factor to be applied to the water quantity.
2. Water quantities in header sections.

Solution:

1. Pump A supplies north and west exposure but diversity can be applied to north exposure only. The total gpm in pump A circuit is 280 gpm and the accumulated gpm in the north exposure is 160 gpm. The ratio of accumulated gpm to the total water quantity in the circuit is:

$$\frac{160}{280} = .57$$

Enter *Chart 6* at the ratio .57 and read the diversity factor .785.

Pump B circuit has a ratio for the east exposure of:

$$\frac{120}{280} = .43$$

Entering *Chart 6* at the ratio of .43, the diversity factor is read as .725.

2. The following table illustrates how the diversity factors are applied to the maximum water quantities to obtain the design water quantities.

PUMP "A" CIRCUIT			
Section	Max Quantity (gpm)	Diversity Factor	Design Quantity (gpm)
A-R1	280	.785	220
R1-R2	260	.785	204
R2-R3	240	.785	188
R3-R4	220	.785	173
R4-R5	200	.785	157
R5-R6	180	.785	141
R6-R7	160	.785	126
R7-R8	140	.785	(110)120*
R8-R9	120	1.00	120
R9-R10	100	1.00	100
R10-R11	80	1.00	80
R11-R12	60	1.00	60
R12-R13	40	1.00	40
R13-R14	20	1.00	20

PUMP "B" CIRCUIT			
Section	Max Quantity (gpm)	Diversity Factor	Design Quantity (gpm)
B-R28	280	.725	203
R28-R27	260	.725	188
R27-R26	240	.725	174
R26-R25	220	.725	160
R25-R24	200	.725	(145) 160*
R24-R23	180	.725	(130) 160*
R23-R22	160	1.00	160
R22-R21	140	1.00	140
R21-R20	120	1.00	120
R20-R19	100	1.00	100
R19-R18	80	1.00	80
R18-R17	60	1.00	60
R17-R16	40	1.00	40
R16-R15	20	1.00	20

*When applying diversity, the design water quantity in the last section of the exposure is usually less than the water quantity in the first section on the adjoining exposure. When this occurs, the water quantity in the last section or last two sections is increased to equal the water quantity in the first section of the next exposure.

CHART 6—DIVERSITY FACTORS

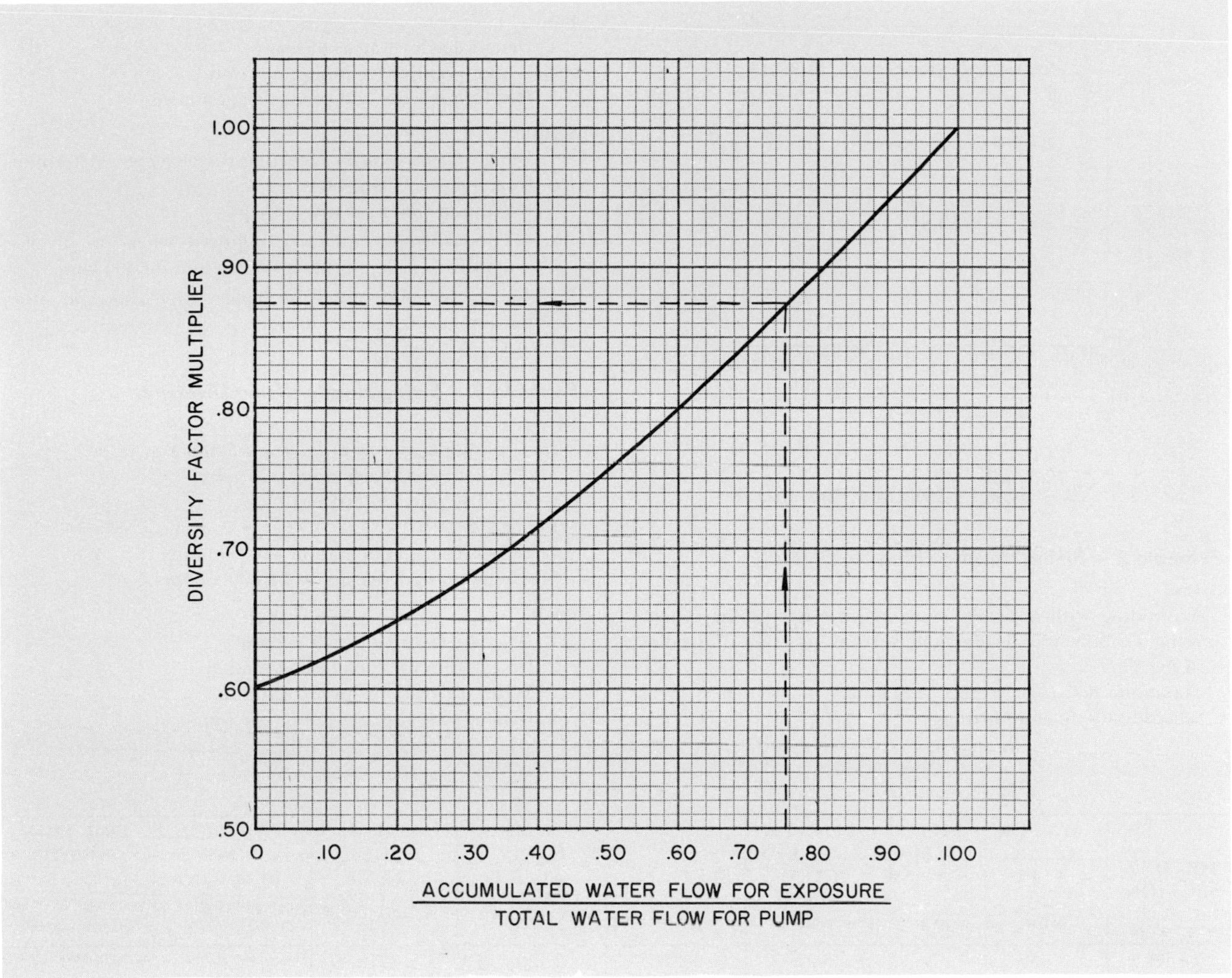

DIVERSITY FACTOR MULTIPLIER (vertical axis: .50, .60, .70, .80, .90, 1.00)

ACCUMULATED WATER FLOW FOR EXPOSURE
TOTAL WATER FLOW FOR PUMP (horizontal axis: 0, .10, .20, .30, .40, .50, .60, .70, .80, .90, .100)

In *Example 1* pump "A" is selected for 220 gpm and pump "B" is selected for 203 gpm. The pipe sizes in the north and east exposures are reduced using the design gpm, whereas the pipes in the south and west exposures are selected full size.

Example 2 and 3 illustrate the economics involved when applying diversity. *Example 2* shows a typical header layout with one pump serving all four exposures. The header is sized without diversity.

Example 3 is the same piping layout but diversity is used to size the header.

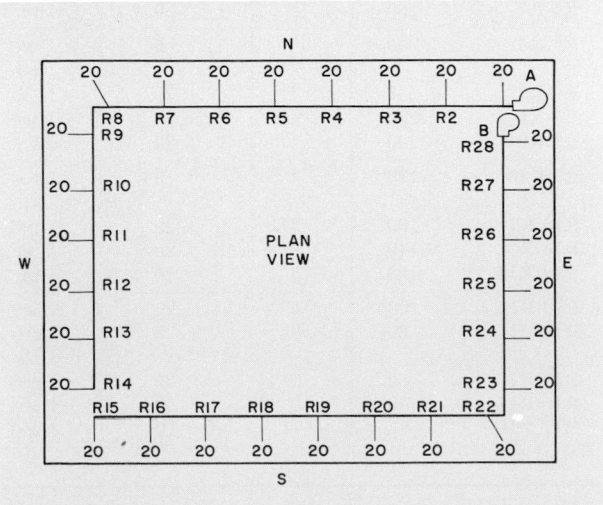

FIG. 27 — SUPPLY WATER HEADER

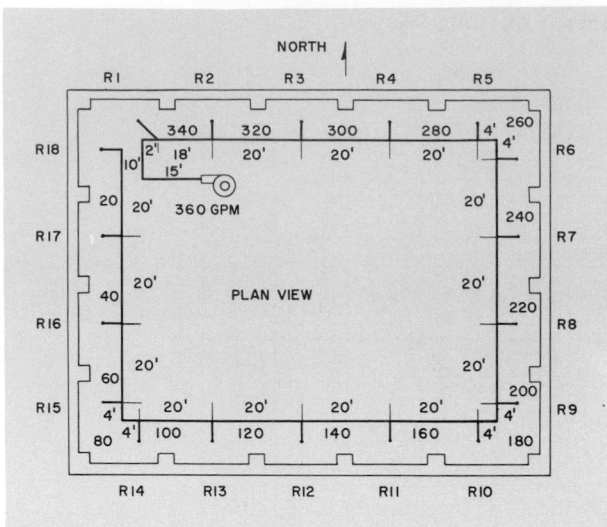

FIG. 28 — SUPPLY HEADER PIPE SIZING

Example 2 — Sizing Header Using No Diversity

Given:

A building with a closed recirculation water piping system using a horizontal header and vertical risers as illustrated in *Fig. 28*.

Maximum flow to each riser — 20 gpm

Schedule 40 pipe and fittings.

Elbows, R/D — I

Expected length of operation — 6000 hours

Find:

1. Design header water velocity
2. Water quantity for pump selection
3. Header pipe size and pump friction head

Solution:

1. Design water velocity for sizing the headers is determined from *Tables 13 and 14*.

   Maximum water velocity — 7 fps

2. Maximum water quantity required when no diversity is applied is 360 gpm. Pump is selected for 360 gpm.

3. The table below gives the header pipe sizes and pump friction head when no diversity is applied.

Example 3 — Sizing Header Using Diversity

Given:

Same piping layout as in *Example 2* and *Fig. 28*.

Maximum flow to each riser — 20 gpm

Schedule 40 pipe and fittings

Elbows, R/D = 1

Expected length of operation — 6000 hours

Maximum design velocity — 7 fps *(Example 2)*

Find:

1. Diversity factor for each exposure
2. Design gpm for each header section
3. Water quantity for pump selection
4. Header pipe size and pump friction head

HEADER SECTION	WATER QUANTITY (gpm)	PIPE SIZE† (in.)	LENGTH BETWEEN TAKE-OFFS (ft)	FITTINGS	FITTING EQUIVALENT LENGTH* (ft)	TOTAL EQUIVALENT LENGTH (ft)	FRICTION LOSS† (ft of water per 100 equiv ft)	FRICTION HEAD (ft of water)
To R1	360	5	27	2-ells	26	53.0	2.3	1.22
R1-R2	340	5	18	1-tee	8.2	26.2	2.0	.53
R2-R3	320	5	20	1-tee	8.2	28.2	1.8	.51
R3-R4	300	5	20	1-tee	8.2	28.2	1.6	.45
R4-R5	280	4	20	1-red. tee	12.0	32.0	4.4	1.41
R5-R6	260	4	8	1-tee	6.7			
				1-ell	10.0	24.7	3.8	.94
R6-R7	240	4	20	1-tee	6.7	26.7	3.2	.85
R7-R8	220	4	20	1-tee	6.7	26.7	2.7	.72
R8-R9	200	4	20	1-tee	6.7	26.7	2.3	.62
R9-R10	180	4	8	1-tee	6.7			
				1-ell	6.0	20.7	2.1	.43
R10-R11	160	3	20	1-red. tee	9.0	29.0	5.5	1.59
R11-R12	140	3	20	1-tee	5.0	25.0	4.6	1.15
R12-R13	120	3	20	1-tee	5.0	25.0	3.2	.80
R13-R14	100	3	20	1-tee	5.0	25.0	2.5	.63
R14-R15	80	3	8	1-tee	5.0			
				1-ell	7.5	20.5	1.6	.33
R15-R16	60	2	20	1-red. tee	7.0	27.0	6.8	1.84
R16-R17	40	2	20	1-tee	3.3	23.3	3.2	.75
R17-R18	20	1¼	20	1-red. tee	5.0	25.0	6.5	1.62

Pump friction head‡   16.39

*Fitting losses are determined from *Table 11*. For reducing tees enter *Table 11* at the larger diameter.

†Friction rate and pipe size are determined from *Chart 3* not exceeding the maximum design water velocity (7 fps).

‡Pump friction head does not include losses for valves, strainers, etc., which must be included in the actual design.

Solution:

1. *Chart 6* is used with the ratio of accumulated gpm in the exposure to the total pump gpm, in order to determine the diversity factors. The following table illustrates the method of determining diversity factors. (First exposure listed is always first exposure served by pump.)

EXPO- SURE	EXPOSURE WATER QUANTITY (gpm)	ACCUM. GPM / TOTAL PUMP GPM	DIVER- SITY FACTOR
North	100	100/360 = .28	.67
East	80	180/360 = .50	.76
South	100	280/360 = .78	.89
West	80	360/360 = 1.00	1.00

2. The diversity factor found in *Step 1* is applied to the maximum water quantity in each header section to establish the design gpm for sizing the header. The table at right gives the design water quantity for the various header sections.

3. The design water quantity required for pump selection when diversity is applied is 240 gpm.

4. The design water quantity found in *Step 2* is used in sizing the header pipe and in establishing the pump friction head. The table below illustrates the header pipe sizing:

HEADER SECTION	MAXIMUM WATER QUANTITY (gpm)	DIVER- SITY FACTOR	DESIGN WATER QUANTITY (gpm)
To R1	360	.67	240
R1-R2	340	.67	227
R2-R3	320	.67	214
R3-R4	300	.67	201
R4-R5	280	.67	(187)197*
R5-R6	260	.76	197
R6-R7	240	.76	182
R7-R8	220	.76	167
R8-R9	200	.76	(152)160*
R9-R10	180	.89	160
R10-R11	160	.89	142
R11-R12	140	.89	124
R12-R13	120	.89	106
R13-R14	100	.89	90
R14-R15	80	1.00	80
R15-R16	60	1.00	60
R16-R17	40	1.00	40
R17-R18	20	1.00	20

*When applying diversity, the design water quantity in the last section of the exposure is usually less than the design water quantity in the first section of the adjoining exposure. When this occurs, the water quantity in the last section or last two sections is increased to equal the water quantity in the first section of the next exposure.

HEADER SECTION	DESIGN WATER QUAN- TITY (gpm)	PIPE SIZE† (in.)	LENGTH BETWEEN TAKE- OFFS (ft)	FITTINGS	FITTING EQUIV- ALENT LENGTH* (ft)	TOTAL EQUIV- ALENT LENGTH (ft)	FRICTION LOSS† (ft of water per 100 equiv ft)	FRICTION HEAD (ft of water)
To R1	240	4	27	2-ells	20.0	47.0	3.4	1.60
R1-R2	227	4	18	1-tee	6.7	24.7	3.0	.74
R2-R3	214	4	20	1-tee	6.7	26.7	2.7	.72
R3-R4	201	4	20	1-tee	6.7	26.7	2.3	.61
R4-R5	197	4	20	1-tee	6.7	26.7	2.3	.61
R5-R6	197	4	8	1-ell	10.0	24.7	2.3	.57
				1-tee	6.7			
R6-R7	182	4	20	1-tee	6.7	26.7	2.0	.53
R7-R8	167	4	20	1-tee	6.7	26.7	1.8	.48
R8-R9	160	3	20	1-red. tee	9.0	29.0	5.6	1.62
R9-R10	160	3	8	1-ell	7.5	20.5	5.6	1.15
				1-tee	5.0			
R10-R11	142	3	20	1-tee	5.0	25.0	4.5	1.12
R11-R12	124	3	20	1-tee	5.0	25.0	3.5	.87
R12-R13	106	3	20	1-tee	5.0	25.0	2.7	.68
R13-R14	90	3	20	1-tee	5.0	25.0	2.0	.50
R14-R15	80	3	8	1-ell	7.5	20.5	1.6	.33
				1-tee	5.0			
R15-R16	60	2	20	1-red. tee	7.0	27.0	6.8	1.84
R16-R17	40	2	20	1-tee	3.3	23.3	3.2	.75
R17-R18	20	1¼	20	1-red. tee	5.0	25.0	6.5	1.63

Pump friction head‡ 16.35

*Fitting losses are determined from *Table 11*. For reducing tee enter *Table 11* at the larger diameter.

†Friction rate and pipe size are determined from *Chart 3* with the water velocity not exceeding 7 fps.

‡Pump friction head does not include losses for valves, strainers, etc., which must be included in the actual design.

Examples 2 and 3 indicate that the following reductions in pipe and fitting size can be made when diversity is used:

1. 57 ft of 5 in. pipe replaced with 4 in. pipe.
2. 28 ft of 4 in. pipe replaced with 3 in. pipe.
3. 8 fittings reduced 1 size.

In addition the pump can be selected for 240 gpm instead of 360 gpm which is approximately a ⅓ reduction. Other areas where a reduction in size is possible are:

1. Pipe and fittings in the return piping header.
2. Valves, unions, couplings, strainers and other elements located in the supply and return headers.

PUMP SELECTION

Pumps are selected so that there is no sustained rise in pressure when the water flow is throttled. Systems having considerable throttling have the pump selected on the flat portion of the "head-versus-flow" curve.

Normally, new installed pipe has less than design friction and, therefore, the pump delivers greater gpm than design and requires more horsepower. For this reason a centrifugal pump is always selected for the calculated pump head without the addition of safety factors. If the pump is selected for the calculated head plus safety factors, the pump must handle a larger water quantity. When this occurs and provision is not made to throttle or bypass the excess water flow, the possibility of pump motor overload exists.

Again, if the pump is selected for maximum water quantity and diversity is not applied, the water flow must be throttled. This increases the pump head.

SYSTEM ACCESSORIES AND LAYOUT

This section discusses the function and selection of piping accessories and describes piping layout techniques for coils, condensers, coolers, air washers, cooling towers, pumps and expansion tanks.

ACCESSORIES

Expansion Tanks

An expansion tank is used to maintain system pressure by allowing the water to expand when the water temperature increases, and by providing a method of adding water to the system. It is normally required in a closed system but not in an open system; the reservoir in an open system acts as the expansion tank.

The open and closed expansion tanks are the two types used in water piping systems. Open expansion tanks are open to the atmosphere and are located on the suction side of the pump above the highest unit in the system. At this location the tank provides atmospheric pressure at or above the pump suction, thus preventing air leakage into the system. The static head on the pump due to the expansion tank must be greater than the friction drop of the water in the pipe from the expansion line connection to the pump suction. In *Fig. 29* the static head *AB* must be greater than the friction loss in line *AC*. Adding any accessories such as a strainer in line *AC* increases the friction drop in *AC* and results in raising the height of the expansion tank. To keep the height of the tank at a reasonable level, accessories should be placed at points 1 and 2 in *Fig. 29*. At these designations the friction loss in line AC is not affected.

The following procedure may be used to determine the capacity of an open expansion tank:

1. Calculate the volume of water in the piping, from *Tables 2 and 3, pages 2 and 3.*
2. Calculate the volume of water in the coils and heat exchangers.
3. Determine the percent increase in the volume of water due to operating at increased temperatures from *Table 15.*
4. Expansion tank capacity is equal to the percent increase from *Table 15* times the total volume of water in the system.

The closed expansion tank is used for small or residential hot water heating systems and for high temperature water systems. Closed expansion tanks are not open to the atmosphere and operate above

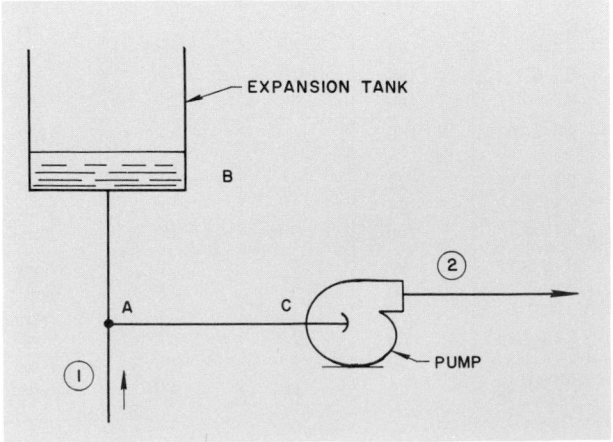

FIG. 29 — STRAINER LOCATION IN WATER PIPING SYSTEM

TABLE 15—EXPANSION OF WATER

(Above 40 F)

TEMP (F)	VOLUME INCREASE (%)	TEMP (F)	VOLUME INCREASE (%)
100	.6	275	6.8
125	1.2	300	8.3
150	1.8	325	9.8
175	2.8	350	11.5
200	3.5	375	13.0
225	4.5	400	15.0
250	5.6		

atmospheric pressure. Air vents must be installed in the system to vent the air. Closed expansion tanks are located on the pump suction side of the system to permit the pump suction to operate at or near constant pressure. Locating the expansion tank at the pump discharge is usually not satisfactory. All pressure changes caused by pump operation are subtracted from the original static pressure. If the pressure drop below the original static is great enough, the system pressure may drop to the boiling point, causing unstable water circulation and possible pump cavitation. If the system pressure drops blow atmospheric, air sucked in at the air vents can collect in pockets and stop water circulation.

The capacity of a closed expansion tank is larger than an open expansion tank operating under the same conditions. ASME has standardized the calculation of the capacity of closed expansion tanks. The capacity depends on whether the system is operating above or below 160 F water temperature.

Water temperatures below 160 F use the following formula to determine the tank capacity:

$$V_t = \frac{E \times V_s}{\dfrac{P_a}{P_f} - \dfrac{P_a}{P_o}}$$

where: V_t = minimum capacity of the tank (gallons).

E = percent increase in the volume of water in the system (*Table 15*).

V_s = total volume of water in the system (gallons).

P_a = pressure in the expansion tank when water first enters, usually atmospheric pressure (feet of water absolute).

P_f = initial fill or minimum pressure at the expansion tank (feet of water absolute).

P_o = maximum operating pressure at the expansion tank (feet of water absolute).

When the system water temperature is between 160 and 280 F, the following equation is used to determine the expansion tank capacity:

$$V_t = \frac{(0.00041\, t - 0.0466)\, V_s}{\dfrac{P_a}{P_f} - \dfrac{P_a}{P_o}}$$

where t = maximum average operating temp (F).

Strainers

The primary function of a strainer is to protect the equipment. Normally strainers are placed in the line at the inlet to pumps, control valves or other types of equipment that should be protected against damage. The strainer is selected for the design capacity of the system at the point where it is to be inserted in the line. Strainers for pump protection should be not less than 40 mesh and be made of bronze. On equipment other than pumps the manufacturer should be contacted to determine the degree of strainer protection necessary. For example, a control valve needs greater protection than a pump and, therefore, requires a finer mesh strainer.

Thermometers and Gages

Thermometers and gages are required in the system wherever the design engineer considers it important to know the water temperature or pressure. The following temperatures and pressures are usually considered essential:

1. Water temperature entering and leaving the cooler and condenser.

2. Pump suction and discharge pressure.

3. Spray water temperature and pressure entering the air washer.

Water thermometers are usually selected for an approximate range of 30 F to 200 F; they should be equipped with separable wells and located where they can be easily read.

Pressure gages are selected so that the normal reading of the gage is near the midpoint of the pressure scale.

Air Vents

Air venting is an important aspect in the design of any water system. The major portion of the air is normally vented thru the open expansion tank.

Air vents should be installed in the high points of any water system which cannot vent back to the open expansion tank. Systems using a closed expansion tank require vents at all high points. Runoff drains should be provided at each vent to carry possible water leakage to a suitable drain line.

PIPING LAYOUT

Each installation has its own problems regarding location of equipment, interference with structural members, water and drain locations, and provision for service and replacement. The following guides are presented to familiarize the engineer with accepted piping practice:

1. Shut-off valves are installed in the entering and leaving piping to equipment. These are normally gate valves. This arrangement permits servicing or replacing the equipment without draining the entire system. Occasionally a globe valve is installed in the system to serve as one of the shut-off valves and in addition is used to balance water flow. Most often it is located at the pump discharge. In a close coupled system the shut-off valves may be omitted if the time and expense required to drain the system is not excessive. This is a matter of economics, the first cost of the valves versus the cost of new water treatment and time spent in draining the system.

2. Systems using screwed, welded or soldered joints require unions to permit removal of the equipment for servicing or replacement. If gate valves are used to isolate the equipment in the system, unions are placed between the equipment and each gate valve. Unions are also located before and after control valves, and in the branch of a three-way control valve. It is good practice to locate the control valve between the equipment and the gate valve used to shut off flow to the equipment. This permits removal of the control valve from the system without draining the system. By locating the control valve properly, it is possible to eliminate the unions required for removal of the equipment. If the system uses flanged valves and fittings, the need for unions is eliminated.

3. Strainers, thermometers and gages are normally located between the equipment and the gate valves used to shut off the water flow to the equipment.

The following piping diagrams are illustrated with screwed connections. However, flanged, welded or soldered connections may be used. These layouts have been simplified to show various principles involved in piping practice.

Water Coils

Figures 30 thru 36 illustrate typical piping layouts for chilled water coils in a closed piping system.

The coil layout illustrated in *Fig. 30* contains a three-way mixing valve. This valve, located at the cooling coil outlet, maintains a desired temperature by proportioning *automatically* the amount of water

flowing thru the coil or thru the bypass. It is regulated by a temperature controller. Gage cocks are usually installed in both the supply and return lines to the coil. This permits pressure gages to be connected to determine pressure drop thru the coil. The plug cock is manually adjusted to set the pressure drop thru the coil.

Figure 31 illustrates an alternate method of piping a water coil. The plug cock shown is used to adjust manually the water flow for a set pressure drop thru the coil. The pressure drop is determined by connecting pressure gages to the gage cocks. In this piping layout control of the leaving air temperature from the coil is maintained within a required range since normally the entering water is controlled to a set temperature. Often an air bypass around the coil is used to maintain final air temperature.

Figure 32 illustrates a multiple coil arrangement. Piping connections for drain and vent lines for the coil are included and should be $\frac{1}{2}$ in. nominal pipe size. The same principles covered in *Figs. 30 and 31* are applicable to multiple coil arrangements.

A globe valve may be substituted for the plug cock and gate valve combination in the return lines in *Figs. 30, 31 and 32*. In this arrangement the globe valve is used to balance the pressure drop thru the coil, and also to shut off the water when servicing is required. However, it has disadvantages that

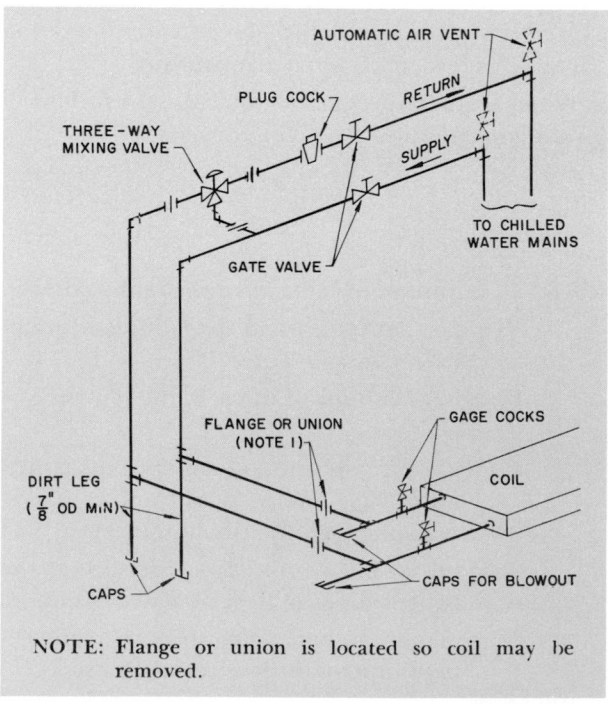

NOTE: Flange or union is located so coil may be removed.

FIG. 30— CHILLED WATER PIPING FOR COILS
(AUTOMATIC CONTROL)

should be considered. First, the valve setting is not fixed and, therefore, can be changed accidentally; secondly, the valve must be reset whenever it is used to shut off the flow.

Capped tees are provided on coils as a means of blowing out the water when the system is drained for freeze protection.

The use of a gate valve with a hose bib should be considered in the dirt leg when floor drains are remote in relation to the unit. The bib makes it possible to connect a hose to the dirt leg for draining the coil. The dirt leg has a 7/8 in. nominal minimum diameter, should be approximately 18 in. long and be located at an accessible servicing point. A gate valve is preferred in the dirt leg because sediment passes thru it more freely than thru a globe valve.

Figure 33 illustrates multiple coil units piped with vertical risers. This is a common approach to air conditioning multi-room, multi-story buildings. The units are connected to common supply and return risers which pass thru the floors of the building. A dirt leg is required at the bottom of each riser as shown in *Fig. 33*.

Gate valves are recommended as illustrated to permit servicing without disturbing the remainder of the system. On very small systems these **valves** may be omitted.

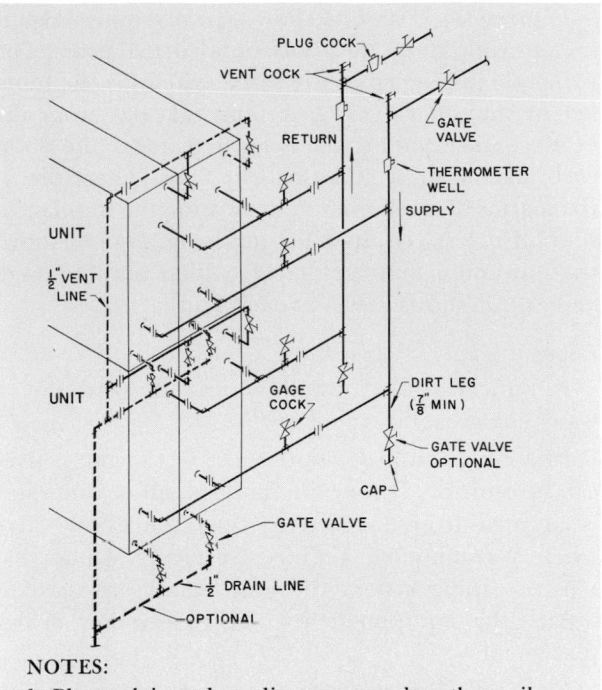

NOTES:
1. Plug cock is used to adjust pressure drop thru coil.
2. All valves shown are gate valves.

FIG. 32 — CHILLED WATER PIPING FOR
MULTIPLE COILS

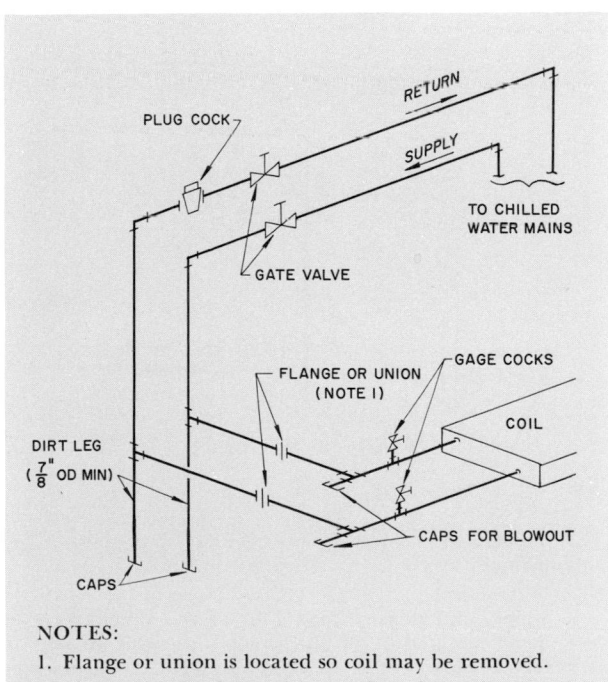

NOTES:
1. Flange or union is located so coil may be removed.
2. Plug cock is used for adjusting flow thru coil.

FIG. 31 — CHILLED WATER PIPING FOR COILS
(MANUAL CONTROL)

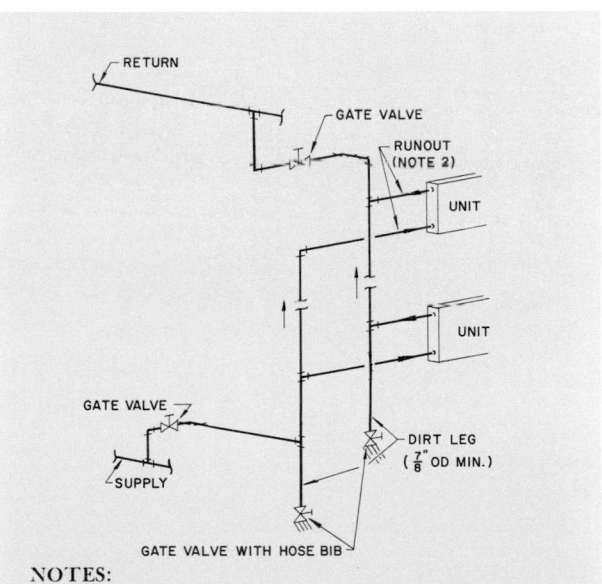

NOTES:
1. Headers are pitched upward in the direction of water flow so that air can be vented thru the expansion tank.
2. Supply and return runouts to the coil should have flared connections if runouts are soft copper. Otherwise unions or flanges are installed to facilitate servicing units.

FIG. 33 — PIPING LAYOUT FOR VERTICAL
MULTIPLE COILS

Figures 34, 35 and 36 show typical piping layouts for multiple units in a horizontal installation. The principle difference in the three systems is the number of shut-off valves (gate) and take-offs from the header. Since the header is located under the floor, each take-off must pass thru the floor. Therefore, it is a matter of economics to determine the number of shut-off valves required for servicing. *Fig. 35* shows the minimum number of valves that may be used, and *Fig. 36* shows valves at each unit.

Cooler

A typical chilled water piping diagram for a water cooler is illustrated in *Fig. 37*.

In a close coupled system most of the gate valves can be omitted. If they are omitted, all of the water is drained from the system thru the drain valve when a component requires servicing. In an extensive piping system the gate valves are used to isolate the equipment requiring servicing or replacement.

Figure 37 illustrates the recommended water piping and accessories associated with a cooler.

Condenser

Figure 38 shows a water-cooled condenser using city, well or river water. The return is run higher than the condenser so that the condenser is always full of water. Water flow thru the condenser is modulated by the control valve in the supply line.

Figure 39 is an illustration of an alternate drain arrangement for a condenser discharging waste water. Drain connections of all types must be checked for compliance with local codes. Codes usually require that a check valve be installed in the supply line when city water is used.

Figure 40 illustrates a condenser piped up with a cooling tower. If the cooling tower and condenser are close coupled, most of the gate valves can be eliminated. If the piping system is extensive, the gate valves as shown are recommended for isolating the equipment when servicing.

When more than one condenser is to be used in the same circuit, the flow thru the condensers must be equalized as closely as possible. This is complicated by the following:

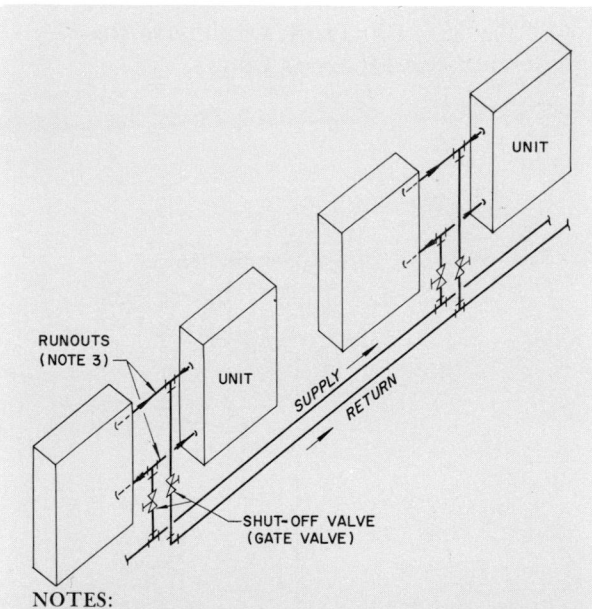

NOTES:
1. Though not shown, control valves (automatic or manual) may be required to control flow thru each unit.
2. A shut-off valve may be installed in the supply and return branch headers when headers serve 3 to 5 units.
3. Supply and return runouts to the coil should have flared connections if the runouts are soft copper. Otherwise unions or flanges are installed to facilitate servicing units.

FIG. 34 – PIPING LAYOUT FOR HORIZONTAL MULTIPLE COILS (4 UNITS – 4 SHUT-OFF VALVES)

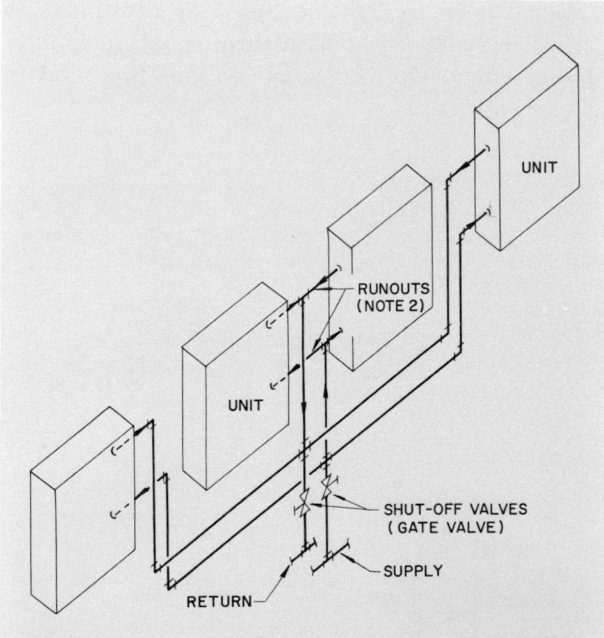

NOTES:
1. Though not shown, control valves (automatic or manual) may be required to control flow thru each unit.
2. Supply and return runouts to the coil should have flared connections if the runouts are soft copper. Otherwise unions or flanges are installed to facilitate servicing units.

FIG. 35 – PIPING LAYOUT FOR HORIZONTAL MULTIPLE COILS (4 UNITS – 2 SHUT-OFF VALVES)

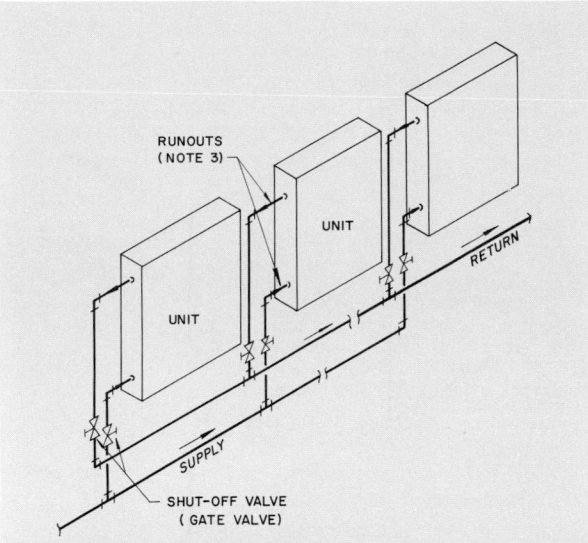

NOTES:

1. Though not shown, control valves (automatic or manual) may be required to control flow thru each unit.
2. A shut-off valve may be installed in the supply and return branch headers when headers serve 3 to 5 units.
3. Supply and return runouts to the coil should have flared connections if the runouts are soft copper. Otherwise unions or flanges are installed to facilitate servicing units.

FIG. 36 — PIPING LAYOUT FOR HORIZONTAL MULTIPLE COILS (3 UNITS — 6 SHUT-OFF VALVES)

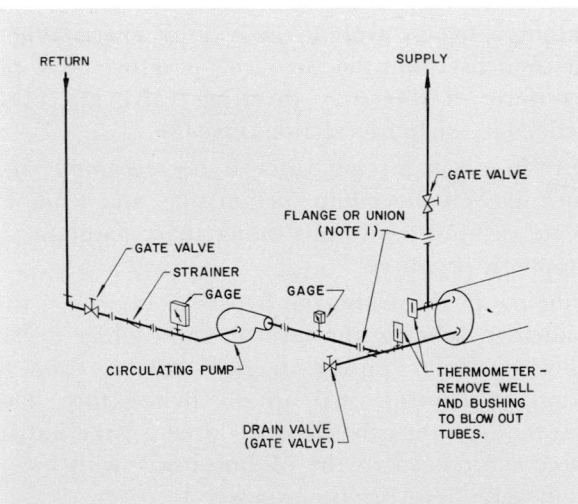

NOTES:

1. Flange or union is located to allow cooler head removal.
2. Gate valves shown may be eliminated in a close coupled system.

FIG. 37 — PIPING AT A WATER COOLER

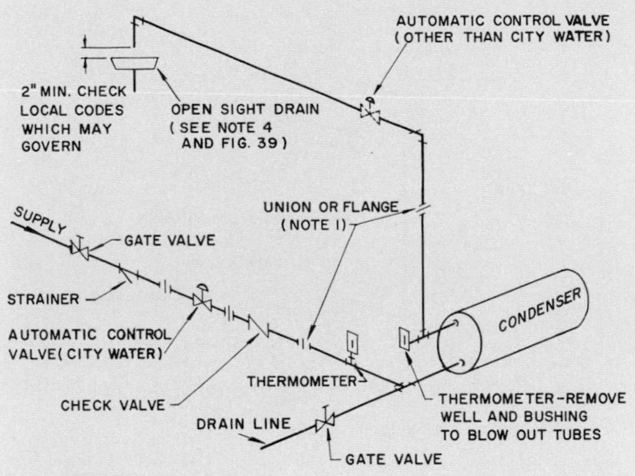

NOTES:

1. Flange or union is located to allow condenser head removal.
2. With outlet at top, condenser will be flooded even though automatic control valve is in modulating position.
3. Check valve is required by most sanitary codes (city water).
4. Required for city water only.

FIG. 38 — CONDENSER PIPING FOR A ONCE-THRU SYSTEM

1. The pressure drops thru the condensers are not always equal.
2. Water entering the branch line and leaving the run of tees seldom divides equally.
3. Workmanship in the installation can affect the pressure drop.

To equalize the water flow thru each condenser, the pipe should be sized as follows:

1. Size the branches for a water flow of 6 fps minimum. The branch connections to each condenser should be identical.

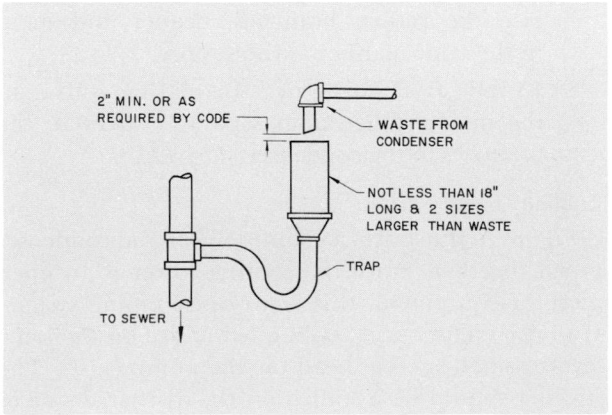

FIG. 39 — ALTERNATE DRAIN CONNECTION

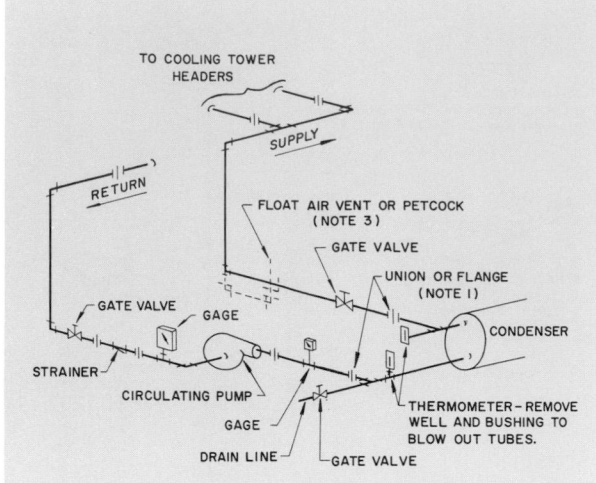

NOTES:

1. Flange or union is located to allow condenser head removal.

2. Gate valves shown may be eliminated in a close coupled system (except drain valve).

3. When water enters bottom of condenser, air will vent naturally thru cooling tower sprays. If it is necessary to drop piping after leaving condenser, install air vent at high point of line before drop. See dotted line in figure.

FIG. 40 — CONDENSER PIPING FOR A COOLING TOWER

2. Size the header for the total required water quantity for all the condensers with a velocity of not more than 3 fps. The header is extended approximately 12 in. beyond the last branch to the condenser.

3. Size the water main supplying the header for a velocity of 5 to 10 fps with 7 fps a good average. The water main may enter the header at the end or at any point along the length of the header. Care should be used so that crosses do not result.

4. Size the return branches, header and main in the same manner as the supply.

5. Install a single water regulating valve in the main, rather than separate valves in the branches to the condenser (Fig. 41).

Cooling Tower

Figure 40 illustrates a cooling tower and condenser piped together. Since the cooling tower is an open piece of equipment, this is an open piping system. If the condenser and cooling tower are on the same level, a small suction head for the pump exists. The strainer should be installed on the discharge side of the pump to keep the suction side of the pump as close to atmospheric as possible.

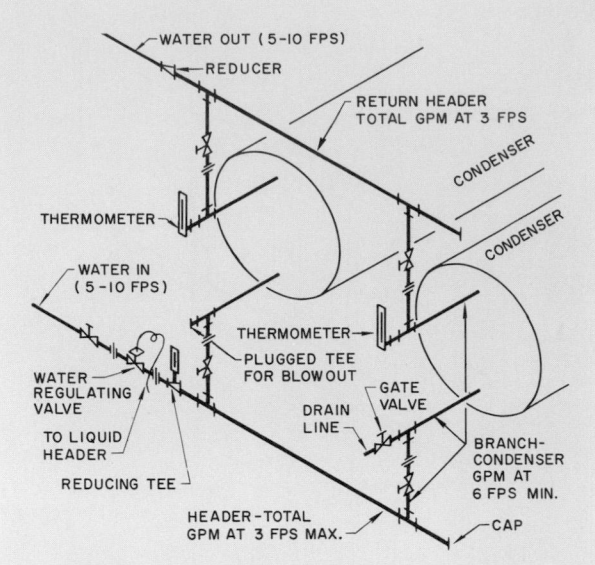

NOTES:

1. Thermometer wells are inserted in tees. Remove wells to blow out coils.

2. A single water regulating valve must be used as shown. If under capacity, install two valves in parallel and connect pressure tube in liquid header.

3. Water supply in or return out can be at any point in the headers.

FIG. 41 — MULTIPLE CONDENSER COOLING WATER PIPING (REFRIGERANT CONNECTIONS PARALLEL)

It is often desirable to maintain a constant water temperature to the condenser. This is done by installing a bypass around the cooling tower. When the condenser is at the same level or above the cooling tower, a three-way diverting valve is recommended in the bypass section *(Fig. 42)*.

A three-way mixing valve is not recommended since it is on the pump suction side, and tends to create vacuum conditions rather than maintain atmospheric pressures.

Figure 43 illustrates the bypass layout when the condenser is below the level of the cooling tower. This particular piping diagram uses a two-way automatic control valve in the bypass line. The friction drop thru the bypass is sized for the unbalanced static head in the cooling tower with maximum water flow thru the bypass.

If multiple cooling towers are to be connected, it is recommended that piping be designed such that the loss from the tower to the pump suction is approximately equal for each tower. *Fig. 44* illustrates typical layouts for multiple cooling towers.

Equalizing lines are used to maintain the same water level in each tower.

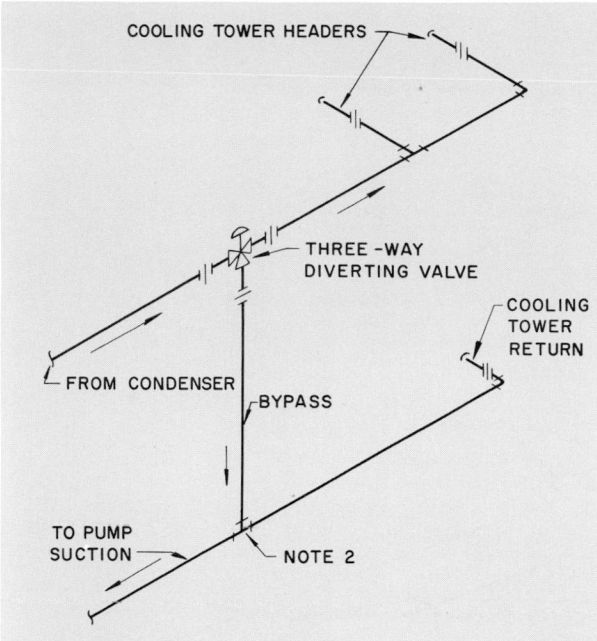

NOTES:

1. A three-way diverting valve is used when the condenser is at the same level as or above the cooling tower. See *Fig. 43* for piping layout when condenser is below the cooling tower.

2. A three-way mixing valve is not recommended at this point as it imposes additional head at the pump suction.

FIG. 42 — COOLING TOWER PIPING FOR CONSTANT LEAVING WATER TEMPERATURE (CONDENSER AND TOWER AT SAME LEVEL)

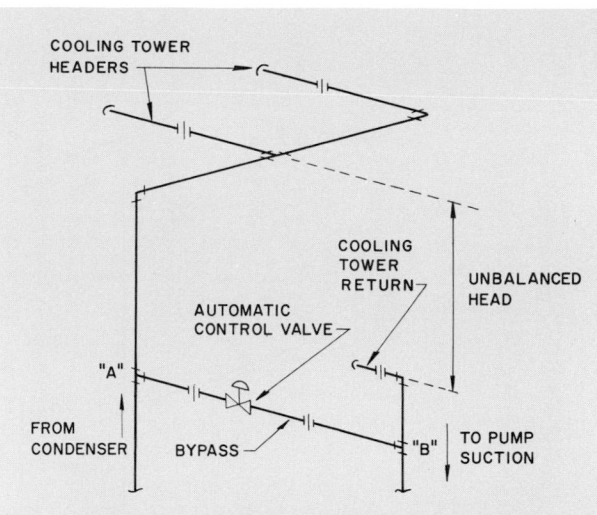

NOTES:

1. A two-way automatic control valve is used when the condenser is below the cooling tower. See *Fig. 42* for piping layout when the condenser is at the same level as or above the cooling tower.

2. The friction loss from "A" to "B" includes the loss thru that section of the pipe and the loss thru the two-way automatic control valve. This friction loss should be designed for the unbalanced head of the cooling tower.

3. Locate the automatic control valve close to the cooling tower to prevent pump motor overload and tower spill-over when valve is in full open position.

FIG. 43 — COOLING TOWER PIPING FOR CONSTANT LEAVING WATER TEMPERATURE (CONDENSER BELOW TOWER)

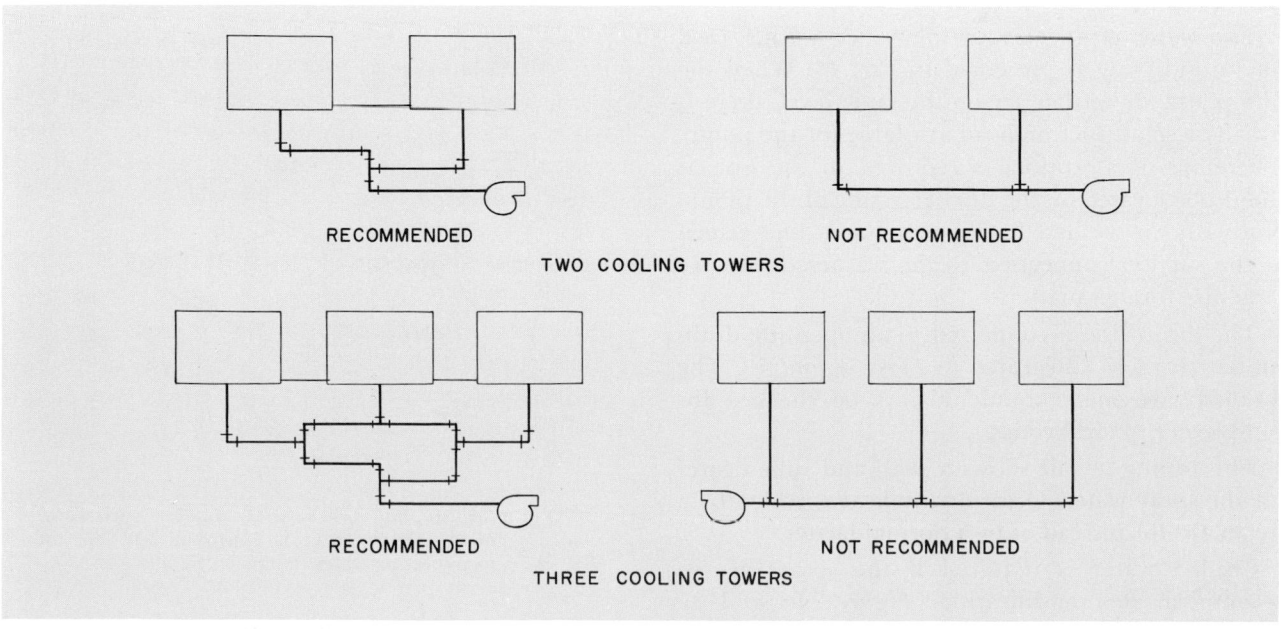

FIG. 44 — MULTIPLE COOLING TOWER PIPING

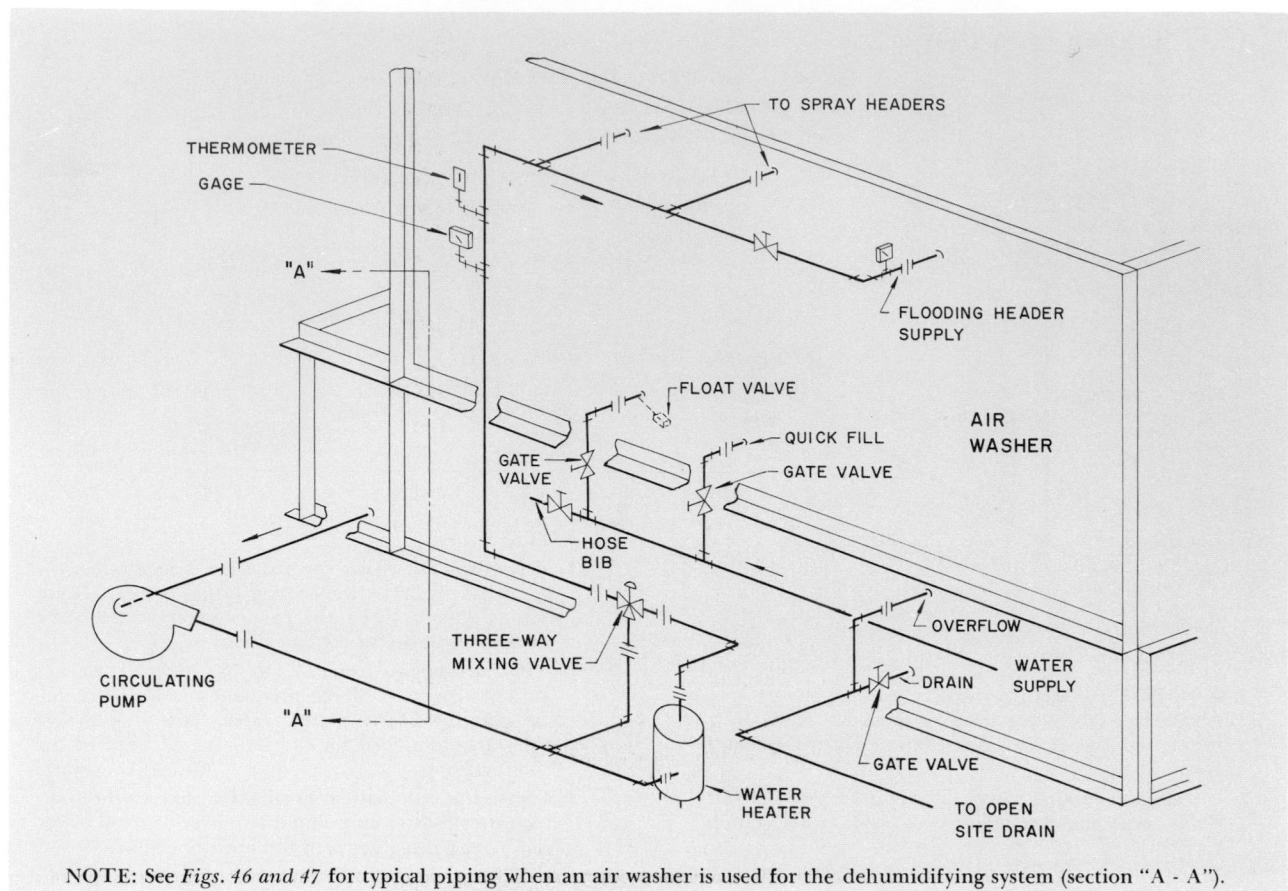

NOTE: See *Figs. 46 and 47* for typical piping when an air washer is used for the dehumidifying system (section "A - A").

FIG. 45 — AIR WASHER PIPING

Air Washer

The water piping layout for an air washer used for humidifying is presented in *Fig. 45*. When the pump and air washer are on the same level, there is usually a small suction head available for the pump. Therefore, if a strainer is required in the line, it should be located on the discharge side of the pump. Normally air washers have a permanent type screen at the suction connection to the washer to remove large size foreign matter.

The drain line is connected to an open-site drain similar to those illustrated in *Figs. 38 and 39*. The drain arrangement should always be checked for compliance to local codes.

The piping layout shows a shell and tube heater for the spray water. Occasionally heat is added by a steam ejector instead of by a normal heater.

Chilled water is required if the sprays are to accomplish dehumidification. *Figures 46 and 47* illustrate two typical methods of connecting the chilled water supply. The plug cock in both dia-

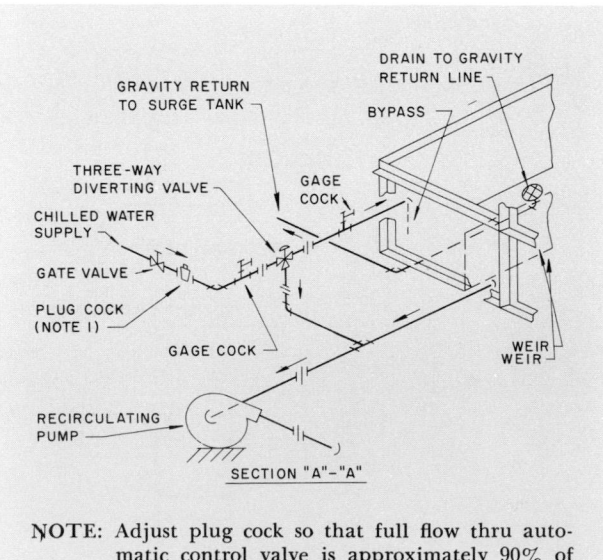

NOTE: Adjust plug cock so that full flow thru automatic control valve is approximately 90% of recirculating water design.

FIG. 46 — AIR WASHER PIPING USING A THREE-WAY CONTROL VALVE

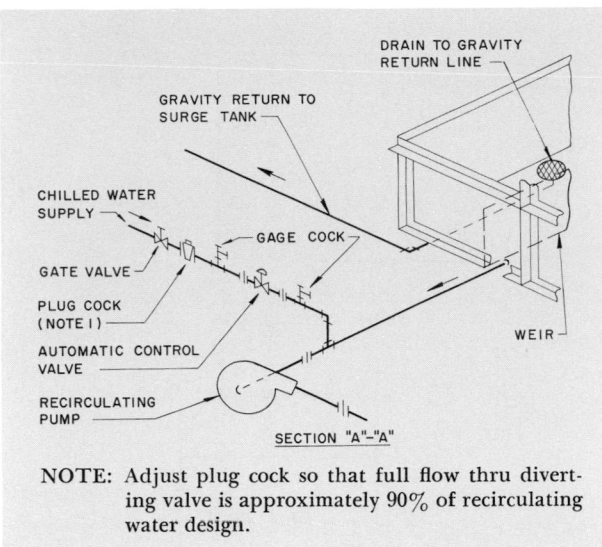

NOTE: Adjust plug cock so that full flow thru diverting valve is approximately 90% of recirculating water design.

FIG. 47 — AIR WASHER PIPING USING A TWO-WAY CONTROL VALVE

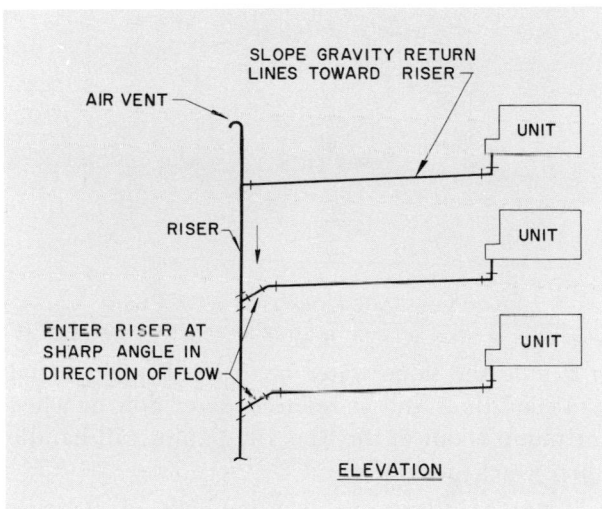

FIG. 48 — AIR WASHER RETURN CONNECTIONS AT DIFFERENT ELEVATIONS

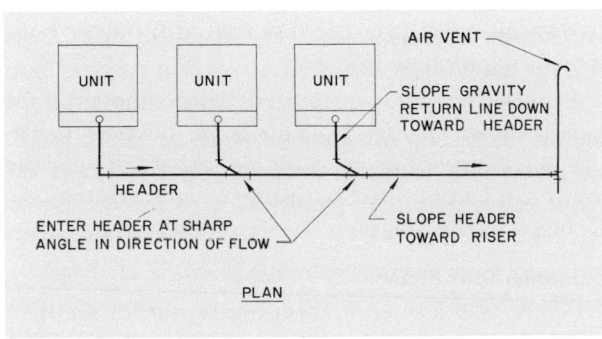

FIG. 49 — AIR WASHER RETURN CONNECTIONS AT THE SAME LEVEL

grams is adjusted so that full flow thru the three-way diverting valve (*Fig. 46*) and thru the automatic control valve (*Fig. 47*) is approximately 90% of the recirculating water design quantity.

Figures 48 and 49 are schematic sketches of multiple air washers with gravity returns piped to the same header.

Sprayed Coil

A typical layout for a sprayed coil piping system is shown in *Fig. 50*. The diagram shows a water heater which may be required for humidification. If a preheat coil is used, the water heater may be eliminated.

The drain line should be fitted with a gate valve rather than a globe valve since it is less likely to become clogged with sediment.

Pump Piping

The following items illustrated in *Fig. 51* should be kept in mind when designing piping for a pump:

1. Keep the suction pipe short and direct.
2. Increase the suction pipe size to at least one size larger than the pump inlet connection.
3. Keep the suction pipe free from air pockets.
4. Use an eccentric type reducer at the pump suction nozzle to prevent air pockets in the suction line.

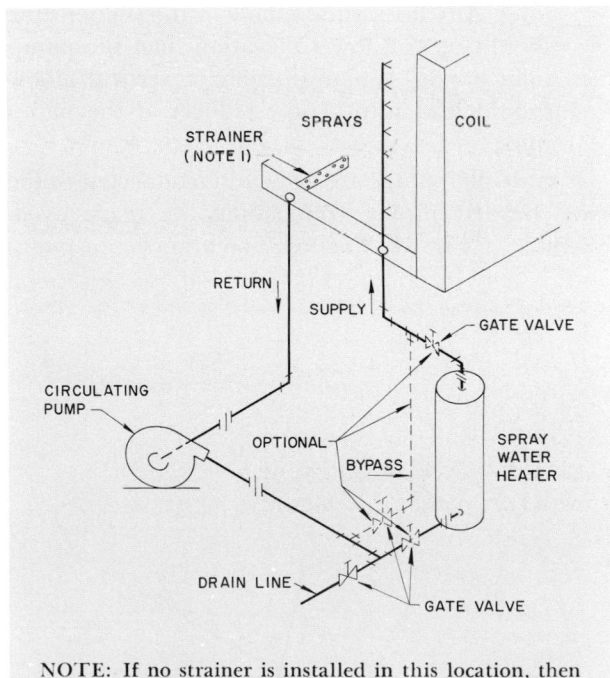

NOTE: If no strainer is installed in this location, then a strainer is recommended on the pump discharge.

FIG. 50 — SPRAY WATER COIL WITH WATER HEATER

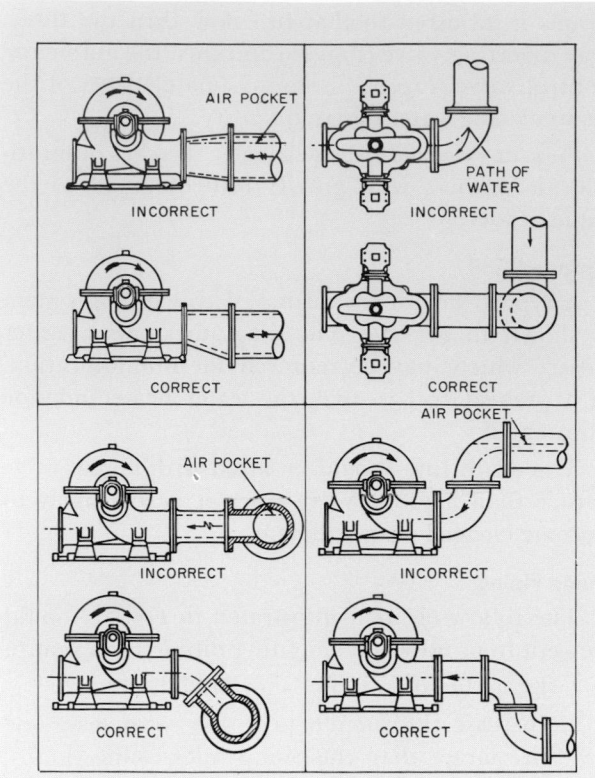

FIG. 51 — PUMP SUCTION CONNECTIONS

5. Never install a horizontal elbow at the pump inlet. Any horizontal elbow in the suction line should be at a lower elevation than the pump inlet nozzle. Where possible, a vertical elbow should lead into a pipe reducer at the pump inlet.

If multiple pumps are to be interconnected to the same header, piping connections are made as illustrated in *Fig. 52*. This method allows each pump

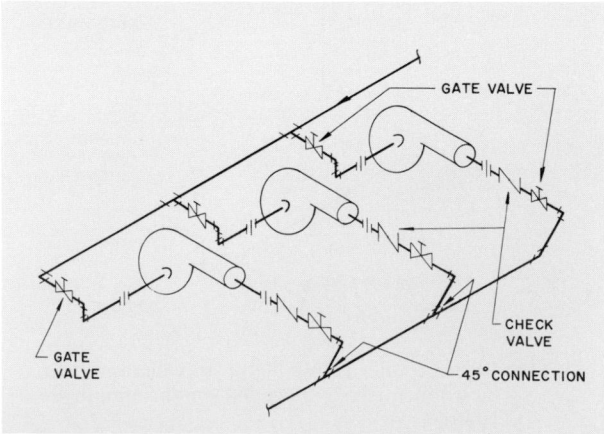

FIG. 52 — MULTIPLE PUMP PIPING

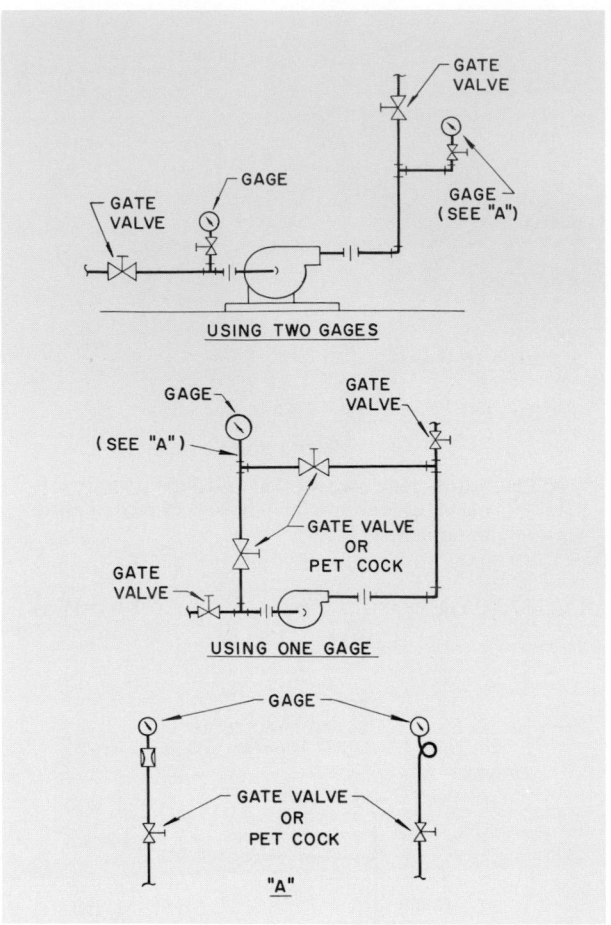

FIG. 53 — GAGE LOCATION AT A PUMP

to handle the same water quantity. Under partial load conditions and at reduced water flow or when one pump is out of the line, the pumps still handle equal water quantities.

Figure 53 illustrates two methods of locating pressure gages at the pump; one method uses two gages and the other uses one. The use of one gage has the advantage of always giving the correct pressure differential across the pump. Two gages may give an incorrect pressure differential if one or both are reading high or low.

A pulsating damper located before the pressure gage is shown in *Fig. 53*. This is an inexpensive device for dampening pressure pulsations. The same result can be obtained by using a pigtail in the line as shown in the diagram.

Expansion Tank Piping

Figure 54 is a suggested piping layout for an open expansion tank. Piping is enlarged at the connection to the expansion tank. This permits air entrained or carried along with the water to separate

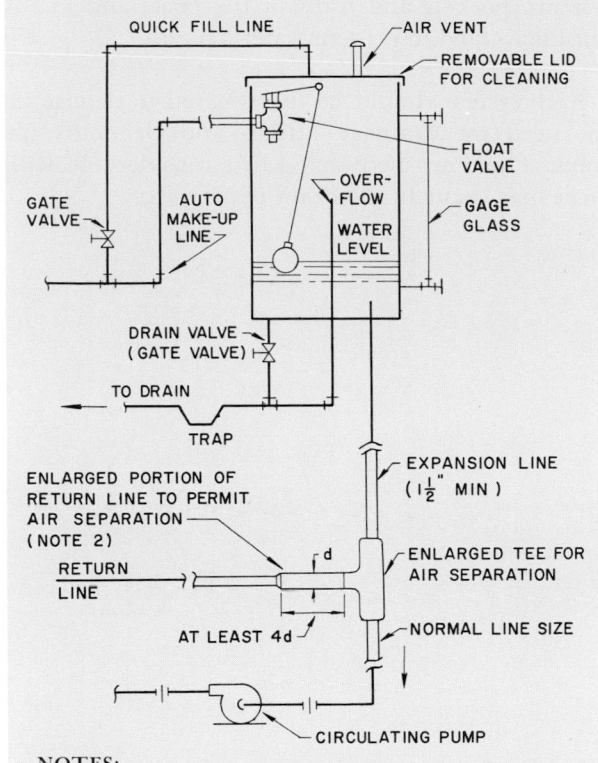

NOTES:
1. Do not put any valve strainer or trap in the expansion line.
2. Enlarged portion of return line and enlarged tee are each two standard pipe sizes larger than return line.

FIG. 54 — OPEN EXPANSION TANK PIPING

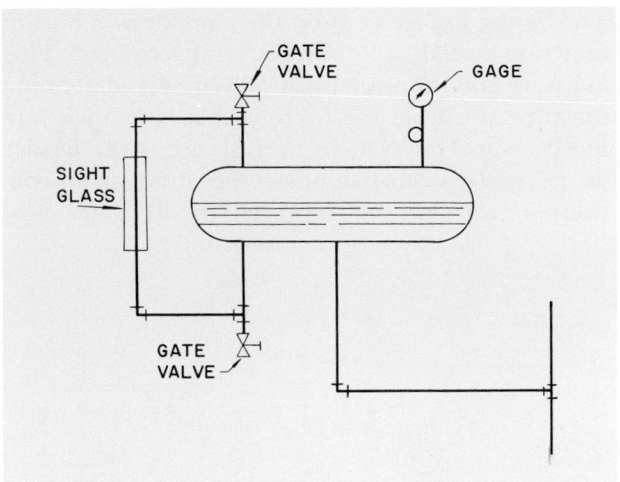

FIG. 55 — CLOSED EXPANSION TANK PIPING

and be vented thru the tank. The expansion tank should be located at the pump suction side at the highest point in the system.

Valves, strainers and traps must be omitted from the expansion line since these may be accidentally turned off or become plugged.

Figure 55 illustrates the piping diagram for a closed tank.

Drain Line Piping

Moisture that forms on the cooling coils must be collected and carried off as waste. On factory fabricated fan-coil units a drain pan is used to collect this moisture. For built-up systems the floor or base of the system (before and after the cooling coil) is used to gather the moisture.

Since, under operating conditions, the drain water is subject to pressure conditions slightly above or slightly below atmospheric pressure, the line used to carry off this water must be trapped. This trap prevents conditioned air from entering the drain line when the drain water is under positive pres-

sure as in a blow-thru fan-coil unit. When the system is under negative pressure as in a draw-thru unit, the trap prevents water from hanging up in the drain pan.

Figure 56 illustrates the trapping of a drain line from the drain pan. The length of the water seal or trap depends on the magnitude of the positive or negative pressure on the drain water. For instance, a 2-inch negative fan pressure requires a 2-inch water seal.

Normally, under-the-window fan-coil units have the drip pan subject to atmospheric conditions only and the drain line from these units is not trapped.

The drain line runout for all systems is pitched to offset the line friction. For a single unit the runout is piped to an open site drain. Local codes and regulations must be checked to determine proper piping practice for an open site drain. The runout is run full size corresponding to the drain pan connection size.

Some applications have multiple units with the drain lines connected to a common header or riser.

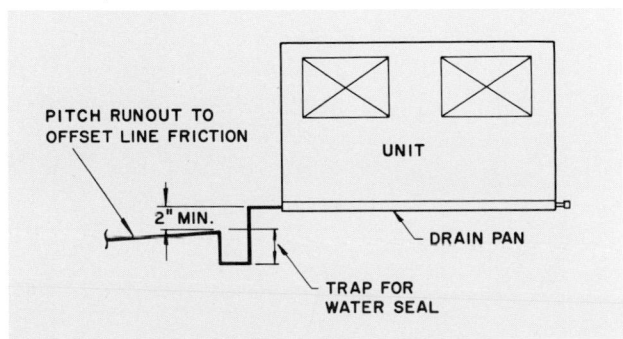

FIG. 56 — PIPING FOR DRAIN PANS

To size the header or riser, the amount of moisture that is expected to form must be determined. This moisture and the available head is used to determine the pipe size from the friction chart for open piping systems. However, in no instance is the header or riser sized smaller than the drain pan connection size. Also, as required in all water flow systems, pockets and traps in the risers and mains must be vented to prevent water hangup.

Each system should be investigated to determine the need for drainage fittings and cleanouts for traps. These are necessary when considerable sediment may occur in the drain pan.

CHAPTER 3. REFRIGERANT PIPING

GENERAL SYSTEM DESIGN

This chapter includes that practical material required for the design and layout of a refrigerant piping system at air conditioning temperature levels, using either Refrigerant 12, 22 or 500.

APPLICATION CONSIDERATIONS

A refrigerant piping system requires the same general design considerations as any fluid flow system. However, there are additional factors that critically influence system design:

1. The system must be designed for minimum pressure drop since pressure losses decrease the thermal capacity and increase the power requirement in a refrigeration system.
2. The fluid being piped changes in state as it circulates.
3. Since lubricating oil is miscible with Refrigerants 12, 22 and 500, some provision must be made to:
   a. Minimize the accumulation of liquid refrigerant in the compressor crankcase.
   b. Return oil to the compressor at the same rate at which it leaves.

Piping practices which accomplish these objectives are discussed in the following pages.

CODE REGULATIONS

System design should conform to all codes, laws and regulations applying at the site of an installation.

In addition the Safety Code for Mechanical Refrigeration (ASA-B9.1-1958) and the Code for Refrigeration Piping (ASA-B31.5-1962) are primarily drawn up as guides to safe practice and should also be adhered to. These two codes, as they apply to refrigeration, are almost identical, and are the basis of most municipal and state codes.

REFRIGERANT PIPING DESIGN

DESIGN PRINCIPLES

Objectives

Refrigerant piping systems must be designed to accomplish the following:

1. Insure proper feed to evaporators.
2. Provide practical line sizes without excessive pressure drop.

3. Protect compressors by —
   a. Preventing excessive lubricating oil from being trapped in the system.
   b. Minimizing the loss of lubricating oil from the compressor at all times.
   c. Preventing liquid refrigerant from entering the compressor during operation and shutdown.

Friction Loss and Oil Return

In sizing refrigerant lines it is necessary to consider the optimum size with respect to economics, friction loss and oil return. From a cost standpoint it is desirable to select the line size as small as possible. Care must be taken, however, to select a line size that does not cause excessive suction and discharge line pressure drop since this may result in loss of compressor capacity and excessive hp/ton. Too small a line size may also cause excessive liquid line pressure drop. This can result in flashing of liquid refrigerant which causes faulty expansion valve operation.

The effect of excessive suction and hot gas line pressure drop on compressor capacity and horsepower is illustrated in *Table 16*.

TABLE 16—COMPRESSOR CAPACITY VS LINE PRESSURE DROP

42 F Evaporator Temperature

SUCTION AND HOT GAS LINE PRESSURE DROP	COMPRESSOR	
	Capacity (%)	Hp/Ton (%)
No Line Loss	100	100
2F Suction Line Loss	95.7	103.5
2F Hot Gas Line Loss	98.4	103.5
4F Suction Line Loss	92.2	106.8
4F Hot Gas Line Loss	96.8	106.8

Pressure drop is kept to a minimum by optimum sizing of the lines with respect to economics, making sure that refrigerant line velocities are sufficient to entrain and carry oil along at all loading conditions.

For Refrigerants 12, 22 and 500, consider the requirements for oil return up vertical risers.

Pressure drop in *liquid lines* is not as critical as in suction and discharge lines. However, the pressure drop should not be so excessive as to cause gas formation in the liquid line or insufficient liquid pressure at the liquid feed device. A system should normally be designed so that the pressure drop in the liquid line is not greater than one to two degrees change in saturation temperature. In terms of pressure drop, this corresponds to about 1.8 to 3.8 psi for Refrigerant 12, 2.9 to 6 psi for Refrigerant 22, and 2.2 to 4.6 psi for Refrigerant 500.

Friction pressure drop in the liquid line includes accessories such as solenoid valve, strainer, drier and hand valves, as well as the actual pipe and fittings from the receiver outlet to the refrigerant feed device at the evaporator.

Pressure drop in the *suction line* means a loss in system capacity because it forces the compressor to operate at a lower suction pressure to maintain the desired evaporator temperature. Standard practice is to size the suction line for a pressure drop of approximately two degrees change in saturation temperature. In terms of pressure loss at 40 F suction temperature, this corresponds to about 1.8 psi for Refrigerant 12, 2.9 psi for Refrigerant 22, and 2.2 psi for Refrigerant 500.

Where a reduction in pipe size is necessary to provide sufficient gas velocity to entrain oil upward in vertical risers at partial loads, a greater pressure drop is imposed at full load. To keep the total pressure drop within the desired limit, excessive riser loss can be offset by properly sizing the horizontal and "down" lines.

It is important to minimize the pressure loss in *hot gas lines* because these losses can increase the required compressor horsepower and decrease the compressor capacity. It is usual practice not to exceed a pressure drop corresponding to one to two degrees change in saturation temperature. This is equal to about 1.8 to 3.8 psi for Refrigerant 12, 2.9 to 6 psi for Refrigerant 22, and 2.2 to 4.6 psi for Refrigerant 500.

REFRIGERANT PIPE SIZING

Charts 7 thru 21 are used to select the proper steel pipe and copper tubing size for the refrigeration lines. They are based on the Darcy-Weisbach formula:

$$h = f \times \frac{L}{D} \times \frac{V^2}{2g}$$

where h = loss of head in feet of fluid
f = friction factor
L = length of pipe in feet
D = diameter of pipe in feet
V = velocity in fps
g = acceleration of gravity = 32.17 ft/sec/sec

The friction factor depends on the roughness of pipe surface and the Reynolds number of the fluid. In this case the Reynolds number and the Moody chart are used to determine the friction factor.

Use of Pipe Sizing Charts

The following procedure for sizing refrigerant piping is recommended:

1. Measure the length (in feet) of straight pipe.
2. Add 50% to obtain a trial total equivalent length.
3. If other than a rated friction loss is desired, multiply the total equivalent length by the correction factor from the table following the appropriate pipe or tubing size chart.
4. If necessary, correct for suction and condensing temperatures.
5. Read pipe size from *Charts 7 thru 21* to determine size of fittings.
6. Find equivalent length (in feet) of fittings and hand valves from *Chapter 1* and add to the length of straight pipe *(Step 1)* to obtain the total equivalent length.
7. Correct as in *Steps 3 and 4* if necessary.
8. Check pipe size.

In some cases, particularly in liquid and suction lines, it may be necessary to find the *actual pressure drop*. To do this, use the procedure described in *Steps 9 thru 11:*

9. Convert the friction drop (F from *Step 3*) to psi, using refrigerant tables or the tables in *Part 4*.
10. Find the pressure drop thru automatic valves and accessories from manufacturers' catalogs. If these are given in equivalent feet, change to psi by multiplying by the ratio:

$$\frac{Step\ (9)}{Step\ (6)}$$

11. Add *Steps 9 and 10*.

In systems in which automatic valves and accessories may create a relatively high pressure drop, the line size can be increased to minimize their effect on the system.

Example 1 — Use of Pipe Sizing Charts

Given:
Refrigerant 12 system
Load — 46 tons
Equivalent length of piping — 65 ft
Saturated suction — 30 F
Condensing temperature — 100 F
Type L copper tubing

Find:
Suction line size for pressure drop corresponding to 2 F.
Actual pressure drop in terms of degrees F for size selected.

Solution:

See *Chart 7*.

1. Line sizes for 40 F saturated suction and 105 F condensing temperature are shown on *Chart 7*. Determine the correction factor for a 30 F suction temperature of 1.19 from table in notes following *Chart 9*.

2. Determine adjusted tons to be used in *Chart 7* by multiplying correction factor in *Step 1* by load in tons:

$$1.19 \times 46 = 55 \text{ tons}$$

3. Enter *Chart 7* and project upward from 55 tons, to a $2\frac{5}{8}$ in. OD pipe size, then to a $3\frac{1}{8}$ in. OD pipe size. At $2\frac{5}{8}$ in. OD, a 2 F drop is obtained with 33 ft of pipe; at $3\frac{1}{8}$ in. OD a 2 F drop is obtained with 71 ft of pipe. Select a $3\frac{1}{8}$ in. OD pipe to obtain less than a 2 F drop.

4. Use the following equation to determine actual pressure drop in terms of degrees F in the $3\frac{1}{8}$ in. OD pipe with a 46 ton load:

Actual pressure drop

$$= \frac{\text{equivalent ft of pipe}}{\text{piping allowed for 2 F drop}} \times 2 \text{ F}$$

$$= \frac{65}{71} \times 2 = 1.8 \text{ F}$$

LIQUID LINE DESIGN

Refrigeration oil is sufficiently miscible with these refrigerants in the liquid phase to insure adequate mixing and oil return. Therefore low liquid velocities and traps in liquid lines do not pose oil return problems.

The amount of liquid line pressure drop which can be tolerated is dependent on the number of degrees subcooling of the liquid. Usually this amounts to 2 F to 5 F as the liquid leaves the condenser. Liquid lines should not be sized for more than a 2 F drop under normal circumstances. In addition, liquid lines passing thru extremely warm spaces should be insulated.

Friction Drop and Static Head

With an appreciable friction drop and/or a static head due to elevation of the liquid metering device above the condenser, it may be necessary to resort to some additional means of liquid subcooling to prevent flashing in the liquid line. Increasing the liquid line pipe size minimizes pipe friction and flashing due to friction drop.

In large systems where the cost is warranted, a liquid pump may be used to overcome static head.

An arrangement shown in *Fig. 57* illustrates a method which may be used to overcome the effect of excessive flash gas caused by a high static head in the system. This arrangement does not prevent the forming of flash gas, but does offset the effect it might have on the operation of the evaporator and valves.

Liquid Subcooling

Where liquid subcooling is required, it is usually accomplished by one or both of the following arrangements:

1. A liquid suction heat interchanger (heat dissipates internally to suction gas).

2. Liquid subcooling coils in evaporative condensers and air-cooled condensers (heat dissipates externally to atmosphere).

The amount of liquid subcooling required may be determined by use of a nomograph, *Chart 22* or by calculation. The following examples illustrate both methods.

Example 2 — Liquid Subcooling from Nomograph

Given:

Refrigerant 12 system
Condensing temperature — 100 F (131.6 psia)
Liquid line pressure drop (incl. liquid lift) — 29.9 psi

Find:

Amount of liquid subcooling in degrees F required to prevent flashing of liquid refrigerant.

Solution:

Use *Chart 22*.

1. Determine pressure at expansion valve:

$$131.6 - 29.9 = 101.7 \text{ psia}$$

2. Draw line from point *A* (100 F cond temp) to point *B* (101.7 psia at expansion valve).

3. Draw line from point *C* (intersection of *AB* with line *Z*) thru point *D* (0% flash gas) to point *E* (intersection of *CD* with liquid subcooling line).

4. Liquid subcooling at point *E* = 18 F. Liquid subcooling required to prevent liquid flashing = 18 F.

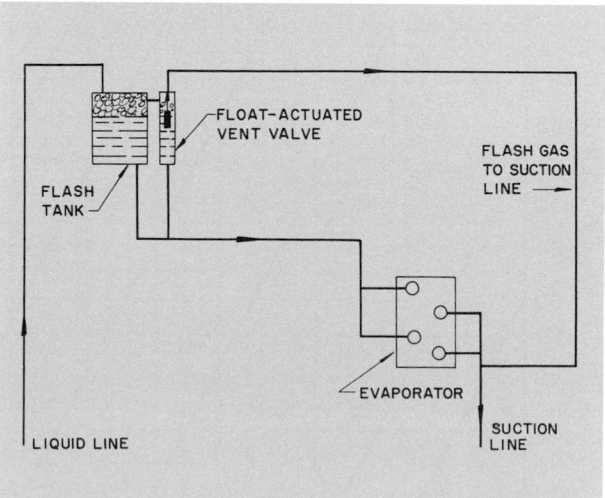

FIG. 57 — METHOD OF OVERCOMING ILL EFFECTS OF
SYSTEM HIGH STATIC HEAD

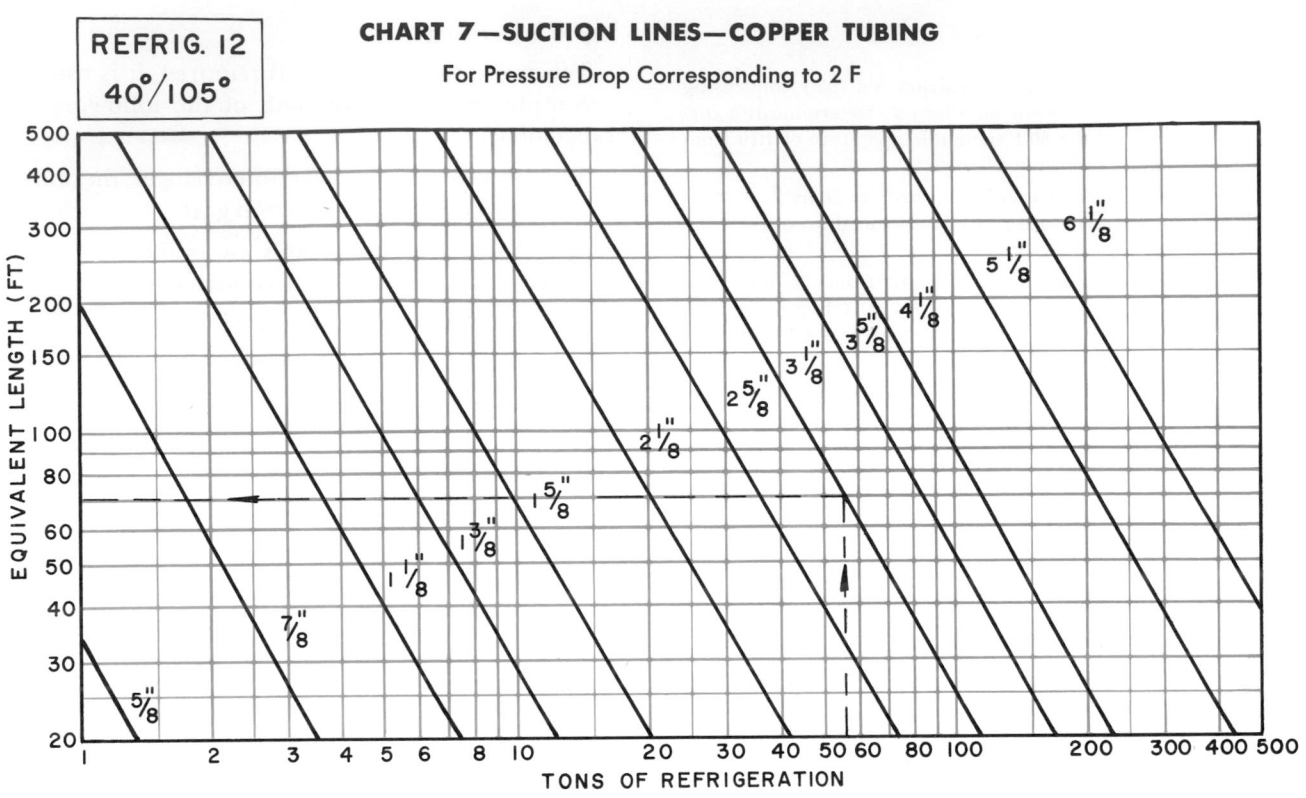

CHART 7—SUCTION LINES—COPPER TUBING

For Pressure Drop Corresponding to 2 F

REFRIG. 12
40°/105°

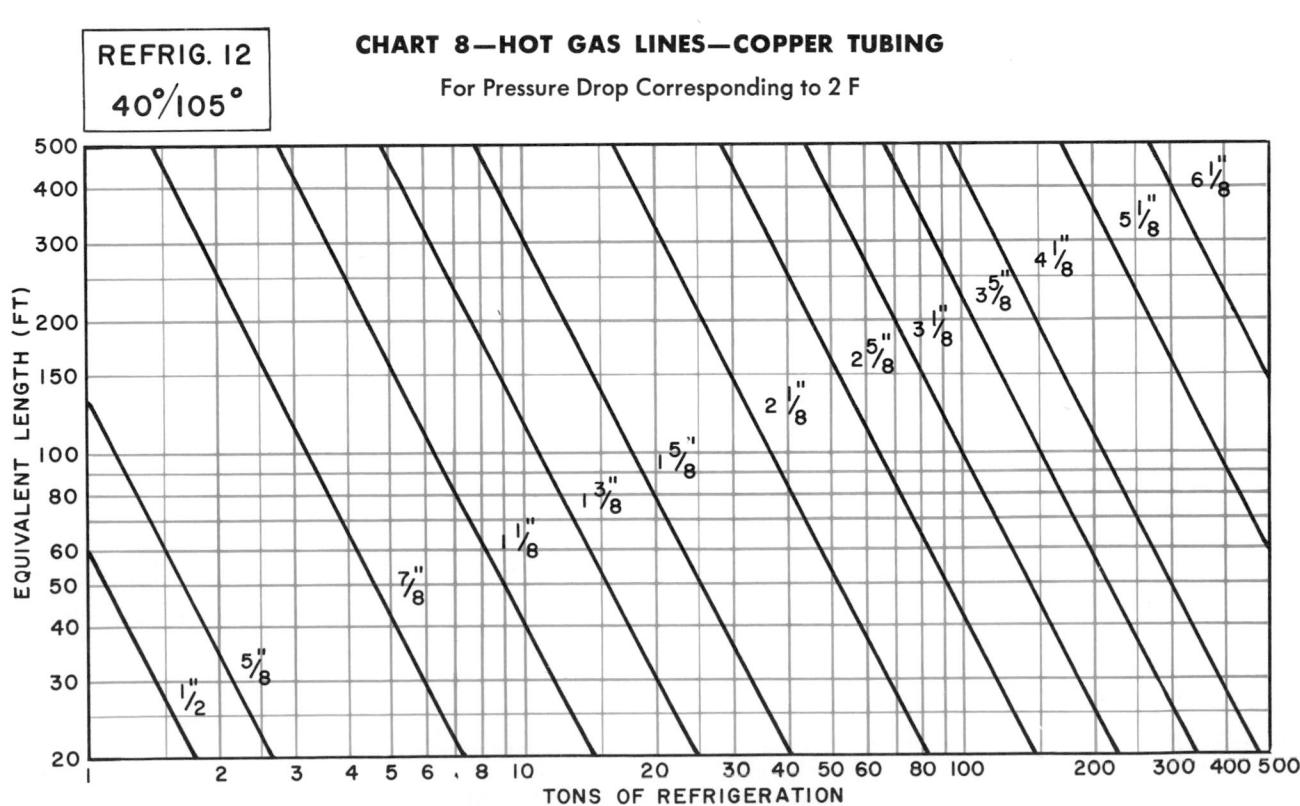

CHART 8—HOT GAS LINES—COPPER TUBING

For Pressure Drop Corresponding to 2 F

REFRIG. 12
40°/105°

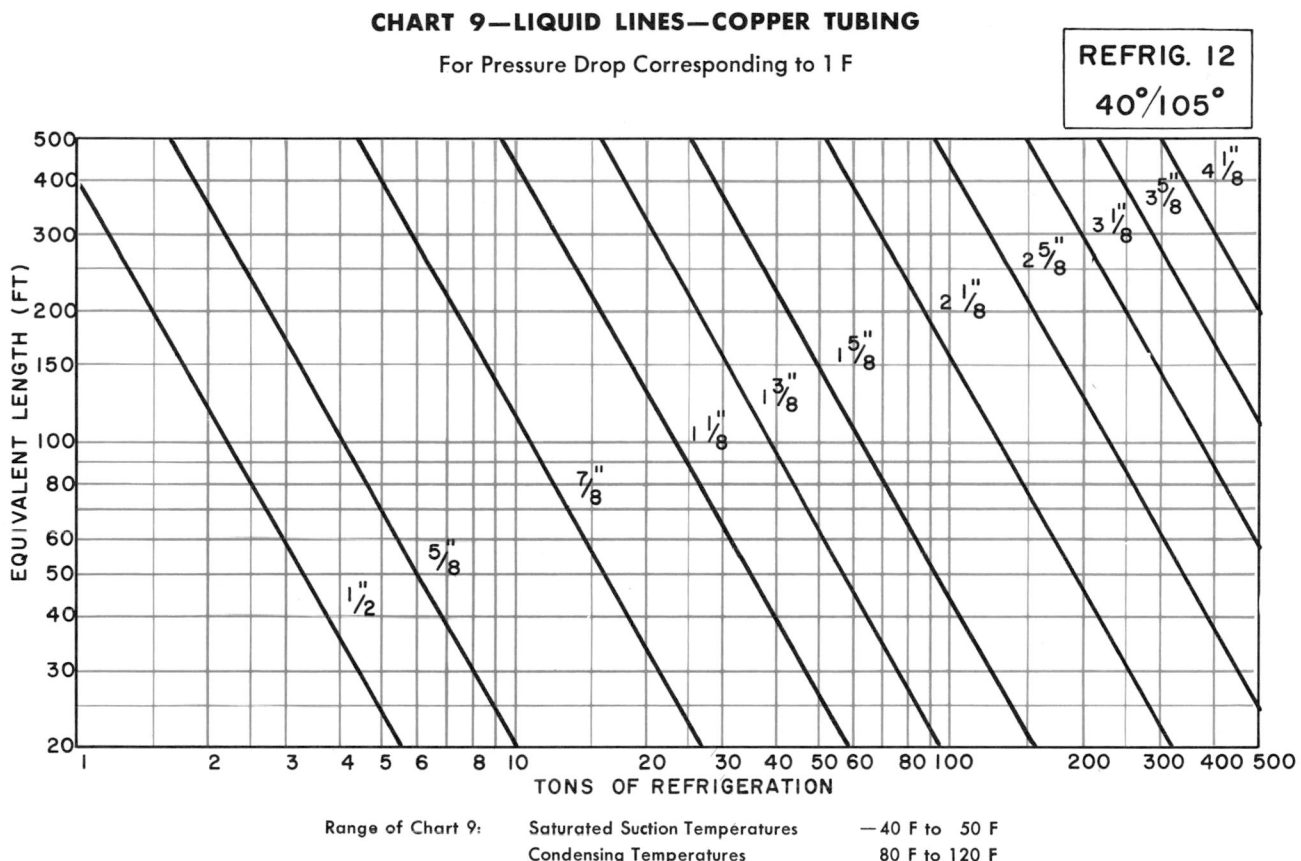

CHART 9—LIQUID LINES—COPPER TUBING
For Pressure Drop Corresponding to 1 F

REFRIG. 12
40°/105°

Range of Chart 9:   Saturated Suction Temperatures   —40 F to 50 F
                    Condensing Temperatures   80 F to 120 F

Pressure drop is given in equivalent degrees because of the general acceptance of this method of sizing. The corresponding pressure drop in psi may be determined by referring to the saturated refrigerant tables.

To use Charts 7 and 8 **for conditions other than 40 F saturated suction, 105 F condensing,** multiply the refrigeration load in tons by the factor below and use the product in reading the chart (S = Suction, HG = Hot Gas).

COND TEMP (F)	SATURATED SUCTION TEMPERATURE (F)																		
	−40		−30		−20		−10		0		10		20		30		40		50	
	TONS MULTIPLYING FACTOR																		
	S	HG	S	HG	S	HG	S	HG	S	HG	S	HG	S	HG	S	HG	S	HG	S	HG
80	4.90	1.47	4.09	1.43	3.04	1.41	2.41	1.37	1.94	1.34	1.60	1.32	1.29	1.29	1.09	1.26	.90	1.24	.77	1.22
90	5.17	1.34	4.30	1.30	3.20	1.28	2.54	1.23	2.00	1.22	1.68	1.19	1.35	1.17	1.12	1.14	.94	1.12	.76	1.10
100	5.45	1.24	4.55	1.21	3.33	1.19	2.69	1.13	2.10	1.12	1.76	1.10	1.41	1.08	1.18	1.05	.98	1.03	.80	1.01
110	5.80	1.17	4.81	1.12	3.56	1.09	2.83	1.04	2.20	1.03	1.86	1.01	1.48	1.00	1.23	.96	1.02	.95	.84	.92
120	6.20	1.09	5.08	1.04	3.78	1.00	3.00	.98	2.37	.97	1.96	.94	1.58	.91	1.30	.89	1.06	.87	.88	.85
130	6.68	1.02	5.50	.98	4.03	.95	3.21	.93	2.54	.90	2.09	.87	1.68	.85	1.39	.83	1.15	.81	.96	.79
140	7.20	.98	5.91	.94	4.35	.91	3.43	.88	2.71	.86	2.24	.83	1.80	.80	1.50	.79	1.22	.77	1.03	.75
150	7.90	.95	6.43	.91	4.74	.88	3.74	.85	2.97	.82	2.41	.79	1.94	.78	1.62	.76	1.31	.73	1.11	.71
160	8.70	.94	7.10	.90	5.20	.87	4.06	.84	3.22	.81	2.62	.78	2.10	.76	1.74	.73	1.41	.71	1.20	.69

NOTES:

1. To use suction and hot gas line charts **for friction drop other than 2 F** or liquid line charts **for friction drop other than 1 F,** multiply equivalent length by factor below and use product in reading chart.

	Liquid Line	0.25	0.5	.75	1.0	1.25	1.5	2.0	2.5	3.0
Friction Drop (F)	Hot Gas Line Suction Line	0.5	1.0	1.5	2.0	2.5	3.0	4.0	5.0	6.0
Multiplier		4.0	2.0	1.3	1.0	0.8	0.7	0.5	0.4	0.3

2. Pipe sizes are OD and are for Type L copper tubing.

CHART 10—SUCTION LINES—STEEL PIPE

For Pressure Drop Corresponding to 2 F

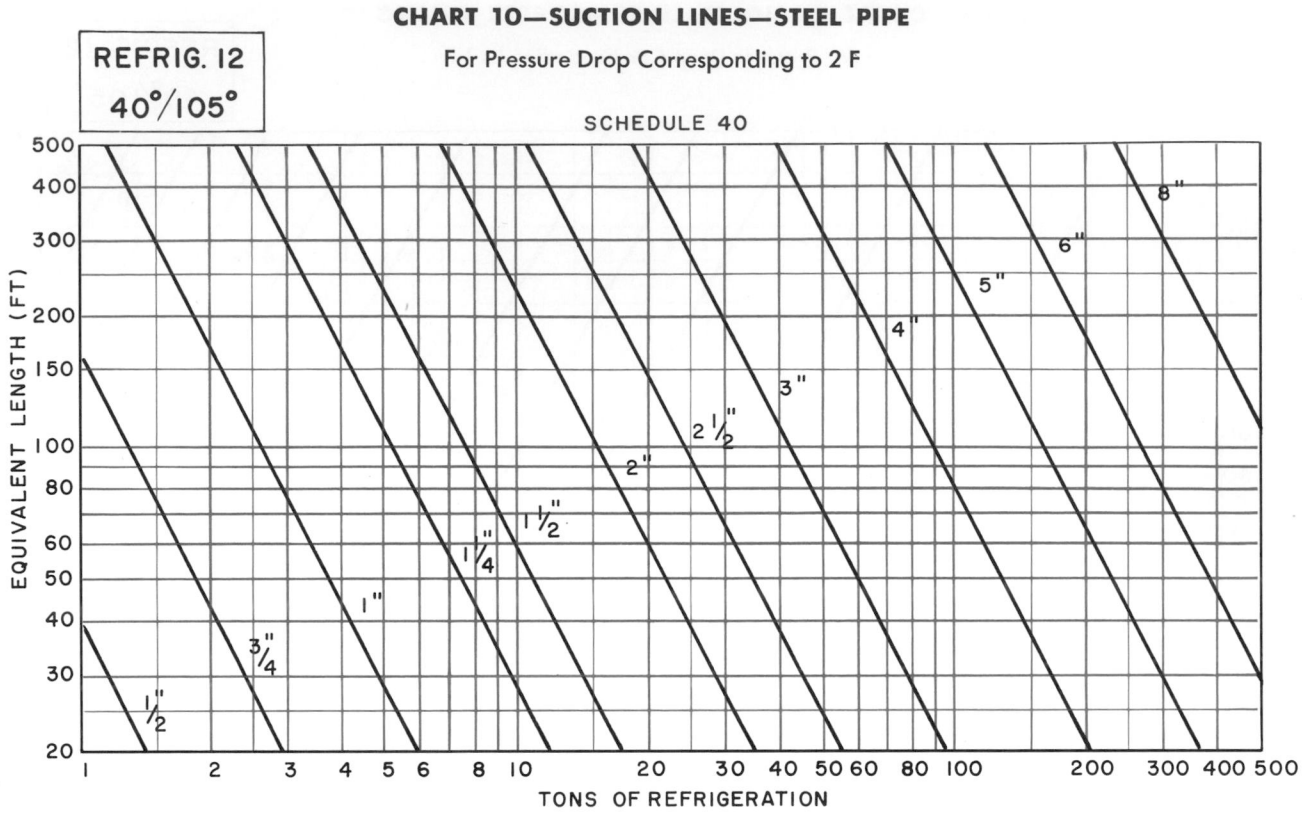

CHART 11—HOT GAS LINES—STEEL PIPE

For Pressure Drop Corresponding to 2 F

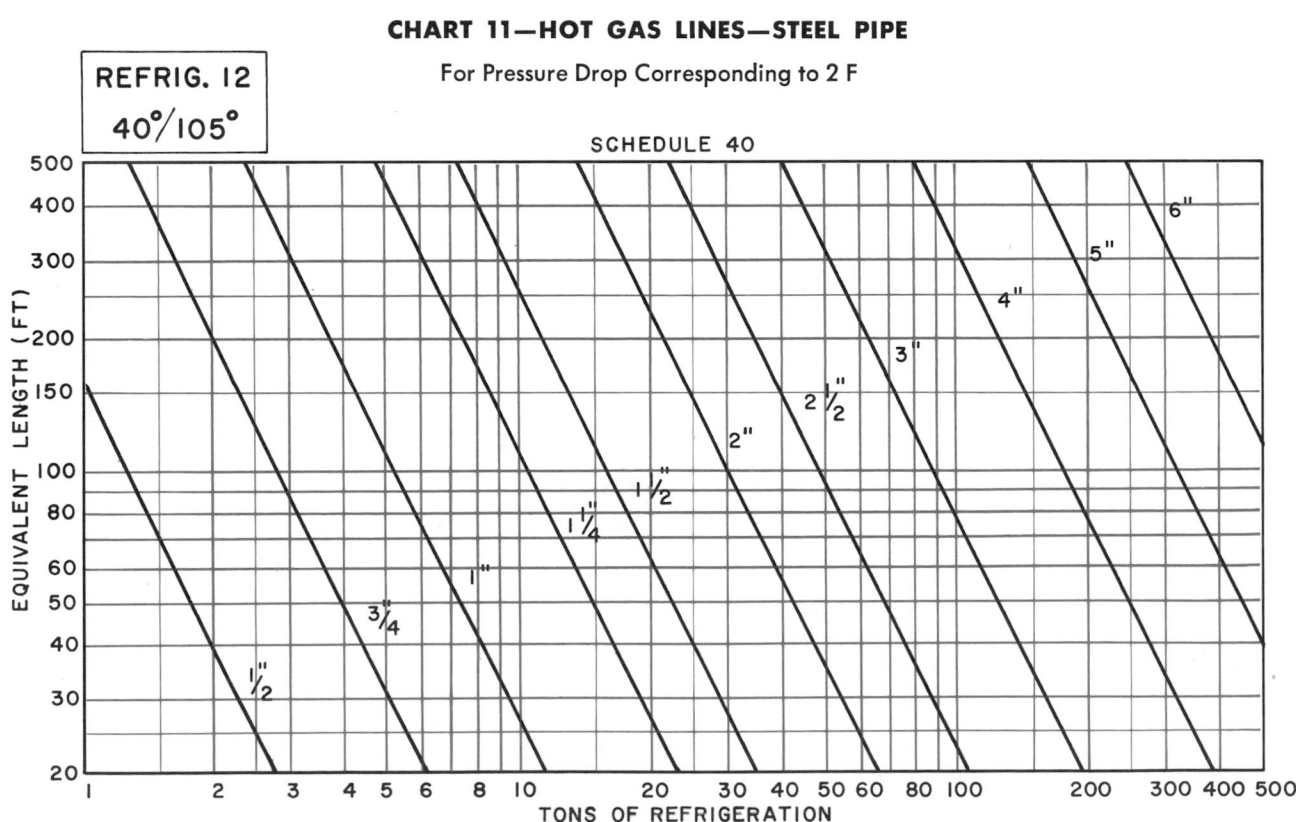

CHART 12—LIQUID LINES—STEEL PIPE

For Pressure Drop Corresponding to 1 F

REFRIG. 12
40°/105°

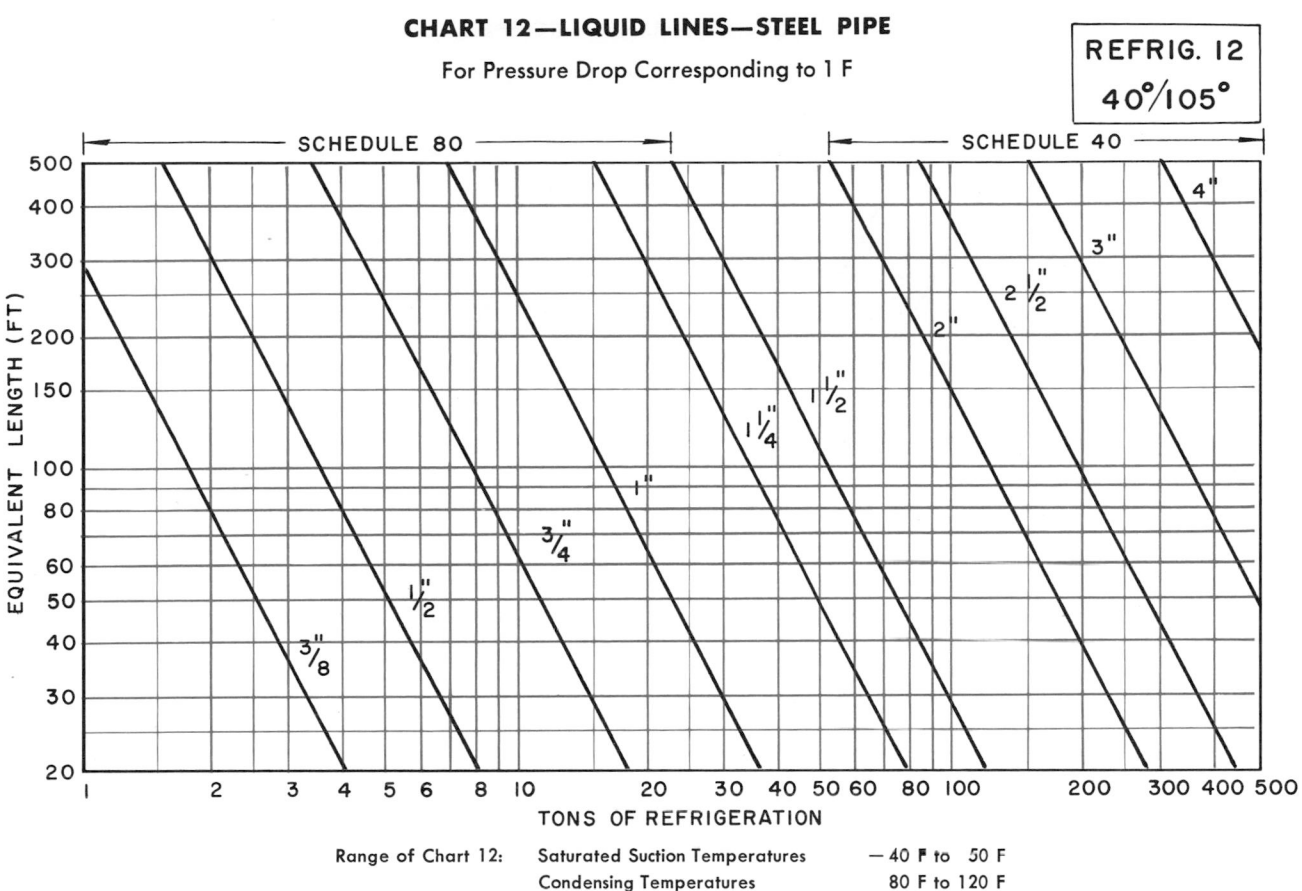

Range of Chart 12:   Saturated Suction Temperatures   − 40 F to 50 F
Condensing Temperatures   80 F to 120 F

Pressure drop is given in equivalent degrees because of the general acceptance of this method of sizing. The corresponding pressure drop in psi may be determined by referring to the saturated refrigerant tables.

To use Charts 10 and 11 **for conditions other than 40 F saturated suction, 105 F condensing,** multiply the refrigeration load in tons by the factor below and use the product in reading the chart (S = Suction, HG = Hot Gas).

COND TEMP (F)	SATURATED SUCTION TEMPERATURE (F)																	
	−40		−30		−20		−10		0		10		20		30		40		50
	TONS MULTIPLYING FACTOR																	
	S	HG	S	HG	S	HG	S	HG	S	HG	S	HG	S	HG	S	HG	S	HG
80	4.56	1.44	3.58	1.41	2.88	1.38	2.26	1.35	1.83	1.32	1.54	1.29	1.29	1.27	1.07	1.24	.90	1.22	.78	1.20
90	4.78	1.33	3.70	1.29	3.00	1.26	2.36	1.20	1.92	1.20	1.59	1.18	1.34	1.16	1.12	1.13	.94	1.11	.79	1.09
100	5.04	1.24	3.90	1.20	3.16	1.16	2.46	1.12	2.00	1.11	1.69	1.08	1.39	1.08	1.17	1.05	.98	1.02	.81	1.00
110	5.38	1.14	4.16	1.11	3.36	1.06	2.55	1.05	2.13	1.02	1.78	1.00	1.47	1.00	1.23	.97	1.02	.94	.87	.92
120	5.72	1.07	4.45	1.03	3.56	1.00	2.74	.98	2.25	.95	1.89	.93	1.56	.92	1.29	.89	1.06	.87	.93	.85
130	6.18	1.00	4.79	.99	3.82	.94	2.95	.91	2.39	.89	2.00	.86	1.65	.84	1.38	.82	1.13	.81	.99	.79
140	6.67	.97	5.19	.95	4.14	.90	3.20	.87	2.58	.83	2.16	.80	1.78	.80	1.46	.79	1.20	.77	1.06	.75
150	7.26	.95	5.65	.91	4.48	.88	3.48	.85	2.78	.82	2.32	.79	1.92	.78	1.59	.76	1.30	.73	1.13	.71
160	8.06	.94	6.20	.90	4.94	.87	3.81	.84	3.04	.81	2.52	.78	2.08	.76	1.72	.73	1.41	.71	1.23	.69

NOTES:

1. To use suction and hot gas line charts **for friction drop other than 2 F** or liquid line charts **for friction drop other than 1 F,** multiply equivalent length by factor below and use product in reading chart.

Friction Drop (F)	Liquid Line	0.25	0.5	.75	1.0	1.25	1.5	2.0	2.5	3.0
	Hot Gas Line Suction Line	0.5	1.0	1.5	2.0	2.5	3.0	4.0	5.0	6.0
Multiplier			4.0	2.0	1.3	1.0	0.8	0.7	0.5	0.4	0.3

2. Pipe sizes are nominal and are for steel pipe.

CHART 13—SUCTION LINES—COPPER TUBING

For Pressure Drop Corresponding to 2 F

REFRIG. 500
40°/105°

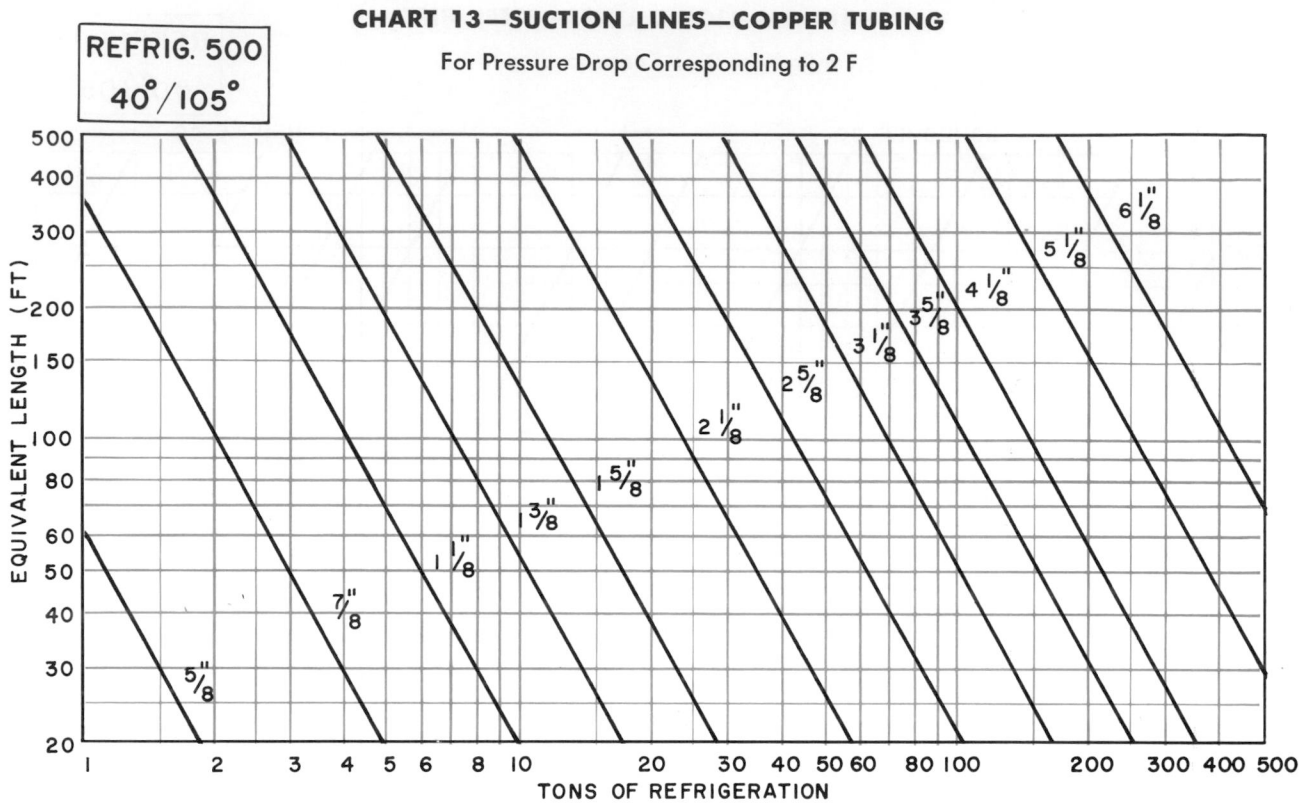

CHART 14—HOT GAS LINES—COPPER TUBING

For Pressure Drop Corresponding to 2 F

REFRIG. 500
40°/105°

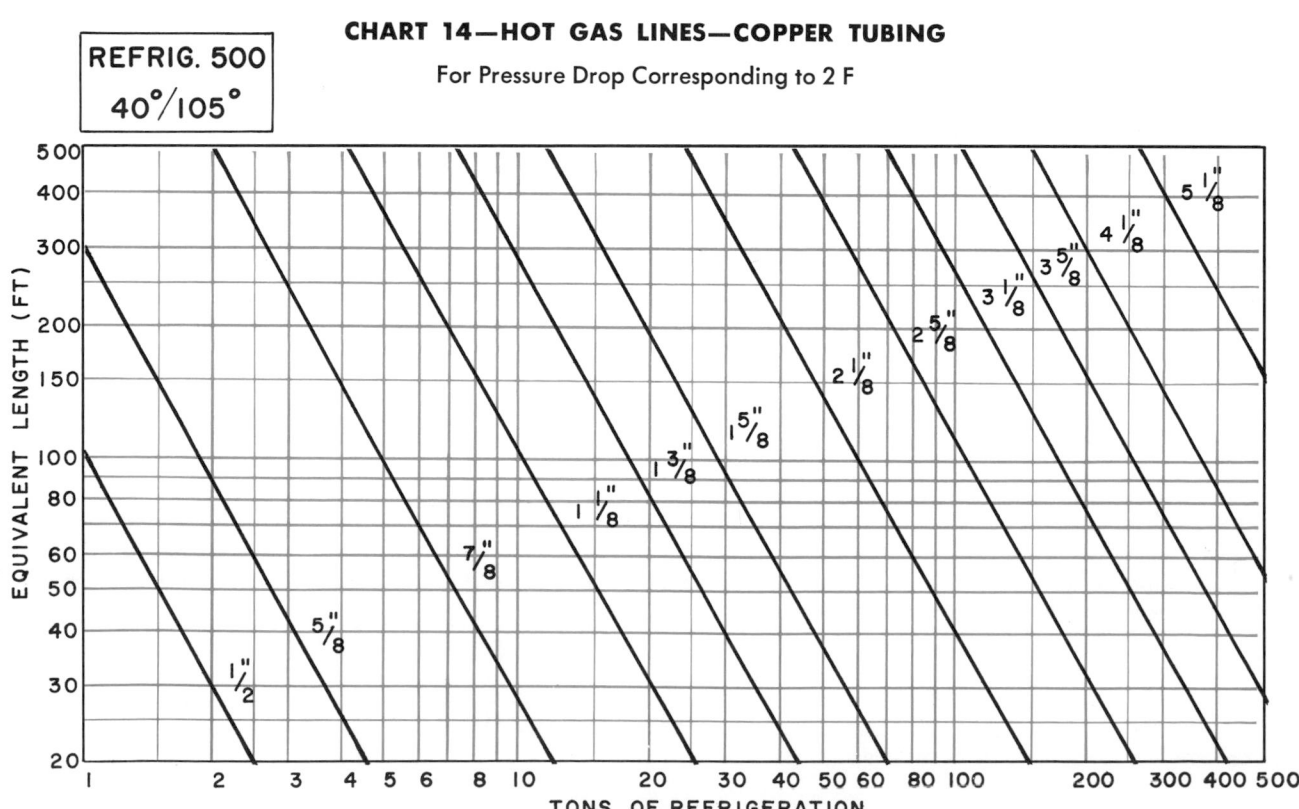

CHART 15—LIQUID LINES—COPPER TUBING

For Pressure Drop Corresponding to 1 F

REFRIG. 500
40°/105°

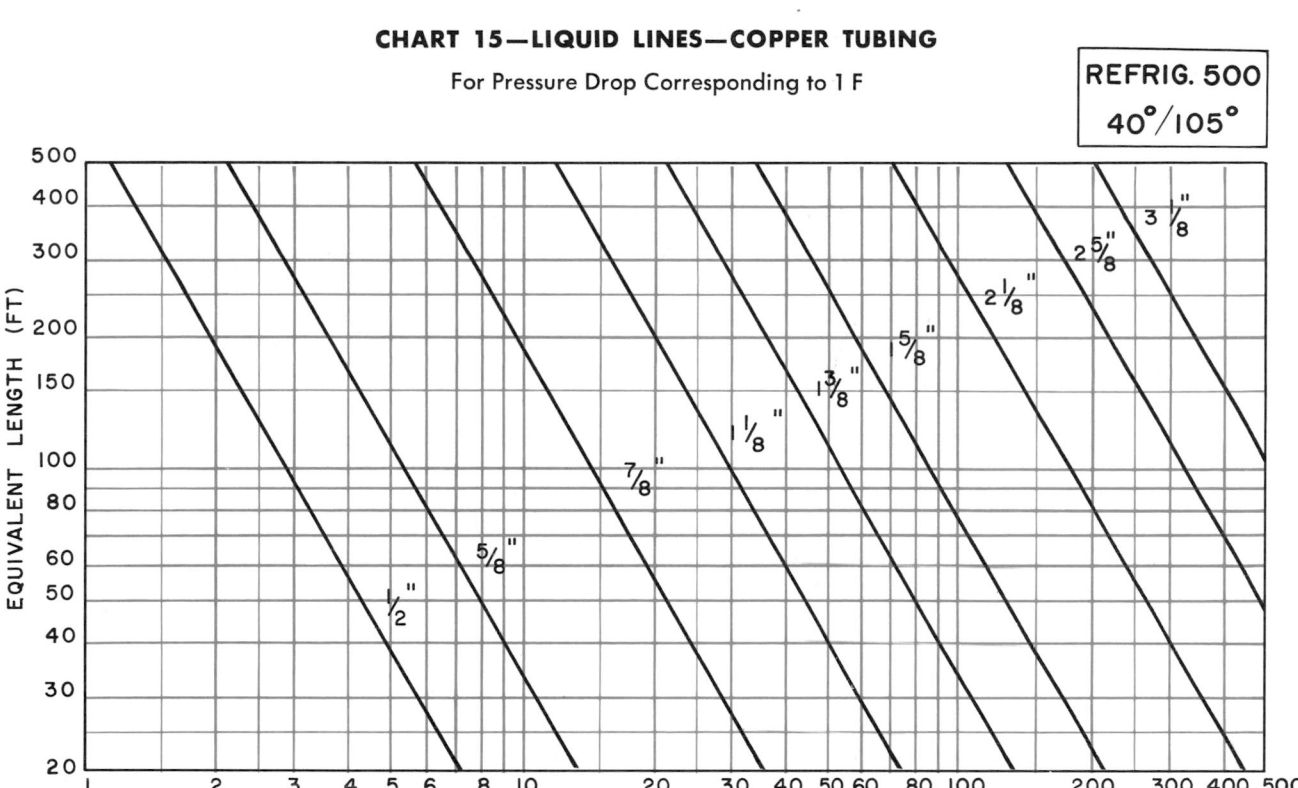

Range of Chart 15:   Saturated Suction Temperatures   —40 F to 50 F
                        Condensing Temperatures   80 F to 120 F

Pressure drop is given in equivalent degrees because of the general acceptance of this method of sizing. The corresponding pressure drop in psi may be determined by referring to the saturated refrigerant tables.

To use Charts 13 and 14 **for conditions other than 40 F saturated suction, 105 F condensing,** multiply refrigeration load in tons by factor below and use product in reading chart. (S = Suction—HG = Hot Gas)

CON-DENSING TEMPERATURE (F)	SATURATED SUCTION TEMPERATURE (F)																			
	—40		—30		—20		—10		0		10		20		30		40		50	
	TON MULTIPLYING FACTOR																			
	S	HG	S	HG	S	HG	S	HG	S	HG	S	HG	S	HG	S	HG	S	HG	S	HG
80	5.4	1.7	4.3	1.6	3.4	1.6	2.7	1.5	2.1	1.4	1.7	1.4	1.4	1.4	1.1	1.3	0.9	1.2	0.8	1.2
90	5.8	1.6	4.6	1.5	3.6	1.5	2.9	1.4	2.3	1.4	1.8	1.3	1.5	1.2	1.2	1.2	0.9	1.1	0.8	1.1
100	6.1	1.4	4.8	1.4	3.7	1.3	3.0	1.3	2.4	1.2	1.9	1.2	1.5	1.1	1.2	1.1	1.0	1.0	0.8	1.0
110	6.5	1.4	5.1	1.3	4.0	1.2	3.1	1.2	2.5	1.2	2.0	1.1	1.6	1.1	1.3	1.0	1.0	1.0	0.9	0.9
120	6.9	1.3	5.4	1.2	4.2	1.2	3.3	1.1	2.6	1.1	2.1	1.0	1.7	1.0	1.3	0.9	1.1	0.9	0.9	0.9

NOTES:

1. To use suction and hot gas line charts **for friction drop other than 2 F** or liquid line chart **for friction drop other than 1 F,** multiply equivalent length by factor below and use product in reading chart.

Friction Drop (F)	Liquid Line	0.25	0.5	.75	1.0	1.25	1.5	2.0	2.5	3.0
	Hot Gas Line Suction Line	0.5	1.0	1.5	2.0	2.5	3.0	4.0	5.0	6.0
Multiplier		4.0	2.0	1.3	1.0	0.8	0.7	0.5	0.4	0.3

2. Pipe sizes are OD and are for Type L copper tubing.

CHART 16—SUCTION LINES—COPPER TUBING

For Pressure Drop Corresponding to 2 F

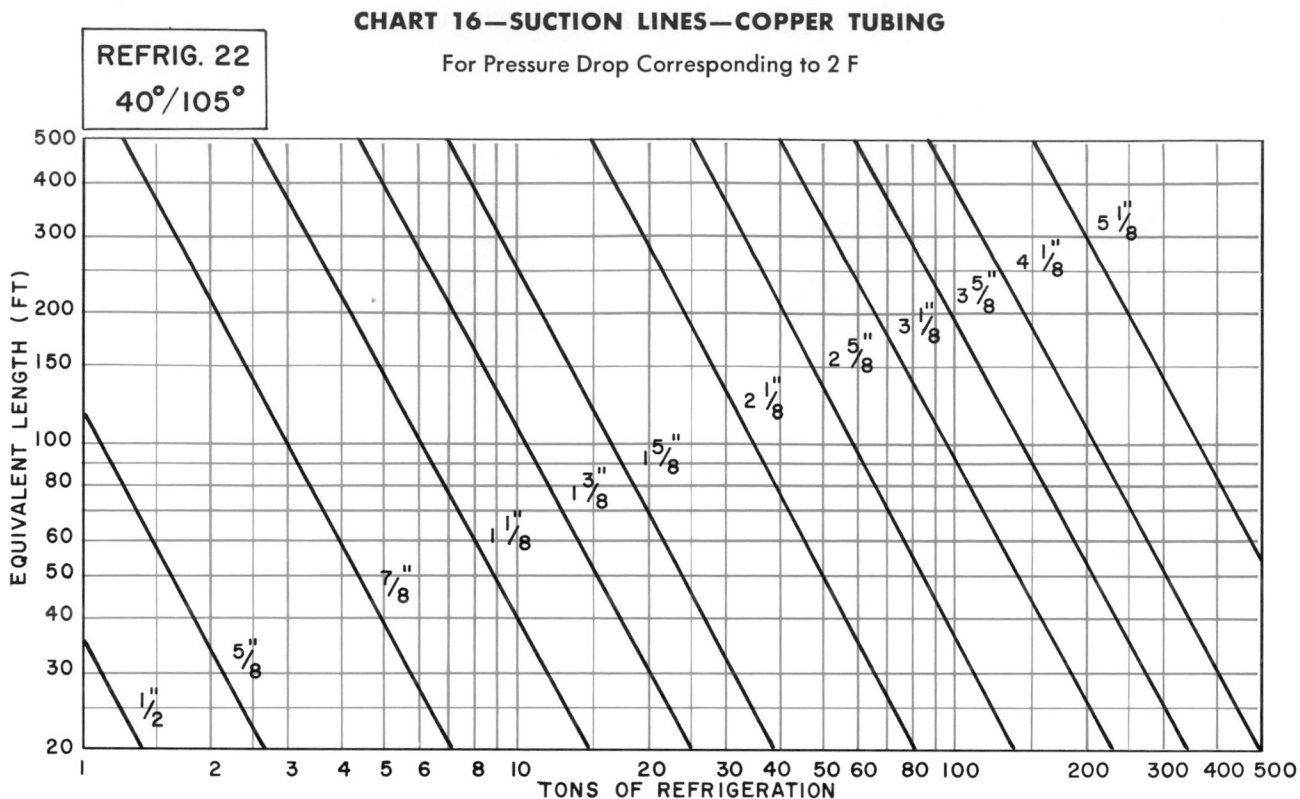

CHART 17—HOT GAS LINES—COPPER TUBING

For Pressure Drop Corresponding to 2 F

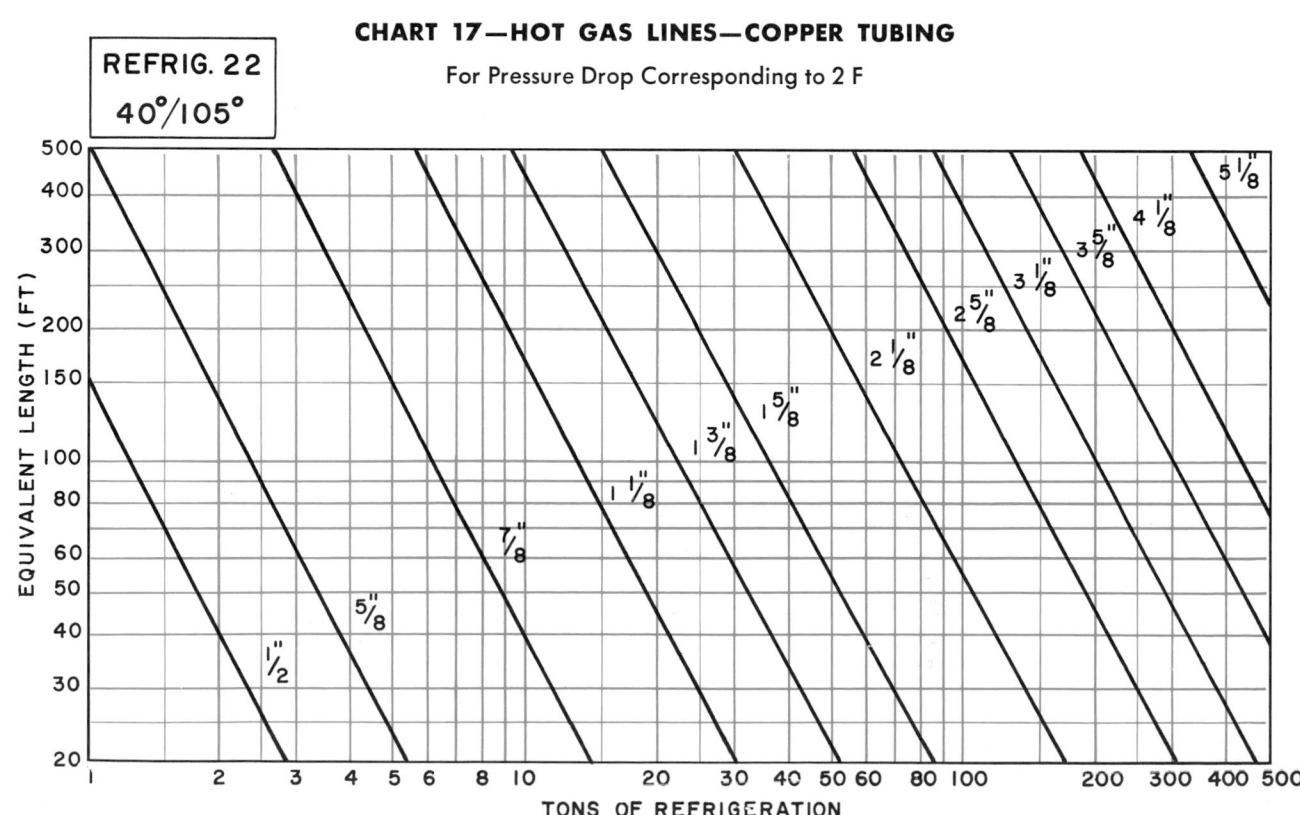

CHART 18—LIQUID LINES—COPPER TUBING

For Pressure Drop Corresponding to 1 F

REFRIG. 22
40°/105°

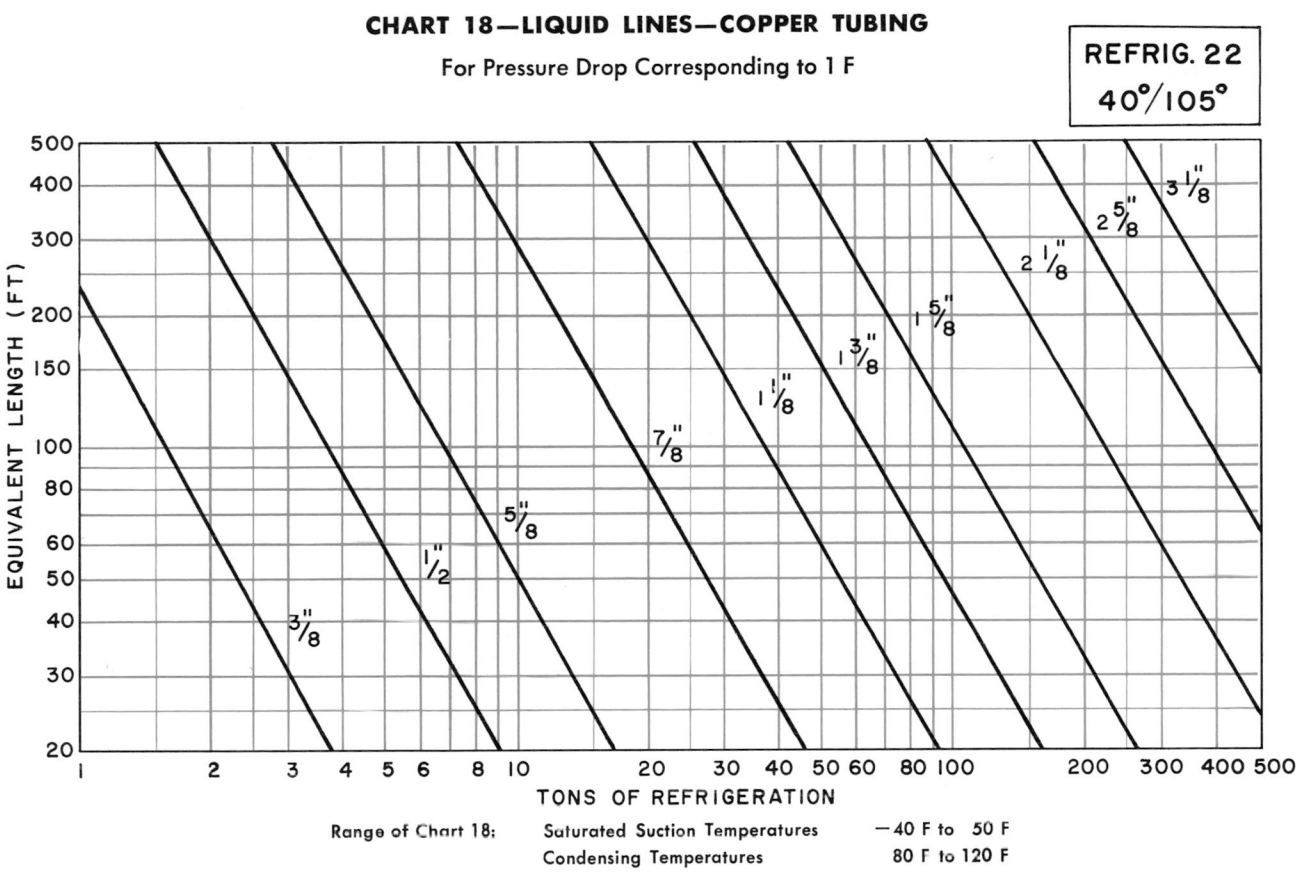

TONS OF REFRIGERATION

Range of Chart 18:   Saturated Suction Temperatures   −40 F to 50 F
                       Condensing Temperatures   80 F to 120 F

Pressure drop is given in equivalent degrees because of the general acceptance of this method of sizing. The corresponding pressure drop in psi may be determined by referring to the saturated refrigerant tables.

To use Charts 16 and 17 **for conditions other than 40 F saturated suction, 105 F condensing,** multiply the refrigeration load in tons by the factor below and use the product in reading the chart (S = Suction, HG = Hot Gas).

COND TEMP (F)	SATURATED SUCTION TEMPERATURE (F)																
	−40		−30		−20		−10		0		10		20		30		40
	TONS MULTIPLYING FACTOR																
	S	HG	S	HG	S	HG	S	HG	S	HG	S	HG	S	HG	S	HG	S	HG
80	4.65	1.40	3.70	1.38	2.93	1.36	2.32	1.34	1.91	1.32	1.54	1.29	1.27	1.28	1.06	1.25	.91	1.24	.81	1.23
90	4.83	1.27	3.87	1.24	3.05	1.23	2.42	1.21	1.98	1.19	1.61	1.17	1.33	1.15	1.11	1.13	.93	1.12	.84	1.10
100	5.12	1.18	4.04	1.16	3.20	1.14	2.56	1.12	2.06	1.10	1.70	1.08	1.39	1.06	1.16	1.04	.97	1.03	.88	1.01
110	5.42	1.08	4.28	1.06	3.39	1.04	2.70	1.02	2.20	1.01	1.80	1.00	1.45	.98	1.21	.97	1.01	.95	.92	.93
120	5.75	1.01	4.55	1.00	3.60	.98	2.85	.95	2.36	.95	1.89	.91	1.53	.91	1.28	.90	1.07	.87	1.00	.87
130	6.20	.95	4.88	.94	3.81	.92	3.00	.88	2.48	.89	2.01	.82	1.63	.83	1.36	.83	1.13	.80	1.00	.81
140	6.68	.90	5.22	.88	4.12	.86	3.21	.82	2.64	.83	2.14	.77	1.73	.78	1.43	.77	1.20	.75	1.07	.75
150	7.21	.85	5.67	.82	4.45	.80	3.45	.77	2.82	.77	2.27	.73	1.85	.73	1.52	.71	1.27	.70	1.16	.69
160	7.90	.80	6.15	.78	4.81	.76	3.78	.73	3.06	.71	2.46	.70	2.02	.68	1.64	.66	1.39	.65	1.27	.64

NOTES:

1. To use suction and hot gas line charts **for friction drop other than 2 F** or liquid line charts **for friction drop other than 1 F,** multiply equivalent length by factor below and use product in reading chart.

Friction Drop (F)	Liquid Line	0.25	0.5	.75	1.0	1.25	1.5	2.0	2.5	3.0
	Hot Gas Line Suction Line	0.5	1.0	1.5	2.0	2.5	3.0	4.0	5.0	6.0
Multiplier			4.0	2.0	1.3	1.0	0.8	0.7	0.5	0.4	0.3

2. Pipe sizes are OD and are for Type L copper tubing.

CHART 19—SUCTION LINES—STEEL PIPE

For Pressure Drop Corresponding to 2 F

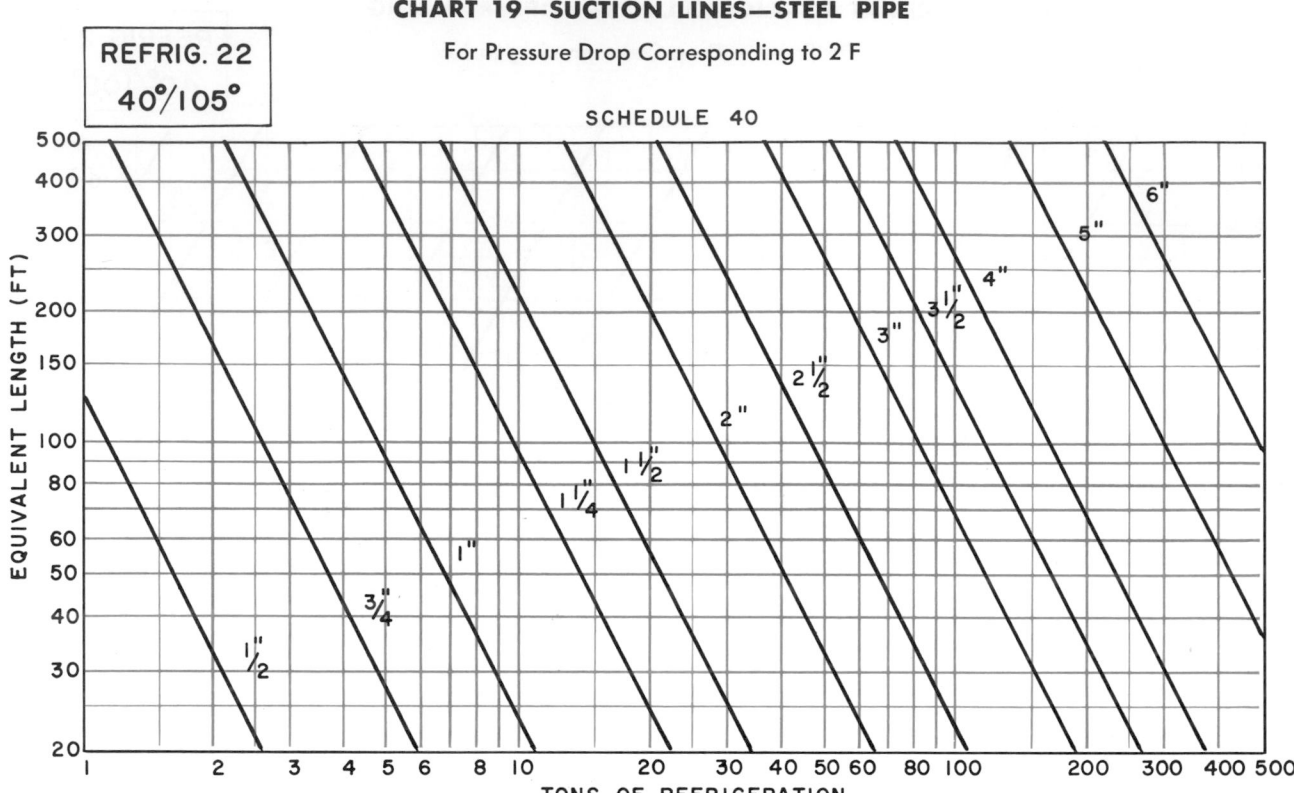

CHART 20—HOT GAS LINES—STEEL PIPE

For Pressure Drop Corresponding to 2 F

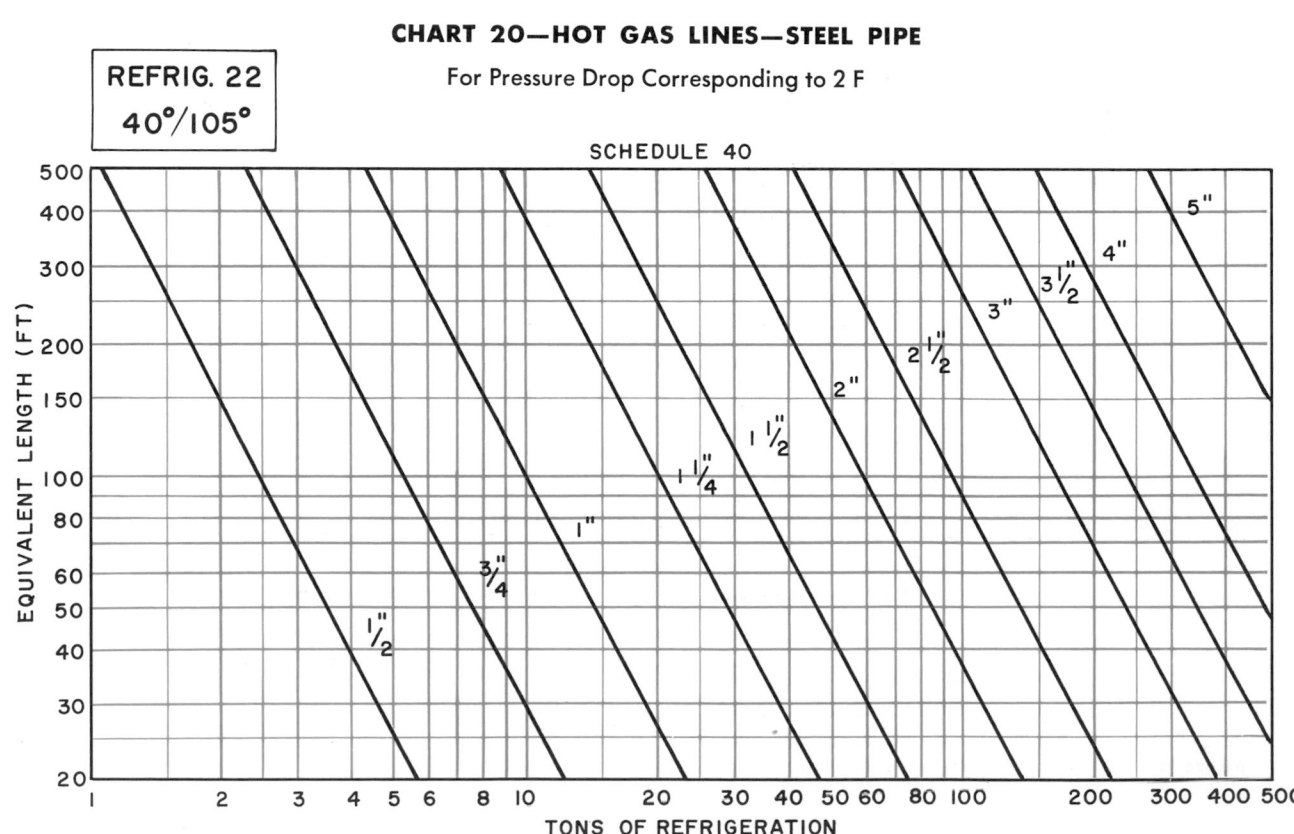

CHART 21—LIQUID LINES—STEEL PIPE

For Pressure Drop Corresponding to 1 F

REFRIG. 22
40°/105°

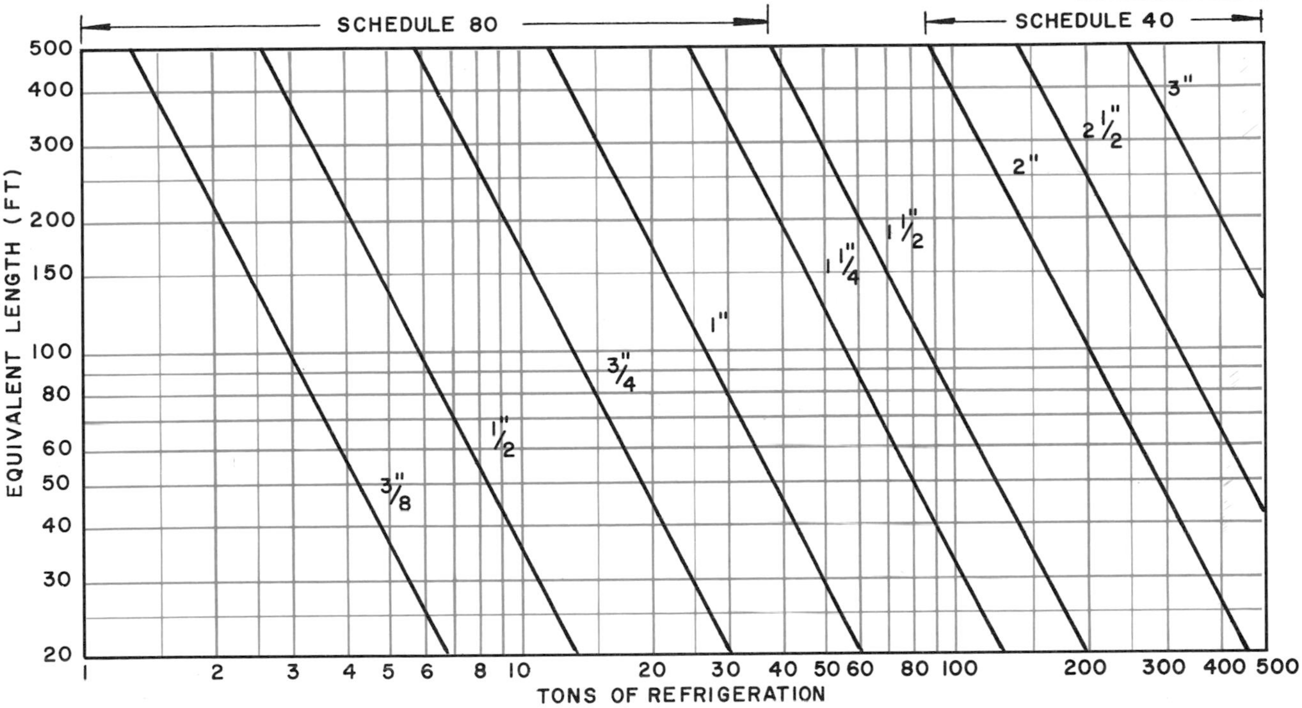

Range of Chart 21:		
	Saturated Suction Temperatures	−40 F to 50 F
	Condensing Temperatures	80 F to 120 F

Pressure drop is given in equivalent degrees because of the general acceptance of this method of sizing. The corresponding pressure drop in psi may be determined by referring to the saturated refrigerant tables.

To use Charts 19 and 20 **for conditions other than 40 F saturated suction, 105 F condensing,** multiply the refrigeration load in tons by the factor below and use the product in reading the chart (S = Suction, HG = Hot Gas).

COND TEMP (F)	SATURATED SUCTION TEMPERATURE (F)																			
	−40		−30		−20		−10		0		10		20		30		40		50	
	TONS MULTIPLYING FACTOR																			
	S	HG	S	HG	S	HG	S	HG	S	HG	S	HG	S	HG	S	HG	S	HG	S	HG
80	4.40	1.38	3.53	1.36	2.80	1.33	2.28	1.31	1.86	1.29	1.54	1.27	1.27	1.26	1.06	1.24	.90	1.22	.81	1.21
90	4.60	1.26	3.71	1.22	2.94	1.23	2.40	1.20	1.93	1.15	1.60	1.17	1.33	1.15	1.10	1.13	.93	1.12	.85	1.11
100	4.88	1.18	3.91	1.14	3.10	1.13	2.52	1.10	2.01	1.07	1.68	1.07	1.39	1.05	1.15	1.03	.97	1.02	.89	1.01
110	5.19	1.09	4.14	1.06	3.26	1.05	2.66	1.01	2.13	1.00	1.77	1.00	1.45	.98	1.21	.95	1.00	.95	.93	.93
120	5.54	1.00	4.40	1.00	3.44	.99	2.81	.94	2.26	.93	1.87	.93	1.54	.91	1.28	.88	1.05	.88	.96	.85
130	5.92	.95	4.71	.94	3.68	.92	3.00	.87	2.40	.86	2.00	.85	1.64	.85	1.37	.81	1.10	.81	1.00	.80
140	6.35	.90	5.05	.90	3.96	.87	3.20	.92	2.58	.80	2.14	.79	1.75	.79	1.46	.77	1.18	.75	1.08	.75
150	6.88	.86	5.45	.86	4.27	.82	3.43	.80	2.75	.77	2.27	.75	1.86	.74	1.54	.73	1.28	.71	1.17	.70
160	7.50	.85	5.95	.83	4.65	.80	3.75	.78	3.01	.76	2.46	.74	2.01	.72	1.67	.71	1.38	.69	1.27	.68

NOTES:

1. To use suction and hot gas line charts **for friction drop other than 2 F** or liquid line charts **for friction drop other than 1 F,** multiply equivalent length by factor below and use product in reading chart.

Friction Drop (F)										
	Liquid Line	0.25	0.5	.75	1.0	1.25	1.5	2.0	2.5	3.0
	Hot Gas Line Suction Line	0.5	1.0	1.5	2.0	2.5	3.0	4.0	5.0	6.0
Multiplier		4.0	2.0	1.3	1.0	0.8	0.7	0.5	0.4	0.3

2. Pipe sizes are nominal and are for steel pipe.

CHART 22—SUBCOOLING TO COMPENSATE FOR LIQUID LINE PRESSURE DROP

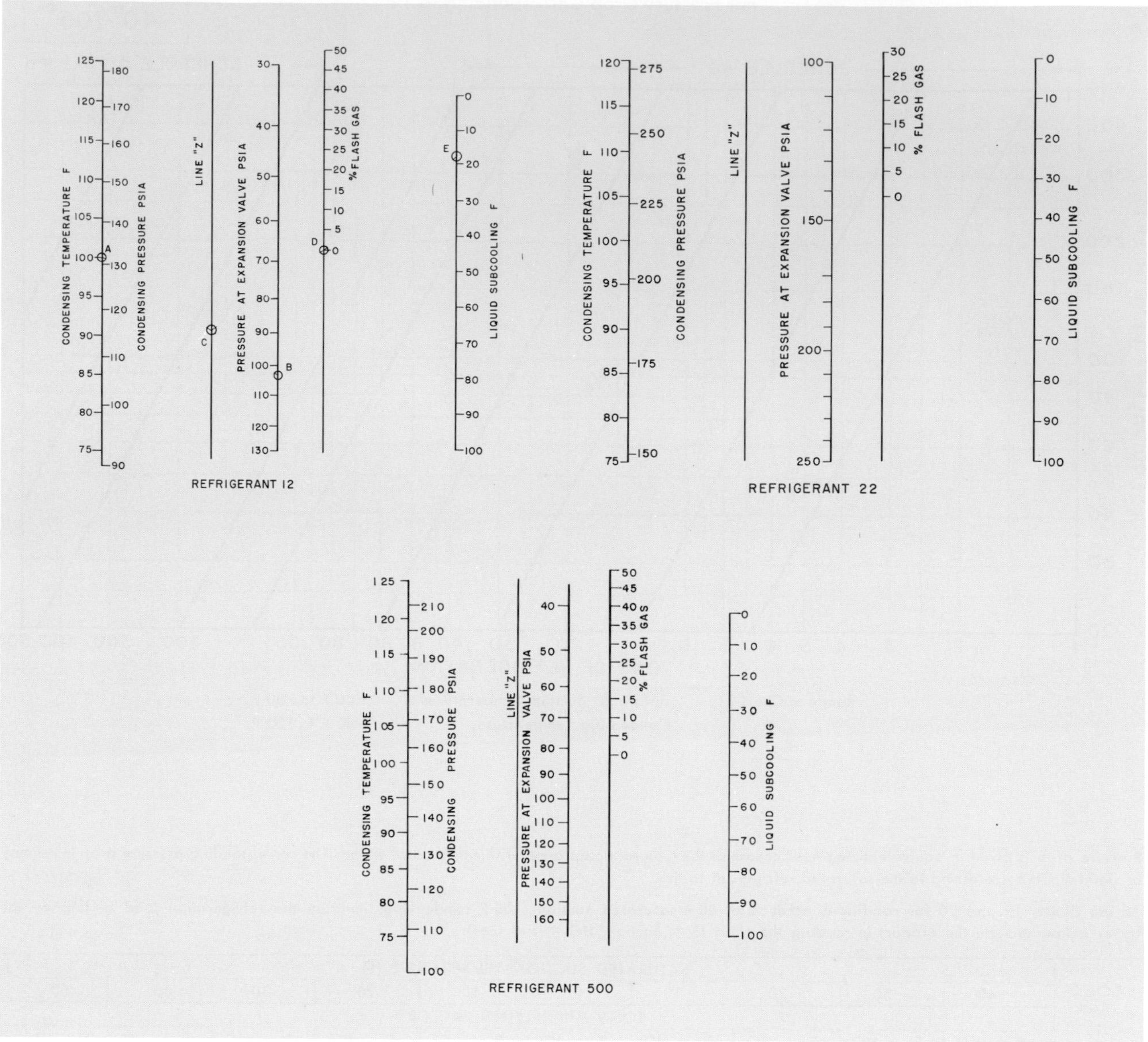

REFRIGERANT 12

REFRIGERANT 22

REFRIGERANT 500

Example 3 — Liquid Subcooling by Calculation

Given:

    Refrigerant 12 system

    Condensing temperature — 100 F

    Liquid lift — 35 ft

    Piping friction loss — 3 psi

    Losses thru valves and accessories — 7.4 psi

Find:

    Amount of liquid subcooling required to prevent flashing of liquid refrigerant.

Solution:

    1. Pressure loss due to pipe friction   = 3.0 psi

       Pressure loss due to valves and accessories = 7.4 psi

       Pressure loss due to 35 ft liquid lift

                  = 35/1.8*   = 19.5 psi

       Total pressure loss in liquid line   = 29.9 psi

2. Condensing pressure at 100 F   = 116.9 psig

    Pressure loss in liquid line   = 29.9 psi

    Net pressure at liquid feed valve = 87 psig

3. Saturation temperature at 87 psig = 82 F

    (from refrigerant property tables)

4. Subcooling required

    = condensing temp — saturation temp at 87 psig

    = 100 − 82 = 18 F

    Liquid subcooling required to prevent liquid flashing

    = 18 F.

*At normal liquid temperatures the static pressure loss due to elevation at the top of a liquid lift is one psi for every 1.8 ft of Refrigerant 12, 2.0 ft of Refrigerant 22, and 2.1 ft of Refrigerant 500.

Sizing of Condenser to Receiver Lines (Condensate Lines)

Liquid line piping from a condenser to a receiver is run out horizontally (same size as the condenser outlet connection) to allow for drainage of the condenser. It is then dropped vertically a sufficient distance to allow a liquid head in the line to overcome line friction losses. Additional head is required for coil condensers where the receiver is vented to the inlet of the coil. This additional head is equivalent to the pressure drop across the condenser coil. The condensate line is then run horizontally to the receiver.

Table 17 shows recommended sizes of the condensate line between the bottom of the liquid leg and the receiver.

SUCTION LINE DESIGN

Suction lines are the most critical from a design standpoint. The suction line must be designed to return oil from the evaporator to the compressor under minimum load conditions.

Oil which leaves the compressor and readily passes thru the liquid supply lines to the evaporators is almost completely separated from the refrigerant vapor. In the evaporator a distillation process occurs and continues until an equilibrium point is reached. The result is a mixture of oil and refrigerant rich in liquid refrigerant. Therefore the mixture which is separated from the refrigerant vapor can be returned to the compressor only by entrainment with the returning gas.

Oil entrainment with the return gas in a horizontal line is readily accomplished with normal design velocities. Therefore horizontal lines can and should be run "dead" level.

Suction Risers

Most refrigeration piping systems contain a suction riser. Oil circulating in the system can be returned up the riser only by entrainment with the returning gas. Oil returning up a riser creeps up the inner surface of the pipe. Whether the oil moves up the inner surface is dependent upon the mass velocity of the gas at the wall surface. The larger the pipe diameter, the greater the velocity required at the center of the pipe to maintain a given velocity at the wall surface.

Tables 18, 19 and 20 show the minimum tonnages required to insure oil return in upward flow suction risers and the friction drop in the risers in degrees F per 100 ft equivalent length.

Vertical risers should, therefore, be given special analysis and should be sized for velocities that assure

TABLE 17—CONDENSATE LINE SIZING

CONDENSER TO RECEIVER

(Based on Type L Copper Tubing)

CONDENSATE LINE SIZE (OD, In.)	REFRIGERATION, MAX. TONS			"X" MIN.*
	Refrigerant 12	Refrigerant 22	Refrigerant 500	
½ ⅝ ⅞	1.2 2.3 6.4	1.4 2.5 7.7	1.2 2.4 6.8	8"
1⅛ 1⅜ 1⅝	13.3 22.5 34.6	15.9 26 41	14.0 23.6 36	15"
2⅛ 2⅝ 3⅛ 3⅝	69.0 119 184 261	83 143 220 312	72 125 194 274	18"

*This is the minimum elevation required between a condenser coil outlet and a receiver inlet for the total load when receiver is vented to coil outlet header (based on 10 ft of horizontal pipe, 1 valve and 2 elbows).

oil return at minimum load. A riser selected on this basis may be smaller in diameter than its branch or than the suction main proper and, therefore, a relatively higher pressure drop may occur in the riser.

This penalty should be taken into account in finding the total suction line pressure drop. The horizontal lines should be sized to keep the total pressure drop within practical limits.

Because modern compressors have capacity reduction features, it is often difficult to maintain the gas velocities required to return oil upward in vertical suction risers. When the suction riser is sized to permit oil return at the minimum operating capacity of the system, the pressure drop in this portion of the line may be too great when operating at full load. If a correctly sized suction riser imposes too great a pressure drop at full load, a double suction riser should be used *(Fig. 58)*.

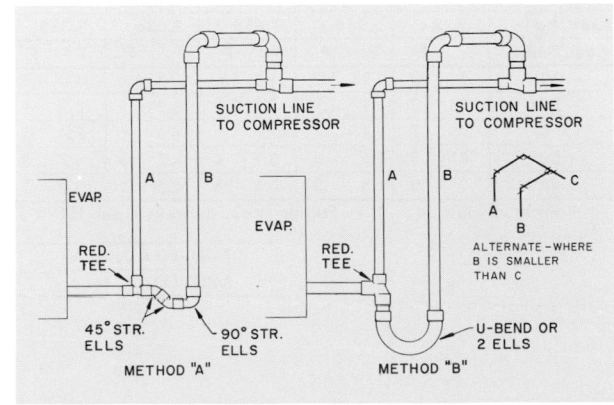

FIG. 58 — DOUBLE SUCTION RISER CONSTRUCTION

TABLE 18—MINIMUM TONNAGE FOR OIL ENTRAINMENT UP SUCTION RISERS

Refrigerant 12

COPPER TUBING—TYPE L																								
Pipe OD	$\frac{7}{8}$		$1\frac{1}{8}$		$1\frac{3}{8}$		$1\frac{5}{8}$		$2\frac{1}{8}$		$2\frac{5}{8}$		$3\frac{1}{8}$		$3\frac{5}{8}$		$4\frac{1}{8}$		$5\frac{1}{8}$		$6\frac{1}{8}$		$8\frac{1}{8}$	
Area, Sq In.	.484		.825		1.256		1.780		3.094		4.770		6.812		9.213		11.97		18.67		26.83		46.85	
Suct Temp	T	F	T	F	T	F	T	F	T	F	T	F	T	F	T	F	T	F	T	F	T	F	T	F
—40	.31	3.1	.61	2.9	1.0	2.7	1.6	2.6	3.2	2.4	5.5	2.2	8.6	2.1	12.5	2.0	17.3	1.9	30.2	1.8	47.4	1.7	95.0	1.6
—20	.40	1.8	.77	1.7	1.3	1.5	2.0	1.5	4.0	1.4	6.9	1.3	10.8	1.2	15.6	1.1	21.8	1.1	37.8	1.0	59.6	1.0	119.1	.9
0	.47	1.1	.93	1.0	1.6	.9	2.4	.9	4.8	.8	8.2	.8	12.8	.7	18.8	.7	26.0	.6	45.4	.6	71.4	.6	143.0	.5
+20	.55	.65	1.1	.6	1.8	.6	2.8	.5	5.6	.5	9.7	.5	14.9	.4	22.0	.4	30.6	.4	53.0	.4	83.1	.4	167.0	.3
+40	.63	.40	1.2	.4	2.1	.4	3.2	.3	6.4	.3	11.1	.3	17.1	.3	25.0	.3	34.8	.2	61.0	.2	95.6	.2	191.5	.2

STEEL PIPE—STANDARD WEIGHT (SCHEDULE 40)																								
IPS	$\frac{3}{4}$		1		$1\frac{1}{4}$		$1\frac{1}{2}$		2		$2\frac{1}{2}$		3		$3\frac{1}{2}$		4		5		6		8	
Area, Sq In.	.533		.864		1.495		2.036		3.355		4.788		7.393		9.89		12.73		20.01		28.99		50.0	
Suct Temp	T	F	T	F	T	F	T	F	T	F	T	F	T	F	T	F	T	F	T	F	T	F	T	F
—40	.35	3.9	.65	3.6	1.3	3.2	1.9	3.1	3.5	2.9	5.5	2.8	9.5	2.6	13.6	2.5	18.7	2.4	33.0	2.3	52.0	2.2	103.8	2.0
—20	.44	2.2	.81	2.1	1.6	1.8	2.4	1.7	4.4	1.7	6.9	1.6	11.8	1.5	17.0	1.4	23.4	1.4	41.3	1.3	65.4	1.3	129.6	1.1
0	.53	1.3	.97	1.2	1.9	1.1	2.9	1.0	5.3	1.0	8.3	1.0	14.2	.9	20.4	.9	28.0	.8	49.0	.8	78.4	.8	152.0	.7
+20	.62	.8	1.1	.7	2.3	.7	3.3	.6	6.2	.6	9.7	.6	16.6	.5	23.9	.5	32.7	.5	57.4	.5	92.0	.4	181.0	.4
+40	.71	.5	1.3	.5	2.6	.4	3.8	.4	7.1	.4	11.2	.4	19.1	.3	27.5	.3	37.7	.3	66.2	.3	105.5	.3	208.0	.3

T = Tons of refrigeration. F = Friction Drop, degrees F per 100 ft equivalent length at tons shown.

$\dfrac{\text{FULL LOAD, tons}}{\text{MIN. LOAD, tons}}$ (%)	FRICTION DROP MULTIPLIER
200	3.5
300	7.0
400	12.0
500	18.0
X	*$(X/200)^{1.8} \times 3.5$

*Solve this equation to determine the friction drop multiplier for any ratio of full load to min. load, tons.

TABLE 19—MINIMUM TONNAGE FOR OIL ENTRAINMENT UP SUCTION RISERS

Refrigerant 22

COPPER TUBING—TYPE L																								
Pipe OD	$\frac{7}{8}$		$1\frac{1}{8}$		$1\frac{3}{8}$		$1\frac{5}{8}$		$2\frac{1}{8}$		$2\frac{5}{8}$		$3\frac{1}{8}$		$3\frac{5}{8}$		$4\frac{1}{8}$		$5\frac{1}{8}$		$6\frac{1}{8}$		$8\frac{1}{8}$	
Area, Sq In.	.484		.825		1.256		1.780		3.094		4.770		6.812		9.213		11.97		18.67		26.83		46.85	
Suct Temp	T	F	T	F	T	F	T	F	T	F	T	F	T	F	T	F	T	F	T	F	T	F	T	F
—40	.45	1.6	.88	1.5	1.5	1.4	2.3	1.3	4.6	1.2	8.0	1.2	12.3	1.1	18.1	1.1	25.0	1.0	43.6	.9	68.5	.9	137.1	.8
—20	.56	1.0	1.1	.9	1.8	.8	2.8	.8	5.6	.7	9.7	.7	15.1	.6	21.9	.6	30.6	.6	53.0	.6	83.6	.5	167.0	.5
0	.66	.6	1.3	.5	2.2	.5	3.3	.5	6.7	.4	11.4	.4	17.9	.4	26.2	.4	36.3	.4	63.3	.3	99.5	.3	199.3	.3
+20	.77	.4	1.5	.3	2.5	.3	3.9	.3	7.8	.3	13.5	.3	20.8	.2	30.8	.2	42.7	.2	74.1	.2	116.2	.2	234.0	.2
+40	.89	.2	1.8	.2	3.0	.2	4.6	.2	9.1	.2	15.8	.2	24.3	.2	35.4	.1	49.4	.1	86.5	.1	135.6	.1	272.0	.1

STEEL PIPE—STANDARD WEIGHT (SCHEDULE 40)																								
IPS	$\frac{3}{4}$		1		$1\frac{1}{4}$		$1\frac{1}{2}$		2		$2\frac{1}{2}$		3		$3\frac{1}{2}$		4		5		6		8	
Area, Sq In.	.533		.864		1.425		2.036		3.355		4.788		7.393		9.89		12.73		20.01		28.99		50.0	
Suct Temp	T	F	T	F	T	F	T	F	T	F	T	F	T	F	T	F	T	F	T	F	T	F	T	F
—40	.51	2.0	.93	1.9	1.8	1.6	2.7	1.6	5.0	1.5	7.9	1.4	13.5	1.3	19.4	1.3	26.7	1.2	47.2	1.2	74.3	1.1	148.1	1.0
—20	.63	1.2	1.1	1.1	2.3	1.0	3.3	.9	6.2	.9	9.7	.9	16.6	.8	23.9	.7	33.0	.7	58.2	.7	92.1	.7	182.4	.6
0	.74	.7	1.4	.7	2.7	.6	4.0	.6	7.4	.5	11.5	.5	19.9	.5	28.5	.5	39.2	.4	68.7	.4	110.0	.4	212.5	.4
+20	.87	.5	1.6	.4	3.2	.4	4.7	.3	8.7	.3	13.6	.3	23.4	.3	33.6	.3	46.0	.3	80.6	.3	129.1	.3	254.5	.2
+40	1.0	.3	1.8	.3	3.6	.2	5.3	.2	10.0	.2	15.7	.2	26.4	.2	38.6	.2	53.0	.2	93.0	.2	148.0	.2	292.0	.1

T = Tons of refrigeration. F = Friction Drop, degrees F per 100 ft equivalent length at tons shown.

$\dfrac{\text{FULL LOAD, tons}}{\text{MIN. LOAD, tons}}$ (%)	FRICTION DROP MULTIPLIER
200	3.5
300	7.0
400	12.0
500	18.0
X	*$(X/200)^{1.8} \times 3.5$

*Solve this equation to determine the friction drop multiplier for any ratio of full load to min. load, tons.

TABLE 20—MINIMUM TONNAGE FOR OIL ENTRAINMENT UP SUCTION RISERS
Refrigerant 500

COPPER TUBING—TYPE L																								
Pipe OD	7/8		1 1/8		1 3/8		1 5/8		2 1/8		2 5/8		3 1/8		3 5/8		4 1/8		5 1/8		6 1/8		8 1/8	
Area, Sq In.	.484		.825		1.256		1.780		3.094		4.770		6.812		9.213		11.97		18.67		26.83		46.85	
Suct Temp	T	F	T	F	T	F	T	F	T	F	T	F	T	F	T	F	T	F	T	F	T	F	T	F
−40	.37	2.6	.72	2.4	1.2	2.2	1.9	2.1	3.8	2.0	6.5	1.8	10.1	1.7	14.8	1.7	20.8	1.6	35.7	1.5	56.0	1.4	112.1	1.3
−20	.45	1.4	.89	1.3	1.5	1.2	2.3	1.2	4.6	1.1	7.9	1.0	12.5	1.0	18.0	.9	25.2	.9	43.5	.8	68.7	.8	137.7	.7
0	.56	.9	1.1	.8	1.8	.7	2.8	.7	5.7	.6	9.7	.6	13.5	.5	22.2	.5	30.8	.5	53.7	.5	84.5	.5	169.2	.4
+20	.65	.5	1.3	.5	2.1	.5	3.3	.4	6.6	.4	11.4	.4	17.6	.4	26.0	.3	36.2	.3	62.7	.3	98.5	.3	197.9	.3
+40	.76	.3	1.5	.3	2.6	.3	4.0	.3	7.8	.3	13.5	.3	20.6	.2	30.2	.2	41.9	.2	73.2	.2	115.1	.2	231.0	.2

STEEL PIPE—STANDARD WEIGHT (SCHEDULE 40)																								
IPS	3/4		1		1 1/4		1 1/2		2		2 1/2		3		3 1/2		4		5		6		8	
Area, Sq In.	.533		.864		1.425		2.036		3.355		4.788		7.393		9.89		12.73		20.01		28.99		50.0	
Suct Temp	T	F	T	F	T	F	T	F	T	F	T	F	T	F	T	F	T	F	T	F	T	F	T	F
−40	.42	3.1	.75	2.8	1.5	2.5	2.2	2.3	4.1	2.2	6.4	2.1	11.0	2.0	15.7	1.9	22.0	1.9	39.2	1.8	60.2	1.7	120.0	1.6
−20	.52	1.8	.96	1.6	1.9	1.5	2.8	1.4	5.2	1.3	8.1	1.3	13.9	1.2	20.0	1.1	27.4	1.1	48.5	1.0	77.0	1.0	158.5	.9
0	.63	1.0	1.2	1.0	2.3	.8	3.4	.8	6.3	.8	9.7	.7	16.9	.7	24.2	.7	33.2	.6	58.3	.6	93.1	.6	180.5	.5
+20	.74	.6	1.3	.6	2.7	.5	4.0	.5	7.3	.5	11.5	.5	19.7	.4	28.3	.4	38.7	.4	68.0	.4	109.0	.4	214.0	.3
+40	.85	.4	1.6	.4	3.1	.3	4.5	.3	8.5	.3	13.3	.3	22.8	.3	32.7	.3	45.0	.3	78.9	.2	122.0	.2	248.0	.2

T = Tons of refrigeration. F = Friction Drop, degrees F per 100 ft equivalent length at tons shown.

$\dfrac{\text{FULL LOAD, tons}}{\text{MIN. LOAD, tons}}$ (%)	FRICTION DROP MULTIPLIER
200	3.5
300	7.0
400	12.0
500	18.0
X	*$(X/200)^{1.8} \times 3.5$

*Solve this equation to determine the friction drop multiplier for any ratio of full load to min. load, tons.

Double Suction Risers

There are applications in which single suction risers may be sized for oil return at minimum load without serious penalty at design load. Where single compressors with capacity control are used, minimum capacity corresponds to the compressor capacity at its minimum displacement. The maximum to minimum displacement ratio is usually three or four to one, depending on compressor size.

The compressor capacity at minimum displacement should be taken at an arbitrary figure of approximately *20 F* suction and not the design suction temperature for air conditioning applications.

Where multiple compressors are interconnected and controlled so that one or more may shut down while another continues to operate, the ratio of maximum to minimum displacement becomes much greater. In this case a double suction riser may be necessary for good operating economy at design load. The sizing and operation of a double suction riser is described as follows:

1. In *Fig. 58* the minimum load riser indicated by *A* is sized so that it returns oil at the minimum possible load.

2. The second riser *B* which is usually larger than riser *A* is sized so that the parallel pressure drop thru both risers at full load is satisfactory, providing this assures oil return at full load.

3. A trap is introduced between the two risers as shown in *Fig. 58*. During partial load operation when the gas velocity is not sufficient to return oil through both risers, the trap gradually fills with oil until the second riser *B* is sealed off. When this occurs, the gas travels up riser *A* only and has enough velocity to carry oil along with it back into the horizontal suction main.

The fittings at the bottom of the riser must be close coupled so that the oil holding capacity of the trap is limited to a minimum. If this is not done, the trap can accumulate enough oil on partial load operation to seriously lower the compressor crankcase oil level. Also, larger slug-backs of oil to the compressor occur when the trap clears out on increased load operation. *Fig. 58* shows that the larger riser *B* forms an inverted loop and enters the horizontal suction line from the top. The purpose of this loop is to prevent oil drainage into this riser which may be "idle" during partial load operation.

Example 4 — Determination of Riser Size

Given:

    Refrigerant 12 system
    Type L copper tubing
    Condensing temperature — 105 F
    Suction temperature — 40 F
    Height of riser — 10 ft
    Equivalent length — 22 ft (10 ft pipe + 2 ells)
    Full load — 98.5 tons
    Minimum load — 8.1 tons
      Two of 16 cylinders operating at 20 F
      2 compressors, 8 cylinders each.

Find:

    Size of riser
    Suction line pressure drop

Solution:

1. From *Table 18* for 8.1 tons minimum load (20 F suction temperature) read $2\frac{1}{8}$ in. OD tubing and .5 F minimum load pressure drop per 100 ft equivalent length of pipe.

2. Calculate minimum load pressure drop:

    Pressure drop for 5.6 ton

$$= \frac{.5}{100} \times 22 = .11 \text{ F}$$

    Minimum load $\left(\dfrac{8.1}{5.6}\right) = 145\%$ of 5.6 tons

    Multiplier $= \left(\dfrac{145}{200}\right)^{1.8} \times 3.5 = 1.96$

    Minimum load pressure drop $= .11 \times 1.96 = .22$ F

3. Determine full load pressure drop:

    Full load $= \dfrac{98.5}{5.6} = 1750\%$ of minimum load

    Multiplier $= \left(\dfrac{1750}{200}\right)^{1.8} \times 3.5 = 174$

    Full load pressure drop $= .11 \times 174 = 19.2$ F

    This full load pressure drop of 19.2 F together with a drop for the remainder of the suction line of approximately 1.5 F results in a 20.7 F total suction line drop. This is obviously too large and consequently a double suction riser must be used.

4. Determine size of smaller riser (same as in *Step 1* for $2\frac{1}{8}$ in. OD) and large riser for suitable pressure drop with a total load of 98.5 tons divided between the two risers.

    Let pressure drop = .5 F.

    Corrected equivalent length
        = equivalent length × correction factor
        (notes under *Chart 9*)
        = 22 × 4 = 88 ft (10 ft of pipe and 2 ells)

    Enter *Chart 7* with a $2\frac{1}{8}$ in. OD pipe and an equivalent length of 88 ft; the capacity is 18.2 tons.

    The capacity for $4\frac{1}{8}$ in. OD pipe at an equivalent length of 88 ft is 102 tons. Therefore, a small riser of $2\frac{1}{8}$ in. OD and a large riser of $4\frac{1}{8}$ in. OD in parallel carry a load of 120.2 tons at a pressure drop of .5 F. However, since the load is only 98.5 tons and a .5 pressure drop is obtained with 125 ft of pipe ($2\frac{1}{8}$ in. pipe, 15 tons; $4\frac{1}{8}$ in. pipe, 84 tons), the actual pressure drop (degrees F) is 88/125 × .5 = .35 F.

    The minimum capacity required to return oil is 34.8 tons for a $4\frac{1}{8}$ in. OD riser and 6.4 tons for a $2\frac{1}{8}$ in. OD

riser at 40 F suction temperature. Therefore, a $2\frac{1}{8}$ in. OD plus a $4\frac{1}{8}$ in. OD pipe are capable of returning oil at maximum load.

DISCHARGE (HOT GAS) LINE DESIGN

The hot gas line should be sized for a practical pressure drop. The effect of pressure drop is shown in *Table 16, page 43*.

Discharge Risers

Even though a low loss is desired in the hot gas line, the line should be sized so that refrigerant gas velocities are able to carry along entrained oil. In the usual application this is not a problem. However, where multiple compressors are used with capacity control, hot gas risers must be sized to carry oil at minimum loading.

Tables 21 and 22 show the minimum tonnages required to insure oil return in upward flow discharge risers. Friction drop in the risers in degrees F per 100 ft equivalent length is also included.

Double Discharge Risers

Sometimes in installations of multiple compressors having capacity control a vertical hot gas line sized to entrain oil at minimum load has an excessive pressure drop at maximum load. In such a case a double gas riser may be used in the same manner as it is used in a suction line. *Fig. 59* shows the double riser principle applied to a hot gas line.

Sizing of double hot gas risers is made in the same manner as double suction risers described earlier.

REFRIGERANT CHARGE

Table 23 is used to determine the piping system refrigerant charge required. The system charge should be equal to the sum of the charges in the refrigerant lines, compressor, evaporator, condenser and receiver (minimum operating charge).

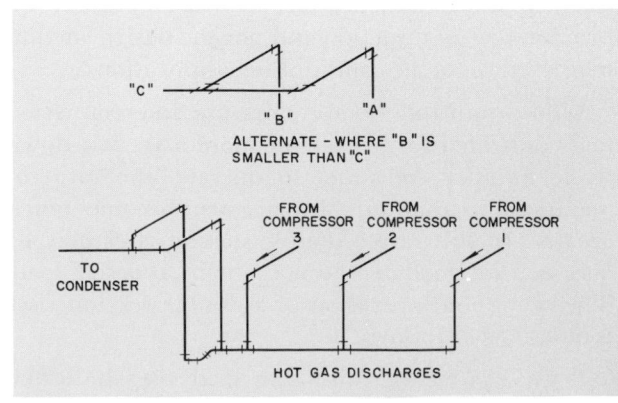

FIG. 59 — DOUBLE HOT GAS RISER

TABLE 21—MINIMUM TONNAGE FOR OIL ENTRAINMENT UP HOT GAS RISERS

Refrigerant 12

	COPPER TUBING—TYPE L																				
Pipe OD	7/8		1 1/8		1 3/8		1 5/8		2 1/8		2 5/8		3 1/8		3 5/8		4 1/8		5 1/8		6 1/8	
Area, Sq In.	.484		.825		1.256		1.78		3.094		4.77		6.812		9.213		11.97		18.67		26.83	
Disch Sat Temp	T	F	T	F	T	F	T	F	T	F	T	F	T	F	T	F	T	F	T	F	T	F
80	.71	.1	1.4	.2	2.3	.2	3.6	.2	7.2	.1	12.3	.1	19.2	.1	28.0	.1	38.6	.1	68.0	.1	106.0	.1
90	.75	.1	1.5	.2	2.5	.1	3.9	.1	7.7	.1	13.3	.1	20.6	.1	30.4	.1	42.0	.1	73.0	.1	114.0	.09
100	.80	.1	1.6	.1	2.6	.1	4.0	.1	8.0	.1	14.0	.1	21.7	.1	31.8	.09	43.7	.09	76.6	.08	120.0	.08
110	.87	.1	1.7	.1	2.9	.1	4.4	.1	8.9	.1	15.0	.09	23.7	.09	34.8	.08	48.0	.08	83.0	.07	131.0	.07
120	.95	.07	1.9	.1	3.1	.09	5.0	.09	9.7	.08	16.9	.08	26.0	.07	38.2	.07	52.5	.06	91.5	.06	143.0	.07

TABLE 22—MINIMUM TONNAGE FOR OIL ENTRAINMENT UP HOT GAS RISERS

Refrigerant 22

	COPPER TUBING—TYPE L																				
Pipe OD	7/8		1 1/8		1 3/8		1 5/8		2 1/8		2 5/8		3 1/8		3 5/8		4 1/8		5 1/8		6 1/8	
Area, Sq In.	.484		.825		1.256		1.78		3.094		4.77		6.812		9.213		11.97		18.67		26.83	
Disch Sat Temp	T	F	T	F	T	F	T	F	T	F	T	F	T	F	T	F	T	F	T	F	T	F
80	1.1	.08	2.1	.11	3.5	.10	5.4	.10	10.6	.09	18.1	.08	28.4	.08	42.0	.08	57.6	.07	101.0	.07	146.0	.06
90	1.2	.07	2.3	.10	3.8	.09	5.8	.09	11.5	.08	19.9	.07	31.2	.07	45.6	.07	62.5	.06	110.0	.06	158.0	.05
100	1.25	.07	2.5	.09	4.1	.08	6.3	.08	12.5	.07	21.6	.07	33.9	.06	49.6	.06	68.5	.06	120.0	.05	173.0	.04
110	1.4	.06	2.7	.08	4.4	.07	6.9	.07	13.7	.06	23.5	.06	38.1	.06	53.7	.05	74.0	.05	129.0	.05	187.0	.04

T = Tons of refrigeration. F = Friction Drop, degrees F per 100 ft equivalent length at tons shown.

$\dfrac{\text{FULL LOAD, tons}}{\text{MIN. LOAD, tons}}$ (%)	FRICTION DROP MULTIPLIER
200	3.5
300	7.0
400	12.0
500	18.0
X	*(X/200)$^{1.8}$ × 3.5

*Solve this equation to determine the friction drop multiplier for any ratio of full load to min. load, tons.

TABLE 23—FLUID WEIGHT OF REFRIGERANT IN PIPING

(Lb/10 ft of length)

PIPE SIZE*		SUCTION LINES 40 F SAT SUCT TEMP			LIQUID LINES 100 F TEMP			HOT GAS LINES 100 F SAT COND TEMP		
Copper OD In.	Steel Nom. In.	R12	R500	R22	R12	R500	R22	R12	R500	R22
1/2	3/8	.013	.013	.016	.80	.70	.72	.032	.032	.047
5/8	1/2	.021	.02	.025	1.28	1.13	1.15	.051	.052	.075
7/8	3/4	.043	.042	.051	2.65	2.33	2.40	.105	.11	.16
1 1/8	1	.073	.072	.087	4.52	3.98	4.09	.18	.18	.27
1 3/8	1 1/4	.110	.11	.13	6.87	6.06	6.22	.27	.28	.40
1 5/8	1 1/2	.16	.15	.19	9.74	8.56	8.81	.39	.39	.57
2 1/8	2	.27	.27	.33	16.9	14.9	15.3	.67	.69	.99
2 5/8	2 1/2	.42	.41	.51	26.1	23.0	23.6	1.04	1.1	1.5
3 1/8	3	.60	.59	.72	37.3	32.9	33.7	1.5	1.5	2.2
3 5/8	3 1/2	.81	.80	.98	50.5	44.3	45.6	2.0	2.0	3.0
4 1/8	4	1.05	1.04	1.27	65.5	57.6	59.3	2.6	2.7	3.8
5 1/8	5	1.64	1.62	1.98	102.0	90.0	93.0	4.1	4.1	6.0
6 1/8	6	2.36	2.33	2.84	147.0	130.0	133.0	5.8	6.0	8.6
8 1/8	8	4.11	4.06	4.96	257.0	227.0	232.0	10.2	10.4	15.0

To Correct for Temperatures Other Than Above, Multiply by the Following Factors:

REFRIGERANT	SUCT LINE—SAT. TEMP F					LIQUID LINE—SAT. TEMP F					HOT GAS LINE—SAT. TEMP F				
	50	30	10	−10	−30	40	60	80	100	120	80	90	100	110	120
12	1.18	.84	.59	.39	.26	1.09	1.06	1.03	1.00	.97	.75	.87	1.00	1.15	1.33
500	1.18	.84	.58	.39	.26	1.10	1.07	1.04	1.00	.96	.74	.86	1.00	1.16	1.33
22	1.18	.84	.58	.39	.25	1.11	1.08	1.04	1.00	.98	.74	.86	1.00	1.16	1.35

*Refrigerants 12, 22 and 500 weights are for OD sizes of Type L copper pipe.

REFRIGERANT PIPING LAYOUT

EVAPORATORS

Suction Line Loops

Evaporator suction lines should be laid out to accomplish the following objectives:

1. Prevent liquid refrigerant from draining into the compressor during shutdown.
2. Prevent oil in an active evaporator from draining into an idle evaporator.

This can be done by using piping loops in the lines connecting the evaporator, the compressor and the condenser. Standard arrangements of suction line loops based on standard piping practices are illustrated in *Fig. 60*.

Figure 60a shows the *compressor located below a single evaporator*. An inverted loop rising to the top of the evaporator should be made in the suction line to prevent liquid refrigerant from draining into the compressor during shutdown.

A *single evaporator below the compressor* is illustrated in *Fig. 60b*. The inverted loop in the suction line is unnecessary since the evaporator traps all liquid refrigerant.

Figure 60c shows *multiple evaporators on different floor levels with the compressor below*. Each individual suction line should be looped to the top of the evaporator before being connected into the suction main to prevent liquid from draining into the compressor during shutdown.

Figure 60d illustrates *multiple evaporators stacked on the same floor level* or may represent a two-circuit single coil operated from one liquid solenoid valve with the compressor located below the evaporator. In this arrangement it is possible to use one loop to serve the purpose.

Where coil banks on the same floor level have separate liquid solenoid valves feeding each coil, a separate suction riser is required from each coil, similar to the arrangements in *Fig. 60c and 60e* for best oil return performance. Where separate suction risers are not possible, use the arrangement shown in *Fig. 60f*.

Figure 60g shows *multiple evaporators located on the same level and the compressor located below the evaporators*. Each suction line is brought upward and looped into the top of the common suction line. The alternate arrangement shows individual suction lines out of each evaporator dropping down into a common suction header which then rises in a single loop to the top of the coils before going down to the compressor. An alternate arrangement is shown in *Fig. 60h* for cases where the *compressor is above the evaporator*.

When automatic compressor pumpdown control is used, evaporators are automatically kept free of liquid by the pumpdown operation and, therefore, evaporators located above the compressor can be free-draining to the compressor without protective loops.

The small trap shown in the suction lines immediately after the coil suction outlet is recommended to prevent erratic operation of the thermal expansion valve. The expansion valve bulb is located in the suction line between the coil and the trap. The trap drains the liquid from under the expansion valve bulb during compressor shutdown, thus preventing erratic operation of the valve when the compressor starts up again. A trap is required only when straight runs or risers are encountered in the suction line leaving the coil outlet. A trap is not required when the suction line from the coil outlet drops to the compressor or suction header immediately after the expansion valve bulb.

Suction lines should be designed so that oil from an active evaporator does not drain into an idle one. *Fig. 60e* shows *multiple evaporators on different floor levels and the compressor above the evaporators*. Each suction line is brought upward and looped into the top of the common suction line if its size is equal to the main. Otherwise it may be brought into the side of the main. The loop prevents oil from draining down into either coil that may be inactive.

Figure 60f shows *multiple evaporators stacked with the compressor above the evaporators*. Oil is prevented from draining into the lowest evaporator because the common suction line drops below the outlet of the lowest evaporator before entering the suction riser.

If evaporators must be located both above and below a common suction line, the lines are piped as illustrated in *Figs. 60a and 60b,* with (a) piping for the evaporator above the common suction line and (b) piping for the evaporator below the common suction line.

Multiple Circuit Coils

All but the smallest coils are arranged with multiple circuits. The length and number of circuits are determined by the type of application. Multiple circuit coils are supplied with refrigerant thru a distributor which regulates the refrigerant distribution evenly among the circuits. Direct expansion coils can be located in any position, provided proper

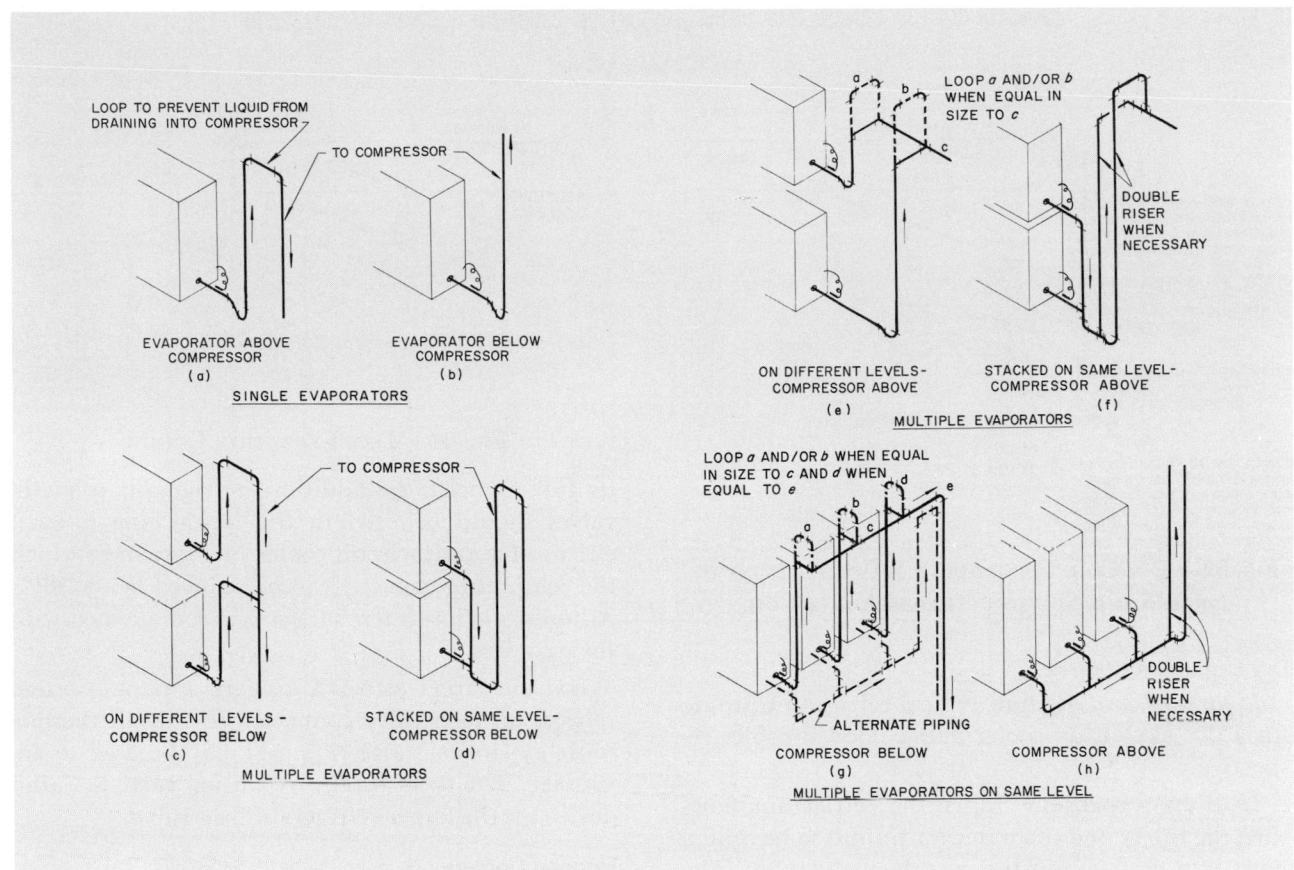

FIG. 60 — STANDARD ARRANGEMENTS OF SUCTION LINE LOOPS (ONE-CIRCUIT COILS SHOWN)

refrigerant distribution and continuous oil removal facilities are provided.

In general the suction line piping principles shown in *Fig. 60* should be employed to assure proper expansion valve operation, oil return and compressor protection.

Figures 61 and 62 show direct expansion air coil piping arrangements in which the suction connections drain the coil headers effectively. *Fig. 61* shows individual suction outlets joining into a common suction header below the coil level. *Fig. 62* illustrates an alternate method of bringing up each suction line and looping it into the common line. The expansion valve equalizing lines are connected at the top of each suction header at the opposite end from the suction connection.

Figure 63 illustrates the use of a coil having connections at the top or in the middle of each coil header, and piped so that this connection does not drain the evaporator. In this case oil may become trapped in the coil. The figure shows oil drain lines from connections supplied for this purpose. The

drain lines extend from the suction connection at the bottom end of each coil header to the common suction header below the coil level.

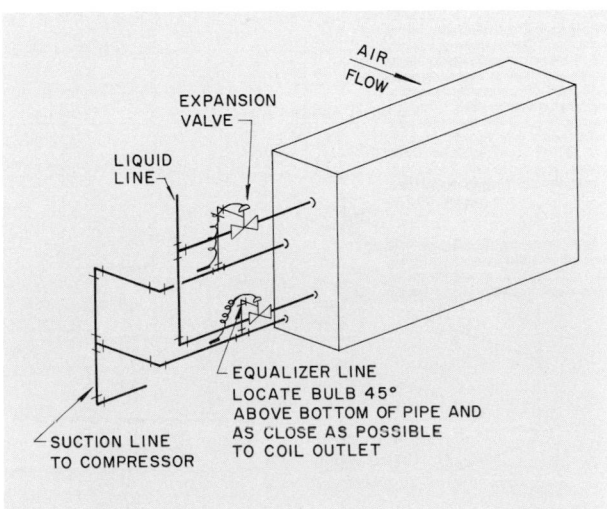

FIG. 61 — DX COIL USING SUCTION CONNECTIONS TO DRAIN COIL, SUCTION HEADER BELOW COIL

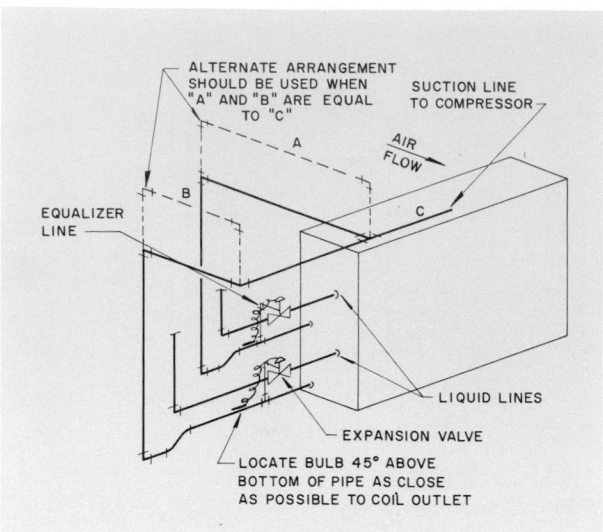

FIG. 62 — DX COIL USING SUCTION CONNECTIONS TO
DRAIN COIL, SUCTION HEADER ABOVE COIL

Dry Expansion Coolers

Figures 64 and 66 show typical refrigerant piping
for a dry expansion cooler and a flooded cooler respectively.

In a dry expansion chiller the refrigerant flows
thru the tubes, and the water (or liquid) to be cooled
flows transversely over the outside of the tubes. The
water or liquid flow is guided by vertical baffles.
Multi-circuit coolers should be used in systems in
which the compressor capacity can be reduced below
50%. This is recommended since oil cannot be properly returned and good thermal valve control cannot
be expected below this minimum loading per circuit.

It is also recommended that the minimum capacity of a single circuit should be not less than 50% of

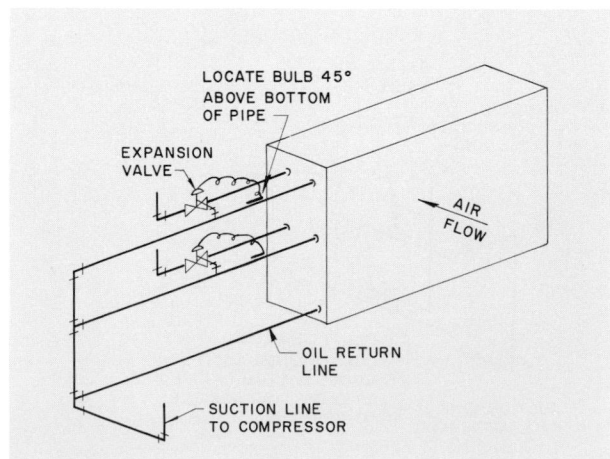

FIG. 63 — DX COIL USING OIL RETURN DRAIN
CONNECTIONS TO DRAIN OIL

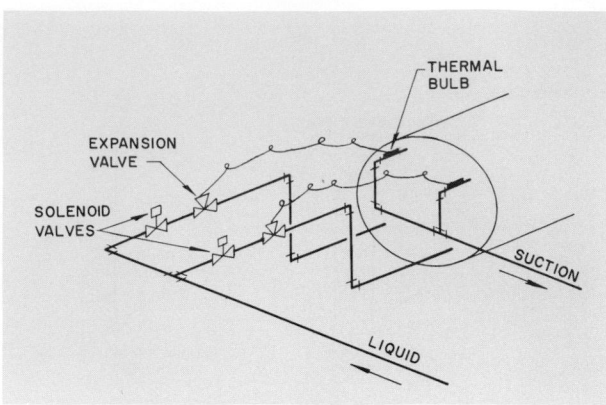

FIG. 64 — DRY EXPANSION COOLER

its full capacity. In addition, refrigerant solenoid
valves should be used in the liquid line to each
circuit of a multi-circuit cooler in a system in which
the compressor capacity can be reduced below 50%.
A liquid suction interchanger is recommended with
these coolers.

For the larger size DX coolers a pilot-operated
refrigerant feed valve connected to a small thermostatic expansion valve *(Fig. 65)* may be used to advantage. The thermostatic expansion valve is a pilot
device for the larger refrigerant feed valve.

Flooded Coolers

In a flooded cooler the refrigerant surrounds the
tubes in the shell, and water or liquid to be cooled
flows thru the tubes in one or more passes, depending on the baffle arrangement.

Flooded coolers require a continuous liquid bleed
line from some point below the liquid level in the
cooler shell to the suction line. This continuous
bleed of refrigerant liquid and oil assures the required return of oil to the compressor. It is usually

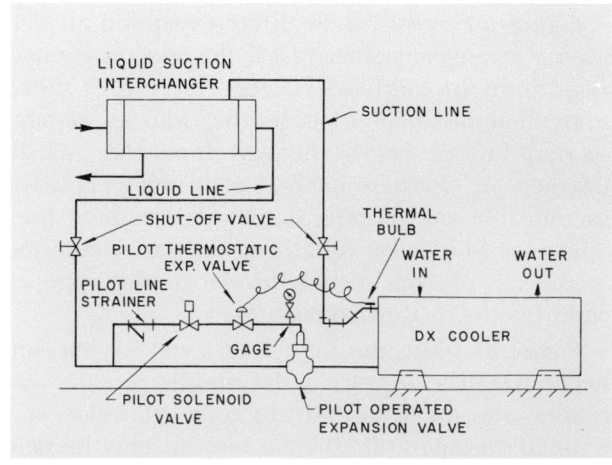

FIG. 65 — HOOKUP FOR LARGE DX COOLERS

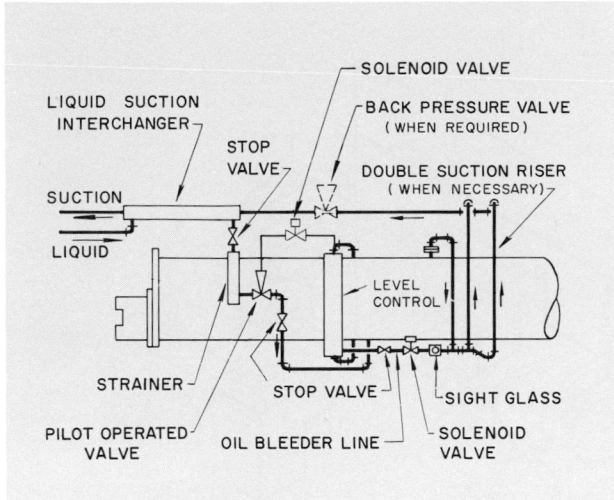

FIG. 66 — FLOODED COOLER

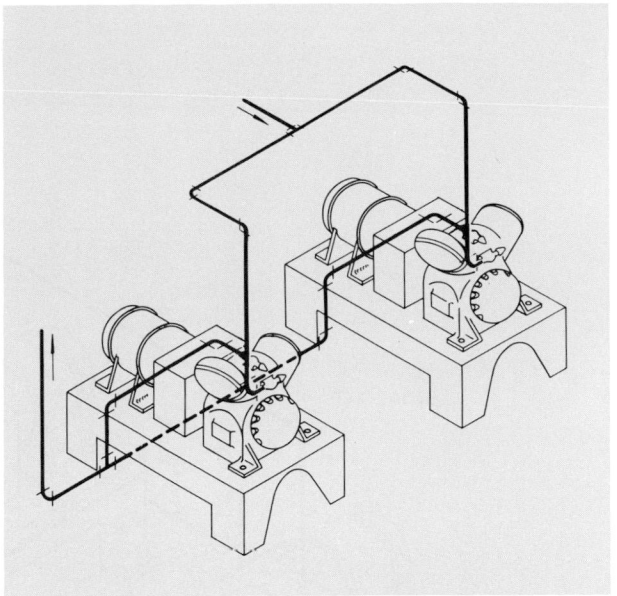

FIG. 67 — LAYOUT OF SUCTION AND HOT GAS LINES FOR MULTIPLE COMPRESSOR OPERATION

drained into the suction line so that the oil can be returned to the cooler with the suction gas. This drain line should be equipped with a hand shut-off valve, a solenoid valve and a sight glass. The solenoid valve should be wired into the control circuit in such a manner that it closes when the compressor stops.

A liquid suction interchanger, installed close to the cooler, is required to evaporate any liquid refrigerant from the refrigerant oil mixture which is continuously bled into the suction line.

Since flooded coolers frequently operate at light loads, double suction risers are often necessary.

To avoid freeze-up the water supply to a flooded cooler should never be throttled and should never bypass the cooler.

COMPRESSORS

Suction Piping

Suction piping of parallel compressors should be designed so that all compressors run at the same suction pressure and so that oil is returned to the running compressors in equal proportions. To insure maintenance of proper oil levels, compressors of unequal sizes may be erected on foundations at different elevations so that the recommended crankcase operating oil level is maintained at each compressor.

All suction lines are brought into a common suction header which is run full size and level above the compressor suction inlets. Branch suction line take-offs to the compressors are from the side of the header and should be the same size as the header. No reduction is made in the branch suction lines to the compressors until the vertical drop is reached.

This allows the branch line to return oil proportionally to each of the operating compressors.

Figure 67 shows suction and hot gas header arrangements for two compressors operating in parallel.

Discharge Piping

The branch hot gas lines from the compressors are connected into a common header. This hot gas header is run at a level below that of the compressor discharge connections which, for convenience, is often at the floor. This is equivalent to the hot gas loop for the single compressor shown in *Fig. 68*.

The hot gas loop accomplishes two functions:

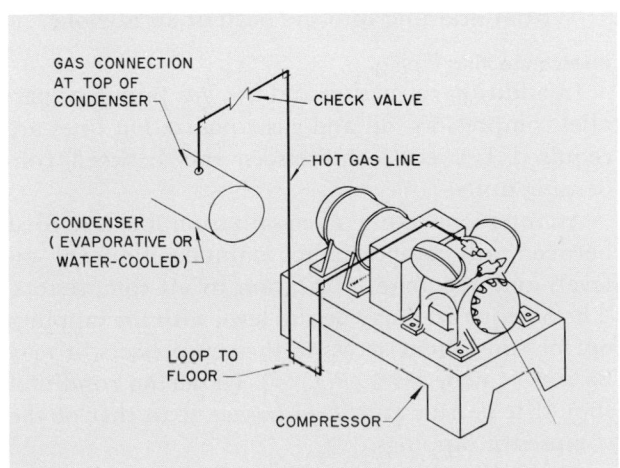

FIG. 68 — HOT GAS LOOP

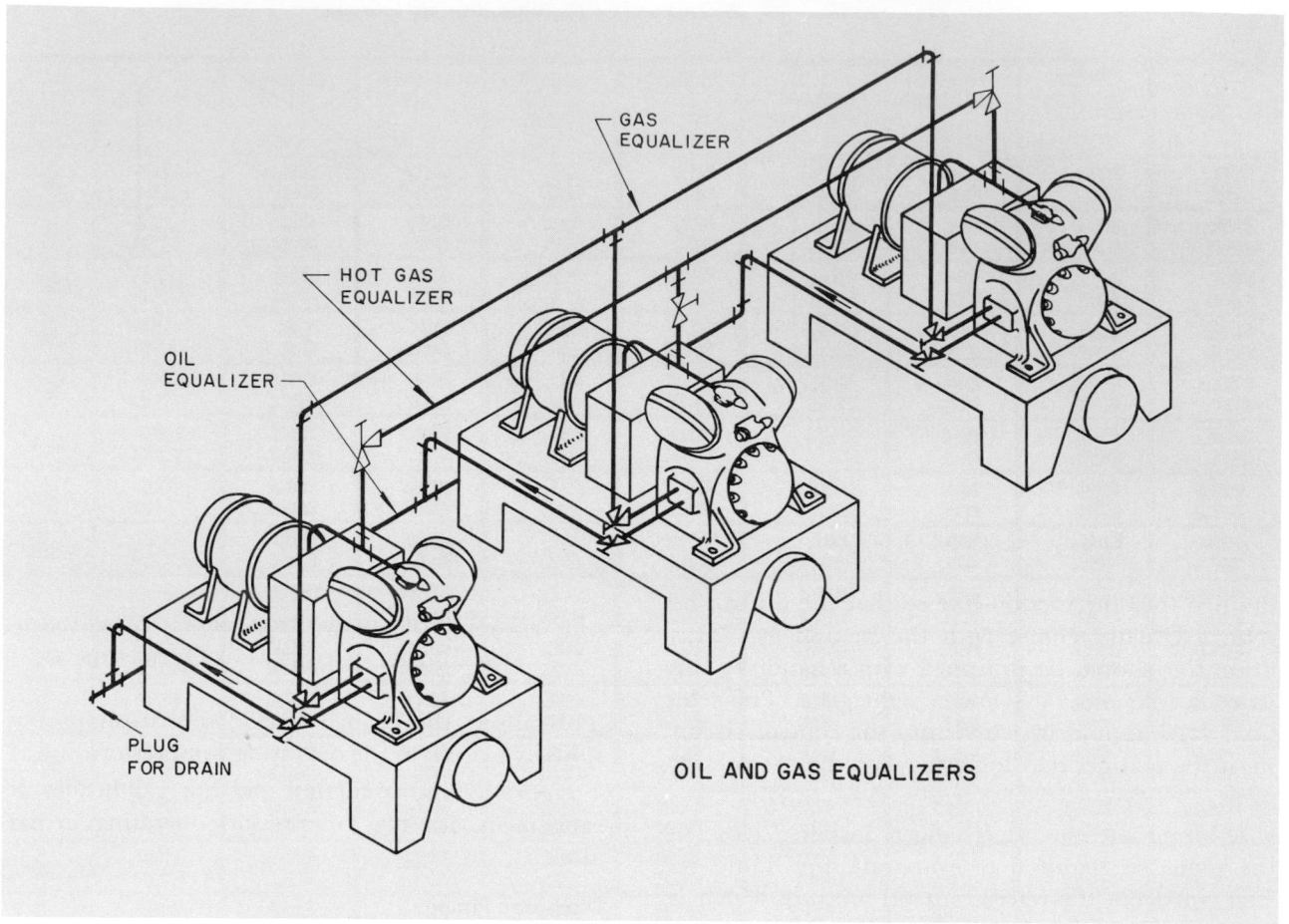

FIG. 69 — INTERCONNECTING PIPING FOR MULTIPLE CONDENSING UNITS

1. It prevents gas, which may condense in the hot gas line during the off cycle, from draining back into the heads of the compressors. This eliminates compressor damage on start-up.
2. It prevents oil, which leaves one compressor, from draining into the head of an idle one.

Interconnecting Piping

In addition to suction and hot gas piping of parallel compressors, oil and gas equalization lines are required between compressors and between condensing units.

An interconnecting oil equalization line is needed between all crankcases to maintain uniform oil levels and adequate lubrication in all compressors. The oil equalizer may be run level with the tappings or, for convenient access to the compressors, it may be run at floor level (*Fig. 69*). Under no condition should it be run at a level higher than that of the compressor tappings.

Ordinarily, proper equalization takes place only if a gas equalizing line is installed above the com-

pressor crankcase oil line. This line may be run level with the tappings, or may be raised to allow head room for convenient access. It should be piped level and supported as required to prevent traps from forming.

Shut-off valves should be installed in both lines so that any one machine may be isolated for repair without shutting down the entire system. Both lines should be the same size as the tappings on the largest compressor. Neither line should be run directly from one crankcase into another without forming a U-bend or hairpin to absorb vibration.

When multiple condensing units are interconnected as shown in *Fig. 69*, it is necessary to equalize the pressure in the condensers to prevent hot gas from blowing thru one of the condensers and into the liquid line. To do this a hot gas equalizer line is installed as shown. If the piping is looped as shown, vibration should not be a problem. The equalizer line between units must be the same size as the largest hot gas line.

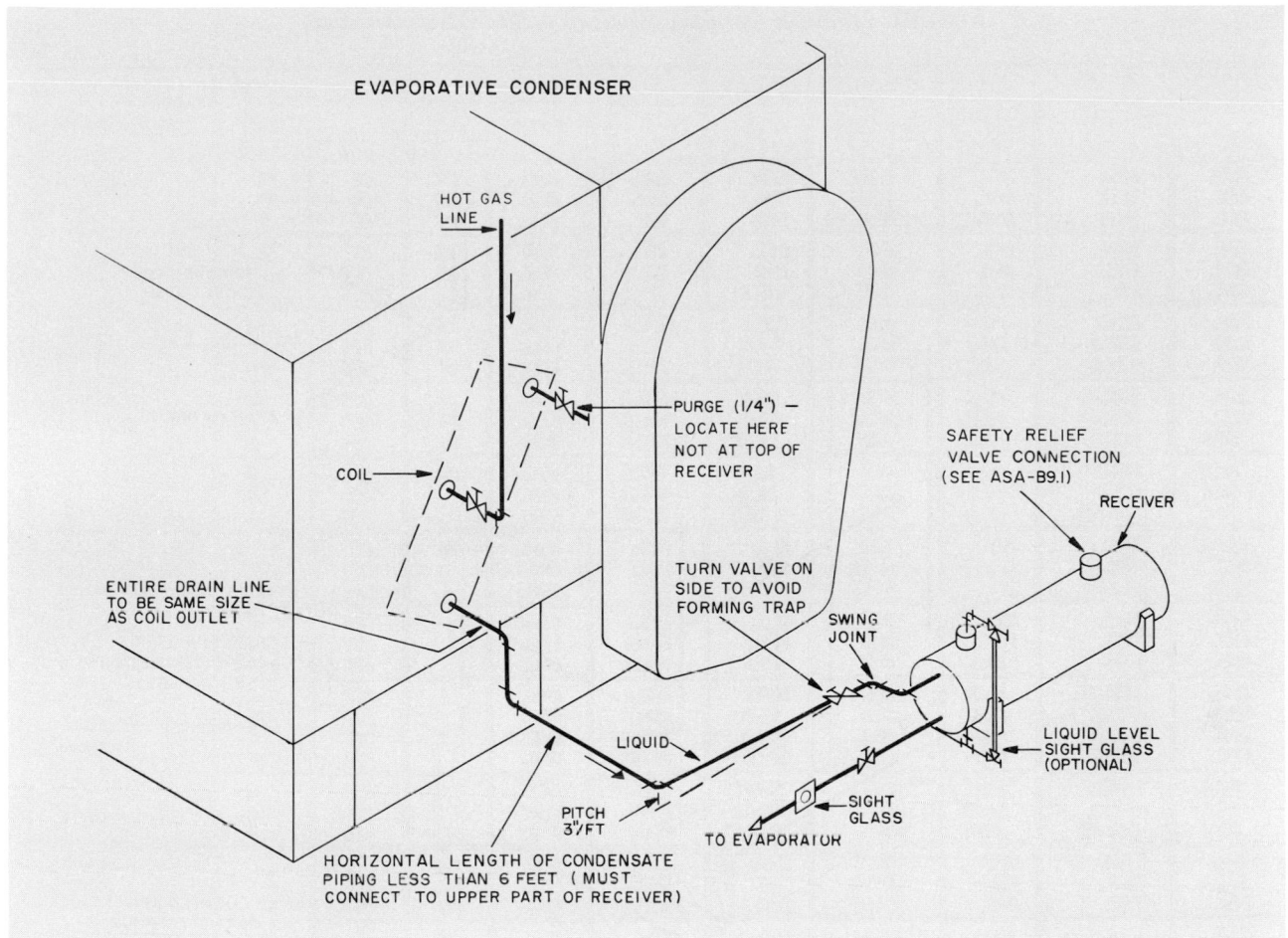

FIG. 70 — HOT GAS AND LIQUID PIPING, SINGLE COIL UNIT WITHOUT RECEIVER VENT

CONDENSERS

Liquid receivers are often used in systems having evaporative or air-cooled condensers and also with water-cooled condensers where additional liquid storage capacity is required to pump down the system. However, in many water-cooled condenser systems the condenser serves also as a receiver if the total refrigerant in the system does not exceed its storage capacity.

When receivers are used, liquid piping from the condenser to the receiver is designed to allow free drainage of liquid from the condenser at all times. This is possible only if the pressure in the receiver is not allowed to rise to the point where a restriction in flow can occur.

Liquid flow from the condenser to the receiver can be restricted by any of the following conditions:

1. Gas binding of the receiver.
2. Excessive friction drop in the condensate line.
3. Incorrect condensate line design.

The following piping recommendations are made to overcome these difficulties.

Evaporative Condenser to Receiver Piping

Liquid receivers are used on evaporative condensers to accommodate fluctuations in refrigerant liquid level, to maintain a seal, and to provide storage facilities for pumpdown. An equalizing line from the receiver to the condenser is required to relieve gas pressure tending to develop in the receiver. Otherwise liquid hang-up in the condenser due to restricted drainage can occur. The receiver can be vented directly thru the condensate line to the condenser outlet, or by an external equalizer line to the condenser.

Figure 70 shows a single evaporative condenser and receiver vented back thru the condensate drain line to the condensing coil outlet. Such an arrangement is applicable to a close coupled system. A separate vent is not required. However, it is limited to a horizontal length of condensate line of less than

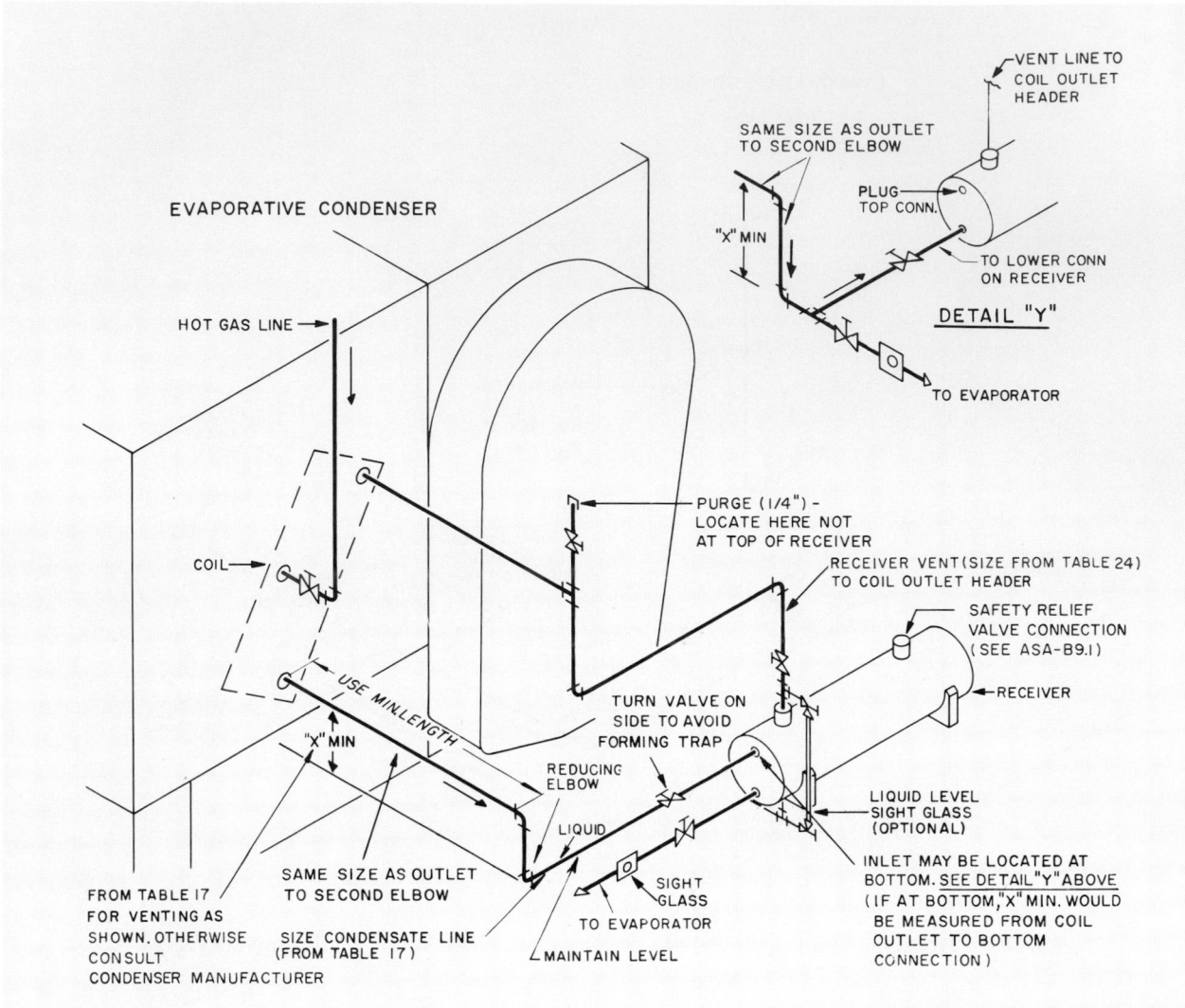

FIG. 71 — HOT GAS AND LIQUID PIPING, SINGLE COIL UNIT WITH RECEIVER VENT

6 ft. The entire condensate line from the condenser to the receiver is the same size as the coil outlet. The line should be pitched as shown.

Figure 71 shows the refrigerant piping for a single unit with receiver vent. Note that the condensate line from the condenser is the full size of the outlet connection and is not reduced until the second elbow is reached. This arrangement prevents trapping of liquid in the condenser coil.

Table 24 lists recommended sizing of receiver vent lines.

There are some systems in current use without a receiver but it must be recognized that problems can occur which can be avoided if a receiver is used.

Such an arrangement is more critical with respect to refrigerant charge. An overcharged system can

waste power and cause a loss of capacity if the overcharge backs up into the condenser. An undercharged system also wastes power and causes a loss of capacity because the evaporator is being fed partially with hot gas. Therefore, if the receiver is omitted, an accurate refrigerant charge must be maintained to assure normal operation.

TABLE 24—RECEIVER VENT LINE SIZING

Receiver to Condenser

VENT LINE SIZE BASED ON TYPE L COPPER TUBING (In. OD)	REFRIGERATION (Tons, Max.)
5/8	to - 40
7/8	40 - 80
1 1/8	80 - 120
1 3/8	Above 120

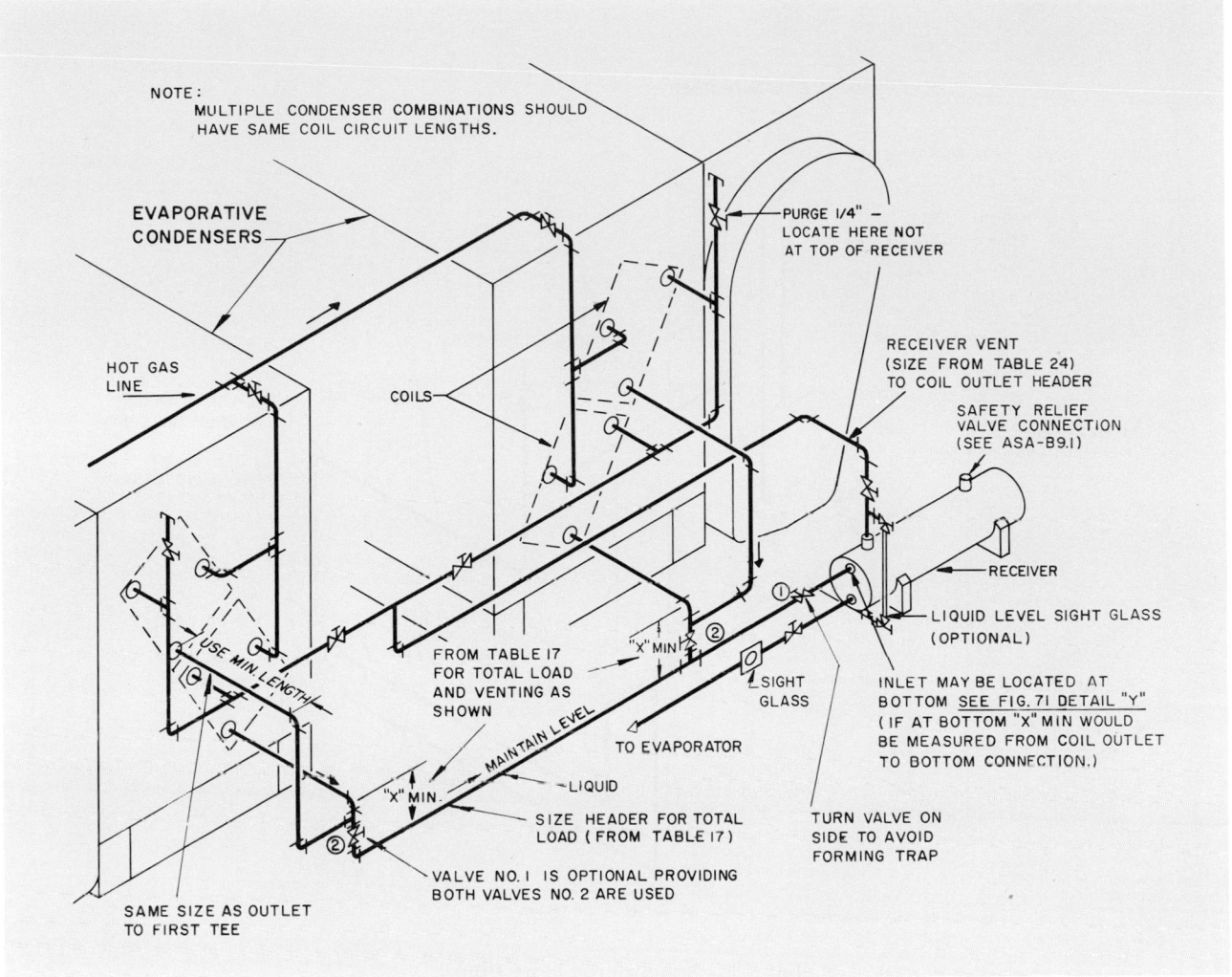

NOTE:
MULTIPLE CONDENSER COMBINATIONS SHOULD
HAVE SAME COIL CIRCUIT LENGTHS.

EVAPORATIVE
CONDENSERS

HOT GAS
LINE

COILS

PURGE 1/4" —
LOCATE HERE NOT
AT TOP OF RECEIVER

RECEIVER VENT
(SIZE FROM TABLE 24)
TO COIL OUTLET HEADER

SAFETY RELIEF
VALVE CONNECTION
(SEE ASA-B9.1)

RECEIVER

LIQUID LEVEL SIGHT GLASS
(OPTIONAL)

USE MIN LENGTH

FROM TABLE 17
FOR TOTAL LOAD
AND VENTING AS
SHOWN

"X" MIN

SIGHT
GLASS

MAINTAIN LEVEL

TO EVAPORATOR

LIQUID

"X" MIN

SIZE HEADER FOR TOTAL
LOAD (FROM TABLE 17)

TURN VALVE ON
SIDE TO AVOID
FORMING TRAP

INLET MAY BE LOCATED AT
BOTTOM SEE FIG. 71 DETAIL "Y"
(IF AT BOTTOM "X" MIN WOULD
BE MEASURED FROM COIL OUTLET
TO BOTTOM CONNECTION.)

SAME SIZE AS OUTLET
TO FIRST TEE

VALVE NO. 1 IS OPTIONAL PROVIDING
BOTH VALVES NO. 2 ARE USED

FIG. 72 — HOT GAS AND LIQUID PIPING, MULTIPLE DOUBLE COIL UNIT

The advantage of such an arrangement is an economic one; equipment cost is lower since receiver and valves are eliminated and the system operating charge is lower if charged accurately.

Figure 72 shows a piping arrangement for multiple units. Note that there are individual hot gas and vent valves for each unit. These valves permit operation of one unit while the other is shut down. These are essential because otherwise the idle unit, at lower pressure, causes hot gas to blow thru the operating unit into the liquid line. Purge cocks are also shown, one for each unit.

The hot gas piping should be such that the pressure in each condenser is substantially the same. To accomplish this the branch connection from the hot gas header into each condenser should be the same size as the condenser coil connection.

Figure 73 shows a subcooling coil piping arrangement. The subcooling coil must be piped between the receiver and the evaporator for best liquid subcooling benefits.

Multiple Shell and Tube Condensers

When two or more shell and tube condensers are applied in parallel in a single system, they should be equalized on the hot gas side and arranged as shown in *Fig. 75*.

The elevation difference between the outlet of the condenser and the horizontal liquid header must be at least 12 in., preferably greater, to prevent gas blowing thru. The bottoms of all condensers should be at the same level to prevent backing liquid into the lowest condenser.

When water-cooled condensers are interconnected as shown, they should be fed from a common water regulating valve, if used.

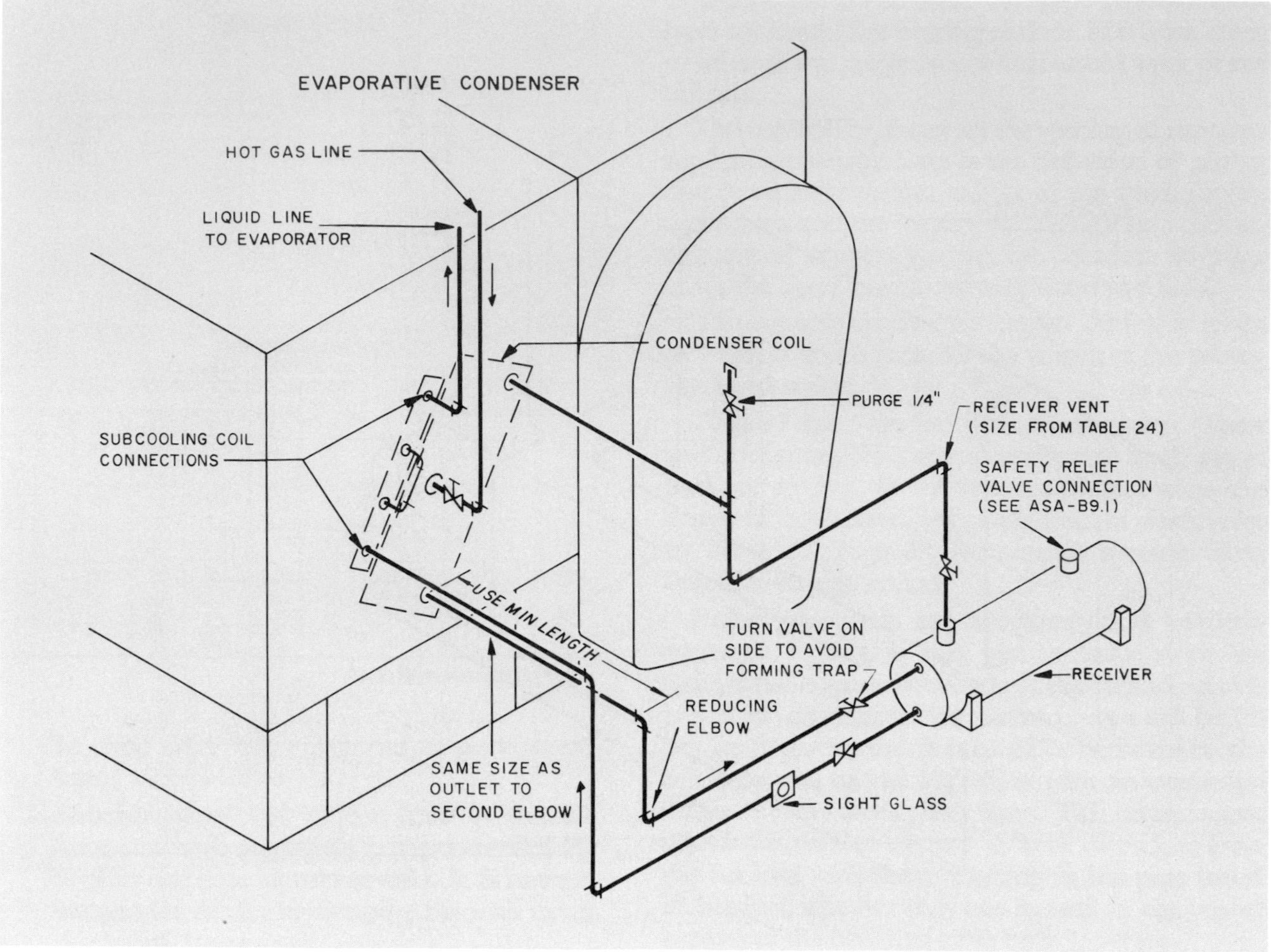

EVAPORATIVE CONDENSER

HOT GAS LINE

LIQUID LINE
TO EVAPORATOR

CONDENSER COIL

SUBCOOLING COIL
CONNECTIONS

PURGE 1/4"

RECEIVER VENT
(SIZE FROM TABLE 24)

SAFETY RELIEF
VALVE CONNECTION
(SEE ASA-B9.1)

USE MIN LENGTH

TURN VALVE ON
SIDE TO AVOID
FORMING TRAP

RECEIVER

SAME SIZE AS
OUTLET TO
SECOND ELBOW

REDUCING
ELBOW

SIGHT GLASS

FIG. 73 — SUBCOOLING COIL PIPING

An inverted loop of at least 6 ft is recommended in the liquid line to prevent siphoning of the liquid into the evaporator (or evaporators) during shutdown. Where a liquid line solenoid valve or valves are used, the loop is unnecessary.

Figure 74 shows a similar loop for a single condenser with tne evaporator below.

Vibration of Piping

Vibration transmitted thru or generated in refrigerant piping and the objectionable noise which results can be eliminated or greatly minimized by proper design and support of the piping.

The best way to prevent compressor vibration from being transmitted to the piping is to run the suction and discharge lines at least 6 pipe diameters in each of three directions before reaching the first point of support. In this manner the piping can absorb the vibration without being overstressed.

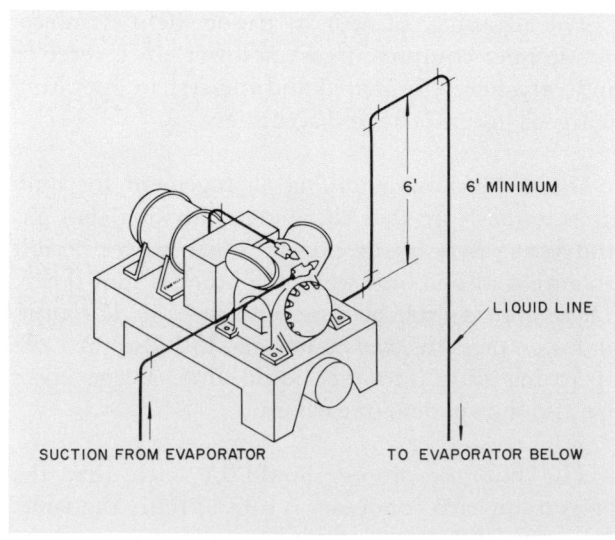

6' MINIMUM

LIQUID LINE

SUCTION FROM EVAPORATOR

TO EVAPORATOR BELOW

FIG. 74 — LIQUID LINE FROM CONDENSER OR RECEIVER
TO EVAPORATOR LOCATED BELOW

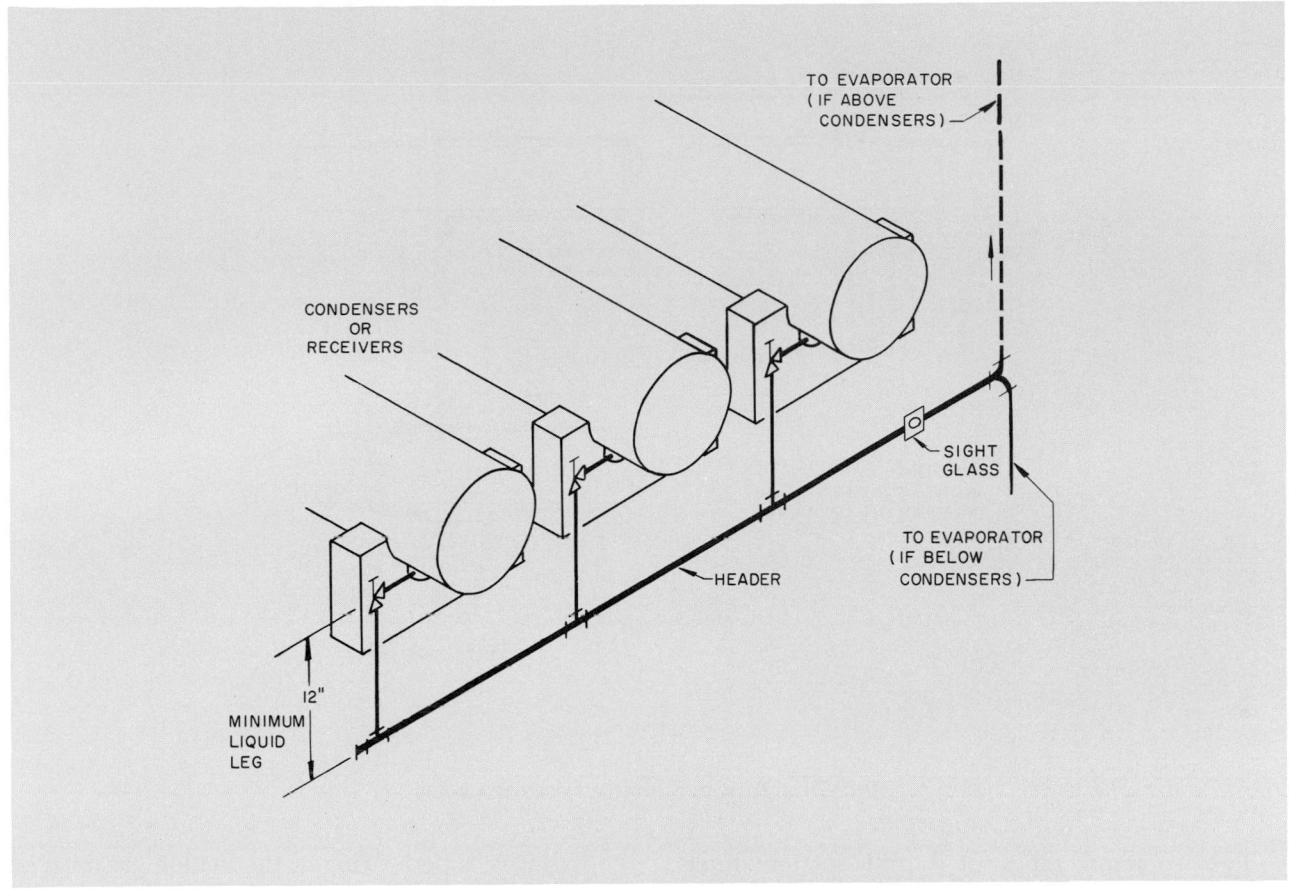

FIG. 75 — LIQUID PIPING TO INSURE CONDENSATE FLOW FROM INTERCONNECTED CONDENSERS

The hot gas loop from the compressor can be attached to the compressor base by means of a bracket if the base is isolated. If there is enough space in the horizontal run of the loop, two brackets are recommended to eliminate excessive rocking movement of the pipe. Brackets should be attached at the point of minimum movement of the compressor assembly. The riser following the loop is supported as close as possible to the compressor.

If the compressor is mounted on a resilient base, the pipe support should have a resilient isolator. The isolator is selected for four times the deflection in the spring support of the compressor base.

See *"Vibration Isolation of Piping Systems"* in *Chapter I* for further discussion of the subject.

REFRIGERANT PIPING ACCESSORIES

LIQUID LINE

Liquid Suction Interchangers

These are devices which subcool the liquid refrigerant and superheat the suction gas. The follow-ing describes four reasons for their use and the best location for each application:

1. To subcool the liquid refrigerant to compensate for excessive liquid line pressure drop. Location — near condenser. Liquid suction interchangers are not recommended for single stage applications using Refrigerant 22 because superheating of the suction gas must be limited to avoid compressor overheating. However, where they are used to prevent liquid slop-over to the compressor, superheating of the suction gas should be limited to 20 F above saturation temperature. A liquid suction interchanger so designed to limit the superheat of the suction gas should have a bypass so that operating adjustments may be made.

2. To act as an oil rectifier. Location — near evaporator.

3. To prevent liquid slop-over to the compressor. Location — near evaporator.

4. To increase the efficiency of the Refrigerant 12 and 500 cycles. Location — near evaporator to avoid insulation of subcooled liquid line.

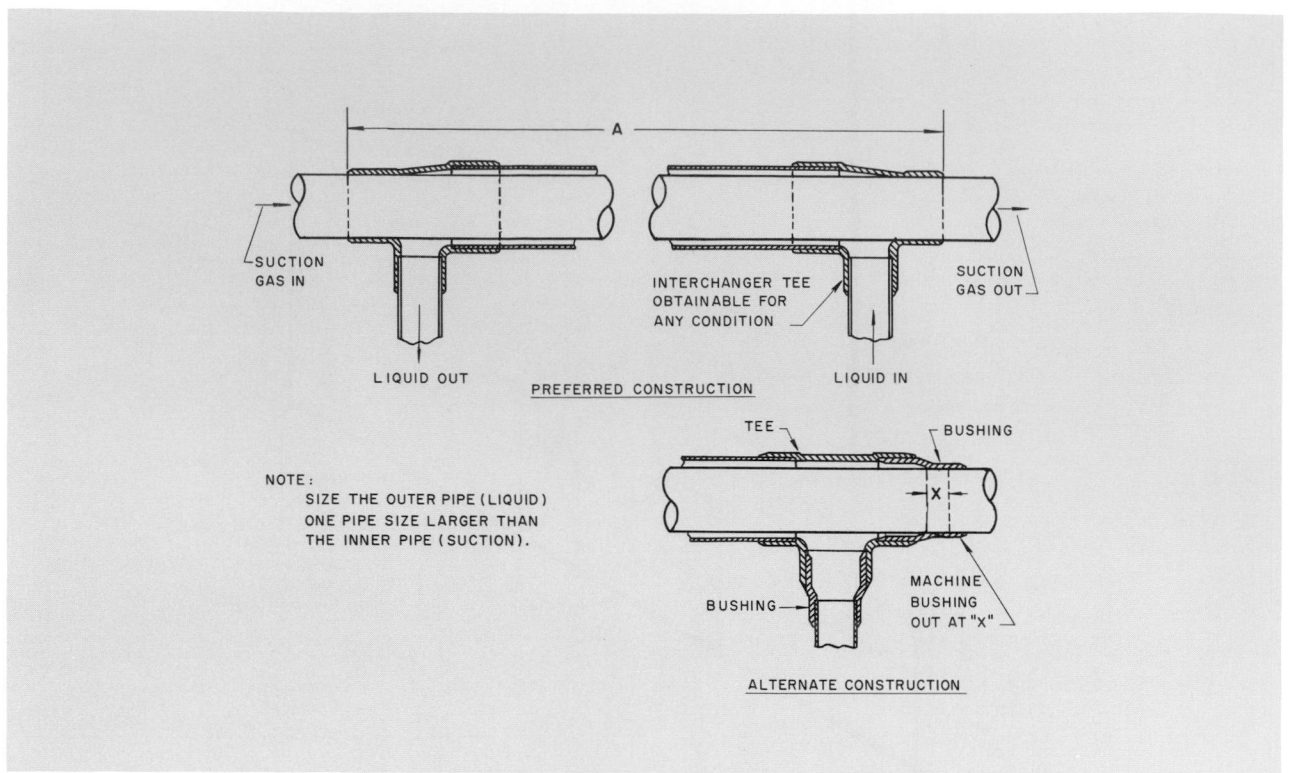

FIG. 76 — LIQUID SUCTION INTERCHANGER

Two common types of liquid suction interchangers are:

1. The *shell and coil* or the *shell and tube exchanger,* suitable for increasing cycle efficiency and for liquid subcooling. This type is usually installed so that the suction outlet drains the shell to prevent oil trapping.

2. The *tube-in-tube interchanger (Figs. 76 and 77),* a preferable type for controlling slop-over caused by erratic expansion valve feed or for "rectifying" lube oil from a refrigerant oil mixture bled from a flooded evaporator.

Excessive superheating of the suction gas must be avoided with heat exchangers since it causes excessive compressor discharge temperatures. Therefore, the amount of liquid subcooling permissible by a liquid suction interchanger is limited to the amount of suction gas superheating that does not cause compressor damage when the gas is compressed to the discharge pressure. Beyond this point additional subcooling should be obtained from other sources.

Charts 23 and 24 are used to determine the length (A) of a *concentric tube-in-tube interchanger (Fig. 76).* The amount of liquid subcooling available is

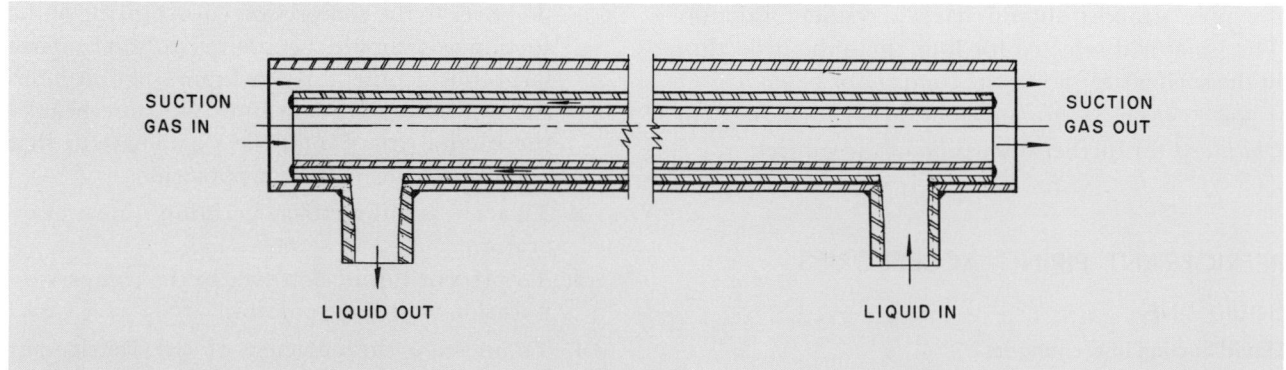

FIG. 77 — ECCENTRIC THREE-PIPE LIQUID SUCTION INTERCHANGER

CHART 23—EFFICIENCY CURVES, DOUBLE TUBE LIQUID SUCTION INTERCHANGER,
Refrigerants 12 and 500

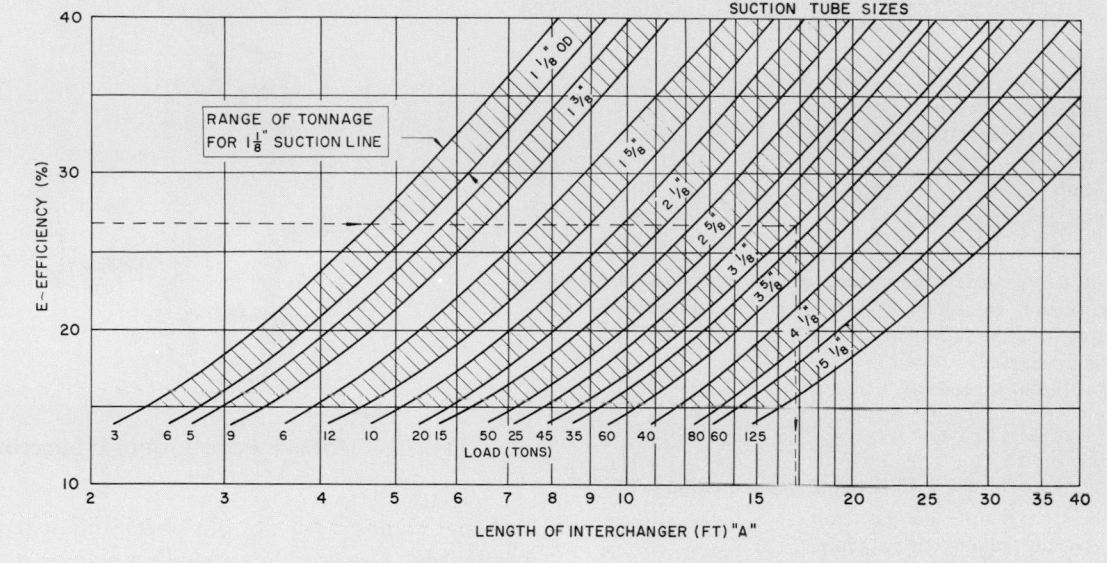

NOTES: 1. E (exchanger efficiency) = $\frac{\text{Leav gas temp} - \text{Ent gas temp}}{\text{Ent liq temp} - \text{Ent gas temp}} \times 100$

       2. Liquid refrigerant sub-cooling multipliers — Refrigerant 12 = .653, Refrigerant 500 = .571

3. Type L tubing for Refrigerants 12 and 500. Liquid in annular space.

4. Liquid tube is next larger size.

CHART 24—EFFICIENCY CURVES, DOUBLE TUBE LIQUID SUCTION INTERCHANGER,
Refrigerant 22

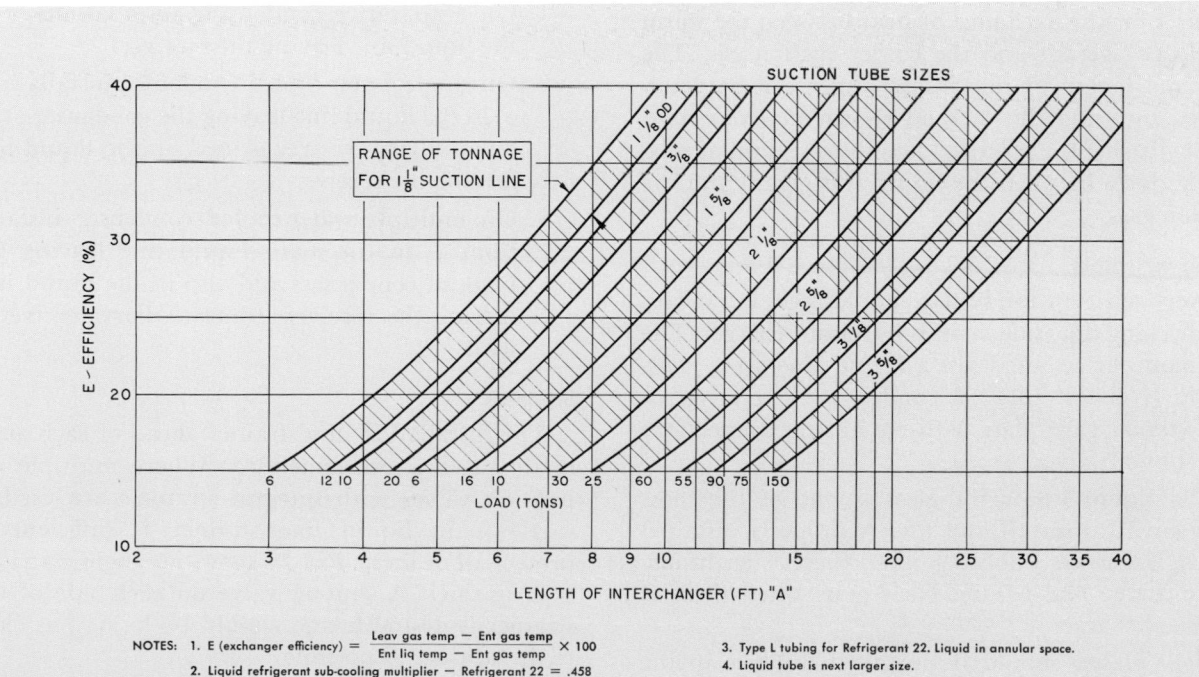

NOTES: 1. E (exchanger efficiency) = $\frac{\text{Leav gas temp} - \text{Ent gas temp}}{\text{Ent liq temp} - \text{Ent gas temp}} \times 100$

       2. Liquid refrigerant sub-cooling multiplier — Refrigerant 22 = .458

3. Type L tubing for Refrigerant 22. Liquid in annular space.

4. Liquid tube is next larger size.

calculated by using the ratio of the specific heats of the suction gas and of the liquid (subcooling multiplier). *Example 5* illustrates the use of these charts.

Example 5 — Determining the Length of a Concentric Tube-in-Tube Interchanger

Given:
   Refrigerant 12 system
   Load — 45 tons
   Suction line — 3⅛ in. OD Type L copper
   Expansion valve — 10 F superheat
   Suction temp — 40 F
   Condensing temp — 105 F

Find:
   Length of a concentric tube-in-tube interchanger to superheat the suction gas to 65 F (suction gas temperature to compressor in accordance with ASRE Standard 23R on rating compressors).
   Amount of liquid subcooling.

Solution:
   See *Chart 23*.

1. Determine interchanger efficiency E from equation

$$E = \frac{\text{leav gas temp} - \text{ent gas temp}}{\text{ent liq temp} - \text{ent gas temp}} \times 100$$

$$= \frac{65 - 50}{105 - 50} \times 100 = \frac{15}{55} \times 100 = 27.2\%$$

2. With an efficiency of 27.2% a 3⅛ in. OD pipe size and a 45 ton load, enter *Chart 23* as indicated per dashed line to determine a length (A) of 17 ft.

3. For Refrigerant 12 the ratio of gas to liquid specific heat is .653. Therefore, subcooling of liquid refrigerant is .653 × 15 F (leav gas temp — ent gas temp) or 9.8 F.

An *eccentric three-pipe interchanger* is shown in *Fig. 77*. The inner pipe and the outer pipe offer two surfaces for the exchange of heat between the warm liquid refrigerant and the colder suction gas. The required length of this interchanger can be determined by using the method shown in *Example 5* and by basing the required length on a ratio of relative surfaces between the liquid refrigerant and the suction gas.

Liquid Indicators

Every refrigeration system should include a means of checking for sufficient refrigerant charge. The common devices used are a liquid line sight glass, liquid level test cock on condenser or receiver, or an external gage glass with equalizing connections and shut-off valves.

The liquid line sight glass is one of the most convenient to install and use. A properly installed sight glass shows bubbling when there is an insufficient charge and a solid clear glass when there is sufficient charge.

Sight glasses should be installed full size in the main liquid line and not in a bypass line that parallels the main line.

FIG. 78 — DOUBLE PORT LIQUID INDICATOR

A sight glass with double ports and seal caps is preferable. The double ports allow a light to be put behind one port so that the state of the refrigerant is easily seen. The seal caps serve as an added protection against leakage or breakage since they are removed only when checking the refrigerant. *Fig. 78* shows a double port liquid indicator with seal caps.

The installation of a double port or see-through sight glass is recommended in each of the following locations:

1. On evaporative condensing installations — in the liquid line leaving the receiver.

2. On single water-cooled condenser installations — in the liquid line leaving the condenser or, if an auxiliary receiver is used, in the liquid line leaving the receiver.

3. On multiple water-cooled condenser installations — in the main liquid line leaving the bank of condensers and also in the liquid line leaving the receiver if an auxiliary receiver is used.

Strainers

The installation of a strainer ahead of each automatic valve is recommended. Where multiple expansion valves with integral strainers are used, a single main liquid line strainer is sufficient to protect all of these. *Fig. 79* shows an angle cartridge type strainer. A shut-off valve on each side of the strainer is desirable and should be located as close to the strainer as possible.

On steel piping systems an adequate strainer should be installed in the suction line and a filter-

FIG. 79 — ANGLE CARTRIDGE TYPE STRAINER

FIG. 80 — ANGLE TYPE DRIER-STRAINER

drier in the liquid line to remove the scale and rust inherent in steel pipe.

Refrigerant Driers

A permanent refrigerant drier is recommended for most systems and is essential for all low temperature systems. It is also essential for all systems using hermetic compressors since the compressor motor winding is exposed to refrigerant gas. If the gas contains excessive moisture, the winding insulation may break down and cause the motor to burn out. A full-flow drier must be used for this type system.

Figure 80 shows an angle type cartridge drier. The drier should be mounted vertically in the liquid line near the liquid receiver. A three-valve bypass *(Fig. 81)* should be used to permit isolation of the drier for servicing and to allow partial refrigerant flow thru the drier.

Reliable *moisture indicators (Fig. 82)* for liquid refrigerant lines are available. These devices indicate the proper time to replace the drier cartridge.

Filter-Driers

Filter-driers *(Fig. 83)* are more commonly used than strainers and driers together. The drier material actually filters the liquid refrigerant.

Solenoid Valves

Solenoid valves are commonly used in the following places:

1. In the liquid line of any system operating on single pump-out or pump-down compressor control.
2. In the liquid line of any single or multiple DX evaporator system.
3. In the oil bleeder lines from flooded evaporators to stop the flow of oil and refrigerant into the suction line when the system shuts down.

In many cases it is desirable to use solenoid valves with opening stems. The opening stem serves as a by-pass so that the system may continue to operate in case of solenoid coil failure.

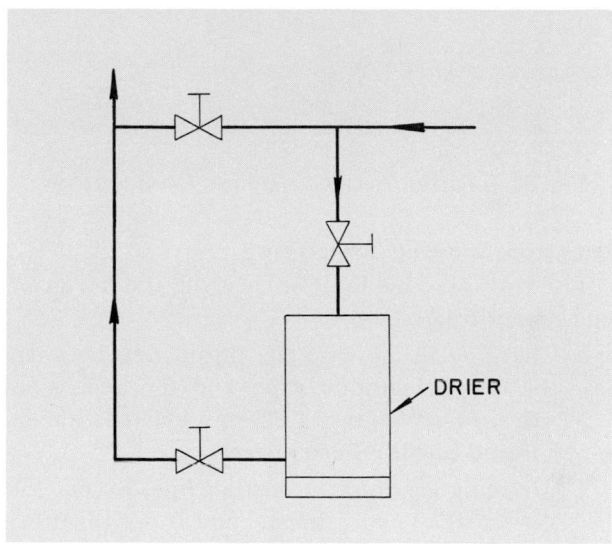

FIG. 81 — THREE-VALVE BYPASS FOR REFRIGERANT DRIER

FIG. 82 — COMBINATION MOISTURE AND LIQUID INDICATOR

FIG. 83 — FILTER-DRIER

Figures 82 and 83, courtesy of Sporlan Valve Co.

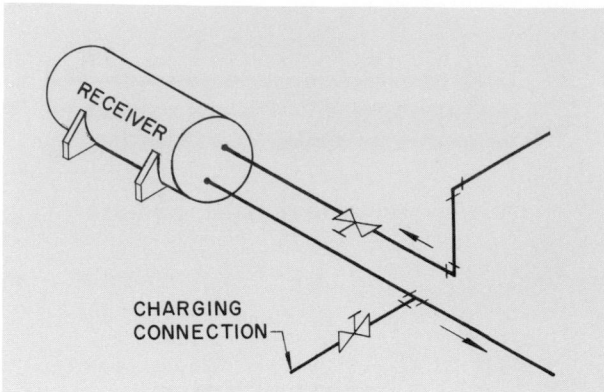

FIG. 84 — REFRIGERANT CHARGING CONNECTIONS

Refrigerant Charging Connections

The two usual methods of charging the refrigeration system are:

1. Charging liquid into the liquid line between the receiver shut-off valve and the expansion valve. *Fig. 84* shows a charging connection in a liquid line leaving a receiver.

2. Charging gas into the suction line. Except for very small systems this method is not practical because of the time required to evaporate the refrigerant from the drum and because of the danger of dumping raw liquid into the compressor.

Expansion Valves

Thermal expansion valves should be sized to avoid both the penalties of being undersized at full load and of being excessively oversized at partial load. The following items should be considered before sizing valves:

1. Refrigerant pressure drop thru the system must be properly evaluated to determine the correct pressure drop available across the valve.

2. Variations in condensing pressure during operation affect valve pressure and capacity. Condensing pressure should be controlled, therefore, to maintain required valve capacity.

3. Oversized thermal expansion valves do not control as well at full system capacity as properly sized valves and control gets progressively worse as the coil load decreases. Capacity reduction, available in most compressors, further increases this problem and necessitates closer selection of expansion valves to match realistic loads.

When sizing thermal expansion valves, make the selection on the basis of maximum load at the design operating pressure and at least 10 degrees super-

heat. Five degrees is the usual change in superheat between a full open and closed position. This is called the operating superheat. Thus a valve which operates at 10 degrees superheat at design load balances out at 5 to 6 degrees superheat at low load. A low superheat setting at design load, therefore, does not provide sufficient margins of safety at low loads because of the 5 degrees necessary for operating superheat.

The expansion valve bulb should be located immediately after the coil outlet on the suction line and 45° above the bottom of the pipe. With this arrangement the coil is the source of superheat for valve operation. The valve should be set so that overfeeding does not occur at times of partial load.

The effect of condensing temperature on the capacity of an expansion valve for two different systems is illustrated in *Example 6*.

Example 6 — Effect of Condensing Temperature on Expansion Valves

Given:

Two refrigeration systems using Refrigerant 500, one operating at 40 F suction and 90 F condensing, the other operating at 40 F suction and 130 F condensing.

	SYSTEM 1 40 F Suction 90 F Condens.	SYSTEM 2 40 F Suction 130 F Condens.
Condensing pressure	121.2 psig	218.2 psig
Liquid line drop	6.2	6.2
Pressure at thermal expansion valve inlet	115.0 psig	212.0 psig
Suction pressure	46.2 psig	46.2 psig
Suction line losses	2.8	2.8
Coil pressure drop	7.0	7.0
Distributor pressure drop	17.0	17.0
Pressure at thermal expansion valve outlet	73.0 psig	73.0 psig
Pressure drop available across valve	42.0 psi	139.0 psi

Assume that a valve of 27.5 ton capacity at 40 F suction and 60 psi differential is selected.

Find:

Capacity at the pressure drop available across the valve of systems 1 and 2.

Solution:

The capacities will vary approximately as the square root of the pressures:

$$\frac{\text{Capacity 1}}{\text{Capacity 2}} = \left(\frac{P_1}{P_2}\right)^{1/2}$$

For system 1　$\dfrac{\text{Cap.}}{27.5} = \left(\dfrac{42}{60}\right)^{1/2}$　Cap. = 23 tons

For system 2　$\dfrac{\text{Cap.}}{27.5} = \left(\dfrac{139}{60}\right)^{1/2}$　Cap. = 42 tons

Note that the expansion valve capacity is nearly double at the higher head pressure.

On certain low temperature applications and on high temperature applications where the design or partial load least temperature difference (L.T.D.) between the refrigerant and air or water is extremely small, it may become necessary to consider the use of the liquid suction interchanger as a source of superheat. This is done to increase the effective evaporator surface by allowing the liquid suction interchanger to supply the superheat function.

If only one liquid suction interchanger is used for the applications just mentioned, it should be an eccentric three-pipe interchanger as shown in *Fig. 77.* This arrangement permits the expansion valve bulb to sense the suction gas temperature from the outside surface of the interchanger. Otherwise two tube-in-tube interchangers should be used with the thermal expansion valve bulb located between the interchangers.

The preferred refrigerant flow in a coil circuit to obtain superheat is illustrated in *Fig. 85.*

SUCTION LINE

Back Pressure Valves

A conventional type back pressure regulating valve is used in a refrigerating system to maintain a predetermined pressure in the evaporator. A conventional type regulator controls the upstream pressure. The regulator has a spring loaded diaphragm designed to actuate a seat pilot valve. The actuating pressure comes from the evaporator or upstream side of the regulator. When the upstream pressure against the diaphragm is greater than that exerted by the spring, the pilot valve opens and a flow of gas is admitted to the power piston. The piston in turn causes the main port to open. This permits a flow of gas from the upstream side of the valve to the downstream side. When the actuating pressure becomes less than that controlled by the spring pressure, the flow of gas to the power piston is stopped and the regulator closes.

There are many variations of the back pressure regulating valve. Several are described in the following:

1. The compensating type, actuated by air or electricity, varies the suction pressure in accordance with temperature or load demand.
2. The dual pressure regulator is designed to operate at two predetermined pressures without resetting or adjustment; by opening and closing a pilot solenoid, either the low pressure or the high pressure head operates.

Figure 86 shows a simple back pressure regulating

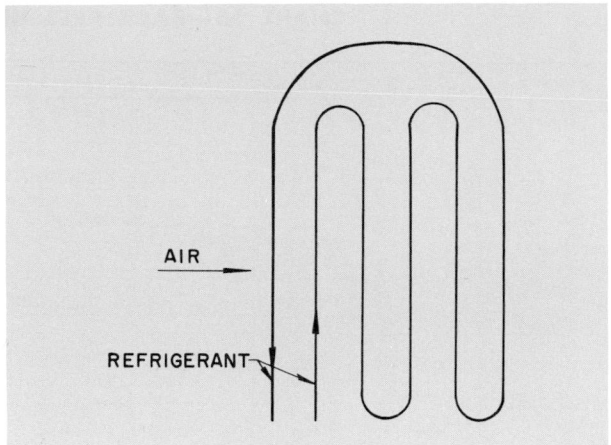

FIG. 85 — PREFERRED REFRIGERANT FLOW IN COIL CIRCUIT TO OBTAIN SUPERHEAT (PLAN VIEW)

valve which is ordinarily used for one of the following purposes:

1. To control evaporator suction pressure in spite of compressor suction pressure variation.
2. To establish evaporator suction pressure when lower compressor suction pressure is demanded by another part of the same system.
3. To prevent evaporator freezing when operating near the freezing temperature.

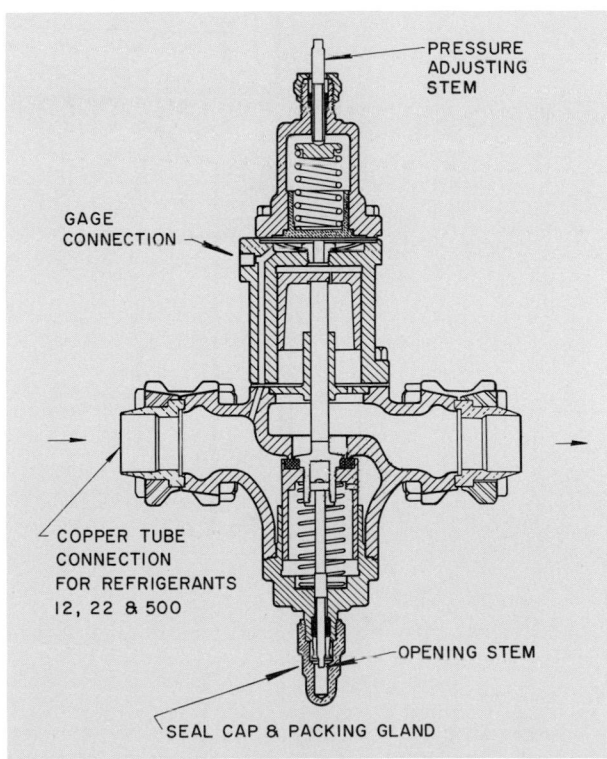

FIG. 86 — BACK PRESSURE VALVE

CHART 25—BACK PRESSURE VALVE APPLICATION CHART

SERVICE	NUMBER OF EVAPORATORS	ROOM CONTROL		COMPRESSOR Reduction	CAPACITY Controlled By	BACK PRESSURE VALVE REQUIRED	REASON	REMARKS
		THERMOSTAT	CONTROLS					
Air Conditioning	Single	1-Step	Liquid Line Solenoid or Compressor Motor	None	— —	Not Usually	See Notes 1, 2	Analyze for Large % OA Jobs
				50%	Pressure	Not Usually	See Notes 1, 2	Analyze for Large % OA Jobs
		2-Step	Compressor Motors	50%	Temperature	Not Usually	See Notes 1, 2	Analyze for Large % OA Jobs
		Modulating	Air Bypass	None	— —	Yes	See Note 1	
			Air Bypass or Modulating Expansion Valve	50%	Pressure	Sometimes	See Note 1	Must Be Analyzed
		Any	Any	3 or more steps	Pressure	Not Usually	See Note 1	Check frosting on last step
	Multiple	1-Step	Liquid Line Solenoid	None	— —	Usually	See Note 1	Check frosting at minimum load
				50%	Pressure	Not Usually	See Note 1	Check frosting at minimum load
		2-Step	2 Liquid Line Solenoids (Rarely Used Control)	None	— —	Usually	See Note 1	Check frosting at minimum load
				50%	Pressure	Not Usually	See Note 1	Check frosting at minimum load
		Modulating	Air Bypass	None	— —	Yes	See Note 1	
			Air Bypass or Modulating Expansion Valve	50%	Pressure	Sometimes	See Note 1	Must be analyzed
		Any	Any	3 or more steps	Pressure	No	See Note 1	
Refrigeration	Single	1-Step	Liquid Line Solenoid or Compressor Motor	None	— —	Sometimes	See Note 3	Used on commercial chill rooms only. Back pressure valve is by-passed during chilling operation.
			Liquid Line Solenoid	50%	Pressure	Sometimes	See Note 3	
				3 or more steps	Pressure	No		
	Multiple	1-Step	Liquid Line Solenoid	None	— —	Sometimes	See Note 3	Used only on rooms where low limit humidity control is desired.
				50%	Pressure	Sometimes	See Note 3	
				3 or more steps	Pressure	No		
Water Cooling (Flooded Coolers)	Single or Multiple	1-Step	Compressor Motor	None	— —	Usually	See Note 4	Check surface temperature at minimum load and on last step.
				50%	Pressure	Sometimes	See Note 4	
				3 or more steps	Pressure	Not Usually	See Note 4	

NOTES:

1: Reason for use of back pressure valve on **any** air conditioning system is to avoid frosting of coil at light load. Chart applies only to normal applications—refrigerant temperatures at full load of 40°F or above. For full load refrigerant temperatures from 35°-40°, frosting at light load should be checked. For full load refrigerant temperatures below 35°, except where certain of constant load, back pressure valve should be used.

2: Except for jobs using large % OA, this control indicates nearly constant load.

3: Reason for use of back pressure on **any** refrigeration system is to set a low limit below which room relative humidity cannot drop at light load.

4: Reason for use of back pressure valve on **any** water cooling system is to avoid freeze-up at light load. Chart applies only to variable loads and to normal applications—leaving water temperatures of 40°F or above. For leaving water temperatures below 40°, analysis should be made to make certain that surface temperature is not lower than 33°, unless certain of constant load, back pressure valve should be used.

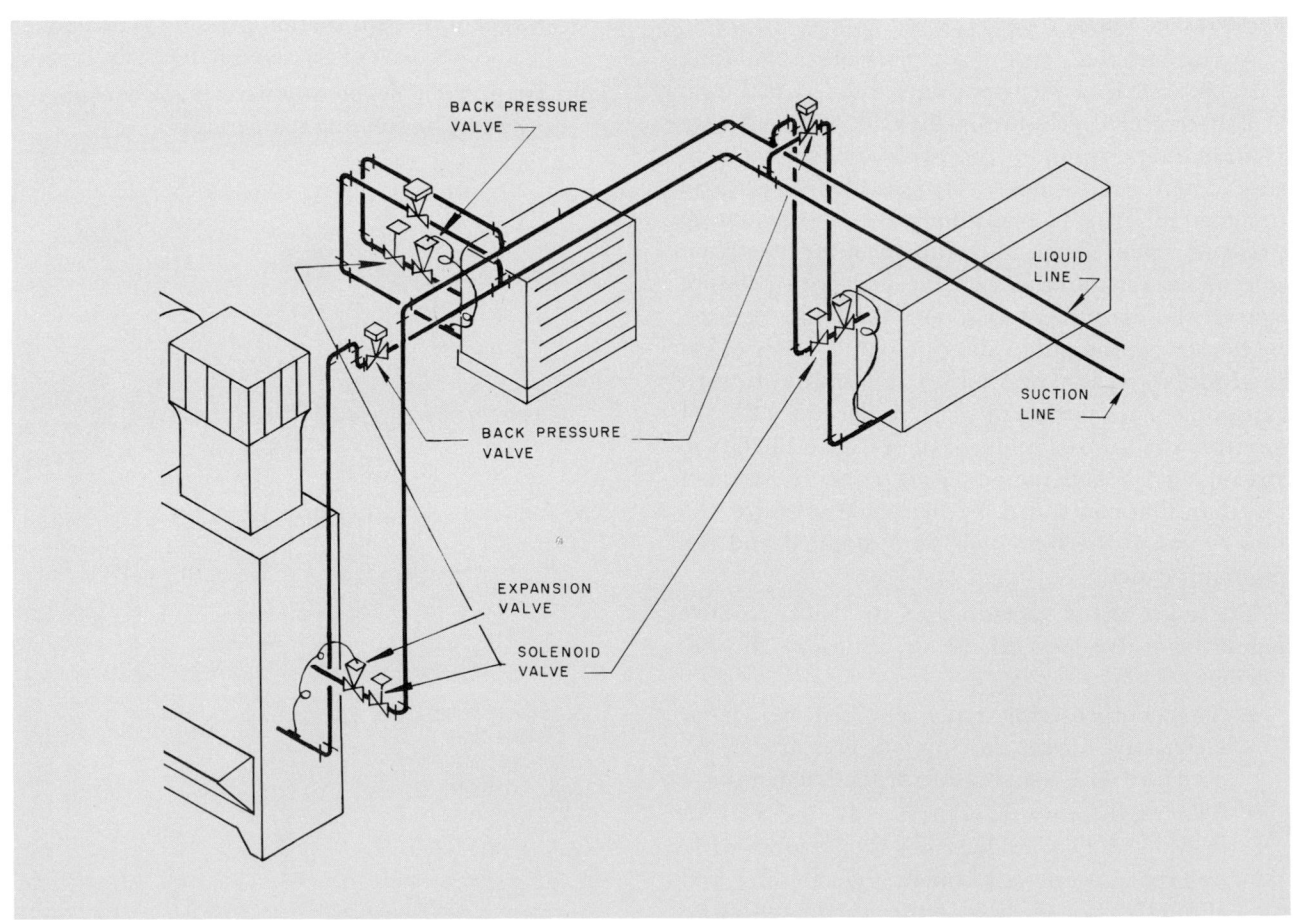

FIG. 87 – INSTALLATION USING BACK PRESSURE VALVES

Chart 25 illustrates the application of back pressure valves for various services such as number and types of evaporators, and types of room and compressor control.

Figure 87 illustrates the location of back pressure valves.

DISCHARGE LINE

Oil Separators

Oil separators reduce the rate of oil circulation. However they are not 100% efficient since some oil always circulates thru the system.

Oil separators are of particular value on certain types of installations such as:

1. Systems requiring a sudden and frequent capacity variation.

2. Systems having extensive pipe lines and numerous evaporators. The large volumes inherent in such systems result in appreciable oil hangup.

There are several objections to oil separators:

1. Oil separators permit some oil to be carried over into the system and, therefore, proper piping design to return oil is still required even though a separator is used.

2. On start-up, gas may condense in the shell of the separator. As a result the separator delivers liquid refrigerant into the crankcase. This in turn increases crankcase foaming and oil loss from the compressor.

During the "off" cycle the oil separator cools down and acts as a condenser for liquid refrigerant that evaporates in the warmer parts of the system. Thus a cool oil separator acts as a liquid condenser during "off" cycles and also on compressor start-up until the separator has warmed up. Large amounts of liquid refrigerant in the crankcase result in poor lubrication and may also result in removing the oil from the crankcase completely.

Figure 88 shows the recommended method for piping an oil separator.

Mufflers

If a muffler is used in the hot gas line, it should be installed in downward flow risers or in horizontal lines as close to the compressor as possible.

The hot gas pulsations from the compressor can set up a condition of resonance with certain lengths of refrigerant piping in the hot gas line. A muffler installed in the compressor discharge aids in eliminating such a condition.

Figure 89 shows a muffler in a hot gas line at the compressor.

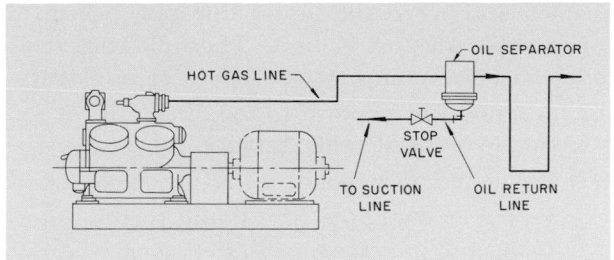

FIG. 88 — OIL SEPARATOR LOCATION

Check Valves

Check valves contribute a relatively large addition to a line pressure drop at full load and must be taken into account in the selection of refrigeration equipment. In addition a check valve cannot be relied upon for 100% shut-off.

Whenever the receiver is warmer than the compressor during shutdown, refrigerant in the receiver tends to boil off and flow back thru the condenser and hot gas discharge line to the compressor where it condenses. If there is sufficient refrigerant in the receiver, liquid refrigerant eventually enters the compressor despite the loop in the hot gas line at the base of the compressor. To prevent this, a check valve should be used *(Fig. 68, page 65)*.

In a non-automatic system a hand valve may be used at the inlet to the condenser to manually shut off the flow on shutdown, in which case the pressure drop involved will be much less than that encountered using a check valve.

REFRIGERANT PIPING INSULATION

Liquid lines should not be insulated if the surrounding temperature is lower than or equal to the temperature of the liquid. Insulation is recom-

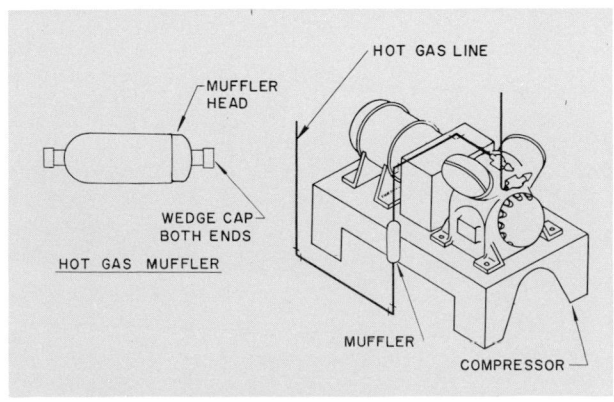

FIG. 89 — HOT GAS MUFFLER AND LOCATION IN
HOT GAS LINE

mended only when the liquid line can pick up a considerable amount of heat. The following areas in a refrigerant piping system should be insulated:

1. A liquid line exposed to the direct rays of the sun for a considerable distance.
2. Piping in boiler rooms.
3. Piping at the outlet of a liquid suction interchanger to preserve the subcooling effect.

Where liquid and suction lines can be strapped together, a single insulating covering can be used over both lines. This induces an exchange of heat and is desirable from the standpoint of the subcooling effect on the liquid. However excessive superheating of the suction gas can result from too much exchange of heat.

Hot gas lines should not be insulated. Any heat lost by these lines reduces the work to be done by the condenser.

Suction lines should be insulated only to prevent dripping where this causes a nuisance or damage. It is generally desirable to have the suction line capable of absorbing some heat to evaporate any liquid which may have entered the suction line from the evaporator. For unusual conditions of high ambient temperatures and simultaneous high relative humidities extra insulation must be applied.

The thickness of insulation required to prevent condensation on the outer surface is that thickness which raises the temperature of the outer surface of the insulation to a point slightly higher than the maximum expected dewpoint of the surrounding air. The external vapor barrier must be made as nearly perfect as possible in order to prevent leakage of vapor into the insulation.

Regular "ice water" thickness moulded cork covering wired on and sealed with asphalt primer is desirable for most work in the air conditioning range. For lower temperatures, "brine" thickness moulded cork covering should be used. Insulation that is not vapor-proof soon becomes saturated with moisture and rapidly deteriorates.

A cellular glass or cellular plastic type of insulation is fast becoming accepted as an ideal insulation. Its cellular structure provides exceptionally high resistance to water and water vapor. The cellular glass, being inorganic, is fire-proof. Cellular plastic which is also available is self-extinguishing.

When located out of doors, insulation must be weatherproofed unless, of course, it is inherently waterproof.

CHAPTER 4. STEAM PIPING

This chapter describes practical design and layout techniques for steam piping systems. Steam piping differs from other systems because it usually carries three fluids — steam, water and air. For this reason, steam piping design and layout require special consideration.

GENERAL SYSTEM DESIGN

Steam systems are classified according to piping arrangement, pressure conditions, and method of returning condensate to the boiler. These classifications are discussed in the following paragraphs.

PIPING ARRANGEMENT

A one- or two-pipe arrangement is standard for steam piping. The one-pipe system uses a single pipe to supply steam and to return condensate. Ordinarily, there is one connection at the heating unit for both supply and return. Some units have two connections which are used as supply and return connections to the common pipe.

A two-pipe steam system is more commonly used in air conditioning, heating, and ventilating applications. This system has one pipe to carry the steam supply and another to return condensate. In a two-pipe system, the heating units have separate connections for supply and return.

The piping arrangements are further classified with respect to condensate return connections to the boiler and direction of flow in the risers:

1. Condensate return to boiler
   a. Dry-return — condensate enters boiler above water line.
   b. Wet-return — condensate enters boiler below water line.
2. Steam flow in riser
   a. Up-feed — steam flows up riser.
   b. Down-feed — steam flows down riser.

PRESSURE CONDITIONS

Steam piping systems are normally divided into five classifications — high pressure, medium pressure,

low pressure, vapor and vacuum systems. Pressure ranges for the five systems are:

1. High pressure — 100 psig and above
2. Medium pressure — 15 to 100 psig
3. Low pressure — 0 to 15 psig
4. Vapor — vacuum to 15 psig
5. Vacuum — vacuum to 15 psig

Vapor and vacuum systems are identical except that a vapor system does not have a vacuum pump, but a vacuum system does.

CONDENSATE RETURN

The type of condensate return piping from the heating units to the boiler further identifies the steam piping system. Two arrangements, gravity and mechanical return, are in common use.

When all the units are located above the boiler or condensate receiver water line, the system is described as a gravity return since the condensate returns to the boiler by gravity.

If traps or pumps are used to aid the return of condensate to the boiler, the system is classified as a mechanical return system. The vacuum return pump, condensate return pump and boiler return trap are devices used for mechanically returning condensate to the boiler.

CODES AND REGULATIONS

All applicable codes and regulations should be checked to determine acceptable piping practice for the particular application. These codes usually dictate piping design, limit the steam pressure, or qualify the selection of equipment.

WATER CONDITIONING

The formation of scale and sludge deposits on the boiler heating surfaces creates a problem in generating steam. Scale formation is intensified since scale-forming salts increase with an increase in temperature.

Water conditioning in a steam generating system should be under the supervision of a specialist.

TABLE 25—RECOMMENDED HANGER SPACINGS FOR STEEL PIPE

NOM. PIPE SIZE (in.)	DISTANCE BETWEEN SUPPORTS (FT)		
	Average Gradient		
	1″ in 10′	½″ in 10′	¼″ in 10′
¾	9	—	—
1	13	6	—
1¼	16	10	5
1½	19	14	8
2	21	17	13
2½	24	19	15
3	27	22	18
3½	29	24	19
4	32	26	20
5	37	29	23
6	40	33	25
8	—	38	30
10	—	43	33
12	—	48	37
14	—	50	40
16	—	53	42
18	—	57	44
20	—	60	47
24	—	64	50

NOTE: Data is based on standard wall pipe filled with water and average number of fittings.

Courtesy of Crane Co.

PIPING SUPPORTS

All steam piping is pitched to facilitate the flow of condensate. *Table 25* contains the recommended support spacing for piping pitched for different gradients. The data is based on Schedule 40 pipe filled with water, and an average amount of valves and fittings.

PIPING DESIGN

A steam system operating for air conditioning comfort conditions must distribute steam at all operating loads. These loads can be in excess of design load, such as early morning warmup, and at extreme

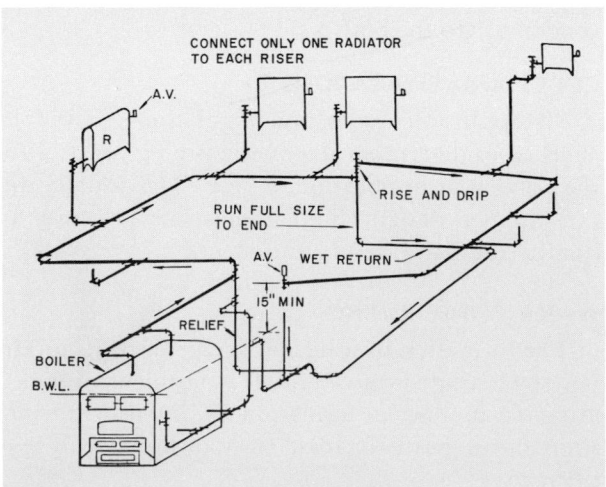

Fig. 90 — One-Pipe, Upfeed Gravity System

partial load, when only a minimum of heat is necessary. The pipe size to transmit the steam for a design load depends on the following:

1. The initial operating pressure and the allowable pressure drop thru the system.
2. The total equivalent length of pipe in the longest run.
3. Whether the condensate flows in the same direction as the steam or in the opposite direction.

The major steam piping systems used in air conditioning applications are classified by a combination of piping arrangement and pressure conditions as follows:

1. Two-pipe high pressure
2. Two-pipe medium pressure
3. Two-pipe low pressure
4. Two-pipe vapor
5. Two-pipe vacuum
6. One-pipe low pressure

ONE-PIPE SYSTEM

A one-pipe gravity system is primarily used on residences and small commercial establishments. *Fig. 90* shows a one-pipe, upfeed gravity system. The steam supply main rises from the boiler to a high point and is pitched downward from this point around the extremities of the basement. It is normally run full size to the last take-off and is then reduced in size after it drops down below the boiler water line. This arrangement is called a wet return. If the return main is above the boiler water line, it is called a dry return. Automatic air vents are required at all high points in the system to remove non-condensable gases. In systems that require long mains, it is necessary to check the pressure drop and make sure the last heating unit is sufficiently above the water line to prevent water backing up from the boiler and flooding the main. During operation, steam and condensate flow in the same direction in the mains, and in opposite direction in branches and risers. This system requires larger pipe and valves than any other system.

The one-pipe gravity system can also be designed as shown in *Fig. 91,* with each riser dripped separately. This is frequently done on more extensive systems.

Another type of one-pipe gravity system is the down-feed arrangement shown in *Fig. 92.* Steam flows in the main riser from the boiler to the building attic and is then distributed throughout the building.

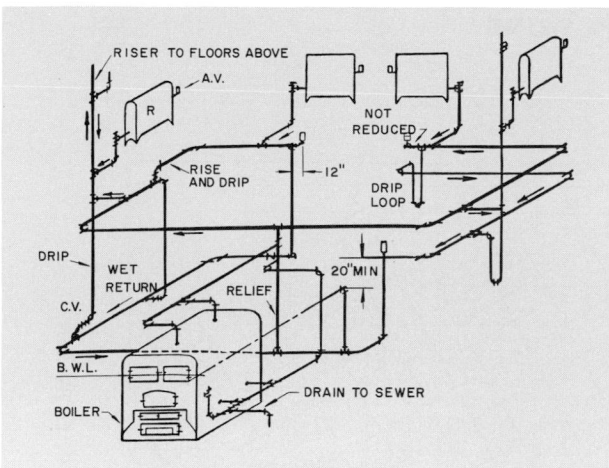

FIG. 91 — ONE-PIPE GRAVITY SYSTEM WITH
DRIPPED RISERS

TWO-PIPE SYSTEM

A two-pipe gravity system is shown in *Fig. 93.* This system is used with indirect radiation. The addition of a thermostatic valve at each heating unit adapts it to a vapor or a mechanical vacuum system. A gravity system has each radiator separately sealed by drip loops on a dry return or by dropping directly into a wet return main. All drips, reliefs and return risers from the steam to the return side of the system must be sealed by traps or water loops to insure satisfactory operation.

If the air vent on the heating unit is omitted, and the air is vented thru the return line and a vented condensate receiver, a vapor system as illustrated in *Fig. 94* results.

The addition of a vacuum pump to a vapor system classifies the system as a mechanical vacuum system. This arrangement is shown in *Fig. 95.*

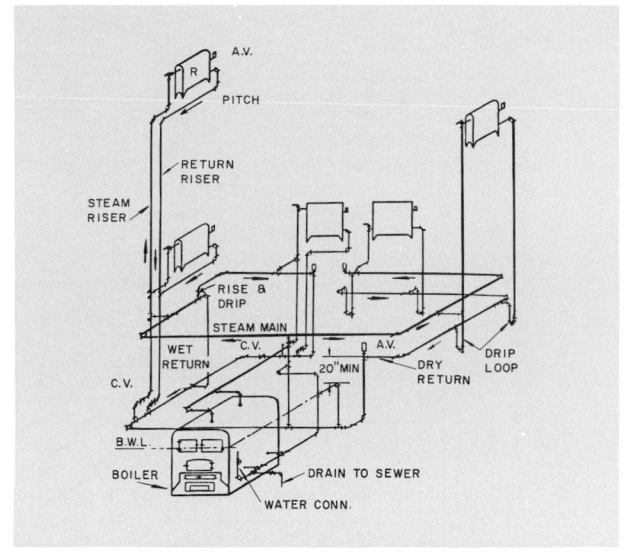

FIG. 93 — TWO-PIPE GRAVITY SYSTEM

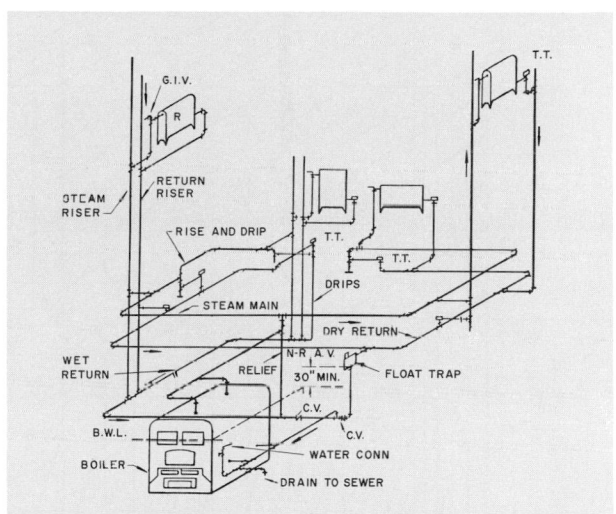

FIG. 94 — VAPOR SYSTEM

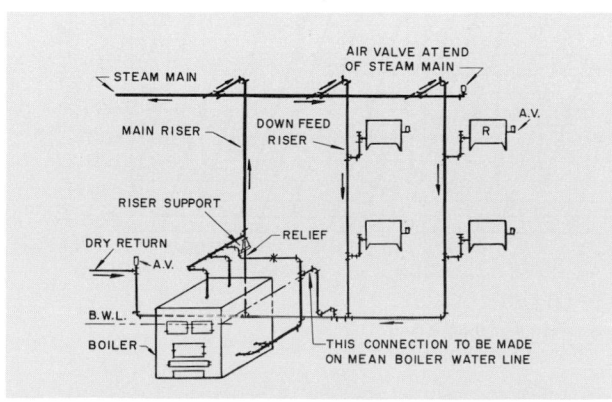

FIG. 92 — ONE-PIPE, DOWNFEED GRAVITY
SYSTEM

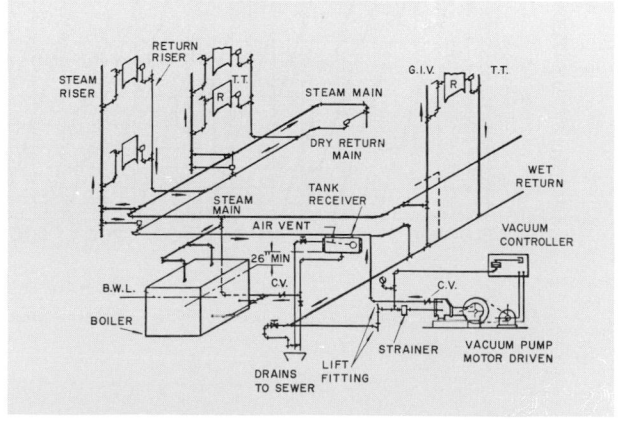

FIG. 95 — MECHANICAL VACUUM SYSTEM

Figures 90 thru 95, courtesy of Kent's Mechanical Engineer's Handbook, Power Volume, Eleventh Edition, John Wiley & Sons Publisher.

CHART 26—PIPE SIZING*

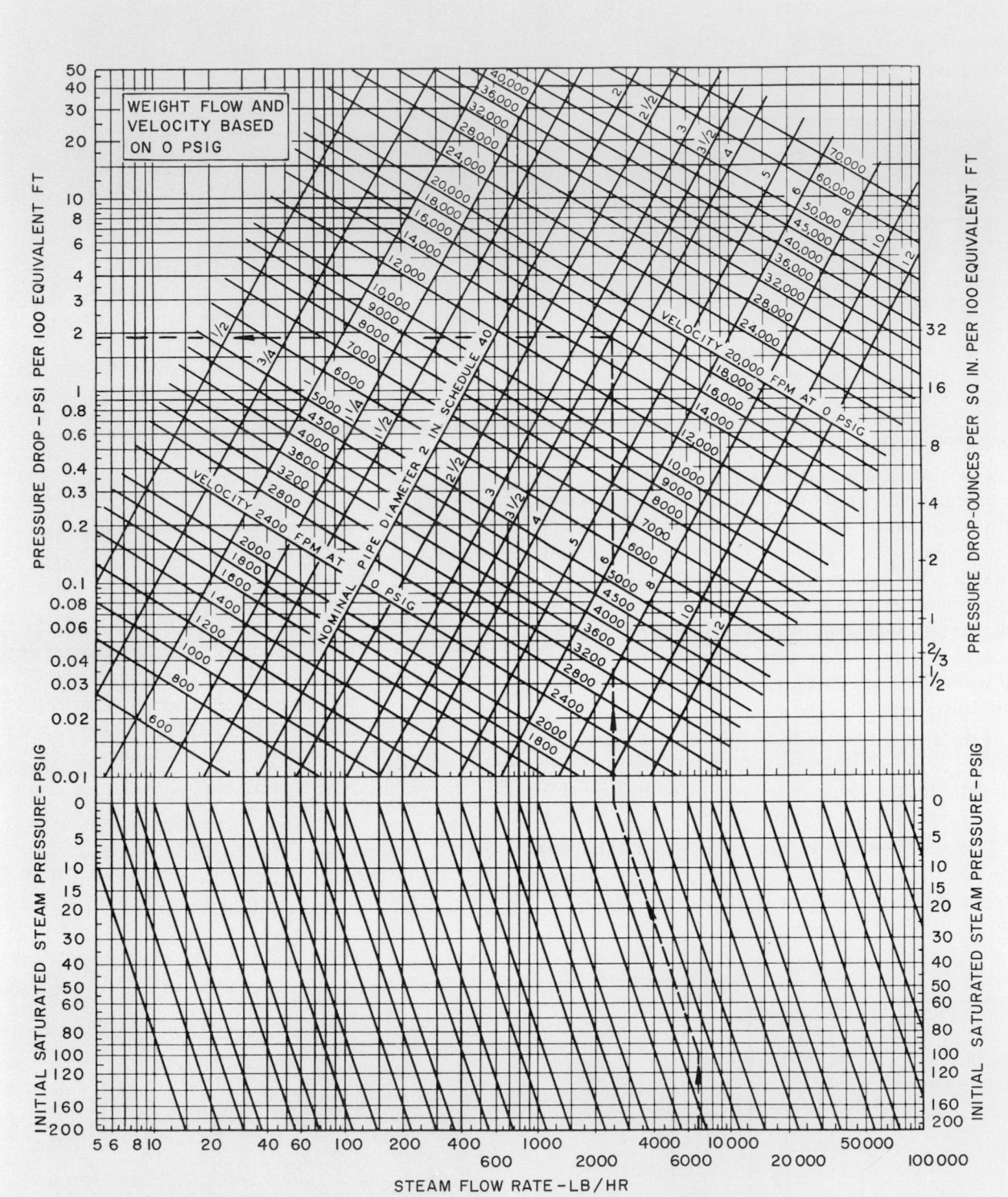

STEAM FLOW RATE – LB/HR

*Use *Chart 27* to determine steam velocity at initial saturated steam pressures other than 0 psig.

CHART 27—VELOCITY CONVERSION*

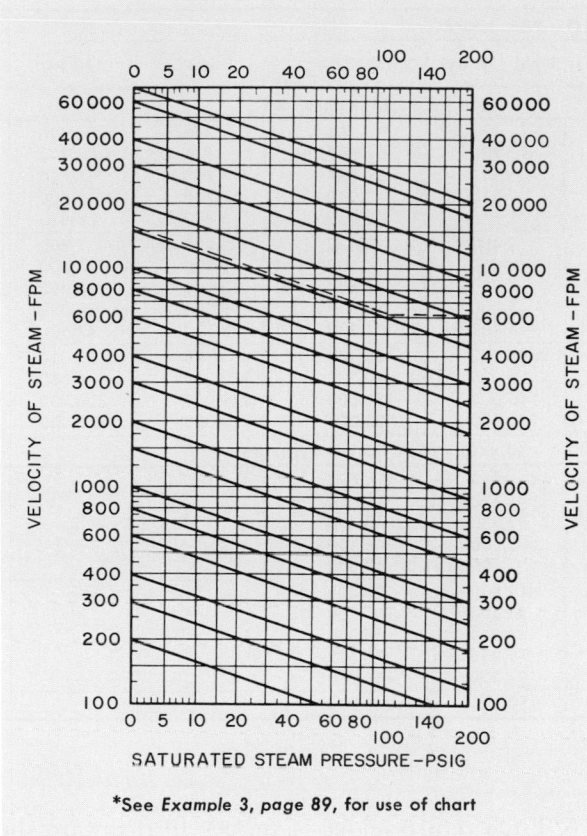

*See *Example 3, page 89,* for use of chart

PIPE SIZING

GENERAL

Charts and tables have been developed which are used to select the proper pipe to carry the required steam rate at various pressures.

Chart 26 is a universal chart for steam pressure of 0 to 200 psig and for a steam rate of from 5 to 100,000 pounds per hour. However, the velocity as read from the chart is based on a steam pressure of 0 psig and must be corrected for the desired pressure from *Chart 27.* The complete chart is based on the Moody friction factor and is valid where condensate and steam flow in the same direction.

Tables 26 thru 31 are used for quick selection at specific steam pressures. *Chart 26* has been used to tabulate the capacities shown in *Tables 26 thru 28.* The capacities in *Tables 29 thru 31* are the results of tests conducted in the ASHAE laboratories. Suggested limitations for the use of these tables are shown as notes on each table. In addition, *Table 31* shows the total pressure drop for two-pipe low pressure steam systems.

RECOMMENDATIONS

The following recommendations are for use when sizing pipe for the various systems:

Two-Pipe High Pressure System

This system is used mostly in plants and occasionally in commercial installations.

1. Size supply main and riser for a maximum drop of 25-30 psi.
2. Size supply main and risers for a maximum friction rate of 2-10 psi per 100 ft of equivalent pipe.
3. Size return main and riser for a maximum pressure drop of 20 psi.
4. Size return main and riser for a maximum friction rate of 2 psi per 100 ft of equivalent pipe.
5. Pitch supply mains ¼ in. per 10 ft away from boiler.
6. Pitch return mains ¼ in. per 10 ft toward the boiler.
7. Size pipe from *Table 26.*

Two-Pipe Medium Pressure System

This system is used mostly in plants and occasionally in commercial installations.

1. Size supply main and riser for a maximum pressure drop of 5-10 psi.
2. Size supply mains and risers for a maximum friction rate of 2 psi per 100 ft of equivalent pipe.
3. Size return main and riser for a maximum pressure drop of 5 psi.
4. Size return main and riser for a maximum friction rate of 1 psi per 100 ft of equivalent pipe.
5. Pitch supply mains ¼ in. per 10 ft away from the boiler.
6. Pitch return mains ¼ in. per 10 ft toward the boiler.
7. Size pipe from *Table 27.*

Two-Pipe Low Pressure System

This system is used for commercial, air conditioning, heating and ventilating installations.

1. Size supply main and risers for a maximum pressure drop determined from *Table 31,* depending on the initial system pressure.
2. Size supply main and riser for a maximum friction rate of 2 psi per 100 ft of equivalent pipe.
3. Size return main and riser for a maximum

TABLE 26—HIGH PRESSURE SYSTEM PIPE CAPACITIES (150 psig)
Pounds Per Hour

PIPE SIZE (in.)	PRESSURE DROP PER 100 FT							
	⅛ psi (2 oz)	¼ psi (4 oz)	½ psi (8 oz)	¾ psi (12 oz)	1 psi (16 oz)	2 psi (32 oz)	5 psi	10 psi
SUPPLY MAINS AND RISERS				130 - 180 psig—Max Error 8%				
¾	29	41	58	82	116	184	300	420
1	58	82	117	165	233	369	550	790
1¼	130	185	262	370	523	827	1,230	1,720
1½	203	287	407	575	813	1,230	1,730	2,600
2	412	583	825	1,167	1,650	2,000	3,410	4,820
2½	683	959	1,359	1,920	2,430	3,300	5,200	7,600
3	1,237	1,750	2,476	3,500	4,210	6,000	9,400	13,500
3½	1,855	2,626	3,715	5,250	6,020	8,500	13,100	20,000
4	2,625	3,718	5,260	7,430	8,400	12,300	19,200	28,000
5	4,858	6,875	9,725	13,750	15,000	21,200	33,100	47,500
6	7,960	11,275	15,950	22,550	25,200	36,500	56,500	80,000
8	16,590	23,475	33,200	46,950	50,000	70,200	120,000	170,000
10	30,820	43,430	61,700	77,250	90,000	130,000	210,000	300,000
12	48,600	68,750	97,250	123,000	155,000	200,000	320,000	470,000
RETURN MAINS AND RISERS				1 - 20 psig - Max Return Pressure				
¾	156	232	360	465	560	890		
1	313	462	690	910	1,120	1,780		
1¼	650	960	1,500	1,950	2,330	3,700		
1½	1,070	1,580	2,460	3,160	3,800	6,100		
2	2,160	3,300	4,950	6,400	7,700	12,300		
2½	3,600	5,350	8,200	10,700	12,800	20,400		
3	6,500	9,600	15,000	19,500	23,300	37,200		
3½	9,600	14,400	22,300	28,700	34,500	55,000		
4	13,700	20,500	31,600	40,500	49,200	78,500		
5	25,600	38,100	58,500	76,000	91,500	146,000		
6	42,000	62,500	96,000	125,000	150,000	238,000		

pressure drop determined from *Table 31,* depending on the initial system pressure.

4. Size return main and riser for a maximum friction rate of ½ psi per 100 ft of equivalent pipe.

5. Pitch mains ¼ in. per 10 ft away from the boiler.

6. Pitch return mains ¼ in. per 10 ft toward the boiler.

7. Use *Tables 28 thru 30* to size pipe.

Two-Pipe Vapor System

This system is used in commercial and residential installations.

1. Size supply main and riser for a maximum pressure drop of 1/16 - ⅛ psi.

2. Size supply main and riser for a maximum friction rate of 1/16 - ⅛ psi per 100 ft of equivalent pipe.

3. Size return main and supply for a maximum pressure drop of 1/16 - ⅛ psi.

4. Size return main and supply for a maximum friction rate of 1/16 - ⅛ psi per 100 ft of equivalent pipe.

5. Pitch supply ¼ in. per 10 ft away from the boiler.

6. Pitch return mains ¼ in. per 10 ft toward the boiler.

7. Size pipe from *Tables 28 thru 30.*

Two-Pipe Vacuum System

This system is used in commercial installations.

1. Size supply main and riser for a maximum pressure drop of ⅛ - 1 psi.

2. Size supply main and riser for a maximum friction rate of ⅛ - ½ psi per 100 ft of equivalent pipe.

3. Size return main and riser for a maximum pressure drop of ⅛ - 1 psi.

4. Size return main and riser for a maximum friction rate of ⅛ - ½ psi per 100 ft of equivalent pipe.

5. Pitch supply mains ¼ in. per 10 ft away from the boiler.

6. Pitch return mains ¼ in. per 10 ft toward the boiler.

7. Size pipe from *Tables 28 thru 30.*

One-Pipe Low Pressure System

This system is used on small commercial and residential systems.

TABLE 27—MEDIUM PRESSURE SYSTEM PIPE CAPACITIES (30 psig)
Pounds Per Hour

PIPE SIZE (in.)	PRESSURE DROP PER 100 FT					
	⅛ psi (2 oz)	¼ psi (4 oz)	½ psi (8 oz)	¾ psi (12 oz)	1 psi (16 oz)	2 psi (32 oz)
	SUPPLY MAINS AND RISERS			25 - 35 psig - Max Error 8%		
¾	15	22	31	38	45	63
1	31	46	63	77	89	125
1¼	69	100	141	172	199	281
1½	107	154	219	267	309	437
2	217	313	444	543	627	886
2½	358	516	730	924	1,033	1,460
3	651	940	1,330	1,628	1,880	2,660
3½	979	1,414	2,000	2,447	2,825	4,000
4	1,386	2,000	2,830	3,464	4,000	5,660
5	2,560	3,642	5,225	6,402	7,390	10,460
6	4,210	6,030	8,590	10,240	12,140	17,180
8	8,750	12,640	17,860	21,865	25,250	35,100
10	16,250	23,450	33,200	40,625	46,900	66,350
12	25,640	36,930	52,320	64,050	74,000	104,500
	RETURN MAINS AND RISERS			0 - 4 psig - Max Return Pressure		
¾	115	170	245	308	365	
1	230	340	490	615	730	
1¼	485	710	1,025	1,285	1,530	
1½	790	1,155	1,670	2,100	2,500	
2	1,575	2,355	3,400	4,300	5,050	
2½	2,650	3,900	5,600	7,100	8,400	
3	4,850	7,100	10,250	12,850	15,300	
3½	7,200	10,550	15,250	19,150	22,750	
4	10,200	15,000	21,600	27,000	32,250	
5	19,000	27,750	40,250	55,500	60,000	
6	31,000	45,500	65,500	83,000	98,000	

1. Size supply main and riser for a maximum pressure drop of ¼ psi.
2. Size supply main and risers for a maximum friction rate of $\frac{1}{16}$ psi per 100 ft of equivalent pipe.
3. Size return main and risers for a maximum pressure drop of ¼ psi.
4. Size return main and risers for a maximum friction rate of $\frac{1}{16}$ psi per 100 ft of equivalent pipe.
5. Pitch supply main ¼ in. per 10 ft away from the boiler.
6. Pitch return main ¼ in. per 10 ft toward the boiler.

TABLE 28—LOW PRESSURE SYSTEM PIPE CAPACITIES
Pounds Per Hour
CONDENSATE FLOWING WITH THE STEAM FLOW

NOM. PIPE SIZE (in.)	PRESSURE DROP PER 100 FT													
	$\frac{1}{16}$ psi (1 oz)		⅛ psi (2 oz)		¼ psi (4 oz)		½ psi (8 oz)		¾ psi (12 oz)		1 psi		2 psi	
	SATURATED PRESSURE (PSIG)													
	3.5	12	3.5	12	3.5	12	3.5	12	3.5	12	3.5	12	3.5	12
¾	9	11	14	16	20	24	29	35	36	43	42	50	60	73
1	17	21	26	31	37	46	54	66	68	82	81	95	114	137
1¼	36	45	53	66	78	96	111	138	140	170	162	200	232	280
1½	56	70	84	100	120	147	174	210	218	260	246	304	360	430
2	108	134	162	194	234	285	336	410	420	510	480	590	710	850
2½	174	215	258	310	378	460	540	660	680	820	780	950	1,150	1,370
3	318	380	465	550	660	810	960	1,160	1,190	1,430	1,380	1,670	1,950	2,400
3½	462	550	670	800	990	1,218	1,410	1,700	1,740	2,100	2,000	2,420	2,950	3,450
4	726	800	950	1,160	1,410	1,690	1,980	2,400	2,450	3,000	2,880	3,460	4,200	4,900
5	1,200	1,430	1,680	2,100	2,440	3,000	3,570	4,250	4,380	5,250	5,100	6,100	7,500	8,600
6	1,920	2,300	2,820	3,350	3,960	4,850	5,700	7,200	7,200	8,600	8,400	10,000	11,900	14,200
8	3,900	4,800	5,570	7,000	8,100	10,000	11,400	14,300	14,500	17,700	16,500	20,500	24,000	29,500
10	7,200	8,800	10,200	12,600	15,000	18,200	21,000	26,000	26,200	32,000	30,000	37,000	42,700	52,000
12	11,400	13,700	16,500	19,500	23,400	28,400	33,000	40,000	41,000	49,500	48,000	57,500	67,800	81,000

The weight-flow rates at 3.5 psig can be used to cover sat. press. from 1 to 6 psig, and the rates at 12 psig can be used to cover sat. press. from 8 to 16 psig with an error not exceeding 8 percent.

Tables 26 thru 28 from Heating Ventilating Air Conditioning Guide 1959. Used by permission.

TABLE 29—RETURN MAIN AND RISER CAPACITIES FOR LOW PRESSURE STEAM SYSTEMS

PIPE SIZE (in.)	1/32 psi (1/2 oz)			1/24 psi (2/3 oz)			1/16 psi (1 oz)			1/8 psi (2 oz)			1/4 psi (4 oz)			1/2 psi (8 oz)		
	Wet*	Dry	Vac	Wet*	Dry	Vac	Wet*	Dry	Vac	Wet*	Dry	Vac	Wet*	Dry	Vac	Wet*	Dry	Vac
RETURN MAINS																		
¾						42			100			142			200			283
1	125	62		145	71	143	175	80	175	250	103	249	350	115	350			494
1¼	213	130		248	149	244	300	168	300	425	217	426	600	241	600			848
1½	338	206		393	236	388	475	265	475	675	340	674	950	378	950			1,340
2	700	470		810	535	815	1,000	575	1,000	1,400	740	1,420	2,000	825	2,000			2,830
2½	1,180	760		1,580	868	1,360	1,680	950	1,680	2,350	1,230	2,380	3,350	1,360	3,350			4,730
3	1,880	1,460		2,130	1,560	2,180	2,680	1,750	2,680	3,750	2,250	3,800	5,350	2,500	5,350			7,560
3½	2,750	1,970		3,300	2,200	3,250	4,000	2,500	4,000	5,500	3,230	5,680	8,000	3,580	8,000			11,300
4	3,880	2,930		4,580	3,350	4,500	5,500	3,750	5,500	7,750	4,830	7,810	11,000	5,380	11,000			15,500
5						7,880			9,680			13,700			19,400			27,300
6						12,600			15,500			22,000			31,000			43,800
RETURN RISERS																		
¾		48			48	143		48	175		48	249		48	350			494
1		113			113	244		113	300		113	426		113	600			848
1¼		248			248	388		248	475		248	674		248	950			1,340
1½		375			375	815		375	1,000		375	1,420		375	2,000			2,830
2		750			750	1,360		750	1,680		750	2,380		750	3,350			4,730
2½						2,180			2,680			3,800			5,350			7,560
3						3,250			4,000			5,680			8,000			11,300
3½						4,480			5,500			7,810			11,000			15,500
4						7,880			9,680			13,700			19,400			27,300
5						12,600			15,500			22,000			31,000			43,800

*Vac values may be used for wet return risers and mains.

From Heating Ventilating Air Conditioning Guide, 1959. Used by permission.

7. Size supply main and dripped runouts from *Table 28*.

TABLE 30—LOW PRESSURE SYSTEM PIPE CAPACITIES

Pounds Per Hour
CONDENSATE FLOWING AGAINST STEAM FLOW

PIPE SIZE (in.)	TWO-PIPE SYSTEM		ONE-PIPE SYSTEM		
	Vertical	Horizontal	Up-feed Supply Risers	Vertical Connectors	Riser Runouts
A	B*	C†	D‡	E	F
¾	8	—	6	—	7
1	14	9	11	7	7
1¼	31	19	20	16	16
1½	48	27	38	23	16
2	97	49	72	42	23
2½	159	99	116	—	42
3	282	175	200	—	65
3½	387	288	286	—	119
4	511	425	380	—	186
5	1,050	788	—	—	278
6	1,800	1,400	—	—	545
8	3,750	3,000	—	—	—
10	7,000	5,700	—	—	—
12	11,500	9,500	—	—	—
16	22,000	19,000	—	—	—

*Do not use Column B for pressure drops less than 1/16 psi/100 ft of equivalent length. Use *Chart 26, page 84.*

†Pitch of horizontal runouts to riser should be not less than ½ in./ft. Where this pitch cannot be obtained, runouts over 8 ft in length should be one pipe size larger than called for in this table.

‡Do not use Column D for pressure drops less than 1/24 psi/100 ft of equivalent run except on sizes 3 in. and larger. Use *Chart 26, page 84.*

From Heating Ventilating Air Conditioning Guide, 1959. Used by permission.

8. Size undripped runouts from *Table 30, Column F.*

9. Size upfeed risers from *Table 30, Column D.*

10. Size downfeed supply risers from *Table 28.*

11. Pitch supply mains ¼ in. per 10 ft away from boiler.

12. Pitch return mains ¼ in. per 10 ft toward the boiler.

Use of Table 31

Example 1 — Determine Pressure Drop for Sizing Supply and Return Piping

Given:
Two-pipe low pressure steam system
Initial steam pressure — 15 psig
Approximate supply piping equivalent length — 500 ft
Approximate return piping equivalent length — 500 ft

Find:
1. Pressure drop to size supply piping
2. Pressure drop to size return piping

Solution:
1. Refer to *Table 31* for an initial steam pressure of 15 psig. The total pressure drop should not exceed 3.75 psi in the supply pipe. Therefore, the supply piping is sized for a total pressure drop of 3.75 or 3/4 psi per 100 ft of equivalent pipe.
2. Although ¾ psi is indicated in *Step 1, Item 4* under the two-pipe low pressure system recommends a maximum of ½ psi for return piping. Therefore, use ½ psi per 100 ft of equivalent pipe.

Return main pressure drop $= \frac{1}{2} \times \frac{500}{100} = 2.5$ psi.

Friction Rate

Example 2 illustrates the method used to determine the friction rate for sizing pipe when the total system pressure drop recommendation (supply pressure drop plus return pressure drop) is known and the approximate equivalent length is known.

Example 2 — Determine Friction Rate

Given:
    Four systems
    Equivalent length of each system — 400 ft
    Total pressure drop of systems — $\frac{1}{2}$, $\frac{3}{4}$, 1, and 2 psi

Find:
    Friction rate for each system.

Solution:

SYSTEM NUMBER	SYSTEM EQUIV. LENGTH (ft)	TOTAL SYSTEM PRESS. DROP (psi)	FRICTION RATE FOR PIPE SIZING (per 100 ft)
1	400	$\frac{1}{2}$	$(400/100)(x) = \frac{1}{2}$ $x = \frac{1}{8}$
2	400	$\frac{3}{4}$	$(400/100)(x) = \frac{3}{4}$ $x = \frac{3}{16}$
3	400	1	$(400/100)(x) = 1$ $x = \frac{1}{4}$
4	400	2	$(400/100)(x) = 2$ $x = \frac{1}{2}$

Use of Charts 26 and 27

Example 3 — Determine Steam Supply Main and Final Velocity

Given:
    Friction rate — 2 psi per 100 ft of equivalent pipe
    Initial steam pressure — 100 psig
    Flow rate — 6750 lb/hr

TABLE 31—TOTAL PRESSURE DROP FOR TWO-PIPE LOW PRESSURE STEAM PIPING SYSTEMS

INITIAL STEAM PRESSURE (psig)	TOTAL PRESSURE DROP IN SUPPLY PIPING (psi)	TOTAL PRESSURE DROP IN RETURN PIPING (psi)
2	$\frac{1}{2}$	$\frac{1}{2}$
5	$1\frac{1}{4}$	$1\frac{1}{4}$
10	$2\frac{1}{2}$	$2\frac{1}{2}$
15	$3\frac{3}{4}$	$3\frac{3}{4}$
20	5	5

Find:
    1. Size of largest pipe not exceeding design friction rate.
    2. Steam velocity in pipe.

Solution:
    1. Enter bottom of *Chart 26* at 6750 lb/hr and proceed vertically to the 100 psig line (dotted line in *Chart 26*). Then move obliquely to the 0 psig line. From this point proceed vertically up the chart to the smallest pipe size not exceeding 2 psi per 100 ft of equivalent pipe and read $3\frac{1}{2}$ in.
    2. The velocity of steam at 0 psig as read from *Chart 26* is 16,000 fpm. Enter the left side of *Chart 27* at 16,000 fpm. Proceed obliquely downward to the 100 psig line and horizontally across to the right side of the chart (dotted line in *Chart 27*). The velocity at 100 psig is 6100 fpm.

Example 4 illustrates a design problem for sizing pipe on a low pressure, vacuum return system.

Example 4 — Sizing Pipe for a Low Pressure, Vacuum Return System

Given:
    Six units
    Steam requirement per unit   72 lb/hr
    Layout as illustrated in *Figs. 96 thru 98*
    Threaded pipe and fittings
    Low pressure system — 2 psi

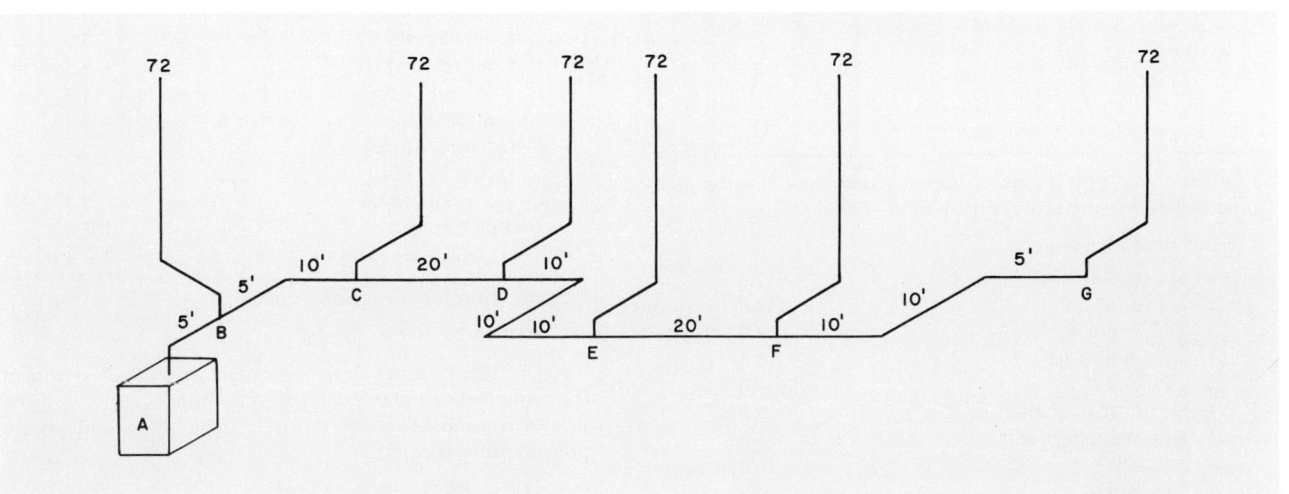

FIG. 96 — LOW PRESSURE STEAM SUPPLY MAIN

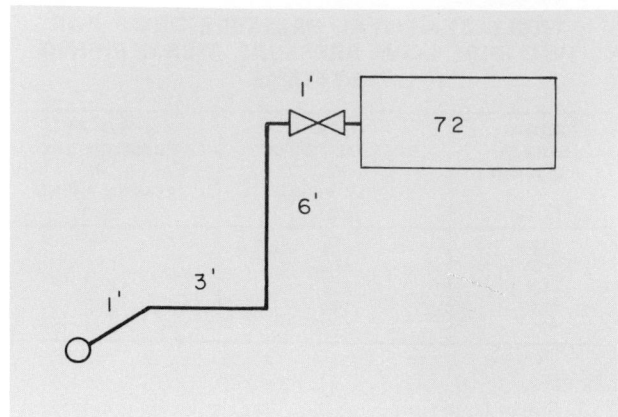

FIG. 97 — LOW PRESSURE RUNOUT AND RISER

Find:

Size of pipe and total pressure drop

Note: Total pressure drop in the system should never exceed one-half the initial pressure. A reasonably small drop is required for quiet operation.

Solution:

Determine the design friction rate by totaling the pipe length and adding 50% of the length for fittings:

$$115 + 11 + 133 = 259$$
$$259 \times .50 = \underline{130}$$
$$\overline{389} \text{ ft equiv length}$$

Check pipe sizing recommendations for maximum friction rate from *"Two-Pipe Vacuum System," Item 2*, ⅛-½ psi. Check *Table 31* to determine recommended maximum pressure drop for the supply and return mains (½ psi for each). Design friction rate = 1/3.89 × (1/2 + 1/2) = 1/4 psi per 100 ft. The supply main is sized by starting at the last unit "G" and adding each additional load from unit "G" to the boiler; use *Table 28*. The following tabulation results:

SECTION	STEAM LOAD (lb/hr)	PIPE SIZE (in.)
F - G	72	1¼
E - F	144	2
D - E	216	2
C - D	288	2½
B - C	360	3
A - B	432	3

Convert the supply main fittings to equivalent lengths of pipe and add to the actual pipe length, *Table 11, page 17*.

Equivalent pipe lengths

1 - 1¼ in. side outlet tee	7.0
2 - 1¼ in. ells	4.6
1 - 2 in. reducing tee	4.7
1 - 2 in. run of tee	3.3
2 - 2 in. ells	6.6
1 - 2½ in. reducing tee	5.6
1 - 3 in. reducing tee	7.0
2 - 3 in. ells	15.0
1 - 3 in. run of tee	5.0
Actual pipe length	115.0
Total equivalent length	167.5 ft.

Pressure drop for the supply main is equal to the equivalent length times pressure drop per 100 ft:

$$167.5 \times .25/100 = .42 \text{ psi}$$

This is within the recommended maximum pressure drop (1 psi) for the supply.

The branch connection for *Fig. 97* is sized in a similar manner at the same friction rate.

From *Table 30* the horizontal runout pipe size for a load of 72 lb is 2½ in. and the vertical riser size is 2 in.

Convert all the fittings to equivalent pipe lengths, and add to the actual pipe length.

Equivalent pipe lengths

1 - 2½ in. 45° ell	3.2
1 - 2½ in. 90° ell	4.1
1 - 2 in. 90° ell	3.3
1 - 2 in. gate valve	2.3
Actual pipe length	11.0
Total equivalent length	23.9 ft.

Pressure drop for branch runout and riser is

$$23.9 \times .25/100 = .060 \text{ psi}$$

The vacuum return main is sized from *Table 29* by starting at the last unit "G" and adding each additional load between unit "G" and the boiler.

Each riser — 72 lb per hr, ¾ in.

SECTION	STEAM LOAD (lb/hr)	PIPE SIZE (in.)
F - G	72	¾
E - F	144	¾
D - E	216	1
C - D	288	1
B - C	360	1¼
A - B	432	1¼

Convert the return main fittings to equivalent pipe lengths and add to the actual pipe length, *Table 11, page 17*.

Equivalent pipe lengths

1 - ¾ in. run of tee	1.4
5 - ¾ in. 90° ells	7.0
1 - 1 in. reducing tee	2.3
1 - 1 in. run of tee	1.7
2 - 1 in. 90° ells	3.4
1 - 1¼ in. reducing tee	3.1
3 - 1¼ in. 90° ells	6.9
1 - 1¼ in. run of tee	2.3
Actual pipe length	133.0
Total equivalent length	161.1 ft.

Pressure drop for the return equals

$$161.1 \times .25/100 = .404 \text{ psi}$$

Total return pressure drop is satisfactory since it is within the recommended maximum pressure drop (⅛ − 1 psi) listed in the two-pipe vacuum return system. The total system pressure drop is equal to

$$.420 + .060 + .404 = .884 \text{ psi}$$

This total system pressure drop is within the maximum 2 psi recommended (1 psi for supply and 1 psi for return).

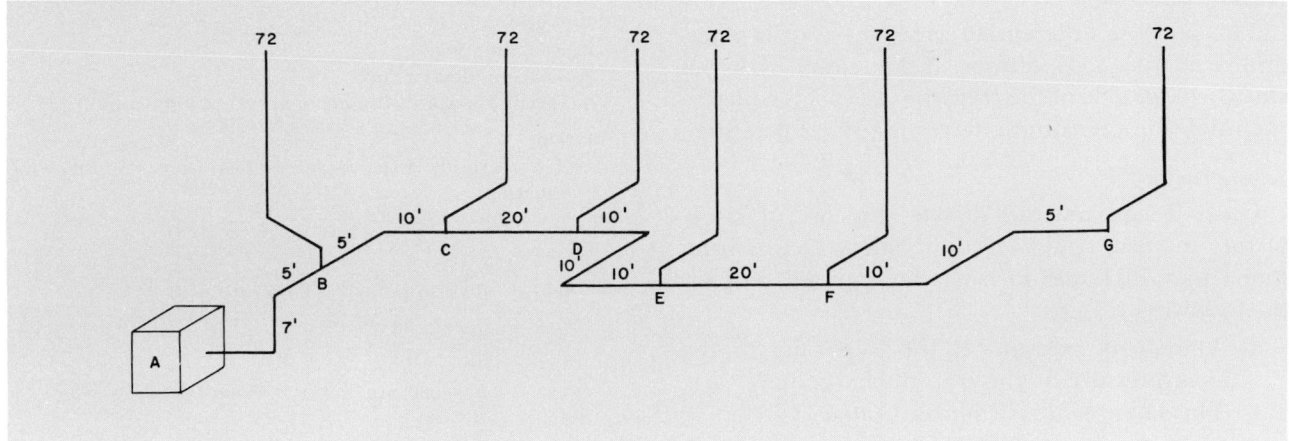

FIG. 98 — LOW PRESSURE VACUUM RETURN

PIPING APPLICATION

The use and selection of steam traps, and condensate and vacuum return pumps are presented in this section.

Also, various steam piping diagrams are illustrated to familiarize the engineer with accepted piping practice.

STEAM TRAP SELECTION

The primary function of a steam trap is to hold steam in a heating apparatus or piping system and allow condensate and air to pass. The steam remains trapped until it gives up its latent heat and changes to condensate. The steam trap size depends on the following:

1. Amount of condensate to be handled by the trap, lb/hr.
2. Pressure differential between inlet and discharge at the trap.
3. Safety factor used to select the trap.

Amount of Condensate

The amount of condensate depends on whether the trap is used for steam mains or risers, or for the heating apparatus.

The selection of the trap for the steam mains or risers is dependent on the pipe warm-up load and the radiation load from the pipe. Warm-up load is the condensate which is formed by heating the pipe surface when the steam is first turned on. For practical purposes the final temperature of the pipe is the steam temperature. Warm-up load is determined from the following equation:

$$C_1 = \frac{W \times (t_f - t_i) \times .114}{h_l \times T}$$

where:

C_1 = Warm-up condensate, lb/hr

W = Total weight of pipe, lb (*Tables 2 and 3, pages 2 and 3*)

t_f = Final pipe temperature, F (steam temp)

t_i = Initial pipe temperature, F (usually room temp)

.114 = Specific heat constant for wrought iron or steel pipe (.092 for copper tubing)

h_l = Latent heat of steam, Btu/lb (from steam tables)

T = Time for warm-up, hr

The radiation load is the condensate formed by unavoidable radiation loss from a bare pipe. This load is determined from the following equation and is based on still air surrounding the steam main or riser:

$$C_2 = \frac{L \times K \times (t_f - t_i)}{h_l}$$

where:

C_2 = Radiation condensate, lb/hr

L = Linear length of pipe, ft

K = Heat transmission coefficient, Btu/(hr) (linear ft) (deg F diff between pipe and surrounding air), *Table 54, Part I*

t_f, t_i, h_l explained previously

The radiation load builds up as the warm-up load drops off under normal operating conditions. The peak occurs at the mid-point of the warm-up cycle. Therefore, one-half of the radiation load is added to the warm-up load to determine the amount of condensate that the trap handles.

Pressure Differential

The pressure differential across the trap is determined at design conditions. If a vacuum exists on the discharge side of the trap, the vacuum is added to the inlet side pressure to determine the differential.

Safety Factor

Good design practice dictates the use of safety factors in steam trap selection. Safety factors from 2 to 1 to as high as 8 to 1 may be required, and for the following reasons:

1. The steam pressure at the trap inlet or the back pressure at the trap discharge may vary. This changes the steam trap capacity.
2. If the trap is sized for normal operating load, condensate may back up into the steam lines or apparatus during start-up or warm-up operation.
3. If the steam trap is selected to discharge a full and continuous stream of water, the air could not be vented from the system.

The following guide is used to determine the safety factor:

DESIGN	SAFETY FACTOR
Draining steam main	3 to 1
Draining steam riser	2 to 1
Between boiler and end of main	2 to 1
Before reducing valve	3 to 1
Before shut-off valve (closed part of time)	3 to 1
Draining coils	3 to 1
Draining apparatus	3 to 1

When the steam trap is to be used in a high pressure system, determine whether or not the system is to operate under low pressure conditions at certain intervals such as night time or weekends. If this condition is likely to occur, then an additional safety factor should be considered to account for the lower pressure drop available during night time operation.

Example 5 illustrates the three concepts mentioned previously in trap selection—condensate handled, pressure differential and safety factor.

Example 5 — Steam Trap Selection for Dripping Supply Main to Return Line

Given:
Steam main — 10 in. diam steel pipe, 50 ft long
Steam pressure — 5 psig (227 F)
Room temperature — 70 F db (steam main in space)
Warm-up time — 15 minutes
Steam trap to drip main into vacuum return line
   (2 in. vacuum gage design)

Find:
1. Warm-up load.
2. Radiation load.
3. Total condensate load.
4. Specifications for steam trap at end of supply main.

Solution:
1. The warm-up load is determined from the following equation:

$$C_1 = \frac{W \times (t_f - t_i) \times .114}{h_l \times T}$$

where: W = 40.48 lb/ft × 50 ft *(Table 2)*
t_f = 227 F
t_i = 70 F
h_l = 960 Btu/lb (from *steam tables*)
T = .25 hr

$$C_1 = \frac{2024 \times (227 - 70) \times .114}{960 \times .25}$$
$$= 150 \text{ lb/hr of condensate}$$

2. The radiation load is calculated by using the following equation:

$$C_2 = \frac{L \times K \times (t_f - t_i)}{h_l}$$

where: L = 50 ft
K = 6.41 Btu/(hr) (linear foot) (deg F diff between pipe and air) from *Table 54, Part I*
t_f = 227 F
t_i = 70 F
h_l = 960 Btu/lb (from *steam tables*)

$$C_2 = \frac{50 \times 6.41 \times (227 - 70)}{960}$$
$$= 52 \text{ lb/hr of condensate}$$

3. The total condensate load for steam trap selection is equal to the warm-up load plus one half the radiation load.

Total condensate load = $C_1 + (\frac{1}{2} \times C_2)$
   = $150 + (\frac{1}{2} \times 52)$
   = 176 lb/hr

4. Steam trap selection is dependent on three factors: condensate handled, safety factor applied to total condensate load, and pressure differential across the steam trap.

The safety factor for a steam trap at the end of the main is 3 to 1 from the table on this page. Applying the 3 to 1 safety factor to the total condensate load, the steam trap would be specified to handle 3 × 176 or 528 lb/hr of condensate.

The pressure differential across the steam trap is determined by the pressure at the inlet and discharge of the steam trap.

Inlet to trap   = 5 psig
Discharge of trap = 2 in. vacuum (gage)

When the discharge is under vacuum conditions, the discharge vacuum is added to the inlet pressure for the total pressure differential.

Pressure differential = 6 psi (approx)

Therefore the steam trap is selected for a differential pressure of 6 psi and 528 lb/hr of condensate.

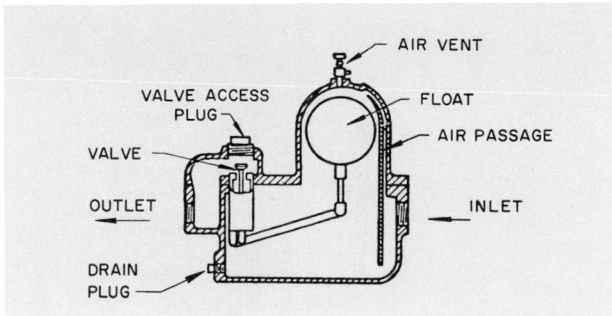

FIG. 99 — FLOAT TRAP

STEAM TRAP TYPES

The types of traps commonly used in steam systems are:

Float   Flash
Thermostatic   Impulse
Float & thermostatic   Lifting
Upright bucket   Boiler return or
Inverted bucket   alternating receiver

The description and use of these various traps are presented in the following pages.

Float Trap

The discharge from the float trap is generally continuous. This type (*Fig. 99*) is used for draining condensate from steam headers, steam heating coils, and other similar equipment. When a float trap is used for draining a low pressure steam system, it should be equipped with a thermostatic air vent.

Thermostatic Trap

The discharge from this type of trap is intermittent. Thermostatic traps are used to drain condensate from radiators, convectors, steam heating coils, unit heaters and other similar equipment. Strainers are normally installed on the inlet side of the steam trap to prevent dirt and pipe scale from

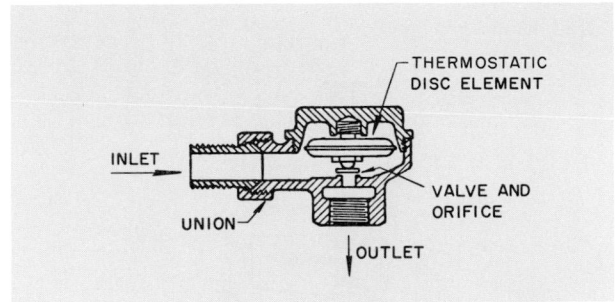

FIG. 101 — THERMOSTATIC TRAP, DISC TYPE

entering the trap. On traps used for radiators or convectors, the strainer is usually omitted. *Fig. 100* shows a typical thermostatic trap of the bellows type and *Fig. 101* illustrates a disc type thermostatic trap.

When a thermostatic trap is used for a heating apparatus, at least 2 ft of pipe are provided ahead of the trap to cool the condensate. This permits condensate to cool in the pipe rather than in the coil, and thus maintains maximum coil efficiency.

Thermostatic traps are recommended for low pressure systems up to a maximum of 15 psi. When used in medium or high pressure systems, they must be selected for the specific design temperature. In addition, the system must be operated continuously at that design temperature. *This means no night setback.*

Float and Thermostatic Trap

This type of trap is used to drain condensate from blast heaters, steam heating coils, unit heaters and other apparatus. This combination trap (*Fig. 102*) is used where there is a large volume of condensate which would not permit proper operation of a thermostatic trap. Float and thermostatic traps are used in low pressure heating systems up to a maximum of 15 psi.

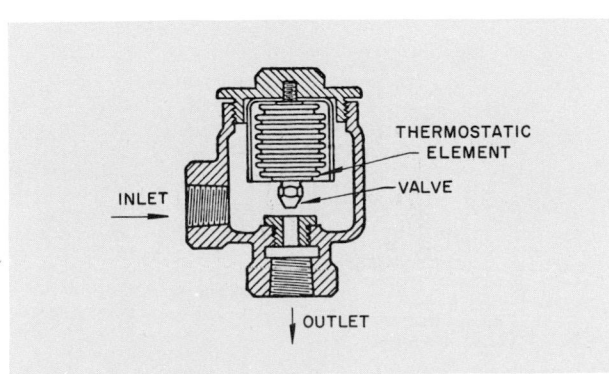

FIG. 100 — THERMOSTATIC TRAP, BELLOWS TYPE

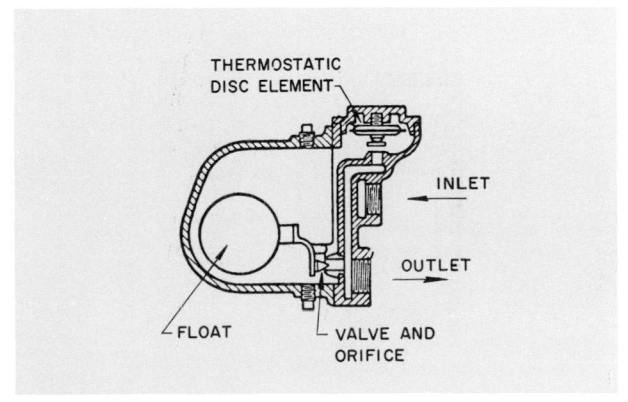

FIG. 102 — TYPICAL FLOAT AND THERMOSTATIC TRAP

Figures 99 thru 102 from Heating Ventilating Air Conditioning Guide 1959. Used by permission.

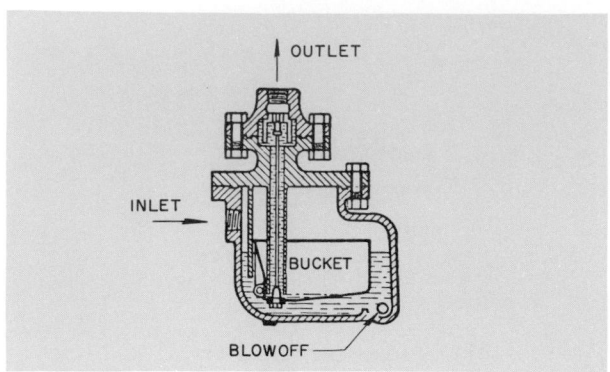

FIG. 103 — UPRIGHT BUCKET TRAP

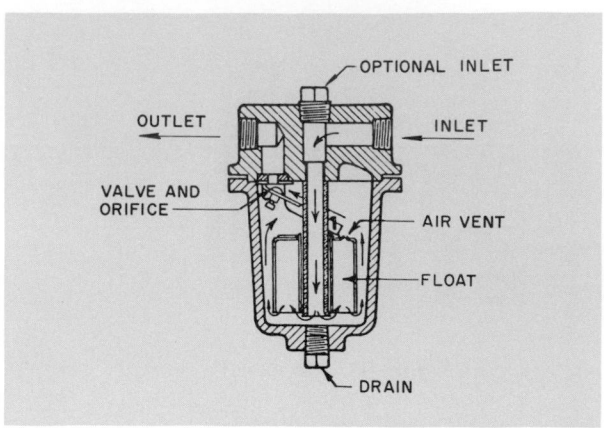

FIG. 105 — INVERTED BUCKET TRAP WITH GUIDE

For medium and high pressure systems, the same limitations as outlined for thermostatic traps apply.

Upright Bucket Trap

The discharge of condensate from this trap *(Fig. 103)* is intermittent. A differential pressure of at least 1 psi between the inlet and the outlet of the trap is normally required to lift the condensate from the bucket to the discharge connection. Upright bucket traps are commonly used to drain condensate and air from the blast coils, steam mains, unit heaters and other equipment. This trap is well suited for systems that have pulsating pressures.

Inverted Bucket Trap

The discharge from the inverted bucket trap *(Figs. 104 and 105)* is intermittent and requires a differential pressure between the inlet and discharge of the trap to lift the condensate from the bottom of the trap to the discharge connection.

Bucket traps are used for draining condensate and air from blast coils, unit heaters and steam heating coils. Inverted bucket traps are well suited for drain-

ing condensate from steam lines or equipment where an abnormal amount of air is to be discharged and where dirt may drain into the trap.

Flash Trap

The discharge from a flash trap *(Fig. 106)* is intermittent. This type of trap is used only if a pressure differential of 5 psi or more exists between the steam supply and condensate return. Flash traps may be used with unit heaters, steam heating coils, steam lines and other similar equipment.

Impulse Trap

Under normal loads the discharge from this trap *(Fig. 107)* is intermittent. When the load is heavy, however, the discharge is continuous. This type of trap may be used on any equipment where the pressure at the trap outlet does not exceed 25% of the inlet pressure.

Lifting Trap

The lifting trap *(Fig. 108)* is an adaption of the upright bucket trap. This trap can be used on all steam heating systems up to 150 psig. There is an

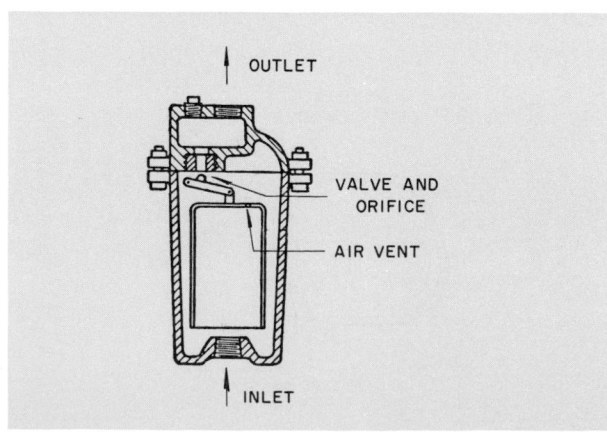

FIG. 104 — INVERTED BUCKET TRAP

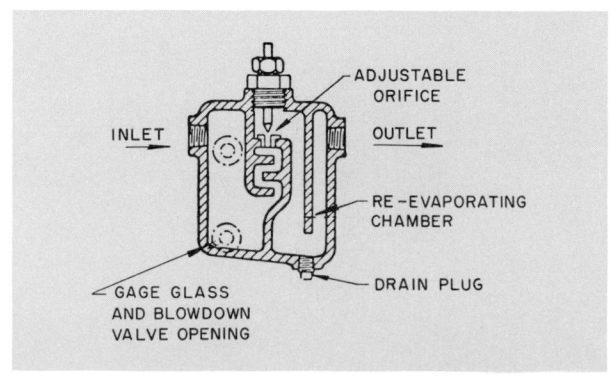

FIG. 106 — FLASH TRAP

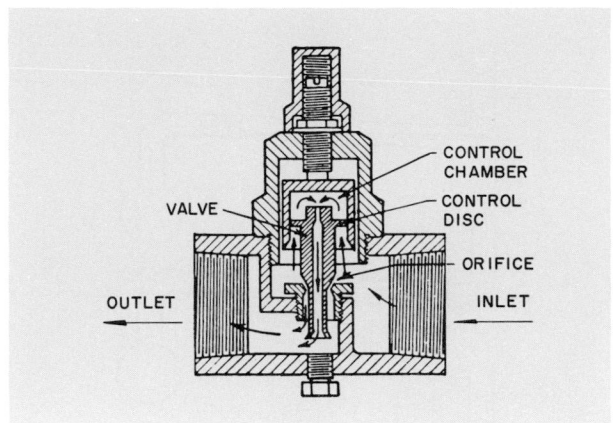

FIG. 107 — IMPULSE TRAP

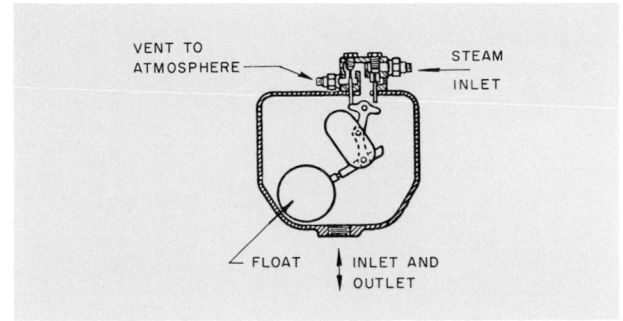

FIG. 109 — BOILER RETURN TRAP OR ALTERNATING
RECEIVER

auxiliary inlet for high pressure steam, as illustrated in the figure, to force the condensate to a point above the trap. This steam is normally at a higher pressure than the steam entering at the regular inlet.

Boiler Return Trap or Alternating Receiver

This type of trap is used to return condensate to a low pressure boiler. The boiler return trap (*Fig. 109*) does not hold steam as do other types, but is an adaption of the lifting trap. It is used in conjunction with a boiler to prevent flooding return mains when excess pressure prevents condensate from returning to the boiler by gravity. The boiler trap collects condensate and equalizes the boiler and trap pressure, enabling the condensate in the trap to flow back to the boiler by gravity.

CONDENSATE RETURN PUMP

Condensate return pumps are used for low pressure, gravity return heating systems. They are normally of the motor driven centrifugal type and have a receiver and automatic float control. Other types

of condensate return pumps are the rotary, screw, turbine and reciprocating pump.

The condensate receiver is sized to prevent large fluctuations of the boiler water line. The storage capacity of the receiver is approximately 1.5 times the amount of condensate returned per minute, and the condensate pump has a capacity of 2.5 to 3 times normal flow. This relationship of pump and receiver to the condensate takes peak condensation load into account.

VACUUM PUMP

Vacuum pumps are used on a system where the returns are under a vacuum. The assembly consists of a receiver, separating tank and automatic controls for discharging the condensate to the boiler.

Vacuum pumps are sized in the same manner as condensate pumps for a delivery of 2.5 to 3 times the design condensing rate.

PIPING LAYOUT

Each application has its own layout problem with regard to the equipment location, interference with structural members, steam condensate, steam trap and drip locations. The following steam piping diagrams show the various principles involved. The engineer must use judgment in applying these principles to the application.

Gate valves shown in the diagrams should be used in either the open or closed position, *never for throttling.* Angle and globe valves are recommended for throttling service.

In a one-pipe system gate valves are used since they do not hinder the flow of condensate. Angle valves may be used when they do not restrict the flow of condensate.

All the figures show screwed fittings. Limitations for other fittings are described in *Chapter 1.*

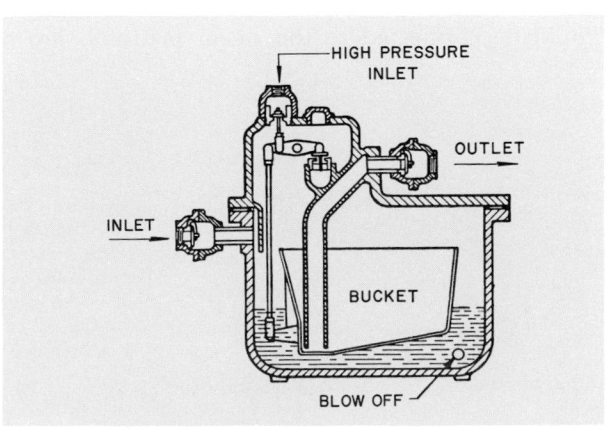

FIG. 108 — LIFTING TRAP

Figures 103 thru 109 from Heating Ventilating Air Conditioning Guide 1959. Used by permission.

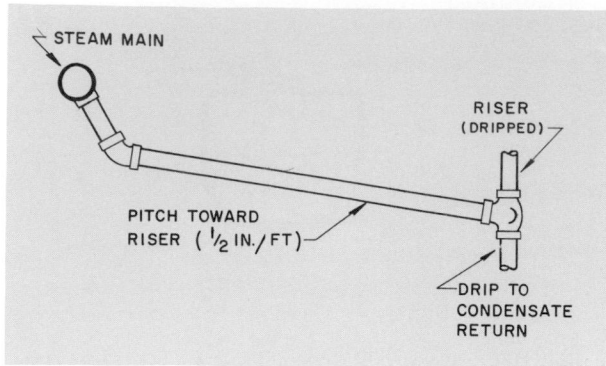

FIG. 110 — CONNECTION TO DRIPPED RISER

Steam Riser

Figures 110 and 111 illustrate steam supply risers connected to mains with runouts. The runout in *Fig. 110* is connected to the bottom portion of the main and is pitched toward the riser to permit condensate to drain from the main. *This layout is used only when the riser is dripped.* If a dry return is used, the riser is dripped thru a steam trap. If a wet return is used, the trap is omitted.

Fig. 111 shows a piping diagram when the steam riser is not dripped. In this instance the runout is connected to the upper portion of the steam main and is pitched to carry condensate from the riser to the main.

Prevention of Water Hammer

If the steam main is pitched incorrectly when the riser is not dripped, water hammer may occur as illustrated in *Fig. 112*. Diagram *"a"* shows the runout partially filled with condensate but with enough space for steam to pass. As the amount of condensate increases and the space decreases, a wave motion is started as illustrated in diagram *"b"*. As the wave or slug of condensate is driven against the turn in the pipe (diagram *"c"*), a hammer noise is

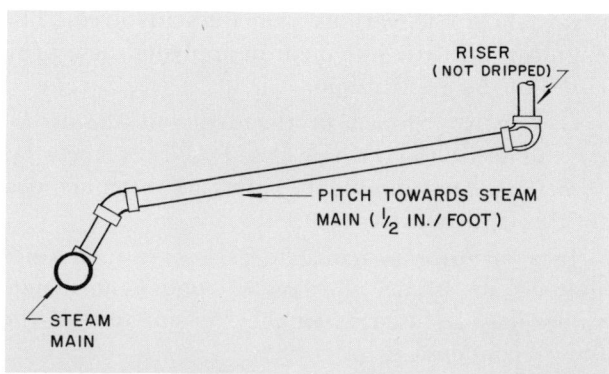

FIG. 111 — CONNECTION TO RISER (NOT DRIPPED)

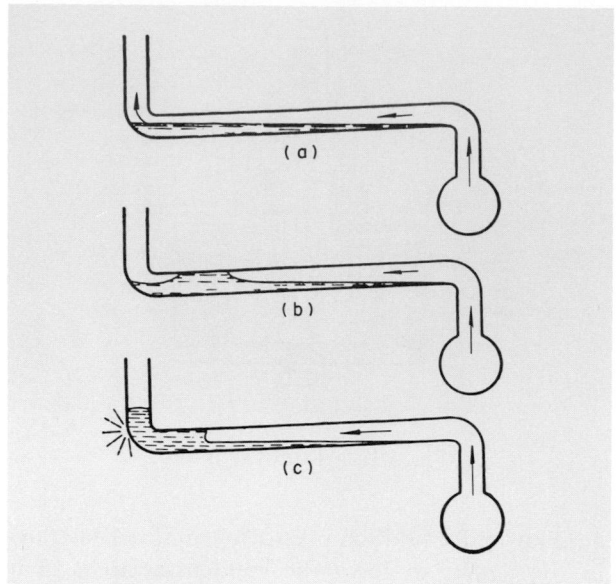

FIG. 112 — WATER HAMMER

caused. This pounding may be of sufficient force to split pipe fittings and damage coils in the system.

The following precautions must be taken to prevent water hammer:

1. Pitch pipes properly.
2. Avoid undrained pockets.
3. Choose a pipe size that prevents high steam velocity when condensate flows opposite to the steam.

Runout Connection to Supply Main

Figure 113 illustrates two methods of connecting runouts to the steam supply main. The method using a 45° ell is somewhat better as it offers less resistance to steam flow.

Expansion and Contraction

Where a riser is two or more floors in height, it should be connected to the steam main as shown

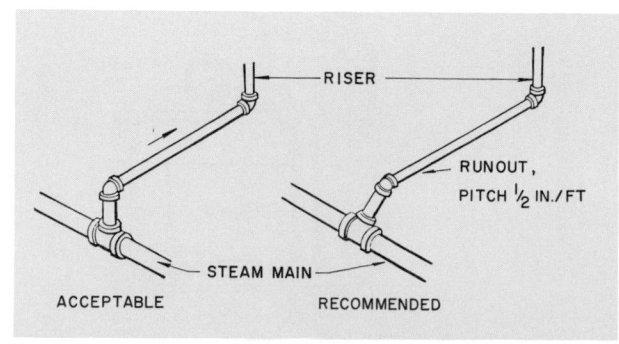

FIG. 113 — RUNOUT CONNECTIONS

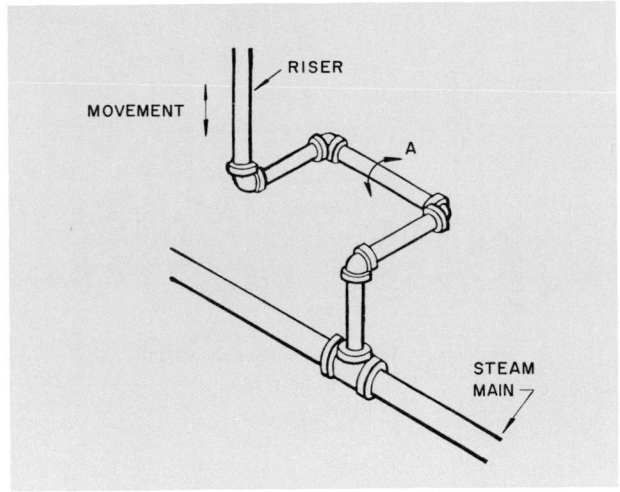

FIG. 114 — RISER CONNECTED TO ALLOW FOR
EXPANSION

in *Fig. 114*. Point *(A)* is subject to a twisting movement as the riser moves up and down.

Figure 115 shows a method of anchoring the steam riser to allow for expansion and contraction. Movement occurs at *(A)* and *(B)* when the riser moves up and down.

Obstructions

Steam supply mains may be looped over obstructions if a small pipe is run below the obstruction to take care of condensate as illustrated in *Fig. 116* The reverse procedure is followed for condensate return mains as illustrated in *Fig. 117*. The larger pipe is carried under the obstruction.

Dripping Riser

A steam supply main may be dropped abruptly to a lower level without dripping if the pitch is downward. When the steam main is raised to a higher level, it must be dripped as illustrated in *Fig. 118*.

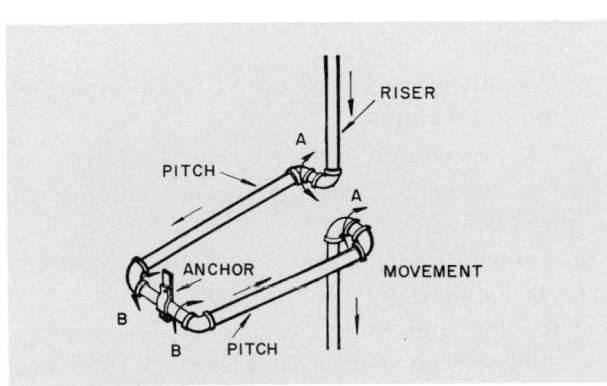

FIG. 115 — RISER ANCHOR

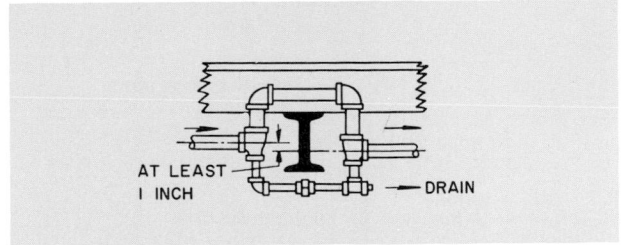

FIG. 116 — SUPPLY MAIN LOOPS

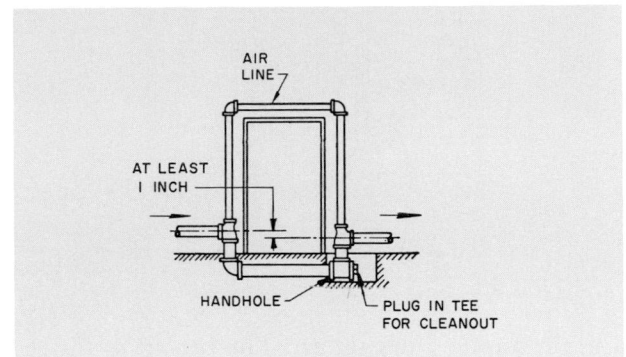

FIG. 117 — RETURN MAIN LOOP

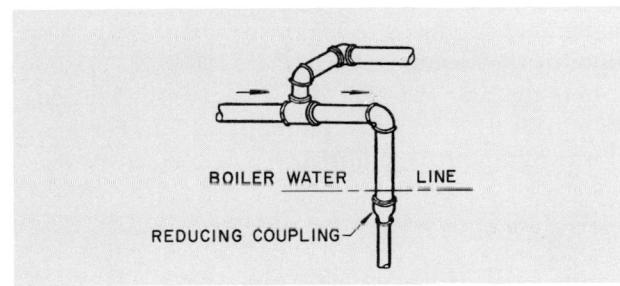

FIG. 118 — DRIPPING STEAM MAIN

This diagram shows the steam main dripped into a wet return.

Figure 119 is one method of dripping a riser thru

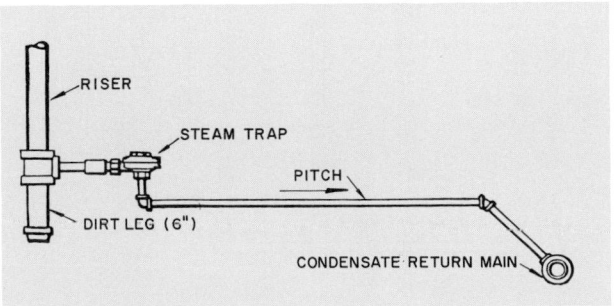

FIG. 119 — RISER DRIPPED TO DRY RETURN

Figures 116 thru 118 from Heating Ventilating Air Conditioning Guide 1959. Used by permission.

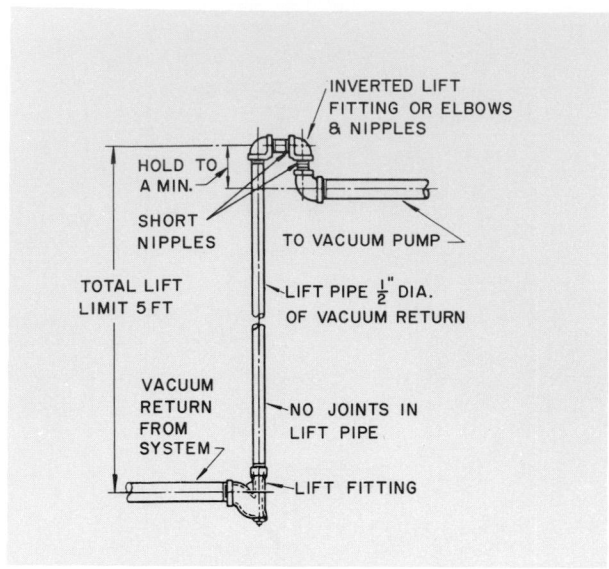

FIG. 120 — ONE-STEP CONDENSATE LIFT

a steam trap to a dry return. The runout to the return main is pitched toward the return main.

Vacuum Lift

As described under vacuum systems, a lift is sometimes employed to lift the condensate up to the inlet of the vacuum pump. *Figs. 120 and 121* show a one-step and two-step lift respectively. The one-step lift is used for a maximum lift of 5 ft. For 5 to 8 ft a two-step lift is required.

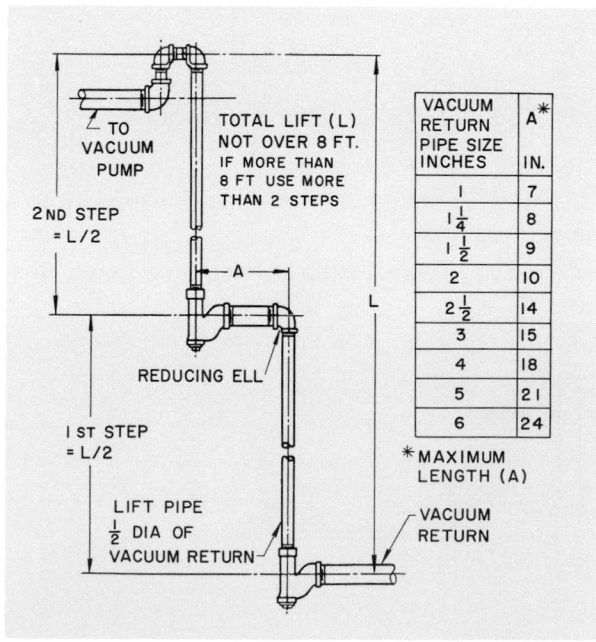

FIG. 121 — TWO-STEP CONDENSATE LIFT

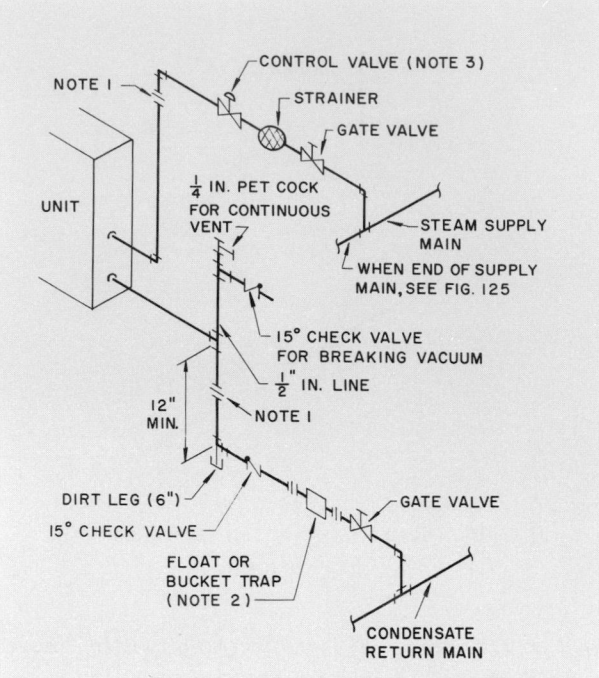

NOTES:
1. Flange or union is located to facilitate coil removal.
2. Flash trap may be used if pressure differential between steam and condensate return exceeds 5 psi.
3. When a bypass with control is required, see *Fig. 126*.
4. Dirt leg may be replaced with a strainer. If so, tee on drop can be replaced by a reducing ell.
5. The petcock is not necessary with a bucket trap or any trap which has provision for passing air. The great majority of high or medium pressure returns end in hot wells or deaerators which vent the air.

FIG. 122 — HIGH OR MEDIUM PRESSURE COIL PIPING

Steam Coils

Figures 122 thru 131 show methods of piping steam coils in a high or low pressure or vacuum steam piping system. The following general rules are applicable to piping layout for steam coils used in all systems:

1. Use full size coil outlets and return piping to the steam trap.
2. Use thermostatic traps for venting only.
3. Use a 15° check valve only where indicated on the layout.
4. Size the steam control valve for the steam load, not for the supply connection.
5. Provide coils with air vents as required, to eliminate non-condensable gases.
6. Do not drip the steam supply mains into coil sections.

Figures 120 and 121 from Heating Ventilating Air Conditioning Guide 1959. Used by permission.

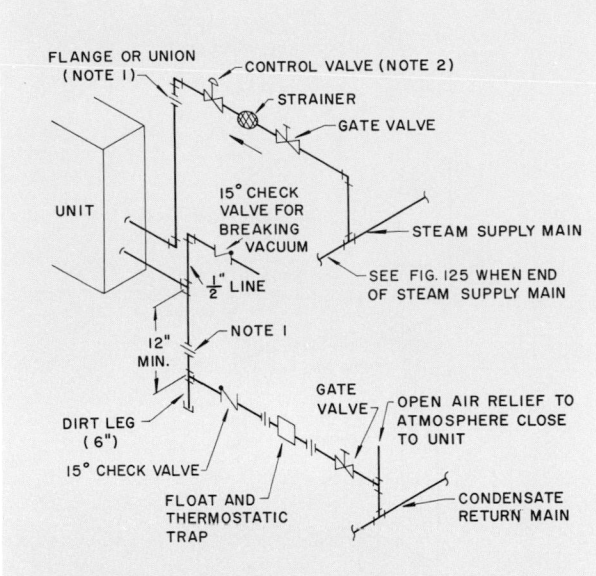

NOTES:

1. Flange or union is located to facilitate coil removal.
2. When a bypass with control is required, see *Fig. 126*.
3. Check valve is necessary when more than one unit is connected to the return line.
4. Dirt pocket is the same size as unit outlet. If dirt pocket is replaced by a strainer, replace tee with a reducing ell from unit outlet to trap size.

FIG. 123 — SINGLE COIL LOW PRESSURE PIPING
GRAVITY RETURN

7. Do not pipe tempering coils and reheat coils to a common steam trap.
8. Multiple coils may be piped to a common steam trap if they have the same capacity and the same pressure drop and if the supply is regulated by a control valve.

Piping Single Coils

Figure 122 illustrates a typical steam piping diagram for coils used in either a high or medium pressure system. If the return line is designed for low pressure or vacuum conditions and for a pressure differential of 5 psi or greater from steam to condensate return, a flash trap may be used.

Low pressure steam piping for a single coil is illustrated in *Fig. 123*. This diagram shows an open air relief located after the steam trap close to the unit. This arrangement permits non-condensable gases to vent to the atmosphere.

Figure 124 shows the piping layout for a steam coil in a vacuum system. A 15° check valve is used to equalize the vacuum across the steam trap.

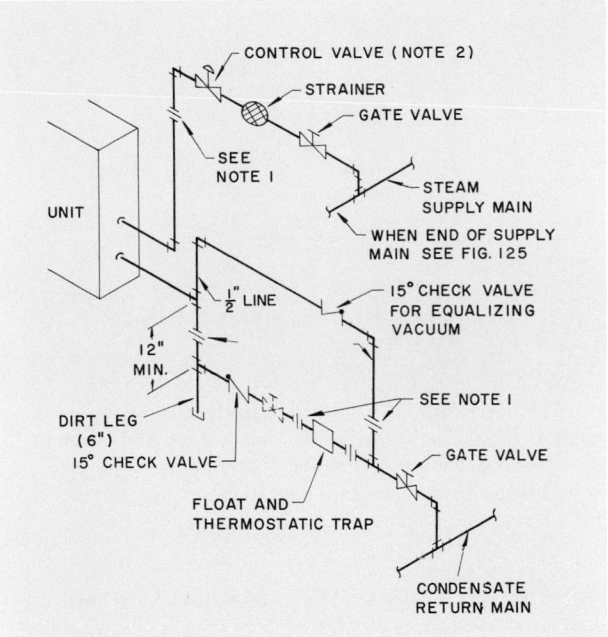

NOTES:

1. Flange or union is located to facilitate coil removal.
2. When a bypass with control is specified, see *Fig. 126*.
3. Check valve is necessary when more than one unit is connected to the return line.

FIG. 124 — VACUUM SYSTEM STEAM COIL PIPING

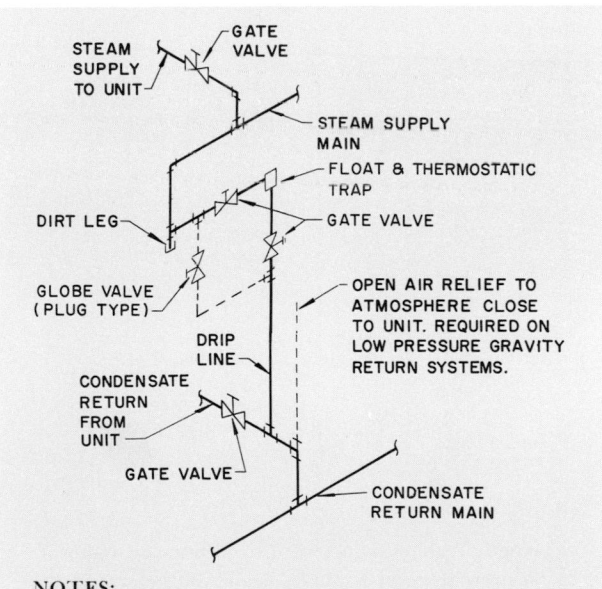

NOTES:

1. A bypass is necessary around trap and valves when continuous operation is necessary.
2. Bypass to be the same size as trap orifice but never less than ½ inch.

FIG. 125 — DRIPPING STEAM SUPPLY TO CONDENSATE
RETURN

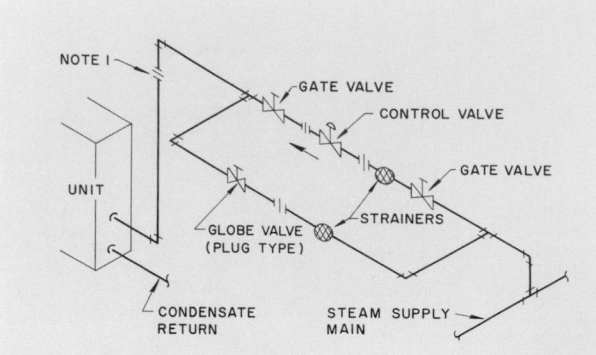

NOTES:

1. Flange or union is located to facilitate coil removal.
2. A bypass is necessary around valves and strainer when continuous operation is necessary.
3. Bypass to be the same size as valve port but never less than ½ inch.

FIG. 126 — BYPASS WITH MANUAL CONTROL

Dripping Steam Supply Main

A typical method of dripping the steam supply main to the condensate return is shown in *Fig. 125*.

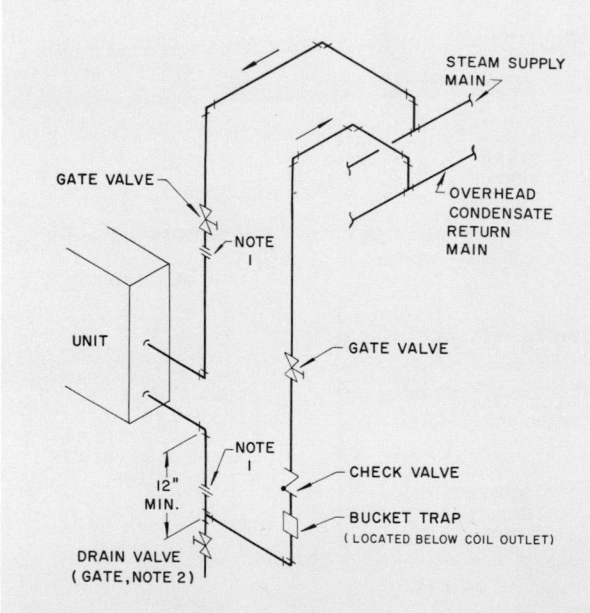

NOTES:

1. Flange or union is located to facilitate coil removal.
2. To prevent water hammer, drain coil before admitting steam.
3. Do not exceed one foot of lift between trap discharge and return main for each pound of pressure differential.
4. *Do not use this arrangement for units handling outside air.*

FIG. 127 — CONDENSATE LIFT TO OVERHEAD RETURN

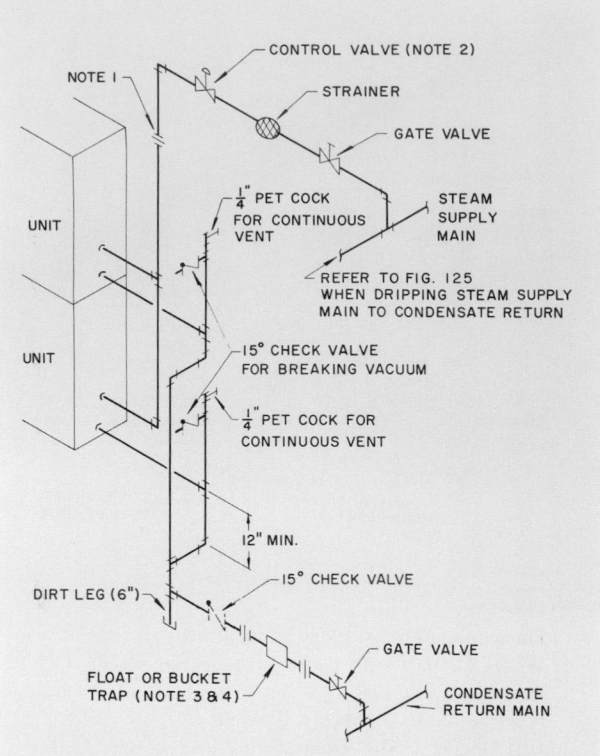

NOTES:

1. Flange or union is located to facilitate coil removal.
2. When bypass control is required, see *Fig. 126*.
3. Flash trap can be used if pressure differential between supply and condensate return exceeds 5 psi.
4. *Coils with different pressure drops require individual traps.*
5. Dirt pocket may be replaced by a strainer. If so, tee on drop can be replaced by a reducing ell.
6. The petcock is not necessary with a bucket trap or any trap which has provision for passing air. The great majority of high pressure return mains terminate in hot wells or deaerators which vent the air.

FIG. 128 — MULTIPLE COIL HIGH PRESSURE PIPING

Steam Bypass Control

Frequently a bypass with a manual control valve is required on steam coils. The piping layout for a control bypass with a plug type globe valve as the manual control is shown in *Fig. 126*.

Lifting Condensate to Return Main

A typical layout for lifting condensate to an overhead return is described in *Fig. 127*. The amount of lift possible is determined by the pressure differential between the supply and return sides of the system. The amount of lift is not to exceed one foot for each pound of pressure differential. The maximum lift should not exceed 8 ft.

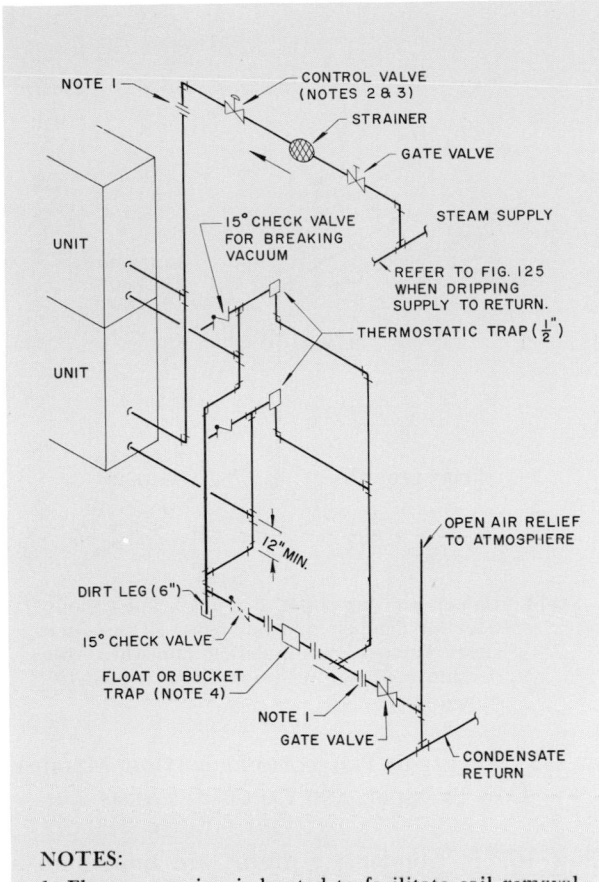

NOTES:
1. Flange or union is located to facilitate coil removal.
2. See *Fig. 131* when control valve is omitted on multiple coils in parallel air flow.
3. When bypass control is required, see *Fig. 126.*
4. *Coils with different pressure drops require individual traps.*

FIG. 129 — MULTIPLE COIL LOW PRESSURE PIPING

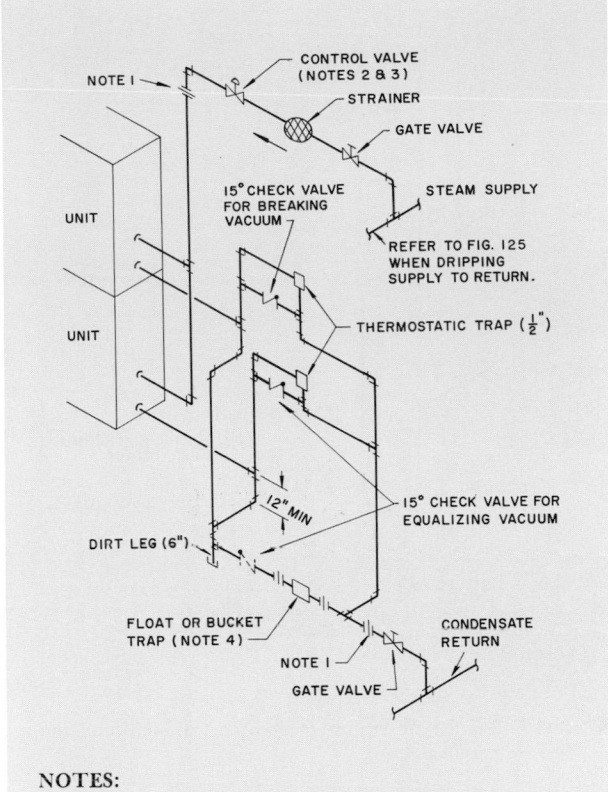

NOTES:
1. Flange or union is located to facilitate coil removal.
2. See *Fig. 131* when control valve is omitted on multiple coils in parallel air flow.
3. When bypass control is required, see *Fig. 126.*
4. *Coils with different pressure drops require individual traps.*

FIG. 130 — MULTIPLE COIL LOW PRESSURE VACUUM
SYSTEM PIPING

Piping Multiple Coils

Figures 128 thru 131 show piping layouts for high pressure, low pressure and vacuum systems with multiple coils. If a control valve is not used, each coil must have a separate steam trap as illustrated in *Fig. 131.* This particular layout may be used for a low pressure or vacuum system.

If the coils have different pressure drops or capacities, separate traps are required with or without a control valve in the system.

Boiler Piping

Figure 132 illustrates a suggested layout for a steam plant. This diagram shows parallel boilers and a single boiler using a "Hartford Return Loop."

FREEZE-UP PROTECTION

When steam coils are used for tempering or preheating outdoor air, controls are required to prevent freezing of the coil.

In high, medium, low pressure and vacuum systems, an immersion thermostat is recommended to protect the coil. This protection device controls the fan motor and the outdoor air damper. The immersion thermostat is actuated when the steam supply fails or when the condensate temperature drops below a predetermined level, usually 120 F to 150 F. The thermostat location is shown in *Fig. 133.*

The 15° check valve shown in the various piping diagrams provides a means of equalizing the pressure within the coil when the steam supply shuts off. This check valve is used in addition to the immersion thermostat. The petcock for continuous venting removes non-condensable gases from the coil.

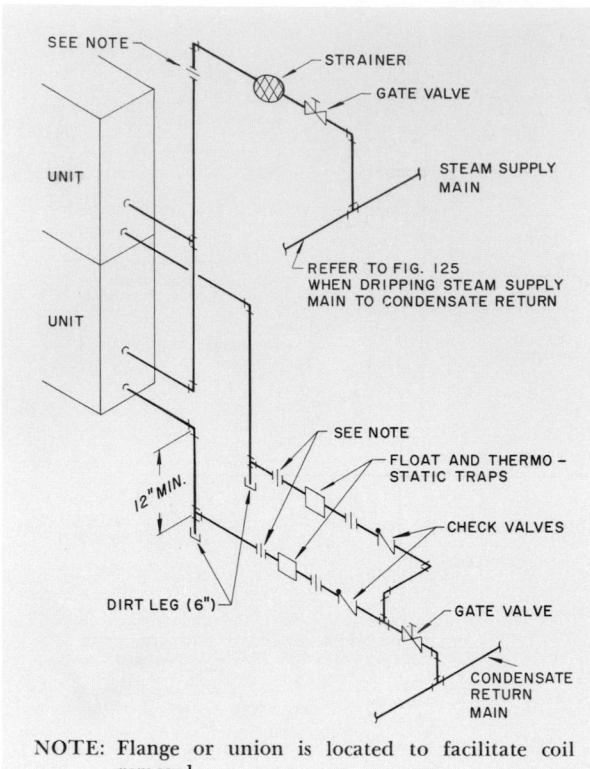

NOTE: Flange or union is located to facilitate coil removal.

FIG. 131 — LOW PRESSURE OR VACUUM SYSTEM STEAM TRAPS

Non-condensable gases can restrict the flow of condensate, causing coil freeze-up.

On a low pressure and vacuum steam heating system, the immersion thermostat may be replaced by a condensate drain with a thermal element (*Fig. 134*). The thermal element opens and drains the coil when the condensate temperature drops

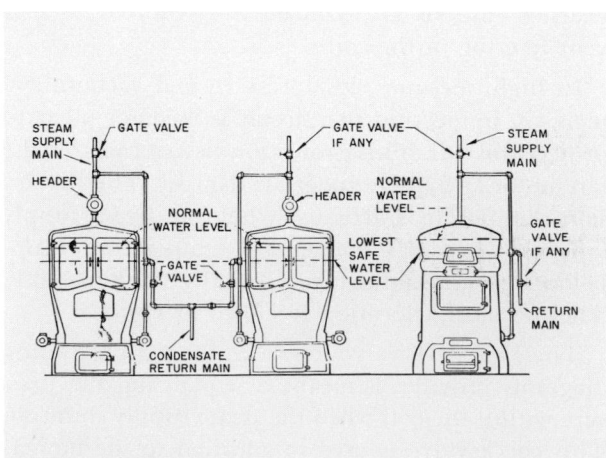

FIG. 132 — "HARTFORD" RETURN LOOP

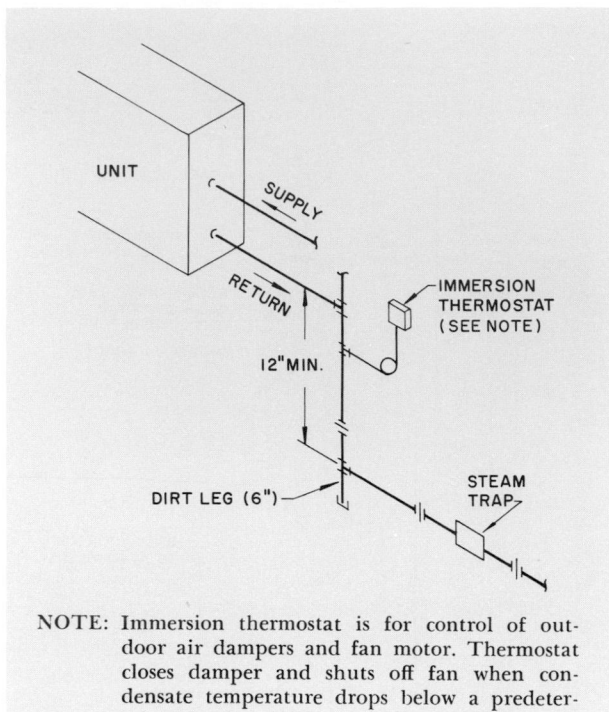

NOTE: Immersion thermostat is for control of outdoor air dampers and fan motor. Thermostat closes damper and shuts off fan when condensate temperature drops below a predetermined level.

FIG. 133 — FREEZE-UP PROTECTION FOR HIGH, MEDIUM, LOW PRESSURE, AND VACUUM SYSTEMS

below 165 F. Condensate drains are limited to a 5 lb pressure.

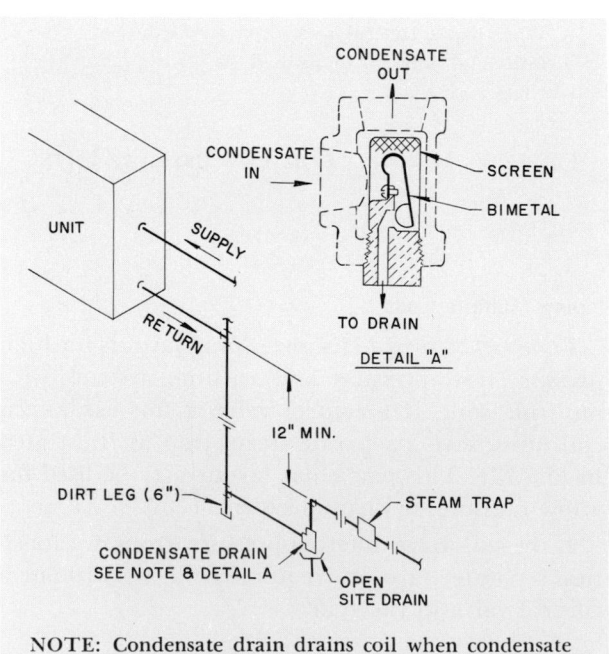

NOTE: Condensate drain drains coil when condensate temperature drops below a predetermined level.

FIG. 134 — FREEZE-UP PROTECTION FOR LOW PRESSURE AND VACUUM SYSTEMS

The following are general recommendations in laying out systems handling outdoor air below 35 F:

1. Do not use overhead returns from the heating unit.

2. Pressure controls are not recommended since they do not necessarily reflect actual conditions. For example, it is possible for the coil to become air bound and have pressure but no steam. Also, the steam trap may be plugged. Pressure controls are slow acting by comparison with thermostatic controls.

3. Use a strainer in the supply line and a dirt leg ahead of the trap.

Part 4
REFRIGERANTS, BRINES, OILS
CHAPTER 1. REFRIGERANTS

This chapter provides information concerning the refrigeration cycles and characteristics of the commonly used refrigerants and their selection for use in air conditioning applications.

To provide refrigeration, a refrigerant may be utilized either:

1. In conjunction with a compressor, condenser and evaporator in a compression cycle, or

2. With an absorbent in conjunction with an absorber, generator, evaporator, and condenser in an absorption cycle.

The refrigerant absorbs heat by evaporation generally at a low temperature and pressure level. Upon condensing at a higher level, it rejects this heat to any available medium, usually water or air.

In a compression system the refrigerant vapor is increased in pressure from evaporator to condenser pressure by the use of a compressor.

In an absorption system the increase in pressure is produced by heat supplied from steam or other suitable hot fluid which circulates thru a coil of pipe. The absorber-generator is analogous to a compressor in that the absorber constitutes the suction stroke and the generator the compression stroke. The evaporator spray header corresponds to the expansion valve. The evaporator and condenser are identical for both compression and absorption systems.

This chapter includes a discussion of the refrigeration cycle, refrigerant selection, and the commonly used refrigerants as well as tables indicating their characteristics and properties.

REFRIGERATION CYCLES

ABSORPTION CYCLE

The absorption refrigeration cycle utilizes two phenomena:

1. The absorption solution (absorbent plus refrigerant) can absorb refrigerant vapor.

2. The refrigerant boils (flash cools itself) when subjected to a lower pressure.

These two phenomena are used in the lithium bromide absorption machine to obtain refrigeration by using the bromide as an absorbent and water as a refrigerant.

Water is sprayed in an evaporator which is maintained at a high vacuum. A portion of the water flashes and cools that which remains. The water vapor is absorbed by a lithium bromide solution in the absorber. The resulting solution is then heated in the generator to drive off the water vapor which is condensed in the condenser. The water is returned to the evaporator, completing the cycle.

Figure 1 illustrates the absorption cycle. *Figure 2* illustrates the cycle plotted on the equilibrium diagram with numbered points representing pressures, temperatures, concentrations in the cycle.

On the lower left side of *Fig. 1* is the absorber partially filled with lithium bromide solution. On the lower right side is the evaporator containing water. A pipe connecting the shells is evacuated so that no air is present. The lithium bromide begins to absorb the water vapor; as the vapor is absorbed, the water boils, generating more vapor and causing the remainder of the water to be cooled.

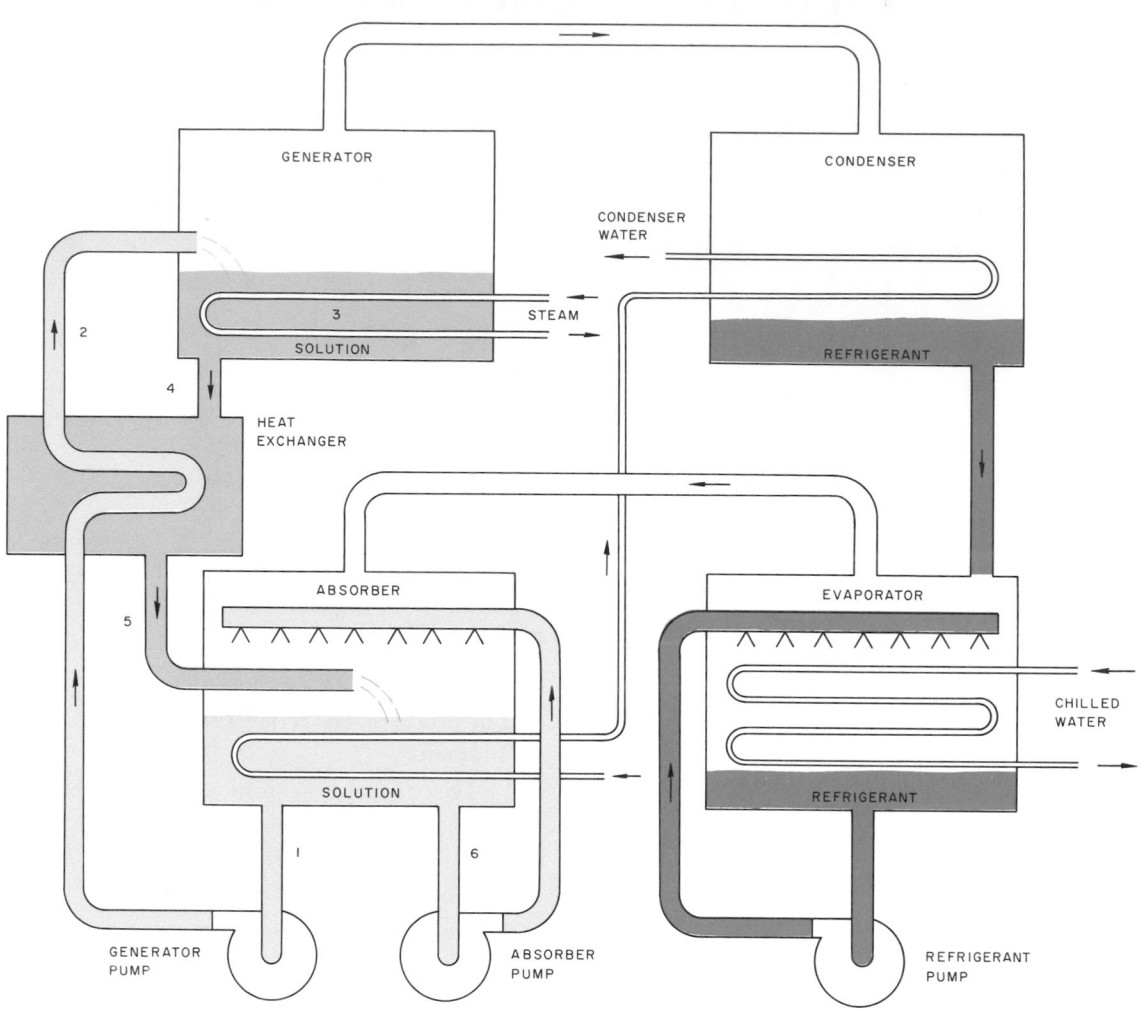

FIG. 1 — ABSORPTION REFRIGERATION CYCLE

Since the water can vaporize more easily if it is being sprayed, a pump is used to circulate the water from the bottom of the evaporator to a spray header at the top. An evaporator tube bundle is located under the evaporator spray header; water inside the tubes, returning from the air conditioning coils or other load, is flash-cooled by the water on the outside of the evaporator tubes. The lithium bromide solution absorbs water vapor easier if it is sprayed; therefore, a pump is used to circulate the solution from the bottom of the absorber to a spray header at the top of the absorber.

As the lithium bromide continues to absorb water vapor, it becomes diluted, and its ability to absorb additional water vapor decreases. The weak solution is pumped to the generator where heat is applied by steam or other suitable hot fluid in the

generator tube bundle to boil off the water vapor. The solution is concentrated and returned to the absorber. Since the weak solution going to the generator must be heated and the strong solution coming from the generator must be cooled, a heat exchanger is used in the solution circuit to conserve heat.

Water vapor boiled from the solution in the generator passes to the condenser to contact the relatively cold condenser tubes. The vapor condenses in the condenser and returns to the evaporator so that there is no loss of water in the cycle. Before the condenser water goes thru the condenser tubes, it passes thru a tube bundle located in the absorber. Here it picks up the heat of dilution and the heat of condensation which is generated as the solution absorbs water vapor.

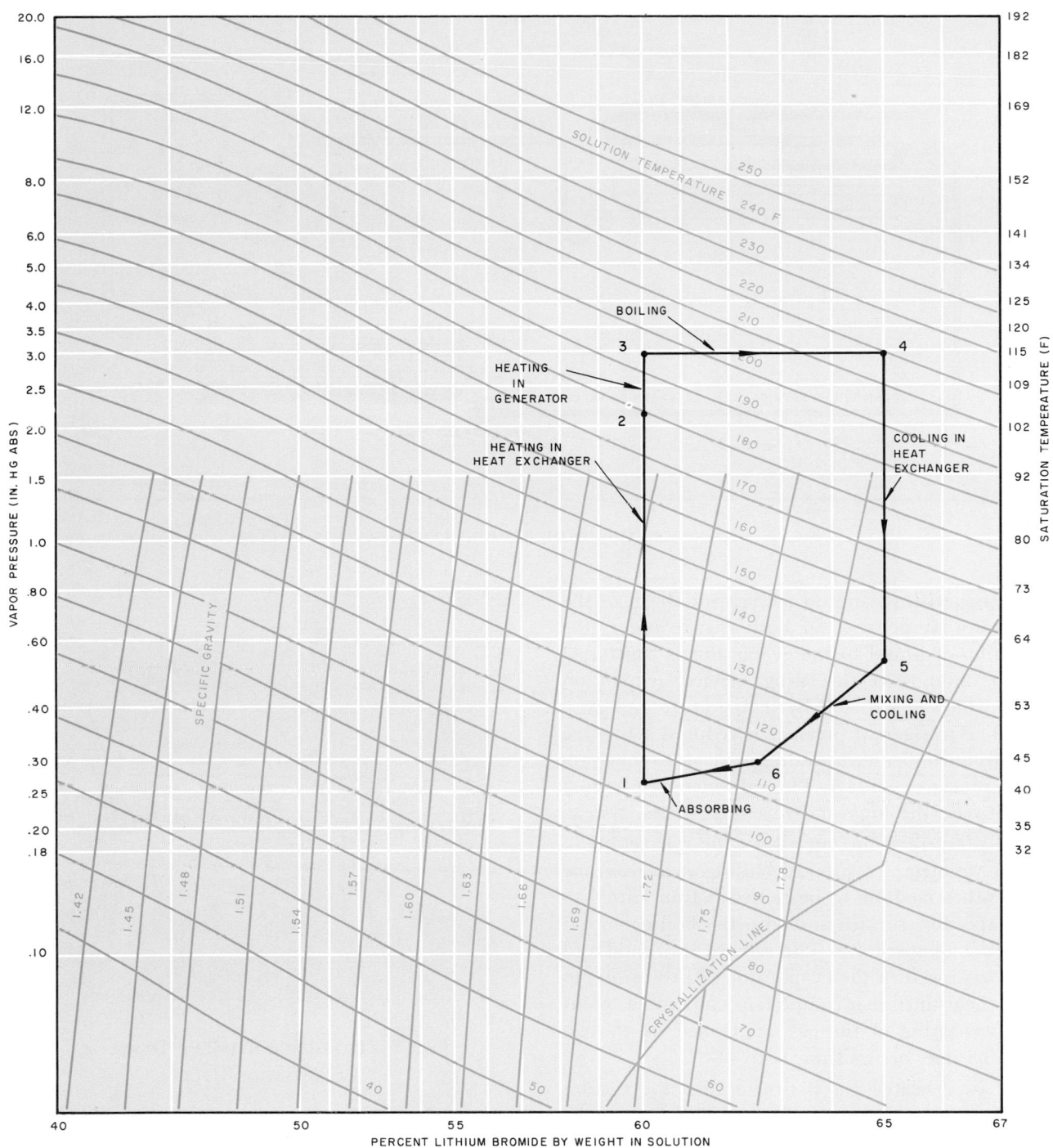

FIG. 2 — EQUILIBRIUM DIAGRAM FOR LITHIUM BROMIDE

COMPRESSION CYCLE

The compression refrigeration cycle utilizes two phenomena:

1. The evaporation of a liquid refrigerant absorbs heat to lower the temperature of its surroundings.

2. The condensation of a refrigerant vapor rejects heat to raise the temperature of its surroundings.

The cycle may be traced from any point in the system. *Figure 3* is a schematic and *Fig. 4* is a pressure-enthalpy diagram of a compression cycle.

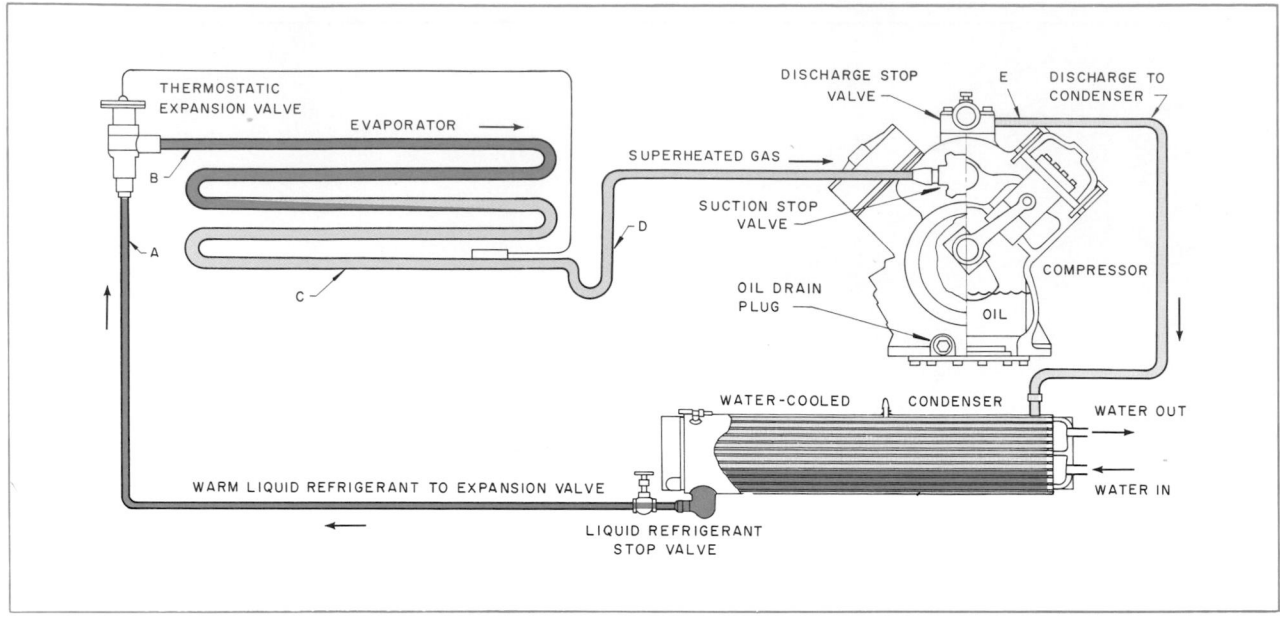

FIG. 3 — RECIPROCATING COMPRESSION REFRIGERATION CYCLE

Starting with the liquid refrigerant ahead of the evaporator at point *A* in both *Fig. 3 and 4*, the admission of liquid to the evaporator is controlled by an automatic throttling device (expansion valve) which is actuated by temperature and pressure. The refrigerant pressure is reduced across the valve from condenser pressure, point *A*, to the evaporator pressure, point *B*. The valve acts as a boundary between the *high* and *low* sides of the system.

The pressure reduction allows the refrigerant to boil or vaporize. To support boiling, heat from the air or other medium to be cooled is transmitted to the evaporator surface and into the boiling liquid at a lower temperature. The refrigerant liquid and vapor passing thru the evaporator coil continues to absorb heat until it is completely evaporated, point *C*. Superheating of the gas, controlled by the expansion valve, occurs from *C* to *D*.

The superheated gas is drawn thru the suction line into the compressor cylinder. The downstroke of the piston pulls a cylinder of gas thru the suction valve and compresses it on the upstroke, raising its temperature and pressure to point *E*. The pressure produced causes the hot gas to flow to the condenser. The compressor discharge valve prevents re-entry of compressed gas into the cylinder and forms a boundry between the *high* and *low* sides.

In the condenser the condensing medium (air or water) absorbs heat to condense the hot gas. Liquid refrigerant is collected in receiver which may be combined with or separate from condenser.

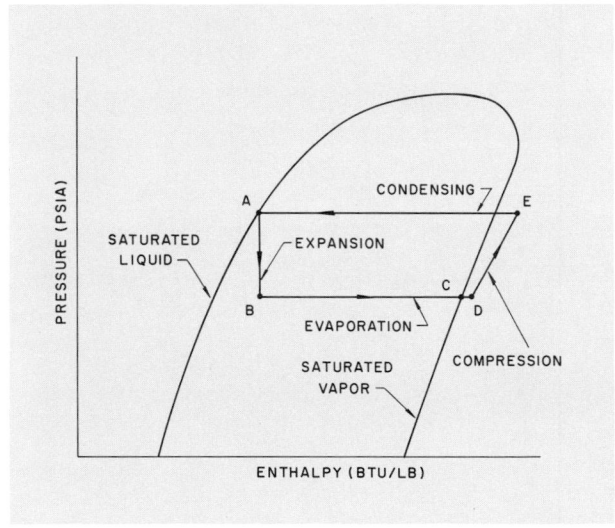

FIG. 4 — PRESSURE-ENTHALPY DIAGRAM,
COMPRESSION CYCLE

The liquid is then forced thru the liquid line to the expansion valve *A* to repeat the cycle.

Liquid-Suction Interchangers

Compressor ratings for Refrigerants 12 and 500 are generally based on 65 F actual suction gas temperatures. When this suction gas temperature is not obtained at the compressor, its rating must be lowered by an appropriate multiplier. To develop the full rating, the required superheat which is over and above that available at the evaporator outlet may be obtained by a liquid-suction interchanger.

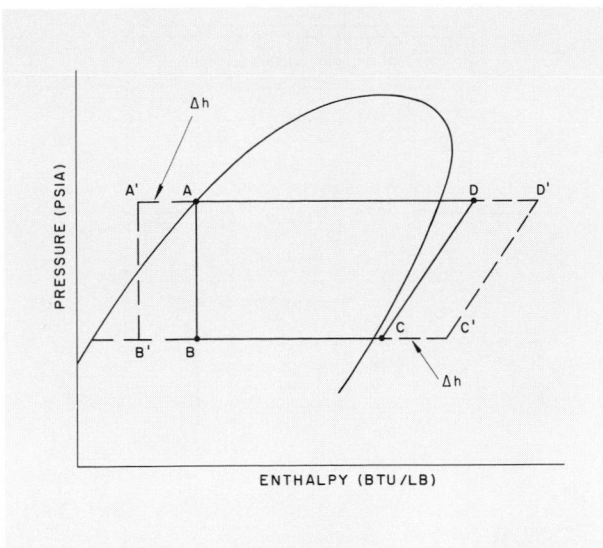

FIG. 5 — EFFECT OF LIQUID-SUCTION INTERCHANGER
ON COMPRESSION CYCLE

The effect of a liquid-suction interchanger on the refrigeration cycle is shown on the pressure-enthalpy diagram *(Fig. 5)*.

The solid lines represent the basic cycle, while the dashed lines represent the same cycle with a liquid-suction interchanger. The useful refrigerating effect with an interchanger is *B'C* rather than *BC* as in the basic cycle.

Superheat increases the specific volume of the suction vapor to reduce the total weight of refrigerant circulated for a given displacement. It also increases the enthalpy of the vapor and may improve compressor volumetric efficiency. Provided the heat absorbed represents useful refrigeration such as liquid subcooling, the refrigerating effect per pound of refrigerant circulated is increased. With Refrigerants 22 and 717 volume increases faster than refrigerating effect; hence, superheating theoretically reduces capacity. With Refrigerants 12 and 500 the reverse is true, and superheating theoretically reduces both the cfm per ton and the power per ton. *Figure 6* illustrates the loss due to superheating of the refrigerant vapor and the gain due to liquid subcooling. Net gain equals the gain minus the loss.

REFRIGERANT PROPERTIES

Refrigerant characteristics have a bearing on system design, application and operation. A refrigerant is selected after an analysis of the required characteristics and a matching of these require-

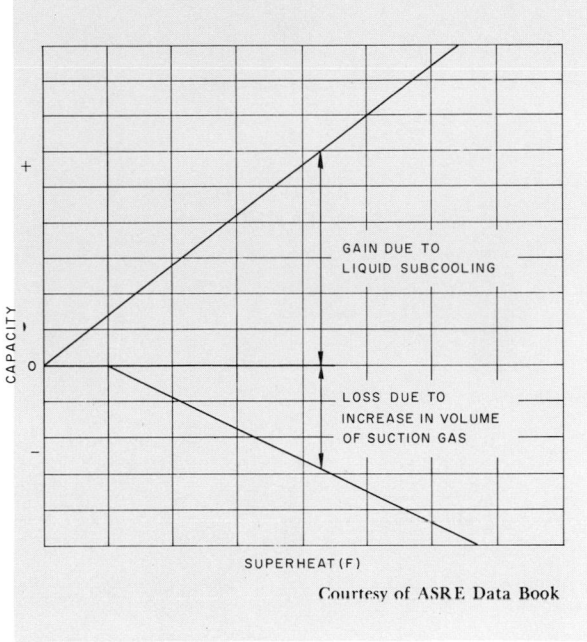

Courtesy of ASRE Data Book

FIG. 6 — EFFECT OF LIQUID-SUCTION INTERCHANGER
ON CAPACITY

ments with the specific properties of the available refrigerants.

Significant refrigerant characteristics are:

1. *Flammability* and *Toxicity* as they pertain to the safety of a refrigerant. The refrigerants treated in this chapter are classified in ASA B9.1 as Group 1, the least hazardous relative to flammability and explosiveness. The Underwriters' Laboratories classification with respect to toxicity puts these refrigerants in Groups 4 to 6. The higher numbered groups in this classification are the least toxic.

   Figure 7 shows the structural formula of the refrigerant compounds treated in this chapter. The chlorine and fluorine elements in these refrigerants make them the least hazardous and least toxic respectively.

2. *Miscibility* of a refrigerant with compressor oil aids in the return of oil from the evaporator to the compressor crankcase in reciprocating machine applications. Centrifugal units have separate oil and refrigerant circuits.

   Some refrigerants are highly miscible with compressor oil. Refrigerants 12 and 500 and lubricating oils are miscible in any proportion; Refrigerant 22 is less miscible. The effect of miscibility in a refrigeration system is illustrated in *Fig. 8*. If Refrigerant 12 is placed

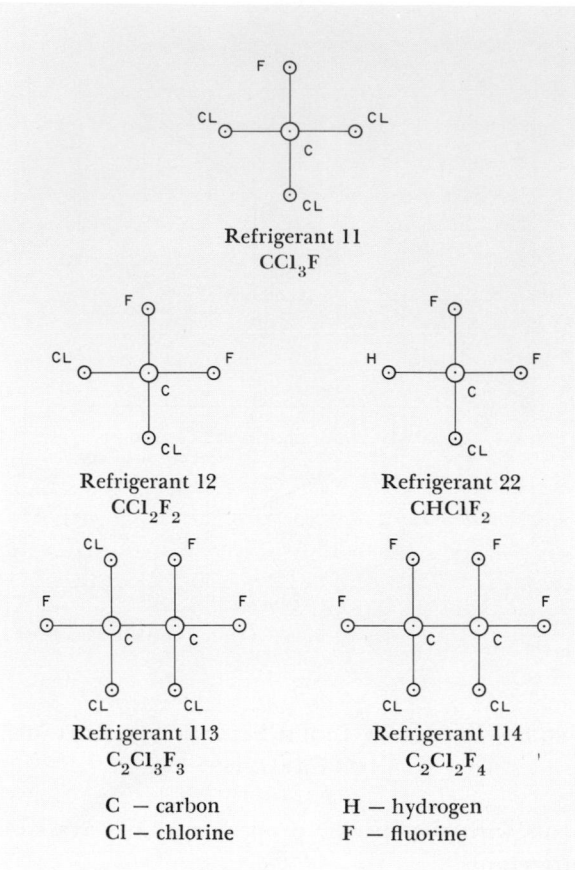

FIG. 7 — STRUCTURAL FORMULAS FOR REFRIGERANTS

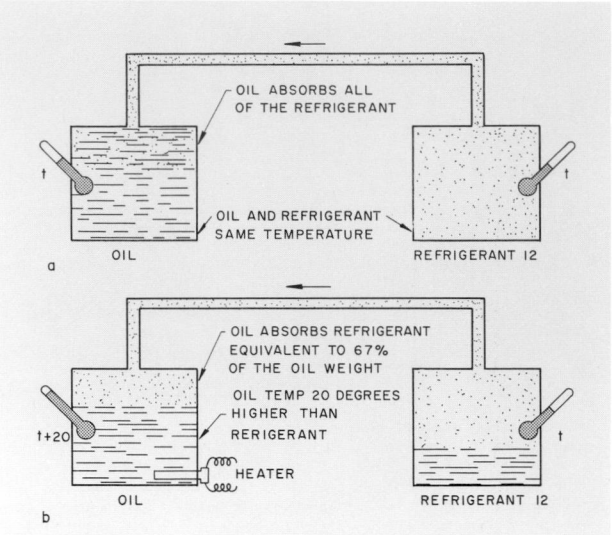

FIG. 8 — MISCIBILITY OF REFRIGERANT 12 AND OIL

in one vessel and lubricating oil in another *(Fig. 8a)*, and if both vessels are placed in a common ambient temperature, all of the refrigerant migrates to the vessel containing oil because of the absorption head of the oil. Raising the oil temperature limits this migration. For instance, if the oil is at an ambient temperature 20 degrees higher than the refrigerant or if the oil is heated by an immersion type heater to 20 degrees higher than the refrigerant temperature *(Fig. 8b)*, only 67% of the oil weight of the refrigerant becomes dissolved in the oil.

3. Theoretical *Horsepower Per Ton* of refrigeration for most refrigerants at air conditioning temperature levels is approximately the same *(Table 1)*.

4. *Rate of Leakage* of a refrigerant gas increases directly with pressure and inversely with molecular weight. The pressure of a refrigerant for a given saturated temperature increases in the following order: Refrigerant 113, 11, 114, 12, 500, 22. The molecular weight of a

refrigerant decreases in the following order: Refrigerant 113, 114, 11, 12, 500, 22. Molecular weight is directly related to vapor specific volume; the higher the molecular weight, the higher the specific volume.

5. *Leak Detection* of the refrigerant should be simple and positive for purposes of maintenance, cost and safety. The use of a halide torch makes it possible to detect and locate minute leaks of the halogen refrigerants.

6. *Vapor Density* influences the compressor displacement and pipe sizing. High vapor density accompanied by a reasonably high latent heat of vaporization (low cfm per ton) is desirable in a refrigerant. A low cfm per ton results in compact equipment and smaller refrigerant piping. Reciprocating refrigeration equipment requires a relatively high vapor density refrigerant for optimum performance. Centrifugal compressors require a low vapor density refrigerant for optimum efficiency at comparatively low tonnages. High vapor density refrigerants are used with centrifugals of large tonnage. The cfm per ton increases in the following order: Refrigerant 22, 500, 12, 114, 11, 113 *(Fig. 9)*.

Cost which is usually a consideration in all selections should not influence the choice of a refrigerant since it generally has little economic bearing on the normal refrigeration system. Although Refrigerant 22 costs approximately twice as much as Refrigerant 12, the compressor required is smaller, tending to offset the additional cost of refrigerant.

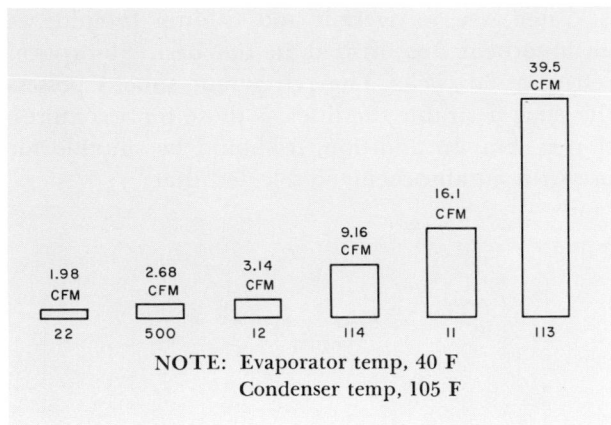

NOTE: Evaporator temp, 40 F
Condenser temp, 105 F

Fig. 9 — Suction Volumes of Refrigerants
(cfm/ton)

HEAT TRANSFER COMPARISONS

The value of the evaporating and condensing film coefficient (Btu/sq ft/F) for Refrigerant 22 is greater than that for Refrigerants 12 and 500. However, it does not follow that cooling coils and condensers can, therefore, be rated for higher capacities with Refrigerant 22. The higher coefficient for Refrigerant 22 does not tell the complete story.

Various other factors should be considered:

1. Whether the heat exchanger is designed for either Refrigerant 22 or Refrigerants 12 and 500.
2. Whether heat transfer is between refrigerant and air or between refrigerant and water.
3. Whether tubes in a heat exchanger are prime surface or extended surface (including the amount of extended surface.)

The evaporating or condensing film coefficient is one of a number of factors which make up the total over-all transfer rate for the heat exchanger. Other factors involved are these:

1. Tube wall resistance (including extended surface, if any).
2. Air or water film coefficient.
3. Refrigerant pressure drop per circuit, which affects the mean effective temperature difference.
4. Surface ratio of tube outside surface to inside surface.
5. Fouling factors (water-cooled condensers).

COOLING COILS

Cooling coils using Refrigerant 22 normally provide a greater capacity than those using Refrigerant 12 or 500. A cooling coil, normally, has considerable extended surface on the air side of the tubes. The higher evaporating film coefficient of Refrigerant 22

combined with the extended surface results in a significant increase in the overall heat transfer rate.

When cooling coil is designed to permit a normal pressure drop with Refrigerant 22, it may have a rather large pressure drop with Refrigerant 12 or 500. In such a case the decreased performance for Refrigerants 12 and 500 is due partially to the difference in their condensing film coefficients. In addition, it is affected by the pressure drop and the resulting lower mean effective temperature difference.

CONDENSERS

Water-cooled condensers using Refrigerant 22 normally provide a greater capacity than those using Refrigerant 12 or 500, depending on the fouling factor used in its selection. There are a number of reasons for this improvement other than the basically higher condensing film coefficient of the refrigerant.

1. The water film coefficient is relatively high as compared to air.
2. The extended surface or refrigerant side of the exchanger tubes assures maximum transfer rate and optimum balance between the inside and outside surfaces.

With Refrigerant 12 or 500 the performance of water-cooled condensers is otherwise not adversely affected since the pressure drop in the shell is not a consideration.

Air-cooled condensers using Refrigerant 22 normally provide a greater capacity than those using Refrigerant 12 or 500.

Air-cooled condensers have considerable extended surface on the air side of the tubes. The higher evaporating film coefficient of Refrigerant 22, combined with the extended surface results in a significant increase in the overall heat transfer rate.

When an air-cooled condenser is designed to allow for a normal pressure drop with Refrigerant 22, it may have a considerable pressure drop when used with Refrigerant 500 or 12. In such a case the decreased performance for Refrigerants 500 and 12 is not due entirely to the difference in their condensing film coefficients, but is affected also by pressure drop and the resulting lower mean effective temperature difference.

Normally, evaporative condensers use prime surface tubing (without extended surface on the outside of tubes). Where the unit is designed for Refrigerant 12 or 500, there is no significant increase in capacity when using Refrigerant 22.

If the unit is designed for Refrigerant 22 (smaller tubing or longer circuits), and is used with Refrigerant 12 or 500, the pressure drop is sufficient to reduce the rating. Such a design may create the impression that the increased rating for Refrigerant 22 is due to the condensing film, whereas actually the performance for Refrigerant 12 or 500 decreases due to coil design.

REFRIGERANT SELECTION

COMPRESSION CYCLE

The choice of a refrigerant for a compression system is limited by:

1. Economics,
2. Equipment type and size
3. Application

The manufacturer of the refrigeration compressor generally preselects the refrigerant to result in optimum owning cost. The specific refrigerant is determined by the type and size of the equipment.

To minimize the number of reciprocating compressor sizes required to fill out a line, the manufacturer rates each size for several refrigerants of relatively dense vapor such as Refrigerants 12, 500 and 22. In effect, this increases the number of units offered without adding sizes.

Centrifugal compressors at comparatively low tonnages require a high vapor volume refrigerant such as Refrigerant 113 or 11 to maintain optimum efficiency. For most sizes Refrigerant 114 or 12 can be used to obtain greater capacities. Refrigerants 500 and 22 are used with specially built centrifugals to obtain the highest capacities.

The refrigerant selected depends on the type of application. Air-cooled condensers may not use certain refrigerants because of the design condensing temperature required and corresponding limitations on compressor head pressure.

The temperature-pressure relationship of a refrigerant is of considerable importance in low temperature applications. If the evaporator pressure is comparatively low for the required evaporator temperature, the volume of vapor to be handled by the compressor is excessive. If the evaporator pressure is comparatively high for the required evaporator temperature, the system pressures are high.

The refrigerants that have been mentioned are the halogens (fluorinated hydrocarbon compounds), except for Refrigerant 500 which is an azeotropic mixture of two fluorinated hydrocarbons. The mixture does not separate into its component refrigerants with a change of temperature or pressure. It has its own fixed thermodynamic properties which are unlike either of its components.

ABSORPTION CYCLE

Water as a refrigerant and lithium bromide as an absorbent are utilized in the basic absorption refrigeration cycle. The refrigerant should possess the same desirable qualities as those for a compression system. In addition, it should be suitable for use with an absorbent, so selected that:

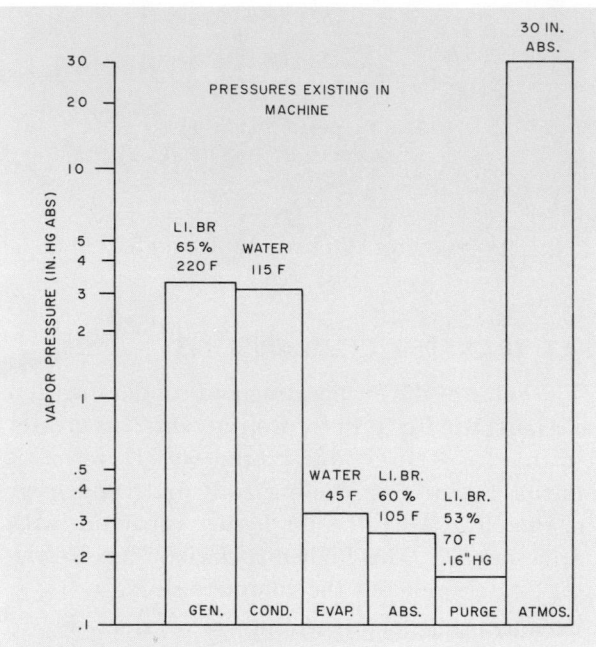

NOTE: Absorption machine at full load, using lithium bromide as absorbent.

FIG. 10 — PRESSURES AND TEMPERATURES OF A TYPICAL ABSORPTION MACHINE

1. The vapor pressures of the refrigerant and absorbent at the generator are different.

2. The temperature-pressure relations are consistent with practical absorber and generator temperatures and pressures. *Figure 10* shows absolute pressures and temperatures existing in a typical absorption machine at full load.

3. The refrigerant has a high solubility in the absorbent at absorber temperature and pressure and a low solubility at generator temperature and pressure.

4. The refrigerant and absorbent together are stable within the evaporator-generator range of temperatures. Normally, the absorbent must remain liquid at absorber and generator temperatures and pressures. It should have a low specific heat, surface tension and viscosity and must be neutral to the materials used in the equipment.

REFRIGERANT TABLES

Table 1 is a tabulation of the various refrigerants and their characteristics. *Tables 2 to 7* list the properties of the refrigerants for various saturated temperatures.

TABLE 1—COMPARATIVE DATA OF REFRIGERANTS

REFRIGERANT NUMBER (ARI DESIGNATION)	11	12	22	113	114	500
Chemical Name	Trichloromono-fluoromethane	Dichlorodi-fluoromethane	Monochlorodi-fluoromethane	Trichlorotri-fluoroethane	Dichlorotetra-fluoroethane	Azeotrope of Dichlorodi-fluoromethane and Difluoroethane
Chemical Formula	CCl_3F	CCl_2F_2	$CHClF_2$	CCl_2F-$CClF_2$	$C_2Cl_2F_4$	73.8% CCl_2F_2 26.2% CH_3CHF_2
Molecular Wt	137.38	120.93	86.48	187.39	170.93	99.29
Gas Constant, R (ft-lb/lb-R)	11.25	12.78	17.87	8.25	9.04	15.57
Boiling Point at 1 atm (F)	74.7	−21.62	−41.4	117.6	38.4	−28.0
Freezing Point at 1 atm (F)	−168	−252	−256	−31	−137	−254
Critical Temperature (F)	388.0	233.6	204.8	417.4	294.3	221.1
Critical Pressure (psia)	635.0	597.0	716.0	495.0	474.0	631.0
Specific Heat of Liquid, 86 F	.220	.235	.335	.218	.238	.300
Specific Heat of Vapor, C_p 60 F at 1 atm	*	.146	.149	*	.156	.171
Specific Heat of Vapor, C_v 60 F at 1 atm	*	.130	.127	*	.145	.151
Ratio $\frac{C_p}{C_v} = K$ (86 F at 1 atm)	1.11	1.14	1.18	1.12	1.09	1.13
Ratio of Specific Heats $\frac{\text{Liquid, 105 F}}{\text{Vapor, } C_p, \text{ 40 F sat. press.}}$	2.04	1.55	2.14	1.47	1.59	1.77
Liquid Head (ft), 1 psi at 105 F	1.61	1.84	2.04	1.51	1.65	2.10
Saturation Pressure (psia) at: −50 F	0.52	7.12	11.74	*	1.35	*
0 F	2.55	23.85	38.79	0.84	5.96	27.96
40 F	7.03	51.67	83.72	2.66	15.22	60.94
105 F	25.7	141.25	227.65	11.58	50.29	167.85
Net Refrigerating Effect (Btu/lb) 40 F-105 F (no subcooling)	67.56	49.13	66.44	54.54	43.46	59.82
Cycle Efficiency (% Carnot Cycle) 40 F-105 F	90.5	83.2	81.8	87.5	84.9	82.0
Solubility of Water in Refrigerant	Negligible	Negligible	Negligible	Negligible	Negligible	Negligible
Miscibility with Oil	Miscible	Miscible	Limited	Miscible	Miscible	Miscible
Toxic Concentration (% by vol)	Above 10%	Above 20%	*	*	Above 20%	Above 20%
Odor	Ethereal, odorless when mixed with air	Same as R 11	Same as R 11	Same as R 11	Same as R 11	Same as R 11
Warning Properties	None	None	None	None	None	None
Explosive Range (% by vol)	None	None	None	None	None	None
Safety Group, U.L.	5	6	5A	4-5	6	5A
Safety Group, ASA B9.1	1	1	1	1	1	1
Toxic Decomposition Products	Yes	Yes	Yes	Yes	Yes	Yes
Viscosity (centipoises) Saturated Liquid 95 F	.3893	.2463	.2253	.5845	.3420	.2150
105 F	.3723	.2395	.2207	.5472	.3272	.2100
Vapor at 1 atm 30 F	.0101	.0118	.0120	.0097	.0108	*
40 F	.0103	.0119	.0122	.0098	.0109	*
50 F	.0105	.0121	.0124	.0100	.0111	*
Thermal Conductivity (k) Saturated Liquid 95 F	.0596	.0481	.0573	.0512	.0435	*
105 F	.0581	.0469	.0553	.0500	.0421	*
Vapor at 1 atm 30 F	.0045	.0047	.0060	.0037	.0056	*
40 F	.0046	.0049	.0061	.0039	.0057	*
50 F	.0046	.0051	.0063	.0040	.0059	*
Liquid Circulated, 40 F-105 F (lb/min/ton)	2.96	4.07	3.02	3.66	4.62	3.35
Theoretical Displacement, 40 F-105 F (cu ft/min/ton)	16.1	3.14	1.98	39.5	9.16	2.69
Theoretical Horsepower Per Ton 40 F-105 F	0.676	0.736	0.75	0.70	0.722	0.747
Coefficient of Performance 40 F-105 F (4.71/hp per ton)	6.95	6.39	6.29	6.74	6.52	6.31
Cost Compared With R 11	1.00	1.57	2.77	2.15	2.97	2.00

*Data not available or not applicable.

TABLE 2—PROPERTIES OF REFRIGERANT 12, LIQUID AND SATURATED VAPOR

TEMP (F) t	PRESSURE (lb/sq in.) Absolute P	PRESSURE Gage p	VOLUME (cu ft/lb) Liquid v_f	VOLUME Vapor v_g	DENSITY (lb/cu ft) Liquid $1/v_f$	DENSITY Vapor $1/v_g$	ENTHALPY (Btu/lb) Liquid h_f	ENTHALPY Latent h_{fg}	ENTHALPY Vapor h_g	ENTROPY (Btu/lb-R) Liquid s_f	ENTROPY Vapor s_g	TEMP (F) t
−100	1.4280	27.0138*	0.009985	22.164	100.15	0.045119	−12.466	78.714	66.248	−0.032005	0.18683	−100
−98	1.5381	26.7896*	.010002	20.682	99.978	.048352	−12.055	78.524	66.469	−.030866	.18623	−98
−96	1.6551	26.5514*	.010020	19.316	99.803	.051769	−11.644	78.334	66.690	−.029733	.18565	−96
−94	1.7794	26.2984*	.010037	18.057	99.627	.055379	−11.233	78.144	66.911	−.028606	.18508	−94
−92	1.9112	26.0301*	.010055	16.895	99.451	.059189	−10.821	77.954	67.133	−.027484	.18452	−92
−90	2.0509	25.7456*	0.010073	15.821	99.274	0.063207	−10.409	77.764	67.355	−0.026367	0.18398	−90
−88	2.1988	25.4443*	.010091	14.828	99.097	.067441	−9.9971	77.574	67.577	−.025256	.18345	−88
−86	2.3554	25.1255*	.010109	13.908	98.919	.071900	−9.5845	77.384	67.799	−.024150	.18293	−86
−84	2.5210	24.7884*	.010128	13.056	98.740	.076597	−9.1717	77.194	68.022	−.023049	.18242	−84
−82	2.6960	24.4321*	.010146	12.226	98.561	.081525	−8.7586	77.003	68.244	−.021953	.18192	−82
−80	2.8807	24.0560*	0.010164	11.533	98.382	0.086708	−8.3451	76.812	68.467	−0.020862	0.18143	−80
−78	3.0756	23.6592*	.010183	10.852	98.201	.092151	−7.9314	76.620	68.689	−.019776	.18096	−78
−76	3.2811	23.2409*	.010202	10.218	98.021	.097863	−7.5173	76.429	68.912	−.018695	.18050	−76
−74	3.4975	22.8002*	.010221	9.6209	97.839	.10385	−7.1029	76.238	69.135	−.017619	.18004	−74
−72	3.7254	22.3362*	.010240	9.0802	97.657	.11013	−6.6881	76.046	69.358	−.016547	.17960	−72
−70	3.9651	21.8482*	0.010259	8.5687	97.475	0.11670	−6.2730	75.853	69.580	−0.015481	0.17916	−70
−68	4.2172	21.3350*	.010278	8.0916	97.292	.12359	−5.8574	75.660	69.803	−.014418	.17874	−68
−66	4.4819	20.7959*	.010298	7.6462	97.108	.13078	−5.4416	75.467	70.025	−.013361	.17833	−66
−64	4.7599	20.2299*	.010317	7.2302	96.924	.13831	−5.0254	75.273	70.248	−.012308	.17792	−64
−62	5.0516	19.6360*	.010337	6.8412	96.739	.14617	−4.6088	75.080	70.471	−.011259	.17753	−62
−60	5.3575	19.0133*	0.010357	6.4774	96.553	0.15438	−4.1919	74.885	70.693	−0.010214	0.17714	−60
−58	5.6780	18.3607*	.010377	6.1367	96.367	.16295	−3.7745	74.691	70.916	−.009174	.17676	−58
−56	6.0137	17.6773*	.010397	5.8176	96.180	.17189	−3.3567	74.495	71.138	−.008139	.17639	−56
−54	6.3650	16.9619*	.010417	5.5184	95.993	.18121	−2.9386	74.299	71.360	−.007107	.17603	−54
−52	6.7326	16.2136*	.010438	5.2377	95.804	.19092	−2.5200	74.103	71.583	−.006080	.17568	−52
−50	7.1168	15.4313*	0.010459	4.9742	95.616	0.20104	−2.1011	73.906	71.805	−0.005056	0.17533	−50
−48	7.5183	14.6139*	.010479	4.7267	95.426	.21157	−1.6817	73.709	72.027	−.004037	.17500	−48
−46	7.9375	13.7603*	.010500	4.4940	95.236	.22252	−1.2619	73.511	72.249	−.003022	.17467	−46
−44	8.3751	12.8693*	.010521	4.2751	95.045	.23391	−0.8417	73.312	72.470	−.002011	.17435	−44
−42	8.8316	11.9399*	.010543	4.0691	94.854	.24576	−0.4211	73.112	72.691	−.001003	.17403	−42
−40	9.3076	10.9709*	0.010564	3.8750	94.661	0.25806	0.0000	72.913	72.913	0.000000	0.17373	−40
−38	9.8035	9.9611*	.010586	3.6922	94.469	.27084	0.4215	72.712	73.134	.001000	.17343	−38
−36	10.320	8.909*	.010607	3.5198	94.275	.28411	0.8434	72.511	73.354	.001995	.17313	−36
−34	10.858	7.814*	.010629	3.3571	94.081	.29788	1.2659	72.309	73.575	.002988	.17285	−34
−32	11.417	6.675*	.010651	3.2035	93.886	.31216	1.6887	72.106	73.795	.003976	.17257	−32
−30	11.999	5.490*	0.010674	3.0585	93.690	0.32696	2.1120	71.903	74.015	0.004961	0.17229	−30
−28	12.604	4.259*	.010696	2.9214	93.493	.34231	2.5358	71.698	74.234	.005942	.17203	−28
−26	13.233	2.979*	.010719	2.7917	93.296	.35820	2.9601	71.494	74.454	.006919	.17177	−26
−24	13.886	1.649*	.010741	2.6691	93.098	.37466	3.3848	71.288	74.673	.007894	.17151	−24
−22	14.564	0.270*	.010764	2.5529	92.899	.39171	3.8100	71.081	74.891	.008864	.17126	−22
−20	15.267	0.571	0.010788	2.4429	92.699	0.40934	4.2357	70.874	75.110	0.009831	0.17102	−20
−18	15.996	1.300	.010811	2.3387	92.499	.42758	4.6618	70.666	75.328	.010795	.17078	−18
−16	16.753	2.057	.010834	2.2399	92.298	.44645	5.0885	70.456	75.545	.011755	.17055	−16
−14	17.536	2.840	.010858	2.1461	92.096	.46595	5.5157	70.246	75.762	.012712	.17032	−14
−12	18.348	3.652	.010882	2.0572	91.893	.48611	5.9434	70.036	75.979	.013666	.17010	−12
−10	19.189	4.493	0.010906	1.9727	91.689	0.50693	6.3716	69.824	76.196	0.014617	0.16989	−10
−8	20.059	5.363	.010931	1.8924	91.485	.52843	6.8003	69.611	76.411	.015564	.16967	−8
−6	20.960	6.264	.010955	1.8161	91.280	.55063	7.2296	69.397	76.627	.016508	.16947	−6
−4	21.891	7.195	.010980	1.7436	91.074	.57354	7.6594	69.183	76.842	.017449	.16927	−4
−2	22.854	8.158	.011005	1.6745	90.867	.59718	8.0898	68.967	77.057	.018388	.16907	−2
0	23.849	9.153	0.011030	1.6089	90.659	0.62156	8.5207	68.750	77.271	0.019323	0.16888	0
2	24.878	10.182	.011056	1.5463	90.450	.64670	8.9522	68.533	77.485	.020255	.16869	2
4	25.939	11.243	.011082	1.4867	90.240	.67263	9.3843	68.314	77.698	.021184	.16851	4
6	27.036	12.340	.011107	1.4299	90.030	.69934	9.8169	68.094	77.911	.022110	.16833	6
8	28.167	13.471	.011134	1.3758	89.818	.72687	10.250	67.873	78.123	.023033	.16815	8
10	29.335	14.639	0.011160	1.3241	89.606	0.75523	10.684	67.651	78.335	0.023954	0.16798	10
12	30.539	15.843	.011187	1.2748	89.392	.78443	11.118	67.428	78.546	.024871	.16782	12
14	31.780	17.084	.011214	1.2278	89.178	.81449	11.554	67.203	78.757	.025786	.16765	14
16	33.060	18.364	.011241	1.1828	88.962	.84544	11.989	66.977	78.966	.026699	.16750	16
18	34.378	19.682	.011268	1.1399	88.746	.87729	12.426	66.750	79.176	.027608	.16734	18
20	35.736	21.040	0.011296	1.0988	88.529	0.91006	12.863	66.522	79.385	0.028515	0.16719	20
22	37.135	22.439	.011324	1.0596	88.310	.94377	13.300	66.293	79.593	.029420	.16704	22
24	38.574	23.878	.011352	1.0220	88.091	.97843	13.739	66.061	79.800	.030322	.16690	24
26	40.056	25.360	.011380	0.98612	87.870	1.0141	14.178	65.829	80.007	.031221	.16676	26
28	41.580	26.884	.011409	0.95173	87.649	1.0507	14.618	65.596	80.214	.032118	.16662	28

*Inches of mercury below one atmosphere.

Courtesy of E. I. du Pont de Nemours & Co.

TABLE 2—PROPERTIES OF REFRIGERANT 12, LIQUID AND SATURATED VAPOR (Contd)

TEMP (F)	PRESSURE (lb/sq in.)		VOLUME (cu ft/lb)		DENSITY (lb/cu ft)		ENTHALPY (Btu/lb)			ENTROPY (Btu/lb-R)		TEMP (F)
t	Absolute P	Gage p	Liquid v_f	Vapor v_g	Liquid $1/v_f$	Vapor $1/v_g$	Liquid h_f	Latent h_{fg}	Vapor h_g	Liquid s_f	Vapor s_g	t
30	43.148	28.452	0.011438	0.91880	87.426	1.0884	15.058	65.361	80.419	0.033013	0.16648	30
32	44.760	30.064	.011468	.88725	87.202	1.1271	15.500	65.124	80.624	.033905	.16635	32
34	46.417	31.721	.011497	.85702	86.977	1.1668	15.942	64.886	80.828	.034796	.16622	34
36	48.120	33.424	.011527	.82803	86.751	1.2077	16.384	64.647	81.031	.035683	.16610	36
38	49.870	35.174	.011557	.80023	86.524	1.2496	16.828	64.406	81.234	.036569	.16598	38
40	51.667	36.971	0.011588	0.77357	86.296	1.2927	17.273	64.163	81.436	0.037453	0.16586	40
42	53.513	38.817	.011619	.74798	86.066	1.3369	17.718	63.919	81.637	.038334	.16574	42
44	55.407	40.711	.011650	.72341	85.836	1.3823	18.164	63.673	81.837	.039213	.16562	44
46	57.352	42.656	.011682	.69982	85.604	1.4289	18.611	63.426	82.037	.040091	.16551	46
48	59.347	44.651	.011714	.67715	85.371	1.4768	19.059	63.177	82.236	.040966	.16540	48
50	61.394	46.698	0.011746	0.65537	85.136	1.5258	19.507	62.926	82.433	0.041839	0.16530	50
52	63.494	48.798	.011779	.63444	84.900	1.5762	19.957	62.673	82.630	.042711	.16519	52
54	65.646	50.950	.011811	.61431	84.663	1.6278	20.408	62.418	82.826	.043581	.16509	54
56	67.853	53.157	.011845	.59495	84.425	1.6808	20.859	62.162	83.021	.044449	.16499	56
58	70.115	55.419	.011879	.57632	84.185	1.7352	21.312	61.903	83.215	.045316	.16489	58
60	72.433	57.737	0.011913	0.55839	83.944	1.7909	21.766	61.643	83.409	0.046180	0.16479	60
62	74.807	60.111	.011947	.54112	83.701	1.8480	22.221	61.380	83.601	.047044	.16470	62
64	77.239	62.543	0.11982	.52450	83.457	1.9066	22.676	61.116	83.792	.047905	.16460	64
66	79.729	65.033	.012017	.50848	83.212	1.9666	23.133	60.849	83.982	.048765	.16451	66
68	82.279	67.583	.012053	.49305	82.965	2.0282	23.591	60.580	84.171	.049624	.16442	68
70	84.888	70.192	0.012089	0.47818	82.717	2.0913	24.050	60.309	84.359	0.050482	0.16434	70
72	87.559	72.863	.012126	.46383	82.467	2.1559	24.511	60.035	84.546	.051338	.16425	72
74	90.292	75.596	.012163	.45000	82.215	2.2222	24.973	59.759	84.732	.052193	.16417	74
76	93.087	78.391	.012201	.43666	81.962	2.2901	25.435	59.481	84.916	.053047	.16408	76
78	95.946	81.250	.012239	.42378	81.707	2.3597	25.899	59.201	85.100	.053900	.16400	78
80	98.870	84.174	0.012277	0.41135	81.450	2.4310	26.365	58.917	85.282	0.054751	0.16392	80
82	101.86	87.16	.012316	.39935	81.192	2.5041	26.832	58.631	85.463	.055602	.16384	82
84	104.92	90.22	.012356	.38776	80.932	2.5789	27.300	58.343	85.643	.056452	.16376	84
86	108.04	93.34	0.12396	.37657	80.671	2.6556	27.769	58.052	85.821	.057301	.16368	86
88	111.23	96.53	.012437	.36575	80.407	2.7341	28.241	57.757	85.998	.058149	.16360	88
90	114.49	99.79	0.012478	0.35529	80.142	2.8146	28.713	57.461	86.174	0.058997	0.16353	90
92	117.82	103.12	.012520	.34518	79.874	2.8970	29.187	57.161	86.348	.059844	.16345	92
94	121.22	106.52	.012562	.33540	79.605	2.9815	29.663	56.858	86.521	.060690	.16338	94
96	124.70	110.00	.012605	.32594	79.334	3.0680	30.140	56.551	86.691	.061536	.16330	96
98	128.24	113.54	.012649	.31679	79.061	3.1566	30.619	56.242	86.861	.062381	.16323	98
100	131.86	117.16	0.012693	0.30794	78.785	3.2474	31.100	55.929	87.029	0.063227	0.16315	100
102	135.56	120.86	.012738	.29937	78.508	3.3404	31.583	55.613	87.196	.064072	.16308	102
104	139.33	124.63	.012783	.29106	78.228	3.4357	32.067	55.293	87.360	.064916	.16301	104
106	143.18	128.48	.012829	.28303	77.946	3.5333	32.553	54.970	87.523	.065761	.16293	106
108	147.11	132.41	.012876	.27524	77.662	3.6332	33.041	54.643	87.684	.066606	.16286	108
110	151.11	136.41	0.012924	0.26769	77.376	3.7357	33.531	54.313	87.844	0.067451	0.16279	110
112	155.19	140.49	.012972	.26037	77.087	3.8406	34.023	53.978	88.001	.068296	.16271	112
114	159.36	144.66	.013022	.25328	76.795	3.9482	34.517	53.639	88.156	.069141	.16264	114
116	163.61	148.91	.013072	.24641	76.501	4.0584	35.014	53.296	88.310	.069987	.16256	116
118	167.94	153.24	.013123	.23974	76.205	4.1713	35.512	52.949	88.461	.070833	.16249	118
120	172.35	157.65	0.013174	0.23326	75.906	4.2870	36.013	52.597	88.610	0.071680	0.16241	120
122	176.85	162.15	.013227	.22698	75.604	4.4056	36.516	52.241	88.757	.072528	.16234	122
124	181.43	166.73	.013280	.22089	75.299	4.5272	37.021	51.881	88.902	.073376	.16226	124
126	186.10	171.40	.013335	.21497	74.991	4.6518	37.529	51.515	89.044	.074225	.16218	126
128	190.86	176.16	.013390	.20922	74.680	4.7796	38.040	51.144	89.184	.075075	.16210	128
130	195.71	181.01	0.013447	0.20364	74.367	4.9107	38.553	50.768	89.321	0.075927	0.16202	130
132	200.64	185.94	.013504	.19821	74.050	5.0451	39.069	50.387	89.456	.076779	.16194	132
134	205.67	190.97	.013563	.19294	73.729	5.1829	39.588	50.000	89.588	.077623	.16185	134
136	210.79	196.09	.013623	.18782	73.406	5.3244	40.110	49.608	89.718	.078489	.16177	136
138	216.01	201.31	.013684	.18283	73.079	5.4695	40.634	49.210	89.844	.079346	.16168	138
140	221.32	206.62	0.013746	0.17799	72.748	5.6184	41.162	48.805	89.967	0.080205	0.16159	140
142	226.72	212.02	.013810	.17327	72.413	5.7713	41.693	48.394	90.087	.081065	.16150	142
144	232.22	217.52	.013874	.16868	72.075	5.9283	42.227	47.977	90.204	.081928	.16140	144
146	237.82	223.12	.013941	.16422	71.732	6.0895	42.765	47.553	90.318	.082794	.16130	146
148	243.51	228.81	.014008	.15987	71.386	6.2551	43.306	47.122	90.428	.083661	.16120	148
150	249.31	234.61	0.014078	0.15564	71.035	6.4252	43.850	46.684	90.534	0.084531	0.16110	150
152	255.20	240.50	.014148	.15151	70.679	6.6001	44.399	46.238	90.637	.085404	.16099	152
154	261.20	246.50	.014221	.14750	70.319	6.7799	44.951	45.784	90.735	.086280	.16088	154
156	267.30	252.60	.014295	.14358	69.954	6.9648	45.508	45.322	90.830	.087159	.16077	156
158	273.51	258.81	.014371	.13976	69.584	7.1551	46.068	44.852	90.920	.088041	.16065	158
160	279.82	265.12	0.014449	0.13604	69.209	7.3509	46.633	44.373	91.006	0.088927	0.16053	160

TABLE 3—PROPERTIES OF REFRIGERANT 500, LIQUID AND SATURATED VAPOR

TEMP (F)	PRESSURE (lb/sq in.)		VOLUME (cu ft/lb)		DENSITY (lb/cu ft)		ENTHALPY (Btu/lb)			ENTROPY (Btu/lb-R)		TEMP (F)
t	Absolute P	Gage p	Liquid vf	Vapor vg	Liquid 1/vf	Vapor 1/vg	Liquid hf	Latent hfg	Vapor hg	Liquid sf	Vapor sg	t
−40	10.84	7.86*	0.0119	4.0757	84.37	0.2454	0.00	89.91	89.91	0.00000	0.21421	−40
−38	11.42	6.68*	.0119	3.8829	84.19	.2575	0.51	89.68	90.19	.00122	.21387	−38
−36	12.03	5.44*	.0119	3.6992	84.00	.2703	1.02	89.45	90.47	.00243	.21352	−36
−34	12.66	4.15*	.0119	3.5276	83.82	.2835	1.54	89.21	90.75	.00365	.21318	−34
−32	13.31	2.83*	.0120	3.3671	83.63	.2970	2.05	88.97	91.02	.00485	.21286	−32
−30	14.00	1.40*	0.0120	3.2121	83.45	0.3113	2.57	88.73	91.30	0.00606	0.21252	−30
−28	14.71	0.01	.0120	3.0674	83.26	.3260	3.09	88.49	91.58	.00725	.21221	−28
−26	15.45	0.75	.0120	2.9302	83.07	.3413	3.60	88.25	91.85	.00845	.21189	−26
−24	16.21	1.51	.0121	2.8020	82.89	.3569	4.13	88.00	92.13	.00963	.21160	−24
−22	17.01	2.31	.0121	2.6788	82.70	.3733	4.65	87.75	92.40	.01083	.21130	−22
−20	17.84	3.14	0.0121	2.5622	82.51	0.3903	5.17	87.50	92.67	0.01202	0.21101	−20
−18	18.70	4.00	.0121	2.4520	82.32	.4078	5.70	87.25	92.95	.01322	.21072	−18
−16	19.59	4.89	.0122	2.3477	82.13	.4260	6.22	87.00	93.22	.01438	.21044	−16
−14	20.51	5.81	.0122	2.2491	81.94	.4446	6.75	86.74	93.49	.01557	.21017	−14
−12	21.47	6.77	.0122	2.1548	81.75	.4641	7.27	86.49	93.76	.01674	.20991	−12
−10	22.46	7.76	0.0123	2.0657	81.56	0.4841	7.80	86.23	94.03	0.01793	0.20965	−10
− 8	23.49	8.79	.0123	1.9807	81.37	.5049	8.33	85.97	94.30	.01909	.20939	− 8
− 6	24.55	9.85	.0123	1.9004	81.17	.5262	8.87	85.70	94.57	.02027	.20914	− 6
− 4	25.65	10.95	.0123	1.8238	80.98	.5483	9.40	85.44	94.84	.02144	.20890	− 4
− 2	26.79	12.09	.0124	1.7507	80.78	.5712	9.94	85.17	95.11	.02259	.20866	− 2
0	27.96	13.26	0.0124	1.6818	80.59	0.5946	10.47	84.90	95.37	0.02376	0.20843	0
2	29.18	14.48	.0124	1.6155	80.39	.6190	11.01	84.63	95.64	.02491	.20819	2
4	30.43	15.73	.0125	1.5530	80.20	.6439	11.55	84.35	95.90	.02608	.20797	4
6	31.73	17.03	.0125	1.4929	80.00	.6698	12.09	84.08	96.17	.02722	.20775	6
8	33.06	18.36	.0125	1.4362	79.80	.6963	12.63	83.80	96.43	.02839	.20754	8
10	34.45	19.75	0.0126	1.3813	79.60	0.7239	13.17	83.52	96.69	0.02954	0.20732	10
12	35.88	21.18	.0126	1.3292	79.40	.7523	13.71	83.24	96.95	.03067	.20711	12
14	37.35	22.65	.0126	1.2796	79.20	.7815	14.26	82.95	97.21	.03182	.20691	14
16	38.86	24.16	.0127	1.2325	79.00	.8114	14.81	82.66	97.47	.03296	.20672	16
18	40.42	25.72	.0127	1.1873	78.80	.8422	15.35	82.37	97.72	.03411	.20652	18
20	42.03	27.33	0.0127	1.1440	78.59	0.8741	15.91	82.07	97.98	0.03526	0.20633	20
22	43.69	28.99	.0128	1.1027	78.39	0.9069	16.45	81.78	98.23	.03638	.20614	22
24	45.40	30.70	.0128	1.0631	78.18	0.9407	17.01	81.48	98.49	.03752	.20595	24
26	47.15	32.45	.0128	1.0254	77.98	0.9752	17.56	81.18	98.74	.03865	.20578	26
28	48.96	34.26	.0129	0.9892	77.77	1.0109	18.12	80.87	98.99	.03978	.20560	28
30	50.82	36.12	0.0129	0.9545	77.56	1.0476	18.67	80.57	99.24	0.04092	0.20542	30
32	52.74	38.04	.0129	.9212	77.35	1.0855	19.22	80.26	99.48	.04203	.20525	32
34	54.71	40.01	.0130	.8894	77.14	1.1244	19.79	79.94	99.73	.04316	.20508	34
36	56.72	42.02	.0130	.8591	76.93	1.1640	20.34	79.63	99.97	.04430	.20491	36
38	58.80	44.10	.0130	.8298	76.72	1.2051	20.91	79.31	100.22	.04542	.20475	38
40	60.94	46.24	0.0131	0.8017	76.50	1.2473	21.47	78.99	100.46	0.04654	0.20458	40
42	63.14	48.44	.0131	.7747	76.29	1.2908	22.03	78.67	100.70	.04764	.20442	42
44	65.39	50.69	.0131	.7490	76.07	1.3352	22.60	78.34	100.94	.04875	.20426	44
46	67.71	53.01	.0132	.7241	75.86	1.3811	23.16	78.01	101.17	.04986	.20410	46
48	70.09	55.39	.0132	.7002	75.64	1.4282	23.75	77.66	101.41	.05099	.20395	48
50	72.52	57.82	0.0133	0.6774	75.42	1.4763	24.31	77.33	101.64	0.05212	0.20380	50
52	75.02	60.32	.0133	.6554	75.20	1.5258	24.88	76.99	101.87	.05321	.20365	52
54	77.57	62.87	.0133	.6343	74.97	1.5764	25.46	76.64	102.10	.05431	.20350	54
56	80.22	65.52	.0134	.6138	74.75	1.6291	26.04	76.29	102.33	.05543	.20335	56
58	82.91	68.21	.0134	.5943	74.52	1.6826	26.59	75.96	102.55	.05649	.20320	58
60	85.66	70.96	0.0135	0.5756	74.30	1.7374	27.19	75.59	102.78	0.05763	0.20306	60
62	88.49	73.79	.0135	.5575	74.07	1.7939	27.76	75.24	103.00	.05872	.20291	62
64	91.39	76.69	.0135	.5400	73.84	1.8518	28.33	74.89	103.22	.05979	.20277	64
66	94.37	79.67	.0136	.5231	73.61	1.9116	28.92	74.51	103.43	.06091	.20262	66
68	97.41	82.71	.0136	.5069	73.38	1.9727	29.52	74.13	103.65	.06203	.20248	68
70	100.51	85.81	0.0137	0.4914	73.14	2.0350	30.10	73.76	103.86	0.06311	0.20234	70
72	103.71	89.01	.0137	.4763	72.91	2.0996	30.68	73.39	104.07	.06419	.20220	72
74	106.97	92.27	.0138	.4618	72.67	2.1655	31.27	73.01	104.28	.06527	.20206	74
76	110.29	95.59	.0138	.4479	72.43	2.2328	31.87	72.62	104.49	.06637	.20192	76
78	113.70	99.00	.0139	.4344	72.19	2.3021	32.47	72.22	104.69	.06749	.20178	78
80	117.20	102.5	0.0139	0.4213	71.95	2.3735	33.06	71.83	104.89	0.06855	0.20164	80
82	120.78	106.1	.0139	.4087	71.70	2.4469	33.63	71.46	105.09	.06959	.20149	82
84	124.40	109.7	.0140	.3967	71.46	2.5210	34.24	71.04	105.28	.07071	.20136	84
86	128.14	113.4	.0140	.3849	71.21	2.5981	34.84	70.63	105.47	.07179	.20121	86
88	131.96	117.3	.0141	.3735	70.96	2.6772	35.44	70.22	105.66	.07288	.20107	88

*Inches of mercury below one atmosphere.

TABLE 3—PROPERTIES OF REFRIGERANT 500, LIQUID AND SATURATED VAPOR (Contd)

TEMP (F)	PRESSURE (lb/sq in.)		VOLUME (cu ft/lb)		DENSITY (lb/cu ft)		ENTHALPY (Btu/lb)			ENTROPY (Btu/lb-R)		TEMP (F)
t	Absolute P	Gage p	Liquid vf	Vapor vg	Liquid 1/vf	Vapor 1/vg	Liquid hf	Latent hfg	Vapor hg	Liquid sf	Vapor sg	t
90	135.86	121.2	0.0141	0.3626	70.70	2.7581	36.04	69.81	105.85	0.07395	0.20092	90
92	139.83	125.1	.0142	.3520	70.45	2.8409	36.66	69.37	106.03	.07505	.20078	92
94	143.90	129.2	.0142	.3418	70.19	2.9261	37.26	68.96	106.22	.07611	.20064	94
96	148.03	133.3	.0143	.3319	69.93	3.0128	37.88	68.51	106.39	.07722	.20049	96
98	152.27	137.6	.0144	.3223	69.67	3.1024	38.47	68.10	106.57	.07826	.20035	98
100	156.61	141.9	0.0144	0.3130	69.41	3.1947	39.09	67.65	106.74	0.07935	0.20019	100
102	161.02	146.3	.0145	.3041	69.14	3.2889	39.71	67.20	106.91	.08042	.20004	102
104	165.55	150.9	.0145	.2953	68.87	3.3860	40.33	66.74	107.07	.08151	.19989	104
106	170.14	155.4	.0146	.2869	68.60	3.4850	40.96	66.27	107.23	.08261	.19974	106
108	174.84	160.1	.0146	.2788	68.33	3.5871	41.59	65.80	107.39	.08367	.19958	108
110	179.62	164.9	0.0147	0.2709	68.05	3.6914	42.22	65.32	107.54	0.08479	0.19942	110
112	184.51	169.8	.0148	.2632	67.78	3.7990	42.82	64.87	107.69	.08482	.19926	112
114	189.47	174.8	.0148	.2558	67.49	3.9086	43.47	64.37	107.84	.08691	.19910	114
116	194.55	179.9	.0149	.2486	67.21	4.0218	44.11	63.87	107.98	.08800	.19893	116
118	199.71	185.0	.0149	.2417	66.92	4.1376	44.72	63.40	108.12	.08905	.19876	118
120	204.99	190.3	0.0150	0.2349	66.63	4.2566	45.35	62.90	108.25	0.09012	0.19859	120
122	210.40	195.7	.0151	.2283	66.34	4.3807	46.02	62.35	108.37	.09124	.19841	122
124	215.88	201.2	.0151	.2219	66.04	4.5068	46.69	61.81	108.50	.09236	.19823	124
126	221.44	206.7	.0152	.2157	65.74	4.6357	47.33	61.29	108.62	.09342	.19805	126
128	227.13	212.4	.0153	.2097	65.43	4.7690	47.98	60.75	108.73	.09451	.19786	128
130	232.89	218.2	0.0154	0.2039	65.13	4.9050	48.65	60.19	108.84	0.09560	0.19767	130
132	238.80	224.1	.0154	.1982	64.81	5.0463	49.28	59.66	108.94	.09666	.19747	132
134	244.84	230.1	.0155	.1926	64.49	5.1924	49.91	59.13	109.04	.09771	.19726	134
136	250.96	236.3	.0156	.1872	64.17	5.3420	50.59	58.54	109.13	.09883	.19705	136
138	257.17	242.5	.0157	.1820	63.85	5.4951	51.30	57.91	109.21	.09996	.19684	138
140	263.52	248.8	0.0157	0.1769	63.52	5.6528	52.02	57.27	109.29	0.10114	0.19663	140
142	269.99	255.3	.0158	.1719	63.18	5.8170	52.73	56.63	109.36	.10228	.19639	142
144	276.55	261.9	.0159	.1670	62.84	5.9878	53.45	55.97	109.42	.10344	.19615	144
146	283.24	268.5	.0160	.1623	62.49	6.1626	54.15	55.33	109.48	.10456	.19591	146
148	290.04	275.3	.0161	.1577	62.14	6.3431	54.86	54.67	109.53	.10569	.19565	148
150	296.97	282.3	0.0162	0.1531	61.78	6.5304	55.62	53.95	109.57	0.10691	0.19539	150
152	304.01	289.3	.0163	.1487	61.42	6.7241	56.34	53.25	109.59	.10806	.19511	152
154	311.18	296.5	.0164	.1444	61.05	6.9253	57.07	52.54	109.61	.10922	.19483	154
156	318.47	303.8	.0165	.1402	60.66	7.1342	57.82	51.80	109.62	.11040	.19453	156
158	325.87	311.2	.0166	.1360	60.28	7.3510	58.61	51.01	109.62	.11163	.19421	158
160	333.40	318.7	0.0167	0.1320	59.88	7.5772	59.35	50.25	109.60	0.11280	0.19389	160

TABLE 4—PROPERTIES OF REFRIGERANT 22, LIQUID AND SATURATED VAPOR

TEMP (F)	PRESSURE (lb/sq in.)		VOLUME (cu ft/lb)		DENSITY (lb/cu ft)		ENTHALPY (Btu/lb)			ENTROPY (Btu/lb-R)		TEMP (F)
t	Absolute P	Gage p	Liquid v_f	Vapor v_g	Liquid $1/v_f$	Vapor $1/v_g$	Liquid h_f	Latent h_{fg}	Vapor h_g	Liquid s_f	Vapor s_g	t
−155	0.19901	29.51*	0.0102	188.1	97.67	0.005316	−29.07	115.85	86.78	−0.0808	0.2996	−155
−150	0.2605	29.39*	.0103	146.1	97.33	.006847	−27.79	115.15	87.36	− .0767	.2952	−150
−145	0.3375	29.23*	.0103	114.5	96.99	.008733	−26.52	114.46	87.94	− .0727	.2912	−145
−140	0.4332	29.04*	.0103	90.61	96.63	.01104	−25.25	113.78	88.53	− .0687	.2874	−140
−135	0.5511	28.80*	.0104	72.33	96.27	.01383	−23.99	113.10	89.11	− .0647	.2837	−135
−130	0.6949	28.51*	0.0104	58.21	95.91	0.01718	−22.73	112.43	89.70	−0.0609	0.2803	−130
−125	0.8692	28.15*	.0105	47.23	95.53	.02118	−21.47	111.76	90.29	− .0571	.2770	−125
−120	1.079	27.72*	.0105	38.60	95.15	.02591	−20.22	111.10	90.88	− .0534	.2738	−120
−115	1.329	27.21*	.0106	31.77	94.76	.03147	−18.98	110.45	91.47	− .0497	.2708	−115
−110	1.626	26.61*	.0106	26.33	94.37	.03798	−17.73	109.80	92.07	− .0461	.2680	−110
−105	1.976	25.90*	0.0106	21.96	93.97	0.04554	−16.48	109.15	92.67	−0.0425	0.2653	−105
−100	2.386	25.06*	.0107	18.43	93.56	.05427	−15.23	108.50	93.27	− .0390	.2627	−100
− 95	2.865	24.09*	.0107	15.54	93.14	.06433	−13.98	107.85	93.87	− .0356	.2602	− 95
− 90	3.417	22.96*	.0108	13.20	92.72	.07578	−12.73	107.20	94.47	− .0322	.2579	− 90
− 85	4.055	21.67*	.0108	11.26	92.29	.08884	−11.47	106.55	95.08	− .0288	.2556	− 85
− 80	4.787	20.18*	0.01090	9.650	91.85	0.1036	−10.22	105.90	95.68	−0.0255	0.2535	− 80
− 78	5.100	19.55*	.01091	9.086	91.67	.1101	− 9.72	105.64	95.92	− .0242	.2526	− 78
− 76	5.430	18.87*	.01093	8.561	91.49	.1168	− 9.21	105.37	96.16	− .0229	.2518	− 76
− 74	5.79	18.14*	.01095	8.072	91.31	.1239	− 8.70	105.10	96.40	− .0216	.2510	− 74
− 72	6.17	17.37*	.01097	7.616	91.13	.1313	− 8.20	104.84	96.64	− .0203	.2502	− 72
− 70	6.57	16.55*	0.01100	7.192	90.95	0.1391	− 7.69	104.57	96.88	−0.0190	0.2494	− 70
− 68	6.99	15.70*	.01102	6.795	90.77	.1472	− 7.19	104.31	97.12	− .0177	.2487	− 68
− 66	7.40	14.86*	.01104	6.426	90.58	.1556	− 6.68	104.04	97.36	− .0164	.2479	− 66
− 64	7.86	13.93*	.01106	6.079	90.39	.1645	− 6.17	103.77	97.60	− .0151	.2472	− 64
− 62	8.35	12.93*	.01109	5.755	90.21	.1738	− 5.67	103.51	97.84	− .0138	.2465	− 62
− 60	8.86	11.89*	0.01111	5.452	90.03	0.1834	− 5.16	103.24	98.08	−0.0126	0.2458	− 60
− 58	9.39	10.81*	.01113	5.166	89.84	.1936	− 4.65	102.97	98.32	− .0113	.2451	− 58
− 56	9.94	9.69*	.01115	4.900	89.65	.2041	− 4.13	102.69	98.56	− .0100	.2444	− 56
− 54	10.51	8.53*	.01118	4.650	89.46	.2151	− 3.61	102.41	98.80	− .0087	.2438	− 54
− 52	11.11	7.31*	.01120	4.415	89.27	.2265	− 3.09	102.13	99.04	− .0075	.2431	− 52
− 50	11.74	6.03*	0.01123	4.192	89.08	0.2386	− 2.58	101.86	99.28	−0.0062	0.2425	− 50
− 48	12.40	4.68*	.01125	3.986	88.88	.2509	− 2.06	101.58	99.52	− .0050	.2418	− 48
− 46	13.09	3.28*	.01128	3.793	88.68	.2636	− 1.54	101.30	99.76	− .0037	.2412	− 46
− 44	13.80	1.83*	.01130	3.611	88.49	.2769	− 1.02	101.02	100.00	− .0025	.2406	− 44
− 42	14.54	0.326*	.01133	3.440	88.30	.2907	− 0.51	100.74	100.23	− .0012	.2400	− 42
− 40	15.31	0.610	0.01135	3.279	88.10	0.3050	0.00	100.46	100.46	0.0000	0.2394	− 40
− 38	16.12	1.42	.01138	3.126	87.90	.3199	0.53	100.17	100.70	.0013	.2389	− 38
− 36	16.97	2.27	.01140	2.981	87.70	.3355	1.05	99.88	100.93	.0025	.2383	− 36
− 34	17.85	3.15	.01143	2.844	87.50	.3517	1.58	99.59	101.17	.0037	.2377	− 34
− 32	18.77	4.07	.01146	2.713	87.29	.3686	2.10	99.30	101.40	.0050	.2372	− 32
− 30	19.72	5.02	0.01148	2.590	87.09	0.3862	2.62	99.01	101.63	0.0062	0.2367	− 30
− 28	20.71	6.01	.01151	2.474	86.89	.4043	3.15	98.71	101.86	.0074	.2361	− 28
− 26	21.73	7.03	.01154	2.365	86.69	.4229	3.69	98.41	102.10	.0086	.2356	− 26
− 24	22.79	8.09	.01156	2.262	86.48	.4421	4.22	98.11	102.33	.0099	.2351	− 24
− 22	23.88	9.18	.01159	2.165	86.27	.4619	4.75	97.81	102.56	.0111	.2346	− 22
− 20	25.01	10.31	0.01162	2.074	86.06	0.4822	5.28	97.51	102.79	0.0123	0.2341	− 20
− 18	26.18	11.48	.01165	1.987	85.85	.5032	5.82	97.20	103.02	.0135	.2336	− 18
− 16	27.39	12.69	.01168	1.905	85.64	.5249	6.40	96.89	103.25	.0147	.2331	− 16
− 14	28.64	13.94	.01171	1.827	85.43	.5474	6.90	96.58	103.48	.0159	.2326	− 14
− 12	29.94	15.24	.01174	1.752	85.21	.5707	7.43	96.27	103.70	.0170	.2321	− 12
− 10	31.29	16.59	0.01177	1.681	84.99	0.5948	7.96	95.96	103.92	0.0182	0.2316	− 10
− 8	32.69	17.99	.01180	1.613	84.78	.6198	8.49	95.65	104.14	.0194	.2312	− 8
− 6	34.14	19.44	.01183	1.549	84.56	.6456	9.02	95.34	104.36	.0205	.2307	− 6
− 4	35.64	20.94	.01186	1.488	84.34	.6723	9.55	95.03	104.58	.0217	.2302	− 4
− 2	37.19	22.49	.01189	1.429	84.12	.6997	10.09	94.71	104.80	.0228	.2298	− 2
0	38.79	24.09	0.01192	1.373	83.90	0.7282	10.63	94.39	105.02	0.0240	0.2293	0
2	40.43	25.73	.01195	1.320	83.68	.7574	11.17	94.07	105.24	.0251	.2289	2
4	42.14	27.44	.01198	1.270	83.45	.7877	11.70	93.75	105.45	.0262	.2285	4
6	43.91	29.21	.01201	1.221	83.23	.8191	12.23	93.43	105.66	.0274	.2280	6
8	45.74	31.04	.01205	1.175	83.01	.8514	12.76	93.11	105.87	.0285	.2276	8
10	47.63	32.93	0.01208	1.130	82.78	0.8847	13.29	92.79	106.08	0.0296	0.2272	10
12	49.58	34.88	.01211	1.088	82.55	0.9191	13.82	92.47	106.29	.0307	.2268	12
14	51.59	36.89	.01215	1.048	82.32	0.9545	14.36	92.14	106.50	.0319	.2264	14
16	53.66	38.96	.01218	1.009	82.09	0.9911	14.90	91.81	106.71	.0330	.2260	16
18	55.79	41.09	.01222	0.9721	81.86	1.029	15.44	91.48	106.92	.0341	.2257	18
20	57.98	43.28	0.01225	0.9369	81.63	1.067	15.98	91.15	107.13	0.0352	0.2253	20
22	60.23	45.53	.01229	0.9032	81.39	1.107	16.52	90.81	107.33	.0364	.2249	22
24	62.55	47.85	.01232	0.8707	81.16	1.149	17.06	90.47	107.53	.0375	.2246	24
26	64.94	50.24	.01236	0.8398	80.92	1.191	17.61	90.12	107.73	.0379	.2242	26
28	67.40	52.70	.01239	0.8100	80.69	1.235	18.17	89.76	107.93	.0398	.2239	28

*Inches of mercury below one atmosphere.

TABLE 4—PROPERTIES OF REFRIGERANT 22, LIQUID AND SATURATED VAPOR (Contd)

TEMP (F)	PRESSURE (lb/sq in.)		VOLUME (cu ft/lb)		DENSITY (lb/cu ft)		ENTHALPY (Btu/lb)			ENTROPY (Btu/lb-R)		TEMP (F)
t	Absolute P	Gage p	Liquid v_f	Vapor v_g	Liquid $1/v_f$	Vapor $1/v_g$	Liquid h_f	Latent h_{fg}	Vapor h_g	Liquid s_f	Vapor s_g	t
30	69.93	55.23	0.01243	0.7816	80.45	1.280	18.74	89.39	108.13	0.0409	0.2235	30
32	72.53	57.83	.01247	.7543	80.21	1.326	19.32	89.01	108.33	.0421	.2232	32
34	75.21	60.51	.01250	.7283	79.97	1.373	19.90	88.62	108.52	.0433	.2228	34
36	77.97	63.27	.01254	.7032	79.73	1.422	20.49	88.22	108.71	.0445	.2225	36
38	80.81	66.11	.01258	.6791	79.49	1.473	21.09	87.81	108.90	.0457	.2222	38
40	83.72	69.02	0.01262	0.6559	79.25	1.525	21.70	87.39	109.09	0.0469	0.2218	40
42	86.69	71.99	.01266	.6339	79.00	1.578	22.29	86.98	109.27	.0481	.2215	42
44	89.74	75.04	.01270	.6126	78.76	1.632	22.90	86.55	109.45	.0493	.2211	44
46	92.88	78.18	.01274	.5922	78.51	1.689	23.50	86.13	109.63	.0505	.2208	46
48	96.10	81.40	.01278	.5726	78.26	1.747	24.11	85.69	109.80	.0516	.2205	48
50	99.40	84.70	0.01282	0.5537	78.02	1.806	24.73	85.25	109.98	0.0528	0.2201	50
52	102.8	88.10	.01286	.5355	77.77	1.868	25.34	84.80	110.14	.0540	.2198	52
54	106.2	91.5	.01290	.5184	77.51	1.929	25.95	84.35	110.30	.0552	.2194	54
56	109.8	95.1	.01294	.5014	77.26	1.995	26.58	83.89	110.47	.0564	.2191	56
58	113.5	98.8	.01299	.4849	77.01	2.062	27.22	83.41	110.63	.0576	.2188	58
60	117.2	102.5	0.01303	0.4695	76.75	2.130	27.83	82.95	110.78	0.0588	0.2185	60
62	121.0	106.3	.01307	.4546	76.50	2.200	28.46	82.47	110.93	.0600	.2181	62
64	124.9	110.2	.01312	.4403	76.24	2.271	29.09	81.99	111.08	.0612	.2178	64
66	128.9	114.2	.01316	.4264	75.98	2.346	29.72	81.50	111.22	.0624	.2175	66
68	133.0	118.3	.01320	.4129	75.72	2.422	30.35	81.00	111.35	.0636	.2172	68
70	137.2	122.5	0.01325	0.4000	75.46	2.500	30.99	80.50	111.49	0.0648	0.2168	70
72	141.5	126.8	.01330	.3875	75.20	2.581	31.65	79.98	111.63	.0661	.2165	72
74	145.9	131.2	.01334	.3754	74.94	2.664	32.29	79.46	111.75	.0673	.2162	74
76	150.4	135.7	.01339	.3638	74.68	2.749	32.94	78.94	111.88	.0684	.2158	76
78	155.0	140.3	.01344	.3526	74.41	2.836	33.61	78.40	112.01	.0696	.2155	78
80	159.7	145.0	0.01349	0.3417	74.15	2.926	34.27	77.86	112.13	0.0708	0.2151	80
82	164.5	149.8	.01353	.3313	73.89	3.019	34.92	77.32	112.24	.0720	.2148	82
84	169.4	154.7	.01358	.3212	73.63	3.113	35.60	76.76	112.36	.0732	.2144	84
86	174.5	159.8	.01363	.3113	73.36	3.213	36.28	76.19	112.47	.0744	.2140	86
88	179.6	164.9	.01368	.3019	73.09	3.313	36.94	75.63	112.57	.0756	.2137	88
90	184.8	170.1	0.01374	0.2928	72.81	3.415	37.61	75.06	112.67	0.0768	0.2133	90
92	190.1	175.4	.01379	.2841	72.53	3.520	38.28	74.48	112.76	.0780	.2130	92
94	195.6	180.9	.01384	.2755	72.24	3.630	38.97	73.88	112.85	.0792	.2126	94
96	201.2	186.5	.01390	.2672	71.95	3.742	39.65	73.28	112.93	.0803	.2122	96
98	206.8	192.1	.01396	.2594	71.65	3.855	40.32	72.69	113.00	.0815	.2119	98
100	212.6	197.9	0.01402	0.2517	71.35	3.973	40.98	72.08	113.06	0.0827	0.2115	100
102	218.5	203.8	.01408	.2443	71.05	4.094	41.65	71.47	113.12	.0839	.2111	102
104	224.6	209.9	.01414	.2370	70.74	4.220	42.32	70.84	113.16	.0851	.2107	104
106	230.7	216.0	.01420	.2301	70.42	4.347	42.98	70.22	113.20	.0862	.2104	106
108	237.0	222.3	.01426	.2233	70.11	4.479	43.66	69.58	113.24	.0874	.2100	108
110	243.4	228.7	0.01433	0.2167	69.78	4.614	44.35	68.94	113.29	0.0886	0.2096	110
112	249.9	235.2	.01440	.2104	69.45	4.752	45.04	68.30	113.34	.0898	.2093	112
114	256.6	241.9	.01447	.2043	69.12	4.896	45.74	67.64	113.38	.0909	.2089	114
116	263.4	248.7	.01454	.1983	68.78	5.043	46.44	66.98	113.42	.0921	.2085	116
118	270.3	255.6	.01461	.1926	68.44	5.192	47.14	66.32	113.46	.0933	.2081	118
120	277.3	262.6	0.01469	0.1871	68.10	5.345	47.85	65.67	113.52	0.0945	0.2078	120
122	284.4	269.7	.01475	.1825	67.75	5.48	48.60	64.97	113.57	.0959	.2076	122
124	291.7	277.0	.01483	.1772	67.40	5.64	49.40	64.21	113.61	.0973	.2073	124
126	299.1	284.4	.01491	.1724	67.05	5.80	50.20	63.45	113.65	.0986	.2070	126
128	306.5	291.8	.01498	.1675	66.70	5.97	50.80	62.89	113.69	.0997	.2067	128
130	314.0	299.3	0.01507	0.1629	66.35	6.14	51.50	62.21	113.71	0.1009	0.2064	130
132	321.8	307.1	.01515	.1585	66.00	6.31	52.30	61.44	113.74	.1022	.2061	132
134	329.9	315.2	.01523	.1538	65.65	6.50	53.10	60.67	113.77	.1035	.2057	134
136	338.3	323.6	.01532	.1492	65.25	6.70	53.80	59.99	113.79	.1046	.2053	136
138	347.0	332.3	.01541	.1449	64.85	6.90	54.60	59.20	113.80	.1059	.2050	138
140	356.0	341.3	0.01551	0.1408	64.45	7.10	55.30	58.51	113.81	0.1070	0.2046	140
142	365.0	350.3	.01561	.1368	64.05	7.31	56.10	57.70	113.80	.1084	.2043	142
144	374.1	359.4	.01571	.1330	63.65	7.52	56.90	56.89	113.79	.1096	.2039	144
146	383.3	368.6	.01581	.1292	63.25	7.74	57.70	56.08	113.78	.1110	.2036	146
148	392.6	377.9	.01591	.1253	62.85	7.98	58.40	55.36	113.76	.1120	.2031	148
150	401.9	387.2	0.01601	0.1216	62.45	8.22	59.20	54.54	113.74	0.1132	0.2027	150
152	411.3	396.6	.01612	.1179	62.02	8.48	60.00	53.71	113.71	.1145	.2023	152
154	420.8	406.1	.01625	.1141	61.58	8.76	60.80	52.87	113.67	.1156	.2018	154
156	430.3	415.6	.01636	.1105	61.13	9.05	61.60	52.02	113.62	.1168	.2013	156
158	439.8	425.1	.01648	.1070	60.67	9.35	62.50	51.06	113.56	.1181	.2008	158
160	449.3	434.6	0.01661	0.1035	60.20	9.66	63.50	50.00	113.50	0.1196	0.2003	160

TABLE 5—PROPERTIES OF REFRIGERANT 11, LIQUID AND SATURATED VAPOR

TEMP (F)	PRESSURE (lb/sq in.)		VOLUME (cu ft/lb)		DENSITY (lb/cu ft)		ENTHALPY (Btu/lb)			ENTROPY (Btu/lb-R)		TEMP (F)
t	Absolute P	Gage p	Liquid v_f	Vapor v_g	Liquid $1/v_f$	Vapor $1/v_g$	Liquid h_f	Latent h_{fg}	Vapor h_g	Liquid s_f	Vapor s_g	t
−40	0.7387	28.42*	0.00988	44.25	101.25	0.02260	0.00	87.53	87.53	0.0000	0.2086	−40
−38	0.7911	28.31*	.00989	41.51	101.10	.02409	0.40	87.37	87.77	.0009	.2081	−38
−36	0.8466	28.20*	.00991	38.97	100.96	.02566	0.79	87.22	88.01	.0019	.2077	−36
−34	0.9053	28.08*	.00992	36.60	100.81	.02732	1.19	87.06	88.25	.0028	.2073	−34
−32	0.9674	27.95*	.00993	34.41	100.66	.02906	1.59	86.91	88.49	.0038	.2070	−32
−30	1.0330	27.82*	0.00995	32.36	100.52	0.03090	1.99	86.75	88.74	0.0047	0.2066	−30
−28	1.1023	27.68*	.00996	30.47	100.37	.03282	2.38	86.60	88.98	.0056	.2062	−28
−26	1.1754	27.53*	.00998	28.70	100.22	.03485	2.78	86.44	89.22	.0065	.2058	−26
−24	1.253	27.37*	.00999	27.05	100.07	.03697	3.18	86.29	89.47	.0074	.2055	−24
−22	1.334	27.20*	.01001	25.51	99.92	.03920	3.58	86.13	89.71	.0084	.2051	−22
−20	1.419	27.03*	0.01002	24.08	99.78	0.04154	3.98	85.98	89.95	0.0093	0.2048	−20
−18	1.509	26.85*	.01004	22.74	99.63	.04398	4.38	85.82	90.20	.0102	.2045	−18
−16	1.604	26.65*	.01005	21.48	99.48	.04655	4.77	85.67	90.44	.0111	.2042	−16
−14	1.704	26.45*	.01007	20.31	99.33	.04923	5.17	85.51	90.69	.0120	.2038	−14
−12	1.809	26.24*	.01008	19.22	99.18	.05204	5.57	85.36	90.93	.0129	.2035	−12
−10	1.918	26.01*	0.01010	18.19	99.03	0.05497	5.98	85.20	91.19	0.0138	0.2032	−10
− 8	2.034	25.78*	.01011	17.23	98.88	.05804	6.38	85.05	91.43	.0147	.2029	+ 8
− 6	2.155	25.53*	.01013	16.33	98.72	.06123	6.77	84.89	91.67	.0155	.2027	− 6
− 4	2.282	25.27*	.01014	15.48	98.57	.06458	7.19	84.74	91.92	.0164	.2024	− 4
− 2	2.415	25.00*	.01016	14.69	98.42	.06807	7.59	84.58	92.17	.0173	.2021	− 2
0	2.554	24.72*	0.01018	13.95	98.27	0.07171	7.99	84.43	92.42	0.0182	0.2018	0
2	2.699	24.42*	.01019	13.25	98.11	.07550	8.39	84.27	92.67	.0191	.2016	2
4	2.852	24.11*	.01021	12.59	97.96	.07945	8.79	84.12	92.91	.0199	.2013	4
6	3.011	23.79*	.01022	11.97	97.81	.08356	9.19	83.96	93.16	.0208	.2011	6
8	3.178	23.45*	.01024	11.38	97.65	.08785	9.60	83.80	93.40	.0217	.2009	8
10	3.352	23.10*	0.01026	10.83	97.50	0.09233	10.00	83.65	93.65	0.0225	0.2006	10
12	3.533	22.73*	.01027	10.32	97.35	.09694	10.41	83.49	93.90	.0234	.2004	12
14	3.723	22.34*	.01029	9.828	97.19	.1018	10.81	83.33	94.15	.0243	.2002	14
16	3.921	21.94*	.01031	9.367	97.04	.1068	11.22	83.18	94.39	.0251	.2000	16
18	4.127	21.52*	.01032	8.932	96.88	.1120	11.62	83.02	94.64	.0260	.1998	18
20	4.342	21.08*	0.01034	8.521	96.72	0.1174	12.03	82.86	94.89	0.0268	0.1996	20
22	4.566	20.62*	.01036	8.133	96.57	.1230	12.43	82.70	95.14	.0277	.1994	22
24	4.799	20.15*	.01037	7.766	96.41	.1288	12.85	82.55	95.39	.0285	.1992	24
26	5.041	19.66*	.01039	7.418	96.25	.1348	13.26	82.39	95.65	.0294	.1990	26
28	5.294	19.14*	.01041	7.090	96.10	.1410	13.67	82.23	95.89	.0302	.1988	28
30	5.556	18.61*	0.01042	6.779	95.94	0.1475	14.07	82.07	96.14	0.0310	0.1986	30
32	5.829	18.05*	.01044	6.484	95.78	.1542	14.48	81.91	96.39	.0319	.1985	32
34	6.112	17.48*	.01046	6.205	95.62	.1612	14.89	81.75	96.64	.0327	.1983	34
36	6.407	16.88*	.01048	5.940	95.46	.1684	15.30	81.59	96.89	.0335	.1981	36
38	6.712	16.25*	.01049	5.688	95.30	.1758	15.71	81.42	97.14	.0344	.1980	38
40	7.030	15.61*	0.01051	5.450	95.14	0.1835	16.12	81.26	97.39	0.0352	0.1978	40
42	7.359	14.94*	.01053	5.223	94.98	.1914	16.54	81.10	97.63	.0360	.1977	42
44	7.700	14.24*	.01055	5.008	94.82	.1997	16.95	80.94	97.88	.0368	.1975	44
46	8.054	13.52*	.01056	4.804	94.66	.2082	17.36	80.77	98.13	.0377	.1974	46
48	8.420	12.78*	.01058	4.609	94.50	.2169	17.77	80.61	98.38	.0385	.1973	48
50	8.800	12.00*	0.01060	4.425	94.34	0.2260	18.19	80.44	98.63	0.0393	0.1971	50
52	9.193	11.20*	.01062	4.249	94.18	.2354	18.61	80.28	98.89	.0401	.1970	52
54	9.600	10.37*	.01064	4.081	94.02	.2450	19.03	80.11	99.14	.0409	.1969	54
56	10.02	9.52*	.01065	3.922	93.86	.2550	19.44	79.94	99.38	.0417	.1968	56
58	10.46	8.63*	.01067	3.770	93.69	.2653	19.86	79.77	99.63	.0425	.1966	58
60	10.91	7.71*	0.01069	3.625	93.53	0.2759	20.27	79.61	99.88	0.0434	0.1965	60
62	11.37	6.76*	.01071	3.487	93.37	.2868	20.69	79.44	100.13	.0442	.1964	62
64	11.85	5.79*	.01073	3.355	93.20	.2981	21.11	79.27	100.38	.0450	.1963	64
66	12.35	4.77*	.01075	3.229	93.04	.3097	21.53	79.10	100.62	.0458	.1962	66
68	12.86	3.73*	.01077	3.109	92.87	.3216	21.95	78.92	100.87	.0466	.1961	68
70	13.39	2.65*	0.01079	2.995	92.71	0.3339	22.37	78.75	101.12	0.0474	0.1960	70
72	13.94	1.54*	.01081	2.885	92.54	.3466	22.79	78.58	101.37	.0481	.1959	72
74	14.51	0.39*	.01083	2.781	92.38	.3596	23.21	78.41	101.61	.0489	.1959	74
76	15.09	0.39	.01084	2.681	92.21	.3730	23.63	78.23	101.86	.0497	.1958	76
78	15.69	0.99	.01086	2.585	92.05	.3868	24.06	78.05	102.12	.0505	.1957	78
80	16.31	1.61	0.01088	2.494	91.88	0.4010	24.48	77.88	102.36	0.0513	0.1956	80
82	16.94	2.25	.01090	2.406	91.71	.4156	24.91	77.70	102.61	.0521	.1955	82
84	17.60	2.91	.01092	2.322	91.54	.4306	25.33	77.52	102.85	.0529	.1955	84
86	18.28	3.58	.01094	2.242	91.38	.4461	25.76	77.34	103.10	.0537	.1954	86
88	18.97	4.28	.01096	2.165	91.21	.4619	26.18	77.16	103.34	.0544	.1953	88
90	19.69	4.99	0.01098	2.091	91.04	0.4781	26.61	76.98	103.59	0.0552	0.1953	90
92	20.43	5.73	.01100	2.021	90.87	.4949	27.03	76.80	103.83	.0560	.1952	92
94	21.19	6.49	.01103	1.953	90.70	.5121	27.46	76.62	104.08	.0568	.1951	94
96	21.97	7.27	.01105	1.888	90.53	.5297	27.89	76.43	104.32	.0575	.1951	96
98	22.77	8.08	.01107	1.826	90.36	.5477	28.32	76.25	104.56	.0583	.1950	98

*Inches of mercury below one atmosphere.

Courtesy of Allied Chemical Corporation

TABLE 5—PROPERTIES OF REFRIGERANT 11, LIQUID AND SATURATED VAPOR (Contd)

TEMP (F)	PRESSURE (lb/sq in.)		VOLUME (cu ft/lb)		DENSITY (lb/cu ft)		ENTHALPY (Btu/lb)			ENTROPY (Btu/lb-R)		TEMP (F)
t	Absolute P	Gage p	Liquid v_f	Vapor v_g	Liquid $1/v_f$	Vapor $1/v_g$	Liquid h_f	Latent h_{fg}	Vapor h_g	Liquid s_f	Vapor s_g	t
100	23.60	8.90	0.01109	1.766	90.19	0.5663	28.75	76.06	104.81	0.0591	0.1950	100
102	24.45	9.75	.01111	1.708	90.02	.5854	29.17	75.88	105.05	.0598	.1949	102
104	25.32	10.62	.01113	1.653	89.85	.6049	29.62	75.69	105.30	.0606	.1949	104
106	26.21	11.52	.01115	1.600	89.68	.6250	30.05	75.50	105.54	.0614	.1948	106
108	27.13	12.44	.01117	1.549	89.51	.6455	30.48	75.31	105.79	.0621	.1948	108
110	28.08	13.38	0.01119	1.500	89.34	0.6667	30.91	75.12	106.03	0.0629	0.1947	110
112	29.05	14.35	.01122	1.453	89.16	.6882	31.34	74.92	106.27	.0637	.1947	112
114	30.04	15.35	.01124	1.408	88.99	.7104	31.78	74.73	106.51	.0644	.1947	114
116	31.06	16.37	.01126	1.364	88.82	.7330	32.21	74.54	106.75	.0652	.1946	116
118	32.11	17.41	.01128	1.322	88.65	.7563	32.65	74.34	106.99	.0660	.1946	118
120	33.18	18.49	0.01130	1.282	88.47	0.7801	33.08	74.14	107.22	0.0667	0.1946	120
122	34.29	19.59	.01133	1.243	88.30	.8045	33.52	73.94	107.46	.0674	.1945	122
124	35.42	20.72	.01135	1.206	88.12	.8295	33.95	73.75	107.70	.0682	.1945	124
126	36.57	21.88	.01137	1.170	87.95	.8551	34.40	73.54	107.95	.0689	.1945	126
128	37.76	23.06	.01139	1.135	87.77	.8812	34.84	73.34	108.18	.0697	.1945	128
130	38.97	24.28	0.01142	1.101	87.60	0.9081	35.28	73.14	108.42	0.0704	0.1945	130
132	40.22	25.52	.01144	1.069	87.42	0.9355	35.72	72.94	108.65	.0712	.1944	132
134	41.49	26.80	.01146	1.038	87.25	0.9636	36.16	72.73	108.89	.0719	.1944	134
136	42.80	28.10	.01149	1.008	87.07	0.9923	36.60	72.52	109.12	.0726	.1944	136
138	44.13	29.44	.01151	0.9788	86.89	1.022	37.04	72.31	109.35	.0733	.1944	138
140	45.50	30.81	0.01154	0.9508	86.68	1.052	37.48	72.10	109.59	0.0741	0.1944	140
142	46.90	32.20	.01156	.9238	86.50	1.082	37.92	71.89	109.82	.0749	.1943	142
144	48.33	33.64	.01158	.8977	86.32	1.114	38.37	71.68	110.05	.0756	.1943	144
146	49.80	35.10	.01161	.8725	86.14	1.146	38.81	71.47	110.28	.0763	.1943	146
148	51.29	36.60	.01163	.8482	85.96	1.179	39.25	71.25	110.51	.0770	.1943	148
150	52.83	38.13	0.01166	0.8247	85.78	1.213	39.70	71.04	110.74	0.0778	0.1943	150
152	54.39	39.71	.01168	.8020	85.60	1.247	40.15	70.82	110.96	.0785	.1943	152
154	55.99	41.31	.01171	.7800	85.41	1.282	40.60	70.60	111.20	.0792	.1943	154
156	57.63	42.94	.01173	.7588	85.23	1.318	41.05	70.38	111.43	.0800	.1943	156
158	59.30	44.61	.01176	.7382	85.04	1.355	41.50	70.16	111.65	.0807	.1943	158
160	61.01	46.31	0.01178	0.7183	84.86	1.392	41.95	69.93	111.88	0.0814	0.1943	160

TABLE 6—PROPERTIES OF REFRIGERANT 113, LIQUID AND SATURATED VAPOR

TEMP (F)	PRESSURE (lb/sq in.)		VOLUME (cu ft/lb)		DENSITY (lb/cu ft)		ENTHALPY (Btu/lb)			ENTROPY (Btu/lb-R)		TEMP (F)
t	Absolute P	Gage p	Liquid vf	Vapor vg	Liquid 1/vf	Vapor 1/vg	Liquid hf	Latent hfg	Vapor hg	Liquid sf	Vapor sg	t
−30	0.2987	29.31*	0.00947	82.26	105.64	0.01216	1.97	72.68	74.65	0.0047	0.1738	−30
−28	0.3214	29.27*	.00948	76.81	105.50	.01302	2.36	72.57	74.93	.0056	.1737	−28
−26	0.3458	29.22*	.00949	71.71	105.37	.01395	2.76	72.45	75.21	.0065	.1736	−26
−24	0.3718	29.16*	.00950	66.99	105.23	.01493	3.16	72.33	75.49	.0074	.1735	−24
−22	0.3995	29.11*	.00952	62.63	105.09	.01597	3.56	72.21	75.77	.0083	.1733	−22
−20	0.4288	29.05*	0.00953	58.61	104.96	0.01706	3.96	72.09	76.05	0.0092	0.1732	−20
−18	0.4600	28.98*	.00954	54.88	104.82	.01822	4.36	71.98	76.34	.0101	.1731	−18
−16	0.4931	28.92*	.00955	51.42	104.68	.01945	4.76	71.86	76.62	.0110	.1730	−16
−14	0.5280	28.85*	.00957	48.23	104.54	.02074	5.16	71.74	76.90	.0119	.1729	−14
−12	0.5652	28.77*	.00958	45.25	104.40	.02210	5.56	71.62	77.18	.0128	.1729	−12
−10	0.6046	28.69*	0.00959	42.48	104.26	0.02354	5.96	71.51	77.47	0.0137	0.1728	−10
−8	0.6462	28.60*	.00960	39.92	104.12	.02505	6.36	71.39	77.75	.0146	.1727	−8
−6	0.6902	28.51*	.00962	37.54	103.98	.02664	6.76	71.27	78.03	.0155	.1726	−6
−4	0.7369	28.42*	.00963	35.31	103.84	.02832	7.17	71.15	78.32	.0164	.1726	−4
−2	0.7860	28.32*	.00964	33.24	103.70	.03009	7.57	71.03	78.60	.0173	.1725	−2
0	0.8377	28.21*	0.00966	31.31	103.56	0.03194	7.98	70.92	78.89	0.0182	0.1725	0
2	0.8924	28.10*	.00967	29.52	103.41	.03388	8.38	70.80	79.18	.0190	.1724	2
4	0.9503	27.99*	.00968	27.84	103.27	.03592	8.78	70.68	79.46	.0199	.1724	4
6	1.011	27.86*	.00970	26.27	103.13	.03806	9.19	70.56	79.75	.0208	.1723	6
8	1.075	27.73*	.00971	24.81	102.98	.04031	9.59	70.44	80.03	.0216	.1723	8
10	1.142	27.60*	0.00972	23.45	102.84	0.04265	10.00	70.32	80.32	0.0225	0.1723	10
12	1.213	27.45*	.00974	22.17	102.69	.04511	10.41	70.20	80.61	.0234	.1722	12
14	1.288	27.30*	.00975	20.97	102.55	.04769	10.81	70.08	80.89	.0242	.1722	14
16	1.366	27.14*	.00977	19.84	102.40	.05040	11.22	69.96	81.18	.0251	.1722	16
18	1.448	26.97*	.00978	18.79	102.25	.05322	11.62	69.84	81.46	.0259	.1722	18
20	1.534	26.80*	0.00979	17.81	102.10	0.05616	12.03	69.72	81.75	0.0268	0.1722	20
22	1.624	26.61*	.00981	16.89	101.96	.05922	12.44	69.60	82.04	.0276	.1721	22
24	1.719	26.42*	.00982	16.02	101.81	.06243	12.85	69.48	82.33	.0285	.1721	24
26	1.818	26.22*	.00984	15.20	101.66	.06579	13.26	69.36	82.62	.0293	.1722	26
28	1.922	26.01*	.00985	14.43	101.51	.06929	13.67	69.24	82.91	.0302	.1722	28
30	2.031	25.79*	0.00987	13.71	101.36	0.07294	14.08	69.12	83.20	0.0310	0.1722	30
32	2.145	25.55*	.00988	13.03	101.21	.07675	14.49	69.00	83.49	.0318	.1722	32
34	2.264	25.31*	.00990	12.39	101.06	.08071	14.91	68.87	83.78	.0327	.1722	34
36	2.388	25.06*	.00991	11.79	100.91	.08483	15.32	68.75	84.07	.0335	.1722	36
38	2.519	24.79*	.00993	11.22	100.76	.08913	15.74	68.62	84.36	.0343	.1722	38
40	2.655	24.52*	0.00994	10.68	100.60	.09361	16.16	68.50	84.65	0.0352	0.1723	40
42	2.797	24.23*	.00996	10.18	100.45	.09826	16.57	68.37	84.94	.0360	.1723	42
44	2.944	23.93*	.00997	9.703	100.30	.1031	16.99	68.25	85.24	.0368	.1723	44
46	3.098	23.61*	.00999	9.253	100.14	.1081	17.41	68.12	85.53	.0377	.1724	46
48	3.258	23.29*	.01000	8.830	99.99	.1133	17.82	68.00	85.82	.0385	.1724	48
50	3.427	22.94*	0.01002	8.426	99.83	0.1187	18.24	67.87	86.11	0.0393	0.1725	50
52	3.602	22.59*	.01003	8.044	99.68	.1243	18.66	67.74	86.40	.0401	.1726	52
54	3.784	22.22*	.01005	7.682	99.52	.1302	19.08	67.61	86.69	.0410	.1726	54
56	3.973	21.83*	.01006	7.342	99.37	.1362	19.50	67.48	86.98	.0418	.1727	56
58	4.170	21.43*	.01008	7.018	99.21	.1425	19.93	67.35	87.28	.0426	.1727	58
60	4.374	21.02*	0.01010	6.713	99.05	0.1490	20.35	67.22	87.57	0.0434	0.1728	60
62	4.586	20.59*	.01011	6.424	98.89	.1557	20.77	67.09	87.86	.0442	.1729	62
64	4.807	20.14*	.01013	6.149	98.73	.1626	21.19	66.96	88.15	.0450	.1729	64
66	5.036	19.67*	.01015	5.889	98.58	.1698	21.62	66.83	88.45	.0459	.1730	66
68	5.275	19.18*	.01016	5.640	98.42	.1773	22.05	66.69	88.74	.0467	.1731	68
70	5.523	18.68*	0.01018	5.404	98.26	0.1851	22.48	66.56	89.04	0.0475	0.1731	70
72	5.780	18.16*	.01019	5.180	98.10	.1931	22.90	66.43	89.33	.0483	.1732	72
74	6.042	17.62*	.01021	4.971	97.93	.2012	23.33	66.29	89.62	.0491	.1733	74
76	6.320	17.06*	.01023	4.769	97.77	.2097	23.76	66.16	89.92	.0499	.1734	76
78	6.607	16.47*	.01025	4.574	97.61	.2186	24.19	66.02	90.21	.0507	.1735	78
80	6.902	15.87*	0.01026	4.392	97.45	0.2277	24.63	65.88	90.51	0.0515	0.1736	80
82	7.208	15.25*	0.1028	4.218	97.28	.2371	25.06	65.74	90.80	.0523	.1737	82
84	7.527	14.60*	.01030	4.051	97.12	.2468	25.49	65.60	91.09	.0531	.1738	84
86	7.856	13.93*	.01031	3.893	96.96	.2569	25.93	65.46	91.39	.0539	.1739	86
88	8.194	13.24*	.01033	3.742	96.79	.2672	26.36	65.32	91.68	.0547	.1740	88
90	8.545	12.53*	0.01035	3.600	96.63	0.2778	26.80	65.18	91.98	0.0555	0.1741	90
92	8.908	11.79*	.01037	3.463	96.46	.2888	27.24	65.04	92.28	.0563	.1742	92
94	9.281	11.03*	.01039	3.333	96.30	.3001	27.67	64.90	92.57	.0571	.1743	94
96	9.668	10.24*	.01040	3.208	96.13	.3117	28.11	64.75	92.86	.0578	.1744	96
98	10.07	9.42*	.01042	3.089	95.96	.3237	28.55	64.60	93.15	.0586	.1745	98
100	10.48	8.59*	0.01044	2.976	95.79	0.3360	28.99	64.46	93.45	0.0594	0.1746	100
102	10.91	7.71*	.01046	2.867	95.63	.3488	29.44	64.31	93.75	.0602	.1747	102
104	11.35	6.82*	.01048	2.762	95.46	.3620	29.89	64.16	94.05	.0610	.1748	104
106	11.81	5.88*	.01050	2.662	95.29	.3756	30.33	64.01	94.34	.0618	.1750	106
108	12.28	4.93*	.01051	2.567	95.12	.3896	30.78	63.86	94.64	.0626	.1751	108

*Inches of mercury below one atmosphere.

Courtesy of E. I. du Pont de Nemours & Co.

TABLE 6—PROPERTIES OF REFRIGERANT 113, LIQUID AND SATURATED VAPOR (Contd)

TEMP (F)	PRESSURE (lb/sq in.)		VOLUME (cu ft/lb)		DENSITY (lb/cu ft)		ENTHALPY (Btu/lb)			ENTROPY (Btu/lb-R)		TEMP (F)
t	Absolute P	Gage p	Liquid v_f	Vapor v_g	Liquid $1/v_f$	Vapor $1/v_g$	Liquid h_f	Latent h_{fg}	Vapor h_g	Liquid s_f	Vapor s_g	t
110	12.76	3.95*	0.01053	2.477	94.95	0.4038	31.22	63.71	94.93	0.0634	0.1752	110
112	13.25	2.95*	.01055	2.391	94.78	.4182	31.67	63.56	95.23	.0641	.1753	112
114	13.76	1.91*	.01057	2.308	94.61	.4333	32.12	63.40	95.52	.0649	.1755	114
116	14.29	0.83*	.01059	2.228	94.43	.4489	32.57	63.25	95.82	.0657	.1756	116
118	14.84	0.14	.01061	2.151	94.26	.4649	33.03	63.09	96.12	.0665	.1757	118
120	15.40	0.70	0.01063	2.078	94.09	0.4813	33.48	62.93	96.41	0.0673	0.1758	120
122	15.97	1.27	.01065	2.008	93.92	.4981	33.93	62.78	96.71	.0680	.1760	122
124	16.56	1.86	.01067	1.941	93.74	.5153	34.38	62.62	97.00	.0688	.1761	124
126	17.17	2.47	.01069	1.876	93.57	.5330	34.83	62.46	97.29	.0696	.1763	126
128	17.80	3.10	.01071	1.814	93.39	.5514	35.29	62.30	97.59	.0704	.1764	128
130	18.45	3.74	0.01073	1.754	93.22	0.5702	35.75	62.14	97.89	0.0712	0.1765	130
132	19.11	4.41	.01075	1.697	93.04	.5894	36.21	61.97	98.18	.0719	.1767	132
134	19.79	5.09	.01077	1.642	92.86	.6091	36.67	61.80	98.47	.0727	.1768	134
136	20.48	5.78	.01079	1.590	92.69	.6290	37.13	61.64	98.77	.0735	.1770	136
138	21.19	6.49	.01081	1.540	92.51	.6494	37.59	61.48	99.06	.0742	.1771	138
140	21.93	7.23	0.01083	1.491	92.33	0.6707	38.05	61.31	99.36	0.0750	0.1773	140
142	22.69	7.99	.01085	1.444	92.15	.6926	38.52	61.13	99.65	.0758	.1774	142
144	23.47	8.77	.01087	1.399	91.98	.7150	38.98	60.96	99.94	.0765	.1775	144
146	24.27	9.57	.01089	1.355	91.80	.7379	39.45	60.79	100.24	.0773	.1777	146
148	25.09	10.39	.01092	1.313	91.62	.7615	39.92	60.61	100.53	.0781	.1778	148
150	25.93	11.23	0.01094	1.273	91.44	0.7856	40.38	60.44	100.82	0.0789	0.1780	150
152	26.79	12.09	.01096	1.234	91.25	.8102	40.85	60.27	101.11	.0796	.1782	152
154	27.67	12.97	.01098	1.197	91.07	.8353	41.32	60.09	101.41	.0804	.1783	154
156	28.56	13.86	.01100	1.162	90.89	.8608	41.79	59.91	101.70	.0812	.1785	156
158	29.48	14.78	.01102	1.128	90.71	.8689	42.26	59.73	101.99	.0819	.1786	158
160	30.44	15.74	0.01105	1.094	90.53	0.9141	42.74	59.55	102.29	0.0827	0.1788	160

TABLE 7—PROPERTIES OF REFRIGERANT 114, LIQUID AND SATURATED VAPOR

TEMP (F)	PRESSURE (lb/sq in.)		VOLUME (cu ft/lb)		DENSITY (lb/cu ft)		ENTHALPY (Btu/lb)			ENTROPY (Btu/lb-R)		TEMP (F)
t	Absolute P	Gage p	Liquid vf	Vapor vg	Liquid 1/vf	Vapor 1/vg	Liquid hf	Latent hfg	Vapor hg	Liquid sf	Vapor sg	t
−80	0.464	28.97*	0.009469	51.26	105.603	0.01951	− 8.73	69.17	60.44	−0.0227	0.1595	−80
−78	0.50	28.90*	.009484	49.83	105.398	.02007	− 8.30	69.01	60.71	− .0215	.1593	−78
−76	0.535	28.85*	.009506	44.87	105.193	.02229	− 7.87	68.85	60.98	− .0203	.1592	−76
−74	0.58	28.73*	.009513	41.69	105.058	.0240	− 7.44	68.69	61.25	− .0191	.1590	−74
−72	0.62	28.66*	.009528	38.92	104.924	.0257	− 7.01	68.53	61.52	− .0179	.1589	−72
−70	0.670	28.59*	0.009543	36.40	104.790	0.02747	− 6.57	68.36	61.79	−0.0167	0.1587	−70
−68	0.72	28.46*	.009558	34.20	104.624	.02925	− 6.14	68.20	62.06	− .0155	.1586	−68
−66	0.775	28.33*	.009573	31.74	104.458	.0315	− 5.71	68.04	62.33	− .0143	.1585	−66
−64	0.833	28.23*	.009589	29.77	104.292	.0336	− 5.28	67.88	62.60	− .0132	.1583	−64
−62	0.895	28.10*	.009604	27.79	104.126	.0360	− 4.84	67.72	62.88	− .0121	.1582	−62
−60	0.959	27.99*	0.009619	26.06	103.960	0.03838	− 4.40	67.56	63.16	−0.0110	0.1580	−60
−58	1.028	27.83*	.009635	24.55	103.792	.04075	− 3.96	67.39	63.43	− .0098	.1579	−58
−56	1.10	27.68*	.009651	22.94	103.622	.0436	− 3.52	67.22	63.70	− .0087	.1578	−56
−54	1.175	27.52*	.009666	21.50	103.452	.0465	− 3.08	67.05	63.97	− .0075	.1577	−54
−52	1.26	27.35*	.009682	20.16	103.282	.0496	− 2.64	66.89	64.25	− .0064	.1576	−52
−50	1.349	27.20*	0.009698	18.96	103.113	0.05274	− 2.20	66.73	64.53	−0.0054	0.1575	−50
−48	1.438	27.0 *	.009714	17.85	102.938	.0560	− 1.76	66.56	64.80	− .0043	.1574	−48
−46	1.535	26.8 *	.009731	16.80	102.766	.0595	− 1.32	66.39	65.07	− .0032	.1573	−46
−44	1.635	26.6 *	.009747	15.75	102.594	.0635	− 0.88	66.23	65.35	− .0021	.1572	−44
−42	1.745	26.4 *	.009764	14.87	102.422	.06725	− 0.44	66.07	65.63	− .0010	.1571	−42
−40	1.866	26.12*	0.00978	14.02	102.25	0.07132	0.00	65.91	65.91	0.0000	0.1571	−40
−38	1.990	25.87*	.00980	13.20	102.08	.07574	0.45	65.74	66.19	.0011	.1570	−38
−36	2.121	25.60*	.00981	12.44	101.90	.08038	0.91	65.56	66.47	.0021	.1569	−36
−34	2.259	25.32*	.00983	11.73	101.72	.08524	1.36	65.38	66.74	.0032	.1568	−34
−32	2.404	25.03*	.00985	11.07	101.55	.09034	1.81	65.21	67.02	.0042	.1567	−32
−30	2.557	24.72*	0.00987	10.45	101.37	0.09568	2.27	65.03	67.30	0.0053	0.1567	−30
−28	2.718	24.39*	.00988	9.877	101.19	.1013	2.72	64.86	67.58	.0063	.1567	−28
−26	2.887	24.04*	.00990	9.338	101.01	.1071	3.17	64.68	67.86	.0074	.1566	−26
−24	3.064	23.68*	.00992	8.833	100.83	.1132	3.63	64.51	68.14	.0084	.1565	−24
−22	3.249	23.31*	.00994	8.362	100.65	.1196	4.08	64.34	68.42	.0095	.1565	−22
−20	3.444	22.91*	0.00995	7.921	100.47	0.1263	4.54	64.16	68.70	0.0105	0.1565	−20
−18	3.648	22.49*	.00997	7.508	100.29	.1332	4.99	63.99	68.98	.0116	.1565	−18
−16	3.862	22.06*	.00999	7.121	100.11	.1404	5.44	63.81	69.26	.0126	.1564	−16
−14	4.085	21.61*	.01001	6.757	99.92	.1480	5.90	63.64	69.54	.0136	.1564	−14
−12	4.319	21.13*	.01003	6.416	99.74	.1559	6.35	63.46	69.82	.0146	.1564	−12
−10	4.564	20.63*	0.01005	6.095	99.56	0.1641	6.81	63.29	70.10	0.0157	0.1564	−10
− 8	4.819	20.11*	.01006	5.794	99.37	.1726	7.26	63.11	70.38	.0167	.1564	− 8
− 6	5.086	19.57*	.01008	5.510	99.19	.1815	7.72	62.94	70.66	.0177	.1564	− 6
− 4	5.365	19.00*	.01010	5.244	99.00	.1907	8.18	62.77	70.94	.0187	.1564	− 4
− 2	5.655	18.41*	.01012	4.992	98.81	.2003	8.63	62.59	71.22	.0197	.1565	− 2
0	5.958	17.79*	0.01014	4.756	98.62	0.2103	9.09	62.42	71.50	0.0207	0.1565	0
2	6.274	17.15*	.01016	4.533	98.44	.2206	9.54	62.24	71.78	.0217	.1565	2
4	6.603	16.48*	.01018	4.322	98.25	.2314	10.00	62.07	72.07	.0227	.1565	4
6	6.945	15.78*	.01020	4.123	98.06	.2425	10.46	61.89	72.35	.0236	.1566	6
8	7.301	15.06*	.01022	3.935	97.87	.2541	10.91	61.71	72.63	.0246	.1566	8
10	7.671	14.31*	0.01024	3.758	97.68	0.2661	11.37	61.54	72.91	0.0256	0.1566	10
12	8.057	13.52*	.01026	3.591	97.48	.2785	11.83	61.36	73.19	.0266	.1567	12
14	8.457	12.71*	.01028	3.432	97.29	.2914	12.29	61.19	73.47	.0275	.1567	14
16	8.873	11.86*	.01030	3.282	97.10	.3047	12.75	61.01	73.75	.0285	.1568	16
18	9.305	10.98*	.01032	3.140	96.90	.3185	13.20	60.83	74.04	.0295	.1568	18
20	9.753	10.07*	0.01034	3.005	96.71	0.3328	13.66	60.65	74.32	0.0304	0.1569	20
22	10.22	9.12*	.01036	2.877	96.51	.3476	14.12	60.48	74.60	.0314	.1569	22
24	10.70	8.14*	.01038	2.756	96.32	.3629	14.58	60.30	74.88	.0323	.1570	24
26	11.20	7.12*	.01040	2.641	96.12	.3786	15.05	60.12	75.17	.0333	.1571	26
28	11.72	6.07*	.01043	2.532	95.92	.3949	15.51	59.94	75.45	.0342	.1571	28
30	12.25	4.99*	0.01045	2.429	95.73	0.4118	15.97	59.76	75.73	0.0352	0.1572	30
32	12.81	3.85*	.01047	2.330	95.53	.4292	16.43	59.58	76.01	.0361	.1573	32
34	13.38	2.69*	.01049	2.236	95.33	.4472	16.89	59.40	76.29	.0370	.1574	32
36	13.98	1.47*	.01051	2.147	95.13	.4658	17.36	59.22	76.58	.0380	.1575	36
38	14.59	0.22*	.01053	2.062	94.93	.4849	17.82	59.04	76.86	.0389	.1575	38
40	15.22	0.52	0.01056	1.982	94.73	0.5047	18.28	58.86	77.14	0.0398	0.1576	40
42	15.88	1.18	.01058	1.905	94.52	.5250	18.75	58.67	77.42	.0408	.1577	42
44	16.56	1.86	.01060	1.832	94.32	.5460	19.21	58.49	77.70	.0417	.1578	44
46	17.26	2.56	.01063	1.762	94.12	.5676	19.68	58.31	77.99	.0426	.1579	46
48	17.98	3.28	.01065	1.695	93.91	.5899	20.14	58.12	78.27	.0435	.1580	48
50	18.73	4.03	0.01067	1.632	93.71	0.6129	20.61	57.94	78.55	0.0444	0.1581	50
52	19.50	4.80	.01070	1.571	93.50	.6365	21.08	57.75	78.83	.0453	.1582	52
54	20.29	5.59	.01072	1.513	93.30	.6609	21.54	57.56	79.11	.0463	.1583	54
56	21.11	6.41	.01074	1.458	93.09	.6859	22.01	57.38	79.39	.0472	.1584	56
58	21.96	7.26	.01077	1.405	92.88	.7117	22.48	57.19	79.67	.0481	.1585	58

*Inches of mercury below one atmosphere.

Courtesy of E. I. du Pont de Nemours & Co.

TABLE 7—PROPERTIES OF REFRIGERANT 114, LIQUID AND SATURATED VAPOR (Contd)

TEMP (F)	PRESSURE (lb/sq in.)		VOLUME (cu ft/lb)		DENSITY (lb/cu ft)		ENTHALPY (Btu/lb)			ENTROPY (Btu/lb-R)		TEMP (F)
t	Absolute P	Gage p	Liquid v_f	Vapor v_g	Liquid $1/v_f$	Vapor $1/v_g$	Liquid h_f	Latent h_{fg}	Vapor h_g	Liquid s_f	Vapor s_g	t
60	22.83	8.13	0.01079	1.354	92.68	0.7383	22.95	57.00	79.95	0.0490	0.1587	60
62	23.72	9.02	.01082	1.306	92.47	.7655	23.42	56.81	80.23	.0499	.1588	62
64	24.64	9.94	.01084	1.260	92.26	.7936	23.89	56.62	80.51	.0508	.1589	64
66	25.59	10.89	.01086	1.216	92.05	.8225	24.36	56.43	80.79	.0517	.1590	66
68	26.57	11.87	.01089	1.174	91.84	.8521	24.83	56.24	81.07	.0526	.1591	68
70	27.57	12.87	0.01091	1.133	91.63	0.8826	25.30	56.04	81.35	0.0534	0.1593	70
72	28.61	13.91	.01094	1.094	91.41	0.9140	25.78	55.85	81.62	.0543	.1594	72
74	29.67	14.97	.01097	1.057	91.20	0.9462	26.25	55.65	81.90	.0552	.1595	74
76	30.76	16.06	.01099	1.021	90.99	0.9793	26.73	55.45	82.18	.0561	.1596	76
78	31.88	17.18	.01102	0.9869	90.77	1.013	27.20	55.26	82.46	.0570	.1597	78
80	33.04	18.34	0.01104	0.9541	90.56	1.048	27.68	55.06	82.73	0.0579	0.1599	80
82	34.22	19.52	.01107	.9226	90.34	1.084	28.15	54.86	83.01	.0587	.1600	82
84	35.44	20.74	.01110	.8923	90.13	1.121	28.63	54.66	83.29	.0596	.1601	84
86	36.69	21.99	.01112	.8632	89.91	1.159	29.11	54.46	83.56	.0605	.1603	86
88	37.97	23.27	.01115	.8353	89.69	1.197	29.58	54.25	83.84	.0613	.1604	88
90	39.29	24.59	0.01118	0.8084	89.47	1.237	30.06	54.05	84.11	0.0622	0.1605	90
92	40.64	25.94	.01120	.7827	89.25	1.278	30.54	53.84	84.39	.0631	.1607	92
94	42.02	27.32	.01123	.7579	89.03	1.320	31.02	53.64	84.66	.0639	.1608	94
96	43.44	28.74	.01126	.7340	88.81	1.362	31.50	53.43	84.93	.0648	.1609	96
98	44.89	30.19	.01129	.7111	88.59	1.406	31.99	53.22	85.21	.0656	.1611	98
100	46.39	31.69	0.01132	0.6890	88.37	1.452	32.47	53.01	85.48	0.0665	0.1612	100
102	47.92	33.22	.01135	.6677	88.15	1.498	32.95	52.80	85.75	.0674	.1614	102
104	49.48	34.78	.01137	.6472	87.93	1.545	33.43	52.59	86.02	.0682	.1615	104
106	51.09	36.39	.01140	.6274	87.70	1.594	33.92	52.37	86.29	.0691	.1617	106
108	52.73	38.03	.01143	.6084	87.48	1.644	34.40	52.16	86.56	.0699	.1618	108
110	54.41	39.71	0.01146	0.5901	87.25	1.695	34.89	51.94	86.83	0.0708	0.1619	110
112	56.14	41.44	.01149	.5724	87.01	1.747	35.38	51.72	87.09	.0716	.1621	112
114	57.90	43.20	.01152	.5554	86.77	1.801	35.87	51.50	87.36	.0725	.1622	114
116	59.70	45.00	.01156	.5389	86.54	1.856	36.35	51.28	87.63	.0733	.1624	116
118	61.55	46.85	.01159	.5230	86.31	1.912	36.84	51.05	87.89	.0741	.1625	118
120	63.44	48.74	0.01162	0.5077	86.08	1.970	37.33	50.83	88.16	0.0750	0.1627	120
122	65.37	50.67	.01165	.4929	85.85	2.029	37.83	50.60	88.42	.0758	.1628	122
124	67.35	52.65	.01168	.4787	85.61	2.089	38.32	50.37	88.69	.0767	.1630	124
126	69.37	54.67	.01171	.4649	85.37	2.151	38.81	50.14	88.95	.0775	.1631	126
128	71.43	56.73	.01175	.4515	85.13	2.215	39.31	49.91	89.21	.0783	.1633	128
130	73.54	58.84	0.01178	0.4387	84.89	2.280	39.80	49.67	89.47	0.0792	0.1634	130
132	75.69	60.99	.01181	.4262	84.65	2.346	40.30	49.44	89.73	.0800	.1635	132
134	77.90	63.20	.01185	.4142	84.41	2.414	40.80	49.20	89.99	.0808	.1637	134
136	80.15	65.45	.01188	.4025	84.16	2.484	41.29	48.96	90.25	.0816	.1638	136
138	82.44	67.74	.01192	.3912	83.91	2.556	41.79	48.72	90.51	.0825	.1640	138
140	84.79	70.09	0.01195	0.3803	83.66	2.629	42.29	48.47	90.76	0.0833	0.1641	140

CHAPTER 2. BRINES

This chapter provides information to guide the engineer in the selection of brines, and includes the properties of the commonly used brines.

At temperatures above 32 F, water is the most commonly used heat transfer medium for conveying a refrigeration load to an evaporator. At temperatures below 32 F, brines are used. They may be:

1. An aqueous solution of inorganic salts, i.e. sodium chloride or calcium chloride. For low temperatures, a eutectic mixture may be used.

2. An aqueous solution of organic compounds, i.e. alcohols or glycols. Ethanol water, methanol water, ethylene glycol and propylene glycol are examples.

3. Chlorinated or fluorinated hydrocarbons and halocarbons.

A solution of any salt in water, or in general any solution, has a certain concentration at which the freezing point is at a minimum. A solution of such a concentration is called a eutectic mixture. The temperature at which it freezes is the eutectic temperature. A solution at any other concentration starts to freeze at a higher temperature. *Figure 11* illustrates the relationship between the freezing point (temperature) of a brine mixture and the percent of solute in the mixture (concentration). *Chart 18* covers a range of temperatures wide enough to reveal the two freezing point curves.

When the temperature of a brine with a concentration below the eutectic falls below the freezing point, ice crystals form and the concentration of the residual solution increases until at the eutectic temperature the remaining solution reaches a eutectic concentration. Below this temperature the mixture solidifies to form a mechanical mixture of ice and frozen eutectic solution.

When the temperature of a brine with a concentration above the eutectic falls below the freezing point, salt crystallizes out and the concentration of the residual solution decreases until at the eutectic temperature the remaining solution reaches a eutectic concentration. Below this temperature the mixture solidifies to form a mechanical mixture of salt and frozen eutectic solution.

This chapter includes a discussion of these brines, also tables and charts indicating properties.

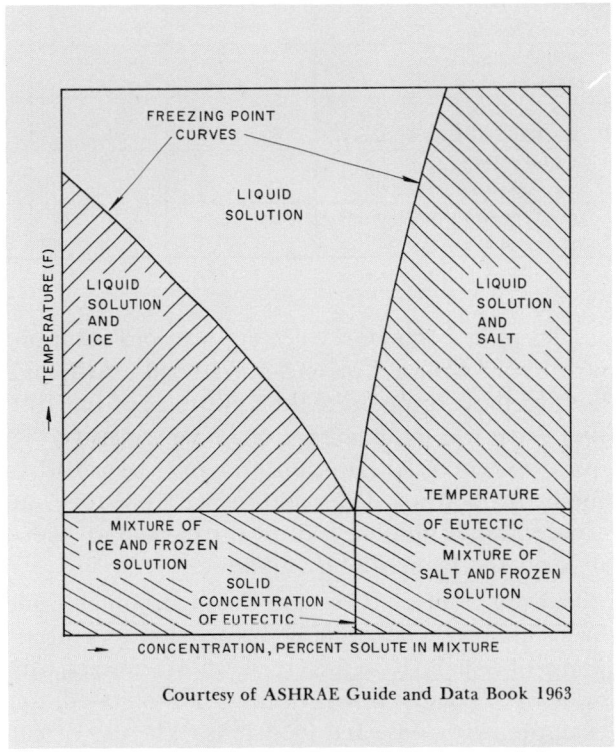

Courtesy of ASHRAE Guide and Data Book 1963

Fig. 11 — Brine Mixture

BRINE SELECTION

The selection of a brine is based upon a consideration of the following factors:

1. *Freezing Point* — Brine must be suitable for the lowest operating temperature.

2. *Application* — When using an open piping system, the possibility of product contamination by the brine should be checked.

3. *Cost* — The initial charge and quantity of make-up required are factors in the determination of costs.

4. *Safety* — Toxicity and flammability of brine.

5. *Thermal Performance* — Viscosity, specific gravity, specific heat and thermal conductivity are utilized to determine thermal performance.

6. *Suitability* — Piping and system equipment material require a stable and relatively corrosive-free brine.

7. *Codes* — Brine must be acceptable by codes, ordinances, regulatory agencies and insuror.

TABLE 8—TYPICAL BRINE APPLICATIONS

Application	Sodium Chloride	Calcium Chloride	Ethylene Glycol	Propylene Glycol	Methanol Water	Ethanol Water	Chlorinated or Fluorinated Hydrocarbons
Breweries				X			
Chemical Plants	X	X	X		X	X	X
Dairies	X	X		X			
Food Freezing	X	X		X		X	
Heat Pumps		X	X				X
Meat Packing	X	X					
Preheat Coils (Air Conditioning Systems)			X				
Skating Rinks		X	X		X		
Snow Melting	X	X	X				
Low Temperature (Special)		X	X				X
Ice Cream		X		X			

The most common brines are aqueous solutions of calcium chloride or sodium chloride. Although both of these brines have the advantage of low cost, they have the disadvantage of being corrosive. To overcome corrosion, an inhibitor may be added to the brine. Sodium dichromate is a satisfactory and economical inhibitor. Sodium hydroxide is added to keep the brine slightly alkaline.

Sodium chloride is cheaper than calcium chloride brine; however, it cannot be used below its eutectic point of −6 F. It is preferred where contact with calcium chloride brine cannot be tolerated, for example, with unsealed foodstuffs. The use of calcium chloride of commercial grade is not satisfactory below −40 F.

Systems using aqueous solutions of alcohol or glycol are more susceptible to leakage than those using salts. A disadvantage of alcohol is its flammability. It is utilized mainly in industrial processes where similar hazards already exist, and in the same temperature range as the salts (down to −40 F). The toxicity of methanol water (wood alcohol) is a disadvantage. Conversely, the nontoxicity of ethanol water (denatured grain alcohol) is an advantage.

Corrosion inhibitors should be used with alcohol type brines as recommended by the manufacturer of the alcohol.

Aqueous solutions of glycol are utilized mainly in commercial applications as opposed to industrial processes. Ethylene and propylene glycol possess equal corrosiveness which an inhibitor can neutralize. Galvanized surfaces are particularly prone to attack by the glycols and should be avoided.

An inhibitor and potable water are recommended for making up glycol brines. The glycol manufacturer should be consulted for inhibitor recommendations. Some manufacturers have a brine sample analysis service to assist in maintaining a satisfactory brine condition in the system. Heat transfer glycols are available with nonoily inhibitors which do not penalize heat transfer qualities (sodium nitrite or borox).

Glycols can be used as heat transfer media at relatively high temperatures. With stabilizers, glycol oxidization in air at high temperatures is eliminated for all practical purposes.

Ethylene glycol is more toxic than propylene glycol, but less toxic than methanol water. Propylene glycol is preferred to ethylene glycol in food freezing for example.

Chlorinated and fluorinated hydrocarbons are expensive and are used in very low temperature work (below −40 F).

Table 8 presents typical applications for the various brines.

Load, brine quantity and temperature rise are all related to each other so that, when any two are known, the third may be found by the formula:

$$\text{Load (tons)} = \frac{\text{gpm} \times \text{temp rise (F)} \times \text{sp gr} \times C_p}{24}$$

where:
sp gr = specific gravity of brine
C_p = specific heat of brine (Btu/lb-F)

PIPING

All materials in the piping system including flange gaskets, valve seats and packing, pump seals and other specialities must be compatible with the brine. Copper tubing (except for the salt brines) and standard steel pipe are suitable for general use.

The pump rating and motor horsepower should be based on the particular brine used and the actual operating temperature.

FRICTION LOSS

To determine the friction loss in a brine piping system, the engineer should first calculate the loss as if water were being used. A multiplier is then used to convert the calculated loss to the actual loss for the brine system. The multiplier is calculated as follows:

$$\text{Multiplier} = \text{sp gr} \times \frac{f_b}{f_w}$$

where:
sp gr = specific gravity of brine
f_b = friction factor for the brine
f_w = friction factor for water at the brine velocity

Friction factor is determined from the Reynolds number. The Reynolds number is:

$$\text{Re} = \frac{7740 \times d \times v \times \text{sp gr}}{\mu'}$$

where:
d = inside pipe diameter (in.)
v = brine velocity (ft/sec)
sp gr = specific gravity of brine = $\dfrac{\text{lb/cu ft}}{62.5}$
μ' = viscosity (centipoises) = $\dfrac{\text{absolute viscosity, lb/(hr) (ft)}}{2.42}$

Example 1 illustrates the use of the multiplier to determine the brine friction loss thru a heat transfer coil.

Example 1 — Friction Loss Multiplier

Given:
A ⅝ in. copper tube coil with a circuit water velocity of 4.29 ft/sec and a pressure drop of 7.5 psi.
Mean water temperature = 55 F.

Find:
Friction loss multiplier and pressure drop when using ethylene glycol at a mean brine temperature of 92.5 F and 41% solution by weight at the same circuit liquid velocity.

Solution:
Refer to *Chart 1*.

$$\frac{\epsilon}{d} = \frac{.00006}{.575} = .000104$$

where:
ϵ = absolute roughness of drawn tubing
d = inside diameter of ⅝ in. copper tubing

Refer to *Chart 19*.
Specific gravity of ethylene glycol at a mean brine temperature of 92.5 F and 41% solution by weight is 1.05.

Refer to *Chart 18*.
Viscosity of ethylene glycol at the same conditions equals 2.1 centipoises.

$$\text{Re} = \frac{7740 \times .575 \times 4.29 \times 1.05}{2.1} = 9520 = 9.52 \times 10^3$$

Refer to *Chart 1*.
For a Reynolds number of 9.52×10^3 and a relative roughness of .000104, the chart indicates friction factor $f_b = .031$.

Specific gravity of fresh water at a mean temperature of 55 F = 1.00

Refer to *Chart 28*.
Viscosity of fresh water at a mean temperature of 55 F = 1.2 centipoises.

$$\text{Re (water)} = \frac{7740 \times .575 \times 4.29 \times 1.00}{1.2} = 15,900 = 1.59 \times 10^4$$

Refer again to *Chart 1*.
For a Reynolds number of 1.59×10^4 and a relative roughness of .000104, the chart indicates a friction factor $f_w = .027$.

$$\text{Friction multiplier} = \text{sp gr (brine)} \times \frac{f_b}{f_w}$$
$$= 1.05 \times \frac{.031}{.027} = 1.21$$
$$\text{Brine friction loss} = 1.21 \times 7.5 \text{ psi} = 9.08 \text{ psi or}$$
$$= \frac{9.08 \times 2.31}{1.05} = 20.0 \text{ ft brine}$$

PUMP BRAKE HORSEPOWER

To determine the horsepower required by a pump with brine, the following formula may be used:

$$\text{bhp} = \frac{\text{gpm} \times \text{total head (ft brine)} \times \text{sp gr}}{3960 \times \text{eff}}$$

where:
gpm = gallons/min. of brine
total head = total pump head (ft brine)
sp gr = specific gravity of brine
eff = pump efficiency

BRINE PROPERTIES

Specific gravity, viscosity, conductivity, specific heat, concentration, and freezing and boiling points are important factors in the consideration of liquids other than water suitable for cooling and heating purposes. High values of specific gravity, conductivity and specific heat, and low values of viscosity, promote a high rate of heat transfer. High values of specific gravity and viscosity result in high pumping head and consequently high pumping costs. High specific heats are desirable in that they reduce the quantity of liquid required to be circulated or stored for a given duty. Low viscosities are desirable from a standpoint of both rate of heat transfer and low pumping costs. They are particularly desirable at the lower temperatures where the viscosity increases.

Table 9 is a tabulation of the various brines covered in this chapter, giving the properties of these brines at different temperatures and suitable concentrations. *Charts 2 to 28* present the viscosity, specific gravity, specific heat and thermal conduc-

TABLE 9—BRINE PROPERTIES

Temp. (F)	Brine	Solution (by wt) (%)	Density (lb/cu ft)	Specific Heat (Btu/lb-F)	Thermal Cond. (Btu/hr -sq ft-F/ft)	Viscosity (centi- poises)	Freezing Point (F)	Boiling Point (F)	Gpm/ton /10 deg rise	h_b *	V_b †	Relative Cost per Gal. of Solution
30	Sodium Chloride	12	68.2	.86	.28	2.2	17.5	215.	2.55	941	1.61	1
	Calcium Chloride	12	69.2	.83	.32	2.4	19.0	213.	2.62	971	1.78	3
	Methanol Water	15	61.5	1.00	.28	3.2	13.5	187.	2.45	781	2.63	13
	Ethanol Water	20	61.0	1.04	.27	5.5	12.0	189.	2.37	621	4.60	20
	Ethylene Glycol	25	64.7	.92	.30	3.7	12.9	217.	2.52	775	2.92	42
	Propylene Glycol	30	64.5	.94	.26	8.0	13.0	216.	2.47	525	6.35	43
15	Sodium Chloride	21	72.8	.80	.25	4.2	1.0	216.	2.57	693	2.90	1
	Calcium Chloride	20	74.8	.72	.31	4.8	1.0	214.	2.77	730	3.28	5
	Methanol Water	22	60.4	.97	.26	5.3	4.5	182.	2.56	599	4.44	19
	Ethanol Water	25	61.0	1.02	.25	8.2	4.5	187.	2.41	504	6.85	25
	Ethylene Glycol	35	66.0	.86	.28	6.8	0.0	219.	2.65	576	5.25	60
	Propylene Glycol	40	65.3	.89	.24	20.0	− 4.2	218.	2.58	103	‡	58
−5	Calcium Chloride	25	78.4	.67	.29	10.3	−21.0	215.	2.85	513	6.75	6
	Methanol Water	35	60.0	.89	.23	9.9	−22.0	176.	2.82	98	8.40	30
	Ethanol Water	36	60.6	.95	.22	13.5	−16.0	183.	2.62	97	‡	35
	Ethylene Glycol	45	67.4	.79	.25	17.2	−15.5	223.	2.82	103	‡	78
	Propylene Glycol	50	66.5	.83	.23	80.0	−29.0	222.	2.72	98	‡	75
−30	Calcium Chloride	30	82.1	.63	.28	27.8	−47.0	216.	2.90	110	‡	8
	Methanol Water	45	60.0	.80	.22	18.0	−45.0	171.	3.13	91	‡	39
	Ethanol Water	52	59.5	.81	.19	20.2	−50.0	179.	3.11	83	‡	50
	Ethylene Glycol	55	69.0	.73	.22	75.0	−43.0	227.	2.98	93	‡	97
	Propylene Glycol	60	67.2	.77	.21	700.0	−55.0	227.	2.90	91	‡	90

*h_b = coefficient of heat transfer between brine and surface (Btu/hr-sq ft-F), at 7 fps velocity for .554 in. ID tubing.

†V_b = minimum brine velocity (ft/sec), at Re = 3500 for .554 in. ID tubing.

‡Above 10 ft/sec.

tivity of the brines for various mean brine temperatures and compositions.

Note that specific gravity for propylene glycol (*Chart 23*) in the composition range of 50% to 100% (same mean brine temperature, F) is the same for two compositions. Specific gravity alone, therefore, is not a reliable method of determining the solution composition of this brine.

CHART 1—FRICTION FACTORS FOR COMMERCIAL PIPE

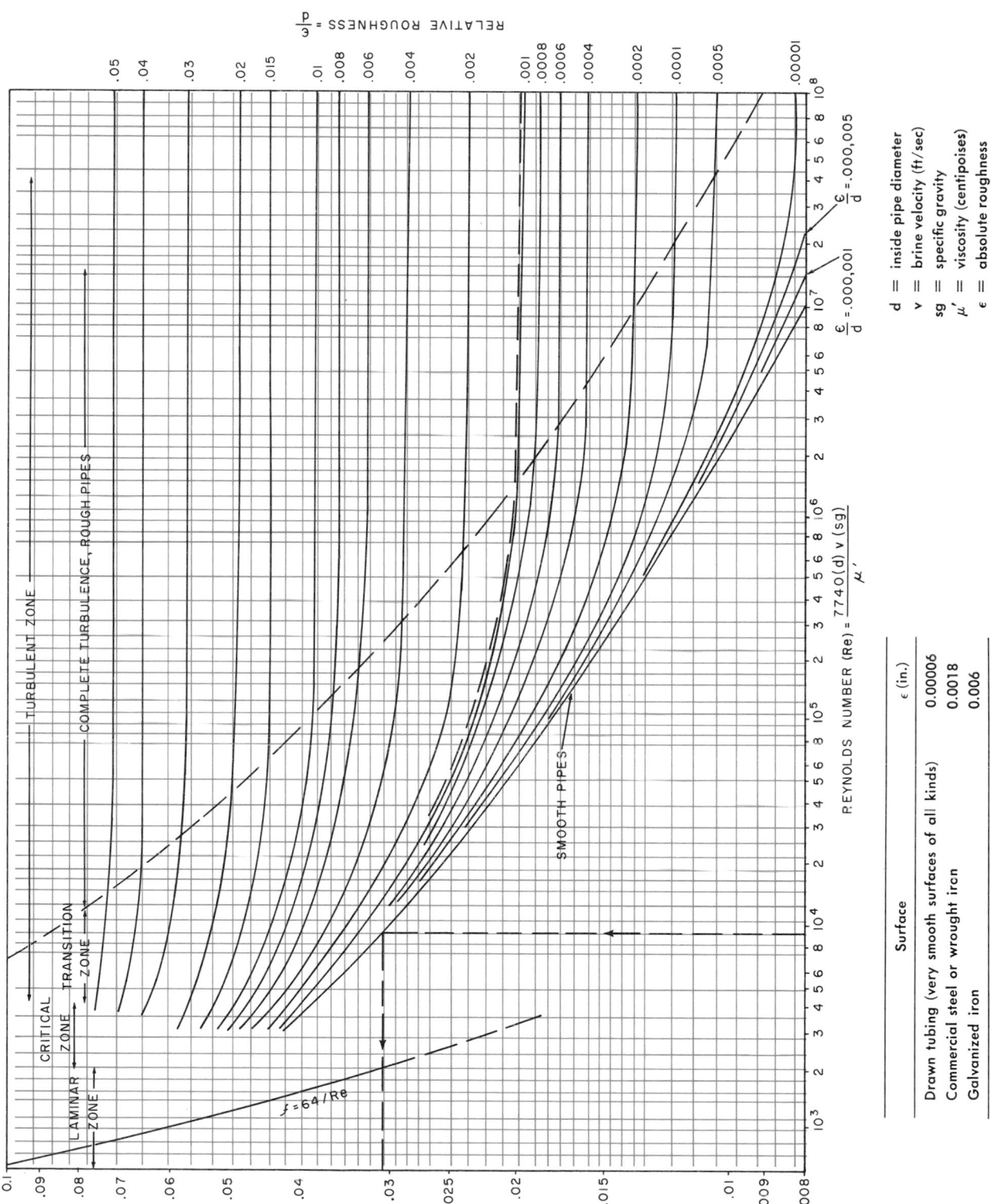

Surface	ϵ (in.)
Drawn tubing (very smooth surfaces of all kinds)	0.00006
Commercial steel or wrought iron	0.0018
Galvanized iron	0.006

d = inside pipe diameter
v = brine velocity (ft/sec)
sg = specific gravity
μ' = viscosity (centipoises)
ϵ = absolute roughness

From *Friction Factors For Pipe Flow*, by L. F. Moody, with permission of the publisher, The American Society of Mechanical Engineers

CHART 2—SODIUM CHLORIDE—VISCOSITY

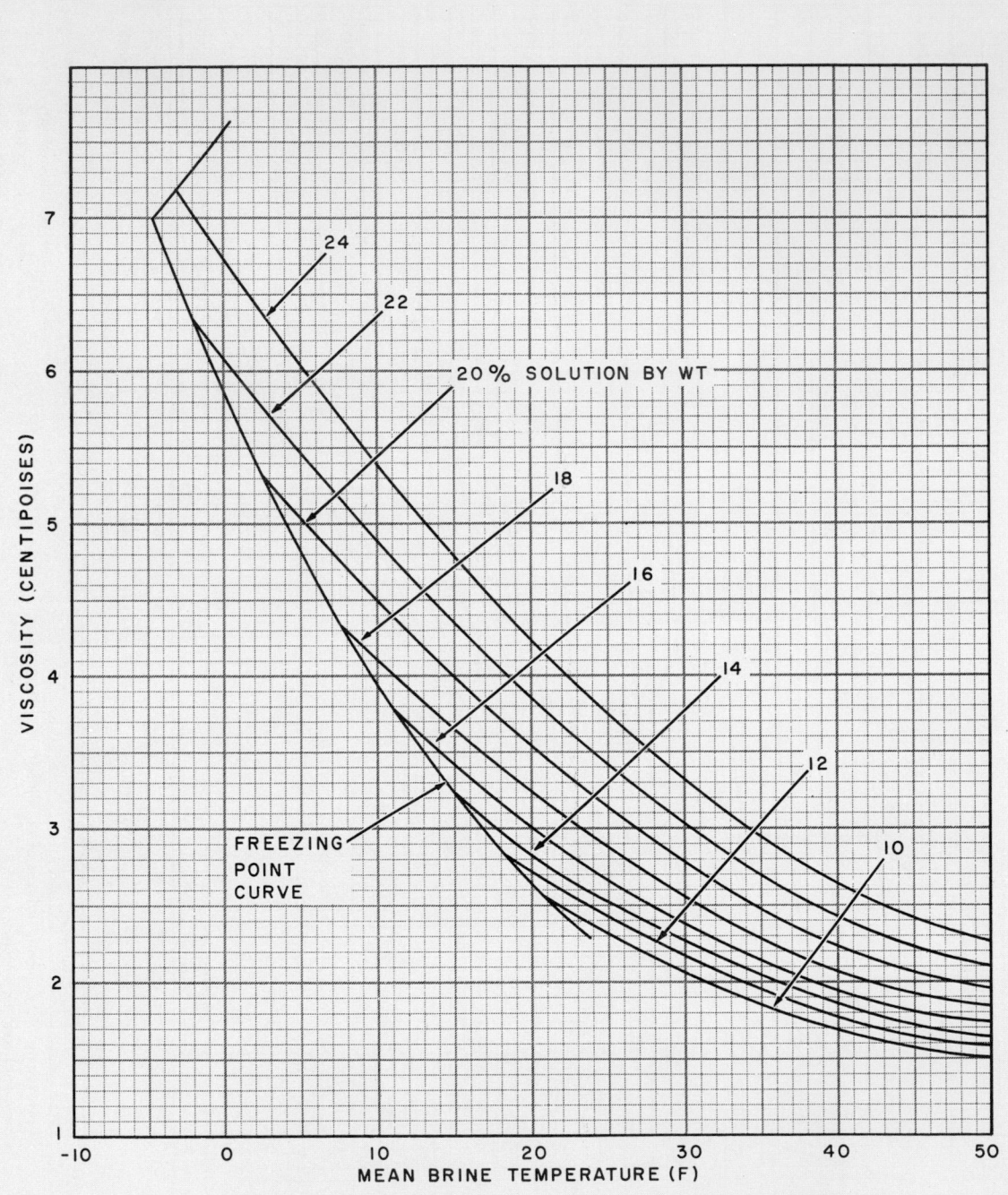

CHART 3—SODIUM CHLORIDE—SPECIFIC GRAVITY

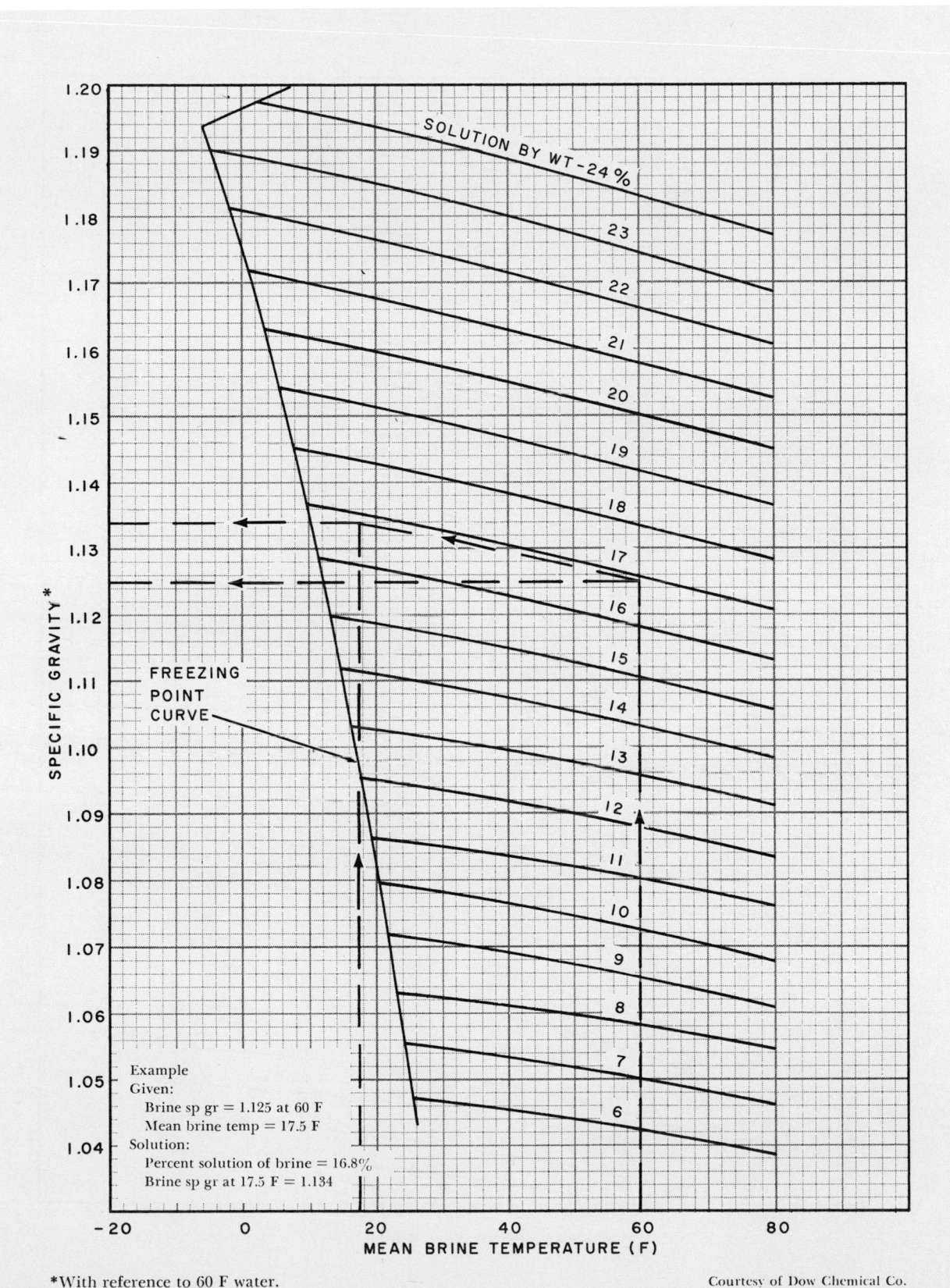

*With reference to 60 F water.   Courtesy of Dow Chemical Co.

CHART 4—SODIUM CHLORIDE—SPECIFIC HEAT

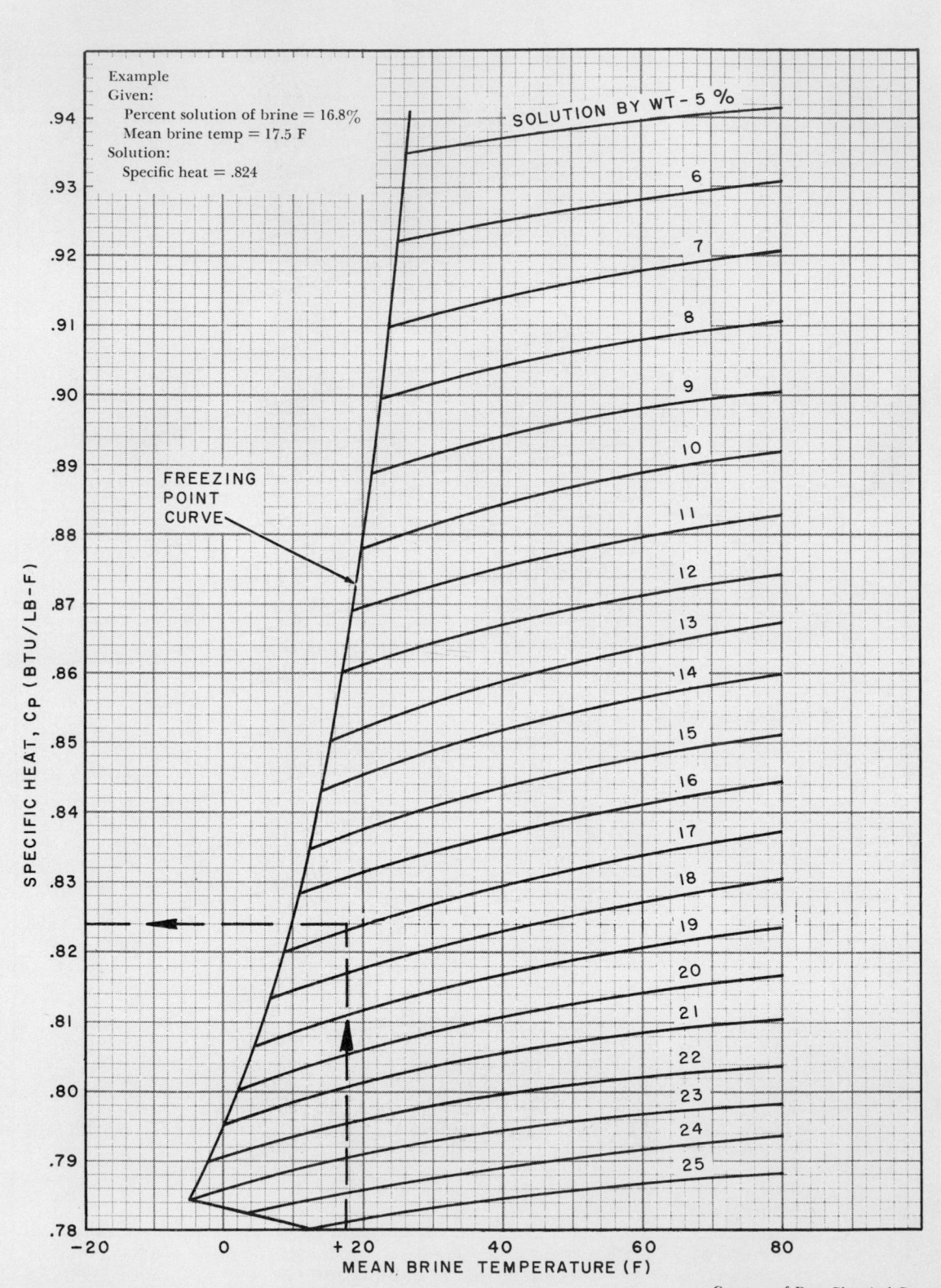

Courtesy of Dow Chemical Co.

CHART 5—SODIUM CHLORIDE—THERMAL CONDUCTIVITY

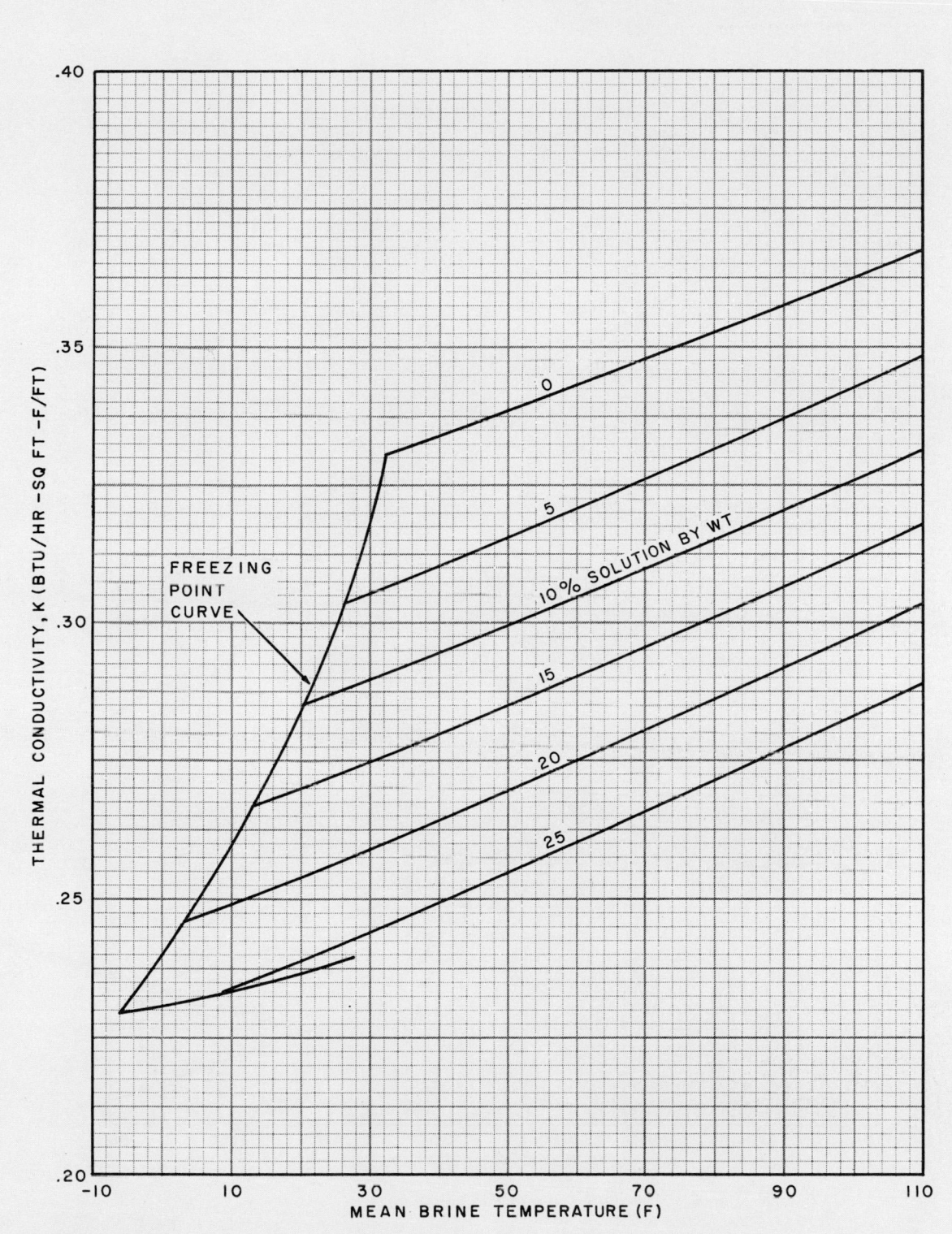

CHART 6—CALCIUM CHLORIDE—VISCOSITY

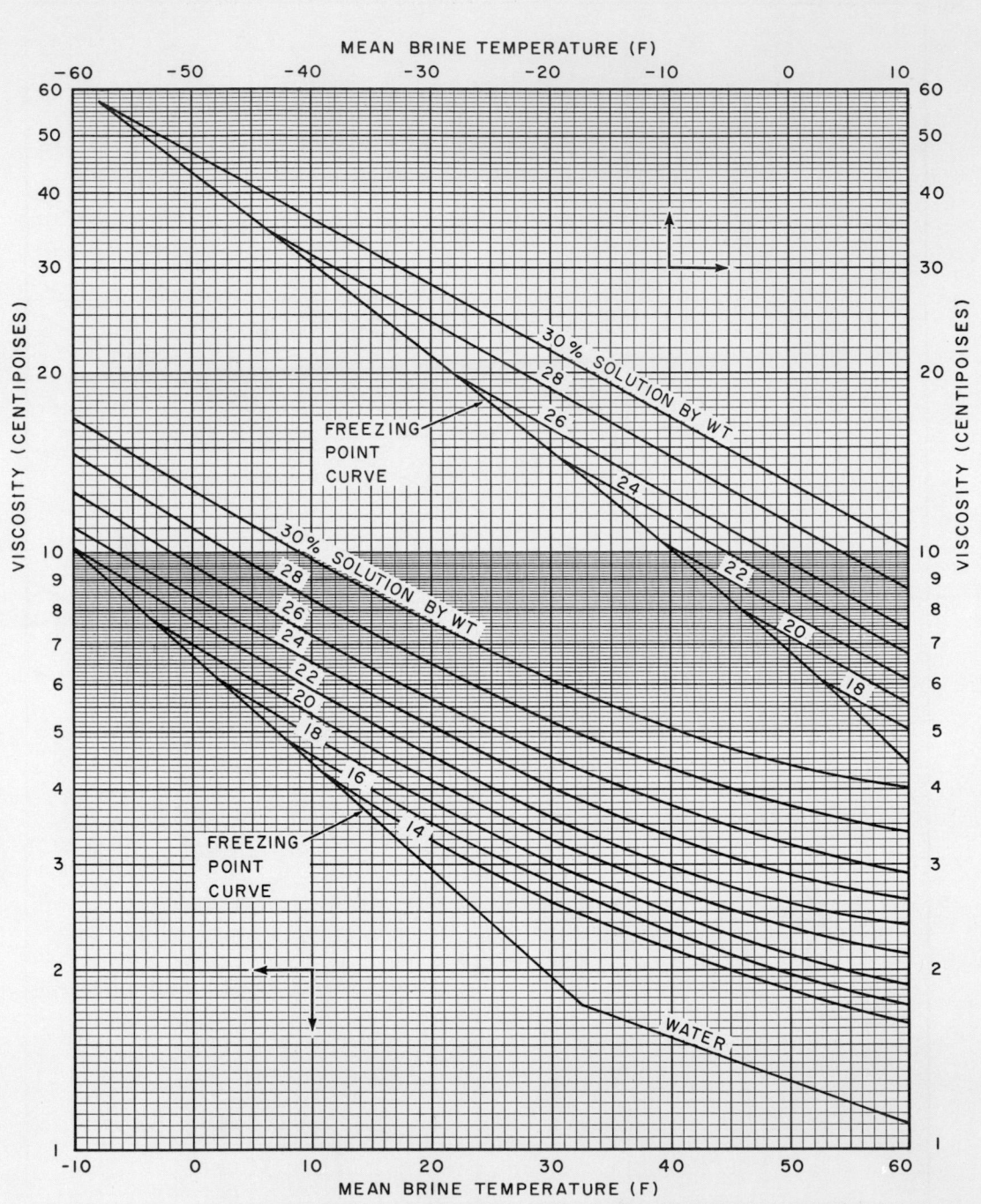

CHART 7—CALCIUM CHLORIDE—SPECIFIC GRAVITY

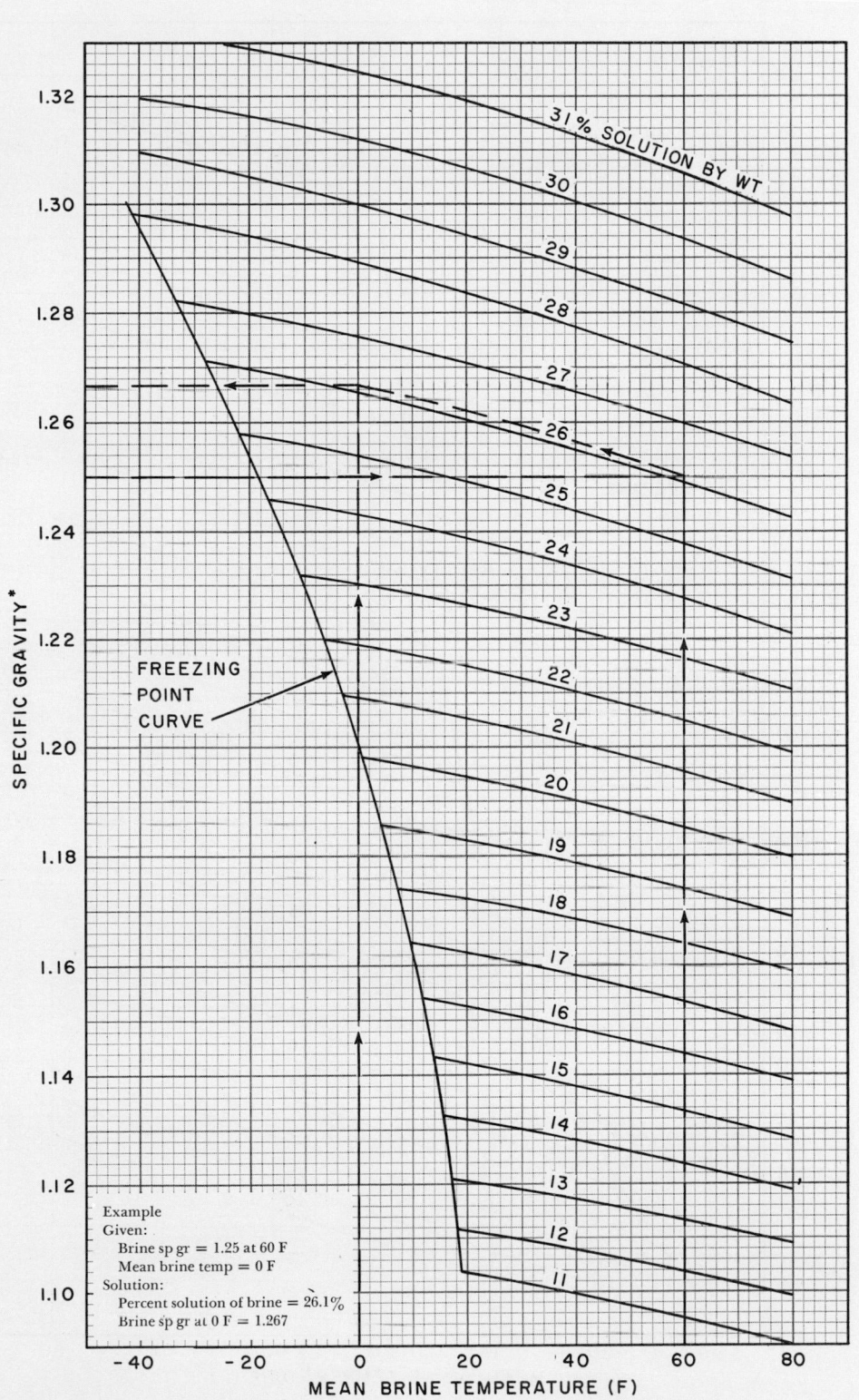

*With reference to 60 F water.   Courtesy of Dow Chemical Co.

CHART 8—CALCIUM CHLORIDE—SPECIFIC HEAT

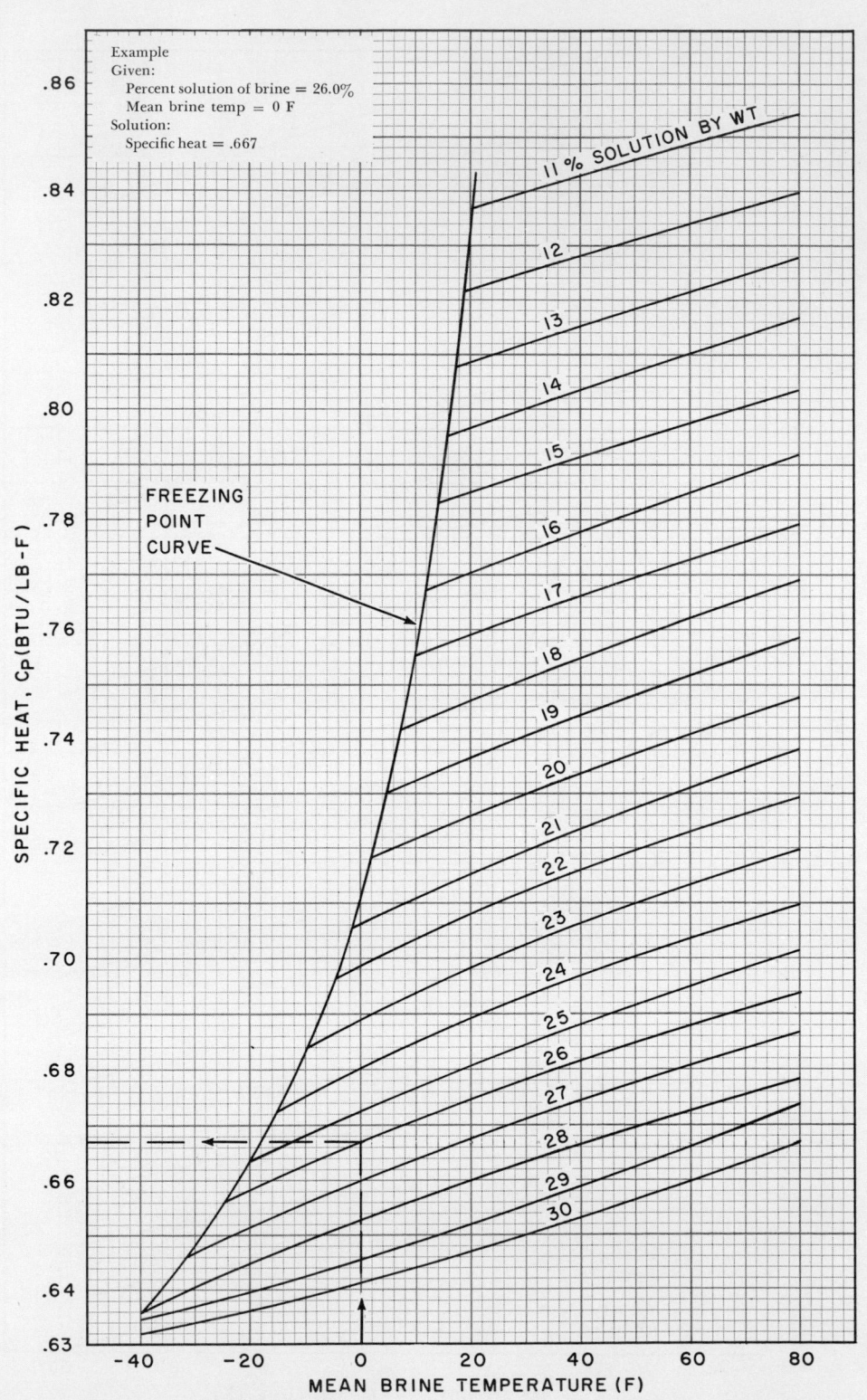

Courtesy of Dow Chemical Co.

CHART 9—CALCIUM CHLORIDE—THERMAL CONDUCTIVITY

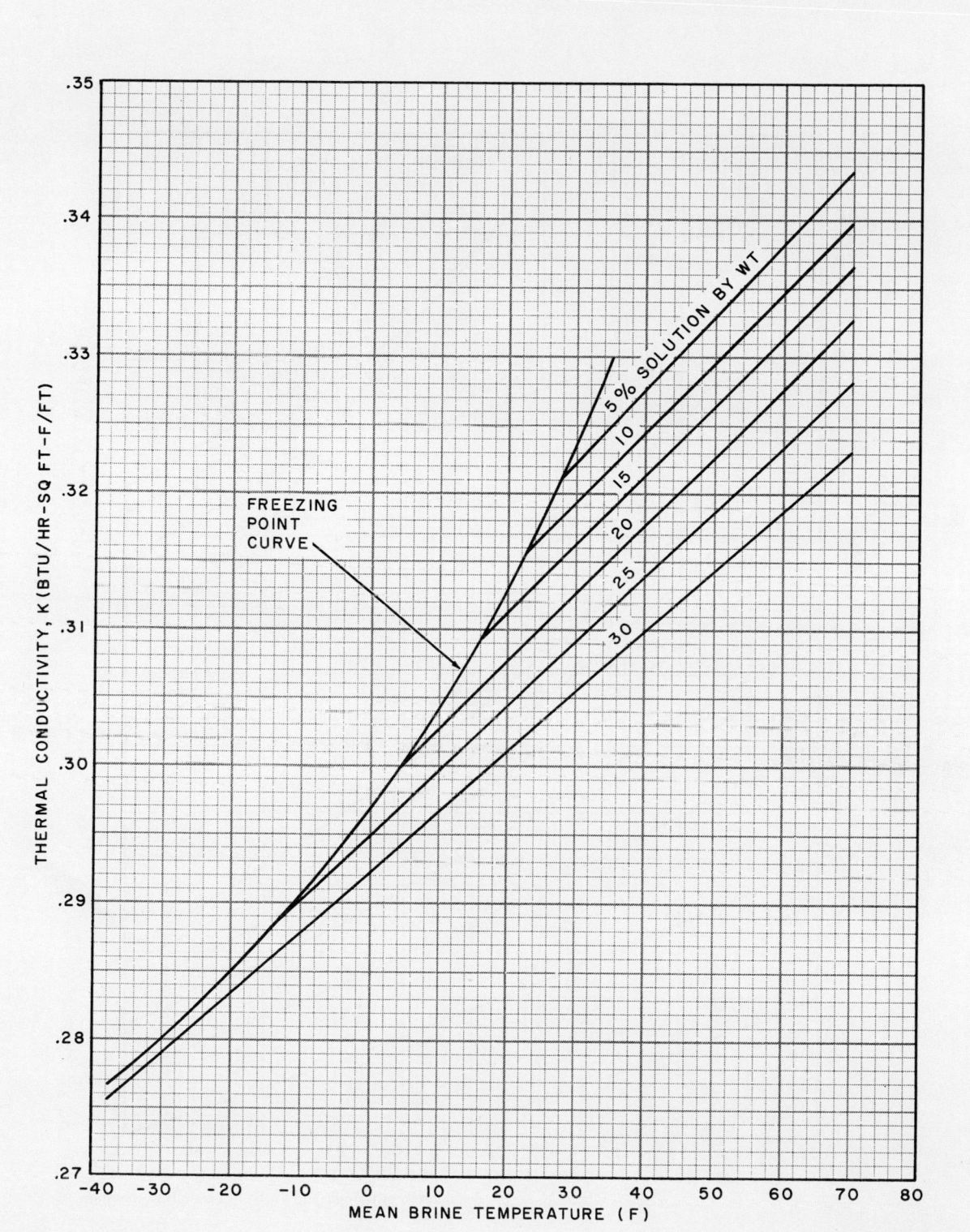

CHART 10—METHANOL BRINE—VISCOSITY

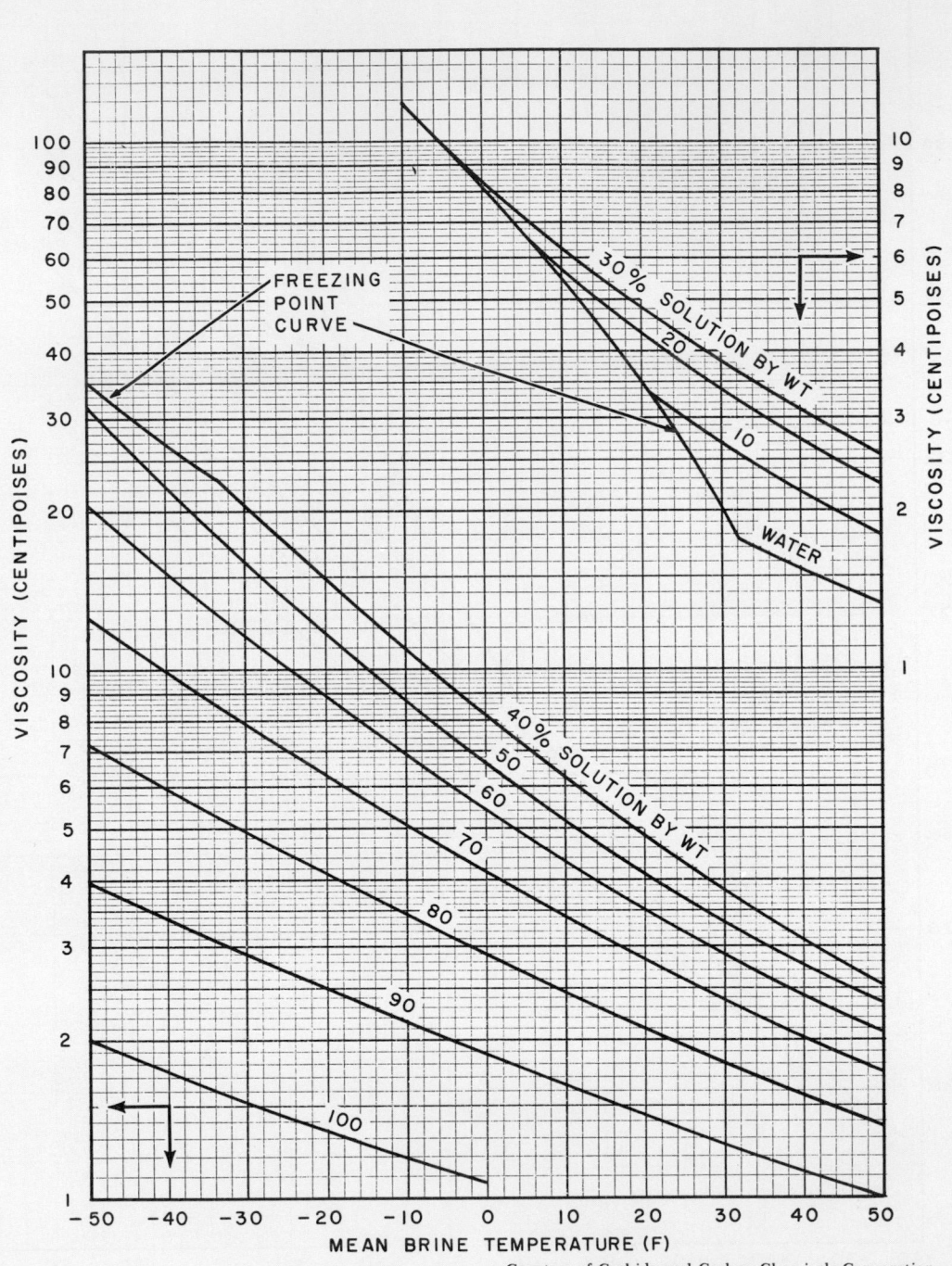

Courtesy of Carbide and Carbon Chemicals Corporation

CHART 11—METHANOL BRINE—SPECIFIC GRAVITY

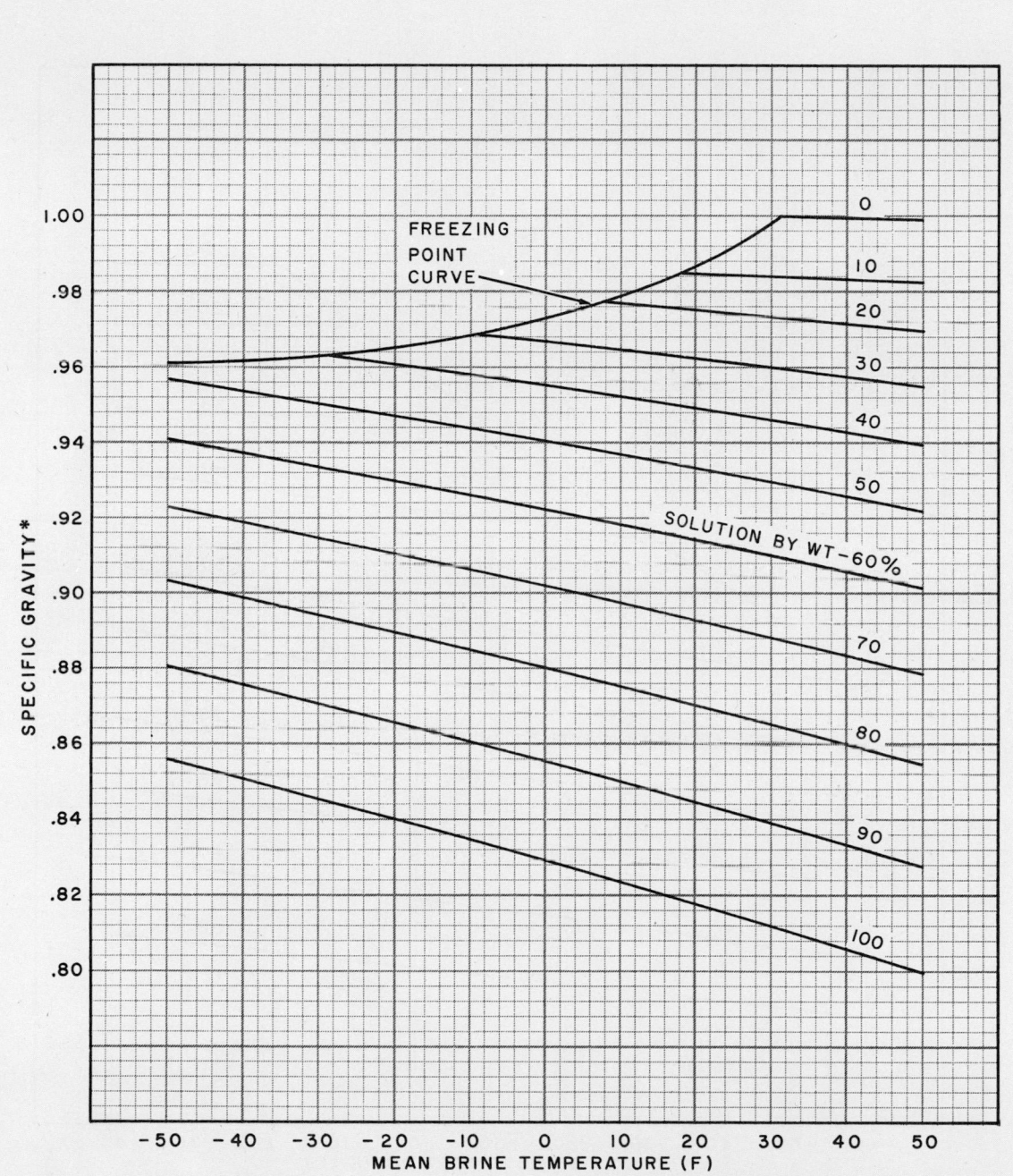

*With reference to 60 F water.   Courtesy of Carbide and Carbon Chemicals Corporation

CHART 12—METHANOL BRINE—SPECIFIC HEAT

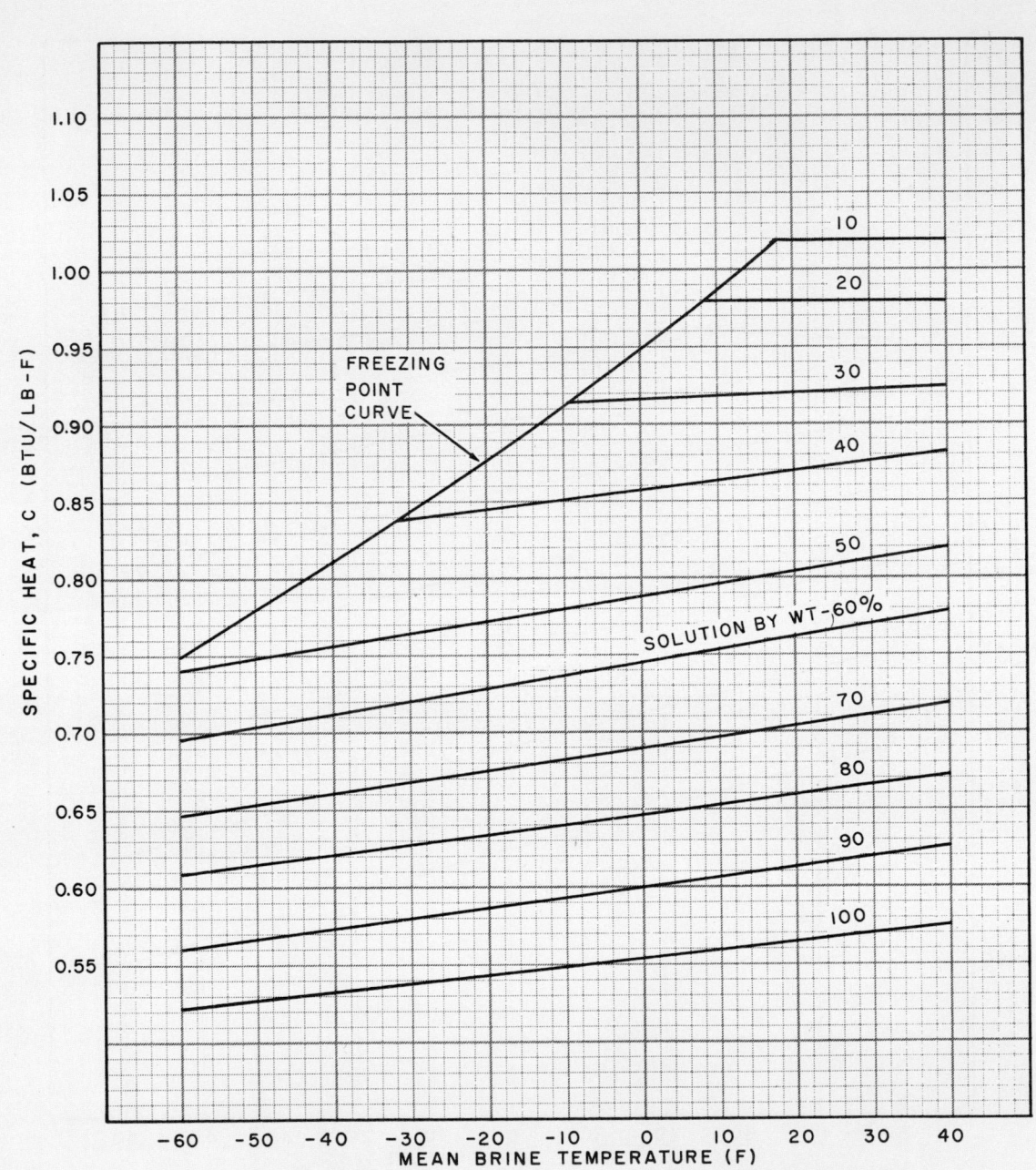

CHART 13—METHANOL BRINE—THERMAL CONDUCTIVITY

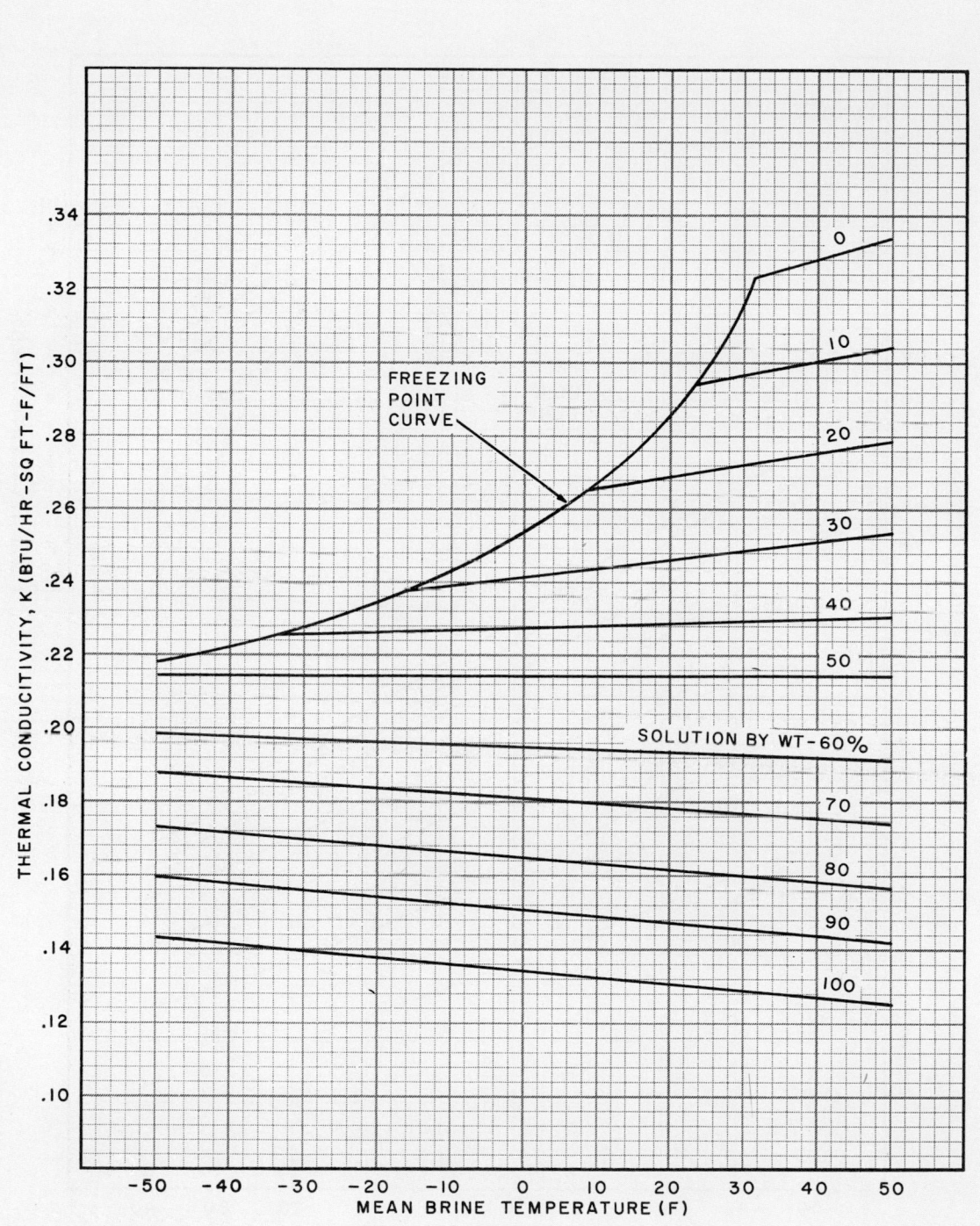

CHART 14—ETHANOL BRINE—VISCOSITY

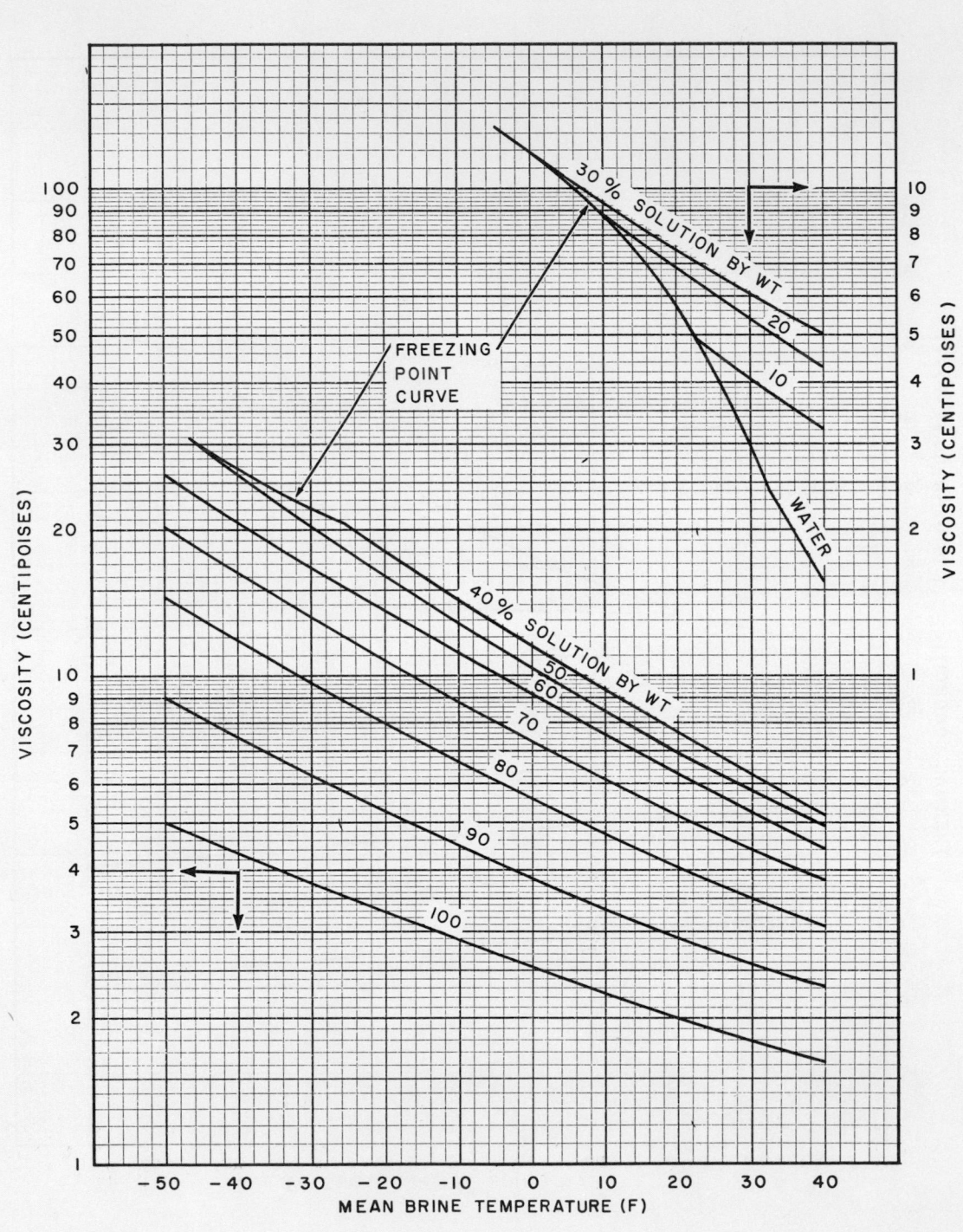

CHART 15—ETHANOL BRINE—SPECIFIC GRAVITY

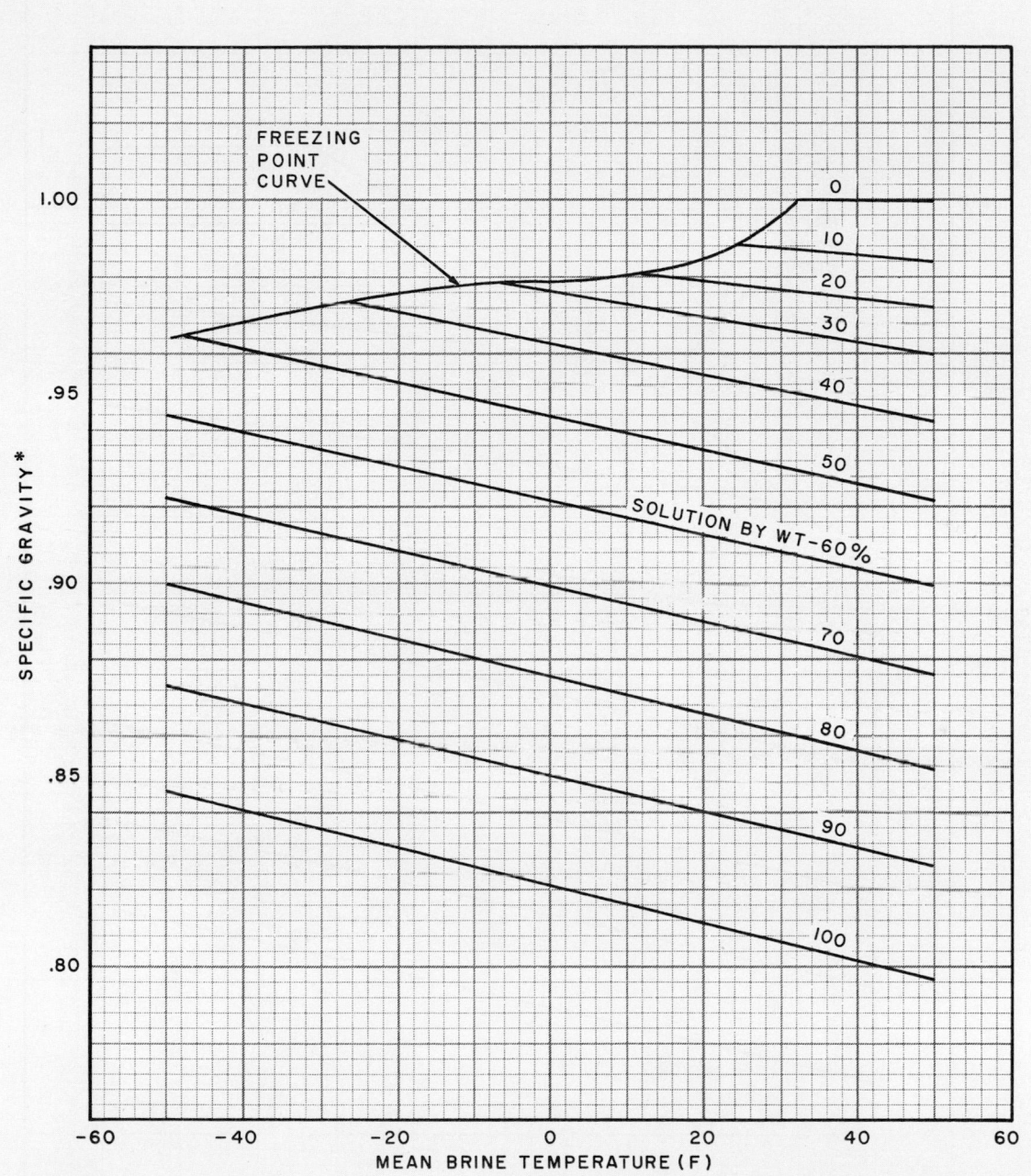

*With reference to 60 F water.   Extrapolated values from International Critical Tables

CHART 16—ETHANOL BRINE—SPECIFIC HEAT

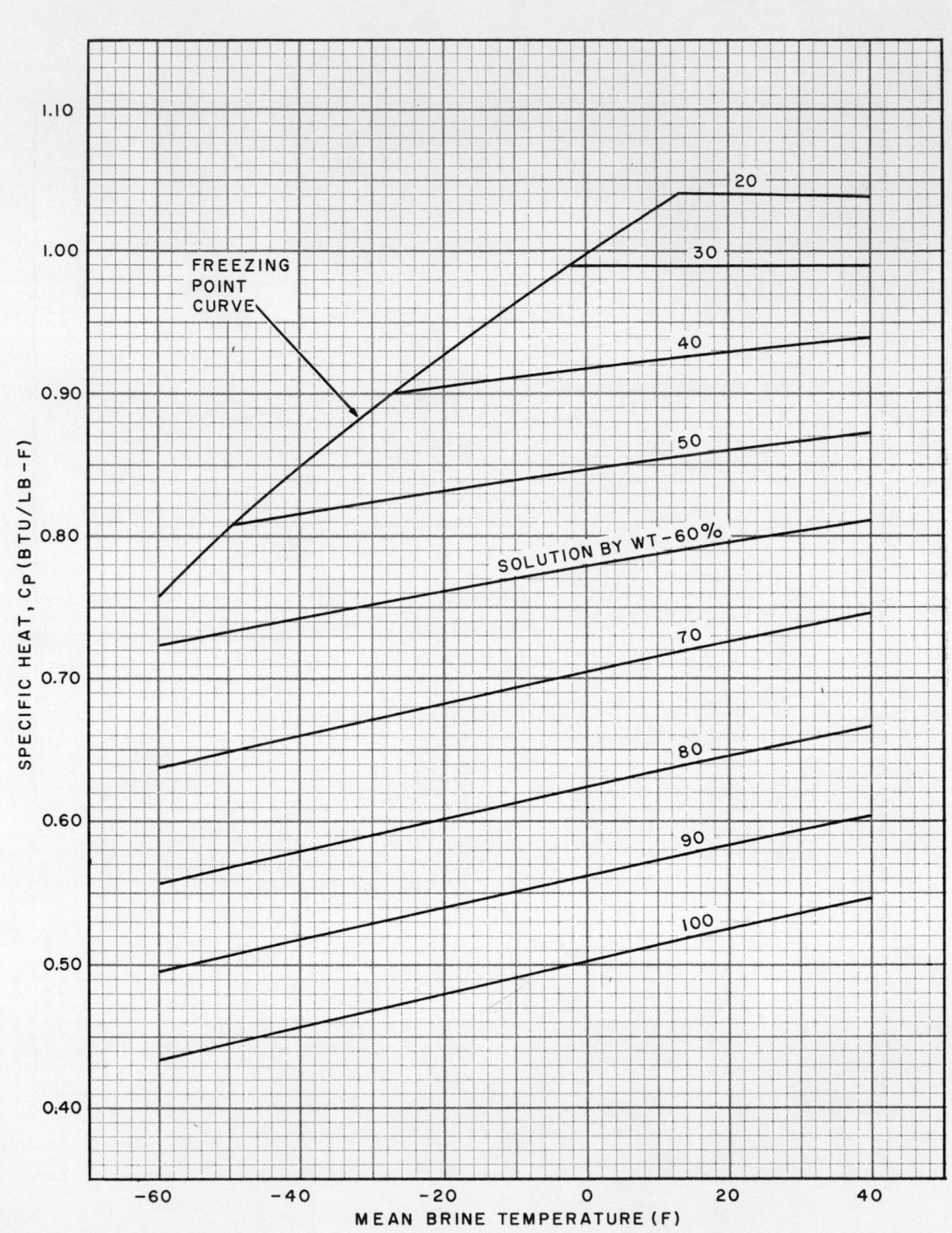

CHART 17—ETHANOL BRINE—THERMAL CONDUCTIVITY

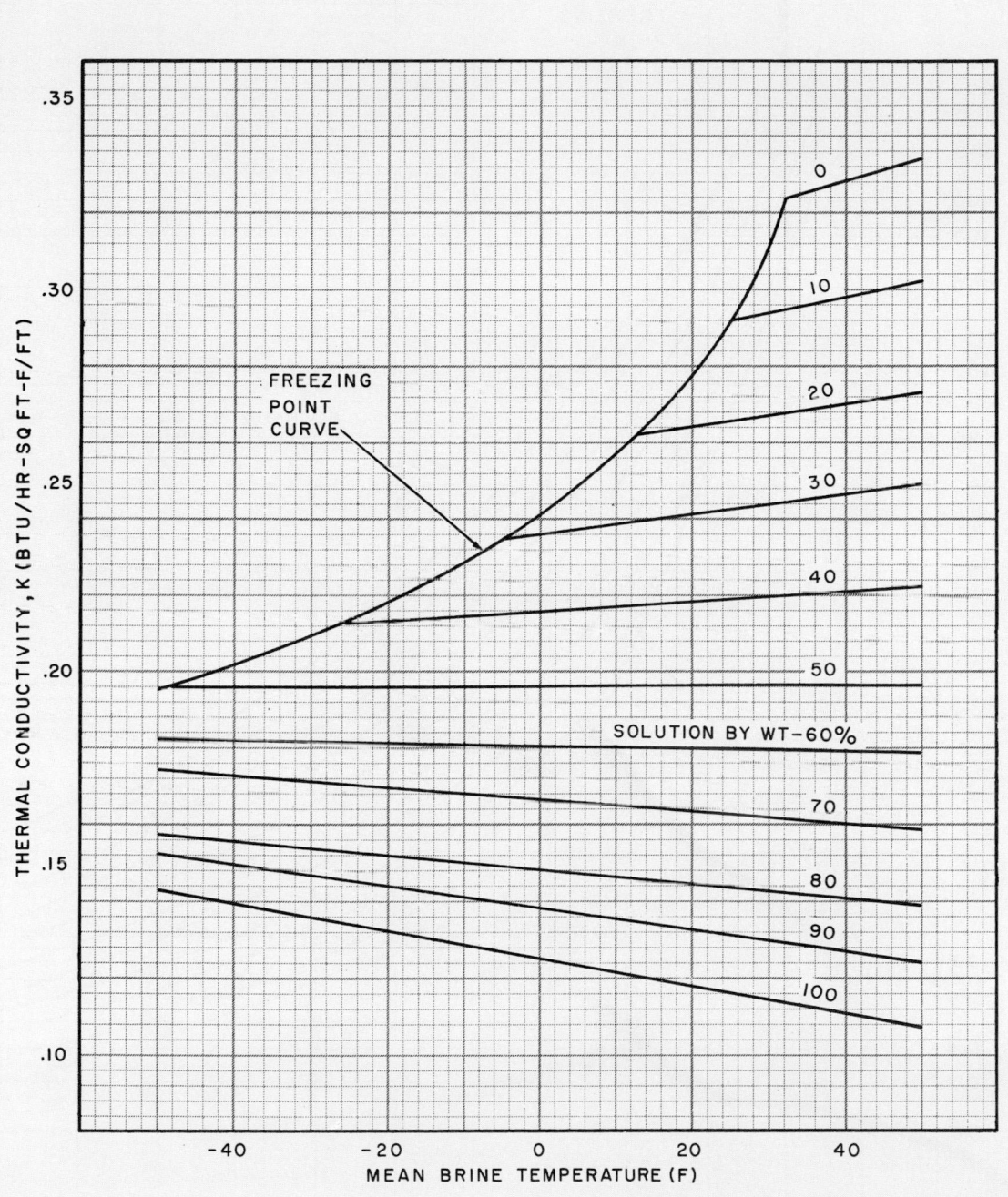

CHART 18—ETHYLENE GLYCOL—VISCOSITY

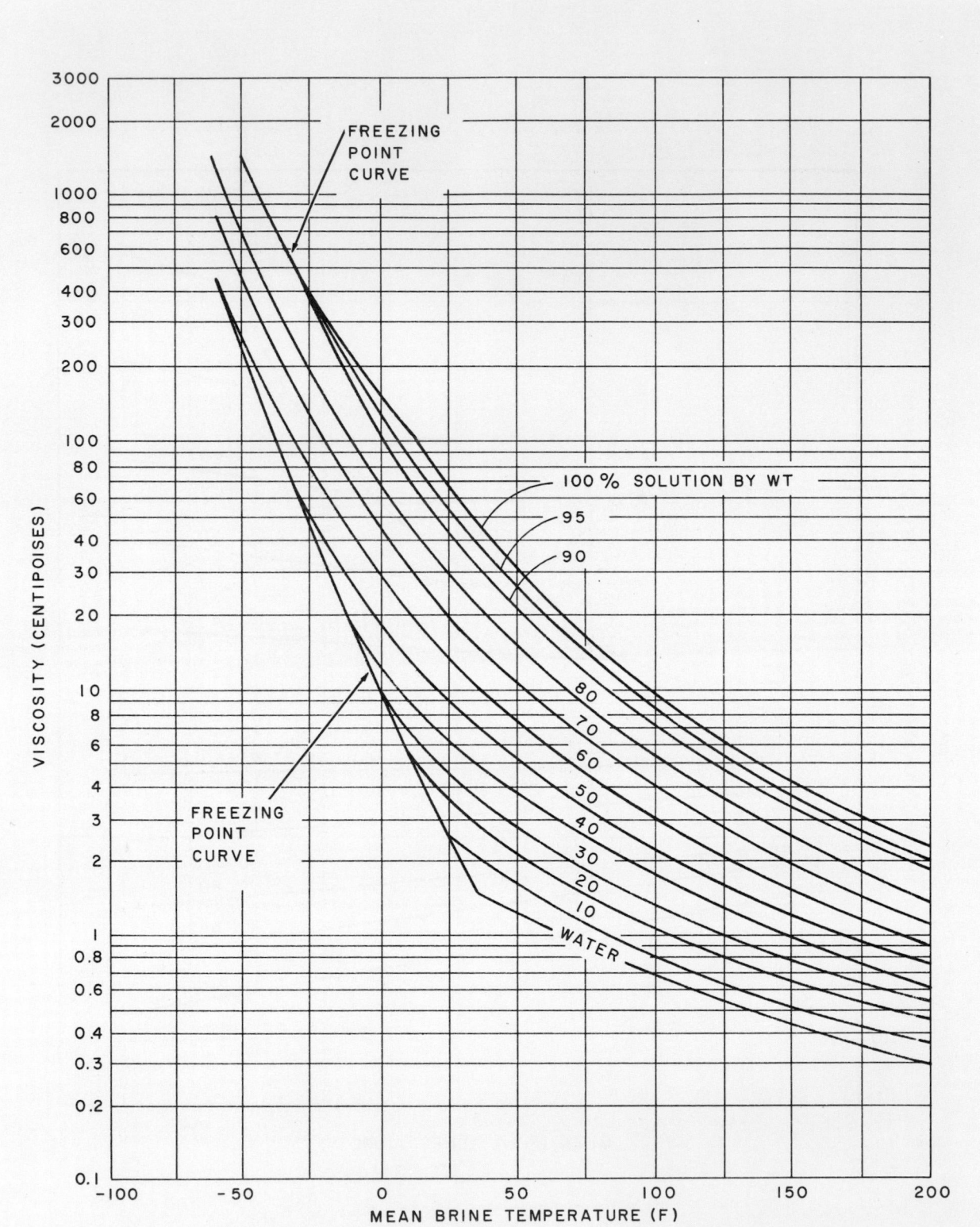

From *Glycols, Properties and Uses*, Dow Chemical Co. 1961

CHART 19—ETHYLENE GLYCOL—SPECIFIC GRAVITY

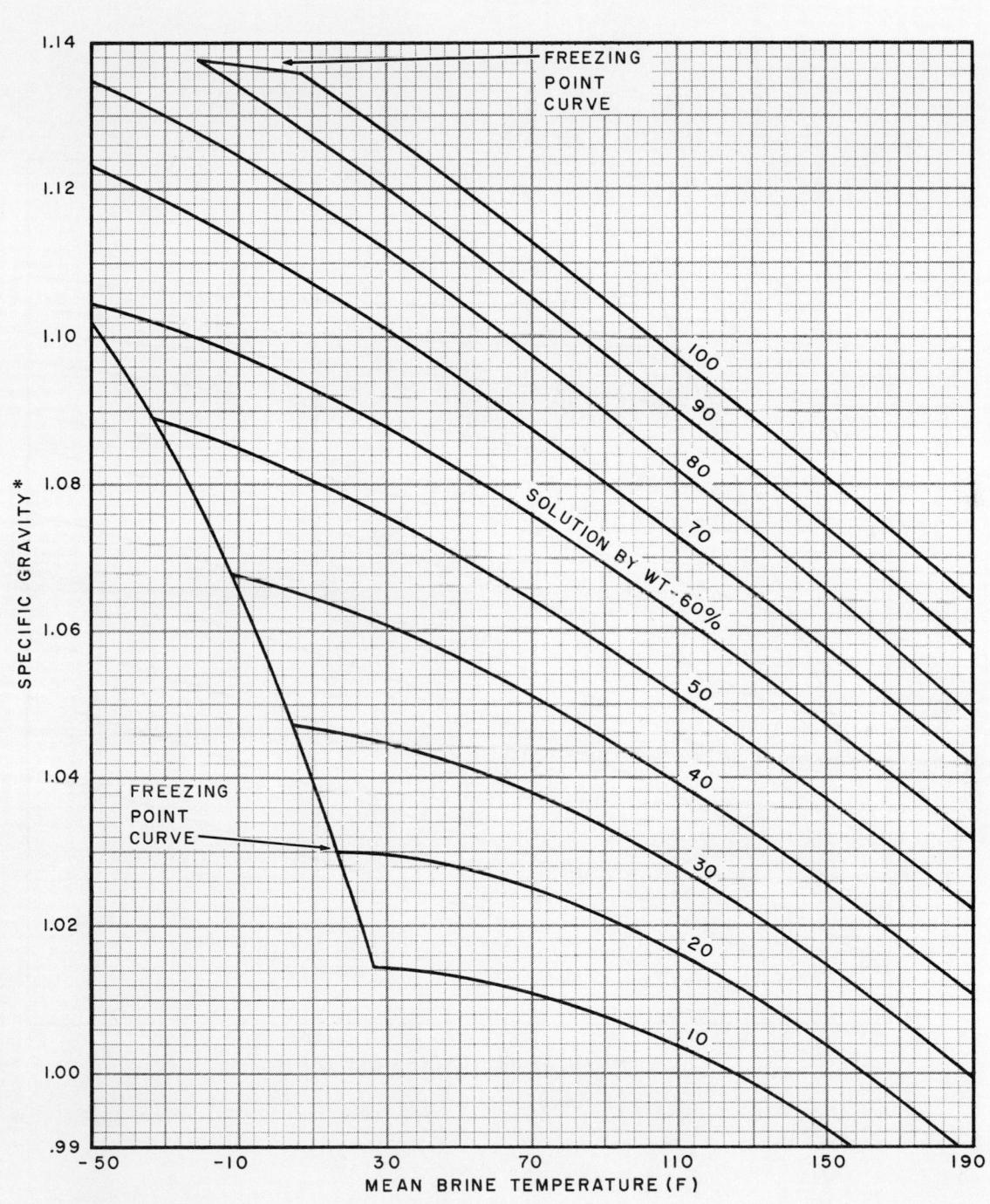

*With reference to 60 F water.

From *Glycols, Properties and Uses*, Dow Chemical Co. 1961

CHART 20—ETHYLENE GLYCOL—SPECIFIC HEAT

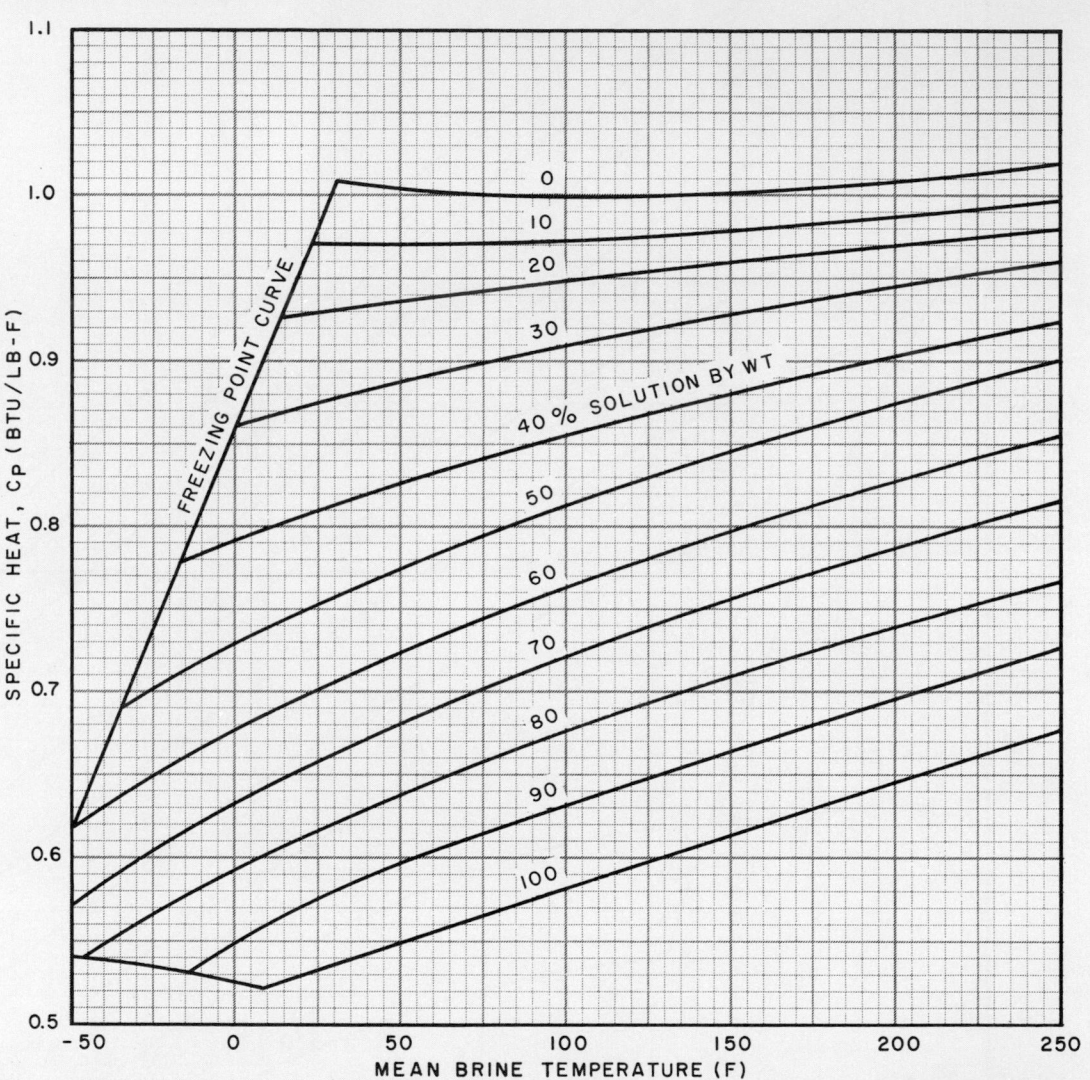

From *Glycols*, Union Carbide Chemicals Co. 1958

CHART 21—ETHYLENE GLYCOL—THERMAL CONDUCTIVITY

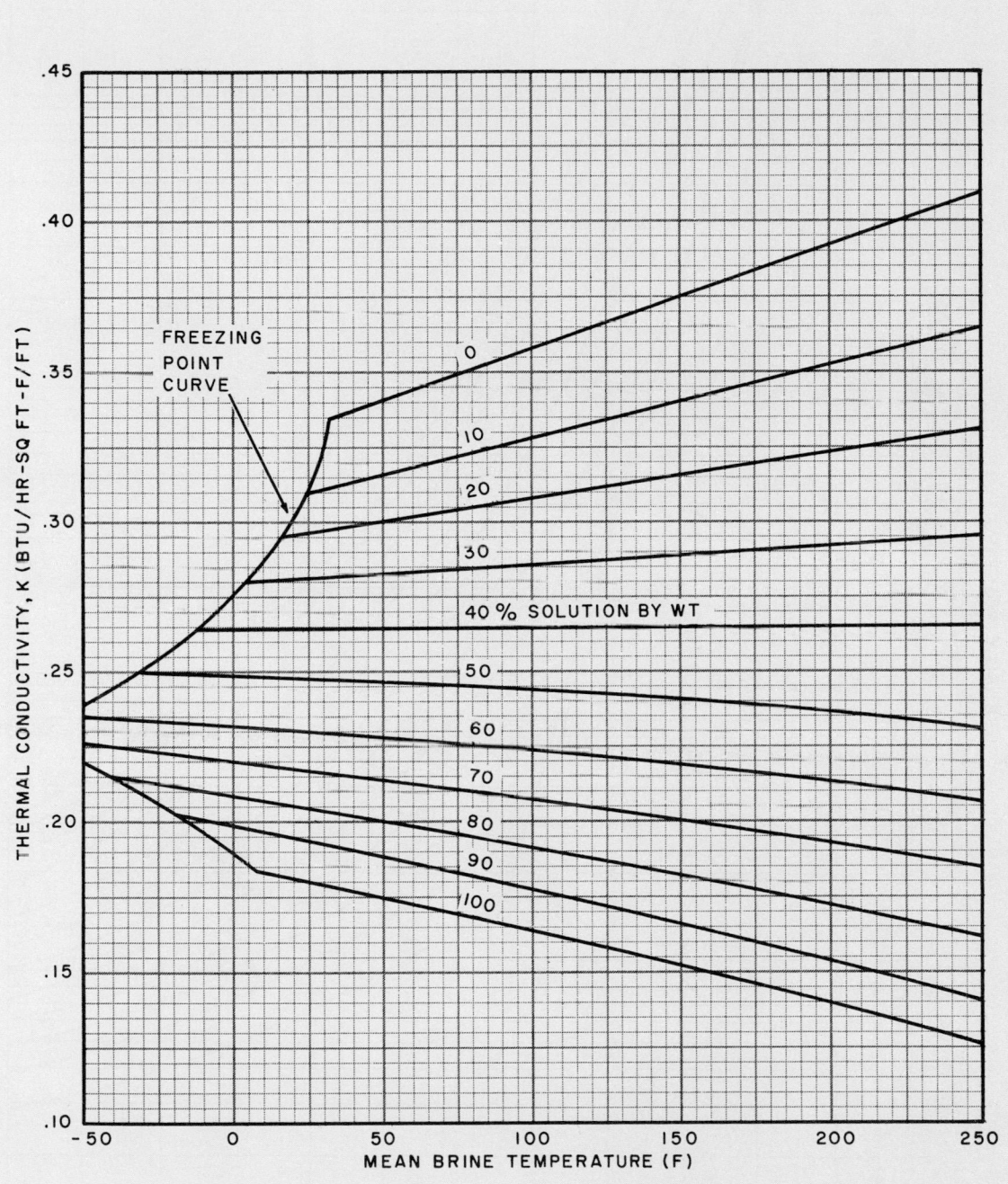

From *Glycols*, Union Carbide Chemicals Co. 1958

CHART 22—PROPYLENE GLYCOL—VISCOSITY

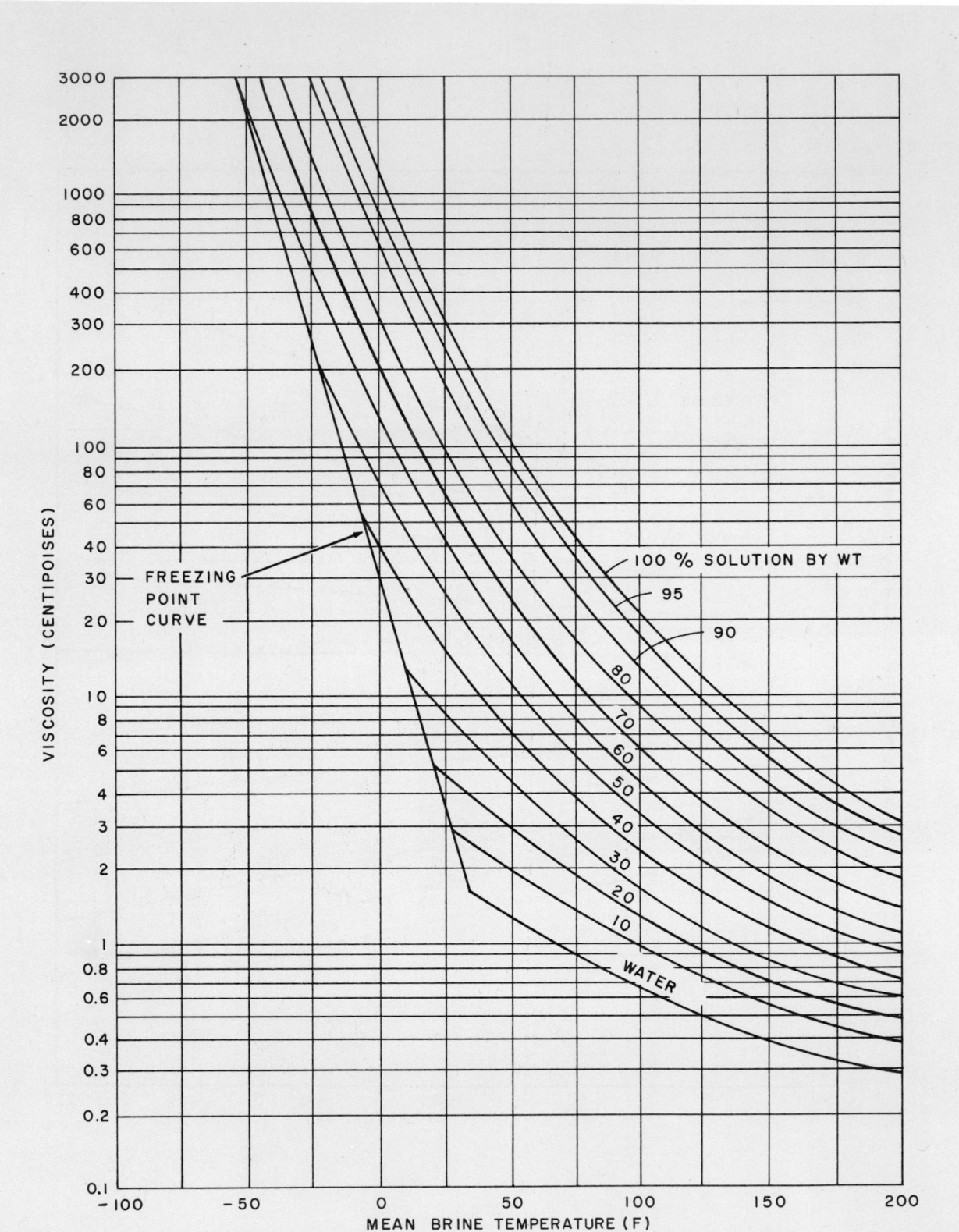

From *Glycols, Properties and Uses*, Dow Chemical Co. 1961

CHART 23—PROPYLENE GLYCOL—SPECIFIC GRAVITY

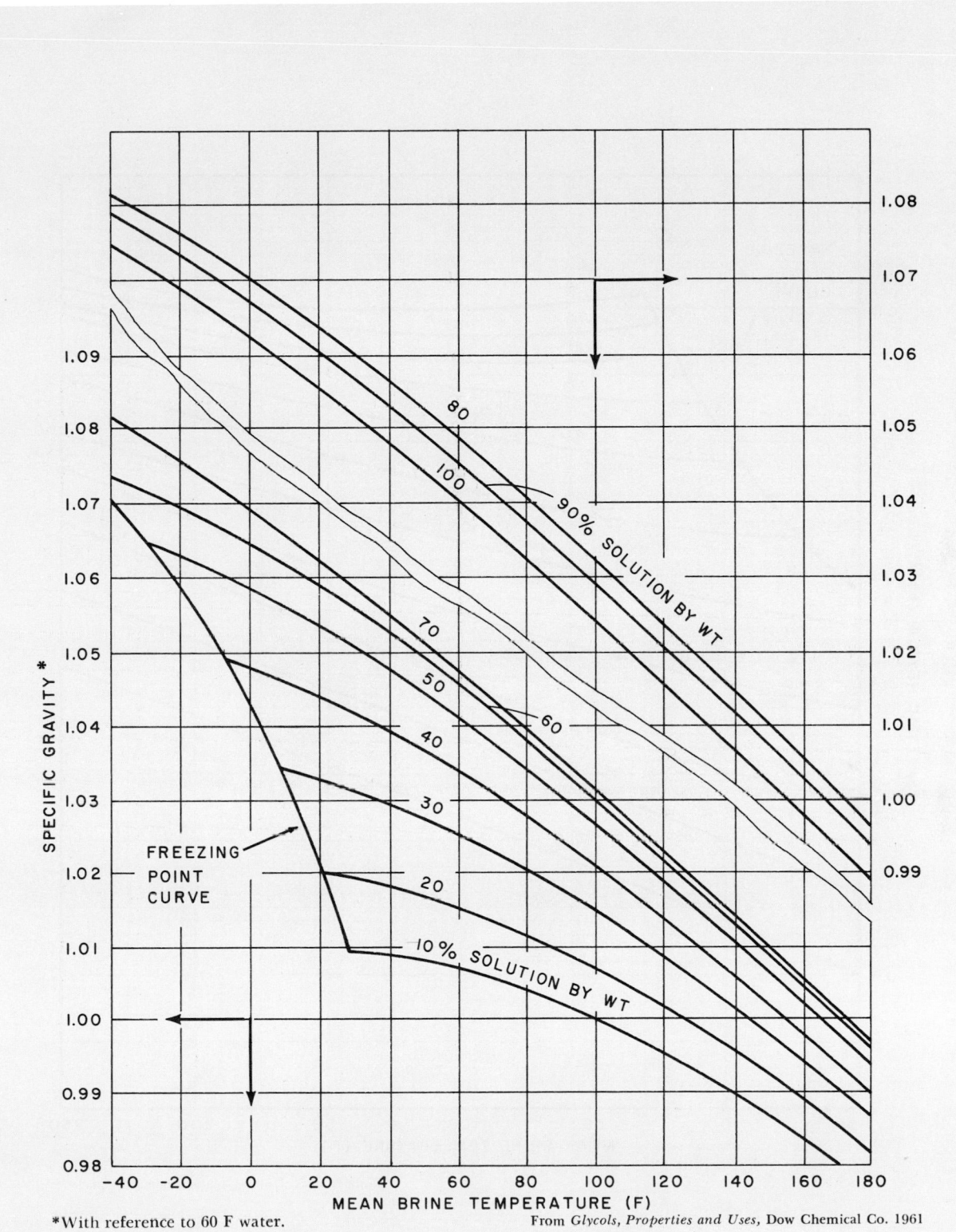

*With reference to 60 F water.

From *Glycols, Properties and Uses*, Dow Chemical Co. 1961

CHART 24—PROPYLENE GLYCOL—SPECIFIC HEAT

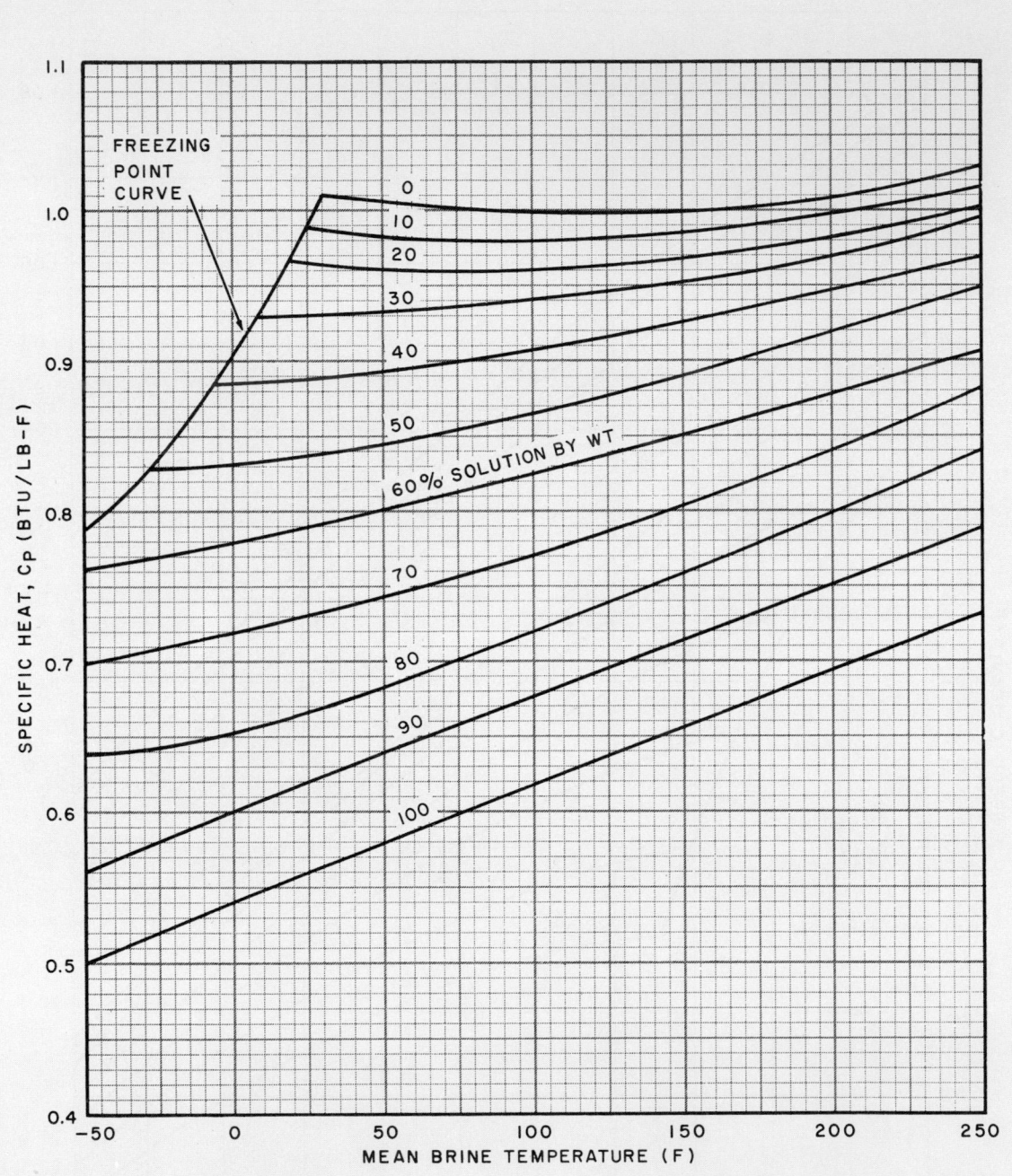

From *Glycols*, Union Carbide Chemicals Co. 1958

CHART 25—PROPYLENE GLYCOL—THERMAL CONDUCTIVITY

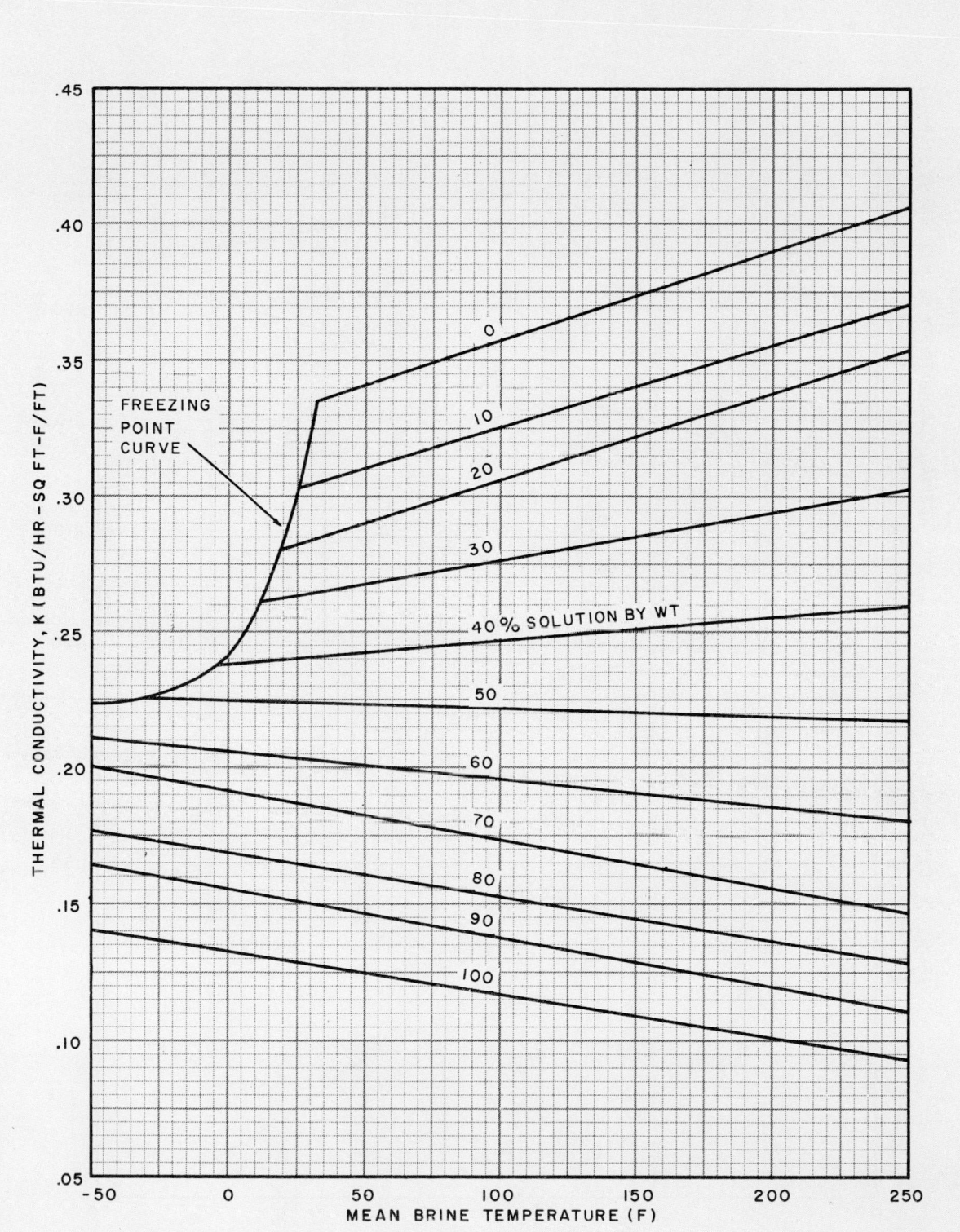

From *Glycols*, Union Carbide Chemicals Co. 1958

CHART 26—TRICHLOROETHYLENE—PROPERTIES

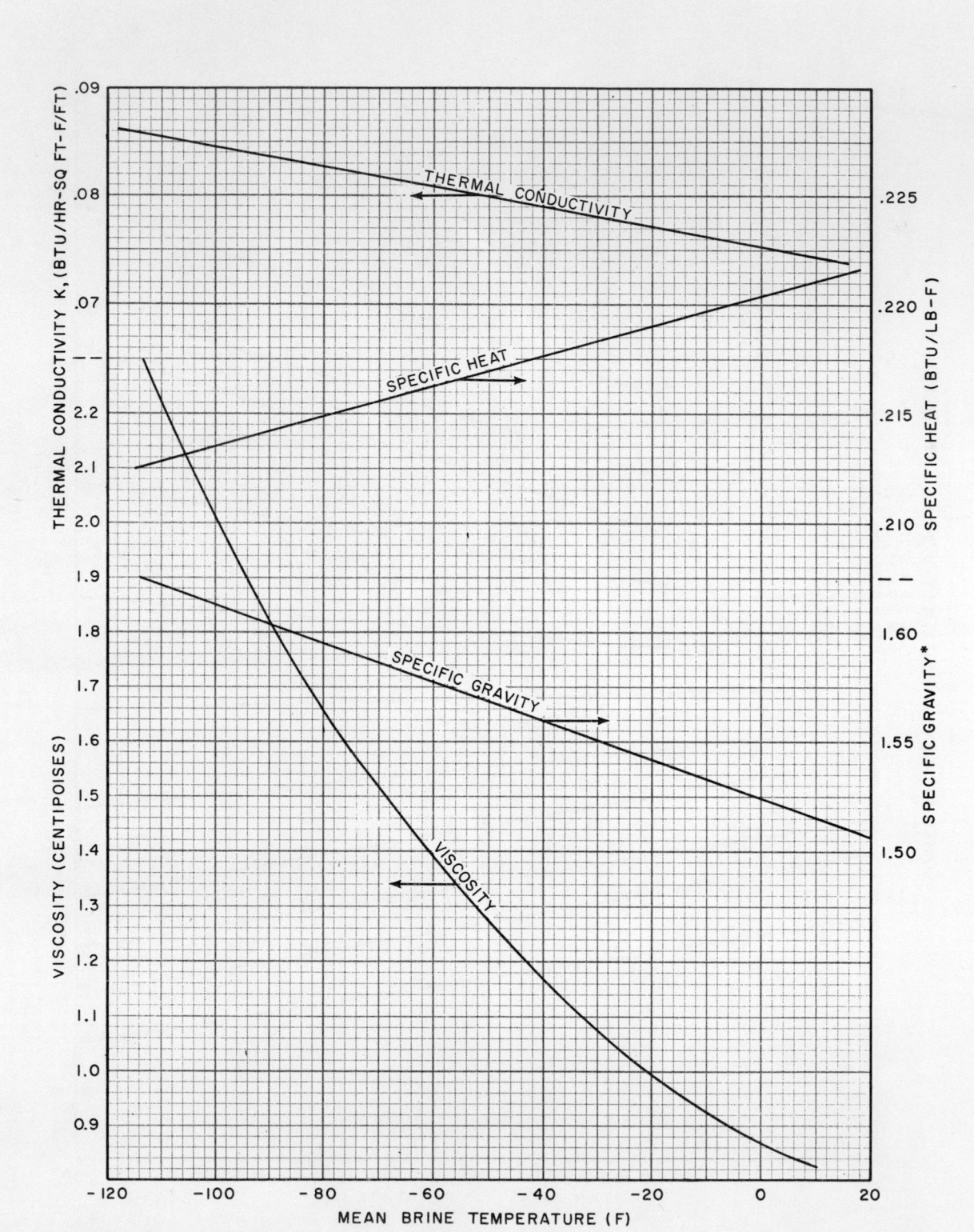

*With reference to 60 F water.

CHART 27—REFRIGERANT 11—PROPERTIES

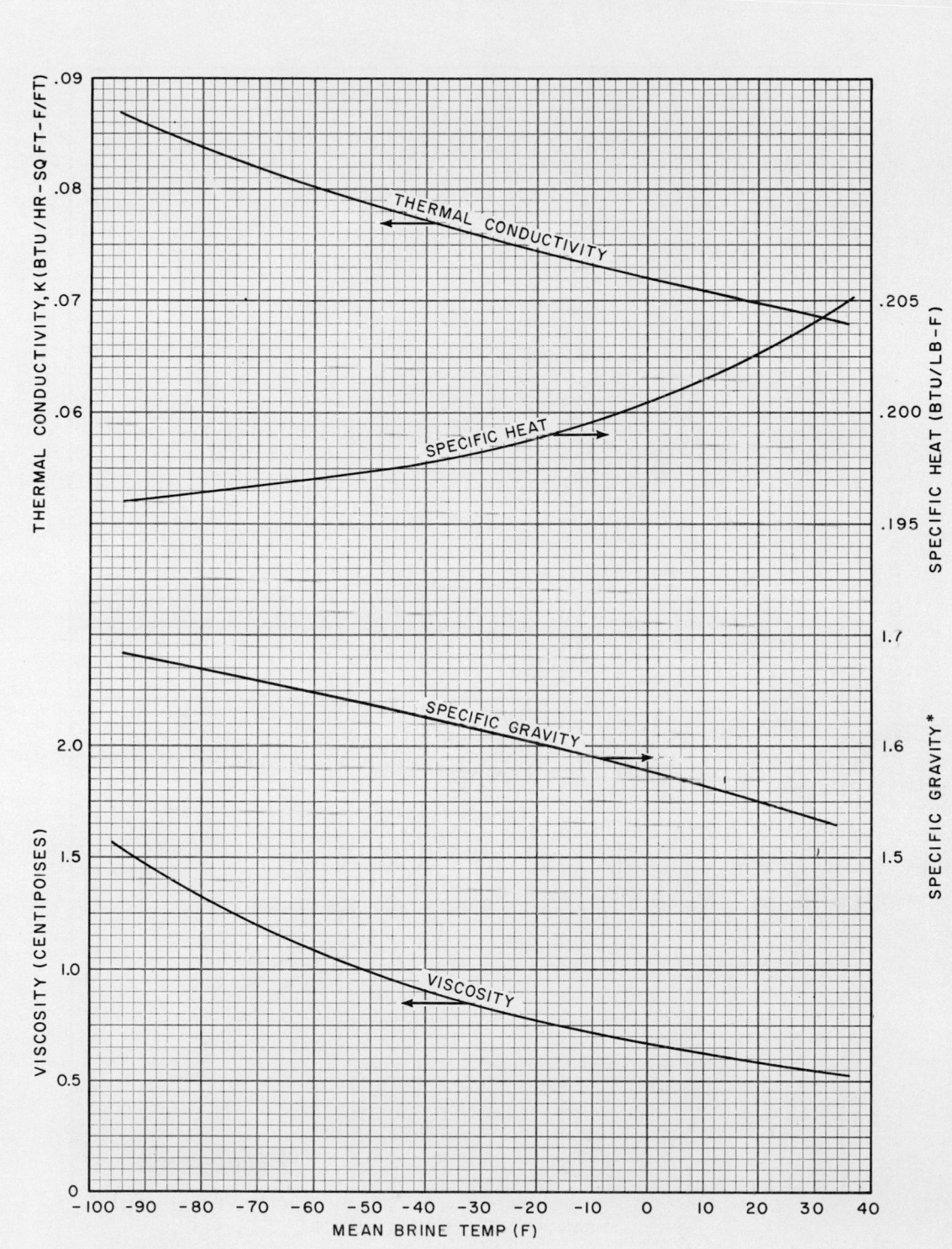

*With reference to 60 F water.

CHART 28—WATER—VISCOSITY

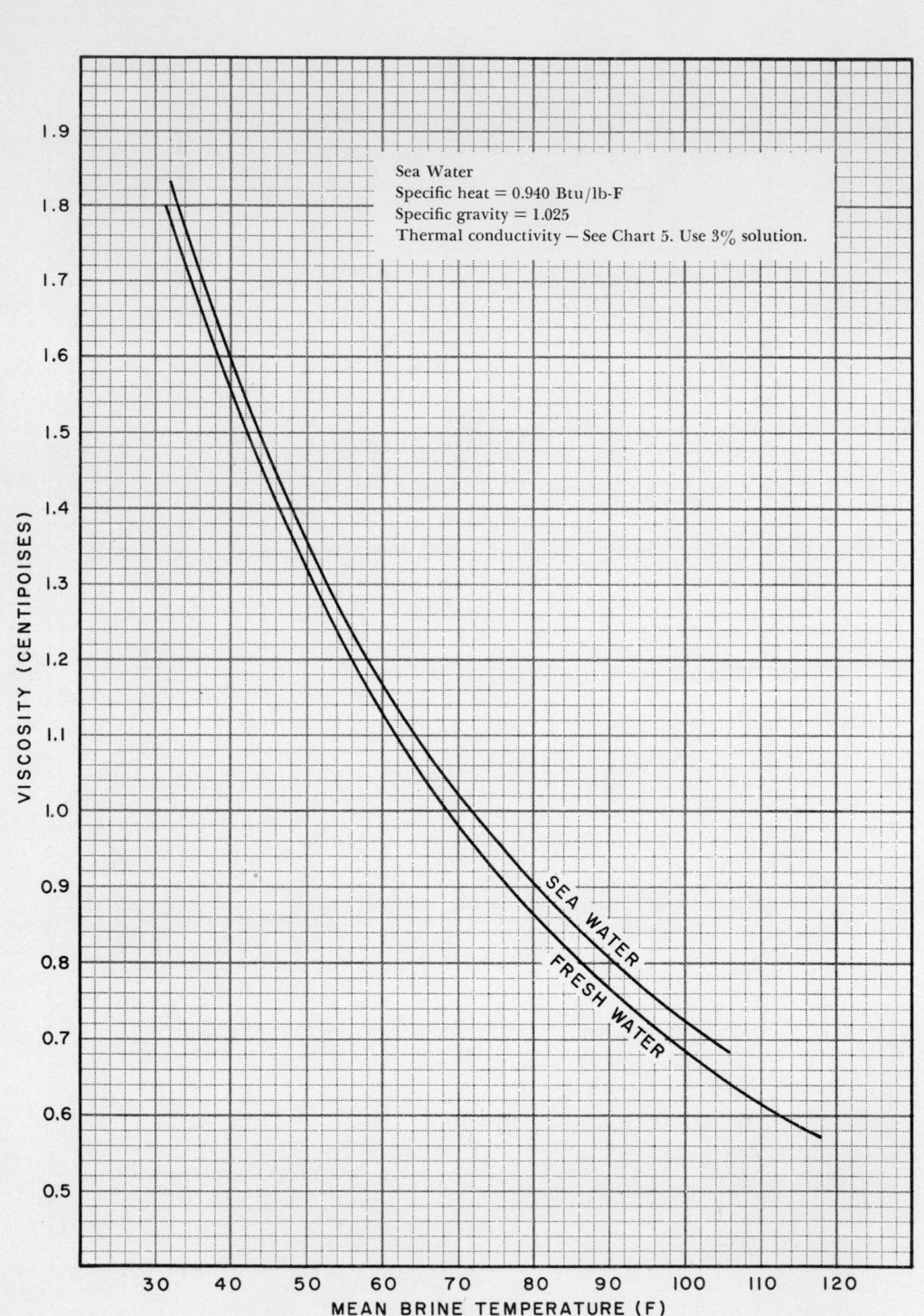

CHAPTER 3. REFRIGERATION OILS

This chapter covers general classifications and quality of lubricating oils that are important in refrigeration. The recommendation of oils to be used in a refrigeration system is primarily the responsibility of the refrigeration system manufacturer. However, it is important for an engineer to understand the basis of the selection of these oils in order to properly apply them in the field.

CLASSIFICATION

Oils classified by source fall in three main groups: animal, vegetable and mineral. Animal and vegetable oils are called fixed oils because they cannot be refined without decomposing. They are unstable and tend to form acids and gums that make them unsuitable for refrigeration purposes.

There are three major classifications of mineral oil: naphthene base, paraffin base, and mixed base. When distilled, a naphthene base oil yields a residue of heavy pitch or asphalt. California oils, some Gulf Coast and heavy Mexican oils are in this class. A paraffin base oil yields a paraffin wax when distilled. The best sources of paraffin base oils are Pennsylvania, Northern Louisiana, and parts of Oklahoma and Kansas. The mixed oils contain both naphthene and paraffin bases. Illinois and some mid-continent oils are in this class.

Experience has shown that the naphthene base oils are more suited for refrigeration work for three main reasons:

1. They flow better at low temperatures.
2. Carbon deposits from these oils are of a soft nature and can easily be removed.
3. They deposit less wax at low temperatures.

When obtained from selected crudes and properly refined and treated, all three classes of mineral oil can be considered satisfactory for refrigeration use.

PROPERTIES

To meet the requirements of a refrigeration system, a good refrigeration oil should:

1. Maintain sufficient body to lubricate at high temperature and yet be fluid enough to flow at low temperature.
2. Have a pour point low enough to allow flow at any point in the system.
3. Leave no carbon deposits when in contact with hot surfaces encountered in the system during normal operation.
4. Deposit no wax when exposed to the lowest temperatures normally encountered in the system.
5. Contain little or no corrosive acid.
6. Have a high resistance to the flow of electricity.
7. Have a high flash and fire point to indicate proper blending.
8. Be stable in the presence of oxygen.
9. Contain no sulfur compounds.
10. Contain no moisture.
11. Be light in color, to indicate proper refining.

As lubricating oils for refrigeration compressors are a specialty product, they require consideration apart from normal lubricants. The emphasis in this chapter is on oil used in refrigeration. Do not consider the emphasis as applicable to lubricants in general.

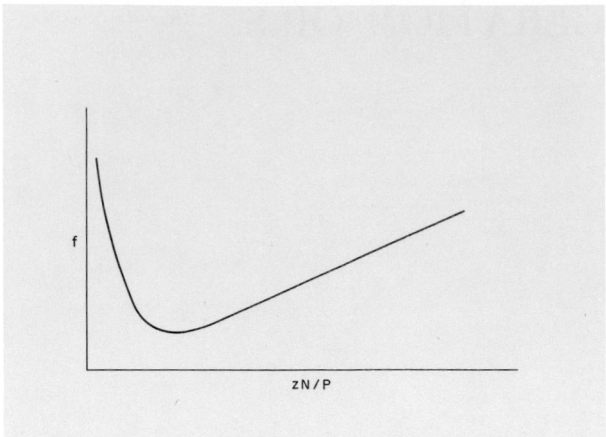

FIG. 12 — LUBRICATION CHARACTERISTICS

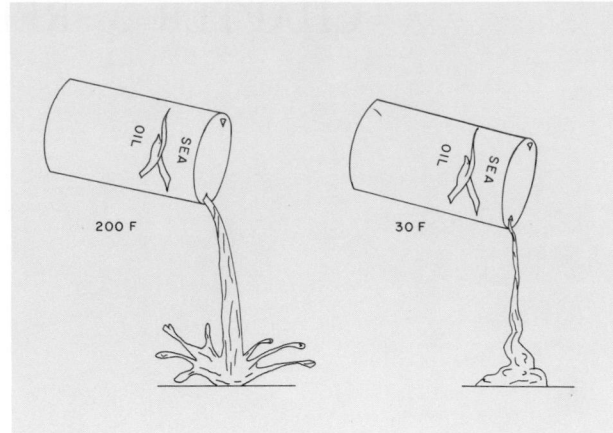

FIG. 13 — EFFECT OF TEMPERATURE ON VISCOSITY

The characteristics of oil for refrigeration (not necessarily in order of importance) are these:

1. Viscosity
2. Pour point
3. Carbonization
4. Floc point
5. Neutralization
6. Dielectric strength
7. Flash point
8. Fire point
9. Oxidation stability
10. Corrosion tendency
11. Moisture content
12. Color

VISCOSITY

Viscosity or coefficient of internal friction is that property of a liquid responsible for resistance to flow; it indicates how thick or thin an oil is.

The purpose of an oil is to lubricate bearing or rubbing surfaces. If the oil is too thin, it does not stay between the rubbing surfaces but is forced out, leaving no protective film. If the oil is too thick, it causes drag and loss of power, and may not be able to flow between the bearing or rubbing surfaces.

Friction loss f is illustrated in *Fig. 12* as a function of viscosity z, speed N in revolutions per unit time, and load P per unit area.

Viscosity is usually measured in terms of Saybolt Seconds Universal (SSU). Under standard temperature conditions, oil is allowed to flow thru a carefully calibrated orifice until a standard volume has passed. The number of seconds necessary for the given volume of oil to flow thru the orifice is the viscosity of the oil in Saybolt Seconds Universal.

The higher the viscosity, the more seconds it takes to pass thru the hole, or the higher the viscosity, the thicker the oil.

Viscosity is affected by temperature (*Fig. 13*), thus making it an important characteristic of refrigeration oils. The viscosity increases as the temperature decreases, or the lower the temperature the thicker the oil. In low temperature applications, this thickening of oil with its increasing resistance to flow is a major problem. As low temperatures occur in the evaporator, an oil that is too viscous thickens and may stay in the evaporator, thus decreasing the heat transfer and possibly creating a serious lack of lubrication in the compressor.

Oil may thin out or become less viscous at high temperatures. Too warm a crankcase may conceivably thin the oil to a point where it can no longer lubricate properly. A refrigeration oil must maintain sufficient body to lubricate at high temperatures and yet to be fluid enough to flow at low temperatures. Oil should be selected which has the lowest viscosity possible to do the assigned job.

Viscosity is also affected by the miscibility of the oil and the refrigerant. The miscibility of oil and refrigerants varies from almost no mixing (with Refrigerant 717, ammonia) to complete mixing (with some halogenated hydrocarbons such as Refrigerant 12).

Refrigerant 717 has almost no effect on the viscosity of a properly refined refrigeration oil. As it is not miscible, there is no dilution of oil and therefore no change in viscosity.

In the case of miscible refrigerants such as Refrigerant 12, the refrigerant mixes with and dilutes the oil, and lubrication must be performed by this mixture. This mixing reduces the viscosity of the oil.

CHART 29—OIL-REFRIGERANT VISCOSITY

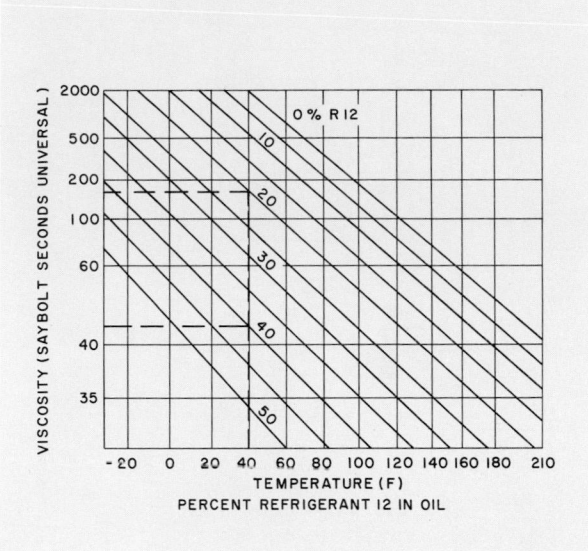

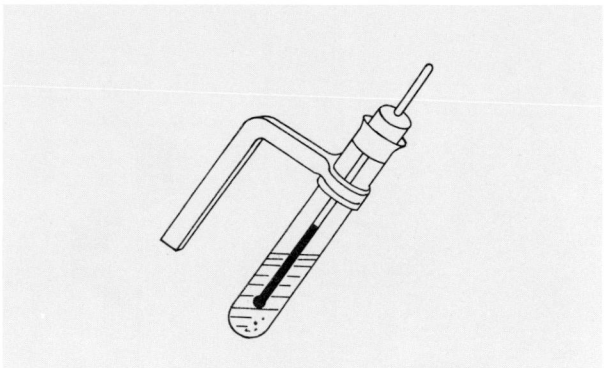

FIG. 14 — POUR POINT TUBE

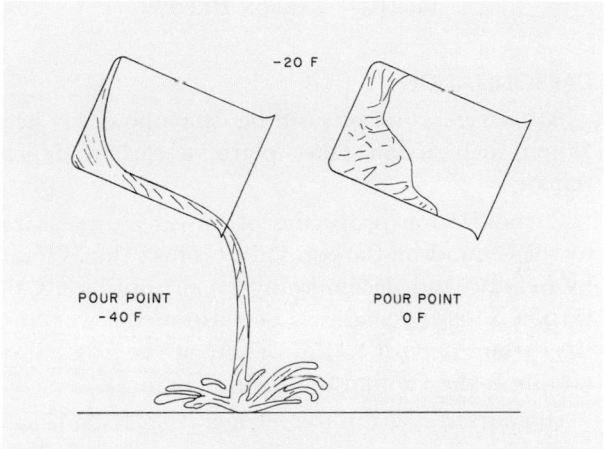

FIG. 15 — POUR POINT

Chart 29 shows viscosity change for mixtures of oil and Refrigerant 12. For example, oil at 40 F containing 20% Refrigerant 12 by weight has a viscosity of approximately 150 SSU.

When the amount of Refrigerant 12 is increased to 40%, the SSU is reduced to 45.

When oil and refrigerants are miscible, the oil is carried thru the system by the refrigerant. It is imperative that the oil be returned to the compressor. Keeping low side gas velocity up assures this proper oil return. With a completely miscible refrigerant, the oil is diluted sufficiently even at low temperatures to keep the viscosity low and to allow the oil to return easily with the refrigerant to the compressor.

There is a third group of refrigerants whose miscibility with oil varies. For example, Refrigerant 22 is completely miscible with oil at high temperatures, but at low temperatures it separates into two layers with the oil on top. When designing or selecting equipment for this group of refrigerants, great care must be taken to allow for low temperature separation of oil and refrigerant.

Oil separation in a flooded cooler necessitates the use of an oil bleed line from the bottom of the cooler to the suction loop. Because the oil is lighter than Refrigerant 22 and floats on top of the liquid refrigerant, an auxiliary oil bleed line from the side of the cooler is required.

POUR POINT

The pour point of an oil is that temperature at which it ceases to flow.

Pour point is simple to determine. Using the apparatus shown in the *Fig. 14*, the selected batch of oil is slowly cooled under test conditions until the oil no longer flows. This temperature is the pour point.

Figure 15 shows two oils with different pour points which have been cooled to the same temperature (−20 F). On the left the oil with a −40 F pour point flows freely. On the right the oil with the 0 F pour point does not flow.

Pour point depends on the wax content and/or viscosity.

With all refrigerants some oil is passed to the evaporator. Regardless of how small an amount, this oil must be returned to the compressor. In order that it may be returned, it must be able to flow thruout the system.

Oil pour point is very important with nonmiscible and partly miscible refrigerants; with the miscible refrigerants the viscoscity of the oil refrigerant mixture assumes greater importance as shown in *Chart 29, Oil-Refrigerant Viscosity.*

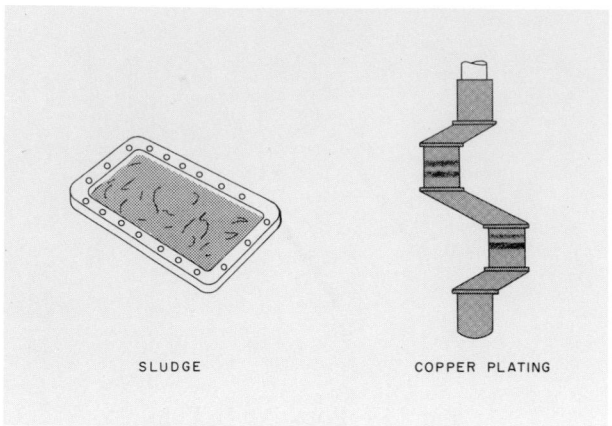

SLUDGE   COPPER PLATING

FIG. 16 — CARBON DEPOSIT

CARBONIZATION

All refrigeration oils can be decomposed by heat. When such action takes place, a carbon deposit remains.

Carbonization properties of an oil are measured by the Conradson Carbon Value. This value is found by heating and decomposing an oil until only the carbon deposit remains. The ratio of the weight of the carbon deposit to the weight of the original oil sample is the Conradson Carbon Value.

Hot surfaces within the refrigeration system sometimes decompose the oil. The carbon remaining is hard and adhesive in paraffin base oils, and forms sludge.

Naphthene base oils form a light fluffy carbon which, though a contaminant, is not as damaging as the hard carbon. However, neither type of carbon deposit is desirable as there is some indication that a relationship exists between oil breakdown, carbonization and copper plating (*Fig. 16*).

A good oil should not carbonize when in contact with hot surfaces encountered in the system during normal operation. A refrigeration oil should have as low a Conradson Carbon Value as practical.

FLOC POINT

All refrigeration oils contain some wax though the amount varies considerably. As the temperature of the oil decreases, the solubility of the wax also decreases. When there is more wax present than the oil can hold, some separates and precipitates.

The method used to determine the waxing tendencies of a refrigeration oil is the floc test (*Fig. 17*). A mixture of 10% oil and 90% Refrigerant 12 in a clear container is cooled until the wax starts to separate, turning the mixture cloudy. As cooling continues, small clusters of wax form. The tempera-

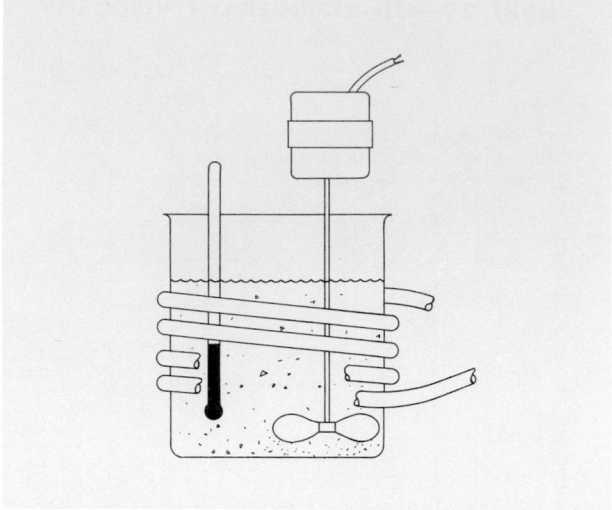

FIG. 17 — FLOC TEST

ture at which these clusters are first noticeable to the unaided eye is the floc point.

The free wax that is formed when a refrigeration oil is cooled can clog metering devices and restrict flow.

Wax normally deposits out in the colder parts of the system such as in the evaporator and its metering device (*Fig. 18*). Wax in the evaporator causes some loss of heat transfer; wax in the metering device can cause restriction or sticking.

A good refrigeration oil should not deposit wax when exposed to the lowest temperatures normally encountered in the refrigeration system.

NEUTRALIZATION

Almost all refrigeration oils have some acid tendencies. Nearly all oil contains material of uncertain composition referred to as organic acids. These are

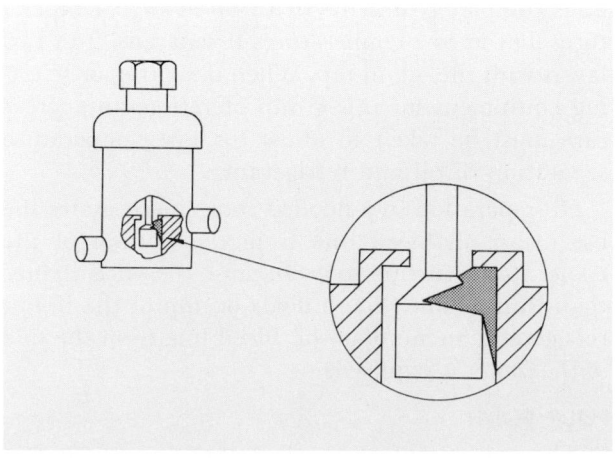

FIG. 18 — WAX DEPOSITS

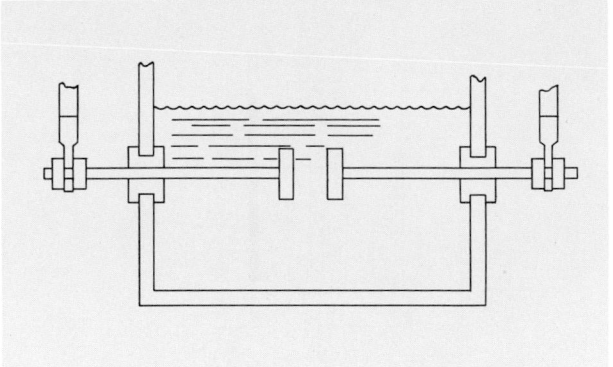

FIG. 19 — DIELECTRIC TESTING

usually harmless and should not be confused with mineral acids which are harmful.

The neutralization number is a measure of the amount of mineral acid, and is determined by measuring the amount of test fluid that must be added to the oil to bring it to a neutral condition. A low neutralization number means that few acids are contained in the oil.

Improper refining may leave a large proportion of corrosive acid present in an oil. A low neutralization number indicates a low acid content. Acids may corrode interior parts of the system; they react with motor insulation and other materials to form sludge which can eventually cause a complete system breakdown. A low neutralization number is highly desirable in refrigeration oils.

DIELECTRIC STRENGTH

Dielectric strength is a measure of the resistance of an oil to the passage of electric current. It is measured in kilovolts on a test cell as shown in *Fig. 19*. The poles in the cell are a predetermined distance apart. They are immersed in the oil so that current must pass thru the oil to flow from one pole to the other. The kilovolts necessary to cause a spark to jump this gap is known as the dielectric rating. Good refrigeration oils normally have a rating of over 25 kilovolts.

A dielectric rating is important because it is a measure of impurities in the oil. If the oil is free of foreign matter, it has a high resistance to current flow. If the oil contains impurities, its resistance to current flow is low.

The presence of foreign matter in a refrigeration system is sufficient reason for considering this test valuable. Hermetic motors make a high kilovolt rating necessary for refrigeration oils since a low kilovolt rating may be a contributing factor to shorted windings.

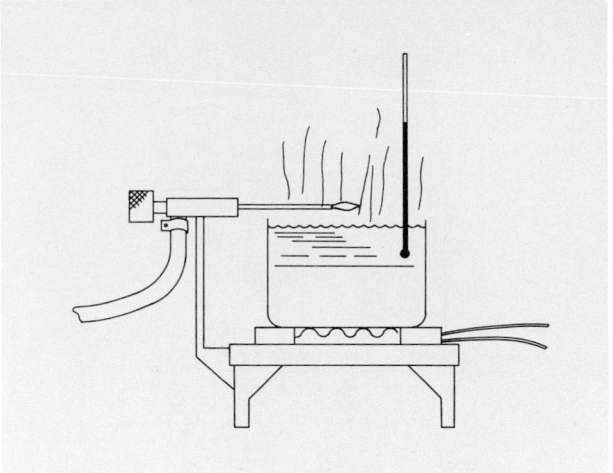

FIG. 20 — FLASH AND FIRE POINT TESTS

FLASH POINT AND FIRE POINT

The flash point of an oil is that temperature at which oil vapor flashes when exposed to a flame. The fire point is that temperature at which it continues to burn.

The apparatus shown in *Fig. 20* heats the oil while a small gas flame is passed closely over the surface of the oil. When a flash of fire is noted at some point on the surface, the flash point has been reached. The apparatus continues to heat the oil until it ignites and continues to burn. This is the fire point.

The flash point of a good refrigeration oil is well over 300 F. Temperatures obtained in the normal refrigeration system rarely reach this point. The test for flash and fire points is important as it is a means of detecting inferior blends.

It is possible to get an acceptable viscosity reading for a refrigerant oil by mixing a small amount of high viscosity oil with a large amount of low viscosity oil. The viscosity of the mixture indicates a satisfactory oil when actually the low viscosity oil is inferior and breaks down under normal use. Fortunately, this can be detected using the flash and fire point test which indicates the inferiority of the low viscosity oil.

OXIDATION STABILITY

Oxidation stability is the ability of refrigeration oil to remain stable in the presence of oxygen. The Sligh Oxidation Test is used to determine this stability (*Fig. 21*).

While exposed to oxygen in the flask, the oil is heated to a high temperature for an extended period of time. The solid sludge that is formed in the flask

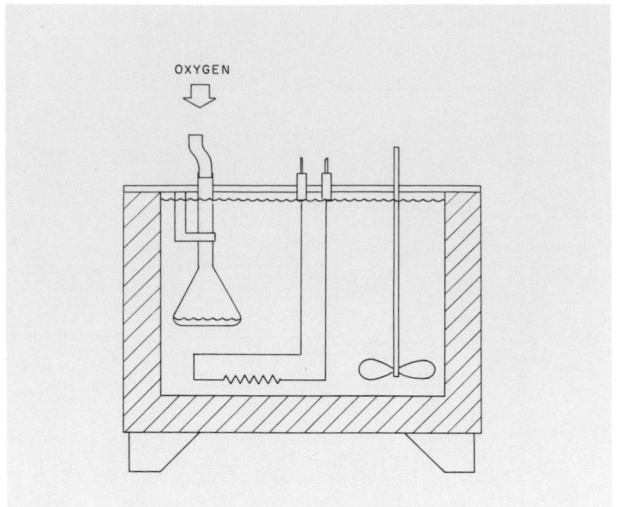

FIG. 21 — SLIGH OXIDATION TEST

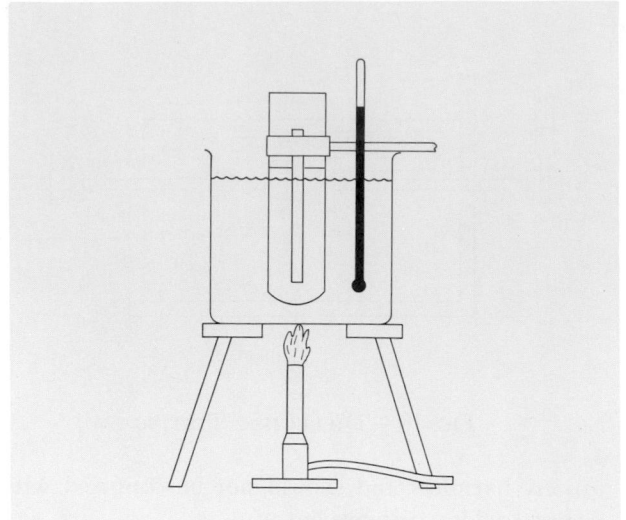

FIG. 23 — CORROSION TEST

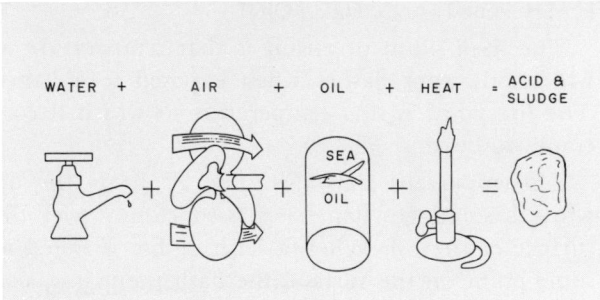

FIG. 22 — OIL BREAKDOWN

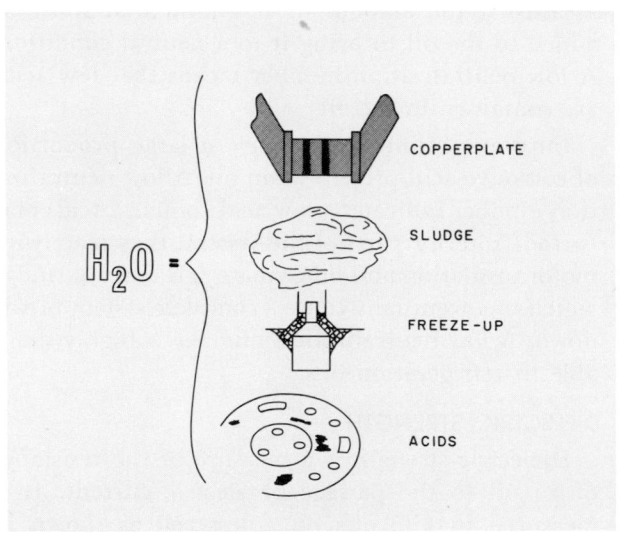

FIG. 24 — MOISTURE AND AIR IN THE SYSTEM

is weighed and reported as the Sligh Oxidation Number.

When air enters a system, some moisture generally accompanies it. The combination of moisture, air, refrigeration oil and discharge temperatures produces acid which creates sludge *(Fig. 22)*. If oil has a low Sligh Oxidation Number, the oil breakdown to acid and sludge is quite slow.

CORROSION TENDENCY

The corrosion tendency of a refrigeration oil is measured by the copper strip corrosion test *(Fig. 23)*. This test is intended to indicate the presence of undesirable sulfur compounds in the refrigeration oil.

A strip of polished copper is immersed in an oil sample in a test tube. This is subjected to temperatures around 200 F. After about 3 hours, the copper is removed from the oil, cleaned with a solvent, and examined for discoloration. If the copper is tarnished or pitted, then sulfur is present in the oil. Well refined oils rarely cause more than a slight tarnishing of copper in this test.

A good refrigeration oil should score negative in the copper strip corrosion test. If it does not, it contains sulfur in a corrosive form.

Sulfur alone is a deadly enemy of the refrigeration system but, in the presence of moisture, surfurous acid is formed, one of the most corrosive compounds in existence. Though the sulfurous acid converts immediately to sludge, this sludge is certain to create serious mechanical problems.

MOISTURE CONTENT

Moisture in any form is an enemy of the refrigeration system; moisture contributes to copper plating, formation of sludges and acids, and can cause freezeup *(Fig. 24)*. No refrigeration oil should contain enough moisture to affect the refrigeration system.

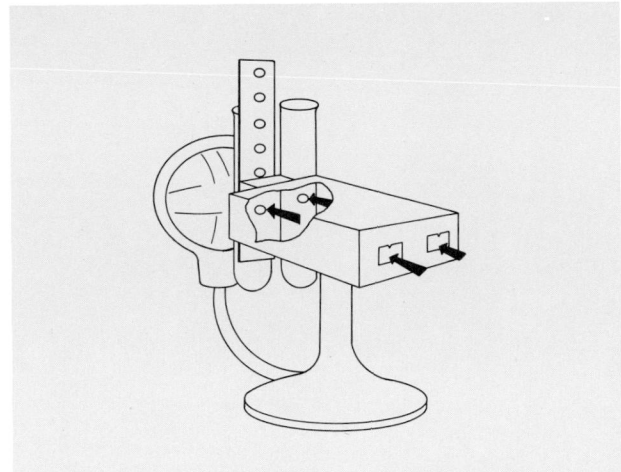

FIG. 25 — COLOR TESTING

COLOR

The color of a refrigeration oil is expressed by a numerical value that is based on comparison of the oil with certain color standards. This is done with the colorimeter shown in *Fig. 25*.

The color of a good refrigeration oil should be light but not water white. Continual refining of a lubricating oil results in a water white color. It also results in poor lubricating qualities.

Under-refining leaves a high content of unsaturated hydrocarbons which darken and discolor an oil. These are believed to be the constituents in oil that act as a solvent for copper. Therefore, the aim is to refine the oil sufficiently to remove these hydrocarbons but not so much as to destroy the lubricating quality.

SPECIFICATIONS

1. For open and hermetic reciprocating compressors at standard air conditioning levels, the following oil characteristics are typical:

   Viscosity, 150 ± 10 SSU at 100 F
        40 to 45 SSU at 210 F
   Dielectric (min.), 25 kv
   Pour Point (max.), −35 F
   Flash Point (min.), 330 F
   Neutralization Number (max.), .05
   Floc Point (max.), −70 F

2. For centrifugal compressors used for water cooling at air conditioning levels, the following are typical properties:

   Viscosity, 300 ± 25 SSU at 100 F
        50 to 55 SSU at 210 F
   Dielectric (min.), 25 kv
   Pour Point (max.), 20 F
   Flash Point (min.), 400 F
   Neutralization Number (max.), 0.1

3. For special applications, consult the equipment manufacturer.

Part 5
WATER CONDITIONING

CHAPTER 1. WATER CONDITIONING—GENERAL

This chapter will help engineers and owners to recognize the value of conditioned water and its effect on the economics of an air conditioning plant. Included are discussions of chemical and physical characteristics of water, properties of water sources in the United States, and characteristics of water circulating systems.

Since there are many companies specializing in the field of water conditioning, it is important to retain an experienced company on a continuous consulting basis to recommend and to administer conditioning for each system. This practice is recommended because the water conditioning program can vary, depending on the season of the year, geographical location of the structure to be air conditioned, and on the changing chemical characteristics of the make-up water source.

Water used in air conditioning systems may create problems with equipment such as scale formation, corrosion, and organic growths. As a universal solvent, water dissolves gases from the air and mineral matter from the soil and rock with which it comes in contact. Environmental conditions at the air conditioning equipment may promote slime and algae. To provide effective control for all components of a system, an appraisal of the system and water source is recommended. A water conditioning program should be considered for all systems, though in the final analysis it may be found unnecessary.

The techniques and equipment required to solve water problems vary with the character of the water, the kind and size of system, and the facilities available. An effective water treatment system can be selected for a particular job, or the program can be designed to treat only some of the water problems.

BENEFITS OF WATER CONDITIONING

A water conditioning program influences the economic aspects of a system; it helps to insure a satisfactory and more continuous operation of the air conditioning system. Water conditioning lowers power and maintenance costs, and increases the equipment life. The following discussion presents some of the reasons why these economic factors are affected.

A good water conditioning program produces high equipment efficiencies.

Hard water in a natural state is scale forming and, when heated, tends to deposit a lime scale. This deposit and other impurities accumulate on tubes, valves, pumps and pipe lines, restricting flow and affecting heat transfer. *Figure 1* shows scale deposit in a brass tube. The finned copper tubes show the effect of corrosion. A very thin film of scale is enough to affect seriously the efficiency of a heat transfer surface.

Chart 1 illustrates the effect of various thicknesses of scale (expressed as fouling factors) on the suction temperature of a typical flooded cooler.

Chart 2 illustrates the effect of scale on the condensing temperature of a typical water-cooled condenser. Since the amount of work accomplished by the refrigeration compressor is a function of the lift (condensing temperature minus suction temperature), minimum fouling factors result in minimum energy requirements.

Chart 3 shows how the compressor horsepower per ton varies with the scale factor, and how only a small amount of scale in the condenser can greatly increase the compressor horsepower.

FIG. 1 — TUBES SHOWING SCALE DEPOSIT AND CORROSION

CHART 1—EFFECT OF SCALE ON SUCTION TEMPERATURE

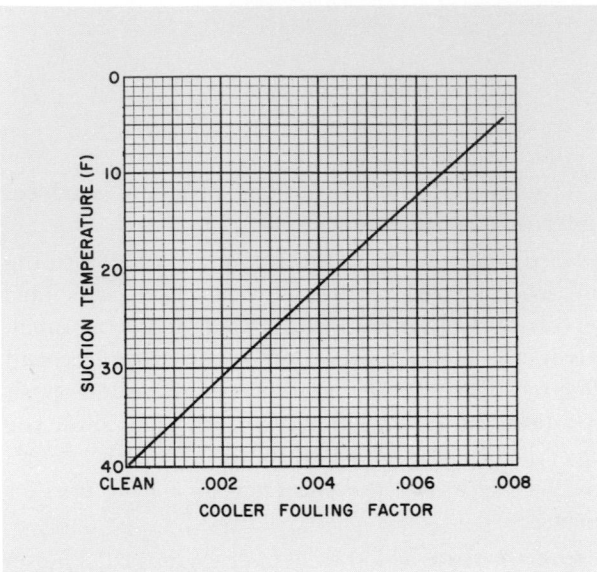

Cleaning may be necessary more frequently if too low a fouling factor is specified without a water conditioning program.

The fouling factor specified for selecting a refrigeration system should be chosen with judgment based on experience. *Table 11* in *Chapter 5* lists suggested fouling factors.

CHART 2—EFFECT OF SCALE ON CONDENSING TEMPERATURE

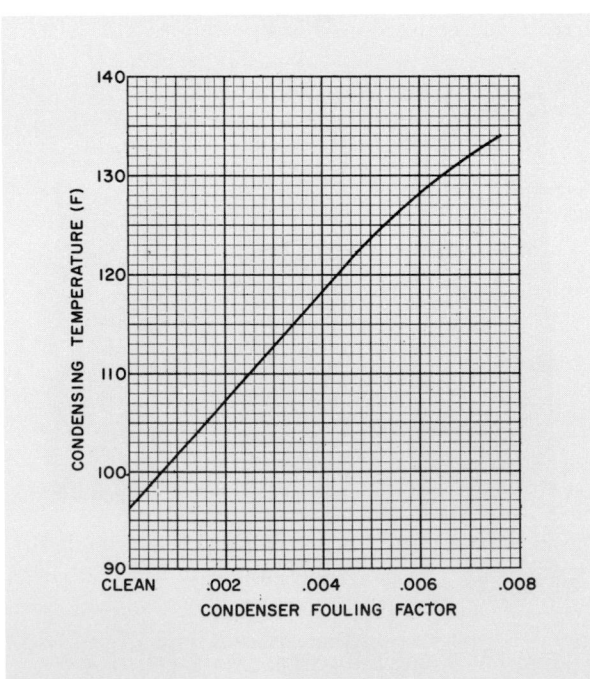

CHART 3—EFFECT OF SCALE ON COMPRESSOR HORSEPOWER
With Clean Cooler Tubes

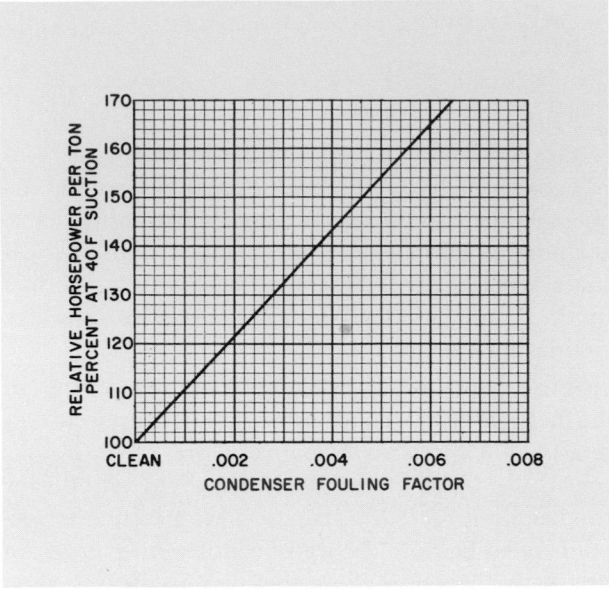

Lower equipment first cost is realized by a proper water conditioning program which allows the selection of a reasonable fouling factor.

Equipment first cost is often affected by the fouling factor which is specified by the engineer. Assume a drive horsepower is specified along with a particular fouling factor. The larger the fouling factor, the more heat exchanger surface is required to accomplish the work load. This increase is illustrated in *Table 1*.

WATER PROBLEMS

Problems produced by water fall into three categories which are discussed in detail in *Chapters 2, 3 and 4* respectively.

1. Scale formation caused by hard water reduces the heat transfer rate and increases the water pressure drop thru the heat exchanger.
2. Corrosion can be caused partly by absorption of gases from the air so that an aggressive condition is created and water attacks the exposed metal. In industrial areas this is frequently a major problem because of heavy pollution of the air by flue gases. Corrosion is prevalent in soft water areas.
3. Organic growths of slime and algae which form under certain environmental conditions can reduce the heat transfer rate by forming an insulating coating or can promote corrosion by pitting.

TABLE 1—HEAT TRANSFER SURFACE REQUIRED TO OFFSET FOULING

FOULING THERMAL RESISTANCE (hr) (sq ft) (deg F temp diff)/Btu‡	OVER-ALL HEAT TRANSFER COEFFICIENT† Btu/(hr) (sq ft) (deg F temp diff)‡	THICKNESS OF SCALE* APPROX. (in.)	INCREASE OF REQUIRED HEAT TRANSFER AREA† (approximate %)
clean tubes	850	.000	0
0.0005	595	.006	45
0.001	460	.012	85
0.002	315	.024	170
0.003	240	.036	250

*Assume a mean value for the thermal conductivity of the scale of 1.0 Btu/(hr) (sq ft) (deg F per ft).

†The over-all heat transfer coefficient U selected for this illustration is typical for a water-cooled refrigerant condenser. However, because it is possible to have different over-all heat transfer coefficients depending on the systems, the effect on the over-all heat transfer by the scale will vary.

‡Sq ft of inside surface of tube in heat exchanger.

WATER SOURCE APPRAISAL

An appraisal of the water supply source should include a chemical analysis and a determination of composition fluctuation over a period of time.

A water supply analysis together with the estimated water temperatures of the system makes possible a determination of the problems that may be encountered on a specific installation.

CONSTITUENTS AND CHARACTERISTICS OF WATER

The constituents of or impurities in water can be classified as dissolved solids, liquids or gases, and suspended matter.

An example of a dissolved solid is sodium chloride or calcium carbonate in solution. Dissolved materials cannot be removed by filtration. Oxygen and carbon dioxide are dissolved in water, but in water conditioning for air conditioning these gases are not normally determined in the water analysis. An example of suspended matter which can be removed by filtration is mud, clay or silt. *Table 2* illustrates the wide range of constituents in typical water sources. However, *pH* is omitted in *Table 2* because it may have a wide range in a particular water.

Several characteristics of water such as *pH* value, alkalinity, hardness and specific conductance are of particular importance in water conditioning and are explained in detail.

pH Value

The *pH* value is one of the most important control factors in water treatment. It is an arbitrary symbol adopted to express the degree of acidity or alkalinity of a water sample. Neutral water has a *pH* of 7.0. Values below 7.0 and approaching 0 are increasingly acid while values from 7.0 to 14.0 are increasingly alkaline. Most natural waters have a *pH* of 6.0 to 8.0. Water containing free mineral acids may have *pH* values below 4.5. A *pH* below 7.0 gives rise to corrosion of equipment with which water comes in contact. When the *pH* is high (above 7.5 or 8.0), calcium carbonate scale is deposited more readily.

TABLE 2—MINERAL ANALYSES TYPIFYING COMPOSITION OF WATERS AVAILABLE AND USED INDUSTRIALLY IN THE U.S.A.

SUBSTANCE	UNIT	LOCATION OR AREA[a,b]								
		(1)	(2)	(3)	(4)	(5)	(6)	(7)	(8)	(9)
Silica	SiO_2	2	6	12	37	10	9	22	14	
Iron	Fe	0	0	0	1	0	0	0	2	
Calcium	Ca	6	5	36	62	92	96	3	155	400
Magnesium	Mg	1	2	8	18	34	27	2	46	1300
Sodium	Na	2	6	7	44	8	183	215	78	11000
Potassium	K	1	1	1		1	1	10	3	400
Bicarbonate	HCO_3	14	13	119	202	339	334	549	210	150
Sulfate	SO_4	10	2	22	135	84	121	11	389	2700
Chloride	Cl	2	10	13	13	10	280	22	117	19000
Nitrate	NO_3	1		0	2	13	0	1	3	
Dissolved Solids		31	66	165	426	434	983	564	948	35000
Carbonate Hardness	$CaCO_3$	12	11	98	165	287	274	8	172	125
Noncarbonate Hardness	$CaSO_4$	5	7	18	40	58	54	0	295	5900

[a] All values are parts per million of the unit cited to nearest whole number (Reference 1).

[b] Numbers indicate location or area as follows:
  (1) Catskill supply — New York City
  (2) Swamp Water (Colored) Black Creek, Middleburg, Florida
  (3) Niagara River (Filtered) Niagara Falls, New York
  (4) Missouri River (Untreated) — Average
  (5) Well Waters — Public Supply — Dayton, Ohio, 30-60 ft
  (6) Well Water — Maywood, Illinois, 2090 ft
  (7) Well Water — Smithfield, Va., 330 ft
  (8) Well Water — Roswell, New Mexico
  (9) Ocean Water — Average

From ASHRAE Guide and Data Book, 1961. Used by permission.

The pH is usually measured by an electrometric pH meter in the laboratory. It can also be determined by the use of color indicators, comparing the solution with standard shades of colors over the range of a particular indicator. There are several standard indicator slides to cover the range from 0–14.0 pH.

The definition of pH is the logarithm of the reciprocal of the hydrogen ion concentration (moles per liter).

$$pH = \log_{10} \frac{1}{H^+}$$

where H^+ is the hydrogen ion concentration.

When pure water, H_2O, ionizes into H^+ and OH^- ions, 0.0000001 grams of hydrogen ions per liter are liberated. This can be written as 1×10^{-7}, or as a pH of 7.0.

Alkalinity

Alkalinity is the most important characteristic of a water when determining the scale forming tendency. It is defined in *Chapter 6*. Generally, alkalinity is a measure of the acid neutralizing power of a water, and is determined by titration with a standard acid to the point of color change of a color indicator.

Alkalinity may be classified in two ways with respect to a pH value as follows:

1. Phenolphthalein alkalinity is a measure of the carbonate and caustic or hydroxyl ions. It is determined by titrating down to a pH of 8.3. In natural waters phenolphthalein alkalinity is usually absent; however, it is sometimes found in water which has been softened by lime or soda ash.

2. Methyl orange (total) alkalinity is a measure of all the alkaline substances, including the phenolphthalein alkalinity. It is determined by titrating down to a pH of 4.3.

The difference between the two kinds of alkalinity is represented by the presence of the bicarbonate ion.

The phenolphthalein and methyl orange tests are commonly used in water conditioning control. *Figure 2* indicates a pH scale showing which constituents can exist at various pH values.

Methyl orange and phenolphthalein alkalinities may be interpreted as follows:

1. When there is no indication of phenolphthalein alkalinity, this means that all of the alkalinity is caused mainly by calcium, magnesium and sodium bicarbonates. This water has a pH value less than 8.5.

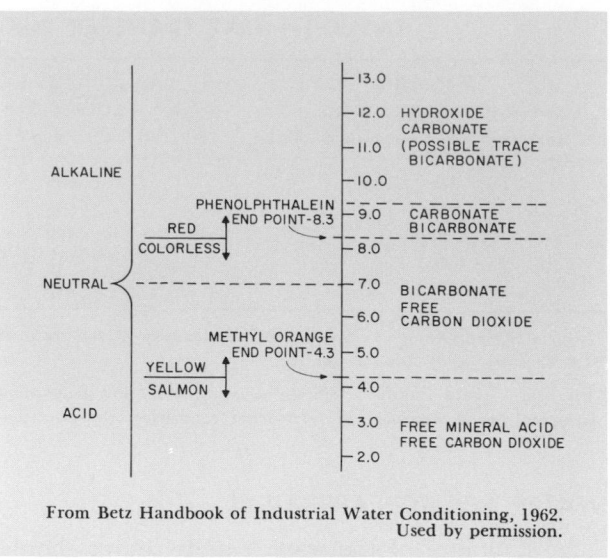

FIG. 2 — pH SCALE FOR CONSTITUENTS EXISTING IN NATURAL AND TREATED WATERS

2. When twice the phenolphthalein alkalinity is less than or equal to the methyl orange alkalinity, the alkalinity is assumed to be caused by calcium, magnesium and sodium carbonates and bicarbonates. This water has a pH greater than 8.5.

3. When twice the phenolphthalein alkalinity exceeds the methyl orange alkalinity, there is no bicarbonate alkalinity. In this case the alkalinity is caused by calcium, magnesium and sodium hydroxides and carbonates. Such water has a pH value greater than 8.5.

The amount of alkalinity and its type in a water determine to a large extent the type of water conditioning required. Examples of how the above rules may be used follow.

Example 1 — Bicarbonate Alkalinity

Methyl orange (MO) alkalinity = 100 ppm as $CaCO_3$
Phenolphthalein (P) alkalinity =   0 ppm as $CaCO_3$

Since this analysis indicates 0 ppm of phenolphthalein alkalinity, all of the alkalinity in the water is caused by bicarbonates.

Example 2 — Bicarbonate and Carbonate Alkalinity

Methyl orange (MO) alkalinity = 100 ppm as $CaCO_3$
Phenolphthalein (P) alkalinity =   5 ppm as $CaCO_3$

Twice the phenolphthalein alkalinity is less than the methyl orange alkalinity. Therefore, the alkalinity is caused by bicarbonates and carbonates.

Two times the phenolphthalein alkalinity $(2 \times 5) = 10$ ppm carbonate alkalinity as $CaCO_3$.

Bicarbonate alkalinity equals total alkalinity minus carbonate alkalinity $(100 - 10) = 90$ ppm as $CaCO_3$.

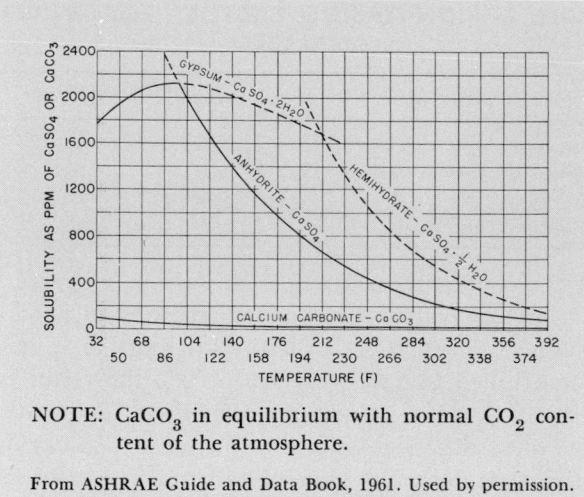

NOTE: $CaCO_3$ in equilibrium with normal CO_2 content of the atmosphere.

From ASHRAE Guide and Data Book, 1961. Used by permission.

FIG. 3 — COMPARISON OF SOLUBILITIES OF CALCIUM SULFATE AND CALCIUM CARBONATE

Example 3 — Carbonate and Hydroxide Alkalinity

Methyl orange (MO) alkalinity = 100 ppm as $CaCO_3$
Phenolphthalein (P) alkalinity = 55 ppm as $CaCO_3$

Since twice the phenolphthalein alkalinity (2×55) exceeds the methyl orange alkalinity, there is no bicarbonate alkalinity present. All of the alkalinity is present as carbonates and hydroxides.

Two times phenolphthalein alkalinity minus methyl orange alkalinity ($110 - 100$) = 10 ppm hydroxide alkalinity as $CaCO_3$.

Methyl orange alkalinity minus hydroxide alkalinity ($100 - 10$) = 90 ppm calcium alkalinity as $CaCO_3$.

Examples 1 and 2 are commonly encountered in both once-thru and recirculating condensing water systems. *Example 3* is seldom encountered.

Hardness

Hardness is represented by the sum of calcium and magnesium salts in water although it may include aluminum, iron, manganese, strontium, or zinc.

Carbonate (temporary) hardness is attributed to carbonates and bicarbonates of calcium and/or magnesium expressed in ppm as $CaCO_3$. The remainder of the hardness is known as noncarbonate (permanent) hardness which is due to the sulfates, chlorides, and/or nitrates of calcium and/or magnesium expressed in ppm as $CaCO_3$.

Noncarbonate hardness is not as serious a factor in water conditioning because it has a solubility which is approximately *70* times greater than the carbonate hardness. In many cases, water may contain as much as 1200 ppm of noncarbonate hardness and not deposit a calcium sulfate scale *(Fig. 3)*.

Water hardness may be classified as follows:

HARDNESS (ppm as $CaCO_3$)	CLASSIFICATION
less than 15	very soft
15 to 50	soft
50 to 100	medium hard
100 to 200	hard
greater than 200	very hard

Hardness is sometimes reduced at the municipal water distributing plant by the use of a cold lime, cold lime soda, or by a sodium cation exchanger (zeolite) water process. The last method is used also in the home for domestic softening.

In domestic use hardness is recognized by the difficulty in obtaining lather without an excessive consumption of soap. The insoluble sticky curd is the result of action of the soap with the hardness. With hard water a scale is formed in vessels when water is boiled.

The most accurate method for determining the hardness of a water is by calculation from complete chemical analysis of the water. Many determinations of hardness made at water treatment plants are made with soap solutions. These results tend to be lower and less accurate than those based on laboratory tests. Alkalinity tests may be used to determine the amount of carbonate and noncarbonate hardness for control purposes as follows:

1. When the methyl orange or total alkalinity exceeds the total hardness, all of the hardness is caused by bicarbonates and carbonates.
2. When the methyl orange or total alkalinity is less than the total hardness:
   a. Carbonate hardness equals the alkalinity.
   b. Noncarbonate hardness equals the total hardness less the methyl orange alkalinity.

Examples of interpretation of hardness are given here.

Example 4 — Carbonate Hardness

Methyl orange (MO) alkalinity = 100 ppm as $CaCO_3$
Total hardness = 95 ppm as $CaCO_3$.

Methyl orange alkalinity in this case exceeds the total hardness; therefore, all of the hardness is present as carbonate hardness.

Example 5 — Carbonate and Noncarbonate Hardness

Methyl orange (MO) alkalinity = 100 ppm as $CaCO_3$
Total hardness = 150 ppm as $CaCO_3$

Total hardness exceeds the methyl orange alkalinity. The carbonate hardness, therefore, equals the alkalinity, and is equal to 100 ppm as $CaCO_3$.

Noncarbonate hardness equals total hardness minus methyl orange alkalinity ($150 - 100$) or 50 ppm as $CaCO_3$.

Specific Conductance

The specific conductance of a water, defined in *Chapter 6*, indicates whether galvanic corrosion may be a problem. See *Chapter 3*.

METHOD OF ANALYSIS AND EXPRESSION OF RESULTS

Water contains a wide variety of dissolved solids or compounds which are mostly metals joined to acid radicals. When a sample is analyzed in the laboratory, it is not feasible to measure the compound. The metal and acid radicals are, therefore, determined separately. Results are expressed in parts per million (ppm).

When a salt, i.e., sodium chloride, $NaCl$, is dissolved in water, the two ions of which it is composed may be considered as existing separately; Na and Cl can thus be measured independently. It is practical to determine the metal (calcium) separately, irrespective of its combination with the bicarbonate radical, HCO_3, forming calcium bicarbonate, $Ca(HCO_3)_2$, or with the sulfate radical, SO_4, forming calcium sulfate, $CaSO_4$. Similarly the acid radicals are determined separately; the bicarbonate, HCO_3, the sulfate, SO_4, and the choride, Cl, are measured irrespective of the metals combined with them. In the ionic form, each element or radical is separated individually. This method is convenient and easy to interpret and use.

Results are expressed in parts per million (ppm), although grains per gallon (gpg) or equivalents per million (epm) are sometimes used. References 2, 3, and 4 describe the entire procedure for making water analyses and several methods of reporting results. When units are expressed in grains per gallon, multiply gpg by 17.1 to obtain ppm (*Table 3*).

It is common practice to report some of the constituents on a common unit or equivalent weight basis. The unit weight of the individual substance is calculated by the unit weight of another substance. For example, calcium is usually expressed in terms of calcium carbonate, $CaCO_3$. When constituents are measured by the same unit weight basis, they can be added or subtracted directly for convenience in calculating the amount of chemicals required for treatment. Another reason to report some of the constituents on a common unit basis is to simplify the calculation in order to obtain the noncarbonate hardness when the total hardness and alkalinity are known.

The equivalent weight method is illustrated here. Basic chemistry states that 20 parts (by weight) of calcium, Ca, combine with 30 parts of carbonate,

TABLE 3—CONVERSION FACTORS FOR WATER ANALYSES

TO CONVERT	INTO	MULTIPLY BY
Grains per U. S. gallon	ppm	17.1
Grains per Imperial gallon	ppm	14.25
Grams per liter	ppm	1000
Milligrams per liter	ppm	1

CO_3, to form calcium carbonate, resulting in a total weight of 20 + 30 or 50. If the calcium ion (ppm) is multiplied by (20 + 30)/20 or 2.5, the result is ppm of calcium as calcium carbonate. Thus, 100 ppm of calcium ion represents the same quantity of calcium as does 250 ppm of calcium expressed as calcium carbonate, $CaCO_3$. In a similar manner, substances other than calcium and carbonate may be expressed as calcium carbonate by dividing the calcium carbonate total equivalent weight (20 + 30) by the equivalent weight of the other substance. For example, 12.16 parts of magnesium combine with the same quantity of any acid radical as does 20 parts of calcium. The factor for magnesium is (20 + 30)/12.16 or 4.12. Thus, 50 ppm of magnesium ion is the same quantity as 206 ppm (4.12 x 50) of magnesium expressed as $CaCO_3$. Factors for determining equivalent weights are found in References 2, 3 and 5. (The unit equivalent weight of an element is numerically equal to the atomic weight divided by the valence of the element. Equivalent weight of an acid radical such as CO_3, HCO_3, etc., is numerically equal to the molecular weight of the radical divided by the valence of the radical.)

Normally the metals and acid radicals are expressed in the ionic form. Hardness (both carbonate and noncarbonate), phenolphthalein alkalinity, and methyl orange alkalinity are normally given as ppm of $CaCO_3$. The undissolved or suspended solids are expressed in ppm. Usually, analysis of dissolved gases is not made in water conditioning but, if required, it is reported as parts per million of the dissolved gas.

WATER ANALYSIS DATA TABLE

Table 13 in *Chapter 5* was compiled from *The Industrial Utility of Public Water Supplies in the United States*, U. S. Geological Survey Water Supply Papers 1299 and 1300. With this information and the expected water temperature encountered in a system, it is possible to arrive at a reasonable guide to the problems that may be expected on a specific installation. Notes accompanying the table explain the source for each column.

APPRAISAL OF WATER SYSTEMS

System appraisal includes the number and types of water circuits, materials of construction, and equipment location. Each type of water circuit requires a different kind of water conditioning.

ONCE-THRU SYSTEM

A once-thru system may pose either a scaling or corrosion problem, but usually not both. If extensive water conditioning is required, it may be more economical either to design equipment with a large scale factor and to clean it frequently or to use expensive corrosion resistant materials rather than use water conditioning. Slime and algae may also be a problem in this system.

CLOSED RECIRCULATING SYSTEM

Conditioning to prevent corrosion is required, but seldom is scaling or slime and algae a problem.

OPEN RECIRCULATING SYSTEM

This system invariably has both a scaling and corrosion problem. During the warmer months slime and algae need control.

CHARACTERISTICS OF OPEN RECIRCULATING WATER SYSTEMS

Although there are no special characteristics of once-thru and closed recirculating systems to interest the water conditioning engineer, there are several characteristics of open recirculating water systems to cause concern. They are evaporation, windage, cycles of concentration, and control by bleed-off.

Evaporation

The evaporation loss in a cooling tower or evaporative condenser is approximately 1% of the water quantity circulated per 10 degree drop in water temperature thru the tower. This is accurate enough for most calculations. The following equation is a more precise method:

$$gpm_e = \frac{tr \times hrf \times 24}{h_{fg}} \qquad (1)$$

where:

gpm_e = water evaporated
tr = refrigerating capacity (tons)
hrf = heat rejection factor of refrigeration machine
h_{fg} = heat of evaporation of water, normally 1050 Btu/lb

Compression refrigeration machines in normal air conditioning evaporates approximately 0.03 gpm per ton for cooling towers and evaporative condenser equipment.

Windage

Windage, a characteristic of spray systems, consists of small droplets of water carried away from the tower, spray pond or evaporative condenser by the wind. Dissolved minerals as well as water are lost by windage. The amount of windage varies with each type of cooling tower, but these percentages may be taken as typical windage losses, based on the rate of circulation.

HEAT REJECTION EQUIPMENT	WINDAGE LOSS (%)
Spray ponds	1.0 to 5.0
Atmospheric towers	0.3 to 1.0
Mechanical draft towers	0.1 to 0.3
Evaporative condensers	0.0 to 0.1

Cycles of Concentration

Cycles of concentration is the ratio of dissolved solids in the recirculating water to the dissolved solids in the make-up water. For example, with three cycles of concentration the dissolved solids in the recirculating water are three times as great as in the make-up water.

The evaporative process, used to cool water in open recirculating systems, concentrates the dissolved solids in the water. This is a characteristic of cooling towers, evaporative condensers and spray ponds. For example, a 50 ton system using make-up water having a hardness of 110 ppm and other scale-forming salts has an evaporation of about 1½ gallons per minute. Approximately 10 grains of mineral salts (equivalent to two aspirin tablets) enter the system each minute. Assuming a 50% load factor, one pound of salt enters the system every 24 hours, or 30 pounds per month. It is important to carry away the scale forming salts before they deposit on the equipment. Therefore, salt concentration is controlled by bleed-off. Windage loss removes some of the dissolved salts.

A typical flow diagram for an open condenser water cooling system is illustrated in *Fig. 4*. The cycles of concentration C may be calculated as follows:

$$C = \frac{gpm_e + gpm_b + gpm_w}{gpm_b + gpm_w} \qquad (2)$$

where:

gpm_b = water lost by bleed-off
gpm_e = water lost by evaporation
gpm_w = water lost by windage

All quantities may be expressed either as shown or as a percent of recirculated water, providing all are represented in the same units.

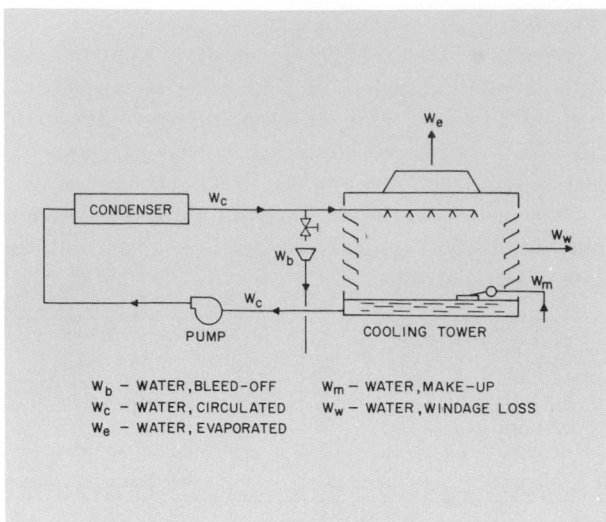

FIG. 4 — TYPICAL FLOW DIAGRAM OF OPEN
CONDENSER WATER COOLING SYSTEM

The equation may be simplified by combining
losses due to bleed-off and windage.

$$C = \frac{gpm_e + gpm_{bw}}{gpm_{bw}} \qquad (3)$$

where:

gpm_{bw} = water lost by bleed-off and windage (gpm)

CHART 4—RELATIONSHIP BETWEEN EVAPORATION LOSS, WINDAGE LOSS AND CYCLES OF CONCENTRATION

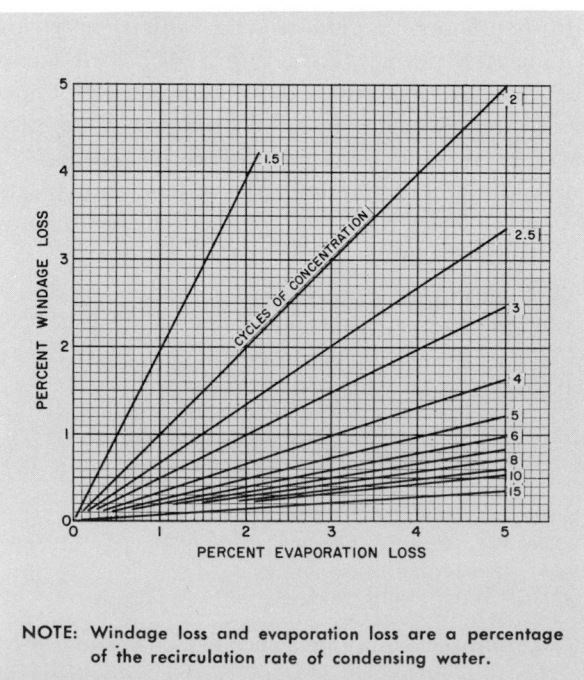

NOTE: Windage loss and evaporation loss are a percentage
of the recirculation rate of condensing water.

The relationship between evaporation loss, windage loss, and cycles of concentration is shown in
Chart 4. For an example, assume a 10 degree drop
in water flow thru the tower and no bleed-off. This
corresponds to an evaporation loss of 1.0% of the
recirculated water quantity. From the typical windage losses (1-5%) illustrated for the spray pond,
assume a loss of 2%. From *Chart 4* the cycles of
concentration is 1.5. If the tower is an atmospheric
type and a 0.6% windage loss is assumed, the cycles
of concentration is 2.7. In the case of a mechanical
draft tower with a windage loss of 0.2% and the
same temperature drop, the cycles of concentration
is 6.0.

The chloride test[2] is used to determine the concentration of solids in both make-up and recirculated water. It is not an effective test for make-up
water when the chloride content is low (0.5-1.0 ppm).

Control of Cycles of Concentration by Bleed-off

Bleed-off is normally required from all open recirculating systems to limit the concentration of solids.
Bleed-off is more readily obtained by calculating the
combined windage and bleed-off, and then subtracting the windage as follows. See *Equations 2 and 3*
for definition of symbols.

$$gpm_{bw} = \frac{gpm_e}{C-1} \qquad (4)$$

$$gpm_b = gpm_{bw} - gpm_w \qquad (5)$$

Bleed-off can be obtained in one step by this
formula:

$$gpm_b = \frac{gpm_e + gpm_w(1-C)}{C-1} \qquad (6)$$

Examples 6 and 7 illustrate the use of cycles of
concentration and evaporation losses to determine
the required bleed-off.

Example 6 — Compressor System

Given:

    Mechanical draft cooling tower
    Refrigeration load   = 100 tons
    Heat rejection factor = 1.25
    Condensing water temperature rise = 10 degrees
    Make-up water methyl orange alkalinity
        = 40 ppm as $CaCO_3$
    Methyl orange alkalinity in recirculated water,
        not to exceed 170 ppm as $CaCO_3$

Find:

    Cycles of concentration
    Recirculated water quantity (gpm)
    Evaporation loss (gpm)
    Windage loss (gpm)
    Bleed-off requirements (gpm)
    Make-up water (gpm)

Solution:

Cycles of concentration, assuming the air does not affect the alkalinity,

$$= \frac{\text{MO alkalinity in recirculated water}}{\text{MO alkalinity in make-up water}} = \frac{170}{40} = 4.25$$

Condensing water quantity

$$= \frac{tr \times hrf \times 24}{\text{temp rise}} = \frac{100 \times 1.25 \times 24}{10} = 300 \text{ gpm}$$

Evaporation loss

$$= \frac{tr \times hrf \times 24}{h_{fg}} = \frac{100 \times 1.25 \times 24}{1050}$$

$= 2.85$ gpm, equivalent to $2.85/300 \times 100$ or 0.95% of the recirculated water flow.

The evaporation loss may also be calculated as 1% of the recirculated water quantity for each 10 degree drop in water temperature at the tower, i.e. $300 \times 0.01 = 3.0$ gpm. This is accurate enough for the usual calculation.

Windage loss $= 0.1\text{-}0.3\%$ of the recirculated water quantity; assume 0.2%.

$0.002 \times 300 = 0.6$ gpm.

Bleed-off requirement (Equations 4 and 5)

$$gpm_{bw} = \frac{gpm_e}{C-1} = \frac{2.85}{4.25-1} = 0.88 \text{ gpm}$$

$$gpm_b = gpm_{bw} - gpm_w = 0.88 - 0.60 = 0.28 \text{ gpm}$$

Make-up water $=$ evaporation $+$ windage $+$ bleed-off
$$= 2.85 + 0.60 + 0.28 = 3.73 \text{ gpm}$$

Example 7 — Absorption System

Given:

Absorption refrigeration machine
Refrigeration load $= 100$ tons
Heat rejection factor $= 2.6$
Condensing water temperature rise $= 15$ degrees
Make-up water methyl orange alkalinity $= 40$ ppm as $CaCO_3$
Recirculated water methyl orange alkalinity, not to exceed 170 ppm as $CaCO_3$.

Find:

Cycles of concentration
Recirculated water quantity (gpm)
Evaporation loss (gpm)
Windage loss (gpm)
Bleed-off requirement (gpm)
Make-up water (gpm)

Solution:

Cycles of concentration
$= 170/40 = 4.25$ (as in Example 6)

Recirculated water quantity
$$= \frac{100 \times 2.6 \times 24}{150} = 417 \text{ gpm}$$

$$gpm_e = \frac{100 \times 2.6 \times 24}{1050} = 5.95 \text{ gpm}$$

$$gpm_w = .002 \times 417 = 0.83 \text{ gpm}$$

$$gpm_{bw} = \frac{gpm_e}{C-1} = \frac{5.95}{4.25-1} = 1.83 \text{ gpm}$$

$$gpm_b = gpm_{bw} - gpm_w = 1.83 - 0.83 = 1.0 \text{ gpm}$$

Make-up water $=$ evaporation $+$ windage $+$ bleed-off
$$= 5.95 + 0.83 + 1.00 = 7.78 \text{ gpm}$$

CHAPTER 2. SCALE AND DEPOSIT CONTROL

This chapter reviews the causes of scale and deposits and methods used to determine scale-forming tendencies and systems to prevent their formation.

When water is heated or evaporated, the formation of insoluble scale causes serious difficulty in air conditioning systems *(Chapter 1)*. Scale forms a protective coating, reducing corrosion, but at the same time reduces heat transfer efficiency and capacity. It also increases the power requirements.

The most common scale deposit found in air conditioning equipment is calcium carbonate, although there may be small amounts of magnesium carbonate and calcium sulfate present. Some waters having a high iron content leave an iron oxide deposit. Polyphosphates which are sometimes used to prevent a calcium carbonate scale deposit may precipitate and form a sludge of calcium phosphate.

TYPES AND CAUSES OF SCALE AND DEPOSITS

Calcium carbonate and calcium sulfate, two of the materials dissolved in water, have decreasing solubilities with rising temperature *(Fig. 3 and 5)*. High temperature surfaces such as condensers are more susceptible to scale formation than the low temperature surfaces in the same water system.

CALCIUM CARBONATE SCALE

The primary factors causing calcium carbonate scale in a water system are:

1. High methyl orange alkalinity, ppm as $CaCO_3$.
2. High calcium content, ppm as $CaCO_3$
3. High pH
4. High temperature
5. High dissolved solids, ppm

Normally, methyl orange alkalinity is a good measure of the amount of calcium bicarbonate in the water. Calcium carbonate is formed by the decomposition of calcium bicarbonate according to the chemical reaction:

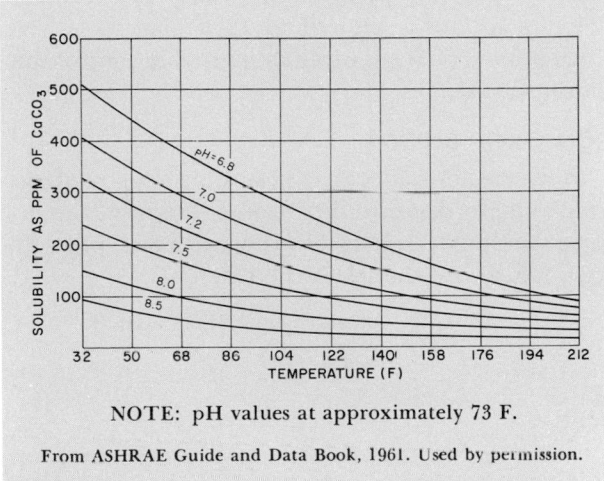

NOTE: pH values at approximately 73 F.

FIG. 5 — SOLUBILITY OF CALCIUM CARBONATE IN DISTILLED WATER CONTAINING CARBON DIOXIDE

$$Ca(HCO_3)_2 + \text{heat} \rightarrow CaCO_3 + CO_2 + H_2O$$

$$\underset{\text{bicarbonate}}{\text{calcium}} + \text{heat} \rightarrow \underset{\text{carbonate}}{\text{calcium}} + \underset{\text{dioxide}}{\text{carbon}} + \text{water}$$

Conversely, carbonates are readily converted to the more soluble bicarbonates by the addition of carbon dioxide or other acidic materials. (Carbon dioxide gas in water forms carbonic acid.) This explains the apparent increase in solubility of calcium carbonate with a lowering of *pH (Fig..5)*, the carbonate going into solution as bicarbonate. Calcium bicarbonate is several times more soluble than calcium carbonate. The addition of acid (explained in this chapter) also produces another effect by changing the bicarbonate and carbonate to a more soluble compound, thus allowing a higher concentration of chemicals in the water. Thus, a high methyl orange alkalinity and a high calcium content aid the deposition of calcium carbonate.

As illustrated in *Fig. 5*, a low *pH* increases the solubility of calcium carbonate, whereas a high *pH* aids the deposition of calcium carbonate.

A high temperature decreases the solubility of calcium carbonate and aids the scale deposit *(Fig. 5)*.

Dissolved solids of high concentration increase the scale deposit; however, the effect is not great.

CALCIUM SULFATE SCALE

Calcium sulfate has a high solubility (*Fig. 3*), and is seldom a problem in water conditioning unless there is a high sulfate in the make-up water. It may be formed (1) by the action of sulfur dioxide gas from the air dissolving in water to form sulfurous, H_2SO_3, or sulfuric, H_2SO_4, acid which in turn combines with calcium carbonate, or (2) by feeding sulfuric acid for calcium carbonate scale control. Calcium sulfate is limited to 1200 ppm as $CaCO_3$ (1630 ppm as $CaSO_4$). Bleed-off is used to control this concentration.

IRON OXIDE DEPOSIT

In the case of well waters possessing a high natural iron content, deposition of iron oxide from the decomposition of ferrous bicarbonate takes place in accordance with the following reaction:

$$4Fe(HCO_3)_2 + O_2 \rightarrow 2Fe_2O_3 + 8CO_2 + 4H_2O$$

$$\underset{\text{bicarbonate}}{\text{ferrous}} + \text{oxygen} \rightarrow \underset{\text{oxide}}{\text{iron}} + \underset{\text{dioxide}}{\text{carbon}} + \text{water}$$

SLUDGE

Sludge may be caused by the products of corrosion in the system. Polyphosphates which are used to prevent calcium carbonate scale formation may cause a precipitation of calcium phosphate whenever the concentration of orthophosphate is too high. (Refer to the discussion of scale inhibitors in this chapter.) A sludge may settle out in the water boxes or tubes. Consequently, a minimum velocity for a solution to prevent sludge settling in the tubes is three fps.

PREDICTION OF SCALE-FORMING TENDENCIES

Water which forms a mild scale can be corrosive, whereas water which forms a heavy scale is less likely to be corrosive. Corrosive waters usually do not form a scale. If a correct balance is found in water conditioning, the water will be neither corrosive nor scale-forming, or only mildly so. Two indexes can be used to predict scale-forming tendencies.

LANGELIER SATURATION INDEX

Professor W. F. Langelier[6] advanced the idea, now widely accepted, of using a calculated saturation index for calcium carbonate for predicting the scaling or corrosive tendencies of a natural water. Calcium carbonate is usually the chief ingredient of scale found on heat transfer surfaces in cooling water systems. The scale that forms under moderate tem-

TABLE 4—PREDICTION OF WATER CHARACTERISTICS BY THE LANGELIER SATURATION INDEX

LANGELIER SATURATION INDEX	TENDENCY OF WATER
+2.0	scale-forming, and for practical purposes noncorrosive
+0.5	slightly corrosive and scale-forming
0.0	balanced, but pitting corrosion possible
−0.5	slightly corrosive and nonscale-forming
−2.0	serious corrosion

perature conditions (50-130 F) is caused by the conversion of a bicarbonate into calcium carbonate upon heating; the scale is also affected by an increase in alkalinity sufficient to cause supersaturation with respect to the calcium carbonate. *pH* has a marked effect on the solubility of calcium carbonate.

The Langelier Saturation Index (I_s) is determined by the following:

$$I_s = pH - pH_s \qquad (7)$$

where:

$pH = pH$ from test
pH_s = calculated pH of saturation of the calcium carbonate.

When I_s equals zero, saturation equilibrium exists; there is no scale formed or little corrosion.

When I_s is negative (pH less than pH_s), corrosion of bare metal occurs and any small scale is dissolved.

When I_s is positive (pH greater than pH_s), there exists a condition of supersaturation which tends to deposit scale on the hottest part of the system. These characteristics are shown in *Table 4*.

When the Langelier Index is +0.5 or greater, scaling usually occurs. The severity of scale formation increases in logarithmic proportion to the increase in Langelier Index values. For example, a water having an I_s of 2.0 theoretically results in approximately 33 times as much scale as an index of +0.5 in the same system. Even though this relation may not be strictly true on a quantitative basis, it indicates that scaling can be very heavy when I_s is 2.0.

Tables prepared by Nordell[5] are more convenient for making repetitive calculations of pH_s in determining the Langelier Index. The method used by Powell[4] is presented in *Chart 5* because the relative effect of the variables is readily apparent by an ex-

CHART 5—LANGELIER SATURATION INDEX

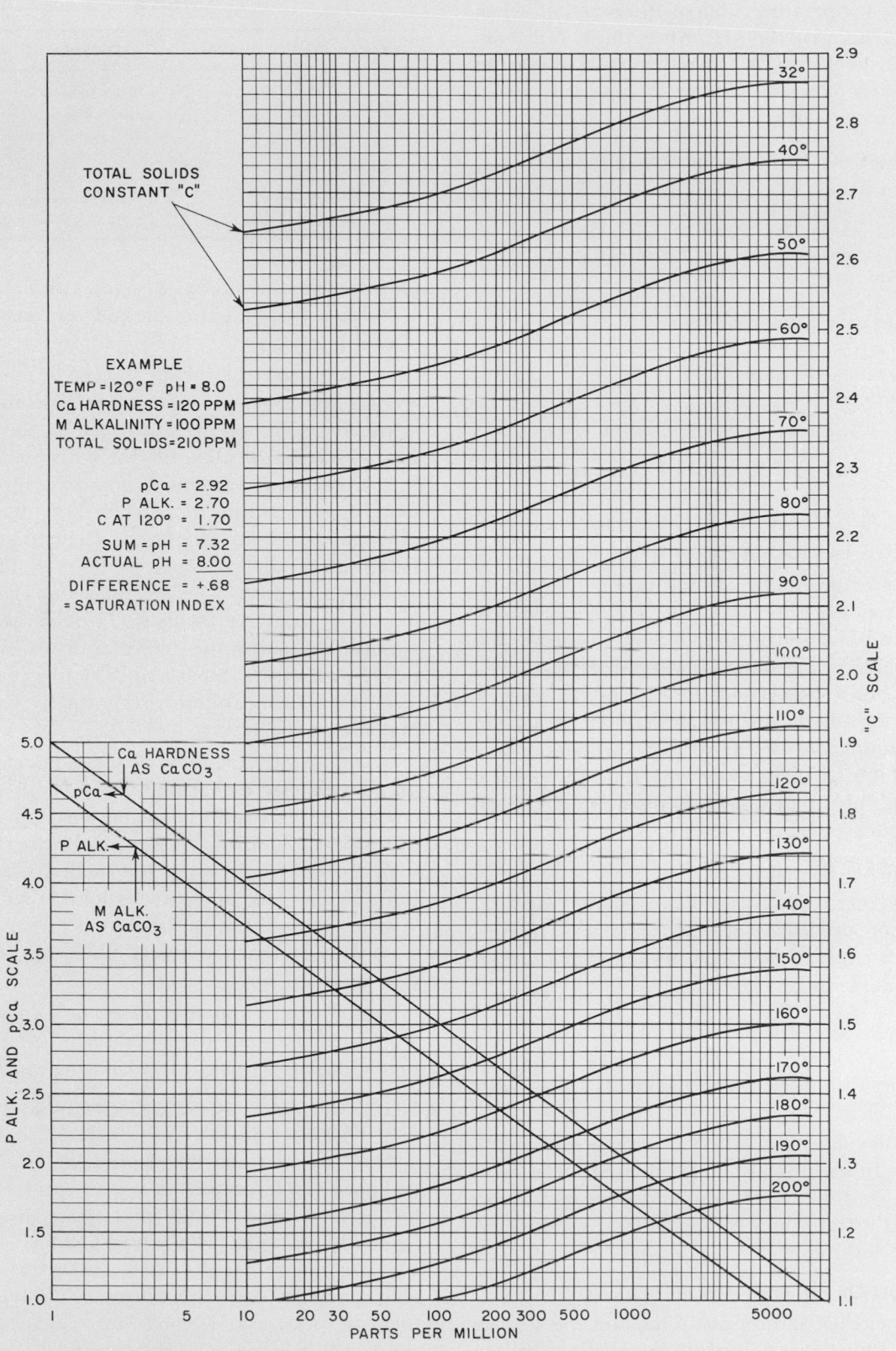

TOTAL SOLIDS
CONSTANT "C"

EXAMPLE
TEMP = 120°F   pH = 8.0
Ca HARDNESS = 120 PPM
M ALKALINITY = 100 PPM
TOTAL SOLIDS = 210 PPM

pCa = 2.92
P ALK. = 2.70
C AT 120° = 1.70
SUM = pH 7.32
ACTUAL pH = 8.00
DIFFERENCE = +.68
= SATURATION INDEX

Ca HARDNESS
AS CaCO₃
pCa
P ALK.
M ALK.
AS CaCO₃

P ALK. AND pCa SCALE

"C" SCALE

PARTS PER MILLION

From *Water Conditioning For Industry*, by Sheppard T. Powell, McGraw-Hill Book Co. Inc., 1954. Used by permission.

amination of the curves. Three curves represent log factors of calcium, alkalinity, and combined total solids and temperature. These three factors when added together give the pH_s. An example is demonstrated on the figure. A second example illustrates the Langelier Saturation Index for a water used in a once-thru condenser.

Example 8 — Langelier Index and Scaling Tendency

Given:
A once-thru condenser
Methyl orange alkalinity	$= 70$ ppm as $CaCO_3$
Calcium hardness	$= 75$ ppm as $CaCO_3$
Total solids	$= 221$ ppm
pH	$= 7.1$
Condensing water temperature	$= 100$ F

Find:
Langelier Saturation Index (I_s)
Scaling tendency

Solution:
From *Chart 5*:
$$pH_s = pALK + pCa + \text{"c" scale}$$
$$= 2.85 + 3.13 + 1.89 = 7.87$$
$$I_s = pH - pH_s = 7.1 - 7.87 = -0.77$$

This water is mildly corrosive and nonscale-forming.

The use of the Langelier Saturation Index indicates the tendency of a water to be either corrosive or scale-forming, and is intended only to serve as a guide in prescribing water conditioning for a given job. It is *not* a measure of the capacity to scale. A water having a high hardness and positive saturation index definitely causes a calcium carbonate scale, whereas a low hardness water with the same positive saturation index may not form any appreciable calcium carbonate scale.

RYZNAR STABILITY INDEX

The Ryznar Stability Index[7] is an empirical method for predicting scaling tendencies, and is based on a study of operating results with water of various saturation indexes.

$$\text{Stability Index} = 2pH_s - pH \qquad (8)$$

where:
$pH_s =$ calculated pH of saturation of the calcium carbonate
$pH = pH$ from test

This index (always positive) is often used with the Langelier Index to assist in predicting more accurately the scaling or corrosion tendencies of a water. *Table 5* illustrates how to predict water characteristics with this index.

PREDICTION OF pH, OPEN RECIRCULATING SYSTEMS

The Langelier Saturation Index and the Ryznar Stability Index are useful in predicting scaling tendencies for once-thru systems. They also are use-

TABLE 5—PREDICTION OF WATER CHARACTERISTICS BY THE RYZNAR STABILITY INDEX

RYZNAR STABILITY INDEX	TENDENCY OF WATER
4.0 - 5.0	heavy scale
5.0 - 6.0	light scale
6.0 - 7.0	little scale or corrosion
7.0 - 7.5	corrosion significant
7.5 - 9.0	heavy corrosion
9.0 and higher	corrosion intolerable

ful for open recirculating systems. However, there is a problem in predicting the pH for various cycles of concentration.

When water recirculates thru a heat exchanger and cooling tower or other aerating device, the pH is usually different from the pH of the make-up water because the alkalinity of the water, aeration in the cooling tower, and contamination of the atmosphere by sulfur dioxide, carbon dioxide and other gases all influence the pH. Generally, contamination from the air tends to lower both the pH and the alkalinity. The pH for a cooling tower or for an evaporative condenser in the majority of systems falls within the shaded area in *Chart 6*. The solid line is used as an average. Cooling towers operating in a clean atmosphere normally have a pH higher than the average line.

In industrial areas there is sometimes enough sulfur dioxide and carbon dioxide gases in the air to neutralize the alkalinity in the make-up water, resulting in a pH down to 4.0 or 5.0 in some cases. This is particularly true when the methyl orange alkalinity in the make-up water is lower than 50 ppm. When this condition exists, there is no scaling; however, corrosion is a problem. Refer to *Chapter 3, Corrosion Control*.

Special curves and data based on experience are available from some water treatment companies for predicting pH values in recirculating water systems.

EFFECT OF SCALE INHIBITORS ON LANGELIER AND RYZNAR INDEXES

The Langelier and Ryznar Indexes are for waters without scale inhibitors. A water with scale inhibitors gives the same index as untreated water. Consequently, the index for a treated water may show heavy scaling although it may not be scale-forming. Ryznar reports that with 1.7 ppm of polyphosphate there is little scale formed with an index of 4.0, whereas this indicates a heavy scale deposit in untreated water.

CHART 6—EXPECTED pH OF COOLING TOWER WATER

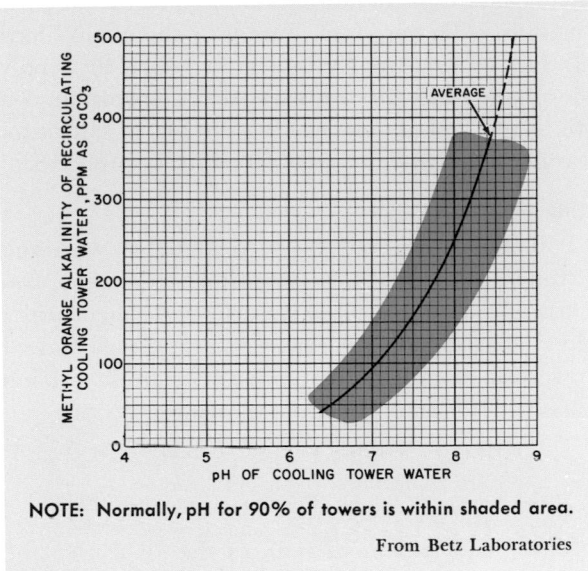

NOTE: Normally, pH for 90% of towers is within shaded area.

From Betz Laboratories

Example 9 — Scaling Tendency of Cooling Tower System

Given:
A cooling tower system
Make-up water methyl orange alkalinity,
ppm as $CaCO_3$ = 94
Calcium hardness, ppm as $CaCO_3$ = 85
Total solids, ppm = 128
Total hardness, ppm as $CaCO_3$ = 109
pH = 7.8
Water temperature (F) = 115

Find:
Scale-forming tendency

Solution:
Assume 1, 1½, 2 and 3 cycles of concentration and determine the Langelier and Ryznar Indexes.

Multiply the concentration of chemicals in the make-up water by the cycles of concentration to obtain the concentration for the various cycles.

Cycles of Concentration	1	1½	2	3
Methyl orange alkalinity, ppm as $CaCO_3$	94	141	188	282
Calcium hardness, ppm as $CaCO_3$	85	128	170	255
Total solids, ppm	128	192	256	384
Total hardness, ppm as $CaCO_3$	109	163	218	327
pH	7.8			
Results				
pH using the average line (Chart 6)	7.0	7.4	7.7	8.1
pH_s (pH of saturation of $CaCO_3$, Chart 5)	7.54	7.22	6.97	6.63
Langelier Index ($pH - pH_s$)	−.54	+.18	+.73	+1.47
Ryznar Index ($2pH_s - pH$)	8.08	7.04	6.24	5.16

The alkalinity may not vary directly as the cycles of concentration because some of the alkalinity is neutralized by acid formed when sulfur dioxide and carbon dioxide gases in the air are absorbed in the water. However, this is accurate enough for an estimate. Note also that the expected pH *(Chart 6)* is a rough approximation.

An examination of the Langelier and Ryznar Indexes in the example and *Tables 4 and 5* leads to the following conclusions.

1. With 1½ cycles of concentration the Langelier Index is +0.18, very slight scaling. The Ryznar Index is 7.04, very little scale or corrosion. Therefore, 1½ cycles may be satisfactory without any conditioning chemicals. However, experience may indicate that a corrosion inhibitor is desirable.

2. With 2 cycles of concentration the Langelier Index is +0.73 which is scale-forming. The Ryznar Index is 6.24 indicating little scale or corrosion. Two cycles may be satisfactory, but may result in some scale deposit.

3. With 3 cycles of concentration the water is definitely scale-forming. Scale may readily be controlled by adding 2-5 ppm of polyphosphate. Since the alkalinity at 4 cycles is only 376, polyphosphate treatment is satisfactory.

SCALE DEPOSIT PREVENTION

Several methods may be used to avoid or lessen the deposition of scale.

1. The increase in total solids caused by evaporation in a recirculating system may be controlled by bleeding off some of the water, accompanied by additional make-up water.

2. The tendency of calcium carbonate to precipitate may be inhibited by the addition of chemicals, i.e. polyphosphates, which tend to hold the calcium carbonate in solution.

3. The pH of the water may be lowered by adding an acid (usually sulfuric) to reduce the alkalinity, but not enough to develop an acid condition which may cause corrosion.

4. The water may be treated before use to remove the elements of calcium, magnesium and iron which form relatively insoluble compounds.

BLEED-OFF METHOD

Bleed-off is used in all open recirculating systems where water is evaporated. This may be sufficient in some cases, but additional conditioning is usually required for either scale or corrosion control, or both.

SCALE INHIBITOR METHOD

Some materials inhibit crystal growth and thereby prevent scale formation. These surface active agents increase the solubility of the scale-forming salts and permit an oversaturated condition to exist without precipitation from solution. Some of these materials

CHART 7—BLEED-OFF REQUIRED TO PREVENT SCALE FORMATION

100 Ton System

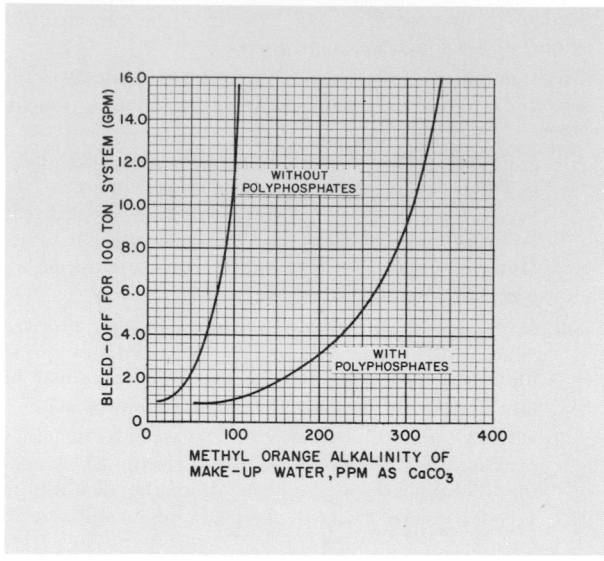

are polyphosphates, tannins, lignins and starches. A combination of several of these surface active agents may be more effective than a single material alone.

Polyphosphates in concentrations of 2-5 ppm are used to prevent or reduce scale formation. *Chart 7* shows the bleed-off required to prevent scale formation, both with and without polyphosphates. The bleed-off in this figure is the sum of bleed-off and windage loss *(Chapter 1)*; subtract the windage loss to obtain the amount of bleed-off wasted to the drain. This is an approximate curve which can be used for rough estimating for the average air conditioning installation where the condensing temperature is between 100 F and 120 F. It is based on a maximum methyl orange alkalinity of 125 ppm as $CaCO_3$ without polyphosphate treatment, and on 400 ppm as $CaCO_3$ when 2-5 ppm of polyphosphates are used. The actual pH, calcium content, and temperature all have a marked effect on the deposit of calcium carbonate. Specialists have developed data and curves from experience which show values both above and below those shown in *Chart 7*. These curves normally include the calcium content and pH in addition to the alkalinity.

Polyphosphates which are molecularly dehydrated phosphates revert eventually to the inaffective orthophosphates when dissolved in water; this reversion takes place in the recirculating water system. For this reason polyphosphates must be stored in the dry form. Too great a concentration of the orthophosphate in the system can result in the deposition of calcium orthophosphate sludge, particularly when high calcium waters are involved. The orthophosphate content of the water is limited to two or three times the polyphosphate content by the use of bleed-off. This is particularly important when high polyphosphate residuals are carried for corrosion control. A method of determining the pH of saturation of calcium phosphate is described in Reference 8.

ACID METHOD

Polyphosphates may be used alone or with acid treatment and corrosion inhibitors. A greater concentration of total solids is permitted when acid is added to the system. The calcium bicarbonate is converted to the more stable and soluble calcium sulfate as illustrated in the following reaction:

$$Ca(HCO_3)_2 + H_2SO_4 \rightarrow CaSO_4 + 2CO_2 + 2H_2O$$

$$\underset{\text{bicarbonate}}{\text{calcium}} + \underset{\text{acid}}{\text{sulfuric}} \rightarrow \underset{\text{sulfate}}{\text{calcium}} + \underset{\text{dioxide}}{\text{carbon}} + \text{water}$$

Thus, acid treatment reduces the alkalinity and, by virtue of this reaction, minimizes the oversaturation of the calcium carbonate. Also, calcium sulfate is much more soluble than calcium carbonate. The use of acid permits a greater concentration of dissolved solids without precipitating calcium carbonate. For estimating purposes in recirculating systems, one ppm of sulfuric acid is required to neutralize one ppm of methyl orange alkalinity as $CaCO_3$ in the make-up water. *Figure 3* illustrates the far greater solubility of calcium sulfate over calcium carbonate.

When sulfuric acid is added to a once-thru system, only enough acid is added to neutralize the methyl orange alkalinity to lower it to the permissible concentration. It is necessary to use only about one-half of the theoretical sulfuric acid for a once-thru system, whereas the full amount is required for an open recirculating system. In the once-thru arangement, the acid is introduced upstream of the condenser where both sulfuric and carbonic acids reduce the alkalinity (carbonic formed by carbon dioxide and water). In the open recirculating system the acid is introduced downstream of the condenser. Carbon dioxide gas is removed by deaeration in the cooling tower.

Occasionally, inhibited sulfamic acid, HSO_3HN_2 is used in place of sulfuric acid. Sulfamic acid (a white crystalline material) is safe to handle when dry, is more convenient for some applications, but is much more costly than sulfuric acid. To prevent formation of ammonium compounds which are very corrosive to copper, the temperature of the acid solution should not exceed 160 F.

REMOVAL OF HARDNESS IN MAKE-UP WATER

Removal of hardness in make-up water for condenser water cooling systems is seldom necessary although this is common practice for steam boiler make-up water. In the ion exchanger (zeolite) softening process, the calcium is replaced by sodium which is very soluble. However, zeolite softening does not decrease the alkalinity of the make-up water; a large bleed-off rate still may be required to prevent scaling. This is one of several types of water softening processes.

WATER CONDITIONING METHOD FOR SCALE CONTROL

The most common practice is to control the chemistry slightly on the corrosive side to prevent scale formation and then add corrosion inhibitors.

The choice of a particular method or methods for water conditioning depends upon the chemical composition of the water, the cost of various methods, and the economics of the various combinations. Sometimes a heavy bleed-off may be the only require-

ment for a recirculating system, but it may be more economical to add water conditioning and save on water cost. Where bleed-off is required, it may be controlled by a meter which measures the conductivity of the water rather than by using a constant rate of bleed-off. The conductivity may be a good indicator of the scaling ability of the water if the majority of the impurities are calcium and bicarbonates. When there is no corrosion problem, a simple or partial treatment to prevent scaling may be selected which may require an occasional acid cleaning of the condenser rather than a complete water conditioning program. This partial treatment is more applicable to the small system when the cost of feeding equipment and consulting services may be large compared to the cost of an occasional cleaning of the condenser tubes. In many locations there is a severe corrosion condition and water conditioning for corrosion prevention is necessary, even though scaling is not a problem.

Chapter 5 gives details for water conditioning to prevent scale formation for the several types of systems.

CHAPTER 3. CORROSION CONTROL

This chapter includes a discussion of the types and causes of water circuit corrosion and methods used to prevent corrosion in a system. Corrosion monitoring and problems external to the conventional water system are also presented.

Corrosion cause and prevention is a large field which is briefly discussed here.

Corrosion in an open water recirculating system in which water sprays come in contact with air is a greater problem than scale deposit. In some industrial areas, it is not uncommon for pipes or parts of a cooling tower system to corrode thru in two to three years; in severely corrosive atmospheres pipes may perforate in less than a year.

Corrosion products reduce the capacity of the lines, increase frictional resistances, and increase pumping costs. The products of corrosion have a volume several times that of the metal they replace. Often they may obstruct or stop flow and clog nozzles in small lines.

TYPES OF CORROSION

An air conditioning system may have several types of corrosion in the water system, as characterized by one or more of the following:

1. Uniform corrosion
2. Pitting corrosion
3. Galvanic corrosion
4. Concentration cell corrosion
5. Erosion-corrosion

UNIFORM CORROSION

Corrosion due to acids such as carbonic or others effect a uniform metal loss. It is the most common type of corrosion encountered in acid environments, and the easiest to predict and control.

PITTING CORROSION

Pitting corrosion is a nonuniform type, the result of a local cell action which is produced when particles are deposited on a metal surface either as flakes, solids or bubbles of gas. The pitting is a local accelerated attack, causing a pit or cavity around which the metal is relatively unaffected. Centralized corrosion in pits creates deeper penetration and more rapid failure at these points. Oxygen deficiency under a deposit sets up anodic areas to cause pitting. Sometimes oxygen causes pitting corrosion as does galvanic cell action.

GALVANIC CORROSION

Galvanic corrosion occurs when dissimilar metals are present in a solution capable of carrying an electric current. It is a form of electrochemical corrosion where a difference in potential associated with the metals themselves causes a small electric current to flow from one metal to the other thru the electrolyte.

Table 6 is a galvanic series of metals and alloys. It is based on corrosion tests in the laboratory under plant operating conditions, and has basic characteristics which make it analogous to the electromotive series in textbooks. However, the metals are not exactly in the same order. When two metals in the same group are joined together, they are relatively safe. Conversely, the coupling of two metals from different groups causes accelerated corrosion in the less noble or the metal higher in the list. The greater the difference in position in the table, the greater the accelerated corrosion. For example, when iron and zinc are joined together, the zinc is corroded and the iron is not affected. Corrosion is accelerated on the zinc, but not on the iron. When copper and steel are joined together, corrosion is accelerated on the

TABLE 6—GALVANIC SERIES OF METALS AND ALLOYS

Corroded end (anodic, or least noble)

Magnesium
    Magnesium alloys
Zinc
Aluminum 2 S
Cadmium
Aluminum 17 ST
Steel or iron
    Cast iron
Chromium-iron (active)
Ni-Resist
18-8 Stainless (active)
    18-8-3 Stainless (active)
Lead-tin solders
    Lead
        Tin
Nickel (active)
    Inconel (active)
Brasses
    Copper
        Bronzes
            Copper-nickel alloys
                Monel
Silver solder
Nickel (passive)
    Inconel (passive)
Chromium-iron (passive)
    18-8 Stainless (passive)
        18-8-3 Stainless (passive)
Silver
Graphite
    Gold
        Platinum
Protected end (cathodic, or most noble)

By permission of International Nickel Co., Inc.

TABLE 7—MAXIMUM WATER VELOCITY TO MINIMIZE EROSION

NORMAL OPERATION (hr)	WATER VELOCITY (fps)
1500	12.
2000	11.5
3000	11.
4000	10.
6000	9.
8000	8.

EROSION-CORROSION

The impingement of rapidly moving water containing entrained air bubbles and/or suspended matter, i.e. sand, may remove or prevent the formation of coatings to permit general corrosion of the metal. This action commonly occurs near the tube inlet ends, or it may extend along the entire tube length. Unfortunately, experimental data is unavailable to determine the maximum allowable water velocity without damage to the protective film. The circulation of solid matter in the system may cause complete deterioration of the tube or pipe walls, particularly on the bottom surface and at the elbows. Abrasive solid matter can be removed by filters.

Since erosion is a function of operating hours, water velocity and suspended materials in the water, the selection of a design water velocity is a matter of judgment. *Table 7* lists the approximate maximum velocity thru copper tubes of a heat exchanger (a cooler or a condenser) to minimize erosion. The values are not intended to be top values. They are based on many years of experience, and permit the attainment of optimum equipment life under normal conditions. Use only 70% of these velocities for heat exchangers where the water temperature ranges from 140-180 F.

Erosion of a centrifugal pump impeller caused by cavitation pitting[9] can occur in a corrosive hot water system when the pump is operated at the bottom end of the pump curve. Although not a normal selection point, this condition may occur when too great a safety factor is used for the estimated pipe friction, particularly since the loss in new pipe is less than the loss obtained in design tables.

OTHER TYPES OF CORROSION

Sometimes corrosion is caused by the dissolution of the more noble metal which is plated out on the anodic member, thus setting up small galvanic cells and causing pitting corrosion. Such corrosion is difficult to control by use of inhibitors. Water tends

steel but not on the copper. The greater the copper area in relation to the steel, the faster is the rate of corrosion. On systems in which all piping and coils are made of copper except for a few steel nipples, the steel parts are subject usually to a rapid corrosive attack and failure in a short time. In such a situation a nonferrous metal should be used instead of steel. Corrosion inhibitors reduce the corrosion rate but do not eliminate galvanic corrosion.

CONCENTRATION CELL CORROSION

Concentration cell corrosion is similar to galvanic action. However, the difference in potential arises from differences in the electrolyte. This potential may be produced by differences in ion concentration, oxygen concentration or pH, or may be caused by dirt, foreign matter or a gas bubble adhering to the metal surface. Each pit in a pitting corrosion represents a concentration cell or anode. The metal surrounding the pit is the cathode.

CHART 8—EFFECT OF OXYGEN CONCENTRATION ON CORROSION AT DIFFERENT TEMPERATURES

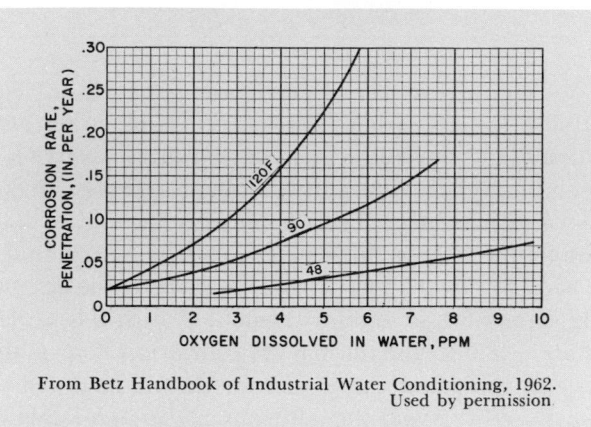

From Betz Handbook of Industrial Water Conditioning, 1962.
Used by permission

to pick up small amounts of copper while passing thru cuprous parts of the system and to deposit this copper on steel parts. Only a small amount of copper, i.e. 0.1 ppm in the water, is sufficient to cause this type of corrosion. It can usually be controlled by maintaining the *pH* above 6.7. An acidic condition or presence of chloride or bromide salts in the water is necessary for this type of corrosion. Copper is satisfactory if the chloride content of the water does not exceed 200 ppm. If 200 ppm is exceeded, the decision of whether to use copper should be determined by experience and consultation with a corrosion engineer. The concentration of the inhibitor should be increased when chlorides are present. Certain inhibitors,[10] i.e. BTT (2-benzothiazolethiol), may be used to prevent dissolution of the copper.

Other types of corrosion[11,12,13,14,15] are dezincification, corrosion fatigue, and stress corrosion. However, since these are uncommon, they are not discussed here.

CAUSES OF CORROSION

Although many factors contribute to corrosion in cooling systems, the chief factor is dissolved oxygen in the cooling water. The reaction of dissolved oxygen with ferrous metal is accelerated with temperature as illustrated in *Chart 8*. The chief variables controlling the corrosive characteristics of a water are:

1. Dissolved oxygen concentration
2. Temperature
3. Carbon dioxide content
4. *pH*
5. Dissolved solids

TABLE 8—DATA TYPIFYING THE DELETERIOUS GAS CONTENT OF DIFFERENT ATMOSPHERES

Gas	Chem. Form.	AIR			
		Rural		Metropolitan	
		% by Volume	Partial Pressure (psia)	% by Volume	Partial Pressure (psia)
Oxygen	O_2	21.	3.143	21.	3.143
Carbon Dioxide	CO_2	0.03	0.004	0.06	0.009
Sulfur Dioxide	SO_2	None	None	0.003	0.004

Gas	Chem. Form.	FLUE GASES					
		Bitum. Coal		Fuel Oils		Natural Gas	
		% by Vol.	Partial Press. (psia)	% by Vol.	Partial Press. (psia)	% by Vol.	Partial Press. (psia)
Oxygen	O_2	2.	0.299	7.	1.048	10.	1.497
Carbon Dioxide	CO_2	15.	2.245	13.	1.946	10.	1.497
Sulfur Dioxide	SO_2	0.07	0.010	0.03	0.004	0.0001	0.0015

From ASHRAE Guide and Data Book, 1961. Used by permission.

6. Suspended solids

7. Velocity

Neutral water (*pH* of 7.0) saturated with air corrodes iron at a rate three times that of air-free water. Hot water containing oxygen corrodes iron three or four times faster than cold water.

Most natural waters contain dissolved substances such as carbon dioxide, oxygen, chlorides and sulfates, all of which corrode metal that is in contact with water. Corrosion affects the heat exchangers, pumps, cooling towers, air washers and piping. In an open recirculating system such as a cooling tower, evaporative condenser or air washer, most of the corrosion is due to the absorption of acidic impurities from the air thru which the water has passed. This is particularly true in a large city with a low alkalinity water supply. Here, the combustion of carbon and sulfur in fuels produces large amounts of sulfur dioxide, SO_2, and carbon dioxide, CO_2, in the flue gases from boiler plants. *Table 8* illustrates the amount of carbon dioxide and sulfur dioxide gases in the atmosphere.

The *pH* is affected directly by the concentration of carbon dioxide which forms carbonic acid. A low *pH* increases the corrosion. Corrosion of iron decreases as the *pH* of water increases, and practically ceases at a *pH* of 11. However, a high *pH* cannot be tolerated because of scale formation in heat exchangers and delignification of cooling tower wood.

Dissolved solids, particularly chlorides and sulfates, intensify the corrosive effect of oxygen and carbon dioxide. Higher solid contents increase the electrical conductivity of water which accelerates corrosion. Corrosion is essentially an electrochemical process; dissolved solids particularly affect corrosion when there are dissimilar metals in the water circuit.

Suspended solids may wear away metal or prevent the maintenance of a protective film by the corrosion inhibitors.

An increase in the flow rate generally increases corrosion.

CORROSION CONTROL

A coating of zinc on steel can be used to prevent the corrosion of steel. Various organic and inorganic coatings can also be used, but not in a piping system because the coating cannot be readily maintained. Certain chemicals in small concentrations protect metal by forming a thin monomolecular film or barrier on its surfaces to prevent the electrochemical process. These corrosion inhibitors greatly reduce the rate of corrosion.

Corrosion may be minimized by several methods:
1. Use of organic or inorganic corrosion inhibitors.
2. Formation of a thin film of calcium carbonate on metal surfaces.
3. Control of *pH* in the range of 7.0 to 8.5.
4. Use of mechanical deaeration of water.

CHROMATE INHIBITORS

The chromates are used extensively to inhibit corrosion and are effective in the water of air conditioning systems in concentrations of 200-500 ppm at a *pH* of 7.0 to 8.5. For corrosion prevention the optimum range in *pH* is 7.5 to 9.5, but there is a greater problem of scaling at the higher *pH*. Consequently, the *pH* is held to the lower side where corrosion protection is excellent. Sodium bichromate, $Na_2Cr_2O_7 \cdot 2H_2O$, is the most commonly used chromate compound because it is more economical. Sodium chromate, Na_2CrO_4, is also used. Chromate concentration is stated in ppm as chromate ion although some treatments may describe concentration in ppm as sodium chromate. The percent of chromate ion in these chemicals is illustrated in the following table:

Chemical	Chromate Ion Percent by Wt
Sodium Bichromate, $Na_2Cr_2O_7 \cdot 2H_2O$	77.9
Sodium Chromate, Na_2CrO_4	71.7

Sodium bichromate is acidic and, when dissolved in pure water at common concentrations, has a *pH* of less than 5.0. Sodium carbonate (soda ash, Na_2CO_3) or sodium hydroxide (caustic soda, NaOH) is added to raise the *pH*. To completely neutralize 100 pounds of sodium bichromate requires 36 pounds of soda ash or 27 pounds of caustic soda; the resulting compound is sodium chromate which has a *pH* of about 8.0 in the normal concentration. On small systems the more costly sodium chromate is sometimes used to avoid the addition of the alkali. Some treatments may require the mixing of soda ash with sodium bichromate in various proportions, depending upon the alkalinity of the treated water. Unless packaged in vapor tight containers, caustic soda absorbs moisture from the air and tends to form lumps if premixed with the bichromate.

Chromate Inhibitors in Hot Water Systems

A greater concentration of chromate is required for high temperatures. Hot water heating systems[16] are treated with 1000-2000 ppm of chromate. In cooling water for natural gas and diesel engines, 800-1200 ppm of chromate are used.

Low Chromates with Other Inhibitors

Much lower concentrations of chromates may be used to obtain a satisfactory over-all corrosion rate, but this may lead to pitting. Conversely, pitting corrosion does not occur when high concentrations are used. A treatment permitting pitting is unsatisfactory. To reduce the cost of conditioning chemicals, other systems[17,18] have been developed, using low concentrations of chromates and high concentrations of phosphates. Concentrations having a total mixture of 30-60 ppm of chromates and polyphosphates have been used to effect a substantial reduction in corrosion and pitting. When high concentrations of polyphosphates are used, there may be some calcium phosphate sludge precipitated as explained in *Chapter 2*.

Low chromate concentrations with other inhibitors are becoming more popular for air conditioning systems. The petrochemical and chemical industries use these inhibitors for large systems where optimum life of equipment may be only seven to ten years due to obsolescence of process. Research is improving this blend of inhibitors.

There are many proprietary formulations of several generic compositions[19] in total treatment concentrations of 20-80 ppm. Usually there is some chromate mixed with one or more other compounds of polyphosphates, zinc, fluorides or organic synergizing agents. A controlled range of pH is required for each formulation; these are sometimes narrow and call for a relatively low pH. The following are examples of the particular ranges of these formulations: 6.0-6.5; 6.4-6.8; 6.8-7.4; 6.7-7.7; 6.5-7.5.

The narrow range indicates the application of an automatic pH controller which is costly and may be justified only on large systems. Pitting corrosion may occur if the pH is not maintained within the recommended limits. Accelerated laboratory tests[19] at 140F show that there is pitting with some formulations in the recommended pH range.

Disadvantages of Chromates

The characteristic yellow color of chromates which stains the sides of the tower, building or cars in the adjacent parking lots is objectionable. However, it is advantageous because it permits an approximate control check on the chromate concentration by a color comparison of a water sample with a set of color standards. Another objection may arise from service operation of control valves or coils in executive offices, hotels or apartments when water spilled on rugs leaves a yellow stain.

Disposal of the treated water to sewage systems or streams may be subjected to ordinances or may be objectionable because of its yellow color.

POLYPHOSPHATE INHIBITORS

A low concentration of polyphosphates (2-4 ppm) is effective in preventing tuberculation. These compounds maintain cleanliness of the pipes although they are not very effective in preventing corrosion, except when used with chromates. Polyphosphates are used for scale control primarily.

NITRITE INHIBITORS

Colorless sodium nitrite has had widespread use as a corrosion inhibitor where chromate staining is objectionable. It is not as effective as the chromates, and it is difficult to maintain effective concentrations because of water infection with organic slimes or bacteria which feed on the nitrite. Periodic treatment of the water with biocides can control the problem of these nitrite-feeding bacteria. The concentration for air conditioning systems is 200-500 ppm of nitrite ion. When the chloride ion is present, nitrite concentration[22] should be maintained at a higher level.[20] Also, sulfates affect the performance of nitrites adversely. Some water treatments recommend a nitrite concentration approximately twice as great as chromates to provide equivalent protection. Generally, the pH should be maintained above 7.0 to make the treatment effective.

INHIBITORS FOR ALUMINUM COOLING TOWERS

Aluminum cooling towers[21] constructed of an alloy such as Alclad 3S or Alclad 4S can be protected adequately by the use of 200-500 ppm chromate accompanied by a pH of 7.0 to 8.5. Contact between dissimilar metals must be avoided. It is particularly important that aluminum be cleaned regularly. Painting of the pan or basin interior and other surfaces on which dirt can settle is necessary to minimize concentration cell corrosion. It is important to keep the pH above 7.0 because traces of copper, iron and other metals in an acidic water can lead to rapid and serious pitting of aluminum. See Reference 21 for detailed recommendations. Another method of treating aluminum towers or piping is the use of a mixture of polyphosphates, citrate and mercaptobenzothiazole.[22]

DEPOSITION OF CALCIUM CARBONATE FILM

The deposition of a controlled thin calcium carbonate film on a heat exchanger and piping can be used to prevent corrosion. However, it is seldom used except in once-thru systems. The amount of deposit is adjusted by control of the Langelier Saturation Index, above +0.5 (Chapter 2). Since temperature is one of the variables, a deposit forms on the surfaces at higher temperatures but not at lower temperatures. An adjustment for pH is made by the use of an inexpensive alkali such as lime, caustic soda, or soda ash to increase the scale formation or by the addition of acid to decrease it.

CONTROL OF pH

In industrial areas the atmosphere contains large amounts of sulfur dioxide and carbon dioxide gases which, when absorbed in the cooling tower water, tend to make it an acidic solution.[23] Although the make-up water may be neutral, the acidity picked up from the air may be sufficient to neutralize the alkalinity in the water and to form an acid condition. The pH of the recirculating water may be as low as 4.0 or 5.0; in extreme cases, even lower. The addition of alkali such as caustic soda or soda ash is used to raise the pH. If the pH is too high, sulfuric acid is normally added to reduce it. The upper limit of pH of a water system is selected to prevent scale formation, not for control of corrosion.

MECHANICAL AND CHEMICAL DEAERATION

Mechanical and chemical deaeration may be applied to once-thru systems to remove oxygen or other corrosive gases; however, these methods are not normally applied to air conditioning systems.

CORROSION MONITORING

Corrosion may be monitored with test coupons, made of the identical materials to be checked and installed in the system. Accurate results are provided by a minimum exposure of thirty days. Corrosion rate is normally expressed as mils per year (mpy) and is calculated on the basis of initial weight, final weight after cleaning and drying, surface area and density of metal.

Thirty days exposure of mild steel coupons in a steel piping system of approximately ¼-inch wall thickness, may be rated as follows:

CORROSION RATE	CORROSION CONTROL
above 5 mpy	poor
2 to 5 mpy	good
0 to 2 mpy	excellent

These ratings apply to uniform corrosion. When pitting attack occurs even with less than a 2 mils per year corrosion rate, there is not sufficient protection. If wall thickness is less than ¼ inch, the acceptable corrosion rates are decreased proportionally.

When copper tubes are used in a system, the corrosion rate on the corresponding copper test coupons must be less than 1 mil per year. Pitting cannot be tolerated. The corrosion rate on copper is generally much lower than on steel in the same system unless chlorides or some compound corrosive to copper is in the system.

INSTALLATION OF TEST COUPONS

Samples of mild steel approximately 3″ x ⅜″ x ⅛″ thick are accurately weighed in the chemical laboratory, then carefully wrapped and placed in tight containers until installed in the water system. *Figure 6* illustrates how the test coupons are installed. They are mounted in plastic holders to insulate the coupon from the pipe; they must not touch the pipe. Flow is parallel to the coupon at a velocity approximately 3 feet per second. If the system is a closed type, or if the required bleed-off is less than that removed by the test installation, the water is pumped back into the system. An alternate piping arrangement used is shown in *Figure 6*. The coupon is made of the identical material which is to be checked in the system. When two metals such as steel and copper are used in the system, a couple or a special coupon can be made by bolting together a strip of steel and copper, using a stainless steel bolt and nut. The coupons are returned to the chemical laboratory where they are cleaned, dried and weighed.

ELECTRICAL RESISTANCE METHOD

Another method of corrosion monitoring which has been used for the past few years is the electrical resistance method. With this method corrosion rates can be measured as the corrosion occurs and a corrosive condition can be detected and corrected before any harm has been done. The corrosion is measured by determining changes in the resistance ratio of a measuring element immersed in the water and a reference element of the same metal protected by a coating. As corrosion occurs the electrical resistance of the measuring element increases. This increase can be measured by a bridge circuit by balancing an adjustable resistance with the measuring element and the reference element. The change in resistance required to rebalance the circuit is an indication of the corrosion rate.

EXTERNAL WATER CIRCUIT CORROSION

Corrosion can also occur on sprayed coils and underground piping unless precautions are taken.

COIL CORROSION OF AIR HANDLING UNITS

Target sprays, sometimes used in a once-thru system for humidification in air handling units, leave no appreciable concentration of chemicals in the condensate pan. Aluminum fins on copper tubes are commonly used for cooling coils. A corrosive attack can occur on the aluminum fins whenever the water sprayed over the coil has a high electrical conductivity; this corrosion is due to galvanic action. Based on several years experience, aluminum fins may be used on copper tubes with once-thru city water sprays when the specific conductance of city water is less than 500 micromhos. The specific conductance for city waters for all major cities is shown in *Table 13, Chapter 5*. If 500 micromhos is exceeded, copper fins on copper tubes are used. Normally, copper fins on cooling coils are dipped in solder.

Aluminum fins on copper tubes are never used when humidification is accomplished by a recirculating spray water system.

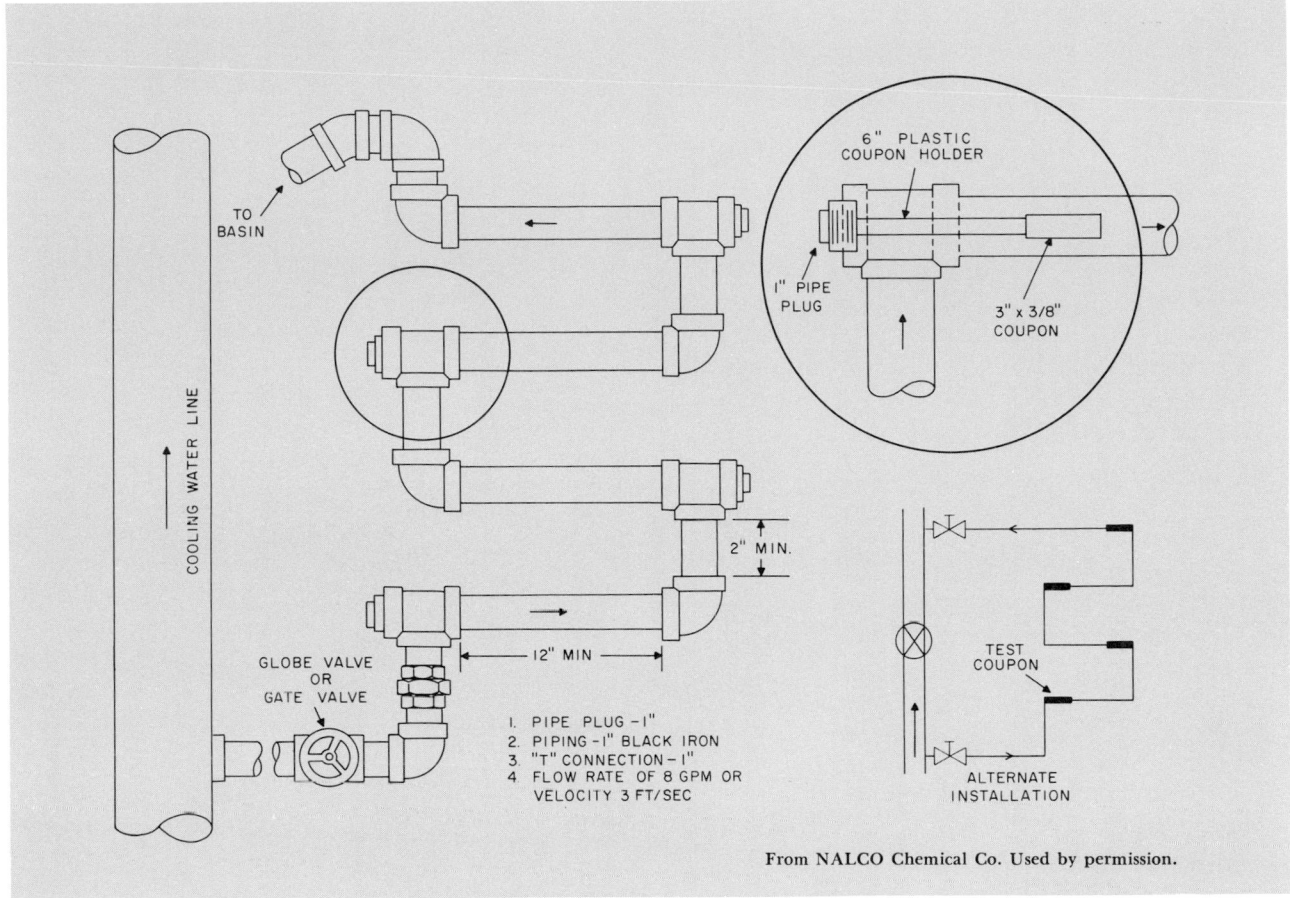

FIG. 6 — INSTALLATION OF CORROSION COUPONS

CORROSION OF PIPES BURIED IN THE GROUND

Chilled water and condenser water lines are sometimes buried in the ground. Chilled water lines are not normally insulated because of insufficient temperature difference between the ground and pipe. Cast iron in the ground has good resistance to corrosion, but steel pipe needs protection. A bituminous coating applied in a molten condition over a prime coat such as red lead provides adequate protection. A wrapping of asbestos fabric saturated with bitumens over this coating gives additional protection because it prevents movement and displacement of the coating. Galvanized wrought iron or steel pipe in the ground is corrosion resistant when the zinc coating is 3 ounces or more per square foot. Zinc is on the electrochemical protective side of iron; as the zinc corrodes, it is changed into zinc compounds before the iron is attacked.

It is good practice to avoid embedding pipe in earth of high salt content.

CHAPTER 4. SLIME AND ALGAE CONTROL

This chapter describes the slime and algae organisms which affect the operation of air conditioning equipment, and discusses the chemicals and methods used to control biological fouling. Wood deterioration and its control is also included.

TYPES OF BIOLOGICAL FOULING

Slime and algae are microorganisms capable of multiplying rapidly and producing large masses of plant material. *Table 9* presents a group listing and description of the slime-forming organisms usually encountered only in open recirculating systems. Slime and algae prevent the maintenance of heat transfer in condensers because they readily attach to surfaces to form slime deposits with a high insulating value.

These organisms are classified with respect to type, as algae, fungi and bacteria.

The species of algae which are of prime importance in cooling water systems are found in locations where they have access to light and air. In the presence of sunlight these microscopic plants carry on processes of photosynthesis to manufacture food and release oxygen. Masses of living algae on metal surfaces can accelerate corrosion in the form of pitting. Dead algae lodged in heat transfer units can cause corrosion cell action to effect severe pitting of the metal.

The second type of biological fouling known as fungi lack chlorophyll so they cannot manufacture their own food. Molds and yeasts are in this group. They are dependent on food to be found in water, and use a wide variety of nitrogenous and cellulose material as food sources.

The third type is bacteria which form slimes, although not all are slime formers. One group reduces the sulfate content of water to the corrosive sulfide ion. Another group utilizes soluble iron, and deposits insoluble iron oxide in a slimy sheath around their cells. Some types of bacteria feed on nitrites used as corrosion inhibitors. The growth of this particular group must be kept under control when sodium nitrite inhibitor is used.

TABLE 9—PRINCIPAL SLIME FORMERS

PHYLA	ROUGH DIVISION OF PHYLA
Algae	Single celled, sometimes forming slimy sheets. Many celled in either sheets or fronds.
Fungi	Bacteria (*Schizomycetes*) frequently forming slimy surface coatings. Slime Molds (*Myxomycetes*) forming slimy sheets as one stage of their life history. Sac fungi (*Ascomycetes*) of which one division, the yeasts, occasionally form slimy aggregates. The alga-like fungi (*Phycomycetes*) and the stalked fungi (*Basidomycetes*) rarely form slimes but their filaments may hold together the slimes of other organisms.

From ASHRAE Guide and Data Book, 1961. Used by permission.

CONTROL OF BIOLOGICAL FOULING

Chemical treatment, rather than removal by mechanical cleaning, is a more satisfactory method of combating these organisms because of the inaccessibility of many of the areas in pipes and equipment. Fungi and bacteria live in dark areas of the system such as heat exchangers and piping, as well as in light areas such as cooling towers or evaporative condensers. If slime and algae are allowed to form an appreciable deposit, it should be removed where practical by mechanical means, and flushed from the system before chemical treatment.

CHEMICALS

Table 10 gives a list of common chemicals used for slime control.

Frequently, microorganisms build up an immunity to a particular algaecide, but not to chlorine. This characteristic makes it necessary to change to other compounds periodically.

Chlorinated Phenols

The most widely used chemicals are the chlorinated phenols, particularly sodium pentachlorophenate, because they are easy and safe to use. These compounds control most slime and algae conditions. Concentrations of 200 ppm of sodium pentachlorophenate are enough for shock feeding. The alter-

TABLE 10—COMMON CHEMICALS USED FOR SLIME CONTROL

CHEMICAL	TRADE NAME	PHYSICAL STATE*
Chlorine Hypochlorites	Chlorine Calcium Hypochlorites Sodium Hypochlorites	Gas Crystalline
Chlorinated Phenols, Sodium —	Chlorophenylphenate Tetrachlorophenate Pentachlorophenate	Briquettes Briquettes Briquettes
Potassium Permanganate Copper Sulfate	Permanganate of Potash Blue Vitriol	Crystalline Crystalline

*As shipped.

From ASHRAE Guide and Data Book, 1961. Used by permission.

nate use of the chlorinated phenols and chlorine is often the best control.

Copper Sulfate

Copper sulfate may be considered a specific algaecide for algae since 0.5 ppm kills most of the common forms. Surface active agents are included to prevent precipitation of the copper ion at a high pH. Wetting agents are also included. Other toxicants are used with copper sulfate for control of bacteria because it alone does not kill many types of bacteria. Copper sulfate is corrosive to steel.

Potassium Permanganate

Potassium permanganate is a powerful oxidizing agent and is highly toxic toward many microorganisms. However, dead as well as live organic matter consumes this compound, thus increasing chemical consumption.

Chlorine

Chlorine is one of the most effective chemicals for control of slime and algae. However, it is not often used in the smaller or medium size installations because of handling difficulties. A residue of only 0.5-1.0 ppm kills bacteria and most microorganisms, but since chlorine acts on all oxidizable matter, organic matter, hydrogen sulfide, ferrous iron, etc., the demand must be satisfied before the required residual can accumulate. When large amounts of chlorine are required, it is obtained in liquid form and fed to the system by a chlorinator. Generally, liquid chlorine with a chlorinator is used only on large systems, 2000 tons and up. Special handling equipment and safety precautions[4,24] are required for liquid chlorine. When small amounts of chlorine are needed, compounds such as calcium or sodium hypochlorite can be used.

When chlorine is used continuously in a system, a concentration of 0.3-0.5 ppm is sufficient. When intermittent chlorination is practical, the concentration is raised to 1.0 ppm for several hours; then feeding is discontinued for several hours. The chlorine is held at a maximum of 1.0 ppm to minimize attack on the cooling tower wood. The pH should not exceed 8.0 during chlorination and should be held preferably in the range of 6.0 to 7.0. Chorine is measured in the recirculated water to the tower, not after the water cascades over the tower. The water passing thru the tower has its chlorine residual decreased or nearly depleted as some of the chlorine is absorbed into the wood.

Quaternary Ammonium Compounds

Some of these compounds in high concentrations are toxic to slime and algae growths. Certain quaternaries react with or are absorbed by organic matter in recirculated water systems, thus losing some of their effectiveness. Volatilization of some compounds occurs when the water passes over a cooling tower.

Other Compounds

There are many proprietary algaecides and biocides available to control biological growths. They may be mixtures of some of the previously mentioned chemicals or other compounds depending on the supplier.

METHODS OF TREATMENT

No toxic agent is effective for the control of all types of biological fouling. Selection of the toxicant is based on the microorganism present and the chemicals used for scale and corrosion control. The methods and frequency of feeding must be varied to suit the individual problem. Compounds may be used to clear the cooling tower of algae, but there may still be active bacteria producing slime in the heat exchanger. Chromates used as corrosion inhibitors also inhibit the growth of some slime-forming organisms. The compatibility of the toxicants with other treatments used for scale and corrosion must be known. This illustrates again the need for the services of a water conditioning specialist.

Feeding of algaecides and biocides in shock doses is more effective and economical than continuous feeding.

Chapter 5 gives some details of water conditioning to prevent biological growth in several types of water systems.

For a comprehensive treatment of slime and algae control, see References 2, 25 and 26.

WOOD DETERIORATION AND ITS CONTROL

Wood in cooling towers is subject to three main types of deterioration: physical, chemical and biological. The three types occur simultaneously.

Wood is composed of cellulose, lignins and natural extractives. The cellulose is similar to cotton fibers and gives wood its strength. The lignins provide the cementing agent which holds the fibers together. The extractives create wood resistance to decay. It is these substances which make redwood so durable. However, the extractives are soluble and are leached from the wood by water. Although this action does not seem to affect the strength of wood, decay may take place more rapidly.

Chemical deterioration usually causes delignification, indicated by the bleached appearance of wood. The chemicals commonly responsible for this attack are oxidizing agents such as chlorine, and alkalies such as calcium bicarbonate, calcium carbonate and sodium carbonate. Chemical attack most frequently takes place in the fill and in the flooded portions of the tower.

Chemical deterioration is controlled by maintaining the pH below 8.0, preferably in the range of 6.0 to 7.0. The free chlorine residue should be maintained at less than 1.0 ppm when using intermittent chlorination.

Biological attack occurs as surface rot and as internal decay. The organisms feed on the cellulose, leaving the lignins; consequently the wood loses much of its strength. Internal decay in cooling towers occurs in the plenum areas, cell partitions, doors and fan housing and supports; surface rot occurs in the flooded area.

The agents which prevent chemical deterioration of wood also minimize biological deterioration of the surface type. Periodic addition of nonoxidizing biocides to the water minimizes the attack. In the areas subject to internal decay, lumber is sprayed with preservatives that are toxic to the organisms. This protective measure must be taken before serious infection starts. A regular spray procedure should be considered for preventative maintenance. A periodic examination of the tower should be made by sending samples of wood to a qualified microbiological laboratory. These studies serve as a basis to determine the beginning of a preventative spraying program.

CHAPTER 5. WATER CONDITIONING SYSTEMS

This chapter describes how water conditioning is applied to the three types of water circulating systems.

Several over-all considerations apply to both large and small air conditioning systems. However, the Guide Specifications presented in this chapter apply primarily to large systems.

The application of water conditioning is classified with respect to refrigeration capacity of the system:

1. Large systems (75 tons and up) for which the customer can afford the services of a water conditioning company for appraisal and administration.

2. Small systems (5 to 75 tons) for which it may not be economical to procure a complete water conditioning service.

   The cost of the complete services and conditioning may be greater than an occasional acid cleaning of the condenser or its replacement.

DESIGN CONSIDERATIONS

Although a complete water conditioning program has a direct bearing on system life, operating efficiency and maintenance, it must be accompanied by good engineering practice to accomplish desirable results. Water conditioning alone is insufficient. Some of these important practices are:

1. Adjustable bleed-off devices with an adequate open site drain to permit easy measuring of the bleed-off flow rate.

2. Access for cleaning and servicing equipment.

3. Use of water velocities to minimize corrosion (*Table 7*).

4. Use of materials in construction compatible with environment and type of water conditioning. Both local experience and the counsel of a corrosion engineer should be considered when specifying the type of pipe for a particular water supply.

5. Application of design practices for dirt traps, valves, etc., described in *Part 3, Piping Design*.

6. Use of sumps large enough to prevent overflow when pumps are shut down.

7. Use of equalizer lines with twin cooling towers or other units connected in parallel to prevent overflow waste to a drain at one unit while make-up water is required at the other.

8. Proper physical location of cooling towers and evaporative condensers. Locations near smoke stacks or other sources of acid gases may cause serious corrosion problems. The exhaust from some photoprinting machines contains ammonia which is very corrosive to copper. In food storage areas, slime and algae problems are more serious.

9. Adequate space for chemical feeding devices, such as pumps and tanks including piping and wiring. Provide piping, or at least tees in the main lines of the piping, for the insertion of test coupons (*Chapter 3*). For large installations a control room with sink is convenient for the making of routine chemical tests.

10. Use of packing glands on pumps for open systems since some bleed-off is necessary.

    Sand or other abrasive materials shorten the life of mechanical seals. Mechanical seals are preferred for closed systems to prevent the loss of conditioned water.

    Chemicals added to a recirculated water system should not create problems with valve packing or pump seals or packing. A check should be made with the pump and valve manufacturers to obtain materials suitable for use with the selected chemical treatment.

FOULING FACTORS

The deposit of scale or other foreign material in the tubes of heat exchangers decreases the heat transfer (*Chapter 1*). Therefore, appropriate fouling factors must be included when equipment is selected if a satisfactory performance is to be maintained over a sustained operating period. Otherwise, excessive system shutdown periods may be required to clean fouled heat exchangers if loss of capacity cannot be tolerated. A fouling factor should be included when selecting equipment for all water systems. Even freshly cleaned equipment can usually have a quick build-up of a heat retardant film. Hence, high fouling factors can usually be expected after the equipment goes into operation. The change may not be a slow steady function of time as many believe.

TABLE 11—EQUIPMENT FOULING FACTORS
TEMP. OF WATER, 35-180 F
TEMP. OF MEDIUM, 0-240 F

SYSTEM OR WATER SOURCE	FOULING FACTORS*
Closed Recirculating Systems	
Chilled Water	
Water chiller	.0005
Coils†	—
Hot Water	
Steam to water heat exchangers	.001
Coils†	—
Open Recirculating Systems	
Air Washers	
Clean conditioned water	.0005-.001
Clean unconditioned water	.001-.002
Oil, lint, etc. in conditioned water	.001-.0025
Cooling Towers	
Conditioned water	.0005
Unconditioned water	.0005-.003
Partially conditioned water	
(small systems, 5-75 tons, see *Chapter 5*)	
Alkalinity, 0-125 ppm make-up water‡	.001-.0015
125-200 ppm make-up water‡	.001-.002
200-300 ppm make-up water‡	.002-.003
Once-Thru Condenser Systems	
Sea Water	.0005-.002
Brackish Water	.0005-.006
Great Lakes Water	.001-.002
River Water	
Minimum	.001
Mississippi	.001-.003
Delaware, Schuylkill	.002-.003
East River and New York Bay	.002-.003
Chicago Sanitary Canal	.006
Muddy or Silty Water	.002-.003
Surface or Well Water, using polyphosphates	
Alkalinity, 0-190 ppm	.001-.0015
190-300 ppm	.001-.002
300-450 ppm	.002-.003

*Fouling factors are for either the cooler or for the condenser.

†Fouling factors are normally not used for hot or chilled water coils because the coil performance is only reduced approximately 1% even with a .001 factor.

‡Using bleed-off of 1½ to 2 times the evaporation rate (cycles of concentration 1.66 to 1.50).

NOTES:

Use for nonferrous and stainless steel tubes, water velocity exceeding 3 feet per second.

Based on Carrier experience, and some data from Standards of TEMA (Tubular Exchanger Manufacturer's Association), Fourth Edition, 1959.

Table 11 contains a range of suggested fouling factors for various types of water systems used with heat exchangers. The lower value shown is for systems on which cleaning on a seasonal basis is not objectionable from a cost standpoint or a drop-off of capacity will not seriously affect desired end results. The higher value shown is suggested for industrial applications, and this value may even be exceeded on systems where excessive fouling may occur or the expense of cleaning is high. For example, systems which require a complete shutdown of plant production to accommodate the normal maintenance may use the higher values for fouling factors.

Fouling can be caused by a scale deposit of calcium carbonate, calcium sulfate, or by the products of corrosion; it can also be caused by an oily film or by a deposit of lint, silt, etc. A film of oil or grease offers greater resistance to heat transfer than does calcium carbonate because the former has a lower thermal conductivity.

Fouling can be significant even with a thin film; after cleaning a heat exchanger with acid and detergents, a film can quickly form on surfaces. Whenever a performance test is run, even directly after cleaning, the fouling factor cannot be assumed to be zero. For field capacity tests, a *minimum* fouling factor of one-half of the value used in selecting the equipment is suggested for test performance. In any instance this value should not be below .00025.

WATER CONDITIONING CHEMICALS

Chemicals introduced into a water system may require special feeding devices and care in handling.

FEEDING EQUIPMENT

The type of equipment used for feeding chemicals can vary greatly, depending on the size of installation, variation in load, variation in the composition of make-up water, and the amount of manual adjustment by operating personnel. Automatic feeding requires more costly equipment.

Large Systems, 75 tons and up

Chemicals packaged dry are usually dissolved in water and introduced as a liquid.

Positive displacement chemical feed pumps of the plunger or diaphragm type producing a uniform flow rate are the most common and dependable means of continuous introduction of chemical treatment solutions. Other types of pumps have either a variable speed drive, variable stroke, or variable speed motor.

Proportional feed can be achieved by means of a special water meter equipped with a contact head which makes an electric contact upon entry of a prescribed quantity of make-up water into the system. A preset timer activates the pump and feeds chemicals for a predetermined period.

A less costly proportional feeder for small installations consists of a measuring column with an electrical probe placed in a pipe between two electric solenoid valves. It is controlled by a timer which permits the treating solution to flow by gravity into the measuring column and discharge a metered amount of solution to the system by gravity flow on a preset time cycle. The time cycle as well as the quantities of solution admitted to the measuring column can be adjusted to vary the quantity of conditioning chemical.

Pumps and devices for feeding acids use corrosion resistant construction such as stainless steel or plastics. Completely automatic control devices for acid feed consist of a *pH* recorder-controller and a variable speed acid pump.

Equipment is available to perform automatically a chemical analysis of any colorimetric determination performed in the laboratory. This equipment can also be made to transmit signals for control of the chemical feeder.

Small Systems, 5 to 75 tons

Chemical feeders used on large systems may be used also on small installations, but the cost cannot normally be justified.

Various types of feeding devices, particularly for the small systems, are available. One type consists of bags of chemicals suspended on the side of the spray chambers above the water. Water flows over the bags whenever the sprays operate, and gradually dissolves the chemicals. Another is a metal can with two openings in the top. The can is placed in the basin of the cooling tower where movement of the water causes flow in one opening and out the other. A third device consists of a metal container with a collecting pan on top which is normally full of water when the sprays are operating. A replaceable orifice in the pan meters a constant flow of water to the chemicals located in the bottom of the container. A fourth device uses a vertical plastic cylinder into which is placed a number of briquettes one on top of another. A controlled amount of water from the pump discharge flows in and out, thru the bottom two or three briquettes. As one dissolves, another drops to take its place.

METHODS OF FEEDING CHEMICALS

The method of feeding chemicals is different for the three classes of water systems. Discussion of both the large and small systems follows.

Once-Thru Systems

Chemical feed (large and small systems) for scale and corrosion control is continuous for once-thru systems. The feeder operates at a constant rate whenever the water pump operates. Batch feed is normally used for slime and algae control.

Open Recirculating Systems

For the small system the chemical feed method for scale and corrosion is the replacement of chemicals or containers periodically, once every two to eight weeks or longer. There is no control check by chemical analysis.

For the large system a daily or regularly scheduled control analysis indicates whether the chemical feed rate should be increased or decreased. Feeding tanks are filled daily or weekly.

Closed Recirculating Systems

Chemicals are added by a batch method for all sizes of closed systems. Chemicals are added periodically, every one to six months or more often, depending on the loss of water from the system.

HANDLING OF CHEMICALS

Some chemicals cause severe skin irritation while others burn the flesh. Precautions should be taken to prevent skin contact with these chemicals; persons handling chromates sometimes complain of chrom-itch. The chlorophenol compounds used in water conditioning (even in low concentration) have been reported to cause dermatitis. Concentrated sulfuric acid, caustic soda and lime will burn the flesh.

Chlorine gas irritates the skin and eyes, and is very dangerous if inhaled; liquid chlorine requires special handling.[24, 26]

Water conditioning chemicals cause injury if taken internally in large doses.

CLEAN SYSTEM REQUIRED

On new systems the piping and units should be chemically cleaned to remove cutting oil, pipe thread compound, and other construction debris before starting the water conditioning. Cleaning prevents mechanical clogging and localized corrosion from dirt left in the system. Cleaning agents commonly used include polyphosphates, synthetic detergents, or combinations of these. Water is circulated at approximately 100F for a day or two, after which the system is drained, flushed and refilled. The maximum water temperature is limited to a temperature safe for the refrigerant contained in the refrigeration equipment. It is important to start the corrosion protection immediately after cleaning since metal surfaces are particularly vulnerable.

If water conditioning is started on an old system not previously treated, the water conditioning specialist usually recommends cleaning before treatment.

CORROSION INHIBITORS, INITIAL CONCENTRATION

Several days may be required to build up a protective film using a normal concentration of inhibitors. This may allow corrosion to start on steel. The rate of film build-up is a function of the concentration. An initial concentration of two to ten times the normal treatment of corrosion inhibitor can be used to form a protective film more quickly. After several hours of operation, the concentration is reduced to normal levels.

WATER CONDITIONING BY NONCHEMICAL METHODS

There are available for water treatment several types of devices which require no technical control and use nonchemical means to prevent corrosion and formation of scale. Some of these devices are made of special processed metal which acts as a catalyst. Others use permanent magnets placed so that water flows thru the magnetic field. An investigation[27] of several of these devices which claim to control corrosion and scale discloses that these particular devices have no significant effect upon scale and corrosion problems.

ORDINANCES AND GOVERNMENT REGULATIONS

In some states water used for humidification is required to meet drinking water standards regarding bacteriological quality. The United States Department of Agriculture Meat Inspection Division prohibits the use of chromates for air washing where the air contacts foodstuffs. Public Health officials are becoming more aware of the need for protection of the public water supplies. An ordinance presently in effect in Detroit states that no physical connection can be made between lines carrying city water and pumps, pipe and tanks from any other source; there must be an atmospheric gap between the city water line and the equipment. This regulation which applies also in other cities is good engineering practice.

WATER CONDITIONING OF LARGE SYSTEMS 75 TONS AND UP

Chapter 1 describes three types of water systems and three kinds of water problems. The various detailed treatments are described in *Chapters 2, 3 and 4*.

Water conditioning for these systems is summarized in *Table 12*. This section discusses the various methods of water treatment utilized in practice and summarized as mentioned. A more general discussion[28] of scale and corrosion emphasizes the need for a water conditioning specialist.

Water conditioning for the three types of systems and three kinds of water problems is discussed under the following headings.

ONCE-THRU SYSTEM

Since no water is recirculated, the cost of water conditioning chemicals for scale and corrosion control can be prohibitive. Therefore, the objective should be to obtain satisfactory protection with low levels of treatment.

Scale Control

The Langelier Saturation Index is most useful in predicting whether water is scale-forming. If the index is positive and less than +0.5, scale is probably not a problem. If greater than +.5, scale may be controlled by the addition of 2 to 5 ppm of polyphosphates unless the index is greater than +1.5 to +2.0, or a limit determined by the water conditioning specialist. If the index is too great, sulfuric acid may be added in addition to polyphosphates to reduce alkalinity and *pH*.

If the Langelier Index is negative (a nonscaling indication), the water is likely to be corrosive. Lime, soda ash or caustic soda may be added to increase the *pH*.

Corrosion may be prevented on heat exchangers by the intentional build-up of a thin film of calcium carbonate. This condition is accomplished by *pH* control with acid or alkali, but is not a common method for air conditioning systems.

Iron (as ferrous bicarbonate) is found in some well water; it decomposes to form a scale of ferrous oxide. Although polyphosphates are used to prevent this type of scale formation, proprietary compounds are available which are more effective.

Corrosion Control

The use of chromates in concentrations of 200-500 ppm is very effective in controlling corrosion, but it is not economical in a once-thru system. Suffi-

TABLE 12—SUMMARY OF WATER CONDITIONING CONTROLS*

	ONCE-THRU	OPEN RECIRCULATING	CLOSED RECIRCULATING

Scale Control	1. Surface active agents, such as polyphosphates
	2. Addition of Acid
	3. pH adjustment
	Other considerations
	  Adequate fouling factor
	  Surface temperature
	  Water temperature
	  Clean system
		1. Bleed-off
		2. Surface active agents such as polyphosphates
		3. Addition of acid
		4. pH adjustment
		5. Softening
		Other considerations
		  Adequate fouling factor
		  Surface temperature
		  Water temperature
		  Clean system
			No control required

Corrosion Control	1. Corrosion inhibitors in low concentrations 2. Deposit of protective scale of calcium carbonate 3. pH control 4. Proper materials of construction	1. Corrosion inhibitors in high concentrations (200 - 500 ppm) 2. Corrosion inhibitors in low concentrations, (20-80 ppm) 3. pH control 4. Proper materials of construction	Corrosion inhibitors in high concentrations (200 - 500 ppm) Proper materials of construction
Slime and Algae Control	Chlorinated phenols Other biocides Chlorine by hypochlorites or by liquid chlorine	Chlorinated phenols Other biocides Chlorine by hypochlorites or by liquid chlorine	No control required

*Abrasive materials must be kept out of the water system, and maximum velocity must not be exceeded in the tube. See *Table 7*.

cient protection at low treatment levels is the chief problem.

Low chromates mixed with other corrosion inhibitors are used in total treatment levels of 20-80 ppm. Control may be required by adding soda ash to raise, or by adding sulfuric acid to lower, the *pH*.

Frequently, well water contains dissolved gases such as carbon dioxide, CO_2, or hydrogen sulfide, H_2S, which may be very corrosive. Iron compounds and chlorides are also troublesome, and require special treatment.

Corrosion can be prevented by the controlled build-up of a thin film of calcium carbonate, as explained under scale control.

Occasionally it may be more economical to use corrosion resistant materials such as cupro-nickel, admiralty brass, stainless steel, or nickel instead of corrosion inhibitors in at least the more critical parts of the system.

Slime and Algae Control

Biological fouling may occur in once-thru systems when the source is polluted surface water. Well water does not normally produce slime formation. Chlorine is commonly used to control slime either by the use of hypochlorites or chlorination. The feed may be either intermittent or continuous.

OPEN RECIRCULATING SYSTEM

Most of the water conditioning problems due to the evaporation of large quantities of water and subsequent concentration of solids are prevalent with this type of system, particularly in cooling towers, evaporative condensers and spray ponds.

Air washers require little water conditioning because in summer there is some water condensed from the air and added to the system. In winter, only a small amount of water is evaporated. Bleed-off is normally required for all open recirculating systems.

Water conditioning is usually designed to be non-scale-forming, a condition which is slightly corrosive. Corrosion inhibitors are added to minimize corrosion.

Scale Control

Three methods to prevent scale formation are described in *Chapter 2*. These involve the use of:
1. Bleed-off.
2. Bleed-off plus scale inhibitors such as polyphosphates.
3. Bleed-off plus scale inhibitors plus acid.

Bleed-off — General rules have been developed for the use of bleed-off to prevent scale formation. When the methyl orange alkalinity of the make-up water does not exceed 100 ppm as $CaCO_3$ and the recirculating water does not exceed a methyl orange alkalinity of 125 ppm, bleed-off alone may be satisfactory. This is illustrated in *Chart 7 (Chapter 2)* which shows the bleed-off rate for a 100 ton air conditioning compressor system based on the above limits of alkalinity.

Accurate determination of the rate of bleed-off is usually accomplished by the use of the Langelier and Ryznar Indexes. These indexes take into consideration the five variables influencing scale formation of which alkalinity is one.

The Langelier Index is determined for several cycles of concentration *(Chapter 2)*. Although bleed-off alone may control scale formation, there is usually a need for corrosion inhibitors.

Bleed-off Plus Scale Inhibitors — Since scale inhibitors such as polyphosphates allow a much higher concentration of solids without scale formation, the amount of make-up water can be greatly reduced, thus saving water costs. Another important factor is the saving of corrosion inhibitors since less chemical is required to maintain the treatment level. The amount of bleed-off required when polyphosphates are used *(Chart 7)* is based on an arbitrary limit of 400 ppm alkalinity. In practice, some systems may have a limit exceeding 400 ppm; others may have a lower limit. Scaling depends on all five properties; alkalinity, calcium, *pH*, temperature and total solids *(Chapter 2)*. All of these factors are evaluated to determine when scaling starts, as scaling cannot be determined by the use of one or even three of these factors alone.

Polyphosphates in concentrations of 2-5 ppm are used to prevent scale formation. Scale inhibitors frequently contain polyphosphates mixed with other compounds which also prevent scale formation.

Polyphosphates prevent the formation of tubercules in the piping system and thus help to keep the water system clean.

Bleed-off Plus Scale Inhibitors Plus Acid — Bleed-off may be reduced by the addition of acid to the system where the water is costly and/or high in concentrations of solids. Sulfuric acid, commonly used because of economics, changes the carbonate ion to the more soluble sulfate ion *(Chapter 2)*. Only enough acid is added to reduce the alkalinity sufficiently to permit scale control by the use of the polyphosphates.

Automatic *pH* control of acid feed, although costly, is recommended because of the hazard involved. A low *pH* due to excess feed of the acid is corrosive.

Air conditioning systems in offices and manufacturing areas of industrial companies often use non-automatic *pH* control. However, daily or even more frequent checks must be made on the chemical control analysis, and adjustment must be made in rate of chemical feed as necessary.

Corrosion Control

System corrosion is controlled by one or both of the following methods.

Corrosion Inhibitors — Chromates are extensively used, and are very effective in controlling corrosion in concentrations of 200-500 ppm at a 7.0-8.5 *pH*.

A mixture of chromates and other corrosion inhibitors with a total treatment in the range of 20 to 80 ppm is sometimes used. Usually the *pH* needs to be held within a narrow range; these are examples of such ranges: 6.0-6.5, 7.0-7.8, 6.5-7.5 *(Chapter 3)*.

When yellow staining due to chromates is objectionable, sodium nitrite can be substituted in the same concentration as the chromates, but the concentration should be approximately twice as great as that of chromates to obtain equal protection, and the *pH* must be maintained above 7.0. Several problems in the use of nitrites are described in *Chapter 3*.

Control of pH — Acid gases in the atmosphere sometimes cause water in cooling towers and air washers to become acidic, even though the make-up water may be neutral. The *pH* in some cooling towers[29] may be as low as 4.0 or 5.0. Addition of an alkali such as soda ash is used to raise the *pH*. If the *pH* is too high, sulfuric acid is commonly added to lower the value. When the *pH* is low, there is no problem with scale.

Slime and Algae Control

Cooling Towers — Slime and algae are usually controlled by chlorinated phenols fed in shot fashion. This may be required at daily, weekly or even longer intervals. Application of chlorine gas by chlorinators or by hypochlorites may be used. Chlorine gas with chlorinator equipment is used only in very large systems.

Air Washers — The more effective substances such as trichlorophenates and chlorine cannot be used because of objectionable odors. Sodium pentachlorophenate may be used. Proprietary biocides have combinations of toxicants for this service. Periodically, the air washers should be sterilized by stopping the air flow and circulating a solution of hypochlorites or a nonoxidizing biocide. The washer is then cleaned with a hose, and slime and dirt removed from the pan.

CLOSED RECIRCULATING SYSTEM

Closed systems require little make-up water unless there is some unusual condition such as a leak at the pump gland or overflow at the expansion tank. Water can also be lost at automatic air vents. Generally, there is no control problem for scale or slime and algae.

Corrosion Control

For chilled water systems chromates in treatment levels of 200-500 ppm at a *pH* of 7.0 to 8.5 are commonly used.

Where yellow staining is a problem, i.e. discoloring of rugs during service on valves and strainers, sodium nitrite may be used in the same concentration as the chromate. Some of the contaminating influences which cause decomposition of sodium nitrite in open systems are absent in closed systems.

In some instances an alkali or acid may be added to keep the *pH* within limits.

Since the cost of the chemicals is small because of little make-up water, the full chromate treatment is used. There is no need to use low chromates with other inhibitors (sometimes utilized in an open system).

The treatment for hot water systems is in the range of 1000-2000 ppm of chromate, the higher concentration being used for water at 212 F. A system which circulates chilled water in summer and hot water in winter should carry a higher concentration of corrosion inhibitors in winter.

WATER CONDITIONING OF SMALL SYSTEMS 5 TO 75 TONS

In small systems where water conditions are poor, an air-cooled condenser is often used to eliminate the water problems. However, it is not always feasible to accomplish this, or the cost of changing to this type of equipment may be too great.

ONCE-THRU SYSTEM

Water conditioning for scale, corrosion, slime and algae is essentially the same as for large systems. Polyphosphate treatment can be used for scale control. Acid treatment is seldom used because of high cost of feeding equipment and controls required.

If corrosion is a problem, use of corrosion resistant materials for the critical components should be considered. Low concentrations of corrosion inhibitors in the range of 20-80 ppm are used, such as mixtures of chromates and other corrosion inhibitors.

CHART 9—BLEED-OFF REQUIRED FOR A 10 TON AIR CONDITIONING SYSTEM

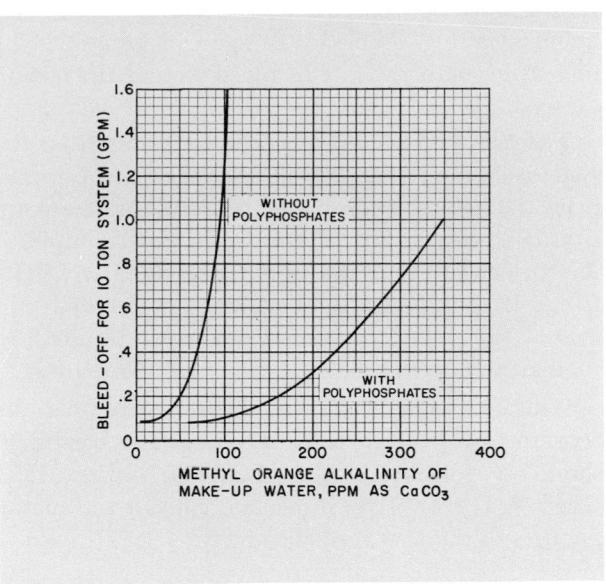

OPEN RECIRCULATING SYSTEM

The open system has all three water problems.

Scale Control

Packages of chemical treatments can be purchased by the consumer thru air conditioning service organizations or directly from the supplier. *Chart 9* illustrates how the bleed-off required varies with the methyl orange alkalinity of the make-up water. This curve is based on a maximum of 125 ppm alkalinity as $CaCO_3$ in the recirculating system without chemical treatment.

The curve for bleed-off with polyphosphates is representative of bleed-off rates recommended by packaged chemical treatments for scale prevention. When the alkalinity of make-up water is in the range of 0-250 ppm as $CaCO_3$, the bleed-off rate is based on maintaining a maximum alkalinity of 400 ppm in the recirculated water. When the alkalinity in the make-up water exceeds 250 ppm, the maximum alkalinity maintained in the recirculated water is 450 ppm as $CaCO_3$.

The use of polyphosphates greatly reduces the bleed-off, and is recommended where the alkalinity in the make-up water exceeds 100 ppm. Generally, when using polyphosphates, a satisfactory scale control can be expected when the methyl orange alkalinity of make-up water does not exceed 150 ppm. When the alkalinity of make-up water is in the range of 150-250 ppm, some scale build-up can be expected. This scale is removed by acid cleaning during equipment shutdown. Depending upon conditions, cleaning may be carried out yearly or once every several years. When the alkalinity exceeds 250 ppm, more scaling may be expected and, therefore, more frequent cleaning. Some advantage can be gained by increasing both the rate of bleed-off and the chemical feed.

The bleed-off (*Chart 9*) is the total lost from the system. The windage loss (*Chapter 1*) is subtracted from the bleed-off to obtain the actual water quantity to be discharged to the drain. For example, if the tower is 30 tons, multiply bleed-off (*Chart 9*) by three. If the alkalinity of make-up water does not exceed 100 ppm, it is possible to adjust bleed-off so no treatment is needed to prevent scale formation.

When packaged treatment chemicals are used, the recommended bleed-off for the particular treatment should be followed, rather than that illustrated in *Chart 9*. If trouble is suspected, consult the manufacturer of the packaged chemicals.

Corrosion Control

Although water conditioning may not be required for scale control, it may be necessary to prevent corrosion or the growth of slime and algae.

Polyphosphates have some effect in reducing corrosion; when scale is not a problem, the concentration may be increased to 10-15 ppm. This is accomplished by increasing the feed rate 2 to 4 times and by reducing the bleed-off to one-half the rate shown in *Chart 9*.

Chromates in concentrations of 200-500 ppm are very effective. Packages of chromates and polyphosphates are available for small systems. Generally, the bleed-off rate is lower where chromates are used because scale is no problem and a lower bleed-off allows higher chromate concentration in the system.

Low level treatment of 20-80 ppm of mixtures of chromates with other inhibitors may be used but, since this generally requires *pH* control, this method is not practical for small installations.

In some areas where the *pH* is too low, an alkali such as soda ash may be mixed with the chromates.

Slime and Algae Control

Shock feeding once a month or oftener is sufficient to control biological growth. Compounds described in this chapter for large systems are commonly used. A continuous drip feed consisting of a plastic bottle and capillary plastic tube is sometimes used.

CLOSED RECIRCULATING SYSTEM

Water conditioning for small closed systems is the same as for large closed systems.

SUGGESTED GUIDE SPECIFICATIONS FOR WATER CONDITIONING

Suggested guide specifications are presented to aid the engineer, contractor and owner to understand what constitutes an adequate water conditioning program.[30] They help evaluate proposals made by water conditioning companies.

GENERAL CONDITIONS

An adequate water conditioning program to control corrosion, scale, algae, slime and sludge shall be provided for a period of _____ years for open and closed water and brine systems, commencing with the start-up of the refrigeration and/or air conditioning equipment. Recirculated water in wooden cooling towers shall be so conditioned as to prevent delignification of wood.

The chemicals used for water conditioning and the methods used to feed chemicals shall comply with local health codes, where applicable.

Chemicals used for water conditioning shall have no detrimental effect on nonmetallic materials such as rubber, etc. used in the system.

The water conditioning program shall be administered by a competent water treatment company acceptable to both the customer and the contractor.

TERMS OF PROPOSAL

The water conditioning company shall submit a complete proposal with its bid.

A complete analysis of the water to be used for make-up in the several systems to be treated shall be included in the proposal.

The chemical names of the water conditioning compounds to be used shall be stated.

The concentration of chemicals to be maintained in the several systems shall be stated.

The estimated amount of continuous bleed-off from recirculating water systems such as cooling towers, evaporative condensers, humidifiers, etc. shall be stated.

Type and frequency of service by the water treatment company shall be stated.

CORROSION CONTROL

Corrosion control shall be provided for all water and brine circuits by using suitable corrosion inhibitors and *pH* control.

Corrosion shall be judged under control when the over-all corrosion rate or maximum depth of pitting does not exceed _____ inches of penetration per year of the thickness of any metal component of the circuit in continuous contact with the conditioned water or brine.*

Over-all corrosion and pitting rates shall be measured by using suitable corrosion test coupons representing the metals in the circuit. ASTM corrosion test procedures D 935-49 and A 224-46 or other equally suitable procedures shall be followed.

Corrosion tests shall be conducted at locations in the circuit as specified by the contractor.†

If over-all corrosion and/or pitting exceeds the maximum allowable rates, an immediate retest shall be made. If the results of the retest are within the maximum allowable rates, corrosion shall be judged under control.

SCALE CONTROL

Formation of adherent mineral deposits which cannot be easily flushed from heat transfer surfaces shall be prevented.

A continuous bleed-off shall be specified for all devices which evaporate water, such as cooling towers, evaporative condensers and humidifiers.†

Internal chemical treatment of water in the circuit and/or external treatment of the make-up water shall be provided when bleed-off alone is inadequate to prevent scale formation or when bleed-off alone is uneconomical because of excessive use of corrosion inhibitors.

ALGAE AND SLIME CONTROL

Algae and slime growths shall be prevented in all circuits by using suitable algaecides. Copper, mercury, or other algaecides which may cause damage to equipment shall not be used.

CHEMICAL FEEDING

Chemicals shall be fed to circuits requiring continuous make-up by automatic proportional feeding devices.†

Acid feeders, when used, shall be controlled by an electronic *pH* controller.†

SERVICE PROCEDURE

The water conditioning company shall provide (complete with alternate a or b, whichever applies):

a. Oral and written instructions and suitable test kits for operating personnel for the maintenance and control of the water conditioning program.

b. The addition of all chemicals and all control testing by its own personnel.

Periodic service calls, water analysis, and corrosion test checks shall be made at intervals as necessary to insure proper control of corrosion, scale, algae and slime.

A written report of each service call, water analysis, and corrosion check test shall be submitted by the water conditioning company (complete with alternate a or b, whichever applies):

a. Upon request.

b. In accordance with a prearranged schedule.

BASIS OF TABLE 13
Water Analysis for Supplies in United States Cities

Data is taken from *The Industrial Utility of Public Water Supplies in the United States*, 1952 Geological Survey Water Supply Paper 1299 (states east of the Mississippi River) and 1300 (states west of the Mississippi River), U.S. Government Printing Office, Washington, D. C.

General

Data is for finished water after treatment at the water plant. When there is one source, averages are given, rather than maximum and minimum. When there is more than one source, the averages are given if available. When the composition is given only for the individual sources, i.e. 6 wells, two figures are included, one for the low level and another for the high level concentration. The water from various sources may be mixed together in some systems; in others, they may be grouped in several systems just like different systems. The composition in one section of the city may, therefore, be considerably different from another section.

Consult the local water companies for more specific data.

Total Hardness

Total hardness is a rough measure of scale-forming properties. It includes both carbonate and non-carbonate hardness (*Chapter 2*). Average values are given, except for those cities having an asterisk (*) in the alkalinity column. For these cities the hard-

---

*Commonly accepted maximum corrosion rate and depth of pitting is .005 inches per year (5 mils per year, mpy). This usually applies to metals having a thickness greater than 0.25 inches. Where metal thickness is less, proportional lower values should be used.

For copper tubing the maximum corrosion rate is .001 inches per year (one mil per year). See *Chapter 3*.

†The contractor responsible for piping and/or electrical work should provide suitable fittings and connections for corrosion testers and continuous bleed-off, and should also provide

necessary facilities for the installation and operation of automatic feeders and their controls.

ness is for a particular day. Hardness may be classified in this manner:

HARDNESS (ppm as $CaCO_3$)	CLASSIFICATION
less than 15	very soft
15 – 50	soft
50 – 100	medium hard
100 – 200	hard
greater than 200	very hard

Bicarbonate (HCO_3)

Concentration of bicarbonate (HCO_3) ion has been multiplied by 0.82 to give the equivalent expressed as $CaCO_3$ (*Table 13*). Generally, this is the chemical analysis taken on a particular day, and is not an average. It may be higher or lower than the average.

Calcium (Ca)

Concentration of calcium (Ca) ion has been multiplied by 2.5 to give the equivalent as $CaCO_3$ (*Table 13*). For example, Ithaca, New York, calcium ion = $22 \times 2.5 = 55$ ppm of calcium as $CaCO_3$. Generally, this is the chemical analysis taken on a particular day, and is not an average. It may be higher or lower than the average.

Total Alkalinity

Total alkalinity (methyl orange alkalinity, *Chapter 2*) is the average, except where indicated by an asterisk (*). No value for alkalinity is given in Papers 1299 and 1300 for these cities. The values shown have been approximated by use of the bicarbonate and carbonate ion; these are given for a particular day and are not average values. Data has been approximated in *Table 13* as follows:

Bicarbonate (HCO_3) ion ppm $\times$ 0.82 =
plus
Carbonate (CO_3) ion ppm $\times$ 1.66 =
equals
Total alkalinity, ppm as $CaCO_3$ =

pH

Values for *pH* are average, except for those cities having an asterisk (*) in the alkalinity column. For these cities the *pH* is for a particular day.

TYPICAL WATER ANALYSIS

An example of water analysis from this publication is shown for Ithaca, New York, one of the less complicated analyses. Many of the cities do not have the regular determinations of alkalinity, *pH* and hardness as shown in the bottom section of the table.

ITHACA, NEW YORK

Ownership: Municipal; also supplies about 2000 people outside the city limits. Total population supplied, about 31,000.

Source: Six Mile Creek impounded. Emergency supply, connection with Cornell University supply.

Treatment: Superchlorination, coagulation with alum, sedimentation, rapid sand filtration, and dechlorination with sulfur dioxide.

Rated capacity, treatment plant: 4,000,000 gpd.
Raw water storage: 316,000,000 gal.
Finished water storage: 4,924,000 gal.

Analysis (ppm) by U. S. Geological Survey

	FINISHED WATER (city tap)		FINISHED WATER (city tap)
Silica, SiO_2	3.5	Carbonate, CO_3	0.
Iron, Fe	.08	Bicarbonate, HCO_3	46.
Manganese, Mn	.00	Sulfate, SO_4	29.
Calcium, Ca	22.	Chloride, Cl	5.8
Magnesium, Mg	2.9	Fluoride, F	.2
Sodium, Na	5.2	Nitrate, NO_3	1.6
Potassium, K	.8	Dissolved solids	96.
Hardness as $CaCO_3$ Total	67		
Noncarbonate	29		
Color	1	Turbidity	0.5
pH	7.4	Temperature (F)	—
Specific Conductance (micromhos at 25 C)	163	Date of collection	

Regular Determinations at Treatment Plant

	ALKALINITY as $CaCO_3$ (ppm)			pH			HARDNESS as $CaCO_3$ (ppm)		
	Avg	Max	Min	Avg	Max	Min	Avg	Max	Min
Raw water	72	113	44	7.5	7.7	7.3	93	129	63
Finished water	66	98	40	6.7	6.8	6.7	92	132	61

TABLE 13—WATER ANALYSES FOR SUPPLIES IN UNITED STATES CITIES

STATE AND CITY	Source of Supply	Total Hardness as CaCO₃ (ppm)	Bicarbonate (HCO₃) as CaCO₃ (ppm)	Calcium (Ca) as CaCO₃ (ppm)	Total Alkalinity as CaCO₃ (ppm)	Dissolved Solids (ppm)	pH	Specific Conductance at 77 F (micromhos)
ALABAMA								
Anniston	S	104	102	55	102*	118	7.5	191
Birmingham	Ri, R, L	27-91	24-73	23-68	24-73*	46-118	8.3-9.0	73-193
Mobile	C	20	10	16	11	38	8.9	54
Montgomery	W	18-67	105-117	12-57	105-117*	154	7.0-7.5	283-257
ARIZONA								
Flagstaff	L	65	33	48	33*	88	8.0	139
Phoenix	Ri, R, W	190-422	126-222	100-205	126-222*	290-887	7.8	473-1490
Tucson	W	94-220	114-208	72-162	114-208*	117-484	8.0	264-736
Winslow†	C	221	172	110	177*	1110	8.4	2110
ARKANSAS								
Fort Smith	Ri, L	48	18	23	21	59	8.9	71
Little Rock	RI, L	20	12	11	14	31	8.3	47
CALIFORNIA								
Bakersfield†‡	W	18-628	48-205	12-320	48-205*	108-1010	7.4	180-1620
Eureka	Ri	60	60-88	50	40	73-134	8.0	—
Fresno	W	72-163	83-165	37-85	83-165*	202-336	7.7	231-451
Long Beach	Ri, W	62	134	48	141*	338	8.5	—
Los Angeles	Ri, W	84-274	113-189	62-192	116-189*	200-478	7.7-8.3	317-724
Oakland	Ri	18-208	20-163	15-120	20-163*	37-316	7.9-9.1	—
Pasadena	Ri, W	65-179	81-177	50-117	81-177*	187-272	7.3-7.8	—
Sacramento	Ri, W	56	15	13	33	37	6.7	55
San Bernadino	W, C	67-169	104-200	60-130	104-200*	197-258	8.0-8.8	319-252
San Diego‡	Ri	170-295	64-193	185-187	79-193*	307-648	7.6-8.4	500-980
San Francisco	Re	59-99	6-86	3-63	6-86*	9-99	6.4-7.7	20-225
San Jose	W	194	182	105	182*	288	8.2	475
Metropolitan Water Dist. So. Cal.‡**	Ri	125	100	78	111	692	8.8	1100
COLORADO								
Denver	Ri, C	48-139	32-92	35-85	35-88	—	7.7-8.0	106-455
Fort Collins	Ri	22	13	30	24	77	7.2	94
Grand Junction	C	84	90	55	90	118	7.8	180
Pueblo‡	Ri	317	74	125	125	329	7.5	491
CONNECTICUT								
Bridgeport	Re	29	14	19	14*	46	7.2	76
Hartford	Re	15	6	9	7	27	6.4	40
New Haven	Re	20-69	5-50	14-57	32	36-111	6.8	64-169
Waterbury	Re	24	14	19	14*	42	7.1	68
DELAWARE								
Wilmington	C	52	22	30	27	89	6.5	137
DIST. OF COLUMBIA								
Washington	Ri	96	65	75	60	151	7.6	236
FLORIDA								
Jacksonville‡	W	274	134	168	134*	461	7.4	—
Key West	W	76	50	53	86	85	8.4	150
Miami	W	85	40	63	43	190	8.8	—
Pensacola	W	20	9	13	9*	56	7.1	87
Tallahasee	W	124-152	120-142	90-102	120-142*	164-173	7.8	287
Tampa	Ri	115	29-57	73	59	123	8.3	197

Sources of Supply — B-brook, C-creek, L-lake, P-pond, Re-reservoir, Ri-river, S-stream, Sp-spring, W-well

Note: See explanation of *Table 13* on page 39.

*Total alkalinity approximated, as described in explanation of *Table 13*.

†Chloride (CI) ion concentration in excess of 100 ppm.

‡Sulfate (SO₄) ion concentration in excess of 100 ppm.

**Metropolitan Water District of Southern California supplies for a population of 3,500,000, to *portions* of Long Beach, Los Angeles, Pasadena, Beverly Hills, Burbank, Compton, Fullerton, Glendale, San Marino, Santa Anna, Santa Monica, Torrance, etc.

TABLE 13—WATER ANALYSES FOR SUPPLIES IN UNITED STATES CITIES (Contd)

STATE AND CITY	Source of Supply	Total Hardness as CaCO3 (ppm)	Bicarbonate (HCO3) as CaCO3 (ppm)	Calcium (Ca) as CaCO3 (ppm)	Total Alkalinity as CaCO3 (ppm)	Dissolved Solids (ppm)	pH	Specific Conductance at 77 F (micromhos)
GEORGIA								
Atlanta	Ri	19	13	15	16	44	8.7	60
Augusta	Ri	20	15	15	15	46	7.5	63
Brunswick	W	196-530	117	98-255	117*	283-1000	7.3	430-1450
Columbus	Ri	18	17	20	14	55	8.2	88
Macon	Ri	32	21	25	25	64	8.5	98
Savannah	W, C	37	28	30	23	67	7.9	108
IDAHO								
Boise	W	85	75	73	75*	157	7.3	228
Lewiston	Ri, W	32-66	35-122	30-43	26-122*	77-200	7.8	109
Pocatello	C, W	102-260	95-246	75-162	95-246*	136-368	7.9-7.5	219-629
Twin Falls	Ri	197	152	127	152*	293	7.4	483
ILLINOIS								
Chicago	L	121-138	113	93	107	148-168	8.1	242
Danville	Ri	183	180	118	144	265	7.3	465
Moline	Ri	150	25	55	84	104	7.6	166
Peoria	W	394-455	292-348	227-275	292-348*	448-583	—	—
Springfield‡	C	94	30	30	36	116	10.5	—
INDIANA								
Evansville	Ri	132	45	75	64	137	8.1	229
Fort Wayne	Ri	76	17	65	30	115	9.8	180
Indianapolis	Ri, C	222-294	155-227	125-190	155-227*	282	7.5	462
South Bend	W	298-544	240-292	180-352	240-292*	318-681	7.5	529-895
Terre Haute	Ri	278	111	140	154	257	7.1	421
IOWA								
Cedar Rapids	Ri	88	36	75	43	182	9.5	278
Davenport	Ri	144	67	90	107	196	7.0	291
Des Moines	Ri	83	4	30	39	212	10.1	341
Dubuque	W	282	264	142	264*	296	7.8	525
Fort Dodge‡	W	466	367	287	367*	692	7.6	1050
Keokuk	Ri	100	94	137	72	246	8.8	362
Sioux City‡	W	472	332	312	332*	672	7.5	982
Waterloo	W	241	193	167	193*	294	7.9	477
KANSAS								
Dodge City	W	248	192	157	192*	368	7.6	528
Salina‡	W	548	370	440	368	800	7.0	1150
Topeka‡	Ri, W	98	35	73	50	390	9.4	638
Wichita	W	106	69	48	136	260	8.0	418
KENTUCKY								
Lexington	C	84	85	90	76	128	7.4	224
Louisville	Ri	97	36	53	40	146	9.0	251
LOUISIANA								
Alexandra	W	4-544	174-490	3-242	174-490*	270-689	6.8-8.1	817-1130
New Orleans	Ri	75	20-30	40-50	35	162	10.1	280
Shreveport	L	63	33	45	32	162	9.0	270
MAINE								
Augusta	L, P	20	9	12	9*	34	6.5	47
Bangor	Ri	33	9	30	18	64	6.7	90
Portland	L	13	7	11	6	23	6.8	31
Presque Isle	S	82	21	40	47	71	7.1	105
Rumford	S	9	5	4	5*	22	6.1	25

Sources of Supply — B-brook, C-creek, L-lake, P-pond, Re-reservoir, Ri-river, S-stream, Sp-spring, W-well
Note: See explanation of *Table 13* on page 39.
*Total alkalinity approximated, as described in explanation of *Table 13.*
†Chloride (Cl) ion concentration in excess of 100 ppm.
‡Sulfate (SO₄) ion concentration in excess of 100 ppm.

TABLE 13—WATER ANALYSES FOR SUPPLIES IN UNITED STATES CITIES (Contd)

STATE AND CITY	Source of Supply	Total Hardness as CaCO₃ (ppm)	Bicarbonate (HCO₃) as CaCO₃ (ppm)	Calcium (Ca) as CaCO₃ (ppm)	Total Alkalinity as CaCO₃ (ppm)	Dissolved Solids (ppm)	pH	Specific Conductance at 77 F (micromhos)
MARYLAND								
Baltimore	Ri, Re	52	40	35	40	81	8.0	—
Cambridge	W	32	384	12	404*	481	8.5	758
Cumberland	C, L	36	53	25	53*	93	7.2	158
Frederick	C	53	21	27	34	78	6.6	131
Salisbury	W	20	18	18	18*	81	7.1	104
MASSACHUSETTS								
Boston	Ri	14	6	10	7	29	6.6	41
Fall River	P	15	8	12	8*	42	7.1	66
Fitchburg	P, Re	6	3	4	3	23	6.2	30
Lowell	W	42	32	33	32	90	6.3	137
New Bedford	P	15	10	11	10*	35	6.5	55
Springfield	Re	12	6	10	6*	30	6.7	39
Worcester	Re	9	3	7	3*	33	6.7	31
MICHIGAN								
Battle Creek	W	201	224	175	222	316	7.0	504
Detroit	Ri	98	76	68	76	132	7.6	—
Escanaba	L, W	132-185	104-164	88-100	88-100*	156-245	7.2	282
Flint	Ri	86	0	58	33*	160	10.2	270
Grand Rapids	L	136	111	87	109	155	7.8	280
Kalamazoo	W	314-381	263-282	202-242	263-282*	357-460	7.6	581-711
Lansing	W	85	3	48	83	177	10.4	—
Marquette	L	46	41	33	41*	51	7.6	94
Saginaw	L	49	33	48	33	111	9.4	194
Sault Ste. Marie	L	43	40	33	40*	52	7.6	94
MINNESOTA								
Duluth	L	46	41	35	43	54	7.6	102
Minneapolis	Ri	77	25	50	42	126	8.0	182
St. Cloud	Ri, L	134	83	88	83*	190	7.0	287
St. Paul	Ri	72	48	53	57	92	8.6	148
MISSISSIPPI								
Jackson	Ri	50	24	38	25	85	9.0	122
Meridian	L, Sp	18	11	14	11*	38	6.8	50
Vicksburg	Ri	110	122	145	90	356	8.4	524
MISSOURI								
Columbia	W	270	302	143	302*	406	8.2	668
Kansas City‡	Ri	84	40	58	40	297	9.4	297
Kirkville	Ri	90	52	127	40	216	7.6	317
St. Joseph	Ri	232	177	177	160	462	7.8	678
St. Louis‡	Ri	97	17	58	37	238	9.2	—
Springfield	Sp, L	178	160	155	157	247	7.4	—
MONTANA								
Billings	Ri	168	120	95	122	257	8.2	407
Butte	Ri, R, C	56	52	40	52*	98	7.4	137
Great Falls	Ri	142	128	102	130	241	7.7	386
Havre‡	W	404	427	207	427*	1120	7.7	1640
Helena	C, Re, Sp	33-190	14-150	23-135	14-150*	76-259	7.1-7.7	93-395
Kalispell	Sp, W	145	148	107	148*	157	7.6	274
Miles City‡	Ri, W	60	70	33	68	422	8.2	663
Missoula	C	16	12	11	12*	22	7.5	27

Sources of Supply — B-brook, C-creek, L-lake, P-pond, Re-reservoir, Ri-river, S-stream, Sp-spring, W-well

Note: See explanation of *Table 13* on *page 39*.

*Total alkalinity approximated, as described in explanation of *Table 13*.

†Chloride (Cl) ion concentration in excess of 100 ppm.

‡Sulfate (SO₄) ion concentration in excess of 100 ppm.

TABLE 13—WATER ANALYSES FOR SUPPLIES IN UNITED STATES CITIES (Contd)

STATE AND CITY	Source of Supply	Total Hardness as CaCO₃ (ppm)	Bicarbonate (HCO₃) as CaCO₃ (ppm)	Calcium (Ca) as CaCO₃ (ppm)	Total Alkalinity as CaCO₃ (ppm)	Dissolved Solids (ppm)	pH	Specific Conductance at 77 F (micromhos)
NEBRASKA								
Grand Island	W	189	136	155	136*	324	7.2	465
Lincoln	W	206-247	174-195	147-172	174-195*	310-396	7.8-7.3	500
Norfolk	W	310	254	234	254*	442	7.8	633
North Platte‡	W	326	206	248	206*	798	7.8	1100
Omaha	Ri	261	139	160	146	448	7.8	682
NEVADA								
Las Vegas	W	229	193	122	193*	266	7.6	439
Reno	Ri, C	40	24	24	24*	91	7.0	110
Winnemucca	W, Sp	224	234	150	234*	483	7.7	759
NEW HAMPSHIRE								
Berlin	B	10	7	8	7*	29	6.6	29
Concord	L	21	7	8	16	25	6.8	38
Keene	L	19	3	5	3*	20	7.2	25
Manchester	L	15	3	9	3*	36	6.5	44
Portsmouth	W	114-139	59-90	65-97	59-90*	153-191	7.1-7.5	204-231
NEW JERSEY								
Atlantic City	W, C	5	3	5	2	41	5.0	52
Bloomfield	Ri, Re	35	11	16	13	39	6.9	65
Camden	W	20-241	1-90	13-97	1-90*	59-460	4.6-6.8	80-444
East Orange	W	139	102	90	102*	194	7.8	304
Jersey City	Ri, Re	44	26	25	24	78	6.8	98
Newark	Ri, Re	33	11	19	14	44	7.3	66
Paterson	Ri, Re	50	23	30	25	86	7.4	—
Trenton	Ri	59	20	43	33	80	7.2	128
NEW MEXICO								
Albuquerque	W	85-116	133	60-85	133*	310-340	7.8	471
Roswell	W	588-664	179-194	420-468	179-194*	886-1160	7.8	1250-1740
Santa Fe	C, W	30-160	25-126	20-140	25-126*	58-217	7.9	72-347
NEW YORK								
Albany	C, Ri, Re	47	34	45	30	70	8.5	122
Binghamton	Ri	57	54	60	54*	91	7.2	153
Buffalo	L	124	89	90	90*	164	7.4	285
Cortland	W	168	139	132	139*	188	7.6	317
Glens Falls	S, Re	16	10	12	10*	29	6.7	41
Ithaca	C	92	38	55	66	96	6.7	163
Jamestown	W	103	82	80	82*	127	8.0	214
New York City	Ri, Re, C	20-50	7-30	13-32	7-30*	43-75	6.7-7.1	44-121
New York City	W	118-282	40-164	—	40-164*	202-392	6.3-7.7	260-432
Ogdensburg	Ri	131	94	95	94*	169	7.7	295
Oswego	L	140	94	105	94*	179	7.9	323
Rochester	L	79-130	48-90	55-95	48-90*	176	7.4	170-295
Schenectady	W	160	135	122	135*	197	7.7	342
Syracuse	L	109	94	85	94*	128	7.8	222
Watertown	Ri	52	58	68	36	124	7.1	201
NORTH CAROLINA								
Asheville	C, Ri	6	5	4	5*	19	5.8	19
Charlotte	Ri	20	15	17	19	45	8.6	75
Greensboro	C, L	33	39	30	36	70	8.6	104
Raleigh	C, L	30	16	24	27	61	9.2	100
Wilmington	Ri	42	25	28	28	68	8.0	116

Sources of Supply — B-brook, C-creek, L-lake, P-pond, Re-reservoir, Ri-river, S-stream, Sp-spring, W-well
Note: See explanation of *Table 13* on *page 39.*
*Total alkalinity approximated, as described in explanation of *Table 13.*
†Chloride (Cl) ion concentration in excess of 100 ppm.
‡Sulfate (SO₄) ion concentration in excess of 100 ppm.

TABLE 13—WATER ANALYSES FOR SUPPLIES IN UNITED STATES CITIES (Contd)

STATE AND CITY	Source of Supply	Total Hardness as CaCO₃ (ppm)	Bicarbonate (HCO₃) as CaCO₃ (ppm)	Calcium (Ca) as CaCO₃ (ppm)	Total Alkalinity as CaCO₃ (ppm)	Dissolved Solids (ppm)	pH	Specific Conductance at 77 F (micromhos)
NORTH DAKOTA								
Bismark‡	Ri	90	39	70	72	244	8.8	338
Fargo	Ri	129	57	75	85	360	9.0	536
Grand Forks‡	Ri	81	37	90	67	248	9.4	363
Williston‡	Ri	137	73	102	55	374	8.9	553
OHIO								
Akron	Ri	119	59	83	69	136	8.3	225
Cincinnati	Ri	111	45	85	39	195	8.5	324
Cleveland	L	125	85	98	82	169	7.5	286
Columbus‡	Re, Ri, W	81	4	48-88	34	181-289	9.5-10.2	181-289
Dayton	W	358	284	226	284*	391	8.2	654
Lima	Ri, W	90	18	45	34	153	9.6	244
Sandusky	L	124	90	90	84	166	7.2	297
Toledo	L	146	113	127	89	228	7.8	378
Youngstown	C, Re	84	0	83	38	156	10.6	275
OKLAHOMA								
Ardmore	C, L	148	133	122	133*	181	7.7	313
Bartlesville	C, L	60	50	54	48	104	8.1	170
Oklahoma City	Ri, L	90	0	45	39	262	10.3	479
Tulsa	C, L	93	84	84	85	114	7.6	191
OREGON								
Baker	L, C	37	39	30	39*	56	7.4	80
Eugene	Ri	17	22	9	22*	48	7.5	51
Medford	Sp	35	46	18	46*	99	6.9	—
Pendleton	Sp, W	95	120	63	120*	203	7.2	—
Portland	Ri, Re	9	12	6	15*	30	7.0	—
Roseburg	Ri	43	27	38	27*	80	7.6	97
PENNSYLVANIA								
Altoona‡	S	228	0	50	0*	306	3.9	398
Bethlehem	C, Re	6	5	4	5	22	6.1	22
Erie	L	129	89	93	89*	177	7.8	300
Harrisburg	C, Re, Ri	33	7	9	11	24	8.1	33
New Castle	Ri	85	35	88	32	171	7.5	274
Oil City	W	83	61	65	61*	140	7.6	242
Philadelphia	Ri	52-152	10-25	35-60	22-31	84-151	6.2-6.6	145-238
Pittsburgh‡	Ri	75	2-38	100-150	7	373-554	6.3	569-822
Reading	C, L	87	39-76	50-87	60	102-152	8.0	204
Scranton	Re	26	5	20	5*	55	6.3	70
Williamsport	C	16	8	10	8*	24	6.7	37
RHODE ISLAND								
Pawtucket	Re	34	14	21	12	54	7.1	85
Providence	Re	27	11	22	15	52	9.4	75
SOUTH CAROLINA								
Charleston	C, Ri	21	19	17	17	68	8.4	119
Columbia	Ri	33	26	21	24	63	8.4	97
Greenville	S	1	15	3	15	30	8.4	45
SOUTH DAKOTA								
Huron	Ri	158	70	95	70*	384	7.3	559
Rapid City	Sp, W	193	173	107	173*	219	7.8	376
Sioux Falls‡	W	484	257	305	257*	646	7.6	921
TENNESSEE								
Chattanooga	Ri	62	30	48	58	88	7.6	138
Johnson City	Sp, C	61	57	33	46	70	8.5	122
Knoxville	Ri	112	71	95	71	188	7.8	275
Memphis	W	32-42	53-69	20-23	53-69*	84-98	7.4	119-145
Nashville	Ri	80	60	63	64	130	7.1	184

Sources of Supply — B-brook, C-creek, L-lake, P-pond, Re-reservoir, Ri-river, S-stream, Sp-spring, W-well

Note: See explanation of *Table 13* on page 39.

*Total alkalinity approximated, as described in explanation of *Table 13*.

†Chloride (Cl) ion concentration in excess of 100 ppm.

‡Sulfate (SO₄) ion concentration in excess of 100 ppm.

TABLE 13—WATER ANALYSES FOR SUPPLIES IN UNITED STATES CITIES (Contd)

STATE AND CITY	Source of Supply	Total Hardness as CaCO₃ (ppm)	Bicarbonate (HCO₃) as CaCO₃ (ppm)	Calcium (Ca) as CaCO₃ (ppm)	Total Alkalinity as CaCO₃ (ppm)	Dissolved Solids (ppm)	pH	Specific Conductance at 77 F (micromhos)
TEXAS								
Abilene	L	194	192	100	185	361	8.2	640
Amarillo	W	258-549	236-360	80-145	236-360	372-530	7.4-7.8	511-799
Austin	Ri	71	14	28	46	220	9.9	412
Brownsville†‡	Ri	286	113	180	146	652	7.8	1,100
Corpus Christi	Ri, W	135	77	95	132	319	7.8	530
Dallas	L	67	14	48	39	184	10.4	331
Del Rio	Sp	218	206	185	206*	254	7.6	448
El Paso†‡	Ri, W	159	37	80	46	788	7.3	1,270
Fort Worth	L	130	129	113	117	203	7.8	374
Galveston†	W	115	275	75	275*	1,010	8.0	1,830
Houston	W	19-131	179-344	310	179-344*	298-576	7.5-8.0	475-1000
Palestine	W	7	185	5	185*	215	8.0	334
Port Arthur	Ri	50	57	25	45	211	8.3	348
San Antonio	W	182-240	144-200	125-162	144-200*	244-296	7.6-8.3	425-498
UTAH								
Logan	Sp	206	187	124	187*	202	8.1	365
Ogden	W, C	156-435	133-222	90-300	133-222*	188-1,120	7.5-8.1	334-2,070
Salt Lake City	W, C	75-233	49-203	55-192	49-203*	101-270	7.4-8.2	169-436
VERMONT								
Bennington	Sp	16	13	8	13*	27	7.2	40
Burlington	L	54	35	38	40	70	7.0	127
Rutland	B	42	33	24	33*	52	7.3	89
VIRGINIA								
Lynchburg	Ri, L	15	13	7	9	39	6.3	43
Norfolk	C, Re	61-78	40-44	65	36	111-134	8.0-8.3	169-205
Richmond	Ri	65	29	58	30	137	8.8	215
Roanoke	Sp, C, Re	18-133	15-124	9-68	12-130	28-127	7.5-9.5	39-228
WASHINGTON								
Seattle	Ri	18	18	16	18*	37	7.3	—
Spokane	W	118-161	103-143	73-97	103-143*	132-195	7.6-8.0	—
Tacoma	Ri, W	64-115	25-46	16-30	25-46*	64-115	7.1-7.8	—
Walla Walla	C, W	38	49	23	49*	89	7.6	102
Wenatchee	Ri	46	59	50	45	100	7.5	143
Yakima	Ri	21	25	15	25*	50	7.1	57
WEST VIRGINIA								
Bluefield	Sp, R	97	105	77	95	124	7.1	217
Charleston	Ri	37	16	24	21	52	9.0	89
Elkins	Ri	45	25	35	29	64	8.2	108
Huntington‡	Ri	70	25	98	30	239	8.4	374
Martinsburg	Sp	264	234	200	234*	279	7.5	468
Parkersburg‡	W	167	64	160	82	366	8.4	554
Wheeling‡	Ri	141	24	100	22	224	9.0	345
WISCONSIN								
Eau Claire	W	54	47	30	47*	86	7.4	122
Green Bay	W	208-272	159-225	125-150	159-225*	296-395	7.7-8.1	473-645
La Crosse	W	236-338	189-290	125-185	189-290*	290-362	7.3-7.8	433-576
Madison	W	300	275-345	137-234	275-345*	272-450	7.5	495-810
Milwaukee	L	128	103	85	104	152	7.5	265
WYOMING								
Casper‡	Ri	329	149	227	149*	615	7.7	869
Cheyenne	C, W	91	86	73	86*	134	7.8	207
Sheridan	C	26	28	17	28*	54	7.2	68

Sources of Supply — B-brook, C-creek, L-lake, P-pond, Re-reservoir, Ri-river, S-stream, Sp-spring, W-well
Note: See explanation of *Table 13* on page 39.
*Total alkalinity approximated, as described in explanation of *Table 13*.
†Chloride (Cl) ion concentration in excess of 100 ppm.
‡Sulfate (SO₄) ion concentration in excess of 100 ppm.

CHAPTER 6. DEFINITIONS

Definitions of the common terms used in water conditioning are presented here.

Algaecide is a substance used to kill algae.

Alkalinity[24] is expressed by the sum of the carbonate, bicarbonate and hydrate ions in water; other ions such as phosphate or silicate may contribute partially to alkalinity. Alkalinity is normally expressed as the number of parts per million (ppm) of calcium carbonate, $CaCO_3$. See further discussion on methyl orange and phenolphthalein alkalinities in *Chapter 1*.

Anode[24] is a positive electrode toward which negatively charged nonmetallic ions migrate and at which reduction occurs in an electrolytic cell. In corrosion processes, the anode is usually the electrode having a greater tendency to go into solution.

Biocide is a substance used to kill living organisms such as algae, bacteria and fungi. It may kill only one of these groups of organisms or only some specific bacteria, algae or fungi in these groups.

Biological Deposits[31] are water-formed deposits of biological organisms or the products of their life processes. These deposits may be microscopic in nature such as slimes, or macroscopic such as barnacles or mussels. Slimes are usually composed of deposits of a gelatinous or filamentous nature.

Bleed-off is the continuous ejection or draining of a portion of the concentrated water from a circulating system, which is replaced with water from the normal supply, thus causing a lowering of concentration. Bleed-off is normally expressed in gallons per minute (gpm).

Cathode[24] is a negative electrode toward which positively charged metallic ions migrate and at which reduction occurs in the electrolytic cell. In corrosion processes, the cathode is usually the electrode tending to resist corrosion.

Chlorinator is a device used to measure, dissolve, and feed liquid chlorine into a water system.

Closed Recirculating System is a system in which water flows in a repetitive (or continuous) circuit thru heat exchangers. There is no evaporation and no make-up except to compensate for leakage.

Corrosion[11] is destruction of a metal by chemical or electrochemical reaction with its environment. In the corrosion process, the reaction products formed may be soluble or insoluble in the contacting environment. Insoluble corrosion products may deposit at or near the attacked area, or be carried along and deposited at a considerable distance away.

Cycles of Concentration is the ratio of dissolved solids in recirculating water to the dissolved solids in make-up water.

Delignification is the deterioration of lignin, the binding material which holds the cellulose fibers together in wood.

Dissolved Solids are in true solution in water and cannot be removed by filtration. Their presence is due to the solvent action of water in contact with the minerals in the earth.

Electrolyte is a solution thru which an electric current flows.

Erosion is the wearing away by action of rapidly moving water, particularly where entrained gas bubbles or suspended abrasive solids are present.

Free Mineral Acid may be sulfuric, nitric or hydrochloric acid, sometimes occurring in acid mine drainage or industrial waste.

Galvanic Corrosion[24] generally results from the presence of two dissimilar metals in an electrolyte. It is characterized by an electron movement from the metal of high potential (anode) to the metal of lower potential (cathode), resulting in corrosion of the anodic metal. Galvanic cell corrosion may also result from the presence of two similar metals in an electrolyte of nonuniform concentration.

Galvanic Series[12] is a list of metals and alloys arranged according to their relative potentials in a given environment.

Hardness is primarily the sum of calcium and magnesium salts in water, although it may include other elements, mainly aluminum, iron, manganese, strontium or zinc. Temporary or carbonate hardness is that portion of total hardness which can combine with the carbonate, CO_3, or bicarbonate, HCO_3, ions. The balance of the hardness is called noncarbonate or permanent hardness. This is principally caused by sulfates, chlorides and/or nitrates of calcium and/or magnesium (*Chapter 1*).

Inhibitor is a chemical substance added to a solution to reduce scale or corrosion or both.

Ion is an electrically charged atom or group of atoms.

Langelier Saturation Index is used to predict the scaling or nonscaling characteristics of a water. A plus value indicates scale-forming tendency; a negative value is nonscaling. The index is the algebraic difference between the actual *pH* from test and the calculated *pH* of saturation for calcium carbonate, $CaCO_3$. (See further explanation in *Chapter 2*.)

Make-Up Water is water from a normal supply which compensates for evaporation, windage and bleed-off losses.

Once-Thru System is a system thru which water passes only once and is discharged. There is no evaporation.

Open Recirculating System is a system thru which water flows in a repetitive circuit thru heat exchangers and reservoirs open to the atmosphere, such as cooling towers or air washers. Water is aerated and evaporated, or water vapor is condensed in the sprays. Some bleed-off is directed to the drain to limit a build-up of solids in the system.

Parts Per Million (ppm) represents the parts of a substance per million parts by weight of the solution. It is normally used to express the results of a water analysis. One ppm is equal to one ten thousandth of one percent (0.0001%).

pH represents the logarithm of the reciprocal of the hydrogen ion concentration of a solution. It denotes the degree of acidity or alkalinity of a solution. A value of 7.0 is neutral; values below 7.0 are increasingly acid; values above 7.0 are increasingly alkaline. Refer to *Chapter 1* for further discussion.

Ryznar Stability Index is a practical extension of the Langelier Saturation Index based on experience. This index (always positive) is equal to two times the calculated *pH* of saturation of $CaCO_3$ minus the actual *pH* from test. Index values above 6.5 indicate a corrosive tendency while values below 6.5 indicate a tendency to scale *(Chapter 2)*.

Scale is a deposit formed from solution directly upon a confining surface. It is usually crystalline and dense, frequently laminated and occasionally columnar in structure.

Sludge[31] is a water-formed sedimentary deposit. It usually does not adhere sufficiently to retain its physical shape when mechanical means are used to remove it from the surface on which it is deposited. Sludge is not always found at the place where it is formed. It may be hard and adherent and baked to the surface upon which it has been deposited.

Specific Conductance is a measure of the ability of water to conduct an electric current. It is expressed as micromhos per cubic centimeter. A micromho is a millionth of a mho which in turn is the reciprocal of an ohm resistance.

Surface Active Agent is a substance that possesses stabilization characteristics tending to minimize deposition of calcium carbonate scale. A typical example are the polyphosphates.

Suspended Solids are solids not in true solution and can be removed from the water by filtration.

Synergizing Agent (water conditioning) is a substance which increases the effectiveness of a corrosion or scale inhibitor.

Titration is a process of adding a liquid of known concentration and of measured volume to a known volume of another liquid until a point is reached at which a definite effect is observed, usually a change in color of an indicator.

Total Solids are the sum of the dissolved and suspended solids.

Tuberculation[12] is the formation of localized corrosion products scattered over a surface in the form of knoblike mounds called tubercules.

Water Analysis is the chemical analysis of the dissolved materials in water. It also includes determination of *pH* and the amount of suspended solids.

Windage Loss is a loss of fine droplets of water entrained by circulating air. The amount varies with different types of equipment. This is a loss of water from the system and is replaced by make-up water. Windage loss tends to limit the cycles of concentration, and is usually expressed as a percentage of the rate of circulation.

BIBLIOGRAPHY

1. Typical Water Analysis For Classification With Reference to Industrial Use, by W. D. Collins; *American Society for Testing Materials Proceedings,* Vol. 44, 1944, p. 1057.

2. *Betz Handbook of Industrial Water Conditioning,* Sixth Edition, Betz Laboratories, Gillingham and Worth Streets, Philadelphia, Pa., 1962.

3. Water Analysis Bulletin No. 11, Fourth Edition, 4th Printing, Allied Chemical Corporation, Solvay Process Division, 61 Broadway, New York, N.Y., 1961.

4. *Water Conditioning for Industry,* by Sheppard T. Powell; McGraw-Hill Book Co. Inc., New York, N.Y., 1954.

5. *Water Treatment for Industrial and Other Uses,* Second Edition, by Eskel Nordell; Reinhold Publishing Corp., New York, N.Y., 1961.

6. The Analytical Control of Anti-Corrosion Water Treatment, by W. F. Langelier, *Journal of the American Water Works Association,* Vol. 28, No. 10, October 1936, p. 1500.

7. A New Index For Determining Amount of Calcium Carbonate Scale Formed by a Water, by John W. Ryznar; *Journal of the American Water Works Association,* Vol. 36, No. 4, April 1944.

8. Calculation of the pH of Saturation of Tricalcium Phosphate, by J. Green and J. A. Holmes, *Journal of the American Water Works Association,* Vol. 39, November 1947, p. 1090.

9. Equipment Room Corrosion Problems Solved by Mechanical Design, by D. B. Gardner; *Materials Protection,* April 1962, p. 54.

10. Control of Couples Developed in Water Systems, by G. B. Hatch; *Corrosion,* Vol. 11, No. 11, November 1955.

11. *Corrosion Handbook,* by H. H. Uhlig; John Wiley & Sons, New York, N.Y., 1948.

12. *Corrosion Causes and Prevention,* Second Edition, by F. N. Speller; McGraw-Hill Book Co. Inc., New York, N.Y. 1935.

13. Corrosion and Its Prevention, Air Conditioning and Refrigeration Institute, 1959 Edition, 1346 Connecticut Avenue N.W., Washington, D.C.

14. Chemistry and Equipment Maintenance, by Charles M. Loucks, *Heating, Piping and Air Conditioning,* November 1962, p. 157.

15. Corrosion and Its Prevention, by Charles M. Loucks, *Heating, Piping and Air Conditioning,* March 1963, p. 145.

16. Corrosion Inhibition With Chromate, Serial No. 55, by Marc Darrin; Allied Chemical Corporation, Solvay Process Division, Mutual Chromium Chemicals, 61 Broadway, New York, N.Y., February 1949.

17. A New Method for the Protection of Metals Against Pitting, Tuberculation and General Corrosion, by H. Lewis Kahler and Charles George; *Corrosion,* Vol. 6, No. 10, October 1950, p. 331.

18. Protection of Metals Against Pitting, Tuberculation and General Corrosion, by H. Lewis Kahler and Philip J. Gaughan; *Industrial and Engineering Chemistry,* Vol. 44, August 1952, p. 1770.

19. Laboratory Evaluation of Fresh Cooling Water Inhibitors for Open Recirculating Systems, by A. Orman Fisher; *Materials Protection,* October 1962, p. 54.

20. Experience With Sodium Nitrite, Unpredictable Corrosion Inhibitor, by Sidney Sussman, Oskar Nowakowski and John J. Constantino; *Industrial and Engineering Chemistry,* Vol. 51, April 1959, p. 581.

21. Corrosion and Its Control in Aluminum Cooling Towers, by Sidney Sussman and J. R. Akers; *Corrosion,* May 1954, p. 151.

22. Proper Materials Selection Can Prevent Cooling Tower Corrosion Problems, by F. C. Risenfield and C. L. Blohm; *Petroleum Engineer,* Vol. 28, No. 3, March 1956, p. D64.

23. Water Composition Changes in Air Conditioning Equipment, by Sidney Sussman and Irving L. Portnoy; *Journal of the American Water Works Association,* Vol. 51, August 1959.

24. *ASHRAE Guide,* Chapter 15, 1961.

25. Biological Fouling in Recirculating Cooling Water Systems, by J. J. Maguire; *Industrial and Engineering Chemistry,* Vol. 48, December 1956, p. 2162.

26. Cooling Water Treatment — A Review, by Maxcy Brooke; *Petroleum Refiner,* February 1957.

27. Experimental Performance of "Miracle" Water Conditioners, by Rolf Eliassen, Rolf Skrinde and William Davis; *Journal of the American Water Works Association,* October 1958, p. 1371.

28. What To Do When Water Changes Its Mind, by F. S. Hodgdon; *Heating, Piping and Air Conditioning,* June 1956, p. 97.

29. How and Why Corrosion Occurs in Air Conditioning Water Circuits, by S. Sussman and I. L. Portnoy; *Heating, Piping and Air Conditioning,* June 1961, p. 134.

30. Your Stake In Water Conditioning, by J. F. Keville and M. A. Scicchitano; *Refrigeration and Air Conditioning Business,* October 1958.

31. Definitions of Terms Relating to Industrial Water and Industrial Waste Water, Book of ASTM Standards, Part 10, American Society for Testing Materials, 1916 Race St., Philadelphia 3, Pennsylvania, 1958, p. 1104.

Part 6
AIR HANDLING EQUIPMENT

CHAPTER 1. FANS

This chapter presents information to guide the engineer in the practical application of fans used in air conditioning systems.

A fan is a device used to produce a flow of air. Use of the term is limited by definition to devices producing pressure differentials of less than 28 in. wg at sea level.

TYPES OF FANS

Fans are identified by two general groups:

1. Centrifugal, in which the air flows radially thru the impeller. Centrifugal fans are classified according to wheel blading; forward-curved, backward-curved and radial (straight).
2. Axial flow, in which the air flows axially thru the impeller. Axial flow fans are classified as propeller (disc), tubeaxial and vaneaxial.

Figures 1, 2a, 2b and 2c show the various types of commonly applied fans.

APPLICATION

When a duct system is needed in an air conditioning application, a tubeaxial, vaneaxial or centrifugal fan may be used. Where there is no duct system and little resistance to air flow, a propeller fan can be applied. However, self-contained equipment often utilizes centrifugal fans for applications without ductwork.

The centrifugal fan is used in most comfort applications because of its wide range of quiet, efficient operation at comparatively high pressures. In addition, the centrifugal fan inlet can be readily attached to an apparatus of large cross-section while the discharge is easily connected to relatively small ducts. Air flow can be varied to match air distribution system requirements by simple adjustments to the fan drive or control devices.

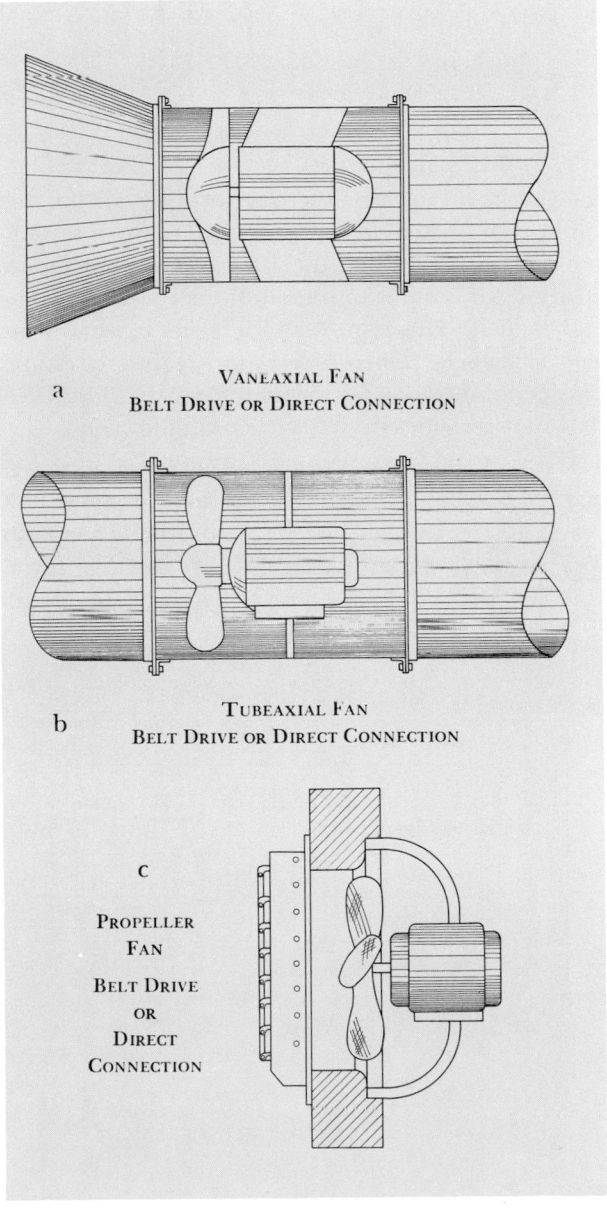

a
VANEAXIAL FAN
BELT DRIVE OR DIRECT CONNECTION

b
TUBEAXIAL FAN
BELT DRIVE OR DIRECT CONNECTION

c
PROPELLER FAN
BELT DRIVE OR DIRECT CONNECTION

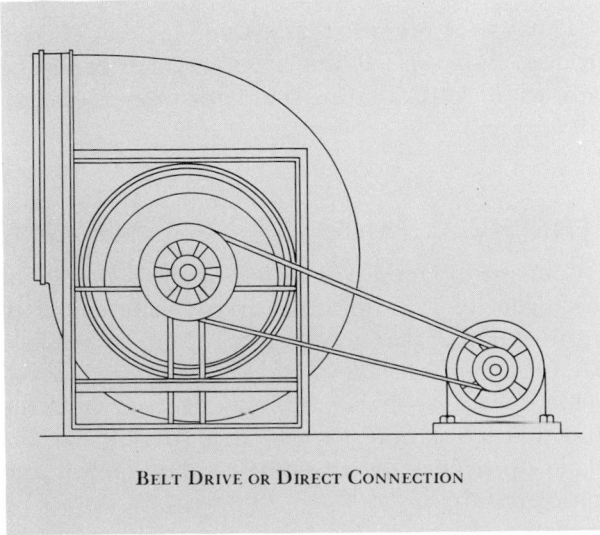

BELT DRIVE OR DIRECT CONNECTION

FIG. 1 — CENTRIFUGAL FAN

FIG. 2 — AXIAL FLOW FANS

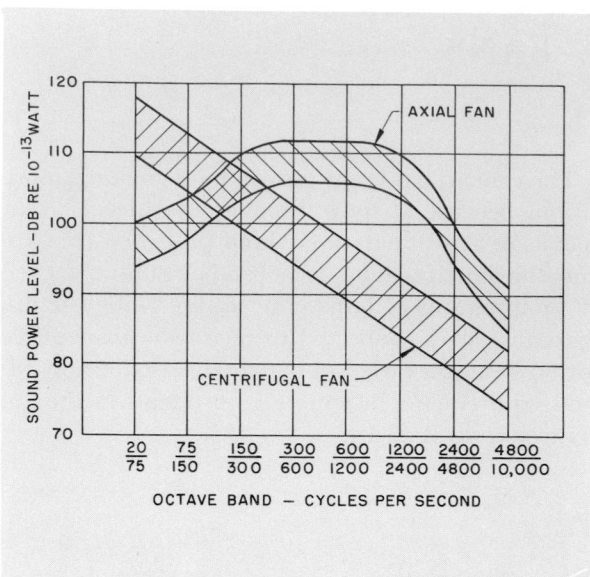

FIG. 3 — SOUND POWER LEVELS

Axial flow fans are excellent for large air volume applications where higher noise levels are of secondary concern. They are, therefore, often used for industrial air conditioning and ventilation. These high velocity fans require guide vanes to obtain the best efficiencies when operating against pressures considered normal for centrifugal fans. However, these fans may be applied without guide vanes.

Figure 3 illustrates the approximate sound power level of a typical centrifugal fan and an axial flow fan. The frequencies detectable by the human ear (300 to 10,000 cycles per second) are the least favorable for the axial flow fan. Therefore, to obtain

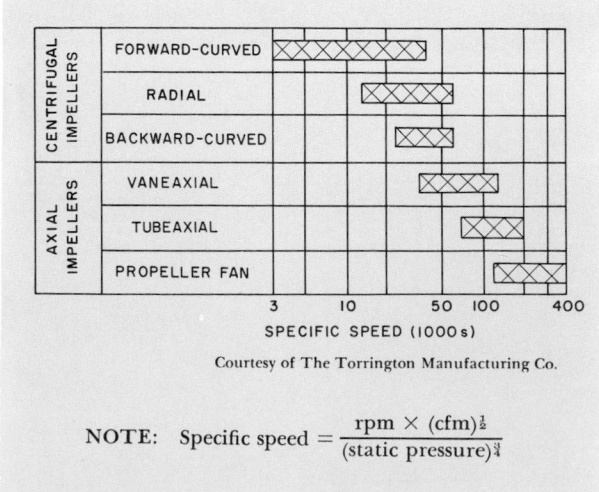

Courtesy of The Torrington Manufacturing Co.

NOTE: Specific speed = $\dfrac{rpm \times (cfm)^{\frac{1}{2}}}{(static\ pressure)^{\frac{3}{4}}}$

FIG. 4 — SPECIFIC SPEED RANGES

acceptable sound levels with the axial flow fan, sound attenuation may be required.

The concept of specific speed is useful in describing the applications of various fan types. Specific speed is a fan performance index based on the fan speed, capacity and static pressure. *Figure 4* shows the ranges of specific speed in which six types of centrifugal and axial flow fans operate at high static efficiencies. This figure indicates that forward-curved blade centrifugal fans attain their peak efficiencies at low speeds, low capacities and at high static pressures. However, propeller fans reach high efficiency at high speeds and capacities and at low static pressures.

The horsepower characteristics of the various fans are such that a type may be overloading or nonoverloading. The backward-curved blade centrifugal fan is a nonoverloading type. The forward-curved and radial blade centrifugal fans may overload. Axial flow fans may be either nonoverloading or overloading.

All fan types may be utilized for exhaust service. Wall fans operate against little or no resistance and therefore are usually of the propeller type. Propeller fans are sometimes incorporated into factory-built penthouses or roof caps. Hooded exhaust fans and central station exhaust fans are typically of the centrifugal type. Axial fans may be suitable for exhaust applications, particularly in factory installations.

STANDARDS AND CODES

Fan application and installation should conform to all codes, laws and regulations applying at the job site.

The AMCA Standard Test Code for Air Moving Devices, Bulletin 210, prescribes methods of testing fans, while AMCA rating standards prescribe methods of rating.

CENTRIFUGAL FANS

Centrifugal fans are identified by the curvature of the blade tip. The forward-curved blade curves in the direction of rotation *(Fig. 5a)*. The radial blade has no curvature *(Fig. 5b)*. The backward-curved blade tip inclines backward, curving away from the direction of rotation *(Fig. 5c)*. The curvature of the blade tip defines the shape of the horsepower and static pressure curves.

The characteristics of the three main types of centrifugal fans are listed in *Table 1*.

FORWARD-CURVED BLADE FAN

A typical performance of a forward-curved blade fan is shown in *Fig. 6*. The pressure rises from 100% free delivery toward no delivery with a characteristic dip at low capacities. Horsepower increases continuously with increasing air quantity.

BACKWARD-CURVED BLADE FAN

A typical performance of a backward-curved blade fan is shown in *Fig. 7*. The pressure rises constantly from 100% free delivery to nearly no delivery. There is no dip in the curve. The horsepower curve peaks at high capacities. Therefore, a motor selected to satisfy the maximum power demand at a given fan speed does not overload at any point on the curve, providing this speed is maintained.

Two modifications of the backward-curved blade fan are the airfoil and backward-inclined blade fans.

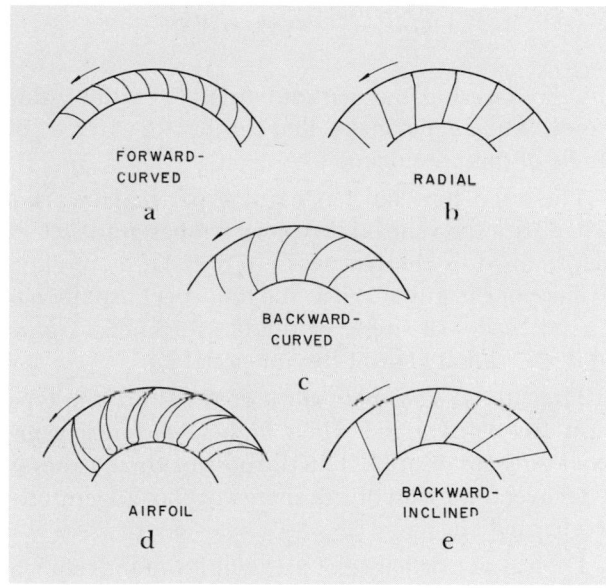

FIG. 5 — FAN BLADES

TABLE 1—CHARACTERISTICS OF CENTRIFUGAL FANS

FAN TYPE	ADVANTAGES
Forward-Curved	1. Runs at a relatively low speed compared to other types for the same capacity. 2. Smaller fan for a given duty, excellent for fan-coil units.
Radial	1. Self-cleaning. 2. Can be designed for high structural strength to achieve high speeds and pressures.
Backward-Curved	1. More efficient. 2. Horsepower curve has a flat peak so that the motor may be sized to cover the complete range of operation from zero to 100% air flow for a single speed. Nonoverloading. 3. Pressure curve is generally steeper than that of the forward-curved fan. This results in a smaller change in air volume for any variation in system pressure for selections at comparable percentages of free delivery. 4. Point of maximum efficiency is to the right of the pressure peak, allowing efficient fan selection with a built-in pressure reserve. 5. Quieter than other types.

These are illustrated in *Fig. 5d and 5e*. Both are nonoverloading types.

The airfoil blade fan is a high efficiency fan because its aerodynamically shaped blades permit smoother air flow thru the wheel. It is normally used for high capacity, high pressure applications where power savings may outweigh its higher first cost. Since the efficiency characteristic of an airfoil blade fan usually peaks more sharply than those of other types, greater care is required in its selection and application to a particular duty.

The backward-inclined blade fan must be selected closer to free delivery; therefore, it does not have as

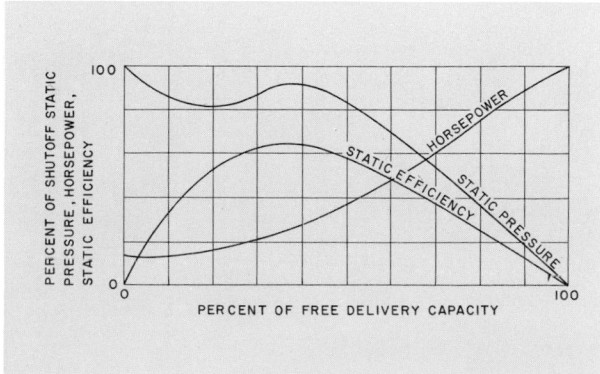

FIG. 6 — FORWARD-CURVED BLADE FAN PERFORMANCE

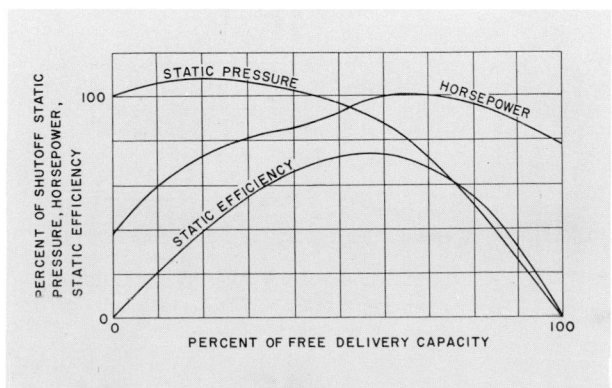

FIG. 7 — BACKWARD-CURVED BLADE FAN PERFORMANCE

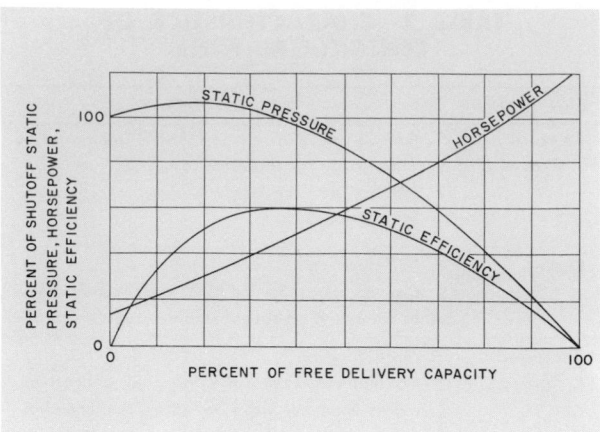

FIG. 8 — RADIAL BLADE FAN PERFORMANCE

great a range of high efficiency operation as does the backward-curved blade fan. Manufacture of an inclined blade is understandably a simpler operation.

RADIAL BLADE FAN

Typical performance of a radial (straight) blade fan is shown in *Fig. 8*. The pressure characteristic is continuous at all capacities. Horsepower rises with increasing air quantity in an almost directly proportional relation. Thus, with this type of fan the motor may be overloaded as free air delivery is approached.

The radial blade fan has efficiency, speed and capacity characteristics that are midway between the forward-curved and backward-curved blade fans. It is seldom used in air conditioning applications because it lacks an optimum characteristic.

AXIAL FLOW FANS

Figure 9 shows a performance characteristic typical of a propeller fan.

The tubeaxial fan is a common axial flow fan in a

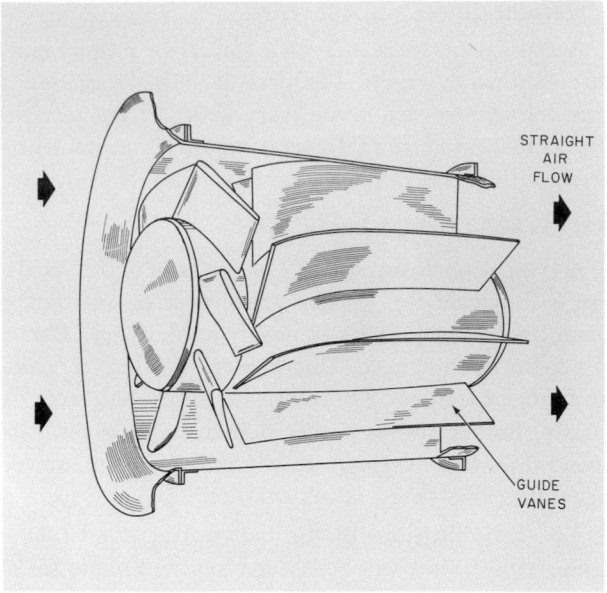

FIG. 10 — VANEAXIAL FAN

tubular housing but without inlet or outlet guide vanes. The blade shape may be flat or curved, of single or double thickness.

The axial flow fan has become particularly associated with the vaneaxial type which has guide vanes before or after the fan wheel. To make more effective use of the guide vanes, the fan wheel usually has curved blades of single or double thickness. *Figure 10* is a sectional view of the vaneaxial fan.

The curved stationary diffuser vanes are the type most frequently used when higher efficiency vaneaxial fans are desired. The purpose of these vanes is to recover a portion of the energy of the tangentially accelerated air.

Typical performance of an axial flow fan is shown in *Fig. 11*.

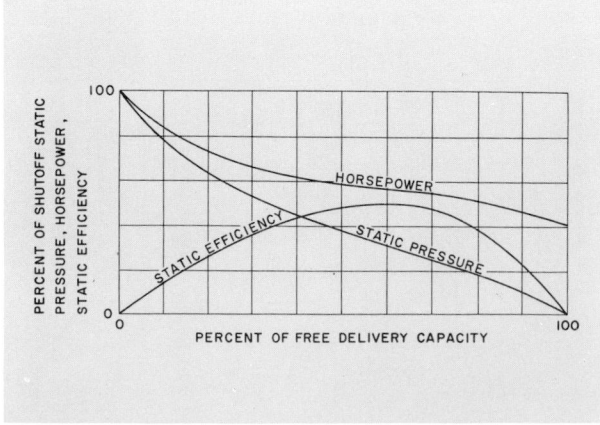

FIG. 9 — PROPELLER FAN PERFORMANCE

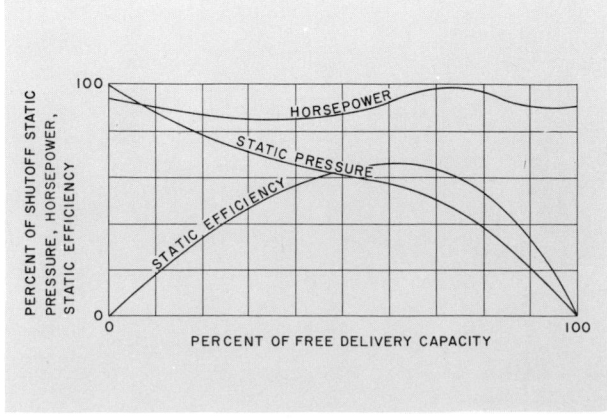

FIG. 11 — AXIAL FLOW FAN PERFORMANCE

FAN DESIGNATION

CLASS OF CONSTRUCTION

The AMCA has developed standards of centrifugal fan construction based on the pressure the fans are required to develop. The four classes of fans appear in *Table 2*. Each of the various fan manufacturers has defined his own maximum wheel tip speed for each class.

The required fan class can be determined from *Chart 1* if outlet velocity and static pressure are known. Calculation of velocity pressure and total pressure is thus eliminated. This chart is based on standard air (29.92 in. Hg barometric pressure and 70 F temperature).

If nonstandard conditions of temperature and altitude are encountered in an application, the calculated static pressure should be corrected before entering *Chart 1*. This procedure is described in the section entitled *Fan Selection*. See *Example 3*.

Minimum first costs can often be achieved by using a larger size fan of a given class than by choosing a smaller fan size of a higher class. If a selection lies on the border line, both alternatives should be considered.

TABLE 2—CLASSES OF CONSTRUCTION
Centrifugal Fans

CLASS	MAXIMUM TOTAL PRESSURE
I	3¾ in. wg — standard
II	6¾ in. wg — standard
III	12¾ in. wg — standard
IV	More than 12¾ in. wg — recommended

Some manufacturers offer packaged fans and motors which are not defined in terms of classes. These packages are made of Class I or II parts, modified slightly to hold the motor within the fan base. The fan package is less expensive than the equivalent Class I or II fan and is satisfactory for most applications. Packaged fans are also offered in construction lighter than Class I. Manufacturers' specifications usually distinguish between light and heavy construction.

A pressure class standard pertaining to centrifugal fans mounted in cabinets has also been published by AMCA. Cabinet fans are commonly used with central station fan-coil equipment. The three classes of such fans are defined in *Table 3*.

CHART 1—CONSTRUCTION CLASS PRESSURE LIMITS

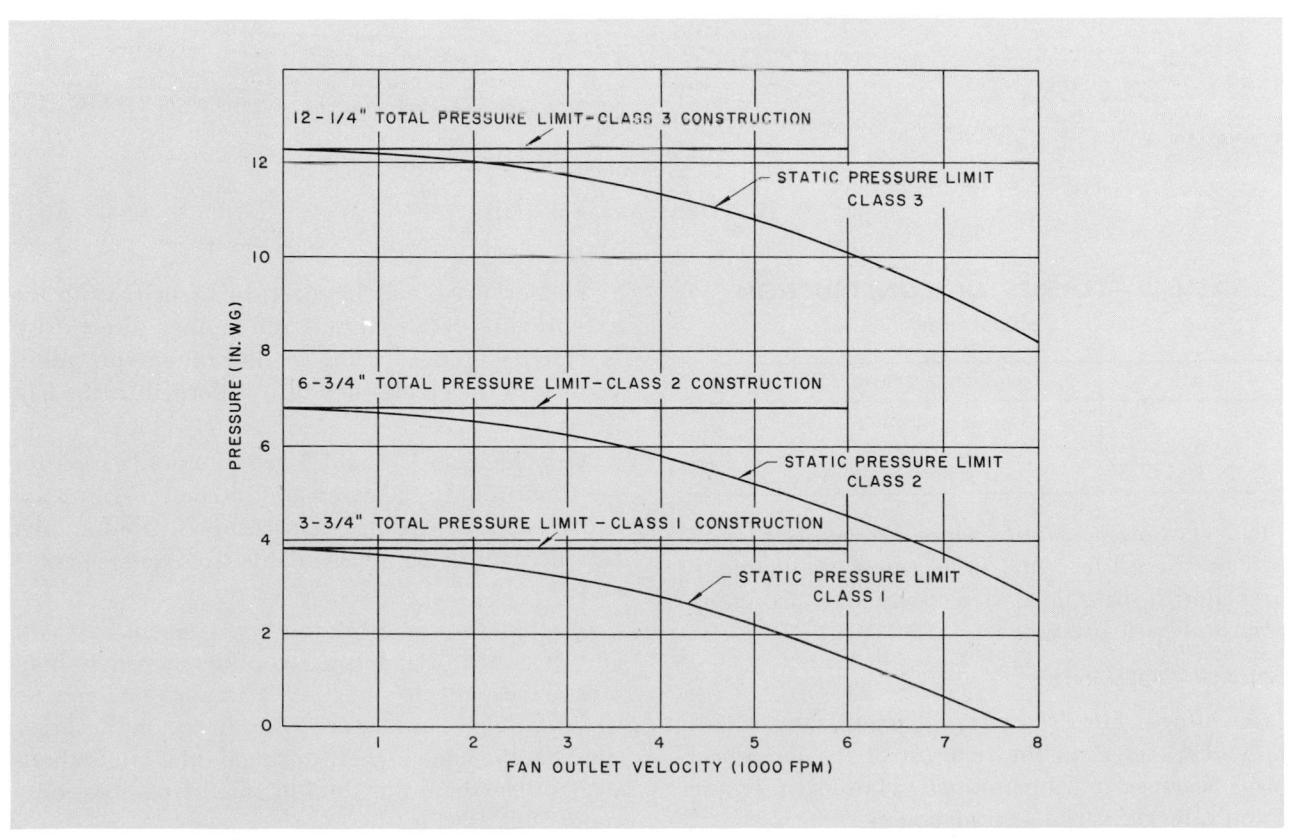

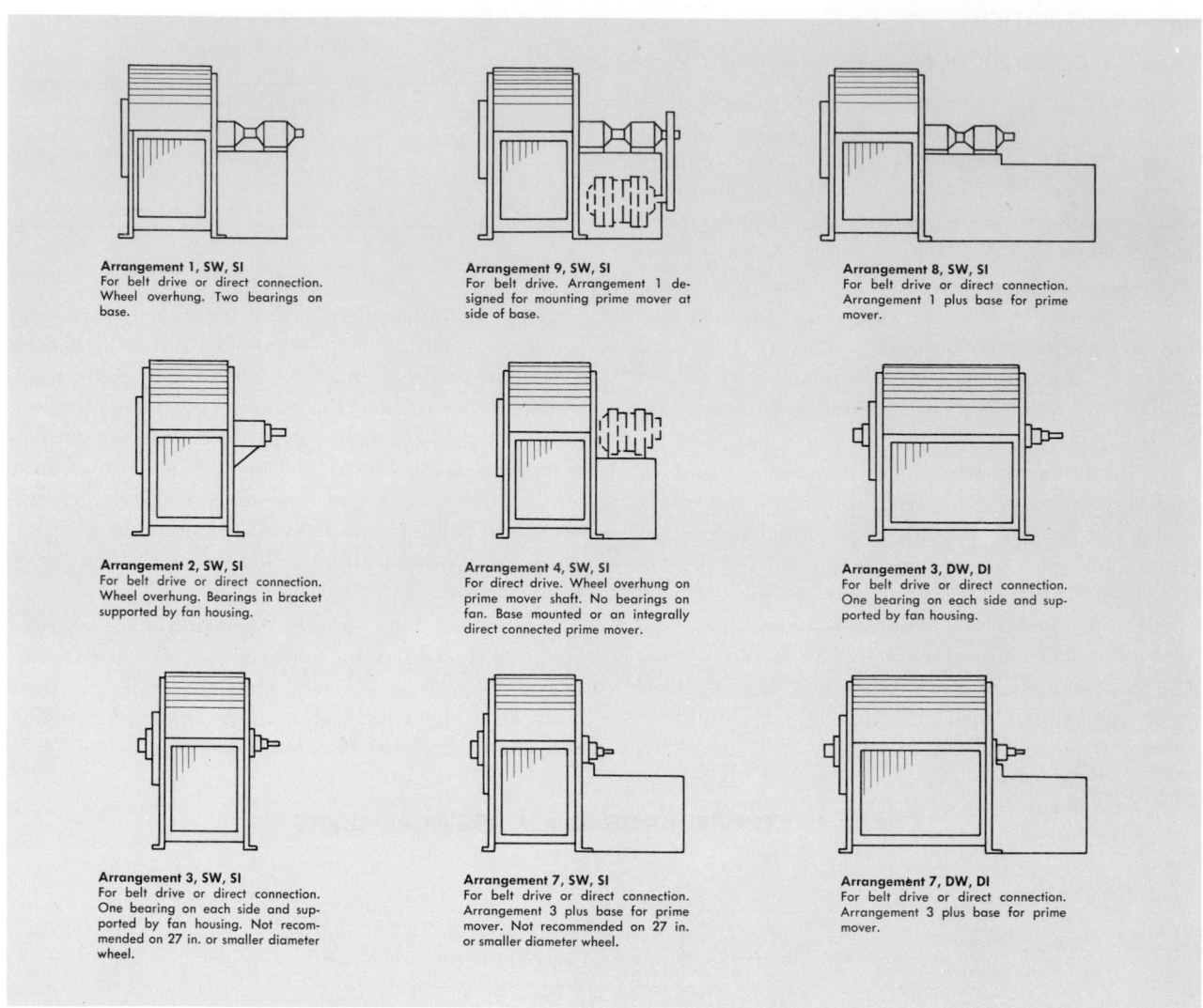

Arrangement 1, SW, SI
For belt drive or direct connection. Wheel overhung. Two bearings on base.

Arrangement 9, SW, SI
For belt drive. Arrangement 1 designed for mounting prime mover at side of base.

Arrangement 8, SW, SI
For belt drive or direct connection. Arrangement 1 plus base for prime mover.

Arrangement 2, SW, SI
For belt drive or direct connection. Wheel overhung. Bearings in bracket supported by fan housing.

Arrangement 4, SW, SI
For direct drive. Wheel overhung on prime mover shaft. No bearings on fan. Base mounted or an integrally direct connected prime mover.

Arrangement 3, DW, DI
For belt drive or direct connection. One bearing on each side and supported by fan housing.

Arrangement 3, SW, SI
For belt drive or direct connection. One bearing on each side and supported by fan housing. Not recommended on 27 in. or smaller diameter wheel.

Arrangement 7, SW, SI
For belt drive or direct connection. Arrangement 3 plus base for prime mover. Not recommended on 27 in. or smaller diameter wheel.

Arrangement 7, DW, DI
For belt drive or direct connection. Arrangement 3 plus base for prime mover.

FIG. 12 — DRIVE ARRANGEMENTS

TABLE 3—CLASSES OF CONSTRUCTION
Cabinet Fans

CLASS	MAXIMUM STATIC PRESSURE
A	3 in. wg
B	5½ in. wg
C	More than 5½ in. wg

Fan class nomenclature does not apply to fans used in fan-coil terminal units where the manufacturer limits such fans to a particular maximum speed and static pressure.

FAN ARRANGEMENTS

Centrifugal fan drive arrangement, standardized by AMCA, refers to the relation of the fan wheel to the bearings and the number of fan inlets. *Figure 12* indicates the various arrangements.

The fan drive may be direct or by belt. With the exception of packaged fans and motors, direct drive is seldom employed in air conditioning applications because of the greater flexibility afforded by the belt drive.

Arrangements 1, 2 and 3 are commonly used for air conditioning. The remaining choices are modified versions of Arrangements 1 and 3. Double inlet fans for belt drive are available in Arrangement 3 and 7.

In selecting a suitable fan arrangement first cost and space requirements are considered. Single inlet fans are usually less expensive in the smaller sizes while double inlet fans are lower in cost in the larger sizes. For the same capacity a single inlet fan is about 30% taller than the double inlet type, but only about 70% as wide.

Arrangement 3 is the most widely used because the bearing location eliminates the necessity for a bearing platform. Cost and required space is therefore minimized.

For single inlet applications Arrangements 1 and 2 are used where the fan wheel is less than 27 inches in diameter. Arrangement 3 is not used since the bearing on the inlet side is large enough, relative to the inlet area, to affect fan performance. Fans of larger sizes and double inlet fans are not limited in this way.

Arrangement 1 is usually more costly than Arrangement 2 because it has two bearings and a base. Where Class III construction is required, Arrangement 1 is preferred over Arrangement 2.

If Arrangement 3 is to be used at air temperatures exceeding 200 F or Arrangement 1 or 2 at temperatures exceeding 300 F, the fan manufacturer should be consulted so that the proper bearing or heat slinger can be specified.

Table 4 compares the costs of fan and drive for several single inlet arrangements. Selections are based on a constant air quantity and static pressure.

TABLE 4—ARRANGEMENT COST COMPARISON

ARRANGEMENTS	MATERIAL COSTS (%)
1	117
2	100
3	100
9	124

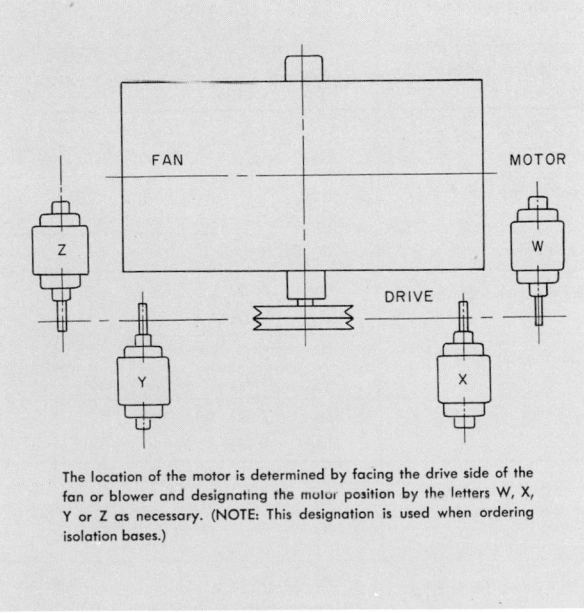

The location of the motor is determined by facing the drive side of the fan or blower and designating the motor position by the letters W, X, Y or Z as necessary. (NOTE: This designation is used when ordering isolation bases.)

FIG. 13 — MOTOR POSITIONS

Figure 13 shows the motor positions possible for a belt-driven fan. Use of Positions W and Z results in the simplest construction of fan base and belt guard.

Figure 14 shows the standard rotation and discharge combinations available.

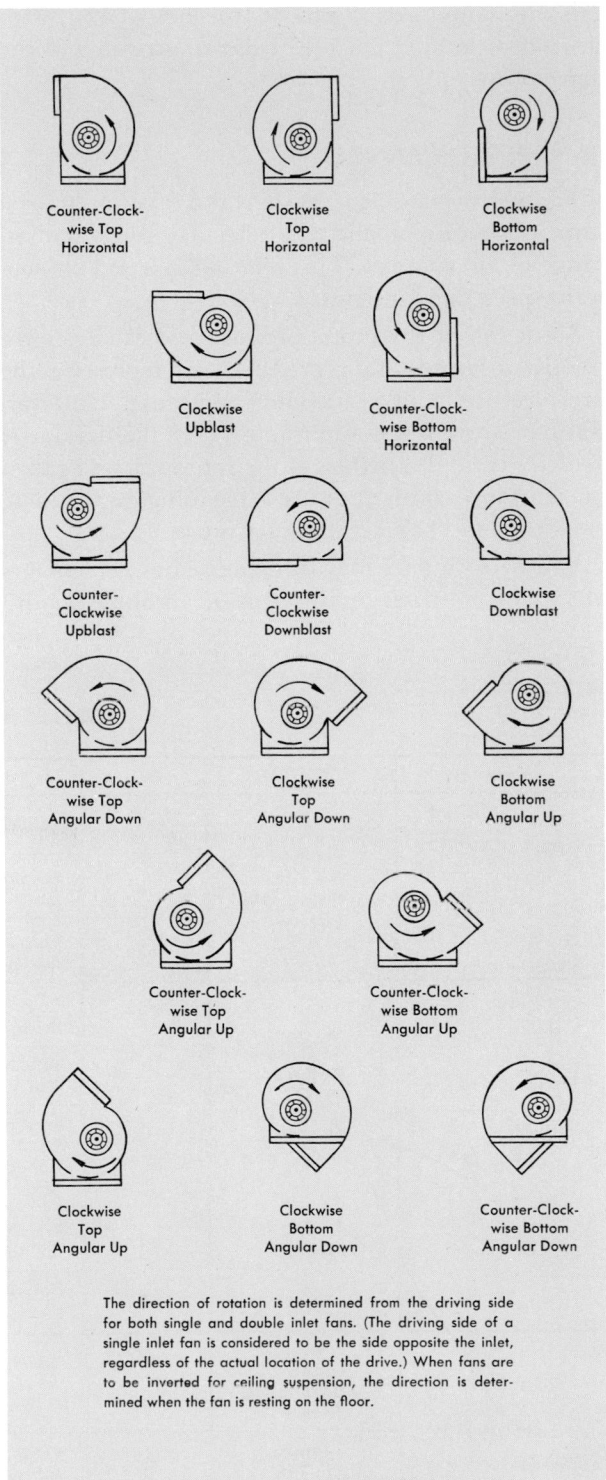

The direction of rotation is determined from the driving side for both single and double inlet fans. (The driving side of a single inlet fan is considered to be the side opposite the inlet, regardless of the actual location of the drive.) When fans are to be inverted for ceiling suspension, the direction is determined when the fan is resting on the floor.

FIG. 14 — ROTATION AND DISCHARGE

Axial flow fans are available for belt drive or direct connections. Therefore, two arrangements have been standardized throughout the industry. Arrangement 4 is driven directly. Since the motor is in the air stream, the application of this arrangement is limited to the handling of air which will not damage the motor. Arrangement 9 is belt-driven, with the motor located outside of the air stream and the drive protected.

FAN PERFORMANCE

Fan performance curves show the relation of pressure, power input and fan efficiency for a desired range of air volumes. This relation is based on constant speed and air density.

Static rather than total pressure and efficiency are usually inferred. Static pressure best represents the pressure useful in overcoming resistance. However, static pressure is less applicable where the fan outlet velocity is high. Further, if the fan operates against no resistance, static pressure is meaningless. In these cases total mechanical efficiency is used.

Fan performance may be expressed as percentages of rated quantities or in terms of absolute quanti-

ties. The former method is illustrated in *Fig. 6, 7, 8, 9 and 11.*

LAWS OF FAN PERFORMANCE

Fan laws are used to predict fan performance under changing operating conditions or fan size. They are applicable to all types of fans.

The fan laws are stated in *Table 5*. The symbols used in the formulas represent the following quantities:

Q — Volume rate of flow thru the fan.

N — Rotational speed of the impeller.

P — Pressure developed by the fan, either static or total.

Hp — Horsepower input to the fan.

D — Fan wheel diameter. The fan size number may be used if it is proportional to the wheel diameter.

W — Air density, varying directly as the barometric pressure and inversely as the absolute temperature.

In addition to the restrictions noted in *Table 5*, application of these laws is limited to cases where fans are geometrically similar and where there is no

TABLE 5—FAN LAWS

VARIABLE	CONSTANT	NO.	LAW	FORMULA
SPEED	Air Density Fan Size Distribution System	1	Capacity varies as the Speed.	$\dfrac{Q_1}{Q_2} = \dfrac{N_1}{N_2}$
		2	Pressure varies as the square of the Speed.	$\dfrac{P_1}{P_2} = \left(\dfrac{N_1}{N_2}\right)^2$
		3	Horsepower varies as the cube of the Speed.	$\dfrac{Hp_1}{Hp_2} = \left(\dfrac{N_1}{N_2}\right)^3$
FAN SIZE	Air Density Tip Speed	4	Capacity and Horsepower vary as the square of the Fan Size.	$\dfrac{Q_1}{Q_2} = \dfrac{Hp_1}{Hp_2} = \left(\dfrac{D_1}{D_2}\right)^2$
		5	Speed varies inversely as the Fan Size.	$\dfrac{N_1}{N_2} = \dfrac{D_2}{D_1}$
		6	Pressure remains constant.	$P_1 = P_2$
	Air Density Speed	7	Capacity varies as the cube of the Size.	$\dfrac{Q_1}{Q_2} = \left(\dfrac{D_1}{D_2}\right)^3$
		8	Pressure varies as the square of the Size.	$\dfrac{P_1}{P_2} = \left(\dfrac{D_1}{D_2}\right)^2$
		9	Horsepower varies as the fifth power of the Size.	$\dfrac{Hp_1}{Hp_2} = \left(\dfrac{D_1}{D_2}\right)^5$
AIR DENSITY	Pressure Fan Size Distribution System	10	Speed, Capacity and Horsepower vary inversely as the square root of Density.	$\dfrac{N_1}{N_2} = \dfrac{Q_1}{Q_2} = \dfrac{Hp_1}{Hp_2} = \left(\dfrac{W_2}{W_1}\right)^{1/2}$
	Capacity Fan Size Distribution System	11	Pressure and Horsepower vary as the Density.	$\dfrac{P_1}{P_2} = \dfrac{Hp_1}{Hp_2} = \dfrac{W_1}{W_2}$
		12	Speed remains constant.	$N_1 = N_2$

change in the point of rating on the performance curves. Because of the latter qualification, fan efficiencies are assumed constant.

Geometrically similar fans are those in which all dimensions are proportional to fan wheel diameter.

The same point of rating for two fans of different size means that for each fan the pressure and air volume at the point of rating are the same fraction of shutoff pressure and volume at free delivery, provided the rotational speed is the same in either case. For example, an operating point on *Fig. 7* will not change with the application of laws 7 thru 9, even though specific values will change.

Example 1 — Use of Laws 1 thru 3

Given:

Air quantity — 33,120 cfm
Static pressure — 1.5 in. wg
Fan speed — 382 rpm
Brake horsepower — 10.5

Find:

Capacity, static pressure and horsepower if the speed is increased to 440 rpm.

Solution:

Capacity $= 33,120 \times (440/382) = 38,150$ cfm
Static pressure $= 1.5 \times (440/382)^2 = 2.0$ in. wg
Horsepower $= 10.5 \times (440/382)^3 = 16.1$ bhp

FAN CURVE CONSTRUCTION

Fan performance is usually presented in tabular form (*Table 6*). However, for a graphic analysis performance curves are more convenient to use. If no curves are available, tabular values of pressure and horsepower may be plotted at constant speeds over the given range of capacities. The resulting curves may then be used as described under *Fan Performance in a System*.

FAN PERFORMANCE IN A SYSTEM

SYSTEM BALANCE

Any air handling system consists of a particular combination of ductwork, heaters, filters, dehumidifiers and other components. Each system therefore has an individual pressure-volume characteristic which is independent of the fan applied to the system. This relation may be expressed graphically on a coordinate system identical to that of a fan performance curve. A typical system characteristic is shown in *Fig. 15*.

System curves are based on the law which states that the resistance to air flow (static pressure) of a system varies as the square of the air volume flowing thru the system. In practice a static pressure is calculated as carefully as possible for a given system at the required air quantity. This establishes one point of the system curve. The remaining points are obtained by calculation from the above law, rather than by further static pressure calculations at other air quantities.

When a fan performance curve for a given fan size and speed is superimposed upon a system characteristic as in *Fig. 15*, there is only one point of intersection. This point is the only possible operating point under the conditions. If the fan speed is increased, the point of operation moves upward toward the right. If the speed is decreased, the operating point moves down and to the left.

Figure 15 illustrates the effect on system performance of operation at other than design conditions. Such a situation could be caused by dirty filters, wet coil versus dry coil operation of the dehumidifier, or the modulation of a damper. Lines of constant brake

TABLE 6—TYPICAL FAN TABLE

| WHEEL DIAM. 44½" | | | | | | | | | INLET AREA = 21.60 | | | | | | | | | | | | BACKWARD-CURVED | |
| TIP SPEED = 11.65 × RPM | | | | | | | | | OUTLET AREA = 20.79 | | | | | | | | | | | Class I Ratings | | |

| CFM | Outlet Veloc. | ¼" SP | | ⅜" SP | | ½" SP | | ⅝" SP | | ¾" SP | | ⅞" SP | | 1" SP | | 1¼" SP | | 1½" SP | | 1¾" SP | |
	FPM	RPM	BHP	RPM	BHP	RPM	BHP	RPM	BHP	RPM	BHP	RPM	BHP	RPM	BHP	RPM	BHP	RPM	BHP	RPM	BHP
12460	600	189	.61																		
14536	700	200	.76	228	1.04																
16613	800	212	.94	237	1.26	264	1.64														
18691	900	228	1.13	250	1.49	272	1.87	292	2.25												
20766	1000	243	1.39	263	1.76	282	2.18	302	2.57	324	3.08										
22841	1100	259	1.66	278	2.09	296	2.52	315	2.95	332	3.44	351	3.98								
24916	1200	276	1.98	294	2.45	310	2.92	327	3.38	343	3.87	359	4.37	377	4.91						
26991	1300	292	2.34	309	2.84	325	3.35	342	3.85	356	4.36	371	4.88	387	5.44	420	6.66				
29065	1400	309	2.77	325	3.29	342	3.83	356	4.37	370	4.91	384	5.47	398	6.03	426	7.22	459	8.62		
31158	1500	326	3.24	342	3.82	357	4.37	371	4.95	385	5.53	398	6.12	411	6.71	437	7.92	465	9.25	492	10.53
33233	1600	344	3.78	359	4.37	373	4.97	387	5.58	400	6.19	413	6.80	425	7.45	449	8.73	473	10.01	500	11.45
35308	1700	361	4.37	375	5.00	390	5.63	403	6.28	416	6.93	427	7.58	440	8.24	463	9.59	486	10.93	509	12.42
37383	1800	379	5.04	394	5.69	407	6.37	420	7.04	431	7.72	443	8.41	454	9.11	477	10.53	498	11.95	520	13.45
39458	1900	397	5.76	411	6.46	423	7.16	436	7.88	447	8.60	459	9.31	470	10.04	492	11.54	512	13.03	533	14.58
41532	2000	416	6.57	428	7.31	441	8.05	452	8.78	464	9.54	475	10.30	486	11.05	507	12.60	526	14.15	546	15.75

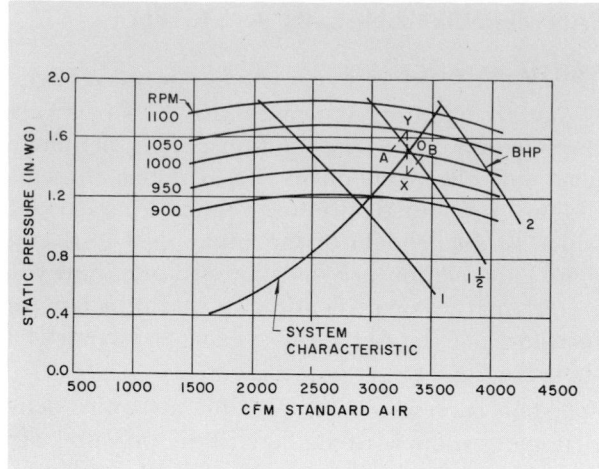

FIG. 15 — EFFECT OF CHANGE IN DESIGN CONDITIONS

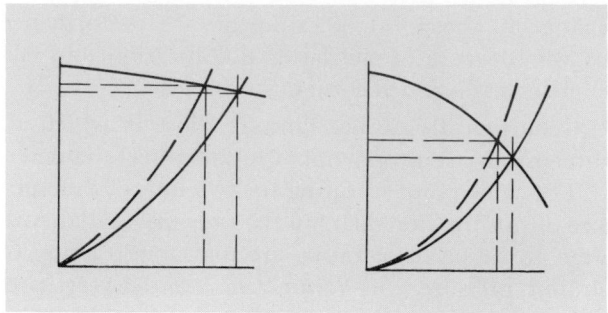

FIG. 16 — EFFECT OF FAN CURVE SLOPE

horsepower have been included for ease of analysis. *Example 2* describes the analysis.

Point *O* is the design point. Points *A* and *B* are new operating points resulting respectively from an increase or decrease in system resistance. Points *A* and *B* are single points each of two new system characteristics.

Example 2 — Operation Above Design Static Pressure

Given:

Air quantity	— 3300 cfm
Static pressure	— 1.5 in. wg
Fan speed	— 1000 rpm
Fan performance	— *Fig. 15*

Find:

Air quantity, static pressure and brake horsepower if the resistance of the filters is 0.15 in. wg greater than estimated for design.

Solution:

1. From design point *O (Fig. 15)* rise vertically to point *Y* at a static pressure of 1.65 in. wg (1.5 + 0.15) and 3300 cfm.

2. Since the fan operates at 1000 rpm, proceed to the 1000 rpm fan curve along a line parallel to the system characteristic. At the new operating point *A* the fan delivers to the system, 3175 cfm at a static pressure of 1.5 in. wg. The required power input at the new conditions is 1.4 bhp.

PRESSURE CONSIDERATIONS

An analysis similar to that of *Example 2* indicates that overestimating the static pressure increases the required horsepower. Operation in this case is at point *B* of *Fig. 15,* rather than at point *O*. Therefore, the addition of a safety factor to the calculated static pressure tends to increase fan horsepower re-

quirements unnecessarily. The static pressure used to select a fan should be that pressure calculated for the system at the design air quantity.

If the static pressure is overestimated, the amount of increase in horsepower and air volume depends upon the steepness of the fan curves in the area of selection. *Fig. 16* shows that volume deviations may be large if the fan curve is relatively flat. With a steep pressure characteristic, pressure differences may have little effect on air volume and horsepower. For this reason a fan with a steep performance curve is well suited to a system requiring an air volume relatively independent of changes in system resistance. An example of such a system is an induction unit primary air system.

Conversely, a variable volume system requires a pressure nearly constant with changes in air volume. Thus, a fan with a comparatively flat pressure characteristic is more appropriate.

STABILITY

Fan operation is stable if it remains unchanged after a slight temporary disturbance or if a slight permanent disturbance produces only a small shift in the operating point.

Instability is a surging or pulsation which may occur when the system characteristic curve intersects the fan curve at two or more points. This is a rare occurrence in a single fan. When two or more forward curved fans are connected in parallel, it is possible that the composite curve have an unstable area such as shown in *Fig 17*. If the operating point falls in this area, either adding or subtracting resistance allows operation at a stable point on either side of this area. When operation occurs such that only one sharp intersection of fan curve and system curve is possible, there is no condition of instability.

System resonance is a rare thing but may occur in systems utilizing high pressure fans with a duct system tuned to a particular frequency like an organ pipe. With operation to the left of the pressure peak, a pressure increase is accompanied by a capacity increase, in turn tending to further increase pressure. This condition may be overcome by altering the system characteristic curve so that the operating point falls between the pressure peak and the free delivery point.

FAN SELECTION

The system requirements which influence the selection of a fan are air quantity, static pressure, air density if other than standard, prevailing sound level or the use of the space served, available space, and the nature of the load. When these requirements are known, the selection of a fan for air conditioning usually involves choosing the most inexpensive combination of size and class of construction with an acceptable sound level and efficiency.

Outlet velocity cannot be used as a criterion of selection from the standpoint of sound generation. The best sound characteristics are obtained at maximum fan efficiency. Fans operating at higher static pressures have greater allowable outlet velocities since maximum efficiency occurs at higher air quantities. Thus, any limits imposed on outlet velocity in relation to sound level depend upon the static pressure in addition to ambient sound levels and the use of the area served. In regard to sound generation a fan should be selected as near to maximum efficiency as is possible and adjacent ductwork should be properly designed, as described in *Part 2*.

The best balance of first cost and fan efficiency usually results with a fan selection slightly smaller than that representing the maximum efficiency available. However, selection of a larger, more efficient fan may be justified in the case of long operating hours. Also, a larger fan may be economically preferable if a smaller selection necessitates a larger motor, drive and starter, or heavier construction.

The selection of a fan and drive can affect psychrometric conditions in the area served. If the combination produces an air quantity below that required at design conditions, the resulting room dry-bulb temperature is higher. When the air quantity is greater than required at design conditions, room controls prevent a fall in temperature.

ATMOSPHERIC CORRECTIONS

Fan sound level does not vary sufficiently with altitude to warrant using sound ratings at conditions other than sea level.

Fan tables and curves are based on air at standard atmospheric conditions of 70 F and 29.92 in. Hg barometric pressure. If a fan is to operate at nonstandard conditions, the selection procedure must include a correction. With a given capacity and static pressure at operating conditions the adjustments are made as follows:

1. Obtain the air density ratio from *Chart 2*.
2. Calculate the equivalent static pressure by dividing the given static pressure by the air density ratio.
3. Enter the fan tables at the given capacity and the equivalent static pressure to obtain speed and brake horsepower. This speed is correct as determined.
4. Multiply the tabular brake horsepower by the air density ratio to find the brake horsepower at the operating conditions.

CHART 2—ATMOSPHERIC CORRECTIONS

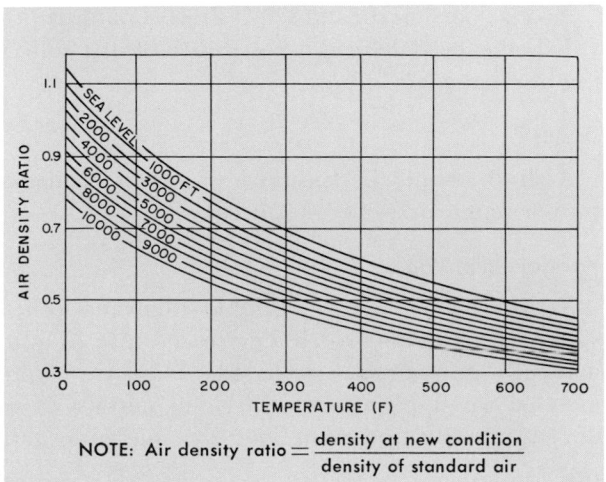

NOTE: Air density ratio = $\dfrac{\text{density at new condition}}{\text{density of standard air}}$

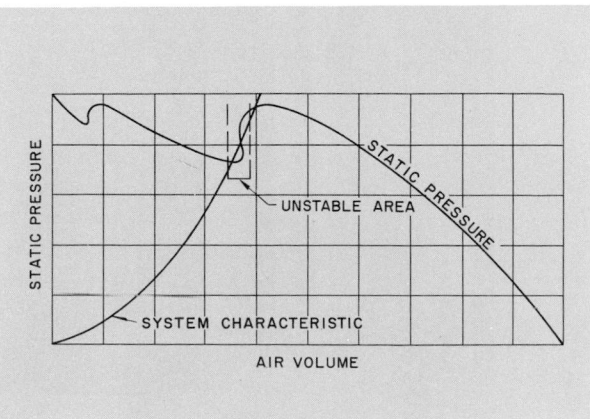

Fig. 17 — System Instability

If atmospheric corrections are ignored in the fan selection, fan speed and air capacity may be too small, and the brake horsepower somewhat high. *Example 3* illustrates a fan selection at high altitude.

Example 3 — Fan Selection at High Altitude

Given:

Air quantity*	— 37,380 cfm
Static pressure*	— 1.45 in. wg
Altitude	— 5000 ft
Air temperature	— 70 F
Fan ratings	— *Table 6*

Find:

Fan speed, brake horsepower and class.

Solution:

1. From *Chart 2* the air density ratio is 0.83.
2. The equivalent static pressure is equal to 1.45/0.83 or 1.75 in. wg.
3. Select from *Table 6* a size 7 double inlet fan, operating at 520 rpm and requiring 13.45 bhp.
4. The design speed at 5000 ft is 520 rpm.
5. The brake horsepower for the less dense air at 5000 ft is 0.83 × 13.45 or 11.2 bhp.
6. At the fan outlet velocity of 1800 fpm and the equivalent static pressure of 1.75 in. wg, enter *Chart 1*. The selection is well within the range of a Class I fan. This is the proper selection.

At altitudes above 3300 feet, fan motor temperature guarantees may not apply. High altitude applications should therefore be brought to the manufacturer's attention.

ACCESSORIES

Fan accessories are available to fulfill specific needs. Where applicable, the following accessories can aid in assuring a satisfactory air conditioning system.

Access Doors

Access doors on the fan scroll sheet should be provided whenever there is a possibility of dirt collecting inside the fan.

Drains

A drain should be specified whenever condensation or water carry-over may occur.

Variable Inlet Vanes

Figure 18 shows a set of variable inlet vanes. These vanes are a volume control device and may be automatically or manually actuated. They are recommended for applications with long periods of reduced capacity operation and for use with static

*Capacity required at 5000 feet. See *Parts 1 and 2* for high altitude load calculations and duct design respectively.

FIG. 18 — VARIABLE INLET VANES

pressure regulators. Use of variable inlet vanes is further discussed under *Control*.

Outlet Dampers

Outlet dampers are a volume control device and may be automatically or manually actuated. They may be used for applications requiring extreme capacity reduction for short periods of time or for small adjustments. These dampers are illustrated in *Fig. 19*. Their use is further discussed under *Control*.

Isolators

In order of decreasing vibration isolation efficiency, steel spring isolators, double rubber-in-shear isolators, and single rubber-in-shear isolators are all used for fan installations. These isolators are normally used in conjunction with steel channel bases so that the fan and the motor may be mounted on

FIG. 19 — OUTLET DAMPERS

an integral surface. For a more complete discussion of vibration isolation, refer to *Chapter 2* of this part.

Bearings

Ball bearings are the most common type of bearing used on fans. The sleeve oil bearing can be provided at an extra cost and is initially a quieter bearing. However, its quietness has been overemphasized since the bearing noise does not materially add to the fan air noise.

CONTROL

Variation of the air volume delivered by a fan may be accomplished by several methods:

1. Variable speed motor control
2. Outlet damper control
3. Variable inlet vane control
4. Scroll volume control
5. Fan drive change

Use of a variable speed motor to control fan capacity is the most efficient means of control and the best from the standpoint of sound level. However, it is the most expensive method.

Use of outlet dampers with a constant speed motor is the least expensive method but the least efficient of the first three mentioned above.

Variable inlet vanes may be used to adjust the fan delivery efficiently over a wide range. This method controls the amount of air spin at the fan inlet, thus controlling the static pressure and horsepower requirement at a given fan speed.

Figure 20 compares variable inlet vane control, outlet damper control and speed control as each affects fan performance. The horsepower curves indicate the power required at various vane settings, damper positions and fan speeds respectively.

The horsepower curve for variable inlet vane control *(Fig. 20)* is based on a fan designed with supplementary fixed air inlet vanes, such that there is no loss in efficiency when variable vanes are used instead. A loss of static efficiency as great as 10% results from the use of variable inlet vanes on a fan designed with an open inlet.

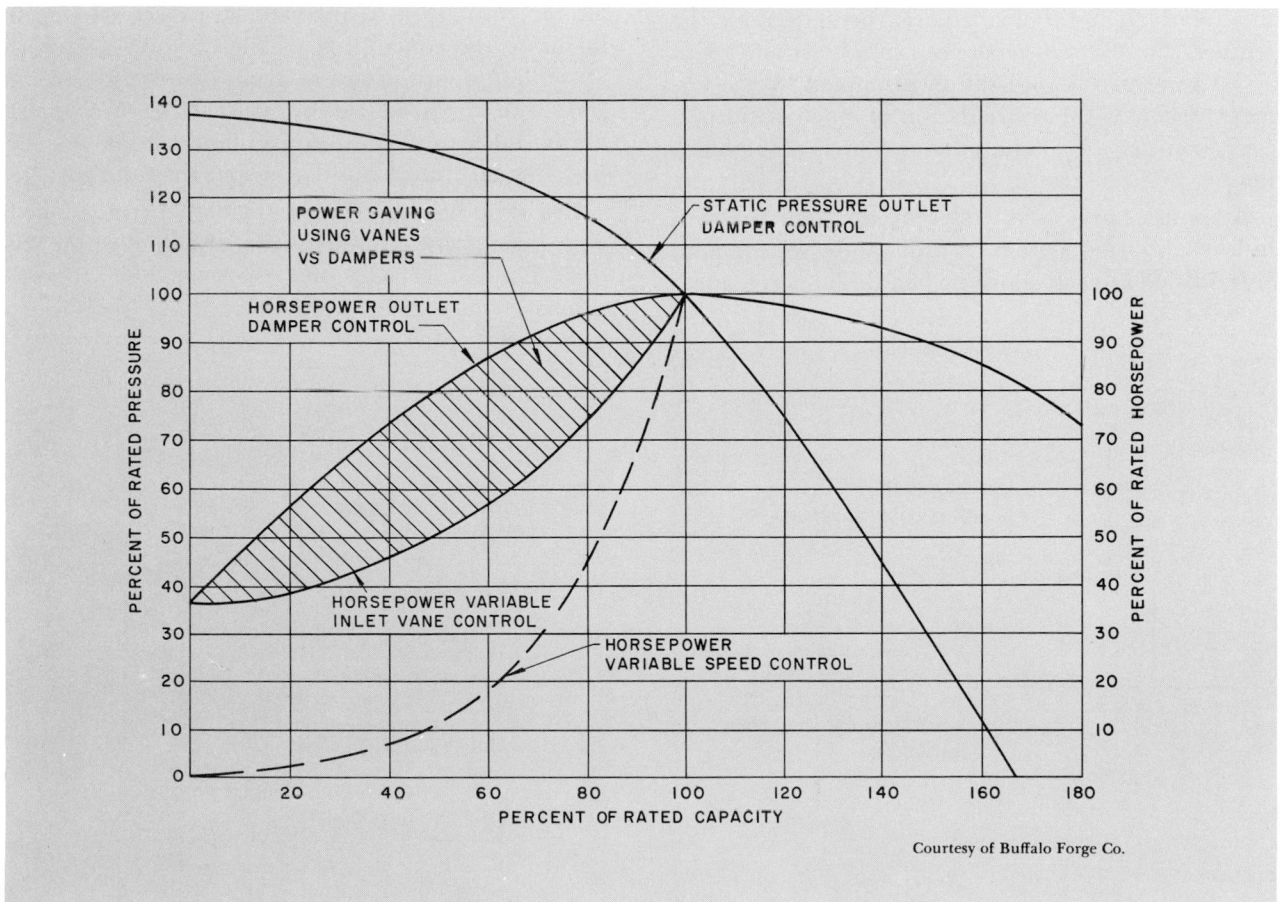

FIG. 20 — COMPARISON OF FAN CONTROL METHODS

Tubeaxial and vaneaxial fans are often equipped with adjustable blades for matching the fan to system requirements.

Propeller fans may be speed-controlled or blade-adjustable.

LOCATION

Refer to *Part 2* for the aspects of fan location. The effect of fan motor location on the system cooling load and air volume is discussed in *Part 1*.

MULTIPLE INSTALLATIONS

Fans may be arranged in series or in parallel to provide for operating conditions not met by the use of a single fan.

Possible series applications include:

1. Recirculating fan
2. Booster fan
3. Return air fan

A recirculating fan increases the supply air to a space without increasing the primary air *(Fig. 21)*. The purpose is to obtain greater air motion, usually in a relatively lightly loaded area, or to decrease the temperature difference between supply air and room air. An industrial application prompted by the former purpose is the recirculation of air in an inspection room served by the same system as a neighboring production area.

A booster fan is used to step up the static pressure in a distribution system in order to serve a remote area, loaded intermittently; when this area is loaded,

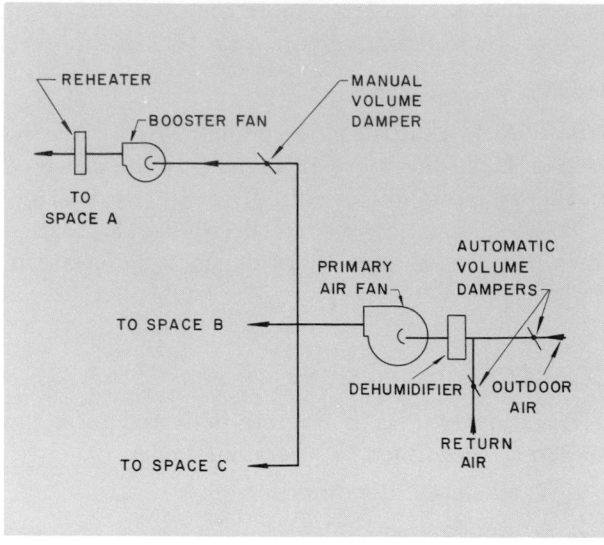

FIG. 22 — BOOSTER FAN

it requires a small air quantity relative to the total primary air *(Fig. 22)*. A conference room (space A) could be conditioned in this manner.

The most common series application is the return air fan, usually used on extensive duct systems to facilitate the controlling of the mixture of return air and outdoor air and to avoid the excessive room static pressures required *(Fig. 23)*. Use of a return air fan also provides a convenient method for exhausting air from a tightly constructed building.

In air conditioning, fans are seldom directly staged, with the outlet of the first being the inlet of the second. The fan efficiency and the operating economy suffers if this method is used for merely obtaining a higher static pressure.

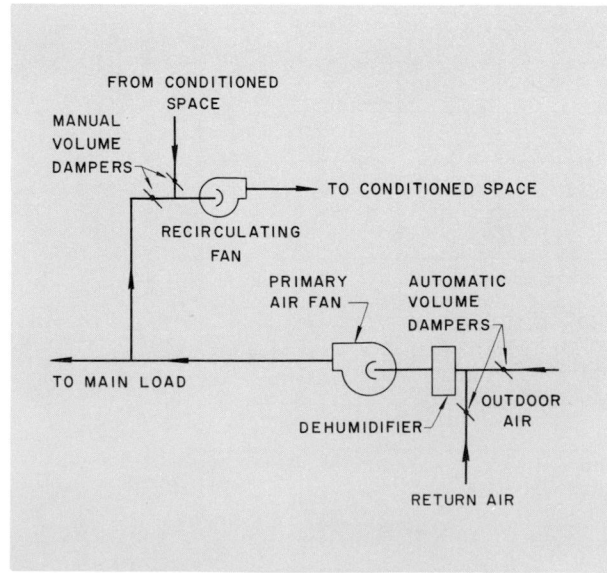

FIG. 21 — RECIRCULATING FAN

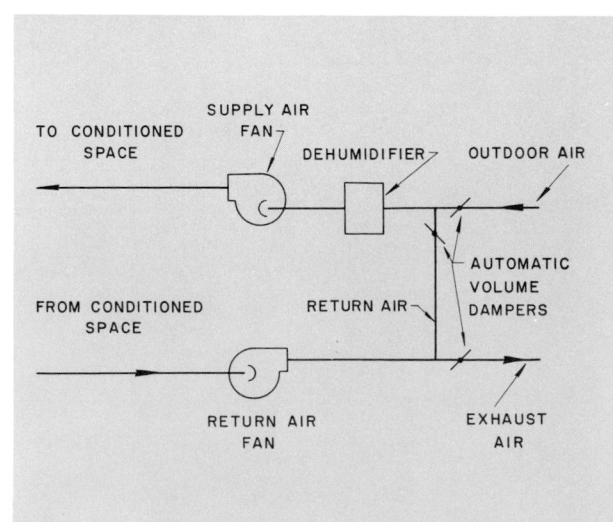

FIG. 23 — RETURN AIR FAN

Fans may be applied in parallel if dictated by space limitations or if provision is to be made for a future addition. Centrifugal fans are available factory-mounted in cabinets for the former reason. Parallel fans provide greater capacity at a common static pressure. However, a parallel design is seldom chosen just to increase capacity since no improvement in fan efficiency occurs and economy is not warranted by the greater first cost of the parallel installation.

CHAPTER 2. AIR CONDITIONING APPARATUS

This chapter contains practical information to guide the engineer in the application, selection and installation of various types of air conditioning apparatus, remote from the source of refrigeration.

Although the concept of air conditioning includes the moving, heating and cleaning of air, this chapter is devoted primarily to cooling, dehumidifying and humidifying equipment. Other types of air handling equipment are discussed in *Chapters 1 and 3* of this part.

TYPES OF APPARATUS

Air conditioning apparatus may be classified into two major groups:

1. Coil equipment in which the conditioning medium treats the air thru a closed heat transfer surface.
2. Washer equipment in which the conditioning medium contacts the air directly.

These two groups may be subclassified as shown in *Chart 3*.

Because of its specialized application, packaged or unitary air conditioning equipment is described in *Chapter 3* of this part. Terminal equipment is discussed in *Parts 10 and 11*.

STANDARDS AND CODES

The application and installation of air conditioning apparatus should conform to all codes, laws and regulations applying at the job site.

Applicable provisions of the American Standard Safety Code B9.1 and ARI, ASHRAE and AMCA Standards govern the testing, rating and manufacture of air conditioning apparatus.

FAN-COIL EQUIPMENT

As the term implies, the primary constituents of a fan-coil unit are a fan to produce a flow of air and a

CHART 3—APPARATUS CLASSIFICATION

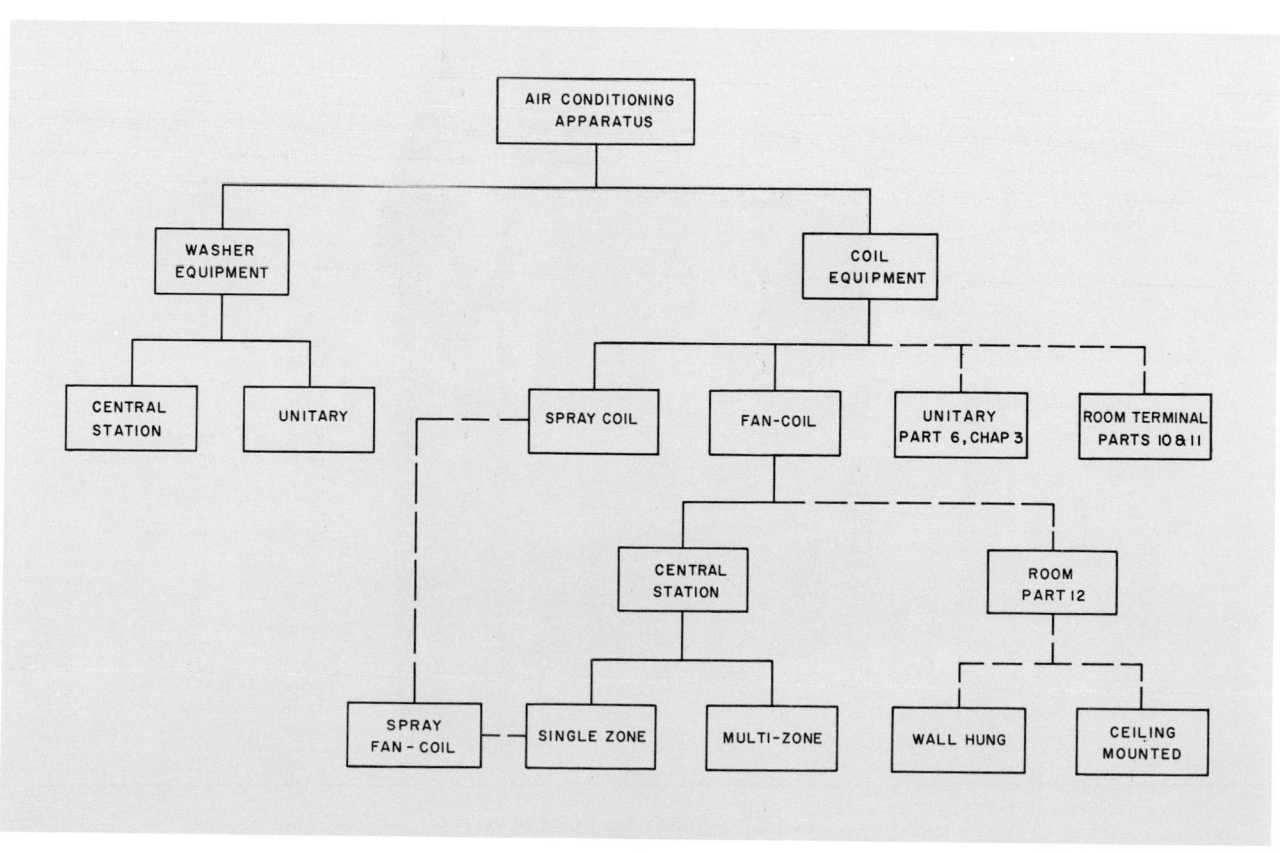

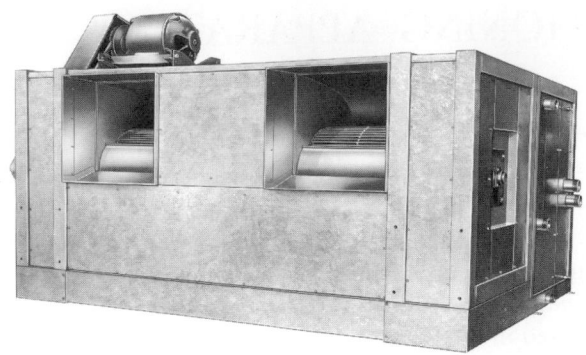

FIG. 24 — SINGLE ZONE FAN-COIL UNIT

FIG. 26 — SPRAY COIL SECTION

FIG. 25 — MULTI-ZONE FAN-COIL UNIT

chilled water or direct expansion coil to cool and dehumidify the air. Accessories such as a heating coil, a humidifier and a filter section are normally available to perform, if necessary, the remaining air conditioning functions. The required components may be assembled into a factory-fabricated, cabinet style package. *Figures 24 and 25* show respectively a single zone and a multi-zone fan-coil unit.

A spray coil section is shown in *Fig. 26*. Since such equipment is intended for incorporation in a built-up apparatus, it is not fan-coil equipment. However, because of the similarity of function, spray coil equipment is discussed in this section. Differences in

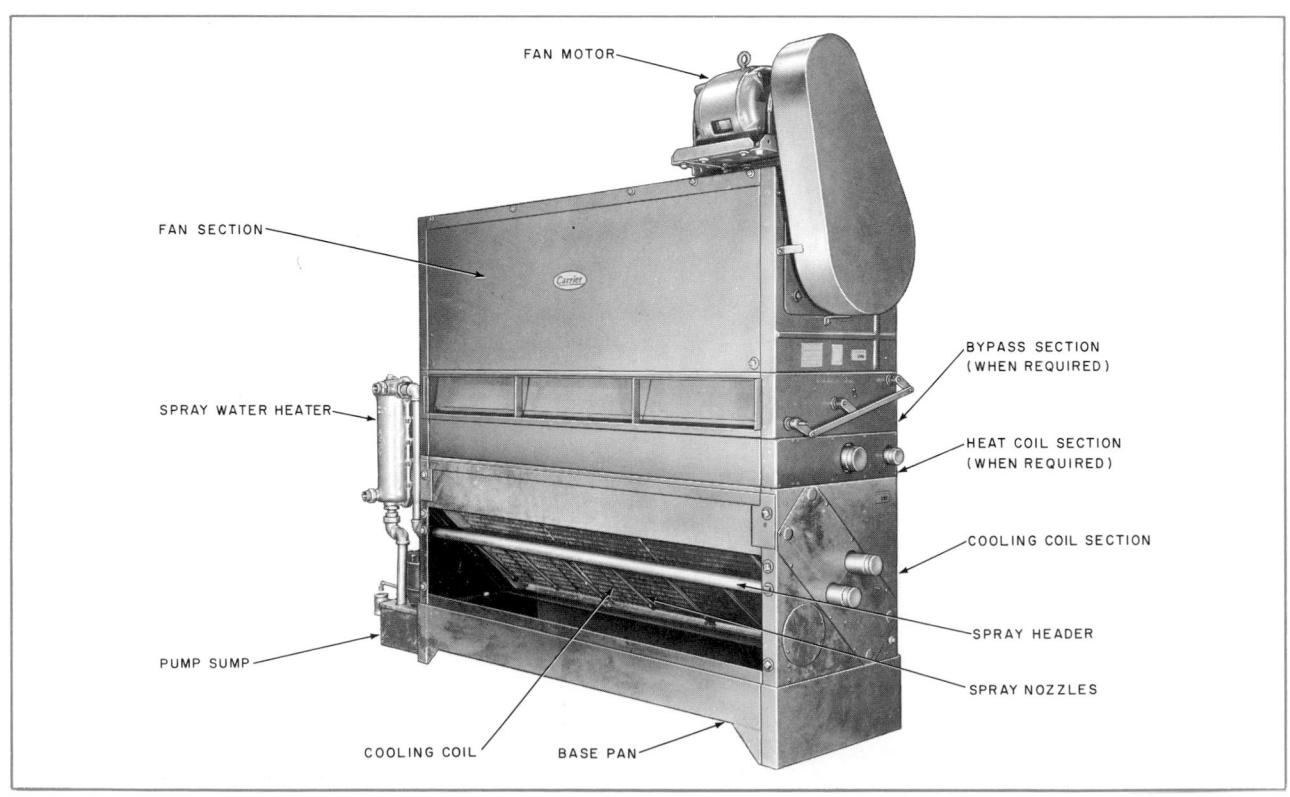

FIG. 27 — SPRAY FAN-COIL UNIT

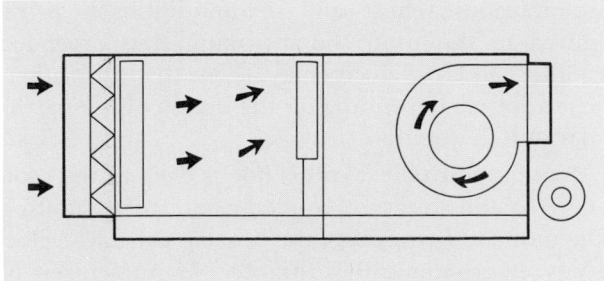

FIG. 28 — AIR FLOW — SINGLE ZONE UNIT

application and layout will be noted as they exist.

Figure 27 illustrates a spray fan-coil unit.

Single zone and multi-zone fan-coil units differ physically in the location of the fan relative to the cooling coil. In a single zone unit, the fan is located downstream of the cooling coil. Therefore, this type of unit is often termed a "draw-thru" unit. A multizone unit may be referred to as a "blow-thru" unit since the fan is located upstream of the coil. *Figures 28 and 29* indicate the flow of air thru the two types of central station fan-coil apparatus.

Typical variations occurring in total pressure, static pressure and velocity pressure, as air passes thru a fan-coil unit, are illustrated in *Fig. 30 and 31*. The use of a fan equipped with a diffuser helps to convert velocity pressure to static pressure with a minimum energy loss.

Fan-coil units are furnished with either forward- or backward-curved blades. Forward-curved blade fans are well suited for such use, since they perform at slower speeds than other types of fans. Fan wheel construction is lighter in weight, more compact and less expensive than with backward-curved blades. Longer fan shafts are permissible because of the slower speeds.

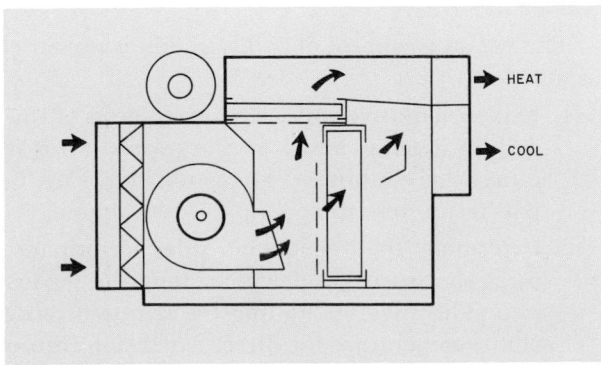

FIG. 29 — AIR FLOW — MULTI-ZONE UNIT

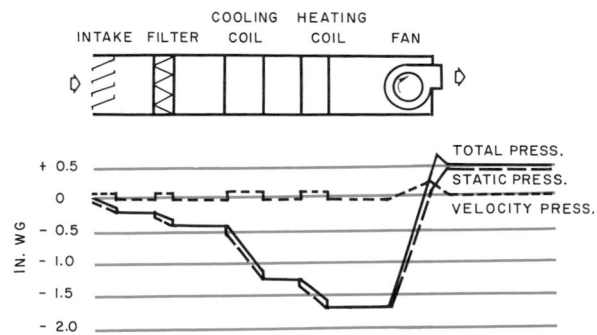

FIG. 30 — PRESSURE VARIATIONS (DRAW-THRU UNIT)

APPLICATION

The application of air conditioning equipment is influenced by the cooling load characteristics of the area to be served and the degree of temperature and humidity control required.

A single zone unit most effectively serves an area characterized by a relatively constant or uniformly varying load. Ideally, this area would be a single large room. However, multi-room applications are practical, provided a given variation in load occurs in all rooms simultaneously and in the same proportion. If required, zoning may be accomplished by reheat or air volume control in the branch ducts.

In a multi-room application where load components vary independently and as a function of time, a multi-zone apparatus provides individual space control with a single fan unit. For this type of load a multi-zone installation is less expensive than a single zone installation with a multiplicity of duct reheat coils.

Since a multi-zone unit permits outdoor air to bypass the cooling coil at partial loads, its use is particularly adapted to applications with high sensible heat factors and a minimum of outdoor air. If humidity control is required with a multi-zone unit, a precooling coil may be installed in the minimum outdoor air duct.

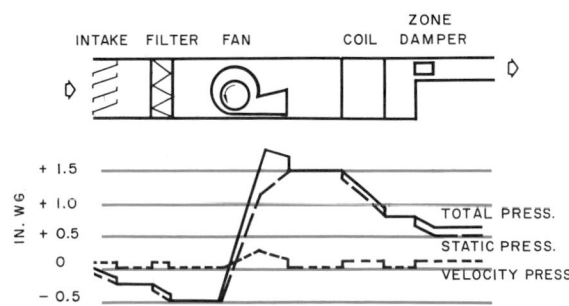

FIG. 31 — PRESSURE VARIATIONS (BLOW-THRU UNIT)

A standard fan-coil unit affords a close temperature control only. A measure of humidity control may be obtained by providing a unit humidifier such as a city water spray package. However, if more certain humidity control is required, a spray coil section or spray fan-coil unit is better suited to the application.

Spray coil equipment may be utilized for summer cooling and dehumidification, winter humidification and evaporative cooling in intermediate seasons. Its use is preferred for applications where humidity control is required, such as in industrial processes, hospitals, libraries and museums. Spray coil equipment may also be equipped with a spray water heater to provide simultaneous cooling or heating and humidification.

Standard fan-coil equipment, both single zone and multi-zone, may be obtained for air deliveries as high as 50,000 cfm. Multiple spray coil sections are available for air quantities exceeding 60,000 cfm. Where the size of available factory-fabricated equipment is exceeded, the apparatus must be constructed of individual cooling coils or spray coil sections.

Static pressure limitations on fan-coil unit fans vary widely with the manufacturer considered. Available cabinet fan pressure classes are defined in *Chapter 1* of this part.

UNIT SELECTION

The selection of fan-coil equipment is a matter of achieving the required performance at the maximum economy. The economic aspect includes not only the particular unit and coil chosen but also the effect of that choice on other system components, such as piping, ductwork and refrigeration equipment.

The selection procedure involves choosing the unit size and the coil. A coil selection includes the determination of the coil depth in rows required, the optimum coil fin spacing and, in the case of chilled water coils, the appropriate circuiting.

Unit Size

With the dehumidified air quantity known, the choice of unit size normally precedes the coil selection. In most cases, the size is determined by the cooling coil face velocity.

When selecting a coil face velocity, it is suggested that the highest allowable face velocity be used in the interest of economy. Manufacturers rate their coils at maximum face velocities proven by tests to be satisfactory, with respect to both the entrainment of moisture droplets and air resistance. However, if simultaneous reheat and dehumidification is required of the unit, the maximum recommended cooling coil face velocity may be less than that otherwise allowed, depending on the design of the particular unit in question.

Since a unit reheat coil is not as deep as the cooling coil and does not condense moisture, limiting the unit size by limiting the heating coil face velocity is not economically justifiable. Manufacturers of fan-coil equipment have designed their internal heating coils to provide optimum performance at recommended cooling coil face velocities.

As explained in *Chapter 1* of this part, fan outlet velocity should not be used as a selection criterion reflecting the intensity of sound generation. Sound characteristics improve with rising fan efficiency, rather than with decreased outlet velocities.

Coil Selection

A particular cooling coil is selected to produce a desired effect on the air passed thru it, in accordance with the sensible, latent and total cooling loads calculated for the space and with the condition of the air entering the coil. However, the final selection defines also the required chilled water flow, the pressure drop at that flow and the required entering water temperature; or in the case of a direct expansion coil, the refrigerant temperature. Therefore, the coil selection should be made with regard to refrigerant side or chilled water side performance as well as to air side performance.

Thus, each coil selection has two facets which may be regarded as independent for the purposes of selection. Air side and refrigerant side performances should be considered separately and then matched to produce the final economically optimum coil selection. The apparatus dewpoint method of coil selection provides means for matching air side and refrigerant side performances. This method is described in *Part 1*.

The two-step concept of coil selection is presented as follows:

1. Make a tentative coil selection in terms of rows and fin spacing, based on the bypass factor required by established air conditions. Coil bypass factor determines apparatus dewpoint.

2. Determine the refrigerant side performance, using the apparatus dewpoint found in the first step. This involves finding the required refrigerant temperature for direct expansion coils or the chilled water quantity, temperature and resulting pressure drop for water coils.

Thus, the coil can be tentatively selected without regard to the final refrigeration machine selection. If the first coil selection does not provide satisfactory refrigerant side performance, another coil with adequate air side performance may be tried. The optimum selection assures proper performance at the least owning and operating cost.

Often in a multi-zone application, the apparatus dewpoints of the various areas differ. Rather than penalizing the cost of the entire system by selecting the lowest room apparatus dewpoint as the coil apparatus dewpoint, a higher, more representative apparatus dewpoint may be chosen, and a compromise accepted in the design relative humidity in the room with the lower apparatus dewpoint. The increased relative humidity is offset by a decrease in dry-bulb temperature. Such a decision may be required in the case of a conference room, with its relatively high latent load. If a compromise is unacceptable for this application, maximum economy may be achieved by furnishing the special area with a separate system.

The various types of coil ratings and selection techniques encountered either use directly, or are derived from, one of two methods. They are the apparatus dewpoint (effective surface temperature) method and the modified basic data method. The latter involves calculating coil performance from basic heat transfer data and equations. It combines air side and refrigerant side performance determination into a single operation. However, the basic data method requires assumptions which are usually modified later in the selection, and is therefore a trial-and-error procedure. Calculated coil depth may be a decimal figure which must be rounded to a whole number, in turn necessitating a recalculation of performance. The apparatus dewpoint method is derived from the two-step concept of coil selection and implements its use. Coil rows are dealt with in terms of standard whole numbers only.

Charts 4 and 5 are conversion charts used to evaluate the air side performance of any cooling coil, with entering and leaving air conditions established. This performance is in terms of coil bypass factor and apparatus dewpoint. A straight edge, fixed at the entering dry-bulb temperature and rotated to pass thru the various intersections of the coil bypass factor and the line connecting entering and leaving wet-bulb temperatures, indicates the coil bypass factor which satisfies the leaving dry-bulb temperature. The apparatus dewpoint can be read at the chosen intersection.

Where the bypass factor for a particular coil is unknown, the coil performance may be plotted on the chart, and the bypass factor may be read at the intersection of the entering-leaving wet-bulb and dry-bulb lines. The bypass factors of various coils may thus be directly compared.

When selecting a cooling coil in conjunction with an air conditioning load estimate form, the bypass factor of the coil selected should agree reasonably with the bypass factor assumed in the estimate. If it does not, the estimate should be adjusted accordingly, as indicated in *Part 1*.

Refrigerant side coil ratings presume a tentative coil selection when based on the apparatus dewpoint. *Chart 6* and *Table 7* illustrate apparatus dewpoint refrigerant side ratings for chilled water and direct expansion coils respectively. Such charts are used in the second step of the two-step approach described above.

Table 8 shows the entering wet-bulb type of rating for direct expansion coils. This method of presentation is used frequently and may or may not be derived from the apparatus dewpoint method.

For a direct expansion coil, optimum coil circuiting is incorporated by the manufacturer into the coil design. A direct expansion coil experiences a decreased capacity with an increased refrigerant pressure drop caused by a greater coil circuit length. This is true even with a given coil surface.

Chilled water coils are usually offered with two or more circuiting arrangements, and the final coil selection prescribes the circuiting. The coil with the least number of circuits has the greatest number of passes back and forth across the coil face and vice versa. The minimum circuited coil has a greater capacity and produces a higher chilled water temperature rise at a given water quantity. However, the greater number of passes of a minimum circuited coil results in a pressure drop higher than that thru a coil of the same size but with more circuits and less passes. Minimum circuited coils are often used on large extensive systems in which the greater pumping head required is more than offset economically by the reduced first cost of piping and insulation.

With the required air side coil performance given, the greater the difference between apparatus dewpoint and entering chilled water temperature the smaller the required water quantity will be. Therefore, the choice of a chilled water temperature may involve an economic analysis of the first costs and operating costs of the refrigeration plant versus the

CHART 4—CONVERSION CHART (48F TO 60F ADP)

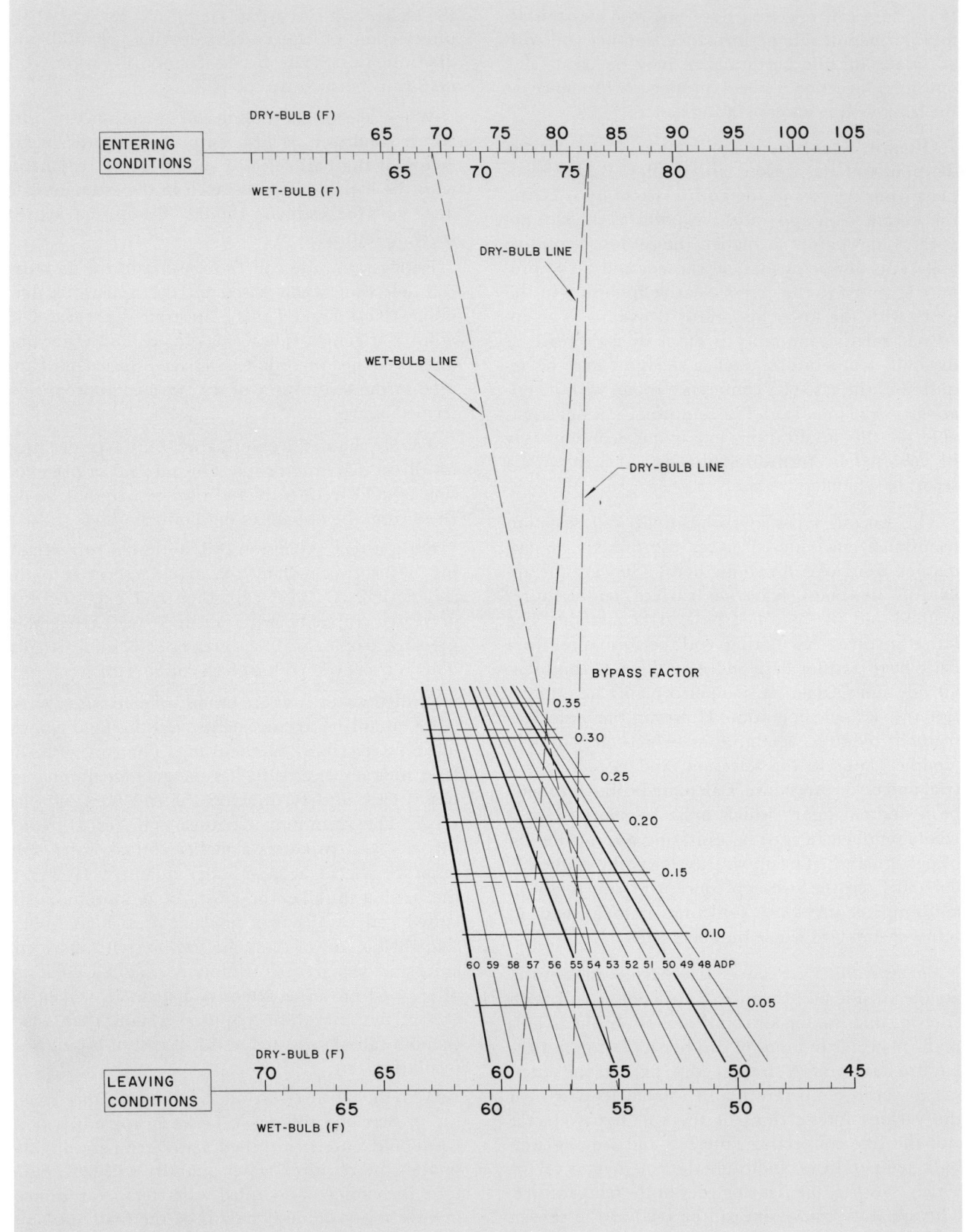

CHART 5—CONVERSION CHART (36F TO 48F ADP)

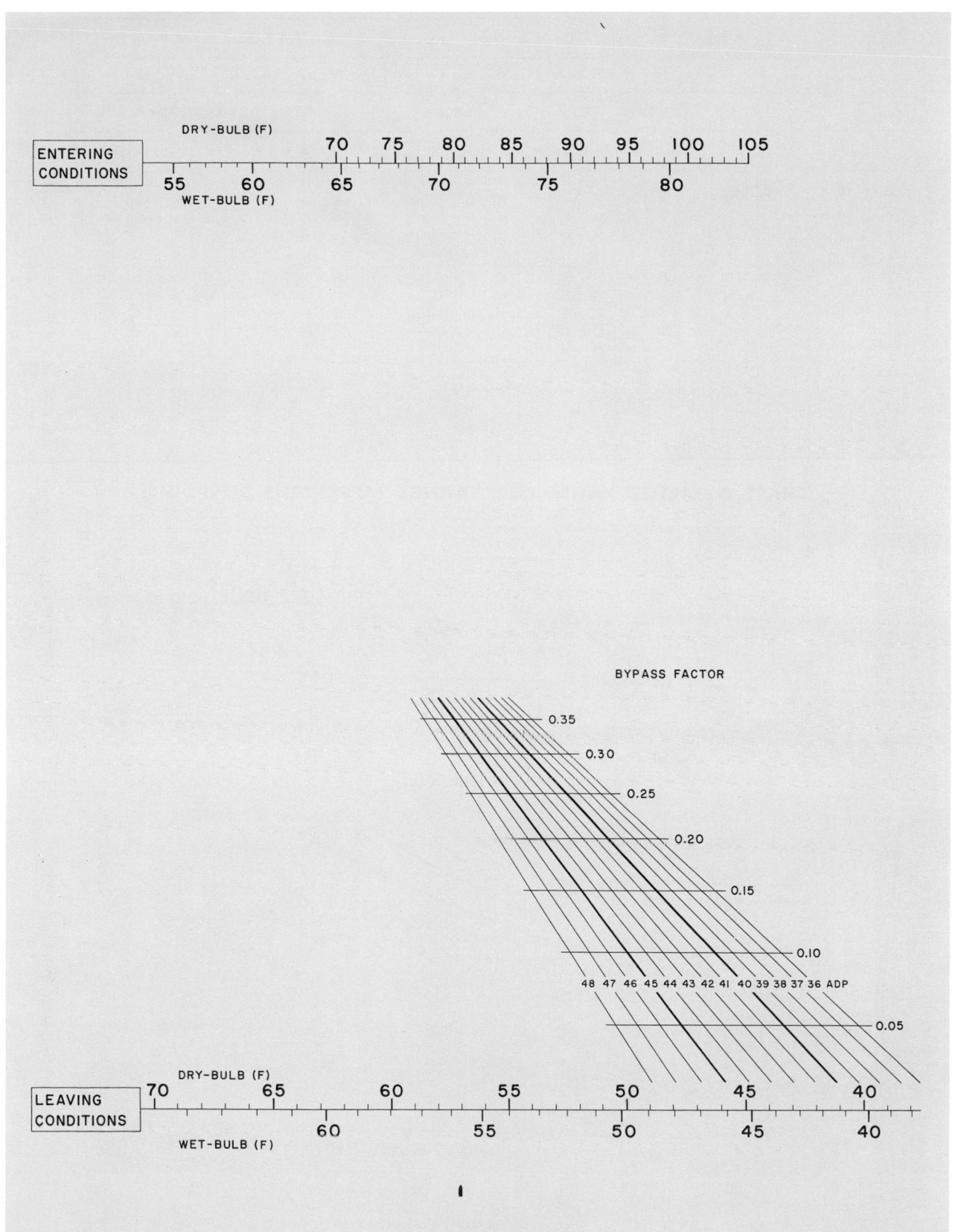

TABLE 7—DIRECT EXPANSION COIL RATINGS (APPARATUS DEWPOINT)

GTH (1000 BTU/HR)	APPARATUS DEWPOINT (F)													APPARATUS DEWPOINT (F)												
	36	38	40	42	44	46	48	50	52	54	56	58	60	36	38	40	42	44	46	48	50	52	54	56	58	60
	REFRIGERANT TEMP (F)													REFRIGERANT TEMP (F)												
	4 ROWS, 8 FINS/INCH													4 ROWS, 14 FINS/INCH												
48		26	28	30	32	34	36	38	40	42	44	46	48	26	28	30	32	34	36	38	40	42	44	46	48	50
65		26	28	30	32	34	36	38	40	42	44	46	48	26	28	30	32	34	36	38	40	42	44	46	48	50
82		26	28	30	32	34	36	38	40	42	44	46	48	26	28	30	32	34	36	38	40	42	44	46	48	50
99		26	28	30	32	34	36	38	40	42	44	46	48	26	28	30	32	34	36	38	40	42	44	46	48	50
115		26	28	30	32	34	36	38	40	42	44	46	48	25	27	29	31	33	35	37	39	41	43	45	47	49
132			26	28	30	32	35	37	39	41	43	45	47	25	27	29	31	33	35	37	39	41	43	45	47	49
149				26	29	31	33	35	37	39	41	43	45		25	27	30	32	34	36	38	40	42	44	46	48
166				25	27	29	31	33	35	37	39	42	44			26	28	30	32	34	36	38	40	43	45	47
183					25	27	29	31	33	35	37	39	42				26	28	30	32	34	37	39	41	43	45
201							26	28	31	33	35	37	39					26	28	30	32	34	37	39	41	43
218								26	28	30	32	35	37						25	28	30	32	34	37	39	41
235									25	28	30	32	34							26	28	30	32	35	37	39
252										25	27	30	32								25	27	30	32	35	37
270												26	29									25	27	30	32	34
288													26										25	27	29	32

CHART 6—CHILLED WATER COIL RATINGS (APPARATUS DEWPOINT)

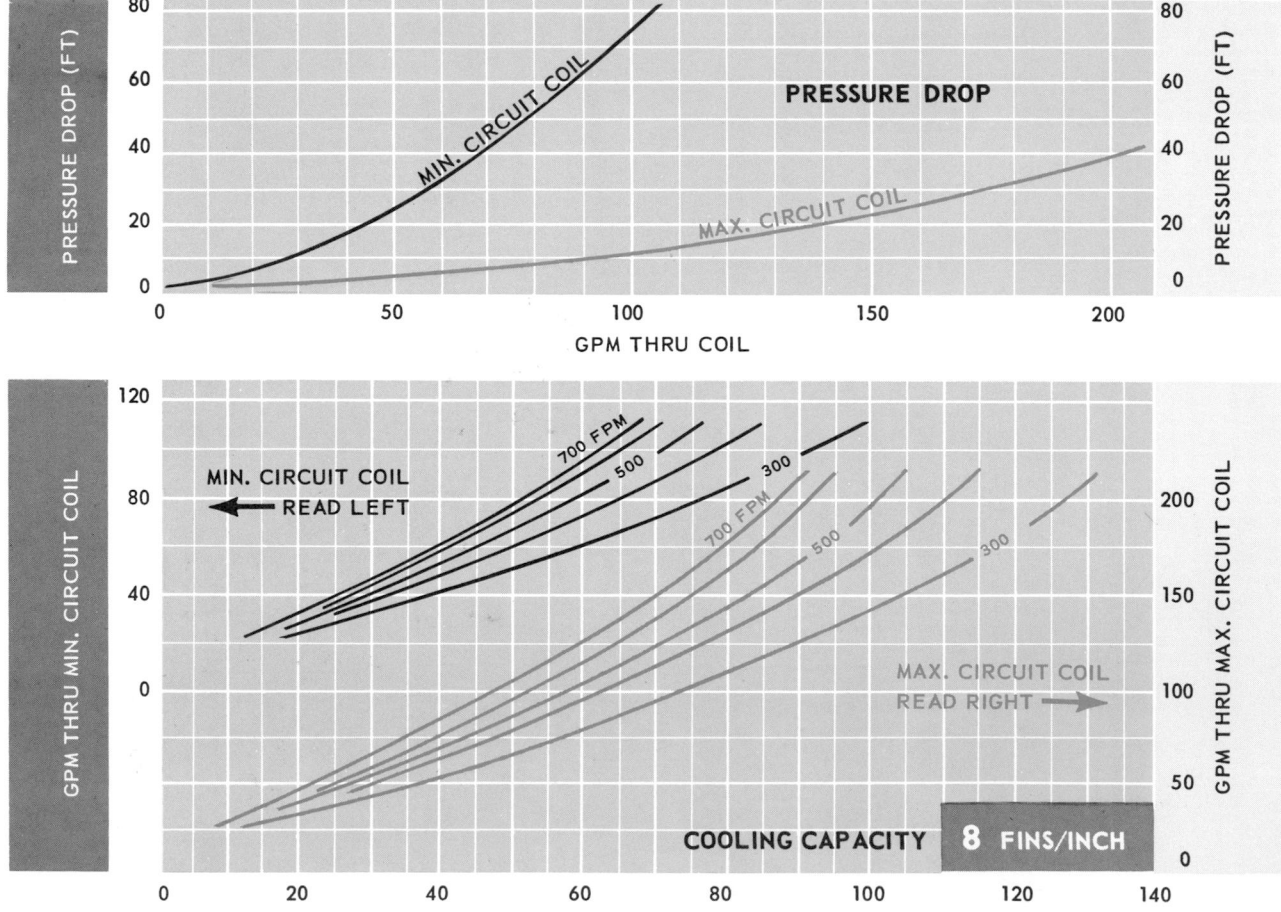

$$Q \text{ (thousands)} = \frac{\text{GRAND TOTAL HEAT (BTU/HR)}}{t_{adp} - t_{ew} \text{ (F)}}$$

TABLE 8—COOLING COIL RATINGS (ENTERING WET-BULB TEMPERATURE)

LEAVING AIR WET BULB TEMPERATURE (F) AND CAPACITY (TONS*)

Cooling Coil Face Velocity (FPM) = 600

Refrig. Temp. (F)	Rows	Fins Per Inch	Ratings*	65	66	67	68	69	70	71	72	73	74	75	76	77	78	79	80
35	4	7	LWB	53.7	54.5	55.2	55.9	56.7	57.4	58.4	59.2	59.9	60.9	61.8	62.6	63.5	64.3	65.3	66.3
			TONS	1.68	1.75	1.83	1.91	1.99	2.07	2.13	2.21	2.30	2.36	2.44	2.52	2.60	2.68	2.75	2.81
	4	14	LWB	50.4	51.0	51.7	52.5	53.2	54.0	54.8	55.7	56.5	57.3						
			TONS	2.12	2.21	2.29	2.38	2.46	2.55	2.63	2.72	2.80	2.89						
	6	7	LWB	50.6	51.6	52.5	53.5	54.4	55.2	56.0	56.9	57.6	58.5	59.5	60.2				
			TONS	2.08	2.13	2.19	2.25	2.32	2.39	2.47	2.55	2.64	2.71	2.78	2.87				
	6	14	LWB	48.5	49.3	50.2	51.0	51.8	52.6	53.6	54.5								
			TONS	2.34	2.42	2.48	2.57	2.64	2.72	2.81	2.89								
40	4	7	LWB	55.3	56.1	56.8	57.4	58.1	59.0	59.8	60.5	61.5	62.2	63.0	63.6	64.6	65.5	66.3	67.2
			TONS	1.47	1.54	1.61	1.70	1.78	1.84	1.92	2.00	2.06	2.17	2.25	2.33	2.41	2.49	2.58	2.65
	4	14	LWB	52.7	53.4	53.9	54.6	55.0	55.9	56.7	57.4	58.2	59.0	59.6	60.4				
			TONS	1.82	1.91	2.00	2.09	2.18	2.28	2.37	2.46	2.56	2.64	2.75	2.84				
	6	7	LWB	53.0	53.6	54.3	55.0	55.7	56.5	57.2	58.0	58.8	59.6	60.4	61.3	62.2	63.0		
			TONS	1.78	1.87	1.95	2.04	2.12	2.21	2.29	2.38	2.46	2.55	2.63	2.72	2.80	2.89		
	6	14	LWB	50.6	51.3	52.0	52.7	53.4	54.1	54.8	55.7	56.5	57.3						
			TONS	2.08	2.17	2.25	2.35	2.45	2.53	2.62	2.71	2.80	2.89						
45	4	7	LWB	57.6	58.2	58.8	59.5	60.1	60.7	61.5	62.2	63.0	63.7	64.5	65.2	66.1	66.8	67.2	68.6
			TONS	1.15	1.23	1.32	1.40	1.49	1.57	1.65	1.74	1.82	1.90	1.99	2.07	2.15	2.24	2.31	2.38
	4	14	LWB	54.8	55.5	56.1	56.8	57.5	58.1	59.0	59.6	60.2	60.9	61.5	62.2	62.8	63.8	64.6	
			TONS	1.53	1.61	1.70	1.78	1.87	1.96	2.05	2.15	2.25	2.36	2.47	2.59	2.76	2.85	2.97	
	6	7	LWB	54.8	55.5	56.1	56.8	57.5	58.2	59.0	59.7	60.5	61.3	62.2	62.8	63.5	64.4	65.2	66.0
			TONS	1.53	1.61	1.70	1.78	1.87	1.96	2.04	2.13	2.21	2.30	2.38	2.47	2.58	2.66	2.76	2.85
	6	14	LWB	52.7	53.4	54.1	54.7	55.4	56.0	56.6	57.5	58.3	58.9	59.6	60.0				
			TONS	1.81	1.90	1.99	2.08	2.17	2.27	2.38	2.45	2.55	2.65	2.76	2.89				
50	4	7	LWB	60.5	61.1	61.5	62.0	62.5	63.2	63.7	64.2	65.0	65.6	66.4	66.9	67.6	68.5	69.2	70.0
			TONS	0.72	0.81	0.92	1.02	1.12	1.20	1.32	1.41	1.50	1.59	1.67	1.78	1.87	1.95	2.06	2.12
	4	14	LWB	57.9	58.3	58.8	59.4	59.8	60.3	61.0	61.8	62.4	63.0	63.6	64.1	65.0	65.8	66.5	67.2
			TONS	1.10	1.22	1.32	1.42	1.53	1.64	1.73	1.83	1.93	2.03	2.12	2.24	2.34	2.43	2.55	2.64
	6	7	LWB	57.6	58.2	58.7	59.3	59.8	60.4	61.0	61.7	62.5	63.0	63.7	64.5	65.2	65.9	66.6	67.2
			TONS	1.13	1.23	1.33	1.43	1.53	1.62	1.72	1.83	1.91	2.01	2.12	2.21	2.30	2.40	2.51	2.64
	6	14	LWB	55.6	56.1	56.6	57.1	57.6	58.2	58.9	59.5	60.1	60.8	61.4	62.1	62.9	63.6	64.2	
			TONS	1.42	1.52	1.63	1.74	1.84	1.95	2.05	2.16	2.27	2.38	2.49	2.59	2.69	2.80	2.92	

LEAVING AIR TEMPERATURE DIFFERENCE (F) (LEAVING DRY BULB MINUS LEAVING WET BULB) (LDB-LWB)

ENTERING AIR TEMPERATURE DIFFERENCE (F) (ENTERING DRY BULB MINUS ENTERING WET BULB) (EDB-EWB)

Rows	Fins Per Inch	Refrig. Temp. (F)	600 FPM Coil Face Velocity											700 FPM Coil Face Velocity										
			10	11	12	13	14	15	16	17	18	19	20	10	11	12	13	14	15	16	17	18	19	20
4	7	35	2.1	2.4	2.7	3.0	3.3	3.6	3.9	4.2	4.5	4.7	5.1	2.6	2.9	3.2	3.5	3.8	4.1	4.4	4.8	5.1	5.4	5.7
		40	2.3	2.6	2.8	3.2	3.5	3.7	4.0	4.3	4.6	4.9	5.3	2.7	3.0	3.3	3.6	3.9	4.2	4.5	4.9	5.2	5.5	5.8
		45	2.6	2.9	3.2	3.4	3.7	4.0	4.3	4.6	4.9	5.2	5.5	2.8	3.1	3.4	3.7	4.0	4.4	4.6	5.0	5.4	5.6	6.0
		50	2.7	3.0	3.3	3.6	3.9	4.2	4.5	4.7	5.0	5.4	5.7	3.0	3.3	3.7	4.1	4.4	4.8	5.1	5.5	5.8	6.2	6.5
	14	35	1.0	1.1	1.2	1.3	1.5	1.6	1.7	1.9	2.0	2.1	2.2	1.0	1.1	1.3	1.5	1.6	1.7	1.9	2.0	2.1	2.3	2.4
		40	1.0	1.1	1.2	1.3	1.5	1.6	1.7	1.9	2.0	2.1	2.2	1.3	1.4	1.6	1.7	1.9	2.0	2.1	2.2	2.4	2.5	2.6
		45	1.1	1.2	1.3	1.5	1.6	1.7	1.8	1.9	2.0	2.1	2.2	1.8	1.9	2.0	2.2	2.3	2.4	2.6	2.7	2.9	3.0	3.2
		50	1.1	1.3	1.4	1.5	1.6	1.7	1.8	1.9	2.1	2.2	2.3	2.0	2.1	2.2	2.4	2.6	2.8	2.9	3.0	3.1	3.3	3.5
6	7	35	1.2	1.3	1.4	1.6	1.8	1.9	2.1	2.3	2.4	2.5	2.7	1.4	1.6	1.9	2.0	2.2	2.3	2.4	2.6	2.8	3.0	3.2
		40	1.2	1.3	1.5	1.7	1.8	1.9	2.1	2.4	2.5	2.6	2.8	1.7	1.9	2.1	2.2	2.4	2.6	2.8	3.0	3.1	3.3	3.5
		45	1.4	1.5	1.7	1.8	1.9	2.1	2.3	2.5	2.6	2.7	2.9	2.0	2.1	2.3	2.4	2.7	2.8	3.0	3.2	3.4	3.6	3.8
		50	1.5	1.6	1.8	1.9	2.1	2.2	2.4	2.5	2.7	2.8	3.0	2.0	2.3	2.5	2.6	2.8	3.0	3.2	3.4	3.6	3.7	3.9
	14	35	0.3	0.4	0.5	0.5	0.6	0.6	0.6	0.6	0.7	0.7	0.8	0.2	0.2	0.3	0.3	0.4	0.5	0.5	0.5	0.6	0.7	0.8
		40	0.3	0.4	0.5	0.5	0.6	0.6	0.6	0.6	0.7	0.7	0.8	0.3	0.3	0.3	0.4	0.5	0.5	0.6	0.6	0.7	0.7	0.8
		45	0.4	0.4	0.5	0.5	0.6	0.6	0.6	0.6	0.7	0.7	0.8	0.7	0.8	0.8	0.9	0.9	0.9	1.0	1.0	1.1	1.1	1.2
		50	0.4	0.4	0.5	0.5	0.6	0.6	0.6	0.6	0.7	0.7	0.8	1.2	1.2	1.3	1.3	1.4	1.4	1.5	1.5	1.6	1.7	1.9

*Tons given per square foot of coil face area.

costs of the piping system. The selection of the water temperature should not be arbitrary; however, experience has shown that a temperature approximately 5 degrees below the apparatus dewpoint is the maximum water temperature that should be used to effect an economical system design. If the resulting water quantities seem to be too high, a lower temperature can be assumed, and its influence on the refrigeration machine size, power input and piping costs should be studied. With a given coil, load and apparatus dewpoint, when the chilled water temperature is reduced, the required water quantity decreases and the temperature rise increases.

By using a coil which requires a smaller water quantity at a higher temperature rise, the following advantages may be realized:

1. a. A smaller refrigeration machine may be selected, *or*

   b. The horsepower requirement may be reduced for the same size machine by operating at an increased evaporator temperature, *or*

   c. The condenser piping or heat rejection equipment may be reduced for the same size machine by operating at a higher condensing temperature with less condenser water.

2. Lower chilled water distribution costs with savings in piping, pump and insulation may be obtained.

A limitation is also imposed on the minimum chilled water quantity by the velocity required for efficient heat transfer. A minimum Reynolds number of 3500 is suggested to insure predictable and efficient performance of a coil. The minimum chilled water flow required to maintain this Reynolds number is approximately 0.9 gpm per circuit for a ⅝ in. OD coil tube diameter. For a ½ in. OD tube diameter, the minimum flow suggested is 0.7 gpm per circuit.

Well water may be circulated thru chilled water coils if it is of sufficient quantity and at a satisfactory temperature and quality. However, well water temperatures are usually low enough to produce sensible cooling only and little or no latent heat removal. In such a case, the well water may be utilized in a precooling coil to remove some of the sensible heat. The remaining cooling load, sensible and latent, is handled by supplementary refrigeration.

Manufacturer's recommendations regarding maximum and minimum direct expansion coil loadings should be followed. Selections at loadings below the minimum may result in unsatisfactory oil return, poor refrigerant distribution and coil frosting.

Atmospheric Corrections

Cooling coil ratings are based upon the standard atmospheric conditions of 29.92 in. Hg barometric pressure. For atmospheric pressures significantly different, such as at altitudes exceeding 2500 feet, a correction should be applied to the air quantity before making the coil selection.

Assuming that the necessary corrections have been made to the load calculation and the sensible heat factor as described in *Part 1,* the following procedure should be applied to the unit selection:

1. Obtain the density ratio from *Chapter 1, Chart 2.*

2. Multiply the calculated dehumidified air quantity by the density ratio to determine the equivalent air flow at sea level.

3. Use this adjusted air quantity, together with the calculated cooling load and sea level refrigerant side coil ratings, to determine the coil water flow and pressure drop or refrigerant temperature.

The calculated dehumidified air quantity is used with no correction to determine unit size and coil face velocity. However, the coil air side pressure drop must be corrected, as described in *Part 2.*

Fan performance is analyzed in *Chapter 1* of this part and motor selection is influenced as outlined in *Part 8.*

ACCESSORIES

Heating Coils

Unit heating coils for fan-coil equipment are available in a variety of depth and fin spacing combinations and in both the nonfreeze steam and U-bend types. The latter type may be used with hot water or steam and may be obtained in different combinations of tube face and fin spacing to produce different rises with the same entering air temperature, face velocity, and steam pressure or hot water temperature. Heating coils are also usually capable of being mounted before or after the cooling coil.

During the intermediate season, any zone served by a multi-zone unit should be able to obtain heating or cooling on demand. Since there is no common supply air duct in which air mixing can occur on a multi-zone installation, the problem of air temperature stratification is of considerable importance. Stratification across the unit heating coil may cause some zones to be denied heat when it is needed.

The throttling of a steam control valve may produce stratification if the steam condenses fully before reaching the end of the tube or circuit. Therefore, it is suggested that full steam pressure be applied to a multi-zone unit heating coil whenever heating may be required by any zone.

In order to provide an air path of approximately equal pressure drop thru either of the two widely differing heat transfer surfaces in a multi-zone fan-coil unit, perforated balancing plates are often used. It may be necessary for the engineer to select such a device, particularly if no heating coil is required.

The application and selection of heating coils is described in detail in *Chapter 4* of this part.

Humidifiers

On a fan-coil unit not equipped with recirculated water sprays, humidification may be obtained by means of a city water spray humidifier, a steam pan humidifier, a steam grid humidifier or a humidifying pack. Spray coil and steam grid equipment provides the most effective humidity control.

A city water spray humidifier consists of a header, spray nozzles and strainer. Either atomizing or nonatomizing sprays are available. The latter type requires a lower water pressure. In either case, the spray density or amount of water circulated per square foot of cooling coil face area is considerably less than that of a recirculated spray coil. Therefore, although lower in first cost, the city water spray humidifier is less efficient than a recirculated spray coil. An eliminator is not usually required for a city water spray.

The use of copper fins on copper tubes with spray humidifiers is suggested when city water has a specific electrical conductance of 500 or more micromhos at 77 F. (See *Part 5* for values of water conductance in various locations.) Aluminum fins may be used if city water is of the proper quality. The use of copper fins should be considered where industrial gases such as hydrogen sulphide, sulphur dioxide or carbon dioxide are present and where salty atmospheres prevail.

If air flow is not maintained thru a spray section when it is operating, wetting of the unit and leakage may result. Therefore, a solenoid valve should be installed or other suitable precautions taken to stop the sprays when the unit fan is not running. To maintain a minimum coil air flow when utilizing face and bypass dampers, a minimum closure device should be provided on the face dampers.

The spraying of heating coils may result in scaling on the coil and the production of odors. This practice should therefore be avoided.

The use of spray humidifiers with a multi-zone unit should be avoided. Since the coil is subjected to a positive static pressure, spray water may leak from the unit cabinet. If sprays are used, they should be of the atomizing type. A grid or pan humidifier is preferred for this usage.

Grid humidifiers are lengths of perforated steam piping wrapped with wicking such as asbestos. The pipe is mounted in an open pan, pitched to facilitate condensate drainage. The condensate drain line from the unit should be trapped to provide a water seal, as described in *Part 3*. Steam pressures should not exceed 5 psig for this application, and the steam used should be free of odors.

The mixing of steam with conditioned air normally produces a negligible increase in the air dry-bulb temperature. This type of humidification, therefore, approximates a vertical line on a psychrometric chart. *Figure 32* illustrates the process. When designing a system using a grid humidifier, the temperature of the air entering the humidifier should be high enough to permit a moisture content at saturation (point *C*) equal to or greater than the desired air moisture content.

Pan humidifiers include a pan to hold water, a steam coil to evaporate the water, and a float valve for water make-up. For this application, a steam pressure of 20 psig is suggested for maximum humidifying efficiency.

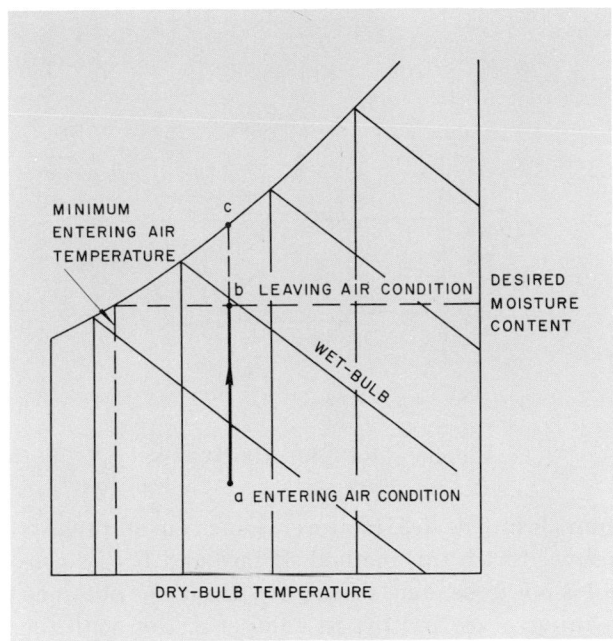

Fig. 32 — Humidification With Steam

Humidifying packs use a fill (often of glass fibers) as an evaporative surface. The pack is located in the air stream and water is sprayed over the fill.

Spray Water Heaters

Spray coil equipment may be provided with a spray water heater to permit simultaneous cooling or heating and humidification. Such flexibility is required during winter operation or where the volume of outdoor air is large in relation to the total air quantity. Typical applications include certain industrial processes or hospital operating rooms. These processes are described in *Part 1*.

Face and Bypass Dampers

On applications employing face and bypass control of coil equipment, the fan selection and air distribution system should be based on an air quantity 10% above the design dehumidified air volume. This additional air quantity compensates for leakage thru a fully closed bypass damper and for air quantity variations occurring when face and bypass dampers are in an intermediate position. With the bypass dampers fully open, system static pressure may be reduced and air quantity and fan brake horsepower increased. Therefore, on face and bypass applications especially, fan motors should be selected so that nominal horsepower ratings are not exceeded.

Where a fixed bypassed air quantity is required, the bypass damper may be provided with a mini-

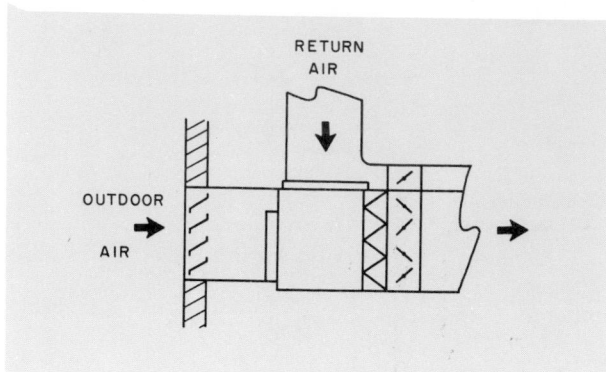

FIG. 33 — RETURN AIR BYPASS

mum closure device. However, some control range is sacrificed with this method. If face and bypass control is not to be used, a fixed bypass may be obtained by using a face and bypass damper section with the face dampers removed.

The bypass of outside and return air mixtures introduces high humidity air directly to the conditioned space. When employing face and bypass control, it is preferable to bypass return air only. This may be accomplished as shown in *Fig. 33.*

Vibration Isolation

Four types of isolators are normally used to absorb the vibrations produced by fan-coil equipment as well as other types of rotating or reciprocating machinery. In order of decreasing effectiveness and first cost, they are:

1. Steel coil springs
2. Double rubber-in-shear
3. Single rubber-in-shear
4. Cork

Steel spring or rubber-in-shear isolators are available for floor-mounted or suspended equipment. Ribbed neoprene pads may be bonded to any of the isolators noted above for floor-mounted units. These pads resist horizontal movement, compensate for slight irregularities in the floor surface, and protect floors from marring.

The proper bearing surface should be provided for cork pad isolators as recommended by the isolator manufacturer. Underloading does not permit the full resiliency to be utilized, while overloading may result in permanent deformation of the cork structure.

Similarly, if spring or rubber isolators are loaded past the point of full compression, binding occurs and there is no isolation.

Vibration isolation efficiency is the percentage of a vibration of a given frequency absorbed by the isolator. Thus, the vibration transmitted beyond the isolator is the difference between 100% and the isolation efficiency.

Isolation efficiency is a function of the isolator deflection when loaded and the disturbing frequency of the machine isolated. For a fan or fan-coil unit the disturbing frequency is the fan speed. *Chart 7* shows the relation between static deflection, disturbing frequency and isolation efficiency for any case of vibration. In addition, *Chart 7* indicates the ranges of deflection for which the various types of isolators are normally obtainable.

As illustrated by *Chart 7,* for a given disturbing frequency, isolator efficiency increases with deflection. Since greater deflections are obtainable with springs than with other types of isolators, springs provide the most effective isolation over all frequencies. Cork is not an effective isolation material for frequencies below about 3000 rpm.

A minimum vibration isolation efficiency of 85% is usually satisfactory for ground floor or basement applications in noncritical buildings. Upper floors may require as high an efficiency as 93%, while critical upper floors usually allow no less than 95%. When several pieces of vibrating equipment are concentrated in one room in a critical upper floor installation, the required efficiency may approach 98.5%, and the transmission of vibration to the floor should be considered in the building design.

Whether floor-mounted or suspended, unit may be mounted on a steel channel base which is then isolated. Units may also be mounted on vibration isolators directly, with no intermediate base. Manufacturers provide support points or hanging brackets for fan-coil equipment, and their recommendations regarding support points and isolator loadings should be followed. Often, larger units of a series or those including the most components require a channel frame base for mounting. Mixing boxes and low velocity filter boxes are usually mounted on their own isolators, so as to prevent a cantilever effect.

If a unit is to be directly isolated with no base employed, the deflection at each support point should be the same. If equal loading is assumed, individual isolators may be overloaded to the point of binding or underloaded, resulting in decreased isolation efficiency. Point loadings for a given unit, coil and components are usually available from the unit manfacturer.

Operating weights should be used in selecting vibration isolators. This is particularly important where water coils are used.

Example 4 — Vibration Isolator Selection

Given:
Fan-coil unit operating weight — 1640 lb equally distributed at four points.
Fan speed — 800 rpm
Design isolation efficiency — 90%

Find:
Required type and characteristics of isolator.

Solution:
1. From *Chart 7* read the required isolator deflection of 0.6 in., within the spring application range.
2. Determine the individual isolator loading; $\frac{1640}{4} = 410$ lb.

3. Select a spring isolator with a maximum characteristic of 410/0.6 or 683 lb per inch. If the characteristic of the spring chosen is less, the deflection is greater than 0.6 in., and the isolation efficiency greater than 90%. However, the spring must not be loaded above its maximum.

Filters

Factory-fabricated filter sections for both high velocity or low velocity filters are normally obtainable from the manufacturer of a fan-coil unit. Either throw-away or cleanable filters can be used. For built-up apparatus field-assembled filter frames are available.

If high velocity filters are to be used in a low velocity filter section, the full area of air flow is not required. Rather than fill up the entire section with

CHART 7—VIBRATION ISOLATOR DEFLECTION

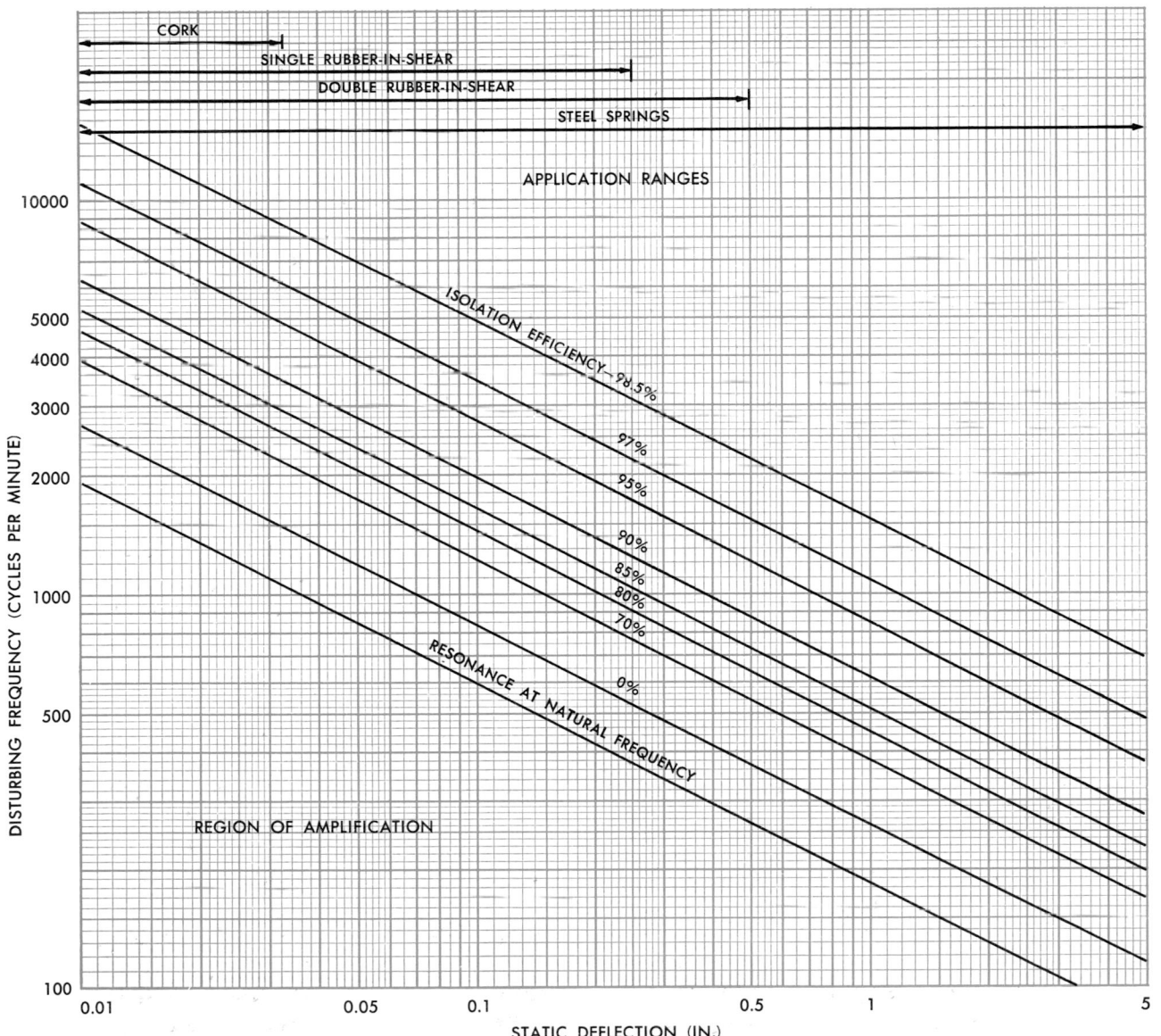

high velocity filters operating at a low velocity, blank-off pieces may be installed, thus lowering the effective area. Blank-offs should be located uniformly across the face of the filter section instead of concentrated in one place.

Filters are discussed in detail in *Chapter 4* of this part.

INSTALLATION

Location

The economic and sound level considerations pertaining to the location of air handling apparatus, as discussed in *Part 2,* are applicable to fan-coil equipment.

Two of the most important factors in the location of air conditioning equipment are the availability of outdoor air and the ease of air return. Outdoor air may be brought to a unit thru a wall, roof or central building chase. It is preferable to locate outdoor air intakes so that they do not face walls of spaces where noise would be objectionable. Air may be returned thru a duct system or directly to the equipment room.

Layout

A fan-coil unit may be of the vertical or horizontal type, depending upon the direction of air flow entering the fan cabinet. It may be floor-mounted or, in the case of a horizontal unit, suspended from above. The choice of unit style and mounting usually depends upon space requirements and optimum duct layout. A support base may be employed, if necessary, as discussed under *Vibration Isolation.*

A practical location recognizes the need for effective servicing of the unit. A minimum of 30 inches is suggested to provide access between the unit and the nearest wall. This facilitates servicing of steam traps, fan bearings, damper motors and fan motor. In addition, service space about the unit must be provided for filter removal, coil removal, fan shaft removal, and the cleaning of cleanable coils.

Suspended units should be accessible from above, if possible. If frequent access is required and space permits, a catwalk may be required.

An access plenum and door should be provided between the filter section and coil section of a spray fan-coil unit. This access permits periodic inspection and cleaning of the sprays and drain pan.

A level unit is necessary to insure proper drainage from coils and drain pan. Manufacturers of spring vibration isolators usually provide leveling devices

in the isolator to compensate for deflection differences.

Units located outdoors require suitable motors and the protection of fan drive and shaft bearings, as well as insulation as noted below.

For information pertaining to the design of air distribution system components and piping at the unit, refer to *Parts 2 and 3.*

Insulation

In a fan-coil unit the casing housing the fan section, cooling coil section and components downstream of the cooling coil are usually internally insulated. This insulation is adequate for normal interior applications. The outdoor air intake duct should be insulated and vapor sealed to prevent condensation on the duct during cold weather. If the intake is kept as short as possible, insulation costs are minimized. Insulation and vapor sealing of the mixing box may be required, depending on the quantity of outdoor air introduced and on the winter design temperature. Intakes for units circulating 100% outdoor air should be insulated up to the preheater.

Units located outdoors should be completely covered and caulked with weatherproofing material. If the outdoor air temperature can fall below the dewpoint of the air within the unit, the unit should be externally insulated, vapor sealed and weatherproofed to prevent interior condensation and to minimize heat losses. The insulation on the top surfaces of the unit should be slightly crowned so that water can run off.

CONTROL

If an air conditioning apparatus is to perform satisfactorily under a partial load in the conditioned area, a means of effecting a capacity reduction in proportion to the instantaneous load is required. The three methods most commonly employed for capacity control are air bypass control, chilled water control and air volume control.

With a drop in room load the sensible heat ratio usually decreases since the room latent load remains constant. This condition commonly occurs in areas where a large proportion of sensible load such as solar heat gain may be decreased with no influence on latent load, as from people or infiltration.

In order to maintain design conditions at partial loads and with decreased sensible heat ratios, the effective coil surface temperature for a given coil must be lower than that surface temperature consistent with full load design conditions. This requirement is illustrated in *Fig. 34*. The relation between effective coil surface temperature and percent of design room sensible heat depends on the volume of outdoor air conditioned by the coil.

However, decreasing the flow of chilled water thru the coil as a means of capacity control causes the effective coil surface temperature to rise as the load decreases. Therefore, room humidity also rises. For this reason it is preferable to maintain the design flow of chilled water thru the coil at all times.

Figure 35a shows a typical cooling coil process at full load for a given set of entering air and water conditions. *Figures 35b, 35c and 35d* depict at half load the three methods of control cited and the influence in each case on effective coil surface temperature. Air volume control is similar in effect to air

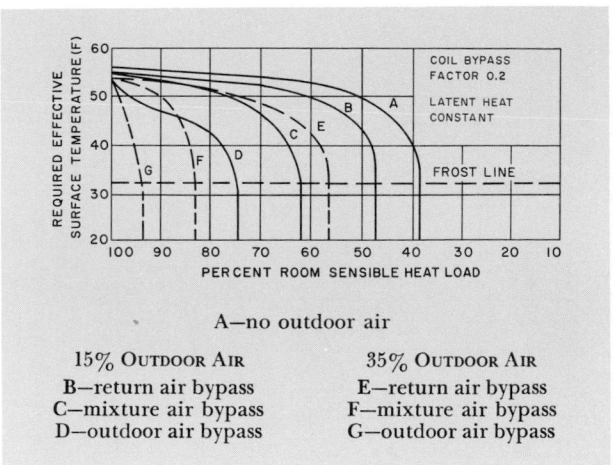

A—no outdoor air

15% Outdoor Air	35% Outdoor Air
B—return air bypass	E—return air bypass
C—mixture air bypass	F—mixture air bypass
D—outdoor air bypass	G—outdoor air bypass

FIG. 34 — REQUIRED COIL PERFORMANCE

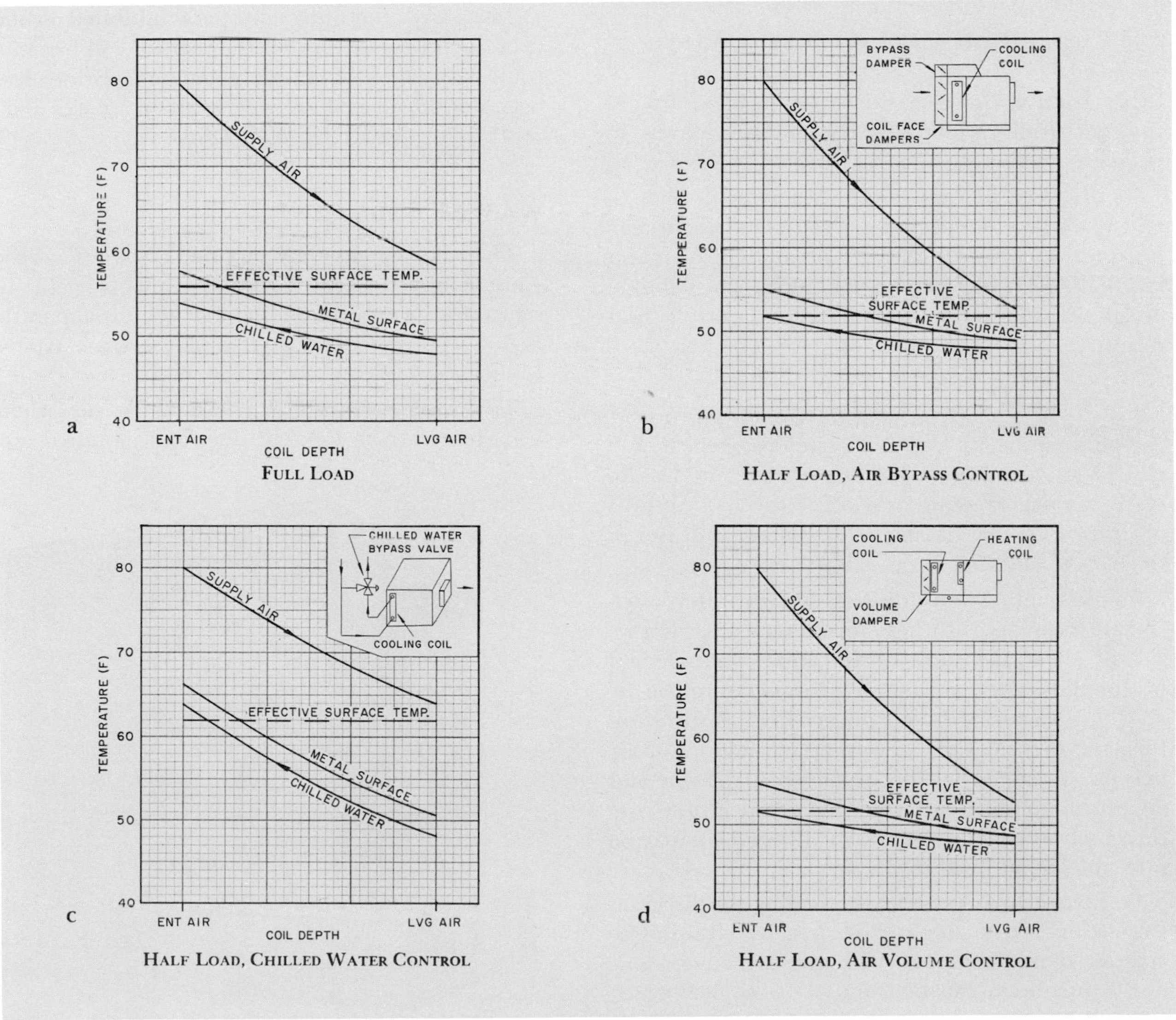

a

FULL LOAD

b

HALF LOAD, AIR BYPASS CONTROL

c

HALF LOAD, CHILLED WATER CONTROL

d

HALF LOAD, AIR VOLUME CONTROL

FIG. 35 — TYPICAL COOLING COIL PROCESSES

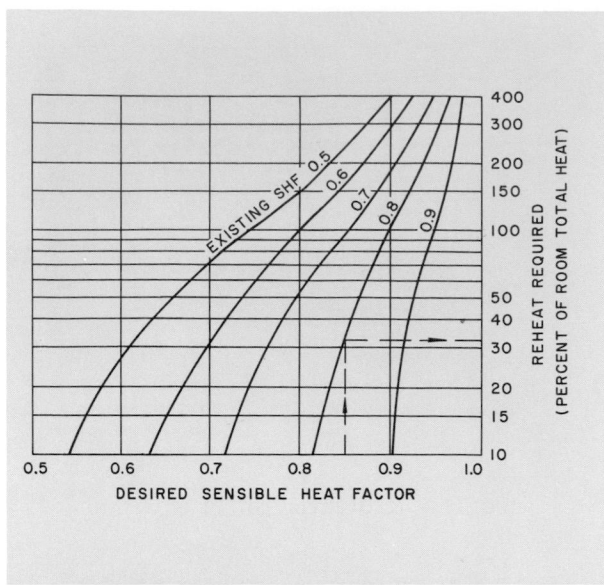

FIG. 36 — REHEAT CONTROL REQUIREMENTS

bypass control. However, the bypassing of air around the coil permits a relatively constant air delivery to be maintained.

As discussed in Part 1, applications with high latent loads may require reheat control of room temperature. *Figure 36* indicates the amount of reheat required, relative to room total heat, to maintain design relative humidity at various sensible heat ratios.

COIL FREEZE-UP PROTECTION

The freezing of water in preheat, reheat and chilled water coils may damage the coils and lead to costly repairs. Freezing may occur not only in coils of units operating during cold weather but also in the coils of units not in operation.

Outdoor air at subfreezing temperatures often comes in contact with heat transfer surfaces as a result of air temperature stratification. Stratification is caused usually by incomplete mixing of return air and outdoor air or by an uneven temperature rise thru the preheat coil. The complete mixing of air may be promoted by the proper arrangement and design of the ductwork. Uneven temperature rises thru preheat coils and heating coil freeze-up may be prevented as outlined in *Chapter 4* of this part.

Coil freeze-up may also be caused by the direct introduction of cold air thru an unprotected coil. Circulation of outdoor air thru an interior fan-coil unit not in operation can be induced by a stack effect, particularly if the unit is on one of the lower floors of a tall building.

In addition to design precautions against stratification, the following methods may be employed to protect a water coil:

1. Remove the water from the coil during the winter.
2. Run the chilled water pump.
3. Decrease the freezing point of the coil water.

Removal of the water from the coil should be accompanied by blowing out the coil with a portable blower to remove residual water. An alternate method of freeze protection is to circulate an inhibited antifreeze solution thru the coil before final drainage.

Operating the chilled water pump during the winter is a costly solution to the problem of freezing. In addition, it is not a certain method since a plugged tube could still freeze.

The practice of using a properly inhibited alcohol brine or antifreeze thruout the year for coil freeze-up protection is becoming more common. Brines have been developed and are now available, particularly for this purpose. Refer to *Part 4*.

WASHER EQUIPMENT

The most commonly applied type of washer equipment is the central station washer (*Fig. 37*), designed for incorporation into a field-built apparatus. *Figure 38* is a cutaway view of the same type of washer and indicates the direction of air flow.

This washer consists of a rectangular steel chamber, closed at the top and sides and mounted on a

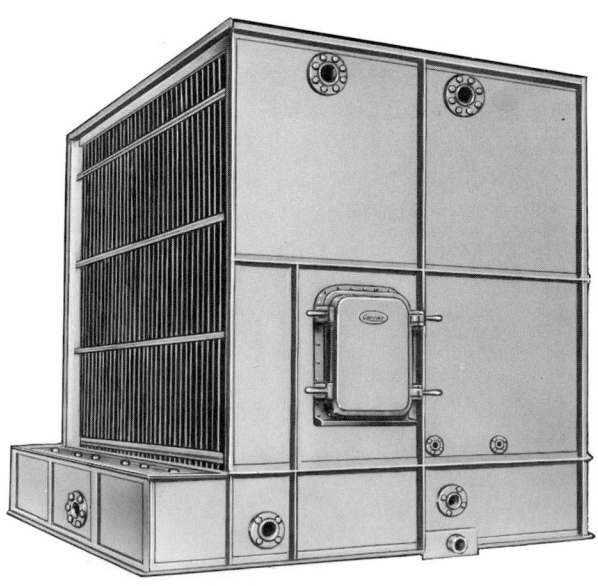

FIG. 37 — CENTRAL STATION WASHER

shallow watertight tank of steel or concrete. Inlet baffles located at the air-entering end of the washer promote uniform air velocities thru the washer and minimize the spraying back of water into the entrance chamber as a result of air eddy currents. At the air-leaving end of the washer, eliminators are provided to remove entrained water droplets.

Within the washer spray chamber two banks of opposing spray nozzles provide finely divided droplets of water uniformly distributed. After contacting the air, the water is collected in the tank and is returned to the sprays by a recirculating pump.

A central station washer may be designed for use as a humidifier or as a dehumidifier. The physical arrangement is the same in either case. A dehumidifier is normally shorter in airway length than a humidifier.

Washers may also be obtained in a unitary design. A unitary spray washer, comparable in design and function to a central station washer, is shown in

Fig. 39. Other types of unitary washers involve the wetting of a fibrous fill or set of pads located in the air stream.

The particular washer shown in *Fig. 39* operates at high spray chamber air velocities and is, therefore, smaller than a central station washer for a given air volume. *Figure 40* indicates the path of the air thru the unit components. The unit includes an inlet air mixing plenum, a vaneaxial fan, a diffuser section, a spray section and a rotating eliminator.

Two to six banks of sprays condition the air and clean it of dirt and other airborne particles. After contact with the air, the water drains from the spray section to a central tank from which it is recirculated.

APPLICATION

Air washers are primarily employed in industrial air conditioning applications. The use of sprays permits humidification, dehumidification or evaporative cooling, as required. In addition, sprays enable

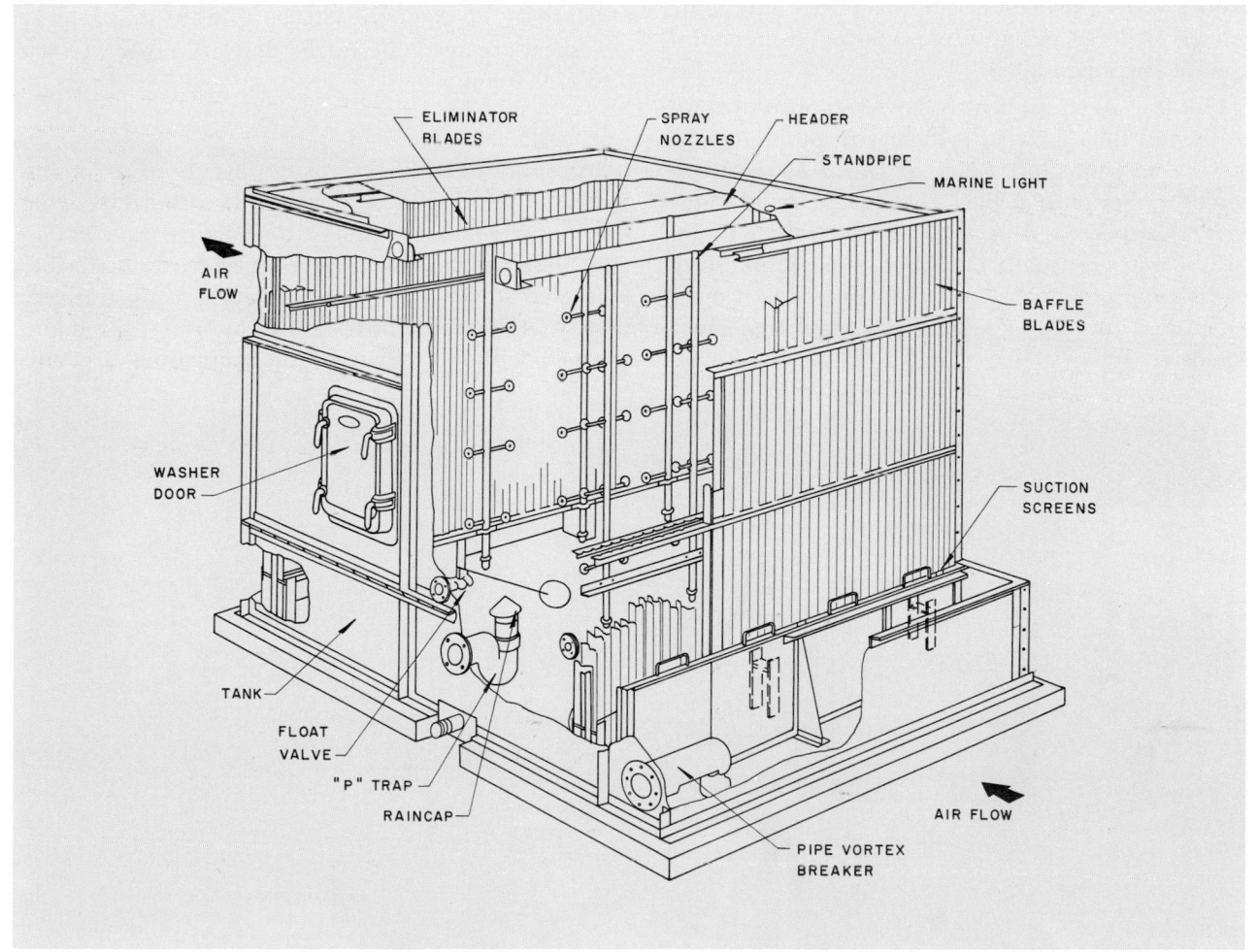

FIG. 38 — CENTRAL STATION WASHER (SECTIONAL VIEW)

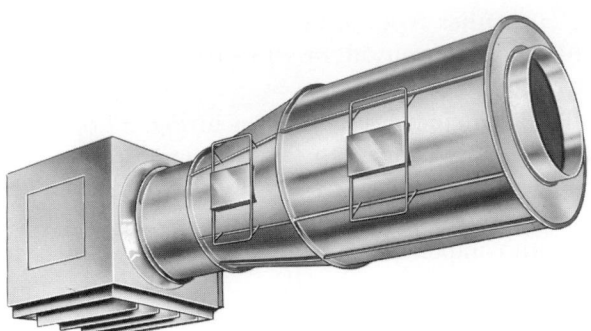

FIG. 39 – HIGH VELOCITY WASHER

a degree of humidity control not possible with coils alone.

Washer equipment is effective in the removal of certain types of odors and dirt from the air. In applications where coils could become clogged with airborne solid particles, washers require a minimum of maintenance.

This flexibility of function is obtained at a relatively low installed cost of equipment per unit of air delivery. A large air capacity is realized from equipment of low weight.

This type of equipment is, however, open hydraulically and thus presents problems in piping design and system balancing. Refer to *Part 3* for a discussion of washer piping. Since the flow of air and water thru the apparatus is parallel and since a gravity return of water is usually employed in a dehumidifier application, pipe sizes tend to be larger in an open system and the piping system and insulation more expensive.

The spraying of water at high pressures such as are required in washer equipment produces a noise level high enough to be objectionable under some circumstances. Sound treatment is not usually required on equipment serving manufacturing areas or areas of high ambient sound levels. For more critical applications the need for sound absorption should be investigated.

The unitary spray washer shown in *Fig. 39* requires considerably less space than a central station type and requires no special apparatus room. It is more flexible in meeting the necessities of plant layout change and is more adaptable to zoning. The salvage value is high and the operating weight low.

The central station washer installation results in lower fan noise levels and lower fan operating costs. Since central station washers are usually fewer in number and more centrally located than unitary washers, they require less piping when used as dehumidifiers for a given installation.

Central station washers may be obtained for air deliveries of 2000 to 336,000 cfm. Unitary spray washers are available in the delivery range of 7800 to 47,000 cfm.

Humidifier

A spray humidifier provides evaporative cooling thruout the year, as required, and heating during the winter season, if necessary. It is particularly suitable to applications where large quantities of sensible heat are to be removed, and where comparatively high relative humidities are to be uniformly maintained without the need for controlling dry-bulb

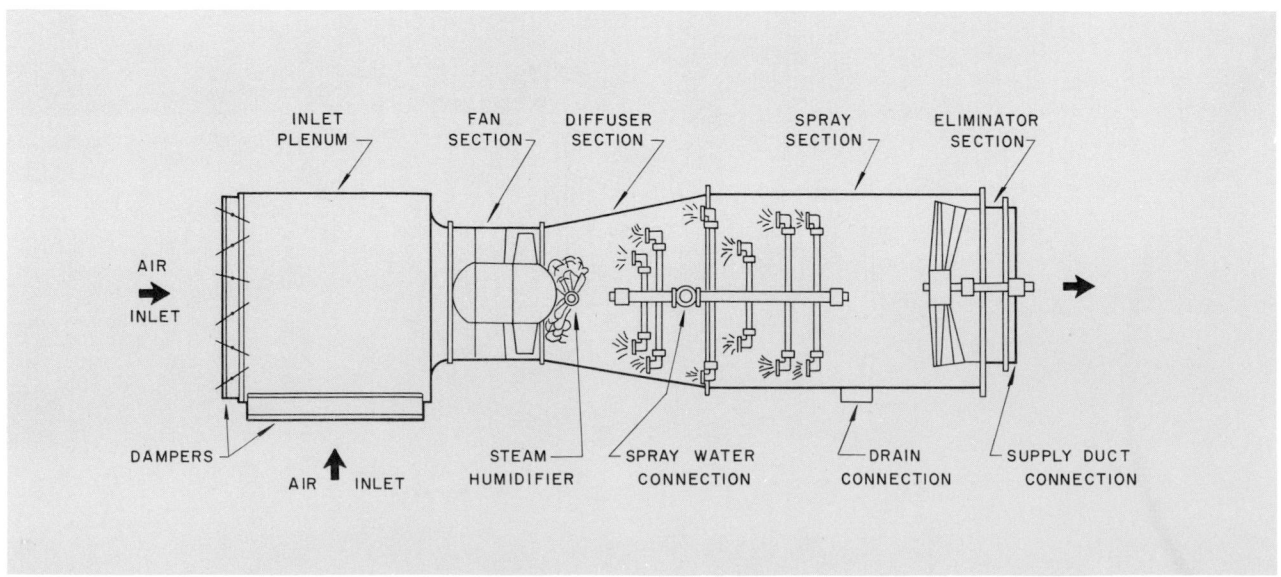

FIG. 40 – HIGH VELOCITY WASHER (SECTIONAL VIEW)

temperature above a prescribed minimum. This type of washer equipment has been used extensively in the conditioning of industrial facilities engaged in the manufacture or processing of hygroscopic materials. Such industries include textiles, paper manufacturing, printing and tobacco processing.

A system of supplementary room atomizers is often used in conjunction with a spray humidifier in order to lower the first cost of the system. The psychrometrics of a combination system are outlined in *Part 1*.

Spray humidifiers require the recirculation of water with no refrigeration. Recirculation occurs at the apparatus in the case of the central station washer. With the unitary washer, the recirculation of the water is produced centrally.

Dehumidifier

A spray dehumidifier provides sensible cooling and dehumidification during the summer season, evaporative cooling during the rest of the year, and heating, if necessary, during the winter. It is used where lower relative humidities are to be uniformly maintained and where dry-bulb temperatures are to be controlled at a comfortable level. A source of chilled water is required for this application.

In a multiple central station system installation, the recirculated water quantity remains constant for each washer, and the chilled water is introduced in varying quantities at the suction of the recirculating pump during the dehumidifying season. See *Part 3*. The excess water returning to the washer tank is either pumped back to a central collection tank or, more commonly, drained from the washer to the central tank by gravity. If a gravity return is employed, a weir is required in the washer tank to maintain the water level in the tank. Rate of return in a pump-back application may be varied by a control valve actuated by the washer tank water level. A return pump should be sized to provide from 10% to 20% more capacity than is required. In either case, the amount of chilled water admitted to the apparatus should be limited to a maximum of 90% of the recirculated water quantity.

Various central station tank arrangements are shown in *Fig. 41*. *Figures 41a and 41b* apply to gravity return dehumidifiers. *Figures 41c and 41d* are typical of pumped return dehumidifiers or evaporative cooling applications.

Although unitary spray washers may be arranged in the same manner as central station washers, they are usually supplied directly with chilled water with no recirculation at the unit. Spray density, therefore, varies with load. Water return is by gravity to a central collection tank.

During the months that refrigeration is not required, the chilled water pump serving the central station spray dehumidifier is idle.

Although efficient heat transfer is promoted by the direct contact of air and spray water in a washer, the parallel flow of air and water is less conducive to heat transfer than the counterflow process possible

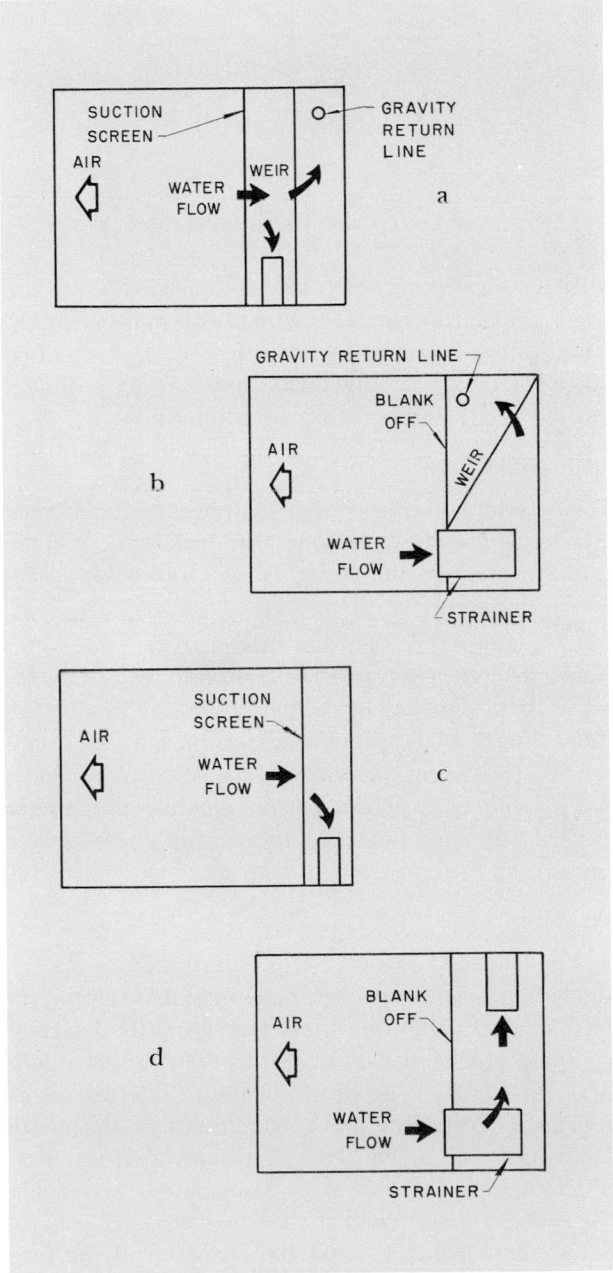

FIG. 41 — WASHER TANK ARRANGEMENTS
(PLAN VIEW)

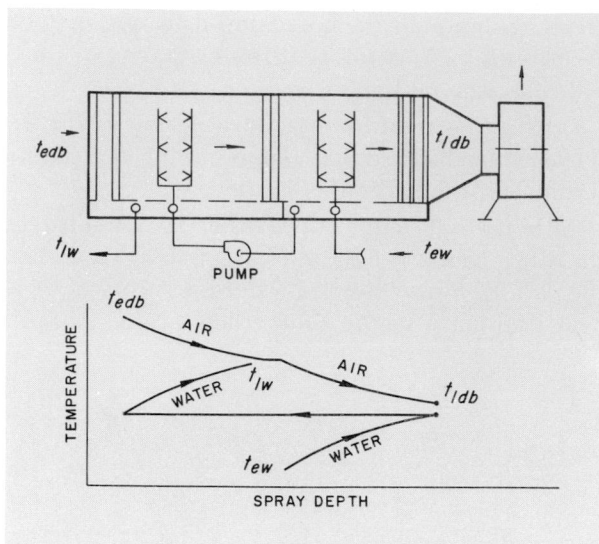

FIG. 42 — TWO-STAGE COUNTERFLOW WASHER

with a coil. *Figure 42* illustrates a method of obtaining a counterflow process with a two-stage spray dehumidifier. Flow is parallel thru each individual stage. Such an arrangement may permit a higher chilled water temperature or a smaller water flow.

UNIT SELECTION

The selection of a washer includes the determination of optimum washer size and dimensions and the establishment of the recirculated spray water quantity and pressure.

In the case of a dehumidifier, a study of the economic effects of a washer selection on other components such as piping or refrigeration equipment may be required. Increasing the recirculated water quantity or decreasing the washer face velocity by selecting a larger washer can permit operation at higher chilled water temperatures or at lower chilled water quantities.

Unit Size

The face area of a washer is determined by the design air quantity and the recommended maximum face velocity. Dehumidifiers are normally designed to operate at velocities of 300 to 650 fpm. Humidifiers are usually selected in the 300 to 750 velocity range. Velocities above or below these limits are not conducive to efficient eliminator performance. Therefore, if a washer must be oversized to provide for future capacity and if the resulting face velocity is less than 300 fpm, a partial blank-off of the face area is suggested to increase the velocity until that time when full capacity is required. Similarly, if volume control is used to maintain space conditions,

the air velocity should not be allowed to drop below 300 fpm.

For maximum economy and flexibility of control, it is suggested that washers be selected at a face velocity as near as possible to the recommended maximum.

With an approximate face area determined, several washers of various heights and widths may be selected. First cost of the washer is usually minimized if it is selected as square as possible, with the height approximately equal to the width. However, at washer heights above a specified maximum the manufacturer may stack eliminators, thus in effect creating two washers. It is preferable economically to select the washer with a height below this maximum, even if the washer width then exceeds the height.

Washer saturation efficiency or contact factor decreases as face velocity increases. Thus, for a required air temperature rise, slightly more air is required at higher washer face velocities. However, the effect is not economically significant enough to justify lower washer face velocities.

The unitary spray washer *(Fig. 39)* operates at velocities up to 2600 fpm with efficient elimination of entrained moisture. This type of washer is rated to handle a nominal air quantity, and selections are made in the range of 75% to 105% of nominal.

Spray Water

Washer saturation efficiency and contact factor are determined by various spray characteristics in addition to face velocity. These characteristics include the number of spray banks and the spray water pressure. At a given spray pressure, spray water quantity may be varied over a relatively wide range with little change in contact factor or saturation efficiency. This can be accomplished with different combinations of spray nozzle orifice size and number of nozzles.

Spray pressures usually lie in the range of 20 to 40 psig, with the higher pressures producing higher saturation efficiencies. Dehumidifiers normally require lower spray pressures than humidifiers.

At a given recirculated water quantity the fewer the number of spray banks, the greater the saturation efficiency since the spray pressure is greater. However, in the design and rating of central station washers, the number of spray banks available is usually standardized and limited.

Optimum dehumidifier efficiency is usually obtained at a spray water quantity of approximately 5 gpm per square foot and a pressure of 25 psig. The spray density may vary from 3 to 11 gpm per square foot without an appreciable effect on the performance, providing the 25 psig nozzle pressure is maintained.

Humidifier spray densities vary from 2.25 to 3.0 gpm per square foot depending on the number and size of nozzles used.

Evaporative cooling applications require only a knowledge of washer size and saturation efficiency to complete the selection. However, for a dehumidifier selection the relation between leaving air wet-bulb temperature and recirculated water temperature after air contact should be known. This information is necessary in order to calculate the quantity of chilled water required at a given temperature. *Chart 8* illustrates such a rating.

The unitary spray washer may be selected at various water quantities. A greater selection of spray banks is therefore available so that a range of contact factors may be obtained. Dehumidifier ratings are based on the apparatus dewpoint concept, as may be fan-coil unit ratings. A typical dehumidifying performance for a given unit size is shown in *Chart 9*.

Recirculating water pump heads for central station washers usually range from 50 to 85 ft wg, provided the pump is close to the washer. The pump head is primarily determined by spray nozzle pressure.

Fouling factors used for selection of refrigeration equipment used with washer equipment should be a minimum of .001. See *Part 5*.

CHART 8—SPRAY DEHUMIDIFIER RATINGS (CENTRAL STATION)

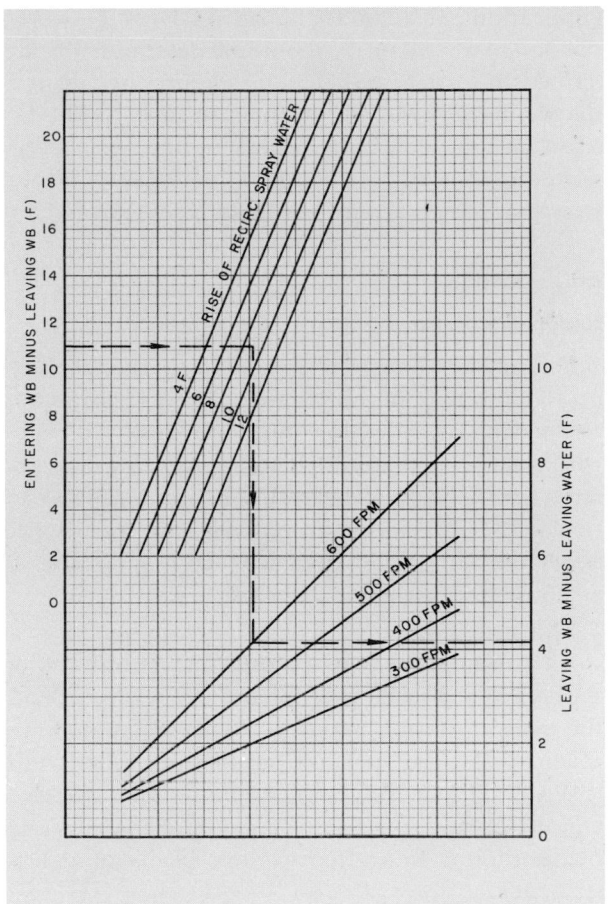

CHART 9—SPRAY DEHUMIDIFIER RATINGS (UNITARY TYPE)

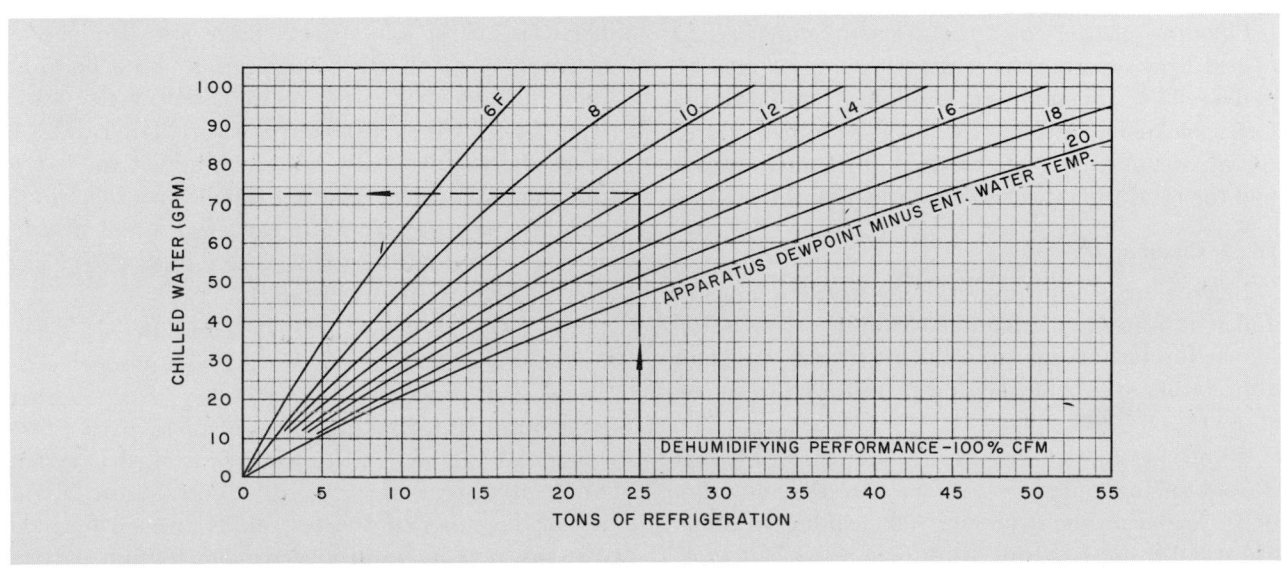

Atmospheric Corrections

No correction to washer ratings is necessary for applications at altitudes above sea level. However, the design air quantity should be determined as described in *Part 1,* and the air side pressure drop of the washer adjusted as outlined in *Part 2.* The fan selection should be in accordance with the suggested procedure found in *Chapter 1* of this part. Motor selection at high altitudes is described in *Part 8.*

ACCESSORIES

Flooding Nozzles

For applications where solid airborne particles may accumulate on eliminator blades, flooding nozzles may be provided to continually flush the blades with recirculated water. Flooding nozzles may also serve the baffles at the entering face of a central station washer. Baffle sprays, however, are usually necessary only in applications with large quantities of airborne lint, such as textile mills.

Eliminator flooding nozzles usually operate at spray pressures of 3 to 10 psig for central station washers and 5 to 20 psig for unitary washers. With the central station type of washer a spray water quantity of 4 gpm per row per foot of washer width is suggested. One row is generally required for each eliminator section. The flooding of eliminator blades should be limited to those blades of at least six bends.

Baffle spray nozzles may be designed for spray pressures of 5 to 15 psig, and should be spaced to provide effective baffle coverage at a spray water quantity of 3 to 6 gpm per foot of washer width per pipe header. Headers should be spaced 2 to 3 feet apart at the entering face.

Flooding nozzle water requirements may be furnished by a separate recirculating pump or may be delivered by the main recirculating pump. If the latter means is chosen, the flooding nozzle water quantity should be added to that of the main sprays and the total then used to select the pump.

Water Cleaning Devices

In order to insure proper spray nozzle operation and a minimum of manual cleaning and maintenance, foreign matter from the air stream and from eliminators and baffles should be removed from the spray water.

Two types of cleaning devices are commonly employed for this purpose: stationary screens and automatic self-cleaning strainers. Self-cleaning strainers are usually the rotating drum or endless belt type.

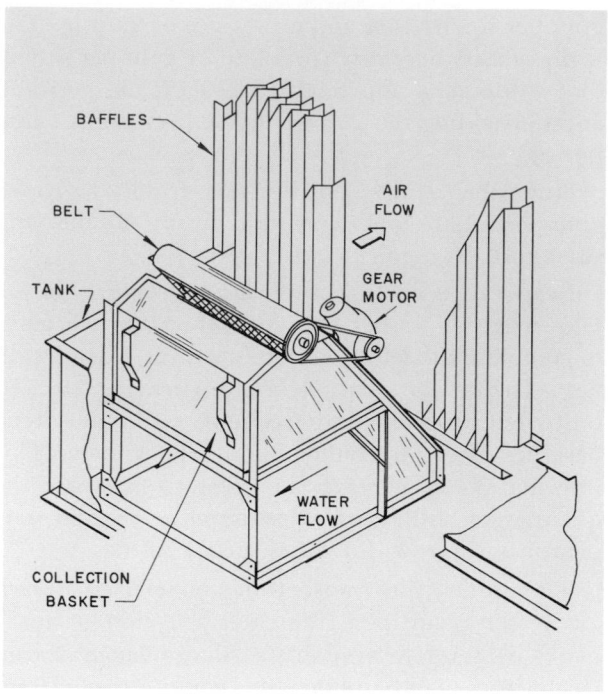

FIG. 43 — BELT TYPE WATER STRAINER

Stationary screens are located in the washer tank so that spray water must pass thru them before being recirculated. Cleaning the screens is a manual operation and can be facilitated by using two screens in series, supported by independent screen guides. The screen openings should be smaller than the spray nozzle orifice size. The washer tanks shown in *Fig. 41a and 41c* should be equipped with stationary screens.

An endless belt self-cleaning strainer may be used with or in place of the stationary screens and is suitable mainly for applications where foreign matter particles are of a relatively large size. It operates continuously, collecting the particles on a belt and then flushing them with recirculated or city water from the belt into a basket. If recirculated water is used, the requirement should be added to that of the main spray and flooding nozzles in order to determine the required pump capacity. A belt strainer can be located within the washer tank *(Fig. 41b and 41d).* A belt strainer is shown in *Fig. 43.*

The rotating drum strainer is installed in a central storage and collection tank. It is a more efficient cleaning device than the stationary screen or belt type strainers. For this reason it is particularly suited for use with a unitary spray washer system where all water is returned to a central location and where the tubes of a water cooler and various control valves must be protected from foreign particle

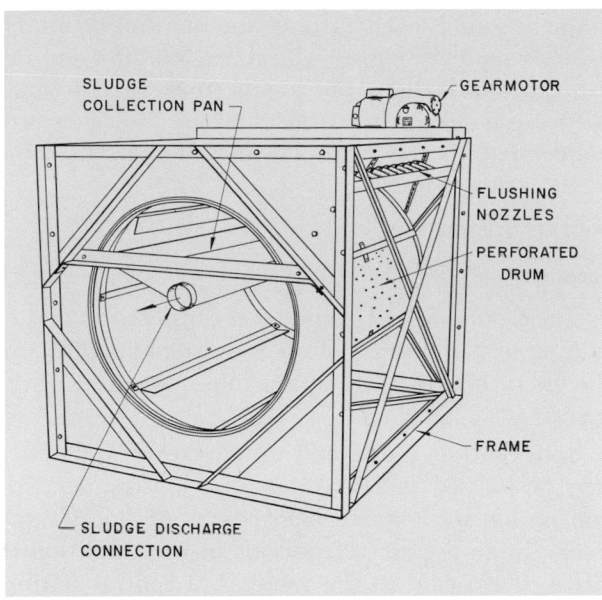

FIG. 44 — ROTATING DRUM WATER STRAINER
(WATER ENTERING SIDE)

accumulations. *Figure 44* illustrates a rotating drum strainer.

With this method of cleaning, water is filtered continuously thru a perforated drum. Accumulations of residue on the drum surface cause the water level to rise in order to seek more perforations. The variations in water level control the periodic drum rotation and flushing required to clean the drum and to remove the waste matter to a collecting basket.

Spray Water Heaters

Spray water heaters may be required in winter when the mixing of outdoor and return air upstream of the washer cannot be controlled to produce the washer entering wet-bulb temperature required to maintain room design conditions with an evaporative cooling process. This condition may occur on very cold days or where minimum outdoor air requirements are relatively high, particularly with high room relative humidities and/or high room sensible heat ratios.

After a shutdown, as over a winter weekend, it may take some time to bring the room humidity up to design conditions when operating on an evaporative cooling cycle, even when the outdoor air quantity is reduced to damper leakage. Therefore, the spray water heater is used to add moisture to the air at approximately the same dry-bulb temperature. The heater thus provides a deviation from an adiabatic saturation process (*Fig. 45*).

Both steam ejector heaters and closed water heaters are available for heating spray water in central station washers. The steam ejector heater is a perforated steel pipe, closed at one end and submerged in the washer tank. Low pressure steam is admitted directly to the washer tank water at a controlled rate. The closed water heater is located on the discharge side of the recirculating pump and is installed in parallel with the main recirculating supply line. It is selected to heat a minimum quantity of spray water and requires suitable service and balancing valves. The closed water heater produces less noise than the steam ejector heater and enables recovery of steam condensate. However, the closed water heater is more costly to purchase and install.

A spray water heater may be sized on the basis of requirements calculated at the particular conditions encountered. Its capacity may also be calculated by determining the steam quantity required to heat and humidify the minimum outdoor air, or outdoor air damper leakage, from outdoor to room design conditions. The latter method provides sufficient capacity to maintain room design conditions during the period following equipment start-up.

The high velocity unitary washer utilizes a steam grid humidifier for humidity control under the conditions described above. Since steam is released directly to the air, relatively little sensible heating is accomplished.

Weirs

A weir is employed in a central station spray dehumidifier tank to insure a minimum submergence of the recirculating pump suction pipe and to maintain a water seal under the eliminators. During the

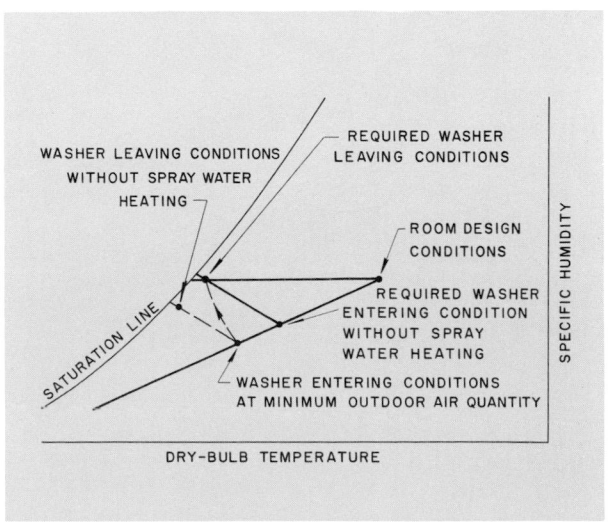

FIG. 45 — EFFECT OF SPRAY WATER HEATER

evaporative cooling season there is no flow over the weir. During the dehumidifying season, however, the flow is equal to the chilled water admitted to the recirculating system plus the moisture removed from the conditioned air.

Weir tanks may usually be obtained from the washer manufacturer. If a concrete tank is to be used, the length of the weir may be calculated from the Francis formula for sharp-crested rectangular weirs.

$$Q = 3.33 \times L \times H^{1.5}$$

The flow Q is expressed in cubic feet per second. The length L and the head H are in feet. If the weir is to have end contractions, the length should be reduced by $0.1 \times H$ for each end contraction, for formula use.

A sharp-crested concrete weir may be obtained by bolting a steel angle to the flat crest.

A flow rate of 5 gpm per foot of weir length is common for dehumidifier tanks.

Isolation

A central station washer requires no vibration isolation. The supply air fan isolation requirements should be investigated, however, with regard to the ambient sound levels thruout the building. Unitary washers seldom require vibration isolation for industrial applications but a vibration analysis may be necessary for critical installations. Isolation recommendations may be found under *Fan-Coil Equipment* in this chapter.

INSTALLATION

Location

The economic and sound level considerations pertaining to the location of air handling apparatus, as discussed in *Part 2,* are applicable to washer equipment.

Both central station and unitary spray apparatus may be located indoors or outdoors, although central station washers are most commonly located indoors, in an apparatus room or in the conditioned space. If exposed to the weather, a central station washer should operate with no water level maintained in the tank, and the fan motor, drive and bearings should be suitably chosen and protected.

Central station equipment is floor-mounted while the unitary washer may be either floor-mounted or suspended from above *(Fig. 46)*.

As with fan-coil equipment the availability of outdoor air and the ease of air return to the apparatus

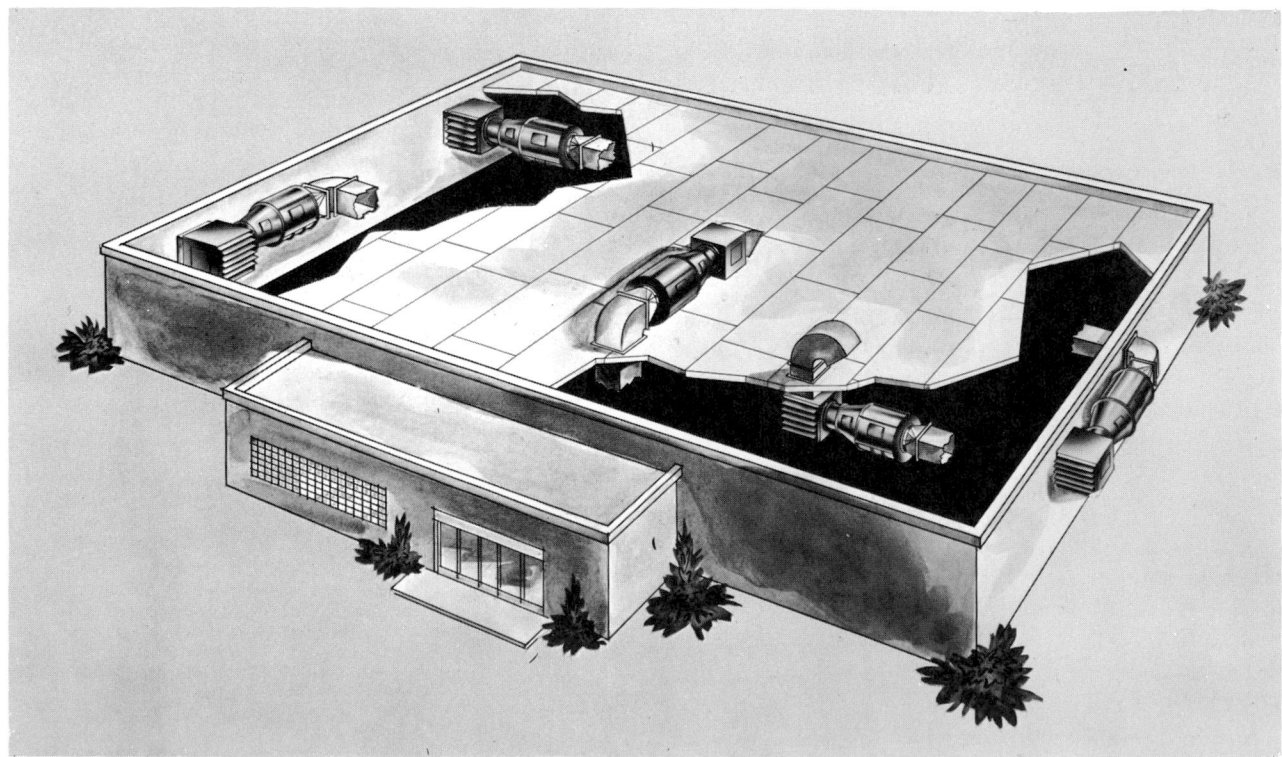

FIG. 46 — UNITARY WASHER INSTALLATION

is of primary concern in selecting a washer location. Outdoor air intakes should, if possible, be located and oriented so that they do not face nearby residential areas or walls of spaces where noise would be objectionable. Air may be returned thru a duct system but, if it is returned directly to the apparatus, the apparatus should be located so as to receive return air from the area it serves.

Another important location consideration is the availability of space. Particularly in industrial production areas space may be difficult to acquire. Limited head room and interferences such as electrical equipment, conveyors or belt drives may also present problems.

In addition, the location and orientation of the washer should be guided by the following considerations:

1. A spray dehumidifier should be located so that the gravity return of water to the central tank is possible. If this condition cannot be met, a pumped return should be employed.

2. Adequate building openings and passages should be available for the admittance of large equipment. If this consideration is overlooked, special openings in the building may later be required.

2. Locating a spray dehumidifier below the refrigeration equipment or pumping return water to a lower elevation may lead to problems of siphoning or overflow at shutdown. If such an arrangement is necessary, consideration should be given to the checking of water backflow tendencies and the breaking of a siphon.

4. The ability of a roof, floor or combination of structural members to withstand the operating weight of a washer should be investigated.

5. A washer should be located and oriented so that the simplest possible duct layout results.

6. Appearance should be considered. For example, a washer mounted on a flat roof may be less noticeable if located some distance from the building perimeter.

Layout

Figure 46 illustrates several layout alternatives for a high velocity unitary washer. A typical central station washer apparatus room layout resembles that shown for coil equipment in *Part 2*.

Central station washers should be provided with inlet plenums of adequate airway depth to promote outdoor and return air mixing and to minimize air eddy currents at the inlet face of the washer. The plenum chamber on the leaving air side of the washer should be large enough to provide unrestricted air flow to the fan at uniform eliminator velocities. Plenums downstream of the washer should also permit easy cleaning and removal of the eliminator blades for both central station and unitary apparatus.

Sufficient space should be provided around the washer for maintenance access, particularly on the side where access doors and piping connections are located. Minimum clearance should be provided on the far side for cleaning, painting and for the application of insulation if required. If suspended well above the floor, unitary washers may require catwalks.

Access doors should be installed between central station apparatus components as required for proper maintenance and service.

A mounting base at least two inches high should be provided for central station equipment. A base provides a level and uniform bearing surface for the tank, prevents damage to the tank or to the insulation under the tank from water seepage, and increases the tank water level available for priming the recirculating pump.

If a concrete tank is designed for a central station washer, it should be of reinforced construction and provided with pipe sleeves, baffle and eliminator supports and anchor bolts, as required. In addition the plenum at the leaving air side of the washer should be provided with a curb at least four inches high.

The recirculating water pump may be located in the air stream entering the washer or outside of the washer casing.

Marine lights should be provided within the washer and between components of a central station washer.

In addition to the washer piping details shown in *Part 3*, the following suggestions apply:

1. Floor drains should be provided in the entering and leaving air plenums, near the recirculating pump, near the outdoor air intake and as required for the cleaning of filters or other components. Usually the drain in the leaving air plenum requires a deep seal trap.

2. If a pumped return is used for a spray dehumidifier, a continuously running $\frac{1}{2}$ inch bleed line from the return pump discharge to the tank prevents overheating of the water in the pump when the return control valve is closed.

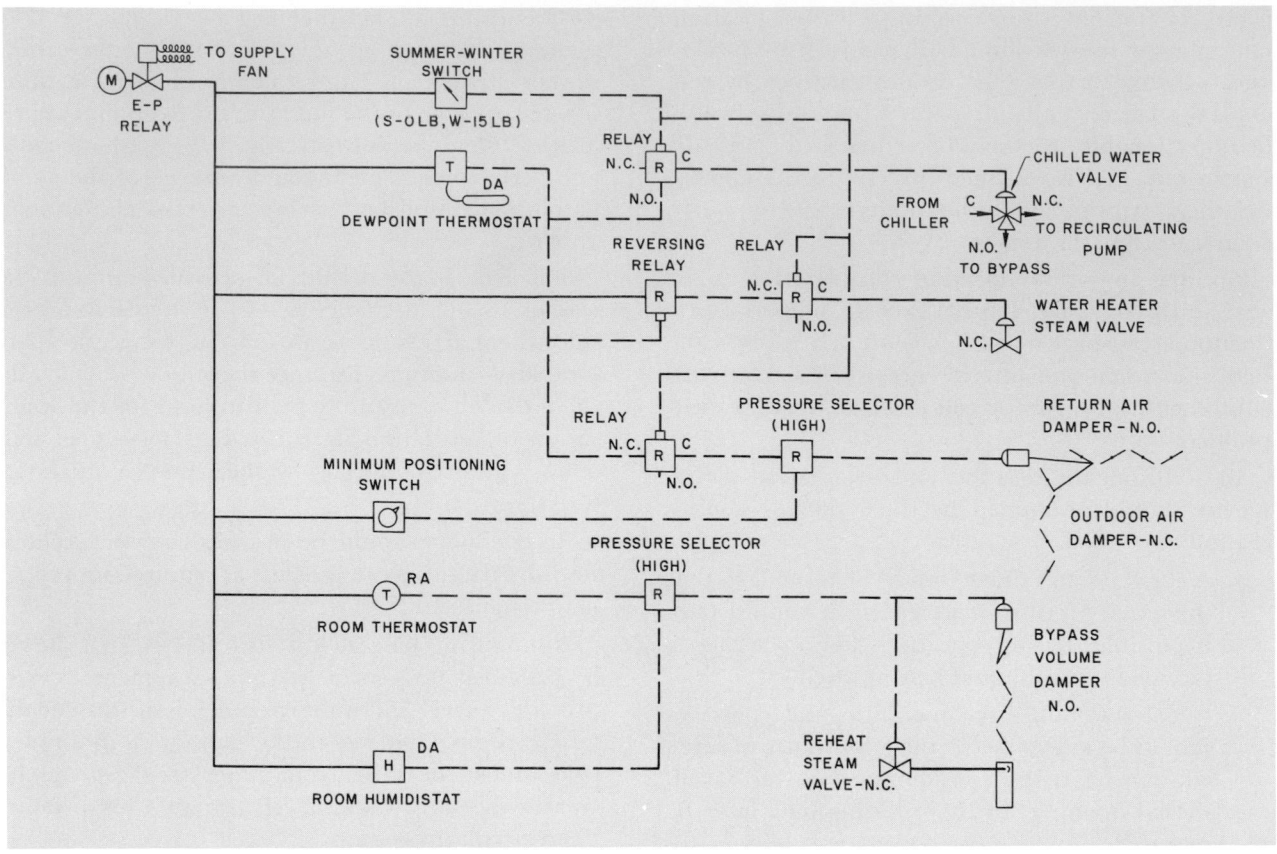

FIG. 47 — CENTRAL STATION DEHUMIDIFIER CONTROL

3. Because the water in a washer tank is relatively shallow, a well designed vortex breaker on the pump suction pipe is required.

Insulation

The top and sides of a spray dehumidifier should be insulated as required to prevent condensation on the apparatus and to minimize heat transfer. A central station humidifier should be similarly insulated if the dewpoint for the return air is higher than the spray water temperature, such as occurs when supplemental atomizer systems are employed.

The high velocity unitary washer described previously should be completely insulated, regardless of the application.

A thickness of cork insulation may be required beneath the washer tank. If used, the cork layer should be coated on both sides with sealing compound and positioned on the tank pad before the unit is installed.

Washers located outdoors should be insulated, vapor sealed and weatherproofed. The insulation on the top surfaces should be slightly crowned so that water will run off.

Water and steam riser insulation in industrial applications is sometimes subject to damage from trucks and material handling equipment. If such is the case, a sheet metal shield around the insulation is suggested, from the floor to a height of several feet.

CONTROL

The function of controls is to produce a balance between the air conditioning load and the apparatus capacity in order to maintain room design conditions.

Apparatus control may be accomplished in one or a combination of two ways:

1. Varying the supply air volume at a given air condition.
2. Varying the air condition with no change in volume.

The condition of the air may be altered by such methods as spray water temperature variation, air reheating, spray water heating, washer bypass, spray throttling, variation of the outdoor and return air mixture proportions, and air humidification, as with a steam grid humidifier or an atomizer system.

A central station air washer operating on a dehumidifying cycle normally utilizes spray water temperature variation, air volume reduction and air reheating. See *Fig. 47* for the simplified control diagram. A dry-bulb thermostat, located on the leaving air side of the washer and set to maintain the leaving dry-bulb condition necessary to achieve the required leaving air dewpoint, controls the chilled water valve supplying the recirculating water system. The spray water temperature is thus controlled.

Dewpoint control is practical because the difference between dewpoint and leaving air dry-bulb temperature is small as a result of the high contact factor of the dehumidifier.

Air volume and reheat control is obtained by a thermostat and humidistat together controlling a reheat coil bypass volume damper and a reheater steam valve in sequence. Closing the volume damper reduces the supply air volume to a predetermined fraction of full load air delivery, usually from 60% to 85%, depending on the fan characteristic and the allowable maximum air pressure drop thru the heating coil. A further reduction in room load causes the reheater valve to begin opening.

Figure 48a illustrates a typical spray dehumidifying process at full load. *Figure 48b* shows the temperature relations at half load. The entering spray water temperature has been increased to maintain a relatively constant apparatus dewpoint.

Chilled water supply to the central station dehumidifier recirculating system can be controlled by a two-way throttling valve or a three-way diverting valve. Use of a two-way valve on a multiple washer system may necessitate a pressure bypass line and

valve at the central collection tank in order to minimize line pressure fluctuations.

Dewpoint control of a central station air washer operating on an evaporative cooling cycle is achieved by controlling the outdoor and return air mixture condition and by operating the spray water heater, if necessary. Zone control may be identical with that control employed when on the dehumidifying cycle.

A measure of humidity control may be obtained during a period of refrigeration shutdown by cycling the recirculating pump and/or the baffle or eliminator spray pump, if any.

In a high velocity unitary washer application, room conditions are controlled directly by a combination of spray throttling, air reheating and, if necessary, humidification. *Figure 49* is a control diagram for units operating with dehumidification control for an all air system. During the dehumidifying season the outdoor air dampers are in a minimum position; the reheat valve is controlled by the room thermostat alone; the spray throttling valve is controlled thru the pressure selector, and the humidifier is controlled by the humidistat, providing the thermostat is satisfied. When operating on the evaporative cooling cycle, the outdoor and return air dampers are controlled by the room thermostat; the reheat valve is controlled thru the pressure selector; the spray throttling is controlled by the humidistat, and the humidifier is controlled as in the dehumidifying cycle.

Since spray throttling is always utilized with this system, a pressure bypass line and valve are usually required at the central water collection tank to minimize fluctuation in line water pressures.

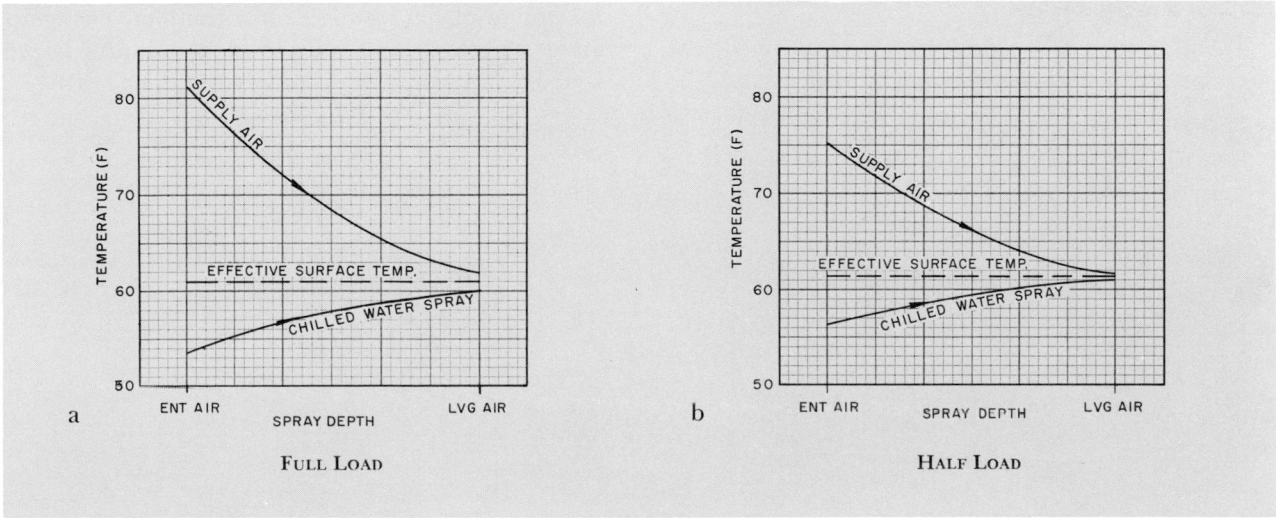

FIG. 48 — TYPICAL SPRAY DEHUMIDIFIER PROCESSES

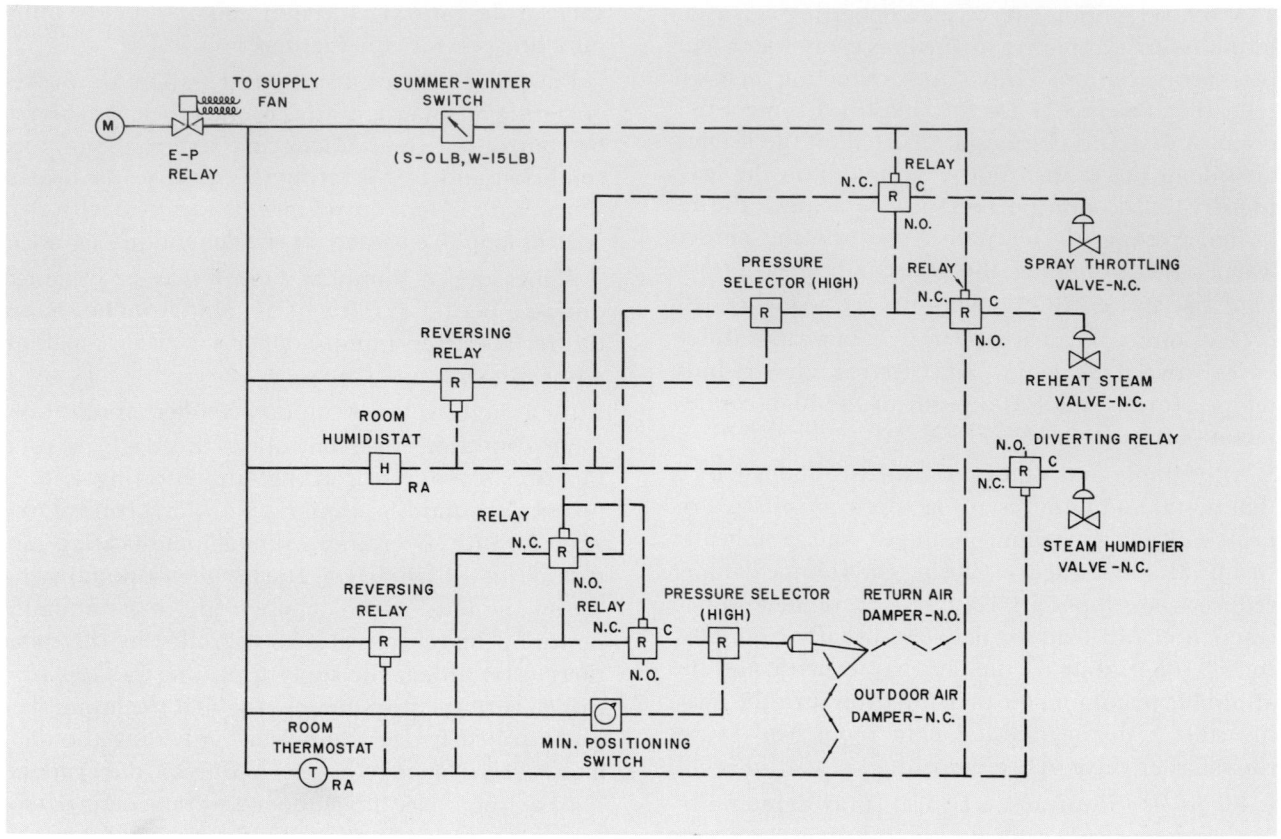

Fig. 49 — Unitary Washer Control (Dehumidifier All Air)

General Control Considerations

For industrial applications, particularly in the case of process air conditioning, room controls are often mounted within a cabinet provided with a small fan. The circulation of room air thru the cabinet provides a constant and positive sampling by the control sensing devices.

If large open spaces are to be conditioned, the area served by each set of room controls should be limited in order to maintain an adequate control response. Maximum areas of 10,000 square feet for a temperature zone and 8000 square feet for a humidity zone are suggested.

Control accuracy and response are also affected by the air circulation in the conditioned space. A maximum of ten minutes for a complete change of air is suggested, and four to eight minutes is preferred.

CHAPTER 3. UNITARY EQUIPMENT

A unitary air conditioning unit, sometimes referred to as packaged equipment, consists of one or more factory-fabricated assemblies designed to provide the functions of air moving, air cleaning, cooling and dehumidification. The functions of heating and humidifying are also usually possible with such equipment. Heat pump versions are available for most types of apparatus.

Unitary equipment includes a direct expansion or chilled water cooling coil and a compressor-condenser combination or water chiller in addition to fans, auxiliaries and internal wiring and piping. If more than one assembly is required, the separate assemblies are designed for use with each other, and combined equipment ratings are based on matched assemblies of equal or differing nominal capacities.

The design of unitary equipment is often styled for installation within the conditioned space.

It is the purpose of this chapter to guide the engineer in the practical application and selection of unitary equipment.

TYPES OF EQUIPMENT

Unitary equipment may be classified as either a self-contained or a split system. A self-contained unit houses all components in a single assembly. Split system equipment incorporates the following assemblies:

1. A coil and compressor combined with a remote condenser.
2. A coil combined with a remote condensing unit.
3. A coil combined with a remote water chiller.

A self-contained unit is illustrated in *Fig. 50*. The self-contained concept is further described in *Fig. 51*. *Figure 52* shows an air-cooled condensing unit — one component of a two-component split system as described in Item 2 above.

FIG. 50 — SELF-CONTAINED UNIT

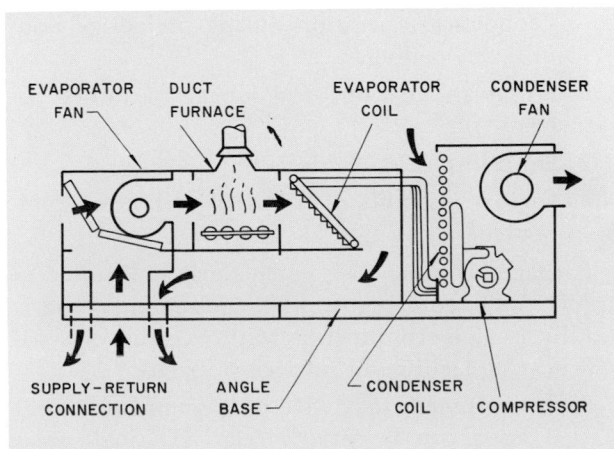

FIG. 51 — SELF-CONTAINED UNIT

FIG. 52 — AIR-COOLED CONDENSING UNIT

The use of matched components differentiates unitary equipment from the fan-coil equipment discussed in *Chapter 2* of this part. Unitary equipment thus affords less flexibility of arrangement and less choice of cooling coil surface. Also, face and bypass control is usually unavailable in packaged equipment.

Split system apparatus provides in packaged form a measure of flexibility not usually obtainable with self-contained equipment.

APPLICATION

The use of unitary equipment should be considered for applications where the following advantages are of primary importance:

1. Low first cost of equipment and installation.
2. Immediate air conditioning benefits and prompt delivery.
3. Ease of installation or removal, if necessary, with a minimum of disturbance.
4. The ability to provide air conditioning in increments without cost penalty.
5. Economical operation during periods of non-uniform loading.
6. High salvage value and longer warrantee periods.
7. Simplified field engineering.
8. Factory assembly of balanced and tested components.

Packaged equipment is particularly well suited to applications requiring summer cooling only, and is readily used in conjunction with existing or separate heating facilities of sufficient capacity.

Such equipment may effectively augment central station apparatus by serving relatively small areas with special design requirements. Typical applications of this nature are laboratories and dining areas.

Applications completely conditioned with unitary equipment include existing office buildings and hotels, motels, shopping center tenant areas, department stores, industrial facilities and residences.

Equipment components are usually matched to provide 300 to 500 cfm per ton of air conditioning at sensible heat ratios of 0.65 to 0.85, in the case of self-contained equipment. Therefore, packaged equipment is most economically applied where these values are specifically required. As mentioned previously some additional application flexibility may be obtained by employing split system equipment. Sensible heat ratios as high as 0.95 are attainable. Such equipment also affords greater choice of location and mounting method.

Self-contained equipment is commonly available in capacities up to 60 tons, while up to 75 tons may be obtained with a split system. The trend has recently been toward larger packaged equipment.

Water-cooled, air-cooled or evaporative condensing may be utilized with unitary apparatus.

STANDARDS AND CODES

Applicable provisions of the American Standard Safety Code B9.1, ARI Standard 210 and Underwriters' Laboratories Standards govern the testing, rating and construction of unitary air conditioning equipment.

The application and installation of such equipment should conform to pertinent government agency regulations and to all codes and laws prevailing at the job site.

UNIT SELECTION

SELECTION RATINGS

Unit size is usually determined by the required cooling capacity and air quantity, adjusted to suit the sensible heat ratio. Cooling ratings present total and sensible heat capacities, based on air quantity, evaporator entering air wet-bulb temperature and, in the case of water-cooled equipment, condensing temperature. A typical cooling rating table is illustrated in *Fig. 53*. Although tabular cooling ratings are most common, some manufacturers present graphical data in place of, or in addition to, tabular ratings.

Cooling ratings *(Fig. 53)* may be expanded to apply to more than one evaporator entering air dry-bulb temperature. If they are not expanded, deviation corrections are usually suggested. Cooling rat-

EVAPORATOR AIR		CONDENSING TEMPERATURE (F)											
		90			100			105			110		
Qty	Ent Wet-bulb	Total Cap.	Sens Heat Cap.*	Compr Motor Power Input	Total Cap.	Sens Heat Cap.*	Compr Motor Power Input	Total Cap.	Sens Heat Cap.*	Compr Motor Power Input	Total Cap.	Sens Heat Cap.*	Compr Motor Power Input
(cfm)	(F)	(1000 Btuh)	(1000 Btuh)	(kw)	(1000 Btuh)	(1000 Btuh)	(kw)	(1000 Btuh)	(1000 Btuh)	(kw)	(1000 Btuh)	(1000 Btuh)	(kw)
4500	72	204	97	12.8	198	95	14.0	194	94	14.6	190	92	15.2
	67	186	119	12.4	181	117	13.7	178	116	14.2	173	114	14.9
	62	169	142	12.2	165	139	13.4	162	138	13.8	158	136	14.4
6000	72	213	108	12.9	207	106	14.2	203	105	14.8	198	103	15.4
	67	196	136	12.7	190	134	13.9	186	133	14.4	182	131	15.1
	62	179	164	12.4	173	162	13.6	170	161	14.0	167	159	14.6
7500	72	220	118	13.0	214	115	14.3	210	114	14.9	204	113	15.6
	67	204	151	12.7	197	150	14.0	192	148	14.6	188	146	15.2
	62	185	184	12.4	180	180	13.7	176	176	14.2	173	173	14.8

*Sensible heat capacity is based on 80 F entering air dry-bulb temperature.

Fig. 53 — Typical Ratings (Water-Cooled Self-Contained Units)

ings also may indicate grand sensible heat factor rather than total sensible heat capacity.

For air-cooled condensing or evaporative condensing, selection ratings are normally based upon condenser entering air dry-bulb temperature or wet-bulb temperature respectively, instead of condensing temperature.

Self-contained equipment is rated as a system with no individual component ratings. A split system apparatus is usually rated both as an individual item of equipment and in combination with its intended components. For example, an air-cooled condensing unit may be rated in terms of cooling capacity available from the package when the condensing unit is used with a particular fan-coil unit at different evaporator wet-bulb and outdoor dry-bulb temperatures.

ECONOMICS

The cooling ratings described above are based upon the capacities of components in balance with each other. It is therefore usually unnecessary to determine component balance capacities when selecting packaged equipment. However, if equipment is to be economically selected for use with unmatched components, the determination of the component balance points and the subsequent selection analysis should be pursued as described in *Part 7*.

The economical balance of component capacities may be upset if the grand sensible heat factor required for an application differs considerably from that characteristic of the package. For example, with relatively high sensible heat ratios the desired air

quantity per ton of capacity is also high, relative to that available from standard packaged equipment. With self-contained equipment, therefore, it may be necessary to provide oversized refrigeration components, in order to acquire the appropriate air delivery. In this case an economical balance can be restored by designing for lower relative humidities, thus permitting a greater air temperature rise and a smaller air quantity. An alternate solution, applicable within limits, is to vary the evaporator air quantity. Split system equipment may be mixed as to nominal component capacities to produce an economical balance.

Economy of equipment selection may also be promoted by the following methods:

1. Select equipment to be fully loaded, taking advantage of room temperature swing, storage effects and reduced safety factors.

2. Avoid the arrangement of unit by zones where selections must be made for peak loads. Diversity benefits may be realized if more than one exposure is served by a unit.

3. Consider operation at relatively high condensing temperatures with possible savings as explained in *Part 7*.

4. Introduce the least outdoor air possible at peak apparatus load.

ATMOSPHERIC CORRECTIONS

Unitary equipment ratings are based on air at standard atmospheric conditions of 70 F and 29.92 in. Hg barometric pressure. For applications deviating significantly from this standard such as at altitudes exceeding 2500 feet, ratings should be adjusted for the difference in air density. The several corrections involved have been described elsewhere and may be summarized as follows:

1. Load calculations should be modified as described in *Part 1*.
2. Air side pressure drop should be adjusted in proportion to the air density ratio as outlined in *Part 2*.
3. If the apparatus includes an evaporator, the unit capacity should be corrected by entering the rating tables at a supply air quantity equivalent to standard atmospheric conditions. This procedure is similar to that described for fan-coil units in *Chapter 2* of this part.
4. Fan speed and brake horsepower should be adjusted as detailed in *Chapter 1* of this part.

The decrease in performance of air-cooled condensers at high altitudes and/or air temperatures, although in itself significant, produces only a slight deviation in the rating of combined components. This deviation may amount to from one to three percent at an altitude of 5000 feet.

COIL FREEZE-UP PROTECTION

The freezing of hot water coils located downstream of the cooling coil may occur during the cooling season. This is particularly possible in packaged fan-coil equipment because of the proximity of the heating and cooling coils and because of the relatively low settings usually employed on the low pressure compressor cutoff switch to prevent excessive cycling. At lower evaporator wet-bulb temperatures, equipment components balance at diminished suction temperatures. Hot water coil freeze-up may occur under these conditions.

Although coil freeze-up may be prevented by the draining of the hot water coil or by the use of a properly inhibited antifreeze solution as described in *Chapter 2* of this part, the use of a protective thermostat to cycle the compressor is suggested instead in the interest of economy. The thermostat should be mounted outside of the air stream with the bulb at the entering face of the hot water coil. An air temperature of approximately 35 F is suggested as a compressor shutoff point.

Coil freeze-up may sometimes be traced to reduced air quantities resulting from dirty filters in the apparatus.

INSTALLATION
LOCATION

Unitary air conditioning equipment may usually be located either outdoors or indoors. Possible specific locations include basements, crawl spaces, attics, garages, roofs, and on the ground as well as within the conditioned space or in an equipment room. Such equipment may be mounted on the floor, suspended from the ceiling or installed in a wall opening, transom or window. Since packaged apparatus is sometimes designed for a specific location such as on a roof or under a window, the manufacturer's literature should be studied for location recommendations. The suggestions in *Part 2, Part 7,* and *Chapter 2* of this part regarding equipment location are applicable to packaged equipment.

Although weatherproofing kits are available for most self-contained equipment, an indoor location is preferred. However, self-contained equipment designed specifically for outdoor use is available. With any equipment featuring outdoor compressors, crankcase heaters should be employed to prevent the migration of refrigerant to the compressor and the damage which may result. For outdoor or indoor locations insulation considerations apply as noted in *Chapter 2* of this part.

LAYOUT

The references noted above should be consulted for suggestions dealing with equipment layout.

Unitary apparatus is available in both horizontal and vertical arrangements and it is usually designed for use with or without duct systems. However, distribution systems should be simple in design and limited in extent.

Often, unitary equipment may conveniently use the same air distribution system as an existing heating system. This is particularly true in residential applications. In such an installation appropriate shutoff or diverting dampers may be required if the heating and cooling functions are in parallel. Existing duct sizes should be checked for adequacy in handling the dehumidified air quantity.

For roof-top installations the roof must be of adequate strength, and the equipment weight should be evenly distributed on the support members. If any doubt exists as to the adequacy of support, a structural engineer should be consulted. Appropriate framing around roof openings, flashing, counter flashing and pitch pockets should be provided.

External vibration isolation of packaged equipment is seldom required because the individual components are usually isolated within the cabinet.

However, for critical installations and light building construction, vibration isolation should be considered for unitary equipment as for any other type of equipment. Vibration isolation is discussed in *Chapter 2* of this part. Isolation recommendations may also be solicited from the manufacturers of vibration equipment.

Layout and location of unitary equipment are influenced by the availability of service facilities such as gas, city water and electrical power.

CONTROL

The reduction of packaged equipment capacity at partial loads is usually effected by cycling the compressor or compressors in accordance with the setting of a room dry-bulb thermostat. Decreasing compressor capacity by the unloading of cylinders is another widely employed method of control.

Unit fans may be operated continuously or cycled with the compressor. Continuous operation of fans provides continuous air circulation. However, alternately condensing moisture from the air and re-evaporating coil moisture produces fluctuations in room humidity conditions. Cycling of fans requires the use of a room thermostat rather than a return air thermostat.

The desirability of controlling equipment capacity and the unavailability of face and bypass control may affect adversely the equipment capacity for latent heat removal. For example, with a single compressor serving a single evaporator coil, cylinder unloading causes an increase in the grand sensible heat ratio and thus a relative decrease in the latent heat capacity. This occurs at a time when the opposite effect is usually desired.

This effect may be overcome by the use of multiple compressors with multiple coils or coil circuits operating on signals from a two-step thermostat. This permits an improvement in latent heat removal at partial loads. Coils may be of equal or unequal capacities. In either case, some additional latent capacity is obtained by the decrease in air quantity over the operative coil as the inoperative coil dries. With unequal coils, where the larger is the first to be removed from operation, additional latent capacity is also obtained thru the lower sensible heat ratio of the smaller coil and the prolonged "on" period of the operating cycle.

Multiple compressors are usually obtainable only on equipment of ten tons capacity, or greater.

Loss of latent capacity at partial loads and the resulting fluctuations in room conditions are intensified by the oversizing of equipment. Fully loaded equipment provides the best insurance of maintaining reasonable humidity conditions at partial loads. Unit air deliveries may also be varied from nominal to obtain the most desirable full load sensible heat ratio, and therefore the best possible latent capacity at partial load.

Under special conditions where accurate control of temperature and humidity is required, packaged equipment may be easily adapted for control by reheat or humidification.

The necessity and means of condensing pressure control is discussed for various condensing methods in *Part 7*.

CHAPTER 4. ACCESSORY EQUIPMENT

This chapter presents practical information to guide the engineer in the application and layout of air cleaning and heating devices, as used in conjunction with air conditioning systems.

AIR CLEANERS

The control of air purity consists of reducing or eliminating unwanted particulate or gaseous matter from the air supplied to a space. This is a function of the air conditioning system. However, normal applications are concerned with particulate matter only.

Effectively applied air cleaners can materially reduce operating expenses and increase productivity. Specific benefits include:

1. The reduction of building cleaning costs — an item otherwise accounting for as much as forty percent of total operating expenses.
2. The reduction of employee absenteeism — a result of the removal of bacteria, viruses and allergens from the air.
3. An increase in employee efficiency.
4. An increase in product quality.
5. An increase in the life of machinery or equipment.

CONTAMINANTS

Air is contaminated in varying degrees by soil, organic matter, spores, viruses, bacteria and allergens, as well as aerosols such as smokes, dusts, fumes and mists. These contaminants may be introduced into the air from outdoors, or they may be returned to the air conditioning apparatus from within the space. The ease and efficiency with which they may be removed depends on the size, shape, specific gravity, concentration and surface characteristics of the particle.

Contaminant characteristics vary widely. Particle diameters range from molecular size up to 5000 microns*. Concentrations as high as 400 grains per 1000 cubic feet may be encountered. However, air conditioning applications usually involve the removal of particles no smaller than 0.1 micron in di-

*One inch equals 25,400 microns.

ameter and as large as 200 microns. Normal concentrations seldom exceed 4 grains per 1000 cubic feet. The specific characteristics of the particles to be removed are determined by the application. Thus, air purity control is a relative concept.

The sizes of common contaminant particles are shown in *Chart 10*. Typical outdoor air dust concentrations for various localities are noted in *Table 9*. Concentrations may increase during the heating season, especially in residential areas.

Particles of an oily nature with irregular surfaces or charged electrostatically tend to agglomerate more readily. The settling and adherence of contaminants is, therefore, affected by other characteristics in addition to size and concentration.

TABLE 9—DUST CONCENTRATION RANGES

LOCALITY	DUST CONCENTRATION (grains per 1000 cu ft)
Rural and suburban districts	0.02 - 0.2
Metropolitan districts	0.04 - 0.4 (0.06 avg)
Industrial districts	0.10 - 2.0
Ordinary factories or workrooms	0.20 - 4.0
Excessively dusty factories or mines	4.0 - 400.

PERFORMANCE CRITERIA

Atmospheric air cleaners (referred to as air filters) are rated in terms of efficiency (arrestance), resistance to air flow, and dust capacity. The three most critical performance factors are the following:

1. The variation of filter resistance with air flow.
2. The variation of filter resistance with dust load at design air flow.
3. The effect of dust loads at design air flow on filter efficiency.

Performance data for a typical unit filter are illustrated in *Fig. 54*. Filter resistance increases with air flow (face velocity) or with dust load at design air flow. The efficiency of a particular filter varies not only with dust load but also with the characteristics of the contaminating particles. For this reason, *Fig. 54* specifies the test procedure used to rate the filter.

The capacity of a filter is a measure of its usable life prior to disposal, renewing or cleaning.

CHART 10—FILTER APPLICATION

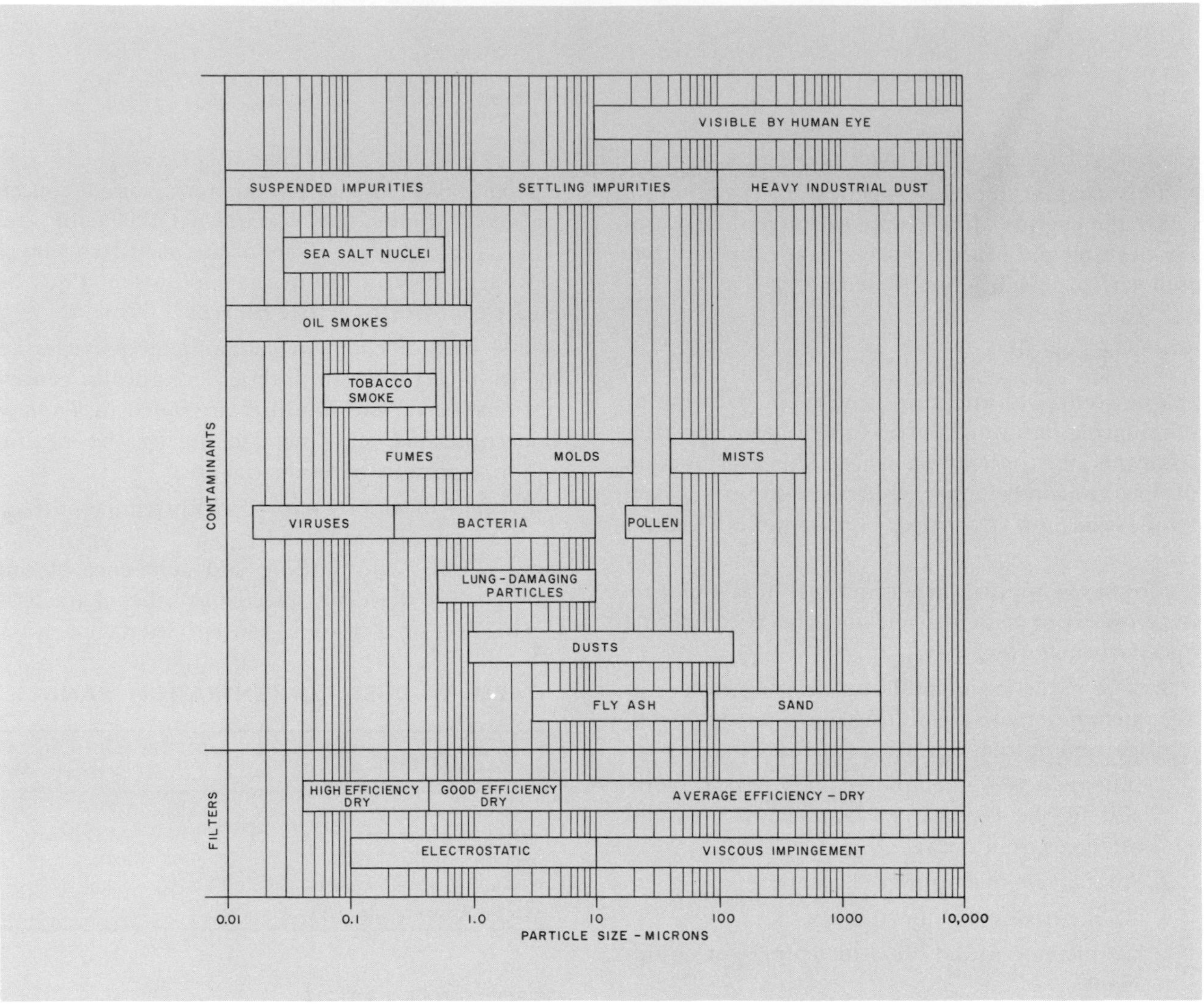

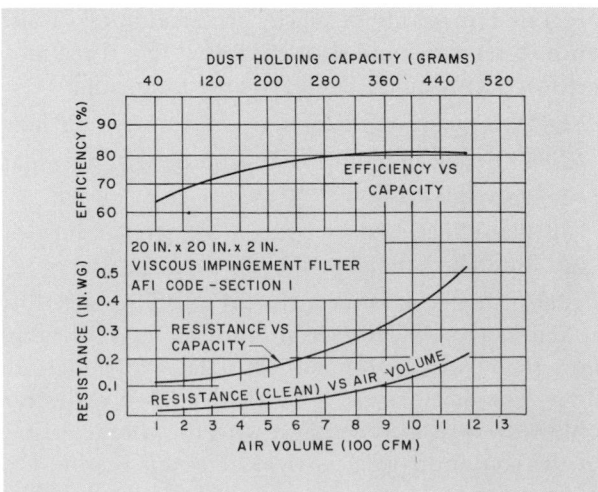

FIG. 54 — TYPICAL PERFORMANCE DATA
(UNIT FILTER)

STANDARDS AND CODES

Air filter manufacture and installation should conform to the recommendations in Pamphlet 90A of the National Board of Fire Underwriters and to all codes and laws applying at the job site.

The efficiency and capacity of air filters are determined by several standardized test methods, differing primarily in the test aerosol used and the method of measuring the amount of dust passed by the filter. The three most common test procedures are these:

1. The weight method, with test aerosols as specified by the Air Filter Institute Code and a modification of the former ASHVE Code.

2. The dust spot method, with procedures as standardized by the Air Filter Institute and the National Bureau of Standards.

3. The D.O.P.* test, a particle count method utilizing a chemical smoke aerosol.

These three methods differ in application and the results are difficult to convert to common terms. In comparing the performance of various filters it is therefore imperative that the test used to obtain the published data be noted in each case.

The weight method expresses filter efficiency in terms of the particle weight removed, relative to the weight introduced to the air stream. It is particularly useful in evaluating the performance of mechanical filters of average efficiency. However, this test overstates filter effectiveness in removing small particles of light weight.

The dust spot test rates filters in terms of the relative opacity of stains on filter paper thru which the air is passed. The optical density of the spots are measured photometrically. This type of test is useful primarily in evaluating air cleaning devices of high efficiency, such as electronic air cleaners. In addition it provides a measure of filter efficiency in removing the sort of dust most likely to cause discoloration of walls and ceilings. Test results are, however, sometimes inconsistent and are difficult to interpret.

The D.O.P. test relates filter performance to the light-scattering tendency of smoke particles approximately 0.3 microns in diameter. Measurements are made photoelectrically. The test is used primarily to determine the ability of filters of very high efficiency to remove specific particles, such as pollen. It requires carefully controlled laboratory conditions and expensive equipment. It cannot be used to determine filter capacity.

TYPES OF AIR CLEANERS

Viscous Impingement

Filters of the viscous impingement type utilize a filtering medium relatively coarse in texture and constructed of fiber, screen, wire mesh, metal stampings or plates. The medium is coated with a viscous substance such as oil or grease. As the many small air streams abruptly change direction thru the filter, contaminating particles are thrown against the medium where they adhere. Efficiencies of 65% to 80%, based on the weight method of testing, are achieved in the case of cleanable media.

This type of filter is available in a throwaway style or may be obtained with a replaceable medium, a manually cleanable medium, or an automatically renewed medium.

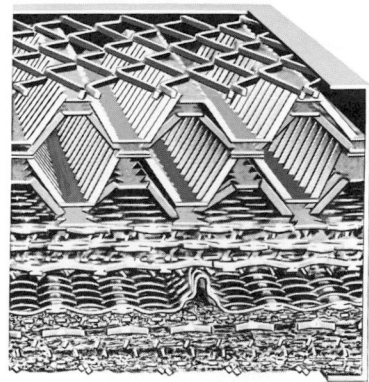

Courtesy of American Air Filter Co., Inc.

FIG. 55 — VISCOUS IMPINGEMENT FILTER SECTION

Media designed for filtering velocities of approximately 300 feet per minute usually increase in density in the direction of the air flow. Thus, the larger particles are the first removed, prolonging filter life. This progressive density is illustrated in *Fig. 55.* High velocity filters operating at approximately 500 feet per minute are normally nondirectional and of uniform density. *Figure 56* shows a cleanable viscous impingement panel filter.

Automatic viscous impingement filters may be of the replaceable media or renewable media type. The former consists of a moving filter roll. The latter is constructed of overlapping filter panels attached to a moving chain and moving thru an oil bath. The self-cleaning filter is shown in *Fig. 57.* In either case the filter curtain may be actuated by a timing mech-

Courtesy of American Air Filter Co., Inc.

FIG. 56 — CLEANABLE VISCOUS IMPINGEMENT FILTER

*Di-Octyl-Phthalate

FIG. 57 — AUTOMATIC VISCOUS IMPINGEMENT FILTER

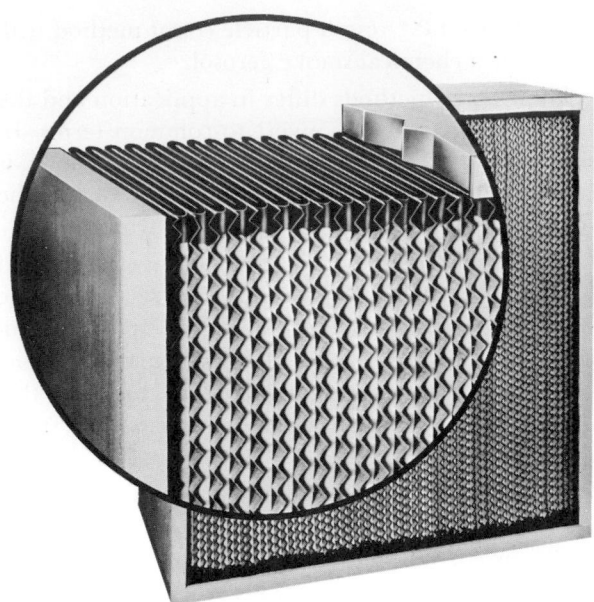

FIG. 59 — HIGH EFFICIENCY DRY FILTER

anism or a pressure sensing device. Automatic filters present a relatively constant resistance to air flow, while panel filter resistance varies considerably as the dust load increases. Automatic filter efficiencies vary from 80% to 90%, based on the weight method.

Dry Media

Dry filters consist usually of a permanent frame and a dry replaceable medium of cellulose, asbestos or glass fibers, specially treated paper, cotton batting, wool felt or synthetic material. The air passages thru the medium are smaller than those of the viscous impingement type filter, and therefore lower air velocities are necessary to avoid excessive resist-

FIG. 58 — DRY FILTER CELL WITH FRAME

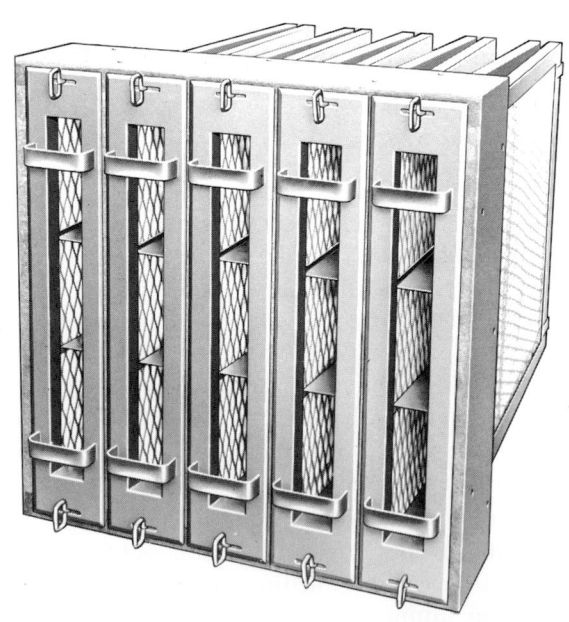

FIG. 60 — DRY FILTER CELL, POCKET TYPE

Courtesy of American Air Filter Co., Inc.

FIG. 61 — AUTOMATIC DRY FILTER

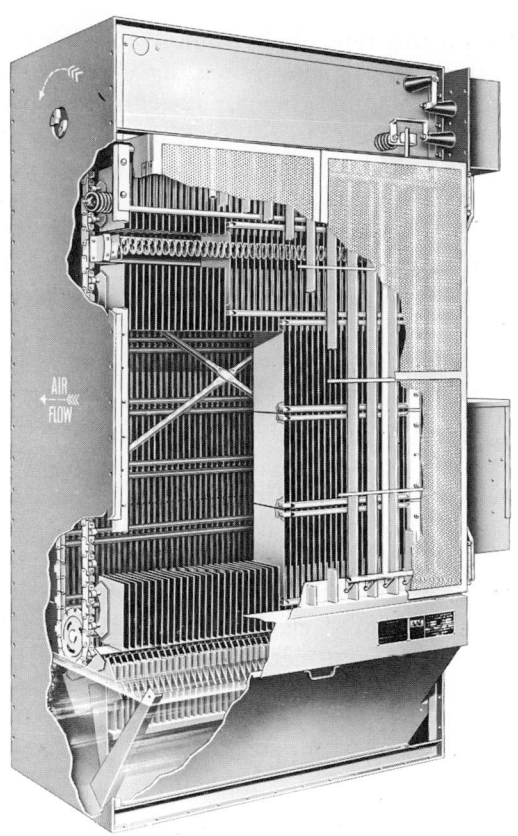

Courtesy of American Air Filter Co., Inc.

FIG. 62 — ELECTRONIC AIR CLEANER, IONIZING TYPE

ances. In order to obtain a large surface area relative to cross-sectional area, the medium is usually pleated in accordion form.

Figure 58 shows a medium efficiency dry filter with an area ratio of 7:1. Such a filter is capable of a wide efficiency range varying from 84% to 95%, based on the AFI weight test, depending on the medium used.

Figures 59 and 60 illustrate very high efficiency filters with area ratios of from 25:1 to 50:1. This type of filter may be obtained with an efficiency as high as 99.97%, D.O.P. test method. D.O.P. efficiencies above 90% are usual.

Dry filters are available in automatic construction, normally utilizing a moving roll of disposable medium *(Fig. 61)*. Movement may be controlled by a differential pressure sensing device. Thus, operating air resistance is maintained relatively constant.

The efficiency of a dry filter depends on the size and spacing of the fibers in the medium used. Media with the smallest, most densely distributed fibers provide the highest efficiencies. High efficiencies, however, are usually associated with high resistance, short life and low dust holding capacity.

Electronic

Electronic air cleaners, often referred to as precipitators, are of two varieties: the ionizing type and the charged media type. They are illustrated respectively in *Fig. 62 and 63*.

Courtesy of American Air Filter Co., Inc.

FIG. 63 — ELECTRONIC CHARGED MEDIA FILTER

The ionizing type of electronic air cleaner ionizes contaminating particles by passing the air thru an electric field of approximately 12,000 volts potential. The particles are then collected on charged plates which are usually coated with an adhesive to prevent re-entrainment of the particles. Efficiencies of 85% to 90%, based on the dust spot test, are achieved. The collecting stage operates at approximately 6000 volts. The high voltages are obtained from rectifiers supplied with 110/120 volt single phase electrical service. Power consumption varies from 12 to 15 watts per 1000 cubic feet per minute, with an additional 40 watts required to energize the rectifier tube heaters.

The charged media electronic air cleaner consists of a panel filter with an electrostatically charged dry medium. It therefore combines the principles of electronic precipitation and dry mechanical filtration. The produced efficiency averages about 60% by the dust spot test method. Approximately 12,000 to 13,000 volts are required to charge the dielectric medium. Power requirements are about 8 watts per 1000 cubic feet per minute.

Although the ionizing type of electronic cleaner may be obtained in replaceable cell construction, the precipitator shown in *Fig. 62* is automatically self-cleaning. Moving collector plates are cleaned and reoiled by the same method as employed with the automatic viscous impingement filter *(Fig. 57)*. Semiautomatic cleaning is also available, utilizing nozzles for water cleaning and reoiling.

The resistance to air flow imposed by an ionizing type electronic cleaner is very small. For this reason the unit may feature screens or perforated plates at the air entering and/or leaving side to promote uniform air flow thru the precipitator.

APPLICATION

The choice of a particular type of air filter for a given application involves the following steps:

1. A determination of the size, concentration and characteristics of contaminants present in both the outdoor air and return air.
2. A decision regarding the size of particles to be removed and the efficiency required for removal.
3. The selection of the filter which will provide most economically the desired efficiency under the prevailing conditions of labor cost, power costs and annual hours of operation.

Air contamination may be appraised by costly laboratory analysis or by an estimation based on past experience and general data. The latter method is preferable in all except highly specialized applications.

Chart 10 and *Table 9* may be used with judgment to determine air contamination. Additional data of local interest may be obtained from the city Bureau of Health or smoke control agency.

The determination of which contaminants are to be removed, and to what degree, should be based on the requirements of the processes, equipment, material or occupants within the conditioned space. For example, a greater filtering efficiency would be required for an electronics laboratory than for a bowling alley. However, for any application certain contaminants should be removed. These contaminants include abrasive dusts, lint, pollen, concentrations of toxic fumes, if present, and carbon, if in appreciable quantities.

Chart 10 also indicates the approximate normal ranges of application of the various filter types based upon particle size only.

Viscous impingement filters efficiently remove contaminating particles larger than 10 microns in diameter, particularly if the particles are oily. The coarse media used are well suited to large particle sizes and concentrations. The capacity and life of such filters are great and therefore maintenance is relatively inexpensive. High velocity unit filters (500 feet per minute) are not suitable to heavy lint applications since the media density is not progressive.

Dry media filters are more efficient than viscous impingement filters in removing particles in the submicron range. However, capacities are smaller with the finer media used. Filters of average to medium efficiency are useful in the collection of lint. High efficiency filters are intended primarily for removing particles of small size and concentration. Dry filter life is relatively short, and therefore, maintenance costs are usually higher than for impingement filters.

Automatic dry filters of the roll medium type are seldom used to remove atmospheric dust. They are well suited to the removal of lint as found in textile mills or dry cleaning establishments, and may be used for the removal of ink mist in the printing industry.

High efficiency dry filters *(Fig. 59 and 60)* are especially effective in the removal of viruses and bacteria, and are therefore useful in hospital air conditioning. They may also be considered for protection from nuclear fallout and agents of chemical and biological warfare.

Electronic air cleaner efficiencies compete with the more efficient dry filters in the submicron particle size range. Only the self-cleaning ionizing type is suited to high contaminant concentrations since collector plates rapidly become less efficient as their dust load increases. Where high particle concentrations are encountered, the use of a viscous impingement prefilter should be investigated.

Ionizing type precipitators are useful on high pressure or high velocity applications for the removal of relatively fine dirt which otherwise tends to accumulate around discharge nozzles. Because of the relatively small maintenance requirements of this type of filter, it may be applied to large air deliveries and installations where equipment is relatively inaccessible or where service is infrequent or incomplete.

The characteristics of charged media air cleaners are similar to those of medium efficiency dry filters. Charged media filters are less efficient than ionizing precipitators but electrical failure of the associated equipment does not completely destroy their usefulness. Operation at relative humidities exceeding 70% may adversely affect the dielectric properties of the medium. Individual resistors may be provided for each filter circuit to limit the current flow thru the medium, should the medium become damp.

Regardless of the type of filter selected, the automatic self-cleaning feature renders servicing less dependent on the human element and provides a relatively uniform air resistance and air flow.

Outdoor air and return air may be separately cleaned with different filter types if the characteristics of the contaminants to be removed are widely different.

Table 10 indicates the relative initial and total annual costs of different types of air cleaning installations.

SELECTION

Filter size is usually determined by the rated air quantity per unit or panel as published by the manufacturer. These rated air deliveries are established with regard to practical air velocities as dictated by the characteristics of the medium employed. An overload of as much as 10% to 15% may be permissible, depending on the medium and on the filter construction.

Table 11 is a tabulation of typical velocities and air resistances for various filter types. The air resistances are based on clean filters. Pressure drops reflecting a partially expended medium should be

TABLE 10—RELATIVE AIR CLEANING COSTS*

TYPE OF AIR CLEANER	RELATIVE COST PER 1000 CFM	
	Initial	Annual Owning and Operating**
Viscous Impingement		
Throwaway (2 in.)	0.55	1.45
Renewable (4 in.)	0.80	2.0
Cleanable (2 in.)†	1.0	1.0
(4 in.)	2.3	1.4
Automatic self-cleaning‡	3.6 - 8.0	1.9 - 3.0
Dry Media		
Cleanable and renewable (2 in.)	0.95	1.1
(8 in.)	3.3	1.5
High efficiency renewable	9.5 - 17.6	5.9 - 7.6
Electronic (ionizing)‡		
Plate or cell	12.4 - 21.3	2.9 - 5.9
Automatic	17.8 - 31.1	4.3 - 7.2
Electronic (charged media)‡	10.7 - 14.2	2.9 - 3.7

*3000 hrs per year.
†Basis of comparison.
‡Assumed minimum size of 10,000 cfm.
**Includes interest and depreciation.

used for fan static pressure calculations, as recommended by filter manufacturers.

The sizes of filter units or panels are normally standardized and limited in number. Installations are then built up from the basic units. Manually serviced viscous impingement filters are usually available in sizes 20 in. x 25 in., 20 in. x 20 in., 16 in. x 25 in. and 16 in. x 20 in. Standard thicknesses are

TABLE 11—OPERATING DATA

TYPE OF AIR CLEANER	Nominal Velocity thru Media (fpm)	Resistance thru Clean Filter (in. wg)
Viscous Impingement		
Throwaway (2 in.)	300	0.06 - 0.12
Renewable (4 in.)	300	0.12 - 0.24
Cleanable (2 in.)	300 - 500	0.04 - 0.12
(4 in.)	300	0.08 - 0.20
Automatic self-cleaning	500	0.30 - 0.50
Dry Media		
Cleanable and renewable (2 in.)	60	0.08 - 0.13
(8 in.)	35	0.10 - 0.12
High efficiency renewable	5 - 20	0.50 - 1.20
Electronic (ionizing)*		
Plate or cell	300 - 400	0.15 - 0.30
Automatic	400 - 500	0.20 - 0.32
Electronic (charged media)	35	0.03 - 0.12

*Includes front and rear screens.

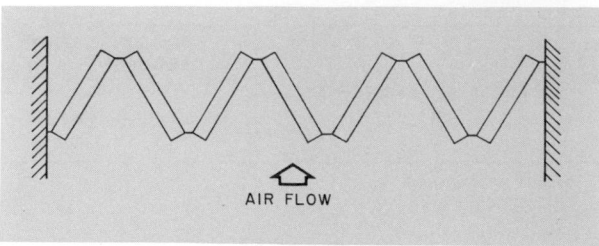

FIG. 64 — UNIT FILTER, "V" BANK

1 in., 2 in. and 4 in. Odd sizes are available only at considerably higher cost unless relatively large quantities are required.

Dry media filters are often available in only one size from each manufacturer, but some may offer a size selection. Charged media electronic filters are also of limited size selection.

Ionizing type electronic cleaners and self-cleaning filters are normally available in height increments of several inches but in standard widths limited to two or three sizes or combinations thereof.

Some flexibility of selection may be exercised in the use of manually serviced mechanical filters by utilizing a "V" bank to obtain a greater ratio of filter area to cross-sectional area. This method is illustrated in *Fig. 64*.

Desirable characteristics for viscous impingement filter adhesive include homogeneity of film, a viscosity relatively constant with temperature change, a resistance to the development of mold spores and bacteria, a high ability to wet and retain dust at all temperatures, minimal evaporation, fire resistance, and freedom from odor.

INSTALLATION

Location

In an air conditioning system, air filters are usually located upstream of the fan, between the cooling coil and preheat coil, if any. This location simplifies duct and casing design for a built-up system, avoids the net static pressure loss associated with an acute transformation downstream of the fan, and produces a more uniform air distribution thru the filters. In addition, a measure of comfort is provided in the winter for the service attendant, and coils are protected from dust deposits and algae formation. Such a location minimizes the possibility of introducing rain or snow, a factor of great importance in the application of electronic air cleaners. In the case of viscous impingement filters, adhesive temperature variations are minimized.

If high efficiency air cleaners are used, the preferred location is downstream of the fan. Any air leakage thru the duct will be outward, and air cleanliness will thus be maintained. With any high efficiency filtering device, mechanical or electronic, located upstream of the fan, the duct or casing between the filter and the fan should be carefully caulked and the connections felted against leakage.

If no preheat coil is used, an electronic cleaning device should be located no closer to the outdoor air intake than the height dimension of the device.

Layout

The unsatisfactory performance of air cleaning devices can often be traced to improper installation practices or the lack of regular maintenance. Therefore, filter installations should be planned to meet engineering requirements and to facilitate service.

An inspection and service area of sufficient depth should be provided before and after the filter bank. A minimum access of two feet for viscous impingement filters and three feet for high efficiency dry filters is suggested. Electronic air cleaners may require five feet on the entering air side to permit the full opening of swinging doors or ionizer panels. The manufacturer's data should be consulted for more detailed information.

Access doors should be installed in the apparatus, upstream and downstream of the filter bank. In addition, ladders or catwalks are required for access to filter tiers at heights above six feet. Electric lights of the marine type facilitate service, and are suggested on both sides of the filter bank.

Viscous impingement and dry media unit filters are usually removed from the entering air side of a bank. However, they may be available for servicing from the leaving air side if such is specifically requested. The air flow arrow should be observed in installing or replacing progressive density filters such as the viscous impingement type.

Duct and apparatus casing, both at the entering and leaving air sides of the filter bank, should be designed and installed to insure even air distribution over the face of the filters. This is especially important in the case of electronic ionizing air cleaners or other low air resistance filters. For this reason perforated plates, grilles or screens are often installed upstream or downstream of electronic ionizing air cleaners. Manufacturers may include such baffling devices with the precipitator unit.

Prefilters may be considered upstream of high efficiency dry media and electronic air cleaners if high dust or lint concentrations are present. These pre-filters also serve to distribute the air uniformly.

Proper provision should be made for the collection and drainage of water if the filters are to be cleaned in place by hoses or nozzles.

Outside air intakes should be located at a height and position such that the introduction of heavy concentrations of surface or roof dirt, automobile fumes and refuse is minimized. Intake screens should be no coarser than 16 mesh. Louvers should be well constructed, particularly if the filter bank is near the intake.

A "V" or staggered filter bank may be employed to increase the ratio of filter surface area to cross-sectional area. In factory-built air conditioning units with filter boxes designed for low velocity filters, it may be necessary to blank off uniformly a portion of the cross-sectional area if high velocity filters are to be used. In this case, the use of a factory-built high velocity filter box is more appropriate.

Operation of electronic filters should be dependent upon electrical interlocks with apparatus access doors in the interests of operator safety. These interlocks interrupt operation as long as an access door remains open.

If sprinkler protection is required, piping must be provided from the building sprinkler system or the city water system. Many air cleaners feature built-in sprinkler provisions requiring only connection.

Mechanical air filter installations should include a draft gage or other differential pressure indicator to signal the need for cleaning or replacing filters or to warn of the failure of automatic filters.

MAINTENANCE

It is difficult to predict, on the basis of filter air resistance, when a manually serviced filter will require cleaning or replacement. Two indicators used to determine the need for servicing are a 10% decrease in air flow or an increase in resistance of two to three times the initial resistance. The intervals between cleanings vary with the application, type of filter and stage of job installation.

The rotation method of cleaning is often used, particularly on extensive installations. Under this method, only certain filter units are cleaned each week. Such a practice insures a more constant work load and a more uniform air resistance at any time.

The size of the installation dictates the most economic manual cleaning means. Filters may be cleaned in place with hoses or fixed nozzles on large

jobs, while smaller installations may favor the use of a filter cleaning tank together with an appropriate number of spare filters.

Self-cleaning filters and precipitators should be observed for the expiration of disposable media or the accumulation of sludge in the collecting pan. Many manufacturers provide signals for their equipment to indicate the need for service.

Manufacturers' recommendations regarding the method and interval of cleaning or replacing filters should be followed.

HEATING DEVICES

The heating devices commonly employed directly with air conditioning systems are designed to heat air under forced convection. They are usually located within the air conditioning apparatus and/or ductwork.

The media used for heating include steam, hot water, electricity and gas flame. In addition, for special applications, glycols and hot refrigerant gas may be used.

STANDARDS AND CODES

Methods of testing and rating forced circulation air heating coils utilizing steam or hot water are prescribed in ASHRAE Standard 33.

Various aspects of the construction and installation of electric heaters are dictated by Underwriters' Laboratories requirements. Installation of such equipment is also governed by the National Electric Code.

The installation and piping of gas-fired duct furnaces is prescribed by the American Standards Association Bulletin 21.30. Installation is also influenced by requirements of the National Board of Fire Underwriters. The manufacture of gas-fired apparatus is directed by standards of the American Gas Association.

The application and installation of all types of heating devices should also conform to local codes and regulations.

TYPES OF EQUIPMENT

Steam Coils

Steam heating coils consist of a series of tubes connected to common headers and mounted within a metal casing. To insure efficient heat transfer, either plate type or spiral fins are bonded to the tubes mechanically or with solder. Tubing is usually constructed of copper, in standard tubing sizes up to and including one inch OD. Fins are often of alumi-

FIG. 65 — STEAM HEATING COIL

FIG. 66 — STEAM DISTRIBUTING TUBE COIL

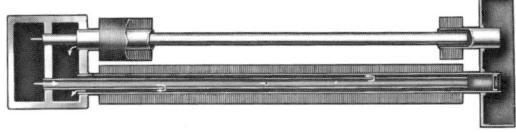

FIG. 67 — STEAM DISTRIBUTING TUBE PRINCIPLE

num with spacings ranging from three to fourteen to the inch. One-row and two-row coils are available with tubes spaced from one to three inches on centers. *Figure 65* illustrates a steam heating coil. The offset tubes provide for changes in length due to temperature variations.

Since the proper performance of steam coils depends on the uniform distribution and condensation of steam in the tube, several methods have been devised to insure this uniformity. Individual orifices may be built into the supply end of each tube, or distributing plates may be installed within the steam header.

Uniform steam distribution and leaving air temperature are also provided with the steam distributing tube type of coil. This design features a tube within a tube, with the inner tube perforated along its length. Steam is supplied to the inner tube and admitted thru the orifices to the outer or condensing tube. Condensate is collected in the return header. A steam distributing tube coil is shown in *Fig. 66*, and the principle is described in *Fig. 67*. In *Fig. 66* note that the tubes are pitched within the casing to promote the rapid return of condensate.

Steam heating coils are available in various tube lengths ranging from one foot to ten feet. Casing widths up to four feet may be obtained for a single coil.

Hot Water Coils

Hot water heating coils are similar in construction, size and appearance to single tube steam coils. Although comfort heating systems seldom require hot water coils of more than two rows, greater depth of surface is available. Fins are usually spaced from a minimum of seven to a maximum of fourteen to the inch. A hot water heating coil is shown in *Fig. 68*.

In order to provide optimum combinations of capacity and water side pressure drop, various circuiting arrangements are employed. On multiple-circuited coils, turbulators are sometimes installed within the tubes to produce the turbulent flow necessary for efficient heat transfer.

Electric Heaters

Electric heating devices are commonly available in either the open type or finned tubular type. These are illustrated in *Fig. 69 and 70*. The open type consists of a series of electrical resistance coils

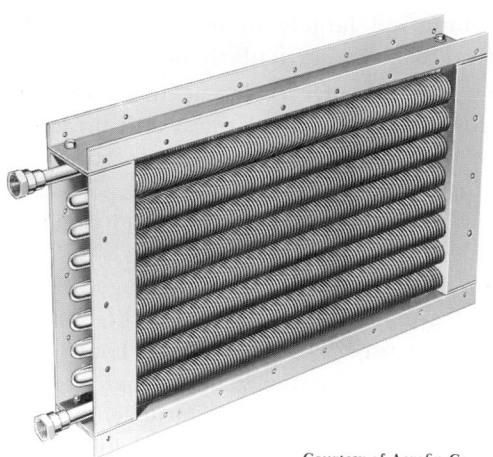

Courtesy of Aerofin Corp.

FIG. 68 — HOT WATER HEATING COIL

Courtesy of Industrial Engineering and Equipment Co.

FIG. 70 — FINNED TUBULAR ELECTRIC HEATER

framed in a metal casing and exposed directly to the air stream. The finned tubular type of heater is made of finned steel sheaths containing resistance wire surrounded by refractory material.

In order to achieve incremental control of heater output, multiple electrical circuits may be obtained. Although normal applications seldom require more than three circuits, as many as required are possible.

Standard voltages include 115 volt, single phase, and 208 and 230 volts in single or three phase. Heaters are also available for operation on 440 and 550 volt service. Direct or alternating current control voltages may be specified.

Duct Furnaces

Gas-fired furnaces are built for installation in air ducts and in some packaged air conditioning units.

Figure 71 shows a duct furnace, and *Figure 51* illustrates the use of a gas-fired furnace in a packaged unit. Such equipment consists of a burner assembly, a heat exchanger, a plenum and controls. Natural, manufactured or liquified petroleum gas may be used.

APPLICATION

Heating devices are used as preheaters and reheaters. A preheater is located upstream of the dehumidifier in an air conditioning apparatus, and is used either to raise the temperature of the entering air to a temperature above freezing or to supply the heat necessary for control of the temperature of the air leaving the dehumidifier. Both functions are often performed by a single heater. A reheater located

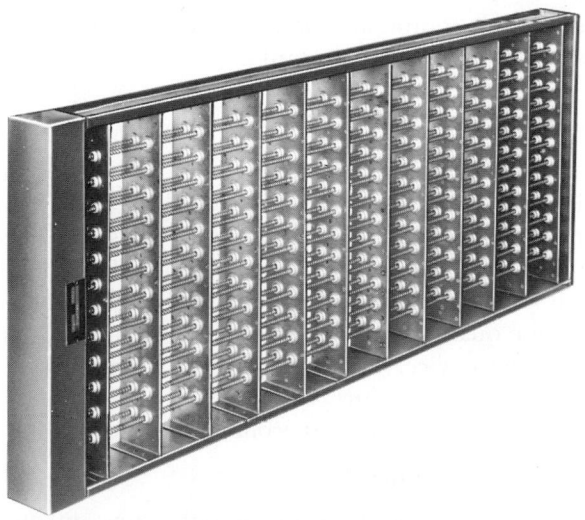

Courtesy of Industrial Engineering and Equipment Co.

FIG. 69 — OPEN ELECTRIC COIL HEATER

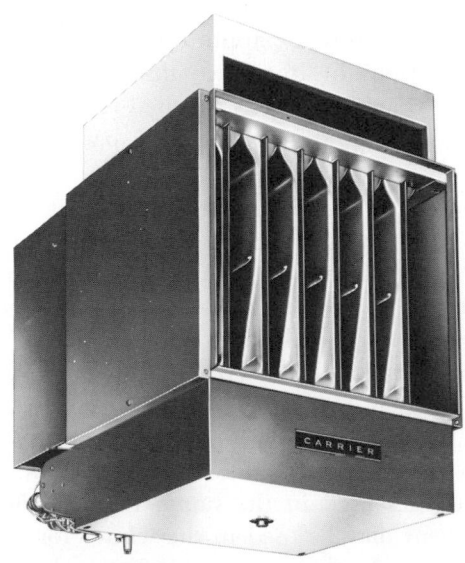

FIG. 71 — DUCT FURNACE

downstream of the dehumidifier is used to control the temperature of the conditioned space when it is subjected to varying cooling loads. *Figure 36* indicates the approximate reheat requirement dictated by a particular design relative humidity and sensible heat ratio. A reheater may also be used as a booster heater, compensating for wide differences in cooling load characteristics between a particular zone and the rest of the space conditioned by an apparatus. If both functions are required, a central reheater may be used to raise the supply air temperature to approximately room temperature or slightly higher. Booster heaters may then be installed in the branch ducts to the various spaces in order to provide control of room temperature.

Steam and hot water heating coils are most commonly employed for the applications cited. Steam coils are normally available for steam pressures up to 200 psig although special coils may be obtained for higher pressures. Hot water heating coils are used on low, medium and high temperature hot water systems. However, applications involving water temperatures exceeding 300 F should be brought to the attention of the manufacturer.

Steam coils of the distributing tube type are preferred over single tube steam coils and hot water coils for service where freezing air temperatures are encountered or where uniform heater leaving air temperatures are mandatory. Single tube steam coils and hot water heaters may, however, be used for preheat service if controlled as described under *Coil Freeze-up Protection*. A minimum entering water temperature of 150 F is suggested for hot water preheat service.

Plate fin steam and hot water coils are preferred to spiral fin coils for applications involving heavy concentrations of lint since they are more easily cleaned. If spiral fin coils are used for such applications, the widest appropriate tube spacing should be chosen.

Where corrosive substances are present in the air, steam or hot water, special coil materials are available. Most steam side corrosion problems may be avoided by the proper trapping and venting of noncondensables.

The advantages of electric heating are low initial equipment and installation cost, a saving of floor space, compactness, simplicity of operation and control, a fast control response and cleanliness. Electrical facilities used in the summer for refrigeration equipment may be used in the winter for the heating system. At the same time electric heating has often proven costly to operate. For this reason it has been employed largely in mild climates or where electrical costs are particularly low.

Since the use of electric heaters eliminates the need for a central heating plant and piping system, electric heating can be applied to tenant areas of shopping centers, department stores, schools, industrial facilities, banks, motels, railroad cars and markets. Electric heaters may be used for churches because of the short duration of usage, the low initial cost and the quick response. Electric heating may also be used in areas such as board rooms or executive offices where occupancy at night or on weekends may be common. It has been used successfully in conjunction with self-contained air conditioning units and as a source of auxiliary heat for heat pump systems.

The open type of electric heater operates at a temperature lower than that of the finned tubular heater, and therefore exhibits a longer life. It is lighter in weight, more rapid in response, offers less air resistance and tends to cycle less. The finned tubular heater is particularly suited to applications where the heater may be subject to mechanical injury or where an explosion hazard exists. Stainless steel fins and sheaths are available for high humidities or corrosive atmospheres.

Gas-fired duct furnaces may be used for preheat and reheat service. The advantages of such equipment are similar to those of electric heaters. Hence, duct furnaces may be used for similar applications in areas of relatively high power cost. As with electric heating the problem of coil freeze-up is not encountered. Gas-fired equipment should never be operated in corrosive atmospheres or in rooms where explosive vapors are present as in paint rooms. This equipment should also be sufficiently removed from acid baths or degreasing tanks.

SELECTION

The selection of a heating device involves a consideration of the heating capacity required, the heating medium available or required and its characteristics, the allowable resistance to the flow of air and/or heating fluid, the entering air temperature, the air quantity to be heated and the air velocity thru the device, dimensional limitations, installation requirements such as the type of control, special design requirements, and economy.

Steam and Hot Water Coils

The capacity of a steam or hot water coil of a given type may be increased not only by increasing the coil surface but also by increasing the coil face velocity thru reducing the face area. Since higher coil face velocities result in higher air side pressure drops, the selection of a coil surface may be more limited than at lower face velocities. Therefore, the size and capacity of a heating coil are interdependent, and each must be determined in relation to the other.

Minimum coil face area is usually determined by the design air quantity and a maximum allowable face velocity. The dimensions of the coil may then be chosen from among those dimensions available with the required face area. For a given face area, coils of greater tube length and smaller tube face are usually the least expensive. However, space requirements may limit both size and dimensions of a coil.

Coils are rated at face velocities of 300 to 1500 feet per minute. The maximum face velocity should be determined by the allowable air side pressure drop and the ambient sound level of the space served by the coil. Air pressure drops of 0.10 to 0.30 in. wg are suggested for preheat applications, while reheat coil friction may range from 0.15 to 0.35 in. wg.

Since heating coils do not condense moisture, and since no entrainment of moisture is possible, the face velocity of a heating coil mounted within a factory-built air conditioning unit should not be limited to the cooling coil face velocity.

The calculated heating load required of a coil is usually the primary determinant of the surface selected. Various combinations of fin spacing, tube spacing and coil depth result in a wide variety of available surfaces. The heat transfer capacity of a given surface varies directly with face velocity, steam pressure, entering water temperature or water tube velocity. It varies inversely with entering air temperature.

Reheat coils are usually oversized. A 15% to 25% safety factor added to the calculated heating load provides for a rapid morning pick-up and compensates for duct heat losses. Steam preheat coils chosen to operate at subfreezing air temperatures with throttling control of steam should be undersized rather than oversized, if the required load cannot be met exactly. This practice reduces valve throttling at air temperatures of 25 F to 32 F, the range where excessive throttling most usually results in the freezing of condensate in the tubes.

When using duct reheat coils for large air quantities, it may be more economical to select a smaller coil to handle only a portion of the air, with the remainder being handled thru a fixed bypass around the coil. The air thru the coil is then heated to a higher temperature so that the mixture air is at the proper temperature. This may require a coil with more heating surface per square foot of face area, but in a smaller casing size. Assuming a coil face velocity, fin spacing and rows of coil, the coil air quantity is determined by dividing the required over-all temperature rise by the coil temperature rise, and multiplying by the total air quantity. The required coil face area can be found from the coil air quantity and the assumed coil velocity. The coil size is then selected to match closely the calculated face area. The coil bypass is sized as described in *Part 2*, and the dimensions are chosen to coincide with the coil casing length.

Figure 72 illustrates a steam coil selection table. Coil capacity may be expressed in terms of steam quantity condensed, heat transferred or final air temperatures alone. Hot water coil ratings may be similarly tabulated, except that, at each entering air temperature, capacities are listed for each surface at various entering water temperatures. Another method of presenting hot water heating coil ratings is illustrated in *Fig. 73*.

Heating coil performance ratings assume a rapid elimination of air and other noncondensables and a uniform distribution of air thru the coil surface.

When steam coils are selected at face velocities exceeding those considered standard by the manufacturer, the amount of condensate per tube should be checked against the maximum recommended by the manufacturer. If the maximum allowable condensate per tube is exceeded, excessive steam pressure drops, water hammer and poor venting may result.

Electric Heaters

In addition to size and capacity an electric heater selection should specify electrical characteristics and the number of circuits required.

Electric heaters are usually chosen to fit a branch duct of given dimensions without requiring entering and leaving transformations. Therefore, face velocity is not the usual determinant of coil size, although for Underwriters' Laboratories approval a minimum face velocity must be maintained and uniform airflow provided. This minimum velocity is a function of entering air temperature and the total watts per square foot of duct area. Velocities may range up to 1800 feet per minute. Air side pressure drops are small compared to steam and water coil pressure drops, seldom exceeding 0.10 in. wg for an open type coil.

All of the electrical energy used in an electric heater is converted to heat. Thus, heater capacities

5 LB STEAM — 227 F

ENT AIR TEMP (F)	COIL SUR-FACE	COIL FACE VELOCITY (FPM)											
		300		400		500		600		700		800	
		Final Temp (F)	Cond*	Final Temp (F)	Cond*	Final Temp (F)	Cond*	Final Temp (F)	Cond*	Final Temp (F)	Cond*	Final Temp (F)	Cond*
0	A	43.5	14.6	39.1	17.5	36.0	20.2	33.8	22.8	31.9	25.0	30.2	27.3
	B	62.5	21.0	56.7	25.5	52.4	29.4	49.7	33.5	47.3	37.2	45.0	40.5
	C	76.0	25.6	68.0	30.6	62.0	35.0	57.5	38.9	54.0	42.5	51.0	46.0
	D	108.9	36.8	100.6	45.4	93.2	52.6	88.1	59.7	84.4	66.6	80.6	72.7
	E	125.0	42.3	114.6	51.8	107.1	60.4	101.2	68.5	96.3	76.0	92.4	83.5
40	A	75.8	12.0	72.2	14.4	69.7	16.6	67.9	18.8	66.3	20.6	64.9	22.1
	B	91.5	17.4	86.7	21.0	83.2	24.2	81.0	27.6	79.0	30.5	77.1	33.3
	C	102.6	21.0	96.0	25.2	91.1	28.8	87.4	32.0	84.5	35.0	82.0	38.0
	D	129.7	30.2	122.9	37.4	116.8	43.4	111.6	48.4	109.6	55.0	106.5	60.0
	E	143.0	34.8	134.5	42.6	128.3	49.9	123.4	56.4	119.3	62.8	116.1	68.6

*Condensate in pounds per hour per square foot face area.

FIG. 72 — TYPICAL STEAM COIL RATINGS

in Btuh are determined by multiplying the kilowatt rating of the heater by 3412.

The number of circuits chosen depends on the degree and period of heating load fluctuations. The amount of control hunting permitted should be weighed against the economics of purchasing and installing the multiple circuit heater.

Duct Furnaces

Gas-fired duct furnaces are chosen according to the output satisfying the heating load. The required air temperature rise thru the furnace determines the air quantity to be handled by the device and the resulting air friction. If a greater branch duct air quantity is required, a fixed bypass may be provided as described under *Steam and Hot Water Coils*.

Atmospheric Corrections

Heating coil ratings are based on the standard atmospheric conditions of 29.92 in. Hg barometric pressure and 70 F. For significantly different air conditions such as at altitudes exceeding 2000 feet or at average air temperatures above 125 F, a correction should be applied to the required air temperature rise and the air quantity upon which the selection is based.

In determining the coil air temperature rise required or the heating load imposed by the admittance of air at temperatures below the design temperature, such as thru ventilation or infiltration, the factor 1.08 should be adjusted in proportion to the ratio of air densities as found from *Chart 2*.

The design air quantity should be multiplied by the density ratio in order to determine the equivalent air flow at sea level. The adjusted air quantity and heating load (or air temperature rise) are then used to select a coil surface.

The coil size and face velocity are determined by the design air flow with no correction applied. However, the coil air side pressure drop should be corrected as described in *Part 2*.

Since the capacity of an electric heater does not depend on air quantity, no correction to the ratings

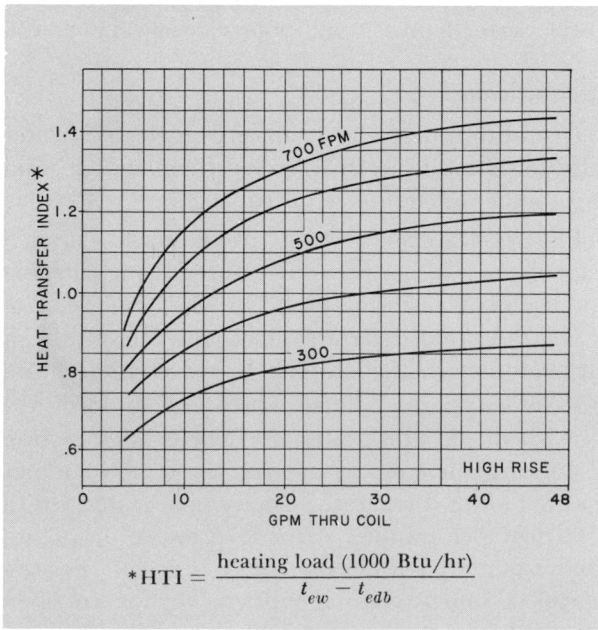

$$*HTI = \frac{\text{heating load (1000 Btu/hr)}}{t_{ew} - t_{edb}}$$

FIG. 73 — TYPICAL HOT WATER COIL RATING CURVES

is required. However, the heating load and air friction should be corrected as necessary.

Gas-fired duct furnaces employed at elevations exceeding 2000 feet should be derated in output by 4% for each 1000 feet above sea level.

COIL FREEZE-UP PROTECTION

The exposure of hot water or steam preheat and reheat coils to subfreezing temperatures, either by accident or intent, makes possible the freezing of water accumulated within the tubes and thus costly damage. The prevention of such occurrences requires consideration of the problem in the apparatus design and layout, in the selection of equipment, and in the choice of control methods.

The primary requirement for positive freeze protection is the assurance of uniform coil leaving air temperatures. Air temperature stratification may be caused by incomplete mixing of outdoor and return air or by an uneven temperature rise thru the coil.

Where the mixing of outdoor and return air takes place upstream of a heating coil, mixing should be promoted by introducing the denser cold air at the top of the plenum and by providing as much airway length as possible. If a steam coil is employed, the steam should be supplied from the naturally colder side of the plenum.

If the mixing of outdoor and return air is to occur downstream of a preheat coil, it is suggested that only the minimum outdoor air be heated and that the maximum outdoor air dampers be maintained closed at subfreezing temperatures. If efficient downstream mixing has been provided, the preheat coil may be used instead to heat the return air to a temperature predetermined to yield the desired mixture air temperature.

Where a steam preheat coil served by a modulating valve tempers outdoor air, freezing of condensate in the tubes occurs most often at entering air temperatures in the range of 25 F to 32 F. Within this range the coil is usually operating under a severe partial load. The relatively small amount of steam admitted to the coil condenses completely before the end of the tube is reached, resulting in stratification. For this reason, if modulating control of steam is required at subfreezing entering air temperatures, the use of the steam distributing tube type of coil is suggested.

Single tube steam and hot water coils may be used for the tempering of subfreezing air, but the heating medium should not be modulated at entering air temperatures below 35 F. However, in cold climates such a design may produce overheating. To provide a degree of control while avoiding stratification, two preheat coils in series, each furnishing a part of the required capacity and controlled in sequence, may be employed. An alternative method consists of the use of face and bypass dampers controlled by a plenum thermostat. The bulb of such an instrument should be located well downstream of the heating coil if space permits. Otherwise, an averaging type bulb should be used.

These same methods of obtaining control without stratification should be considered where steam distributing type coils of large capacity are used for preheat service. However, rather than employing two coils in series, it may prove economically preferable to utilize one coil with two control valves piped in parallel. The first valve to open may be sized to pass the minimum steam quantity necessary for even distribution of steam thru the tubes at a signal from a two-position outdoor air thermostat.

As mentioned previously, a minimum entering water temperature of 150 F is suggested for tempering subfreezing air with hot water. Uniform leaving air temperatures should be insured as described above. In addition, a safety control closing the outdoor air damper at entering water temperatures below 150 F is suggested.

Another method of freeze protection is the circulation of an inhibited glycol solution thru a water coil. A two-row coil with a single circuit is the best protection against stratification. The system should be designed to supply the glycol solution to the coil at a temperature of about 150 F at peak conditions with a high temperature drop of about 50 degrees. The steam valve to the glycol heat exchanger is controlled by the air temperature leaving the coil.

Another requirement for adequate freeze-up protection of steam coils is the positive and complete drainage of condensate from the tubes. Any type of steam coil may be damaged if condensate is allowed to accumulate and freeze thru poor design of the system or coil. For this reason an ideal position for a steam preheat coil is with the tubes vertical and with the condensate header at the bottom. For either horizontal or vertical air flow, steam preheat coils installed with tubes horizontal should be pitched downward toward the condensate header to facilitate drainage. Many steam distributing tube coils feature tubes internally pitched for either horizontal or vertical air flow. In this case installation is simplified, and the only precaution necessary is to make sure that the condensate header is lower than the steam header if air flow is vertical.

Positive condensate drainage is also insured by the proper design of the condensate return system. Adequately sized steam traps and vacuum breakers are among the most important design considerations. Refer to the discussion in this chapter under *Layout* and to *Part 3*.

Larger heater tube diameters provide more positive condensate drainage and more uniform leaving air temperatures.

The outdoor air dampers of an apparatus should be closed whenever the fan is not running. In this way the induction of cold air by stack effect thru an inoperative coil is minimized.

Steam preheat coil control valves, if used, should be of the "normally open" type. Such valves should be sized to provide the maximum required capacity at a large pressure drop. Since valve capacity varies as the square root of the pressure drop, the valve tends to be undersized if steam pressure falls, and freezing is less likely to occur within the coil.

Although occurring less frequently, the freezing of reheat coils may be a problem, particularly if preheaters are not employed. If such is the case, the same provisions as described for preheat coils should be considered if complete mixing of outdoor and return air cannot first be guaranteed. Where face and bypass control of a dehumidifier is employed, the preheat coil should be located so that the air bypassed to the reheat coil is tempered as well as the dehumidified air.

INSTALLATION

Location

In an air conditioning apparatus the preheater is usually located between the outdoor air intake and the filters. Reheaters are mounted downstream of the dehumidifier coil, either within the apparatus or in the ducts. The latter location is often chosen where more than one control zone is served by a single air conditioning unit.

Duct-mounted heating devices may be located outdoors as well as indoors. Heater and duct should be externally insulated and weatherproofed. As much of the steam condensate return system as possible should lie within the heated space. Terminal boxes for electric heaters should be weatherproof.

Layout

Hot water coils and single-pass steam coils of both the single tube and steam distributing types may be installed with tubes horizontal or vertical and used for vertical or horizontal air flow. Multi-pass steam coils designed for use also with hot water are limited to horizontal tube applications. Regardless of the orientation, steam coils should be mounted so that the condensate connection is below the steam connection.

Steam and water coils may be assembled in banks. Coils so mounted should be supported individually in angle iron frames, thus protecting the lower coils from damage and facilitating coil removal.

Sufficient access space should be provided around a heater to permit maintenance and removal. Connections to ductwork should be so constructed to allow coil removal without disturbing the duct. Duct access doors on either side of the coil permit cleaning of the equipment in place. Refer to *Part 2* for a description of the design of ductwork surrounding a heater.

A fixed heater bypass may be located around or below the heating surface. A single-acting bypass damper with blades inclined toward the leaving air side of the heater promotes the mixing of heated and bypassed air.

The design of hot water and steam coil piping is described in *Part 3*. Hot water coils should be piped so that the water enters at the bottom connection, and coil vents should be provided as required. In a steam coil installation where the condensate return main is higher than the coil steam trap, a condensate pump, lift trap or boiler return trap should be used to move the condensate to the main. A minimum of 18 inches should be maintained between a steam coil condensate outlet and the floor to provide space for traps and piping.

In the design of heater installations considerations should be given to the prevention of air temperature stratification. Uneven air temperature rises thru a heating coil may result not only in coil freeze-up problems but also in the supplying of air of nonuniform temperature to branch ducts splitting off downstream of the heating surface. Stratification may be minimized by the proper design of duct splits, by the use of two coils mounted in parallel and supplied from opposite sides of the apparatus and, if necessary, by the use of individual duct heaters. A horizontal supply air split is suggested with single fan air handling units, while a vertical split is more appropriate for multi-fan units. Other measures to reduce stratification such as the use of steam distributing tube coils are outlined in the section dealing with coil freeze-up protection.

The location and layout of electric heaters and gas-fired duct furnaces relative to surrounding com-

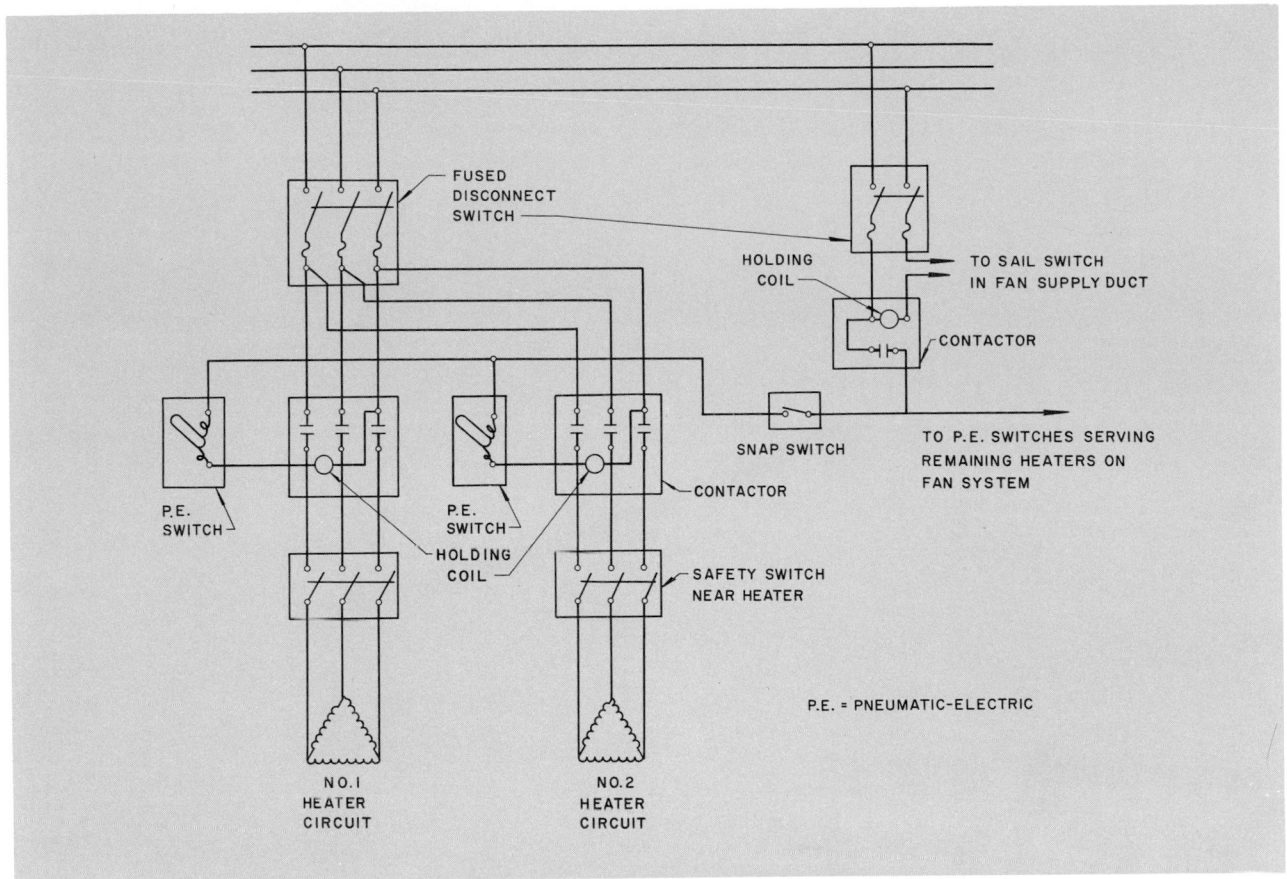

FIG. 74 — ELECTRIC HEATER CONTROL

bustible surfaces is limited by applicable standards and codes.

When locating electric heaters in equipment in front of fan motors, overheating of the motors may result because of the high temperatures that can be obtained.

Duct furnaces may be grouped in series or parallel. Outdoor air of approximately 14 cubic feet of air per cubic foot of gas should be provided. Flue design should be in accordance with American Gas Association standards.

CONTROL

The capacity of a heating coil may be varied in accordance with the load by control of the flow of the heating medium, by air volume control, or by air bypass control. The control of steam or hot water flow is most commonly employed. A multi-zone air conditioning apparatus may utilize air bypass control. If steam is used in this case, "on-off" control of the coil is preferred to minimize stratification.

Where a reheat coil has been selected with excess capacity as suggested above, the use of two control valves mounted in parallel and furnishing respectively one-third and two-thirds of the steam required may improve the accuracy of control at relatively low heating loads.

Figure 74 illustrates the wiring and control of electric heaters. The controlling instrument shown is a pressure-electric switch actuated by a pneumatic zone thermostat. Electric thermostats may also be utilized. The supply duct sail switch insures that the heaters operate only when the fan is running. On a single-heater fan system, a thermal switch may also be used for this purpose. Where multiple circuit heaters are employed, individual pressure-electric switches and contactors are used for each circuit. A step thermostat may be used in place of the pressure switches.

Gas-fired duct furnaces require safety controls such as gas valve low voltage control, a normally closed gas valve, a high bonnet temperature cutout, a pilot safety control, and a gas pressure regulator for other than LP gas.

Part 7
REFRIGERATION EQUIPMENT

CHAPTER 1. RECIPROCATING REFRIGERATION MACHINE

This chapter presents data to guide the engineer in the practical application of reciprocating refrigeration machines when used in conjunction with air conditioning systems.

The main component of these machines is the reciprocating compressor which is a positive displacement device employing the vapor compression cycle, and which is applied with refrigerants having low specific volumes and relatively high pressure characteristics.

A reciprocating refrigeration machine may be classified as one of the following:

1. A compressor unit consisting of a compressor, motor, and safety controls mounted as a unit.
2. A condensing unit consisting of a compressor unit plus an interconnected water-cooled or air-cooled condenser mounted as a unit.
3. A water-chilling unit consisting of either a compressor unit or a condensing unit, plus an interconnected water cooler and operating controls mounted as a unit.

Figures 1, 2 and 3 show a compressor unit, condensing unit, and water-chilling unit respectively.

TYPES OF COMPRESSORS

Compressors may be classified as either open or hermetic.

OPEN COMPRESSOR

An open compressor requires an external drive (*Fig. 1*) and may be direct driven thru a coupling or belt driven to operate at a specific speed, depending on load requirements. The type of the drive (electric motor, internal combustion engine or steam turbine) may be selected to provide sufficient horsepower to match the job requirement. This compressor may use any type of electric motor.

HERMETIC COMPRESSOR

A hermetic compressor has an electric motor and a compressor built into an integral housing (*Fig. 4*). The motor and compressor utilize a common shaft and bearings. The motor is generally cooled by suction gas passing thru the windings but may, in some cases, be water-cooled. These compressors eliminate problems of motor mounting, coupling alignment, motor lubrication, and refrigerant leakage at the shaft seal.

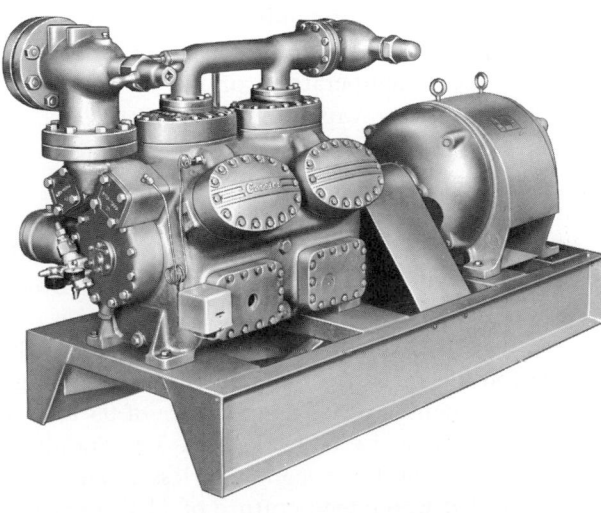

FIG. 1 — OPEN COMPRESSOR UNIT

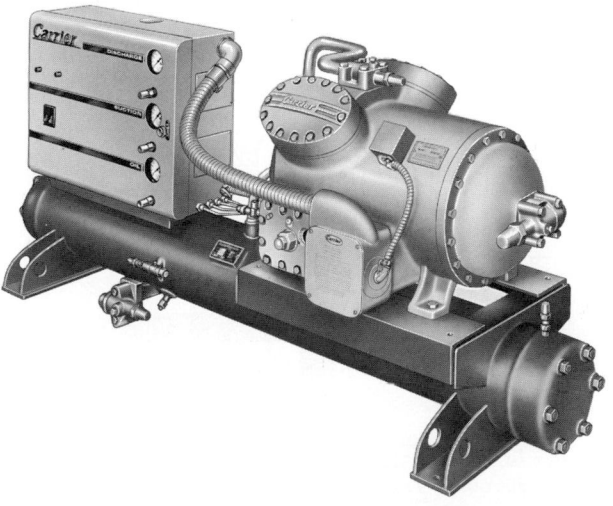

FIG. 2 — HERMETIC CONDENSING UNIT

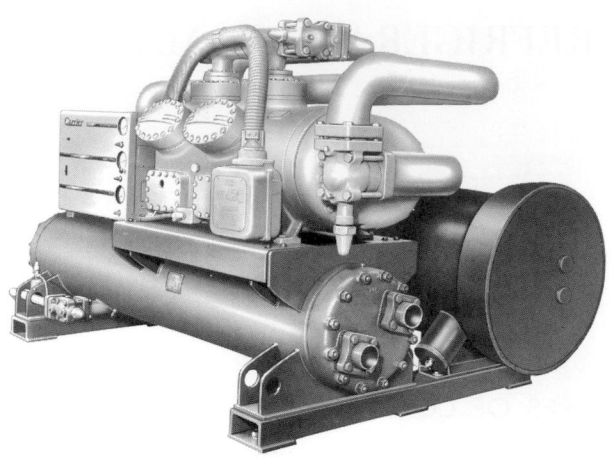

FIG. 3 — WATER CHILLING UNIT

The compressor operating limits depend on the refrigerant used and the horsepower output of the motor. Generally, the motor horsepower is matched to the compressor and a refrigerant so that the motor does not overload when the unit operates within normal air conditioning levels. Hermetic compressors may be classified as either (1) sealed (requiring factory service for repairs) or (2) accessible (permitting service on the job).

APPLICATION

Refrigeration applications up to 60 tons generally utilize reciprocating compressors. From 60 to 200 tons either reciprocating compressors or other types such as centrifugal water chillers or absorption machines are used. Above 200 tons centrifugal water chillers or absorption machines are normally used.

A compressor unit must be combined with a device such as an air-cooled, water-cooled or evaporative condenser to condense the refrigerant. In a field built-up system this combination or a factory-assembled condensing unit can be applied with

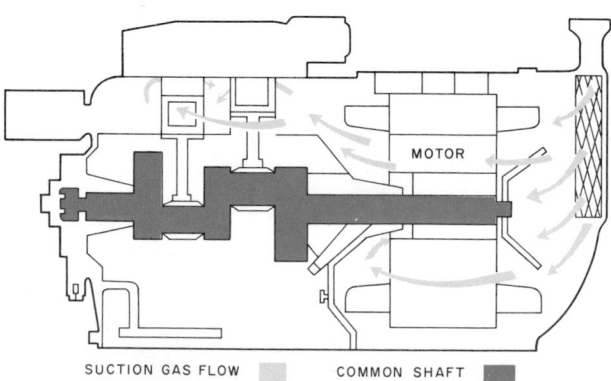

SUCTION GAS FLOW   COMMON SHAFT

FIG. 4 — CUTAWAY VIEW OF HERMETIC COMPRESSOR

direct expansion evaporator coils in fan-coil equipment or a built-up apparatus. It may also be applied to a water or brine chiller or for any other type of refrigeration duty.

A water-chilling unit may be applied to an air conditioning system or to any process requiring chilled water. Package water-chilling units can be obtained complete with or without a water-cooled condenser so that an air-cooled or evaporative condenser can be utilized if desired. When two units are required, they may be applied with their coolers piped in parallel or series water flow. Series connected coolers can be used on high rise water systems to effect a saving in the over-all horsepower per ton required for the system.

STANDARDS AND CODES

The location and installation of a reciprocating compressor should be in accordance with local and other code requirements.

The equipment should be manufactured to conform to the ASA B9.1 Safety Code for Mechanical Refrigeration.

The cooler, condenser and accessories of the system should be built to conform to the ASME Unfired Pressure Vessel Code. This code covers the minimum construction requirements for design, fabrication, inspection and certification of unfired pressure vessels.

ARI Standards for open and sealed compressors establish recommended specifications for (1) standard equipment, (2) methods of testing and rating, including Standard Rating Conditions, and (3) provisions for safety. The Standard Rating Condition usually published by the manufacturer for a compressor unit used for air conditioning duty is Group IV which is based on an entering saturated refrigerant vapor temperature of 40 F, an actual entering refrigerant vapor temperature of 55 F, a leaving saturated refrigerant vapor temperature of 105 F, an ambient temperature of 90 F, and no liquid subcooling.

ARI Standards for a Reciprocating Liquid Chilling Package establish a Standard Rating Condition for a water-cooled model of a leaving chilled water temperature of 44 F, a chilled water range of 10 F, a .0005 fouling factor in the cooler and condenser, a leaving condenser water temperature of 95 F, and a condenser water rise of 10 degrees. The Standard Rating Condition for a condenserless model is a leaving chilled water temperature of 44 F, a chilled water range of 10 degrees, a .0005 fouling factor in the cooler, and a condensing temperature of 105 F or 120 F.

These Standard Rating Conditions can be used to make comparisons between compressors. When comparing catalog ratings of compressors of different manufacturers, the rating conditions must be known, particularly the amount of subcooling and superheating needed to produce the capacities shown.

Specifications should call for conformance to these standards and codes to assure a high quality product.

UNIT SELECTION

The selection of a reciprocating refrigeration machine is influenced by the economic aspects of the complete system; a balance between first cost and operating cost should be considered. The evaporator as well as the heat rejection equipment should be included in the analysis. Refer to *Chapter 5* for the economic consideration of the heat rejection equipment. Refer to *Part 6* for a discussion of dehumidifiers.

COMPRESSOR UNIT

Factors which influence the selection of a compressor unit include the following:

1. *Capacity* — The amount of heat to be exchanged by the refrigeration system in the evaporator. This heat depends on refrigerant weight flow and on entering and leaving enthalpy of the refrigerant at the evaporator.

2. *Evaporator Temperature* — The refrigerant temperature required to absorb heat from the medium being cooled. Lowering the evaporator temperature 10 degrees from a base of 40 F and 105 F reduces the capacity about 24%, and at the same time increases the compressor horsepower per ton about 18%.

3. *Condensing Temperature* — The refrigerant temperature required to reject heat to the condensing medium. Increasing the condensing temperature 15 degrees from a base of 40 F and 105 F reduces the capacity about 13%, and at the same time increases the compressor horsepower per ton about 27%.

4. *Refrigerant* — The three main refrigerants used in reciprocating compressors are Refrigerants 12, 22 and 500. *Table 1* indicates some comparative data on these three; refer to *Part 4* for more information.

5. *Subcooling of the Condensed Refrigerant* — Subcooling increases the potential refrigeration effect by reducing the percentage of liquid flashed during expansion. Subcooling may be accomplished in the condenser, in an external subcooler, or in a liquid suction heat exchanger. For each degree of subcooling, the compressor capacity is increased about 0.5% due to the increased refrigeration effect per pound of refrigerant flow.

6. *Superheating of the Suction Gas* — Superheating can occur by heat pickup in piping outside the cooled space, in a liquid suction heat exchanger, or in an evaporator within the cooled space. Superheating increases the compressor capacity slightly (0.3-1.0% per 10 degrees) when using Refrigerants 12 and 500, providing the heat absorbed by the vapor represents useful refrigeration such as coil superheat, not superheating from a liquid suction heat exchanger. Superheating from a liquid suction heat exchanger increases the capacity of the compressor by the subcooling effect on the condensed liquid. Compressor ratings for Refrigerants 12 and 500 are generally published

TABLE 1—COMPARATIVE DATA OF REFRIGERANTS

Refrigerant No. (ARI Designation)	12	22	500
Chemical Name	Dichlorodifluoro-methane	Monochlorodifluoro-methane	Azeotrope of Dichlorodifluoro-methane and Difluoroethane
Chemical Formula	CCl_2F_2	$CHClF_2$	73.8% CCl_2F_2 and 26.2% CH_3CHF_2
Boiling Point at 1 atmosphere (F)	−21.62	−41.4	−28.0
Saturation Pressure (psig) at: 40 F / 105 F	51.67 / 141.25	83.72 / 227.65	60.94 / 167.85
Net Refrigerating Effect (Btu/lb) 40F to 105F (no subcooling)	49.13	66.44	59.82
Displacement (cfm/ton), 40 F to 105 F (no subcooling, no superheating)	3.14	1.98	2.69

with an actual suction gas temperature of 65 F as a maximum.

Although tests using Refrigerant 22 in a compressor indicate a negligible increase in capacity due to improvement in the volumetric efficiency, superheating is not recommended because of the possibility of compressor overheating. Therefore, compressor ratings for Refrigerant 22 are generally published with only the superheat obtained with the normal expansion valve action and line transmission losses. Superheat is generally limited to 15 to 20 degrees between 40 F and 20 F saturated refrigerant temperatures respectively. Liquid suction heat exchangers are not used with Refrigerant 22 except where necessary to evaporate liquid refrigerant in the oil returning to the compressor and to eliminate liquid slugging to the compressor.

7. *Refrigerant Line Pressure Drops* — The operation of the reciprocating compressor in a refrigeration system is similar to that of a pump in a water system. It must be selected to overcome the system resistance and to produce the required refrigerant flow. At normal air conditioning levels, a piping loss equivalent to approximately 2 degrees is allowed in the suction piping, and a loss equivalent to 2 degrees is allowed in the hot gas discharge piping. Thus, when an evaporator requires a refrigerant temperature of 42 F to handle a required load, the compressor must be selected for a 40 F suction temperature (42 − 2 piping loss). Correspondingly, if the condenser requires 103 F to reject the proper amount of heat, the compressor must be selected for a 105 F condensing temperature (103 + 2 piping loss).

At lower suction temperatures it may be necessary to allow a larger difference between the evaporator and compressor suction temperatures because of the pressure-temperature relationship of the refrigerant. For example, with Refrigerant 12 a pressure change of 1 psi at 0 F is equivalent to a temperature change of 2 degrees, and at +50 F equivalent to 1 degree. For Refrigerant 22 a change of 1 psi at 0 F equals a change of 1¼ degrees, and at +50 F equals ⅔ degree. For Refrigerant 500 a change of 1 psi at 0 F equals a change of 1⅔ degrees, and at +50 F equals ¾ degree.

8. *Operating Limits* — The manufacturer of a compressor unit generally states the operating limits of his compressor. Capacity limits may be shown in the rating tables with a note stating that extrapolation of the ratings is not permitted. There may be limitations on suction temperature, superheat, compression ratio, discharge temperature, compressor speed, horsepower, or motor cooling. As examples of these limitations, most manufacturers limit (1) the saturated suction temperature to a maximum of 50 F, (2) the compression ratio for Refrigerant 22 compressors to 5, and (3) the discharge temperature at the discharge valve to 275 F. This compression ratio limitation may be exceeded if water-cooled heads are used. The horsepower or kilowatt input may be limited by the size of motor available with the compressor. Generally, this limitation occurs with hermetic compressors which have built-in motors of specific sizes. If 50-cycle motors are used, the compressor speed is reduced and the capacity has to be adjusted accordingly. For all practical purposes the capacity is approximately proportional to the speed.

9. *Heat Rejection* — In order to select a condenser to match the compressor, the heat rejection of the compressor must be known. This is usually given in the ratings by the manufacturer, or may be approximated by multiplying the capacity by a given heat rejection factor. The heat rejection is dependent upon the compressor brake horsepower for the conditions involved, less the heat transferred to ambient air, jacket water or oil cooler during the compression of the refrigerant vapor. For an open type compressor the heat rejection can be approximated by adding the required brake horsepower converted to tons to the refrigeration capacity. (To convert brake horsepower to tons, multiply by 0.212.) For a gas-cooled hermetic type compressor the heat rejection can be approximated by adding the kilowatt input in tons (or Btu) to the refrigeration capacity. (To convert kilowatts to tons, multiply by 0.285.)

A typical rating of an open type compressor unit is shown in *Table 2*.

Example 1 — Selection of a Compressor Unit

Given:
Evaporator conditions

Load	= 36 tons
Evaporator temperature	= 30 F
Coil superheat	= 20 deg
Suction line pressure drop	= 2 psi

TABLE 2—OPEN COMPRESSOR RATINGS

REFRIGERANT 12   8-CYLINDER COMPRESSOR

Suction Temp (F)	CONDENSING TEMPERATURE (F)														
	90			100			105			110			120		
	Cap. (tons)	Power Input (bhp)	Heat Rej (tons)	Cap. (tons)	Power Input (bhp)	Heat Rej (tons)	Cap. (tons)	Power Input (bhp)	Heat Rej (tons)	Cap. (tons)	Power Input (bhp)	Heat Rej (tons)	Cap. (tons)	Power Input (bhp)	Heat Rej (tons)
−40	5.1	15.3	7.9	4.3	14.3	6.7	—	—	—	—	—	—	—	—	—
−30	8.0	20.9	11.7	6.9	20.3	10.5	6.3	19.9	9.8	5.8	19.4	9.1	—	—	—
−20	11.8	26.1	16.8	10.4	26.0	15.3	9.8	25.9	14.5	9.1	25.7	13.8	7.9	25.1	12.3
−10	16.7	30.8	22.9	15.0	31.4	21.2	14.2	31.5	20.3	13.4	31.6	19.4	11.8	31.6	17.8
0	22.7	35.0	30.1	20.7	36.2	28.1	19.6	36.7	27.1	18.6	37.1	26.1	16.6	37.8	24.1
10	29.9	38.6	38.3	27.5	40.5	36.0	26.2	41.3	34.9	25.0	42.1	33.7	22.5	43.5	31.4
20	38.4	41.5	47.5	35.5	44.1	44.9	34.0	45.3	43.6	32.5	46.5	42.3	29.5	48.7	39.6
30	48.2	43.5	57.7	44.8	47.0	54.7	43.0	48.6	53.2	41.2	50.2	51.7	37.7	53.3	48.6
40	59.5	44.7	68.8	55.4	49.0	65.5	53.3	51.1	63.7	51.3	53.2	62.0	47.1	57.1	58.5
50	72.2	44.9	80.8	67.5	50.1	77.1	65.1	52.7	75.1	62.7	55.3	73.2	57.9	60.2	69.1

NOTES:

1. Where values are not shown for capacity and power input, the operating conditions are beyond the operating limits of the compressor.

2. Capacities are based on liquid subcooling of 15 degrees in the system.

3. Although interpolation is permitted, do not extrapolate. Operation outside limits of the table is not allowed.

4. The refrigerant temperatures shown are the saturation temperatures corresponding to the pressures indicated at the compressor. The actual gas temperatures are higher because of superheat.

5. Ratings are based on operation at 1750 rpm. See table below for multiplying factors for other speeds.

Multiplying Factors for Other Speeds

Rpm	1450	1160
Capacity	0.835	0.674
Bhp	0.798	0.602

RATING BASIS AND CAPACITY MULTIPLIERS FOR REFRIGERANT 12 AND REFRIGERANT 500

Saturated Suction Temp (F)	Rated Suction Gas Temp (F)	CAPACITY MULTIPLIERS									
		Actual Suction Gas Temperature to Compressor (F)									
		−20	−10	0	10	20	30	40	50	60	65
−40	35	0.927	0.940	0.953	0.967	0.980	0.993				
−30	45	0.922	0.934	0.946	0.958	0.970	0.982	0.994			
−20	55	0.920	0.931	0.941	0.952	0.963	0.973	0.984	0.995		
−10	65		0.930	0.939	0.949	0.958	0.967	0.977	0.986	0.995	1.000
0	65			0.940	0.949	0.958	0.968	0.977	0.986	0.995	1.000
10	65				0.950	0.959	0.968	0.977	0.986	0.995	1.000
20	65					0.960	0.969	0.978	0.987	0.996	1.000
30	65						0.970	0.979	0.987	0.996	1.000
40	65							0.987	0.992	0.997	1.000
50	65								0.997	0.999	1.000

Example 1 (contd)

Condensing temperature = 105 F
Compressor speed = 1750 rpm
Subcooling (water-cooled condenser), assume = 5 deg
Refrigerant 12

Find:

Compressor size, horsepower, heat rejection

Solution:

Suction line loss = approximately 1.4 deg per psi at 30 F

Suction temperature = $30 - (1.4 \times 2) = 30 - 2.8$
= 27.2 F or 27 F

Compressor capacity correction factors

Superheat correction, 27 F suction, 20 deg superheat
= .985

Subcooling correction = $1 - .005 (15 - 5) = 1 - .05 = .95$

Equivalent capacity = $\dfrac{36}{.985 \times .95} = 38.5$ tons

Select an 8 cylinder compressor. By interpolation from *Table 2*, at 27 F suction and 105 F condensing,

Compressor capacity = 40.3 tons
Power input = 47.6 bhp
Heat rejection = 50.3 tons

CONDENSING UNIT

There are two types of condensing units, water-cooled and air-cooled.

The selection factors stated under *Compressor Unit* also apply to condensing units with the addition of the following.

For *Water-Cooled Condensing Units:*

1. Condenser Water Source — Water for condensing may be available from such sources as city water, well water, river water, sea water, cooling tower, or spray ponds. If a choice is available, the selection of a water source is generally a matter of economics. The cost of city water is a factor, and also there may be sewage charges for disposing of water used in a once-thru system. The cost of the tower or spray pond as well as the water conditioning cost also influences the choice of a source.

2. Fouling Factor — Fouling factors constitute the thermal resistance to heat flow introduced by scale and other water impurities. Normally, manufacturers rate a water-cooled condenser for various values of water side fouling. Nothing less than a .0005 factor should be used when selecting a condenser, even when good quality water is available, because some surface fouling is present from the beginning of operation. Fouling factors have only a small influence on the capacity of reciprocating compressor equipment as compared with other types of refrigeration equipment. An increase in a scale factor of .0005 reduces the capacity of a condensing unit only about 2%. Ranges of fouling factors used for equipment selection are shown in *Part 5, Water Conditioning.*

3. Entering Condenser Water Temperature — If city water or well water is used for the condensing media in a once-thru system, the maximum water temperature prevailing at the time of maximum refrigeration load is used for the selection. This temperature must be obtained locally from the water company or other local sources. If a cooling tower is used to cool the water, the temperature is based on the design wet-bulb temperature and on the approach which is the difference between the wet-bulb temperature and the water temperature leaving the tower. For further data on tower selection and temperature levels, refer to *Chapter 5.*

4. Water Quantity — The required water quantity may be found from the condenser ratings or may be given as an available quantity. It

TABLE 3—MAXIMUM SUGGESTED WATER VELOCITIES THRU COOLERS AND CONDENSERS

Normal Operation (hr)	Water Velocity (fps)
1500	12.
2000	11.5
3000	11.
4000	10.
6000	9.
8000	8.

may be limited by the suggested maximum water tube velocities for various total operating hours per year (*Table 3*). City water quantities generally run from 1-2 gpm per ton. Cooling tower water quantities are usually selected for 3 gpm per ton.

A typical rating of a hermetic water-cooled condensing unit is shown in *Table 4.*

For *Air-Cooled Condensing Units:*

1. Entering Air Temperature — The normal summer outdoor design dry-bulb temperature is used as the temperature of the air entering the condenser.

2. Air Flow — The unit must be located so that the flow of air to and from the condenser coil is not impeded. There must be enough space surrounding the unit to prevent recirculation of air. Units with direct-driven propeller fans should not use ductwork for the condenser air because the capacity is lowered and the condensing temperature is raised. Ductwork can be used on some units which use belt-driven centrifugal condenser fans.

When selecting condensing units, consideration should be given to the variation expected in the load and the types and steps of capacity control available on the unit.

Example 2 — Selection of a Condensing Unit

Given:

Refrigeration load	= 10.0 tons
Saturated suction temperature	= 40 F
Entering condenser water temperature	= 75 F
Fouling factor	= .0005
Refrigerant 12	

Find:

Condensing unit size
Condensing temperature
Power input
Condenser water quantity
Circuiting

Solution:

There are two possible selections depending on the number of condenser passes.

TABLE 4—WATER-COOLED CONDENSING UNIT RATINGS

CAPACITY, POWER INPUT, HEAT REJECTION, CONDENSER WATER   REFRIGERANT 12, COMP MODEL 40, COND MODEL 40

Cond Temp (F)		Passes	Ent Cond Water Temp (F)	Suction Temperature (F)				
				50	40	30	20	10
				Suction Pressure (psi g)				
				46.7	37.0	28.5	21.1	14.6
90	Capacity (tons)			14.4	11.6	9.2	7.2	5.6
	Power Input (kw)			8.1	8.0	7.7	7.3	6.7
	Heat Rejection (tons)			16.7	13.9	11.4	9.3	7.5
	Gpm	8	50	12	10			
		8	60	18	14	11		
		4	60	22	17			
		4	65	30	22			
		4	70	46	32	23	17	
100	Capacity (tons)			13.4	10.8	8.6	6.7	
	Power Input (kw)			9.0	8.7	8.3	7.7	
	Heat Rejection (tons)			16.0	13.3	11.0	8.9	
	Gpm	8	60	11	9			
		8	70	17	13	10		
		4	70	21				
		4	75	28	21			
		4	80	42	30	22		
110	Capacity (tons)			12.4	10.0	8.0	6.2	
	Power Input (kw)			10.0	9.5	8.9	8.2	
	Heat Rejection (tons)			15.3	12.8	10.5	8.6	
	Gpm	8	70	11	9			
		8	80	16	12	10		
		4	80	20				
		4	85	26	20			
		4	90	38	28	21		

NOTES:

1. Where values are not shown for capacity and power input (kw), the operating conditions are beyond operating limits of the compressor.
2. Where values for gpm are not shown, the operating conditions require a water quantity outside the condenser operating limits.
3. The condenser water quantities shown are based on a .0005 fouling factor.
4. Although interpolation is permitted, do not extrapolate. Operation outside the compressor limits of the table is not allowed.
5. For 50-cycle operation, multiply the capacity and power input by 0.83. The condenser water quantities are based on 60-cycle operation. For 50-cycle operation, use the condenser ratings.

6. Condenser leaving water temperature

$$= \text{entering water temp} + \frac{\text{heat rejection (tons)} \times 24}{\text{gpm}}$$

where:

heat rejection (tons) = unit heat rejection (rating tables)
24 = conversion constant
gpm = gal/min (rating table)

7. The refrigerant temperatures shown are the saturation temperatures corresponding to the pressure indicated at the compressor. The actual gas temperature is higher because of superheat.

Example 2 (contd)

	Selection 1	Selection 2
Condensing unit size	40	40
Condensing temperature	100 F	110 F
Power input	8.7 kw	9.5 kw
Condenser water quantity	21 gpm	10.5 gpm
Circuiting	4-pass	8-pass

Four-pass selection is usually used with cooling tower water and eight-pass selection with city water.

WATER-CHILLING UNIT

The factors influencing the selection of a water chilling unit are:

1. *Capacity, Chilled Water Quantity, Temperature Range* — These are related to each other and, when any two are known, the third can be found by the formula:

$$\text{Capacity (tons)} = \frac{\text{gpm} \times \text{temperature range}}{24}$$

Temperature range is the difference between the water temperature entering and leaving the chiller. Capacity is the total load of the chiller, and the water quantity is the design flow; these are generally determined from the dehumidifier(s) selection.

2. *Water Temperature Levels* — The leaving chilled water temperature is usually selected as the entering water temperature required at the load source. The proper determination of the water temperature required for chilled water coils and spray washers is discussed in *Part 6*.

The entering condenser water temperature is determined by the source of the water, i.e. city water used in a once-thru system, or cooling tower water used in a recirculating system.

3. *Fouling Factors* — The same discussion included under water-cooled condensing units applies to the condenser of a water-chilling unit. The fouling factor used for the chiller selection depends on the water conditioning and the system to which the chiller is applied, either an open or closed recirculating system. Water chillers applied to open recirculating chilled water systems should be selected with a minimum fouling factor of .001 in the cooler. For closed recirculating systems, a minimum fouling factor of .0005 should be used. Refer to *Part 5* for recommended fouling factors for different applications and also for information on water conditioning.

A typical rating of a water-cooled water-chilling unit is shown in *Table 5*.

Example 3 — Water-Chilling Unit Selection

Given:

Chilled water quantity	= 200 gpm
Leaving chilled water temperature	= 44 F
Chilled water temperature rise	= 10 degrees
Entering condenser water temperature	= 85 F
Fouling factor (cooler and condenser)	= .0005

Find:
Unit selection
Condensing temperature
Power input
Condenser water flow rate
Pressure drops thru cooler and condenser

Solution:

$$\text{Load} = \frac{200 \times 10}{24} = 83.3 \text{ tons}$$

Select Model 100. By interpolation:

Condensing temperature	= 107.4 F
Power input	= 82.0 kw
Condenser load	= 106.5 tons

$$Q_c = \frac{106.5}{107.4 - 85} = 4.75$$

For a 3 pass condenser, gpm = 170

Condenser pressure drop	= 8 psi
Cooler pressure drop	= 10.5 psi

DRIVE SELECTION

Drive selection involves a consideration of types, sizes, starting torque, overload, and starting requirements.

DRIVE TYPE

For an open compressor almost any type of driver can be used. The most common is the polyphase squirrel cage induction motor. Other types used in special cases are d-c motors, wound rotor induction motors, or single phase motors. There may be a need to apply 2-phase motors or 25 or 50 cycle motors for compressors in regions where these electrical conditions exist.

Another type of driver that can be applied to an open compressor is an internal combustion engine, either diesel or natural gas driven. These may be used when an economic analysis of the relative costs of fuel and electrical energy, maintenance and the relative investment required is made. Engines are available commercially in sizes from 10 horsepower and up; they usually are direct-connected.

Hermetic compressor equipment is normally supplied with a squirrel cage induction motor. It may be wound for part-winding starting to provide a method of reducing current inrush; it may be furnished in some sizes with a single phase or two-phase motor, and may be available for 50-cycle duty.

SIZE

Driver selection based on the brake horsepower required at the design operating condition is usually satisfactory for an open compressor application for air conditioning duty. For selections at low design suction temperatures, it is usually necessary to consider the initial pulldown operating condition. The compressor operates at a higher suction temperature at start-up, requiring a higher input during the pulldown. This consideration frequently dictates the size of the motor required rather than the brake horsepower required at the design operating condition.

The motor for a hermetic compressor is selected by the manufacturer to prevent overload when the compressor is running at its normal operating condition.

STARTING TORQUE

It should be noted in the selection of a drive that the required starting torque available by the driver must equal the compressor starting torque only when the compressor is selected to operate at the driver speed (direct-driven). If the design compressor speed is less than the driver speed (as on

TABLE 5—WATER CHILLING UNIT RATINGS

MODEL 100, 60 CYCLE   ARI Standard Ratings, 85.8 tons, 78.8 kw, 257 gpm condenser

Leaving Chilled Water Temperature (F) Fouling Factor: .0005 in Cooler Cooling Range: 5 F to 15 F		WATER-COOLED MODEL based on 5 F subcooling					CONDENSERLESS MODEL based on 15 F subcooling				
		Compressor Saturated Discharge Temperature (F)					Compressor Saturated Discharge Temperature (F)				
		90	95	100	105	110	105	110	120	130	135
40	Capacity (tons)	85.3	82.8	80.3	78.1	75.8	81.6	79.1	73.6	67.6	64.4
	Power Input (kw)	71.2	74.0	76.4	79.0	81.9	78.9	81.5	86.6	93.8	97.4
	Condenser Load (tons)	105.5	103.8	102.0	100.6	99.1	104.1	102.3	98.3	94.3	92.2
42	Capacity (tons)	88.6	86.1	83.6	81.4	79.0	84.6	82.2	76.5	70.3	66.9
	Power Input (kw)	71.5	74.2	77.1	79.6	82.5	79.5	82.4	87.6	95.2	99.0
	Condenser Load (tons)	108.9	107.2	105.5	104.0	102.5	107.2	105.6	101.4	97.4	95.1
44	Capacity (tons)	92.0	89.4	87.0	84.6	81.9	87.9	85.3	79.6	73.1	69.6
	Power Input (kw)	71.8	74.7	77.5	80.4	83.4	80.2	83.2	88.6	97.0	100.9
	Condenser Load (tons)	112.4	110.6	109.0	107.5	105.6	110.7	109.0	104.8	100.7	98.3
46	Capacity (tons)	95.5	92.6	90.3	87.9	85.0	91.3	88.4	82.5	76.0	72.4
	Power Input (kw)	72.1	75.1	78.1	81.0	84.3	80.9	84.1	89.7	98.4	102.5
	Condenser Load (tons)	116.0	114.0	112.5	110.9	109.0	114.2	112.3	108.0	104.0	101.6
48	Capacity (tons)	98.8	96.1	93.5	90.9	88.1	94.5	91.6	85.6	78.9	75.1
	Power Input (kw)	72.1	75.3	78.4	81.6	84.9	81.4	84.6	90.7	99.7	104.0
	Condenser Load (tons)	119.3	117.5	115.8	114.1	112.3	117.7	115.7	111.4	107.3	104.7
50	Capacity (tons)	102.1	99.4	97.1	94.1	91.3	97.6	94.7	88.8	81.9	78.0
	Power Input (kw)	72.2	75.5	78.8	82.1	85.5	81.9	85.3	91.5	100.9	105.6
	Condenser Load (tons)	122.6	120.9	119.5	117.5	115.6	120.8	118.9	114.8	110.6	108.1

ARI STANDARD RATINGS

ARI Standard 590 for Reciprocating Liquid Chilling Packages requires that Standard Ratings be published for established operating conditions.

The ARI rating shown above the rating table applies to water-cooled models only, and is based on 54 F to 44 F chilled water and 85 F to 95 F condenser water with a .0005 fouling factor in both exchangers.

ARI ratings for the condenserless models are indicated in the rating table by heavy lined boxes and are based on a .0005 fouling factor in the cooler.

COOLING FOULING FACTOR ADJUSTMENTS

Fouling Factor	Capacity	Power Input (kw)
Clean	1.02	.98
.0005	1.00	1.00

Condenser fouling factors are covered in the curves below.

RATINGS IN THE TABLES are based on the following:
1. Cooling range: 5 to 15 degrees
2. Fouling factor: .0005 in the cooler
3. Liquid subcooling: water-cooled models, 5 degrees in condenser
   condenserless models, 15 degrees in condenser

(If a condenser is selected for less than 15 degrees of subcooling, adjust as follows: Multiply tabular capacity rating by 0.94. Then adjust this result upward by 0.4 percent for each one degree of available subcooling.)

FORMULAS

$$\text{Capacity (tons)} = \frac{24}{\text{gpm x temp drop}}$$

$$Q_c = \frac{\text{condenser load (tons)}}{\text{condensing temp} - \text{entering condenser water temp}}$$

$$\text{Condenser water rise} = \frac{24 \times \text{condenser load}}{\text{gpm}}$$

CONDENSER GPM

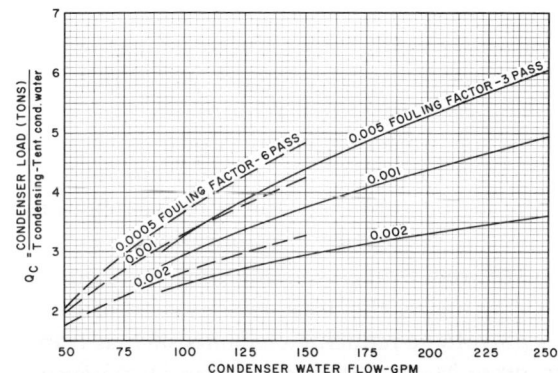

PRESSURE DROP

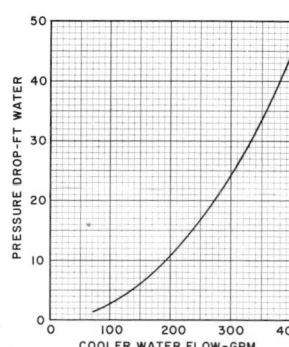

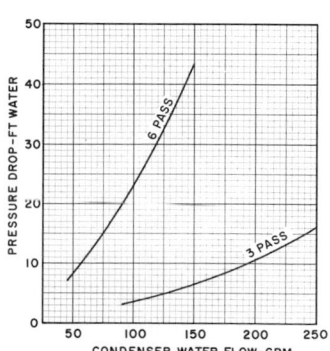

belt-drive units), the starting torque requirements are reduced in proportion to the speed ratio between the compressor and drive, because of the mechanical advantage available to the driver.

The following formulas are useful in analyzing motor performance with respect to starting torque requirements:

1. Motor full load torque (lb ft) $= \dfrac{5250 \times \text{hp}}{\text{motor rpm}}$

2. Starting torque available at motor (lb ft)

$$= \frac{5250 \times \text{motor hp} \times \text{percent starting torque} \times 0.81}{\text{motor rpm} \times 100}$$

3. Starting torque available at compressor (lb ft)

$$= \frac{5250 \times \text{motor hp} \times \text{percent starting torque} \times 0.81}{\text{compressor rpm} \times 100}$$

4. Minimum motor horsepower required to start compressor

$$= \frac{\text{comp starting torque (lb ft)} \times \text{comp rpm} \times 100}{5250 \times \text{percent starting torque} \times 0.81}$$

(Incorporation of the 0.81 factor in formulas 2, 3 and 4 allows up to 10% drop in voltage at the motor terminals during the starting periods.)

The starting torque which a motor develops is proportional to the square of the voltage at the motor terminals. Thus, at half voltage a motor develops only one-quarter of the torque developed at full voltage. To make starting possible, the voltage at the motor terminals must be high enough to provide the required starting torque.

A compressor equipped with a capacity control may be supplied with a normal starting torque motor (NEMA design B) when it is started partially loaded. A compressor not equipped with capacity control should be provided with a high starting torque motor (NEMA design C). The compressor manufacturer should be consulted as to the starting torque requirements of the compressor.

With reduced voltage starting, it is usually necessary to use a high starting torque motor when the starting inrush must be kept to the lowest possible value coincident with actual starting of the compressor. *Table 6* shows NEMA starting torque and locked rotor current values for both normal and high torque motors in the 5-200 horsepower range. The full voltage starting torque (locked rotor torque) is usually expressed in percent of the torque delivered by the motor at full load, full speed and at rated voltage and frequency. For convenience, the actual torque (pound feet) of the motor is also shown in the table.

OVERLOAD

NEMA standards permit continuous overloading of 40 C rise open squirrel cage motors up to 15% over nameplate rating when operated at full rated voltage, frequency and ambient temperatures not exceeding 40 C. Whether or not part of this 15% service factor should be used in making an initial motor selection depends on having available accurate information concerning local voltage and frequency variation, ambient temperature, compressor speed, and maximum suction and condensing pressures at which the compressor operates. Where conditions of voltage, frequency, and ambient temperatures are comparatively unknown, *motors should positively not be selected overloaded*. Selection for operation above full load current may shorten the life of a motor as much as 50%.

BELT DRIVE

When using a belt drive for a compressor, it is recommended that 3% be added to the direct drive brake horsepower ratings to obtain the approximate motor brake horsepower required.

For more complete information on motors and their characteristics, refer to *Part 8*. For information on other types of drives, refer to the manufacturer.

STARTING EQUIPMENT

HERMETIC COMPRESSOR

Normally, hermetic compressor equipment is completely wired at the factory and supplied with either a standard motor and across-the-line starter or a part-winding motor and increment type starter, designed specifically for use with the compressor. The increment type starter can be augmented with a resistance type accessory to make it a three-step starter with a lower inrush current than with a part-winding type alone.

OPEN COMPRESSOR

Whenever possible, across-the-line starting is most desirable because it is less costly and less troublesome than the complicated reduced voltage equipment. However, limitations of the power distribution system capacity often require the use of a reduced voltage starter for compressor motors above a certain size. In each case the power company concerned should be consulted and a rating obtained with respect to the particular application.

When reduced voltage starting must be used, increment starting is the least expensive because it requires no voltage-reducing elements such as transformers or resistors. Although the most expensive

TABLE 6—NEMA STARTING TORQUE VALUES

STANDARD OPEN SQUIRREL CAGE MOTORS, 60 CYCLE, 3 PHASE

MOTOR HP	FULL LOAD RPM	MINIMUM STARTING TORQUE (At Full Voltage)				LOCKED ROTOR CURRENT AT 220 VOLTS (amps)
		Design "B" Normal Starting Torque Normal Starting Current		Design "C" High Starting Torque Normal Starting Current		
		Percent of Full Load	Pound Feet	Percent of Full Load	Pound Feet	
5	1750	185	27.8	250	37.4	90
7½	1750	175	39.4	250	56.2	120
10	1750	175	52.5	250	74	150
15	1750	165	74.2	225	101	220
20	1750	150	90	200	120	290
25	1750	150	112	200	150	365
30	1750	150	135	200	180	435
40	1750	150	180	200	240	580
50	1750	150	225	200	300	725
60	1750	150	270	200	360	870
75	1750	150	337	200	450	1085
	1160	135	458	200	680	1085
	870	125	565	200	906	1085
100	1750	125	375	200	600	1450
	1160	125	565	200	906	1450
	870	125	755	200	1205	1450
125	1750	110	410	200	750	1815
	1160	125	705	200	1130	1815
	870	125	940	200	1510	1815
150	1750	110	495	200	900	2170
	1160	125	850	200	1355	2170
	870	125	1130	200	1810	2170
200	1750	100	600	200	1200	2900
	1160	125	1130	200	1810	2900
	870	125	1510	200	2410	2900

type, auto transformer starters obtain the most effective reduction of inrush current drawn from the line. Primary resistance starters are less expensive and obtain a smaller degree of reduction of inrush current. Refer to *Part 8* for data on starters.

CONTROLS

Compressor control consists mainly of capacity control, safety controls, and the method of compressor operation.

COMPRESSOR CAPACITY CONTROL

Various methods of capacity control are available from different compressor manufacturers. A brief description and some of the advantages or disadvantages of these methods are included here.

Suction-Valve-Lift Unloading

This unloading is accomplished by unseating the suction valves of certain cylinders in the compressor so that compression cannot take place. This is inherently the most efficient method of capacity control since passage of the refrigerant vapor in and out of the cylinder thru the suction valves without compression involves smaller losses than other methods. The compressors are generally controlled in steps down to one fourth or one third of full load capacity, depending on the number of cylinders. The cylinders may be unloaded internally under the control of suction pressure or externally from a thermostat or pressurestat. *Figure 5* illustrates a method of this type of capacity control.

Cylinder Head Bypass

On multi-cylinder compressors one or more cylinders or banks of cylinders may be made ineffective by a bypass of gas from the cylinder discharge to the intake port (*Fig. 6*). A check valve is installed to separate the inactive cylinders from active ones. A solenoid valve can be installed to operate the by-

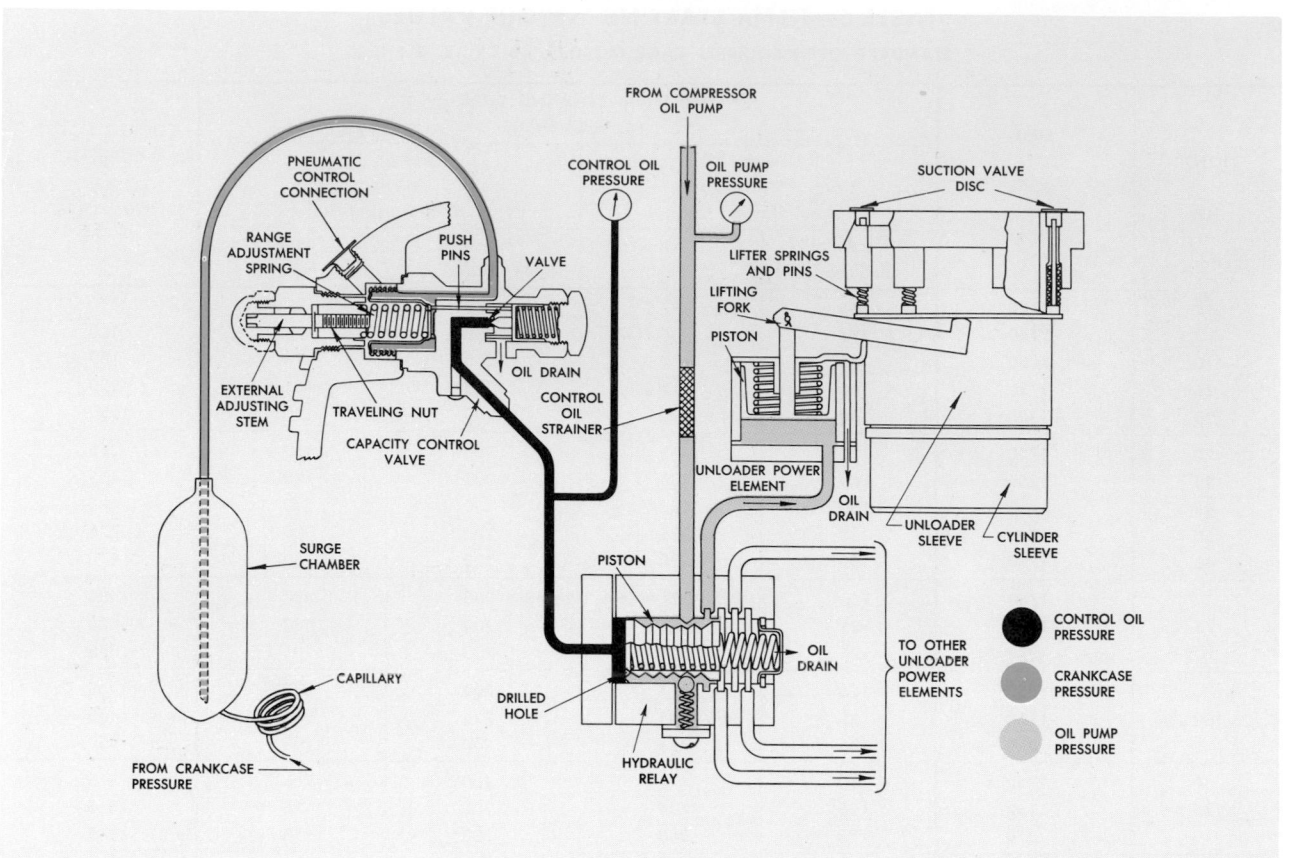

Fig. 5 — Suction-Valve-Lift Unloading

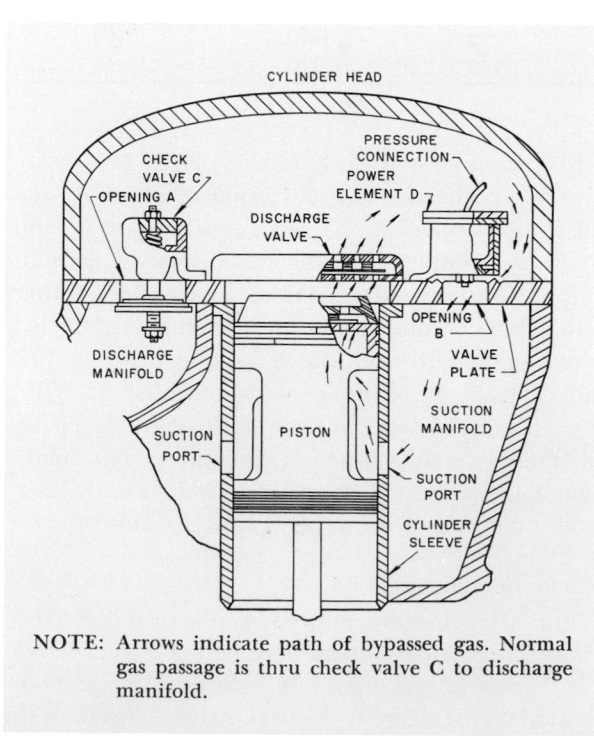

NOTE: Arrows indicate path of bypassed gas. Normal gas passage is thru check valve C to discharge manifold.

Fig. 6 — Cylinder Head Bypass

pass, thus permitting automatic control by a thermostat or pressurestat. Gas passes thru the inactive cylinders but is not compressed; therefore, the only loss is thru the valves, cylinders, and connections. Because of these losses, the power requirement does not decrease proportionally with the capacity.

Speed Control

Compressor capacity is almost directly proportional to the speed while the compressor brake horsepower is proportional to the ratio of the speeds raised to a power of 1.0 to 1.3, depending on the compressor design. The speed control can be obtained by using a multi-speed motor which provides two or three speeds, or by using an internal combustion engine which can produce multiple speeds. Care must be taken that the compressor is not operated at a speed below the range for proper operation of the lubricating system.

Multiple Units

The use of multiple compressors for capacity control (*Fig. 7*) has the following advantages: (1) single speed motors may be selected and operated con-

Fig. 7 — Multiple Compressors

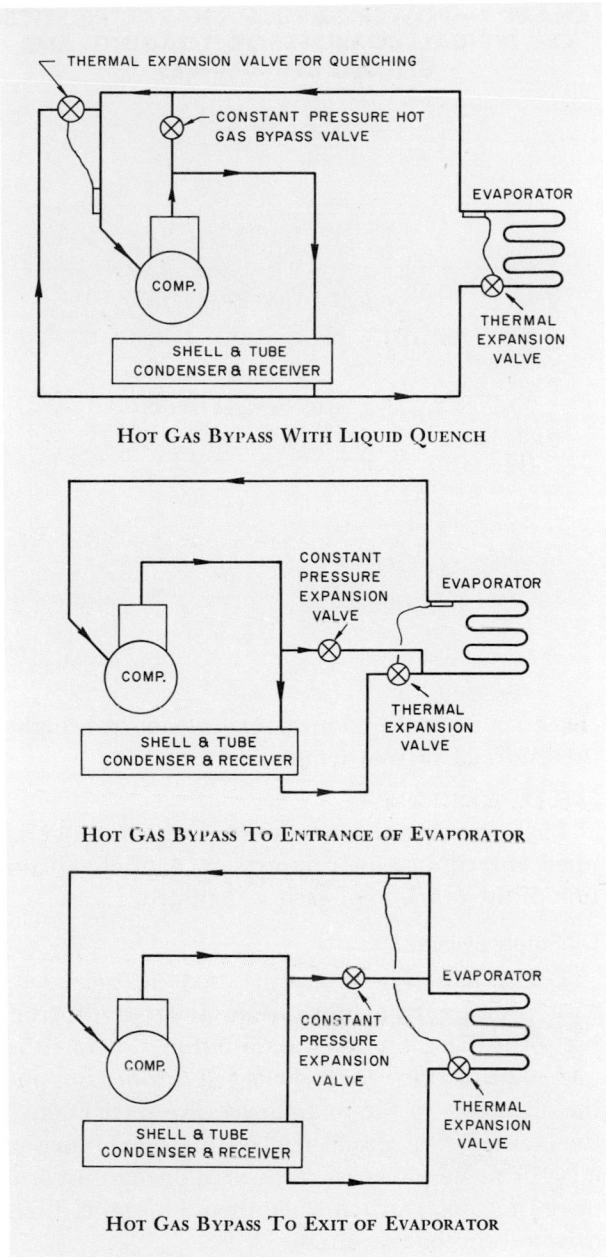

Hot Gas Bypass With Liquid Quench

Hot Gas Bypass To Entrance of Evaporator

Hot Gas Bypass To Exit of Evaporator

Fig. 8 — Hot Gas Bypasses

tinuously at their best efficiency; (2) stand-by equipment is available which allows partial load operation if one of the machines breaks down; (3) compressors may be started in sequence to limit the current inrush if time-delay devices are employed. The compressors may or may not be interconnected as conditions require. Thermostats or pressurestats may be used to start and stop compressors in accordance with load demands.

Hot Gas Bypass

Another method of capacity control is to load the compressor artificially. This can be accomplished by transferring heat to the suction gas in the form of a hot gas bypass. The discharge is connected to the low side thru a constant pressure valve which admits hot gas to the low side as the evaporator pressure tends to drop, thus maintaining a constant suction pressure. *Figure 8* shows three arrangements of hot gas bypasses. Since this is a method of loading and not unloading the compressor, the compressor brake horsepower remains fairly constant.

Though not a method of capacity control, a back pressure valve does permit a constant speed compressor to operate at lower capacities and horsepowers while maintaining a constant evaporator temperature. As the load is decreased, the evaporator temperature is reduced. This causes the back pressure valve to start to close, creating a restriction between the evaporator and the compressor suction, which causes a reduction in suction pressure while the evaporator pressure remains close to design. As the suction pressure decreases, the density of the suction gas decreases, and the weight flow of refrigerant is reduced, thus lowering the capacity of the compressor.

Chart 1 shows the power-saving characteristics of typical compressor loading and unloading devices.

Compensation of the compressor capacity from a room thermostat or a chilled water thermostat is used when control of the temperature must be closer than the 8-10 degree swing normally obtained when using only suction-valve-lift unloading. This compensation is accomplished by resetting the control point at which the compressor unloads. For example, as the temperature at the room thermostat or chilled water thermostat decreases, the control point is in-

CHART 1—POWER-SAVING CHARACTERISTICS OF TYPICAL COMPRESSOR LOADING AND UNLOADING DEVICES

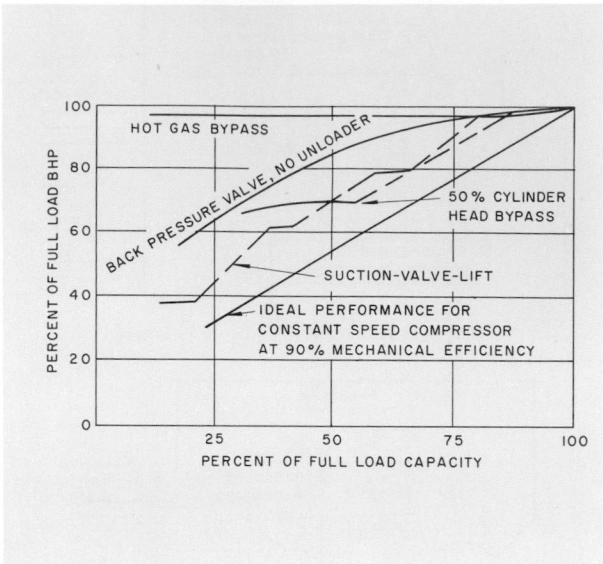

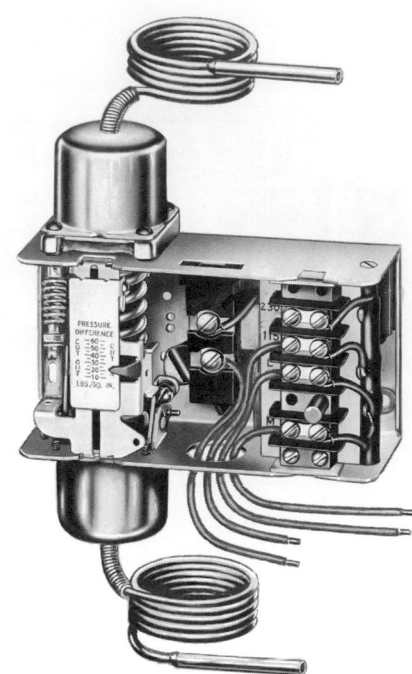

FIG. 9 — OIL SAFETY SWITCH

creased so that the compressor unloads at a higher than normal suction temperature.

SAFETY CONTROLS

There are several safety controls which can be applied to reciprocating compressors. A brief explanation of the function of each is included.

Oil Safety Switch

This switch (*Fig. 9*) may be used on compressors with pressure type lubrication. It stops the compressor if there is a lubrication failure due to either oil leaks from the system, a clogged strainer stopping the oil intake to the pump, excessive refrigerant in the crankcase, or insufficient oil pressure. Provision must be made to bypass this switch on start-up or to use a time delay switch when the oil pump is direct-driven from the compressor shaft.

Low Pressure Switch

This switch is used to stop the compressor when the suction pressure is reduced to a point which could produce freezing in a water chiller, or which would permit operation beyond the prescribed operating limits of the compressor.

High Pressure Switch

This switch is used to stop the compressor when the discharge pressure rises above the prescribed limits because of either inadequate condensing, an overcharge of refrigerant air in the system, or any other reason. It is usually combined with the low pressure switch into one control called a dual-pressure switch (*Fig. 10*).

Chilled Water Safety Switch

This switch is used with water-chilling units to stop the compressor when the water temperature in the chiller approaches the freezing point.

Time Delay Relay

This relay should be used on hermetic compressors, and can be used on open compressors to prevent short cycling of the unit. The relay should be set to prevent the starting of the compressor until an elapse of time (such as 5 minutes) after the compressor has been shut down, due to action of one of the safety or operating controls. This reduces the chances of overheating the motor, and eliminates a possible burnout because of frequency of starting.

Motor Temperature Switch

This switch is used on hermetic compressors to stop the compressor when the temperature in the motor windings becomes excessive.

Motor Overloads

These overloads are included in the wiring of the compressor circuit to stop the compressor when the motor draws excessive current.

COMPRESSOR OPERATION

Reciprocating compressor operation must prevent excessive accumulation of liquid refrigerant in the crankcase during off cycles. This minimizes the rapid evaporation of the refrigerant on start-up, resulting in foaming of the oil and loss of lubrication.

Courtesy of Penn Controls, Inc.

FIG. 10 — DUAL PRESSURE SWITCH

Methods of Control

Either of the following methods of control may be used to prevent this accumulation of excess refrigerant:

1. *Automatic Pumpdown Control (DX Systems)*— The most effective and most common means of keeping liquid out of the crankcase during system shutdown periods is to operate the compressor on automatic pumpdown control. It is most practical on small systems using a single DX evaporator. A typical wiring diagram for this control is shown in *Fig. 11.* The recommended control arrangement involves the following devices and provisions:

   a. A tight-closing solenoid valve in the main liquid line or in the branch to each evaporator.

b. Compressor operation thru a low pressure switch providing for pumpdown whenever the valve closes, whether the balance of the system is in operation or not.

c. Electrical interlock of the liquid solenoid valve(s) with the evaporator fan or water chiller pump, so that the refrigerant flow is stopped when either the fan or the pump is out of operation.

d. Electrical interlock of the refrigerant solenoid valve(s) with the safety devices (high pressure cutout, oil safety switch and motor overloads) so that the valve(s) closes when the compressor stops, due to any one of these safety devices.

e. Low pressurestat settings, such that the cut-in point corresponds to a saturated refrigerant temperature lower than any ambient air temperature to which the compressor may be subjected.

2. *Crankcase Oil Heater With Single Pumpout At The End of Each Operating Cycle (DX Systems)* — This arrangement is not as positive in keeping liquid refrigerant out of the crankcase as automatic pumpdown control, but it is a substitute where pumpdown control (resulting in compressor cycling) meets with customer objections. A typical schematic wiring diagram for single pumpout control with crankcase heater is shown in *Fig. 12.* The use of this method at the end of each operating cycle requires the following:

   a. A tight-closing solenoid valve in the main liquid line or in the branch to each evaporator.

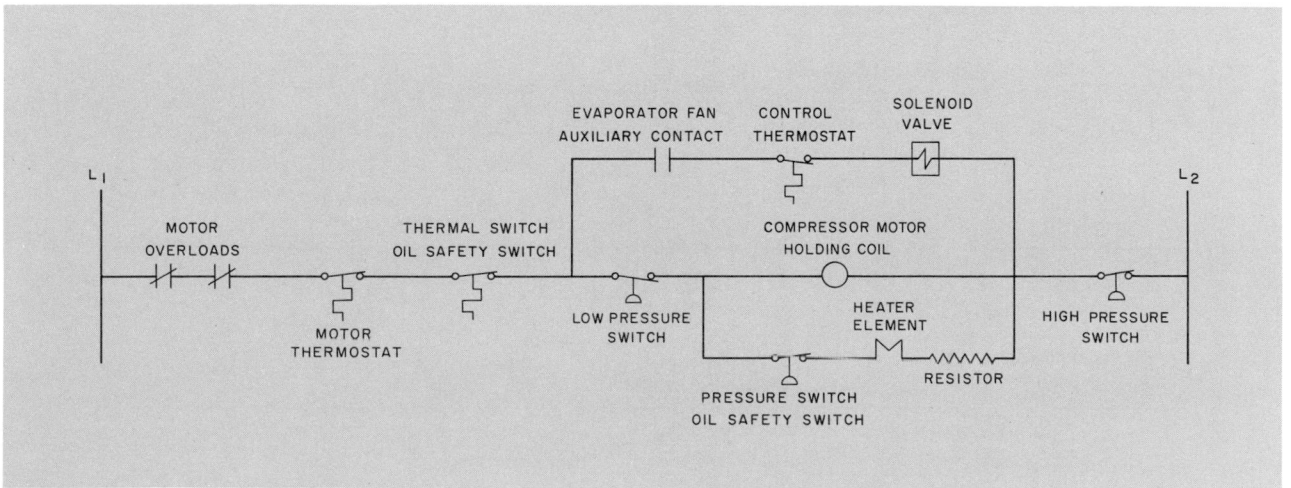

FIG. 11 — TYPICAL AUTOMATIC PUMPDOWN, WIRING DIAGRAM

b. A relay or an auxiliary contact of the compressor motor starter to maintain compressor operation until the low pressure switch opens.

c. A relay or auxiliary contact for energizing the crankcase heater during the compressor off cycle and de-energizing it during the compressor on cycle.

d. Electrical interlock of the refrigerant solenoid valve(s) with the evaporator fan or chilled water pump, so that the refrigerant flow is stopped when either the fan or pump are out of operation.

e. Electrical interlock of the refrigerant solenoid valves(s) with the safety devices, (high pressure cutout, oil safety switch and motor overloads), so that the valve(s) closes when the compressor stops, due to any one of these safety devices.

3. *Compressor Control With Flooded Evaporators* — Neither automatic pumpdown control nor single pumpout operation is practical in systems employing flooded evaporators unless suction line solenoid valves are added to the system. Therefore, with flooded evaporators the following arrangements are often used:

a. Manual operation (Item 4). No crankcase heaters are required.

b. Automatic control from temperature controllers or other devices, provided crankcase heaters are used and energized on the off cycles, and the liquid solenoid valve is closed whenever the compressor is stopped.

Where water cooling of the compressor head is employed, a solenoid valve in the water supply line closes whenever the compressor is stopped.

c. Item b, with the added precaution of a single pumpout of the compressor for night or weekend shutdowns. This can be accomplished by manually closing the compressor suction stop valve.

4. *Manual Compressor Operation* — Compressors may be controlled manually without the use of automatic pumpdown control, or by single pumpout and a crankcase heater, provided the system is under the control of a qualified operator at all times. The operator pumps down the system by use of the manual valves, and keeps the liquid, suction and discharge valves closed when the machine is not operating.

Effect of a Short Operating Cycle

It is characteristic of the reciprocating compressor operation that oil leaves the crankcase at an accelerated rate immediately after starting. Therefore, each start must be followed by a sufficiently long operating period to permit the regain of the oil level. Operation under control of a room thermostat generally provides enough operating time in most cases. However, if the compressor is controlled in response to a thermostat located in the supply air or in the water leaving a water chiller, a rapid cycle may result. This thermostat should have a differential wide enough so that the running cycle is not less than seven or eight minutes.

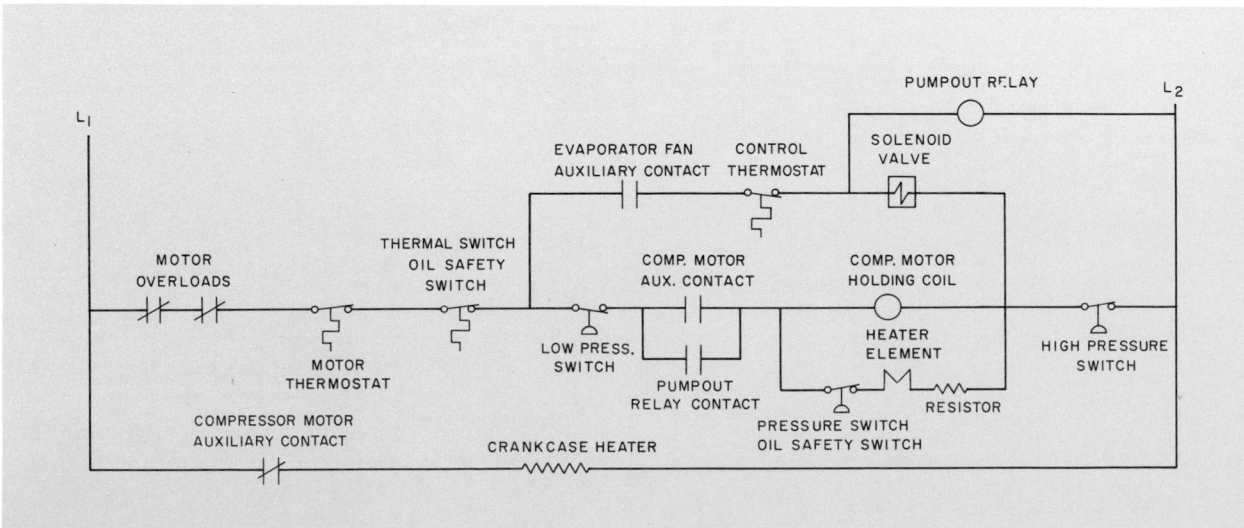

FIG. 12 — TYPICAL SINGLE PUMPOUT WITH CRANKCASE HEATER, WIRING DIAGRAM

ACCESSORIES

The following accessories may be required and can be supplied with the compressors.

1. *Coupling* — Used with an open compressor when driven at motor speed.

2. *Belt Drive Package* — Used with an open compressor utilizing a fly wheel and motor pulley for driving the compressor at any specific speed.

3. *Vibration Isolators* — Used to isolate compressor units, condensing units or water chilling units to reduce transmission of noise and vibration to the floor or building structure.

4. *Crankcase Heaters* — Used to keep the oil warm in the crankcase when the compressor is not running. This heating prevents the oil from absorbing refrigerant to an excessive degree, thus maintaining its full lubricating and protective qualities.

5. *Water-Cooled Compressor Heads* — Used to prevent excessive temperatures at the discharge valve. They are usually required when a compression ratio of 5:1 is exceeded when using Refrigerant 22. To prevent condensation of refrigerant in the cylinders, water flow must be stopped when the compressor stops. The maximum leaving water temperature should be 100 F.

6. *Suction Strainer* — Used to prevent foreign particles from entering the compressor. It is important that the system be thoroughly cleaned before start-up.

7. *Structural Steel Base* — Used to mount the compressor unit, condensing unit, or water chilling unit as a completely fabricated assembly, for ease of installation.

8. *Crankcase Connections* — Used to interconnect two or more compressors (connected to the same system) to return oil equally to all compressors.

9. *Muffler* — Used to minimize refrigerant noise. It should be installed so oil is not trapped.

INSULATION

Cold surfaces such as the cooler shell of a water chilling unit and the suction pipe should be insulated to prevent dripping where this condition creates a nuisance or causes damage. The thickness should be such that the temperature of the outer surface is slightly higher than the expected dewpoint of the surrounding air. An external vapor barrier should be used to prevent leakage of vapor into the insulation. Cellular plastic or cellular glass type of insulation can be used since they have a high resistance to water and water vapor, and are good insulators.

Hot gas lines are not insulated unless there is some danger of receiving burns by contact with the lines. If this is to be prevented, the hot gas lines may be insulated up to 5 feet from the floor with a high temperature insulation such as magnesia.

Liquid lines should not be insulated unless heat can be picked up from the surrounding air, i.e. where they are installed exposed to the direct sunlight for a considerable distance or installed in boiler rooms. Insulation may also be used at the outlet of a liquid suction interchanger to preserve the subcooling effect.

LOCATION

The location of a reciprocating refrigeration machine should be carefully planned; it directly influences the economic and sound level aspects of any system.

In general, the compressor should be located in a clean, dry, well-ventilated space. Cleanliness and absence of dampness insure long life to motors and belts, and reduce the necessity of frequent painting of exposed piping. If natural ventilation is inadequate or cannot be supplied thru windows and doors, forced ventilation thru ductwork should be provided. It is essential that the starter and open motors have adequate ventilation to prevent overheating of the starter and overloading of the motor.

Space should be available at the end of all replaceable tube coolers and/or condensers so that the tubes can be cleaned or replaced. Adequate space should be left around and over the compressor for servicing; it should be accessible from all sides. Sufficient space should be left above the unit for removing cylinders, and on either side to permit removal of the flywheel and crankshaft.

The unit should be protected so that the water-cooled condenser, water lines and accessories are not subject to freezing during winter shutdown periods.

The machine should be located near the equipment it serves to effect a minimum first cost system. However, there may be cases where the machine must be located elsewhere because of space, structural, or sound considerations. The machine should be located where moderate sound levels can be accepted; otherwise special sound proofing may be required when adjacent to low ambient sound level areas such as executive offices or conference rooms.

In new construction the floor framing of the equipment room should be laid out by the architect or consulting engineer to match equipment supports, and should be designed for weights, reactions, and speeds furnished by the equipment supplier. This framing transfers equipment loads to the building columns.

In existing buildings, use of existing floor slabs should be carefully studied. Any deflection in the floor due to weight of the equipment together with vibrations transmitted thru equipment isolation may result in magnifying the vibration in the building structure. Supplementary steel framing to transfer all equipment loads to building columns may be required by the architect or consulting engineer.

LAYOUT

In the layout of a reciprocating refrigeration machine, consideration should be given to foundations and electrical connections.

FOUNDATIONS

Where a foundation is required for a machine, it must be of ample size, have proper proportions, and be constructed of first class materials.

The functions of a foundation are the following:

1. Support and distribute the weight of the machine over a sufficient area so that it is rigidly located.

2. Absorb the forces produced by the rotating and reciprocating parts of the machine. Forces produced by the reciprocating parts act along the center line of the piston. Those produced by rotating parts act radially in all directions from the center of the crank. The magnitude of these forces depends on the weight of the parts and the speed of the machine. These forces produce perceptible and perhaps objectionable vibration of the machine and foundation if the latter does not have sufficient mass.

3. Hold the machine rigid against unbalanced forces produced by the pull of the belts or other sources.

ELECTRICAL CONNECTIONS

Rigid conduit should never be fastened directly to the compressor or base because it can transmit vibration. Instead, flexible conduit should be used.

UNIT ISOLATION

Isolators are of value not only on upper floors in the prevention of transmission of vibration to the

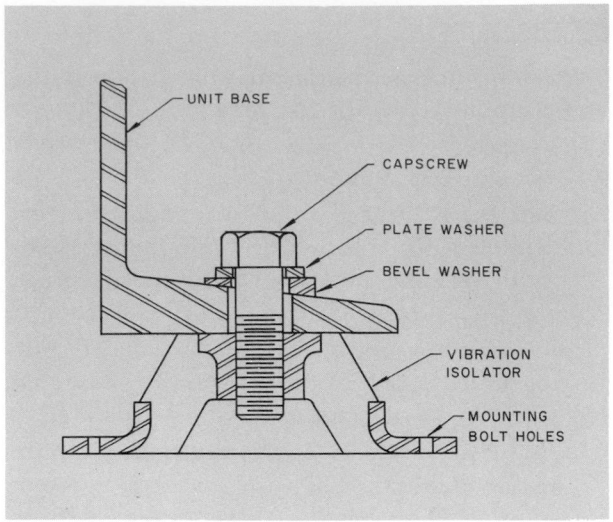

FIG. 13 — TYPICAL VIBRATION ISOLATOR MOUNTING

building structure, but also on concrete basement floors.

Two common ways in which vibration is transmitted from reciprocating refrigeration compressors to building structures are the following:

1. Thru the compressor base directly to the building structure.

2. Thru refrigerant and condenser water piping directly to the building structure (*Part 3*).

Figure 13 shows a typical vibration isolator mounting.

When using belt drive compressors, a greater deflection of the isolator is required to maintain the same effectiveness of isolation, as the compressor rpm decreases.

While the standard isolator package is suitable for most applications, for example, on ground or basement floors, a superior isolation may sometimes be required particularly for upper floor installations where complete freedom from vibration transmission and unusual quietness is a prerequisite.

Where the isolation problem is critical as in upper floor equipment rooms, spring mountings are recommended. They should be used in conjunction with a stabilizing mass such as a concrete foundation or steel base, and should be selected for the lowest disturbing frequency, which is the compressor speed.

The important considerations in selecting such isolating equipment are these:

1. The isolator must permit sufficient deflection under a load to produce high isolating efficiency.

2. The isolator must retain its resilience; that is, it must not become permanently deformed as otherwise its efficiency would be impaired.

3. The isolator must be structurally suitable to the load imposed and must be applied to a base which distributes the load effectively. Inequality of weight distribution, i.e. a flywheel projecting beyond the base, often necessitates various sizes of isolator units under a common base to produce the required deflection at all points. Manufacturers of isolating equipment publish ratings and physical details of these units. Ratings are published in terms of deflection under load and maximum permissible loading. With an understanding of the problem the designer can use such data to select the necessary equipment for his particular application.

CHAPTER 2. CENTRIFUGAL REFRIGERATION MACHINE

Centrifugal refrigeration equipment is built for heavy duty continuous operation and has a reputation for dependability in all types of commercial and industrial applications.

This chapter presents data to guide the engineer in the practical application and layout of centrifugal refrigeration machines used for cooling water or brine at comfort air conditioning temperature levels.

A centrifugal refrigeration machine consists basically of a centrifugal compressor, a cooler and a condenser. The compressor uses centrifugal force to raise the pressure of a continuous flow of refrigerant gas from the evaporator pressure to the condenser pressure. A centrifugal compressor handles high volumes of gas and, therefore, can use refrigerants having high specific volumes. The cooler is usually a shell-and-tube heat exchanger with the refrigerant in the shell side. The condenser is also a shell-and-tube type utilizing water as a means of condensing; it may be an air-cooled or evaporative condenser for special applications.

TYPES OF CENTRIFUGAL REFRIGERATION MACHINES

Centrifugal refrigeration machines may be classified by the type of compressor:

1. Open compressors have a shaft which projects outside the compressor housing, requiring a seal to isolate the refrigerant space from the atmosphere.
2. Hermetic compressors have the driver built into the unit, completely isolating the refrigerant space from the atmosphere.

OPEN MACHINE

Open type equipment may be obtained for refrigeration duty in single units up to approximately 4500 tons capacity at air conditioning temperature levels. The compressor is normally designed with one or two stages, and is driven by a constant or variable speed drive. Compressors are usually driven at speeds above 3,000 rpm and may operate up to 18,000 rpm.

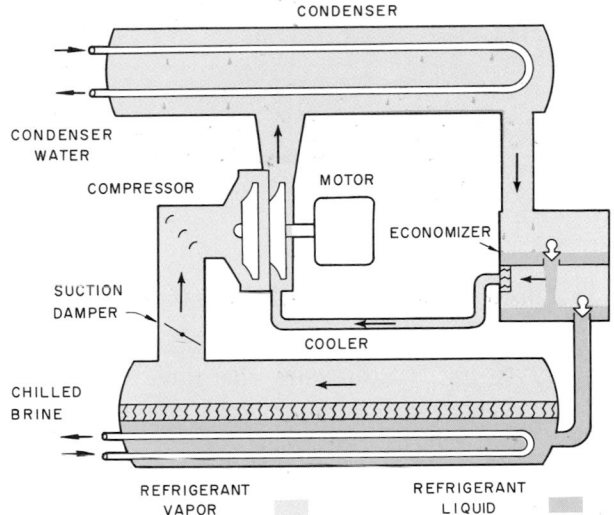

FIG. 14 — OPEN CENTRIFUGAL MACHINE

The centrifugal drive may be an electric motor, steam turbine, gas engine, gas turbine or diesel engine. An electric motor, gas engine or diesel engine usually requires a speed-increasing gear between the drive and the compressor. Gas turbines operating at high speeds may require a speed-decreasing gear between the turbine and the machine. Steam turbines are usually directly connected to the compressor.

Figure 14 illustrates the three basic components, compressor, cooler and condenser, as well as the refrigerant cycle.

Capacity can be varied to match the load by means of a constant speed drive with variable inlet guide vanes or suction damper control, or a variable speed drive with the suction damper control.

HERMETIC MACHINE

Standard hermetic equipment may be obtained in single units up to approximately 2000 tons capacity. They are normally designed with either one or two stages and are driven at a single speed. The drive motor may be either refrigerant- or water-cooled.

A hermetic machine may be driven at motor speed or, by means of a speed-increasing gear between the motor and compressor, at a single higher

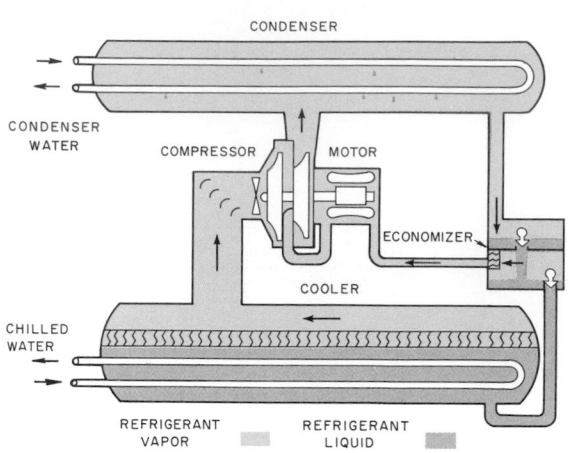

FIG. 15 — HERMETIC CENTRIFUGAL MACHINE

FIG. 16 — THE ORIGINAL CENTRIFUGAL MACHINE
(1922)

speed. *Figure 15* illustrates the three basic components, compressor-motor, cooler and condenser, as well as the refrigerant cycle.

Most machines use variable inlet guide vanes for capacity control.

APPLICATION

Centrifugal refrigeration machines were developed to fill the need for single refrigeration units of large capacity. A single centrifugal machine can be used in place of many reciprocating units.

Since the original one was installed (*Fig. 16*), centrifugal refrigeration machines have been known for:

1. Reliability
2. Compactness
3. Low maintenance costs
4. Long life
5. Ease of operation
6. Quietness

Open centrifugal machines are essentially multi-purpose machines. They are used in special and industrial applications requiring higher temperature lifts than normally encountered at air conditioning levels. They are flexible in regard to speed selection and staging, and are used for standard water chilling applications where one or more large capacity machines are required, or where a steam turbine, gas engine, gas turbine, diesel engine or special motor drive is desired.

The application of a gas engine or gas turbine drive to a centrifugal machine is particularly attrac-

tive when the engine or turbine exhaust gases can generate steam in a waste heat boiler to produce additional refrigeration from absorption machine equipment.

Hermetic centrifugal machines are single purpose machines and are generally used for water chilling applications. They are low in first cost because they are a factory package. They can be installed easily and quickly with a minimum of field problems involving motor mounting, coupling and alignment.

STANDARDS AND CODES

Equipment installation should conform to all codes, laws and regulations applying at the site.

The equipment should be manufactured to conform to the ASA B9.1 Safety Code for Mechanical Refrigeration. This safety code requires conformance to the ASME Unfired Pressure Vessel Code.

Specifications should call for conformance to these standards and codes to assure a high quality product. Pressure vessels are ASME stamped when required by the code.

UNIT SELECTION

The factors involved in the selection of a centrifugal machine are load, chilled water or brine quantity, temperature of the chilled water or brine, condensing medium to be used, quantity of the condensing medium and its temperature, type and quantity of power available, fouling factor allowance, amount of usable space available, and the nature of the load, whether variable or constant.

The final selection is usually based on the least expensive combination of machine and heat rejection device as well as a reasonable machine operating cost.

Load, chilled water or brine quantity, and temperature rise are all related to each other such that, when any two are known, the third can be found by the formula:

$$\text{Load (tons)} = \frac{\text{quantity (gpm)} \times \text{temp rise (F)} \times \text{sp ht} \times \text{sp gr}}{24}$$

where:
sp ht = specific heat (1.0 for water)
sp gr = specific gravity (1.0 for water)

Table 7 illustrates typical hermetic centrifugal machine chilled water ratings. Ratings in tons based on various leaving chilled and condenser water temperatures are given for a particular machine size. The ratings in bold face type require rated kilowatt input while those in italics require less input.

Brine cooling normally requires special selection by the manufacturer.

The choice of a chilled water temperature for air conditioning applications should be carefully considered as pointed out in *Part 6*. The selection is an economic one since it involves the analysis of the owning and operating costs of several systems to determine the optimum chilled water temperature.

The selection of multiple machines for a common load is normally based on availability, reliability and/or flexibility: availability because of limitations to the physical size it is economical to produce; reliability because of the need to handle a portion of the load when one machine may be down for service; flexibility because of the ability to more efficiently match compressor capacity to partial load requirements. As a general rule, seldom are multiple machine applications made on normal air condition loads less than about 400 tons.

When multiple machines are considered, series water flow thru the coolers may be advantageous (*Fig. 17*). Generally, the longer the piping distribution system, the higher the over-all chilled water rise. For instance, close-coupled chilled water coils and coolers normally have an economic optimum rise of about 8-10 degrees. Conversely, chilled water distribution systems for a campus type operation would normally have an economic optimum rise of about 15-20 degrees. For the higher rises, series water flow thru the chillers may offer an operating cost savings. The first machine operates at a higher suction temperature which requires less power.

TABLE 7—TYPICAL HERMETIC CENTRIFUGAL MACHINE RATINGS

REFRIGERATION CAPACITY (TONS)
Italic Rating Requires Less than 330 kw Input

Leaving Chilled Water Temp (F)	Leaving Condenser Water Temperature (F)			
	85	90	95	100
40	409	405	399	386
41	417	414	406	391
42	426	422	413	396
43	434	429	419	401
44	442	435	424	406
45	450	441	430	412
46	457	447	435	417
47	461	453	442	423
48	465	459	449	429
49	470	465	456	435
50	474	472	462	442

NOTE: Ratings are based on a 2-pass cooler using 380-1260 gpm and on a 2-pass condenser using 430-1430 gpm.

CAPACITY ADJUSTMENT

	Water Flow (gpm)	Pass	Nominal Capacity Adjustment to 2-Pass Rating	
Cooler	190 to 630	4	ADD	3%
	250 to 840	3	ADD	1½%
	755 to 2520	1	DEDUCT	4%
Condenser	215 to 715	4	ADD	3%
	285 to 955	3	ADD	1½%
	860 to 2860	1	DEDUCT	2½%

The optimum machine selection involves matching the correct machine and cooling tower as well as the correct entering chilled water temperature and water rise. A selection of several machines and cooling towers often results in finding one combination having a minimum first cost. In many instances it is possible to reduce the condenser water quantity and increase the leaving condenser water temperature, resulting in a smaller tower.

The use of an economizer can effect a power reduction for the compressor of as much as 6% for the same cooler and condenser surface. This same power reduction can be obtained by adding 15% to 30% more surface in the heat exchangers. The method of saving this power is a machine design consideration and becomes a matter for the manufacturer to decide which is the least expensive way to accomplish this reduction. Economizers can only be used with multi-stage compressors.

On low temperature, industrial applications where four or more stages are used, a two-stage economizer is generally justified.

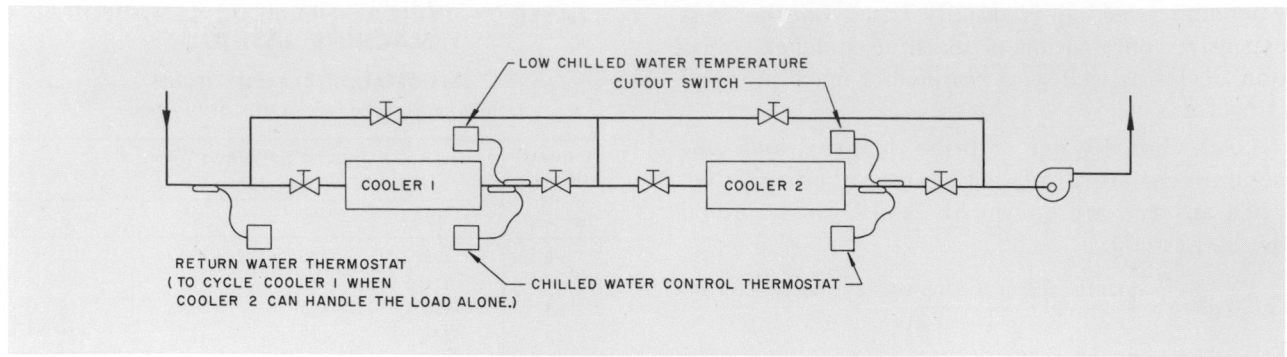

FIG. 17 — SERIES ARRANGEMENT OF TWO COOLERS

Fouling factors used when the cooler and condenser are selected have a direct bearing on the system economics. Too conservative a factor results in a high first cost while too low a factor increases operating costs by increasing the frequency of tube cleaning or increasing the costs of the water conditioning to maintain the low factor. *Part 5* includes detailed effects of fouling on equipment selection, and suggests various fouling factors based on equipment and systems.

Table 8 lists the relative costs and resistance to corrosion of various metals and alloys for special cases where unusual water conditions require other than the standard copper tubing for either the cooler or the condenser.

OPERATING COSTS

The power operating costs of an electrically driven centrifugal machine may be realistically determined by a mechanical integration of power cost increments for the total operating hours of the machine as follows:

$$C = c\,(p_1 h_1 + p_2 h_2 + p_3 h_3 + \dots\dots + p_n h_n)$$

where:

C = annual power costs

c = cost per kilowatt hour, including demand and energy charges

p = power consumption for incremental percentage of nominal full load, expressed as

(1) motor kilowatt input, OR

(2) $\dfrac{\text{motor output brake horsepower} \times .746}{\text{motor efficiency}}$

(See *Chart 2* for a typical hermetic centrifugal performance.)

h = hours of machine operation during the year at above percentage of nominal full load. (See *Chart 3* for a typical graph showing percentages of full load vs operating hours.)

TABLE 8—RELATIVE COSTS AND RESISTANCE TO CORROSION OF METALS AND ALLOYS

Material	Relative Costs per Tube	Sea Water or Brackish Salt Water	Soft Fresh Water high oxygen and carbon dioxide content	Soft Fresh Water low oxygen and carbon dioxide content	Water of High Hardness Content tendency to scale
		➝ CORROSION TENDENCY DECREASES			
Steel (SAE 1010)	1.6	N.R.	N.R.	A	A
Copper	1.0	N.R.	B	A	A
Nickel	—	A	A	A	A
Red Brass	1.9	N.R.	B	A	A
Admiralty Brass (inhibited)	2.0	B	A	A	A
Cupro-Nickel 70/30	2.6	A	A	A	A
90/10	1.9	A	A	A	A
Aluminum Brass	2.5	A	A	A	A
Stainless Steel (304L)	6.0	N.R.	A	A	A
Nickel Steel (3½%)	2.6	N.R.	N.R.	A	A
Aluminum	—	N.R.	B	B	B

NOTE: The relative resistance to corrosion from fresh water increases as the tendency for scaling increases. Water that causes scaling usually does not cause corrosion.

Symbols in Table: A — Generally acceptable for use.
B — Used under certain conditions, when experience shows metal is acceptable.
N.R. — Not recommended.

CHART 2—TYPICAL HERMETIC CENTRIFUGAL PERFORMANCE

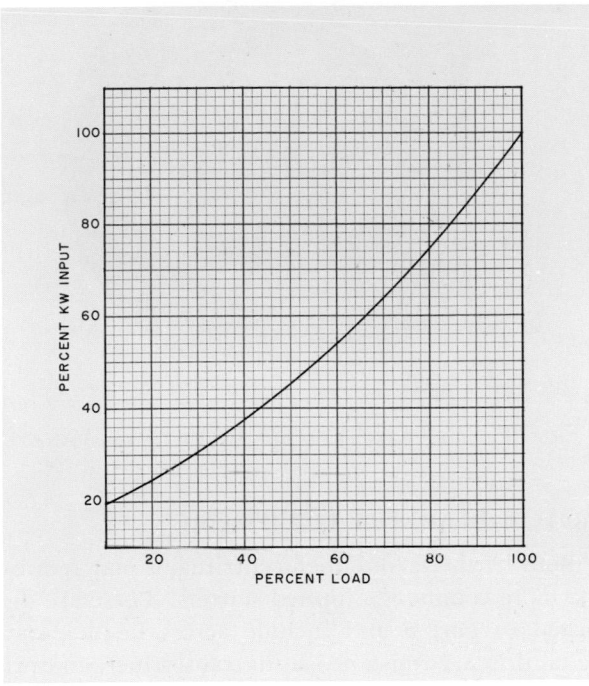

The annual power costs of auxiliary equipment may be calculated as follows:

$$\text{Power costs} = \frac{.746 \times \text{bhp} \times \text{hr} \times \text{cost/kw-hr}}{\text{motor efficiency}}$$

DRIVE SELECTION

There are four types of drives in general use for centrifugal compressors.

1. Steam turbine
2. Variable speed motor
3. Constant speed motor
4. Constant speed engine

Steam turbines are ideally suited for centrifugal compressors. They afford variable rpm, permitting the compressor to operate at a minimum speed and brake horsepower. They usually have a good efficiency characteristic over the required speed range with economy of operation. Refer to *Part 8* for additional information on drives.

Variable speed motors of the wound rotor type are used for open centrifugal machine applications because of the favorable starting inrush characteristics and the range of speed regulation. Capacity can be controlled by varying the speed manually or automatically. *Figure 18* indicates that a rapid decrease in power input results when the speed is reduced.

CHART 3—TYPICAL GRAPH, PERCENT FULL LOAD VS OPERATING HOURS

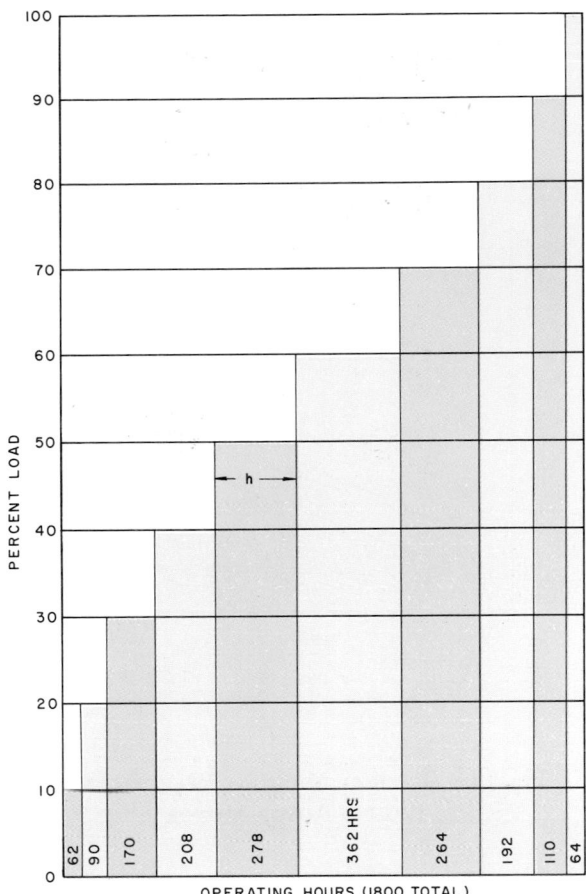

Motors normally used for a constant speed drive are the squirrel cage induction or synchronous type. It is sometimes possible to obtain a motor that has low starting current features and uses an across-the-line starter, thus saving on starting equipment cost when power company limitations on inrush current are satisfied.

Hermetic compressors use only induction type motors since they usually operate in a refrigerant atmosphere and do not require brushes or commutators which may cause a breakdown of the refrigerant due to arcing.

Synchronous motors may be applied to advantage if a power factor correction is desired. Another method is to use a standard induction motor plus the necessary capacitor.

Natural gas engines may also be applied as drives for centrifugal machines. Engine speeds normally range from 900-1200 rpm, the lower speeds being used on applications having longer annual operating hours. Speed increasing gears are used between the engine and the centrifugal machine.

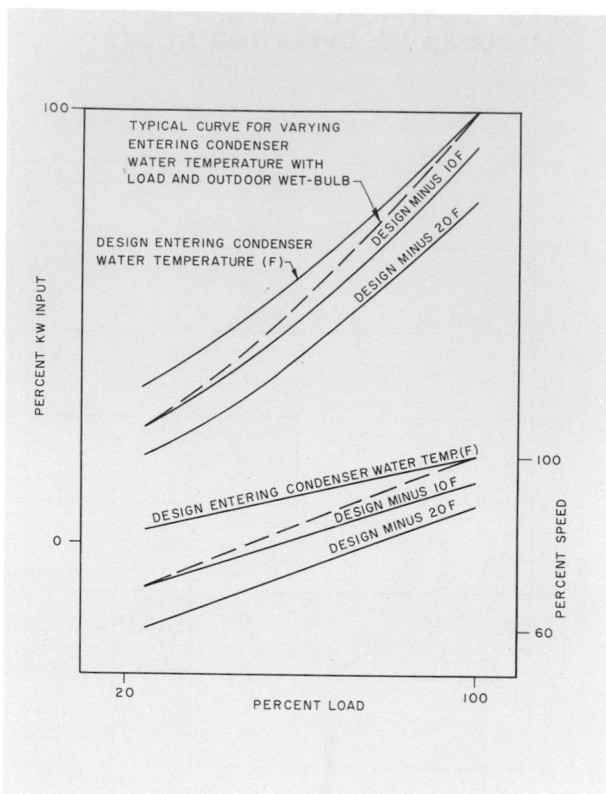

FIG. 18 — TYPICAL MACHINE PERFORMANCE,
WOUND ROTOR MOTOR

Centrifugal compressors have low starting torques; therefore, most drives can easily be matched with these machines.

However, not only must starting torque be checked, but also acceleration time required to bring the centrifugal up to speed. Too fast an acceleration time is not desirable because design stresses for keyways may be exceeded and lubrication problems may be created. Minimum recommended acceleration times for open machines are available from the manufacturer.

GEARS

Speed increasing gears used for an open centrifugal compressor drive are usually the double helical (herring-bone) type. The horsepower loss in the gear must be included with the actual compressor brake horsepower to determine the motor horsepower.

The selection of the proper gear for a particular centrifugal application depends on motor horsepower, motor rpm and compressor rpm. Water-cooled oil coolers are normally included with the gear.

FIG. 19 — TYPICAL VARIABLE INLET GUIDE VANES

MOTOR STARTING EQUIPMENT

Both hermetic and open centrifugals may require the more commonly applied starters. These are discussed in *Part 8* and include across-the-line, star-delta, primary resistance, auto transformer, and primary reactor starters.

CONTROLS

CAPACITY CONTROL

If a centrifugal machine is to perform satisfactorily under a partial load, a means of effecting a capacity reduction in proportion to the reduction of the instantaneous load is required.

Hermetic Centrifugal

Water temperature control is obtained by means of variable inlet guide vanes (*Fig. 19*) at the suction inlet to the compressor.

This control reduces capacity by varying the angle at which the suction gas is directed into the eye of the impeller. It also conserves power because it promotes aerodynamic gas flow thru the compressor. At low flow the change of inlet gas direction has little effect on capacity and the control operates primarily as a suction damper. The minimum partial load capacity of the machine is based upon the amount of gas leakage thru the fully closed capacity regulating vanes. *Chart 2* shows a typical power input curve for a hermetic centrifugal machine operating with condenser water supplied from a cooling tower when the refrigeration load closely follows the outdoor wet-bulb temperature. The curve is based on the design water flow being maintained at a constant rate for both the cooler and condenser.

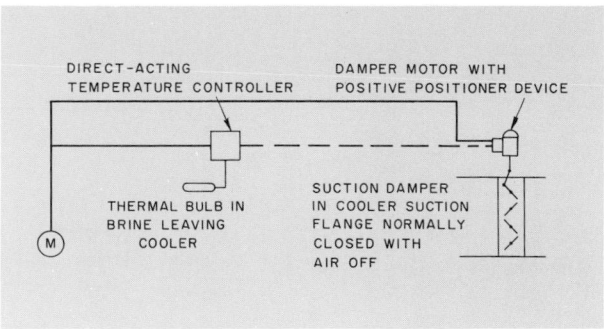

FIG. 20 — SUCTION DAMPER CONTROL SYSTEM

A chilled water control thermostat automatically controls the leaving chilled water temperature. When the temperature changes, the thermostat signals the chilled water control to reposition the capacity regulating vanes which change the capacity of the machine to maintain the desired temperature. When the vanes reach the closed position and the leaving temperature continues to decrease to a predetermined minimum, the low chilled water temperature cutout switch stops the machine.

Open Centrifugal

Capacity control on an open centrifugal machine may be obtained with a suction damper (*Fig. 20*), variable inlet guide vanes, or variable speed drive (steam turbine, gas turbine, gas engine or wound rotor motor).

A suction damper is controlled by a thermostat in the leaving chilled water to reduce the capacity of the compressor by throttling the suction gas. The variable inlet guide vane control is identical to that discussed under *Hermetic Centrifugal*.

Variable speed drives may be controlled manually when the change in loading is gradual or when a suction damper is used for automatic control. Automatic speed control is used with steam turbine, gas turbine or gas engine drives. Automatic speed control provides very economical operation and requires less input than other methods of control. Automatic speed control of wound rotor motors is expensive and is seldom used.

Chart 4 shows the comparative performances of different methods of centrifugal compressor capacity control.

CONTROL OF SURGE

Surge is a characteristic of centrifugal compressors which occurs at reduced capacities. This condition is a result of the breakdown in flow which occurs in the impeller. When this happens, the impeller can no longer maintain the condenser pressure, and

CHART 4—COMPARATIVE PERFORMANCES, CENTRIFUGAL COMPRESSOR CAPACITY CONTROL

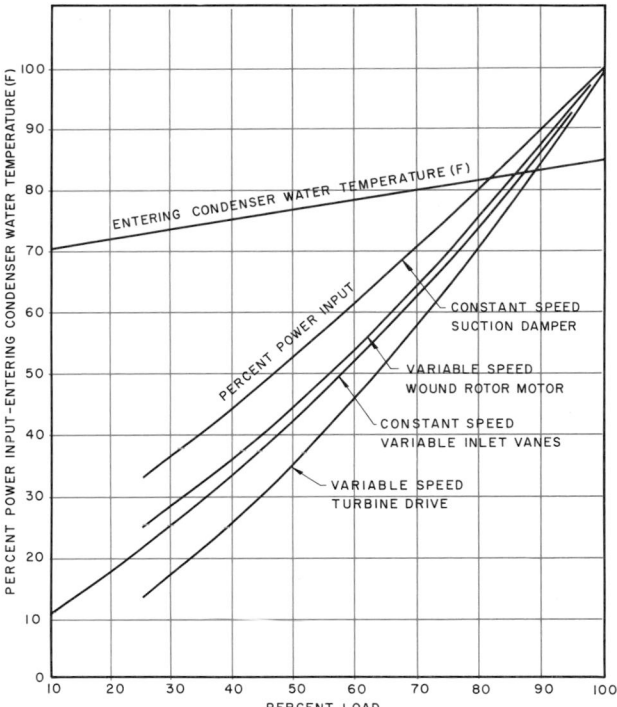

there occurs a momentary reversal of flow which is accompanied by a lowering of condenser pressure. This allows the impeller to function normally again, and the gas flow returns to its normal direction. The operation is stable until the condenser pressure builds up and surge occurs again. Surging can be detected primarily by the change in sound level of the machine.

Surge in a centrifugal machine does not occur at partial loads if the head or lift decreases sufficiently with the load.

Chart 5 shows a typical lift versus load diagram for a hermetic centrifugal. A series of curves is plotted indicating the compressor operating curves at different positions of the inlet guide vanes. Line *B* represents a series of operating points of a machine when the characteristic of the loading is such that the condensing temperature or the total lift of the compressor reduces as the loading also reduces. An example is a comfort air conditioning job using a cooling tower to provide the condenser water. As the outdoor wet-bulb decreases, the refrigeration load decreases and the condenser water temperature is reduced, allowing the condensing temperature to drop. Line *A* represents a series of operating points of a machine when the condensing temperature or lift remains almost constant or decreases only

CHART 5—TYPICAL LIFT-LOAD DIAGRAM, HERMETIC CENTRIFUGAL

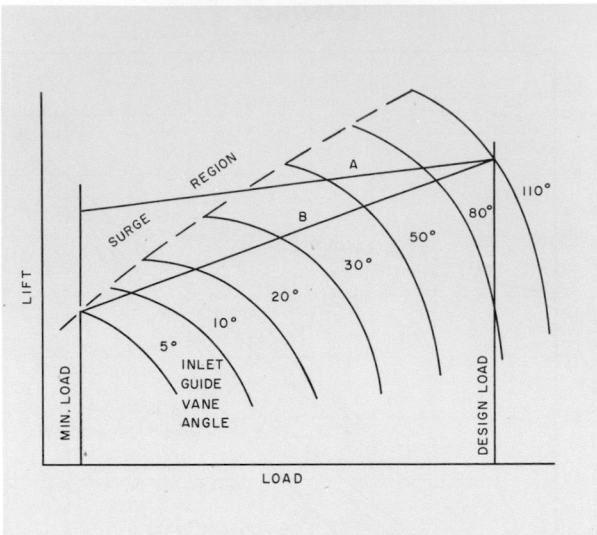

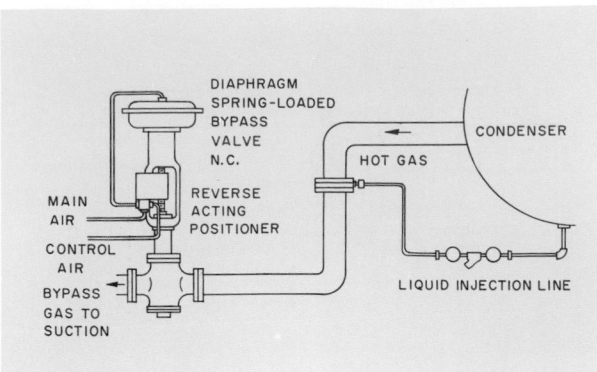

Fig. 21 — Automatic Hot Gas Bypass Valve

slightly. An example is the condition in which a fixed temperature of condenser water is available all year, or the condenser water temperature is still at the design temperature and a partial load condition exists as in a process application.

It can be seen from the chart that the line *B* does not enter the surge region until the loading is under the minimum load (approximately 10%). The line *A* enters the surge region above the minimum loading and, therefore, the machine needs some method of maintaining the loading above this point, such as a hot gas bypass.

To control surge occurring at partial load for either an open or hermetic centrifugal machine, a valved gas connection between the condenser and cooler is normally used to load the compressor artificially. The valve may be either manual or automatic (*Fig. 21*). As applied to open centrifugal machines, the automatic hot gas bypass valve is usually controlled in sequence with the automatic suction damper or with the speed of the compressor so that the valve starts to open just before the suction damper position or speed of the compressor indicates surge.

SAFETY CONTROL

Hermetic Centrifugal

The variable inlet guide vanes control capacity and are used to prevent motor overload in two ways:

1. When starting, the capacity regulating vanes remain closed until the motor is connected across the line at full voltage and the current drawn is below full load current.

2. Motor overload control overrides the chilled water temperature control to prevent further opening of the vanes at 100% motor load. If the current drawn continues to rise above 100%, the vanes begin to close, reducing the motor load.

Similar controls may be obtained for an open centrifugal machine driven by a constant speed motor.

A typical ladder diagram of the safety controls for a hermetic centrifugal is shown in *Fig. 22*. Safety controls common to both the hermetic and open type of centrifugal machine are described as follows:

1. Condenser high pressure cutout switch stops the compressor when the condenser pressure becomes too high due to a condenser water stoppage, excessive condenser scaling or air in the system.

2. Low refrigerant temperature cutout switch stops the compressor when the evaporator pressure becomes too low due to a chilled water stoppage, excessive cooler scaling, or insufficient refrigerant charge.

3. Low oil pressure cutout switch stops the compressor when the oil pressure drops below the required minimum and prevents either starting (on compressors with external oil pumps) or operating (on compressors with shaft-driven oil pumps) the compressor motor before the oil pressure is up to the minimum.

4. Low brine or chilled water temperature cutout switch stops the compressor when the leaving brine or chilled water temperature drops below the minimum allowable temperature.

5. Chilled water flow switch stops the compressor when chilled water ceases to flow, and prevents a start-up of the compressor motor until chilled water flow is established (optional).

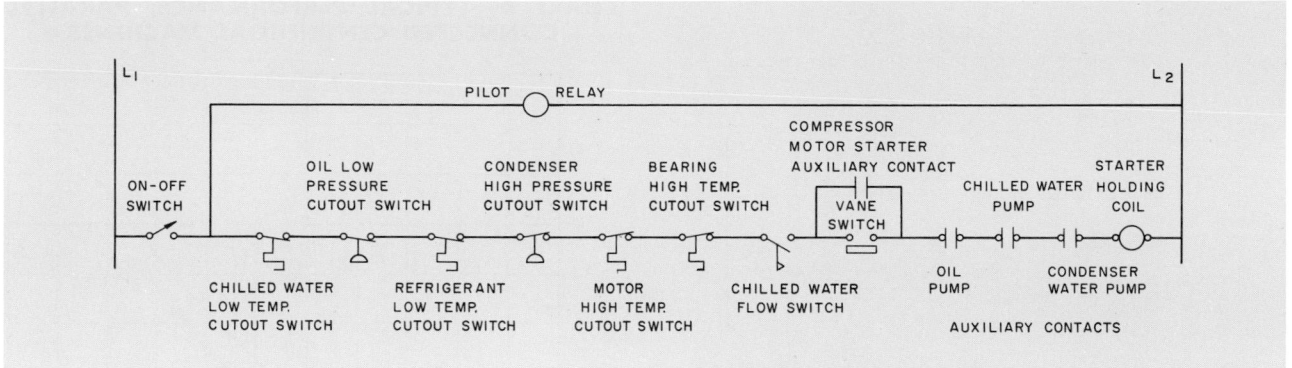

FIG. 22 — TYPICAL SAFETY CONTROL SYSTEM, HERMETIC MACHINE

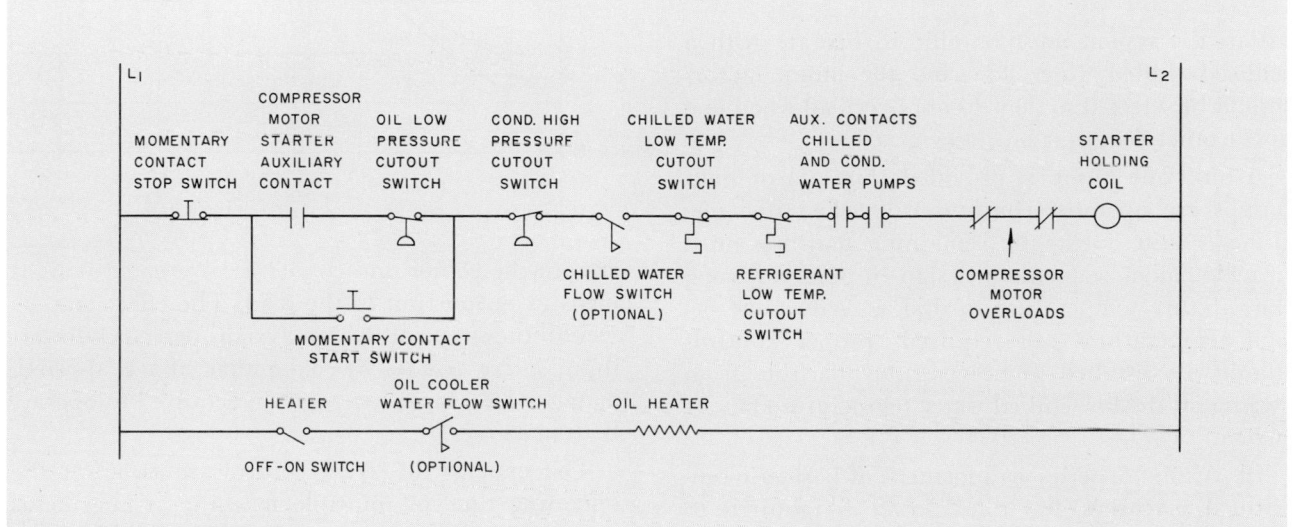

FIG. 23 — TYPICAL SAFETY CONTROL SYSTEM, OPEN MOTOR-DRIVEN MACHINE

Open Centrifugal

A ladder diagram of the safety controls for a motor-driven centrifugal is shown in *Fig. 23*.

ELECTRICAL DEMAND CONTROL

This control can override the capacity control to limit the current drawn during off-season operation. This allows operation of the machine without creating high electrical demand charges during months when full load capacity is not required. The control can be set to reduce the amount of current which can be drawn by the motor down to as low as 40% of the full load amperage.

MULTIPLE MACHINE CONTROL

When two or more centrifugal machines are required to handle a load, they may be applied in a parallel or series arrangement of coolers. The arrangements are controlled in a manner similar to single machines.

Installations with machine coolers in parallel may utilize two or more machines. With series chilled water flow, the cooler pressure loss is cumulative and may become excessive if more than two machines are installed in series.

Parallel Arrangement

When two or more machines are installed with the coolers connected in parallel in the chilled water circuit (*Fig. 24*), each machine may control its own leaving chilled water at design temperature as in a single machine installation. The same throttling range should be used for each machine. As the system load is reduced, each machine reduces capacity simultaneously, thus individually producing the same leaving chilled water temperature.

When each cooler is provided with a separate chilled water pump, the pump and cooler may be shut down during partial load operation. This

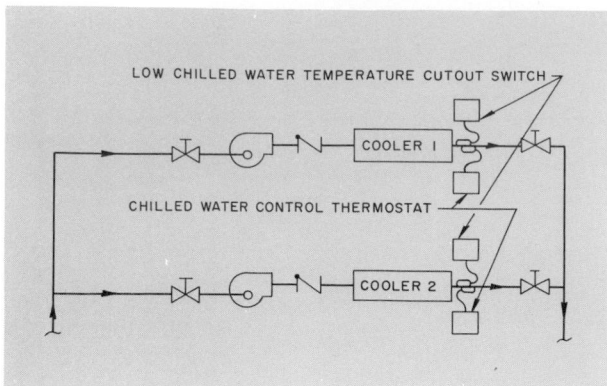

FIG. 24 — PARALLEL ARRANGEMENT OF TWO
MACHINES (TWO PUMPS)

CHART 6—TYPICAL PERFORMANCE, PARALLEL
CONNECTED CENTRIFUGAL MACHINES

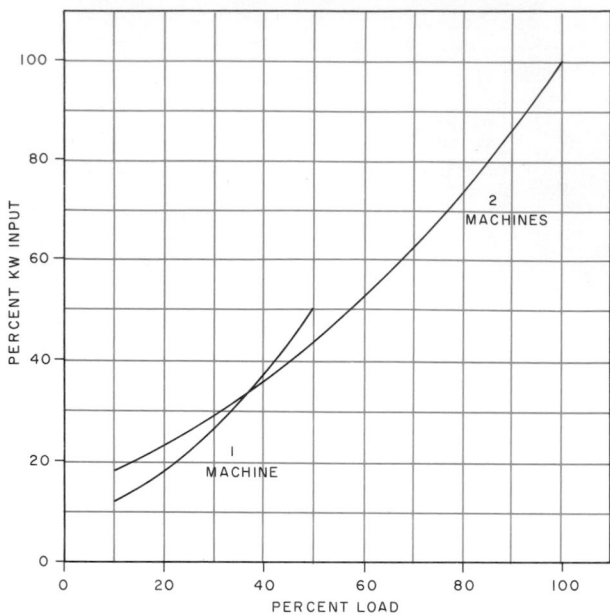

means the system must be able to operate with a reduced chilled water flow and the pump motors should be selected so they do not overload when one of the other pumps is shut down.

If only one pump is provided (*Fig. 25*) or both pumps are operated continuously, when one machine is shut down, the remaining machine must provide colder water than design in order to bring the mixture temperature to design. When low design temperatures are required, proper controls should be installed to prevent the machine from cycling on the low chilled water temperature cutout switch.

In parallel or series arrangement of hermetic centrifugal machines, less total power is required to operate both machines simultaneously down to approximately 35% load than to run only one, throttling it to the load. This occurs because the surface

area in the cooler and condenser is greater at light loads in proportion to the load. The effect may be seen in the shape of the load versus percent kilowatt input curve (*Chart 6*). Note that above approximately a 35% load, less power is required to operate both machines.

The expense of extra controls to equalize the operating time of multiple machine arrangements is not required. When using reciprocating equipment, changing the order of starting and stopping multiple compressor arrangements is sometimes justified. Due to the absence of wearing parts in a centrifugal machine, this changing is seldom used.

Series Arrangement

When coolers are connected in series, equal reduction of loading of each machine produces the best power consumption. The throttling range of the high stage machine must be adjusted to insure that each machine handles the same percentage of the system load, both at design and at partial load conditions.

In any series selection the throttling range required on the high stage machine equals the chilled water temperature drop thru the low stage machine plus the throttling range of the low stage machine.

Figure 17 shows a series cooler arrangement of two hermetic centrifugal machines and controls. The extra thermostat (return water thermostat) is used to cut in and cut out the first machine at light loads.

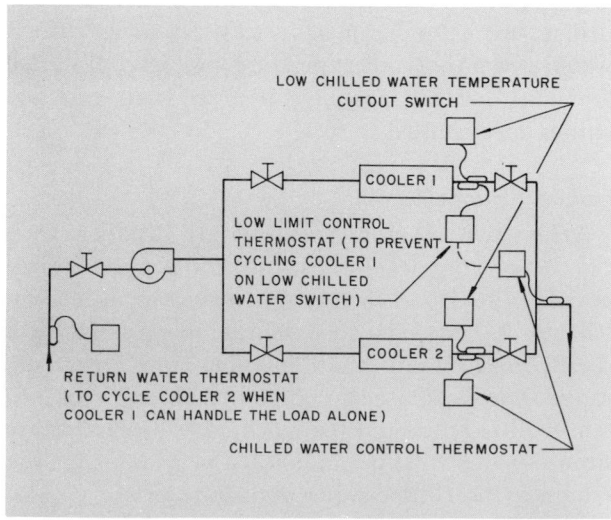

FIG. 25 — PARALLEL ARRANGEMENT OF TWO
MACHINES (ONE PUMP)

TABLE 9—NORMAL REFRIGERANT LOSS EXPECTED WITH CENTRIFUGAL MACHINES

Installation Size (tons)	125-175	175-250	250-350	350 or more
Refrigerant Loss (lb/yr)	75-125	100-150	100-150	125-175

*Based on comfort conditioning 120 days per year, 10 hours per day (1200 hour season).

NOTE: Factors for conditions other than above:

Time factors — 2500 hr/yr		1.25
— continuous for year		1.50
Application factor, low temp		1.20

PURGE UNIT

A purge unit for a centrifugal machine may be either a thermal or a compressor type.

The purpose of the unit is to evacuate air and water from the centrifugal machine and to recover and return refrigerant which is mixed with the air. Even though a machine may be perfectly airtight, it may develop a water leak which is detected only by operation of the purge system. If water is allowed to remain in the machine, serious damage to tubes and other internal parts can result.

The compressor type purge unit operates independently by means of a small reciprocating compressor to remove the air, moisture and a small quantity of refrigerant from the condenser.

The thermal type purge unit operates on a pressure differential principle, and a reciprocating compressor is not required.

Although the purge unit performs a highly efficient job of removing refrigerant from the air being purged, it is physically impossible to recover all the entrained refrigerant; some is always lost.

Table 9 shows an approximation of the normal refrigerant loss to be expected with a centrifugal machine. It must be realized that the actual loss varies widely from one installation to the next; this variation is based on machine tightness, frequency of purging and other factors.

INSULATION

The cooler, suction piping and other cold surfaces should be insulated to prevent sweating. Float valve chambers, water boxes and other parts of the machine which may require servicing should be provided with a removable type of insulation, such as sheet metal covers filled with granulated cork.

Various types of insulation can be used, such as vegetable cork, closed cell foamed plastic and expanded polystyrenes.

LOCATION

Machine location and layout should be carefully studied when applying a centrifugal machine. The location of the machine directly influences the economic and possibly the sound level aspects of any system.

The construction of the room where the machine is located should contain mass to reduce transmissibility of noise to surrounding spaces and should also provide acoustical treatment to maintain reasonable sound levels in the room.

A floor adequately strong and reasonably level is all that is required for the location of a hermetic centrifugal machine. However, though the foregoing statement may be taken literally, it is to the engineer's advantage to consider other aspects relative to the location of the machine.

1. It should be located so that the installation costs of the piping between the unit and the equipment it supplies and the costs of the wiring or piping of the services to the unit are at a minimum.

2. There should be sufficient space near the machine for auxiliary equipment such as chilled and condenser water pumps and piping.

3. There should be adequate clearances around the machine for access and servicing.

In new construction on upper floors, steel floor framing should be laid out by the architect or consulting engineer to match machine supports in order to transfer loads to the building columns. On upper floors in existing buildings, the use of existing floor slabs should be avoided. Supplementary steel framing for transferring all machine loads to building columns should be designed by the structural engineer.

LAYOUT

In the layout of centrifugal refrigeration machines, consideration should be given to nozzle arrangements.

NOZZLE ARRANGEMENTS

Cooler — When arranged for multi-pass, water should enter the bottom tubes and leave thru the top tubes. This method gives the best efficiency and promotes venting of any air trapped inside the tubes.

Condenser — When arranged for multi-pass, water should enter the top tubes first. This provides the coldest surface at the top of the condenser shell to stratify noncondensables for proper purging.

The nozzle arrangement chosen for both cooler and condenser should result in a minimum number of chilled water and condenser water pipe fittings, the optimum access to the centrifugal and auxiliary equipment, and a neat appearance.

OUTDOOR INSTALLATION

A hermetic centrifugal machine is basically designed for indoor operation. Outdoor installation is usually not encouraged. The machines should not be located outdoors when they may be subjected to freezing temperatures.

A simple heated structure enclosing the machine is preferred since complete protection for the machine, instruments, starter and auxiliary equipment is provided. Erection of such an enclosure may be less costly than the outdoor installation precautions which may be required. If it is necessary to install the machine outdoors, refer to the manufacturer for recommendations and precautions.

UNIT ISOLATION

Normally, only the hermetic compressor assembly is isolated from the floor with moulded grooved neoprene isolation pads. For upper story installations, isolation pads may also be required under the feet of the cooler. In the case of highly critical installations, spring isolators may be required under the compressor assembly and cooler, in which case auxiliary pumps and piping should be isolated.

The base which holds the open compressor and its driver is designed for each individual job. With a steel base design, the complete unit is mounted on an independent fabricated steel foundation. With a concrete type of base, the various components are mounted on individual steel plates which are anchored to the concrete.

Figure 26 shows a depressed concrete base isolated from the adjoining floor with mastic.

Cork is not a satisfactory isolation material for most applications. Under no conditions should it be considered for isolation on an upper floor of a building where the least amount of vibration would be objectionable. However, four-inch thick cork pads may be used in noncritical locations.

The machine foundation should be located away from building column footings. In order to do a first class job such as may be required on an upper floor, spring mounting should be used. Sandwich type spring mountings, although nonadjustable, may also be used under a metal pan into which the concrete is poured.

When spring isolation is used, flexible rubber connections are recommended at the points where the chilled and condenser water piping is connected to the cooler and condenser to take up the movement of machine and base when starting and stopping.

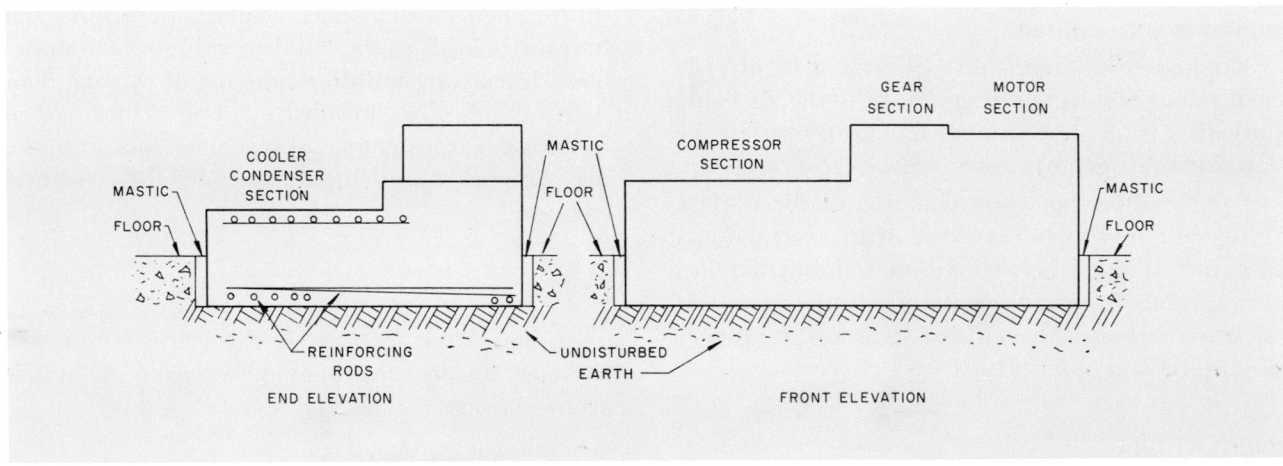

FIG. 26 — DEPRESSED CONCRETE BASE (ISOLATED FROM FLOOR)

CHAPTER 3. ABSORPTION REFRIGERATION MACHINE

The absorption refrigeration machine is a water chilling package which uses heat directly without the use of a prime mover, thus utilizing the heating facilities on a full time, year-round basis. Because of its compactness and vibrationless operation, it can be installed anywhere space and a heat source is available, from basement to roof. It uses the cheapest, safest and most available of all refrigerants, ordinary tap water. Its absorbent is a simple salt.

This chapter presents data to guide the engineer in the practical application and layout of absorption refrigeration equipment when used for comfort air conditioning systems.

APPLICATION

Since heat in the form of steam or hot water is generally the operating force of an absorption machine, the following situations are favorable to the application of absorption refrigeration machines:

1. Where low cost fuel is available, as in natural gas regions.

2. Where electric rates are high. Whenever the cost of steam in dollars per thousand pounds is less than fifty times the cost of electricity in dollars per kilowatt, a lower operating cost can be expected for the absorption machine. This is approximately the break-even point in operating cost (at design) between this machine and the electrically-driven compressor. The cost of steam is shown graphically in *Chart 7* for different fuels. Comparison curves in *Chart 8* indicate the operating cost of refrigeration for various costs of steam and electricity when applied to an absorption and centrifugal machine respectively. Demand charges should be included in the average electric cost. When comparing the operating cost of a steam turbine-driven centrifugal and an absorption

CHART 7—COMPARATIVE COSTS OF STEAM

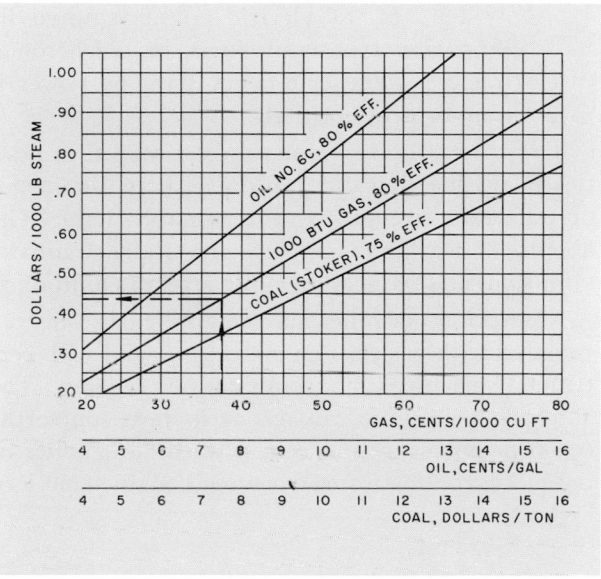

CHART 8—STEAM COSTS VS POWER COSTS

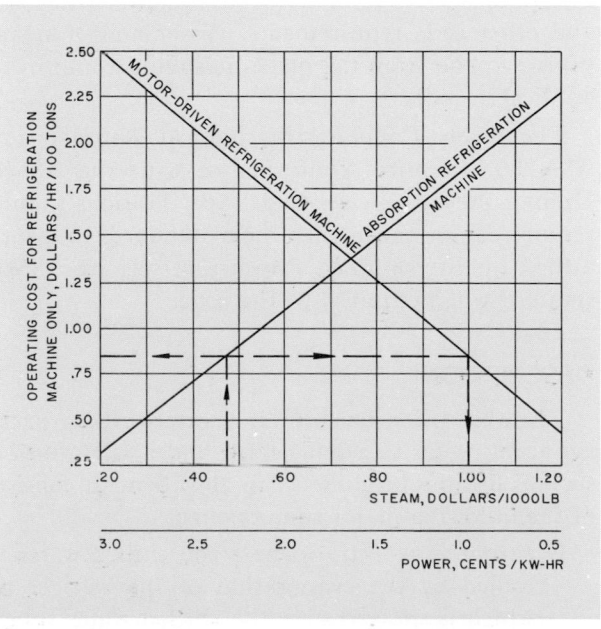

machine, a straight steam rate per ton of refrigeration is not a proper criterion. To obtain a correct analysis, the total system heat input should be used.

3. Where steam or gas utilities are desirous of promoting summer loads.

4. Where low pressure heating boiler capacity is largely or wholly unused during the cooling season.

5. When waste steam is available.

6. Where there is a lack of adequate electric facilities for installing a conventional compression machine. Since the absorption machine uses only 2-9% of the electric power required by compression type equipment, its use becomes attractive where emergency stand-by power is required, as in hospitals.

The absorption machine can be installed in practically any location in a building where the floor is of adequate strength and reasonably level. The absence of heavy moving parts practically eliminates vibrations and reduces the noise level to a minimum.

Absorption machines may be applied also in conjunction with gas engines or turbines and with centrifugal machines as combination systems. The absorption machine can use as its heat source the steam or hot water made in a waste heat boiler or the jacket cooling water from a gas engine (250 F or higher).

STANDARDS AND CODES

The location and installation of absorption machines should be made in accordance with local and other code requirements. Water and/or steam piping to and from the machine should conform to applicable codes.

The Safety Code for Mechanical Refrigeration ASA B9.1 requires conformance with the ASME Unfired Pressure Vessel Code. Specifications should require conformance with these standards to assure a high quality product. Pressure vessels are ASME stamped when required by the code.

DESCRIPTION

The absorption machine is a water-chilling package using water as a refrigerant and a salt solution such as lithium bromide as an absorbent. It consists of the following major components:

1. *Evaporator Section* where the chilled water is cooled by the evaporation of the refrigerant which is sprayed over the chilled water tubes.

2. *Absorber Section* where the evaporated water vapor is absorbed by the absorbent. The heat of absorption is removed by condenser water circulated thru this section.

3. *Generator Section* where heat is added in the form of steam or hot water to boil off the refrigerant from the absorbent to reconcentrate the solution.

4. *Condenser Section* where the water vapor produced in the generator is condensed by condenser water circulated thru this section.

5. *Evaporator Pump* which pumps the refrigerant over the tube bundle in the evaporator section.

6. *Solution Pumps* which pump the salt solution to the generator and also to the spray header in the absorber.

7. *Heat Exchanger* where the dilute solution being pumped to the generator from the absorber is heated by the hot concentrated solution which is returned to the absorber.

8. *Purge Unit* which is used to remove noncondensables from the machine and to maintain a low pressure in the machine.

Figure 27 shows a schematic of an absorption cycle. The machine may be constructed in one, two or more shells or sections depending on the manufacturer or the application.

UNIT SELECTION

The factors influencing the selection of an absorption machine are load, chilled water quantity, temperature of the chilled water, condenser water source, condenser water temperature, condenser water quantity, fouling factor allowance, and heat source. The final selection is usually based on the least expensive combination of machine and cooling tower as well as a reasonable machine operating cost. The absorption machine can be utilized with any conventional open or closed circuit chilled water system.

Load, chilled water quantity, and temperature rise are all related to each other so that, when any two are known, the third can be found by the formula:

$$\text{Load (tons)} = \frac{\text{water quantity (gpm)} \times \text{temp rise}}{24}$$

Table 10 illustrates typical absorption machine chilled water ratings using steam as the energy source. Ratings in tons based on various leaving chilled water temperatures, entering condenser

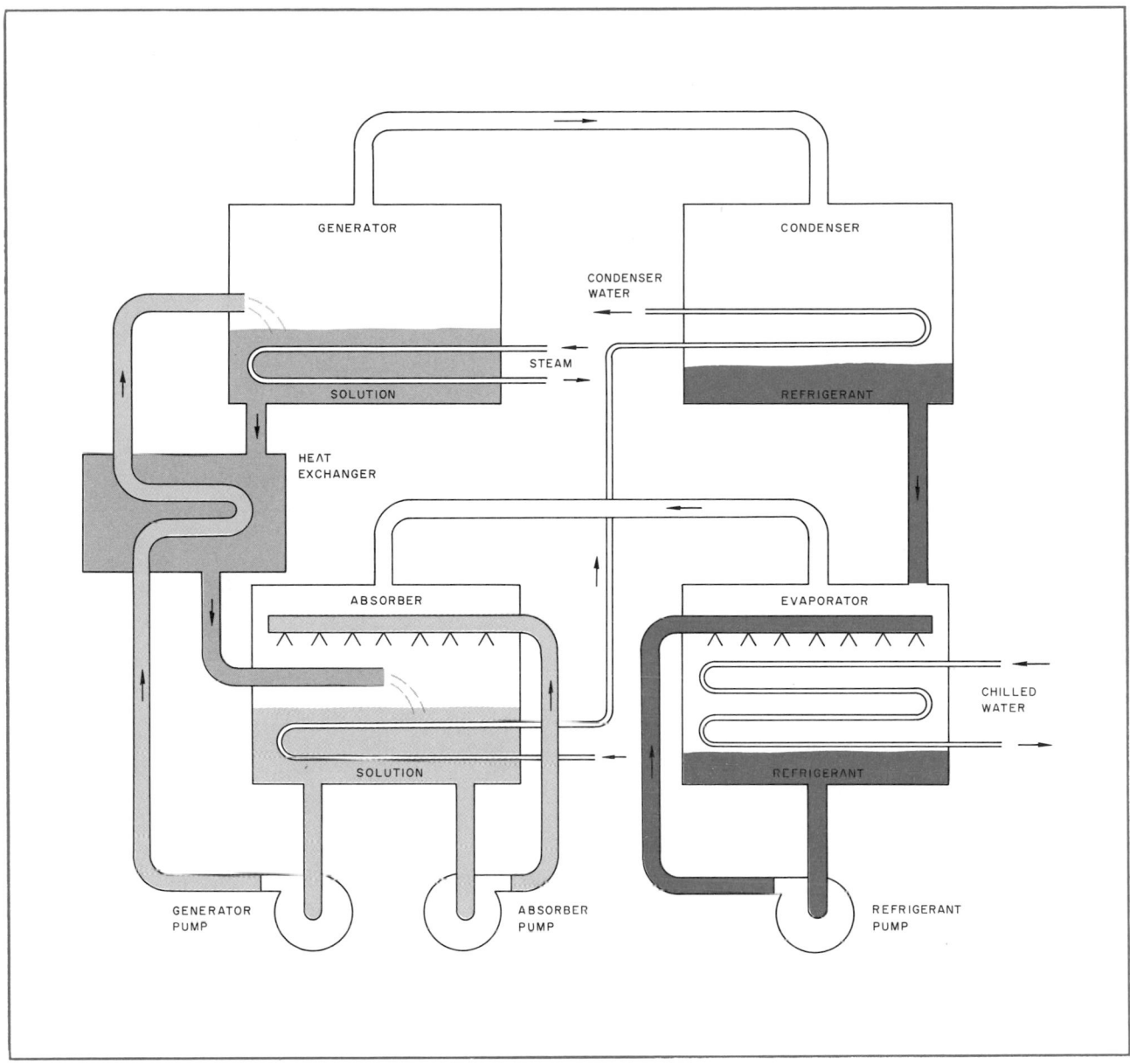

FIG. 27 — SCHEMATIC OF BASIC ABSORPTION CYCLE

water temperatures and steam pressures are given for a particular machine size.

The chilled water temperature should be carefully chosen rather than casually selected or assumed. The proper determination of water quantity and temperature for chilled water coils is discussed in *Part 6*. When low water quantities and a high rise (15 to 20 degrees) are required for the chilled water system, the use of two machines piped in series may be an economic advantage since one machine operates at a higher level and requires less heat input.

Almost any source of condenser water is suitable for use in an absorption machine, provided it is of a good quality. Cooling towers are generally used, but river, lake or well water can also be used when available in sufficient quantity and temperature.

If the condenser water source is a lake, river, well or existing process water, the maximum expected water temperature should be used in selecting the machine. The water quantity required depends on the temperature and load. When a cooling tower is to be used in conjunction with the absorption machine, the tower selection should be matched to the machine selection to provide the most economical combination. In many cases the optimum tower selection will indicate a condenser water tempera-

TABLE 10—TYPICAL ABSORPTION MACHINE RATINGS

REFRIGERATION CAPACITY (TONS)

(Italic Ratings Require Less Than Nominal Condenser Water Flow)

LEAVING CHILLED WATER TEMP. (F)	Entering Condenser Water Temperature (F)											
	75			80			85			90		
	STEAM PRESSURE (PSIG)											
	12	10	8	12	10	8	12	10	8	12	10	8
40	766	759	723	707	695	664	639	628	598	545	529	482
42	824	805	773	761	749	714	702	682	648	605	592	551
44	870	856	818	817	797	760	750	737	695	663	643	603
45	889	878	840	838	823	782	773	760	717	688	673	630
46	889	889	858	858	843	802	800	784	737	710	700	654
48	889	889	889	889	883	840	840	825	778	758	745	696
50	889	889	889	889	889	877	872	857	814	800	782	728
	STEAM PRESSURE (PSIG)											
	6	4	2	6	4	2	6	4	2	6	4	2
40	700	659	631	635	593	556	562	505	454	426	–	–
42	748	711	676	685	637	595	608	557	500	495	411	–
44	791	756	718	730	686	637	652	602	539	553	480	–
45	814	776	739	748	709	662	674	624	559	576	519	385
46	832	797	761	770	731	682	697	647	580	597	539	406
48	871	838	803	808	771	723	736	686	622	636	578	459
50	889	876	840	843	811	765	771	725	665	674	610	496

ture higher than the temperature normally estimated, which is usually 7 to 10 degrees above the design wet-bulb temperature. This may mean a considerable saving in cooling tower cost by reducing the size of the tower. Since this is a heat-operated machine, the heat rejection to the cooling tower is approximately two times that of a motor-driven refrigeration machine. The cooling tower used with the absorption machine is usually about 75% larger than that used with the motor-driven machine. The condenser water temperature drop thru the tower is usually about 17 to 20 degrees. Refer to *Chapter 5* for details on the economics of the cooling tower selection.

Typical fouling factor allowances which should be used for the chilled water and condenser water systems in the selection of the machine are given in *Part 5, Water Conditioning*. Generally a *minimum* factor of .0005 is used for both a closed recirculating chilled water system and an open recirculating condenser water system with conditioned water.

Absorption machines normally use either low pressure steam or high temperature water as an energy source. The pressure or temperature limits are usually defined by the manufacturer, although 12 psig steam pressure or a leaving hot water temperature of 240 F and a temperature drop of 160 degrees are usually considered as maximum values.

When the temperature or pressure (energy source) exceeds the machine design limits, methods such as a steam pressure reducing valve, a steam-to-hot water converter, a hot water-to-steam converter, a water-to-water heat exchanger, or a run-around system, blending return water with supply water can be used to reduce the energy source to the acceptable limits. Other energy sources that may be adapted for use in the absorption machine are hot chemical solutions or petroleum.

Whenever the capacity requirements are less than the capacity of the machine, a lower operating steam pressure should be considered. This usually permits a lower steam rate and a lower total steam consumption. The condenser water quantity must be maintained at full nominal flow for this condition. This also may allow the machine to operate, using very low pressure steam manufactured in a waste heat boiler.

OPERATING COSTS

An important aspect of the economics of an absorption machine other than the first cost is the operating cost.

The annual steam costs may be accurately determined by means of a mechanical integration of steam cost increments for the total operating hours of the machine as follows:

CHART 9—STEAM RATES FOR VARIOUS METHODS OF CONTROL

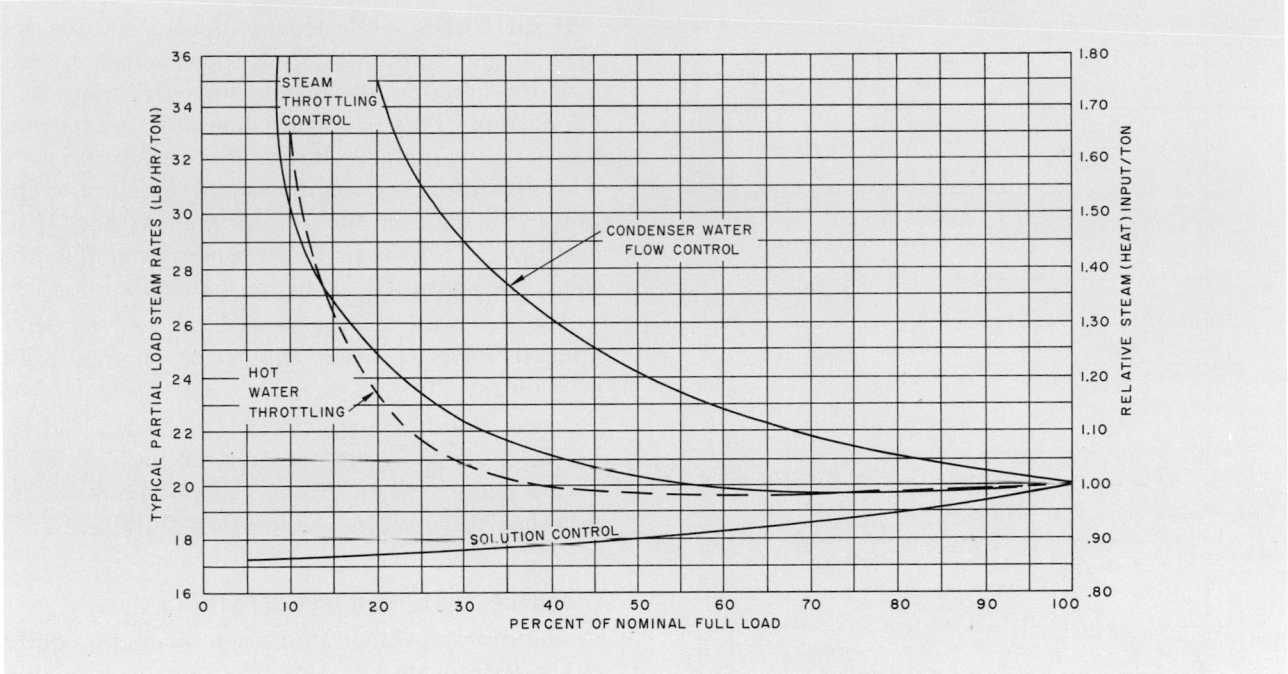

$$C = c \, (s_1 h_1 + s_2 h_2 + s_3 h_3 + \ldots\ldots + s_n h_n)$$

where:

C = annual steam costs

$c = \dfrac{\text{cost per 1000 lb steam}}{1000}$

s = steam consumption for incremental percentage of nominal full load (pounds of steam/hour). Steam consumption for each percentage increment is found by multiplying the steam rate (lb/hr/ton) at each percentage increment (*Chart 9*) by the load (tons) at each increment.

h = hours of operation during the year at percentage of nominal full load. (See *Chart 10* for typical graph showing percentage of full load vs. operating hours.)

When comparing differences in operating costs between absorption machines and turbine drive centrifugals, steam rates can be misleading unless the proper steam costs are used for each machine. The amount of heat used or fuel consumed may be identical although the steam rates are considerably different. *Example 1* illustrates that the steam rates for each machine may be different, but the total heat required is the same.

Example 1 — Comparison of Heat Requirements

Given:

Chilled water temperature from chiller = 45 F
Available condenser water temperature = 85 F
Absorption machine:

    Steam supply   = 12 psig
    Steam rate   = 19 lb/hr/ton

Turbine driven centrifugal:

    Steam supply   = 125 psig
    Condensing pressure = 26 in. vacuum
    Steam rate   = 17 lb/hr/ton

Find:

Amount of heat used for each machine (Btu/hr/ton).

Solution:

Absorption machine

    Total heat of steam at 12 psig = 1161.7 Btu/lb
    Heat of liquid at 212 F leaving machine = 180.0 Btu/lb
    1161.7 − 180.0 = 981.7 Btu/lb steam
    981.7 Btu/lb × 19 lb/hr/ton = 18,700 Btu/hr/ton

Turbine driven centrifugal

    Total heat of steam at 125 psig = 1192.4 Btu/lb
    Heat of liquid at 26 in. vacuum or 125 F = 92.9 Btu/lb
    1192.4 − 92.9 = 1099.5 Btu/lb steam
    1099.5 Btu/lb × 17 lb/hr/ton = 18,700 Btu/hr/ton

This indicates that the amount of heat used for each type of machine is the same even when the steam rates are different.

Because a correct analysis of owning and operating costs uses total system heat input as a criterion rather than a straight steam rate per ton of refrigeration, steam costs must be properly calculated and weighted.

The annual power costs of the auxiliary equipment may be calculated as follows:

$$\text{Power costs} = \frac{.746 \times \text{bhp} \times \text{hr} \times \text{cost/kw-hr}}{\text{motor efficiency}}$$

CHART 10—TYPICAL GRAPH, PERCENT FULL LOAD VS OPERATING HOURS

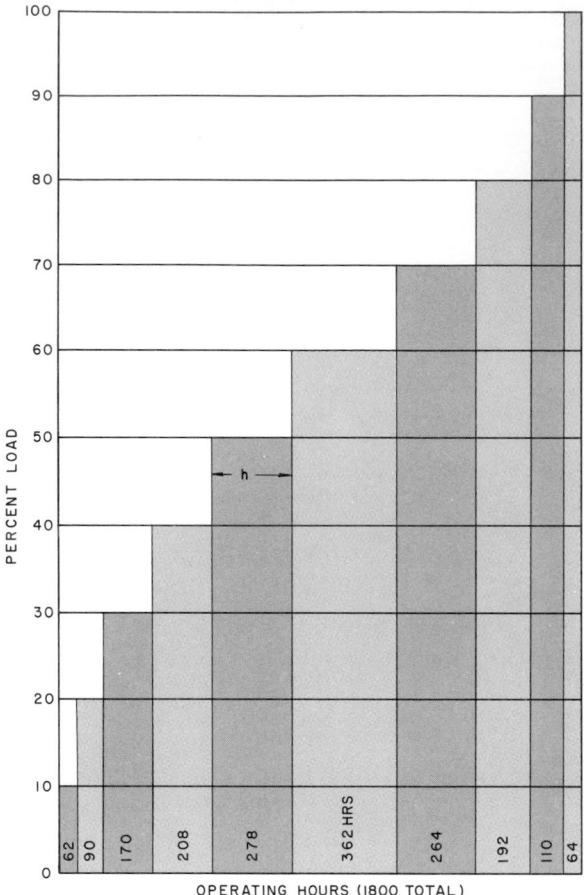

OPERATING HOURS (1800 TOTAL)

STEAM BOILER SELECTION

Any boiler capable of modulating its input to maintain design operating steam pressure within plus or minus one pound is suitable for absorption machine application. This includes:

1. All gas- and oil-fired boilers, since their control is flexible enough to meet this requirement.

2. Coal-fired boilers, when the absorption machine never represents more than 15% of the operating load on the boiler. This is because of the slow build-up and shutdown characteristics which limit their flexibility to adjust to the load. Therefore, these boilers are generally limited to large industrial jobs where the steam is being generated in large quantities year-round for other processes.

If the job conditions require that the absorption machine pick up the load rapidly at start-up, it is recommended that the boiler capacity be based on the start-up steam demand of the machine. This demand is the maximum amount of steam that the

machine can condense at start-up, and must be obtained from the manufacturer.

If the boiler is selected to supply only the full load steam consumption as determined by the machine selection, the boiler temporarily overloads on start-up. This overloading usually affects most boilers by temporarily lowering the steam pressure. This condition is generally not detrimental to low pressure boilers or the absorption machine. If an overload is anticipated, the boiler manufacturer should be consulted for his recommendations.

The net boiler rating should be used to determine its capacity when applied to an absorption machine.

Steam shutoff valves are not necessary for the proper operation of the absorption machine. However, a manual steam shutoff valve is recommended to isolate the machine during long shutdowns.

CONDENSATE RETURN SYSTEMS

The steam-operated absorption machine requires either a steam trap or a direct return to the boiler thru a wet return arrangement.

If single traps of adequate capacity are not available, multiple traps in parallel should be used. Either an inverted bucket or float and thermostatic trap may be used. The operating steam pressure should be used as the inlet pressure to the trap, neglecting the small pressure loss in the generator tubes. The trap discharge pressure depends on the type of return system, and must be determined for the individual application.

A properly sized condensate receiver permits variation of the condensate quantity in the return system from maximum to minimum, with an adequate reserve for the maintenance of boiler feed water requirements.

The most common condensate return system is the steam trap vented receiver type (*Fig. 28*). The steam trap insures condensation of all the steam in the absorption machine.

A condensate piping arrangement generally known as a wet return (*Fig. 29*) is preferable whenever possible. When used under the proper conditions, steam can be returned by gravity from the absorption machine to the boiler without the use of a steam trap. A wet return should not be used if the cost of installation is unreasonably high as compared to the cost of a steam trap discharging into an existing condensate return system. An existing wet return condensate system should be checked for adequate capacity before using.

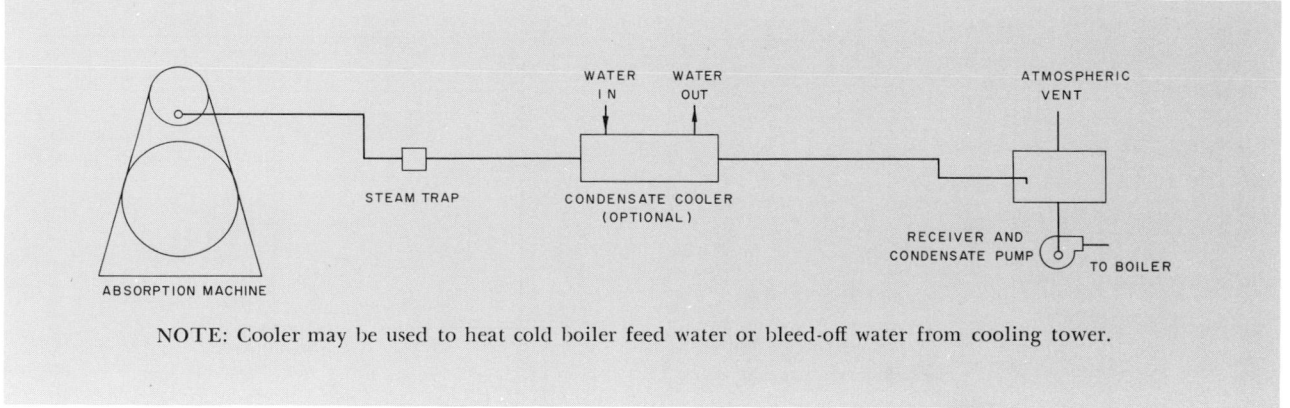

NOTE: Cooler may be used to heat cold boiler feed water or bleed-off water from cooling tower.

FIG. 28 — SCHEMATIC OF CONDENSATE RETURN USING A STEAM TRAP WITH VENTED RECEIVER

It is generally not practical to utilize an existing vacuum pump condensate return system for an absorption machine because the condensate is far higher in temperature than that for which the return pump was originally selected. This hot condensate flashes and causes vapor binding of the piping and/or vacuum return pump. A separate wet return system is recommended where possible. Where impossible, the condensate may be discharged thru a steam trap to an atmospheric vent receiver and then thru a second trap into the vacuum return system. The condensate may also be cooled to an acceptable level in a heat exchanger and discharged to the vacuum pump condensate system. Any cold water source which may be benefitted by the heat rejected can be used.

CONTROLS

Items which require control are:
1. Condenser water temperature
2. Chilled water temperature
3. Energy source
4. Multiple machines

CONDENSER WATER TEMPERATURE CONTROL

Normally a wide range of condenser water temperatures can be used in the selection of an absorption machine. However, once a particular inlet temperature is established, this must be maintained within definite limits.

A bypass type control may be required to maintain inlet temperature. The need for bypass control

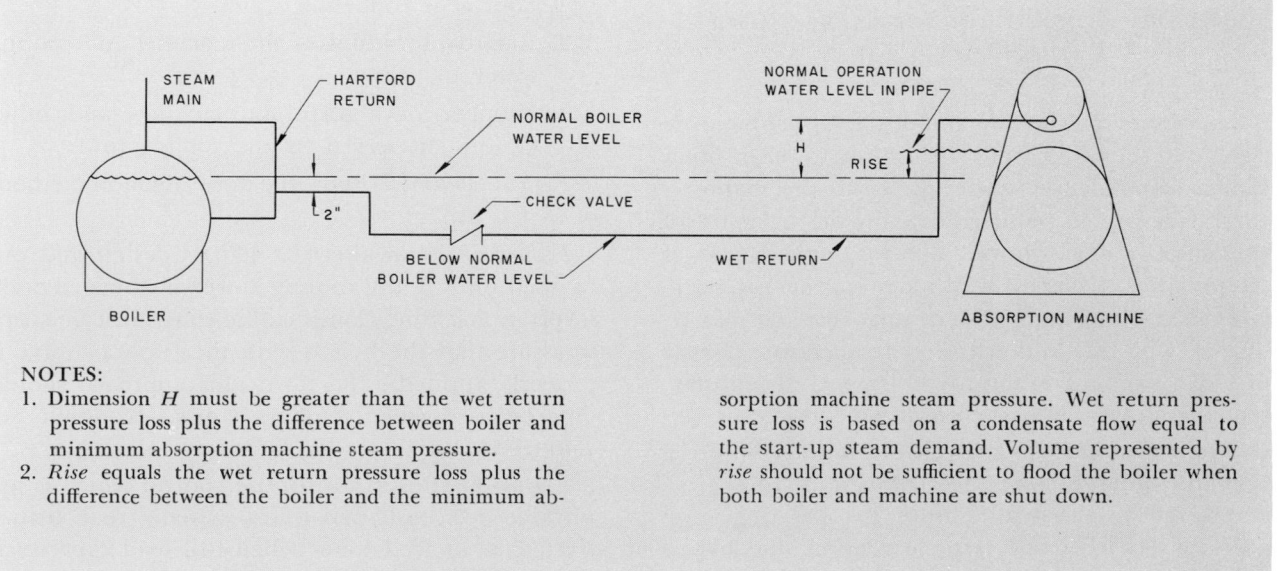

NOTES:
1. Dimension H must be greater than the wet return pressure loss plus the difference between boiler and minimum absorption machine steam pressure.
2. *Rise* equals the wet return pressure loss plus the difference between the boiler and the minimum ab-

sorption machine steam pressure. Wet return pressure loss is based on a condensate flow equal to the start-up steam demand. Volume represented by *rise* should not be sufficient to flood the boiler when both boiler and machine are shut down.

FIG. 29 — SCHEMATIC OF CONDENSATE PIPING USING A WET RETURN

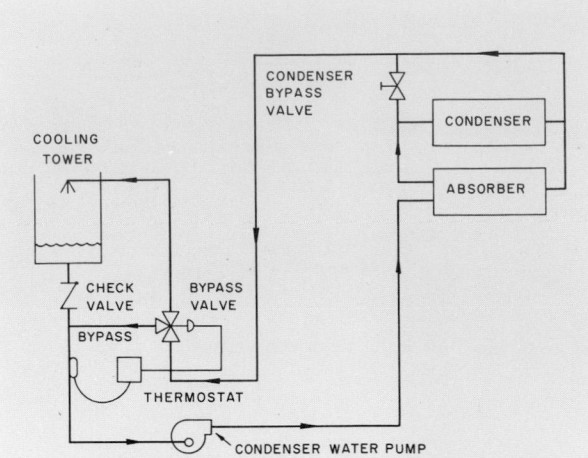

NOTES:
1. Size the bypass 3-way diverting valve and pipe for 100% condenser water flow. Valve pressure loss may be as great as desired.
2. Locate the bypass line and valve close to the cooling tower inlet connection to minimize variations in pump head as the valve position changes.
3. When two absorption machines are installed with a common (or individual) condenser water pump and a common cooling tower, the 3-way diverting valve should be sized for the combined flow of both machines.
4. To keep the system from draining when shut down, shut off the bypass valve when the condenser pump is not operating. Bypass valve port to the cooling tower inlet should be normally closed. Check valve should be installed in the cooling tower drain to prevent water from flowing back into the tower thru the drain.
5. Install the thermostat adjacent to the bypass valve, and the thermal bulb in mixed water adjacent to the bypass line rather than close to the machine.

FIG. 30 — SCHEMATIC OF BYPASS PIPING USED WITH A COOLING TOWER

is determined by the rate and degree of temperature change of the water from the cooling tower or other source of condenser water. Refer to the manufacturer for specific requirements on the necessity of condenser water control. The rate and degree of temperature change of well water is generally negligible; therefore, a bypass control may not be required. The rate and degree of temperature change of water from a cooling tower is generally substantial; therefore, a bypass control is necessary.

The bypass must be capable of limiting the variation in condenser water temperature to 10 degrees and bringing the temperature to operating level quickly. To meet the latter condition, the bypass must always be designed and sized to bypass the total condenser water flow.

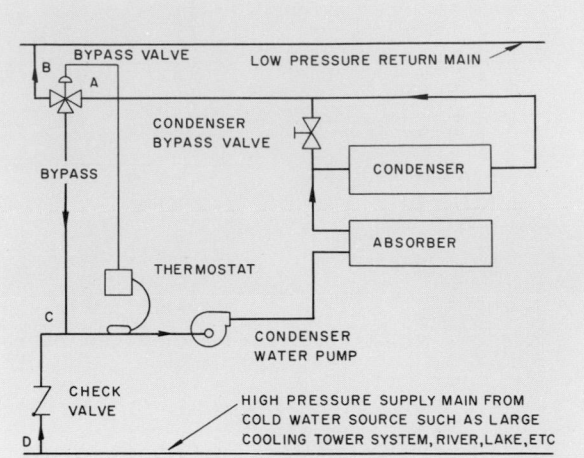

NOTES:
1. Size the bypass 3-way diverting valve for 100% condenser water flow.
2. When two absorption machines are installed, the use of separate condenser water pumps and bypass 3-way diverting valves is recommended. Otherwise, size the bypass valve for combined flow to insure a pressure at point A that is always higher than the pressure at point D plus the pressure differential between points A and C, even when one machine is valved out of the circuit.
3. Install the thermostat with thermal bulb in the mixed water adjacent to the bypass line rather than close to the machine.

FIG. 31 — SCHEMATIC OF BYPASS PIPING USED WITH A CENTRAL WATER SOURCE

The control system design and valve selection are determined from various combinations of the following:
1. Source of condenser water.
2. Relative locations of the machine and cooling tower.
3. Number of absorption machines and other equipment served by the cooling tower.

Figures 30 and 31 show the most common methods of bypass piping.

Figure 32 is an alternate bypass design and can be used only if the cooling tower is above the absorption machine. Considerable time must be spent to assure that the bypass with the two-way valve is properly applied. Therefore, this approach should only be used when it offers a great economic advantage over the smaller three-way valve.

Figure 33 illustrates throttle control which is applicable to condenser water systems that utilize river, lake or well water when full load capacity is not required as the condenser water temperature drops.

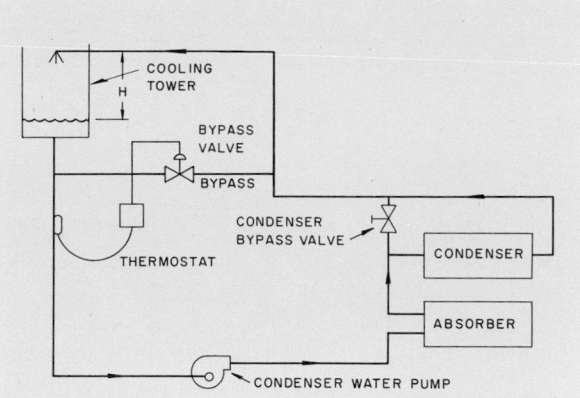

NOTES:

1. Size the bypass valve and pipe for the unbalanced static head of the cooling tower (dimension *H*) and 100% condenser water flow.
2. Locate the bypass line and valve next to the cooling tower base level.
3. If the bypass line and valve cannot be located next to the cooling tower base level, use the arrangement shown in *Fig. 30*.
4. When two absorption machines are installed with a common condenser water pump and a common cooling tower, one bypass line and valve should be installed and sized for the combined flow of both machines and the unbalanced static head of the tower. Locate the valve next to the tower base level.
5. When two absorption machines are installed with individual condenser water pumps and a common cooling tower, individual bypass lines and valves may be installed, each sized and located as specified for a single machine.
6. Install the thermostat adjacent to the bypass valve, and the thermal bulb in mixed water adjacent to the bypass line rather than close to the machine.

FIG. 32 — ALTERNATE SCHEMATIC OF BYPASS PIPING USED ONLY WITH A COOLING TOWER LOCATED ABOVE ABSORPTION MACHINE

These figures illustrate the proper location of the condenser water temperature control valve and schematically show the related condenser water piping. The actual piping layout should be made in accordance with *Part 3, Piping Design*.

When the cooling tower or open drain is below the machine, the piping should contain a loop above the outlet of the condenser nozzle. This prevents the water from draining out of the condenser at shutdown or low flow conditions encountered in throttle type control applications. A vacuum breaker should be installed at the high point of the loop to prevent siphoning of the line.

The bypass valve is usually either a three-way diverting or two-way throttling valve of the globe body type with a linear flow characteristic.

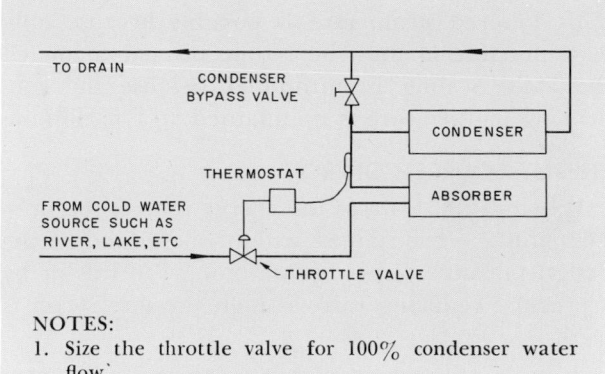

NOTES:

1. Size the throttle valve for 100% condenser water flow.
2. Install the thermostat with the thermal bulb in the condenser water leaving the absorber.

FIG. 33 — THROTTLE CONTROL FOR ONCE-THRU CONDENSER WATER SYSTEMS*

The bypass should be sized and located, and the condenser water pumps selected so that, when water is being bypassed, the flow thru the machine is not increased more than 10%; this prevents overconcentration in the generator and minimizes any increase in pump brake horsepower.

CHILLED WATER TEMPERATURE CONTROL

If an absorption machine is to perform satisfactorily under partial load, a means of effecting a capacity reduction in proportion to the instantaneous load is required. Capacity reduction may be accomplished by steam throttling, control of condenser water flow, or control of reconcentrated solution. For some hot water machines capacity reduction may be accomplished by means of hot water throttling.

These various methods are all used as means to control the ability of the machine to reconcentrate the solution which is returned to the absorber. The more dilute the concentration in the absorber, the less capacity the machine has to chill the water.

Chart 9 shows the comparative performances of these four types of absorption machine capacity control. It is seen that the solution control gives the best steam rate at partial loads; this is where the

*Used if full load capacity is not required when the temperature drops at the cold water source. The ability of the machine to produce full load capacity is not affected when this type of control is applied to condenser water systems where the temperature at the cold water source is constant, as in ground wells. The throttle valve in such systems is used to conserve water rather than maintain condenser water temperature control.

machine is operated most of the operating season. This lowered steam rate is possible because only enough solution must be reconcentrated to match the load. Scaling is minimized because the condensing temperature is maintained at a minimum.

ENERGY SOURCE CONTROL

When using steam as the energy source, the pressure must be maintained within one pound of the design pressure, either by the boiler controls or by a pressure reducing valve if high pressure steam is used.

A back pressure regulator valve to limit steam demand on start-up is rarely used. It may be required if the absorption machine represents most of the load, and if a temporary lowering of the boiler pressure affects operation of the other equipment operated from the boiler. It may also be required where a sudden loss of boiler pressure results in a loss of water in the boiler and/or causes other detrimental effects on the boiler priming.

When high temperature hot water is the energy source, a control valve is usually required to control the hot water flow thru the machine. A two-way throttling or three-way mixing valve is controlled either by a thermostat located in the hot water leaving the machine or by a chilled water thermostat thru a high limit thermostat located in the leaving hot water. The two-way valve should only be used if it does not adversely affect the hot water boiler circulation or the circulating pump. The three-way valve provides a constant flow and is the one most often used.

MULTIPLE MACHINE CONTROL

Absorption machines may be applied to parallel and series arrangements of coolers.

Installations with machine coolers in parallel may utilize two or more machines. With series chilled water flow the cooler pressure loss is cumulative and may become excessive if more than two machines are installed in series.

Parallel Arrangement

When two or more machines are installed with the coolers connected in parallel in the chilled water circuit (*Fig. 34*), each machine should control its own leaving chilled water at design temperature as in a single machine installation. The same throttling range should be used for each machine. As the system load is reduced, each machine automatically reduces capacity simultaneously, thus individually producing the same leaving chilled water temperature.

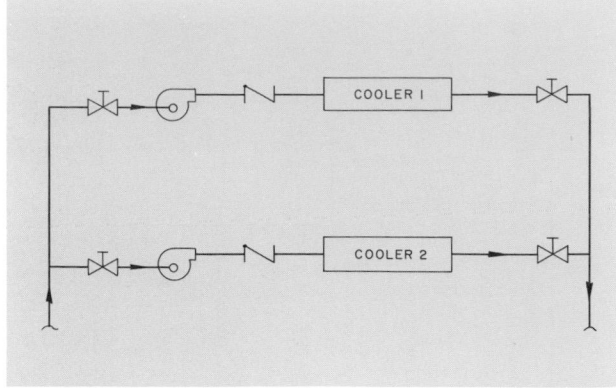

Fig. 34 — Parallel Arrangement of Two Machines

Operating all machines simultaneously down to minimum load provides the best total steam consumption and the most economical operation when using solution control. There is no economic advantage in shutting down any of the machines at partial load since the steam consumption for two machines operating at partial load is less than for one machine operating at full load. Since there is a minimum of moving parts, there is no reason to shut down a machine to prevent its wearing out.

It is recommended that each machine cooler be provided with a separate chilled water pump on the normal air conditioning application. The pump and pump motors should be selected so that the pump motor is not overloaded if one or more machines and their pumps are shut down.

If separate pumps are not provided and a machine is required to be shut down, provision should be made to shut off the chilled water flow and condenser water flow after the shutdown cycle is completed.

Series Arrangement

When the coolers are connected in series (*Fig. 35*), equal reduction of loading of each machine produces the best steam consumption. The throttling

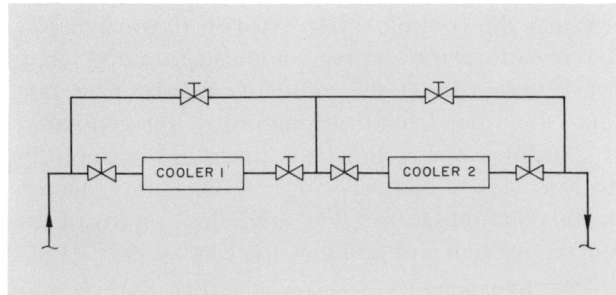

Fig. 35 — Series Arrangement of Two Machines

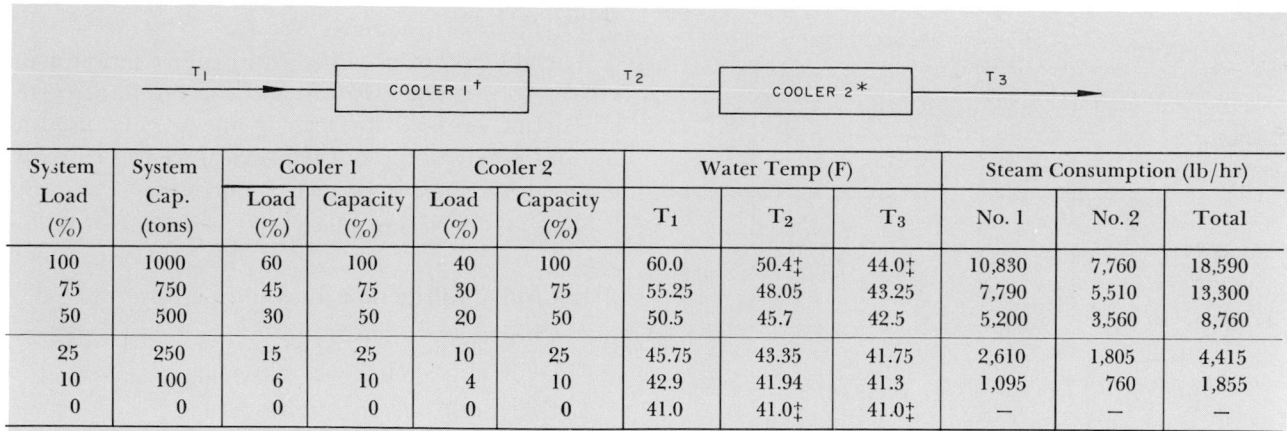

System Load (%)	System Cap. (tons)	Cooler 1		Cooler 2		Water Temp (F)			Steam Consumption (lb/hr)		
		Load (%)	Capacity (%)	Load (%)	Capacity (%)	T_1	T_2	T_3	No. 1	No. 2	Total
100	1000	60	100	40	100	60.0	50.4‡	44.0‡	10,830	7,760	18,590
75	750	45	75	30	75	55.25	48.05	43.25	7,790	5,510	13,300
50	500	30	50	20	50	50.5	45.7	42.5	5,200	3,560	8,760
25	250	15	25	10	25	45.75	43.35	41.75	2,610	1,805	4,415
10	100	6	10	4	10	42.9	41.94	41.3	1,095	760	1,855
0	0	0	0	0	0	41.0	41.0‡	41.0‡	—	—	—

*Same size as Cooler 2 in parallel flow.

†Smaller size than Cooler 1 in parallel flow.

‡Difference in temperature from full load to no load equals the throttling range.

FIG. 36 — CHILLED WATER TEMPERATURE DATA FOR TYPICAL SERIES-CONNECTED COOLERS

range of the high stage machine must be adjusted to insure that each machine handles the same percentage of the system load, both at design and at part load conditions.

In any series selection, the range required on the high stage machine equals the chilled water temperature drop thru the low stage machine plus the throttling range of the low stage machine.

Figure 36 shows the chilled water temperature data from a typical two-machine installation with coolers in series flow. Note that, while the throttling range on the low stage machine (No. 2) is the generally recommended 3 degrees, the throttling range on the high stage machine (No. 1) must be adjusted to 9.4 degrees if both machines are to be proportionally reduced to zero load.

Figure 37 shows the chilled water temperature data from a typical two-machine installation with coolers in parallel flow. In this case the throttling range on both machines is identical.

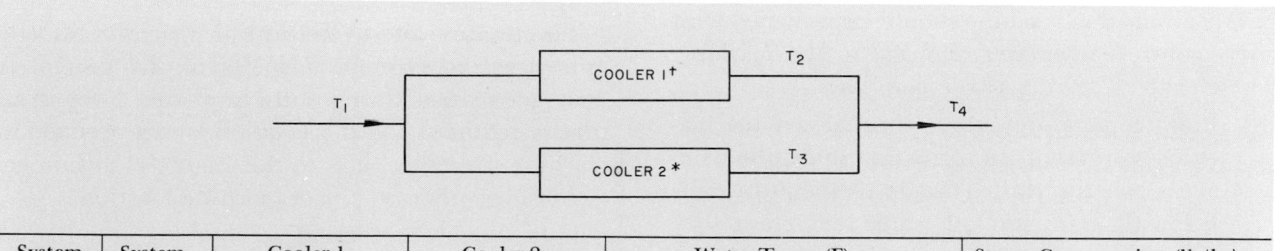

System Load (%)	System Cap. (tons)	Cooler 1		Cooler 2		Water Temp (F)				Steam Consumption (lb/hr)		
		Load (%)	Capacity (%)	Load (%)	Capacity (%)	T_1	T_2	T_3	T_4	No. 1	No. 2	Total
100	1000	60	100	40	100	60.0	44.0‡	44.0‡	44.0	11,400	7,760	19,160
75	750	45	75	30	75	55.25	43.25	43.25	43.25	8,120	5,400	13,520
50	500	30	50	20	50	50.5	42.5	42.5	42.5	5,300	3,530	8,830
25	250	15	25	10	25	45.75	41.75	41.75	41.75	2,620	1,760	4,380
10	100	6	10	4	10	42.9	41.3	41.3	41.3	1,080	755	1,835
0	0	0	0	0	0	41.0	41.0‡	41.0‡	41.0	—	—	—

*Same size as Cooler 2 in series flow.

†Larger size than Cooler 1 in series flow.

‡Difference in temperature from full load to no load equals the throttling range.

FIG. 37 — CHILLED WATER TEMPERATURE DATA FOR TYPICAL PARALLEL-CONNECTED COOLERS

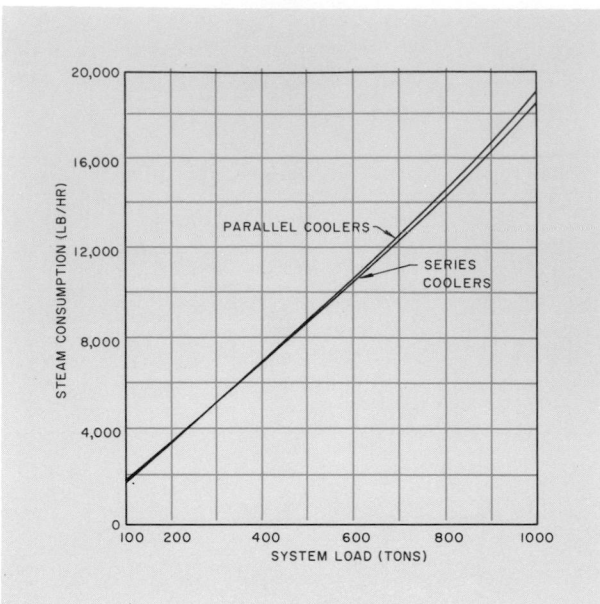

FIG. 38 — PERFORMANCE OF MULTIPLE ABSORPTION MACHINES

Figure 38 indicates the steam consumption for a 1000 ton load for the two types of systems and shows that series operation above about 350 tons uses less steam than parallel operation. The series arrangement has a lower operating cost and generally permits the use of a smaller machine so that the first cost is less.

SAFETY CONTROL

The absorption machine should be provided with safety controls to prevent damage to the machine. These controls are described as follows:

1. Low temperature cutout shuts down the machine to prevent ice formation and tube damage when the chilled water temperature falls below the minimum allowable temperature.

2. Solution pump and evaporator pump auxiliary contacts shut down the machine when either of these pumps become inoperative.

3. Chilled water or condenser water flow switches or their pump's auxiliary contacts shut down the machine when water flow is interrupted in either circuit.

PURGE UNIT

A purge unit is required to remove all noncondensables and to maintain a low pressure in the absorption machine. The purge unit must be able to maintain a pressure below the pressure in the absorber.

INSULATION

The absorption machine requires insulation principally to prevent sweating and the resultant corrosive action on cold surfaces. It may also be used to minimize machine room temperatures and to cover exposed hot lines in or near traffic areas.

Some of the items which may require insulation are:

1. Chilled refrigerant lines and pump
2. Chilled water boxes
3. Generator shell
4. Generator nozzles and headers
5. Solution heat exchanger
6. Hot solution piping.

The cold surfaces may be insulated with flexible fiberglass, closed cell foamed plastic, expanded polystyrenes, plaster or plaster tape, and should include a vapor seal. Water boxes which require removal should be insulated with a removable type of insulation such as sheet metal covers with a granulated fill.

The hot surfaces such as the generator shell may be insulated with a blanket type or low pressure boiler insulation. The generator nozzles and headers should use a removable insulation such as a granulated fill in a sheet metal cover. The hot solution piping requires flexible insulation similar to the types used on the cold piping.

LOCATION

The location of the absorption machine directly influences the economic aspects of the system. A floor adequately strong and reasonably level is all that is required for the location of an absorption machine. However, it is to the engineers advantage to consider other aspects of machine location.

1. It should be located so that the installation costs of the piping between the unit and the equipment it supplies and the wiring and piping of the services to the unit are at a minimum.

2. There should be sufficient space near the machine for auxiliary equipment such as chilled water and condenser water pumps and piping.

3. There should be adequate clearances around the machine for access, servicing, and tube pulling or cleaning.

Many absorption machines along with their boilers and auxiliaries are installed on upper floors or roofs of buildings because the location has many advantages.

1. It allows the basement areas normally used for such mechanical equipment to be available for profitable use.

2. It eliminates many of the pipes and shafts thruout the building. The only services required thru the building are a small fuel line to the boiler, an electric feeder and the normal drain lines.

3. Equipment room ventilation is simplified.

4. All mechanical equipment can be located in the same relative area, providing for better maintenance and supervision.

5. It eliminates a long boiler stack and long steam relief lines.

6. Pumps and water boxes do not have to be designed for high pressures, as may be expected in tall buildings.

LAYOUT

NOZZLE ARRANGEMENT

The nozzle arrangement chosen for chilled and condenser water should result in a minimum number of chilled and condenser water pipe fittings, should allow proper access to the machine and auxiliary equipment, and should present a neat and attractive appearance.

OUTDOOR INSTALLATION

The absorption machine is designed for indoor operation. Outdoor locations are usually not encouraged.

A simple heated structure enclosing the machine is generally preferred, and erection of the enclosure may be less costly than the precautions that may be required. If it is necessary to install the machine outdoors, refer to the manufacturer for his recommodations and precautions.

UNIT ISOLATION

A rubber isolator is normally used under the leg supports of the machine. Such isolation together with the required isolation of the chilled water and condenser water pumps and the piping to and from the machine is usually sufficient for a satisfactory installation.

CHAPTER 4. COMBINATION ABSORPTION-CENTRIFUGAL SYSTEM

A combination refrigeration system is well suited to many large tonnage air conditioning systems where operating economy is important and medium or high pressure steam is planned as the driving force.

This chapter includes System Description, System Features, Engineering Procedure and Controls.

As with all applications an owning and operating cost analysis should be made before selecting any specific refrigeration equipment. However, the combination system is generally most attractive on large tonnage applications. These applications include large buildings, big building complexes such as college campuses, and industrial processes requiring water at air conditioning temperature levels.

Also, on existing systems where more air conditioning is planned, this increased load can often be added without increasing refrigeration energy requirements by either of the following alternatives:

1. If about 1/3 more load* is planned, a back pressure steam turbine driven centrifugal refrigeration machine can be added to existing absorption equipment.

2. If about 2/3 more load* is required, absorption equipment can be added to centrifugal refrigeration equipment. Normally this entails the substitution of a back pressure turbine for a condensing turbine.

As mentioned, medium or high pressure steam must be planned or be available. This can be self-generated or district heating steam.

*As explained more fully under *Apportionment of Chilled Water Cooling*, the best operating economy is normally accomplished when the air conditioning load is divided approximately one-third centrifugal refrigeration and two-thirds absorption refrigeration.

SYSTEM FEATURES

The following are some of the features offered by a combination system:

1. *Minimum Energy Requirements* — Less heat input is required for a combination system than for a condensing turbine-driven centrifugal or an absorption machine alone. A correct analysis of owning and operating costs uses total system heat input as a criterion rather than a straight steam rate per ton of refrigeration. *Chapter 3* discusses this point at greater length.

2. *Minimum Heat Rejection* — Less heat is rejected from a combination system than from a condensing turbine-driven centrifugal or an absorption machine. Condensation of steam from a back pressure steam turbine in an absorption machine eliminates the need for a more expensive condensing turbine and steam condenser. A lower heat rejection to the cooling tower may permit a smaller size tower to be selected.

Table 11 compares typical heat input and heat rejection per ton of refrigeration for the three different systems. *Figure 39* is a graphic analysis of *Table 11*.

SYSTEM DESCRIPTION

The following system description and control arrangement chosen for the description is one of many that can be designed for the system. This particular arrangement is presented because it offers control simplicity and a minimum steam rate when operating from full load down to approximately 15-35% load.

TABLE 11—COMPARISON OF STEAM-OPERATED SYSTEMS

	Combination System*	Condensing Turbine Driven System†	Absorption System
Heat Rejection (Btu/ton)	24,600	28,600	30,600
Heat Input (Btu/ton)	12,600	16,600	18,600

*Back pressure turbine   125 psig inlet, 13 psig exhaust, no moisture
†Condensing turbine   125 psig inlet, 4 in. Hg abs. exhaust

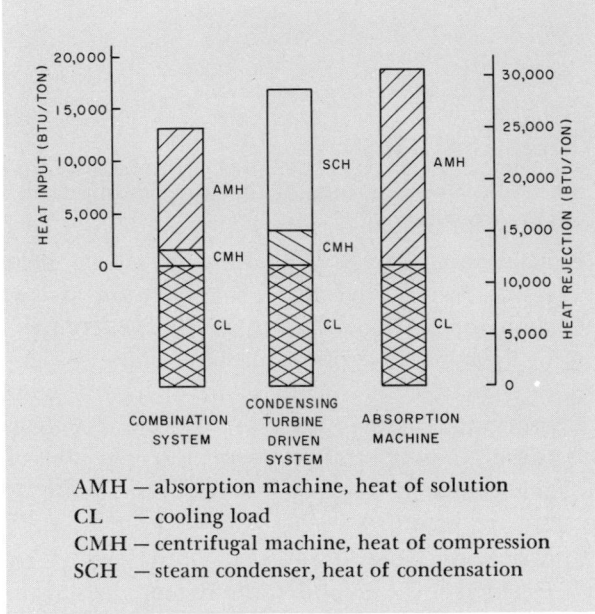

AMH — absorption machine, heat of solution
CL   — cooling load
CMH — centrifugal machine, heat of compression
SCH   — steam condenser, heat of condensation

FIG. 39 – GRAPHIC ANALYSIS OF TABLE 11

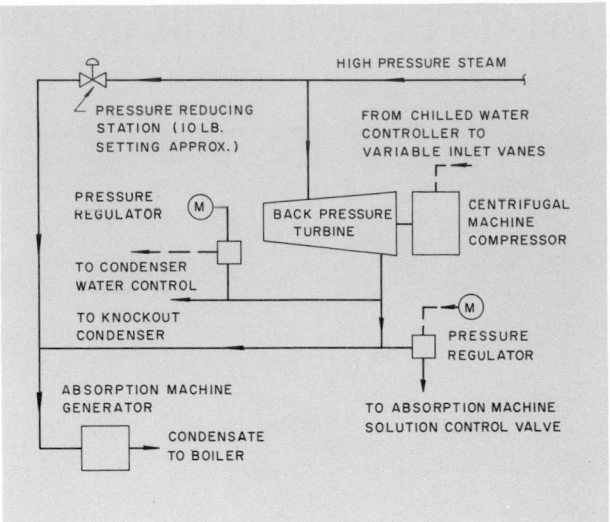

FIG. 40 – SCHEMATIC OF COMBINATION SYSTEM
STEAM PIPING

Although only one centrifugal machine and one absorption machine are shown in *Fig. 40,* a combination of one centrifugal and two or more absorption machines can be used as well. It is shown this way for clarity.

Figures 40 and 41 show the suggested arrangement. Chilled water returns from the load thru the centrifugal machine, to the absorption machine and then back to the load. The condenser water circuit is piped in parallel to the refrigeration equipment with individual pumps for each circuit. This not only allows the versatility of independent machine operation; it also gives good operating economy.

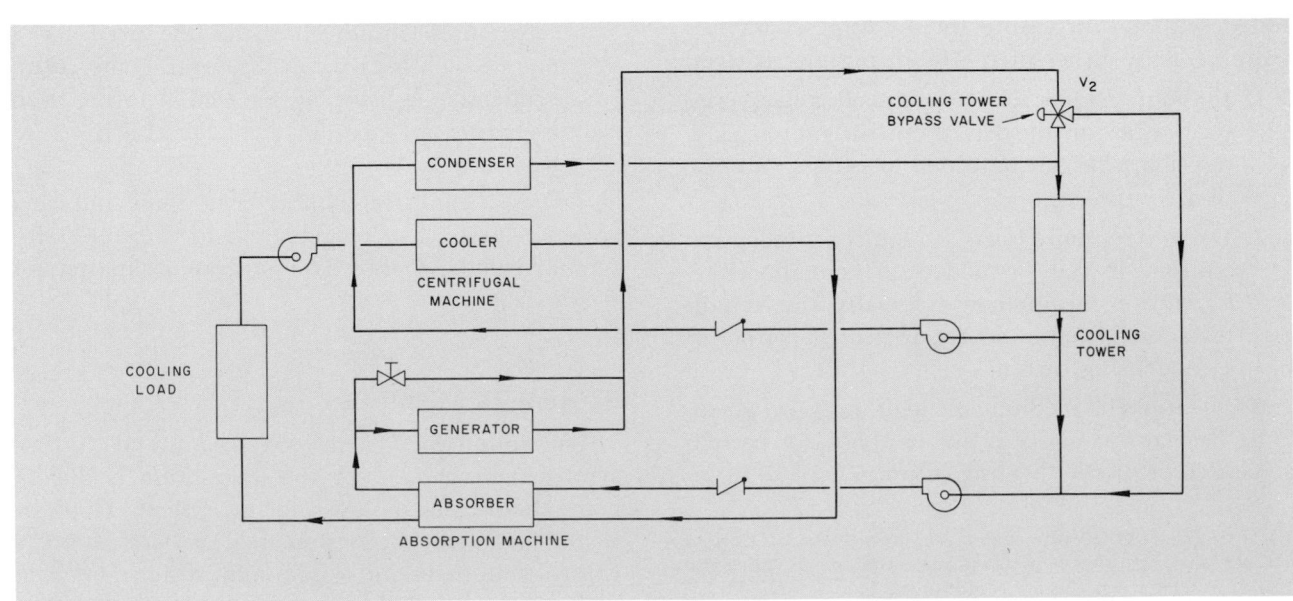

FIG. 41 – SCHEMATIC OF COMBINATION SYSTEM WATER CIRCUITING

TABLE 12—MINIMUM SYSTEM STEAM RATE AND CENTRIFUGAL MACHINE PROPORTION OF SYSTEM LOAD

Turbine Inlet Steam Rate (lb/bhp/hr)	Centr. Power Req.* (bhp/ton)	Ratings†	Turbine Exhaust Steam Quality									
			1.00					0.95				
			Absorption Machine Steam Rate at Design (lb/hr/ton)‡									
			17	18	19	20	21	17	18	19	20	21
44	.7	Rate	11.0	11.4	11.8	12.1	12.5	11.3	11.7	12.1	12.5	12.9
		Percent	36	37	38	39	41	37	38	39	41	42
	.8	Rate	11.5	11.9	12.3	12.8	13.2	11.9	12.3	12.8	13.2	13.6
		Percent	33	34	35	36	38	34	35	36	38	39
	.9	Rate	11.9	12.4	12.8	13.3	13.7	12.3	12.8	13.3	13.7	14.2
		Percent	30	31	32	34	35	31	32	34	35	36
	1.0	Rate	12.3	12.8	13.3	13.8	14.2	12.7	13.2	13.8	14.2	14.7
		Percent	28	29	30	31	32	29	30	31	32	33
42	.7	Rate	10.8	11.2	11.5	11.9	12.3	11.1	11.5	11.9	12.3	12.6
		Percent	37	38	39	40	42	38	39	40	42	43
	.8	Rate	11.3	11.7	12.1	12.5	12.9	11.7	12.2	12.5	12.9	13.3
		Percent	34	35	36	37	38	35	36	37	38	40
	.9	Rate	11.7	12.2	12.6	13.1	13.5	12.2	12.6	13.1	13.5	14.0
		Percent	31	32	33	35	36	32	33	35	36	37
	1.0	Rate	12.1	12.6	13.1	13.6	14.0	12.5	13.1	13.6	14.0	14.5
		Percent	29	30	31	32	33	30	31	32	33	35
40	.7	Rate	10.6	11.0	11.3	11.7	12.0	10.9	11.3	11.7	12.0	12.4
		Percent	38	39	40	42	43	39	40	42	43	44
	.8	Rate	11.1	11.5	11.9	12.3	12.7	11.5	11.9	12.3	12.7	13.1
		Percent	35	36	37	38	40	36	37	38	40	41
	.9	Rate	11.5	12.0	12.4	12.9	13.3	12.0	12.4	12.9	13.3	13.7
		Percent	32	33	34	36	37	33	34	36	37	38
	1.0	Rate	11.9	12.4	12.9	13.3	13.8	12.4	12.9	13.3	13.8	14.2
		Percent	30	31	32	33	35	31	32	33	35	36
38	.7	Rate	10.4	10.7	11.1	11.4	11.7	10.7	11.1	11.4	11.7	12.1
		Percent	39	40	42	43	44	40	42	43	44	45
	.8	Rate	10.9	11.3	11.7	12.1	12.4	11.3	11.7	12.1	12.4	12.8
		Percent	36	37	38	40	41	37	38	40	41	42
	.9	Rate	11.4	11.8	12.2	12.6	13.0	11.7	12.1	12.6	13.0	13.4
		Percent	33	35	36	37	38	34	35	37	38	39
	1.0	Rate	11.7	12.2	12.7	13.1	13.5	12.2	12.6	13.1	13.5	14.0
		Percent	31	32	33	34	36	32	33	34	36	37
36	.7	Rate	10.2	10.5	10.8	11.2	11.5	10.5	10.8	11.2	11.5	11.8
		Percent	40	42	43	44	46	42	43	44	46	47
	.8	Rate	10.7	11.1	11.4	11.8	12.1	11.0	11.4	11.8	12.2	12.5
		Percent	37	39	40	41	42	38	40	41	42	43
	.9	Rate	11.2	11.6	12.0	12.4	12.7	11.5	12.0	12.4	12.8	13.1
		Percent	35	36	37	38	39	36	37	38	40	40
	1.0	Rate	11.5	12.0	12.4	12.9	13.3	12.0	12.4	12.9	13.3	13.7
		Percent	32	33	34	36	37	33	34	36	37	38

*A function of the chilled water temperature leaving the centrifugal which is directly affected by the system chilled water rise and leaving temperature (centrifugal on high temperature side).

†Rate is minimum system steam rate in lb/hr/ton. Percent is centrifugal machine proportion of system load.

‡A function of the chilled water temperature leaving the absorption machine (at 100% machine capacity).

The condenser water temperature is maintained at its required 85 F for the absorption machine, whereas the water entering the centrifugal is allowed to drift downwards at partial loads, thus improving centrifugal operating economies.

The minimum system steam rate occurs when a balanced steam flow condition exists. That is when the absorption machine utilizes exactly all the steam discharged from the turbine. Normal procedure is to accomplish this initial balanced condition at 100% load. The control arrangement (described later) accomplishes the balanced steam flow condi-

tion at partial loads. *Table 12* shows the minimum system steam rate and the centrifugal portion of the system load for various turbine inlet and absorption machine steam rates.

Control is accomplished as follows. A thermostat in the chilled water circuit leaving the combination system controls the inlet guide vanes or a suction damper in the centrifugal. The absorption machine is controlled by a pressure regulator sensing the steam pressure in the header between the turbine exhaust and the absorption machine. The thermostat maintains a constant chilled water temperature

CHART 11—CENTRIFUGAL COMPRESSOR POWER REQUIREMENTS
Refrigerant 11

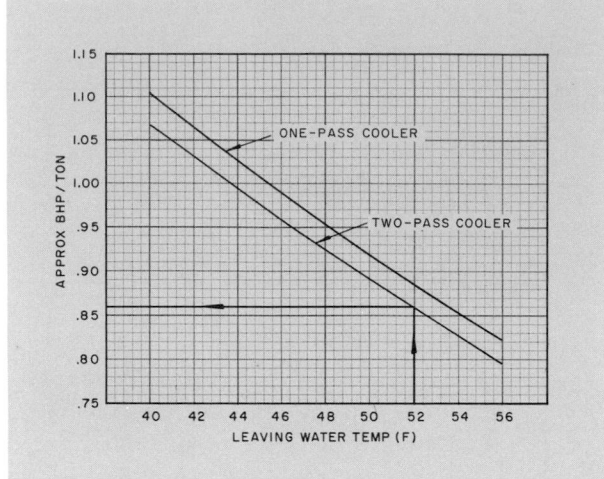

CHART 12—TYPICAL PERFORMANCE DATA FOR STEAM TURBINES

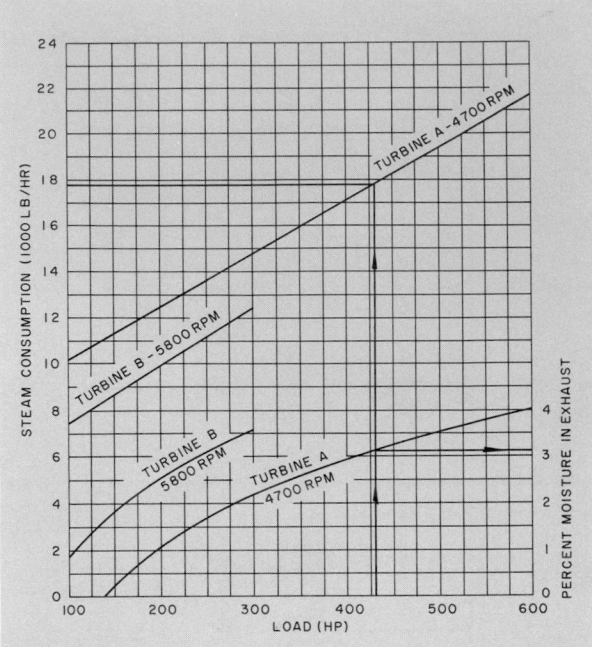

NOTE: Design steam: 125 psig, 0 superheat, 13 psig exhaust.

and the pressure regulator holds the steam pressure constant ahead of the absorption machine.

Essentially, the chilled water thermostat effects a capacity reduction of the combination system at partial load in proportion to the instantaneous load, while the steam pressure regulator maintains a steam flow balance between the centrifugal and absorption machines and consequently a minimum system steam rate.

ENGINEERING PROCEDURE

Selecting the equipment for a combination system is a matter of achieving the required performance with the minimum owning and operating costs. Equipment selection is essentially a trial-and-error process. The following method is recommended to provide a satisfactory system.

PERFORMANCE

The required performance consists of load or tonnage, chilled water temperature and quantity or rise (F). The load or tonnage is determined by using normal methods. A chilled water rise of 15-20 F is suggested for minimum system steam rate and pressure drop thru the chillers. Chilled water temperature selection should not be arbitrary, but should be as high as design permits.

Recommended condenser water quantities are approximately 3 gpm per ton for the centrifugal and 3.5 gpm per ton for the absorption machine.

TURBINE STEAM RATE

Determine the expected single stage turbine steam rate and exhaust steam quality for the specified inlet steam conditions (pressure and superheat) and 13 psig back pressure. (This allows 1 psi pressure loss for the steam piping between the turbine exhaust and the absorption machine inlet, assuming the absorption machine and turbine are close-coupled.)

Base the determination of steam rate upon an assumed one-third system load on the centrifugal and a two-thirds system load on the absorption machine. With the centrifugal on the high temperature side a leaving water temperature may be determined. From *Chart 11* an approximate brake horsepower per ton of refrigeration may be assumed for the centrifugal.

The required turbine performance is obtained from the turbine supplier and includes the following:

1. Curves relating speed, horsepower and steam consumption from full to minimum load of the centrifugal.

2. Curves relating turbine exhaust steam quality to turbine horsepower, speed or steam consumption.

There are no turbine performance characteristics which are typical of all turbines. *Chart 12* shows the performance of two typical back pressure turbines for centrifugal refrigeration machine duty.

CHART 13—TYPICAL ABSORPTION MACHINE STEAM RATES

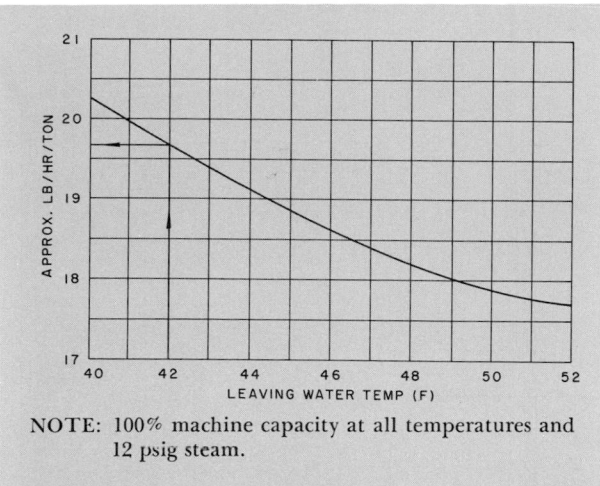

NOTE: 100% machine capacity at all temperatures and 12 psig steam.

DISTRIBUTION OF DESIGN CAPACITY AND MINIMUM SYSTEM STEAM RATE

Calculate the centrifugal machine proportion of the system design load. This is determined from the formula:

$$\text{Load} = \frac{SR_a}{SR_a + (SR_t \times \text{bhp/ton} \times X_{te})} \tag{1}$$

where:

Load	= centrifugal machine proportion of system design load.
SR_a	= absorption machine steam rate (lb/hr/ton) at design load.*
SR_t	= turbine inlet steam rate (lb/bhp/hr) at design load.
X_{te}	= turbine exhaust steam quality.
bhp/ton	= centrifugal machine power requirement.†

The system design load minus the centrifugal proportion equals the absorption machine proportion of the load.

Calculate the minimum system steam rate determined from the formula:

$$SR_{min} = \frac{SR_t \times \text{bhp/ton} \times SR_a}{SR_a + (SR_t \times \text{bhp/ton} \times X_{te})} \tag{2}$$

where:

SR_{min} = minimum system steam rate (lb/hr/ton) of system load.

Example 1 illustrates how these two formulas may be utilized.

---

*For initial determination, use approximate lb/hr/ton from *Chart 13* for chilled water temperature leaving the system.

†Use approximate bhp/ton figures from *Chart 11* for chilled water temperature leaving the centrifugal machine.

Example 1 — Minimum Steam Rate At System Design Load

Given:

2400 gpm of water to be chilled from 57 F to 42 F or a 1500 ton design load.

Centrifugal machine power requirement
= .86 bhp/ton (*Chart 11*) × 500 tons = 430 bhp.

Turbine inlet steam rate at design load

$$\frac{17,700}{430} = 41.2 \text{ lb/bhp/hr } (\textit{Chart 12})$$

Absorption machine steam rate at design load
= 19.65 lb/hr/ton (*Chart 13*)

Turbine exhaust steam quality = .97

Find:

Apportionment of system tonnage
Minimum system steam rate

Solution:

Using formula 1, the centrifugal machine proportion of system design load

$$= \frac{19.65}{19.65 + (41.2 \times .86 \times .97)} = .364$$

then, .364 × 1500 tons = 545 tons

Chilled water temperature range thru the centrifugal
= .364 × 15 F = 5.45 F.

Using formula 2, the minimum system steam rate

$$= \frac{41.2 \times .86 \times 19.65}{19.65 + (41.2 \times .86 \times .97)} = 12.9 \text{ lb/hr/ton}$$

The absorption machine is selected to handle the remainder of the system design load.
1500 − 545 = 955 tons.

APPORTIONMENT OF CHILLED WATER COOLING

Determine the entering and leaving chilled water temperatures for the centrifugal and absorption machine portion of the system design load based on series chilled water flow with the chilled water passing first thru the centrifugal machine. The total required chilled water temperature range is proportioned between the centrifugal and absorption machines in the same ratio as the tonnage. *Example 1* illustrates how the apportionment is determined.

CENTRIFUGAL MACHINE

Select the centrifugal machine for an entering chilled water temperature equivalent to that returning from the load and the leaving chilled water temperature determined previously.

Select the condenser using the available cooling tower water temperature. *Chapter 2* can be used as a guide for the selection of the machine.

ABSORPTION MACHINE

Select the absorption machine for the entering chilled water temperature equal to that leaving the centrifugal and the leaving chilled water temperature required by the design load. Base the machine

CHART 14—TYPICAL STEAM CONSUMPTION, COMBINATION SYSTEM

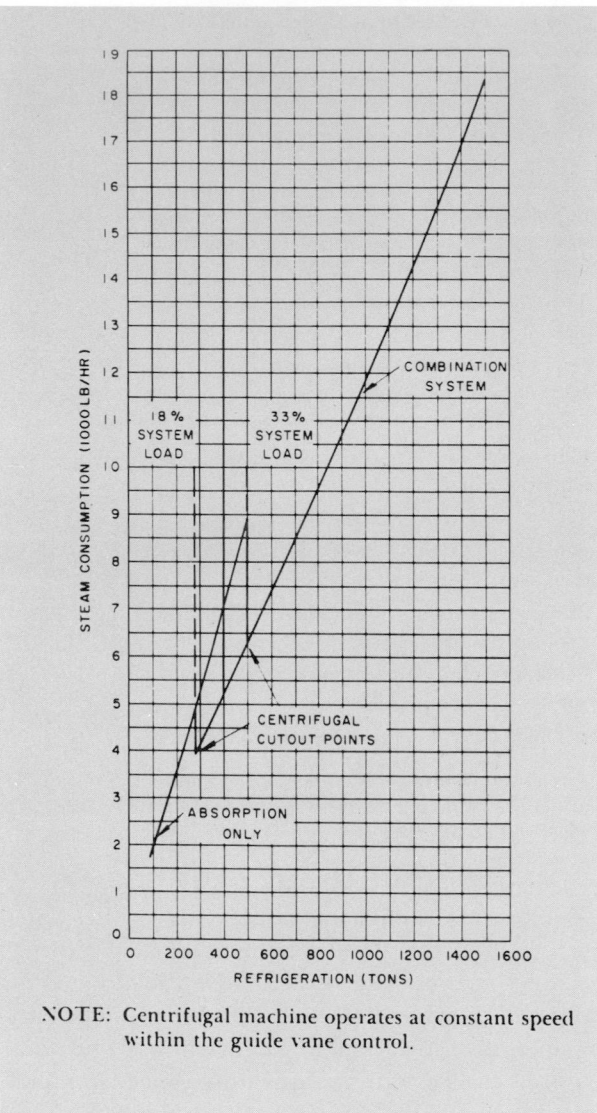

NOTE: Centrifugal machine operates at constant speed within the guide vane control.

CHART 15—TYPICAL STEAM CONSUMPTION, INDIVIDUAL MACHINES

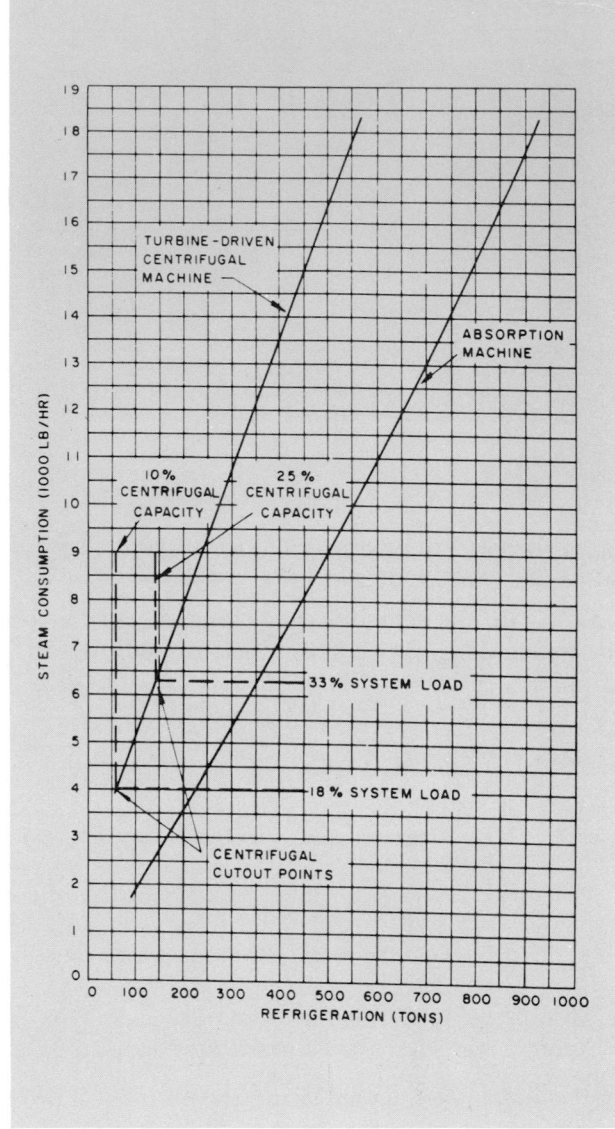

selection on a 12 psig steam supply and the available cooling tower water temperature. *Chapter 3* can be used as a guide for the selection of the absorption machine.

STEAM TURBINE

Select the back pressure steam turbine for the speed and brake horsepower required by the centrifugal at design load. Often a 5% safety factor is added to the required centrifugal brake horsepower before selecting the steam turbine horsepower, to allow for poor entering steam conditions to the turbine.

A single stage turbine is normally used. For equal cost the turbine having the best steam rate should be selected. Two single stage turbines in tandem are recommended in preference to a single, multi-stage turbine. A hydraulic direct-acting governor is recommended for use as the turbine (manual) speed control.

Determine the total required turbine steam flow and exhaust steam quality, and check the balance of steam flow between the turbine and absorption machine. If the quantities determined do not agree with those estimated originally and/or a balance does not exist, further adjustments may be required in the selection of equipment.

Charts 14 and 15 indicate the steam consumption at partial load for a typical combination system and its individual machines respectively.

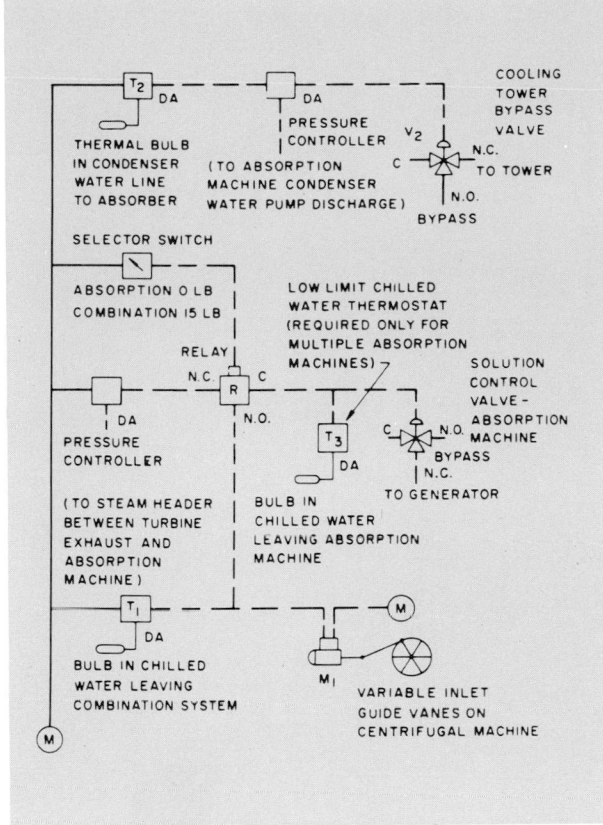

FIG. 42 — TYPICAL CONTROL DIAGRAM,
COMBINATION SYSTEM

CONTROLS

A typical control diagram for a combination system is shown in *Fig. 42*. Either electrical or pneumatic controls may be used; however, pneumatic controls are most commonly utilized. The sequence of operation is the same regardless of which is used.

Capacity control by means of variable inlet guide vanes or a suction damper governs the operation of the centrifugal machine. A thermostat with its thermal element in the chilled water line leaving the absorption refrigeration equipment positions the throttling device at the compressor.

A steam pressure regulator controls the operation of the absorption machine. The sensing element of the pressure regulator is piped to the steam header between the steam turbine exhaust and the absorption machine. The regulator positions the solution control valve at the absorption machine in accordance with the steam pressure to control its capacity and maintain a constant pressure. Minimum system steam rate is thus maintained at both full and partial load because the absorption machine utilizes exactly the steam discharged from the turbine.

A selector switch permits operation of the absorption machine alone below the centrifugal cutout point.

COMBINATION OPERATION

With the selector switch in the Combination position the centrifugal machine is under the control of the chilled water controller, and the absorption machine is put under the control of the pressure regulator thru a low limit thermostat (if required).

The cooling tower bypass valve is interconnected with the absorption machine condenser water pump so that the valve is controlled by the condenser water thermostat when the pump is started.

When the temperature falls, the chilled water controller acts to throttle the variable inlet guide vanes at the centrifugal compressor. If the resulting quantity of exhaust steam from the turbine is insufficient to supply the absorption machine in accordance with its solution control valve position, the steam supply pressure drops. The steam pressure regulator senses the pressure drop and throttles the solution control valve to bring the system into balance. An increase in chilled water temperature causes a reverse action.

The low limit chilled water thermostat is a safety device to prevent a freeze-up when more than one absorption machine is required in the design.

ABSORPTION MACHINE OPERATION

When the system load drops below approximately 15-35% of design, the centrifugal machine is shut down. With the selector switch in the Absorption position the chilled water thermostat controls the solution control valve at the machine.

Steam is supplied to the absorption machine thru the pressure reducing station in the bypass line around the turbine. The steam also maintains the turbine hot for easy start-up.

CHAPTER 5. HEAT REJECTION EQUIPMENT

In order for the refrigeration cycle to be complete, the heat absorbed in the evaporator and the heat equivalent of the work required to raise the pressure of the refrigerant must be removed and dissipated. This is the function of heat rejection equipment. Heat may be dissipated by sensible heat transfer or by a combination of sensible heat transfer and latent heat transfer (mass transfer). The means of heat rejection is the basis of equipment classification.

This chapter contains practical information to guide the engineer in the application and layout of heat rejection equipment.

TYPES OF EQUIPMENT

There are three types of heat rejection equipment commonly used. They are:

1. Air-cooled condenser, in which heat is rejected directly to the air by sensible heat transfer.

2. Evaporative condenser, in which sprayed coils are used to dissipate heat to the air by sensible and latent heat transfer.

3. Water-cooled condenser, in which heat is sensibly transferred to water. Although this water may then be wasted, it is usually conserved by a process of sensible and latent cooling in a cooling tower. The water is then recirculated

to the condenser. For this reason, the water-cooled condenser and the cooling tower should be examined together as a single heat rejection device.

Figures 43, 44, 45 and 46 show an air-cooled condenser, an evaporative condenser, a water-cooled condenser, and a cooling tower respectively.

APPLICATION

An evaluation of owning and operating costs is usually the basis for selection of a means of heat rejection. Customer preference and provision for future conditions may influence the choice as well. Local design factors such as air and water conditions and the system application affect the selection insofar as they affect the economics.

In an economic analysis, system size is important since the installed costs per ton of the various condensing methods decrease at different rates with increasing size. All factors being equal, air-cooled condensing is often chosen for capacities up to 75 tons. Evaporative condensing is a primary alternative in the 50-150 ton range. Above 100 tons, water-cooled condensing, in conjunction with a mechanical draft cooling tower, is the most common choice. There are many applications where well, river or lake water is used for water-cooled condensing purposes. The installed cost per ton of once-thru water-cooled condensing where the water is wasted remains constant with system size.

In the capacity range where all three condensing methods are alternatives, air-cooled condensing is

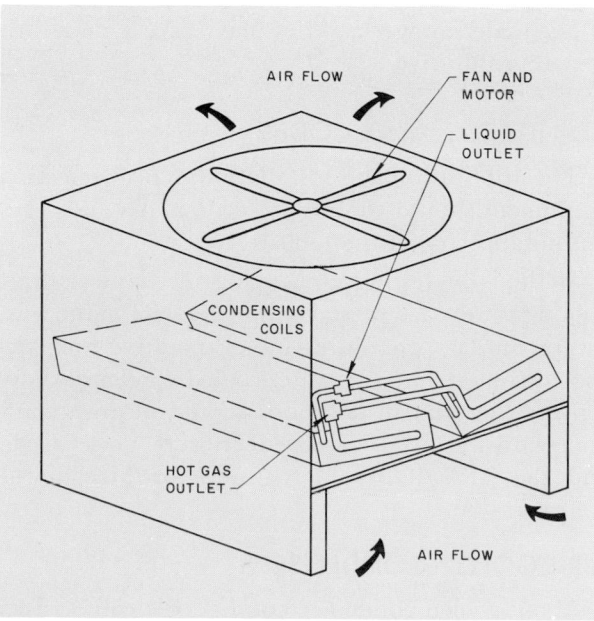

FIG. 43 — AIR-COOLED CONDENSER

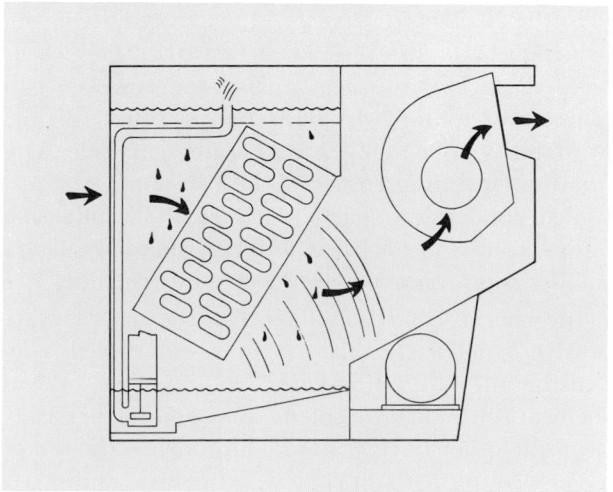

FIG. 44 — EVAPORATIVE CONDENSER

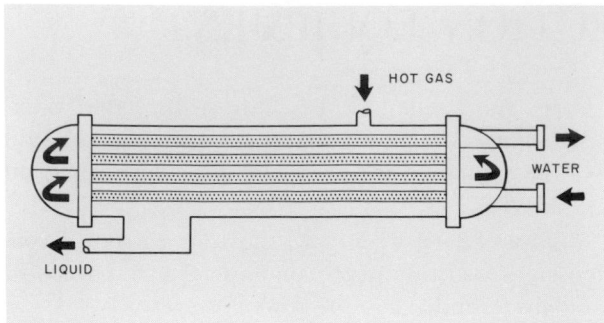

FIG. 45 — WATER-COOLED CONDENSER

usually the highest in first cost. However, maintenance costs for air-cooled condensers are considerably lower for a given capacity. Therefore, air-cooled condensing is well suited to systems where service is infrequent or incomplete. Similarly, long operating hours at light loads favor air-cooled condensing. Over-all operating costs of this method for the commonly applied capacity range are less than for water-cooled condensing, and compare favorably with evaporative condensing.

Other factors supporting the choice of air-cooled condensing are the lack of make-up water or drainage facilities, the availability of only foul water, high summer wet-bulb temperatures, relatively low summer dry-bulb temperatures or high water costs. Installations featuring many independent compressors may possibly be served more satisfactorily by multiple air-cooled condensers or by multiple circuiting in a single air-cooled condenser than by a single evaporative condenser or cooling tower. Also, if operation is required at low outdoor temperatures, air-cooled condensing presents no water freezing problems.

In the 50-150 ton capacity range, evaporative condensing usually has the lowest first cost. Some other factors which encourage its use are low wet-bulb temperatures, high dry-bulb temperatures, or the availability of inexpensive water of adequate quality. Operating costs may be below those of air-cooled condensing, particularly if the condensing temperature considered is lower, with consequently smaller compressor power input requirements.

In general, conditions favoring the use of evaporative condensing also favor water-cooled condensing in combination with a cooling tower. When the heat rejection equipment is located further away from the other refrigeration components, the use of a close-coupled water-cooled condenser and a remote cooling tower becomes economically more attractive. This is because the refrigerant piping

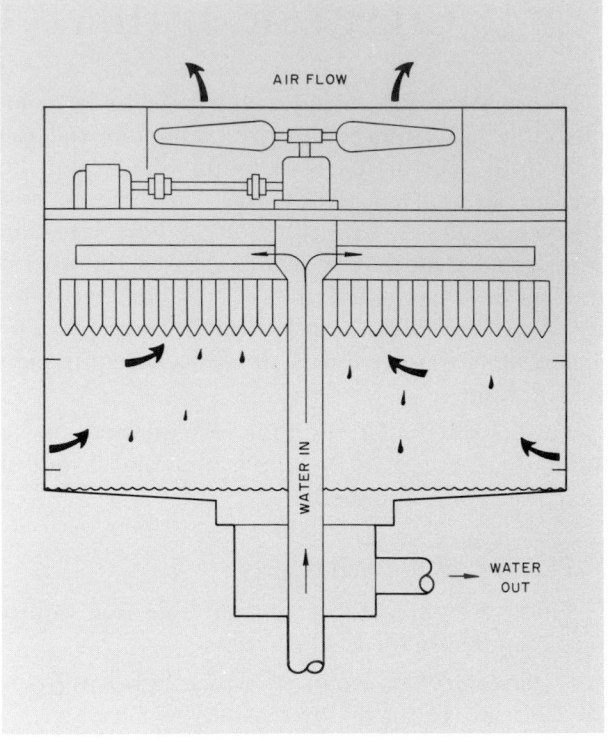

FIG. 46 — COOLING TOWER

necessary with air-cooled or evaporative condensing is more costly than water piping for a given capacity.

Once-thru water-cooled condensing may be the most practical and economical choice if there is a nearby supply of water of adequate temperature and quality such as a river, lake or well. Otherwise, city water costs, local codes for the use of water, or lack of adequate sewage facilities may make a once-thru system prohibitive.

STANDARDS AND CODES

The application and installation of heat rejection equipment should conform to codes, laws and regulations applying at the job site.

Methods of testing and rating mechanical draft cooling towers are prescribed in the ARI Standards, as are similar procedures for evaporative and air-cooled condensers. For water-cooled condensers, design, testing and installations should be in accordance with the ASME Unfired Pressure Vessel Code and the ASA B9.1. Safety Code for Mechanical Refrigeration.

AIR-COOLED CONDENSERS

An air-cooled condenser consists of a coil, casing, fan and motor. It condenses the refrigerant gas by means of a transfer of sensible heat to air passed

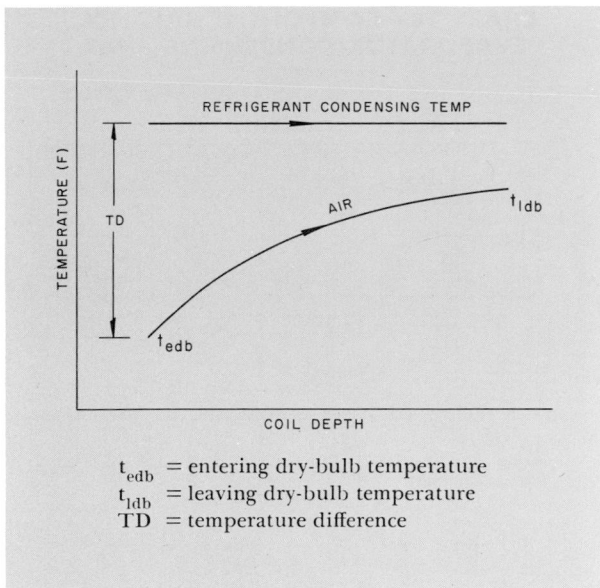

t_{edb} = entering dry-bulb temperature
t_{ldb} = leaving dry-bulb temperature
TD = temperature difference

FIG. 47 — AIR-COOLED CONDENSING PROCESS

CHART 16—COMPONENT COMBINATIONS AIR-COOLED CONDENSING

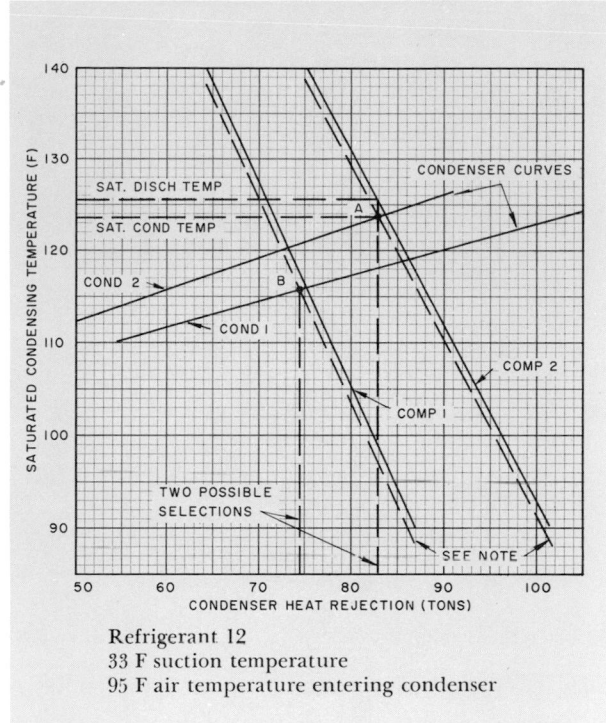

Refrigerant 12
33 F suction temperature
95 F air temperature entering condenser

over the coil. The relation between condensing temperature and air temperature is shown in *Fig. 47*.

For a given surface and air quantity, the capacity of an air-cooled condenser varies, for practical purposes, in direct proportion to the difference (TD) between the condensing temperature and the entering air dry-bulb temperature. Therefore, assuming the heat rejection requirement is constant, a fall or rise in entering air temperature results in an equal decrease or increase in condensing temperature.

Values for TD range from 15-35 degrees, with condensing temperatures between 110 F and 135 F. In desert areas these temperatures may reach 140 F.

UNIT SELECTION

Economics

Air-cooled condensers are most commonly applied to relatively small refrigeration systems. First cost usually dictates the selection of the compressor-condenser combination at conventional condensing temperatures. Condenser first cost advantages may be realized with higher condensing temperatures. However, it must be recognized that, as the chosen condensing temperature is increased, compressor power input increases also. Higher compressor power requirements may be partially or fully offset by decreased condenser fan motor horsepower. Additionally, common practice is to build most air-cooled condensers with subcooling circuits. This has the effect of increasing total system capacity with a slight increase of compressor power input requirements.

Component Balance

The capacities of a condenser and a compressor operated in combination will balance at some final condensing temperature. Rather than determining this balance point by trial and error, it is preferable to calculate it graphically. This is done by plotting condensing temperature versus heat rejection for both the compressor and the condenser on the same set of coordinates. Two or more condensers and compressors can be plotted so that the various combinations can be analyzed for performance, first cost and operating cost.

Chart 16 is a graphical selection for a nominal 60 ton design load. Points *A* and *B* represent two possible combinations. Combination *B* has the smaller compressor while combination *A* has the smaller condenser. Compressor power input requirements are higher with combination *A*, but installed first cost is lower.

In plotting the air-cooled condenser capacities, it is preferable to use condenser ratings given in terms of total heat rejected rather than in evaporator tons. Compressor heat rejection requirements vary not only with suction and condensing temperature but also with the compressor selected. In addition, heat rejection requirements for hermetic refrigeration machines vary with size. The engineer should refer to specific catalogs for the total heat rejection for

CHART 17—COMPONENT BALANCE, AIR-COOLED CONDENSING UNIT

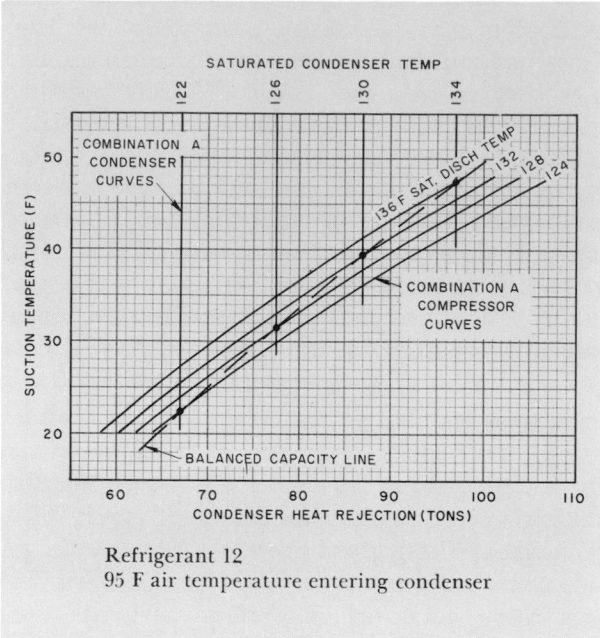

Refrigerant 12
95 F air temperature entering condenser

CHART 18—COMPONENT BALANCE, EVAPORATOR-CONDENSING UNIT

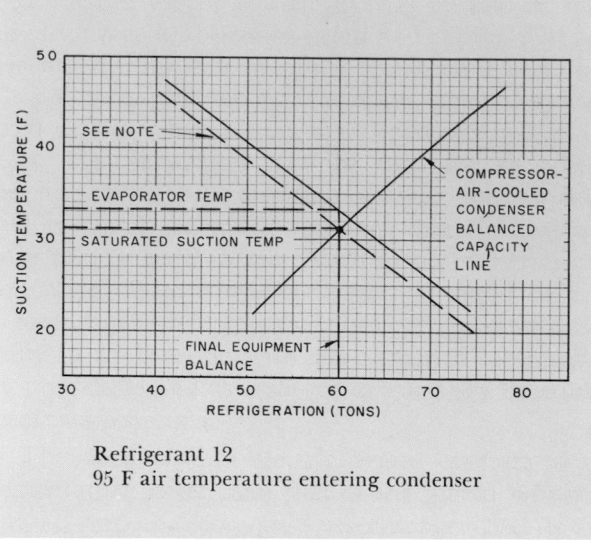

Refrigerant 12
95 F air temperature entering condenser

each individual compressor for plotting purposes.

The compressor capacity line (shown broken) has been corrected for an assumed 2 degree discharge line loss. Thus, for a given design suction temperature, a point on the corrected line represents a condensing temperature 2 degrees less than the compressor saturated discharge temperature. When designing hot gas lines, it is common practice not to exceed a pressure drop corresponding to a 2 degree change in saturation temperature.

An analysis of combination *A* is given in *Charts 17 and 18*. *Chart 17* describes the heat rejection capacity of compressor-condenser combination A at varying suction and condensing temperatures. The balance line (shown broken) indicates the heat rejection possible with this combination at various suction temperatures, assuming a 2 degree line loss.

Chart 18 includes the balance line shown in *Chart 17* in terms of evaporator tons and, in addition, relates evaporator capacity and suction temperature. The broken line on this chart is the evaporator capacity curve corrected for an assumed 2 degree suction line loss. This standard is not usually exceeded in sizing the suction line.

Compressor Limitations

The selection and application of air-cooled condensers may be limited by the restrictions imposed by the manufacturer on operation of the compressor. These restrictions define maximum

compressor saturated suction temperatures and saturated discharge temperatures. Specific data should be obtained from the catalog of the manufacturer involved.

Subcooling

The use of an integral subcooler in an air-cooled condenser provides these operating advantages:

1. An increase in system capacity.
2. A method of offsetting the effects of moderate liquid lifts.
3. Reduced power input per ton of refrigeration.

The subcooler is installed in series with the condensing coil, and the condensed liquid from all circuits of the condensing coil is combined before passing thru the subcooler.

Because of the reduced enthalpy of the subcooled liquid, each pound of refrigerant evaporated can absorb more heat, thus increasing the system capacity (*Fig. 48*).

Flashing of liquid refrigerant due to pressure drops from moderate liquid lifts is offset by subcooling.

The power input per ton is reduced because more capacity is available without an increase in the required horsepower. Line *CD* in *Fig. 48* represents the work input which is not changed due to the subcooling.

A receiver (if used) should be valved out of any system using a subcooling coil because the subcooling effect is often offset by the liquid flashing in the receiver.

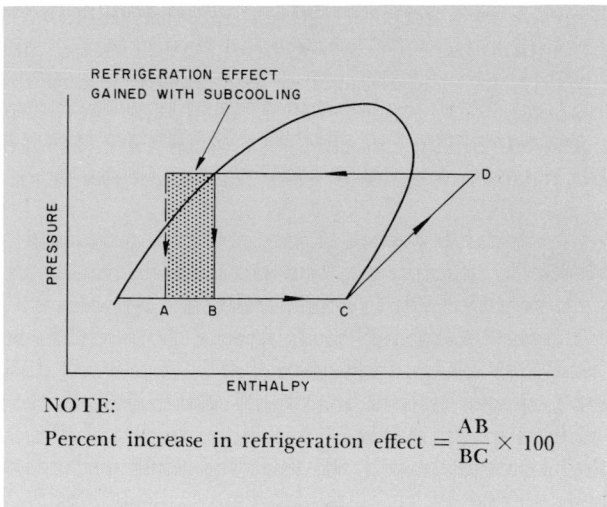

NOTE:

Percent increase in refrigeration effect $= \dfrac{AB}{BC} \times 100$

Fig. 48 — Effect of Liquid Subcooling

Atmospheric Corrections

Air-cooled condenser ratings are based on air at standard atmospheric conditions of 70 F and 29.92 in. Hg barometric pressure. If a condenser is equipped with a direct drive and is to operate at an altitude above sea level, heat rejection ratings should be corrected for the change in air density. This correction amounts to a decrease in capacity of approximately 9% at an altitude of 5,000 feet. However, it should be pointed out that this reflects only about a 3% loss in total system capacity.

On condensers equipped with belt-driven equipment, fan speed can be increased to offset this correction.

Capacity corrections for air temperature deviations are unnecessary unless temperatures exceed 125 F.

Multiple Circuiting

Some installations may include several independent refrigeration systems, each operating at the same or different suction or condensing temperatures. These may be supplied with a single condenser which has a multiple of circuits, each operating with a separate refrigeration system.

CONDENSER CONTROL

There are two basic methods of controlling the capacity of air-cooled condensers. They are air side and refrigerant side control. The air side controls utilize methods of varying the air flow thru the condenser. The refrigerant side controls vary the amount of available condensing surface by flooding portions of the condenser with liquid refrigerant.

Air-cooled condenser controls are needed to maintain a sufficient pressure differential across the refrigerant expansion device to provide the required refrigerant flow to offset the load. The control should operate satisfactorily at any outdoor temperature at which refrigeration is required. The minimum outdoor temperature above which the control system operates satisfactorily must be obtained from the manufacturer of the control. This minimum temperature may depend on whether the refrigeration system can operate unloaded some of the time or if it operates loaded all of the time. The application of the refrigeration system, whether for liquid chilling, direct expansion cooling or a heat pump, may also influence the minimum temperature.

Most manufacturers of air-cooled condensers offer patented methods of control which can only be used with their condensers. Refer to these manufacturers for their recommendations concerning operating limits and applications for their controls.

Refrigerant migration to the condenser on shutdown may be prevented or its effect on a proper start-up may be lessened by the condenser control. Refrigerant migration occurs because the condenser is in a colder atmosphere than the other parts of the system. This may create trouble on start-up since the condenser may be full, or partially full, of subcooled liquid which is at or close to the outdoor temperature. There may be very little pressure difference across the expansion valve and the system may cycle on the low pressure switch when attempting to start.

Several methods are available to aid in providing for winter start-up. One method is to shunt out the low pressure switch at start up with a time delay relay until the system gets up to pressure. Another method is to locate the low pressure switch in the liquid line and put a defrost thermostat on the coil. When single pump out control is used, a low pressure stat is also required in the suction line.

Feeding subcooled liquid to a liquid chiller may cause frost pinching of the tubes or a freeze-up of the chiller due to erratic operation of the expansion valve. With a direct expansion cooling coil, the subcooled liquid may not allow proper distribution of the refrigerant thru the parallel circuits in the evaporator. Frost may deposit on the coil and can build up and block off the air flow.

Proper selection of the expansion valve and the selection of an adequate receiver to hold any excess refrigerant may also be required in connection with the condenser control.

Air Side Control

Examples of the air side controls include the cycling of fans in sequence when multiple fans are used on a single coil system; the modulation of a volume damper installed in the fan discharge or as a face damper on the coil; the modulation of a by-pass damper installed to bypass air around the coil; a variable speed fan; or possibly a combination of these controls.

Refrigerant Side Control

Types of refrigerant side controls include an electrically heated surge type receiver to maintain a minimum liquid temperature in the receiver; a by-pass valve to bypass hot discharge gas around the condenser to maintain a minimum downstream or receiver pressure; a pressure regulating valve in the drain line from the condenser to maintain a minimum pressure in the condenser; or variations and combinations of these controls.

LOCATION

An air-cooled condenser may be located indoors or outdoors. It may be remote or near, above or below the compressor. The greater the distance separating the condenser and compressor, the greater is the first cost and operating cost. Specific location recommendations include the following:

1. Locate the unit so that air circulates freely and rapidly without recirculating.
2. Locate the unit away from areas continuously exposed to loose dirt and foreign matter.
3. Locate the unit away from occupied spaces with low ambient sound levels.

Condenser location with respect to the evaporator can have an influence on the liquid line size. Since most air-cooled condensers are manufactured with a coil designed to give liquid subcooling, the liquid line can be designed for a much higher pressure drop when the evaporator is below the condenser than when the evaporator is located at the side of or above the condenser. It is suggested that this subcooling effect and pipe sizing be studied because of the increased cost of pipe installation and additional refrigerant required for an oversized liquid line.

It may be possible to utilize building exhaust air as part or all of the supply to the condenser, thus reducing the size of the condenser or lowering the head pressure.

When air-cooled condensers are manufactured with the subcooling coil integral with the condensing coil, liquid receivers are not normally used.

The use of a receiver often completely eliminates the subcooling effect if the liquid condenser drains directly into the receiver before passing on to the evaporator. Receivers (when used) are normally bypassed during normal operation and are restricted for pumpdown when service on the system is required.

LAYOUT

Air-cooled condensers are manufactured for both vertical and horizontal air flow. Vertical coils can be affected by condenser orientation. On an outdoor installation the prevailing winds should blow toward the air intake of the unit. If this is impossible, a discharge air shield is suggested. Also, special intake and discharge hoods may be necessary if snow can accumulate in the fan section. Orientation has no effect on the performance of air-cooled condensers equipped with horizontal coils.

For indoor installations, fresh air intakes and discharge ducts to the outdoors must be provided. It is necessary to arrange these ducts so that recirculation does not occur.

Refer to *Part 3, Chapter 3* for refrigerant piping details.

EVAPORATIVE CONDENSERS

An evaporative condenser consists of a condensing coil, fan and motor, water distribution system, sump, recirculating pump, and casing. It condenses the refrigerant gas by means of a combination sensible and latent heat transfer process. Rejected heat is dissipated by water diffused over the coil surface. It is then transferred to the air passing over the coil. Latent heat transfer is more effective as a means of heat dissipation and, therefore, permits a unit with less cubage than an equivalent air-cooled condenser. The relation between condensing temperature, air enthalpy, and coil surface temperature is shown in *Fig. 49.*

The capacity of an evaporative condenser may be increased either by lowering the entering air wet-bulb temperature or by increasing the condensing temperature. Condensing temperatures normally range from 100-115 F. Increasing the air quantity beyond the design has little effect on capacity.

An evaporative condenser may be used to cool other liquids such as oil and water instead of a refrigerant. Most manufacturers provide ratings for such applications. For optimum performance, piping must be designed so that the flow of water thru the condenser coil is opposite in direction to the flow of air.

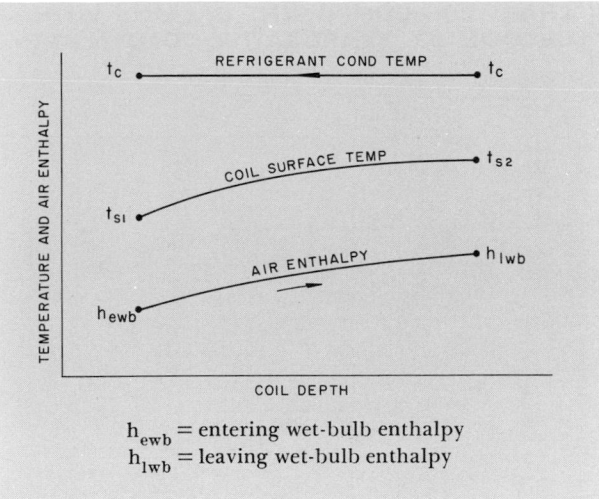

h_{ewb} = entering wet-bulb enthalpy
h_{lwb} = leaving wet-bulb enthalpy

FIG. 49 — EVAPORATIVE CONDENSING PROCESS

UNIT SELECTION

The normal design wet-bulb temperature of a locale should be the basis of an evaporative condenser selection since higher wet-bulb temperatures seldom occur, and then only briefly. If design conditions must be maintained at all times, such as in some industrial process applications, the maximum wet-bulb temperature should be used to insure adequate equipment capacity.

Economics

As in the case of air-cooled condensers, first cost usually dictates the selection of an evaporative condenser-compressor combination. Operating at relatively high condensing temperatures with a heavily loaded condenser lowers the installed cost per ton but may increase the operating cost per ton. For over-all maximum economy of operation, a difference of 25-30 degrees between condensing and entering air wet-bulb temperatures is suggested.

Component Balance

The balanced capacity of an evaporative condenser-compressor combination may be determined graphically as described under *Air-Cooled Condensers*. *Chart 19* illustrates such a graphical analysis at a given wet-bulb temperature and air quantity. As mentioned previously, it is preferable to base the condenser operating characteristic on a known condenser heat rejection performance rather than on condenser ratings in tons of refrigeration effect.

Subcooling

The use of an integral or accessory subcooling coil with an evaporative condenser provides the following operating advantages:

CHART 19—COMPONENT BALANCE, EVAPORATIVE CONDENSING

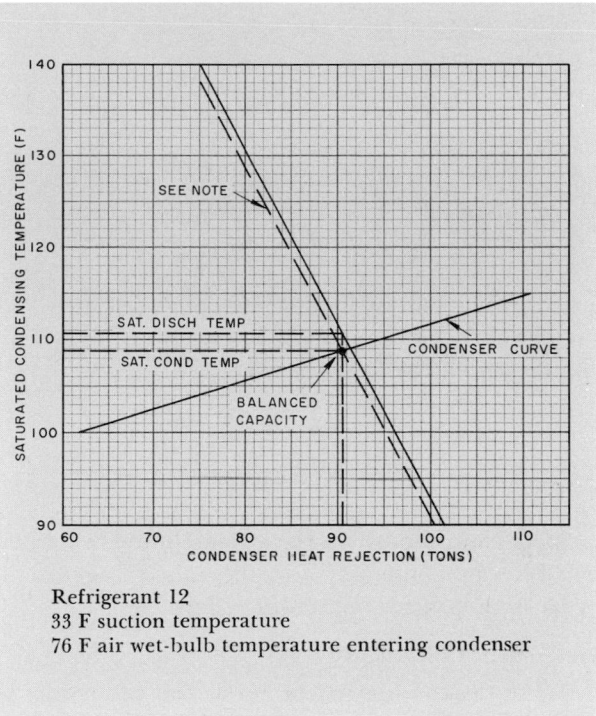

Refrigerant 12
33 F suction temperature
76 F air wet-bulb temperature entering condenser

1. An increase in system capacity.

2. An offset to the effect of moderate liquid lifts.

3. Reduced power input per ton of refrigeration.

The subcooling coil is installed in series with the condensing coil and is the first coil contacted by the entering air.

When the balanced capacity of a condenser-compressor combination is slightly below design requirements, it is usually more economical to add a subcooling coil than to select the next larger combination.

Figure 50 is a pressure-enthalpy diagram illustrating the source of the additional capacity realized from the use of subcooling. The capacity of the condenser-compressor combination is increased because with subcooled liquid each pound of refrigerant evaporated does more work. *Chart 20* shows the effect of a subcooling coil on the same combination as is shown in *Chart 19*.

The amount of increase in system refrigeration capacity as a result of subcooling is obtained from the manufacturers' evaporative condenser ratings.

Liquid subcooling cannot be obtained if a receiver follows the subcooling coil in the refrigeration system. If a receiver is used and if it must be so located, it should be used as a storage vessel only and valved out of the circuit during operation. A

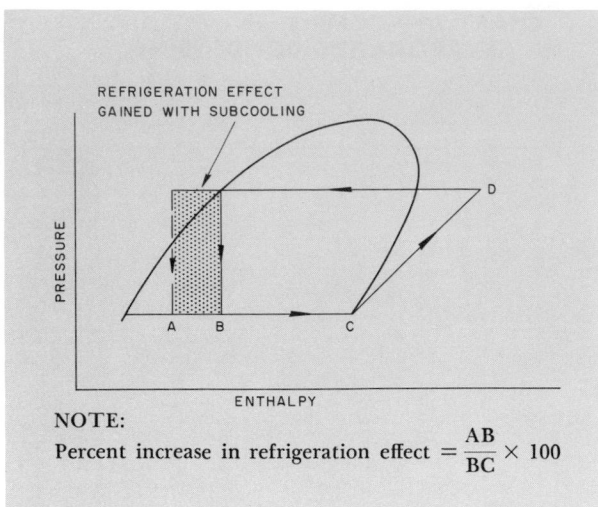

NOTE:

Percent increase in refrigeration effect $= \dfrac{AB}{BC} \times 100$

FIG. 50 — EFFECT OF LIQUID SUBCOOLING

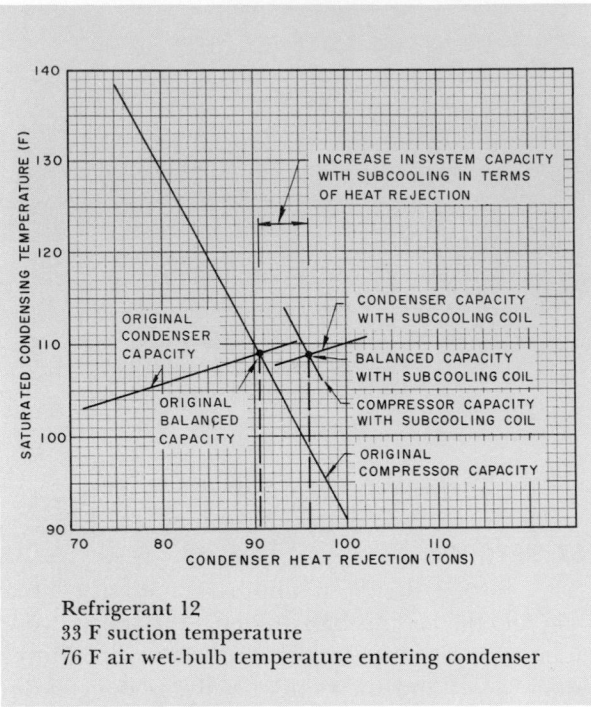

CHART 20—COMPONENT BALANCE WITH SUBCOOLING, EVAPORATIVE CONDENSING

Refrigerant 12
33 F suction temperature
76 F air wet-bulb temperature entering condenser

receiver continuously in the system should be located between the condenser and the subcooling coil in order to obtain the subcooling effect.

Atmospheric Corrections

At elevations above sea level, the reduction in weight flow of air due to the decrease in density of the air at high altitude is offset by the greater latent heat absorbing capacity of the high altitude air.

Fan motor selections made on a sea level basis are conservative.

To determine actual motor brake horsepower at the design fan speed at any elevation, multiply the power requirement at sea level by the ratio of the air density at the design altitude to the air density at sea level.

Multiple Circuiting

For applications in which one condenser serves several separate refrigeration systems, evaporative condensers are available with headers divided to form two or more independent refrigerant circuits. The number and relative capacities of these individual circuits depend on the circuit design of the condenser coil. Individual circuits may be operated at the same or different suction or condensing temperatures.

CONDENSER CONTROL

Operation of an evaporative condenser at low outdoor temperatures requires special consideration in order to prevent the freezing of recirculation water and to insure proper functioning of the thermostatic expansion valve.

The following methods may be used to maintain condensing pressure in an evaporative condenser:

1. An automatic discharge damper which varies the flow of air across the coil.

2. An automatic air recirculating assembly (*Fig. 51*) which controls the entering air wet-bulb temperature.

3. A fan motor having two or more speeds to vary the air flow across the coil.

Automatic discharge dampers or air recirculation damper assemblies are the most satisfactory solutions

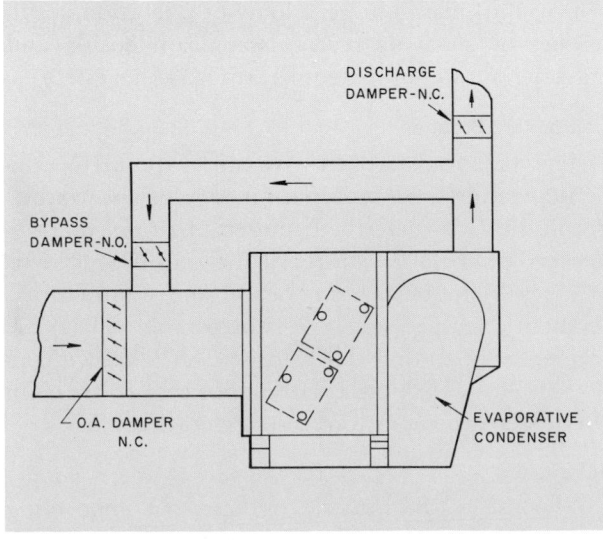

FIG. 51 — AIR RECIRCULATING ASSEMBLY

to the problem of positive condensing pressure control. These dampers are often sold as accessories by the condenser manufacturer. The dampers may be actuated in response to a condensing pressurestat or, in the case of recirculation control, in response to a recirculation water thermostat.

Condensing pressure control methods involving a cycling of the recirculation pump are not recommended; cycling of the pump produces a rapid accumulation of scale deposits.

WINTER OPERATION

When year-round operation is required, it is advisable to locate the unit indoors but, if an outdoor installation is needed, the unit should be operated dry with the sprays shut off and the water drained from the pan. This generally reduces the capacity of the unit, but full capacity is usually not required at this time.

LOCATION

An evaporative condenser may be located indoors or outdoors. An indoor location requires ductwork which increases first cost and fan power requirements; it is recommended if year-round operation is planned; it minimizes problems of unreliable start-up and the freezing of recirculated water. An indoor condenser operated in the summer only should be provided with manual intake and discharge dampers in the duct in order to prevent the introduction of cold air and attendant problems of moisture condensation.

A condenser mounted outdoors usually requires no protective covering, but should be provided with drainage facilities for the tank, pump and water piping.

Other location recommendations are listed under *Location* for air-cooled condensers.

LAYOUT

Whether indoors or outdoors, the unit should be installed elevated above the floor, roof or ground. This may be accomplished by suspending the unit or by providing a mounting pad.

MAKE-UP WATER

Provision must be made for city water make-up for evaporation and bleed-off water losses. Evaporation occurs at the rate of approximately 1.8 gallons per hour per ton of refrigeration. Bleed-off varies from 50-200% of evaporation, depending on water treatment recommendations. Refer to *Part 5, Water Conditioning.*

Refer to *Part 3* for refrigerant piping details.

WATER-COOLED CONDENSERS

A water-cooled condenser consists of heat transfer tubes mounted within a steel shell. Condenser water passes thru the tubes, and the condensing refrigerant occupies the shell surrounding the tubes. The shell is equipped with a hot gas inlet, a liquid sump, purge connections, water regulating valve connection, and a pressure relief device. The shell-and-tube condenser has a supporting tube sheet at each end and removable heads while the shell-and-coil condenser has a spiral or trombone tube bundle, accessible from one end only. Tubes must be cleaned chemically instead of by reaming or brushing. Shell-and-coil condensers are of relatively low cost and are often used in air conditioning applications.

At a given entering water temperature and condensing temperature, condenser capacity is decreased by decreasing the water flow rate, thus increasing the water temperature rise. Condensing temperatures usually range from 100-110 F, but may be as low as 80 F for once-thru city water condensing.

UNIT SELECTION

The water-cooled condenser and the cooling tower should be considered as a single heat rejection device for the purposes of selection and application. Therefore, the economics of condenser selection are reviewed under the subject of *Cooling Towers.*

In a once-thru application, the entering water temperature used for selection should be the maximum water temperature prevailing at the time of maximum refrigeration load.

Selection of the number of water passes should be made with regard to the temperature and pressure of the water available. The low pressure and higher temperature water available from cooling towers generally dictates the least number of passes.

Normally, manufacturers base water-cooled condenser ratings on various conditions of the tube scaling on the water side. A fouling factor represents the resistance to heat flow presented by scaling. Since some surface fouling is present on tube surfaces from the beginning of operation, a minimum fouling factor of .0005 is suggested for selection. Scale factors for various types of condensing water systems are shown in *Part 5.*

The factors shown should be tempered by operating conditions. A reduction in the factor is justified in the case of frequent cleaning, an unusually low condensing temperature, or when operation is less than 4000 hours annually.

CONDENSER CONTROL

Control of water flow thru condensers may be required to limit the condensing pressure to a predetermined minimum. Two methods of restricting this flow are commonly used, a two-way throttling valve or a three-way diverting valve.

The two-way valve is useful in maintaining condensing pressure in once-thru applications utilizing city, well, lake or river water. In the case of city water, a prime objective may be to minimize water costs.

The three-way valve is most often used with a cooling tower. It operates to direct water around the condenser as the condensing temperature is lowered. This allows the pump to maintain its flow and reduces problems of water distribution with multiple unit applications.

COOLING TOWERS

Atmospheric water cooling equipment includes spray ponds, spray-filled atmospheric towers, natural draft atmospheric towers, and mechanical draft towers. Except for relatively small installations on which the spray-filled atmospheric tower may be used, the mechanical draft tower is the most widely used for air conditioning application. Of the types of equipment available, the mechanical draft tower is the most compact, the lowest in silhouette, the lightest and the best suited to meet exacting conditions of water temperature.

Air flow thru a mechanical draft tower may be forced or induced. Referring to the direction of air flow relative to the water flow thru the fill, a tower may be classified as counter-flow, cross-flow or parallel flow. The towers commonly used are induced draft, counter-flow or cross-flow.

A cooling tower consists of a casing, basin and sump, water distribution system, fill, fan, motor and drive.

The relation between the enthalpy of the air and the water temperature is illustrated in *Fig. 52* for a counter-flow tower. The rate of heat transfer from the water to the air depends on the enthalpy of the air which is represented by wet-bulb temperature. This rate is independent of the air dry-bulb temperature. For a given air and water quantity thru a tower, the rate of heat transfer, or rated tower capacity, is increased by lowering the entering air wet-bulb temperature requirement or by raising the temperature of the water entering the tower.

Tower performance is specified in terms of water range and approach. Cooling range is the difference

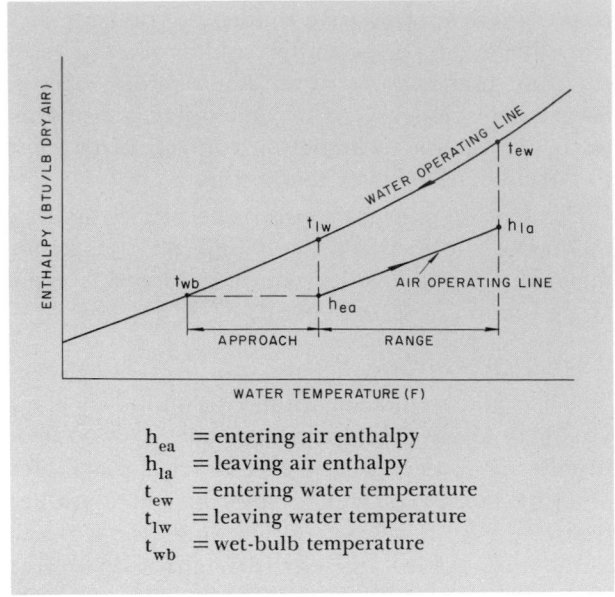

h_{ea} = entering air enthalpy
h_{la} = leaving air enthalpy
t_{ew} = entering water temperature
t_{lw} = leaving water temperature
t_{wb} = wet-bulb temperature

FIG. 52 — WATER COOLING PROCESS, COUNTERFLOW COOLING TOWER

between water entering and leaving temperatures and is equal to the temperature rise thru the condenser. Approach is the difference between the water temperature leaving the tower and the entering air wet-bulb temperature.

UNIT SELECTION

A cooling tower should be selected for the normal design wet-bulb temperature of the locale. If design conditions must be maintained at all times such as in some industrial process applications, the maximum wet-bulb temperature should be used to insure adequate equipment capacity.

Economics

The selection of refrigeration equipment and a cooling tower is influenced primarily by the condensing temperature chosen and its effect on tower range and approach. Relatively small increases in condensing temperature can produce large economies in tower size and cost, weight, space required and grillage or foundation costs. Further first cost savings result from:

1. Reduction in condenser water pump size and power input because of the lower water quantities accompanying higher ranges.
2. Reduction in piping costs with lower water quantities.
3. Reduction in tower fan power input.

At a given wet-bulb temperature, an increased condensing temperature can be obtained by one or both of two methods:

1. Increasing the condenser entering water temperature and, therefore, the approach.
2. Reducing the condenser water quantity.

With no change in the refrigeration equipment, a higher condensing temperature does result in a higher compressor power input requirement and, therefore, higher operating costs and motor and drive first costs. This effect can be countered by accepting a larger condenser size. With increasing condensing temperature, installed cooling tower costs usually fall more rapidly than refrigeration first costs rise.

Although the compressor full load power requirements are greater with increased condensing temperatures, the compressor motor is only partially loaded most of the time. Power savings from the smaller condenser water pump and tower fan motors which run continuously may more than pay for increased compressor operating costs.

Codes and manufacturers' recommendations should be referred to for limitation on maximum condensing temperatures.

Increasing condenser water temperatures may necessitate a more complete and thorough water treatment program than required at lower temperatures.

Atmospheric Corrections

At elevations above sea level, the reduction in weight flow of air due to the decrease in density of the air at high altitude is offset by the greater latent heat absorbing capacity of the high altitude air. Therefore, no correction need be made to cooling tower ratings as a result of altitude effects. Fan motor selections made on a sea level basis are conservative.

Fan Drives

Cooling towers are usually obtainable with either gear or belt drives. Gear drives are recommended on large cooling towers since malfunctions are infrequent.

CONDENSER WATER TEMPERATURE CONTROL

Operation of refrigeration equipment at low outdoor temperatures necessitates a control of condensing pressure. Maintenance of a minimum condensing pressure insures proper operation of the thermostatic expansion valve or the refrigerant float valves. A fall in evaporator temperature to the safety control setting is also prevented.

When water-cooled condensing is used, condensing pressure control is achieved thru condenser water temperature control. It is usually not ad-

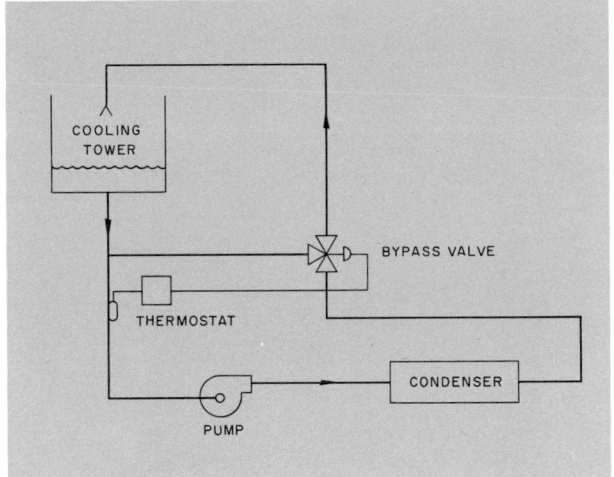

FIG. 53 — CONDENSER WATER TEMPERATURE CONTROL

visable to operate a constant speed centrifugal refrigeration machine at condensing temperatures below 80 F. Variable speed centrifugal machines may be operated at condensing temperatures as low as 50 F. Condensing temperatures required for expansion valve operation for normal reciprocating equipment should be maintained at about 90 F.

One or more of the following methods may be used for controlling condenser entering water temperatures:

1. Cycle the cooling tower fan.
2. Employ a two-speed fan motor to permit a capacity reduction.
3. Stop in sequence the fans on a multi-cell tower.
4. Bypass the cooling tower fill thru a control valve (*Fig. 53*).

Each of the above solutions may be manually or automatically controlled. The control valve may be snap-acting or may modulate the water bypassed, but should be snap-acting at subfreezing temperatures.

WINTER OPERATION

Winter operation of a cooling tower introduces the problems of water freeze-up in the basin and ice formation on the fan blades and louvers.

The use of a depressed or auxiliary sump within the heated space is one solution to the freezing of water in the basin (*Fig. 54*). In this way, the tower basin is dry during periods of shutdown and whenever condenser water bypasses the fill. This auxiliary sump should be sized to provide sufficient storage space for all the water in the tower and basin and to provide a suction head on the pump

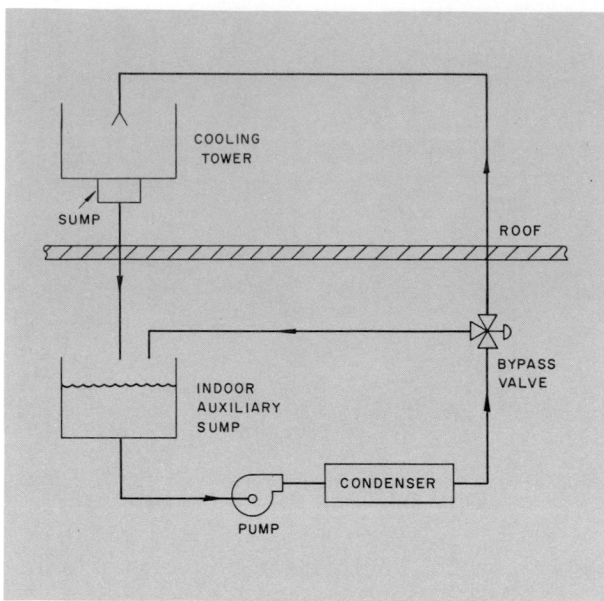

Fig. 54 — Condenser Water Temperature Control, Auxiliary Sump

when the tower is operating. An alternate solution provides for heating the basin water with steam or hot brine passing thru coils.

Ice formation on the fan blades results in excessive vibration which may lead to blade breakage and damage to the tower. Two-speed operation may be employed at tower leaving air temperatures approaching 32 F. If ice continues to form at low speed operation, a vibration switch may be utilized to stop the fan.

Air flow thru the tower can be restricted by an ice build-up on the louvers. Complete prevention of ice formation may be difficult, but small deposits can be melted by reversing the fan motor and operating at top speed. This reversing is usually accomplished manually.

MAKE-UP WATER

A cooling tower loses water by evaporation, drift and bleed-off. Evaporation amounts to about one percent of the total condenser water circulated for each 10 degrees of range. Drift loss is constant at all ranges and is approximately 0.2% of the water circulated. Bleed-off varies with water conditions and should be established by the water treatment program as explained in *Part 5*. Where water conditions are not severe, bleed-off is approximately 0.3% for each 10 degrees of range.

The amount of make-up water required is established by the total of these losses. Water may be made up on demand thru a mechanical float-oper-

ated valve or by a pair of electric level probes used with a relay and a water solenoid valve.

LOCATION

The selection of a tower site and the orientation of the tower should be guided by the following considerations:

1. Locate the tower so that air circulates and diffuses freely and rapidly without recirculating. Manufacturers usually publish recommendations on this subject.

2. Locate the tower away from sources of heat or contaminated air such as smoke stacks.

3. Locate the tower away from areas where wetting or freezing from normal drift is objectionable. Drift may be carried several hundred feet downwind from the tower if the wind is strong.

4. Locate the tower above or remote from occupied spaces or surroundings with low ambient sound levels. The ideal location is on the roof of a tall building. A tower location at a low level between tall buildings is very undesirable with regard to both tower performance and sound level.

5. Condenser water piping may be simplified and its cost reduced if the tower is located immediately adjacent to or above the refrigeration room.

6. Locating a tower below the refrigeration equipment may lead to problems of siphoning or overflow at shutdown. If water is siphoned from the condenser, the condenser may be damaged by water shock when the pump is started again. If such a location is planned, consideration should be given to reliable checking of backflow at shutdown and to breaking any siphon which might form.

7. Where the tower is to be mounted on a roof, the ability of the roof to withstand the added weight should be checked. The tower should be located to afford the most even distribution of weight on the structural members.

LAYOUT

If the tower is remote from the refrigerating machine or at a lower elevation, the condenser water pump may be located adjacent to the tower. With a tower at grade level, a vertical pump may be used instead of the more conventional horizontal pump.

A steel or wood cooling tower mounted on a roof is elevated and supported by a steel grillage. A tower located on the ground may be so mounted, or may be furnished without a basin and mounted on a concrete basin. Grillages and concrete basins should be designed in accordance with manufacturers' recommendations. The design of a concrete basin should include suction screens. Wood and steel basins usually feature sump screens.

On some towers the manufacturers offer a vertical supply line to the water distribution system located at the center of the cell. For certain layouts, this feature can improve the appearance of the installation and reduce the complexity and cost of the piping.

For details of condenser and cooling tower piping, refer to *Part 3, Chapter 2*. The following are further specific recommendations.

1. Tower overflows should be piped to a drain and should not be valved.

2. Provision should be made for draining the tower and equipment.

3. A city water hose bib should be provided at the tower to facilitate cleaning.

4. A tower should be provided with a city water fill line in addition to a make-up line. A tower located above the refrigeration equipment affords an ideal location for filling the entire condenser water system.

5. If the fan motor starter is remote from the tower, a disconnect switch should be provided at the tower to insure safety while servicing.

Part 8
AUXILIARY EQUIPMENT

CHAPTER 1. CENTRIFUGAL PUMPS

This chapter sets forth a review of the selection and application of centrifugal pumps used with air conditioning and refrigeration systems.

There are two major categories of pumps:

1. Positive displacement — reciprocating, rotary and screw.
2. Centrifugal — with a variety of impeller designs classified as plain (radial) flow, mixed flow and axial flow, each within a volute casing; also turbine (diffuser) type pumps.

Figure 1 illustrates the two types of centrifugal pumps as well as the four basic types of impellers. The plain impeller has single curvature vanes always curved backwards. Wider impellers have vanes of double curvature with the suction ends twisted. These vanes are called mixed flow (Francis type) vanes. The extreme mixed flow (minimum of radial element) and axial flow impellers have propeller type vanes. While having axial flow, this class of pumps represents hydraulically an extreme in a continuous series of centrifugal pumps.

It is the plain flow impeller centrifugal pump that is used most frequently in air conditioning and refrigeration. They are used to move chilled, warm, hot and refrigerant condensing water, steam condensate, brine, lubricant oil or refrigerant.

A centrifugal pump is distinguished by a continuous steady flow and characteristic performance curves with a smooth rising head and falling power from maximum capacity to shutoff *(Fig. 2)*. The pump presents an easy load for a driver. The starting torque is small and operating load is constant. As a rule a constant speed squirrel-cage induction electric motor (NEMA Design B) with normal starting torque *(Chapter 2)* is applicable to drive the pump although steam turbines, gasoline or steam engine, and belt and motor drives may also be used.

The centrifugal pump is rated on the basis of capacity, i.e. volume of liquid per unit time or gallons per minute, gpm (1 cu ft/min = 7.48 gpm), against a head, i.e. feet of water required by the fluid transmission system, and the energy required at a given speed.

There are two types of liquid circuits, open and closed. In the open system the pump moves a liquid from a source located above or below the pump but open to atmospheric pressure *(Fig. 3a and 3b)*. A closed system is one in which the liquid circuit is not open to the atmosphere *(Fig. 3c and 3d)*. The most common application in air conditioning and refrigeration systems is the closed water circuit. This is the major subject of the following text.

The rudiments of centrifugal pump behavior are found in the section on *Centrifugal Pump Fundamentals*.

STANDARDS AND CODES

Standards of the Hydraulic Institute (an organization of leading pump manufacturers) define the product, material, process and procedure in the design and testing of any type of pump.

Pump installation should conform to all local codes, rules and regulations.

CENTRIFUGAL PUMP

The centrifugal pump has a unique distinction of simplicity in construction, yet critical in application. There are two major elements in a centrifugal pump assembly — an impeller rotating on a shaft supported in a packed or mechanical seal and bearings, and a casing that is the impeller chamber (volute). The impeller imparts the principal force to the liquid, and the volute guides the liquid from inlet to the outlet, at the same time converting the kinetic (velocity) energy into pressure.*

---

*In a turbine type centrifugal pump the diffusers perform the major part of the energy conversion task.

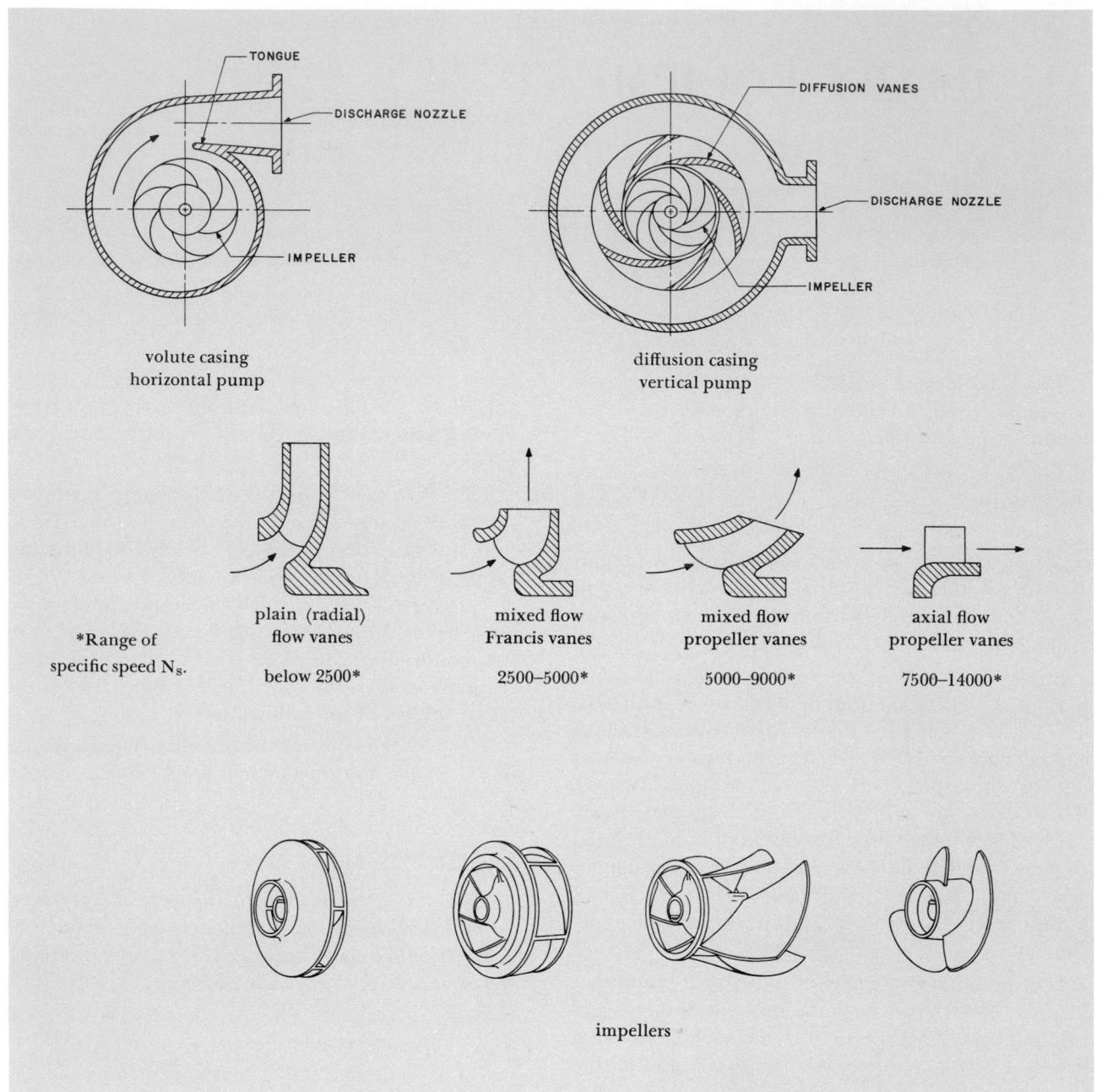

FIG. 1 — CENTRIFUGAL PUMP CASINGS AND IMPELLERS

ADVANTAGES

The centrifugal pump is favored because of the following characteristics:

1. Simplicity of construction.
2. Absence of valves and reciprocating parts.
3. Fewer moving parts.
4. Absence of close clearances.
5. Minimum of power transmission losses.
6. Steady, nonsurging flow.
7. Operation at shutoff condition without excessive build-up in pressure.
8. Absence of contact between liquid pumped and lubricant.
9. Compactness; light in weight.
10. Adaptability for direct connection to standard motors (major types of drive).
11. Long life.
12. Ease of maintenance and minimum repair.
13. Reasonable cost.

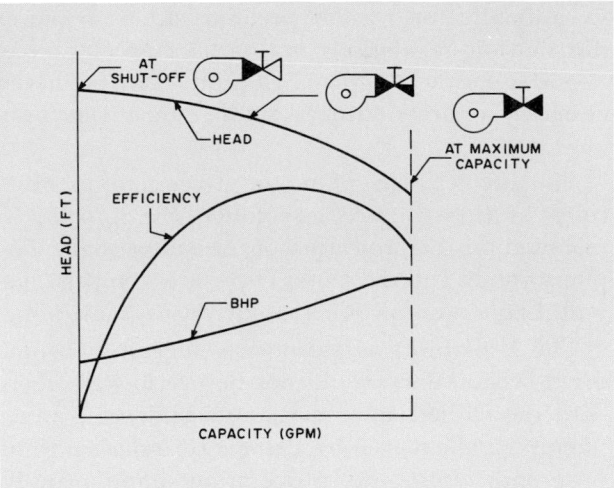

FIG. 2 — TYPICAL PERFORMANCE, PLAIN FLOW VANE
CENTRIFUGAL PUMP

There are two shortcomings in a centrifugal pump:
1. It is not self-priming unless specially equipped with a priming device (or a foot valve).
2. It is inefficient for capacities smaller than 10 gpm at heads higher than 30 feet.

CLASSIFICATION AND DESIGNATION

The manner of liquid flow within the impeller and the casing of centrifugal pumps has already been outlined. The plain radial flow impeller in a volute casing centrifugal pump is the one usually applied in air conditioning and refrigeration applications.

Impeller

The impellers are constructed in three arrangements:
1. Enclosed (vanes within an impeller shroud or side walls).
2. Semi-enclosed (vanes assembled with one side wall).
3. Open (no side walls, casing serving as side walls).

Suction

The liquid approach into the pump may be either:
1. Thru a single inlet with end suction to impeller.
2. Thru a single inlet with double suction, liquid flowing into the impeller along the shaft on two sides (*Fig. 4*).

Casing

The volute casing (*Fig. 5*) may be split axially (horizontally, usually with double suction pumps), or radially (vertically, usually with single suction, end inlet pumps).

Stages

The single stage pump is one with a single impeller; it may have a single or double suction. If the required head is too high for a single impeller to develop, two or more single stage pumps may be used in series, or a set of impellers in series may be put into a single casing. The latter assembly is designated a multi-stage pump.

Assembly

With reference to the axis of rotation of the shaft, centrifugal pumps and drive arrangements are either horizontal or vertical (at times inclined). Horizontal pumps are arranged either with end or side suction inlets; top and bottom suctions are also available. Double suction pumps are usually built with side discharge nozzles (*Fig. 5*).

The single suction pumps are usually made with the end suction inlet (also available in other arrangements) and a variety of discharge nozzle positions (*Fig. 6*). The discharge nozzle is a size or two smaller than the suction. Centrifugal pumps are often identified by a number corresponding to the size of its discharge; however this does not define its capacity which must be stated concurrently.

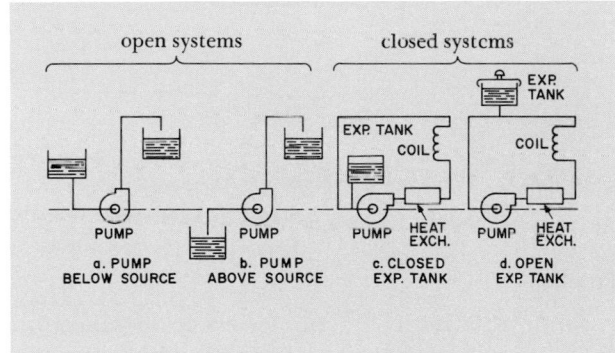

FIG. 3 — LIQUID FLOW SYSTEMS

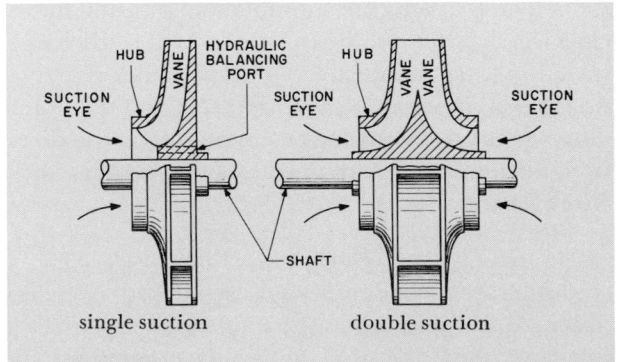

FIG. 4 — IMPELLERS

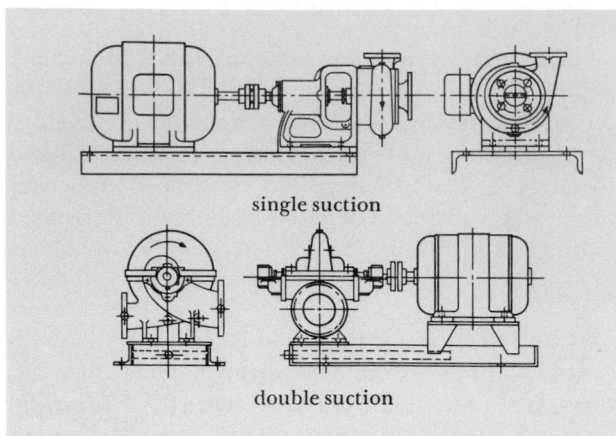

FIG. 5 — SINGLE AND DOUBLE SUCTION
CENTRIFUGAL PUMPS

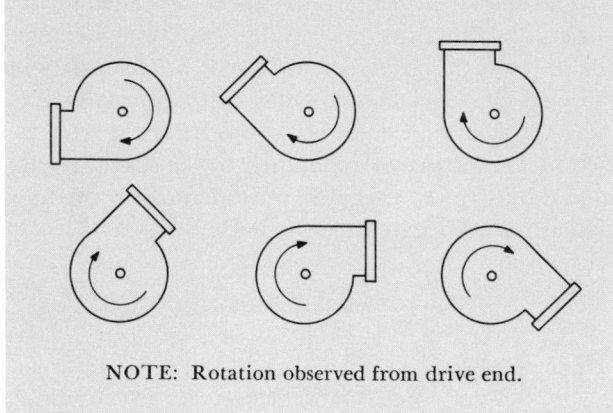

NOTE: Rotation observed from drive end.

FIG. 6 — DISCHARGE NOZZLE ARRANGEMENTS,
SINGLE SUCTION CENTRIFUGAL PUMP

Rotation

Pump rotation is determined when looking from the drive toward the pump. Thus straddling the drive of a horizontal pump or looking down at the motor end of a vertical pump, if the liquid is entering the suction on the right side (double suction) and moves clockwise within the casing towards the discharge, such a pump is designated as having clockwise rotation. With the suction on the left side and the liquid moving within the casing in counterclockwise direction, the pump is designated as having counterclockwise rotation (Fig. 7).

Drive

Motors are the major drivers used to supply energy to centrifugal pumps. Of the single-phase motors the capacitor type are used for small pumps. Of the multi-phase motors the standard squirrel-case induction type (NEMA Design B) are the most popular.

Occasionally for reasons peculiar either to power distribution regulations or to the customer's economic situation (Chapter 2) a pump may be driven either by a part-winding, wound rotor or synchronous motor.

The smaller sizes of motor-driven pumps often come in close-coupled assemblies; the impeller is mounted on the projection of the motor shaft. The pump volute and the motor enclosures comprise one unit. Large pumps connect to drivers by a coupling.

The availability of steam may suggest a turbine drive. A situation should not be overlooked where both the chilled and refrigerant condensing water pumps may be driven by a single thru-shaft turbine since both pumps are in operation simultaneously. High speed (3500 rpm and above) pumps are particularly adaptable for direct connection to turbines.

In critical and emergency situations auxiliary drives may be provided; this is another means to drive a pump in case of failure of the regular drive.

SUPPLEMENTARY COMPONENTS

Several components supplementing the impeller shaft and casing are required to complete the centrifugal pump assembly and to provide various protections and accommodations in order to:

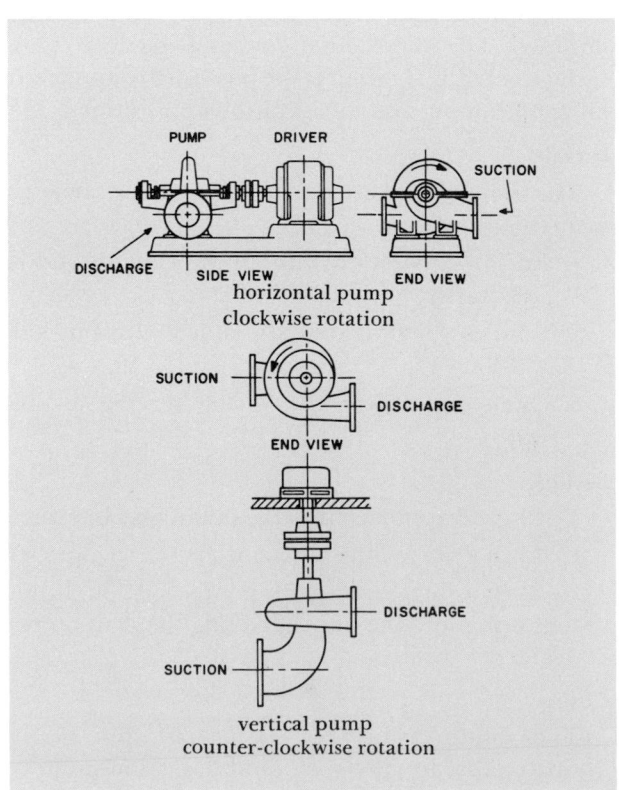

FIG. 7 — ROTATION DESIGNATION, CENTRIFUGAL PUMP

1. Avoid destructive wear to either the total unit of the impeller or the casing; at the same time, provide close-running clearance between the lower pressure inlet and the higher pressure discharge region in the casing (wearing rings).

2. Prevent leakage outward or inward between the inside of the pump and the ambient outside (seal). The direction of leakage depends on whether the pressure within the pump casing is higher or lower than the ambient atmospheric pressure.

3. Support and align the rotating impeller shaft with the casing stationary (bearings).

4. Connect the pump shaft to the driver shaft (coupling) unless the pump impeller is mounted on the extension of the driver shaft as in the end inlet, single suction pumps.

5. Support the total pump-drive assembly (bedplate).

Wearing Rings

To achieve the first protective provision the impeller hub outer surface at the impeller eye and the adjacent casing surface are variously equipped with wearing rings (Fig. 8). They are respectively the impeller ring and casing ring. When necessary, only the rings are replaced, rather than the total impeller or total casing. A variety of designs and combinations of wearing rings and labyrinth arrangements are available.

Shaft

The shaft, a separate carefully designed entity, is treated in this text together with the impeller as a single rotating element. The standard shaft is protected from wear, corrosion and erosion within a stationary support by a sleeve designed and fitted in many forms. This sleeve covers the shaft thru a stuffing box or a mechanical seal. Very small pumps are frequently built with special wear-resisting shafts to avoid the disadvantages of the diameter enlargement of the fitted shaft.

Stuffing Box

The second protective provision (against leakage between the inside of the pump casing and the ambient atmosphere) is achieved either by a stuffing box (Fig. 9) or by a mechanical seal. With the stuffing box the sealing between the rotating shaft or shaft sleeve and the stationary support is achieved by rings of specially lubricated materials such as asbestos or metal packing, held tight by a gland. When the leakage becomes more apparent, the gland is tightened

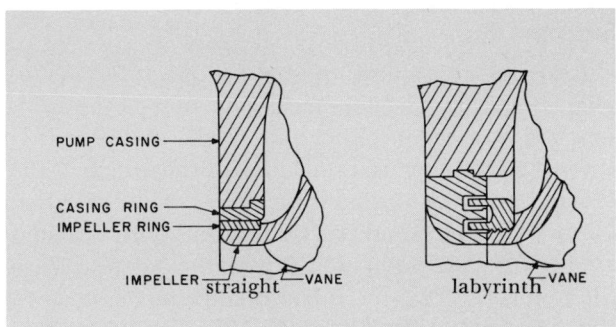

FIG. 8 — WEARING RINGS

(within limits). The sealing, lubricating and cooling liquid is supplied either from the high pressure region of the casing or from external sources.

Mechanical Seal

When handling expensive volatile or high temperature liquids at varying pressures or when attempting to provide a truly positive seal, a single or double mechanical seal is provided. The mechanical seal differs from the stuffing box with its packing by the orientation of the sealing. The stuffing box packing seals axially along the shaft (Fig. 9); the mechanical seal is formed by contact of two highly polished surfaces of dissimilar materials set perpendicular to the shaft. One spring-held inner surface is attached to the shaft, rotating with it; the outer surface is attached to the stationary part of the pump.

It is very important that there is a liquid film between the surfaces to provide lubrication and cooling. Mechanical seals are available in numerous designs that are constantly improved and reduced in cost. They require practically no maintenance.

The zero-leakage wet-winding or "canned" motor-pumps do not need a stuffing box or mechanical seals. They are leakless and sealless close-coupled motor-pump assemblies.

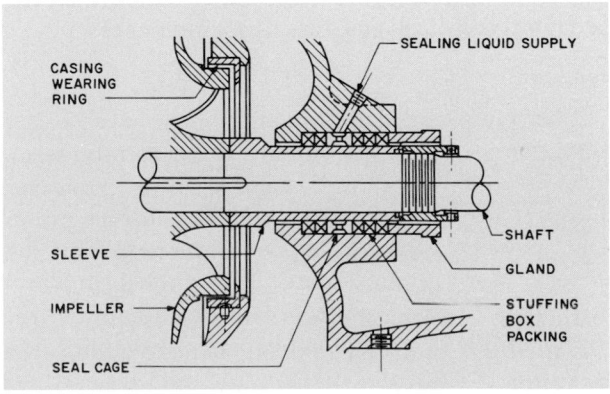

FIG. 9 — SLEEVE, STUFFING BOX, PACKING AND GLAND

Bearings

Bearings are points of shaft support serving to align the shaft. On double suction pumps the bearings are located on either side of the pump casing; the outer bearing is called the outboard, and the bearing between the pump and the driver is called the inboard. On single suction pumps both bearings are between the pump and driver; the one nearest to the pump is called the inboard, and the one nearest the coupling or driver is called the outboard. The bearings are either the sleeve type (minor usage) or the frictionless ball or roller type. The ball bearings are used most frequently. Bearings are often designed to take up thrust resulting from various unbalanced forces exerted within the pump.

Couplings

Except for close-coupled assemblies, pumps are usually connected to the drive thru a coupling. There are two basic classes of couplings, rigid and flexible. The rigid coupling does not permit axial or radial motion. It is a solid connection providing a continuous shaft. The rigid coupling is used for vertical pumps.

While transferring power from driver to pump, the flexible coupling allows a transverse adjustment for a very minor misalignment. However the pump alignment must not be misused; it must be rigidly enforced. Misalignment causes a whipping of the shaft; it adds to pump and driver bearing thrust and may result in excessive maintenance. Misalignment must not be tolerated.

Flexible couplings are effective in providing also lateral adjustments (along the length of the shaft) for either or all of the following: thermal changes, hydraulic float, or shifting of the magnetic center of the motor. A variety of flexible couplings as well as adaptation combinations are available to resolve any particular requirement to accommodate either the behavior peculiarity or ease of maintenance.

Bedplate

The pump and drive assembly must have perfect alignment. Close-coupled pumps are naturally assembled into balanced and aligned units. However pumps that are combinations of drive-coupling-pump units must be assembled either in the field or at the factory. Many pumps come preassembled and prealigned on a cast iron bedplate or structural steel support. These pump-on-bedplate assemblies are ready for bolting and dowling for level installation on a foundation. This does not mean that a factory-assembled pump-coupling-drive unit is inviolably perfect; accidents may happen in transit. Therefore, during installation the pump must be rechecked for alignment and level position. Cast iron bedplates are often equipped with a rim for containing and draining the pump leakages. Otherwise, separate provisions must be made for collecting and disposing of leakage.

MATERIALS

Centrifugal pumps used in air conditioning and refrigeration are usually made of standard materials except in special instances of pumping sea water or corrosive or highly electrolitic brines. The pumps are built of special materials for the case of strenuous hydraulic circumstances or for handling extremely low temperature liquids; in the latter instance the strength and brittleness of standard materials should be examined. For pumps used with high temperature water (HTW) up to 300–350 F, a standard cast iron casing is applicable. With either high temperatures (above 250 F) or low temperatures (below 50 F), the choice of materials for the impeller-shaft-supplementary components assembly becomes critical. The materials must be chosen such that the thermal expansion and contraction of these parts are equal.

According to Hydraulic Institute terminology for a standard fitted pump the standard materials used are: cast iron casing, steel shaft, bronze impeller as well as wearing rings and shaft sleeve (when used). A pump so constructed is termed bronze fitted. When all parts (casing, impeller, various rings, shaft) of the pump that come in contact with the liquid to be pumper* are made of bronze, such a pump is termed all bronze; in the case of all iron parts, the pump is termed all iron. There are many varieties of material deviations to fit the specific needs. *Figure 10* shows the major parts of a bronze-fitted pump.

There are two basic approaches in the selection of pump materials:

1. If the engineer is thoroughly experienced for the given application, he dictates the specifications.
2. If the manufacturer has wide experience in selecting the appropriate materials, then the engineer furnishes the manufacturer comprehensive data on the liquid pumped, including the operating temperature, the physical characteristics at this temperature and any peculiarities of the operation.

---

*Bearings, bearing housings and other parts of the pump external to the liquid passage are not in contact with the liquid pumped, and are made of appropriate industry standard materials.

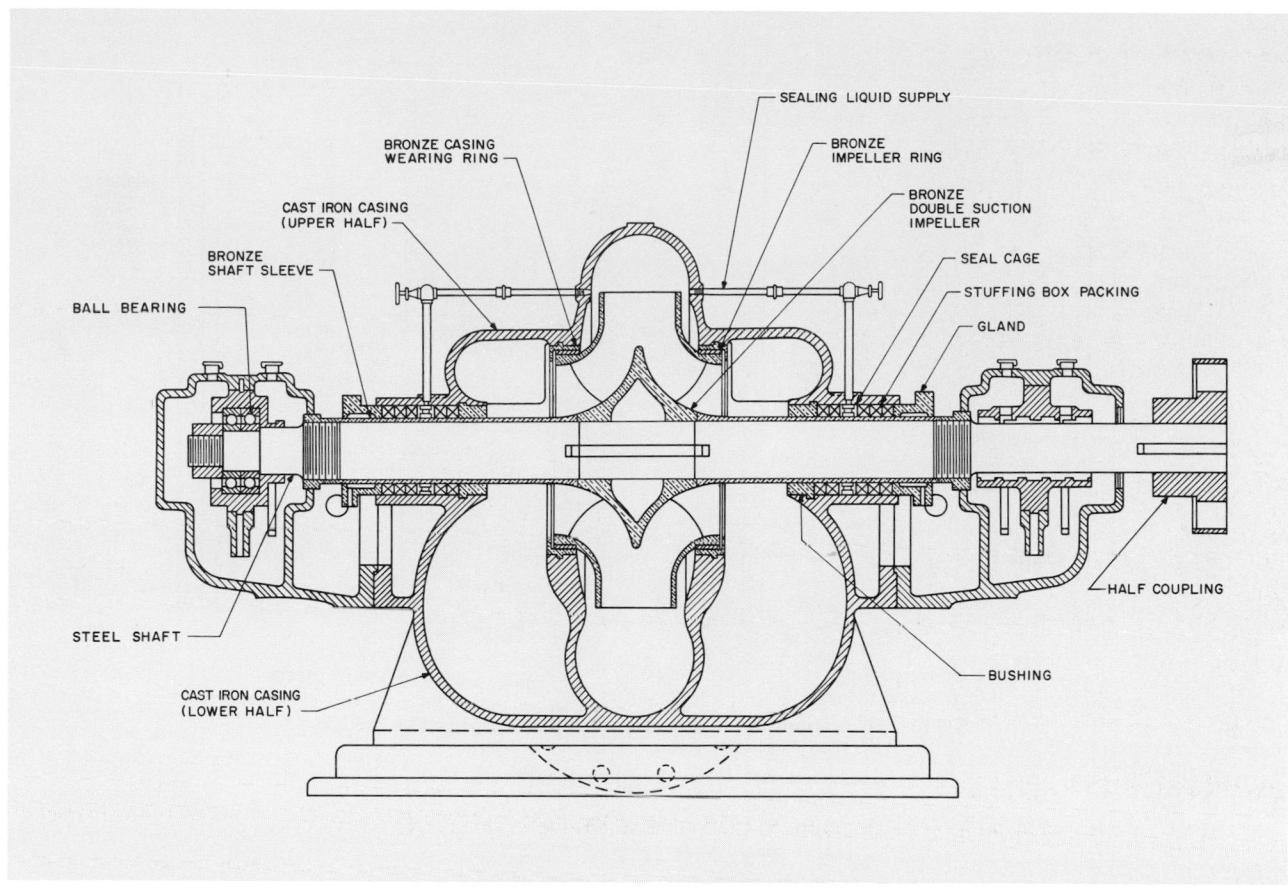

FIG. 10 — MAJOR PARTS OF A PUMP

CENTRIFUGAL PUMP FUNDAMENTALS

Having covered the mechanical aspects of the centrifugal pump and before considering it as part of the liquid circulating system, a discussion of pump behavior is presented.

BASIC THEORY

The rotating impeller imparts to a fluid a centrifugal force, kinetic energy in the form of velocity. The volute converts about 50 percent of the kinetic energy into the pressure head, potential energy measured in feet of fluid handled. As the fluid flows thru the impeller vanes, a reduced pressure zone is created at the inlet to the vanes. The atmospheric or system pressure and the static head of the fluid as available act on the pump suction inlet and force the liquid into the pump. This pressure at the pump suction plus the pressure developed by the rotating impeller in the volute produces the flow of the liquid. This is fundamental to the application of the centrifugal pump.

NET POSITIVE SUCTION HEAD

If the pumping is limited only to that normally applied in the air conditioning cold water closed circuit systems, there is no need to be concerned with sufficiency of the suction pressure to force the liquid into the pump suction. However, the various liquids at any given temperature have a definite saturation pressure at which they turn to vapor. In the field of air conditioning and refrigeration situations exist to handle water, brines and refrigerants at any temperature and pressure level. It is the problem of the process and pump application engineer to be sure that under any set of circumstances there is sufficient pressure on the liquid fed to the pump to prevent the liquid from flashing into vapor.

Between the pump suction nozzle and the minimum pressure point within the pump impeller, there exists in addition to the suction velocity head a pressure drop. This pressure drop is due to velocity acceleration, friction and turbulence losses. The suction head (feet of liquid absolute) determined at the suction nozzle and referred to a datum line less the vapor pressure of the liquid (feet absolute) is called the net positive suction head or NPSH. The suction head necessary to keep liquid flowing into the pump

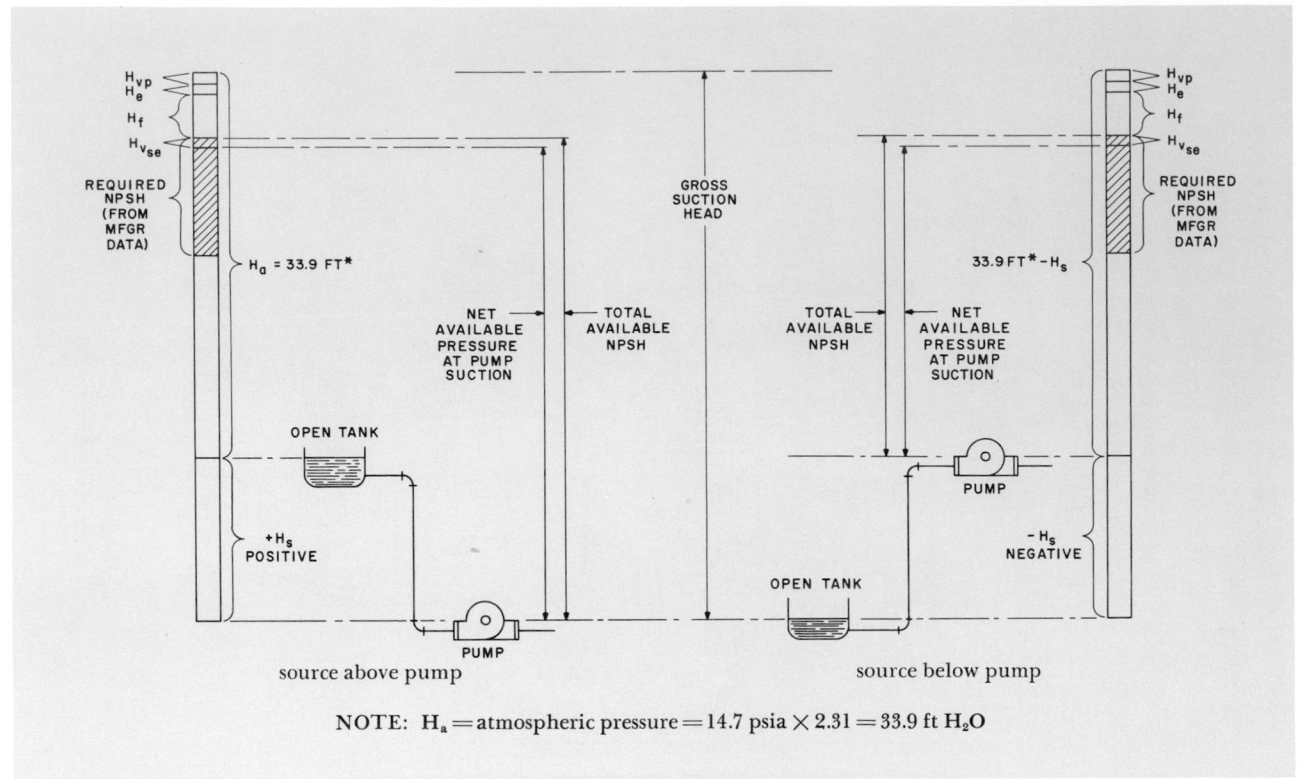

NOTE: H_a = atmospheric pressure = 14.7 psia × 2.31 = 33.9 ft H_2O

FIG. 11 — NET POSITIVE SUCTION HEAD, OPEN SYSTEMS (COLD WATER)

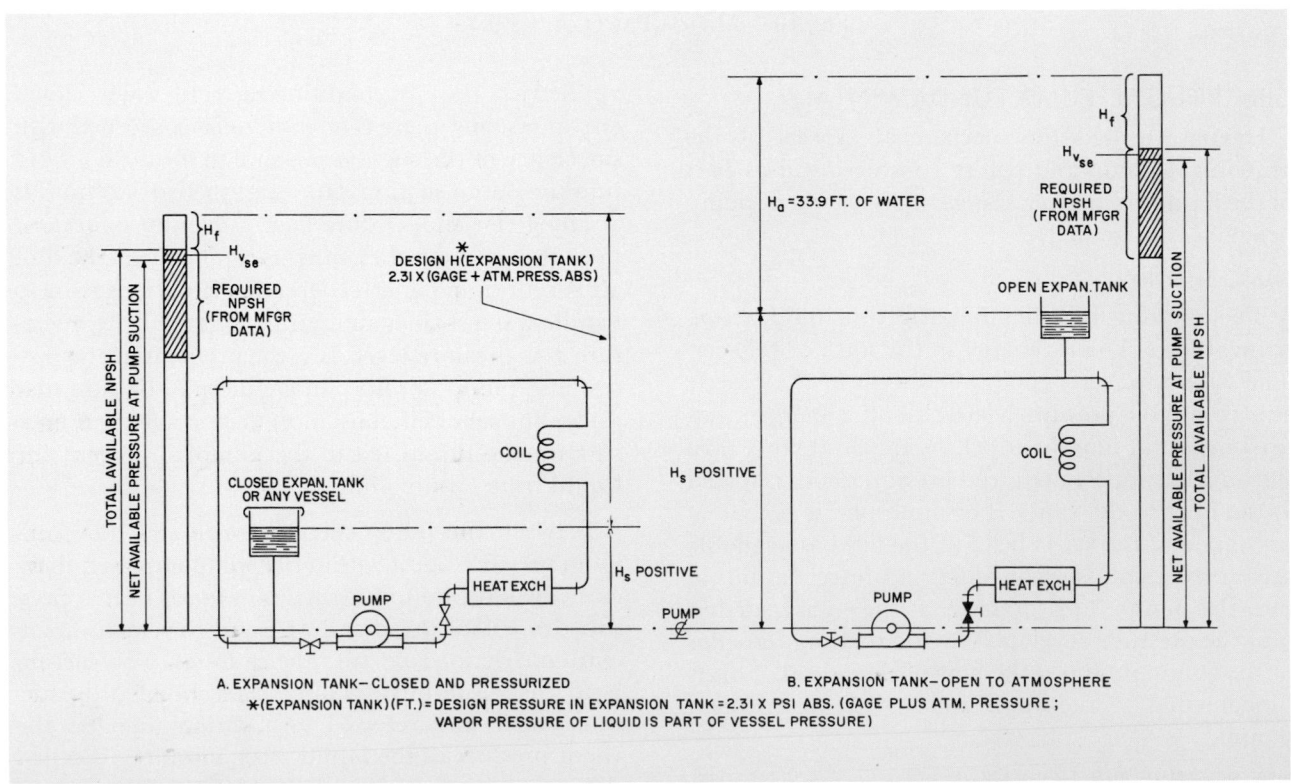

FIG. 12 — NET POSITIVE SUCTION HEAD, CLOSED SYSTEMS (COLD WATER)

and to overcome the pump internal pressure losses is the *required NPSH* of the pump.

The required NPSH of a pump is part of the standard design performance data furnished by the manufacturer or of a design specific to a given process pump.

The net positive suction head (pressure in feet of liquid) of the process liquid system as it exists within the system complex at the entering (suction) side of the pump is called the *available NPSH*. It must be at least equal to or greater than the required NPSH in order to produce a flow thru a pump. A safety factor should be considered to cover a possible excess of required NPSH.

The available NPSH is the algebraic sum determined by the formula:

$$\text{Available NPSH} = \frac{2.31 \, (P_a - P_{vp})}{\text{sp gr}} + H_s - H_f$$

where:

NPSH = net positive suction head (absolute pressure, ft)

2.31 = conversion factor to change one pound pressure at a specific gravity of 1.0 to pressure head in feet of water (1 inch Hg = 1.134 ft of water).

P_a = atmospheric pressure (absolute pressure, psia) in an open system; or pressure (absolute, psia) within a totally closed system.

P_{vp} = vapor pressure (psia) of the fluid at pumping temperature; in a totally closed system it is part of the total pressure P_a.

H_s = elevation head, static head (ft) above or below the pump center line. If above, positive static head; If below, negative static head, sometimes termed suction lift.*

H_f = friction head (ft) on the suction side of the system including piping, fittings, valves, heat exchangers at the design velocity (V_s in ft per sec) within suction system.

sp gr = specific gravity of liquid handled at operating temperature (Fig. 14).

Figures 11 and 12 illustrate the application of the calculation of available NPSH to the variety of open and closed circuits. Three additional terms are introduced in these figures:

H_{vp} = vapor pressure (ft) of the fluid at pumping temperature.

H_e = entrance head (ft), suction pipe entrance loss in open systems.

$H_{V_{se}}$ = pump suction eye velocity head (ft), $(V_{se})^2/2g$. This term is usually very small as shown in the following tabulation:

Velocity (ft/sec)	3	4	5	6	7
Velocity head (ft)	.14	.25	.39	.56	.76

Velocity (ft/sec)	8	9	10	11	12
Velocity head (ft)	.99	1.26	1.55	1.88	2.23

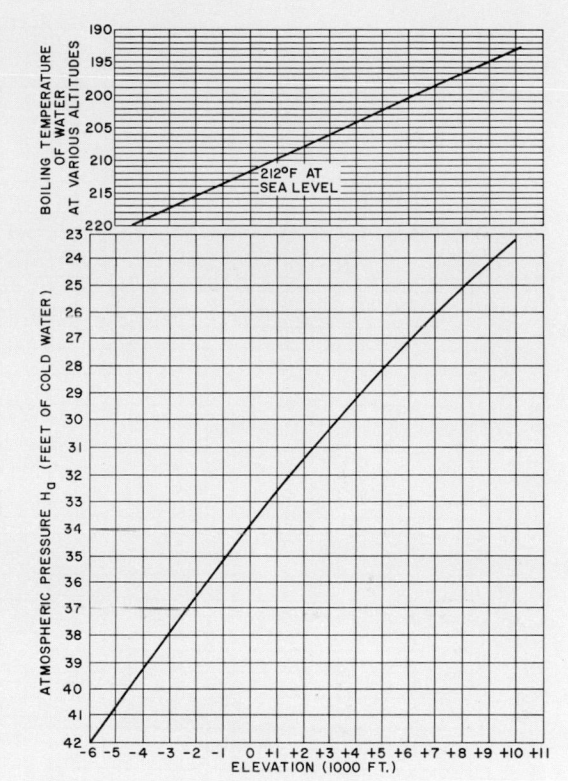

FIG. 13 — EFFECT OF ALTITUDE ON ATMOSPHERIC PRESSURE

A pressure selected to be maintained above atmospheric pressure in the top circuit of a closed piping system determines the design $H_{(expansion \; tanks)}$ pressure *(Fig. 12)*.

On examining *Fig. 11 and 12*, it is evident that the available NPSH may vary, especially with critical fluids. The variables that may be either fixed or adjusted are:

1. Altitude of the system location above or below sea level; *Fig. 13* shows the change of atmospheric pressure (feet of cold water) with the altitude. The greater the altitude, the lower is the available atmospheric pressure (P_a in psia or H_a in ft) which influences an open system. The totally closed system pressure P_a may be regulated.

2. Vapor pressure of the liquid *(Fig. 14)* pumped at operating temperature P_{vp} (psia) or H_{vp} (ft); *Figure 14* shows the vapor pressure of water at various temperatures. This pressure may or may not be adjusted.

3. Friction losses of the pump suction piping system; the larger the pipe, the less are the friction losses H_f (ft) for a given fluid flow.

*It must be remembered that a pump does not lift the liquid it moves; a pump must have pressure to produce the flow.

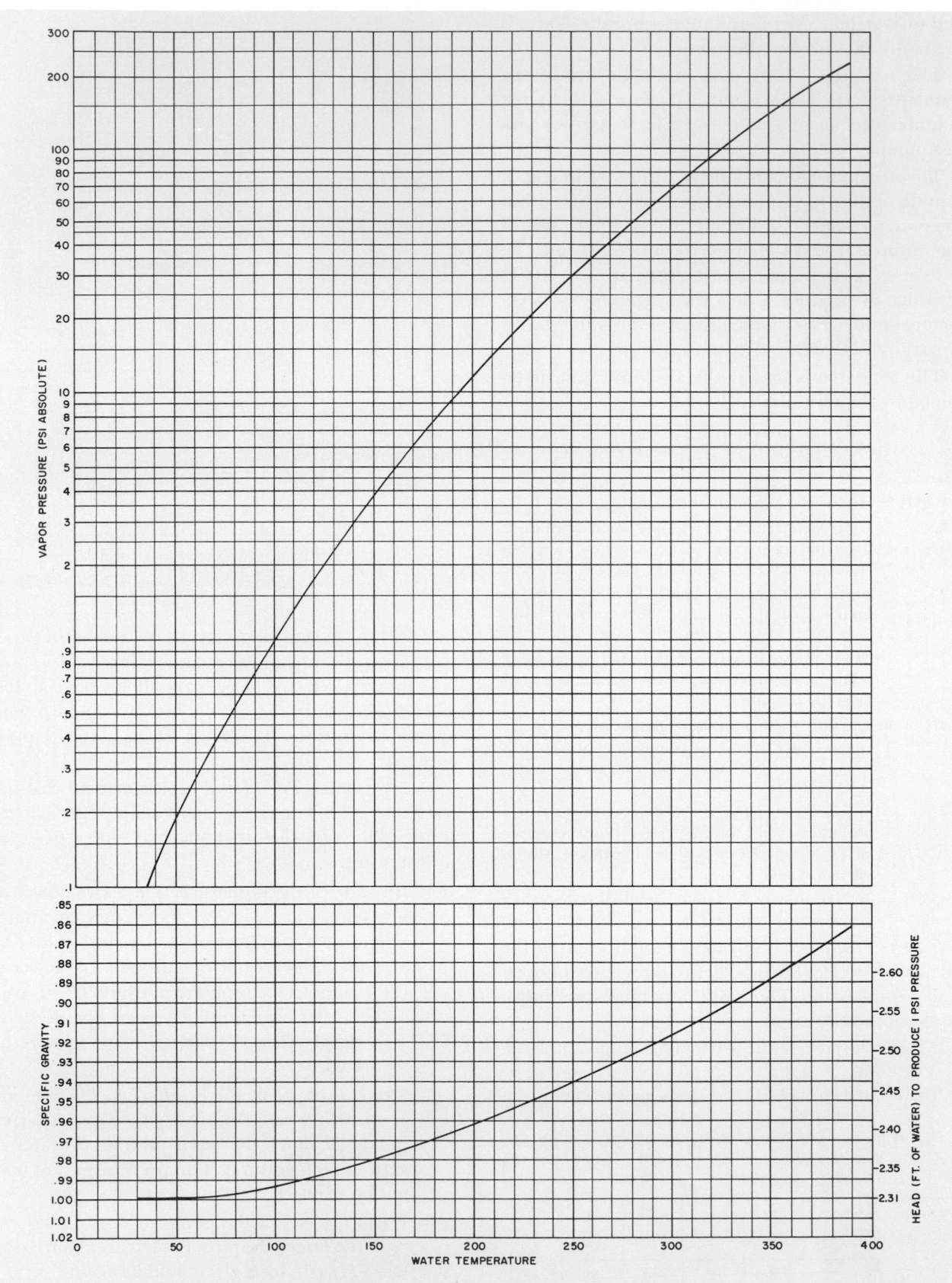

FIG. 14 — PROPERTIES OF WATER AT VARIOUS TEMPERATURES

4. Elevation of the source of liquid static head H_s (ft, positive or negative); pump location may be altered to increase or reduce static head. There are limitations to the negative head (suction lift).

SUCTION LIFT

The suction lift of open systems is not too frequent a factor in the design of air conditioning and refrigeration systems. Basically a pump does not lift; to operate, it must have pressure at its suction. Thus the maximum lift is determined by the required NPSH and limited by the available NPSH pressure.

The atmospheric pressure at sea level is 33.9 feet of water. For cold water at normal temperatures H_{vp} equals approximately one foot; therefore the gross suction pressure may be considered as 33 feet. Based on an operating rule that available NPSH should at least equal the required NPSH, normal suction lift should be limited. The limitation is the amount of available NPSH remaining after deducting from the gross suction pressure the H_e, H_f and the required NPSH (Fig. 11b) of the pump plus a safety factor, to allow for possible vagaries and to prevent appearance of cavitation and consequent vibration of the pump. There must be pressure at the suction for the pump to do normal work.

To quote from paragraph B-44 of the Hydraulic Institute Standards, "Among the more important factors affecting the operation of a centrifugal pump are the suction conditions. Abnormally high suction lifts (low NPSH) beyond the suction rating of the pump usually cause serious reductions in capacity and efficiency, often leading to serious trouble from vibration and cavitation."

SPECIFIC SPEED

Paragraph B-45 of the Hydraulic Institute Standards states: "The effect of suction lift on a centrifugal pump is related to its head, capacity and speed. The relation of these factors for design purposes is expressed by an index number known as the specific speed."

This number is used by pump designers to arrive at an optimum efficiency, expressed as follows:

$$\text{Specific speed } N_s = \frac{\text{gpm}^{0.5}}{H^{0.75}} \times \text{rpm}$$

where:

   H = head (ft) * based on the maximum diameter impeller at the design capacity.

   gpm = capacity at best efficiency.

   rpm = mechanical speed at which the gpm and head are obtained.

*In case of a multi-stage pump, head of each stage.

The specific speed may be defined as the rpm at which a pump of a particular design would have to operate to deliver one gpm against a head of one foot. Specific speed is an index to the type of impeller (Fig. 1). The lower the specific speed, the more the blades of the impeller are of an arrangement to deliver a strictly radial flow; the higher the delivery head, the smaller the required NPSH. Excessive reduction in required NPSH may lead to cavitation. The radial flow impellers afford more regulated flow thru the impeller vanes.

CAVITATION

The lack of available NPSH shows up particularly in pump cavitation. If the pressure at any point inside the pump falls below the operating vapor pressure of the fluid, the fluid flashes into a vapor and forms bubbles. These bubbles are carried along in the fluid stream until they reach a region of higher pressure. Within this region the bubbles collapse or implode with a tremendous shock on the adjacent surfaces. Cavitation accompanied by low rumbling or sharp rattling noise and even vibration causes mechanical destruction in the form of pitting or erosion.

Remedies to eliminate cavitation are apparent from the tabulation of variable elements in an evaluation of available NPSH. The first two factors are fixed; the system is installed at a definite altitude, and the temperature of the fluid is fixed by the process. Therefore, only the two remaining adjustments can be made; decrease the friction loss and/or change the elevation of the pump to increase the static head.

Do not tamper with the pump suction inlet; do not request the pump manufacturer to enlarge the pump suction in order to decrease the required NPSH. The pump efficiency falls off and the whole performance of the impeller is upset.

VORTEX

A whirling fluid forming an area of low pressure at the center of a circle is called a vortex. This is caused by a pipe suction placed too close to the surface of the fluid. Such a vortex impairs the performance of a pump and may cause a loss of prime.

In the case of pump suction in a shallow water sump such a vortex may be prevented by placing a plate close to the intake at a distance of one-third diameter from the suction inlet. The plate should extend $2\frac{1}{2}$ diameters in all directions from the center of the inlet.

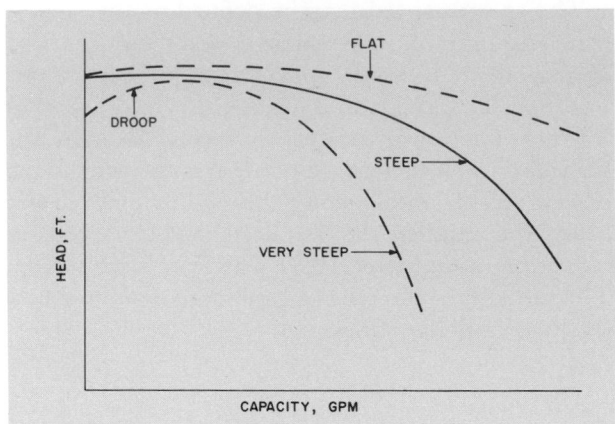

Fig. 15 — Desirable Head-Capacity Curve

PERFORMANCE

When a pump designer has established the specific speed of a pump, its capacity-head curve is then defined. The pump operates on this curve *(Fig. 2 and 22)* unless some physical change is effected.

Head-capacity curves should not droop at shutoff conditions; this leads to a surging operation when the flow is throttled into this range. Neither should the curves be too flat. The steepness in the pump head-capacity curve *(Fig. 15)* most desirable for an air conditioning and refrigeration application is shown by the solid line.

The performance characteristics of a centrifugal pump as expressed by a head-capacity curve are influenced in several ways:

1. Variation in speed — proportionally raises or lowers the head and capacity. The whole head-capacity curve shifts up or down.

2. Varying impeller diameter — varies the capacity and head proportionally, as in Item 1.

3. Varying impeller width — proportionally varies the capacity.

4. Varying the pitch and number of vanes within the impeller changes the shape of the head-capacity curve. Spoke-like vanes or more vanes usually produce a flat curve.

5. Varying impeller and vane designs produce variations in head-capacity relationships. Narrow impellers with larger impeller-to-eye diameter ratios develop larger heads. The wide impellers with low diameter ratios are used for large flows at low heads.

The changes in speed and impeller diameters are reflected in pump performance as follows:

$$\frac{rpm_1}{rpm_2} \text{ or } \frac{impeller\ dia_1}{impeller\ dia_2} = \frac{gpm_1}{gpm_2} = \left(\frac{head_1}{head_2}\right)^{1/2} = \left(\frac{bhp_1}{bhp_2}\right)^{1/3}$$

or

Capacity varies directly,
Head varies as the square,
Bhp varies as the cube of speed or impeller diameter change.

The performance of a centrifugal pump is affected when handling viscous fluids. The effects are a marked increase in brake horsepower and decrease in head, capacity and efficiency *(Fig. 16)*.

POWER AND EFFICIENCY

In pump operation two power requirements may be evaluated, liquid power and actual power (brake horsepower that takes into account the pump efficiency). Liquid power is the product of the weight of the liquid pumped (gpm), pump head (ft), and the conversion factors. Brake horsepower is the actual power output of the driver, pump input, or liquid power divided by pump efficiency. Pump efficiency is the relation between the liquid (theoretical) power and the actual mechanical power input (a greater amount due to machine losses). The efficiency is expressed as a decimal. This should not be confused with driver efficiency since the latter is the relation between the output of the driver and the energy input to produce the power to drive the pump and compensate for losses within the driver.

$$Bhp = \frac{U.S.\ gpm \times pump\ H\ (ft) \times sp\ gr}{3960 \times percent\ pump\ eff}$$

where:

3960 = 33,000 (ft lb) /8.33 (lb/gal water at 1.0 sp gr), used to convert to horsepower.

sp gr = specific gravity of liquid.

Viscosity* of the liquid pumped affects the friction losses and therefore the pump horsepower requirements.

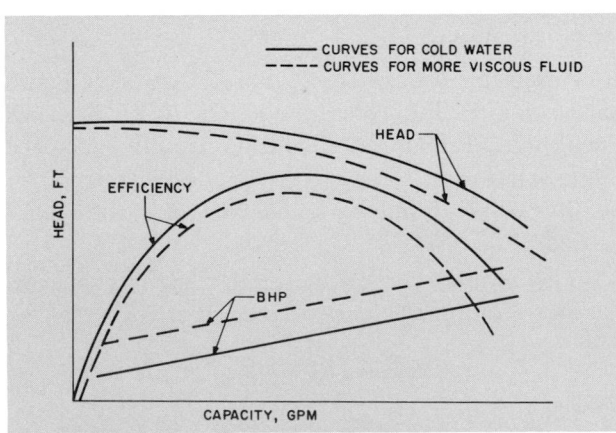

Fig. 16 — Influence of Viscosity on Pump Performance

*Viscosities and specific gravities for various brines at mean brine temperatures are found in *Part 4*. Viscosities must be expressed in consistent units.

CENTRIFUGAL PUMP AND SYSTEM

SYSTEM HEAD

A flow of liquid within any system of piping including fittings, valves and heat exchangers requires a system head consisting of a velocity head (usually insignificant) and friction head, and must overcome a static head. Thus in any piping system the system head is the algebraic sum of the static head on the pump discharge minus the static head on the pump suction plus the friction losses thru the entire system of fluid flow. With an increase in flow the friction losses increase approximately as the square of the flow; when plotting head against capacity flow, a parabolic head curve is formed *(Fig. 17)*.

OPERATION IN A SYSTEM

A given centrifugal pump operates along its own head-capacity curve. At full capacity flow the operating point falls at the crossing of the pump head-capacity curve and the system head curve *(Point 1, Fig. 17)*. If the pump is throttled, the operating point moves up the head-capacity curve *(Point 2)*; if it is desired to obtain greater flow to operate down the head-capacity curve *(Point 3)*, the path of flow in the system must be eased to reduce the friction losses. Otherwise the pump must be either speeded up or the impeller increased in diameter. Then a new head-capacity curve is established *(Point 4)*. The engineer must carefully analyze the system and select the pump from the manufacturer's performance head-capacity curves.

If the system head is overestimated and the pump is selected with a high head-capacity curve, unfortunate results may follow. The pump will operate on its head-capacity curve to produce an increased flow

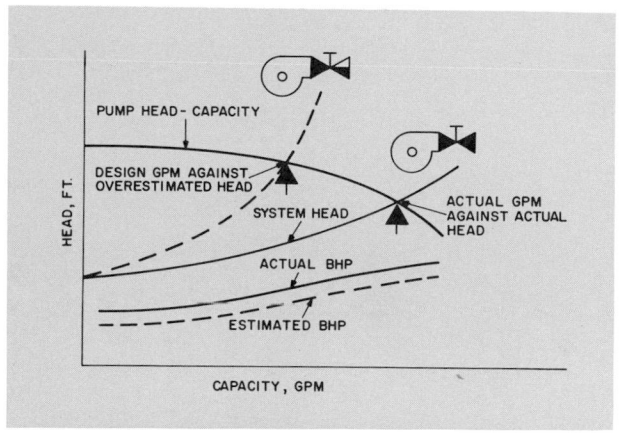

FIG. 18 — EFFECT OF OVERESTIMATING PUMP HEAD

at decreased head and increased horsepower demand *(Fig. 18)*. The system head should always be calculated without undue safety factor extention or as close as practical to the true values to eliminate possible waste of horsepower or possible overload of pump motor with an unvalved system.

The true evaluation of system head is specially important when designing a system with pumps in parallel or in series.

PARALLEL OPERATION

The operation of pumps in parallel results in multiple capacity against a common head *(Fig. 19)*. This type of application is for a system requiring high capacity with a relatively low head or for variable systems where a number of small pumps handle the load with one or more pumps shutting down as required. The pumps should have matched characteristics. Drives should have ample power to avoid overloading when operated singly.

SERIES OPERATION

The operation of pumps in series results in multiple head with a common capacity *(Fig. 20)*. This type of application is for systems requiring a high head and a relatively low capacity. Careful consideration of fluid flow must be made to safeguard the booster pump. Normally a series flow is provided for by a multi-stage pump.

HIGH BUILDINGS

The operation of pumps at the base of high buildings requires an analysis of pressures on the discharge and suction sides of the pump. The static head of the fluid in the piping system plus the head developed in the pump may necessitate the use of 250 lb pipe and fittings and even a reinforced pump casing.

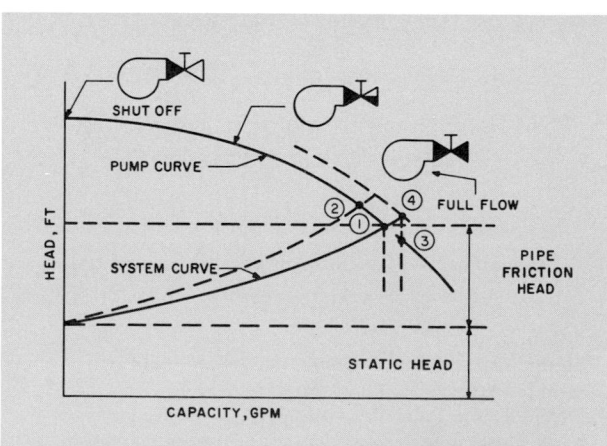

FIG. 17 — CROSSOVER POINT OF PUMP AND SYSTEM
CURVES

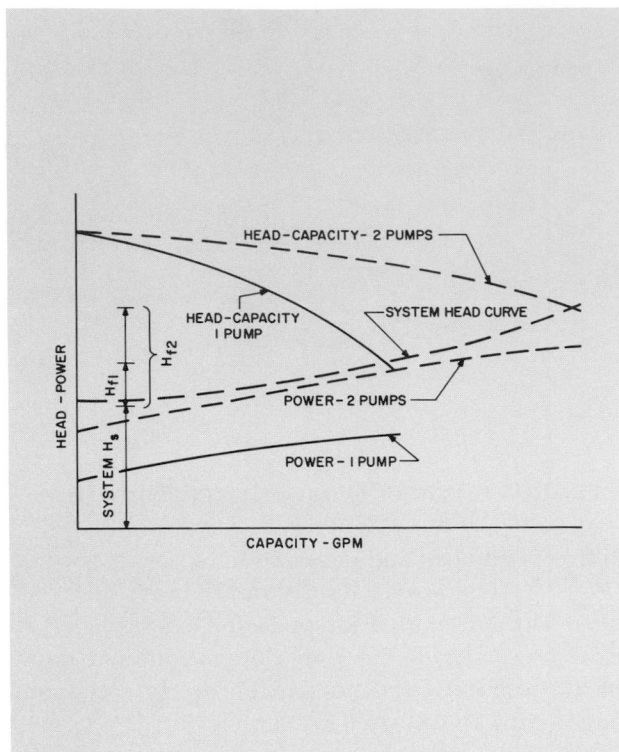

FIG. 19 — TWO EQUAL CAPACITY PUMPS, PARALLEL
OPERATION

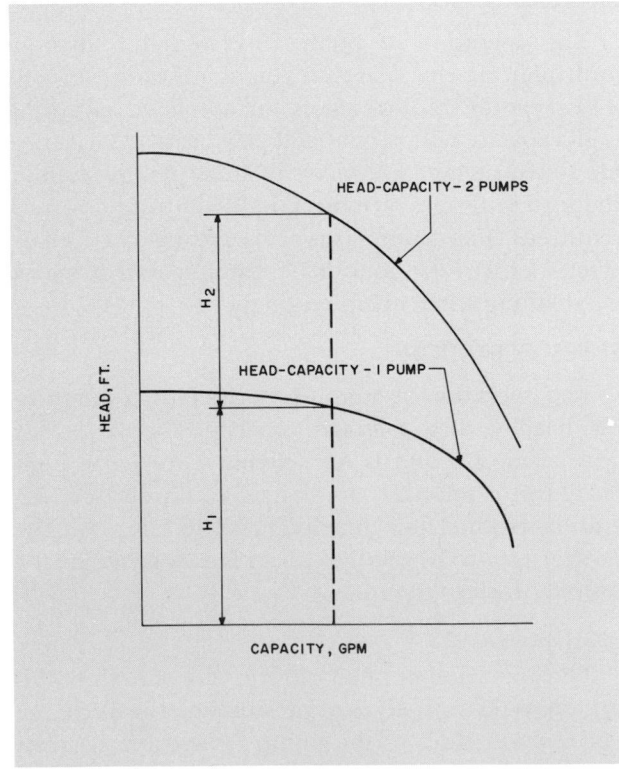

FIG. 20 — TWO EQUAL CAPACITY PUMPS, SERIES
OPERATION

WORKING PRESSURE

Pump casing working pressure is the total head developed by the pump to overcome friction losses of the system plus the suction static head minus the friction losses in the pump suction line, from the junction of the expansion tank line to the pump suction. The problem outlined in *Fig. 21* and its solution serve as an example.

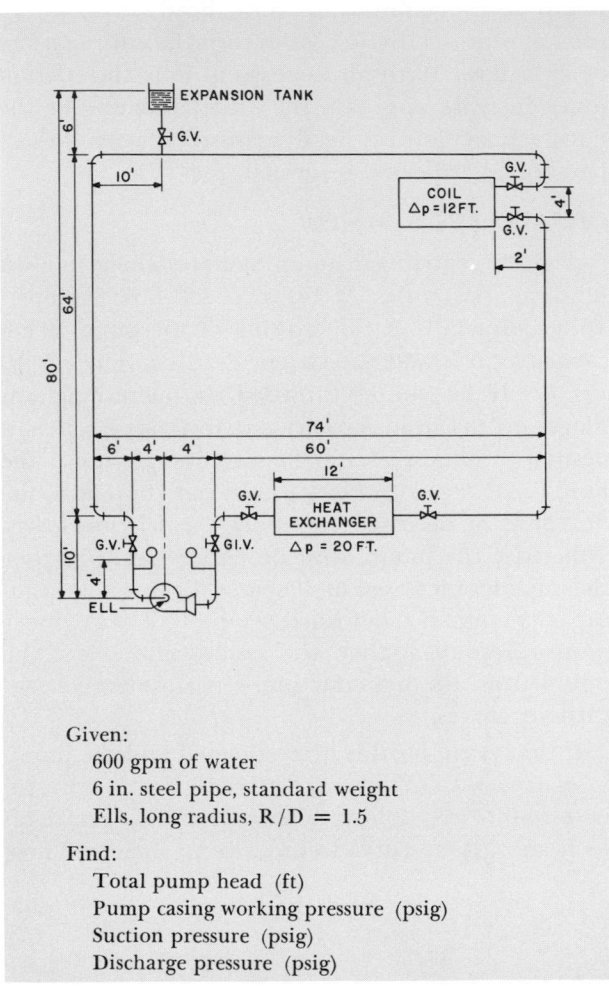

Given:
   600 gpm of water
   6 in. steel pipe, standard weight
   Ells, long radius, R/D = 1.5

Find:
   Total pump head (ft)
   Pump casing working pressure (psig)
   Suction pressure (psig)
   Discharge pressure (psig)

FIG. 21 — PUMP SYSTEM TOTAL HEAD AND PUMP
CASING WORKING PRESSURE

Example 1 — Calculation of Total Head and Working Pressure

Solution:
(based on data in *Part 3, Chapter 2, Water Piping*)
   Friction head H_f, suction line
   (from expansion tank to pump)

Straight pipe	= 94 ft
Five ells	= 50 ft
One gate valve	= 7 ft
Suction head total	= $\overline{151}$ ft × 2.35 ft H_f/100 ft
	= 3.54 ft

Friction head H_f, discharge line
(from pump to expansion tank)

Straight pipe	=	190 ft
One ⅔ enlargement	=	9 ft
Six ells	=	60 ft
Four gate valves	=	28 ft
One globe valve	=	170 ft
Discharge head total	=	$\overline{457}$ ft × 2.35 ft H_f/100 ft
		= 10.75 ft

Pump system total friction losses H_f

Suction piping	=	3.54 ft
Discharge piping	=	10.75 ft
Heat exchanger	=	20.00 ft
Coil	=	12.00 ft
Pump head total	=	$\overline{46.29}$ ft

Pumps casing working pressure

Static head	=	80.00 ft
Less suction line H_f	=	3.54
Subtotal	=	$\overline{76.46}$ ft
Plus system H_f	=	46.29 ft
Working pressure	=	$\overline{122.75}$ ft × 2.31 ft per psi
		= 53.1 psi

Suction pressure gage reading

Static head = 80 − 4		= 76.00 ft
Less H_f of straight pipe*	= 86 ft	
3 ells	= 30 ft	
1 gate valve	= $\underline{7}$ ft	
86 + 30 + 7 = 123 ft		
Less suction head = 123 ft × 2.35 ft H_f/100 ft		
	= −2.89 ft	
	Net = $\overline{73.11}$ ft	

Suction pressure = 73.11 ft/2.31 ft per psi = 31.7 psi

Discharge pressure gage reading

Suction pressure gage plus pump system total H_f		
= 73.11 + 46.29	= 119.40 ft	
Less H_f of straight pipe†	= 16 ft	
1 enlargement	= 9 ft	
3 ells	= $\underline{30}$ ft	
16 + 9 + 30 = 55 ft		
Less discharge head = 55 ft × 2.35 ft H_f/100 ft =		
	−1.29 ft	
	Net = $\overline{118.11}$ ft	

Discharge pressure = 118.11 ft/2.31 (ft per psi) = 51.2 psi

NOISE

The centrifugal pump is inherently a relatively quiet machine. However, in the case of a motor driven pump there are possible motor disturbances such as motor fan, bearings and magnetic noise (*Chapter 2*), in addition to the normal hydraulic and mechanical disturbances originating in the pump.

A fixed frequency vibration (rpm times the number of blades divided by 60) may be set up by a pump using too large an impeller. Sometimes, to produce quiet operation a recommendation is made to use an impeller diameter 10–15% smaller than the largest size that will fit in a given pump casing.

*Distance between expansion tank and gage.
†Distance between gages.

Operation of the pump under conditions of insufficient NPSH must be avoided to preclude formation or aggravation of cavitation and noise. However a small amount of cavitation in a condensate pump operation is permissible.

Well designed pumps operating at either 1750 or 3500 rpm rotative speed may be used.

A pump frequency may coincide with a corresponding frequency in the piping system or building structure; such telegraphing noise must be avoided.

PUMP SELECTION

Pumps are selected from manufacturer's performance curves (*Fig. 22*). Most standard pumps are designed to operate at maximum efficiency about midway on the head-capacity curve. Selecting a pump at the maximum efficiency point or slightly to the left materially assists in minimizing problems of noise and vibration. A selection too far to the right of the efficiency point may lead to cavitation due to an increase in required NPSH.

Pump efficiency is not the only selection criterion; quiet operation, lowest first and operating costs and close conformance to actual needs are parallel objectives.

MOTOR SELECTION

The horsepower of the motor selected to drive a given pump must be equal to or greater than the brake horsepower called for at the operating point of the head-capacity curve. There is always the danger of a pump running away from the selected operating point and overloading the motor. In case of a nonoverloading pump-motor combination the

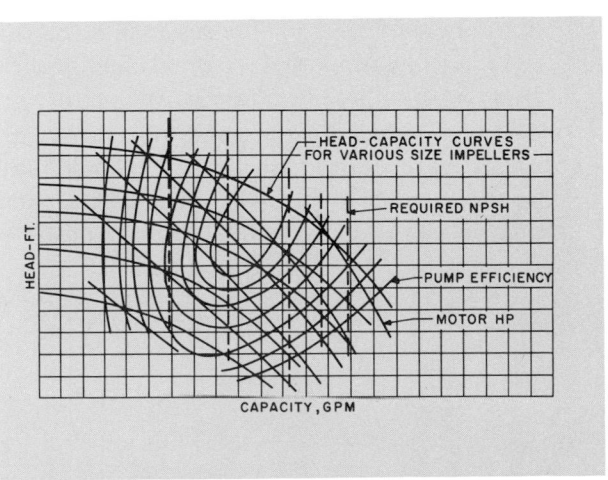

FIG. 22 — TYPICAL PUMP SELECTION CURVES

selected motor horsepower is always larger than the required brake horsepower; a safety margin is provided. If the pump is fitted with a nonoverloading impeller, it may be possible to select a motor of smaller horsepower. In either case the brake horsepower is the same.

CENTRIFUGAL PUMP INSTALLATION

INSTALLATION

There are several aspects in the application of a centrifugal pump that are external to the pump itself, yet important in the installation.

The suction piping at the centrifugal pump must be designed with care to avoid possible malfunctioning. The precautions to be taken (Part 3) are summarized:

1. The suction line approach to the pump should be as straight as possible and all elbows should have large radii.
2. A straight section of pipe should be attached to the suction inlet to allow the fluid to straighten out before entering the pump; this is especially true of double suction pumps.
3. The suction line should be one or two sizes larger than the pump inlet.
4. With an oversized suction line an eccentric reducer must be used, keeping the pipe flat on top.
5. Suction line should be airtight, with no high spots where air or gases may separate out of the fluid.
6. A check valve and a gate valve should be installed at the pump discharges of a multi-pump system. These should be installed in the order named to enable the check valve to be serviced without draining the discharge line.
7. Both the suction and discharge pipe connections must be supported separately and in such a way as to impose no strain on the pump.
8. The suction line for the pump operating with a negative static head (suction lift) should have no valves other than a foot valve. The suction line should be large and as direct as possible.

INSULATION

It is advisable not to insulate the pumps intended for chilled water (brine) service or hot water service. The refrigerant condensing water pumps need not be insulated. If a pump is insulated, the insulation should be applied in a form which permits disassembling of the pump for servicing without wrecking the insulation.

ISOLATION

Cork is not an effective isolation material for rotative speeds below 2000 rpm. Rubber-in-shear or corrugated rubber is useful on the ground floor installations. In more critical installations, on floors above occupied areas (especially those of executive offices, board rooms, libraries, hospital areas) steel spring isolation is recommended for isolation effectiveness approaching 100%. The concrete foundation of one to two times the machinery weight serves as a dampening mass and must be of reinforced construction. For piping isolation refer to Part 3.

FOUNDATION

Where requirements for isolation of a centrifugal pump are at a minimum (basements, outdoors, any remote location), a foundation is desirable to keep the pump off the floor or ground level.

STARTING

Unless the pump is self-priming, it must be primed before starting.

When starting the pump, the discharge valve is usually closed, then gradually opened so as not to run the risk of overloading the drive motor.

For more information there is a vast experience of pump manufacturers recorded in their catalog and handbook data and in innumerable authoritative articles written by engineers from these manufacturing concerns. A classical book on pump design and application is *Centrifugal and Axial Flow Pumps*, 2nd edition, by A. J. Stepanoff, 1957.

CHAPTER 2. MOTORS AND MOTOR CONTROLS

This chapter presents the characteristics of various motors that drive the equipment normally applied in air conditioning and refrigeration systems, the functions of motor controllers, and a brief discussion of the behavior of electric energy to produce mechanical power.

EQUIPMENT SERVED

The air conditioning and refrigeration systems include fans, pumps, and reciprocating, rotary and centrifugal compressors. To effect the transfer or compression of various liquids and gases such as air, water, brine or refrigerants, it is necessary to put this equipment into motion by prime movers, in this instance, electric motors. Apart from the factors (source of electric energy, speed, power) which are used in the selection of particular motor-starter combinations, a knowledge of the load torque characteristic for a particular driven equipment is the most fundamental requirement. The operational torque characteristics include those required for starting from rest, acceleration, and for full load running. The starting or locked rotor torque is the initial turning effort for bringing the driven equipment from standstill into motion; the acceleration torque is the developing of this motion into operating speed in an alloted time. The full load torque is the sustained effort by the motor to maintain the driven equipment in motion under the work load. The driven equipment torque requirements must be matched with a drive complex (motor and its control) of the proper torque and current characteristics.

NORMAL OR HIGH TORQUE MOTORS

With the exception of the reciprocating and rotary compressors, the equipment considered is of the centrifugal type (fans, refrigeration compressors, pumps) operating at starting within a system of high and low sides that are equalized. Ordinarily no specific requirement of starting torque other than normal is needed. At times with large centrifugal compressors or fans there exists a pull-up torque problem because of the large rotational inertia

(Wk^2)* of the massive impellers or wheels. The pull-up minimum torque must exceed Wk^2. A possible problem may exist when small equipment is driven by an oversized motor with overpowering torque; this equipment may be damaged because of excessive acceleration or torque applied.

The duration of acceleration to full speed at full voltage is usually 1 to 3 seconds. With open centrifugal compressors using a standard integral oil pump, it should be at least 8 seconds, permitting oil to reach all the lubricated surfaces before high speed is developed. When an auxiliary oil pump is used, it should be at least 5 seconds to prevent excessive stress concentration on the keyway of the compressor shaft.

The present methods of starting the larger reciprocating compressors with cylinders either fully or partially unloaded permit the use of normal starting torque motors. The smaller, hermetically driven compressors are started fully loaded with a requirement of high starting torque.

The application of rotary compressors as low pressure differential boosters in a refrigeration system does not present any unusual requirement of starting torque; therefore, a normal starting torque motor is applicable.

Centrifugal, propeller or axial fans may be either belt or direct connected to electric motor drives. In all cases the torque requirements are normal. The fans should start smoothly and without undue noise.

The fan cfm is directly proportional to speed; the system resistance varies as the square of speed; and the horsepower required varies as the cube of speed.

Because the gas density at the suction of a centrifugal compressor usually increases during shutdown, the torque requirements of centrifugal compressors do not necessarily follow the fan laws. Normal torque motors are usually used for driving centrifugal compressors; however, they must be started either with an almost completely closed suction damper or pre-rotation guide vanes to prevent the increased gas density from imposing an excessive overload at full speed.

*In this expression W = weight of the body; k = radius of gyration.

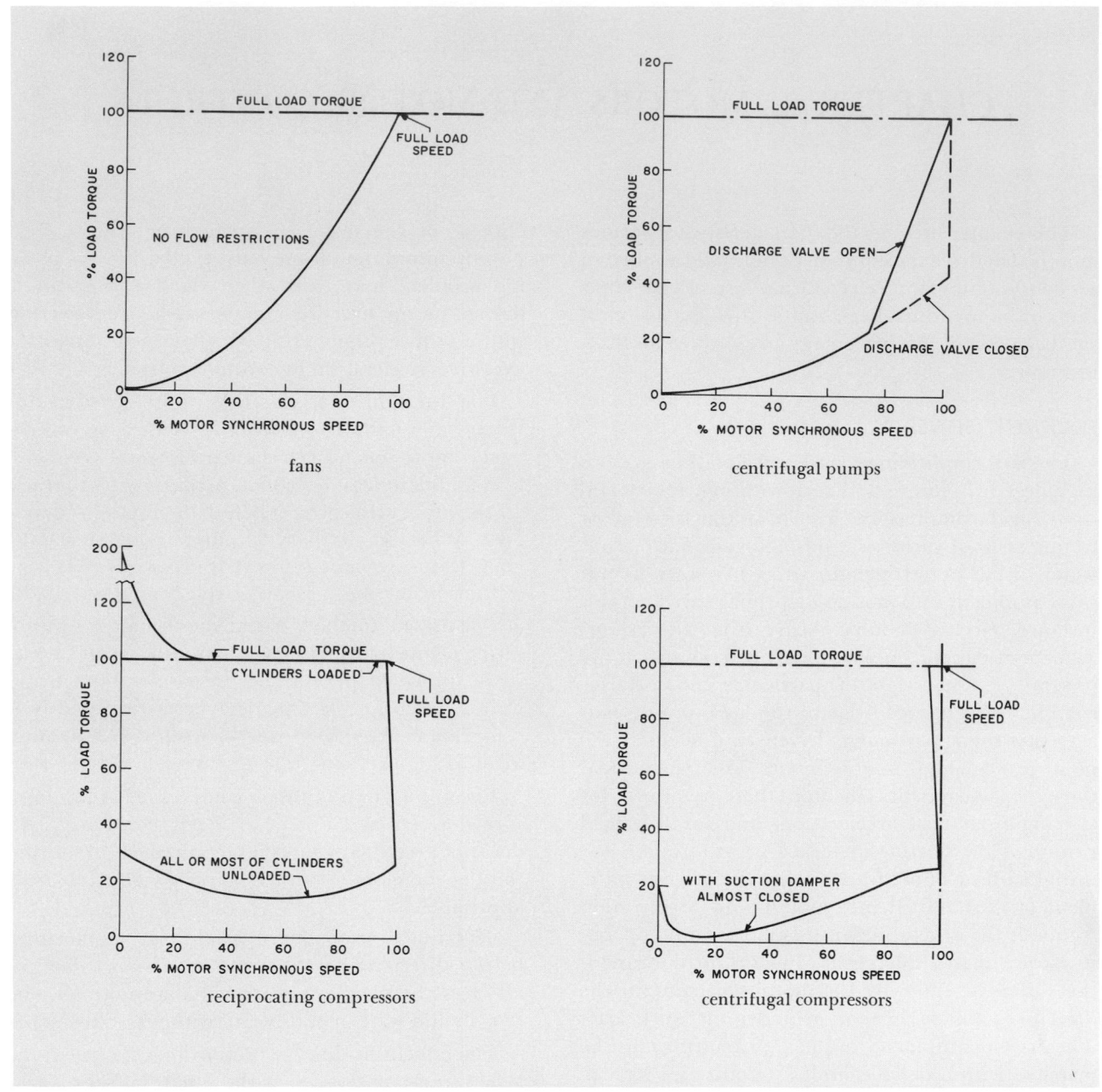

FIG. 23 — TORQUE CHARACTERISTICS

Figure 23 illustrates typical torque requirements of fans, pumps, and reciprocating and centrifugal compressors. Pumps usually have a check valve in the discharge; therefore, the break exists in the torque requirement curve. There is a decided difference in torque requirements of reciprocating compressors, whether they are started with cylinders loaded or unloaded. The breakaway and acceleration torques shown for centrifugal compressors are determined by the friction and the Wk^2.

MOTORS OF MATCHING CHARACTERISTICS

The energy necessary to operate motors is available in two services: (1) direct current (d-c), unidirectional and at constant pressure, and (2) alternating current (a-c), alternating in pressure and direction. Since direct current is used infrequently except in certain industrial processes, special purposes, or in some remote communities, this text concentrates only on a-c electrical equipment.

TABLE 1 — SINGLE AND POLYPHASE A.C. MOTOR CHARACTERISTICS

| Motor Type | HP Rating | Speed Characteristics | Full Voltage | | Remarks |
			Starting Torque	Starting Current	
POLYPHASE					
Squirrel-cage induction	Small to large	Constant and multi-speed	High to normal	Low to normal	Most widely used for constant speed service
Wound-rotor	All	Constant or variable	High	Low	For applications requiring high starting torque and low starting current, or limited variation in speed control
Synchronous	Medium to large	Strictly constant	Normal to low	Low to normal	For constant speed service and where power factor correction is required
SINGLE-PHASE					
Capacitor-start, induction-run	Small*	Constant	High	Normal	General purpose
Capacitor-start, capacitor-run	Small*	Constant	High	Low	High efficiency
Split-phase	Fractional	Constant	Normal	Normal	Least expensive of higher starting torque types, general purpose
Permanent split-capacitor	Fractional and small integral	Constant or adjustable varying	Low	Normal	Quiet, efficient; low running current; poor starting torque
Shaded-pole	Fractional	Constant or adjustable varying	Low	—	Inexpensive; poor starting torque; least efficient; high running current

*Up to 7.5 hp.

This chapter describes a-c motor drives and motor controllers which match the torque characteristics of the driven equipment. The motor sizes discussed are fractional (up to 1), integral (above 1) and medium to large horsepower. The types of motors include single phase capacitor, polyphase squirrel-cage induction, and synchronous motors. However most applications use the simple polyphase squirrel-cage induction motor. Major characteristics of a-c motors are shown in *Table 1*.

Each type in *Table 1* offers specific motor characteristics of starting torque, starting current and operating speed to meet the requirements of various industrial applications. The motors used in air conditioning systems fall into two groups: (1) single-phase motors for small systems, and (2) polyphase squirrel-cage induction motors for large systems. Occasionally wound-rotor and synchronous motors may be used for refrigeration centrifugal compressors.

In the medium sizes of squirrel-cage induction motors Type B (*Table 3*) is preferred, having normal starting torque and low starting current; these characteristics conform to normal equipment torque requirements and to the regulations of power distribution by the utilities. These regulations are directed toward leveling the demand for current by the consumers so that at any one moment of power flow there is no extreme dip that might cause flickering of the lights and other anomalies along the electric system.

EQUIPMENT AND MOTORS

The purpose of a motor is to supply mechanical power to drive equipment. The preceding discussion states that the motor must inherently possess necessary torque, and must not affect the power line adversely by its current requirements. *Table 2* lists the major varieties of equipment and motors that are applied in each case.

TABLE 2 — EQUIPMENT AND MOTORS

EQUIPMENT	MOTORS	
	Single Phase	**Approximate Horsepower Range**
Room fan-coil units	Shaded-pole (air-over)	Fractional up to ⅓ hp
Small fans with any equipment and **Hermetic compressors**	Permanent, split-capacitor (air-over, self-ventilated or refrigerant-cooled) Capacitor-start, capacitor-run (air-over, self-ventilated or refrigerant-cooled)	Fractional and integral up to 5 hp
	Polyphase	**Approximate Horsepower Range**
Fans, pumps and centrifugal compressors	Squirrel-cage induction (constant speed)*	Integral above 1 hp
Open reciprocating compressors	Squirrel-cage induction (constant speed)**	Integral above 1 hp
Centrifugal compressors	Wound rotor (varying speed)	Large
Centrifugal compressors, Pumps	Synchronous (constant speed)	Large, medium
Hermetic reciprocating compressors	Hermetic (refrigerant-cooled)	Small to medium
Hermetic centrifugal compressors	Hermetic (refrigerant-cooled)†	Medium to large
Absorption machine solution pumps	Hermetic (solution-cooled)††	Small

*NEMA Design B, Insulation Class A (Tables 3 and 4).
**NEMA Design C, Insulation Class A (Tables 3 and 4).
†At times wound for star-delta starting.
††Insulation Class F (Table 4).

Fans may be either included with a built-up system or coupled directly in any type of equipment complex containing coils, filters and spray chambers. The composite equipment may include fan-coil units, self-contained packages, condensing units, heat pumps, unit heaters and cooling towers.

The motors may be encased in any enclosure (open type to explosion — or weatherproof type), depending on the design of equipment, application and customer desire.

STANDARDS AND CODES

Motor manufacturers are guided by the Standards of the National Electric Manufacturers Association (NEMA). Motor installation should conform to the utility regulations and local codes and ordinances. NEMA standards for motors cover frame sizes and dimensions, horsepower ratings, service factors, temperature rises, and performance characteristics.

Reference is also made to the National Electric Code (NEC) sponsored by the National Fire Protection Association, Underwriters' Laboratories, Inc., AIEE Standards, and federal and military standards when applicable.

The standard ambient conditions for normal

TABLE 3 — CHARACTERISTICS OF SQUIRREL-CAGE INDUCTION MOTORS, NEMA DESIGN A, B AND C

NEMA DESIGN	TORQUE (Percent of Full Load)		STARTING CURRENT	SLIP	APPLICATION
	Starting	**Breakdown**			
A (Rarely used)	Normal 100 to 275	Higher than Design B	Normal	Normal	On moderately easy-to-start loads requiring slightly more than full load starting torque, low slip and moderately high breakdown torque to sustain occasional overloads (Design A higher than Design B).
B	Normal 100 to 275	200 to 300	Normal	Low	For fans and blowers, centrifugal pumps and compressors, reciprocating compressors (started unloaded).
C	High 200 to 250	190 to 225	Normal	Low	On hard-to-start loads requiring high starting torque but not high overload demands. Used with open reciprocating compressors (started loaded), rotary pumps.

motor operation are assumed to be a location with an unobstructed circulation of clean, dry air at a temperature of 40 C (104 F) and at an altitude not exceeding 3300 feet (1000 meters). At higher altitudes the rarefied air produces insufficient cooling.

Special insulation must be provided for motors applied in the tropics because they may be exposed to excessive ambient temperatures, humidity and fungus.

The motors and motor controls utilized to drive the equipment used in air conditioning and refrigeration systems are described in the following sections.

POLYPHASE A-C MOTORS

The operating characteristics inherent in various polyphase alternating current motors (squirrel-cage, wound-rotor and synchronous) are shown in *Table 1*. The text that follows describes the mechanical construction, electrical behavior and application.

Fundamental relations and terminology used in the field of electricity are to be found in *Fundamental Relations*.

SQUIRREL-CAGE INDUCTION MOTOR

The most widely used motor is the squirrel-cage induction type. It is simple in construction, easy to start, and in combination with the starting equipment it is the least expensive in dollars per horsepower; it is efficient and has a reasonably good power factor. *Figure 24* illustrates a representative performance of a standard squirrel-cage induction motor.

The two basic components of a squirrel-cage induction motor (*Fig. 25*) are the stator (stationary part) and the rotor (rotating part) reminiscent of a squirrel cage.

A stator consists of a laminated iron core within and around the inner rim of which are distributed insulated windings. These primary windings are three in number for three-phase current and two for two-phase current. The arrangement of primary windings is controlled by the line voltage, number of phases, and number of poles. The power source is connected to the primary windings.

The rotor consists of a laminated iron core with bar windings in various shaped slots around the periphery of the core. The bar windings at a design resistance specific to various NEMA design (*Table 3*) are interconnected (short-circuited) with rings. The rotor mounted on a shaft is in turn mounted in bearings. There are no direct power connections to the

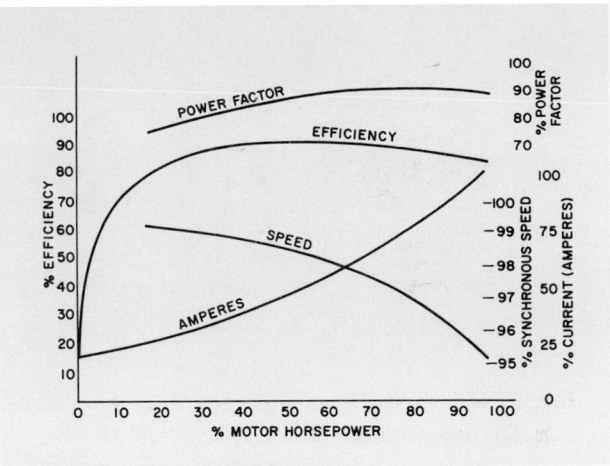

FIG. 24 — TYPICAL PERFORMANCE, STANDARD SQUIRREL-CAGE INDUCTION MOTOR (MEDIUM SIZE)

secondary windings. The current in the rotor is induced, therefore the name induction motor.

Thru a specific design arrangement of windings within the stator and with the positive and negative alternations of the current imposed, a magnetic field known as flux is established. This magnetic field has a definite polarity changing with each alternation in current. With an arrangement of windings to form two poles within the stator, the resultant flux leaves the stator at one point (north) and re-enters at an opposite point (south) (*Fig. 26*). During the next alternation the polarity changes; thus the flux is rotated. In a four-pole windings arrangement the flux leaves the stator at two opposite points (both north) and re-enters at two opposite points (both south). This flux pulsates with the rise and fall of current; these pulsations and alternating polarity in the stator poles produce the rotating magnetic field. As the stator-created flux revolves, it cuts the bars or coils of the rotor; in so doing, it induces a voltage within the rotor circuits (transformer-like action).

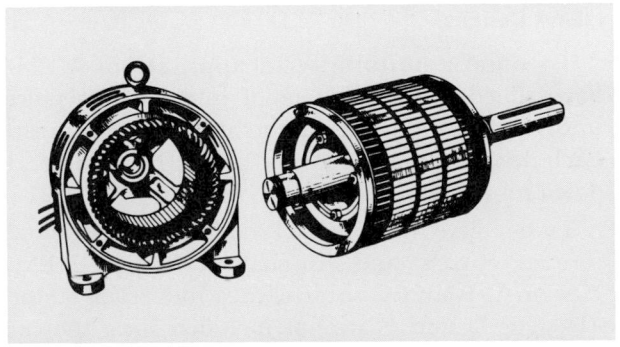

FIG. 25 — STATOR AND ROTOR, SQUIRREL-CAGE INDUCTION MOTOR

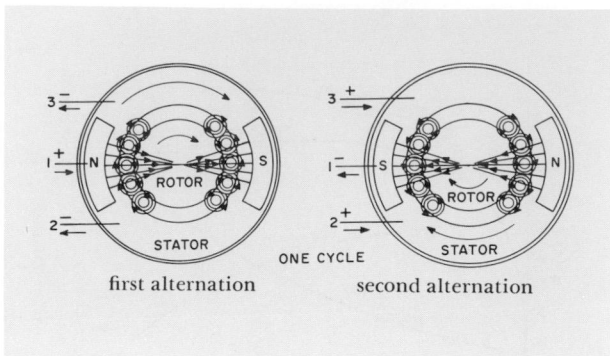

FIG. 26 — INITIAL FLUX IN STATOR, INDUCED FLUX
IN ROTOR, SQUIRREL-CAGE INDUCTION MOTOR
(2-POLE, 3-PHASE)

Current generated in the rotor bars sets up a magnetic field of its own *(Fig. 26)*. The interaction of forces created produces the turning torque that accelerates the rotor and in turn puts in motion the external load to which the motor is connected.

The speed of the stator-created rotating flux is the synchronous speed (rpm) of the motor. If a current is impressed on the stator and if the rotor is held at a standstill, a magnetic field revolves past the rotor conductors at a synchronous speed, generating a maximum current in rotor conductors. If the rotor is driven at the same synchronous speed, then its conductors do not cut the stator flux, and no current is generated within the rotor. Since there is no current in the rotor conductors, there is no torque, no turning effort developed by the rotor. However the windage, friction and applied load create a slowing down of the rotor below the synchronous speed of the flux, allowing the motor to develop torque. The difference (by design) in rotor speed is the slip of the motor. The current generated within the decelerated rotor is sufficient to produce the torque necessary to rotate the driven equipment.

NEMA Designs

To obtain uniformity in applications NEMA has defined specific designs of integral horsepower squirrel-cage induction motors up to 200 hp in size. Each design conforms to specific starting and breakdown torque, starting current and slip.

Table 3 gives the ranges of starting and breakdown torques, current and slip characteristics of NEMA Design A, B and C squirrel-cage induction motors. (Designs D and F are not included since they are not used in air conditioning applications.) The table also cites the equipment to which the motor designs A, B and C are applicable.

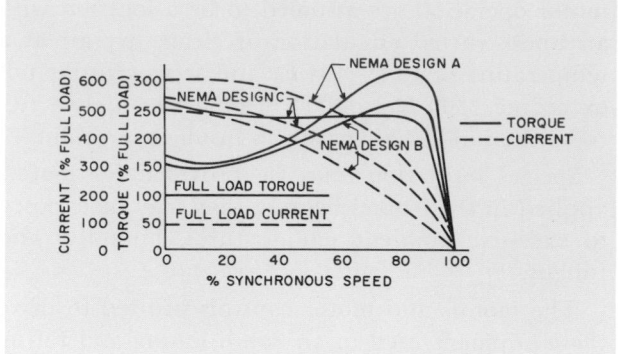

FIG. 27 — TYPICAL CURRENT AND TORQUE CURVES FOR
NEMA DESIGNS, SQUIRREL-CAGE INDUCTION MOTOR

NEMA Design C motors usually have double windings in the rotor. The outer slots are utilized to provide high resistance at starting, creating the high locked rotor torque and moderate starting current. The low resistance inner slots carry most of the induced current during full load, thus offering low slip and high efficiency.

Figure 27 demonstrates the shape of torques developed and current used by NEMA Design A, B and C squirrel-cage induction motors. The significance of the magnitude of starting torques is their ability to overcome driven equipment inertia; the significance of starting currents is their capacity to affect adversely the power supply. *Chart 1* illustrates approximate efficiencies of these motors.

Manufacturers must be consulted for motor data for a given application.

Special Winding Arrangements

To accommodate a power company requirement of reduced current draw on the power line at start-up, there are two varieties of squirrel-cage induction motors. These have a special arrangement of primary stator windings, namely part-winding *(Fig. 44)* and star-delta motors *(Fig. 45)*.

**CHART 1 — APPROXIMATE FULL LOAD
EFFICIENCIES, SQUIRREL-CAGE INDUCTION
MOTORS**

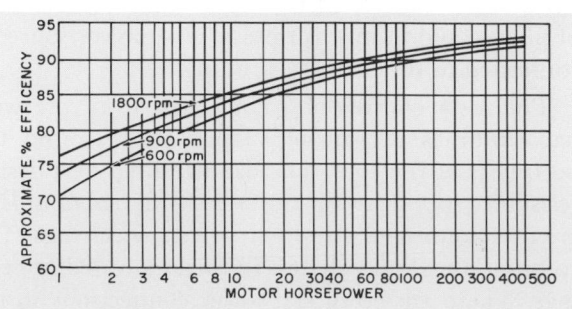

The part-winding type is a polyphase motor with two or more circuit windings. It may be a dual voltage (110/220 or 220/440) motor; the suitability should be checked with the manufacturer (see also *Motor Controls*). The starter is arranged to start the motor on one set of windings and after a time delay to apply all windings across the line. Such a combination provides motor starting on about 65% of full voltage locked-rotor amperes while the locked-rotor torque available is about 48% of full voltage starting torque. This arrangement may be less expensive than the use of a standard motor and a reduced voltage starter. A reciprocating compressor, either open or hermetic, may be equipped with part-winding motors.

The star-delta type is a motor with stator windings in delta arrangement with additional leads to circuit the windings in star arrangement; the leads brought out from each end of each winding may be connected either in star or delta by the starter. The motor is started on star arrangement of windings and is fully operated on delta arrangement of windings. The starting locked-rotor current is about 33% of the maximum, and the torque is about 33% of full voltage starting torque; these conditions are suitable for low starting torque applications such as centrifugals with inlet closed. However the star-delta (wye-delta) arrangement has been adopted for hermetic centrifugal compressors.

Multi-Speed Motors

At times there is a need for two or more fixed steps in speed change for the operation of driven equipment (fans). The polyphase squirrel-cage induction motors may be obtained in two-, three- (rarely used) or four-speed arrangements. The multi-speed motor operation is obtained by either multiplying or rearranging the stator windings. Depending on the complexity of these design requirements, the multi-speed motor may be contained either in the same frame size as the single-speed motor of the same horsepower rating or in a frame larger by at least one size (contributing to an increase in cost).

The two-speed motors have either single consequent windings (separate leads) or two independent windings (double superimposed) in the stator to obtain respectively 2:1 or 3:2 ratios in speed change, and variable or constant torque characteristics.

The four speed motors are usually in 1800/1200/900/600 rpm combinations for 1800 rpm synchronous speed, and 1200/900/600/450 rpm combinations for 1200 rpm synchronous speed. Three- and four-speed motors usually have consequent pole separate windings.

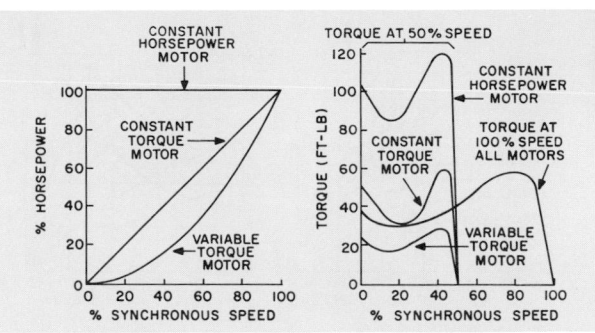

FIG. 28 — TYPICAL HORSEPOWER AND TORQUE CURVES, MULTI-SPEED SQUIRREL-CAGE INDUCTION MOTOR

The multi-speed polyphase squirrel-cage induction motors are constructed according to these standards:

1. Motors producing variable torque, that is decreasing torque with decreasing speed. Horsepower delivered varies as the square of the speed. These motors may be used to drive centrifugal pumps or fans.

2. Motors producing constant torques thruout the range of speeds. Horsepower varies directly as speed. These motors are suitable for reciprocating compressors.

3. Motors producing constant horsepower at all speeds. Torque increases with a decrease in speeds. These motors have no particular application with equipment in air conditioning systems.

Figure 28 illustrates the horsepower and torque delivery of three types of multi-speed induction motors.

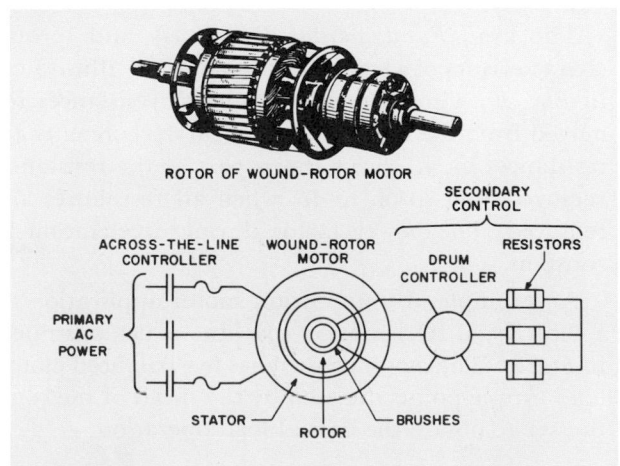

FIG. 29 — WOUND-ROTOR MOTOR ROTOR AND CONTROL ASSEMBLY

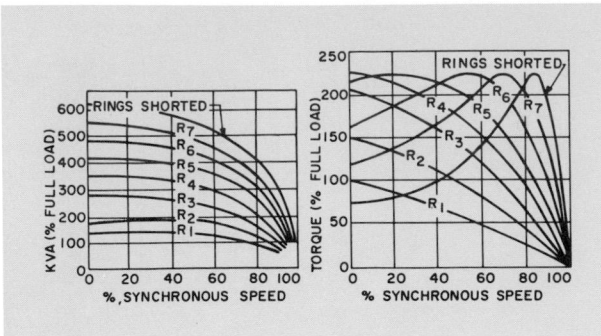

FIG. 30 — TYPICAL CURVES OF KVA AND TORQUE,
WOUND-ROTOR MOTOR

WOUND-ROTOR MOTOR

Another variety of squirrel-cage induction motor that provides higher torque at start-up and a range of adjustable reduced speed operation is the wound rotor or slip-ring motor. The reduced speed operation is of particular interest since it permits part load operation of refrigeration centrifugal compressors. Next to the turbine drive with its infinite speed variations, the wound-rotor motor speed changing is an efficient capacity control *(Part 7, Chapter 2, Fig 4)*.

The wound-rotor motor is constructed with three-phase windings in the rotor. One end of each rotor phase is brought out to a slip ring on the rotor shaft. Stationary brushes in contact with the slip rings are connected to an external secondary circuit into which any desired amount of resistance may be introduced to obtain the needed speed *(Fig. 29)*. With the slip rings shorted (external resistance totally excluded) the wound-rotor motor has speed and torque characteristics of the standard NEMA Design B squirrel-cage induction motor.

The kva *(Fundamental Relations)* and torque characteristics of a wound-rotor motor are illustrated in *Fig. 30*, with succeeding external resistances removed from the circuit. The R_1 curve represents all resistances in; R_2 curve shows part of the resistance removed, and so on to R_7 when all resistances are removed. The power factor during acceleration is constant.

An example of wound-rotor motor application to a centrifugal refrigeration machine is demonstrated in *Fig. 31*. The motor control has five balanced motor load torque points dictated by the speed of the compressor to obtain the partial load operation.

SYNCHRONOUS MOTOR

Synchronous motors are inherently and strictly constant speed motors. Their application is characterized by the high efficiency of conversion of electric energy into mechanical energy and by operating at either unity power factor or leading power factors, for example 0.9, 0.8. Their speed is unaffected by changes in voltage or load. Large horsepower synchronous motors at low speeds are simple and compact, less costly than the squirrel-cage induction motors of equivalent rating.

In construction the synchronous motor consists of a stator to which a-c power is applied, producing the primary revolving magnetic field and the rotor spider that contains field poles and amortisseur (damper) windings; these windings are similar to those used in induction motors. The amortisseur windings develop most of the starting and accelerating torque. At start-up the synchronous motor simulates the squirrel-cage induction motor operation; it depends on the arrangement of slots and windings.

There are several varieties of synchronous motors. This text discusses only the direct connected exciter motors *(Fig. 32)*. On the rotor shaft in addition to the two windings, there are collector rings and a d-c exciter. Another method of excitation is by a current furnished by a separate motor-generator set external to the motor. As soon as the rotor comes up to speed and runs with a slip of about 2–3%, a d-c field current from the exciter is applied to the field windings on the salient (projecting) rotor poles. The torque developed to pull the motor in step is called pull-in

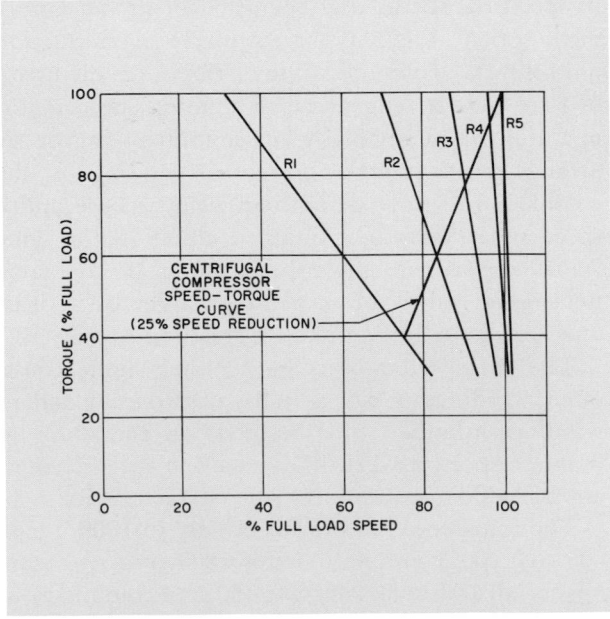

FIG. 31 — SPEED-TORQUE CURVE WITH FIVE BALANCED
POINTS, WOUND-ROTOR MOTOR

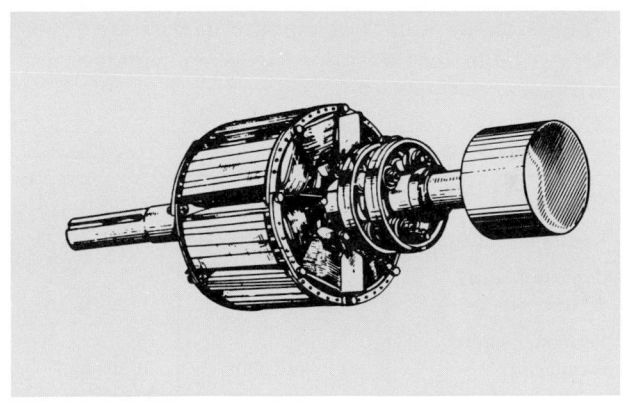

FIG. 32 — SYNCHRONOUS MOTOR ROTOR

torque. The magnetized rotor poles lock in step with the stator revolving magnetic field and the rotor revolves at synchronous speed. However if d-c field current is applied before the rotor reaches 97–98% of synchronous speed, the rotor may not synchronize, resulting in severe vibration and high pulsating input current.

When a mechanical load (resisting torque) is applied to the shaft of a synchronous motor operating at synchronous speed, a balancing counter-torque is developed and the rotor field poles tend to lag the stator magnetic field. Any increase in load is accompanied by an increase in lag angle, resulting in the motor developing its maximum pull-out torque. Any further increase in the mechanical load stops the motor. The normal pull-out torque is usually 150% of full-load torque with unity power factor motors and 200–250% with 0.8 leading power factor motors.

While driving its load a synchronous motor can have the a-c current input into its stator varied by changing the strength of the field excitation. This varies the power factor; it adjusts the armature current to be in phase with or leading the voltage at any given load. With a weaker field strength the power factor is less and the armature current lags the voltage. This is an abnormal operation of the synchronous motor. When the field is overexcited, the synchronous motor provides magnetization (kvar) in excess of its own requirements. This extra margin is fed into the power supply system.

This latter feature (Fig. 33) is especially desirable in systems where a considerable number of squirrel-cage induction motors are operating. The excess kvar produced by a synchronous motor is consumed by the induction equipment in the plant, thus correcting the electric system power factor.

Synchronous motors may be useful because of their inherent tendency to regulate the voltage of the power system. With a fall in line voltage the leading kvar of a synchronous motor is increased, raising the supply line voltage by the improved power factor of the line. Rising voltage in a line reverses the processes. This voltage regulation may be useful on the ends of long transmission lines, especially if a large inductive load is present.

HERMETIC MOTOR

Both centrifugal and reciprocating compressors are available in hermetically closed arrangements including motors. The refrigeration machine hermetic motors are in a separate class since they are cooled by either liquid or vapor refrigerant at temperatures much lower than the air used for cooling open motors. Such motors may operate with a higher temperature rise without exceeding the maximum temperature on which the rating of general purpose squirrel-cage induction motors is based.

Since their application is quite different, hermetic motors are usually not rated on a horsepower basis. They are identified by the full-load and locked-rotor currents; the significance of this identification becomes apparent when selecting controls.

Hermetic motor manufacturers furnish only the matched polyphase squirrel-cage induction motor less shaft, end shields and bearings (no enclosure). The compressor manufacturer assembles these stator-rotor matches with compressors within the same enclosures using proper bearings. The windings are properly insulated and well bonded, specially for larger size motors. The small integral motors are occasionally single-phase and, since sparking contacts cannot be used, they are the capacitor or resistance split-phase types with capacitors and switches mounted outside the compressor assembly.

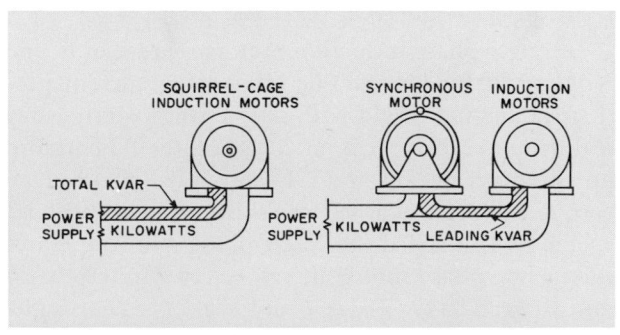

FIG. 33 — POWER FACTOR IMPROVEMENT WITH
SYNCHRONOUS MOTOR

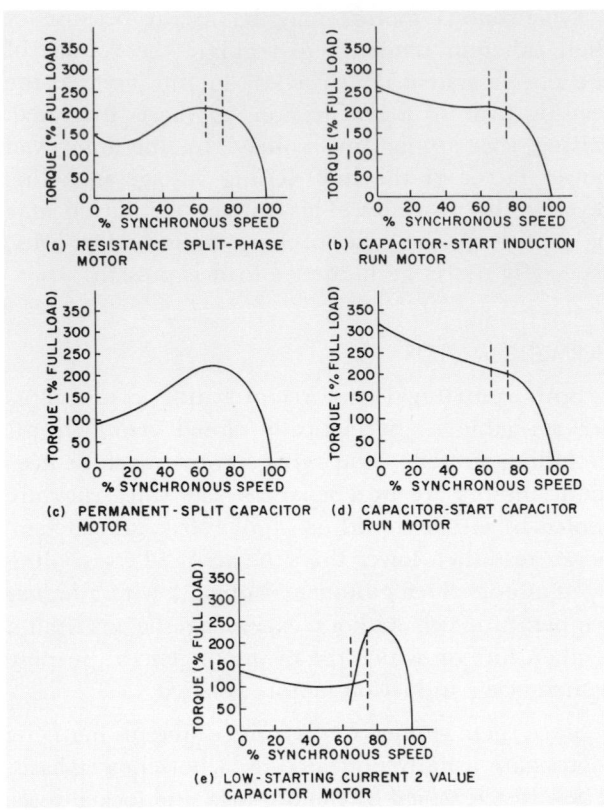

FIG. 34 — TYPICAL SPEED-TORQUE CURVES, SINGLE-
PHASE MOTOR

SINGLE-PHASE MOTORS

FRACTIONAL AND SMALL INTEGRAL HORSEPOWER MOTORS

Small integral motors are generally the polyphase type; however in many areas of extensive use power is available only in single phase. Polyphase motors have two or three separate and uniformly distributed windings in the stator, one for each phase. The current in these windings alternates continually and progressively to produce a revolving magnetic field, resulting in torque that turns the rotor.

In single-phase induction motors there is only one winding in the stator. The alternating current produces a magnetic field with alternating polarity, but it does not revolve; it is on "dead center." Therefore an auxiliary means must be provided to produce torque to start and accelerate the motors to full load speed. At full speed the single-phase motor operates like a polyphase motor in respect to slip, efficiency and power factor (*Table 1 and Fig. 34*). The single-phase motors are limited in overload capacity, and are confined to fractional and small integral sizes (1/100 to 6 horsepower).

The most popular single-phase motors are generally available in the following types, voltages and horsepower ratings:

MOTOR TYPE	VOLTAGE	HORSEPOWER
Shaded pole	115 and 230	1/100 to 1/4
Resistance split-phase	115 and 230	1/20 to 1/3
Capacitor-start, Induction-run	115 and 115/230	1/20 to 1.2
Permanent split-capacitor	115 and 230	1/20 to 5
Capacitor-start, capacitor-run	115 and 230	1/20 to 6

Resistance Split-Phase Motor (Fig. 35)

The resistance split-phase motor is the oldest arrangement of single-phase motors. The rotor has a squirrel-cage winding similar to a conventional polyphase motor. There are two stator windings, main and auxiliary. The main stator winding is made of heavy wire to provide low resistance and high reactance; the auxiliary torque-producing winding is made of fine wire to provide high resistance and hence a higher power factor. The latter winding is magnetically displaced from the main or running winding. This arrangement causes a displacement between the current in the main and phase windings to simulate a revolving magnetic field. With the current on, the rotor is caused to rotate. When a speed approximately 70% of full load speed is reached, a centrifugal switch mounted on the rotor and in series with the auxiliary winding removes this winding from the circuit. With this impetus the rotor comes up to speed and the motor operates as a regular squirrel-cage induction motor. The resistance split-phase motor has a low starting torque and takes

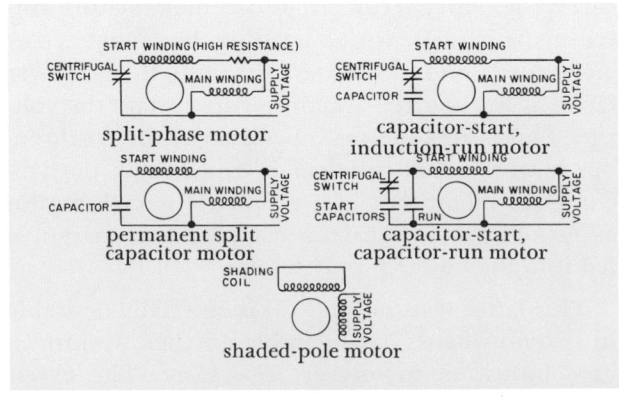

FIG. 35 — CONNECTION DIAGRAM,
SINGLE-PHASE MOTORS

a relatively large starting current, causing a flicker in the lights. The switch is sensitive to heat and must be protected whenever exposed to hot air or radiant heat. These motors are classed as general purpose motors and are applied to small propeller fans.

Capacitor-Start, Induction-Run Motor (Fig. 35)

To improve on the characteristics of the split-phase motor, there is a variety of capacitor motors that have higher starting torques. The capacitor-start, induction-run motor usually has an electrolytic capacitor added in the auxiliary winding circuit; the centrifugal switch is eliminated. As with split-phase motors, when the motor has come up to approximately 70–75% of full load speed, the capacitor and the phase winding are removed from the circuit by a voltage relay and the motor runs as a regular squirrel-cage induction motor, hence the name capacitor-start, induction-run. The displacement between the current in the main and phase windings is increased. As compared to resistance split-phase motors this feature produces increased starting and accelerating torques; the latter enables these motors to come up to speed faster. The increase in starting torque is due to the use of a low impedance capacitor. These motors may be used to operate small fans and blowers of heavier load requirements.

Permanent Split-Capacitor Motor (Fig. 35)

The permanent split-capacitor motor is similar to the capacitor-start, induction-run motor except that the capacitor and the auxiliary winding remain permanently in the circuit with the main winding after the motor is started. The capacitor may be oil-filled for continuous operation. This motor has no switch and therefore is simpler to operate. Its starting torque is low (about 45%) since the value of the capacitance is constant. The capacitance is higher than the normal compromise value; it is selected for running operation rather than for starting needs. The breakdown torque is high. Because of starting torque limitations these motors are not used in belt-driven applications. However they are efficient, quiet, and low in current requirements.

CHART 2 — POWER FACTORS OF ELECTRICAL EQUIPMENT AT FULL LOAD

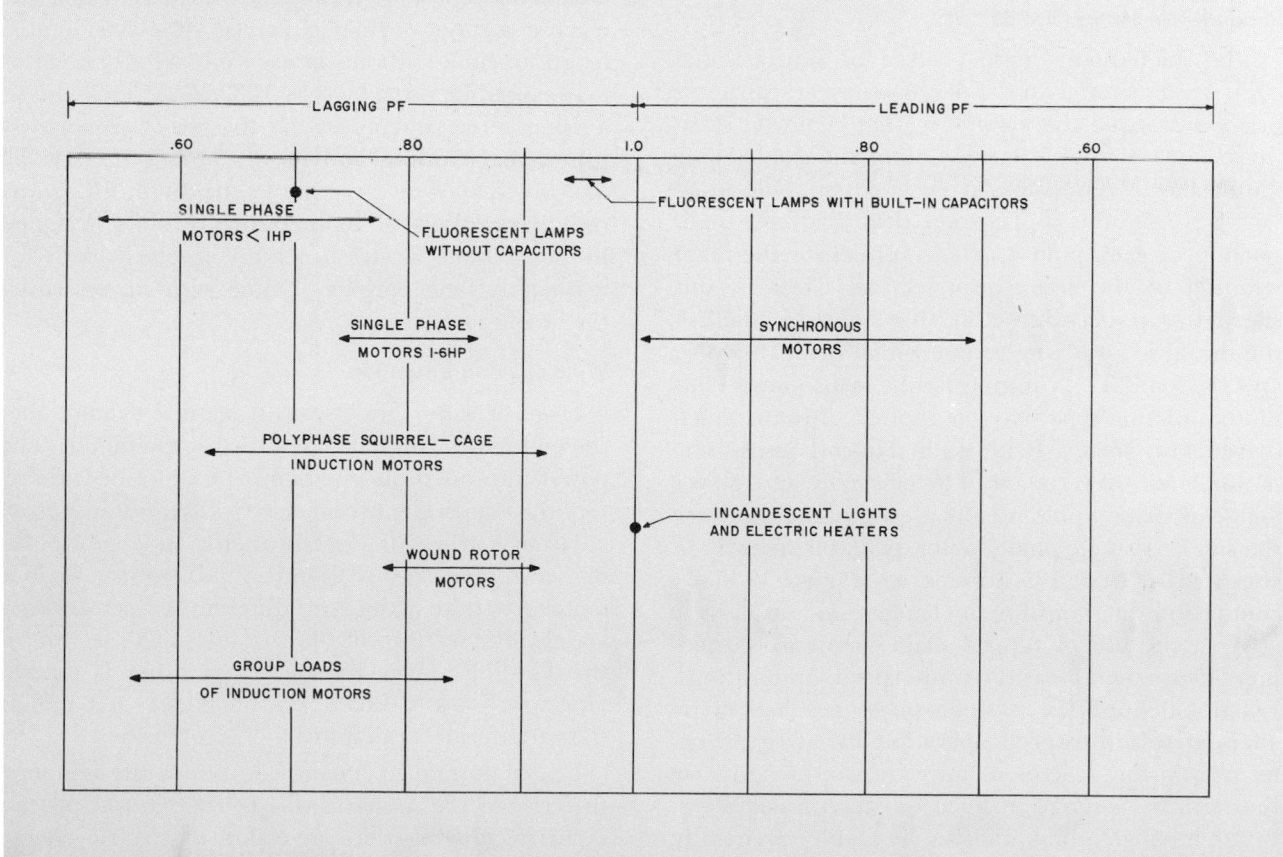

Capacitor-Start, Capacitor-Run Motor (Fig. 35)

The capacitor-start, capacitor-run motor has two capacitors in series with the auxiliary winding and a transfer switch which cuts out the low impedance starting capacitor after the rotor reaches two-thirds to three-quarters full load speed. The running capacitor of high impedance and low capacitance stays in the circuit with the auxiliary winding. This motor has the highest starting torque. *Figure 34* demonstrates the torque characteristics of various single-phase motors (split-phase and capacitor types).

The capacitor motors are suited to applications with normal starting torque such as fans, blowers and centrifugal pumps. Motors of higher starting torque may be applied to reciprocating compressors; these motors have a rather high lagging power factor while running *(Chart 2)*.

The capacitors are either electrolytic or oil-filled types; performance of both may be affected by low ambient temperatures. The electrolytic capacitors fall in performance (about 15%) at a temperature of 0–15 F. Similarly the oil-filled capacitors fall in performance at a temperature approaching −10 F. The motor torque performance must be reduced accordingly at a slightly higher rate.

Shaded-Pole Motor (Fig. 35)

The shaded-pole motor (often of subfractional size) is very similar to the permanent split-capacitor motor in torque characteristics, that is, low at starting. In place of auxiliary winding the shaded-pole motor has a depression at each salient pole filled with a continuous copper loop, thus shading a small portion of each pole. Current applied to the main winding of the stator produces an effect in this shorted loop (shading coil) that helps to establish the initial low starting torque which turns the rotor and the load. The running torque is also low. Thus motors of this type may be applied only to direct-driven fans such as with room fan-coil units. The motor is air-over cooled. The efficiency and power factor of shaded pole motors are very low. They are the smallest single-phase motor available, usually in sizes smaller than 1/5 horsepower. *Figure 35* shows connection diagrams for single-phase motors.

With the use of tapped main windings shaded-pole motors may satisfy multi-speed applications. Fractional horsepower, single-phase motors can be adapted to multi-speed operation by using an external resistor, reactor or auto transformer to vary the terminal voltage in fixed, predetermined steps.

Other single-phase motors available (generally not in use with air conditioning equipment) are the repulsion-induction and repulsion-start, induction-run motors. These are the commutator type. This construction utilizes an arrangement in which the main field winding is connected in series with the compensating winding and the brushes are short-circuited. These motors have been developed to produce a particularly high starting torque. A repulsion motor has the variable speed characteristics while the repulsion-start, induction-run motor has constant speed characteristics of the regular squirrel-cage induction motor.

Fractional horsepower series motors that are adapted for use on either d-c or a-c circuits of a given voltage are called universal motors.

MOTOR MECHANICS AND ENVIRONMENT

This section is devoted to the physical-mechanical aspects of motors in relation to power impressed, full load operation and the environmental conditions to which motors are subjected.

INTERNAL OVERHEATING OF MOTOR

The motor rating is an arbitrarily specified safe operating limit for the machine determined in accordance with certain accepted standards. It is intended to represent the operating limit which the machine cannot ordinarily exceed for a considerable length of time without damage to itself. The motor may exceed its rated load by 10%, 25%, 50%, but at a risk of a rise in temperature that may permanently injure the winding and its insulation; in fact the motor will stall on reaching its maximum in torque rating regardless of temperature. Motors designed for continuous service can carry specified loads for reasonably long periods of time without exceeding the heating limits.

Windings and Insulation

From previous discussions it is quite evident that the motor windings are the heart of the motor. The power impressed on them must be contained; therefore the winding must be electrically insulated from adjacent parts. An electric motor in operation is higher in temperature than the ambient; the various motor parts are actually at different temperatures as are also the sections of the windings. The section of the winding at the highest temperature is termed "hot spot" and is usually on the axial center line of the core in one of the slots.

Under normal operating conditions the temperature rise of the motor is due to the natural process occurring during the conversion of electric energy into mechanical energy and the rotation of parts.

There are three sources of energy losses appearing as heat that raise the motor temperature:

1. Windings — heat produced by a flow of current against resistance and equal to the product of current squared and resistance (I^2R). (With a motor design resulting in lower current and/or lower resistance less heat is produced and the motor is more efficient.)

2. Iron core — heat produced by hysteresis* and eddy current losses set up by the magnetic field in the stator and rotor.

3. Mechanical parts — bearings, fans, brushes (when used). Proper control of the driven load or number of motor starts required can also influence the winding losses.

The losses occurring during full load operation of the motor can be divided into two groups: (1) fixed losses, running light losses (I^2R no-load current losses), iron losses, bearing friction and internal fan (when used), and (2) applied losses of the driven load.

The major heat losses are in the windings. There is a definite maximum temperature which the windings can withstand under a given load and with a given insulation without undue deterioration either within themselves or in the insulation. In order that the maximum output of the motors may be secured without overheating, it is necessary to keep the heat losses to a minimum. Thus the insulation of the motor windings performs a dual function, that of an electric insulator and also a controlled heat dissipator.

NEMA has established six classes of insulation designed for various loads and for keeping the hot-spot temperatures within safe limits. *Table 4* lists class designations, description of insulation materials, and the limiting safe hot-spot temperatures (C).

The limiting hot-spot temperatures shown are determined by adding together:

1. Ambient environment temperature, normally 40 C (104 F).

2. Hot-spot temperature allowance, ranging from 5–15 C (usually 10 C).

3. Service factor, normally 15 C.

4. Allowable design motor temperature rise.

*Hysteresis is the conversion of electrical energy into heat energy due to molecular friction opposing magnetic polarity changes, friction that opposes the turning about of atoms.

TABLE 4 — CLASSIFICATION OF MOTOR-INSULATING MATERIALS

Insulation Class	Description	Open or Dripproof Guarded Motor Temp Rise (C)	Limiting Safe Hot Spot Temp (C)
0	Cotton, silk, paper and similar organic materials neither impregnated with insulating compounds* nor immersed in a liquid dielectric.	40†	90 (194 F)
A	Same materials as in Class 0 but impregnated or immersed in liquid dielectric*; also enamel coated on conductors.	40	105 (221 F)
B	Mica, glass fiber asbestos and other inorganic or organic materials in built-up form using suitable binders.	70	130 (266 F)
C	Entirely of mica, porcelain, glass, quartz or similar inorganic materials.		
F	Same as Class B, using modified organic binders.	90†	155 (311 F)
H	Same as Class B, using silicone resin binders.	110	180 (356 F)

*An insulation is considered to be impregnated when a suitable substance replaces the air between its fibers.

†Approximate temperature rise.

Motors used in air conditioning and refrigeration systems normally use Class A insulation that permits a 40 C temperature rise for the windings. With present standards the life of the motor winding is approximately 35,000 hours when operated at rated temperature and subjected to normal dielectric and mechanical stresses and humidity.

For open motors applied in the tropics or similar environments, special insulation must be provided because the motors may be exposed to excessive ambient temperatures and humidity.

TABLE 5 — SERVICE FACTOR, A.C. INDUCTION MOTORS

Motor Horsepower	Service Factor
⅛ and below	1.40
⅙, ¼, ⅓	1.35
½, ¾, 1	1.25
1½, 2	1.20
3 and larger	1.15

Service Factor

Standard open type motors of NEMA Design A, B and C (also Design F) carry a service factor, an allowable continuous overload above the rated nameplate horsepower, without causing a dangerous temperature rise because of overload. *Table 5* lists the service factors for fractional and integral horsepower induction motors.

Service factors apply only when the voltage and frequency are held at the rated value. When a motor is operated continuously overloaded, the motor naturally has a higher temperature rise and, therefore, may have an efficiency, power factor and speed different than the rated load. The locked-rotor torque, current and breakdown torque remain the same. It must be remembered that the temperature rise caused by operating a motor at a continuous service factor overload shortens the life of insulation and therefore the life of the motor. As a rule each temperature rise of 10 C halves the life of the insulation.

EXTERNAL OVERHEATING OF MOTOR

The temperature rise previously discussed is the result of current flow occasioned by the load applied at the rated voltage and frequency. However there are other considerations under which motor overheating may result:

1. Obstruction to heat dissipation due to improper ventilation.
2. Obstruction to heat dissipation due to physical debris inside or outside the motor.
3. Rise in ambient temperature above 40 C (104 F).
4. Unbalance in voltage.
5. Voltage and/or frequency variation from rated.
6. Other unpredictable misbehaviour of power transmission component parts affecting the winding performance.
7. Failure to start, stalling.

Thermal Protection—Overcurrent

The fundamental requirements for thermal protection of electric motors are contained in the National Electric Code (NEC), Article 430 of Motors and Motor Controllers. There are two methods of motor protection, one external to and separate from the motor, one internal within the motor. The former is an over-current protection in either or all of the following: the power feeder line, motor branch power line, and motor starter, but positively in the motor starter. The internal motor protection is a device responsive to the motor current and temperature.

Overcurrent protection devices guard the motor against excessive current up to and including the locked-rotor current. The branch power line protection guards against a possible short circuit or ground currents in the conductors. (See *Motor Controls*.)

The well-designed internal motor protection allows the motor to carry any load including an overload just long enough so that the motor does not overheat. Only small motors, particularly fractional horsepower, are equipped with inherent overheating protective devices which are imbedded in the motor windings and which respond directly to the heat generated within the motor.

Motor Enclosures

Environmental conditions generally refer to the ambient temperature, humidity, altitude and access to ample dry and clean ventilation air. The enclosure of the stator and rotor as a support and container is also a protector against the following environmental conditions:

1. Moisture — dripping, splashing, corrosive, exceedingly damp or even steaming.
2. Gases (fumes) — corrosive or explosive.
3. Dust — gritty, combustible (explosive) or conductive.
4. Outdoor installation — rain, wind, sun, etc. with above items; also insects, birds and small animals.
5. Temperature below 10 C (50 F).

All items except the last are quite obvious in their physical aspect. Operation below 10 C (50 F) reduces conductor resistance reflected in a small increase in starting current and a decrease in starting torque. Another danger of operation in low ambient is the probable moisture condensation on motor winding insulation. A thermal stress deterioration of the insulator may occur due to changes in temperature from cold standstill to warm running operation.

Motors used in air conditioning and refrigeration systems are normally the standard open type, simplest in construction and lowest in cost. An open machine has ventilating openings which permit the passage of external cooling air over and around the windings of the machine.

NEMA has defined many different types of open enclosures in its publication MG1-1.20, Open Machine. The most common are (1) drip-proof enclosures protecting a motor from solid or liquid drops falling on the motor at any angle not greater than 15 degrees from vertical and (2) splash-proof enclosures protecting from solid or liquid particles falling on the machine or coming towards it in a straight line at any angle not greater than 100 degrees from the vertical. Additional protection is also defined for open machines as semi-guarded, guarded, and drip-proof fully guarded.

There are also the totally-enclosed motors described in NEMA MG1-1.21 Totally-Enclosed Machine. These motors are machines so enclosed as to prevent the free exchange of air between the inside and the outside of the enclosure, but not sufficiently enclosed to be termed air-tight. The totally-enclosed motor range covers nonventilated, fan-cooled, explosion-proof, dust-ignition-proof, waterproof, externally and pipe-ventilated, water-cooled, water-air-cooled, air-to-air-cooled, and fan-cooled guarded and weather-protected machines.

In addition to the NEMA standardized enclosures listed, other motors exist such as lint free (textile systems), sanitary (dairy and food industries), encapsulated (sealed enclosure), and canned pump motors used particularly in nuclear applications. NEMA standards and motor manufacturers should be consulted for specific data on any special motor enclosure for a given application.

Bearings

Next to the stator, rotor and the enclosure, the most important part of the motor is the rotor shaft support, the bearings. There are two major types: (1) anti-friction ball or roller bearings and (2) sleeve bearings. The latter are either waste-packed (used mostly in fractional horsepower motors), oil ring lubricated, or pressure oiled on large motors. Ball bearings are either grease- or oil-lubricated. Most ball bearings are grease-packed, either the pre-lubricated (sealed) or relubrication type. Oil-lubricated ball bearings are applied usually on large size motors; these bearings require more complex housing and careful control of oil level, and must be mounted in a prescribed position. The bearing should be equipped with a sight gage for observing a proper oil level.

The grease-packed ball bearings must have grease that is quiet, have low friction and oxidation rates, and must be clean. For applications where motors are exposed to winter ambient or various low temperature conditions special low temperature greases must be utilized. Excess grease may lead to overheating of the bearing.

NOISE

The noise level is increasingly important in many motor applications. Motors produce airborne sound and physical vibration as unavoidable byproducts of the conversion of electrical to mechanical energy. The undesirable manifestation of sound and vibration is termed noise.

Airborne Noise

Airborne noise is produced by all the vibrating parts of the motor. The initial sources are magnetic, mechanical and windage.

Magnetic noise is produced by magnetic forces (flux) in the air gap and other parts of a magnetic circuit. The frequency (cycles per second), usually twice the line frequency and its harmonic is a function of either the number of slots and rps (varying) or the line frequency (constant). The air gap forces are to be considered only in relation to the stator. The rotor (usually quite rigid) may be a source of noise in the case of hermetic motors and close-coupled motor pumps. In the latter case the shaft vibration noise may be transduced to the water.

Mechanical noise may result from disrepair, unbalance or bearing disorders. The first two are abnormalities that should not exist in a well-constructed, balanced motor. Bearing noise may be differentiated between that of a sleeve type or ball type of bearing. The former is inherently quiet with a few distinguishable sound frequencies. Ball bearings with numerous component parts moving relative to each other produce many sound frequencies. Rigidity of bearing support is very important. The noise level of both types increases with lubrication impurity and surface roughness appearing in the course of wear and tear. Therefore care must be exercised in the choice of motor enclosure in relation to ambient environment.

Although rarely encountered in the air conditioning field, another mechanical noise is brush noise, resulting from the sliding contact of brushes against a slip ring or commutator. In the case of a slip ring the noise is less than that produced by brushes sliding over a segmented commutator. In either case the brush noise is characterized by high frequencies. Since brush noise is a function of surface finish, it also varies during motor operation because of wear.

The passing of ventilation air thru the motor together with the propulsory elements creates only airborne windage noise. Its pulsations contribute to stator vibrations. The windage noise is of broad band characteristics. Motors having open enclosures are principal generators of windage noise. The totally-enclosed fan-cooled motor equipped with an external fan may at times have higher level windage noise.

Totally-enclosed nonventilated motors with internal fan air circulation have subdued noise levels. Windage noise from high speed motors dominate over noises from other sources.

Of all the motors compared on the basis of equal horsepower and speed, synchronous motors are the most quiet.

The single-phase fractional motors have a terminal 120 cps vibration (60 cycle current) caused by the pulsating, 120 alternation power supply and transmitted to driven apparatus and motor supports. Since their bearings are the sleeve type, the bearings seldom contribute to noise.

Mountings and Isolation

The second aspect of undesirable noise is the vibration coupling between the motor and its support, transmitting unwanted vibration noise to building structure. Careful attention must be given to the motor mounting and support isolation. A standard rigid base is the simplest and least expensive normal motor mounting. To reduce vibration and noise either from the motor or motor-driven machine assembly, various resilient mountings are available. Resilient elements are used either under motor feet (where applicable) or under the base of the total assembly. Fractional horsepower motors used on fan-coil units are often isolated by rubber rings around the bearing supports. There exist also flange or face mountings used on such apparatus as close-coupled motor-pumps. These assemblies may be isolated from the floor by resilient mounts.

Motors can be installed in any position, horizontal, vertical, upside-down or sideways, provided they are equipped with proper bearings and lubrication.

The motor mechanical power may be transmitted to driven equipment thru (1) a direct shaft such as with hermetic assemblies or small fans, (2) couplings with pumps and reciprocating compressors, (3) matched V-belts with fans, (4) step-up gears and couplings with centrifugal compressors, or (5) hydraulic and magnetic couplings with fans or centrifugal compressors.

MOTOR CONTROLS

Motor characteristics and the requirements imposed by the equipment used in air conditioning and refrigeration systems are set down in the preceeding section on motors. These requirements and characteristics and the rules imposed by the power companies constitute the guides for selecting motor controls.

In order to obtain adequate motor performance electric energy must be regulated. The controller may be a simple on-off toggle switch or a combination of complex automatic equipment.

PURPOSE

The purpose of the motor controller is to:
1. Admit electric energy to the motor at a proper rate.
2. Protect against any fault that may occur in the electric system which may cause a sudden inrush of current.
3. Prevent overheating of the motor while operating.
4. Regulate the motor speed.
5. Withdraw electric energy when the need ceases.

This discussion gives an insight into the functions of motor starting and protective equipment. The detailed selection of equipment to satisfy given requirements is the responsibility of the electrical engineer.

STANDARDS AND CODES

The National Electric Code, Article 430, Motors, Motor Circuits and Controllers covers basic minimum provisions and rules for the use of the subject equipment. The provisions are a guide to the safeguarding of persons and of buildings and their contents from hazards arising from the use of electricity. This code is not a design manual. The Underwriters Laboratories, Inc. provide standards for Industrial Control Equipment (#508) and for Temperature-Indicating and Regulating Equipment (#873).

The local, city and state codes must also be followed as well as the regulations of the local power company.

CONTROL ELEMENTS

Figure 36 is a schematic diagram of the various possible elements of a power supply circuit to the motors.

Protective equipment that prevents major malfunctioning of the power supply system such as a short circuit, reverse-phase and open-phase operation, voltage variation or stoppage, and interlocking of controls is discussed later. Most starters for motors of larger than one horsepower provide motor overload protection either firmly fixed in the starter or in an assembly of contactor and overload protection. Fractional and small integral horsepower motors often have overload protection as part of the motor. Without this basic protection the starters are switches or contactors.

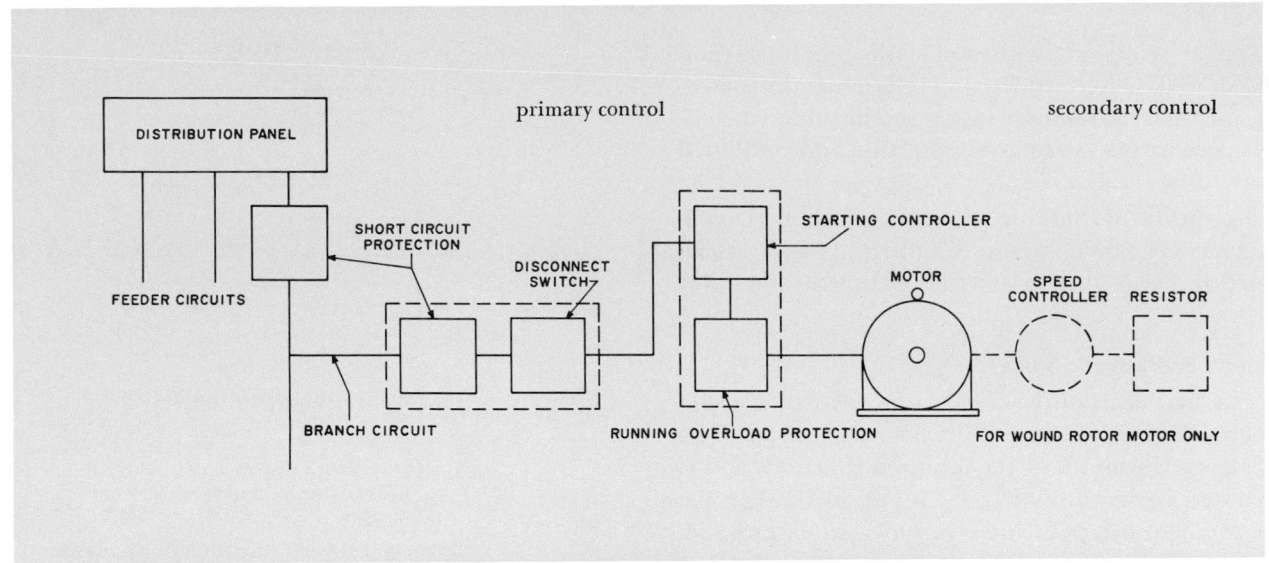

FIG. 36 — MOTOR CONTROL CIRCUIT

Various possible components of motor control are:

1. Switching mechanisms such as manual switches or magnetic contactors which open or close the power circuits.

2. Power-absorbing or transforming devices such as resistors or reactors which absorb part of the power applied to the motor; and auto-transformers which reduce the line voltage before application to the motor.

3. Protective devices which are motivated by temperature or voltage.

4. Pilot or initiating devices such as push buttons, float switches and thermostatic switches.

The push buttons are either a part of the starters or are installed remotely on a separate panel for the convenience of the operator in charge of the air conditioning and refrigeration system. Other initiating devices are discussed with their applied equipment.

The motor controller furnishes a means for a motor to:

1. Start and accelerate.

2. Operate the load.

3. Regulate its speed.

4. Protect itself, including controller and wires.

5. Stop.

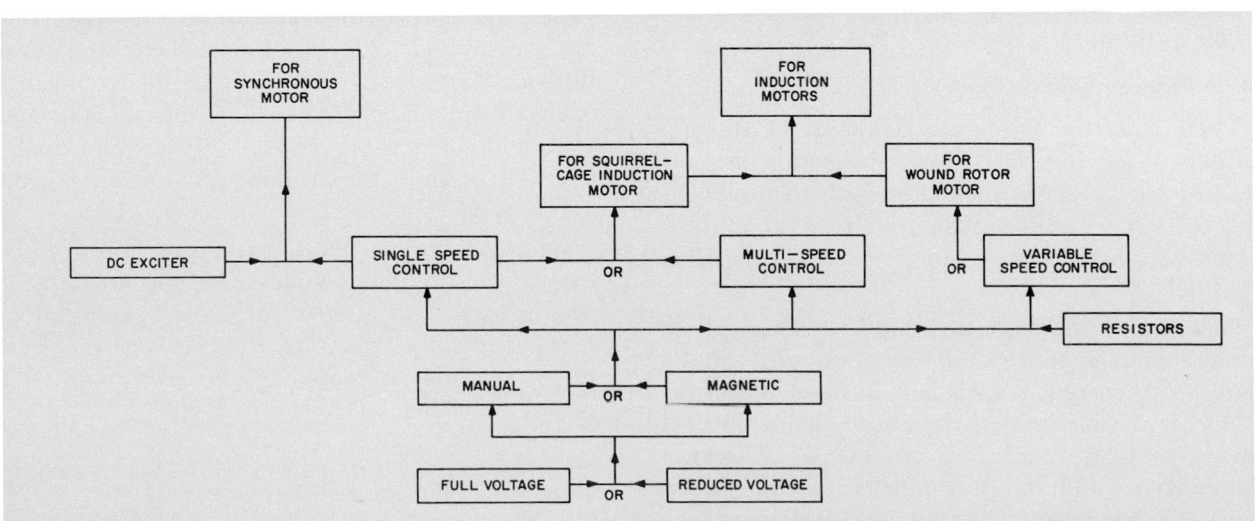

FIG. 37 — STARTER SELECTION GUIDE

STARTERS

Starter selection is integral with motor selection and should be so considered in relation to the following factors: horsepower rating, permissible current, desirable torque, necessary protection and combined economics.

An outline of the basic process of selection of available starters is given in *Fig. 37*. There are many variations and combinations with protective and pilot devices.

Manual or Magnetic Starter

The first decision is whether the particular application calls for a manual or magnetic starter. The former performs all of the required functions and is operated by hand. The latter is automatic thru the use of electromagnets, and is initiated either by hand, push button, or by some starting device involved in the process control; from this point on, the starter is sequenced thru the steps designed into the control circuit.

The choice of manual or magnetic starter is influenced by the size of the motor and frequency of operation. Infrequently operated motors may often have manual starters. Most motors should be started magnetically, either because of the size of the motor or because of the convenience of remote starter operation. Magnetic starters are more expensive but their higher cost is offset by lower maintenance costs and greater safety. Another advantage of a magnetic starter is the fact that it is automatic; once committed to certain duty the possibility of introducing human error is eliminated. Because of the possibility of including a variety of pilot relays, automation has the flexibility to achieve any desired end to operate and protect the motor. Magnetic starters include undervoltage protection.

Full or Reduced Voltage Starter

There are two fundamental classes of starting equipment serving squirrel-cage induction motors: (1) full voltage, across-the-line and (2) reduced voltage, reduced current inrush starters. Wound-rotor motor and synchronous motor controls are discussed separately.

Four factors influence a choice between a full or reduced voltage starter: (1) cost, (2) size of the motor, (3) current inrush and starting torque behavior of the motor with reduced voltage, and (4) power company restrictions on the use of electric energy (in relation to the requirements of the driven machine). These factors are discussed under a specific class of starters.

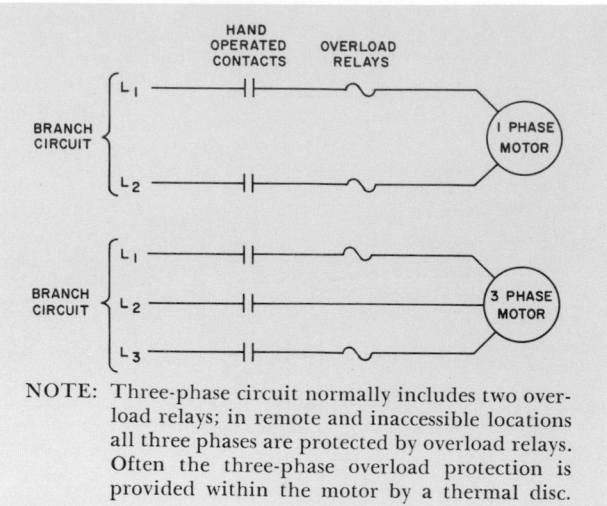

NOTE: Three-phase circuit normally includes two overload relays; in remote and inaccessible locations all three phases are protected by overload relays. Often the three-phase overload protection is provided within the motor by a thermal disc.

FIG. 38 — ACROSS-THE-LINE MANUAL STARTERS

Full Voltage, Across-the-Line Starter

The least expensive are the manual full voltage starters *(Fig. 38)*.* These are applicable especially to small size motors, single phase up to 5 hp and three-phase up to 7½ hp; they consist of switching contacts and overload relay trips. They do not provide automatic undervoltage protection. Motors restart on reinstatement of voltage on the power line.

When using small switches for fractional horsepower motors, protection is provided by inserting fuses in circuits.

The magnetic full voltage starters provide convenience, flexibility and safety that is greater than with manual starters; they include undervoltage protection.

A push-button relay that continues to maintain contact on voltage failure *(Fig. 39a)*† illustrates one variant for safeguarding against low voltage. However with voltage reduction the magnetic strength of the starter coil weakens, permitting the contacts to open. When voltage is reinstated, the coil is re-energized and contacts are closed. This is a low voltage release that allows the equipment to operate on voltage recovery.

In cases where such automatic procedure is undesirable, a momentary contact push button is used in combination with a set of normally open auxiliary contacts in the starter *(Fig. 39b)*. When voltage returns, the motor cannot restart and requires a manual resetting.

*Figures 38, 39, 41-45, 47 and 49 are diagrammatic only. No attempt is made to illustrate the actual starters.

†Control circuits shown here are basic. Further starter diagrams do not show control circuits.

Reduced Voltage Starters

Reduced voltage on the motor reduces locked-rotor current inrush and torque as well as accelerating torque *(Fig. 40)*. The reduced torque produced is generally greater than that required by the equipment used for air conditioning systems *(Fig. 23)*. Thus a reduction in current inrush in accordance with local power company current limitations is the primary reason for the use of reduced voltage starters.

The common practices by power companies are to:

1. Limit starting current to a fixed percentage of locked-rotor current inrush.
2. Limit starting current to certain increments at fixed intervals of time using closed transition between successive steps, thus helping the network to adjust itself to a gradually imposed load.

All deviations from across-the-line starting methods are grouped together under reduced voltage types. However some types actually reduce voltage; others reduce current inrush directly. Both methods result in the reduction of current and torque. Reduction of current is of concern here.

The general class of reduced voltage starters divides into two groups:

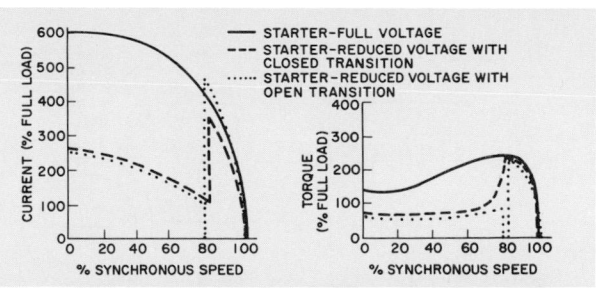

Fig. 40 — Current and Torque Characteristics Squirrel-Cage Induction Motor with Auto-Transformer Reduced Voltage Starter

1. Reduced voltage starters applied to any motor.
   a. Auto-transformer
   b. Primary-resistor
   c. Primary-reactor
2. Reduced current inrush starters applied to specially wound motors.
   a. Part-winding
   b. Star-delta (wye-delta)

Table 6 lists locked-rotor current and torque characteristics of various starters.

TABLE 6 — COMPARISON OF STARTING METHODS

Method of Starting	Inrush Current (percent full voltage locked-rotor current)	Starting Torque (percent full voltage locked-rotor torque)
Across-the-line	100	100
Auto-Transformer		
80% tap	71	64
65% tap	48	42
50% tap	28	25
Primary Resistor or Reactor		
80% applied voltage	80	64
65% applied voltage	65	42
58% applied voltage	58	33
50% applied voltage	50	25
Star-Delta	33	33
Part-Winding	60	48
Part-Winding with Resistors	60–30	48–12
Wound Rotor (approximate)	25	150

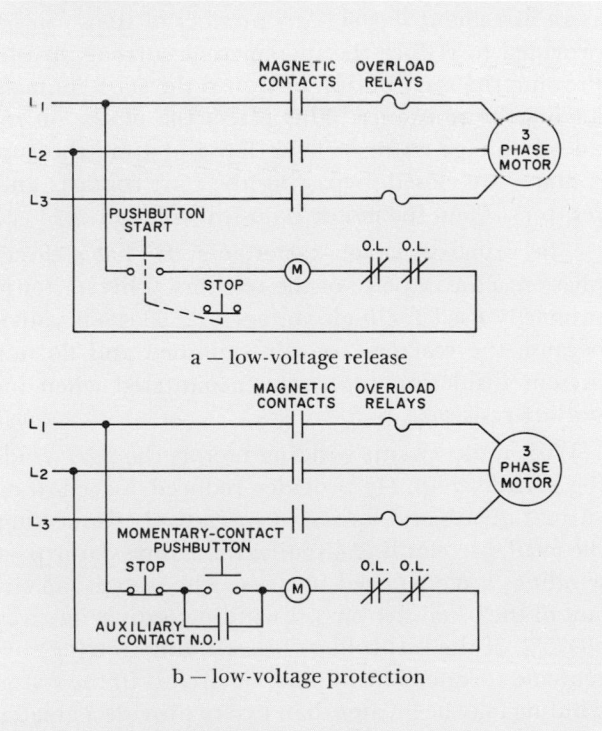

a — low-voltage release

b — low-voltage protection

Fig. 39 — Across-The-Line Magnetic Starters with Low Voltage Safeguard

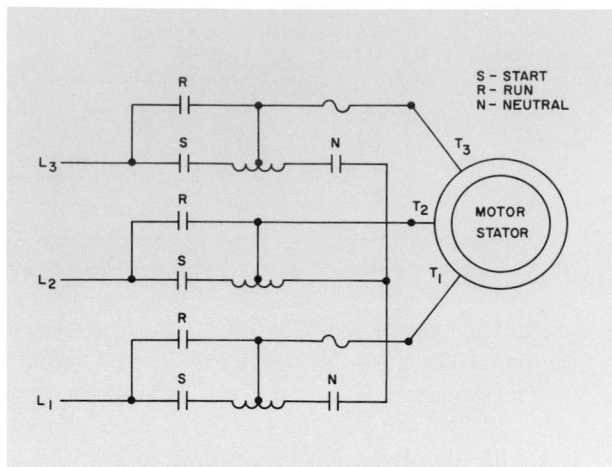

FIG. 41 — AUTO-TRANSFORMER REDUCED VOLTAGE
STARTER, CLOSED TRANSITION

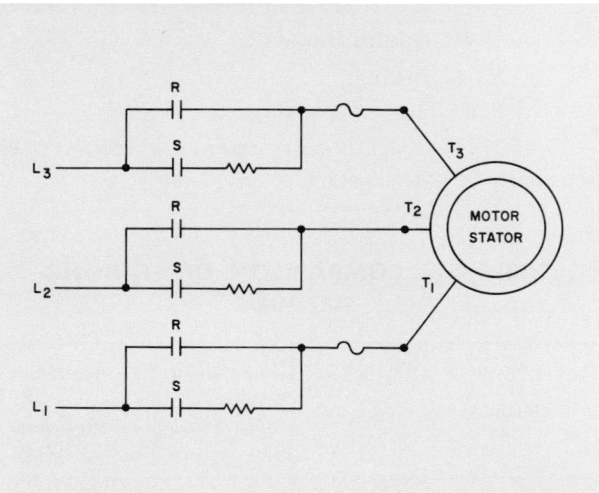

FIG. 42 — REDUCED VOLTAGE STARTER, PRIMARY-
RESISTER

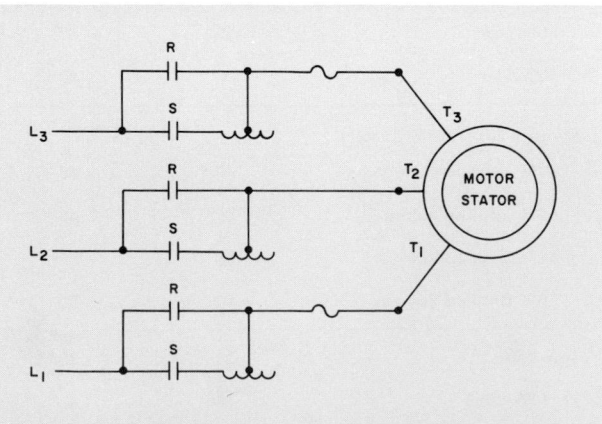

FIG. 43 — REDUCED VOLTAGE STARTER, PRIMARY-
REACTOR

Some of these starters are manual. However since the majority of applications are of the magnetic type, only this variety is discussed. The schematic illustrations are intended to show starting sequence only and do not represent the actual wiring of any starter. Refer to starter manufacturers' catalogs and data for wiring diagrams and complete details.

The auto-transformer starter usually has three voltage taps (50%, 65% and 80%). On closing the circuit for a starting contactor the motor is connected to the power line thru the design voltage tap. Simultaneously a timing unit is energized. After a brief lapse of time the auto-transformers are removed and the motor is connected to full voltage. In effect there is a brief lull, then a "bump"* on the power line. This is an open-circuit transition version of starting. *Figure 41* shows one method used by manufacturers to achieve closed circuit transition starting. Before the run contacts R are closed, the neutral contacts N are open and the auto-transformers operate briefly as reactors in series with starter windings. Then the run contacts are closed. The flow of current is not interrupted.

The primary-resistor starter *(Fig. 42)* with closed circuit transition limits voltage and locked-rotor current by inserting an external resistance in series with the stator windings. This starter is frequently used as an increment starter. Any number of steps can be provided to reduce the incremental current inrush. Pressing the start button energizes the start contacts having the resistance. This places the motor on reduced voltage. After a brief lapse of time the run contacts are closed, bypassing the start contacts and resistors. Then the motor is on full voltage.

The primary-reactor starter *(Fig. 43)* has a three-phase reactor in place of the resistors. This starter is primarily used for high voltages (2300–4800 volts) because the reactors are self-contained and do not present insulation problems encountered when installing resistors.

With multi-circuit winding motors the part-winding starter *(Fig. 44)* provides reduced locked-rotor current inrush and torque by successively connecting the available winding circuits; the motor stator part windings are energized in steps. The use of one circuit of the usual two-circuit winding ordinarily gives 60–75% of the full-voltage full-winding starting current and torque. The number of circuits in the stator winding may be greater than two to provide a greater number of increments in starting. The part-winding

---

*There is also a "bump" in motor torque; this imposes stresses on motor windings, shaft and coupling.

starter is not a reduced-voltage starter, but a reduced-current starter.

Part-winding reduced-current starting has certain advantages. It is simple and less expensive than most reduced-voltage methods because it requires no voltage-reducing elements such as transformers, resistors or reactors; it uses smaller contactors. It is inherently a closed circuit transition starter.

The part-winding starter also has an advantage in that it is not adversely affected by high voltages. Continuous high voltages (not high voltage surges) such as 250 or 260 on a nominal 220-volt system can result in motor burnout when using an auto-transformer starter. The auto-transformer is normally rated for short-time duty and is therefore rather small. If an over-voltage of about 15% is applied, the transformer will saturate and allow very high currents to pass thru the starting contactor which, in turn, welds shut and puts the motor on single phase the next time it is started. The part-winding starter does not impose a limit on the starting duty cycle as does the auto-transformer since there is no insulated voltage reducing equipment which may overheat.

The part-winding motor starter is almost always an increment-start device. Not all motors can be part-winding started; it is quite important that the motor manufacturer be consulted before this type of starting is applied. Some motors are wound sectionally with part-winding starting in mind; indiscriminate application to any dual voltage motor (for example, a 220/440 volt motor which is to run at 220 volts) can lead to excessive noise and vibration during starting, to overheating, and to extremely high transient currents upon switching.

For the delta wound motor provided with six leads (3-phase motor only) there is the star-delta or wye-delta starter that provides reduced locked-rotor current inrush and torque. This behavior is achieved by connecting first the motor windings in star arrangement, and then on the second step by rearranging the windings in delta arrangement. The essential difference is that, for the same motor winding, the star connection draws only one third (33%) as much current as the normal delta connection, and gives one third as much torque.

Figure 45 shows a closed transition arrangement in which additional protection in the form of a resistor is provided against high inrush current during the switch-over period from star to delta winding. Initial inrush is the same as in the open transition arrangement; however, depending on the time lapse the incremental inrush is reduced with closed transition

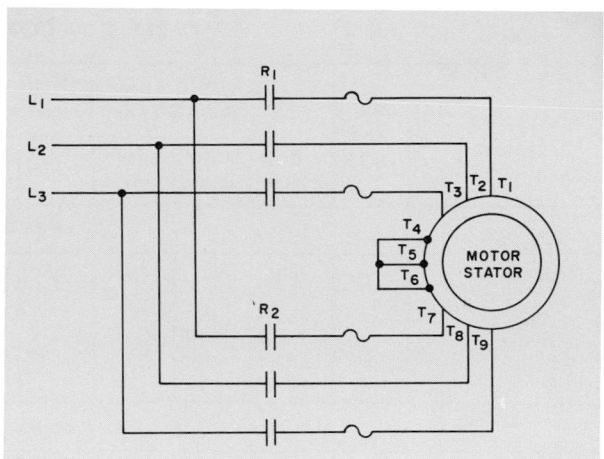

FIG. 44 — PART-WINDING STARTER

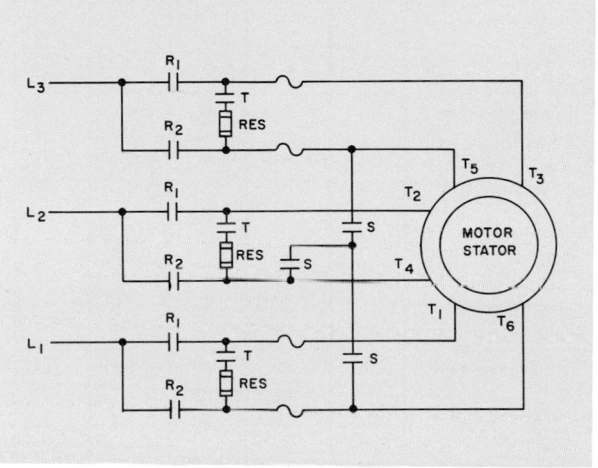

FIG. 45 — STAR-DELTA STARTER, CLOSED TRANSITION

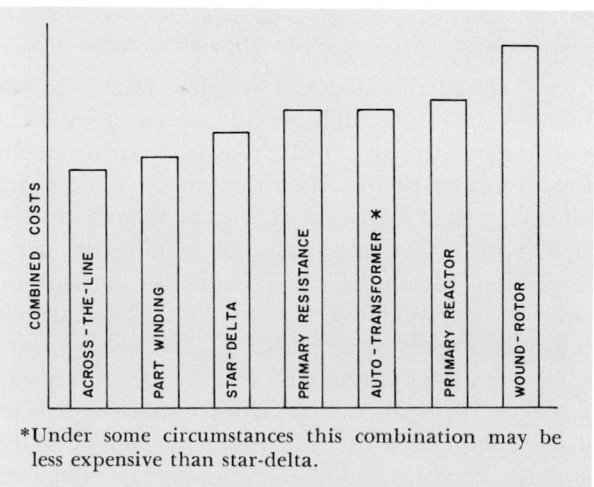

*Under some circumstances this combination may be less expensive than star-delta.

FIG. 46 — COMPARATIVE COMBINED COSTS OF MOTOR AND STARTER FOR VARIOUS STARTING METHODS

TABLE 7 — COMPARISON OF STARTERS

TYPE OF STARTER	PER-CENT TAP	STARTING CHARACTERISTICS (percent of rated value)					ADVANTAGES	LIMITATIONS
		Voltage at motor	Motor current	Line current	Torque	Torque efficiency		
FULL-VOLTAGE								
Magnetic	—	100	100	100	100	100	1. Lowest cost 2. Least complex 3. Least maintenance 4. Suitable for low or high voltage	1. Draws highest current from line during starting 2. Starting torque may be too high for some applications
REDUCED-VOLTAGE								
Auto-transformer (Closed-transition standard)	80 65 50	80 65 50	80 65 50	64* 42* 25*	64 42 25	100 100 100	1. Starting characteristics easily adjusted 2. Provides maximum torque per line ampere for reduced voltage starters (high torque efficiency) 3. Closed-transition starting	1. Additional external voltage-reducing component may impose additional limits to duty cycle
Primary resistor (2-step)	—	80†	80	80	64	80	1. Increment starting can be constructed in any desired number of steps to meet any inrush limitation 2. Improved power factor 3. Inherently closed transition	1. Unavoidable power loss in resistor 2. Low torque efficiency 3. Introduces external voltage-reducing component 4. Not readily adapted to high voltage 5. Duty cycle limited by thermal capacity of resistor
Primary reactor	80 65 50	80 65 50	80 65 50	80 65 50	64 42 25	80 65 50	1. Least complex method to obtain reduced-voltage start for high voltage motors 2. Starting characteristics easily adjusted 3. Inherently closed transition	1. Lower torque per unit of line amperes than auto-transformer-type 2. Additional external voltage-reducing device (reactor) may impose additional limits to duty cycle 3. Low-starting power factor

*Does not include autotransformer magnetizing current which is usually less than 25 percent of motor full-load current.
†If application dictates, other voltage increments can be furnished.

because the current does not first drop to zero before stepping up to the higher value provided by delta connections. In the open transition arrangement there is a brief period when the stator windings are not energized. This may result in a momentarily high current inrush at the instant of making the delta connection. In some power systems this "bump" inrush is objectionable and therefore a closed transition arrangement is preferred.

The star-delta motors and starters are widely employed with hermetic centrifugal machines. The primary appeal of this starting arrangement is the absence of voltage-reducing equipment. Voltage reduction is inherent in the star connection of the delta-wound motor.

Of all the starters discussed here the reduced-voltage stepless resistance starter gives a smoother start for the squirrel-cage induction motor. However in the case of severe limitations of starting current, a resistance starter may not be applicable because of a possible severe torque reduction below a point of positive starting; other starters may have to be used.

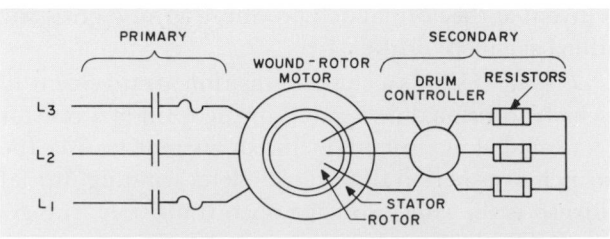

FIG. 47 — WOUND-ROTOR MOTOR CONTROL

TABLE 7 — COMPARISON OF STARTERS (Contd)

TYPE OF STARTER	PER-CENT TAP	STARTING CHARACTERISTICS (percent of rated value)					ADVANTAGES	LIMITATIONS
		Voltage at motor	Motor current	Line current	Torque	Torque efficiency		
		REDUCED CURRENT INRUSH						
Part Winding 1/2–1/2 Winding	—	100	65	65	50	71	1. Reduced starting torque always provided 2. Closed-circuit transition-type provided	1. Possibility of motor not accelerating due to severe torque dip at half speed 2. Limitation in motor selection
2/3–1/3 Winding	—	100	65	65	40	—	1. Reduced starting torque always provided 2. Closed-circuit transition-type provided 3. Motor will usually accelerate to full speed in one step	1. Limitation in motor selection
Star-delta (Open or closed transition)	—	100	33	33	33	100	1. Starting duty cycle usually limited by motor heating only 2. High torque efficiency for all speeds 3. No torque dips or unusual stresses because full-winding energized 4. Closed-circuit transition-type eliminates line surges during transition from start to run	1. Starting characteristics not adjustable 2. Requires motor with a normally delta-connected winding with all leads brought out for connection to control

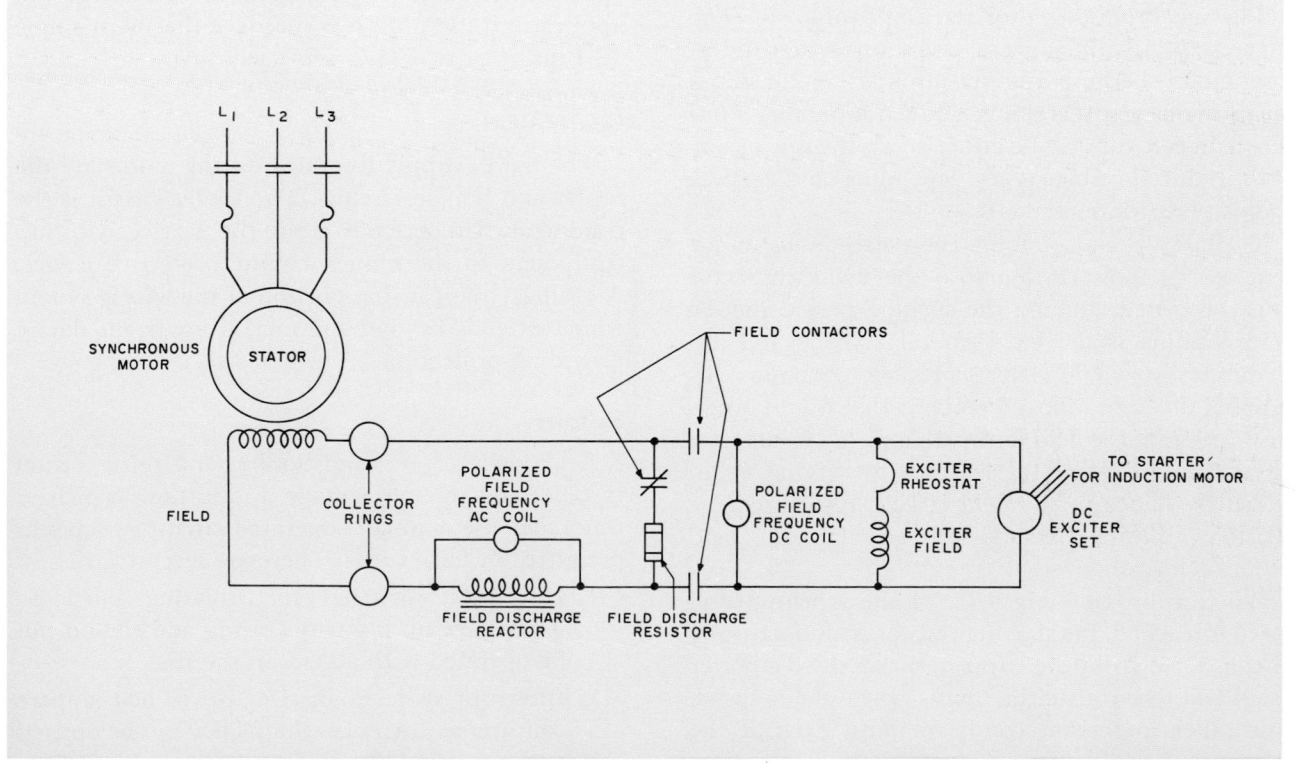

FIG. 48 — TYPICAL SYNCHRONOUS MOTOR CONTROLLER

Across-the-line starters are always lowest in cost. *Figure 46* shows comparative costs of motor-starter combinations. *Table 7* gives comparative data on various starters.

Wound-Rotor Motor Starter

The wound-rotor motor variable secondary resistance starting equipment *(Fig. 47)* is the most expensive motor-starter combination and consists of:

1. Primary across-the-line starter with major protective elements, overload and undervoltage.
2. Secondary resistors and drum controller which acts to insert into the motor rotor windings the resistance required to provide the design speed control. The drum controller may be manually or automatically controlled.

The change in rotor resistance provided by the controller results in a change of the motor speed-torque curve; thus this method can be used as a means of speed control. It also results in a change of the motor speed-current curve so that this method is used to limit the starting current drawn from the power line.

A safety relay does not permit motor starting unless all the resistance is in the circuit.

Synchronous Motor Starter

The synchronous motor starting equipment *(Fig. 48)* is essentially the same as for a squirrel-cage induction motor because the synchronous motor starts operating as a squirrel-cage induction motor. Thus the main control may be either across-the-line or any of the reduced-voltage types, depending on the power company regulations.

To start up and operate, the synchronous motor requires a-c power; therefore the main contactor must be closed. During the starting period the d-c field winding is shorted thru the starting and discharge resistor. The latter serves two purposes; (1) it limits the high induced voltages that would otherwise appear at the field terminals and (2) it increases motor pull-in torque at start-up by serving as an added resistance in the field circuit. As a discharge resistor it limits the field voltage when the field supply is disconnected.

When approximately 97% of the synchronizing speed is reached, the starting resistor is automatically disconnected from the slip rings, and the d-c power is applied to stabilize the motor synchronous speed. One other important condition must exist at the moment of application of d-c excitation; a correct relationship between the rotor field and the stator revolving magnetic field which also contributes to developing maximum pull-in torque.

If a sudden overload or voltage dip occurs, the required load torque can exceed the maximum pull-out torque of the motor. Thus the required torque pulls the motor out of its synchronous speed; the motor is pulled out of step. The d-c field.is disconnected immediately; otherwise severe vibration due to torque pulsations may develop, and stator pulsating current inrush may rise to dangerous levels.

The polarized field relay operates at the instant of proper motor slip, bringing in d-c excitation, and thus automatically resynchronizing the motor if there is sufficient torque left in the motor, after removing the disturbance which caused the pull-out. Otherwise the motor is automatically taken off the line. A series of protective devices are included to provide a proper sequence of operation.

Multi-Speed Motor Controllers

The control of multi-speed motors is achieved thru separate starters (in one enclosure) acting either individually on each winding with separate-winding motors *(Fig. 49b)* or in interlocked manner with the consequent-pole motors *(Fig. 49a)*. No attempt is made to analyze the multiplicity of speed-winding combinations or control circuits that may apply under various circumstances. Most large fan applications use only two speeds, either with single (2:1 ratio) or two (3:2 ratio) windings.

PROTECTION

The power supply lines nearest the motor are the feeder and branch circuits. The feeder circuit is the conductor that extends from the service entrance equipment to the branch circuit protective device. A branch circuit is that portion of the wiring system which extends beyond the final overcurrent device providing protection.

Switches

A majority of air conditioning and refrigeration installations are low voltage applications which at times may use manually-operated circuit switches as permitted by local codes. There are several varieties:

1. Disconnect switches for isolating purposes. They have no interrupt rating and should not be operated with a load on the line.
2. Interrupt switches in sizes up to 600 ampere continuous current rating. They can be opened with a load on the line, and are used mostly as service entrance equipment.

3. Enclosed safety switches, fused or unfused, available for light duty a-c service up to 600 ampere and 240 volt rating; for normal duty service (general purpose) up to 1200 ampere and 600 volt rating; and for heavy duty industrial service up to 1200 ampere and 600 volt rating.

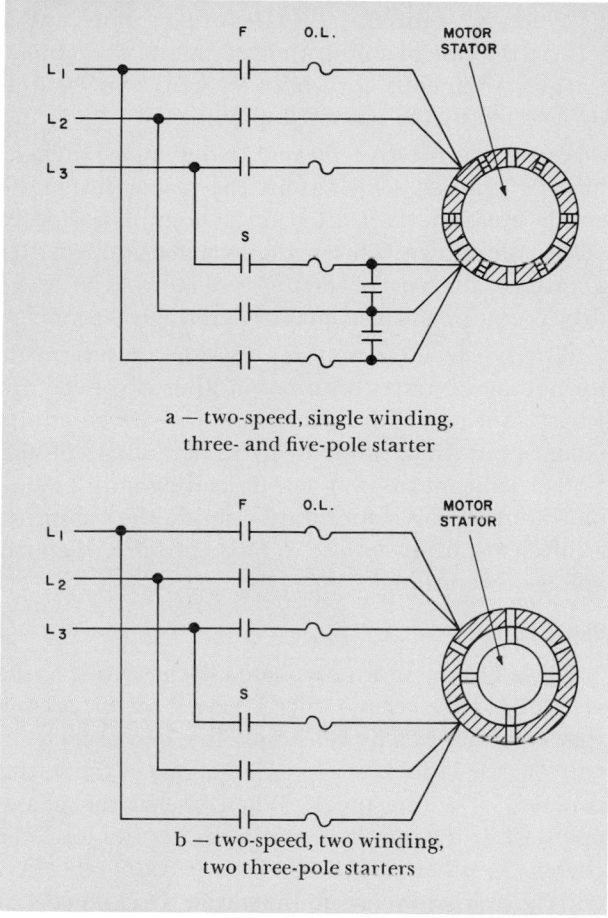

a — two-speed, single winding, three- and five-pole starter

b — two-speed, two winding, two three-pole starters

FIG. 49 — MULTI-SPEED CONTROL

Fuses and Circuit Breakers

The design performance of a motor is delivered under a normal supply of electric energy. Any disturbance to normal flow leads either to overheating and eventual destruction of the motor, or to non-delivery of the required mechanical power.

The greatest hazard to any electric service is a short circuit, a flow of an enormously excessive current (running into thousands of amperes) caused by some fault either in the power line or at the motor. Sufficient protection must be designed into the system to safeguard the feeder and branch circuits as well as to isolate the feeder system from a fault in the individual branch circuit.

In an air conditioning and refrigeration system it is important to maintain full system capacity. Thus protection of the branch circuits that lead to the motors driving various fans and refrigeration machines is very important. Isolation of a defective branch circuit affords immediate attention for diagnosis and repair of the fault while the rest of the system is still able to deliver its partial air conditioning capacity. This does not obviate centralization of the system because present day electrical equipment is well designed and dependable; nevertheless, careful attention must be paid to the selection of electrical equipment appropriate to needs.

The electric system protection is twofold, (1) from a fault in the power supply and (2) from a fault occurring at the motor. A fuse or circuit breaker current interruption in the branch circuit must be instantaneous before the effect is cumulated back to the feeder circuit breaker.

An a-c squirrel-cage induction motor has three levels of electric current usage:

1. Fundamental or operating full-load (100% load) level.
2. Code permissible over-load level (service factor generally 115%, *Table 5*).
3. Almost instantaneous level* of starting current inrush (600%).

Neither the motor nor the starter is able to continuously withstand a short-circuit without damage. The starter (controller) is equipped with normal overload relays; therefore, the time-current characteristics of the short-circuit protection must be carefully coordinated with that of the normal overload protection so that the short-circuit overcurrent protection does not operate under any of the three levels of usage but operates instantaneously under short-circuit conditions. Fuses and circuit breakers are used for short-circuit interruption. When fuses or simple circuit breaker are installed in the same enclosure with the motor starter, the assembly is called a combination starter.

A fuse is a low cost short-circuit protection device. It may be a plug or cartridge type. The plug type is available in ratings up to 30 amperes and is used on lines not exceeding 125 volts. There are two basic types of fuses, one with a single fusible zinc element and one with a dual-element time delay fuse. The latter has the ability to open the circuit on either the overload or short circuit.

---

*The almost instantaneous nature of starting current inrush does not provide sufficient temperature rise in a protective element for it to act.

Commercial or industrial cartridge type fuses are available in either single or dual-element varieties. There are also renewable or nonrenewable designs. Fuses of renewable design are more expensive, but the cost of replacing fuse links is less than that of fuses of a nonrenewable design.

A circuit breaker functions both as a circuit protector and as a branch circuit disconnect switch. Its advantage is that upon being opened by a short circuit it may be reset without the necessity of replacement, as in the case of a fused disconnect switch. With fused polyphase circuits there exists another danger, namely single-phasing in case only one fuse blows. A circuit breaker disconnects all three phases.

Overload Relays

The line overcurrent protection of the motor and starter is provided by varieties of switches, fuses and circuit breakers, or combinations of these. The motor overheating protection is accomplished by overload relays in the starter itself.

Overload relays are either a melting alloy or bimetallic type. The latter may be the compensating type, that is, compensating for ambient temperature difference between motor and starter locations. The compensated overload relay thus safeguards the motor from unnecessary outage and acts to trip only on motor overcurrent conditions.

A third type of overload relay is the magnetic induction type, the Heineman relay; it is non-sensitive to ambient temperature and is instantaneously resetting.

Undervoltage

There are many types of relays that may protect against any failure or malfunction. Only the most applicable are described.

Power systems are subject to occasional voltage fluctuations of varying magnitude and duration. Lightning, accidental short circuits or line overloads may create undervoltage dips and failures; motors may slow down and even stop. Minor voltage dips below 10% or frequency variations below 5% may be tolerated. More extensive variations may overheat the motor causing it to stop. There are three methods of protection that deal with line undervoltage:

1. Instantaneous and complete stoppage — the protective device trips the starter and the motor stops. The motor can be reinstated in operation only manually.
2. Time-delay — the motor is not shut down, but remains connected to the line for a brief duration, i.e. 2 seconds. Beyond this time interval

the motor is disconnected either to be restarted manually or automatically with the voltage on again.
3. Time-delay and automatic reconnection — the starter is instantaneously disconnected from the power line and restarts the motor with voltage restoration. This protection is permissible if automatic reinstatement of equipment operation is required and is safe for personnel, and if in the case of a multitude of motors a combined instantaneous inrush of current can be tolerated by the power supply line.

Another protective means, undervoltage release, may be applied which stops the motor instantaneously but restarts it after any indefinite period of voltage shutdown. This method may be applied after a careful analysis of motor restarting relative to driven equipment and plant operating personnel.

With synchronous motors provisions must be made for automatic resynchronization after a voltage dip where continuity of operation of the driven equipment is required. Time delay undervoltage protection or some means of motor unloading must be provided during resynchronization unless the motor has a sufficient pull-in torque to resynchronize when the voltage is reinstated.

Phase Failure

Phase failure in a polyphase system must be prevented. It may occur under a varied set of circumstances, either in a power supply line or a branch circuit — a fuse failure, malfunction of one of the starter contacts, or a line break. When one of the phases opens while the motor is running under full load, the current in other phases increases in value and may trip the overload relays in the starter. Occasionally at certain partial loads the overload relays may not be tripped, yet current in the circuit of low impedance may cause overheating. Damage to fractional horsepower motors is rare; since this protection is an extra expense, it is usually considered only with larger, more costly motors.

Reversing of Phases

Reversing of phases results in reverse rotation of the motor. Such a situation completely upsets the flow of fluid handled. Usually such a fault is rare since great care is exercised in wiring a motor. For instance, the fans rotating in reverse are unable to deliver the design air quantity. In both instances, either open or reverse phase, there are available relays that stop the motor or prevent backward operation respectively.

Interlocking

One form of protection is interlocking of the equipment or functions within the equipment. Examples of equipment interlock are relays that prevent starting of the refrigeration compressor until (1) the condenser or chilled water pump or both are started, (2) oil pressure is up or oil pump is started (centrifugal compressor), or (3) prerotational vanes are closed (hermetic centrifugal). An example of interlocking functions within a given equipment is an interlock that prevents wound-rotor motor starting until the drum controller is in the starting position, that is, with all the resistors in the circuit.

ENCLOSURES FOR STARTERS

Starters, either singly or in combination or in multiples thereof, are usually encased in general-purpose sheet metal enclosures furnished with a dowelled or hinged access panel and doors. The small enclosures are not ventilated but are inherently protected from dust and light indirect splashing; they also protect the operator from accidental contact with the live parts.

Multiples of individual starters or in combination with protective devices may be assembled in vertical multiple cubicle control centers and may be prewired. Two NEMA types are:

Type A — contains no terminal boards (blocks) for either load or control connections.

Type B — contains terminal boards for load and control connections at each starter (most popular).

There are also dusttight and watertight enclosures that do not admit dust or water; the watertight enclosures may be used for outdoor installations.

For hazardous areas there are Class 1 and 2 enclosures. The Class 1 enclosures for flammable and corrosive vapor areas are made for both air-break and oil-immersed control. The air-break controls are heavy enough to withstand an internal explosion and to prevent hot gases from escaping to the outside. Oil-immersed controls are also used where corrosive atmospheres are present; oil protects the metal parts. Class 2 enclosures are made dusttight for locations where air-dust explosive mixtures may form.

NEMA classification of the motor control enclosures for mechanical and electrical protection of the operator and equipment are listed as follows:

General Purpose NEMA 1 — indoor, atmospheric conditions normal, sheet metal, does not prevent entry of liquid or airborne foreign particles, conduit entrances standard.

General Purpose NEMA 1 Gasketed — indoor, designed to exclude dust and other foreign airborne particles, does not meet dusttight requirements, sheet metal, conduit entrances standard.

Driptight NEMA 2 — indoor, general purpose enclosure with shields for protection from dripping liquids, conduit gasketed entrances may require special glands or hubs.

Weather Resistant (Raintight) (Weatherproof) NEMA 3 — outdoor, controls operate satisfactorily exposed to rain or sleet, sheet metal, special conduit hub or entrance to maintain weather resistant characteristics.

Watertight NEMA 4 — outdoor, water or moisture excluded from splashing or direct stream, meets specific hose test requirements, sheet metal or cast construction, special hubs or glands required for conduit entrance, no conduit entrance knockouts, external mounting feet.

Dusttight NEMA 5 — indoor, prevents entry of dusts, nonhazardous locations, sheet metal or cast construction, gasketed or equivalent, no conduit knockouts, conduit entrances predrilled, sealtight bushings required, external mounting feet.

Hazardous Locations (Gas) NEMA 7D — meets requirements of NEC for Class 1 Group D hazardous locations, cast enclosure bolted or threaded, conduit entrances threaded, special hubs or glands required, external mounting feet.

Hazardous Locations (Dust) NEMA 9E-F-G — meets requirements of NEC for Class II hazardous locations, cast construction, bolted or threaded, conduit entrances threaded, special hubs or glands required, external mounting feet.

Industrial NEMA 12 — indoor, meets JIC electrical standards for industrial equipment, excludes dust, lint, fibers, flyings and oil or coolant seepage, sheet metal gasketed, no conduit entrances, sealtight bushings required, external mounting feet.

Oiltight — no specific NEMA number designation, indoor, designed to exclude entrance of oils or coolant, used in applications similar to NEMA 12.

HAZARDOUS LOCATIONS

The NEMA classifications and definitions of hazardous locations are:

Class I, Group A — atmospheres containing acetylene.

Class I, Group B — atmospheres containing hydrogen or gases or vapors of equivalent hazard such as manufactured gas.

TABLE 8 — STANDARD NEMA SIZES AND MAXIMUM HORSEPOWERS

Size	Maximum Horsepower				
	2- or 3-phase			Single-phase	
	Volts				
	110	208/220	380,440,550	115	230
0	1½	3	5	1	2
1	3	7½	10	2	3
1½				3	5
2	7½	15	25		
3	15	30	50		
4	25	50	100		
5		100	200		
6		200	400		
7		300	600		
8		450	900		

Class I, Group C — atmospheres containing ethyl ether vapor.

Class I, Group D — atmospheres containing gasoline, petroleum, naphtha, alcohols, acetone, lacquer solvent vapors and natural gas.

Class II, Group E — atmospheres containing metal dust.

Class II, Group F — atmospheres containing carbon black, coal or coke dust.

Class II, Group G — atmospheres containing grain dust.

NEMA SIZES FOR STARTERS

Standard NEMA sizes and corresponding maximum horsepowers are listed in *Table 8*.

FUNDAMENTAL RELATIONS

Electric motors consume electric energy. Energy is utilized in various ways by the specific designs of motors. It is the purpose of this section to review briefly the fundamental elements of electric energy and their interplay affecting the performance of a motor drive to provide mechanical power for the operation of various equipment.

The text is elementary; however the correct application of motors and motor control depends on a thorough understanding of the rudiments of electric current phenomena.

The standard alternating current service is generated usually at medium voltage, distributed at very high voltage using smaller size conductors, and transformed at points of use to single, two- or three-phase, 60-cycle current of 120, 240, 480 or 600 voltages. The a-c current is also available in 50 cycles (used primarily in foreign countries) and 25 cycles. The 60-cycle frequency is most utilitarian for power and lighting applications. Medium high voltages (2300 to 4160 and 4800 volts) are used for large motors (200–250 horsepower and up) to achieve whenever possible a lower over-all installation cost of motor and auxiliaries.

Alternating current voltage alternates regularly in value and direction. *Figure 50a* illustrates a single-phase wave. Frequency or cycle is the number of complete 360 degree cycles per second (two alternations per second). If the electric power is supplied over two circuits in one of which the voltage reaches zero and other corresponding values 90 degrees later than in the other circuit, the service is two-phase (*Fig. 50b*). If the power is supplied over three circuits with corresponding current values reached at 120 degree intervals, the service is three-phase (*Fig. 50c*).

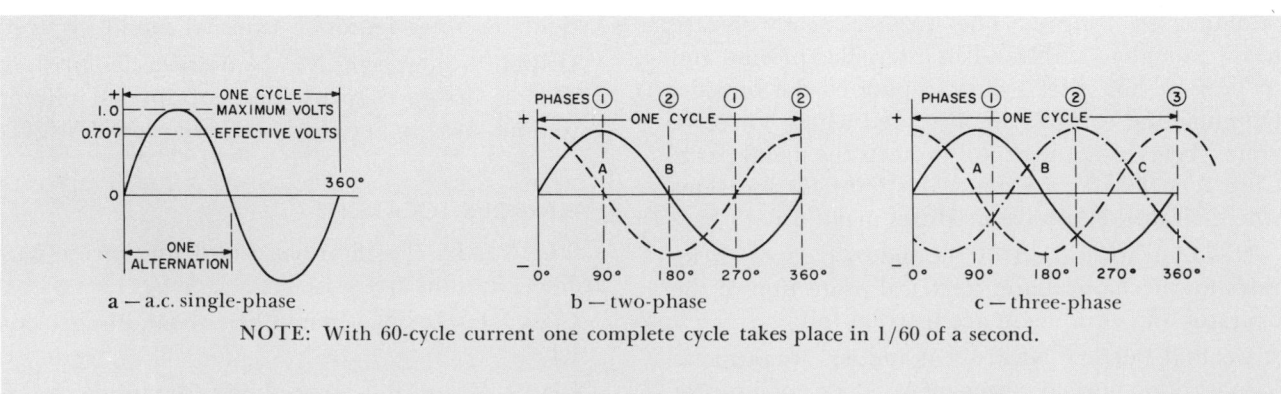

a — a.c. single-phase   b — two-phase   c — three-phase

NOTE: With 60-cycle current one complete cycle takes place in 1/60 of a second.

Fig. 50 — Electric Service

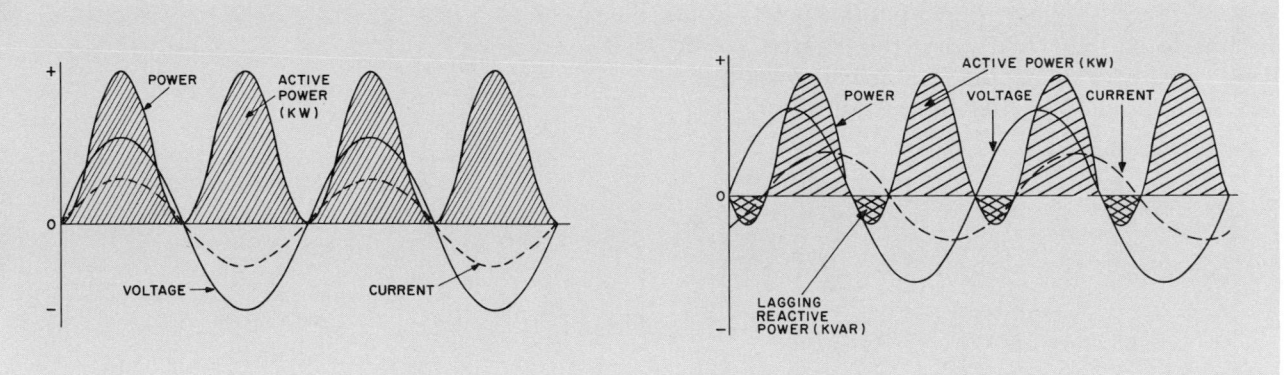

FIG. 51 — CURRENT AND VOLTAGE IN PHASE, POSITIVE ACTIVE POWER ONLY

FIG. 52 — CURRENT LAGS VOLTAGE, POSITIVE AND NEGATIVE POWER

PRESSURE, INTENSITY, RESISTANCE AND ACTIVE POWER

Flow of electricity is caused by electro-motive force (emf); the unit volt (E) is the common measure of electric pressure *(Fig. 50a)*. The top of the sine curve is the maximum line pressure. The actual effective average voltage (root-mean-square, rms, voltage) is 0.707 of the maximum, and is measured by the a-c voltmeter.

The unit ampere (I), the intensity of current, is the measure of rate of flow of electric current. The effective rms value is the value indicated on a common a-c ammeter.

A conductor of electricity inherently has resistance to flow. For a given resistance expressed in ohm units (a flow of one ampere under pressure of one volt) the emf has to be varied to change the rate of flow. With reduced resistance a given pressure (emf) increases the rate of flow.

Electric power (W) is measured in watts.

$$kw = \frac{W}{1000}$$

A watt (EI) is the product of one ampere effective current flowing at a pressure of one effective volt, in a circuit that does not contain either inductance or capacitance, for instance, as an incandescent light bulb or a heating device.

In these noninductive a-c circuits the voltage and current are in phase, reaching the maximums and minimums at the same time *(Fig. 51)*. When voltage and current are in phase, the power is the total active power consumed:

1. On single-phase service, $kw = \dfrac{EI}{1000}$

2. On two-phase service, $kw = \dfrac{2\,EI}{1000}$

3. On three-phase service, $kw = \dfrac{1.73\,EI}{1000}$

CURRENT-VOLTAGE INTERACTIONS

The relationship between current and voltage is an important aspect of motor design for two reasons: (1) power factor and (2) electromagnetism, the life blood of transformers, motors and other electric apparatus (solenoid valve) with magnetic effects created within iron cores. The magnetic lines crossed by a conductor induce a current within the conductor. This is the foundation of the motor concept.

The magnetic effects produced by electric current in an electric circuit containing coils or windings react in turn upon the current. The magnetic effects retard (check back) the current, causing it to lag behind the voltage; the current still flows in the circuit even if the voltage is zero *(Fig. 52)*; the magnetic reaction is called inductance. A condenser in an electric circuit causes current to lead ahead of the voltage. This reaction is called capacitance and tends to counteract the inductance.

APPARENT POWER

In electric circuits containing inductance (induction motors) with a continuous flow of current the product of effective current and effective voltage is greater than the actual power used to drive the motor. The cumulative apparent power is measured in volt-amperes, or usually in kilovolt-amperes (kva).

1. On single-phase service, $kva = \dfrac{EI}{1000}$

2. On two-phase service, $kva = \dfrac{2\,EI}{1000}$

3. On three-phase service, $kva = \dfrac{1.73\,EI}{1000}$

Figure 52 is the power plot, usually positive; when the current or voltage is negative, the product EI is negative power. A motor draws positive power from the line to do the actual work; the negative power (kvar) goes back to the line. The actual power is the net flow of power measured on the wattmeter.

The kvar may be evaluated as follows:

1. On single-phase service, kvar $= \dfrac{EI \sqrt{1 - PF^2}}{1000}$

2. On two-phase service,   kvar $= \dfrac{2\,EI \sqrt{1 - PF^2}}{1000}$

3. On three-phase service, kvar $= \dfrac{1.73\,EI \sqrt{1 - PF^2}}{1000}$

(PF = power factor.)

POWER FACTOR

The negative aspect of the power is the result of magnetism, the reactive current that does no actual work; however it provides the necessary magnetic field. The ratio of real active power to apparent power (watts/volt-ampere, kw/kva) is the power factor. A power factor of 1.0 (unity) all positive (*Fig. 51*) is ideal; this exists in circuits having resistance only such as incandescent lights and electric heaters. A lagging power factor less than unity is undesirable in circuits having squirrel-cage induction motors. A decrease in power factor means an increase in kvar for a given kw load. *Chart 2* shows the range of power factor effects of various electrical equipment on the power circuit.

There are various methods of controlling a power factor as close to unity as possible. The kvar inherent in induction motors is to be contained within the plant which uses the actual power (*Fig. 33*). Without control of the power factor the effects of its lagging are deleterious; they are of vital interest to the power supplier as well as the power consumer. The reasons are three-fold:

1. Low power factor means more current per kilowatt used; hence it costs more to transmit the useful power being dragged by apparent power. The consumer may have to pay more for the real power he uses.

2. Low power factor reduces the capacity of the power system to carry the real power of the total system from generator to switches at the motors; the whole system must be larger to transmit a given kilowatt load. The power producer's investment per kilowatt of load is higher.

3. Low power factor may depress the voltage, resulting in detrimental reduction of the output of electric apparatus. This lowers the performance of the consumer's plant. *Figure 53* shows the effect of a low power factor on voltage.

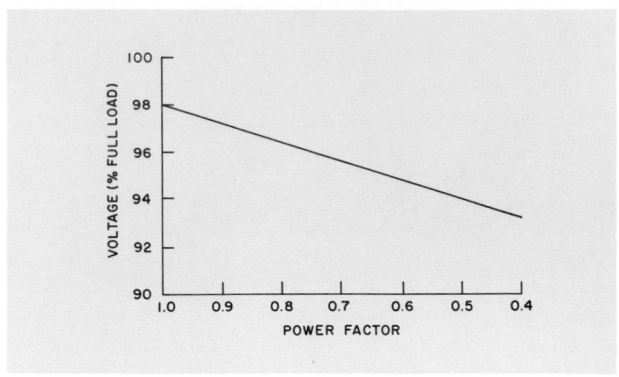

FIG. 53 – APPROXIMATE DROP IN VOLTAGE WITH LAGGING POWER FACTOR

VOLTAGE VARIATION

A variation of the line voltage affects the induction motor power factor and motor efficiency as shown in *Fig. 54*. The motor torques are raised or lowered from their design value proportionally to the square of voltage; for 90% of line voltage (10% drop), there is available only 81% (.9 x .9) of design torque. *Table 9* lists the effects of voltage variation on the elements of motor behavior. *Figure 55* presents graphically the effect of voltage and frequency variation on the two basic characteristics of a motor, the starting torque and starting current.

A decrease in voltage increases the full load current and thus increases the full load temperature rise. An increase in voltage may have the physical effects of shearing off couplings and even producing some damage to the driven equipment itself because of a sharp rise in starting and running torque.

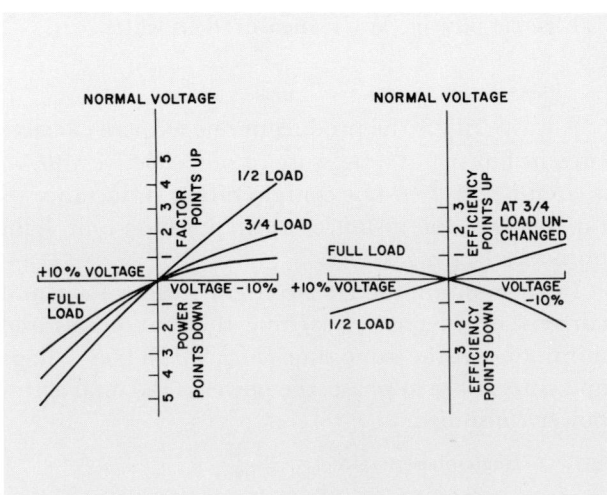

FIG. 54 – EFFECT OF LINE VOLTAGE VARIATION ON MOTOR POWER FACTOR AND EFFICIENCY

TABLE 9 — VARIATION IN MOTOR CHARACTERISTICS WITH CHANGE IN VOLTAGE

Motor Characteristic		Percent Variation with Change in Voltage of		
	Function of Voltage	+20%	+10%	−10%
Starting and Maximum Running Torque	(Voltage)²	Increase 44%	Increase 21%	Decrease 19%
Synchronous Speed	Constant	No Change	No Change	No Change
Percent Slip	1/(Voltage)²	Decrease 30%	Decrease 17%	Increase 23%
Full-Load Speed	(Syn. Speed Slip)	Increase 1.5%	Increase 1%	Decrease 1½%
Efficiency				
Full-Load	—	Small Increase	Increase ½ to 1 point	Decrease 2 points
¾ Load	—	Decrease ½ to 2 points	Practically no change	Practically no change
½ Load	—	Decrease 7 to 20 points	Decrease 1 to 2 points	Increase 1 to 2 points
Power Factor				
Full-Load	—	Decrease 5 to 15 points	Decrease 3 points	Increase 1 point
¾ Load	—	Decrease 10 to 30 points	Decrease 4 points	Increase 2 to 3 points
½ Load	—	Decrease 15 to 40 points	Decrease 5 to 6 points	Increase 4 to 5 points
Full-Load Current	—	Decrease 11%	Decrease 7%	Increase 11%
Starting Current	Voltage	Increase 25%	Increase 10 to 12%	Decrease 10 to 12%
Temperature Rise, Full Load	—	Decrease 5 to 6 C	Decrease 3 to 4 C	Increase 6 to 7 C
Maximum Overload Capacity	(Voltage)²	Increase 44%	Increase 21%	Decrease 19%
Magnetic Noise — No Particular Load	—	Noticeable Increase	Increase Slightly	Decrease Slightly

NOTE: Applicable to NEMA Design A, B and C motors.

MOTOR TORQUE

The load torque of a driven machine must be matched by the torque characteristics of the motor. The relationship between torque, power and speed is given by:

$$T = \frac{P \times 5252}{S}$$

where T = torque, pound-feet
      P = power, horsepower
      S = speed, rpm
      $5252 = \dfrac{33000}{2\pi}$

For example, a 500 hp, 1750 rpm (full load rpm) motor has a full load torque of

$$\frac{500 \times 5252}{1750} = 1500 \text{ pound-feet}$$

Motor torque is created by the interaction of a rotating magnetic field (*Fig. 26*) and the induced voltage in the rotor coils.

Figure 56 illustrates a torque curve characteristic of a squirrel-cage induction motor. It also points out the approximate locked rotor, pull-up breakdown and full load torque values against the percent motor synchronous speed.

Definitions of various torques follow:

1. Locked-Rotor* (starting, static, breakaway) Torque is developed at the instant of starting for any angular position of the motor rotor when the rated voltage is applied at rated frequency. It is the turning effort applied to the load at rest.

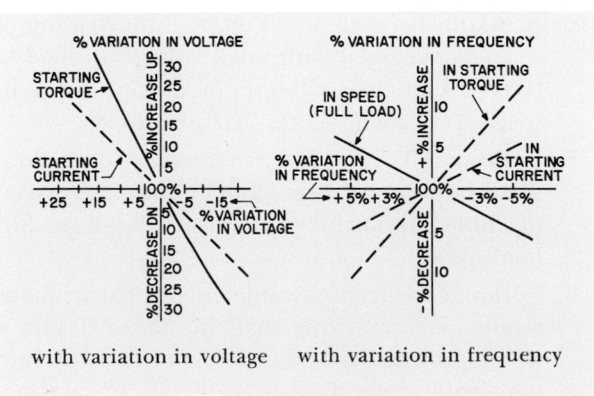

with variation in voltage   with variation in frequency

FIG. 55 — MOTOR PERFORMANCE

*Locked-rotor nomenclature is derived from the fact that in measuring starting torque the rotor is locked in position and is motionless when the current is applied to the rotor.

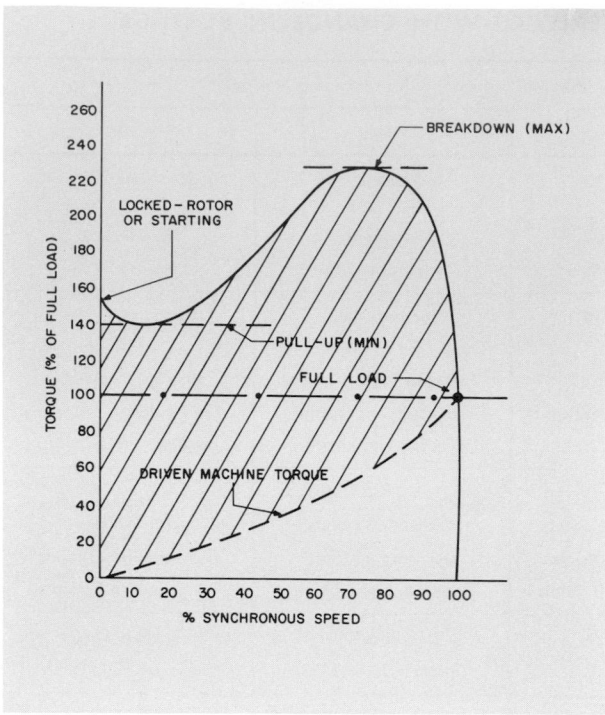

FIG. 56 — SPEED-TORQUE RELATIONSHIP, INDUCTION
MOTOR

2. Pull-Up Torque is the minimum torque developed by a motor at rated voltage during the acceleration from rest to the speed at which the breakdown torque occurs, or in general the minimum torque developed up to rated speed. A drop in minimum torque below the load torque prevents acceleration or stalls the motor.

3. Acceleration (net) torque is the margin by which the motor torque exceeds the load torque from standstill to full speed (*Fig. 56*).

4. Breakdown (pull-out) Torque is the maximum torque developed with rated voltage applied at rated frequency without an abrupt drop in speed. It accommodates brief overloads.

5. Full-Load Torque is that necessary to produce rated horsepower needed to operate the driven machine, performing in turn its task as full load speed.

6. Pull-in Torque, applicable only to synchronous motors, is the torque that the motor develops when pulling its connected inertia load into synchronism upon application of d-c excitation.

SYNCHRONOUS SPEED

The theoretical speed of a rotating magnetic field (flux) produced by electric current in the primary windings of the starter of an a-c induction motor is the synchronous speed of the motor. The synchronous motor is the only one that operates at 100% synchronous speed at full load. Other types of motors have a speed that lags the synchronous speed by a difference called slip.

$$\text{Synchronous speed S} = \text{theoretical motor rpm} = \frac{120\,f}{p}$$

where:

f   = frequency, cycles per second
p   = number of starter poles
120 = number of alternations per second for 60-cycle current

Thus theoretically a two-pole motor rotates at 3600 rpm with 60 cycle current (3000 rpm with 50 cycle, 1500 rpm with 25 cycle current). A sixteen (16) pole motor operates at 450 rpm with 60 cycle current.

SLIP

The difference between the synchronous and the operating speed of a motor is called slip (*Fig. 57*). It is expressed in percent of synchronous speed. The greater the load, the greater the slip; that is, the slower the motor runs. But even at full load the slip generally is below 5%; the motor is still considered a constant-speed motor.

$$\text{Percent slip} = \frac{\text{synchronous speed} - \text{full load speed}}{\text{synchronous speed}} \times 100$$

A motor rated at 1800 rpm and running at 1750 rpm when fully loaded has

$$\frac{1800 - 1750}{1800} \times 100 = 2.77\% \text{ slip.}$$

The significance of low slip is in the utilization of the motor synchronous speed and its maximum efficiency.

MOTOR CURRENT

Because of usually low resistance in the motor circuit when the motor is at rest, the locked-rotor

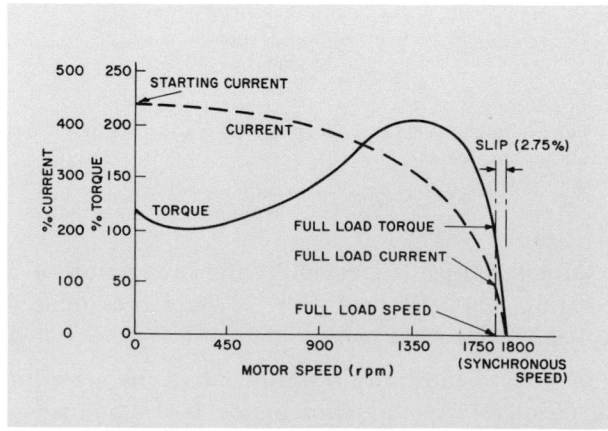

FIG. 57 — TYPICAL CURVES, MOTOR TORQUE AND
CURRENT

TABLE 10 — STANDARD EQUATIONS

DESIRED DATA	ALTERNATING CURRENT		
	Single-phase	2-Phase	3-Phase
Horsepower (output)	$\dfrac{\text{volts} \times \text{amps} \times \text{eff} \times \text{PF}}{746 \times 100}$	$\dfrac{2 \times \text{volts} \times \text{amps} \times \text{eff} \times \text{PF}}{746 \times 100}$	$\dfrac{1.73 \times \text{volts} \times \text{amps} \times \text{eff} \times \text{PF}}{746 \times 100}$
Amperes (when horsepower is known)	$\dfrac{\text{hp} \times 746 \times 100}{\text{volts} \times \text{eff} \times \text{PF}}$	$\dfrac{\text{hp} \times 746 \times 100}{2 \times \text{volts} \times \text{eff} \times \text{PF}}$	$\dfrac{\text{hp} \times 746 \times 100}{1.73 \times \text{volts} \times \text{eff} \times \text{PF}}$
Amperes (when kilowatts are known)	$\dfrac{\text{kilowatts} \times 1000}{\text{volts} \times \text{PF}}$	$\dfrac{\text{kilowatts} \times 1000}{2 \times \text{volts} \times \text{PF}}$	$\dfrac{\text{kilowatts} \times 1000}{1.73 \times \text{volts} \times \text{PF}}$
Amperes (when kva is known)	$\dfrac{\text{kva} \times 1000}{\text{volts}}$	$\dfrac{\text{kva} \times 1000}{2 \times \text{volts}}$	$\dfrac{\text{kva} \times 1000}{1.73 \times \text{volts}}$

NOTE: Equations for 2-phase are set up for 4-wire circuits. In 3-wire circuits the current in the common conductor is 1.41 times that in either of the other two conductors. Efficiency is expressed in an integral number (90%). Power factor is expressed as a decimal (0.85). Refer to Charts 1 and 2.

momentary inrush of current (starting current) is four to six (up to 10) times greater than the full load *(Fig. 27)*. The low figure is the standard for the majority of motors specifically designed to satisfy the power companies' concern with taxing the transmission lines with disturbances that affect the power transmission performance; these disturbances are due to possible large low power factor starting currents. The customer charges normally are affected only slightly since the inrush is momentary in nature; this inrush may have only slight effect on the demand charge which is based on continuous use of current for at least fifteen minutes. However in the case of frequent starting or long acceleration periods, the demand is integrated over a period of time (usually fifteen minutes) and may up the demand charge. The same amount of current is drawn at starting in case a motor is stalled because of mechanical overload.

EFFICIENCY

Motor efficiency is a measure of the motor capacity to convert electric energy input to mechanical horsepower output and is expressed in percent of kw input:

$$\text{Percent efficiency} = \frac{\text{hp output} \times .746}{\text{kw input}} \times 100$$

Approximate comparative efficiencies of standard squirrel-cage induction motors at three different synchronous speeds are shown in *Chart 1*.

HORSEPOWER AND CURRENT

To facilitate the evaluation of horsepower output and current consumption in amperes, standard equations are presented in *Table 10*.

The objective of this chapter has been to provide an outline or an introduction to the great wealth of material available on motors and motor controls. The manufacturers' data and catalogs provide the necessary details. The proper selection of motors and motor controls is an inseparable part of the design of air conditioning and refrigeration systems that are part of the total over-all mechanical equipment. The performance and acceptance of a system may well rest on the electrical equipment selected to operate it.

CHAPTER 3. BOILERS

This chapter presents information to guide the engineer in the practical application and layout of boilers when used in conjunction with air conditioning and refrigeration systems.

The scope of this chapter is limited to packaged boilers which have capacities to cover the applications with which the engineer is concerned. Steam generation as well as water heating may be effected by the utilization of such boilers up to comparatively high steam pressures and water temperatures respectively.

TYPES OF BOILERS

Boilers may be classified in two general groups:
1. Sectional cast iron boilers.
2. Steel firebox boilers, fire-tube or water-tube.

Cast iron boilers may be rectangular (or square) with vertical sections or round with horizontal sections. These boilers are usually shipped in sections and assembled at the place of installation.

Some smaller boilers are factory-assembled. Some have water-filled spaces completely surrounding the combustion chamber. Cast iron boilers are normally limited to 15 psig steam pressure and 30 psig water pressure (274 F) with IBR* net load ratings ranging up to approximately 2,500,000 Btu/hr output.

Fire-tube steel boilers have their combustion gases passing thru tubes surrounded by circulating water. A packaged steam or hot water fire-tube boiler is a modified Scotch type boiler† having all components in an assembled unit. Components include burner, boiler, controls and auxiliary equipment. Most modern fire-tube units operate at or below 250 psig and below about 20,000 lb steam/hr. Fuels for packaged fire-tube units may be oil, gas or a combination of these.

Water-tube steel boilers have their combustion gases circulating around the tubes and water passing thru the tubes. Most modern water-tube packaged units have capacities ranging up to 60,000 lb steam/hr and pressures up to 900 psig. Capacity is limited by shipping clearances. Water-tube units are designed principally for oil, gas or a combination of these fuels. These boilers can be adapted to solid fuels more readily than can fire-tube steel boilers.

APPLICATION

Cast iron boilers are suitable for steam generation or hot water heating where low pressures are used. They may be applied in commercial and industrial buildings within the capacity range available.

The capacities of cast iron and steel boilers overlap. Where this occurs, any comparison should include the following:
1. Steel boilers in the larger sizes are more efficient.
2. With proper maintenance and use, a cast iron boiler outlasts any steel boiler made. However, where the character of maintenance is apt to result in neglect, the serviceability of the steel boiler is of marked advantage.
3. A skilled steam fitter is required to assemble the heating sections of cast iron boilers (when field-erected). The steel boiler has only to be placed into position. However, since the sections of a cast iron boiler are so designed as to be readily carried thru doors or windows, ease of installation generally favors the cast iron boiler.
4. The relative cost of steel boilers in the smaller sizes is greater than that of cast iron boilers of the same capacity.
5. When more boiler capacity is required in the extension of a system, additional sections may be added to a cast iron boiler, whereas a separate or replacement steel boiler of a larger size must be considered.

*Institute of Boiler and Radiator Manufacturers.
†Type of boiler evolved to meet space and weight requirements of the merchant marine. It is self-contained, requires no brick setting, and can be operated at high ratings without damage.

Fire-tube steel boilers are used principally in small heating and industrial plants. Its popularity is growing in the industrial, commercial and institutional fields. Shell diameters are limited to about 96 inches. Cost of installation including setting is considerably less than that of a corresponding water-tube boiler. With a water-filled cylindrical shell housing an internally fired furnace, a relatively long gas-travel path yields high efficiency in a compact unit.

Water-tube steel boilers pick up in capacity near the upper end of the fire-tube range to extend the availability of the packaged concept. Pressure and size limitations of fire-tube boilers do not exist in water-tube units; these boilers require no prepared setting other than a floor of sufficient strength and no skilled labor for assembly prior to operation.

Fire-tube and water-tube steel boilers may be used for hot water heating.

STANDARDS AND CODES

Boiler installation should conform to applicable national, state, local, ASME, utility and insurance code requirements. The ASME boiler and pressure vessel code (Sections I and IV) prescribe methods of boiler design, construction and installation. The Mechanical Contractors Association of America prescribe boiler testing and rating procedures.

STEAM BOILERS

Where applicable, low pressure steam generators are recommended because boilers operated at more than 15 psig pressure must, generally, be tended by a licensed operator.

LOW PRESSURE BOILERS

Cast iron or fire-tube steel boilers may be applied to low pressure steam generation.

Hartford Return Loops illustrated in *Part 3* are recommended to prevent the loss of boiler water by backward flow into the return mains.

Cast Iron Boilers

Cast iron boilers are designed for oil, gas or coal fuel.

Figure 58 shows a typical gas-fired, cast iron steam generator. Good boiler design incorporates methods of breaking up the hot gases leaving the firebox and, by means of passes or baffles, promotes contact with the heating surface at a high gas velocity but with a reasonable resistance.

Courtesy of Bryant Manufacturing Co.

FIG. 58 — TYPICAL GAS-FIRED CAST IRON GENERATOR

Fire-Tube Steel Boilers

Practically all packaged fire-tube units are available for low pressure heating service (15 psig and below). They are relatively inexpensive compared to the corresponding water tube boilers. *Figure 59* shows a typical fire-tube steam generator.

The design and construction of a fire-tube boiler presents definite limits to the sizes which can be built. Practical limits of size and pressure result from the fact that the entire steam-making process takes place inside a shell. Since shell strength to resist rupture is proportional to pressure times diameter, high pressures and large diameters lead to prohibitively thick shells.

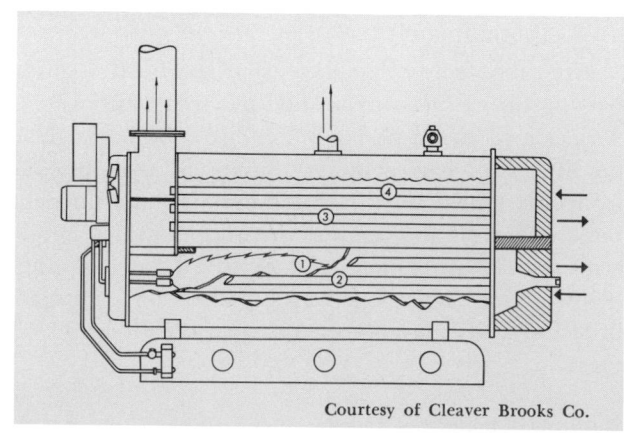

Courtesy of Cleaver Brooks Co.

FIG. 59 — TYPICAL FIRE-TUBE STEAM GENERATOR

The fire-tube boiler has a large water-storage capacity; thus the wide and sudden fluctuations in steam demand are met by only a slight change in pressure. Because of the large water content, a longer time is required to bring the boiler up to operating pressure. Overload capacity is limited and exit gas temperatures rise rapidly with increased output. Oil- or gas-fired, the modern packaged fire-tube boiler operates with efficiencies of about 80% over a wide load range. Coal, stoker-fired boilers operate from 60–75% efficiency.

The fire-tube design is not readily adaptable to the installation of soot blowing equipment. However, with relatively large tube diameters compared with water-tube boilers and with the products of combustion confined within the tubes, the turbulent high speed gas tends to produce a scrubbing action and maintain tube surfaces relatively free of combustion deposits. The fire-side surfaces of the tubes may require brushing at periodic intervals, the length of the intervals depending on the cleanliness of the combustion process and the type of fuel used.

HIGH PRESSURE BOILERS

Both fire-tube and water-tube steel boilers may be applied to high pressure steam generation.

Water-Tube Boilers

Water-tube boilers *(Fig. 60)* have compact and efficient heating surface layouts by combining water-wall and convection surfaces; they are well suited for low head, limited space applications.

Waterwalls handle the bulk of the heat absorption in most packaged water-tube units. Exposed to radiant heat the waterwall heat transfer rate is high. In nearly all designs the drum is arranged with its long axis parallel to the furnace length.

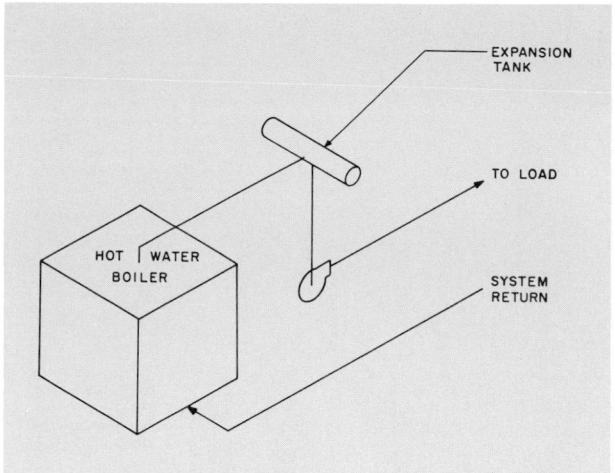

FIG. 61 — DIRECT SYSTEM,
HIGH TEMPERATURE WATER

Practically all water-tube boilers are equipped with soot blowers when delivered, or have provision for easy installation of soot blowing elements. These are usually of the steam type.

HOT WATER SYSTEMS

In hot water applications the range of temperatures involved is from 180 F (conventional gravity hot water heating system) to 400 F (accepted practical upper limit for industrial applications). Two basic types of hot water systems are the indirect cascade and the direct systems.

A direct system *(Fig. 61)* generally has a separate tank which provides expansion as the water temperature changes. If forced circulation is used, a centrifugal pump draws water from the tank, circulates it thru the system, sends it to the boiler for reheating, then returns it to the tank to complete the cycle.

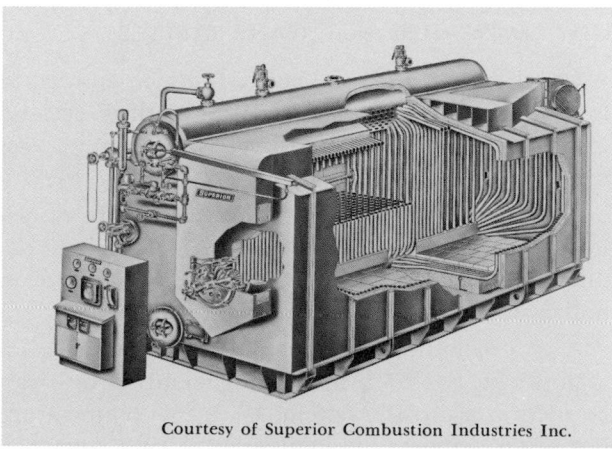

Courtesy of Superior Combustion Industries Inc.

FIG. 60 — TYPICAL WATER-TUBE STEAM GENERATOR

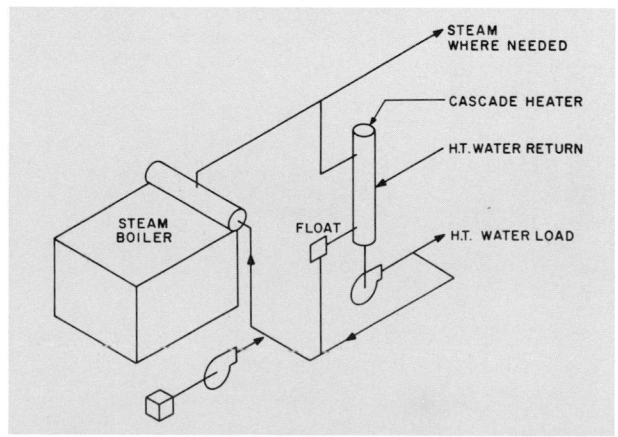

FIG. 62 — INDIRECT SYSTEM,
HIGH TEMPERATURE WATER

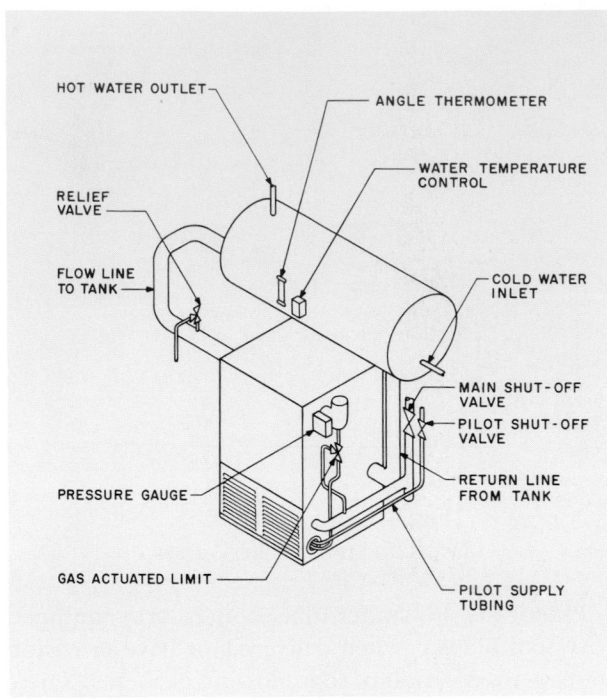

FIG. 63 — DIRECT SYSTEM,
LOW TEMPERATURE HOT WATER

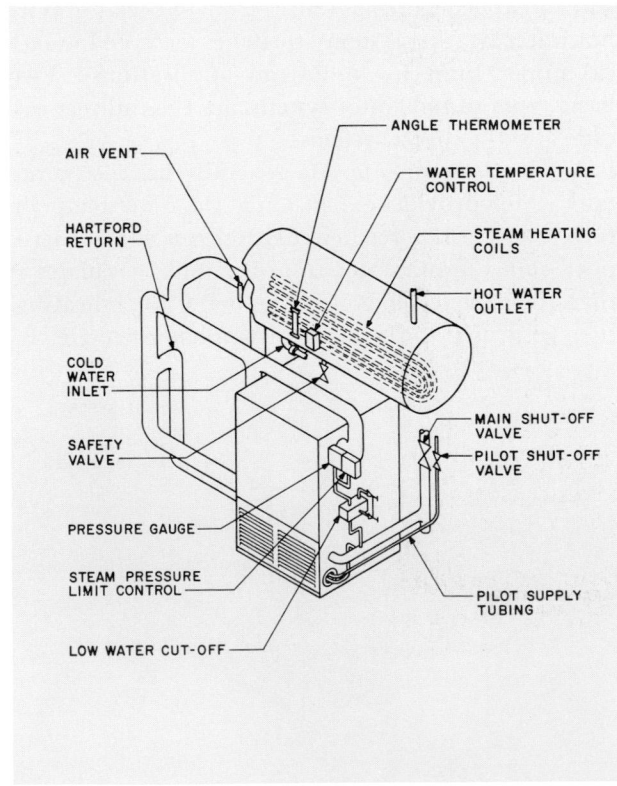

FIG. 64 — INDIRECT SYSTEM,
LOW TEMPERATURE HOT WATER

An indirect system (*Fig. 62*) takes steam from a boiler and carries it to a direct contact heater that raises the water temperature to within about two degrees of the entering steam temperature. From the heater, pumps circulate the high temperature water to heat exchangers at points of use. Condensate is generally returned to a hot well. Feed water may be supplied from both the hot well and water portion of the direct contact heater.

The direct type of system is generally employed except on installations where existing steam generators are used, when a source of exhaust steam is used with a cascade heater to produce high temperature water, or where specific limitations on the use of a cast iron boiler prevent it from being used on a direct system. Specific limitations on a cast iron boiler are these:

1. Not recommended where the hardness of the water produces lime deposits in the boiler sections.

2. Not recommended where the city water mains are used as an expansion tank and the city water supply pressure is greater than the boiler maximum working pressure.

High temperature water applications may be divided into three categories:

1. Low temperature (LTHW) range is from 180–250 F with corresponding saturation pressures from 0–15 psig.

2. Medium temperature (MTW, sometimes called intermediate) range is from 250–300 F with saturation pressure ranging from 15–52 psig.

3. High temperature (HTW) range is from 300–400 F with saturation pressure readings of 52–233 psig.

LOW TEMPERATURE HOT WATER SYSTEMS

Cast iron and fire-tube boilers may be applied to low temperature hot water heating with direct and indirect systems. A typical direct system using a cast iron boiler is shown in *Fig. 63*.

With suitable provision for introducing the return low temperature water without thermal shock to the unit, the conventional fire-tube packaged steam boiler may be adapted to the direct system.

A typical indirect system using a cast iron boiler is shown in *Fig. 64*. Any steam generating unit can be used in a cascade system. With low temperature hot water applications temperature rises up to 50 degrees may be used.

MEDIUM AND HIGH TEMPERATURE WATER SYSTEMS

Water temperatures greater than 274 F can be secured with some cast iron boilers utilizing a direct system. Boilers may also be water-tube or fire-tube, and may be equipped with any conventional fuel firing apparatus.

Since there is a partial correlation between pressure and capacity, maximum pressure and temperature seldom prove economical except in the higher capacity range. Water-tube boilers are seldom designed for pressures below 150 pounds; they are preferred for the higher pressure and capacity ranges.

With medium and high temperature hot water applications, advantage may be taken of high temperature rises up to 100 F and 200 F respectively.

Figure 65 shows a typical fire-tube hot water boiler. *Figure 66* shows the flow pattern in a hot water boiler.

BOILER PERFORMANCE

The term performance refers to the rate of output, efficiency of heat transfer, and draft and pressure requirements of the unit or any of its component parts.

CAPACITY

The capacity or output of a boiler is expressed in many ways. The most accurate method of rating is in terms of total heat transferred per hour to the water or steam as it passes thru the unit. Capacity may also be expressed in equivalent direct radiation, boiler horsepower or actual evaporation.

However such a rating reflects only the boiler output under laboratory test conditions and does not provide for piping loss and starting load allowances; when selecting a boiler, its net load rating (IBR, SBI*, ABMA†) should be equal to or exceed the calculated heat requirements of the building. The net load rating varies from 75% for oil, gas and automatic coal-fired boilers to approximately 40% of the gross output for small hand, coal-fired boilers. When required by job conditions, the calculated heat requirements of the building should take into consideration the startup steam demand of the miscellaneous equipment supplied by the boiler.

Equivalent Direct Radiation

An equivalent square foot of steam radiation surface (EDR) is defined as the amount of surface which emits 240 Btu/hr, with a steam temperature of 215 F and a room air temperature of 70 F. With hot water

*Steel Boiler Institute.

†American Boiler Manufacturers Association.

Courtesy of Superior Combustion Industries Inc.

FIG. 65 — TYPICAL FIRE-TUBE HOT WATER BOILER

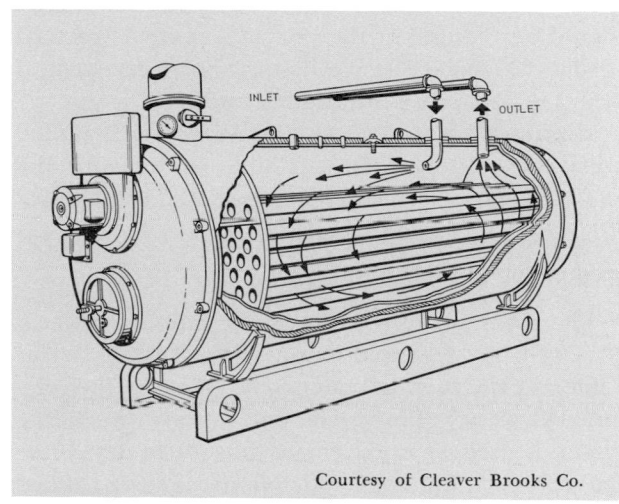

Courtesy of Cleaver Brooks Co.

FIG. 66 — INTERNAL FLOW PATTERN, HOT WATER BOILER

the value of 150 Btu/hr may be used for a 20 F drop. However, the EDR unit is being replaced by the more universal Btu/hr rating.

Boiler Horsepower

A boiler horsepower (BHP) is defined as the evaporation of 34.5 lb of water per hour from a temperature of 212 F into dry saturated steam at the same temperature. This is equivalent to 33,475 Btu/hr or 139.5 sq ft of equivalent direct steam radiation or 223.1 sq ft of hot water radiation.

Actual Evaporation

The term most commonly used is actual evaporation, i.e. pounds of steam generated per hour (lb steam/hr) at the given steam temperature and pressure. This does not offer an accurate comparison between one unit and another since the heat transferred per pound of steam generated may vary widely, depending on steam pressure, temperature and feed-water temperature.

Percent of Rating

It has been customary to rate boilers on the basis of 10 square feet of heating surface per boiler horsepower. However, as boiler designs and firing methods have improved, boilers can now develop several times the capacity based on former methods of rating. The ratio of actual to nominal capacity has been stated as percent of rating. Although the term has become obsolete, it is still used occasionally with reference to standardized boilers of low capacity.

Corrections to Ratings

For elevations above 2000 feet, boiler ratings should be reduced at the rate of 4 percent for each 1000 feet above sea level unless the boiler mechanical draft fan capacity is adjusted accordingly.

Minimum required gas pressures for gas-fired boilers should be adjusted upward for altitudes above 700 feet. Correction factors should be obtained from the boiler manufacturer.

EFFICIENCY OF HEAT TRANSFER

The efficiency of a boiler is the ratio of the heat absorbed by water and steam to the heat (calorific value) in the fuel. In commercial practice, the combined efficiency of the boiler and furnace (including grate) is used. It is extremely difficult to determine the actual efficiency of a boiler alone, as distinguished from the efficiency of the combined apparatus.

Stack losses (sensible heat lost in flue gas) are almost always the most serious source of furnace inefficiency. A maximum stack temperature of 500–600 F is considered good practice by many engineers. Exit gas temperatures of 100–150 degrees above saturated steam temperatures are typical. *Figure 67* shows the typical performance of an oil-fired fire-tube boiler at a constant pressure.

DRAFT AND PRESSURE

The various items included in the pressure differential across the convection surface of a boiler are:
1. Friction due to flow across tubes.
2. Loss in head due to turns.
3. Friction due to flow thru or parallel to tubes.
4. Stack effect.

Turns in boiler passages are usually of the severest

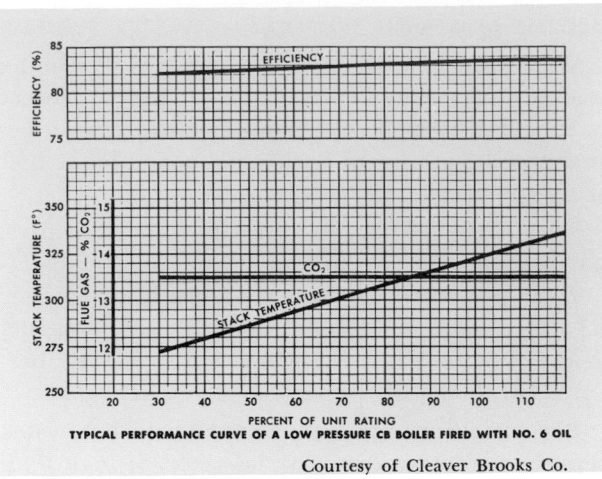

Courtesy of Cleaver Brooks Co.

FIG. 67 — TYPICAL PERFORMANCE CURVES, OIL-FIRED FIRE-TUBE BOILER

type in that they are generally 180 degrees and are very sharp.

Because of the great difference in the coefficient of heat transfer between cross and parallel flow, lower velocities are usually used for cross flow in order to obtain reasonable draft losses.

The flow of air thru the fuel bed and the products of combustion thru the boiler breeching and stack result in a pressure drop. In order to keep the gases moving at the rate required to maintain combustion, either mechanical or natural draft is required to overcome this pressure difference.

Mechanical draft implies the use of forced or induced draft fans whose characteristics and selection are similar to those in ventilating work, except for a heavier construction.

Natural draft is produced by chimneys or stacks which serve to discharge the gaseous combustion products at an elevation sufficiently high to avoid pollution of the immediate surroundings. Their selection involves the determination of (1) the amount of draft required, (2) the stack height needed to produce this draft, (3) the weight rate of flow of flue gases and (4) the stack section area necessary to accommodate this flow.

Chimneys produce a draft as a result of the difference in density between the column of hot gases inside and the air outside. The net useful draft is the difference between this theoretical static head and the resistance of the chimney itself due to gas flow.

Mechanical draft fans for packaged boilers are designed in accordance with the combustion requirements of the burner. Natural stack draft is not required with the packaged automatic boiler. Stack construction and maintenance costs are eliminated. Only a small vent is needed to carry flue gases outdoors.

FUELS

The principal fuels used for combustion in a boiler are coal, oil and gas; the choice of fuel is usually based on availability, dependability, cleanliness, economy, operating requirements and control.

AVAILABILITY

The unavailability of a particular type of fuel may preclude its use. Reasonably certain long term availability of a fuel and less likely interruption of supply in the event of any emergency should be considered.

Local codes may prohibit the use of certain types of a fuel.

DEPENDABILITY

Dependability may be measured in terms of quantity and quality.

An interruptable service rate to consumers of gas fuel is lower than the normal service rate but at the same time permits the gas company to interrupt the service during times of greatest demand, such as may occur under extreme weather conditions. To accept such an arrangement for gas service when available means that the fuel is not dependable as to quantity. However, if the firing equipment of the boiler is a combination gas-oil burner, oil can be used during short periods of interrupted service to provide an optimum arrangement.

It is desirable that the fuel used be of a consistent quality. A varying fuel quality can prevent optimum economy due to decreased efficiency and increased maintenance.

CLEANLINESS

General cleanliness is inherent in oil- and gas-fired boilers. However, there is an increasing demand by both government agencies and private industry for coal-fired packaged boilers. As a result over the past few years there has been a concentrated effort on the part of the coal-fired boiler and firing equipment designers to approach the cleanliness of the oil- and gas-fired equipment.

ECONOMY

Relative economy of various fuels does not depend on the heating value of the fuel itself as much as on the conditions attending its use. The final cost of steam or hot water which determines the most economical fuel depends on (1) charges for operation and maintenance, (2) cost of fuel, (3) charges for handling fuel, (4) cost of operation and maintenance of auxiliary equipment and (5) fixed charges for standby capacity. Item 1 is common to all fuels; items 2 to 5 vary with each fuel.

The average cost of burning coal in this country exclusive of fixed charges is about 5% of its cost; for fuel oil approximately 1.5% of the equivalent coal cost, and for natural gas 0.5%.

Storage and handling problems assume greater proportions with the solid fuels. They are essentially nonexistent with gas and are easily handled with oil. While a somewhat higher initial investment may be required for a coal-fired unit than for a gas- or oil-fired unit, other factors tend to have a balancing influence on a long term basis. In many areas coal offers a lower cost per Btu than other fuels delivered to the plant.

The advantages of fuel oil over coal are these:

1. Weight 30% less and space occupied 50% less than coal of equivalent heat content.
2. Freedom from spontaneous combustion.
3. Storage may be distant from furnace.
4. Fuel immediately available, stored or removed with practically no labor.
5. High combustion rates per cubic foot of combustion space.
6. Great flexibility in furnace to carry peak and valley loads readily and economically.
7. Low labor cost to handle oil at the furnace and to clean boiler tubes.
8. No labor for cleaning fires or removing ashes.
9. High efficiency and relatively no smoke.
10. Absence of wear on machinery due to ash and dust.
11. Low pressure drop thru the furnace.
12. Minimum of excess air required for complete combustion.

The advantages of gas over coal are these:

1. Burned in furnaces where the supply can be varied almost instantaneously between wide limits by manual or automatic control.
2. Complete combustion obtained with low excess air; flue losses low and operation smokeless.
3. Furnace can be maintained with an oxidizing or reducing atmosphere with ease and little reduction in efficiency.
4. No storage facilities needed on the premises of the customer.

OPERATING REQUIREMENTS

With most coal and some gas boilers, when a fuel is burned continuously at a rate to match the load, the maximum load rarely occurs. For this reason poor efficiency can be tolerated at full load if the bulk of operation occurs at good efficiency. Conversely, boilers with intermittent oil or gas burners operating at maximum load at all times during the on cycle, should be selected for good efficiency at this load.

CONTROLS

Controls are a consideration in the choice of a fuel. For equivalent control, coal-fired boiler control is more complicated and expensive than that for oil or gas. Similarly, control for oil is more complex and costly than that for gas.

BOILER SELECTION

Factors which exert the greatest influence on the selection of boilers are fuel characteristics, capacity and steam conditions, space conditions, cost and individual preference.

FUEL CHARACTERISTICS

Prior to a preliminary selection of equipment, complete information should be available concerning fuels on which boiler types and predicted performance are based. Numbers 5 and 6 fuel oil require preheating equipment before the boiler. It is also desirable to determine a secondary fuel supply for emergency use when the primary fuel supply is interrupted or when changes in price make the secondary fuel more economical.

Where possible, equipment selection should be such that secondary fuel performance is equivalent to that of the primary fuel. However, if any interruption in primary fuel supply is only temporary and if the price differential between primary and secondary fuels is fairly stable, it may be more economical to design for maximum efficiency with the primary fuel and to accept some compromise in performance and maintenance costs with the secondary fuel. Packaged boilers are currently offered for gas firing with oil standby. They operate with equal efficiency on oil or gas.

CAPACITY AND STEAM CONDITIONS

Capacity is one of the most important factors in determining the type of unit to be selected. There is a partial correlation between steam conditions and capacity. Maximum pressures and temperatures seldom prove economical except in the higher capacity ranges. Limitations imposed by steam pressure and temperature are predominantly structural. They affect the weight of steel required, hence the cost; temperature affects the space required by the superheater and adaptability of the boiler to provide that space.

SPACE CONDITIONS

In an existing building both shape and volume of the space available have a marked effect on the capacity of the unit to be installed, type of firing, and possibly the range of fuels which can be fired at a given capacity.

COST

Caution should be exercised in the degree to which first cost is allowed to influence the equipment selected. A complete economic study should be made by considering the load factor of the installation, the cost of fuel, and the efficiency of the installation as a whole rather than the boiler equipment alone.

A small plant with an ample supply of low priced fuel and a seasonal load of a few months each year can justify a standard boiler and natural draft. However, a plant with a load factor approaching 100% using a high priced fuel can readily justify an efficient fuel burning system, high steam pressure and temperature, and induced draft fans. The cost of the fuel burned during the life of such a unit may be many times the initial investment. Even a small advantage in reliability, efficiency or flexibility gives economic justification for the relatively small additional first cost necessary to provide the better unit.

INDIVIDUAL PREFERENCE

Individual preference should be considered if the plant personnel is familiar with the operation of a given type of equipment, or if the plant is designed for specific equipment and is unsuitable for other equipment without expensive changes. However the improvements in design and the higher efficiency or capacity that may be obtained within the same space at reduced cost for labor and maintenance should not be overlooked.

LAYOUT

Considerations in the layout of a boiler installation are location, vent or chimney, air supply and water treatment.

Most boiler manufacturers publish information relating to the specific details of their boilers and the requirements of auxiliary equipment used in connection with boiler plant design.

LOCATION

Boilers should be located at a central point with respect to the heat transfer equipment it serves and in a space provided with maximum natural light. For example, a gas fired boiler with mechanical draft located in a roof penthouse with the central station air conditioning equipment may prove economically attractive. The greater part of the piping normally required is that needed to supply gas to the boiler and interconnect the boiler and air conditioning equipment. Minimum gas pressure requirements at the boiler should be checked and compared with the available pressure. Oversized gas trains at the boiler may be used to reduce gas pressure requirements. Only a small vent is needed to carry flue gas from the boiler to the outdoors.

An adequately strong and level floor is required for the location of a packaged boiler. If the floor is not level, a concrete pad should be constructed. This also provides inspection accessibility to the piping beneath the boiler, and added height for washing down the area beneath the boiler. A boiler should not be installed on combustible floors unless so approved. Isolation of the boiler may be necessary in low sound ambient areas.

There should be adequate clearances around the boiler for access and service. Manufacturers should be consulted for recommended clearances. The space in front of the boiler should be sufficient for firing, stoking, ash removal, and cleaning or renewal of flue tubes. Space should be allowed on at least one side of every boiler for convenience of erection and for accessibility to the various dampers, cleanouts and accessories. Space at the rear of the boiler should be ample for vent or chimney connection and cleanouts. A service trench in the boiler room floor for fuel and miscellaneous piping is recommended for optimum room appearance. Boiler room height should be sufficient for the location of boiler accessories and for proper installation of piping. Room height varies directly with:

1. Height and size of boiler.
2. Steam header size and location.
3. Breeching size and location.
4. Local and insurance code requirements.

While more boilers are installed in boiler rooms completely protected from the elements, such housing is not entirely essential to their operation when built especially for permanent outdoor installation. To cover the vital working parts, a special housing called a "dog house" is available or is included in the design. While providing the required protection, this

FIG. 68 — BOILER FOR OUTDOOR INSTALLATION

housing provides access to working parts. *Figure 68* shows a packaged boiler for outdoor service.

VENT OR CHIMNEY

Natural stack draft is not required with mechanical draft packaged boilers. Only a small vent the size of the boiler vent outlet is needed to carry flue gases outdoors. On multiple boiler installations when building conditions permit, the simplest and most efficient method of venting the flue gas is the use of individual stacks. To minimize steel breeching and stack condensation, insulation is used to lower heat losses.

Stack heights in excess of 150 feet or extremely large breeching and stack combinations may cause excessive draft. A barometric damper located close to the stack in the breeching should be considered only after serious burner adjustment problems.

Considerations relating to a chimney for a natural draft boiler are these:

1. Recirculation effect decreases with chimney height and an increase in flue gas velocity.
2. Two or more chimneys (large or small) should be used separately, never connected.
3. Excessive height in a chimney does no harm but means are necessary for controlling the induced draft.

AIR SUPPLY

All rooms or spaces containing boilers should be provided with a constant supply of combustion (and ventilation) air at adequate static pressure to insure proper combustion in the fuel burners. The importance of providing proper combustion air should not be underestimated and a failure to do so may result in erratic or even dangerous operating conditions for the equipment. Rules for providing air supply openings are found in technical standards, in state and municipal building codes, and in service and installation bulletins published by manufacturers of fuel burning equipment.

Approximately 1 cfm of combustion air per 4200 Btu/hr boiler gross output and 1 cfm of ventilation air per 17,000 Btu/hr boiler gross output should be provided for the boiler room for oil- or gas-fired boilers for altitudes up to 1000 feet. At higher elevations three percent more air per each 1000 feet should be provided if the boiler rating is not reduced.

WATER TREATMENT

If the boiler water is scale- or sediment-forming or corrosive, measures should be taken to correct this condition. Consult a water treatment specialist and make necessary piping arrangements to provide such treatment. Refer to *Part 5*.

Bottom blowdown helps remove impurities. A continuous blowoff system should be considered whenever the percentage of raw make-up water is 50% or more or when the raw water contains a high amount of impurities.

CHAPTER 4. MISCELLANEOUS DRIVES

This chapter presents practical information to guide the engineer in the application and layout of steam and gas turbines and gas and diesel engines used with air conditioning systems.

These drives may replace electric motor drives when there is a lack of electric power or when there is an economic advantage. Gas may be used where utility companies offer favorable rates for off-season users. Steam may be available as waste from a high-pressure source.

Steam turbines are used for driving refrigeration machines (centrifugal or reciprocating), fans and pumps. Gas turbines, gas engines and diesel engines are usually used only for driving refrigeration machines.

STEAM TURBINE DRIVE

A steam turbine drive is usually chosen to improve a heat balance where exhaust or high pressure steam is available. When a centrifugal refrigeration machine, fan or pump is fitted into a heat balance, operating economies can be obtained.

TYPES OF STEAM TURBINES

Steam turbine drives are available as single stage and multi-stage.

Single Stage Turbine

In a single stage turbine steam expands from the initial to the final exhaust pressure in one nozzle or set of nozzles (all working at the same pressure), and the energy is absorbed in one or more rows of revolving blades.

Multi-Stage Turbine

In a multi-stage turbine the expansion of steam from the initial to the exhaust pressure is divided into two or more drops thru a series of sets of nozzles. Each set is followed by one or two rows of revolving blades which absorb the energy of each pressure drop.

A typical steam turbine is shown in *Fig. 69*.

APPLICATION

A steam turbine may be operated as a condensing or noncondensing (back pressure) turbine. When steam is not otherwise required for process heating or other application, a condensing turbine drive produces power for the least amount of steam. All the steam used is chargeable to power costs. When steam is required for other applications, a noncondensing turbine provides power at the lowest cost because the exhaust steam from the turbine can be used for other applications.

Condensing Turbine

Condensing turbine drives may be used with either high or low pressure steam. This turbine is higher in first cost than a noncondensing unit because of the additional condensing equipment required.

High pressure condensing turbines are used to produce power with a minimum amount of steam when the exhaust steam cannot be utilized or to secure a maximum amount of power with limited boiler capacity.

Low pressure condensing turbines utilize exhaust steam from existing equipment, producing power from steam that would otherwise be wasted. They

Courtesy of Elliott Co.

Fig. 69 — Typical Steam Turbine

are often utilized in summer when available exhaust steam cannot be used for heating.

Noncondensing Turbine

Noncondensing turbine drives are particularly economical when the demand for process or heating steam is sufficient to utilize all of the turbine exhaust steam. Under such circumstances the turbine acts as a reducing valve and produces power at a very low cost. It is also used when very low cost steam is available and the exhaust from the turbine is wasted or when no condensing water is available.

STANDARDS AND CODES

Steam turbine application and installation should conform to all codes, laws and regulations applying at the job site.

TURBINE SELECTION

The system requirements influencing the selection of a steam turbine drive are type of driven equipment, governor, maximum horsepower and rpm, available turbine inlet steam pressure, superheat and design turbine exhaust pressure. When these requirements are known, the selection of a steam turbine drive usually involves the choice of the most inexpensive combination of frame size and stages with an acceptable steam rate.

The selection of a turbine having a horsepower rating 5% greater than the design horsepower required for the refrigeration machine is recommended for comfort cooling applications. A minimum 10% safety factor is recommended for industrial refrigeration applications at air conditioning temperature levels. These recommendations are based on a minimum scale factor of .0005 for both the cooler and condenser.

Multi-stage turbines are more efficient than single stage turbines and more expensive. However, for the same operating conditions the exhaust steam is of lower quality. For certain applications a lower quality exhaust steam may be undesirable.

STEAM RATE

An approximate turbine steam rate may be determined from this formula:

$$\text{Approximate steam consumption (lb/hp/hr)} = \frac{\text{theoretical steam rate (lb/hp/hr)}}{\text{approx. over-all efficiency}}$$

Theoretical steam rates and approximate over-all efficiencies may be obtained from *Table 11* and *Chart 3* respectively. Actual turbine efficiencies and steam rates should be obtained from the turbine manufacturer.

CONDENSER

A condenser is required to condense the exhaust steam from a condensing turbine. A shell-and-tube type is used with the steam condensed in the shell and condenser water from the refrigeration machine pumped thru the tubes.

Water required for condensing should be piped first thru the refrigeration condenser and then in series thru the steam condenser. This arrangement results in minimum piping, power requirements and steam consumption. A higher fouling factor than that used in the selection of the refrigeration machine condenser is recommended when selecting the steam condenser because the higher condensing temperature causes greater fouling.

CHART 3 — APPROXIMATE OVER-ALL EFFICIENCIES OF STEAM TURBINES

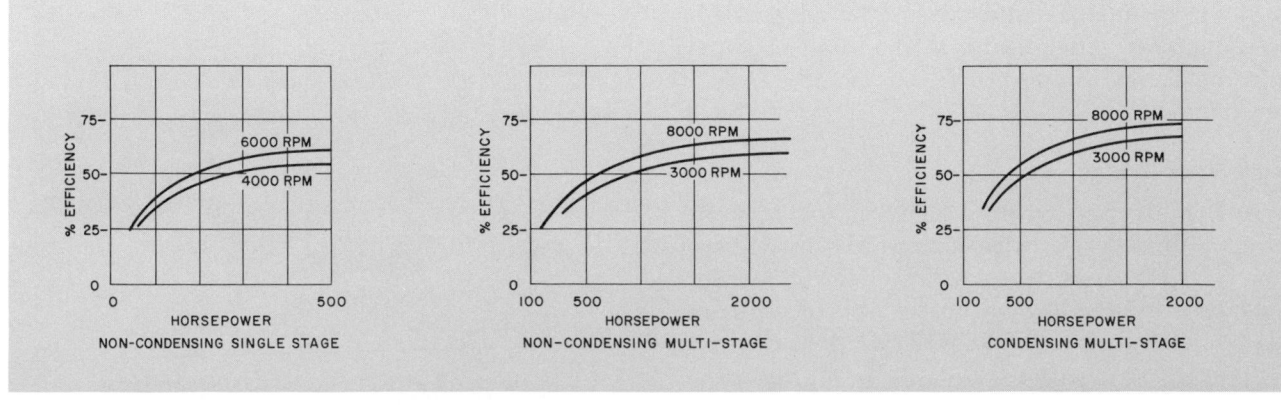

TABLE 11 — TURBINE THEORETICAL STEAM RATE (lb/hp/hr)

		INITIAL PRESSURE (PSIG)														
EXHAUST PRESSURE		60			100			125			150			175		
		INITIAL SUPERHEAT (F)														
		0	42.7	92.7	0	62.1	112.1	0	97.2	122.2	0	84.2	134.2	0	76.2	122.6
Condensing	3.5 in. Hg abs.	10.4	10.1	9.8	9.3	9.0	8.7	8.9	8.4	8.2	8.6	8.1	7.9	8.3	7.9	7.7
	4.0 in. Hg abs.	10.7	10.4	10.1	9.5	9.2	8.9	9.1	8.6	8.4	8.8	8.3	8.0	8.5	8.1	7.8
	4.5 in. Hg abs.	11.0	10.7	10.4	9.8	9.4	9.1	9.3	8.7	8.6	9.0	8.5	8.2	8.7	8.3	8.0
Noncondensing	5 lb/sq in. gage	25.6	24.9	23.9	19.4	18.6	17.8	17.5	16.2	15.9	16.2	15.2	14.5	15.2	14.4	13.8
	10 lb/sq in. gage	30.5	29.6	28.2	22.0	21.0	20.1	19.5	18.1	17.7	17.9	16.7	16.0	16.7	15.7	15.1
	15 lb/sq in. gage	36.2	35.0	33.2	24.8	23.6	22.4	21.6	20.0	19.5	19.6	18.3	17.5	18.1	17.1	16.4

		INITIAL PRESSURE (PSIG)														
EXHAUST PRESSURE		200			250			300			400			600		
		INITIAL SUPERHEAT (F)														
		0	62.2	137.2	0	94	169	0	78.3	153.3	0	101.9	201.9	86.2	186.2	261.2
Condensing	3.5 in. Hg abs.	8.1	7.8	7.4	7.8	7.3	7.0	7.6	7.1	6.8	7.2	6.7	6.2	6.3	5.9	5.6
	4.0 in. Hg abs.	8.3	8.0	7.6	7.9	7.5	7.1	7.7	7.3	6.9	7.3	6.8	6.4	6.4	6.0	5.7
	4.5 in. Hg abs.	8.5	8.1	7.7	8.1	7.6	7.2	7.8	7.4	7.0	7.5	6.9	6.5	6.5	6.1	5.8
Noncondensing	5 lb/sq in. gage	14.4	13.7	12.9	13.3	12.3	11.6	12.5	11.7	11.0	11.5	10.5	9.7	9.5	8.8	8.2
	10 lb/sq in. gage	15.7	15.0	14.1	14.4	13.3	12.5	13.5	12.6	11.9	12.2	11.2	10.3	10.0	9.2	8.7
	15 lb/sq in. gage	17.0	16.2	15.2	15.4	14.3	13.4	14.4	13.5	12.6	13.0	11.9	10.9	10.6	9.7	9.1

$$\text{Theoretical steam rate (lb/hp/hr)} = \frac{2544.1}{h_1 - h_2}$$

where:

h_1 = enthalpy of initial steam (Btu/lb)

h_2 = enthalpy of exhaust steam at the entropy of the initial steam (Btu/lb)

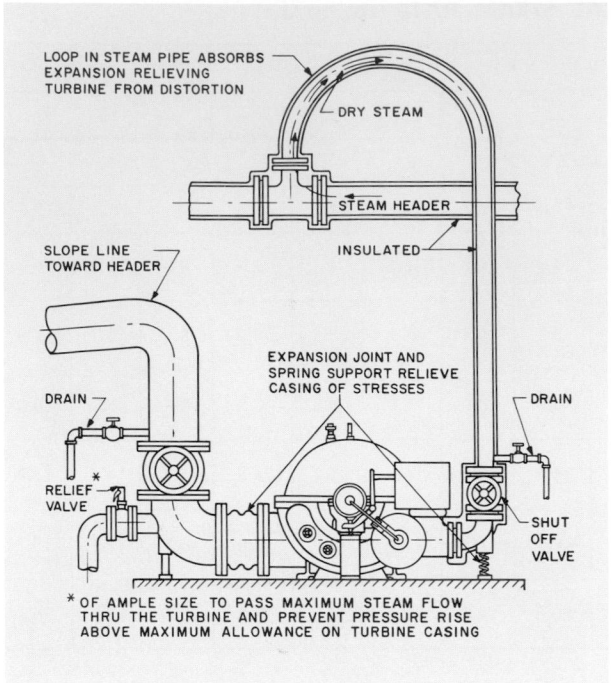

FIG. 70 — RECOMMENDED STEAM AND EXHAUST PIPING ARRANGEMENT

The condenser water rise in the steam condenser may be determined from this formula:

$$\text{Rise} = \frac{\text{Btu/hr}}{500 \times \text{gpm}}$$

where:

Btu/hr = turbine steam rate (lb/bhp/hr)
         $\times$ centrifugal bhp $\times$ 950
gpm   = condenser water flow

PIPING DESIGN

No part of the turbine installation is more important for successful operation than well-designed, installed piping.

Steam Piping

There are two objectives for good piping:
1. Prevent the heated piping from imposing strains on the turbine casing, thus affecting alignment.
2. Connect and drain the turbine inlet and exhaust lines so that dry steam is furnished to the turbine and water accumulation in these lines is prevented.

A shutoff valve is recommended in the steam line (preferably at a convenient accessible location in the turbine room) between the steam header and the turbine inlet to allow service on the turbine without shutting down the boiler.

The turbine casing must be protected from piping weight or expansion strains. Piping weight should be carried separately by suitable supports. Expansion joints or bends should be provided adjacent to the turbine connections. Connections between piping and turbine should be made without forcing the pipe line in any direction in order to make a satisfactory joint. *Figure 70* illustrates the recommended steam and exhaust piping arrangement.

A receiver type separator with ample drains should be provided ahead of the shutoff valve in the steam supply to prevent slugs of water from damaging the turbine. When a separator is not provided, a blow-off valve or continuous drain should be connected to the lowest point of the steam inlet piping. It is imperative that feed water treatment and boiler operation be carefully controlled to insure a supply of clean steam at all times.

Piping must be designed in accordance with the turbine-exhaust hand selected. *Figure 71* shows the available hand for single and multi-stage turbines.

Miscellaneous Piping

Properly planned miscellaneous piping gives a workmanlike appearance to an installation. Drain lines should be grouped and brought to a common drain arrangement. All open drain connections should be brought into a common closed collector box with a glass window for visual checking by the operator. Water cooling connections should be connected to a water supply at a maximum of 85 F. *Figure 72* shows the miscellaneous piping connections to a typical single stage turbine.

INSULATION

All heated surfaces of steam turbines such as casings and chests, connections, flanges and valves should be insulated to prevent heat loss and condensation in the turbine. Wet steam results in power losses, unnecessary wear, and possible damage to the turbine. To protect the insulation, metal lagging is fitted closely over the surfaces of the insulation. Insulation and jacketing can normally be provided by the turbine manufacturer.

CONTROLS

The function of the controls is to adjust the horsepower output of the drive to the horsepower requirement of the load. The speed of the turbine must also be controlled either at a constant speed or variable speed depending on the load requirements.

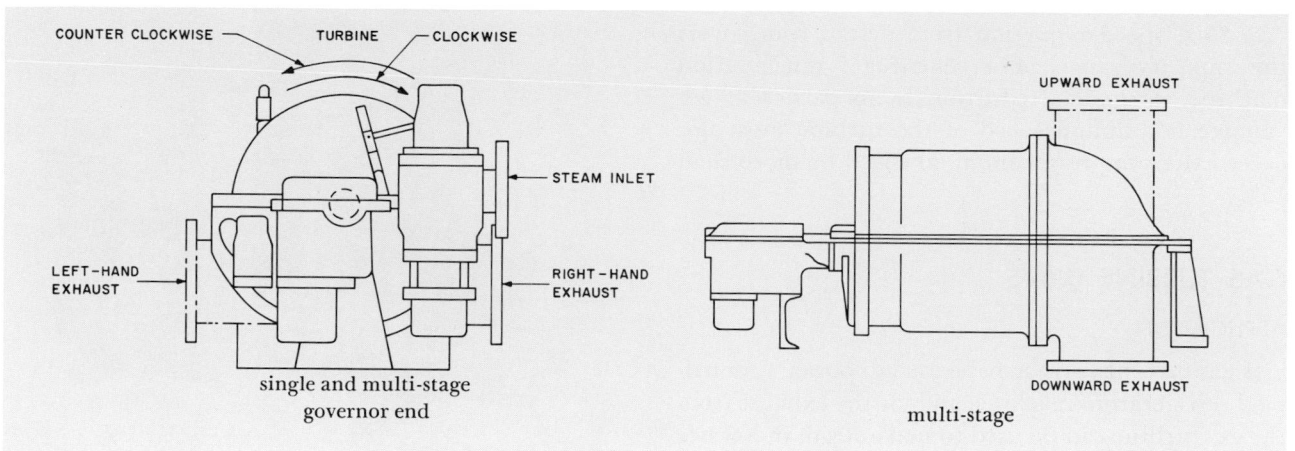

FIG. 71 — TURBINE EXHAUST HAND AND ROTATION

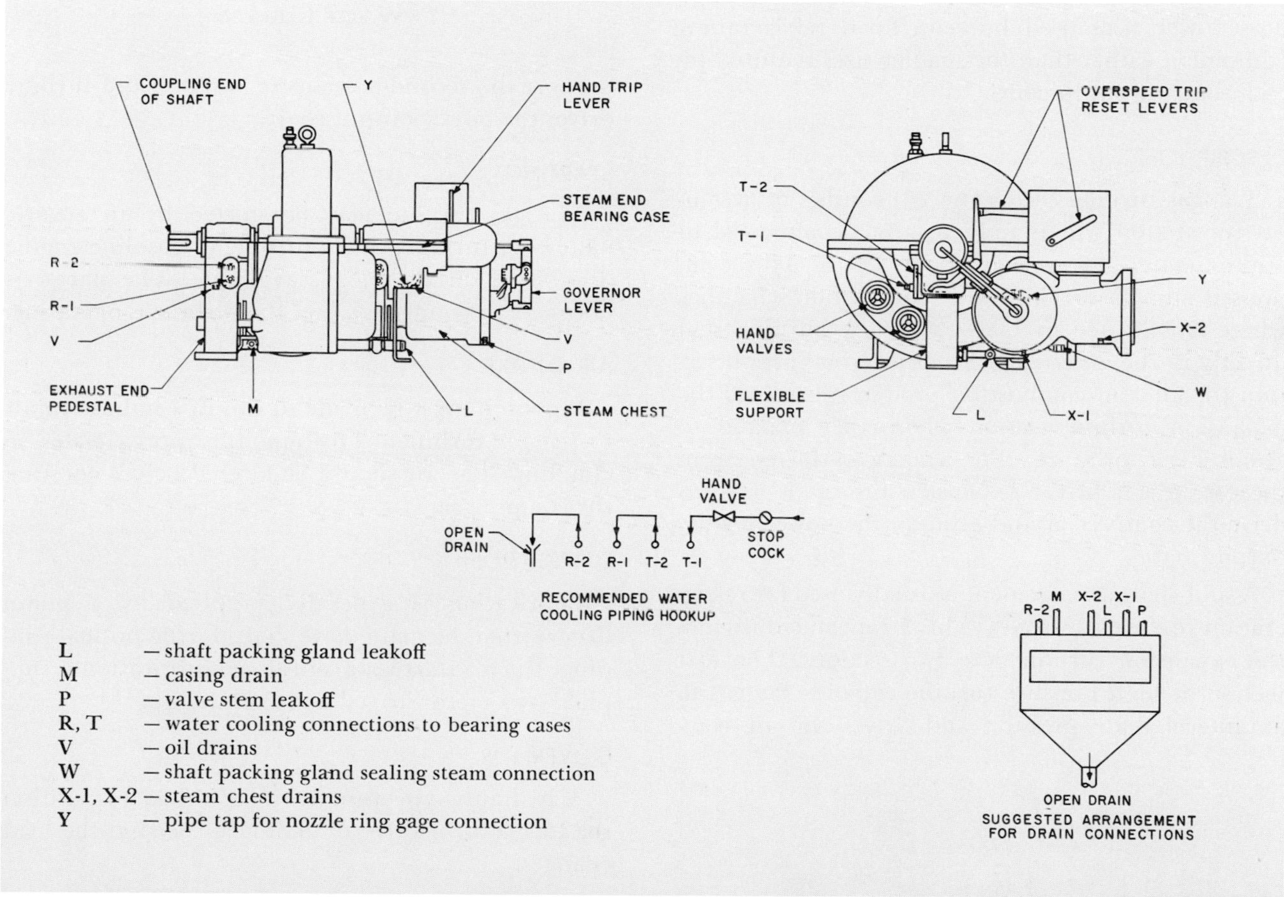

FIG. 72 — MISCELLANEOUS PIPING CONNECTIONS, TYPICAL SINGLE STAGE TURBINE

L — shaft packing gland leakoff
M — casing drain
P — valve stem leakoff
R, T — water cooling connections to bearing cases
V — oil drains
W — shaft packing gland sealing steam connection
X-1, X-2 — steam chest drains
Y — pipe tap for nozzle ring gage connection

GOVERNOR

A speed governor must be employed to maintain or vary the speed. They are of two basic types, mechanical (fly-ball) or hydraulic (oil pump). This classification indicates the type of speed sensitive element. Each may be either direct-acting or controlled by a relay to indicate the means of speed control.

The turbine manufacturer should recommend the type of governor for specific conditions. In general, a direct acting hydraulic governor is used for a constant or variable speed drive. However, above approximately a 5-inch governor valve size or for speeds in excess of 7000 rpm, an oil relay hydraulic governor is used.

A 35% speed reduction from design rpm covers the capacity range of a centrifugal refrigeration machine. A smaller reduction is no particular advantage. Maximum speed of the turbine must not exceed the compressor nominal speed by more than 15%.

GAS TURBINE DRIVE

APPLICATION

A gas turbine drive may be used to power a centrifugal refrigeration machine, and/or the exhaust from the gas turbine can be used to make steam in a waste heat boiler to operate an absorption machine or a steam-driven centrifugal machine.

Gas turbines are usually available in the large horsepower sizes used by centrifugal refrigeration equipment rather than the smaller sizes required by reciprocating equipment.

DESCRIPTION

The gas turbine cycle *(Fig. 73)* consists of a compressor section where ambient air is compressed to approximately 60 psia at about 350 F. This compressed air passes into the combustion chamber where it is heated to 1350–1500 F by burning fuel directly in the air stream. From the combustion section the air and combustion products flow into the expansion turbine section where they expand to atmospheric pressure. The energy extracted from the gas stream in the expansion process is used to drive the compressor and produce the power for the output shaft.

A split-shaft arrangement is usually used for refrigeration compressor drives. This arrangement divides the expansion turbine into two sections. The first section or high pressure turbine expands the gas to an intermediate pressure and drives the air com-

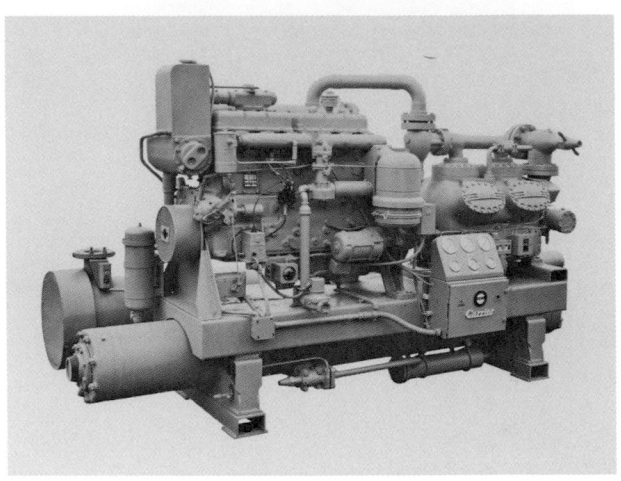

Fig. 74 — Gas Engine Driven Reciprocating Water Chiller

pressor; the second section or low pressure turbine drives the power output shaft.

STARTING

The gas turbine may be started by an electric motor, air turbine, steam turbine or gasoline engine depending on the means available. The starter is normally disconnected after the turbine is operating.

AIR INTAKE

Provision must be made to supply combustion air to the gas turbine: This amount approximates 15 cfm/bhp. The air should be filtered before entering the compressor.

LUBRICATION

Lubrication is generally supplied by a pump driven from the main drive shaft during normal running. During startup or shutdown lubrication is supplied by a motor-driven auxiliary pump.

GOVERNOR

A hydraulic governor is usually used to position the fuel control valve to maintain speed as the load changes.

SAFETY CONTROLS

Safety controls are provided to shut down the unit for the following causes:
1. Low oil pressure
2. Overspeeding of the unit
3. Low fuel pressure
4. High bearing temperature
5. Loss of flame

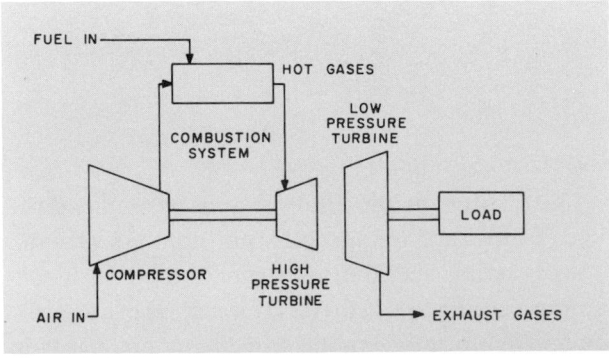

Fig. 73 — Split-Shaft Gas Turbine Cycle

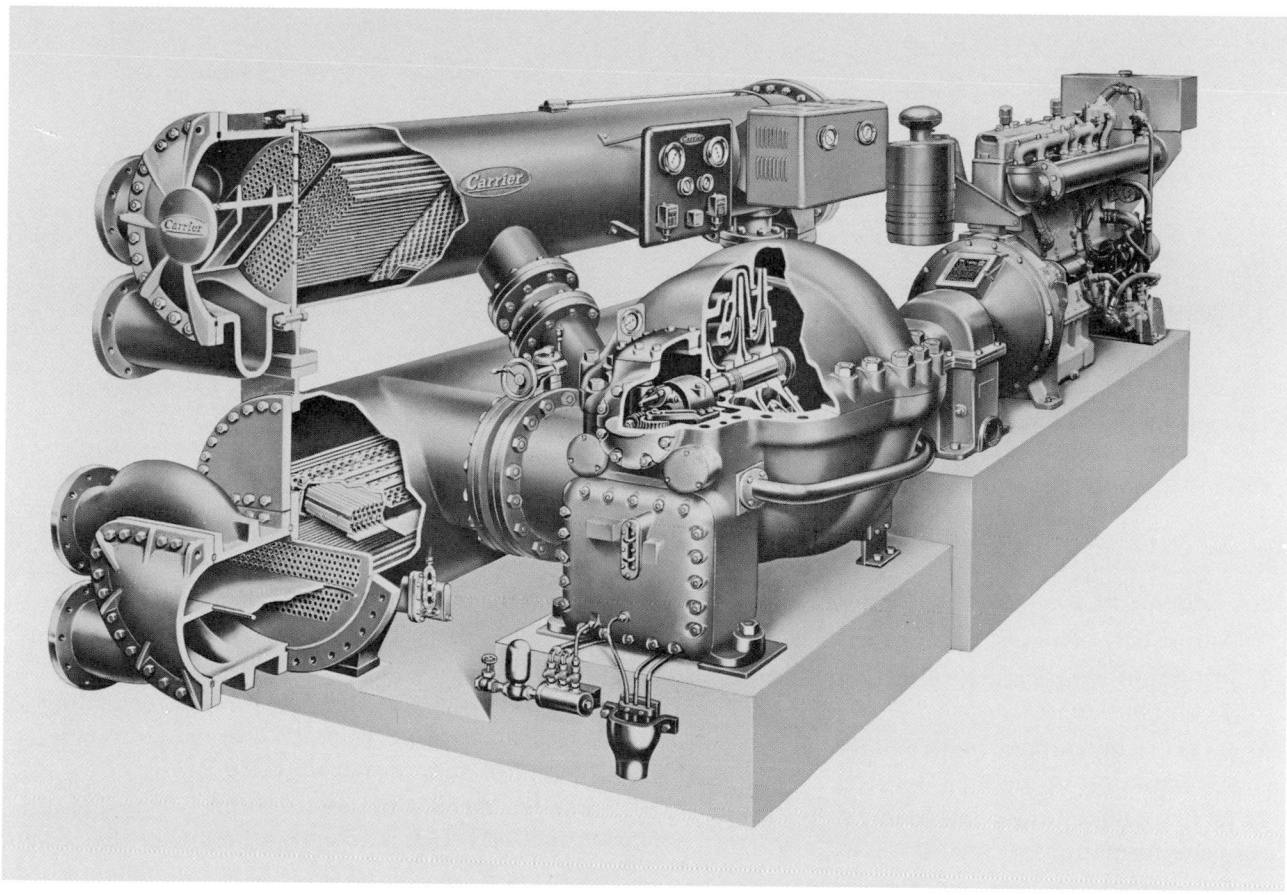

FIG. 75 — GAS ENGINE DRIVEN CENTRIFUGAL REFRIGERATION MACHINE

GAS ENGINE DRIVE

APPLICATION

A gas engine drive may be used when gas is available at a cost which provides a saving in owning and operating costs. Gas engines are used to provide power to drive reciprocating (*Fig. 74*) or centrifugal refrigeration machines (*Fig. 75*) and may also indirectly supply steam to operate an absorption refrigeration machine by using the heat rejected from the engine cooling system and exhaust system.

STANDARDS AND CODES

Gas engines should be installed to conform to all codes, standards and regulations concerning internal combustion engines.

SELECTION

Gas engines used for driving refrigeration equipment should be selected for continuous duty service. This means that the unit should be selected to operate at 80% of the maximum corrected horsepower. Thus, if a compressor requires 100 bhp, the gas engine is selected for 100/.80 or 125 maximum rated horsepower. The reduced output and lower speeds are a major consideration in attaining a longer engine life.

ATMOSPHERIC CORRECTIONS

The maximum rated horsepower of a gas engine is given for an air temperature of 60 F and an air pressure of 29.92 in. Hg (sea level average).

Deduct 1% from the maximum rated horsepower for every 10 degree increase in ambient temperature above 60 F. For every 1000 feet in elevation above sea level, deduct 3% from the maximum rated horsepower.

HEAT REJECTION

The heat rejection of a gas engine to the water jacket circuit is approximately 50–60 Btu/hp/min. This represents about 30% of the input to the engine. Another 30% of the input is given up to the exhaust system, and about 10% is given up as radiation losses.

COOLING SYSTEMS

Any type of cooling system must meet the follow-

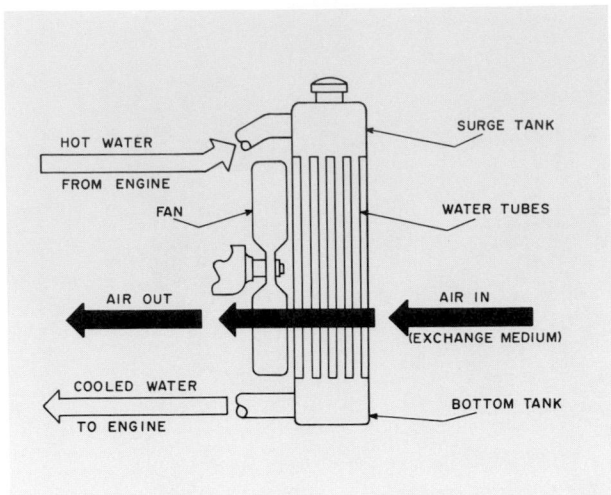

FIG. 76 — FAN AND RADIATOR COOLING

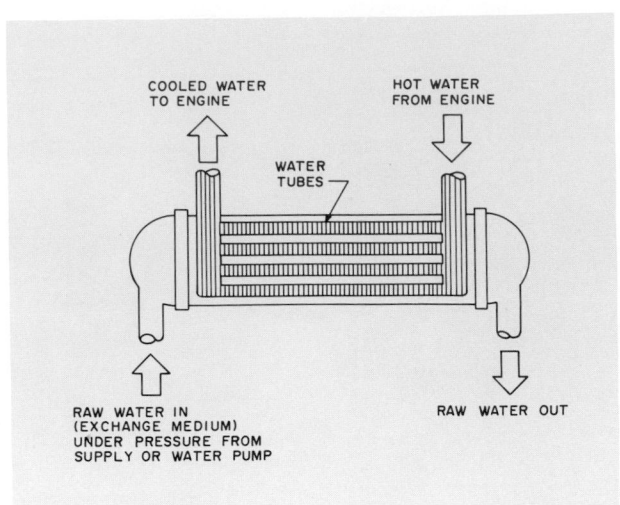

FIG. 77 — SHELL-AND-TUBE COOLING

ing requirements for satisfactory engine operation:

1. Ample flow of water.
2. Minimum temperature differential between inlet and outlet.
3. Jacket temperature high enough to prevent condensation inside the case.
4. Jacket temperature low enough to prevent steam formation.
5. Soft water to prevent scale formation.
6. Clean water to prevent clogging of the engine jacket passages.
7. Positive pressure on the entire system to prevent entry of air.

Fan and Radiator

This system depends on cooling the engine jacket water by an engine-driven fan creating a flow of air over a finned tube radiator to dissipate the heat to the atmosphere *(Fig. 76)*.

The advantage of this system is that it is self-contained, and does not depend on external water sources.

Ductwork must be provided for discharge of the hot air and openings provided for the entry of cool air. An extra 10–30 hp must be supplied by the engine to power the fan. The jacket water is pumped to the radiator by a pump driven by the engine.

Shell-and-Tube Heat Exchanger

This system uses a heat exchanger to cool the jacket water with a separate water source which may be wasted or cooled in a cooling tower *(Fig. 77)*. When the engine is applied to a refrigeration machine, the same water used for condensing the refrig-

erant can be used to cool the jacket water. The condenser and heat exchanger are generally connected in parallel.

The water temperature range for jacket cooling is recommended to be 10 to 12 degrees with an inlet temperature of approximately 180 F.

Another method of jacket cooling is called ebullition cooling which is actually a boiling of the jacket water *(Fig. 78)*. The jacket is kept completely filled with water which circulates because of differences in temperature. The jacket water temperatures are in the boiling range. The concept of this cooling is

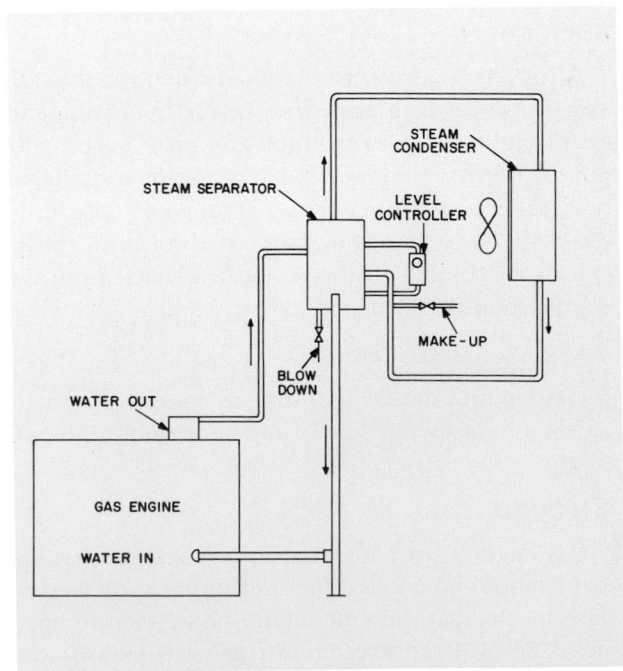

FIG. 78 — EBULLIENT COOLING SYSTEM

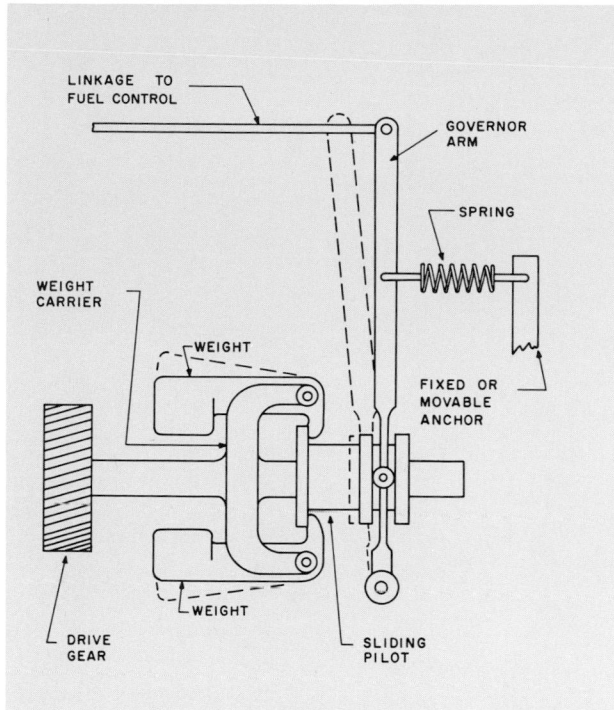

FIG. 79 — SCHEMATIC OF CENTRIFUGAL GOVERNOR

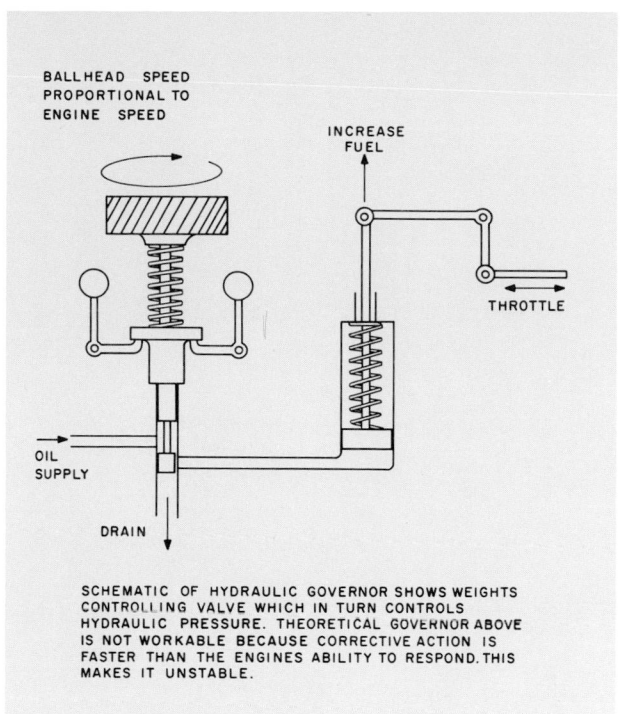

FIG. 80 — SCHEMATIC OF HYDRAULIC GOVERNOR

that a large amount of heat can be picked up at the surface of the metal without increasing the engine temperature because 970 Btu are required to evaporate one pound of water at 14.7 psia. The engine heat is removed in the form of steam which is passed into a separator where the water is removed and any entrained solids are allowed to settle. The steam passes on to a steam condenser or some other equipment. The condensate then returns to the separator and the engine.

AIR INTAKE SYSTEM

The gas engine requires two to five cubic feet of air per minute per horsepower for combustion. This is important because this amount of air may upset the heating and ventilation calculations of an air conditioned building. An intake should be located to provide the cleanest air possible with the least contaminants, especially flammable vapors, tank vents, or other explosive wastes because the natural induction of this material may cause a runaway engine.

EXHAUST SYSTEM

Provision must be made to dispose of the exhaust gases from the engine with a minimum of restriction or back pressure. Excessive back pressure causes a loss of power, poor fuel economy, excessive valve temperatures and jacket water overheating. The

exhaust pipes should be independently supported to prevent strains on the engine manifolds. Exhaust pipes can get red hot when the unit is heavily loaded; thus expansion must be provided as well as provision for disposing of the radiant heat given off by the exhaust pipe. One method of removing the heat is to install sheet metal ductwork around the pipe with an inch or two of space between them to create a chimney effect. Sometimes water-cooled exhaust piping is used for this purpose.

Attention should also be given to adequate silencing of the engine to prevent objectionable noise.

GOVERNOR

Governors are used for adjusting the power output of the unit to match the load and to maintain the speed. The governor measures the engine speed and provides power to move the throttle. The two main types are centrifugal governors *(Fig. 79)* and hydraulic governors *(Fig. 80)*.

DIESEL ENGINE DRIVE

This drive is similar to the gas engine drive; the difference is the fuel used. The compression ratio of a diesel engine is greater than that of a gas engine; the engine may be either a four cycle or a two cycle.

Selection, type of cooling systems, exhaust systems, governors, etc., are identical to the gas engine drive.

Part 9
SYSTEMS AND APPLICATIONS _____

CHAPTER 1. SYSTEMS AND APPLICATIONS

This text introduces the engineer to the preliminary considerations and project aspects for determining the selection of an air conditioning system. The part sketches the systems available and compares their performance and applicability to various spaces and buildings. It also indicates the particular problems of selected applications, stressing the important functions of air conditioning. It presents a broad review only; details of various systems and their design are presented in other parts of this manual.

SCOPE AND INTENT

Selection of the correct air conditioning system for a particular space or building is a very critical decision facing the design engineer. On this decision rests the satisfaction of the customer and occupant and the system fitness to the building it serves. Many factors must be analyzed, judged, screened and coordinated. The desires of the investor and the economic aspects are the foremost considerations.

CUSTOMER AND OBJECTIVE

There is a big gap between a customer who thinks in terms of simple relief cooling in a room or small establishment and the one who builds a monumental building representing the epitome of an integrated concept of a structure and the environment within, whether architectural, acoustical or air conditioning.

Complete air conditioning provides an environment of correct temperature, humidity, air movement, air cleanliness, ventilation and acoustical level. Anything less is a compromise and is not termed air conditioning. Therefore the particular system involved should be identified with the function which the system design is to accomplish, whether heating, cooling, humidity control or complete year-round air conditioning.

There must be complete fusion of system and building to lead to a natural and normal behavior as a whole in disposing of the heat gains or offsetting the heat losses.

ECONOMICS

These considerations affect the type of equipment and the whole system that is offered to the prospective owner. The economic factors are foremost. They originate from the owner's desire and capacity to invest in an installation that is intended to provide either a minimal or a maximum benefit. It must be determined whether the project is an investment for a quick write-off, a resale or a long term investment. The investor may seek either the lowest first cost, the lowest first cost in balance with the economical owning and operating costs, or strictly the lowest owning and operating costs. Above all the investor is interested in a profitable return on his investment.

PROBLEM

To realize a substantial advantage from an operation of air conditioning in a space or a complete building, the design engineer must consider fundamental situations. He must first define the problem. He must be able to anticipate the behavior of the contemplated air conditioning system. For given conditions of external environment and internal load, the system must integrate with the space or the building which it is to serve. The system must satisfy the maximum instantaneous or actual thermal load* and be operable at any partial load conditions.

The fundamental diagnosis should consider:

1. Investor's financial capacity and the investment objective.
2. Space or building
   a. Purpose
   b. Location
   c. Orientation and shape
3. Coincidental occurrence in external environment of
   a. Temperature
   b. Humidity
   c. Wind
   d. Exposure to sun or other heat exchanges
   e. Shade

*Thruout this discussion load shall mean the thermal load of heat gain or loss to or from a building and its content.

4. Diversity of internal load
   a. Occupancy
   b. Lighting
   c. Other heat exchanges
5. Capability for storage of heat gains
6. Necessity and capacity for precooling
7. Physical aspects of space or building to accommodate
   a. Equipment
   b. System
   c. Balanced operation at partial load
8. Customer's concept of environment desired

INTEGRATION

Each space or building presents an individual problem to resolve. There is no universal solution to a system selection even after it is defined, the physical circumstances evaluated, and the actual load of heating and cooling requirements established. The design engineer must have an appreciation of the structure, its thermal capacity behavior, and the response accorded it by the contemplated system. He must understand the interaction of space or building with external and internal thermal loads and the cancellation of these loads by the system. There should be a full realization that the equipment installed, the control of the air conditioning system, and the building are irrevocably integrated into one whole. To be successful these elements must be coordinated to operate as an entity.

The discussion that follows is a résumé of factors that constitute the preliminary qualifications of a project. It offers a guide to the selection of a system best suited to a given circumstance surrounding an application. All types of systems are reviewed briefly, from a self-contained room unit to an elaborate central station system.

The range of systems applications for human comfort is covered from a residence to a high rise apartment building, from the smallest commercial application to a skyscraper or a factory.

SPACE AND BUILDING

OCCUPANCIES

The wide range of spaces and buildings for application of air conditioning for human comfort may be divided into two fields, single-purpose and multi-purpose occupancies.

Single-Purpose Occupancies

Single-purpose occupancies involve either an individual or a multitude of individuals gathered for a common purpose of work, prayer, relaxation or amusement. The predominant characteristic is the presence of a single environmental control zone. Examples include a room, residence, or large open area with or without low partitions.

The large area may be an office space, restaurant, beauty salon, etc., at times set in an individual small building. The larger structure may be a church, theater, auditorium or pavilion. The common feature is a building with one or more large open spaces as a major area to be air conditioned.

Multi-Purpose Occupancies

Multi-purpose occupancies involve a multitude of people gathered for various purposes in one or more multi-room, multi-story buildings. These buildings may serve a single purpose: a sale of goods, department store; book lending or collection, a library; a collection of articles of special interest, museum; research, laboratory; learning, school; manufacturing, factory; etc. Generally the multi-room, multi-story buildings may be office buildings, hotels, apartment buildings or hospitals. To the multi-purpose occupancies also belong building compounds of apartment houses, schools, colleges, medical and shopping centers, and factories.

The major characteristic of these occupancies is a multiplicity of environmental control zones served by a single or multiple, preferably central, air conditioning system. With increases in size and number of central station units in a single system, sources of refrigeration and heating are consolidated into one or more large refrigeration and boiler plants.

THERMAL LOAD

When an engineer is faced with an existing space or building, little can be done to alter these structures to aid either the reduction of the air conditioning load or system accommodation unless a major alteration is embarked upon. Definite limiting circumstances may exist.

On the other hand a new building provides freedom and challenge to the architect-engineer team. They may devise a structure that is architecturally and acoustically acceptable and pleasing, at the same time one that incorporates all possible forethought to minimize the air conditioning load. Proper orientation in regard to exposures and analysis of shading (external or internal) is essential.* Space is required for air conditioning equipment and for transmission and distribution of heating-cooling effects.

*Sun gain thru 150 sq ft of unshaded west glass requires approximately one ton of cooling capacity as against only one tenth of a ton required for the glass with north exposure.

Building Shell and Outdoors

A building or a space is a thermal container, an enclosure. Within it an air conditioning environment for human comfort is to be maintained, regardless of seasons and outdoor climatic conditions. The considerations involving the construction of a building shell are: thin panel vs massive wall and partial vs total glazing. Other considerations are: glass and wall shading, orientation of the building (simple or complex architecture), height and shape, predominance of peripheral or internal areas, and single zone vs multi-zone application.

These considerations are pertinent to evaluating the external influences on the air conditioning load of simultaneous occurrence of temperature, humidity, wind and solar conditions. These constitute the outdoor design conditions.

Internal Elements

The selection of the inside design conditions for space or building establishes the thermal head against which the air conditioning system will operate at any load condition. The internal load behavior is determined by the diversity factor that can be applied to the population, lights and any heat-producing or extracting equipment or situation. The smaller the space, the less the diversity; an air conditioning cycle applied to an individual space takes full account of instantaneous load. However, with an increase in the size of a project, the requirements for refrigeration capacity grow. A larger diversity factor may be applied. In most cases the possibility of application of either the heat storage principle or precooling effect should be considered in order to reduce the cooling-heating load (approach the actual load) or reduce the size of equipment.

Part 1 contains information on the evaluation of an air conditioning load. It must be re-emphasized that *an estimate of the actual cooling-heating load is the most fundamental step* before selecting an air conditioning system and equipment. This must be preceded by a thorough survey.

Partial Load

A necessary corollary to the evaluation of actual load is an appraisal of load behavior at partial conditions: possible variations in the internal load; effect of change in external weather elements; reaction of the thermal enclosure and barrier, the structure. The shell's thermal capacity and physical porosity may have great influence on the amplitude of the peaks and valleys in the daily load curves. The system picked for the particular situation must have certain flexibility. Thus the load, the enclosure and the air conditioning equipment in a complete system (total complex) regulates the space environment.

SPACE FOR EQUIPMENT AND SYSTEM

The air conditioning equipment and system auxiliaries require space. The industry is continually devising means and methods to reduce the size of the equipment, the elements of the total system, and their costs, yet produce the same total capacities of cooling and heating at a reasonable investment. Until some radically new approach is discovered, the present means to provide comfort require space.

Self-Contained Units

The extent of space requirements may be small enough to take care of a room unit or a self-contained apparatus within the air conditioned area. Both are types of miniaturized central station plants of small capacity. Such packages contain all the elements to provide complete air conditioning in one enclosure, yet convenient to handle as a unit. The spaces required affect directly the conditioned area; however

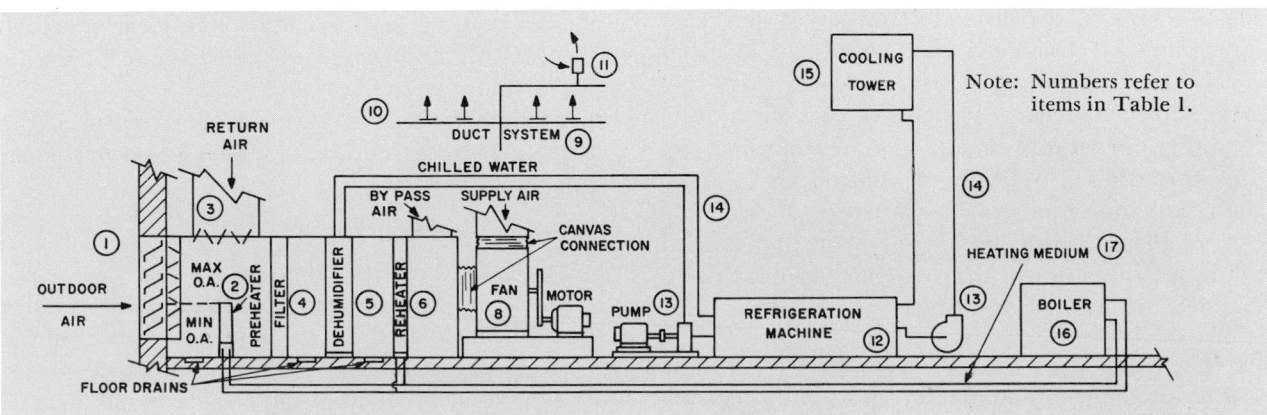

FIG. 1 — AIR CONDITIONING PLANT

there may be cases where the units are installed external to the conditioned space.

Central Stations

For a large central station system space has to be provided for conditioning, heating and refrigeration machinery. The cooling-heating media requires space to be transmitted to the conditioning apparatus and then to the areas to be conditioned. There are terminals within the conditioned areas for the final delivery of the cooling-heating effects.

Thus there is a conditioned air path from the outdoor air intake thru the apparatus, thru ducts to terminals within the conditioned area; there are refrigeration and boiler plants and interconnecting piping to air handling apparatus and in some systems to terminals *(Fig. 1)*. Except for within-the-room terminals the system space required is external to conditioned areas. At times areas that can be utilized more profitably may have to be surrendered for use by some elements of the system.

BASIS FOR DESIGN

The design engineer must make a correct appraisal of the building or space to be conditioned. He must define the problem. He must evaluate the characteristics of the space or building, the climatic environment and the internal heat gains or losses. The evaluation of maximum actual load, its behavior at partial conditions, and the thermal capacity of the enclosing structure are the elements and the foundation to guide the selection of the proper air conditioning system.

QUALIFYING SYSTEM SELECTION

ECONOMICS

Having pointed out the physical aspects and requirements of air conditioning of space and building, a review of the intangible aspects is necessary. These are the customer's concepts of (1) financial involvement and realization of income gain, (2) achievement of the environment of temperature, humidity, air cleanliness, air motion and quietness, (3) realization of flexibility of controls, and (4) insight into limitations of the structure. By clarification of these aspects proper judgment and mutual appreciation between the customer and architect-engineer team is accomplished.

Buyer

The air conditioning market is vast and very competitive. The ultimate user-buyer may be a shrewd investor looking for a quick turnover of capital or a customer who is satisfying a need to neutralize competition that surrounds him, or an owner creating the utmost in monumental progressive design and application.

There are three factors in the economics of system selection: (1) first cost, (2) operating and maintenance costs, and (3) return on an investment. The buyer may be looking for either extreme, the lowest first cost or the lowest operating cost. He may rely on the design engineer to arrive at a balanced proposal. He may desire a monument irrespective of any costs. The design engineer must weigh not only the engineering considerations but the customer's financial attitude and desire for return on the investment.

Investment

One aspect in the economics of selection of an air conditioning system is the actual longevity of equipment and auxiliary components, that is, the write-off life span permitted by the government for depreciation for tax purposes. The net effect is apparent in the analysis of owning and operating costs.

Owning Costs

The owning element of costs is influenced by the price of the equipment, materials, labor and services for an installation. These components must be weighed carefully in arriving at a selection of a system. With an existing building there is an additional element, the interference with regular business.

Operating Expense

Operating costs are influenced by the consumption of energy, whether electrical, steam, gas or other; this is the bulk of operating costs. There is also the item of maintenance consisting of the operating personnel and upkeep of equipment together with supplies of oil, filters and other materials. The equipment working condition is its condition approaching that of the original installation except for the normal duty wear.

The otherwise rentable space given up to accommodate the air conditioning system is also an element in operating costs.

Return on Investment

In the final analysis the rent or any income producing increment is of major interest to the investor. Investment analysis is a gauge for determining whether or not money spent on a proposed project will be wisely invested. It can be used to establish the merits of air conditioning as a sound investment compared with money invested in any other manner.

Investment analysis is the owning and operating cost analysis to determine on a unit basis the incremental cost due to the addition of air conditioning. To the prospective owner this cost is an added cost per square foot, per room, per apartment unit, per hospital bed, or per factory worker. This added cost is compared to the expected incremental income. Such a tally determines whether an investment in this or another system is profitable, and which system is more profitable than another.

Budget Analysis

During the preliminary budget stages approximations based on experienced judgment should be made to determine the thermal load and the cost of various systems. This procedure provides the basis and affects the results of the qualifying investment analysis. That air conditioning can offer attractive investment potential is easily established by a budget analysis. Since investment return varies with each system and with the length of elapsed time before the owner plans to liquidate the investment, two or more analyses must be made to determine which of several applicable systems is the best investment.

ENVIRONMENT

Temperature and Humidity

The foremost prerequisite of comfort air conditioning is temperature control and the regulation of humidity is of secondary importance. All systems have the facility of controlling the comfort range of temperature. The human tolerance of humidity variation is considerable.

Air Cleanliness

There are considerable variations in air cleaning requirements; they depend on the location of the particular project, the fastidiousness of the customer, and the costs of cleaning. The latter include the initial investment and maintenance costs. Odor and bacteria control may also be required; at times this is an absolute necessity. Air must be clean.

Air Movement

With any air conditioning application a system may have the required cooling-heating and cleaning capacity but, if the air is not properly distributed and not in effective motion, all the requisites of proper air conditioning are not fulfilled. Air movement depends on the amount of air in circulation and direction of air delivery; both are determined by the type of air distribution employed. Air movement definitely assists in the sensation and appreciation of comfort.

Acoustics

Part of the work produced by equipment and cooling-heating media in motion is always converted into sound energy. When reaching the conditioned areas, this energy may be a welcome masking which contributes a desirable amount of ambient noise, to make the space "quieter" by making it noisier. On the other hand it may be objectionable and annoying. The designer must evaluate the location of the equipment and system elements in relation to surrounding occupancies, the mass of the structure, the air conditioned space, and the location of the building (busy city or peaceful country). Then the designer must establish the desired sound level within the air conditioned space relative to the terminal or terminals in that space and the type of occupancy. This is part of the total comfort environment.

The achievement of this objective involves the cooperation of the customer and architect-engineer team. Proper design must be applied to the building, the distribution of spaces that are critical soundwise, and to the location and soundproofing of various machinery and system elements. The team must be concerned with the evaluation of sound level within air conditioned spaces.

CONTROL REQUIREMENTS

Choice of an air conditioning equipment and system depends on the character, nature and behavior of the cooling-heating load under partial conditions.

Nature of Thermal Load

It is necessary to determine whether the load is mostly sensible or latent, highly concentrated or light, uniformly distributed or variable within the area, continuously constant or varying in a regular or haphazard pattern. All these factors are involved in a decision, one zone or a multi-zone system. These considerations are necessary for selecting an air conditioning system and its control which maintains the desired conditions under partial load. This means achievement of controlled conditions under broad or exacting requirements.

EXISTING VS. NEW STRUCTURES

There is a distinct difference between systems available for existing or new buildings. In existing enclosures or structures heating and at times ventilation is already supplied. Therefore the additional system is for cooling only, adapting and integrating the existing heating-ventilation into a year-round cycle. The air conditioning systems to be selected are limited to those operable from overhead and sometimes to systems that can be applied on an individual space-to-space or floor-to-floor progressive basis.

In new structures or in those where the existing heating systems are discarded, the exterior areas may be treated by complete year-round systems. In exterior areas there are air conditioning systems specifically designed for highly variable peripheral thermal loads, systems that have many different arrangements. For interior areas, depending on the type of arrangement, their use and thermal loads, there are also available a wide variety of systems.

Having determined the economic aspects of the project, that is, the profit realization by an investor, the extent of the requirements of environmental conditions, and their control, the engineer is ready to select the system.

SYSTEMS

A review of the fundamental evolvement of a variety of systems is presented here.

OBJECTIVE

The objective of an air conditioning system is to provide a comfortable environment for an occupant or occupants of a residential, public, medical, factory or office building. It may be for a number of transient occupants in a commercial establishment such as a store, a bowling alley, a beauty salon, a restaurant or others. It may be for an assembly of occupants gathered in a large space such as a church, theater, auditorium, pavilion, or a factory loft or floor.

The comfort environment is the result of simultaneous control of temperature, humidity, cleanliness and air distribution within the occupant's vicinity that includes the proper acoustic level.

The final media used to achieve a comfortable environment are air and surfaces surrounding the occupant. Both the air and surfaces of the enclosure are sinks for the metabolic heat evolved by the occupant (*Fig. 2*).

Air circulates around the occupant and the surfaces. The occupant also has radiant heat exchange with the surrounding surfaces. Air is brought into motion within a given space either thermally or by force.

Thermal air motion usually occurs over heating convectors within the space or along heating-cooling panels applied to surfaces enclosing the space.

The forced air motion is affected by air delivered thru a diffuser outlet installed in the proper location on a wall, ceiling or at a mixing terminal. Air is

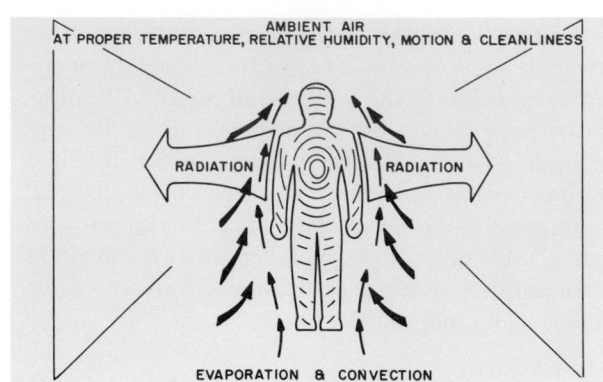

FIG. 2 — HEAT DISSIPATION

brought to the diffusers thru ducts from an apparatus where it is cleaned and passed over heat exchangers within which circulate the primary heating or cooling media. Some induction terminals within the room have secondary coils that provide supplemental heating or cooling. Examples of primary and secondary cooling media are a direct expansion refrigerant, chilled water or brine. The heating media are usually steam, hot water, gas or electricity.

The temperature of the surfaces surrounding the occupant are subject to temperatures prevalent within and outside the structure. The surface temperature may be regulated by thermal panels applied to the ceiling, floor or walls. The secondary media circulated thru these panels is usually either hot or cold water.

To supply and circulate the media, there are available boiler and refrigeration plants complete with necessary auxiliaries and interconnecting piping.

SYSTEM COMPONENTS

The essential elements of an air conditioning system are illustrated in *Figure 1*.

The basic elements, optional components and their functions are listed in *Table 1*.

TYPES OF SYSTEMS

Air conditioning systems are generally divided into four basic types determined by the methods thru which the final within-the-space cooling and heating are attained. The air surrounding the occupant is the end medium which is conditioned; in some systems most of the thermal effect is radiant.

The basic types are:

Direct expansion *(Part 12)*
All-water *(Part 12)*
All-air *(Part 10)*
Air-water *(Part 11)*
Heat pump

TABLE 1 — SYSTEM COMPONENTS

SYSTEM COMPONENTS	FUNCTIONS PERFORMED
Air Side	
1. Outdoor Air Intake (screen, louvers, dampers)	Path for outdoor air used for ventilation and marginal weather cooling
2. Preheater	Preheats air
3. Return Air Intake (dampers)	Path for return and/or recirculated air to apparatus
4. Filter	Removing contaminants from air
5. Dehumidifier (direct spray washer or cooling coils; DX, water, brine, with or without sprays)	Cooling and dehumidifying (air washing with sprays)
6. Heating Coil	Heating in winter and reheat for temperature and/or humidity control
7. Humidifier	Humidifying
8. Fan	Air propulsion
9. Duct System	Path for air transmission
10. Air Outlet	Air distribution within air conditioned space
11. Air Terminal (with outlet)	Enclosure for air handling; may be equipped with air mixing chamber, heating coil, heating and/or cooling coil, acoustic treatment, and outlet
Refrigerant Side	
12. Refrigeration machine (compressor, condenser, cooler and refrigerant piping)	Means for cooling
Water Side	
13. Pump	Water or brine propulsion
14. Water or Brine Piping	Path for transmission of water or brine between heat exchangers
15. Cooling Tower	Heat disposal from water used in condensing refrigerant
Heating Side	
16. Boiler and Auxiliaries	Provides steam or hot water
17. Piping	Path for transmission of steam or hot water

A self-contained compact unit located within or next to the air conditioned space and consisting of the minimum elements essential to producing the cooling effect is a direct refrigerant or *direct expansion (DX) system (Fig. 3)*. Heating may be either included with the unit or separate.

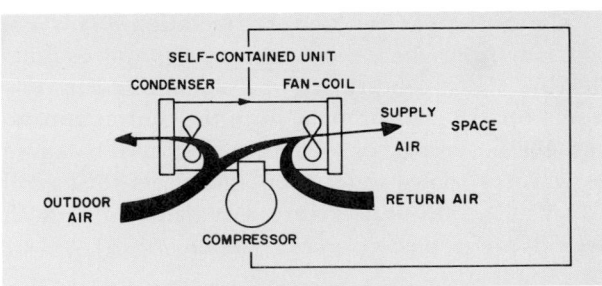

FIG. 3 — DX SYSTEM

A cooling medium (chilled water or brine instead of direct refrigerant) may be supplied from a remote source and circulated thru coils of an air terminal unit within the conditioned space; the cooling medium is warm in winter to provide heating. A system of these terminal units is called an *all-water system (Fig. 4)*.

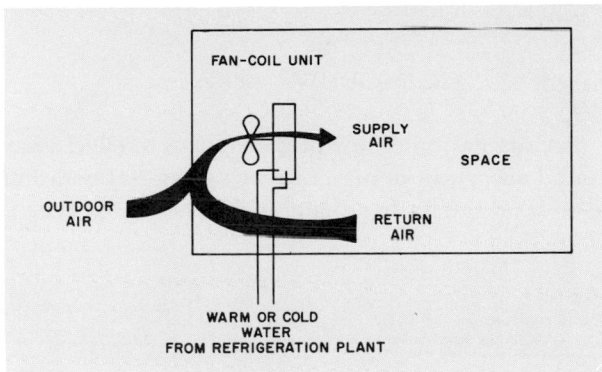

FIG. 4 — ALL-WATER SYSTEM

Both the air treating and refrigeration plants may be located some distance from the conditioned space in a central station apparatus arrangement. Only the final cooling-heating medium (air) is brought into the conditioned space thru ducts and distributed within the space thru outlets or mixing terminal-outlets. Such a system is termed an *all-air system (Fig. 5)*.

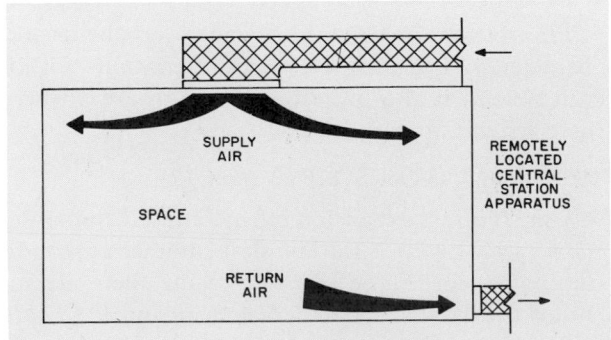

FIG. 5 — ALL-AIR SYSTEM

The air apparatus and refrigeration plant are separate from the conditioned space; the cooling-heating of the conditioned space is affected in only a small part by air brought from the central apparatus. The major part of room thermal load is balanced by warm or cooled water circulated either thru a coil in an induction unit or thru a radiant panel. Such systems are termed *air-water systems (Fig. 6)*.

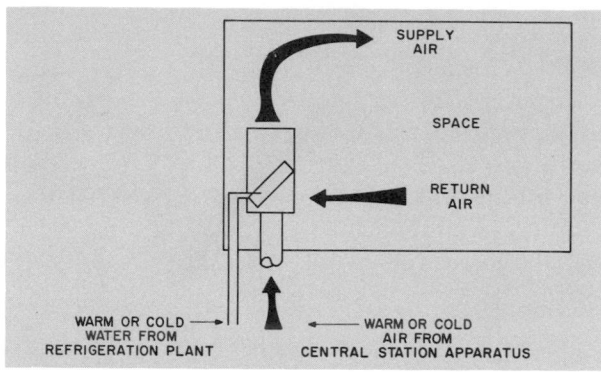

FIG. 6 — AIR-WATER SYSTEM

An adaptation of any of the systems to effect year-round air conditioning utilizing system refrigeration as a heat generating plant is a *heat pump system (Fig. 7)*.

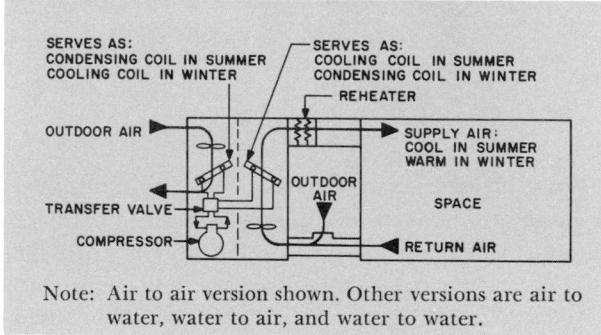

Note: Air to air version shown. Other versions are air to water, water to air, and water to water.

FIG. 7 — HEAT PUMP SYSTEM

SYSTEMS AND APPLICATIONS

This section discusses the specific components and characteristics of each system and variations within each system; it also indicates the major areas where these systems are applicable.

DIRECT EXPANSION SYSTEMS *(Part 12)*

A small direct expansion refrigerant self-contained *room cooling unit* is the simplest summer air conditioning system *(Fig. 8)*. In one casing there are included the basic elements 1, 3, 4, 8, 10 and 12 *(Table 1)*. With the addition of element 6 or adapting the unit to a heat pump arrangement, this self-contained

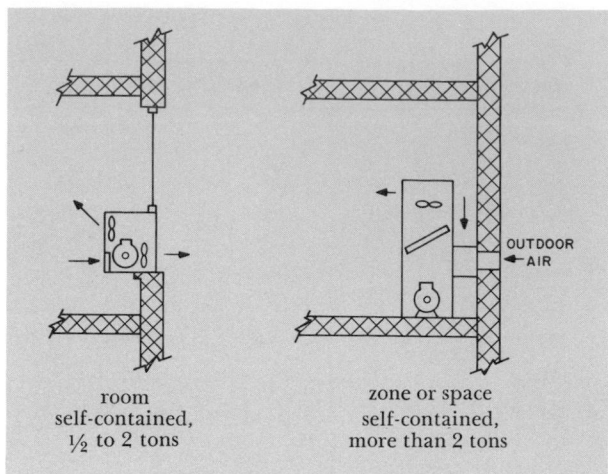

room
self-contained,
½ to 2 tons

zone or space
self-contained,
more than 2 tons

FIG. 8 — DX REFRIGERANT SYSTEMS

unit is converted into a year-round system. A *large capacity self-contained assembly* may in addition incorporate elements 2 and 9 to serve a larger space. Control of the self-contained units is usually the on-off control of refrigeration equipment with step control in the larger sizes. The fan operates continuously in most cases.

The self-contained units are applicable to individual small or large rooms and segregated zones. Such spaces may be oriented to service an individual occupant or a group of occupants. These may exist in a private residence, office, commercial establishment or in a group of offices in a single zone.

ALL-WATER SYSTEMS *(Part 12)*

Maintaining the individual room air conditioning aspect while verging into the central station system of refrigeration is the all-water system of *room fan-coil units (Fig. 9)*. Each unit within the system contains elements 1, 3, 4, 5, 8 and 10 *(Table 1)*. Outdoor air is introduced thru the wall to each unit. The individual room self-contained refrigeration assemblies (element 12) are combined into one or more large capacity remote central source water chilling refrigeration plants with the addition of elements 13, 14 and 15. Room temperature is normally controlled by regulating a water valve at the coil within the room fan-coil unit.

The fan-coil unit all-water system with a central refrigeration and heating source can be converted to an air-water system by centralizing the primary air ventilation supply. This eliminates the individual unit thru-the-wall air intakes. They are combined into a central station air supply system. Air is delivered either to the fan-coil units or directly to the

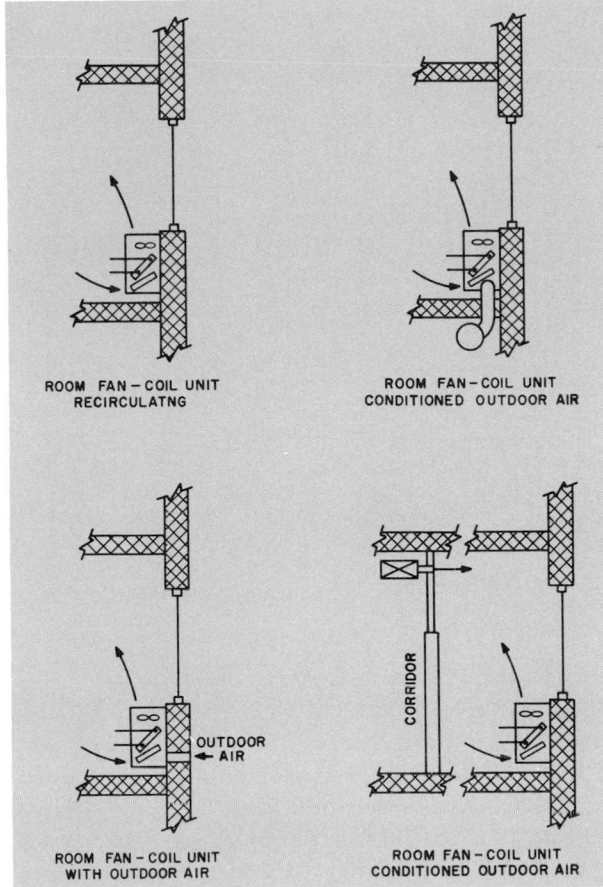

FIG. 9 — ROOM FAN-COIL SYSTEMS

room thru a corridor duct system and separate outlets. The central ventilation air apparatus contains elements 1, 2 and 4 (6 and 7 optional). A return air system is not generally used. The ventilation air reheat coil assists in heating during marginal weather. During the true summer and winter weather the space cooling and heating respectively are provided by the fan-coil unit.

The room fan-coil unit systems are applicable in smaller multi-room buildings such as motor-hotels, factory office buildings and small medical centers.

The air-water system alternate is particularly adaptable where it is necessary to avoid undependable and architecturally unsightly thru-the-wall outdoor air intakes, but there is a need for defined and controlled ventilation as in office or apartment buildings.

ALL-AIR SYSTEMS (Part 10)

Central station systems are conventional all-air systems. Basically the all-air systems are a form of reheat system; space conditions are maintained by various make-ups for load changes.

Volume Control

One way to compensate for a varying load is by regulating the volume of cooling air *(Fig. 10),* that is, without any make-up for room load variation. This all-air *variable volume system* has very limited application, only when the load varies less than 20%. Otherwise, if the air volume varies by more than 20%, the air motion within the space may be upset and create a drafty condition. If an outlet can maintain the space air motion regardless of volume reduction, volume control may be used more broadly. Then the *constant temperature, variable volume system (Part 10)* may be applied.

In the all-air *dual-conduit* system* there are two streams of air. One is permanently cold and varies in volume to compensate for the fluctuating internal and sun load. The other is cold in summer and varies in temperature during marginal or winter weather to compensate for the varying building transmission loss. This is an all-air system which parallels the high pressure air-water induction unit system discussed later in this text.

Bypass Control

Another method to compensate for a varying load is by reducing the amount of cooling air while maintaining the full quantity of supply air by including neutral air recirculated from the conditioned space

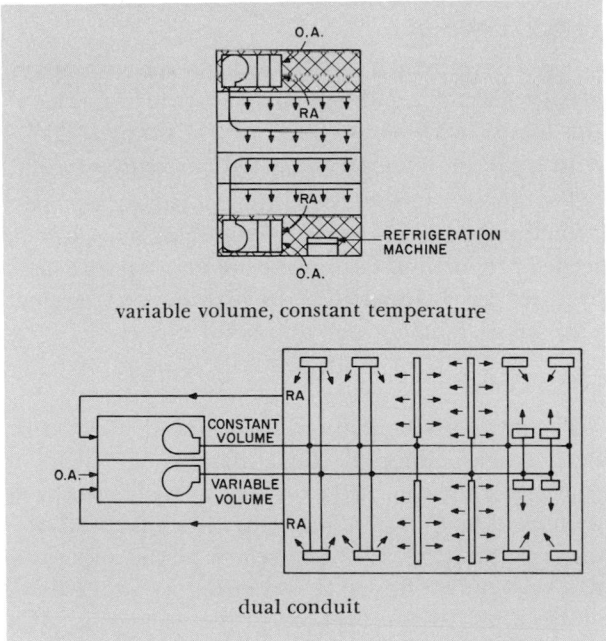

variable volume, constant temperature

dual conduit

FIG. 10 — ALL-AIR VOLUME CONTROL SYSTEMS

*A Carrier patented system.

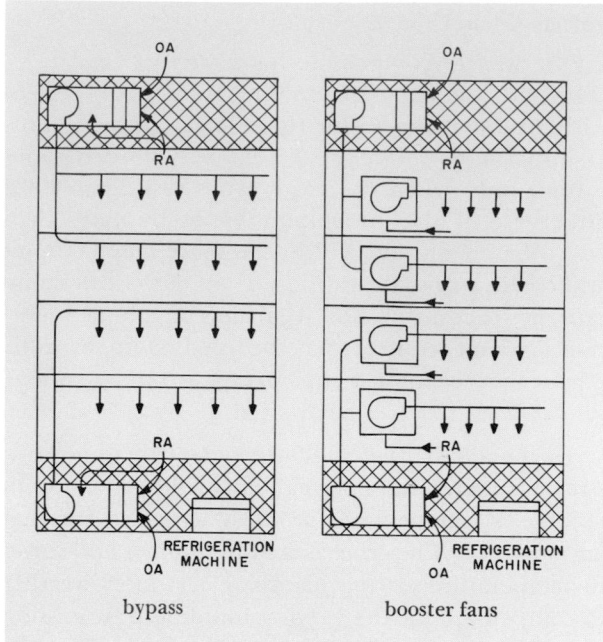

FIG. 11 — ALL-AIR BYPASS SYSTEMS

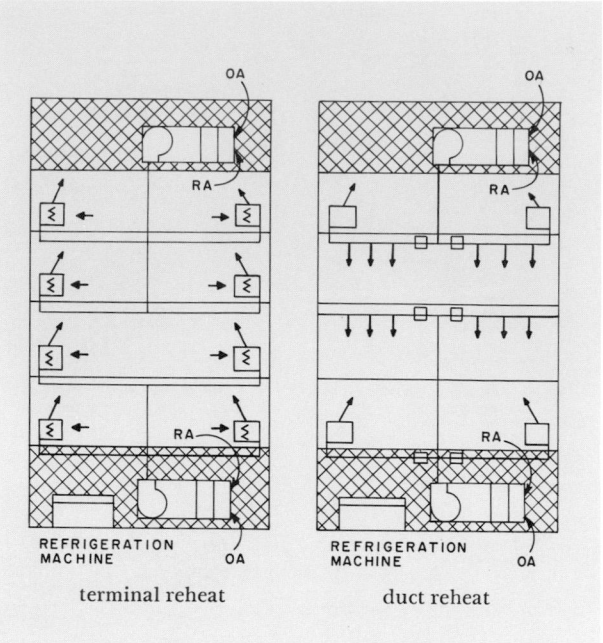

FIG. 12 — ALL-AIR REHEAT SYSTEMS

This is the conventional face and bypass control system *(Fig. 11)*, a variety of the *constant volume, variable temperature systems (Part 10)*. The multi-zone single fan unit single duct system may be at times a bypass system. The air bypass may be remote from the conditioned space (at the central apparatus or unit) or within the room terminal unit (in the dual-duct system).

A system in which a single central station apparatus provides dehumidified air to a number of booster fans located on separate floors of a building *(Fig. 11)* or in separate zones is also classed as a bypass system.

The booster fans pick up the cooled air and pass it on mixed with recirculated (bypassed) air in needed proportions to compensate for a varying load. Bypassed air is often used to furnish a permanent make-up to augment the volume of supply air.

Reheat Control

The make-up for reduced load may be affected by using reheat either at the apparatus or within a terminal in the conditioned space, as with a *constant volume induction system,* or within overhead terminal-outlets *(Fig. 12)*. The reheat at the apparatus may be supplied by a reheater either in a zone duct, within a multi-zone unit, or in the warm duct of a dual-duct system.

The *multi-zone blow-thru unit system (Fig. 13)* is a variation of the bypass and reheat system. Parallel heating and cooling coils provide an operation in

which the cold air is mixed in needed proportions with recirculated air passing thru heating coils; the recirculated air is either unheated (bypass) or heated (reheat). The air mixtures proportioned by several pairs of dampers (located within the unit) are transmitted thru single ducts to individual zones. Thermostats located in these zones control the corresponding mixing dampers.

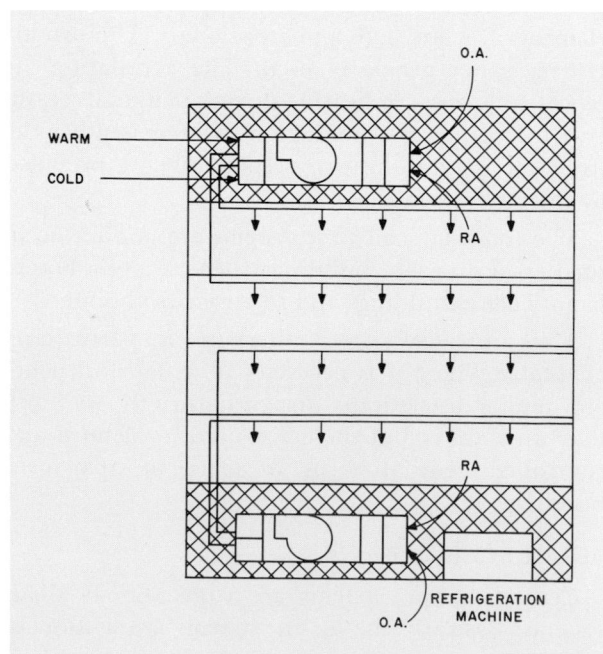

FIG. 13 — ALL-AIR MULTI-ZONE SYSTEM

The all-air *dual-duct system (Fig. 14)* supplies the mixing terminal units with air from two ducts, air streams at two temperature levels; one is cold, the other warm. The mixing terminal unit proportions the cold and warm air in response to a thermostat located in its respective space or zone.

The all-air systems ranging from the conventional bypass system thru the induction terminal reheat unit or zone reheat, multi-zone single duct, dual conduit and dual-duct systems have wide applications. These include the field of small single occupancy, single purpose buildings and multi-story buildings.

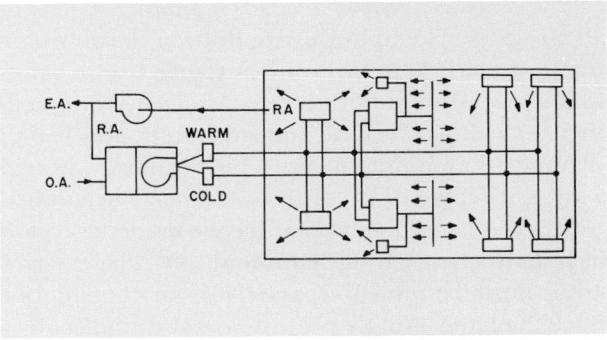

Fig. 14 — All-Air Dual-Duct System

AIR-WATER SYSTEMS (Part 11)

The air-water systems are practical for reducing the floor areas occupied by the terminals, the cooling-heating media transmission system and the central apparatus. Most of the space load (internal and sun heat gains) is balanced by a water coil included in the room terminal. Small water pipes and small high velocity air conduits replace the bulky air ducts of conventional all-air systems.

The original air-water system is the high velocity high pressure *induction unit system (Fig. 15)*. Primary air reduced to 20–25% of the conventional all-air system quantity serves to balance the building shell heat gains or losses, to satisfy the ventilation requirements, and to provide the humidity control and motive power to induce room air across the secondary cooling-heating coil within the terminal. The secondary coil produces year-round cooling or seasonal cooling and heating. In the latter case the primary air is cold year-round except during marginal weather when it is reheated according to a set schedule of temperatures.

The induction unit system is specially adaptable to load characteristics of the perimeter areas of multi-story multi-room buildings. This system results in lower owning and operating costs.

A variant of the high velocity high pressure induction unit system is the *three-pipe induction unit system (Fig. 15)*. Here the air stream (primary air) mentioned previously is not related to the structure shell transmission load; it is divorced from the heating of the building. It does provide the ventilation, humidity control and motive power for secondary air across the terminal coil. The three-pipe induction unit system is a year-round system that provides a simultaneous choice of either cooling or heating as required. One pipe supplies chilled water, the second pipe warm water, and the third pipe serves as a common return.

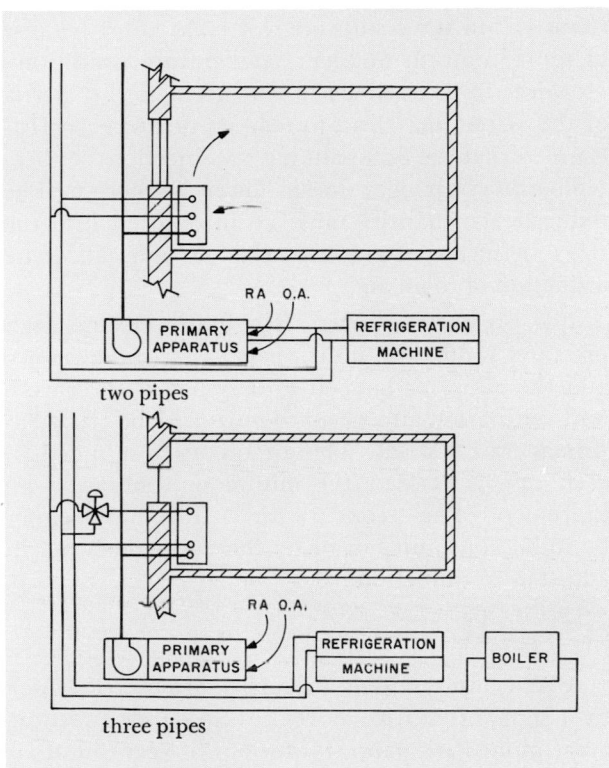

two pipes

three pipes

Fig. 15 — Air-Water Induction Systems

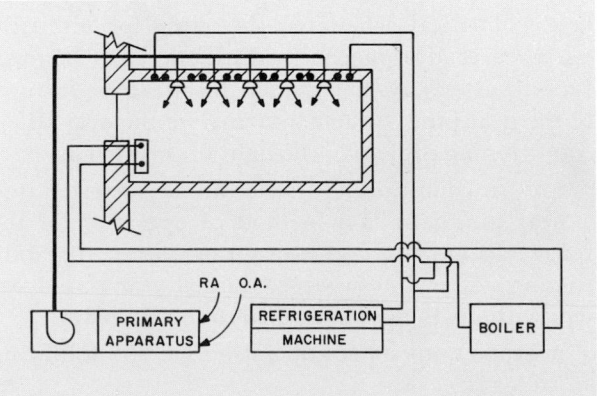

Fig. 16 — Air-Water Panel-Air System

The *panel-air system (Fig. 16)* is another variety of an air-water system. In this application the air quantity is similarly kept to an amount sufficient for ventilation and dehumidification requirements. The ceiling and wall panels have imbedded in or attached to them small pipe circuits thru which chilled or heated water is circulated. Panel-air systems may be applied with advantage to areas that have considerable gain of radiant energy from lights and sun. The exterior areas must be zoned separately from the interior since the latter usually presents a year-round cooling load.

System design and control arrangements are rather sensitive but the results are worth the effort because of more uniform ambient temperature conditions. However in areas of sharp load changes the inertia of the system may lead to temperature swings. It is claimed that the occupant has no objections to these temperature changes due to lower air motion. The reduced air quantity must be introduced into the space by means of an outlet that provides adequate induction of room air.

The air-water systems with high velocity, high pressure induction units, panel-air arrangements, and the all-water fan-coil unit system with centralized ventilation air ducted to units may be classed as primary air systems. The air quantity supplied in each case is at least the minimum ventilation requirement. The secondary air is introduced either by induction, unit fan or by thermal circulation in addition to induction.

HEAT PUMPS

A heat pump system is a refrigeration cycle which by design and control moves heat in either direction. Heat pumps are natural solutions of year-round air conditioning in buildings with a favorable balanced ratio of cooling-heating load, particularly where the two load values are almost equal. Heat pump systems are economical whenever the rates for electrical energy are advantageous against the costs of fossil fuels. It often pays to consider the attractive features of the heat pump system and analyze the over-all air conditioning project in the light of this system.

Any air conditioning system can be converted into a heat pump. It is a method of operation of the refrigeration plant that transforms a given air conditioning system into a self-contained year round system without the use of a separate boiler plant.

A heat pump system operation has the following advantages:

1. Over-all first cost saving — For many new buildings a single system can be installed to serve both cooling and heating. The cost of a boiler plant and auxiliary equipment including a smoke stack can be saved.

2. Space saving — Thru the elimination of boiler plant, smoke stack, fuel storage, etc.

3. Nuisance elimination — Does away with ash removal, smoke, soot and ash dust damage.

4. Single energy source — Electric power, simplifies accounting and service problems.

5. Increased safety — Thru elimination of handling of fossil fuels.

6. Fire insurance rate reduction — By elimination of fire hazards.

A refrigeration cycle can be applied to provide heating only, with future conversion to complete year-round air conditioning.

APPLICATIONS

The descriptions of individual applications that follow point out only the salient features that influence the air conditioning load calculations and selection of the most suitable system. The text is oriented mainly to summer air conditioning because many applications to existing buildings revolve around addition of summer cooling to the existing heating facility or its complete replacement by a year-round system. New buildings are assumed to incorporate year-round air conditioning at the inception of the project design. However even new buildings may include only the heating facility, with provision for future addition of cooling.

Any humanly conceived plan has room for improvement. The design engineer must apply his ingenuity in view of the wide variety of applications and idiosyncrasies within each application. Various technical publications contain articles and comprehensive texts on design guides and solutions for various applications. This text contains only the applications that have some characteristic prevalent to them. Other applications not discussed here either bear some resemblance to one of the described or are so individual that they have to be resolved by a separate unbiased analytical approach.

SINGLE-PURPOSE OCCUPANCIES

In this section are described system applications that are unidirectional, to serve either a home for a single human unit or a building serving an activity accommodating a large group of people. Thus one common temperature and humidity condition is to be maintained within a given space.

Residences

The architecture of residences is extremely varied, ranging from a standard repetitive package within large developments to an individual design for fashionable suburbia. Whatever the design there is a multitude of structural, orientation and shading devices by which the air conditioning load may be reduced. The complete appreciation and coordination of the data concerning the investor's aims, building structure, living habits of the occupant, and outdoor conditions should result in the most economical selection of equipment and its operating costs. Most important is the satisfaction of the owner and the occupant.

Residences represent the greatest single block of individual buildings. Of all the markets and applications of air conditioning, the residential market is the major one and is highly competitive. Special efforts are expended in defining the problem of air conditioning a residence (a 24-hour operation). At present there exists a unified industry-accepted method of load evaluation. Many manufacturers produce heating-air conditioning-heat pump equipment for residences. Dependability and ruggedness of equipment is their aim. Self-contained room units can be applied on an individual room basis in conjunction with the existing heating systems. Solar energy heating and air conditioning of residences may be considered.

Facilities for Dining and Wining

This classification covers restaurants, cafeterias, lunchrooms, lunchwagons, bars, cocktail lounges and night clubs. The outstanding characteristic of these applications is that the air conditioning load experiences sharp peaks at certain times of day and night. These peaks must be coordinated with the coincident outdoor conditions to arrive at a true cooling load. An absolute prerequisite is good ventilation with proper control of exhaust to neutralize odors of food and heavy smoking. This ventilation requirement is not only for the comfort of the occupant but also for preventing odor absorption by the walls and furnishings.

Another facet of load evaluation in this application is concentration of sensible and latent heat in dancing and entertainment areas.

Restaurants range from large dining spaces separated from kitchen and pantry areas to rooms adjoining the food preparing areas or where the cooking and pantry area is located in front of the service counters. In the latter case extreme care should be taken in the exhaust-make-up system. Such a system should trap the heat gains and odors over the cooking area. There must always be an air movement away from the dining area.

Dining areas should be under positive pressure. Occasionally a certain amount of make-up air for kitchen exhausts may come from the dining area ventilation unless it is heavy with tobacco smoke. Kitchen and pantry areas should be under negative pressure. Whether the kitchen is to be air conditioned or ventilated, in either case all the heat and vapor producing equipment must be carefully scrutinized. The most offensive equipment should be equipped with efficient exhaust hoods to eliminate heat and moisture gains. Kitchen and pantry odors present a dual problem. The odors retained by various utensils, boards and other surfaces including the room itself may backfire by rendering otherwise fresh foods undesirable and unusable.

The size of the particular food dispensing establishment dictates the choice of an air conditioning system. The choice may also be directed by relation of the establishment to the over-all building complex if this happens to be either a tenant or an intrinsic part of the building. In some cases the same central station system serving the total building may be operated on a partial load basis during the weekend to serve only the restaurant. This can be done by re-circulating at the apparatus the conditioned air assigned to the rest of the building. The building requirements of conditioned air are shut off while the restaurant continues to operate.

Variety and Specialty Shops

In this classification are shops and stores. These may be dress, shoe, fur, drug, candy or variety stores. The latter include sophisticated "five and ten" stores, supermarkets, or "country fair" type of small department stores. These are characterized by an 8–12 hour (at times 24-hour) operation cycle, varying population, high lighting intensity, and a considerable stock of goods or furnishings.

These areas call for special treatment of heat gains or losses and ventilation considerations: soda fountains; lunch and food counters; luncheonettes; pet sales; photo, candy, cigarette and food dispensers; open frozen food displays; beauty salons with their heat and odor producing equipment. There may be storage areas for candies, furs and other critical commodities that require special temperature and humidity conditions.

The stock and furnishings within a variety store can be utilized in load reduction by taking advantage of precooling effect. Cooling stored in the mass of stock to take care of peak loads helps to minimize the size of equipment. With high ceiling structures advantage may be taken of heat stratification using either natural venting or forced exhaust.

In most cases unitary self-contained equipment is applicable, either floor-mounted, ceiling-suspended or roof-mounted. At times the individual or combined projects may require either field- or factory-assembled fan-coil central stations with extensive ductwork. In conjunction with the air units larger refrigeration plants may be required, either direct expansion or water-chilling packages.

Bowling Alleys

The peculiar aspect of this application is that it is similar to spot or man air conditioning. Only the spaces occupied by bowlers, spectators and food dispensing are air conditioned. The alleys in front of the foul lines and the pin setting areas are not air conditioned. Because of high concentration of people (8 to 19 per alley) strenuous exercise and heavy smoking, ventilation must be generous and positive. The exhaust should be fan operated with the grilles located above the spectators and bowlers, beyond the foul line above the alleys. This prevents the spreading of tobacco smoke haze over the alleys. The room air returns should be numerous and at a distance from the food dispensing areas. Grilles, coffee urns and other equipment should be provided with exhaust hoods. The supply of conditioned air should be engineered to achieve draftless distribution. The equipment applicable is usually of the self-contained type; bypass and multi-zone units may also be used.

Radio and TV Studios

Radio and television studios vary considerably in size, occupying one or more floors of a building or a complete building devoted entirely to a network. Also there are small or large studios devoted to rebroadcasting. Large studios (especially color TV) have tremendous and variable loads of lights, cameras and equipment. Therefore, it is necessary to take full account of the loads and take all advantages to minimize these loads by utilizing every possible method of circumvention. The resulting air conditioning equipment and system should be economical and easy to control. The controls must be individual for each space, whether a studio, auditorium, control room, office or reception area. Electronic racks must have clean ventilating air and a direct exhaust; air necessary for this ventilation may come from surrounding studios, control rooms and office spaces.

Sound is an important element. Vibration and unwanted sound must be carefully controlled.

Individual applications dictate the choice of air conditioning system and equipment as long as the temperature and ventilation are satisfied without creating drafts or sound nuisance. With small installations multi-zone units are applicable.

Country Clubs

These oversize residences are usually located in open country, often on a hilltop completely open to climatic elements. The varying population is a problem, people often remaining late at night. The various activities and manner of occupancy dictates a careful analysis of individual room loads and overall refrigeration needs. The spaces range from social lounges, dining areas, cocktail lounges or bars to locker rooms and individual guest rooms. Some clubs have various recreational areas devoted to ping-pong tables, billiards and bowling.

This application suggests the use of any kind of equipment from room air conditioners to central station fan-coil assemblies serviced by water chilling refrigeration within the range of factory-assembled package equipment.

Funeral Homes

The problem of a large occupancy load for a short duration can be resolved economically by application of various expedients to reduce the load. The occupancy may also vary in size. The presence of flowers requires larger ventilation to dilute the odor effusion. A preparation room should have a good exhaust without any diffusion to the rest of the premises. Equipment selection is subject to the investor's capacity, method of operation and services provided. Quietness, ventilation ability and flexibility of control of capacity together with pre-cooling are the main attributes in the selection of a proper system.

Beauty Salons

Heat and odor emitting equipment (hair dryers, curlers, etc.) and compounds used are the special aspects of this application. An accurate evaluation of the actual load and arrangement for positive ventilation and exhaust are absolutely necessary. Both large beauty salons in individual buildings and small neighborhood shops fall into this category.

An additional problem exists in beauty salons located within department stores, hotels or other buildings. They must have negative pressure to prevent the spread of odors to adjoining areas.

Air distribution should be carefully designed to avoid drafts. There is no need for spot cooling.

Usually beauty salons present large areas with booths that have low partitions. There may be separate rooms for beauty treatments and massages. The air distribution in large areas is usually accomplished with ceiling diffusers, and the individual rooms are supplied by separate outlets.

The common single duct bypass or reheat system is adequate. Return air should incorporate an odor absorber to relieve the system of the requirement of an excessive outdoor air load.

Barber Shops

Air distribution that does not aim at a customer's neck or flow down mirrors should be a prerequisite. A proper ventilation rate should be chosen. Barber shops are often part of a large building unit; therefore they can be conditioned by the building system. Otherwise unitary self-contained equipment can satisfy the requirements.

Churches, Theaters and Auditoriums

These applications have one common feature, large population. A congregation of people may consist of a group quietly relaxed at prayer or an active crowd at a political rally. Judgment should be exercised in applying proper factors to provide sufficient capacity for varying latent and transmission heat gains. The load in ordinary cases may be standardized at 5 to 6 tons of refrigeration per one hundred people. However structures may vary from a large room or space as a part of a large building to cathedrals, opera and symphony halls, and large arenas. Often the cooling and heating air distribution cycles must be analyzed separately.

The major factors to ascertain for summer cooling loads are:

1. Pattern of patronage — One or many services in church, time of services or continuous; matinee and evening performances in theaters and auditoriums; nature of various sport or public events in arenas; coincidence of use of auditorium within a building with the operation of the rest of the building.
2. Ventilation — Possibly reduced to 5 cfm per person. Certain sport events and political rallies with heavy smoking require more ventilation.
3. Stratification — Application to loads, particularly in large structures.
4. Precooling — To reduce the size of air conditioning equipment, particularly refrigeration.
5. Climatic effects — Careful analysis of outdoor and solar design conditions coincident with peak patronage.

6. Inside design conditions — 77 F db and 60 percent rh quite acceptable.

The systems applicable to churches, theaters and auditoriums range from self-contained equipment to large field-assembled central station systems. Either water or ice refrigeration storage systems may be applied. Many of these structures may incorporate extensive auxiliary services for business, social and residential purposes. Use of these facilities should be analyzed as to coincidence of loads with the major auditorium load. This analysis may help in the overall economics of system and equipment selection.

Dance and Roller Skating Pavilions

A large active population characterizes these applications. The sensible heat factor is quite low; therefore some sort of reheat must be added or more air handled to absorb the excess moisture. The active and spectator population must be considered. Inside design conditions should be coordinated with air movement around the active people. With higher air distribution velocities higher temperatures may be selected. Most of the factors pertaining to auditoriums are applicable. However ventilation requirements must be raised due to an excessive odor problem, 15–25 cfm per person (reduced to 10–15 cfm if activated carbon is used on the return air). Coincidence of lighting effects and activity must be analyzed. In roller skating rinks the winter heating requirements should be coordinated with the radiant effect of cold walls and roof. The cooling effect of the enclosure is not objectionable to skaters but may be unpleasant for spectators.

Systems applicable to these structures are judged primarily by the investor's attitude. Systems of self-contained equipment or central station fan-coil assemblies and refrigeration should be judged on a purely economic basis. The air distribution may be designed around a limited number of air diffusers or with extensive ductwork.

Factories

A distinction should be made between factory human comfort and industrial product air conditioning. The primary purpose of the former is comfort of the worker. In the latter case the ambient conditions of the product are of foremost concern. There are cases when the air conditioning is used for the benefit of both the worker and the product as in clean rooms or with precision products. The economics is the deciding factor, to have the worker efficient and satisfied resulting in a production of goods with minimum rejects.

The optimum indoor design conditions of temperature, humidity and air velocity are the keys to the efficiency of the worker. The indoor design conditions for nonfactory comfort application is established at 76 F and 45–50% relative humidity; this is the condition for efficient performance of light activity. Most factory workers can perform at 80–85 F with dewpoints at 55–70 F and ambient air velocities from 25–300 fpm. In a few instances workers may tolerate temperatures as high as 90 F. It must be remembered that it is the combined effect of temperature, humidity and air velocity ambient to worker location that influence the environment for factory worker efficiency. The ambient air velocity is the velocity sweeping the worker. The relative humidities may range from 35–80%.

If the design conditions and air movement for an entire project prove to be uneconomical, consideration should be given to individual cooling for a worker in a given spot or limited area.

In certain factories the mean radiant temperature may be significantly higher than the desired design dry-bulb temperature. Consequently it may be necessary to lower the design ambient air temperature and its dewpoint, at the same time increasing the air movement. If this solution cannot be achieved economically, shielding of the radiation should be considered.

To achieve minimum costs a careful analysis must be made of loads from lights, roof, walls, equipment, processes and workers to arrive at the actual operating load. The influence of these loads may be reduced by the application of diversity factors and principles of stratification and flotation of heat gains *(Part 1)*. These two principles are aided by radiant heat exchange, natural tendency of heat to rise, exhausting of the warm air and an economic balance between the outdoor air taken thru the air conditioning equipment, the outdoor air introduced for heat flotation, and the exhausted air.

Spot or man cooling and standard compact single duct all-air systems are appropriate for applications in factory air conditioning. For widely scattered spot cooling stations it may prove more economical to convey chilled or evaporatively-cooled water thruout the factory instead of air. In such cases an air handling unit can be used directly as the supply air terminal or in conjunction with short duct runs.

MULTI-PURPOSE OCCUPANCIES

Under this heading are described applications of air conditioning systems that are practical for large multi-room multi-story buildings or compounds of buildings. These are designed primarily for a single function common to all occupants, either work, residence, medical, or teaching and learning. Buildings are occupied by multitudes of people heterogeneous in their individual comfort desires; buildings vary in exposure to sun, wind and shading. This diversity requires dividing buildings into individual air conditioned zones. Zones range in size from large areas to a small room. Individual desires and a variety of coincident load conditions within a conglomerate of zones must be satisfied.

Multi-room multi-story buildings include offices and apartment buildings, hotels, motels, dormitories and hospitals. There are some basic problems common to all of these:

1. Simultaneous occupancy on all sides of a building with daily cycle of solar effects on east-south-west exposures.

2. Preference for individual control of space conditions under severe variations of load.

3. Necessity for sound and odor isolation of individual spaces.

4. Limitation in available space to house air conditioning equipment and pipes, conduits and ducts.

The following discussion outlines these basic problems in various types of buildings. It also points out the specific aspects that affect preference of one system or another.

Office Buildings

Office buildings are mostly multi-tenant; at times they may be occupied by a single tenant, a single organization. This aspect varies the approach to a solution of the problems. The multi-tenant building requires more exacting applications. Regardless of the occupancy most existing and some new buildings have the following basic areas to consider — the interior and the peripheral exterior areas. The interior areas are within the center of a building not influenced by outdoor elements, except the top story under a roof. Areas around the periphery of a building (exterior zone) may extend from 12 to 20 feet inward from the outside wall. This zone is exposed to sun, wind, outdoor temperature and shading of the structural elements or neighboring buildings. There is an evident need for two different air conditioning systems to handle the two areas marked by loads of different behavior.

An interior zone has a relatively constant load of lights and people. Therefore a single all-air system is the most applicable. However with complications of scattered addition of electronic equipment and provisions to switch off lights within a fully partitioned space, the system grows more complex. There may appear a need for either a terminal reheat, volume control, or dual-stream all-air system. Occasionally there may be an application for a primary air-secondary water system, particularly if the wattage is high.

Exterior zones have extensively varying load characteristics, from an extreme combination of sun gain thru glass, maximum outdoor heat transmission, lights and people to no load during marginal weather and to a maximum negative transmission load in winter. The exterior zone is also subject to moving shadows of building structural elements (overhangs, reveals, fins), neighboring buildings and clouds. These elements together with the desires of occupants require a very flexible air conditioning system. It must be capable of balancing loads which may change from cooling to heating or vice versa on different exposures and even in adjacent spaces on the same exposure.

Added to this is the requirement peculiar to a south exposure due to the fact that sun gain peaks in winter months when the sun is at a low altitude. Cooling may be required while the rest of the zones are on heating. The south zone MUST be given special treatment in selecting air conditioning terminals or even a separate system with a winter source of cooling.

Another peculiarity of a peripheral zone is the winter behavior of the building shell; it produces downdrafts along the outside walls. The latter also act as sponges for the radiant heat of occupants, making people quite uncomfortable unless compensated for by a higher ambient temperature.

Two additional characteristics of office or any multi-room, multi-story buildings are the fenestration and general architecture. These also influence the choice of air conditioning systems. The fenestration may vary from 25 to 75% of the outdoor wall area. In some buildings fenestration is eliminated. The buildings may be squat, may occupy large areas, and have a predominently interior zone. Buildings may be tall and narrow, consisting of only peripheral zones. Buildings with the same total areas devoted to air conditioning, one squat another tall, may have the refrigeration load vary only 5–10%, but there can be a large deviation in air handling equipment requirements.

To meet the widely fluctuating conditions of the exterior zones, the air conditioning system must have two fluids available for conditioned spaces, one warm or hot and one cold. The dual-stream all-air systems and primary air-secondary water systems can be applied to the exterior zones; the all-air systems require more space for apparatus and supply and return air ducts. Where the necessary space is available or can be sacrificed at the cost of some other gain, the all-air systems are appropriate. These systems are excellent from a ventilation point of view. They provide economy in refrigeration during marginal weather thru outdoor air cooling.

The primary air-secondary water systems are space savers. The primary air either takes care of the building shell transmission losses, minimum ventilation, dehumidification and provides motive power for the induction units, or it provides ventilation and dehumidification only, as with three-pipe induction unit, panel-air or room fan-coil unit systems. Water carries the major burden of heat gain neutralization; space used by pipes is much smaller since water is over 200 times more efficient as a heat energy-carrying fluid than air.

Although office buildings are occupied mainly for 8–10 hour periods with some offices busy into the evenings, the air conditioning equipment should usually operate for at least 16 hours. During peak design conditions the air conditioning system should operate for 24 hours. This contributes to a more economical selection of equipment.

For some buildings the simplest and best system from a performance standpoint is the single duct all-air terminal reheat system. All the cooling is done with the supply air, and any variations in heat gain or loss are satisfied by a hot water, steam or electric reheat coil within the terminal unit.

It is necessary to repeat here the original premise expressed at the opening. There must be an appreciation of coincident elements of outdoor conditions, the building shell behavior and the inside conditions of load behavior. The result can be the proper solution of a given problem, subject to architect-engineer-owner-investor team cooperation.

There are some critical areas in each office building and these must be given special attention. Such spaces are conference rooms, special occupancies such as medical, laboratory, barber shops, beauty salons, restaurants, etc., or a concentration of electronic equipment or special rooms devoted solely to computer services. All of these are to be treated preferably as separate spaces to attend to their individual requirements of ventilation, odor and special thermal and sound problems. Keep equipment areas away from areas of high calibre occupancy and conference rooms since the best techniques in vibration elimination and sound attenuation may not completely satisfy the situation.

Hotels, Apartment Buildings, Dormitories

Much that has been said about office buildings and systems used can be applied to these buildings. However it must be observed that these buildings have mostly exterior zones. They are operated on a 24-hour basis thruout the year. Dormitories are an exception for the present but in the not-too-distant future they may also operate on a year-round basis.

Hotel guest rooms have a transient population, absent most of the day. Coincidentally the lighting load is also reduced. Considering these facts the individual room loads should be determined on the basis of exposure and either for early morning or late afternoon occupancy. The total refrigeration load should be based on the maximum occupancy coincident with the predominant exposure. With predominant east-west exposures the maximum load is either in the morning or in late afternoon with the corresponding occupancy, lights and sun effect; with predominant north-south exposures it is in the late afternoon with the corresponding occupancies, lights and sun effect.

Quick responding room temperature control and draftless, quiet air distribution are essential. Independent corridor ventilation is desirable. The toilet exhaust should approximate the room ventilation requirements. There should be independent systems for public spaces such as lobbies, restaurants, cocktail lounges or any services that a hotel may offer. All systems must be vibration- and sound-isolated from occupied areas. Practically any two-media system may be applied that confines air circulation to individual rooms. However this does not exclude the use of single-duct bypass systems in small hotels and where the local regulations permit the use of corridors as return air plenums.

Closely related to hotel applications are the modern multi-story motels that emerge from widely spread single-two-story motels. The application of various systems to the multi-story motels is similar to that for hotels. All-water 2- or 3-pipe room fan-coil systems or individual self-contained room units are more applicable to the single-two-story motels. In the latter case sometimes the direct use of outdoor air is excluded, relying instead on infiltration (natural or caused by toilet exhaust) for odor dilution. The excess infiltration should be estimated as part of the system load. The all-water system should be designed around a high temperature rise water circulation. Heat pump systems also are applicable.

Apartment houses and buildings range from 2–3 story garden type low-rise to high-rise multi-story complexes. Apartments range from multi-room luxury class to 2–3–4 room size to single room efficiency class. The occupancy of these apartments varies from fully filled during the day to an occupancy much like that of hotels, that is, occupied by professional people who are absent during most of the day. The occupancy diversity factor may vary from 75–80% down to 40–50%. The load calculations must recognize a 24-hour operation. In fact the 24-hour residential load estimate approach is applicable for arriving at the cooling-heating requirements.

Many systems may be applied to this application, depending on the size and configuration of buildings and the arrangement of apartments. These systems include individual self-contained units and heat pumps as well as all-air and air-water central systems. The investor and the size and type of project are prime factors in the selection of a particular system. Large apartment buildings often have auxiliary services such as restaurants, shops and office space rentals; these spaces should be handled by independent systems.

Residential hotels and apartment houses may have kitchen exhausts in addition to bathroom exhausts. This must be taken into account in balancing outdoor air intake with exhaust.

Modern dormitories are very similar to large size apartments; daytime occupancy may be taken at 50%. Building orientation and class schedules must be considered. There are no problems of kitchen exhaust; the interiors are rather austere. This means that the sound problem approaches that of an office building application.

Hospitals

The patient rooms of hospital buildings should be considered like hotels except that the load requirements are based on 100% occupancy with 24-hour, year-round operation. Depending on the predominant exposure the lighting load can be subjected to a diversity analysis. The important factor is that the air circulation must be contained within each room. Corridors, nurses stations and serving areas must have a separate supply. Each room must have an exhaust creating negative pressure. There should be no cross-communication (contamination) between various areas. The special treatment, therapeutic, maternity, surgical, morgue and other service areas usually have individual temperature, humidity and ventilation requirements.

Patient areas are best handled by primary air-secondary water induction systems. In a long building oriented east-west, with predominant north-south exposures, all-air reheat systems are applicable. The individual facility spaces are handled by independent self-contained or central station fan-coil assemblies. Absolute cleanliness from dust, odors and bacteria (along with rigid hospital housekeeping) is essential. This objective can be achieved easily with standard air conditioning systems and equipment. These must be simple to operate and particularly to maintain. Equipment that presents complexity soon loses proper attention.

Schools and Colleges

The need for year-round thermal environment in a schoolhouse is necessitated by concentrated schoolroom occupancy, elevated lighting intensity, sun effect in case of peripheral classrooms with windows and recognized improvement in performance of teacher-student instruction-absorption. Another element for the same need is the present tendency toward extension of the 9-month school year to a full 12 months. Even without the summer months the solution of cooling needs on the 9-month operation by use of ventilation is inadequate 25–90% of the time depending on location. Ventilation for odor dilution is an extremely important function. Depending on room cubage per pupil and student living habits the outdoor air requirement may vary 5–20 cfm per pupil. Thus the load due to outdoor air is significant.

The architecture of primary and secondary school buildings varies from one- or two-story standard arrangement single buildings to windowless highly utilized designs, to a compound of buildings devoted to individual functions. Whatever the arrangement the type of systems applicable to school air conditioning are extensive. Since the main prerequisite is introduction of sufficient outdoor air, all-air systems of any arrangement are the best choice. They operate economically by using outdoor air cooling any time the outdoor temperature drops below 60 F. The package equipment and all-water room fan-coil units have limited applications. The air-water system has extremely limited use.

The previous text has been oriented essentially toward primary and secondary schools. Institutions of higher learning, colleges, institutes and universities, present similar problems which extend to a larger variety of occupancies. The buildings are often multi-story structures for single- and multi-purpose occupancies; therefore the system selection has greater latitude. The refrigeration needs are often supplied by a central plant paralleling the arrangement of central heating.

Department Stores

The more extensive the department store, the more attention must be paid to problems of each floor or to departments within a particular floor. The nature of each floor is very individual. On the other hand a department store may be quite uniform in singularity of purpose and therefore may be handled as a unit.

Care should be exercised in evaluating the population and local shopping habits. The location of a department store is critical: serving a large city (purely urban population) or a small town (catering to farming communities); as part of a shopping center or in a deluxe setting. The patronage pattern is individual in each case.

The air cycle of a department store air conditioning system should correspond to the load of the specific area or floor or at times the whole store. However the refrigeration plant is sized on the basis of a net instantaneous load taking into account the population diversity factor of the entire store. Full advantage should be taken of outdoor air for cooling during the winter. Outdoor air preheaters should be designed to temper the maximum outdoor air quantity.

A variety of air conditioning systems may be applied to department stores; self-contained units or multiple central station fan-coil assemblies with minimum ductwork may be used. A main dehumidifier central station apparatus may supply chilled air to recirculating booster fans in various areas or floors. One central station assembly with extensive ductwork may serve a department store as a unit. Individual conditions can be maintained in a manner suitable to the circumstances. It may be either a total apparatus bypass control or an individual area volume control.

Shopping Centers

The trend toward concentration of shops, variety and department stores in outlying areas has led to the development of shopping centers. Problems of load calculations are stated under individual categories. However the design of air conditioning systems is dominated by the economics of investment and service to the operators of stores or by agreement between the lessees and lessor. The major choice is either individual self-sustaining treatment of each store unit or a central refrigeration along with a boiler plant. In the latter case the cooling and heating services are distributed to each store on an equitable basis. Stores are usually equipped with individual air cycles as part of the unit design.

An additional feature of shopping centers is a covered mall connecting the stores and plazas between the stores. These mall areas are usually air conditioned at somewhat higher temperatures than the surrounding stores. Doors from stores to the malls usually stay open. Ventilation air from the stores is usually exhausted thru the malls. At times there are openings above small central plazas within a store compound; these openings serve as exhaust relief.

Libraries and Museums

This application belongs to the industrial type since it is primarily for the benefit of various products. The products are on permanent display; although the design conditions are strict, this application is similar to one for human comfort.

Libraries and museums housing collections of books, fine and technical arts, historical and natural sciences are often located in large cities. This location exposes the collections to a polluted, destructive atmosphere. Air conditioning with special attention to filtration and elimination of atmospheric pollutants *must maintain year-round constant temperature and humidity.* Fortunately the indoor design conditions are within the human comfort range. A thorough air movement must be established to avoid stagnant corners.

The control of space relative humidity and the object humidity, that is, the humidity of the thin film in close contact with the object surface, is very essential. The space relative humidity affects the equilibrium of moisture content between the object and the ambient atmosphere. The object humidity affects the surface of the object. In both cases excessive moisture acts as a destructive agent.

The equipment for conditioning the air may be any kind within the economics of the given project. In contaminated areas considerations should be given to the removal of acidic fumes and vapors by employing direct spray dehumidifiers or sprayed coils with spray alkalinity maintained at a pH value of 8.5 to 9.0.

The ideal system is one that is well zoned with central dehumidifying equipment and booster fan stations using bypass or steam, hot water or electric reheat control. In all cases the automatic control should incorporate a safety device to prevent over-humidification. Equipment vibration isolation and acoustic attenuation must be part of the design. The system must operate 24 hours year-round.

Laboratories

The use of air conditioning systems in laboratories provides one or more of the following services:

1. Control of regain in hygroscopic materials.
2. Influence of physiological reactions (comfort).
3. Control of chemical reactions.
4. Control of biological reactions.

With regard to their functions laboratories can be grouped in the following types:

1. Research
2. Development
3. Test
4. Calibration
5. Pilot plants

Individual laboratories (singly or collectively) are usually designed for a specific condition or for a range of conditions of temperature, humidity and cleanliness (white or clean rooms). These functions must be maintained to very exacting requirements. The controls and safeguards must often be quite elaborate.

The discussion that follows is confined to laboratory buildings, where modules, research personnel offices and conference rooms, libraries and work shops are combined under one roof. Chemical research laboratories are equipped with exhaust hoods that present specific problems.

In the design of air conditioning systems for laboratory buildings the application of standard design patterns and practices is very limited. Each laboratory and building is a problem in itself that requires the following considerations:

1. Exact room conditions.
2. Specific ventilation, often oriented to the exhaust requirements.
3. Separation of general occupancy spaces.
4. Orientation of heavy load laboratories away from the additional burden of sun gain.
5. High load variation in each laboratory
6. Diversity in use of laboratories.
7. Diversity in loads thruout the building.
8. 24-hour operation of laboratory spaces.
9. Constant or variable exhausts.
10. Concentrations of sensible or latent heat, requiring either special exhausts or spot cooling or both.
11. Corrosive effects of fumes on parts of the air conditioning, ventilation and exhaust systems.
12. Explosion hazards.

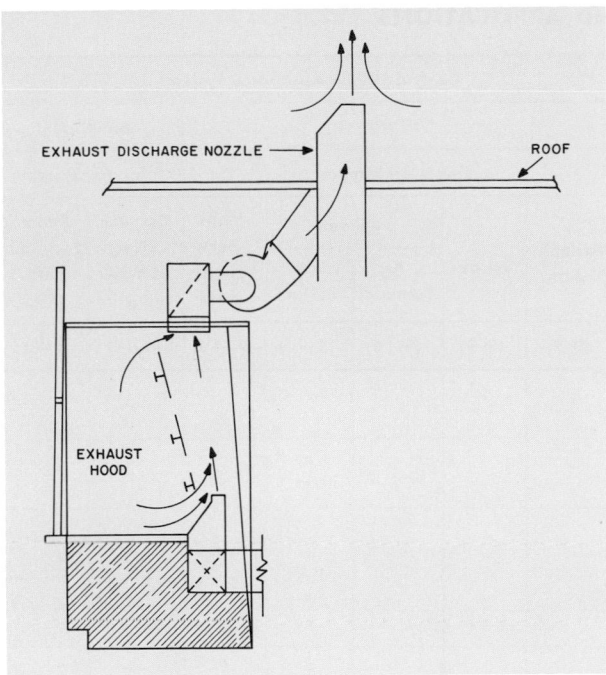

EXHAUST DISCHARGE NOZZLE →   ROOF

EXHAUST
HOOD

FIG. 17 — INDUCTION HOOD

The load calculations are made in accordance with *Part 1,* taking account of any special heat gains and customer specification of indoor design conditions. The recommended ventilation is 15–20 cfm per person unless the amount of exhaust demands larger quantities.

The selection of an air conditioning system is dictated by the grouping of similar spaces (zoning) and individual requirements of separate spaces. The ordinary comfort applications are open to any type of system; the variable load spaces are best treated with multi-zone reheat systems such as a constant volume, variable temperature system; the spaces with concentrated and special loads may need spot treatment of either temperature, humidity and/or cleanliness provisions. Standard equipment is applicable except for explosion hazards and corrosive fumes requirements where applicable.

Fume exhausts may present unusual problems of both functional and economic nature. The conditioned make-up air necessary to fill the exhaust air requirements is a major burden on the system and increases costs. The intermittent and varying use of exhaust hoods presents a difficult air balancing problem, with special attention to selecting proper air handling equipment. One solution of fume exhaust is the application of self-ventilated hoods *(Fig. 17);* the exhaust hood requirements are divorced from the air conditioning system. Untreated outdoor air (heated in winter) is introduced directly to the hood

and exhausted either from a group of hoods or individually. The latter method is followed when there is a danger of mixing exhausts because of an explosive or corrosive nature.

In this connection it is necessary to discharge the exhausts forcibly upward in order to fully diffuse the fumes to protect possible recirculation into the outdoor air intakes, whether for ventilating the exhaust hoods or the air conditioned spaces.

Marine

The air conditioning of ships, whether a luxury passenger liner or a cargo vessel, follows the patterns on land. However, in defining the outdoor design conditions it is necessary to determine the zone of sailing coverage. Normally it is recommended that the summer outdoor design conditions are 95 F db and 82 F wb unless sailing areas are predominately tropical. Then the average warm port of call defines the summer outdoor design condition. The coldest port of call pinpoints the winter outdoor design condition. Indoor design conditions are similar to applications on land. The sun effect thru glass must be evaluated on a maximum basis because of no definite orientation. There is an added factor of sun radiant energy diffusely reflected from the water surface. Ventilation needs aboard ship are larger due to the limited volume of quarters and general malodors specific to ships. A minimum of 12.5 cfm per person or 2.5 air changes per hour, whichever is greater, should be used to estimate ventilation. Deluxe passenger vessels should have more ventilation.

The compactness of the hull of a ship completely utilizing each cubic foot of space leads to specific considerations in the selection and design of air conditioning systems and equipment. Generally systems for staterooms and crew quarters are limited to air-water induction, all-air reheat and dual-duct systems. The public spaces are air conditioned by field- or factory-assembled conventional all-air central station fan-coil systems.

The equipment installed aboard ship should be selected to satisfy the following requirements: a minimum of space; minimum noise and vibration; operation under conditions of ship's roll and pitch; use of materials able to withstand the corrosive effects of sea air and salt water; and sufficiently maintenance-free for the duration of a voyage. To be prepared for emergencies the systems should be adequately supplied with spare parts and maintenance materials.

Table 2 presents a summary of the various applications and air conditioning systems.

TABLE 2 — SYSTEMS AND APPLICATIONS

APPLICATIONS	Page	DX Self-Contained — Room — 1/3 to 2 tons (9–8)	DX Self-Contained — Zone — 2 tons and over (9–8)	All-Water Room Fan-Coil — Recir. Air (9–8)	All-Water Room Fan-Coil — With Outdoor Air (9–8)	All-Air Single Air Stream — Variable Volume (9–9)	All-Air Single Air Stream — Bypass (9–9)	Reheat — At Terminal (9–10)	Reheat — Zone in Duct (9–10)	Multi-Zone Single Duct (9–10)	Air-Water Prim. Air — Secndry. Water H-V H-P Induction (9–11)	Room Fan-Coil with O.A. (9–8)
Single-Purpose Occupancies												
Residential Medium (9–13)		x										
Residential Large			x	x						x		
Restaurants Medium (9–13)			x						x			
Restaurants Large			x				x	x	x	x		
Variety & Spctly. Shops (9–13)			x									
Bowling Alleys (9–14)			x				x					
Radio and TV Studios Small (9–14)			x				x	x	x			
Radio and TV Studios Large			x					x	x			
Country Clubs (9–14)			x				x		x	x		
Funeral Homes (9–14)			x							x		
Beauty Salons (9–14)		x	x									
Barber Shops (9–15)		x	x									
Churches (9–15)			x				x			x		
Theaters (9–15)							x					
Auditoriums (9–15)							x					
Dance and Roller Skating Pavilions (9–15)			x			x	x					
Factories (comfort) (9–15)			x				x		x			
Multi-Purpose Occupancies												
Office Buildings (9–16)						x				x	x	
Hotels, Dormitories (9–18)				x	x						x	x
Motels (9–18)				x								
Apartment Buildings (9–18)					x						x	x
Hospitals (9–18)					x				x		x	
Schools and Colleges (9–19)					x	x	x	x	x			
Museums (9–20)									x	x		
Libraries Standard (9–20)			x				x		x	x		
Libraries Rare Books			x						x			
Department Stores (9–19)							x					
Shopping Centers (9–19)			x				x			x		
Laboratories Small (9–20)			x			x			x	x		
Laboratories Lge Bldg						x		x			x	
Marine (9–21)								x			x	

NOTES:
1. Systems checked for a particular application are the systems *most commonly used*. Economics and design objectives dictate the choice and deviations of systems listed above, other systems as listed in Note 2, and some entirely new systems.

2. There are several systems used on many of these applications when higher quality air conditioning is desired (often at higher expense). They are Dual-Duct (9–11), Dual Conduit (9–9), 3-pipe Induction and Fan-Coil (9–11), 4-pipe Induction and Fan-Coil, and Panel-Air (9–12).

3. Numbers in parentheses are page numbers of the text describing the particular system or application.

Part 10
ALL-AIR SYSTEMS

CHAPTER 1. CONVENTIONAL SYSTEMS

The conventional all-air systems are ordinary single duct air transmission arrangements with standard air distributing outlets, and include direct control of room conditions. Such systems are applied within defined areas of usually constant but occasionally variable occupancies such as stores, interior office spaces and factories, where precise control of temperature and humidity is not required. However, these systems can be arranged to satisfy very exacting requirements.

The conventional systems are classified in two major categories: constant volume, variable temperature and variable volume, constant temperature systems. The first category has the greater flexibility to control space conditions, extending from on-off refrigeration capacity control to exacting reheat control.

The conventional systems and their methods of room temperature control are listed as follows:

1. Constant volume, variable temperature systems with
   a. On-off or variable capacity control of refrigeration.
   b. Apparatus face and bypass damper control.
   c. Air reheat control.
2. Variable volume, constant temperature systems with supply air volume control.

The conditioned area may include either a single zone or several zones, the latter consisting of two or more individually controlled zones. Single zones are usually served by using refrigeration capacity or face and bypass control, and at times reheat control. The multi-zone applications require reheat control or varying volume control systems.

Maintenance of uniform conditions depends on a balanced design of air distribution and matching of design space load with refrigeration capacity.

This chapter includes Systems Features, Systems Description, Controls and Engineering Procedure for designing these conventional systems.

SYSTEM FEATURES

Some of the features of the conventional systems are the following:

1. *Simplicity* — All the systems described are easy to design, install and operate.
2. *Low Initial Cost* — The general simplicity of system design, rudimentary requirements and minimum physical make-up lead to a low initial cost.
3. *Economy of Operation* — Since the systems are the all-air type, the outdoor air may serve as a cooling medium during marginal weather, thus conserving the use of refrigeration. In most cases the areas served by the systems are of limited size; therefore, the operation of the systems may be limited to periods when their use is of maximum benefit.
4. *Quiet Operation* — All mechanical equipment can be remotely located.
5. *Centralized Maintenance* — All elements of the air handling and refrigeration apparatus are in one location, limiting centralized services and maintenance to apparatus rooms.

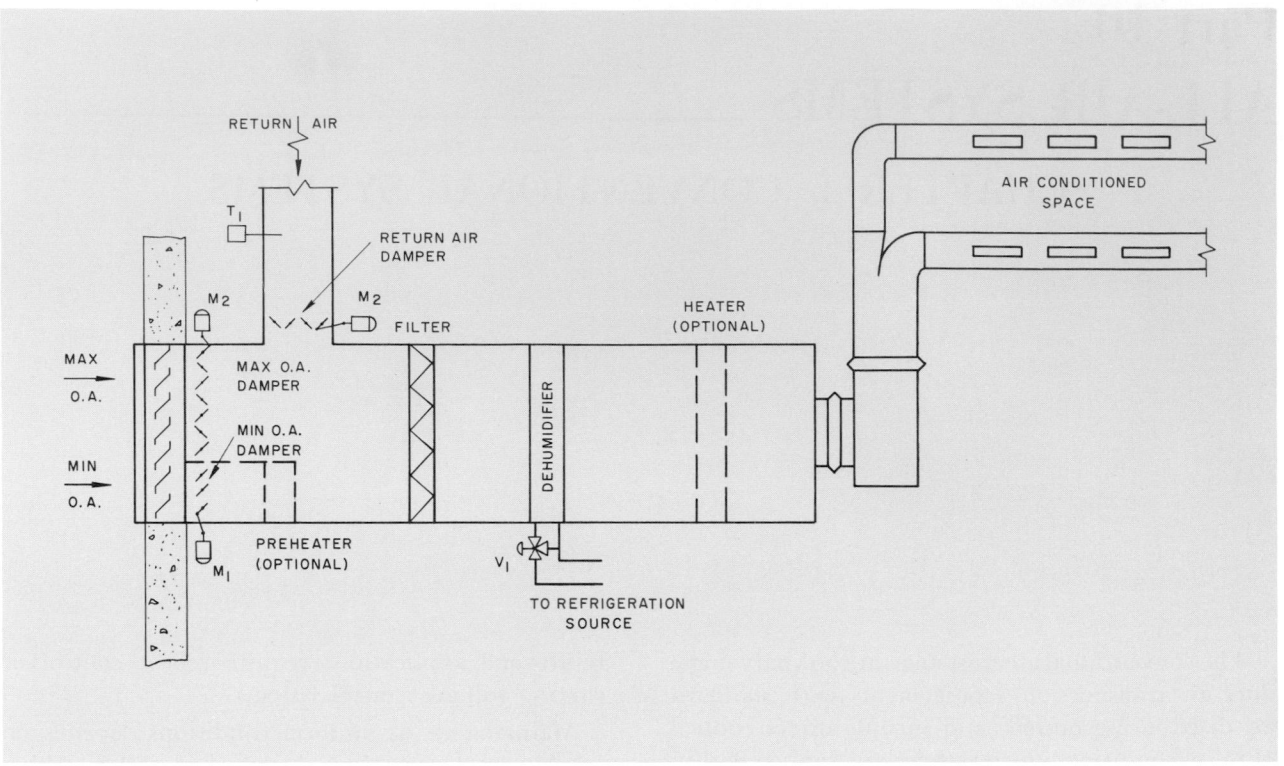

FIG. 1 — BASIC ELEMENTS OF A CONVENTIONAL AIR CONDITIONING SYSTEM

SYSTEM DESCRIPTION

CONSTANT VOLUME, VARIABLE TEMPERATURE SYSTEMS

Figure 1 shows the basic parts of a conventional system required for summer air conditioning: outdoor and return air connections, filter, dehumidifier, fan and motor, and supply air ducts and outlets. The optional elements provide preheating of outdoor air and space heating when required.

Refrigeration Capacity Control

In a summer air conditioning system, a thermostat located in the space return air path is set at the desired room temperature. It controls directly the refrigeration capacity of the dehumidifier, either by on-off, step or modulation controls. The choice of a specific control method depends on the size and type of the refrigeration plant. The resultant temperature and humidity conditions are only relatively constant, since the refrigeration machine capacity does not always match the load. The on-off control of space conditions is intermittent, since humidity conditions can rise during off-cycles because the supply air consists of an unconditioned mixture of return and outdoor air.

The refrigeration plants are either the small-to-medium size direct expansion type or the medium-to-large size water chilling type. Accordingly, the control applied can be either an on-off liquid solenoid valve, a step operation of compressor(s), or a valve to modulate the water flow thru the dehumidifier(s). In marginal weather the space thermostat controls the return and maximum outdoor air dampers to provide cooling from outdoor air.

A heating coil is added if the system is designed for year-round operation to provide winter ventilation and heating. A preheating coil is added at the minimum outdoor air intake when the mixture temperature of minimum outdoor and return air is below the required supply air temperature.

The supply air is transmitted thru low velocity air ducts and distributed in the space by standard outlets or diffusers. Although they are a conventional type, the air ducts and outlets must be engineered carefully to avoid the generation of disagreeable noise.

This type of conventional system is used in many different applications; however, the performance is best in spaces with loads that have relatively stable characteristics and minimum ventilation requirements. For this application equipment selected to match the load is economical to operate and, being fully loaded most of the time, maintains the space conditions at nearly constant level.

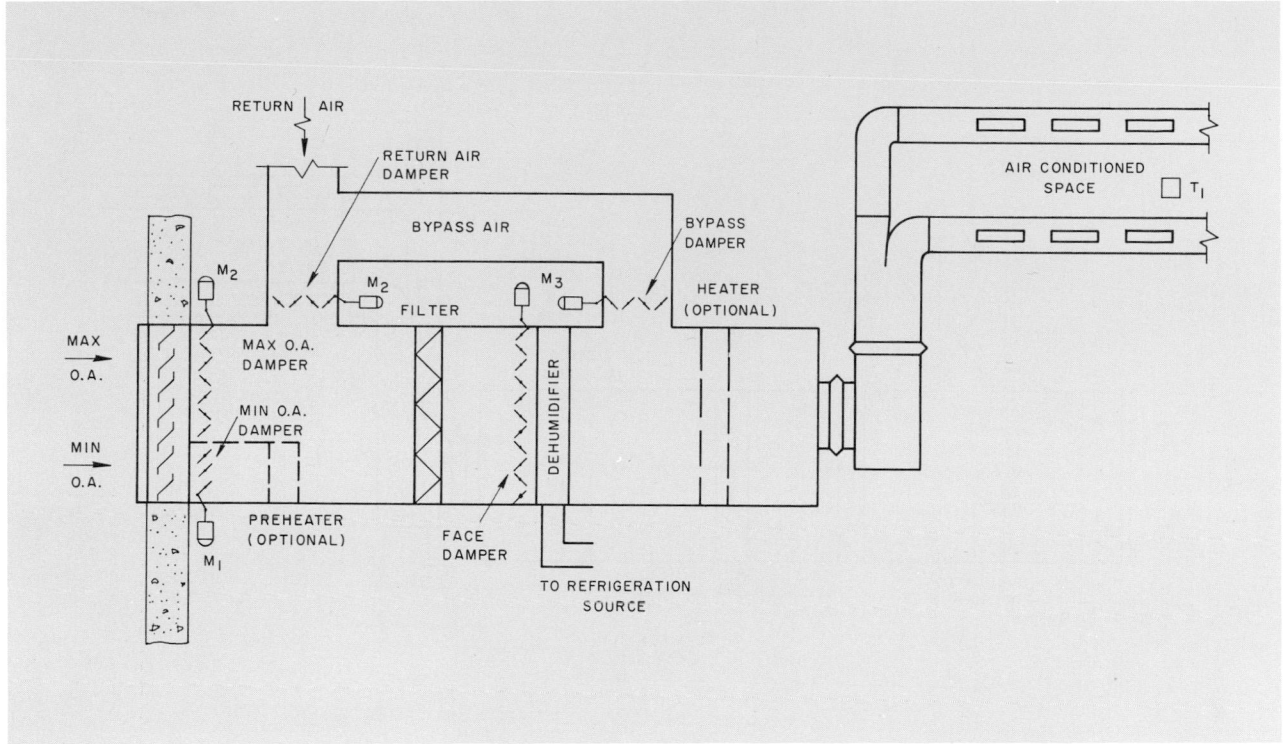

FIG. 2 — TYPICAL CONVENTIONAL SYSTEM WITH FACE AND BYPASS CONTROL

Face and Bypass Control

A variation of the preceding arrangement to improve the control of space conditions and to allow a more economical selection and utilization of the refrigeration plant is the use of an air connection between the return air and fan intake to allow a bypass of air around the dehumidifier (*Fig. 2*). This arrangement for mixing the bypassed return air with the dehumidified air improves the control of space conditions. Space temperature is more constant. Space humidity is still subject to variations though much smaller than with the original system. Care must be exercised to exclude the possibility of short circuiting ventilation outdoor air thru the bypass connection.

The refrigeration capacity is indirectly controlled by the falling temperature of the cooling medium as the dehumidifier face dampers close and the load on the dehumidifier falls off. This has a beneficial effect on the humidity since the temperature leaving the dehumidifier tends to fall with the decreasing air bypass factor and the falling temperature of water. When the face dampers are closed, the refrigeration equipment is stopped. In marginal weather the refrigeration equipment is shut down. The face and bypass dampers are set open and closed respectively. Space conditions are controlled by mixing the outdoor and return air to utilize the cooling available in outdoor air.

Air Reheat Control

The best control of space conditions relative to both temperature and humidity can be obtained by means of the reheat system (*Fig. 3*). Close temperature control is obtained by adding heat to neutralize excess cooling to maintain a constant space temperature.

Space humidity conditions are achieved by maintaining the supply air at a constant dewpoint temperature (constant moisture content). During hours of partial sensible and latent heat loads the space humidity is lowered. This lowering may be considerable if, in the case of applications using water chilling cycles, water is circulated continuously, resulting in lower apparatus dewpoint and supply air temperature.

The capacity of the refrigeration plant is controlled from either the return or supply water temperature. Generally, the dehumidifier capacity is controlled either from a dewpoint thermostat located at the dehumidifier outlet or by a thermostat located in the fan discharge. The setting of a fan discharge thermostat must compensate for the heat gain between the dehumidifier and the fan dis-

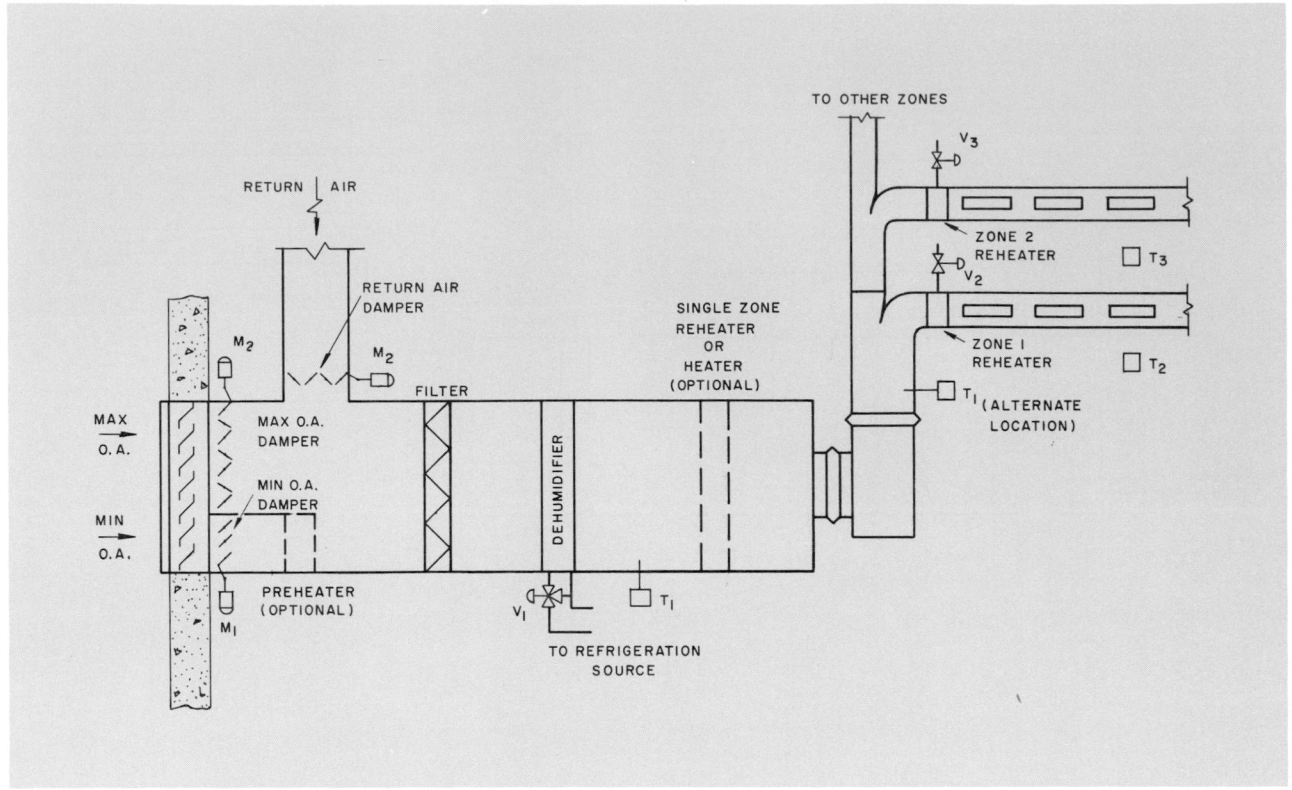

FIG. 3 — TYPICAL CONVENTIONAL SYSTEM WITH AIR REHEAT CONTROL

charge (fan horsepower and duct heat gains). In marginal weather, either of these thermostats controls the return and maximum outdoor air dampers to utilize the cooling effect of outdoor air.

Room conditions are maintained by controlling either the apparatus reheater in the case of a single zone or the duct reheaters in the case of a multi-zone application. The reheaters may serve also to provide winter heating as required.

VARIABLE VOLUME, CONSTANT TEMPERATURE SYSTEM

Variable Volume Control

The variable volume, constant temperature system *(Fig. 4)* parallels the reheat system, except (1) the dehumidifier is sized for instantaneous peak load of zones involved, and (2) individual reheaters are replaced by air volume control applied to either the individual branch ducts or the individual outlets. The dewpoint thermostat controls the dehumidifying capacity in summer and the return and outdoor air dampers in marginal weather. Preheating and heating elements may be added when required. The space conditions are maintained by room thermostats controlling the volume of supply air to the individual space. At partial loads the humidity may

rise because the supply air is not at the lower dewpoint needed by the lower room sensible heat factor (SHF). A lower dewpoint may be achieved with a system of uncontrolled chilled water flow.

This system is applied in multi-zone areas. However, to be fully effective over the complete range of load variations, the supply air terminal must be able to vary the air volume without condensation occurring at outlets or causing noise, and to maintain reasonable air circulation within a space. Such a system is described in *Chapter 5*.

The variable volume, constant temperature system with conventional outlets, particularly the sidewall type, must limit the variation in air volume to 75-80% of the full quantity. The lower volume of air may cause a draft due to incomplete throw of the air stream; thus the load fluctuations within a given zone must be small. The variable volume, constant temperature system is primarily applied to internal areas; it is seldom used in external areas because the solar radiation load constitutes a major portion of the system load.

Efficiency of the various conventional systems described previously is reflected in the pattern of indicated relative humidity profiles resulting during the partial load conditions *(Fig. 5)*.

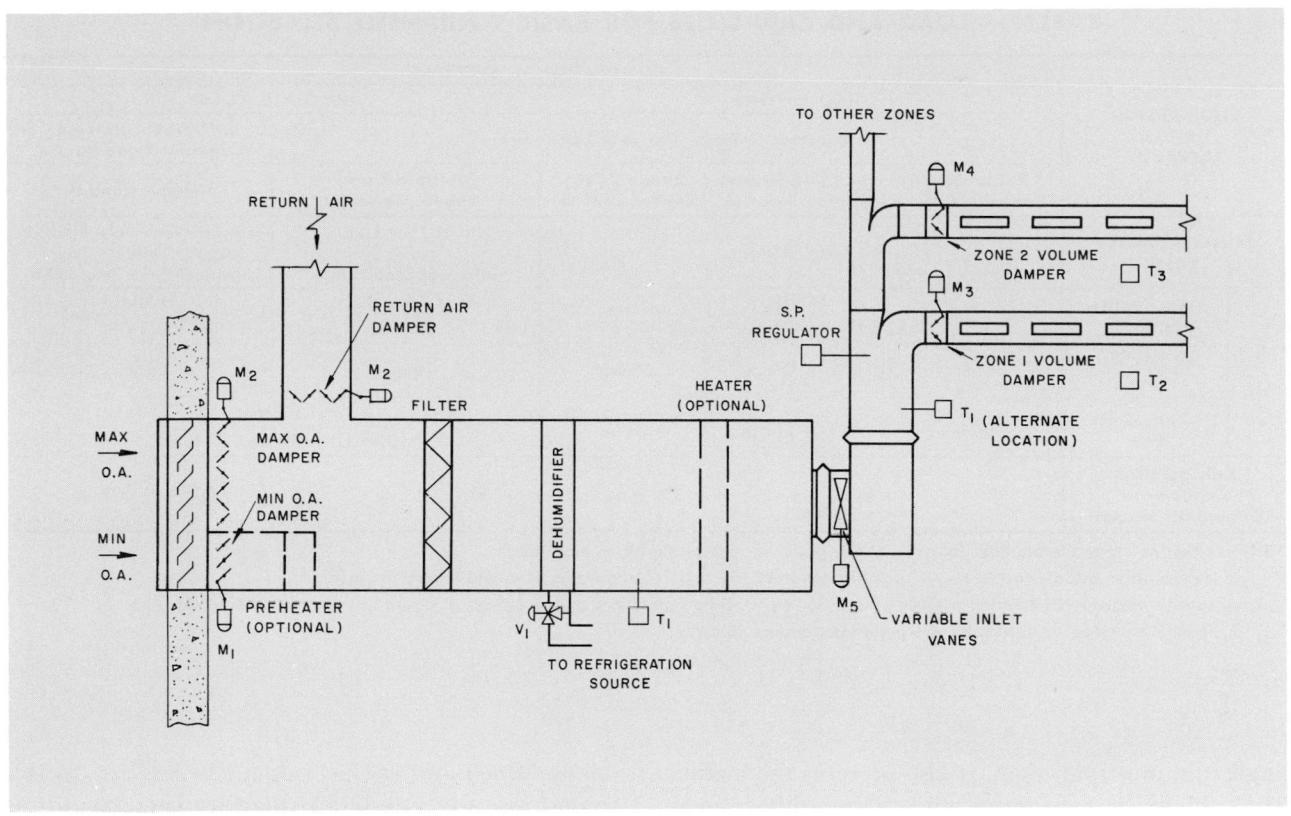

FIG. 4 — TYPICAL CONVENTIONAL SYSTEM WITH AIR VOLUME CONTROL

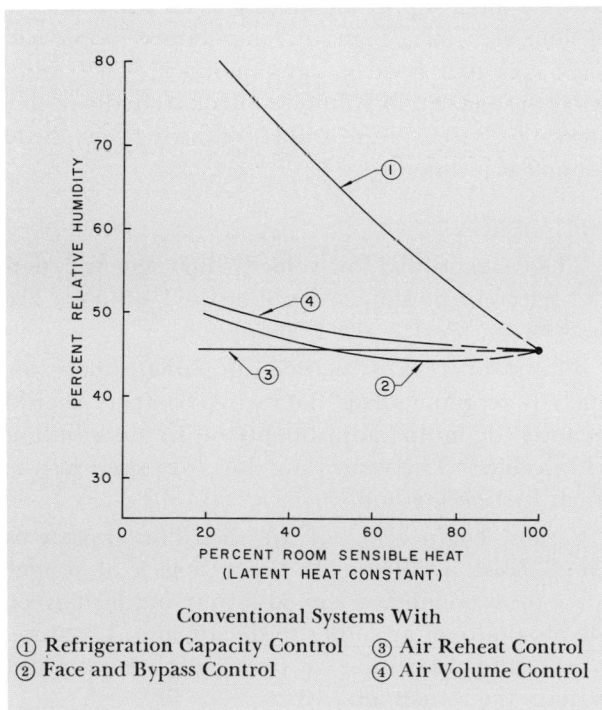

Conventional Systems With

① Refrigeration Capacity Control ③ Air Reheat Control
② Face and Bypass Control ④ Air Volume Control

FIG. 5 — COMPARISON OF RELATIVE HUMIDITY
BEHAVIOR FOR VARIOUS TYPES OF CONTROL,
CONVENTIONAL SYSTEMS

ENGINEERING PROCEDURE

The following design rules are offered to guide an engineer in achieving a practical design. *Part 1* contains data on the initial survey, preliminary layout and load calculation.

COOLING LOAD

Both the sensible and latent loads are calculated for each zone. The sensible heat factor determines the apparatus dewpoint temperature.

In the case of a multi-zone application, a judicious selection of apparatus dewpoint temperature must be made to avoid penalizing the system by using the lowest apparatus dewpoint required by any one of the zones. The apparatus dewpoint may be the one resulting from a block load estimate or one arbitrarily selected to produce acceptable variations in relative humidity in the zones involved.

When calculating the load for systems designed to apply face and bypass damper control, if the outdoor air can be bypassed around the dehumidifier, the Btu calculations should be adjusted by increasing the outdoor air bypass factor by 0.1.

Table 1 summarizes the cooling load requirements of various conventional systems applied to

TABLE 1—LOAD AND CAPACITIES FOR BASIC EQUIPMENT SELECTION

ENGINEERING DESIGN ASPECTS		CONVENTIONAL SYSTEMS				
		SINGLE ZONE			MULTIPLE ZONE	
		Constant Volume, Variable Temperature				Variable Volume Constant Temperature
		Refrigeration Capacity Control	Face and Bypass Damper Control	Single Zone Reheat Control	Multiple Zone Reheat Control	Volume Control
Sensible Cooling ERSH		Zone ERSH$_z$			Sum of Zone ERSH$_z$ and Individual Zone ERSH$_z$	Instantaneous Block ERSH$_{bk}$ and Individual Peak Zone ERSH$_z$
Air Quantity	Dehumidifier Cfm$_{da}$	$\dfrac{\text{ERSH}_z}{1.08\,(1-\text{BF})\,(t_{rm}-t_{adp})}$			$\dfrac{\text{Sum of Zone ERSH}_z}{1.08\,(1-\text{BF})\,(t_{rm}-t_{adp})}$	$\dfrac{\text{ERSH}_{bk}}{1.08\,(1-\text{BF})\,(t_{rm}-t_{adp})}$
Air Quantity	Supply Fan Cfm$_{sa}$	Cfm$_{da}$	1.1 × Cfm$_{da}$	Cfm$_{da}$	Cfm$_{da}$	Cfm$_{da}$
Air Quantity	Zone Duct Cfm$_{sa}$	Cfm$_{da}$	1.1 × Cfm$_{da}$	Cfm$_{da}$	$\dfrac{\text{ERSH}_z}{1.08\,(1-\text{BF})\,(t_{rm}-t_{adp})}$	
Refrigeration Capacity or Dehumidifier Load		Cfm$_{da}$ × 4.45 × (1 − BF) $(h_{ea}-h_{adp})$ = GTH$_z$				GTH$_z$ or GTH$_{bk}$

ERSH = effective room sensible heat (Btu/hr). Subscripts: z = zone peak; bk = block peak.

BF = dehumidifier bypass factor; t_{rm} = room temperature (F); t_{adp} = apparatus dewpoint temperature (F).

h_{ea} = specific enthalpy of entering mixture of outdoor air at design conditions and return air at average system conditions (Btu/lb).

h_{adp} = specific enthalpy at apparatus dewpoint temperature (Btu/lb).

single and multiple zones. It also presents the methods applied in calculating the dehumidified and supply air quantities as well as the refrigeration or the dehumidifier load, and defines the fan and zone supply air quantities.

HEATING LOAD

When heating is required, the load for each zone is calculated to offset the transmission loss plus infiltration. The capacities of the heating or reheating coils should be capable of both raising the supply air temperature to room conditions and offsetting the zone heating load (*Part 2*).

If a preheater is required, it may be selected to temper the minimum outdoor air to 40 F or to heat the mixture of outdoor and return air to the required dewpoint temperature.

SUPPLY AIR

The supply air for the various types of conventional systems may either equal the dehumidified air quantity (*Table 1*) or be increased to maintain the proper air circulation within the conditioned space. If it is increased, it may be accomplished either by the addition of a permanent bypass of untreated recirculated space air to be mixed with the dehumidified air, or by the selection of a larger dehumidified air quantity, using higher apparatus dewpoint temperature but having the same capacity to absorb the space moisture. In the first case the supply air fan quantity is equal to the sum of the

dehumidified and the permanent bypass air. In the second case the supply air quantity is equal to the increased quantity of dehumidified air.

With a variable volume, constant temperature system a large supply air quantity (1-2 cfm per sq ft of floor area) at a high air temperature (approaching 65 F) is a good design approach. This minimizes the amount of volume control. Actually it may make the system quite stable, requiring very little volume adjustment.

DUCT DESIGN

The conventional low velocity duct system design and selection of standard outlets and diffusers are described in *Part 2, Air Distribution*.

The static regain method of sizing supply air ducts is recommended. Balancing dampers should be used for minor adjustments of air distribution within ducts. The return air ducts are sized by the equal friction method.

Careful engineering of air distribution systems avoids noise problems. At times, a lack of proper space to accommodate a good layout and fittings, or the proximity of an outlet to the apparatus, may require sound absorption treatment of the supply and perhaps the return air paths.

PIPING DESIGN

Factors affecting the design of refrigerant, chilled water and steam piping are described in *Part 3*.

CENTRAL APPARATUS

General guidance for the design and arrangement of the various components of the apparatus is found in *Part 2, Air Distribution*.

The engineering procedures point out a specific basis for selecting the dehumidifier, supply air fan, and heating coils for any system. Filters are selected for the required supply air quantity to meet the needs appropriate to the application.

The simplest arrangement for the small conventional system is a prefabricated package or an assembly of fan and coil central station equipment with separate refrigeration plant (*Part 2*).

REFRIGERATION LOAD

Refrigeration capacity is estimated as shown in *Table 1* with the particular type of machinery determined by the size of the load.

CONTROLS

Controls for conventional systems are simple, and can be either electric or pneumatic.

There are several control elements which regulate the functioning of conventional air conditioning systems; five are basic and two are optional.

These are the basic elements.

1. A relay energized by the fan starter opens the minimum outdoor air damper as the fan is started. This provides ventilation in all seasons.

2. A space, dewpoint or fan discharge thermostat controls the dehumidifier cooling capacity and indirectly the refrigeration plant. This provides cooling in summer.

3. A space, dewpoint or fan discharge thermostat controls the cooling capacity by the use of outdoor air. This provides cooling in marginal weather.

4. A summer-winter switch for seasonal change of control cycles.

5. A space or zone thermostat(s) maintains space conditions controlling:

    a. Cooling source directly as indicated in Items 2 and 3 for the basic conventional system (*Fig. 6*).

    b. Face and bypass dampers in summer, and cooling source in marginal weather, as indicated in Item 3 for the face and bypass damper control system (*Fig. 7*).

    c. Zone reheaters in all seasons, for reheater control systems (*Fig. 8*).

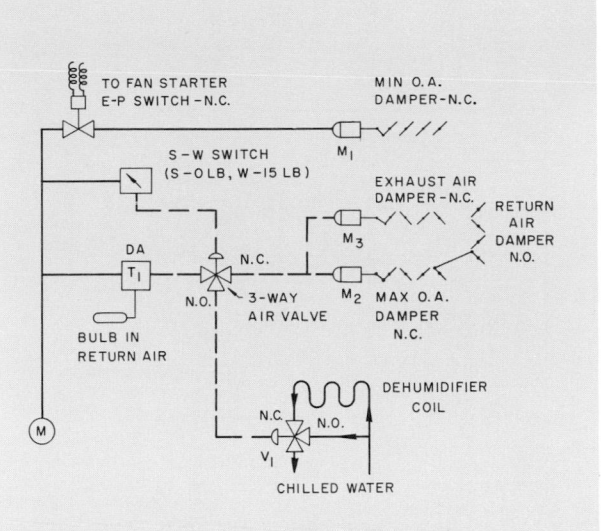

NOTE: Chilled water control is shown; other controls may be arranged as described in the text.

FIG. 6 — REFRIGERATION CAPACITY CONTROL, TYPICAL PNEUMATIC ARRANGEMENT*

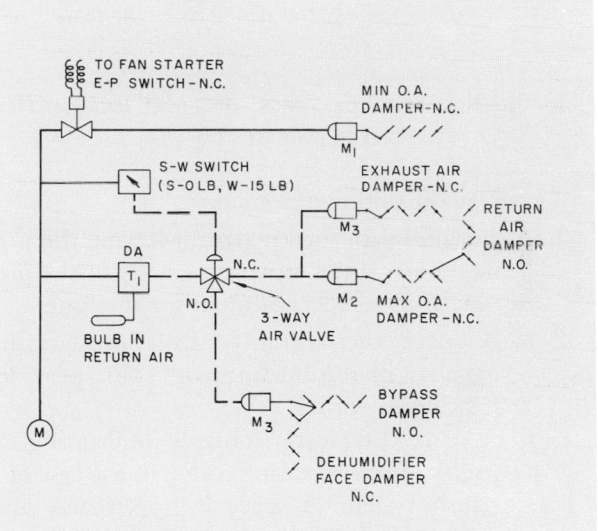

NOTE: Refrigeration control as described in the text.

FIG. 7 — FACE AND BYPASS DAMPER CONTROL, TYPICAL PNEUMATIC ARRANGEMENT*

    d. Volume dampers in all seasons, for variable volume, constant temperature systems (*Fig. 9*).

*Figures 6, 7, 8 and 9 are schematic and for guidance only; they do not include the optional elements: preheater and reheater for winter heating. The design engineer must work out a control diagram for his specific application.

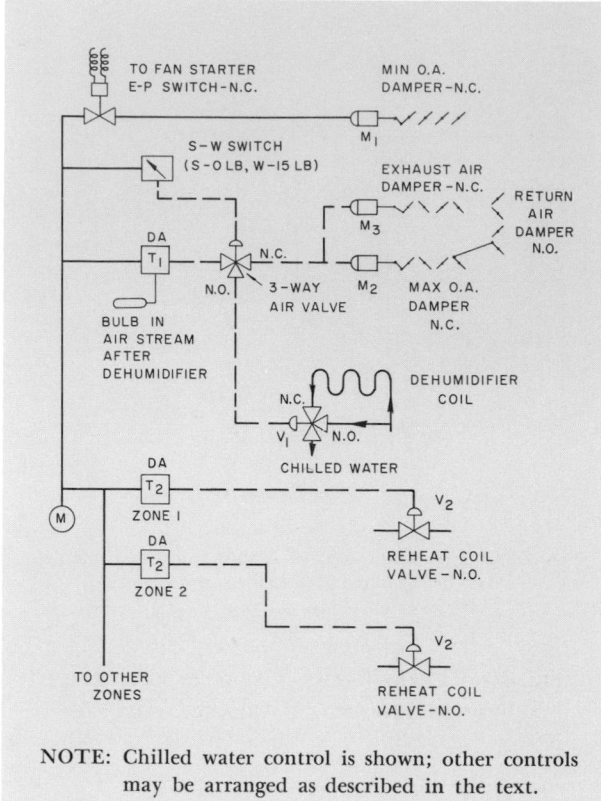

FIG. 8 — REHEATER CONTROL, TYPICAL PNEUMATIC
ARRANGEMENT*

These are the optional elements:

1. A thermostat in the air stream having the pre-heater controls the heating capacity of the pre-heater. This tempers outdoor air in winter.

2. a. A space thermostat controls the heating capacity of the heating coil. This provides heating in winter.

   b. A space thermostat controls the heating capacity of the reheating coil(s) in the case of a reheat system. A space hygrostat may also control the reheater, particularly in the case of a reheat system applied to a single zone.

MODIFICATIONS

This chapter has outlined the basic arrangements of conventional systems. Numerous variations may be devised to suit a design engineer.

One particular modification is an arrangement in which the main apparatus is a source of dehumidi-

*Figures 6, 7, 8 and 9 are schematic and for guidance only; they do not include the optional elements: preheater and reheater for winter heating. The design engineer must work out a control diagram for his specific application.

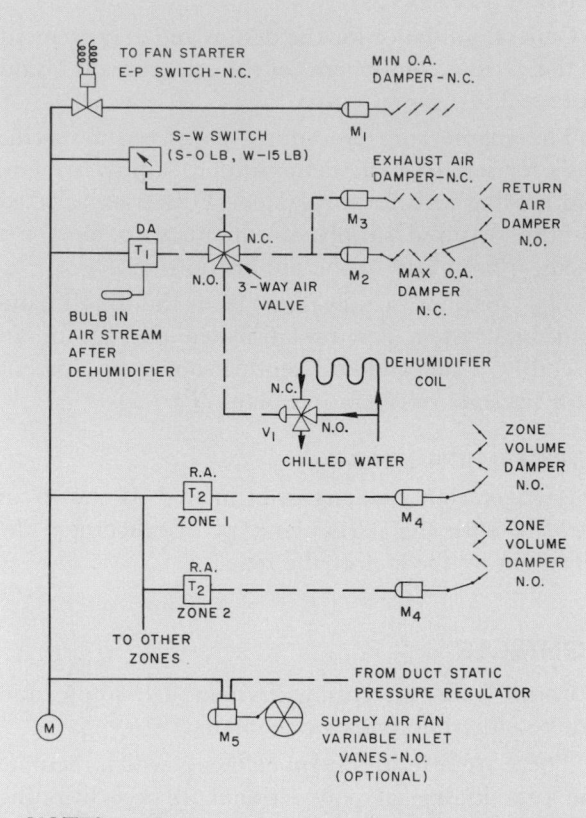

NOTES:
1. Chilled water control is shown; other controls may be arranged as described in the text.
2. System fan must have inlet vanes controlled by a static pressure regulator to provide total volume control where required.

FIG. 9 — VARIABLE VOLUME CONTROL, TYPICAL
PNEUMATIC ARRANGEMENT*

fied air to be distributed to several booster fan stations which have mixing dampers. The zone control mixes dehumidified and room air in proper proportions to maintain the zone temperature. The main dehumidified air fan must have inlet vanes controlled by a static pressure regulator to provide volume control. Such a system may be economically applied to a large building. The main apparatus and services are concentrated in one location with booster fans usually suspended from the ceilings of the floors served. The design of the duct distributing dehumidified air to the booster fans may at times use high velocity principles, while the design of ducts transmitting the supply air to the rooms is usually in accordance with low velocity principles.

An exhaust system should be used to remove the excess air that is brought into the building during marginal weather.

CHAPTER 2. CONSTANT VOLUME INDUCTION SYSTEM

The all-air Constant Volume Induction System is well suited for many applications, particularly medium and small multi-room buildings where individual rooms as well as large spaces may be air conditioned from one central air conditioning plant. It is often applied to buildings having a large ratio of floor area to height, indicating a need for horizontal ductwork and piping.

This system is particularly suited to high latent load applications such as schools and laboratories, as well as existing hotels in which the design sensible cooling load is low and where a serviceable steam or hot water system is available. Hospitals, motels, apartment houses, professional buildings, and office buildings are other applications.

An exceptional application of this system is a school in which heating and ventilation are required at present and conversion to full air conditioning may be required at a future date. In this instance, equipment, air quantities and layout are based on the air conditioning calculations. Future conversion is easily accomplished by adding a refrigeration machine, cooling coils and piping.

This chapter includes System Features, System Description, Controls and Engineering Procedure for designing a complete constant volume induction system.

SYSTEM FEATURES

The constant volume induction system offers many features favorable for its application to medium and small multi-room buildings. Some of these features are:

1. *Individual Room Temperature Control* — Zoning problems are solved without the expense of multiple pumps or zoned piping and ductwork since each room is a zone.

2. *Flexible Air System Design* — The choice of low or high velocity air distribution can be made on the basis of economics and building requirements, since units are designed to handle either type of distribution.

3. *Centralized Primary Air Supply* — One central station apparatus can serve both interior and exterior rooms of the building, since the constant volume, constant temperature characteristic of the primary air is suitable for zones of this type.

4. *Simplified Control System* — A single nonreversing thermostat and control valve or a self-contained valve is the only requirement for each room.

5. *Economy of Operation* — The refrigeration machine is not required during the intermediate season when the outdoor air is at the proper temperature to handle the cooling load; that is, equal to or below the supply air temperature.

6. *Controlled Ventilation, Odor Dilution and Constant Air Motion* — The system provides positive ventilation to each space to dilute odors. In addition, room air motion remains uniform since this is a constant volume system.

7. *Quiet Operation* — All fans and other rotating equipment are remotely located.

8. *Centralized Maintenance* — Since service is required only in the machine room, maintenance is easier to accomplish, with less distraction and in a more orderly manner.

9. *Filter Efficiency* — Since filtration is accomplished at a single location, higher efficiencies to meet the desired requirements are attainable.

10. *Central Outdoor Air Intake* — This central location allows a more desirable architectural treatment. Wind direction has little or no effect on ventilation. Building damage caused by rain leakage thru numerous intakes is eliminated.

11. *Convector Heating* — Night, weekend, and holiday heating is easily accomplished by operating a single hot water pump or a steam system.

12. *High Temperature Differential* — Supply air temperatures may be 25 degrees below room temperatures since room air is mixed with the primary air before the total air stream is discharged into the room. This feature makes possible smaller air quantities at lower temperatures than with a conventional system. Also, this means smaller duct sizes and smaller central station apparatus.

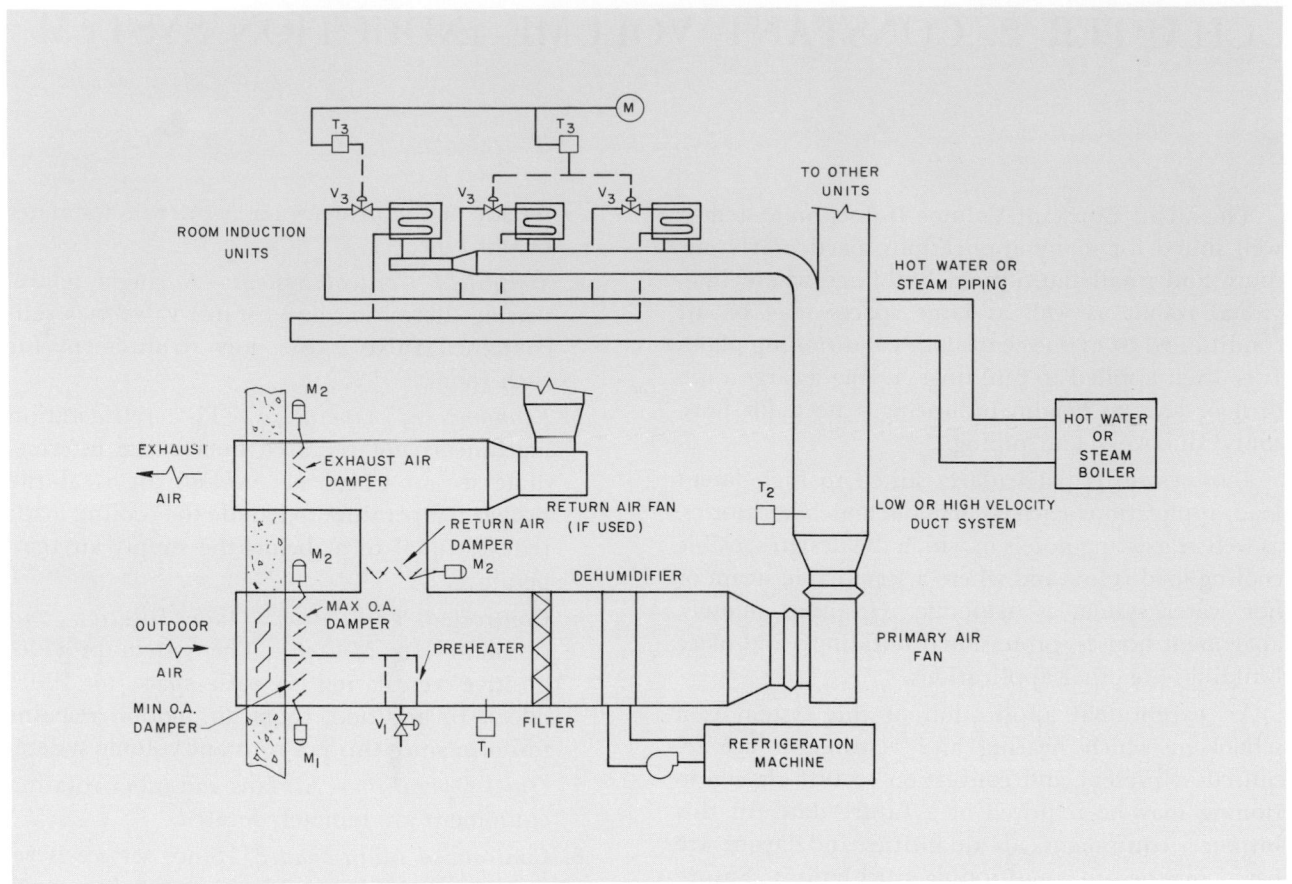

FIG. 10 — TYPICAL CONSTANT VOLUME INDUCTION SYSTEM

13. *Centralized Dehumidification* — Since all dehumidification occurs at the central station, no condensation occurs in the room. Thus, drain lines, drain pans and cleaning of these items are eliminated.

SYSTEM DESCRIPTION

Figure 10 is a schematic diagram of the system.

CENTRAL STATION APPARATUS

The central station apparatus conditions the air and supplies either a mixture of outdoor and return air or 100% outdoor air to the room unit. The apparatus contains filters to clean the air, preheat coils (if required) to temper cold winter air, and a dehumidifier to cool and remove excess moisture from warm humid air or to add winter humidification.

A relatively constant supply air temperature is maintained at the fan discharge, normally from 50-55 F.

A high or low velocity air distribution system is used to move the air from the central station to the room units. A sound absorber (when required)

located downstream from the fan discharge is used to reduce the noise generated by the fan.

Chilled water is circulated or refrigerant is evaporated in the coils of the dehumidifier to remove excess moisture and cool the air. Hot water or steam is supplied to the unit heating coils.

INDUCTION UNIT

The induction unit is designed for use either with a complete air conditioning system or with a system providing heating and ventilating only. *Figure 11* shows the unit elements which include the air inlet, sound attenuating plenum, nozzle and heating coil.

A constant volume of cool conditioned air is supplied to the unit. This air, designated as primary air, handles the entire room requirements for cooling, dehumidification or humidification, and ventilation. The primary air induces room air which is heated by the coil to provide summer tempering (when needed) and winter heating.

Room temperature control is achieved by adjusting the flow of hot water or steam thru the coil by a manual or an automatic control valve.

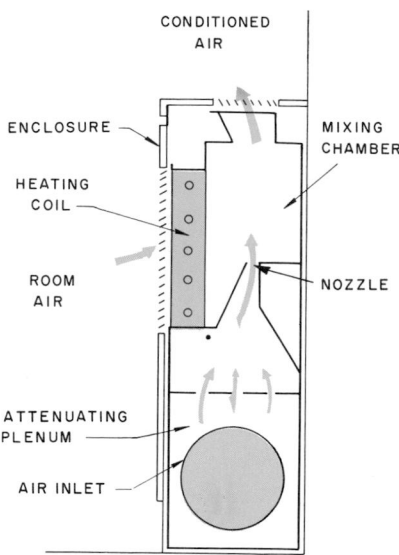

FIG. 11 – TYPICAL INDUCTION UNIT

TABLE 2—TYPICAL COMPARISON OF ROOM LOAD CHARACTERISTICS*

ROOM NO.	EXPOSURE	CONDITIONS AT ROOM PEAK LOAD		
		ESHF	Room Temp (F)	t_{adp} (F)
1	NE	.82	78	51.5
2	E	.86	78	52.0
3	SE	.86	78	52.0
4	S	.86	78	52.0
5	SW	.95	78	54.0
6	W	.95	78	54.0
7	NW	.95	78	54.0
8	N	.81	78	50.0

*Based on maximum design conditions of 78 F, 45% rh, thermostats assumed set at 75 F.

ENGINEERING PROCEDURE

The following procedure is offered to assure a practical operating air conditioning system. As in all design work, a survey and preliminary layout are required, as explained in *Part 1*. Room loads and associated air quantities are determined by using load factors and methods described in *Part 1*.

ROOM COOLING LOAD

Calculate the load for all typical exposures: east, west, north, south and any space that has unusual loads. Some flexibility may be allowed in these calculations to provide for future partition changes, depending on the type of application. In most multi-room applications, 8 to 16 room load calculations for typical sampling may be required. This includes both room sensible and latent load requirements.

AIR QUANTITIES

Calculate the air quantity required for each room. This is determined from the following formula:

$$cfm_{da} = \frac{ERSH}{1.08 \times (1 - BF)(t_{rm} - t_{adp})}$$

where:

cfm_{da} = dehumidified air quantity
ERSH = effective room sensible heat
BF = dehumidifier coil bypass factor
t_{adp} = apparatus dewpoint temperature
t_{rm} = room temperature

The air quantity determined from this formula is used for two purposes: unit selection and design of the air distribution system. All of the values used in this formula are explained in *Part 1, Chapter 8, Psychrometrics*. A short discussion is included here for economic guidance in the selection of apparatus dewpoint.

On installations where the relative humidity should not exceed the design conditions for any reason, the lowest apparatus dewpoint determined from the cooling load estimates must be used in the formula. However, on most installations there may be several rooms which require an apparatus dewpoint lower than the rest of the building. In these instances a compromise value is often used, recognizing that these spaces may have a relative humidity that exceeds average design conditions.

As an illustration, calculations are made for a one-story office building *(Fig. 12)*. The resulting apparatus dewpoints for the various rooms are shown in *Table 2*. Note that the lowest apparatus dewpoint (50.0 F) occurs in the north exposure. If 50.0 F is selected for use in determining the air quantity for all the rooms, the relative humidity will be below the room design condition in all spaces other than the north exposure.

Conversely, if 54 F is used for determining the air quantity, the southwest, west and northwest rooms will have a satisfactory relative humidity, and the remaining rooms will have a relative humidity that exceeds the design condition. The suggested apparatus dewpoint to be used in this instance is 52 F. If 52 F is used, the northeast and north rooms will have a relative humidity slightly higher than design.

Use of the compromise apparatus dewpoint results in a practical system that gives excellent results most of the time, and loses relative humidity control only in a few spaces having a maximum latent load when the complete building is at peak load.

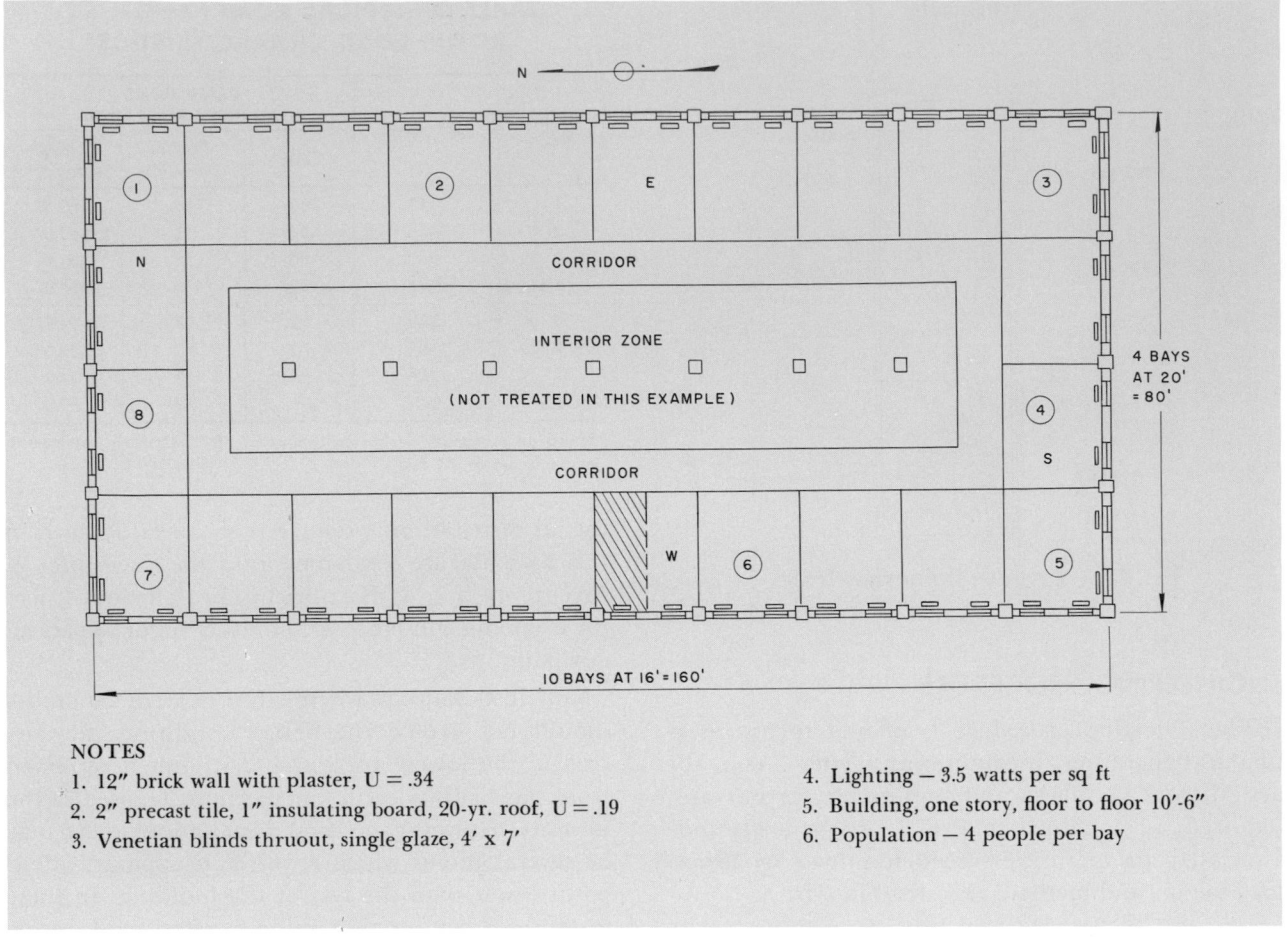

NOTES
1. 12″ brick wall with plaster, U = .34
2. 2″ precast tile, 1″ insulating board, 20-yr. roof, U = .19
3. Venetian blinds thruout, single glaze, 4′ x 7′

4. Lighting — 3.5 watts per sq ft
5. Building, one story, floor to floor 10′-6″
6. Population — 4 people per bay

FIG. 12 — TYPICAL OFFICE BUILDING

ROOM HEATING LOAD

Calculate the room heating load for these two conditions:

1. With the primary air fan operating. This gives the required heating coil capacity for room units with induction.

2. With the primary air fan not operating. This gives the required heating coil capacity when the room induction unit is used as a convector.

The first condition includes heating requirements to offset transmission and infiltration, and to temper the primary air from its entering temperature to the room winter design temperature. The second condition includes heating requirements to offset transmission and infiltration only.

Building type and planned system operating periods can influence the gravity heating calculations. Many applications are designed for a night, weekend and holiday set-back temperature. During these unoccupied periods, the room temperature may be allowed to drop to a range of 60-65 F. This can result in a lower operating cost and, at times, a smaller room unit.

UNIT SELECTION

Select the room units to satisfy the following requirements:

1. Primary air quantity.

2. Room heating load (coil capacity).

3. Sound level appropriate to the application.

4. Space limitations.

Once these items are established, unit selections can be made from the manufacturer's catalog.

Often, the gravity heating requirements (with fan not operating) may indicate a larger unit than that required to satisfy the cooling requirements. It is more economical to operate a fan for limited periods of time during extreme winter weather than to select a larger unit. This is easily accomplished automatically by installing in a typical room a night thermostat which starts the primary air fan when the room temperature drops below the thermostat setting.

The design water flow rate used in selection of the unit can influence the total system cost. The lower the flow rate, the lower the first cost of system piping and pump. However, a check should be made to determine if turbulent flow conditions exist in the unit coil. For a ½ inch OD tube, the minimum flow is approximately 0.7 gpm for turbulent conditions.

CENTRAL APPARATUS

Select the central air handling apparatus for the sum of the air quantities supplied to each space.

Two methods may be used to determine the dehumidifier load. The first *(Example 1)* results in a smaller refrigeration load and, therefore, lower owning and operating costs, but requires more time to calculate. Since all rooms are not at peak load or design temperature simultaneously, the air entering the dehumidifier is at a lower temperature than the mixture of room design temperature and outdoor ventilation air temperature. This condition occurs in systems using return air.

Example 1 — Calculation of the Dehumidifier Load

Given:
 Building shown in *Fig. 3*
 Outdoor design $= 95$ F, 75 F wb, $h_{oa} = 38.6$
 Room design $= 78$ F, 45% rh, $h_{rm} = 29.0$
 Apparatus dewpoint $= 52.0$ F, $h_{adp} = 21.4$
 4-row coil BF $= .20$

	Supply Air Quantity	Ventilation Air Quantity
West exposure =	5600 cfm	920 cfm
East exposure =	4800 cfm	790 cfm
South exposure =	2000 cfm	330 cfm
North exposure =	1280 cfm	210 cfm
Total =	13680 cfm	2250 cfm

Return air at 78 F, 45% rh, $h_{rm} = 29.0$
West exposure $= 5600 - 920$ $= 4680$ cfm

Return air at 75 F, 50% rh, $h_{rm} = 28.2$
East exposure $= 4800 - 790$ $= 4010$ cfm
South exposure $= 2000 - 330$ $= 1670$ cfm
North exposure $= 1280 - 210$ $= 1070$ cfm
Total $= 6750$ cfm

Find:
 Dehumidifier load

Solution:
 Basic equation is
  Load $= 4.45 \times cfm_{da} \times (1 - \text{BF})(h_{ea} - h_{adp})$
 where:
  cfm_{da} = dehumidified air quantity
  h_{ea} = entering air enthalpy
  h_{adp} = apparatus dewpoint enthalpy
  BF = bypass factor

Outdoor air at 95 F, 75 F wb, $h_{oa} = 38.6$
 Load $= 4.45 \times 2250 \times (1 - .2)(38.6 - 21.4)$
  $= 137,800$ Btu/hr

Return air at 78 F, 45% hr, $h_{rm} = 29.0$
 Load $= 4.45 \times 4680 \times (1 - .2)(29.0 - 21.4)$
  $= 126,500$ Btu/hr

Return air at 75 F, 50% rh, $h_{rm} = 28.2$
 Load $= 4.45 \times 6750 \times (1 - .2)(28.2 - 21.4)$
  $= 163,300$ Btu/hr

Total dehumidifier load
 $= 137,800 + 126,500 + 163,300 = 427,600$ Btu/hr

The second method of determining the dehumidifier load is less complex, but results in a load larger than required. It consists of adding the total loads for each of the spaces. Also, this can be determined by using the same formula shown in *Example 1*, and by assuming all of the rooms are at peak load simultaneously.

Using the values shown in *Example 1*, the dehumidifier load in this instance is:

Outdoor air at 95 F, 75 F wb, $h_{oa} = 38.6$
 Load $= 4.45 \times 2250 \times (1 - .2)(38.6 - 21.4)$
  $= 137,800$ Btu/hr

Return air load at 78 F, 45% rh, $h_{rm} = 29.0$
 Load $= 4.45 \times 11,430 \times (1 - .2)(29.0 - 21.4)$
  $= 309,000$ Btu/hr

Total dehumidifier load $= 446,800$ Btu/hr

When the loads calculated by methods 1 and 2 are compared, method 1 represents a 4.3% saving over method 2. This saving is reflected in the cost of the refrigeration system, dehumidifier coil and the interconnecting piping system.

A preheater may be required when the mixture temperature of the minimum outdoor air and return air is below the desired supply air temperature. It may be selected to temper the minimum outdoor air to 40 F or to heat the mixture of outdoor and return air to the required dewpoint temperature.

REFRIGERATION LOAD

The refrigeration load is determined by the dehumidifier load. When more than one dehumidifier is used, the total load is the sum of all of the dehumidifier loads. This assumes that all dehumidifiers are operating normally at the same time with no additional diversity factors.

DUCT DESIGN

Methods described in *Part 2* should be used in designing the air distribution system. Since this is a constant volume system, no special precautions are required to account for variations in air quantities caused by changing load conditions.

Low velocity duct systems are normally preferred since they are simpler to design and result in lower owning and operating costs. However, they do require more space and are more difficult to balance.

In many buildings the amount of space available for ducts is limited and, therefore, the use of a higher velocity system is required. Usually, Class II fans are required for the increased static pressure in a high velocity system, and extra care must be taken in duct layout and duct construction. Particular care must be given to the selection and location of fittings to avoid excessive pressure drop and possible sound problems. Ducts must be carefully sealed to prevent air leakage. Round duct is preferred to rectangular ductwork because of its greater rigidity.

Although other methods of duct sizing such as equal friction or velocity reduction may be used, the static regain method is preferred. A system designed by the static regain method tends to be self-balancing since it is designed for the same static pressure at each terminal. This minimizes field balancing and results in a system that is quieter and more economical to operate.

PIPING DESIGN

Design the piping system in the normal manner. Either a hot water or steam distribution system may be used to supply the unit coils. Although steam is acceptable and has been extensively used in the past, hot water is currently the normally preferred heating medium. It provides quieter operation, easier and more uniform control of room temperature; it requires a simpler and less complicated piping system with a minimum of mechanical specialties.

Regardless of whether steam or hot water is used, normal design practice should be followed in laying out the system as shown in *Part 3*. For hot water, either a direct return or a reverse return system may be used. However, a reverse return system is preferred and should be used wherever practical, since it provides an inherently balanced system.

CONTROLS

A basic pneumatic control arrangement is shown in *Fig. 13*.

UNIT CONTROL

Control of steam or hot water flow to the coil is the only control necessary at the unit. This may be accomplished either manually or automatically.

Manual control is accomplished by operating a hand valve to vary the flow of steam or hot water thru the coil. While satisfactory control of unit capacity can be obtained in this manner, it is in-

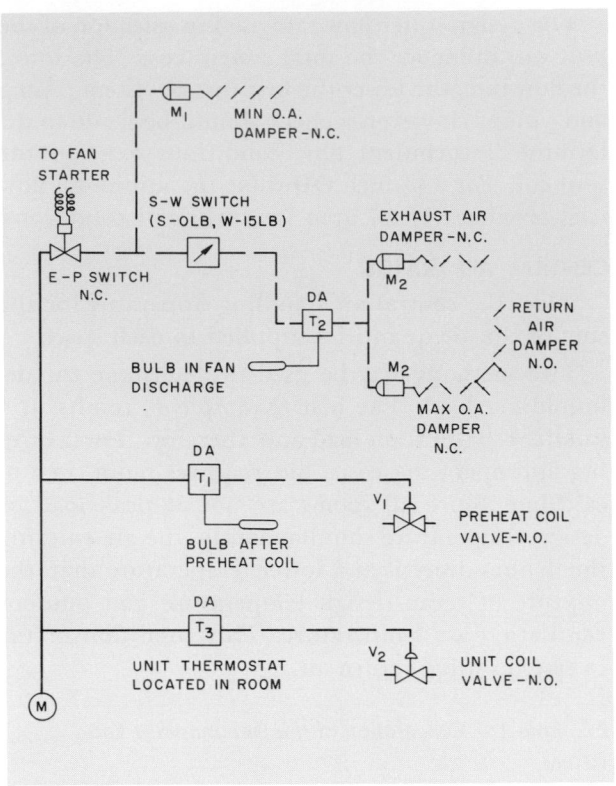

FIG. 13 — CONSTANT VOLUME INDUCTION SYSTEM CONTROL, TYPICAL PNEUMATIC ARRANGEMENT

convenient because the room occupant must adjust the capacity to meet variations in room load caused by such factors as changing outdoor temperatures or sun load. With a hot water system, some of these adjustments may be minimized by varying the supply water temperature depending on the outdoor temperature, and by zoning the piping to supply different temperature water to exposures with different sun loads.

Most installations are supplied with automatic controls to maintain constant room temperatures, regardless of changing load conditions. Either pneumatic, electric or self-contained controls may be used. Since the fluid is always hot, only a nonreversing thermostat is necessary to control the valve.

A direct-acting thermostat and a normally open valve are usually selected with pneumatic or electric controls so that the valve opens when the air supply or electric control circuit is shut off. This is particularly useful for pneumatic systems because gravity heating can be obtained for night and weekend operation without the expense of running the air compressor. In addition, it provides a safety feature in that heating is still available in the event of failure of the air system or electric control circuit.

With pneumatic or electric controls, the thermostat may be located on the wall or conveniently mounted on the unit within the enclosure. Self-contained controls are always mounted within the enclosure because the thermostat and valve are an integral unit.

With unit mounted controls, the temperature sensing bulb of the thermostat is placed in the induced air stream between the recirculating grille and the coil. It may be very close to the coil; however, metal-to-metal contact between the coil and the bulb must be avoided to assure proper control.

Several units may be controlled with one thermostat and one valve. When this is done, the thermostat should be centrally located to assure that the temperature sensed by the thermostat is representative of average room conditions. Consideration should be given to possible future relocation of the partitions.

CENTRAL APPARATUS CONTROL

Either electric or pneumatic controls may be used for the central apparatus. The sequence of operation is the same regardless of which is used.

Summer Operation

The minimum outdoor air damper is interconnected with the fan starter so that the damper opens as the fan is started. With the summer-winter switch in the summer position, the maximum outdoor air damper is closed and the return air damper is wide open. Normal procedure is to maintain a constant leaving water temperature from the chiller. Thus, with a constant entering water temperature to the dehumidifier, the primary air temperature is at its design maximum during the peak load condition, but drops as the load on the dehumidifier decreases. Increased flexibility in room control is therefore provided during off-peak operation.

Winter Operation

When the outdoor air temperature is below the design primary air temperature (50-55 F), the refrigeration machine is shut off and the summer-winter switch is set in the winter position. This allows the thermostat in the fan discharge to modulate the outdoor and return air dampers to maintain the desired temperature. Thus, the cool outdoor air is used as a source of free cooling.

If a preheater is used in the minimum outdoor air, a thermostat located downstream of the coil is set at a minimum of 40 F. If a preheater is used in the mixture of outdoor and return air, a thermostat located downstream of the coil is set for about 5 degrees below the fan discharge thermostat, but not lower than 40 F. The preheater is inoperative until only the minimum outdoor air damper is open.

CHAPTER 3. MULTI-ZONE UNIT SYSTEM

The all-air Multi-Zone Blow-Thru Unit System that has heating and cooling coils in parallel is a constant volume, variable temperature system. It is applied to areas of multiple spaces or zones which require individual temperature control.

This system is considered when one or more of the following conditions exist:

1. The area consists of several large or small spaces to be individually controlled — a school, a suite of offices, an interior zone combining several individual open floors of a multi-story building.

2. The area includes zones with different exposures and different characteristics of internal load — a bank floor of a building, a large open multi-exposure office space.

3. The area combines a large interior zone with a relatively small group of exterior spaces.

4. The area consists of interior spaces with individual load characteristics — radio and television studios.

Examples of these conditions are shown in *Fig. 14.*

The blow-thru system is essentially applicable to locations and areas having high sensible heat loads and limited ventilation requirements. Applications of high ventilation requirements need a dehumidifying coil in the minimum outdoor air with heating available at the heating coil at all times. This is

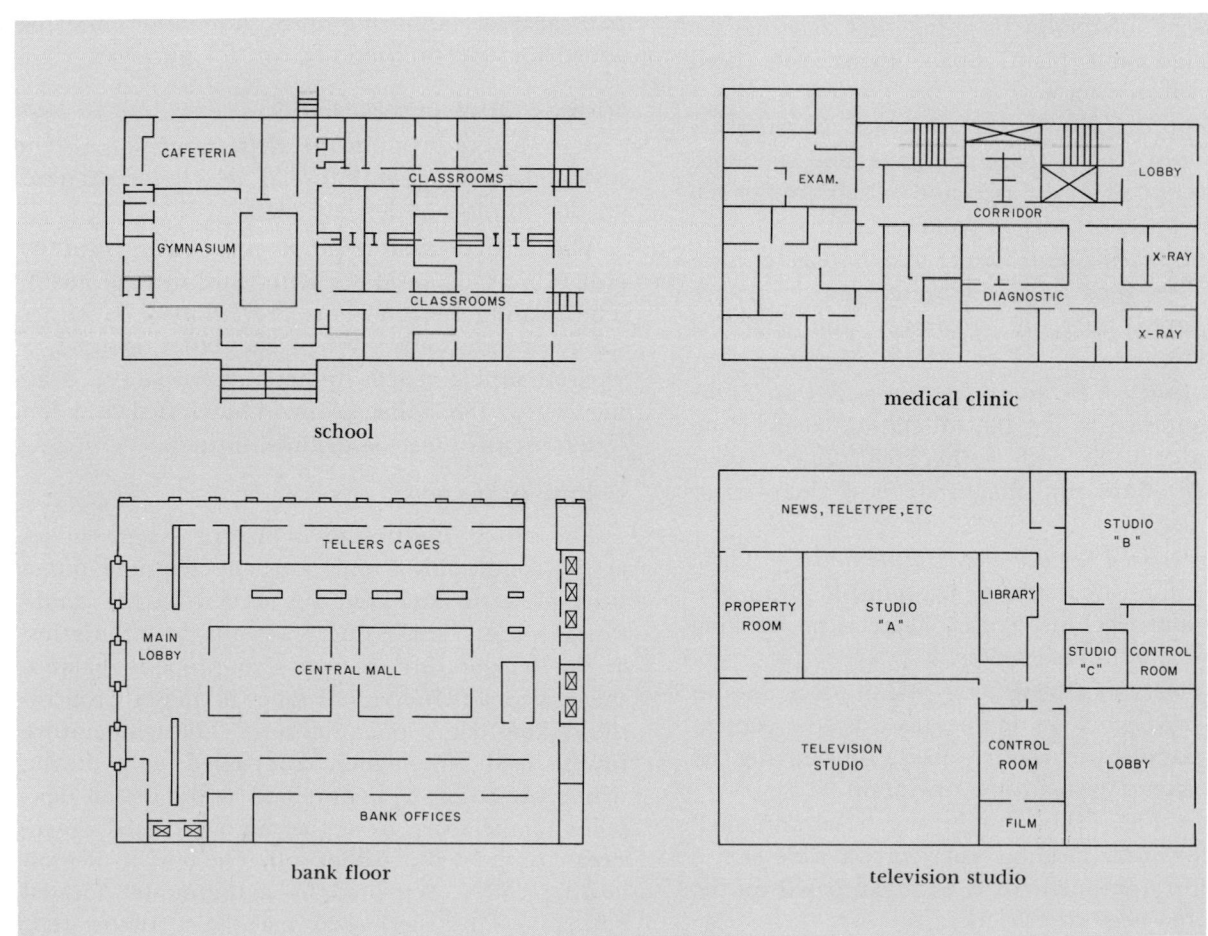

school

medical clinic

bank floor

television studio

FIG. 14 — TYPICAL AREAS SERVED BY A MULTI-ZONE UNIT(S)

necessary to prevent the bypass of humid outdoor air around the cooling coil.

This chapter includes System Features, System Descriptions, Controls, and Engineering Procedure to aid in designing a complete multi-zone unit system.

SYSTEM FEATURES

The following are some of the features of a properly designed multi-zone unit system:

1. *Individual Space or Zone Temperature Control* — Zoning problems are solved since each space is treated as a zone and is supplied with air quantities at the proper temperature.

2. *Individual Zoning from Minimum Apparatus*— Central station zoning is facilitated by having available those sizes of prefabricated units which are most frequently required. The field-assembled apparatus can be fitted to any requirements.

3. *Simple Nonchange-over Operation* — Change-over from summer to winter or vice versa consists of manually stopping and starting the refrigeration plant. Space thermostats need be set only once.

4. *Simplified Air Transmission and Distribution* — Only single air flow ducts and standard selection of air diffusers and outlets are needed. The system is easy to balance.

5. *Centralized Conditioning and Refrigeration* — Services such as power, water and drains are required only in apparatus and machine rooms.

6. *Centralized Dehumidification* — All air is dehumidified at the central station; there is no condensation within the conditioned space, thus eliminating the need for drain pans or piping.

7. *Centralized Service and Maintenance* — These functions are easier to accomplish in apparatus and machine rooms. There is no tracking thru conditioned spaces.

8. *Economy of Operation* — All outdoor air can be used when its temperature is low enough to handle the cooling load, thus saving on refrigeration machine operation.

9. *Filter Flexibility* — Filtration is acomplished at a central location; therefore, a wide choice of filtration methods is afforded, based on the desired need or efficiency.

10. *Quiet Operation* — All fans and other rotating equipment are remotely located.

SYSTEM DESCRIPTION

CENTRAL APPARATUS

A multi-zone unit system is shown in *Fig. 15*. This apparatus may be a factory-assembled unit, or it may be field-assembled. However, a majority of the applications use one or more factory-assembled units, each of which consists of a mixing chamber, filter, fan, a chamber containing heating and cooling coils, warm and cold air plenums, and a set of mixing dampers. The mixing dampers blend the required amounts of warm and cold air to be transmitted thru a single duct to outlets in the zones.

OPTIONAL EQUIPMENT

A system may incorporate a preheat coil for minimum outdoor air when system design may require the maintaining of a higher design temperature of the incoming outdoor air. For applications where more exacting humidity control is required, a dehumidifying coil may be incorporated in the minimum outdoor air. An exhaust air fan may be added if a positive removal of air is required. A steam pan, grid or atomizing spray humidifier may be added for winter humidity control purposes.

OTHER SYSTEM COMPONENTS

For economic reasons the air transmission system is designed using conventional velocities. Standard air distribution outlets are used.

The refrigeration requirements are satisfied by either direct expansion or water chilling equipment. The heating requirements are fulfilled by either steam or hot water. When the latter is used, a separate piping system connects the apparatus heating coil to the boiler plant. The chilled and hot water circuits must be distinct entities.

SYSTEM OPERATION

The all-air multi-zone blow-thru system mixes at the conditioning apparatus the required quantities of warm and cold air needed by the conditioned space. A single duct transmits the air mixture at the temperature necessary to properly balance the space load. Individual zone thermostats control the mixing dampers at the unit. The temperature in the cold air plenum, controlled only during winter operation, is maintained at the design dew-point temperature by a thermostat located downstream of the dehumidifier coil. The hot air plenum heating coil is activated by a thermostat located outdoors. This thermostat may be a master type to reset a control thermostat in the hot air plenum, due to changing outdoor temperature.

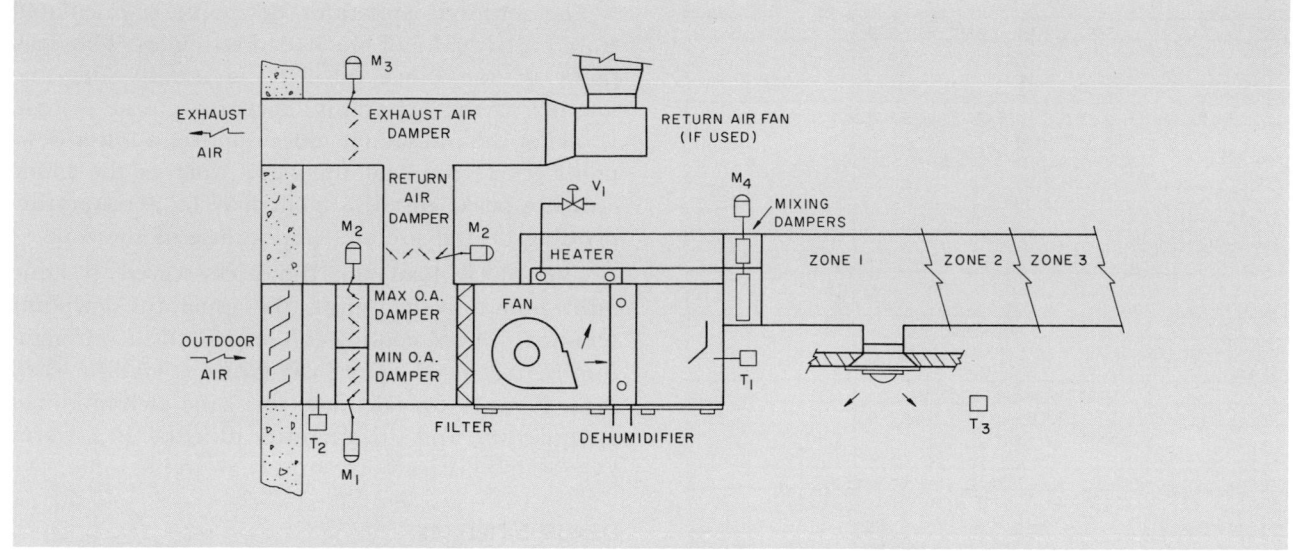

FIG. 15 — TYPICAL MULTI-ZONE UNIT SYSTEM

ENGINEERING PROCEDURE

The design of a multi-zone unit system generally follows conventional practices. *Part 1, Load Estimating*, should be referred to for all information on survey, preliminary layout and load calculations. Particulars are discussed in the following text.

ZONING

In dividing an area into zones, similarities of exposures, internal loads and occupancy must be considered. Also, the grouping of spaces into zones should be determined by physical size, arrangement of constituent spaces, and uniformity of control requirements. In other words, all exposures and interior zones should be grouped individually. The character of occupancy, whether executive, supervisory or general, may also govern the zoning. For a successful zone control, the requirement of cooling and heating, both hourly and seasonal, must be consistant thruout the spaces constituting a zone. A careful analysis of zoning should permit a uniform summer dewpoint selection and a consistent winter heating requirement thruout each zone.

Figure 16 illustrates the floor plan of an office suite. The multiple exposures and the occupancy pattern indicate varying loads both hourly and seasonal. To maintain the desired temperature conditions in each space or group of spaces, it is necessary to shift a portion of the cooling effect from one space to another as the solar load moves around the building with the sun. The east exposure has its maximum in the morning while the west exposure reaches its peak in the late afternoon.

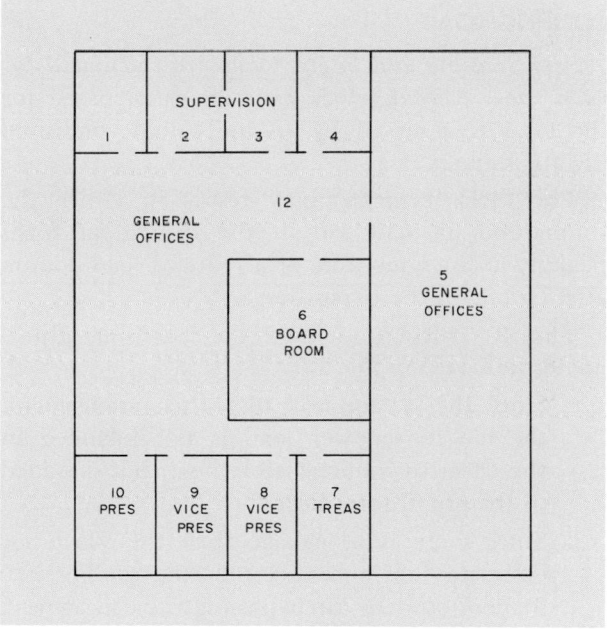

FIG. 16 — TYPICAL OFFICE SUITE

The board room and part of the general offices constitute the interior zone. The general offices, unless directly under a roof, have a constant load thruout the day. The board room may be occupied or empty at any time of the day. The interior spaces usually require cooling during both summer and winter to offset the major load of lights, people and perhaps office equipment. Thus, a careful analysis is indicated, and the only limitation to the number of zones is the first cost and the size(s) of multi-zone units required to perform the task.

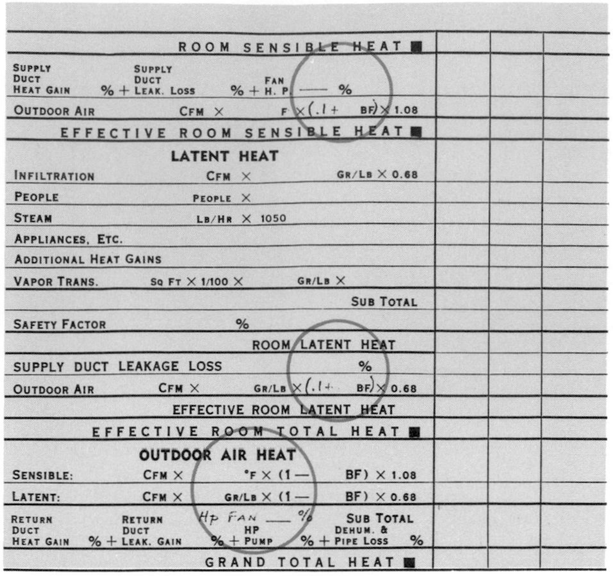

ROOM SENSIBLE HEAT ■			
SUPPLY DUCT HEAT GAIN % + SUPPLY DUCT LEAK. LOSS % + FAN % + H. P. ——— %			
OUTDOOR AIR CFM × F × (.1 + BF) × 1.08			
EFFECTIVE ROOM SENSIBLE HEAT ■			
LATENT HEAT			
INFILTRATION CFM × GR/LB × 0.68			
PEOPLE PEOPLE ×			
STEAM LB/HR × 1050			
APPLIANCES, ETC.			
ADDITIONAL HEAT GAINS			
VAPOR TRANS. SQ FT × 1/100 × GR/LB ×			
SUB TOTAL			
SAFETY FACTOR %			
ROOM LATENT HEAT			
SUPPLY DUCT LEAKAGE LOSS %			
OUTDOOR AIR CFM × GR/LB × (.1 + BF) × 0.68			
EFFECTIVE ROOM LATENT HEAT			
EFFECTIVE ROOM TOTAL HEAT ■			
OUTDOOR AIR HEAT			
SENSIBLE: CFM × °F × (1 — BF) × 1.08			
LATENT: CFM × GR/LB × (1 — BF) × 0.68			
RETURN DUCT HEAT GAIN % + RETURN DUCT LEAK. GAIN % + HP FAN ——— % HP PUMP % + SUB TOTAL DEHUM. & PIPE LOSS %			
GRAND TOTAL HEAT ■			

FIG. 17 — ADJUSTMENTS TO BTU CALCULATIONS

COOLING LOAD

Peak sensible and latent loads are calculated for each zone. A peak block estimate is prepared for the total area served by an individual conditioning apparatus.

The block estimate may result in either the summation of loads of all the individual zones peaking at the same time or a reduced load dominated by one of the exposures.

The Btu calculations have the following adjustments, indicated in *Fig. 17:*

1. Since the system is a blow-thru arangement, the fan horsepower load is not included in the effective room sensible heat, but is added to the grand total heat.

2. Since there is a leakage thru the warm air dampers, it is necessary that the cooling estimate of outdoor air bypassing the coil at peak cooling must be increased approximately 10%. This means that the bypass factor used in determining the outdoor air portion of the effective room sensible and latent heat must be increased by 0.1.

3. If the heating coil is operated in summer, there is an additional heat gain in the bypass air. This is accounted for as a part of the effective room sensible heat load by the use of a supply duct heat gain factor which is approximately 2-4% of the room sensible heat. This bypass air heat gain is in addition to the actual duct heat gain which must also be included.

The required apparatus dewpoint is calculated from individual and block load estimates. The dewpoint of the block estimate is usually the one selected as the apparatus dewpoint of the system. If one of the individual zones requires a lower dewpoint temperature at the same time as the block estimate peak, an adjustment may be necessary depending on the size and importance of the zone.

The block load for the area served by one multi-zone unit determines the apparatus dewpoint temperature, the cooling coil load, and the refrigeration requirements. The instantaneous load for each zone is used to calculate the zone dehumidified air quantity, and this quantity plus the 10% warm bypass air constitutes the supply air to the zone.

DEHUMIDIFIED AIR

The dehumidified air quantity for each zone or the total area is calculated from the peak effective room sensible heat load using the calculated apparatus dewpoint and the estimated coil bypass factor selected from *Part 1.* The applicable formula is as follows:

$$cfm_{da} = \frac{ERSH}{1.08\,(1 - BF)(t_{rm} - t_{adp})}$$

where:

cfm_{da} = dehumidified air quantity

ERSH = effective room sensible heat (Btu/hr), for individual zone or block estimate of all zones

BF   = bypass factor

t_{rm} = room temperature (F)

t_{adp} = apparatus dewpoint temperature (F)

The sum of the zone air quantities may be used to select the apparatus cooling coil. If the block estimate indicates a wide diversity in the load resulting in a smaller quantity of dehumidified air, the cooling coil size may be reduced to handle this smaller quantity.

SUPPLY AIR

The unit supply air quantity is equal to the sum of the dehumidified air quantities supplied to individual zones plus the suggested 10% bypass thru the warm air plenum dampers.

$$\text{zone } cfm_{sa} = cfm_{da} + cfm_{ba} = 1.1 \times cfm_{da}$$

$$\text{unit } cfm_{sa} = 1.1\,(cfm_{da_1} + cfm_{da_2} + \ldots + cfm_{da_n})$$

where:

cfm_{sa} = supply air quantity

cfm_{da} = dehumidified air quantity. (Subscripts of *da*, as in da_1, indicate individual zone dehumidified air quantities.)

cfm_{ba} = warm bypass air quantity

The sum of the zone supply air quantities is the basis for the selection of the fan. The individual zone supply air quantities are used to design the air duct transmission system and select the outlet terminals.

HEATING LOAD

Winter heating load to offset transmission and infiltration is calculated for each individual zone. The total requirements and the highest temperature called for in any one zone form the basis for selecting the heating coil.

$$t_{waw} = t_{rm} + \frac{\text{individual zone heat loss}}{1.08 \times cfm_{sa}}$$

where:

cfm_{sa} = individual zone summer supply air quantity

t_{rm} = room temperature (F)

t_{waw} = winter warm air temperature (F)

The heating coil load must take into account the total area heating requirements and heating of the leakage thru the cold air dampers. The temperature of the winter supply air used for cooling approaches the temperature of the summer apparatus dewpoint which is approximately 55-60 F. Thus, the heating coil capacity is determined as follows:

$$\text{coil heating load} = \text{unit } cfm_{sa} \times 1.08 \times (t_{waw} - t_{edb})$$

where:

cfm_{sa} = total fan supply air quantity

t_{waw} = winter warm air temperature (F), highest of all zones

t_{edb} = coil entering air temperature (F), the lowest temperature of outdoor and return air mixture, or approximately 55-60 F

DUCT DESIGN

For design details of the air transmission system and of air distribution methods, refer to *Part 2, Air Distribution*. For economic reasons, the air ducts are normally designed for conventional velocities which also permit the use of standard air outlets.

Since each duct feeds one or a small number of outlets, the only requirement is to size the parallel ducts such that each has an equal or nearly equal pressure drop. The duct that has a number of outlets must be designed to provide equal pressure at all outlets. This means that the particular run of the duct is sized independently by the static regain method. Balancing dampers may be required and their use should be investigated. The return air duct is usually short and, therefore, may be sized by the equal friction method. Where space is restricted, the air transmission and distribution may be designed using high velocity and high pressure principles.

Under such circumstances, the multi-zone unit system may be converted and adapted to a dual-duct system.

A properly engineered selection of a multi-zone unit and a good design of air distribution system minimizes the need for sound treatment. Normal precautions (*Part 2*) should be followed.

CENTRAL APPARATUS

Refer to *Parts 2 and 6* for details of equipment. Normally, the air handling apparatus is a factory-fabricated multi-zone unit (*Fig. 15*). Occasionally, for special or larger requirements, the apparatus is assembled in the field from separate component parts. Thruout this text, guidance has been indicated concerning the capacities required of the major components. However, a few additional details are necessary.

The fan is sized for an air quantity to satisfy all zones at their peaks, and is selected to operate at a static pressure sufficient to overcome the resistance of all system elements. Specific attention should be given to possible differences in pressure drops of the parallel cooling and heating coils; the larger pressure drop should be used. The zoning dampers should be proportioned on the basis of air quantities for each zone at fairly uniform velocities.

To minimize stratification and to assure that all zones have access to uniform heating, cooling and ventilation, the outdoor and recirculated air is brought to the unit across the full width of the unit. Connecting air streams to either side (or ends) of the unit produces harmful inequality of air temperatures and quality. Under these circumstances, the zone control is utterly defeated.

REFRIGERATION LOAD

Select the refrigeration plant to match the dehumidifier load or, in the case of more than one dehumidifier, the sum of the dehumidifier loads. Allowance should be made for diversity of loading if diversity can occur.

PIPING DESIGN

Design of water, steam and refrigerant piping is discussed in *Part 3, Piping Design*.

If hot water is used for winter heating, separate cold and hot water piping systems should be installed for the cooling and heating coils. It is extremely important that both heating and cooling are available during marginal weather conditions. At these times, while the interior zones call for cooling, some of the exterior zones may call for heating.

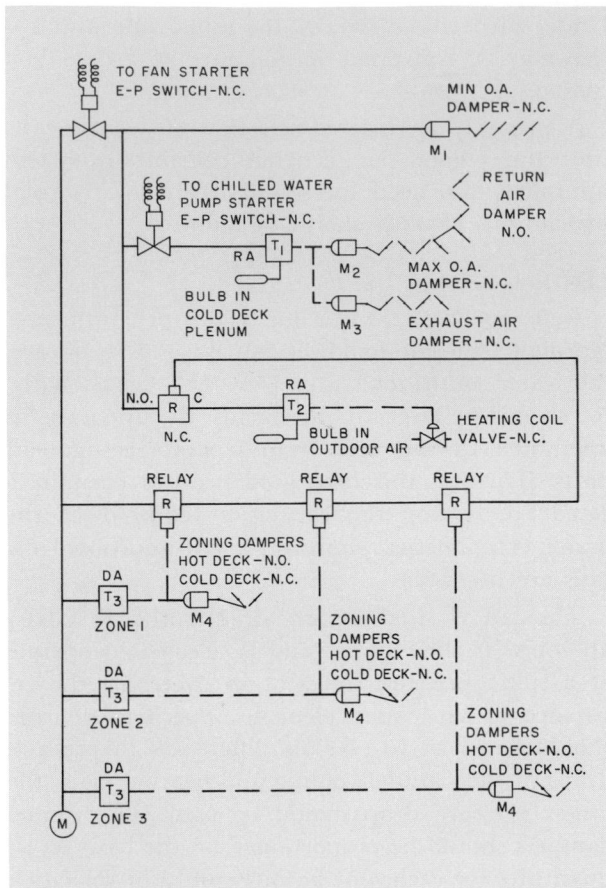

FIG. 18 — MULTI-ZONE UNIT SYSTEM CONTROL,
TYPICAL PNEUMATIC ARRANGEMENT

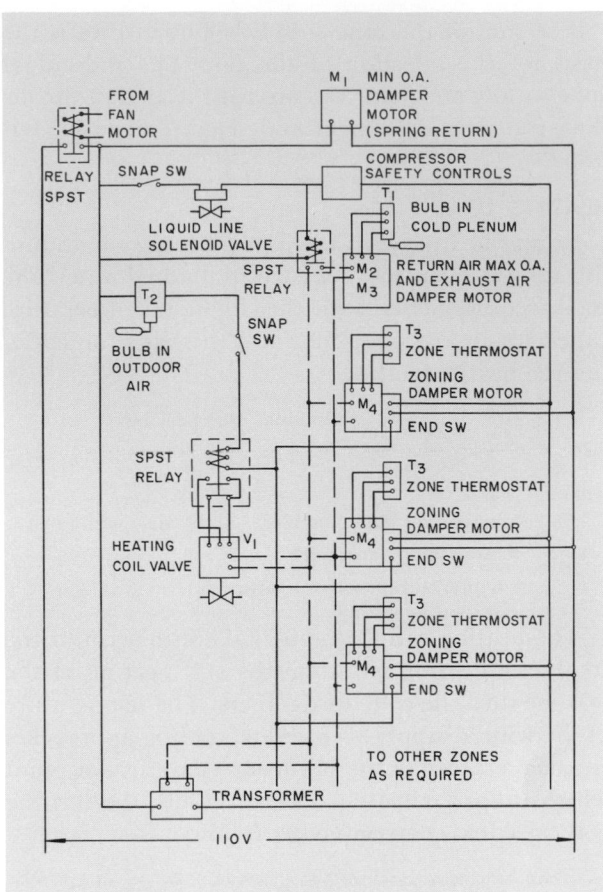

FIG. 19 — MULTI-ZONE UNIT SYSTEM CONTROL,
TYPICAL ELECTRIC ARRANGEMENT

CONTROLS

PNEUMATIC AND ELECTRIC

Either a pneumatic or electric system may be applied to regulate the operation of the multi-zone unit and to control the space or zone temperature with either chilled water or direct expansion refrigeration. *Figures 18 and 19* illustrate respectively the pneumatic and electric arrangements of controls. The sequence of operation is the same with either combination.

With direct expansion cooling coils, special attention should be given to provide flexibility for partial load control. If the capacity control of a single compressor does not satisfy the situation, multiple compressors should be considered in order to prevent compressor cycling at a minimum load.

SUMMER OPERATION

The minimum outdoor air damper motor M_1 *(Fig. 18)* is interconnected with the fan starter. As the fan starts, the minimum outdoor air damper

opens. When the refrigeration is on, the maximum outdoor and exhaust air dampers close and the return air damper opens (motors M_2 and M_3).

Normally, the chilled water temperature is maintained at a constant level and the design quantity of the water is continuously circulated thru the cooling coils. This arrangement allows the apparatus dewpoint temperature to fall at partial load conditions; this reduced temperature helps to maintain better humidity conditions as the zone loads and sensible heat ratios fall.

Each zone thermostat T_3 controls a warm and cold plenum mixing damper motor M_4.

WINTER OPERATION

With outdoor air temperatures below the design cold plenum temperature, the refrigeration source is shut down. At this time thermostat T_1 modulates the outdoor, return and exhaust air dampers to maintain the desired cold plenum temperature. The cool outdoor air is utilized to provide cooling during marginal and winter weather.

Below a predetermined outdoor temperature, the outdoor thermostat T_2 allows full steam pressure or full hot water flow to the heating coil.

VARIATIONS

Depending on the climate, a preheat coil may be added to heat the minimum outdoor air; this coil is under control of the cold plenum thermostat.

If it is desired to precool the minimum outdoor air, a precooling coil may be added and controlled from the thermostat located on the leaving side of the coil; this thermostat is set at the room dewpoint temperature.

If it is desired to hold humidity at a lower level within zones during the summer, this may be achieved by turning on the heating coil in the unit; the added heat calls for an equivalent amount of cooled and dehumidified air, lowering the room humidity.

If it is desired to add a humidifying effect, the means of humidification should be controlled from a humidistat located in the return air path.

CHAPTER 4. DUAL-DUCT SYSTEM

The all-air Dual-Duct System is well suited to provide temperature control for individual spaces or zones. This temperature control is achieved by supplying a mixing terminal unit with air from two ducts with air streams at two different temperature levels; one air stream is *cold* and the other is *warm*. The mixing terminal unit proportions the cold and warm air in response to a thermostat located in its respective space or zone.

The multi-room building is a natural application for this system. Many systems are installed in office buildings, hotels, apartment houses, hospitals, schools and large laboratories. The common characteristic of these multi-room buildings is their highly variable sensible heat load; a properly designed dual-duct system can adequately offset this type of load.

This chapter includes System Features, System Description, Engineering Procedure, Controls, and System Modifications.

SYSTEM FEATURES

The dual-duct system offers many features that are favorable for application to multi-room buildings where individual zone or space temperature control is desired. Some of these features are:

1. *Individual Temperature Control* — Flexibility and instant temperature response are achieved because of the simultaneous availability of cold and warm air at each terminal unit at all times.

2. *Individual Zoning From Minimum Apparatus* — Central station zoning is minimized by having available at each terminal both heating and cooling at the same time.

3. *Simple Nonchange-over Operation* — The space or zone thermostats may be set once, to control year-round temperature conditions. Starting and stopping of the refrigeration machine and boiler is the only requirement for extreme changes in outdoor temperatures.

4. *Centralized Conditioning and Refrigeration* — Services such as power, water and drains are required only in apparatus and machine rooms, and are not needed thruout the building.

5. *Centralized Servicing and Maintenance* — These are accomplished more easily and efficiently in apparatus and machine rooms. Less dust and dirt are tracked thruout the building.

6. *Central Outdoor Air Intakes* — Building stack effect and leakage of wind and rain are minimized. More desirable architectural treatment may be achieved.

7. *Economy of Operation* — All outdoor air can be used when the outdoor temperature is low enough to handle the cooling load, thus saving on refrigeration machine operation.

8. *Filter Efficiency* — Since filtration is accomplished at a central location, higher filtration efficiencies are economically attainable to match the requirements.

9. *Quiet Operation* — All fans and other rotating equipment are remotely located.

10. *Flexible Air System Design* — The choice of medium or high velocity air transmission can be made on the basis of economics and building requirements.

SYSTEM DESCRIPTION

THREE BASIC ARRANGEMENTS

Figure 20 shows three basic arrangements of a dual-duct system, in each of which the two ducts conveying the warm and cold air streams and the air terminal units are of common design.

However, the arrangements of the central station apparatus differ, depending on the degree of precision desired in humidity control. In Arrangement 1, during summer partial load conditions minimum outdoor air may bypass the cooling coil and travel directly into the warm duct. Thus, the

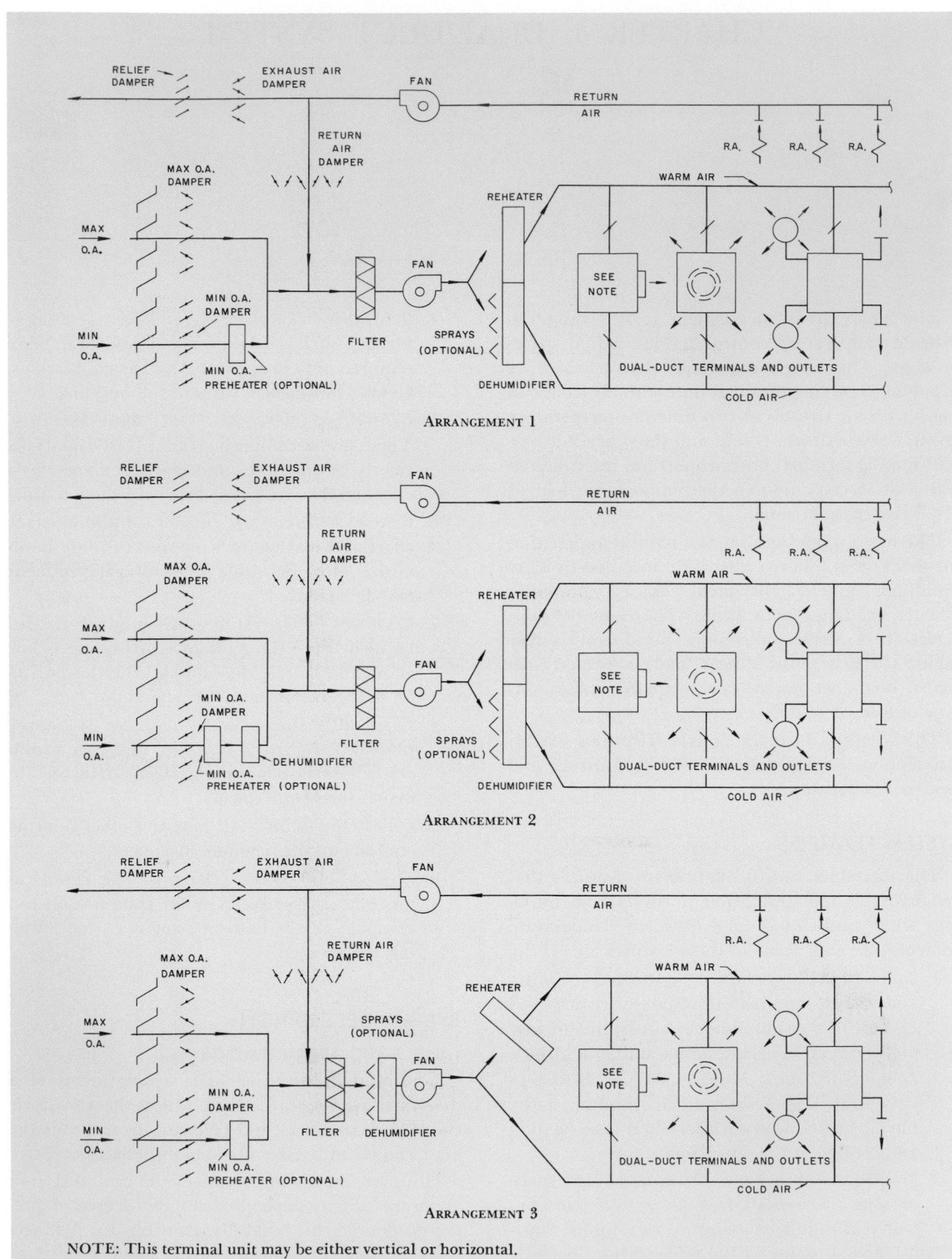

ARRANGEMENT 1

ARRANGEMENT 2

ARRANGEMENT 3

NOTE: This terminal unit may be either vertical or horizontal.

FIG. 20 — TYPICAL DUAL-DUCT SYSTEM ARRANGEMENTS

space or zone relative humidity may rise above design if heat is not applied to the warm duct. The addition of heat in summer does add to operating costs. *Figure 21* shows relative humidity at partial loads with various warm air temperatures when Arrangement 1 is used.

In Arrangement 2 a precooling coil cools and dehumidifies the minimum outdoor air. Therefore, the problem of bypassing unconditioned outdoor air thru the warm duct is eliminated. Operation of both Arrangements 1 and 2 is similar to the operation of either a single-duct face and bypass system or a multi-zone unit system, except that in a dual-duct system the bypass warm air and the cold air are mixed at the terminal unit. It should be noted also that in both Arrangements 1 and 2 the dehumidifier and fan are applied in a blow-thru arrangement.

In Arrangement 3 the dehumidifier and the fan are shown for a draw-thru arrangement; the total air quantity is dehumidified before heat is applied to the warm air stream. Thus, Arrangement 3 is similar in operation to a straight reheat system. It is used primarily to satisfy exacting humidity requirements.

CENTRAL STATION APPARATUS

As seen in *Fig. 20*, there are several variations of the central station apparatus. Generally, the diagrams show that:

1. Whatever the arrangement, the dual-duct system is capable of utilizing 100% outdoor air for cooling purposes during intermediate seasons.

2. A combination return-exhaust air fan is used to exhaust excess air to the outdoors and to return air to the central apparatus in balance with maximum outdoor air required.*

3. The total supply air is always filtered.

4. Minimum ventilation air may be preheated if required.

5. The degree of dehumidification is determined by the apparatus arrangement.

6. Sprays are optional, and may be included where shown in *Fig. 20*.

Standard methods of refrigeration and sources of heating are employed to provide the cooling and heating required to condition the spaces. For ordi-

*On very small systems it is possible to omit the return air fan, provided there is a provision to dispose of the minimum ventilation air as well as the total outdoor air when used for cooling.

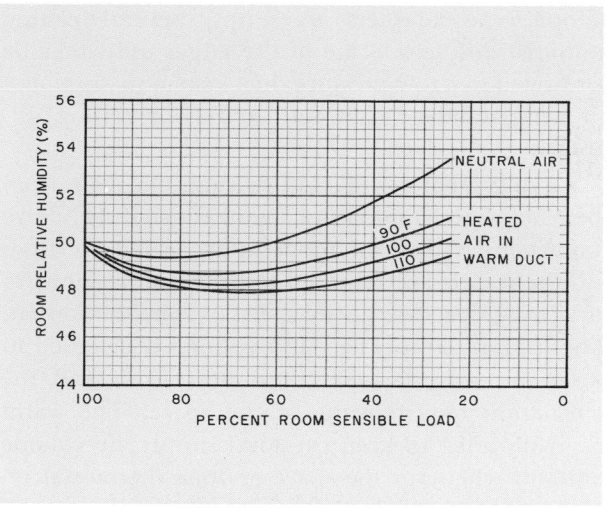

FIG. 21 — PARTIAL LOAD PERFORMANCE OF A DUAL-DUCT SYSTEM, ARRANGEMENT 1

nary comfort applications, precise control of humidity is not essential. However, for economic and comfort reasons during summer operation, the humidity variations should be limited to a range of approximately 45-55% rh. During winter operation lower relative humidities (10-30% rh) are usually maintained to prevent moisture condensation on cold surfaces.

The dual-duct system may be designed by using either a high or medium velocity air transmission system connecting the central apparatus and the terminal units. On both the blow-thru and draw-thru arrangements, care should be exercised to design the apparatus-to-main-duct transitions for a minimum pressure loss and noise generation. Although the terminal units are sound treated, it may be necessary to add additional sound treatment after the fan to reduce the noise generated by the fan.

DUAL-DUCT TERMINAL

The dual-duct terminal unit is designed to:

1. Supply the correct proportions of the cold and warm air streams thru thermostatically controlled air valves.

2. Mix the two air streams and discharge them at an acceptable sound level.

3. Provide a constant volume of discharge air with varying duct static pressures.

The individual terminal units are available in arrangements suitable for either a vertical or horizontal installation, and may be used with an under-the-window grille, a side wall grille, or a ceiling diffuser. The horizontal units are also available with octopus-type adapters to supply several ceiling mounted diffusers. Some of the larger units may be connected to a low pressure duct system to distribute air thru standard side wall grilles or ceiling diffusers.

Varying amounts of cold dehumidified air from the cold duct and either neutral (slightly above room temperature) or moderately heated air from the warm duct are supplied to the terminal units to satisfy the demands of the space or zone thermostat. The cold air supply at 100% volume is designed to offset the sensible and latent heat loads and the ventilation requirements of the space. The warm air is supplied to keep the total supply air volume constant whenever the space or zone thermostat reduces the cold air flow.

There are two commonly used methods of operating the warm air portion of the system. The warm air temperature may be maintained slightly above that of the space or zone; it may also be controlled by a return air hygrostat which raises the temperature as the relative humidity increases. As the warm air temperature is raised, the space or zone thermostat calls for less warm air and more dehumidified air, thus reducing the rising relative humidity. Another method is to maintain the warm air temperature at a higher level constantly.

In winter the warm air duct may supply all the heating requirements of the space or zone served. If the building has, or is designed to have, peripheral heating, the warm air temperature is maintained close to the room temperature.

The dual-duct terminal units are equipped with dampers (valves), damper actuators, and volume compensators to provide constant volume, regardless of varying pressure within the cold and warm air supply ducts. The terminal unit warm air damper is normally open. Space temperature is controlled by a thermostat operating the cold and warm air damper actuators to effect a proper mixture of the two streams to satisfy the load.

ENGINEERING PROCEDURE

BUILDING SURVEY

Before the cooling-heating load can be estimated, a comprehensive survey of the building must be made. This assures an accurate evaluation of the building characteristics, nature and extent of loads, and factors that will direct the establishment of space and zone combinations and selection of the system arrangements.

TABLE 3—HEAT GAINS REFLECTED IN LOAD CALCULATIONS

HEAT GAIN LOAD FROM	ARRANGEMENT			
	Blow-Thru		Draw-Thru	
	Warm Duct (heating)	Cold Duct (dehumid.)	Dehumid.	Warm Duct (reheat)
Supply Duct	heat loss	yes	yes	heat loss
Supply Duct (leakage loss)	yes	yes	yes	——
Supply Fan Motor	yes*	yes	yes	——
Minimum Outdoor Air	yes	yes	yes	——
Return Duct	yes	yes	yes	——
Return Duct (leakage gain)	yes	yes	yes	——
Pump	——	yes	yes	——
Dehumidifier and Pipe Loss	——	yes	yes	——

*Whether the minimum outdoor air affects the warm air temperature, precooled or not, depends on the apparatus layout.

COOLING AND HEATING LOADS

Since a dual-duct terminal unit supplies a constant volume of air to each space or zone, the specific air quantity for a space or zone is determined by either the cooling load, the heating load, or the ventilation needs, whichever requires the maximum air volume. Therefore, it is necessary to estimate the cooling load, the heating load, and the ventilation requirements for each space and zone.

The peak building load is used to determine the apparatus dewpoint, which in turn determines the cooling air requirements. This dewpoint must be checked against individual space or zone apparatus dewpoints for possible deviations, and occasionally some compromise adjustment may be made.

Either the heating load, cooling load, or ventilation air quantity determines the maximum total air supply to each space. However, it is necessary to maintain uniform air change thruout all the spaces in the building. If the air volume as determined by the cooling load is insufficient, then the heating load will determine the air volume.

The over-all refrigeration and heating requirements are established by the totals of the cold and warm air quantities respectively.

Since the dual-duct system central station apparatus may be either a blow-thru or draw-thru arrangement, a different accounting appropriate to each arrangement should be made of the supply duct heat gain, fan horsepower heat gain, and duct leakage (*Table 3*).

Summarizing the preceding, the load estimating procedure should:

1. Establish system zones.
2. Calculate the peak room sensible load for each space (including items from *Table 3*).
3. Calculate the latent load for each space.
4. Calculate the required apparatus dewpoint for each space.
5. Select the desired apparatus dewpoint.
6. Calculate the heating load for each space (including items from *Table 3*).

The data in *Part 1, Load Estimating*, should be used for guidance in survey and load calculations.

AIR QUANTITIES

Before establishing a procedure for estimating the volume of cold and warm air streams, certain aspects of the dual-duct system must be understood. Essentially, this system is a high pressure system because of high air transmission velocities and moderately high pressure drop thru the terminal units. Regardless of whether they are at the peak of their cooling or heating demand, all the spaces and zones receive a constant volume of air, all cold, all warm, or a mixture of each. In summer the warm air duct supplies air above room temperature. This condition occurs because of the return air pick-up (duct and return air fan horsepower). In winter the cold air temperature is slightly warmer than it is in summer, and is maintained between 55 F and 60 F.

In summer the space or zone terminal unit which is not at the peak of its cooling demand admits some air that is warmer than the room air. This bypass air has to be cooled. Therefore, the terminal rebalances the amount of cold air. In winter the reverse is true. Within both ducts there are also variations of flow and pressure which must be compensated for by the terminal unit.

The effects of a terminal unit which create additional bypass are (1) a variable pressure behind the damper, (2) an internal volume compensation, (3) a temperature difference between the two streams of air, and (4) construction characteristics of the terminal itself. This bypass air is either warm at the peak of summer cooling requirements or cold at the peak of winter heating requirements.

The following procedure establishes the various air quantities involved in the design of the dual-duct system:

1. Individual space or zone air quantity (cfm_{sa}) is established on an individual maximum of cooling or heating load, or ventilation requirements.

$$cfm_{sa} = \frac{\text{ERSH}}{1.08\,(t_{rms} - t_{sa})} \quad \text{OR}$$

$$= \frac{\text{heat loss}}{1.08\,(t_{waw} - t_{rmw})} \quad \text{OR}$$

$$= cfm_{oa}$$

These cfm air quantities, at their maximum values for each space and zone, are the basis for the total supply air required by this dual-duct system.

Terminals must be selected for individual air quantities determined, plus 10-20% margin to allow for leakage.

2. Total air supply (cfm_{ta}) is the sum of all maximum air quantities to effect individual space or zone cooling, heating or ventilation. This is the fan design air volume. This total cfm quantity is also the sum of the cold and warm air quantities arrived at in *Steps 3 and 4*, plus respectively the summer or winter bypass air discussed previously.

$$cfm_{ta} = \text{sum of individual maximums} \quad \text{OR}$$
$$= cfm_{ca} + cfm_{ba} \;(\text{summer}) \quad \text{OR}$$
$$= cfm_{wa} + cfm_{ba} \;(\text{winter})$$

3. Cold air quantity (cfm_{ca}) is based on the sum of the peak cooling sensible heat gain and the heat load of the bypass air.

$$cfm_{ca} = \frac{\text{ERSH} + [cfm_{ba} \times 1.08\,(t_{was} - t_{rms})]}{1.08\,(t_{rms} - t_{udp})(1 - \text{BF})}$$

Combining with the summer equation in Step 2 results in:

$$cfm_{ca} = \frac{\text{ERSH} + [cfm_{ta} \times 1.08\,(t_{was} - t_{rms})]}{1.08\,[t_{was} - t_{rms}\,\text{BF} - t_{adp}\,(1 - \text{BF})]}$$

where:
  ERSH = building peak gain
  BF = bypass factor of a dehumidifier, generally assumed between 0.03 and 0.10

This is the air quantity for which the dehumidifying coil is selected and the cold air duct transmission system is designed.

4. Warm air quantity (cfm_{wa}) is based on the sum of the peak heating sensible heat loss and the heat load of the bypass air.

$$cfm_{wa} = \frac{\text{heat loss} + [cfm_{ba} \times 1.08\,(t_{rmw} - t_{caw})]}{1.08\,(t_{waw} - t_{rmw})}$$

Combining with the winter equation in *Step 2* results in:

$$cfm_{wa} = \frac{\text{heat loss} + [cfm_{ta} \times 1.08\,(t_{rmw} - t_{caw})]}{1.08\,(t_{waw} - t_{caw})}$$

where t_{waw} = winter warm supply air temperature assumed between 120 F and 140 F.

This is the air quantity for which the heating coils for the warm air duct may be selected and the warm air duct transmission system is designed. Refer to *Duct Design* for other possible warm air quantities that may be substituted in the equation for cfm_{wa}; this may result in a different design warm supply air temperature (t_{waw}).

5. Minimum outdoor air *(min cfm_{oa})* quantity is the design ventilation air. This air quantity is the basis for sizing the minimum outdoor air dampers, preheaters (when used), and precooling coils (when desired).

6. Maximum outdoor air *(max cfm_{oa})* and return air *(cfm_{ra})* are the same quantities.

$$max\ cfm_{oa} = cfm_{ra} = cfm_{ta} - min\ cfm_{oa}$$

This air quantity is used to size the maximum outdoor air dampers and the return air duct system, together with its fan and exhaust and return air dampers.

NOTE: The various terms used are defined as follows:

cfm_{sa} = individual supply air quantity (space or zone)

cfm_{ca} = cold supply air quantity

cfm_{wa} = warm supply air quantity

cfm_{ba} = bypass supply air quantity

cfm_{ta} = total supply air quantity

$min\ cfm_{oa}$ = minimum outdoor air quantity (ventilation)

$max\ cfm_{oa}$ = maximum outdoor air quantity

t_{adp} = apparatus dewpoint temperature

t_{sa} = supply air temperature (apparatus dewpoint plus heat gain to terminal)

t_{rms} = room temperature in summer (at design condition)

t_{rmw} = room temperature in winter (at design condition)

t_{was} = warm air temperature in summer (at room design temperature plus return air heat gains or an assumed design value)

t_{waw} = warm air temperature in winter (120-140 F)

t_{caw} = cold air temperature in winter (55-60 F)

BF = dehumidifier bypass factor (0.03 to 0.10)

ERSH = effective room sensible heat

DUCT DESIGN

The supply (cold and warm) duct air transmission systems are generally designed using a medium or high velocity air transmission system. Since the dual-duct system is a variable volume system, the method of duct design is not critical. The static

regain method is used for the branch risers or headers which feed zones on the same exposure or with similar loadings. The main headers or branches feeding zones with divergent loading may be sized by either the static regain or equal friction method. Ductwork systems sized by the static regain method usually require lower fan horsepower, and maintain better system stability at all times.

The warm duct air quantity *(cfm_{wa})* is greatest when the need for cold air is the least. This condition occurs during marginal weather in the exterior areas and at no load in the interior areas. The warm air duct is usually sized to handle 80-85% of the cold air quantity *(cfm_{ca})* determined in *Step 3*. At times the warm air duct is sized for 50-60% of the cold air quantity *(cfm_{ca})*, but at the expense of high temperature operation.

The return air duct for this system is usually based on low or medium velocities and sized by the equal friction method.

Part 2 should be used as a guide to duct sizing; specifically, a high velocity dual-duct air transmission system should adhere to the following rules:

1. No dampers or splitters are to be used.

2. All rectangular elbows are to be vaned.

3. Number of offsets is to be at a minimum.

4. Sufficient lengths of straight duct runs are to be allowed between flow disturbances caused by fittings or offsets.

5. Conical take-offs are to be used when higher velocities are applied.

INSULATION

For applications that have basically a constant load, normal practice is to determine whether duct insulation is required from supply heat gain calculations. On variable load applications the amount of insulation required is determined by making the heat gain check when a partial load exists. At this time the supply air volume is reduced with a corresponding decrease in air velocity, resulting in higher duct heat gains or losses. Insulation may be applied on the inside of the duct to increase sound attenuation. Normal considerations apply when insulating other elements of the mechanical equipment.

CENTRAL APPARATUS

The central air conditioning apparatus is selected for the sum of the maximum air quantities supplied to each space or zone. This total supply air is used to select the supply air fan which operates at a static pressure sufficient to overcome the resistance of the apparatus and air transmission components.

TABLE 4—FUNCTION AND LOCATION OF CENTRAL APPARATUS COMPONENTS

APPARATUS COMPONENT	FUNCTION AND LOCATION	
	Arrangements	
	Blow-Thru	Draw-Thru
Preheater (if required)	Minimum outdoor air	Minimum outdoor air
Precooler (if desired)	Minimum outdoor air	—
Fan	Discharge into plenum (70% discharge velocity press. loss)	Discharge into supply duct Fan to duct conversion loss or gain
Dehumidifier	Cold air duct volume	Total supply air volume
Reheater	On return air discharge from supply fan Dual purpose: summer humidity winter heating	On dehumidifier cooled air from supply fan Dual purpose: summer reheat (bypass) winter heating
Sprays (optional)	Humidifying at dehumidifiers	Winter-summer humidifying at dehumidifiers

Included in this resistance are pressure drops from the outdoor air intake to the beginning of the supply air duct, the critical run of the supply air duct, and the terminal unit and outlet combination. The duct connection loss between the terminal unit and outlet is also included.

Since the dual-duct system may be designed in either blow-thru or draw-thru arrangements, several special aspects in equipment selection peculiar to these arrangements must be considered. *Table 4* shows the apparatus components, including their function and location in either one of these arrangements.

The fans should be equipped with variable inlet vanes to assist in system air balancing. Fans in a blow-thru arrangement should have perforated plates placed in front of the air discharge and at the heating and cooling coils to distribute the air evenly. To regulate the fan discharge air pattern in a more efficient manner, an *evasé* section should be used. Its length should be 1½ to 2 times the fan discharge equivalent diameter.

The dehumidifiers, usually a dry coil type (using sprays if needed or desired), should be selected to cool and dehumidify the air mixture from entering conditions to the apparatus dewpoint determined previously. At times it may be desirable to operate the selected dehumidifier at a lower dewpoint to compensate for possible irregularities of apparatus arrangement that may contribute to excessive by-passing of outdoor humid air.

The dehumidifier capacity is calculated by the following formulas:

For a blow-thru arrangement

$$\text{Dehumidifier load} = cfm_{ca} \times 4.45 \, (h_{ea} - h_{adp})(1 - \text{BF})$$

For a draw-thru arrangement

$$\text{Dehumidifier load} = cfm_{ta} \times 4.45 \, (h_{ea} - h_{adp})(1 - \text{BF})$$

where:

cfm_{ca} = cold supply air quantity
cfm_{ta} = total supply air quantity
h_{ea} = enthalpy of the entering air mixture
h_{adp} = enthalpy of the apparatus dewpoint
BF = dehumidifier bypass factor

In either a blow-thru or a draw-thru arrangement, the reheater is selected to heat the warm air quantity from the design winter cold air temperature to the required warm air temperature. This selection provides enough capacity in a draw-thru arrangement to reheat the air in summer from the fan discharge temperature (apparatus dewpoint plus supply fan horsepower heat gain) to room temperature.

It is recommended that reheaters be selected with about 15-25% excess capacity to provide for morning pick-up and duct heat losses.

The minimum outdoor air precooling coils (*Fig. 20, Arrangement 2*) should be designed to cool the minimum outdoor air from outdoor design conditions to the room dewpoint level.

The outdoor air intake screen, louvers, minimum and maximum dampers, minimum outdoor air preheaters (if required), return air fan and dampers, and air filters are selected for both blow-thru and draw-thru arrangements, using standard procedures outlined in *Parts 2 and 6*. The degree of filtration desired determines the filter selection.

REFRIGERATION LOAD

The refrigeration load is determined by the sum of the requirements called for by dehumidifiers and precooling coils (if used). This sum assumes that the component cooling equipment is operating at its peak. The refrigeration machines may be the reciprocating, absorption or centrifugal type.

CONTROLS

The diagram shown in *Fig. 22* illustrates the control elements required for the dual-duct system

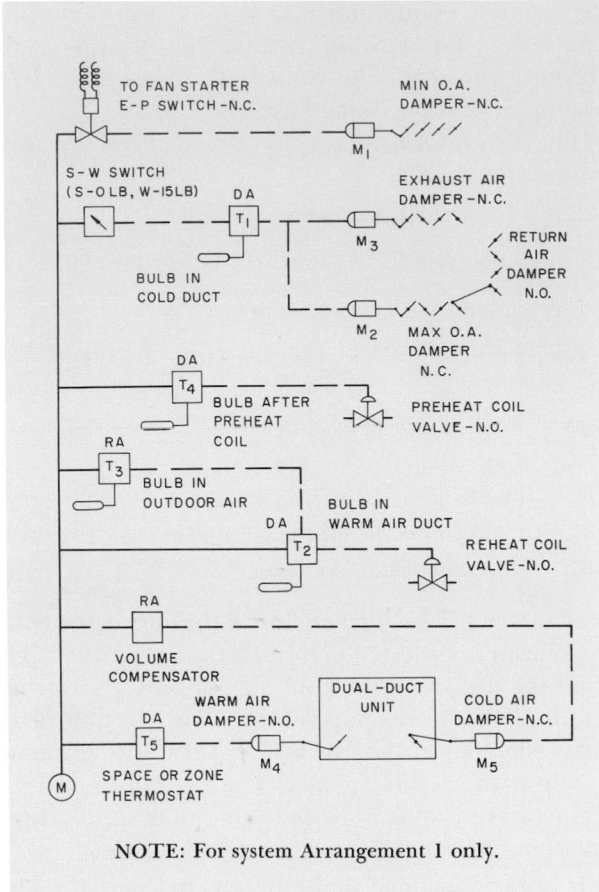

FIG. 22 — DUAL-DUCT SYSTEM CONTROL,
TYPICAL PNEUMATIC ARRANGEMENT

shown in *Fig. 20*, Arrangement 1. Either electric or pneumatic (shown) type may be used. The sequence of operation is the same, regardless of which one is used.

CENTRAL APPARATUS CONTROL

Summer Operation

When the summer-winter switch is in the summer position, the cold air duct thermostat T_1 is inactivated, closing the exhaust air and maximum outdoor air dampers and opening the return air dampers. The temperature leaving the dehumidifier is not controlled, except for maintaining either the entering chilled water temperature of the water coil or the evaporator temperature of the direct expansion coil. This limits the supply air temperature during peak load conditions, but allows it to decrease as the load on the dehumidifier decreases; thus, the room relative humidity may be slightly improved under partial load conditions. The electric-pneumatic switch on the minimum outdoor air damper is interconnected with the fan starter so that

the damper opens as the fan is started. Since the outdoor temperature is above their set point, the preheater and reheater controls do not normally function.

Marginal Weather and Winter Operation

When the outdoor air temperature is below the design coil air supply temperature, the refrigeration machine is shut off and the summer-winter switch is set in the winter position. This allows the cold air duct thermostat T_1 to modulate the maximum outdoor and return air dampers in conjunction with exhaust air dampers, to maintain the desired cold air duct temperature. Thus, cool air is used as a source of free cooling. The submaster thermostat T_2 after the reheater coil is reset by the master outdoor air thermostat T_3 and maintains the desired schedule of temperatures in the warm air duct. If a preheater is required, the thermostat T_4 after the preheater coil is set at a minimum of 40 F.

UNIT CONTROL

Each space or zone thermostat T_5 controls the warm air damper in each terminal to maintain the desired space or zone temperature. A constant volume compensator in the terminal unit maintains a constant supply air quantity by controlling the cold air damper.

SYSTEM MODIFICATIONS

TWO FANS

Both Arrangements 1 and 2 *(Fig. 20)* may be modified by the use of two fans instead of a single fan, each handling about 50% of the total supply air required by the system *(Fig. 23)*. This modification has two bypasses, one before the fans and one after the fans. The latter is a plenum into which both fans discharge, supplying air into the cold and warm air ducts. One fan handles the minimum outdoor air at all times and is supplemented by all outdoor air, all return air, or a mixture of outdoor and return air. The other fan handles primarily the return or outdoor air, or a mixture of both. The room humidity conditions are improved and the bypassing of the minimum outdoor air is least when the warm duct air is less than half of the total air. The advantage of the two fan modifications is that in winter only one fan needs to operate to supply heating under load conditions when only heating is required, i.e. at night and over weekends. The terminal unit volume compensator must be inactivated; then the space or zone thermostat operates the warm air damper (valves).

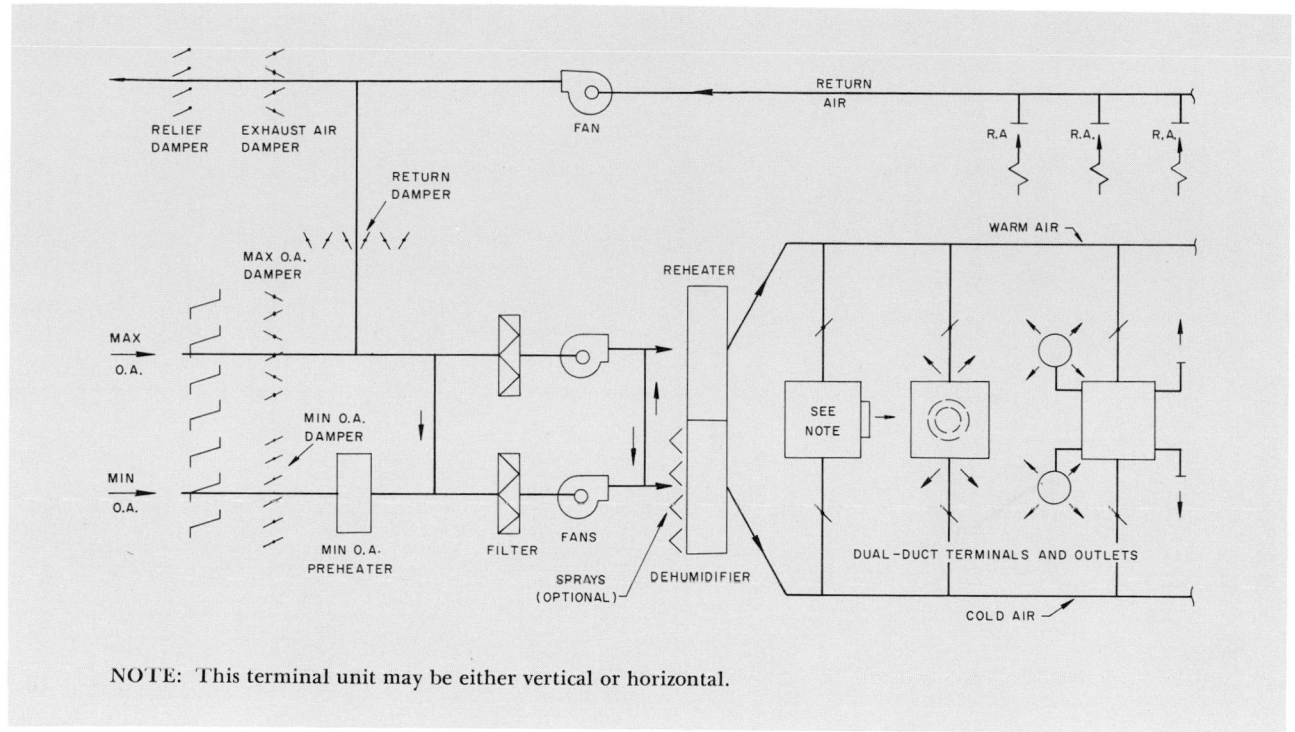

NOTE: This terminal unit may be either vertical or horizontal.

FIG. 23 — TYPICAL DUAL-FAN, DUAL-DUCT SYSTEM ARRANGEMENT

HUMIDITY CONTROL

Another minor modification that may be incorporated is the activating of the reheater in summer to improve room humidity. Under this arrangement the reheater is partially controlled by a hygrostat located in the return air duct. By admitting warmer air, the terminal must compensate by increasing the amount of cold dehumidified air.

A further modification is the decreasing of the warm air quantity by reheating it in summer to higher than room temperature, permitting closer humidity control but at an economic disadvantage.

CHAPTER 5. VARIABLE VOLUME, CONSTANT TEMPERATURE SYSTEM

The all-air Variable Volume, Constant Temperature System is well suited for many applications. Among these are applications for which a relatively constant cooling load exists year round, i.c. interior zones of an office building and department stores.

Other applications for which this system should be considered are those with variable loads having a serviceable steam or hot water heating system and for which only summer cooling is desired. Examples are existing buildings such as office buildings, hotels, hospitals, apartments and schools.

This chapter includes System Features, System Description, Controls, System Modifications and Engineering Procedure for designing a complete variable volume, constant temperature system.

SYSTEM FEATURES

The variable volume, constant temperature system offers many features that are favorable for its application to interior zones and where only summer cooling is required. Some of these features are:

1. *Economical Operation* — Since the volume of air is reduced with a reduction in load, the refrigeration and fan horsepower follows closely the actual air conditioning load of the building. All outdoor air is available during intermediate seasons for free cooling.

2. *Individual Room Temperature Control* — A nonreversing thermostat and volume damper controls the flow of supply air to match the load in each space, giving simplified control. The flow of air literally follows the load around the building.

3. *Simple Operation* — Change-over from summer to winter or winter to summer operation is obtained by stopping or starting the refrigeration equipment manually.

4. *Minimum Apparatus* — Zoning by areas is not required because each space supplied by a controlled outlet is a separate zone.

5. *Low First Cost* — This system is extremely low in first cost when compared to other systems that provide individual space control, because it requires only single runs of duct and a simple control at the air terminal. Also, where diversity of loading occurs, smaller equipment can be used.

6. *Centralized Conditioning and Refrigeration* — Services such as power, water and drains are required only in apparatus and machine rooms, and are not necessary thruout the building.

7. *Centralized Service and Maintenance* — These services are more easily accomplished in apparatus rooms, resulting in more efficient maintenance and service with less dust and dirt tracked thruout the building.

8. *Central Outdoor Air Intake* — Leakage of wind and rain and building stack effect are minimized. This allows a more desirable architectural treatment.

SYSTEM DESCRIPTION

There are many variations that can be applied to this system. The following is a description for a system that may be applied to interior zones where the load is fairly constant.

The room outlet delivers completely filtered, humidity-controlled air during all seasons. Individual space temperature control is accomplished by modulating the air quantity to match the required space load.

The air handling apparatus conditions the air and supplies either a mixture of outdoor and re-

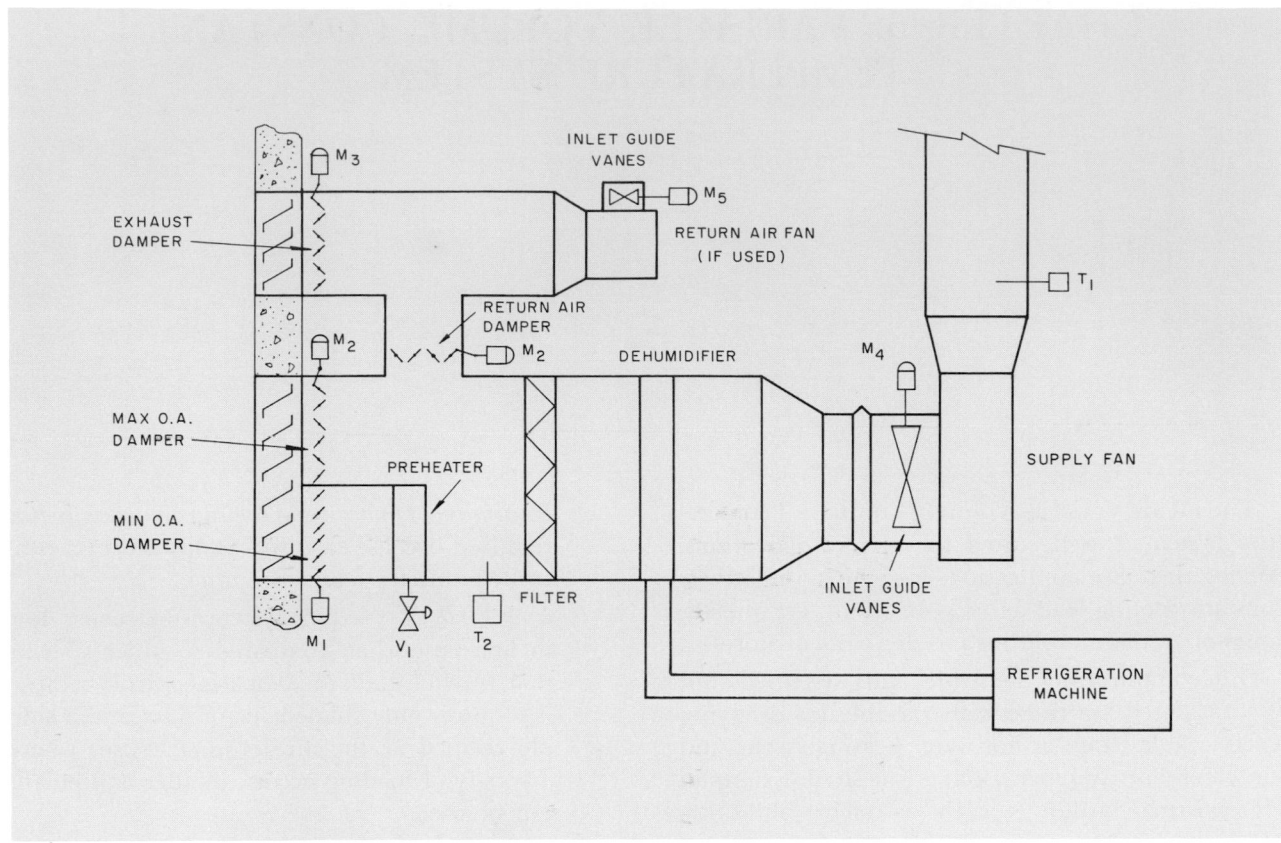

FIG. 24 — TYPICAL VARIABLE VOLUME, CONSTANT TEMPERATURE SYSTEM

turn air or 100% outdoor air to the terminal unit. The apparatus contains filters to clean the air, preheaters (if required) to temper cold winter air, and a dehumidifier to remove excess moisture and cool the supply air. For a typical variable volume, constant temperature system, see *Fig. 24.*

A constant leaving temperature is maintained in the fan discharge during intermediate and winter seasons when the refrigeration machine is not operating.

A high or low velocity air distribution system is used to move the air from the apparatus to the room terminal units. When required, a sound absorber is used to reduce the noise generated by the fan.

The dehumidifier may be supplied by either a direct expansion or chilled water refrigeration system.

ENGINEERING PROCEDURE

The procedure for designing a variable volume system is similar to the procedure for designing any all-air system. *Part 1* can be used as a guide for the survey and preliminary layout and for determining cooling loads and air quantities.

COOLING LOAD

The module concept is often used for determining the area to be conditioned by each terminal unit. This allows for flexibility in relocating partitions (if and when required). It is common practice to design for future modernization of existing buildings and possible shifting of loads in both new and existing buildings.

Both sensible and latent load calculations are made for each space. On interior applications the load is often determined on a square foot basis and multiplied by the number of square feet in the interior zone. Exterior applications require a minimum of one peak load estimate for each exposure and an additional estimate for unusual spaces.

A block load estimate is made for the entire area that is to be supplied from each fan system. This is made at the peak cooling load condition and includes factors for diversity of lights and people (if applicable).

AIR QUANTITIES

The air quantity required is calculated for each typical space by using this formula:

CHART 1—SYSTEM PERFORMANCE AT PARTIAL LOAD

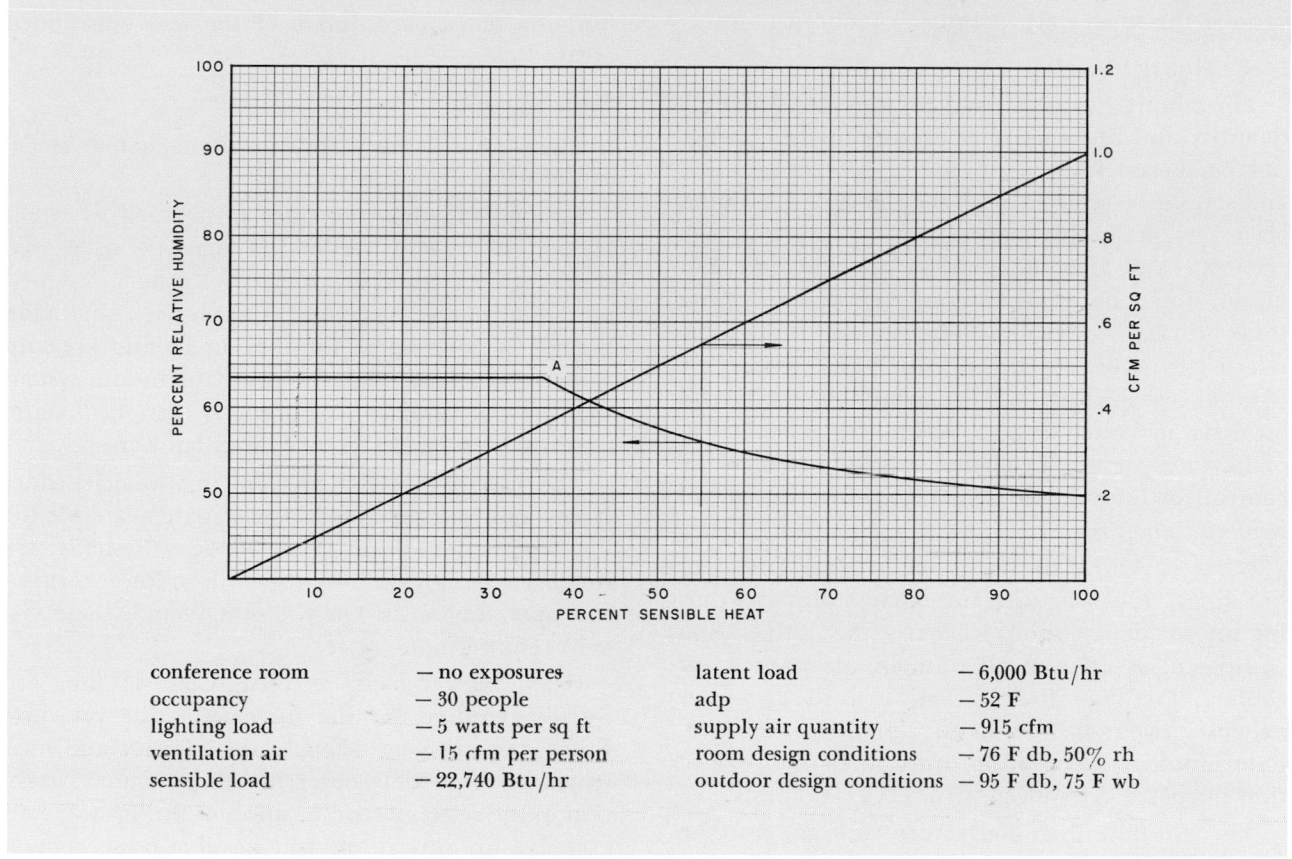

conference room	— no exposures	latent load	— 6,000 Btu/hr
occupancy	— 30 people	adp	— 52 F
lighting load	— 5 watts per sq ft	supply air quantity	— 915 cfm
ventilation air	— 15 cfm per person	room design conditions	— 76 F db, 50% rh
sensible load	— 22,740 Btu/hr	outdoor design conditions	— 95 F db, 75 F wb

$$cfm_{da} = \frac{ERSH}{1.08\,(1 - BF)(t_{rm} - t_{adp})}$$

where:

cfm_{da} = dehumidified air quantity

ERSH = effective room sensible heat

BF = dehumidifier coil bypass factor

t_{rm} = room temperature

t_{adp} = apparatus dewpoint temperature

The air quantity determined from this formula is used for outlet selections and duct sizing. The apparatus dewpoint temperature used in the formula is selected as being the most representative of the majority of the spaces. ERSH is obtained from the load estimates for each typical space.

The fan and dehumidifier air quantity is calculated using the same formula. ERSH is found from the block estimate, and the apparatus dewpoint temperature is that which has been selected previously. A factor of 5% is added to the calculated air quantity to allow for space overcooling, since in some areas temperature settings are below room design; this means that zones which peak at

the time of maximum loading do not receive full cooling unless there is excess air available.

As with other all-air systems, the air quantity supplied to each terminal must have sufficient capacity to offset the sensible and latent load. Since temperature control is maintained by varying the air volume, partial load characteristics of the space should be analyzed for the resulting relative humidity and reduced air quantity.

Chart 1 illustrates the maximum expected relative humidity at different load conditions. In this conference room, as the sensible load drops, a condition is reached eventually where the only load in the room is from people. This is noted on the chart as point A. To obtain a further reduction in load, it is necessary for the people to leave the room. Therefore, to the left of point A the latent load is decreasing as is the sensible load, and the relative humidity remains practically constant at the maximum of 63% for this particular application.

Chart 1 shows also the reduction in supply air quantity as the sensible load is reduced. To maintain a reasonable room air motion, it is desirable to use an outlet which maintains a high induction ratio as the supply air quantity is reduced.

FAN SELECTION

The supply fan is selected for the calculated air quantity and the static pressure required. This fan can be picked from performance curves or tables, and should be selected near the point of maximum efficiency, preferably between the points of peak efficiency and free delivery. In addition, the fan motor brake horsepower must be obtained from these curves or tables. The motor should be selected to allow the fan to supply 20% excess air without overloading; this can avoid motor overload on early morning start-up. With either backward- or forward-curved fan blades, it is advisable to use controlled inlet guide vanes to improve partial load efficiencies.

When a return air fan is required (as in larger buildings), it can be used for exhaust purposes during intermediate seasons whenever the refrigeration equipment is off and all outdoor air is used for cooling. The fan either returns air to the system or exhausts air to the outdoors. Normally, the minimum outdoor air quantity introduced for ventilation purposes is removed thru service exhausts.

The air flow thru the return air fan must be balanced with the air flow thru the supply fan. When the supply fan is throttled, the quantities of return and outdoor air are reduced proportionally; therefore, the inlet vanes of the return air fan must be throttled simultaneously in the same proportion. Thus, the return air fan has inlet vanes acting in unison with the inlet vanes of the supply fan.

DEHUMIDIFIER LOAD

The dehumidifier load is calculated by using this formula:

$$\text{Load} = 4.45 \times cfm_{da} \times (1 - \text{BF})(h_{ea} - h_{adp})$$

where:

cfm_{da} = dehumidified air quantity
h_{ea} = entering air enthalpy
h_{adp} = apparatus dewpoint enthalpy
BF = bypass factor

The entering air condition is generally a mixture of the minimum outdoor air and the return air. Outdoor air is assumed to be at the maximum design temperature. Return air is assumed to be at a temperature equal to the inside design temperature plus any temperature rise due to return duct and fan heat gains and return duct leakage.

REFRIGERATION LOAD

The refrigeration load is determined by the peak building or block estimate of the air conditioned areas.

DUCT DESIGN

Ductwork for the variable volume system is designed using *Part 2* as a guide.

Although methods of duct sizing such as equal friction or velocity reduction may be used, the static regain method is preferred. A system designed by the static regain method is almost self-balancing because it is designed for the same static pressure at each terminal. This helps to maintain system stability. In addition, a properly designed static regain system results in a reduced fan horsepower.

The use of either a low or high velocity duct system can be determined by the space available for the supply air ducts. Low velocity systems are simpler to design and usually result in lower owning and operating costs. On the other hand, these systems require more space.

When high velocity is used, Class II fans are usually required for the increased static pressure. Extra care must be taken in duct layout and construction. A good layout requires particular attention to the selection and location of fittings to avoid excessive pressure drops and possible noise generation problems.

Since this is a variable volume system and the supply air quantity varies directly with the load, duct construction is important. In areas where there is a complete absence of load, duct pressure can build up almost to the fan discharge pressure and, consequently, it is necessary to construct the ducts for that pressure. The ducts must not only withstand a variable pressure, but must also be sealed to prevent air leakage.

The outlets should be selected in conjunction with the fan so that they do not create an objectionable noise when they throttle against the maximum static pressure developed by the fan.

INSULATION

For applications such as interior zones that have a constant load, normal practice is to determine from the calculation of the supply air heat gain whether insulation is required. On variable load applications the amount of insulation required is determined by making the heat gain analysis when a partial load exists since, at this time, the supply air volume is reduced with a corresponding decrease in air velocity.

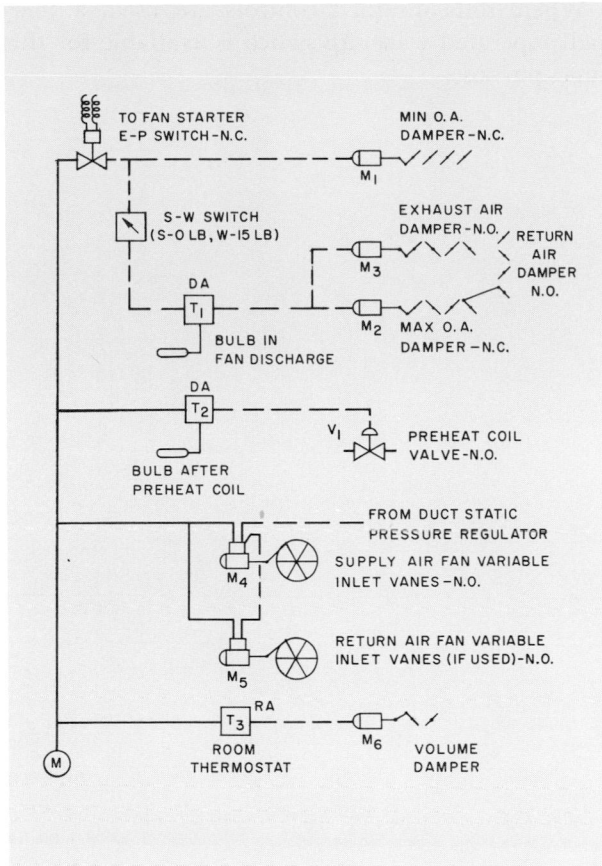

FIG. 25 — VARIABLE VOLUME, CONSTANT TEMPERA-
TURE CONTROL, TYPICAL PNEUMATIC ARRANGEMENT

CONTROLS

A suggested control diagram for a variable volume system is shown in *Fig. 25* and a discussion of controls for the unit and air handling apparatus follows.

UNIT CONTROL

Although either manual or automatic control can be used for the air terminal, automatic control is preferred. With automatic control, a constant room temperature may be maintained, regardless of changing load conditions, when the system is applied to spaces such as interior zones that always require cooling. In such cases, since the air is always cool, a simple nonreversing thermostat is used to control the air flow volume regulator. Duct pressure is controlled by the fan inlet vanes to prevent overblow and excessive throttling at the terminal unit.

Several air terminals may be controlled from one thermostat centrally located to insure that the temperature sensed is representative of the average room temperature. Also, consideration should be given to possible future relocation of partitions.

CENTRAL APPARATUS CONTROL

Either electric or pneumatic control may be used for the central apparatus; the sequence of operation is identical.

Summer Operation

The minimum outdoor air damper is interconnected with the fan starter so that the damper opens as the fan is started. With the summer-winter switch in the summer position, the maximum outdoor air damper is closed and the return air damper is wide open. Normal procedure is to maintain a constant leaving water temperature from the refrigeration plant. This limits the supply air temperature during peak load conditions, but allows it to decrease as the load on the dehumidifier decreases; thus, increased flexibility in room control is provided during off-peak operation.

Winter Operation

When the outdoor temperature is below design supply air temperature, the refrigeration machine is shut off and the summer-winter switch is set in the winter position. This allows the thermostat in the fan discharge to modulate the outdoor and return air dampers in conjunction with the exhaust dampers, in order to maintain the desired leaving air temperature. Thus, cool outdoor air is used as a source of free cooling. If a preheater is used in the minimum outdoor air, the thermostat after the preheater is set at a minimum of 40 F.

SYSTEM MODIFICATIONS

REHEAT COIL

For applications in which the space temperature is allowed to drop below normal design at night or over weekends, and a separate source of heat in the space is not available, a reheater should be installed in the central apparatus. If only part of the building is without heat, the reheater should be installed in the ductwork supplying that part of the building.

The reheater should be designed to heat the required air quantity from the normal supply air temperature (50-55 F) to 15 degrees above the room design temperature. The reheater can be controlled by opening a manual valve in the steam line for a short time following start-up. The fan discharge thermostat must be deactivated during operation of the reheater in the central apparatus.

Provision must be made in the controls to allow the room control damper to open when warm air is being supplied. With pneumatic controls the main line control pressure to the reverse acting room thermostats is reduced to zero so that the dampers assume their normally open position.

Where self-contained controls are used, a thermally-operated warm-up switch is available for this same purpose.

CHAPTER 6. DUAL CONDUIT SYSTEM

The all-air Dual Conduit System* is a modern central station system that can be applied to multi-zone buildings such as schools, offices, apartments and hospitals, for areas that have a reversing transmission load and require individual room temperature control. It can be adapted easily to areas that have variable cooling and heating requirements caused by sun, outdoor temperature and internal loads. Generally, its application is similar to the dual-duct system, but with a more economical first cost.

This chapter includes System Features, System Description, Engineering Procedure, Controls, System Modifications and Air Terminal Units.

SYSTEM FEATURES

The Dual Conduit System offers many features that are favorable for its application to multi-zone buildings where individual room temperature control is desired. Some of these features are:

1. *Smaller Duct Sizes* — Both primary and secondary air streams are used to offset the summer peak load. Therefore, the duct sizes are smaller because the combined areas of the two ducts are utilized for summer cooling, instead of one duct supplying cold air and the other supplying neutral air.
2. *Flexible Air Distribution* — The supply air (*Fig. 32*) can be distributed from many locations: under the window, from the ceiling, or from the side wall. In addition, the two air streams can be separated so that the primary

air is distributed from under the window, at the ceiling, or from the side wall, while the secondary air is distributed from another of these locations.

3. *Economical Operation* — During night and weekend operation in winter, only the small primary air fan is operated. As in most all-air systems, outdoor air is available to provide free cooling during intermediate season operation.
4. *Centralized Conditioning and Refrigeration Equipment* — Services such as power, water and drains are required only in the apparatus rooms and not thruout the building.
5. *Centralized Service and Maintenance* — These functions are more easily accomplished in apparatus rooms where maintenance and service are more efficient. This means there is less dirt and dust tracked thruout the building.
6. *Central Outdoor Air Intake* — Building stack effect and leakage of wind or rain are minimized. This allows a more desirable architectural treatment.
7. *Simple Operation* — Change-over from summer to winter or winter to summer consists of stopping or starting the refrigeration plant manually.
8. *Individual Room Temperature Control* — A nonreversing thermostat and a volume damper are used to control the flow of secondary air to maintain the desired room temperature.
9. *Quieter Rooms* — Mechanical equipment is remotely located; therefore, vibration is easier to control.

*Dual Conduit System is a Carrier patented system.

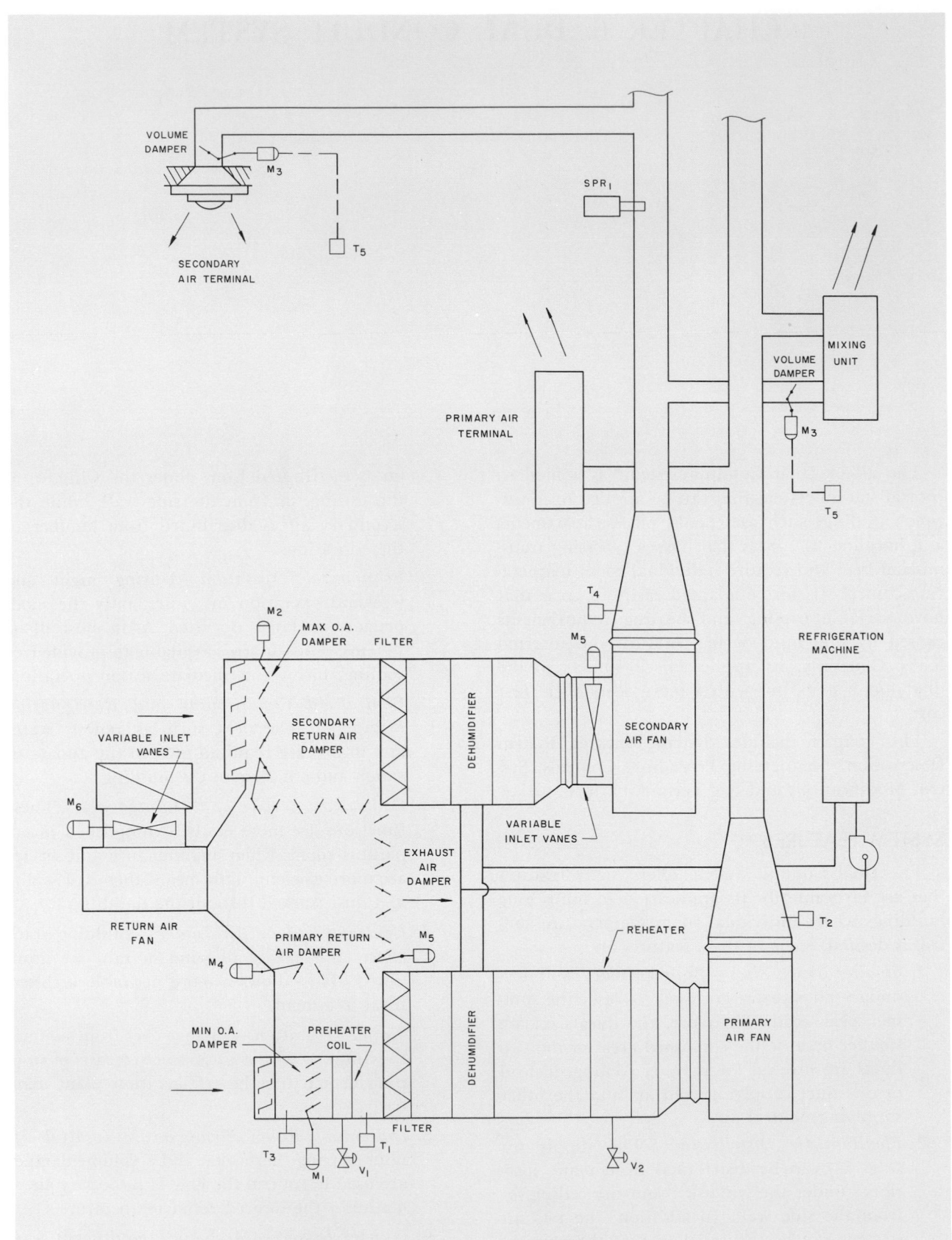

FIG. 26 — TYPICAL DUAL CONDUIT SYSTEM, DUAL-FAN, DUAL APPARATUS

SYSTEM DESCRIPTION

The system is designed to supply two air streams to exposures that have a reversing transmission load.

One air stream called the secondary air is cool the year round, and is constant in temperature and variable in volume to match the capacity required for the changing cooling load caused by sun, lights and people. Therefore, the secondary air is a constant temperature, variable volume air stream.

The other air stream called the primary air is constant in volume, and the air temperature is varied to offset transmission effects; it is warm in winter and cool in summer. The primary air is, therefore, a constant volume, variable temperature air stream.

Various central station arrangements can be used to provide the air temperatures and volumes required for practical temperature control. Two of these are described under *System Modifications*.

A dual fan, dual apparatus system is described here and illustrated in *Fig. 26*.

The primary air apparatus conditions the air and supplies a mixture of outdoor and return air to the room terminals. The apparatus contains filters to clean the air, preheat coils (as required) to temper cold winter air, a humidifier (if desired) to add winter humidification, and a dehumidifier to remove excess moisture and cool the supply air. The primary air stream contains a reheat coil controlled by a master-submaster thermostat arrangement, the function of which is to adjust the air temperature to match the building transmission affects. Outdoor air is admitted to the apparatus thru a rain louver and screen.

The secondary air apparatus conditions the air and supplies all return air, a mixture of outdoor and return air, or all outdoor air, depending on the season. The apparatus contains filters to clean the air and a dehumidifier to remove excess moisture and cool the supply air. A thermostat located in the fan discharge modulates the outdoor and return air dampers to maintain a constant leaving temperature during seasons of nonoperation.

Air from both the primary and secondary apparatus is delivered to the room terminal units thru ductwork. Normal practice requires the use of a high velocity air distribution system for the primary air and either high or medium velocities for the secondary air.

TABLE 5—SCHEDULE OF PRIMARY AIR TEMPERATURES

OUTDOOR DRY-BULB TEMP. (F)	PRIMARY AIR TEMPERATURE (F)												
	A/T Ratio												
	0.2	0.4	0.6	0.8	1.0	1.2	1.4	1.6	2.0	2.5	3.0	3.5	4.0
100	56	56	56	56	56	56	56	56	56	58	61	63	64
95	56	56	56	56	56	56	56	56	57	60	62	64	66
90	56	56	56	56	56	56	56	56	59	62	64	66	67
85	56	56	56	56	56	56	57	59	62	64	66	67	68
80	56	56	56	56	56	58	60	62	64	66	67	69	70
75	56	56	56	57	60	63	64	65	67	68	69	70	71
70	56	56	60	63	65	67	68	69	70	71	71	72	72
65	58	65	68	70	71	71	72	72	72	73	73	73	74
60	80	78	77	76	76	76	75	75	75	75	75	75	75
55	102	90	85	82	81	80	79	78	78	77	76	76	76
50	125	103	93	89	86	84	82	81	80	79	78	78	78
45	147	116	102	95	91	88	86	85	83	81	80	79	79
40		128	110	101	96	92	90	88	85	83	82	81	80
35		140	119	108	101	97	93	91	88	85	83	82	81
30			127	114	106	101	97	94	90	87	85	84	83
25			136	121	111	105	101	98	93	90	87	85	84
20			145	127	117	110	105	101	96	92	89	87	85
15				134	122	114	109	104	99	94	91	88	87
10				140	127	118	112	107	101	96	92	90	88
5				147	132	123	116	111	104	98	94	91	89
0					137	127	120	114	106	100	96	93	91
−5					143	131	124	117	109	102	98	94	92
−10						136	127	121	112	105	100	96	93
−15						140	131	124	114	107	101	97	95
−20							135	127	117	109	103	99	96

NOTE: These temperatures are required at the units, and thermostat settings must be adjusted to allow for duct heat gains or losses.

A refrigeration and heating plant is necessary to complete the system.

ENGINEERING PROCEDURE

The following is a guide for designing a Dual Conduit System for air conditioning the exterior zones of a multi-room building. Several methods can be used in the design of this system; one is presented here and supplemental ideas are presented under *System Modifications*.

The first method is similar to the design of an air-water induction system and is based on the principle that the primary air system offsets the transmission gains or losses, handles the latent load, and supplies the ventilation air requirements. In this instance, the secondary air offsets the sensible heat loads of the sun, lights and people in the space.

Part 1 contains information for making a survey and preliminary layout and for obtaining factors required to determine heating and cooling loads.

AIR QUANTITIES

The following procedure is suggested to aid in determining the primary and secondary air quantities:

1. Divide the area to be conditioned into modules that can be supplied by one or more terminal units. These modules may be as small as a one-man office in a building or as large as a forty pupil classroom in a modern school.

2. Calculate the cooling load for a typical module on each exposure and for nontypical spaces such as top floor and corner rooms.

3. Calculate the A/T ratios *(Example 1)* for the area to be conditioned. These are based on design temperatures, and are found by applying the following formulas. The largest A/T ratio is selected for the design. This ratio is used to select the reheat schedule from *Table 5*.

$$\text{A/T ratio (summer)} = \frac{t_{oas} - t_{rms}}{1.08\,(t_{rms} - t_{pas})}$$

$$\text{A/T ratio (winter)} = \frac{t_{rmw} - t_{oaw}}{1.08\,(t_{paw} - t_{rmw})}$$

where:

t_{oas} = summer outdoor air temperature

t_{rms} = summer room temperature

t_{oaw} = winter outdoor air temperature

t_{rmw} = winter room temperature

t_{pas} = summer primary air temperature, usually taken as 56 F (based on an apparatus dewpoint of 48 F and a temperature rise of 8 degrees for fan and duct heat gain).

t_{paw} = winter primary air temperature, usually taken as 125 F (a suggested limit because of excess duct heat losses at higher temperatures).

Example 1 — Calculation of A/T Ratio

Given:

t_{oas} = 95 F	t_{rmw} = 72 F
t_{oaw} = 0 F	t_{pas} = 56 F
t_{rms} = 75 F	t_{paw} = 125 F

Find:

Design A/T ratio

Solution:

$$\text{A/T ratio (summer)} = \frac{t_{oas} - t_{rms}}{1.08\,(t_{rms} - t_{pas})}$$

$$= \frac{95 - 75}{1.08\,(75 - 56)} = 0.98$$

$$\text{A/T ratio (winter)} = \frac{t_{rmw} - t_{oaw}}{1.08\,(t_{paw} - t_{rmw})}$$

$$= \frac{72 - 0}{1.08\,(125 - 72)} = 1.26$$

The larger design A/T ratio (1.26) is selected.

Example 2 — Calculation of Transmission Per Degree

Given:

Typical modules as shown in *Fig. 27*
U wall = 0.30 Btu/ (hr) (sq ft) (deg F temp diff)
U glass = 1.13 Btu/ (hr) (sq ft) (deg F temp diff)

Find:

Transmission per degree for each of the typical module rooms 1, 2 and 3.

Solution:

Room 1:

glass area = 5 × 8 = 40 sq ft
wall area = (10 × 10) − 40
      = 60 sq ft

transmission per degree
    = (wall area × U wall) + (glass area × U glass)
    = (60 × 0.30) + (40 × 1.13) = 18 + 45.2
    = 63.2 Btu/(hr)(deg F temp diff)

Room 2:

glass area = 5 × 8 = 40 sq ft
wall area = (15 × 10) − 40
      = 110 sq ft

transmission per degree
    = (110 × 0.30) + (40 × 1.13) = 33 + 45.2
    = 78.2 Btu/(hr)(deg F temp diff)

Room 3:

glass area = (10 × 8) + (5 × 8) = 120 sq ft
wall area = [(15 + 15) × 10] − 120
      = (30 × 10) − 120
      = 180 sq ft

transmission per degree
    = (180 × 0.30) + (120 × 1.13) = 54 + 135.5
    = 189.5 Btu/(hr)(deg F temp diff)

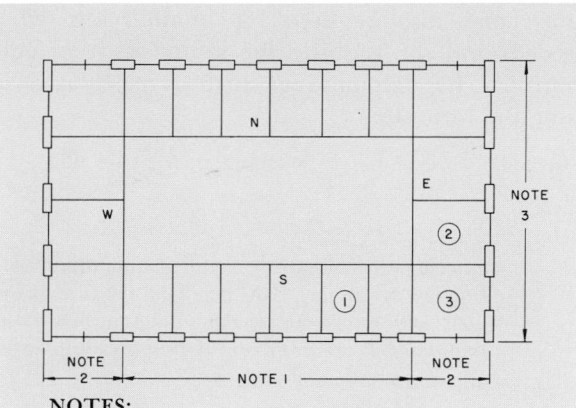

NOTES:
1. Six offices, each 10' wide, with 5' x 8' windows.
2. Office, 15' wide, with 10' x 8' windows.
3. Four offices, each 15' wide, with 5' x 8' windows.

FIG. 27 — TYPICAL FLOOR PLAN

4. Calculate the transmission per degree for each typical module (*Example 2*).
5. Find the primary air quantity for each module (*Example 3*) by multiplying the transmission per degree by the A/T ratio from *Step 3*.
6. Calculate the peak sensible load for each typical module, neglecting the transmission (*Ex-*

ample 4). This is called the secondary load and includes the solar heat gain and the internal loads composed mainly of people and lights.

7. Calculate the secondary air quantity for each module (*Example 5*) by using the formula:

$$cfm_{seca} = \frac{\text{secondary load}}{1.08 \, (t_{rms} - t_{seca})}$$

where:

t_{rms} = **summer room temperature**

t_{seca} = secondary air temperature, usually selected as 55 F. (This allows 5 degrees for duct heat gain and fan heat, using 50 F as the apparatus dewpoint.)

8. The air quantity used to select the primary air fan and apparatus is determined by adding together the primary air quantities for each individual space.

9. The total secondary air quantity used to select the secondary air fan and apparatus is determined by calculating the block estimate secondary load for the entire conditioned area, adding 5% to provide capacity for space overcooling, and substituting this load in the formula from *Step 7*.

Example 3 — Calculation of Primary Air Quantities

Given:

Typical floor plan (*Fig. 27*)

A/T ratio = 1.26

Transmission per degree

Room 1:   63.2 Btu/(hr)(deg F temp diff)

Room 2:   78.2 Btu/(hr)(deg F temp diff)

Room 3: 189.5 Btu/(hr)(deg F temp diff)

Find:

Primary air quantities for each typical space

Solution:

cfm_{pa} = A/T ratio × transmission per degree

Room 1: cfm_{pa} = 1.26 ×   63.2 =   80 cfm

Room 2: cfm_{pa} = 1.26 ×   78.2 =   99 cfm

Room 3: cfm_{pa} = 1.26 × 189.5 = 239 cfm

Example 4 — Calculation of Secondary Loads

Given:

Typical floor plan (*Fig. 27*)

People   load = 100 sq ft per person

Lighting load = 4 watts/sq ft of fluorescent lights

Solar heat gain from *Part I*

Find:

Secondary loads for Rooms 1, 2 and 3

Solution:

Room 1:

people load = people × Btu/(hr)(person)

= 1 × 215 = 215 Btu/hr

light load   = watts/sq ft × sq ft × 1.25 × 3.4

= 4 × 150 × 1.25 × 3.4 = 2550 Btu/hr

solar load   = window area × Btu/(hr)(sq ft)

× storage factor × shade factor

× steel sash factor

= 40 × 166 × 0.79 × 0.56 × 1/0.85

= 3460 Btu/hr

total secondary load

= 215 + 2550 + 3460 = 6225 Btu/hr

Room 2:

people load = 2 × 215 = 430 Btu/hr

light load   = 4 × 225 × 1.25 × 3.4

= 3830 Btu/hr

solar load   = 40 × 164 × 0.73 × 0.56 × 1/0.85

= 3150 Btu/hr

total secondary load

= 430 + 3830 + 3150 = 7410 Btu/hr

Room 3:

people load = 2 × 215 = 430 Btu/hr

light load   = 4 × 225 × 1.25 × 3.4

= 3830 Btu/hr

solar load, south glass

= 80 × 162 × 0.79 × 0.56 × 1/0.85

= 6750 Btu/hr

solar load, east glass

= 40 × 122 × 0.29 × 0.56 × 1/0.85

= 935 Btu/hr

total secondary load

= 430 + 3830 + 6750 + 935 = 11,945 Btu/hr

CENTRAL APPARATUS
Fans

The primary air fan is selected to supply the air quantity at a constant volume and at a static pressure to overcome the duct, apparatus and terminal unit friction losses of the primary air system.

The secondary air fan is a variable volume fan selected to supply the air quantity at a static pressure sufficient to overcome the duct, apparatus and terminal unit friction losses of the secondary air system. The fan should be equipped with variable inlet vanes to throttle the air flow efficiently as the building load is reduced. The fan selected should have a performance curve with a flat, stable portion, and should operate within this portion of its curve.

The fan should be selected so that the discharge static pressure is low enough that no noise is created in a throttled outlet.

The return air fan (if used) is also a variable volume fan and should be equipped with variable inlet vanes operated in conjunction with the volume control of the secondary air fan. This fan must be sized to handle the total return air to the primary and secondary fans. Normally, a return air fan is required in larger buildings, and is ideal for use as a combination return air and exhaust fan.

Dehumidifiers

The primary air dehumidifier is sized to handle the primary air quantity. It may be selected as a spray coil dehumidifier with a 6- or 8-row coil. The

---

Example 5 — Calculation of Secondary Air Quantity

Given:
   Typical floor plan *(Fig. 27)*
   Secondary air load
      Room 1 — 6,225 Btu/hr
      Room 2 — 7,410 Btu/hr
      Room 3 — 11,945 Btu/hr
   Room temperature = 75 F
   Secondary supply air temperature = 55 F

Find:
   Secondary air quantities for Rooms 1, 2 and 3.

Solution:

Room 1: $cfm_{seca} = \dfrac{\text{secondary load}}{1.08\,(t_{rms} - t_{seca})}$

$= \dfrac{6225}{1.08\,(75 - 55)} = 288$ cfm

Room 2: $cfm_{seca} = \dfrac{7410}{1.08\,(75 - 55)} = 343$ cfm

Room 3: $cfm_{seca} = \dfrac{11,945}{1.08\,(75 - 55)} = 553$ cfm

---

sprays may also be used for humidifying (when needed) and for washing the air to assist in odor control. The dehumidifier load is calculated by using the formula:

Dehumidifier load $= cfm_{pa} \times 4.45\,(h_{ea} - h_{adp})(1 - \text{BF})$

where:
cfm_{pa} = primary air quantity
h_{ea} = entering air enthalpy to the dehumidifier on a summer design day. This may be a mixture of outdoor and return air if the minimum ventilation requirements are less than the primary air quantity.
h_{adp} = apparatus dewpoint enthalpy
BF = bypass factor

The secondary air dehumidifier is sized to handle the maximum secondary air quantity. It is usually selected with a bypass factor of approximately 0.1. The dehumidifier load is calculated by using the formula:

Dehumidifier load $= cfm_{seca} \times 4.45\,(h_{ea} - h_{adp})(1 - \text{BF})$

where:
cfm_{seca} = secondary air quantity
h_{ea} = entering air enthalpy
h_{adp} = apparatus dewpoint enthalpy
BF = bypass factor

Filters

Any commercially acceptable filter may be used with these systems, depending on the degree of filtration required. The filters selected for the secondary air system must operate with a varying velocity and volume of air.

Heating Coils

The primary air preheater is sized to handle the minimum ventilation air quantity and must have a capacity equal to that found by the formula:

Preheater capacity $= cfm_{oa} \times 1.08\,(50 - t_{oaw})$

where:
cfm_{oa} = outdoor air quantity required for ventilation
t_{oaw} = winter outdoor air temperature

The primary air reheater is sized to handle the primary air quantity and must have a capacity equal to that found by the formula:

Reheater capacity $= cfm_{pa} \times 1.08\,(t_{sa} - t_{ea} + 15)$

where:
cfm_{pa} = primary air quantity
t_{sa} = supply air temperature determined from *Table 1* at the minimum outdoor air temperature.
t_{ea} = entering air temperature to the reheater
15 = an allowance for duct heat loss and for a quick warm-up after a prolonged shutdown, such as occurs on weekends or nights.

Air Louvers, Screens and Dampers

Primary outdoor air louvers, screens and dampers are sized for minimum ventilation air requirements. Primary return air dampers are sized for the primary air quantity. The secondary outdoor air louvers, screens and dampers are sized for the maximum secondary air quantity. The secondary return air dampers are sized for the maximum secondary air quantity.

REFRIGERATION LOAD

Any of the three basic refrigeration cycles, absorption, centrifugal or reciprocating, may be considered for the refrigeration systems. Either chilled water or direct expansion cooling can be used. When direct expansion is used, the possibility of operating two separate systems at different temperature levels can be considered, one for the primary air and one for the secondary air system.

The refrigeration load is the sum of the primary and secondary dehumidifier loads. In addition, other loads, i.e. from interior zones, must be included.

DUCT DESIGN

Generally the supply ducts of both the primary and secondary systems are designed as high velocity systems. Static regain is used for sizing ducts to provide a more stable system and to reduce fan horsepower. Sound absorbers in the fan discharge are normally required to reduce noise generated by the fan. When selecting and locating fittings, care must be taken to avoid excessive pressure drops and noise generation. Round duct is generally used for the duct systems. *Part 2* should be consulted for design of the duct distribution system.

PIPING DESIGN

There is very little piping design connected with this system since the piping is concentrated in the apparatus room. *Part 3* may be used as a guide for the details and sizing.

INSULATION

Insulation is recommended for both air systems to prevent excessive heat gain or loss. Vapor sealed insulation is used on ducts located outside the conditioned areas. Duct insulation without a vapor seal is used within conditioned areas. Weatherproofing is required on ducts exposed to the outdoors.

CONTROLS

A basic pneumatic control arrangement is shown in *Fig. 28.*

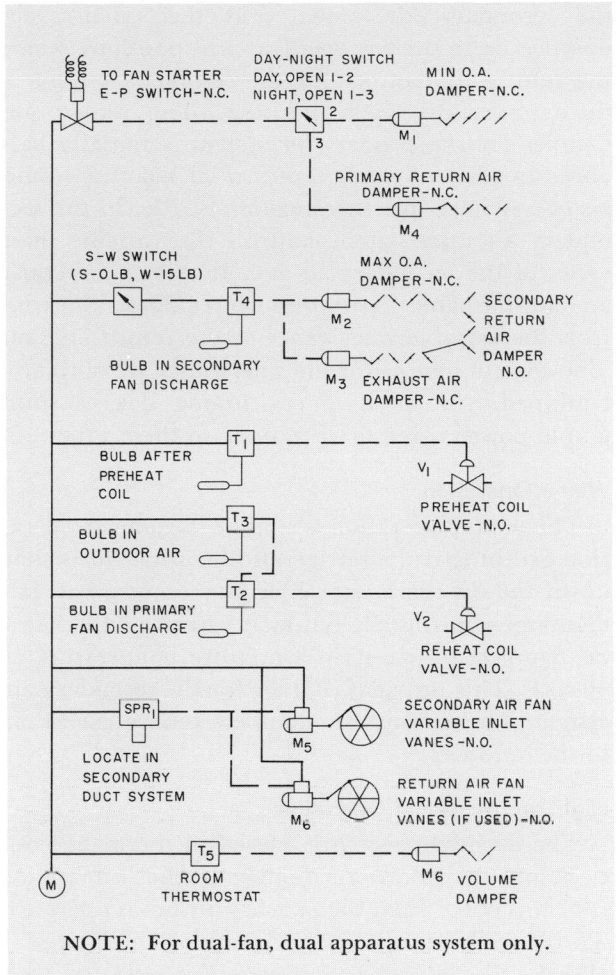

NOTE: For dual-fan, dual apparatus system only.

FIG. 28 — DUAL CONDUIT SYSTEM CONTROL
TYPICAL PNEUMATIC ARRANGEMENT

CENTRAL APPARATUS CONTROLS

Either electric or pneumatic controls may be used for the central apparatus and the sequence of operation is identical.

Summer Operation

During summer the day-night switch is left in the day position. The outdoor air damper M_1 in the primary air system opens when the primary air fan is started. The secondary air fan and return air fan are then started, as well as the refrigeration cycle which provides cooling to the dehumidifiers. The primary air reheat coil valve V_2 is controlled by a fan discharge thermostat T_2 which is reset from an outdoor air thermostat T_3 located outside the outdoor air dampers but protected from the sun. The preheat coil valve V_1 in the primary air system is controlled by a thermostat T_1 located immediately after the preheat coil. The outdoor air damper for the secondary air system and the exhaust air damper are in their normally closed positions, while the return air damper for the secondary system is in its normally open position. The return air damper for the primary air system is initially balanced to admit the design return air quantity to the system. A static pressure regulator SPR_1 in the secondary air duct system controls the variable inlet vanes of the secondary air fan. If there is a return air fan, the same static pressure regulator also controls the variable inlet vanes of the return air fan. The control motors for the inlet vanes are normally equipped with positive positioning devices since ample power is required to operate these vanes.

Winter Operation

The winter operation is similar to summer operation except that the refrigeration equipment is shut down and the thermostat T_4 in the secondary air fan discharge controls the outdoor, return and exhaust air dampers to maintain a mixture temperature of 50-55 F. This provides cool air for the secondary air system. The exhaust air dampers relieve excess air to the outdoors.

Night and Weekend Operation

The day-night switch is placed in the night position, and the return air damper in the primary air system is open when the primary air fan is operated. To reduce heating costs, the outdoor air damper is closed and the primary air system is operated with only return air. The secondary air fan is not operated during these periods. Economical operation can be obtained by controlling the primary air fan from a thermostat which is located in a typical space and which cycles the fan to maintain a minimum temperature in the building. See *Fig. 29* for control diagram.

UNIT CONTROLS

Either electric, pneumatic or self-contained controls may be used for the unit control.

The only control required in the room is a thermostat to operate the damper in the secondary system to modulate the secondary air as the load changes.

SYSTEM MODIFICATIONS

This section enumerates certain variations that can be incorporated in a Dual Conduit System. These variations adapt the system for specific applications or for certain requirements such as a low first cost or low operating cost.

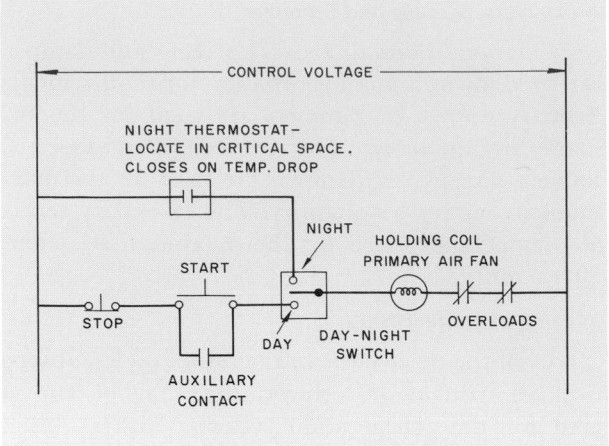

Fig. 29 — Electrical Control Diagram, For Night Fan Control

OTHER APPARATUS ARRANGEMENTS

An arrangement with one apparatus and two fans is shown in *Fig. 30*, and another with one apparatus and one fan is shown in *Fig. 31*. These arrangements have a lower first cost; however, they generally sacrifice ventilation air at partial loads, and have slightly higher operating costs. Their operation is similar to the arrangement shown in *Fig. 26*.

INTERIOR SPACES

In addition to serving as the source of secondary air for the Dual Conduit System, the secondary air apparatus can be increased in size to serve the interior zone. This can be arranged with either a variable volume air system or a constant volume air system. When used with a constant volume system, space terminal units must be carefully selected since the air quantity serving the interior zone is increased at partial load on the exterior spaces. This arrangement can be applied to buildings having many open areas with few private offices.

SERIES WATER FLOW

When two separate dehumidifiers are operated at different apparatus dewpoints, it may be practical to connect their water sides in series. This may mean a saving in pipe and pump size as well as a reduction in first cost of the refrigeration machine required.

AIR TERMINAL UNIT LEAKAGE

An examination of the secondary air terminal unit is required to determine if a tight shutoff can be accomplished. Some terminal units do not close off tightly and, therefore, tend to overcool the space at minimum loads.

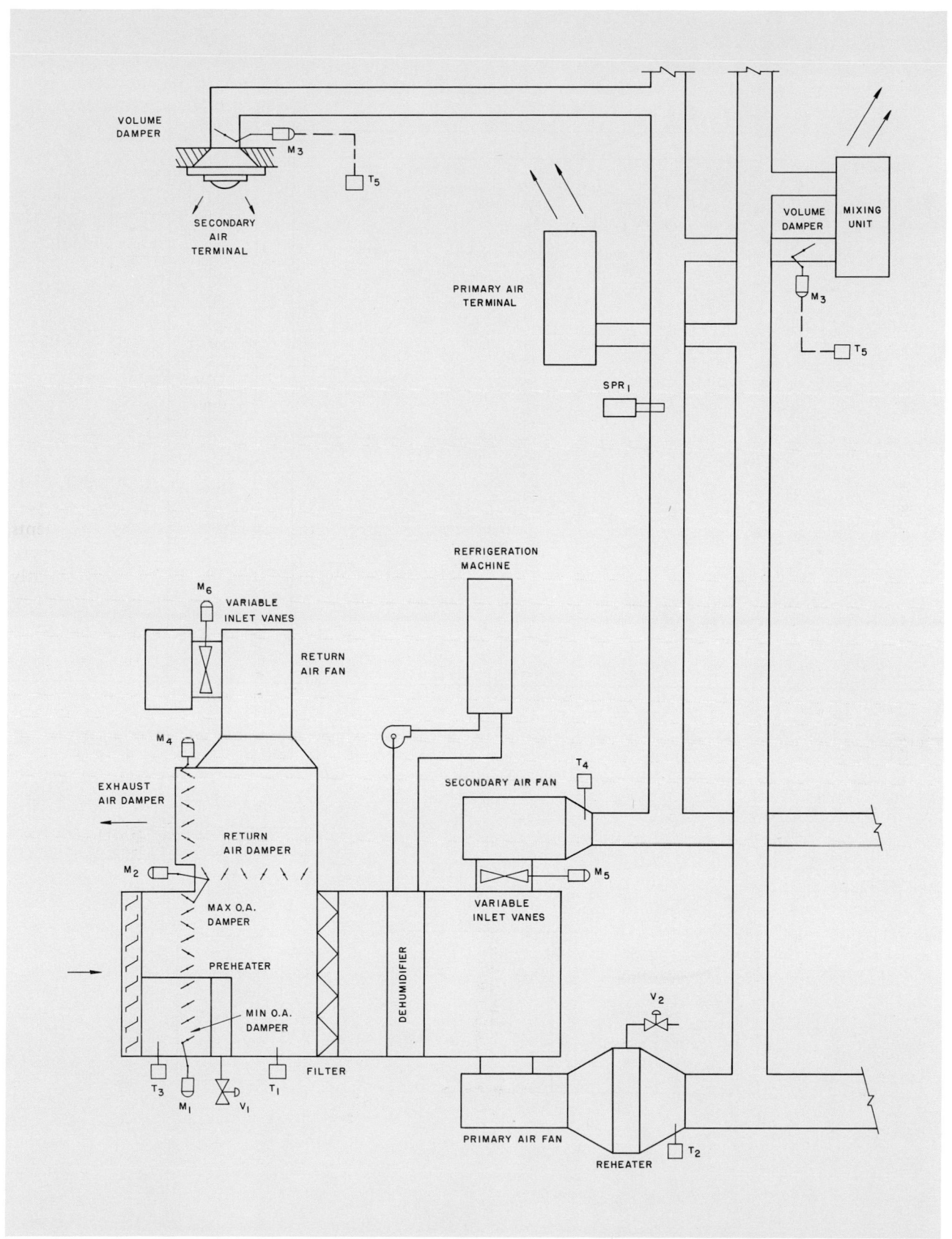

FIG. 30 — TYPICAL DUAL CONDUIT SYSTEM, DUAL-FAN, SINGLE APPARATUS

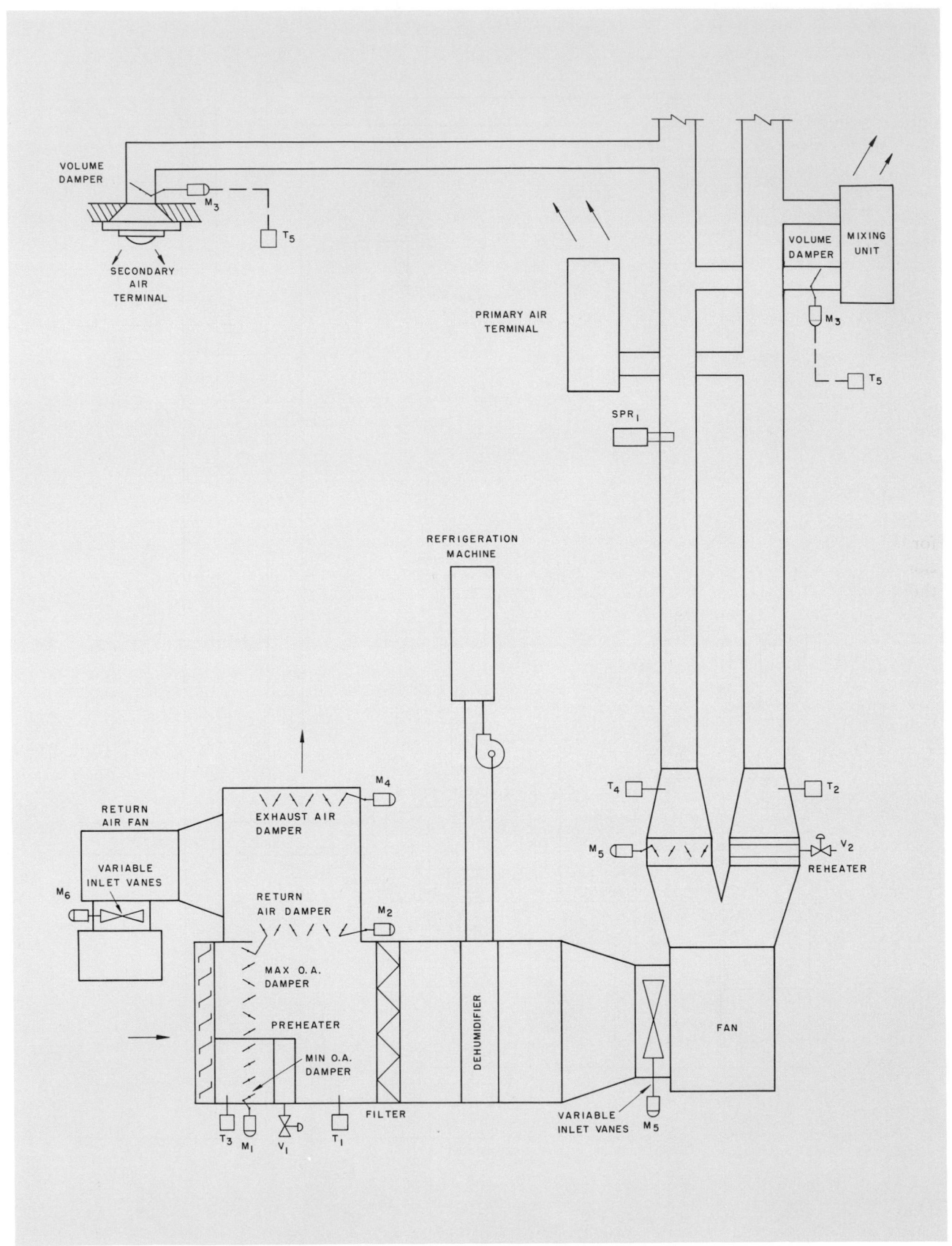

FIG. 31 — TYPICAL DUAL CONDUIT SYSTEM, SINGLE FAN, SINGLE APPARATUS

For example, if the secondary air quantity is 85% of the total air supplied to the space and the secondary terminal unit has a minimum leakage of 15%, the minimum cooling available from the secondary terminal unit is 15% of 85% or 12.8%. This 12.8% plus the primary air capacity of 15% means that the minimum load in the room must not be below 27.8% (12.8 + 15) of full load if overcooling is to be prevented. This condition may occur when the outdoor temperature is near the indoor temperature, when the space is unoccupied with the lights off, and when there is reheat in the primary air system. However, if this condition occurs in an isolated case, transmission thru the walls, floor, etc. to surrounding spaces tends to offset the overcooling capacity of the terminal units. If overcooling becomes a problem, it may be prevented by keeping the lights on when the air conditioning is operating during these periods.

VENTILATION THRU THE SECONDARY AIR SYSTEM

The system may be designed to use all return air for the primary air apparatus and to supply ventilation air thru the secondary air system. This permits the use of only one outdoor air intake, and saves on costs for heating the primary air.

OPERATING ECONOMY

In winter, greater economy may be obtained by operating the primary apparatus with return air only to reduce the energy required to heat the air.

Outdoor air for ventilation is available from the secondary air system because it is necessary to provide the cooling capacity for the internal loads by using outdoor air when the refrigeration equipment is shut down.

RETURN AIR IN A CEILING PLENUM OF SINGLE STORY BUILDINGS

The roof transmission load in an interior space of a single story building with air returned thru a ceiling plenum requires special attention. Since the return air passing thru the plenum above the room can be a variable amount due to throttling of the secondary air quantity, and since it can vary because of the location of the room in relation to the apparatus, the amount of roof transmission load which is picked up by the return air may vary between rooms. Also, portions of the solar gain and light load (if recessed lights are used) may be offset in varying amounts by this return air.

When determining the primary air quantity, only 33% of the roof transmission is considered as an effective load. Therefore, the transmission per degree used for finding the primary air quantity is only one third of the actual calculated value. This transmission per degree is multiplied by the calculated A/T ratio to find the primary air quantity.

The secondary air load which is used to calculate the secondary air quantity is determined by adding 33% of the solar load, the light load (reduced somewhat if recessed lights are used), and the people load, all obtained in the same manner as for the basic system. The remaining portions of these loads must be added to the dehumidifier, resulting in no savings in refrigeration load but a reduction of required air quantities.

DIRECT EXPANSION COOLING

The primary and secondary air apparatus may be serviced by separate direct expansion refrigeration systems for cooling and dehumidifying the air. This arrangement allows the greatest operating economies when the secondary dehumidifier is selected at a higher apparatus dewpoint than that of the normal system. However, secondary air quantities are larger than on the normal system, thus requiring larger ductwork. This feature of separate refrigeration systems offers the additional advantage of operating only the primary direct expansion system when the outdoor temperature is below that required for the secondary air system.

AIR TERMINAL UNITS

These units can be located as illustrated in *Fig. 32*, depending on individual building requirements.

SEPARATE AIR TERMINAL UNITS

Primary Air Terminal Unit

These units may be any conventional or high pressure outlet complete with balancing damper, sound absorbing lining, and pressure reducing device. Air may be distributed from under the window, the ceiling or side wall. In northern climates where the winter design temperature is less than 20 F, air should be discharged from under the window to offset downdrafts.

Secondary Air Terminal Unit

This unit may distribute air from the ceiling, side wall or under the window. It must be capable of controlling the volume of conditioned air supplied to the space while at the same time maintaining a reasonably uniform and draftless air distribution. It should be complete with means for sound attenuation and a method of volume regulation. The damper may operate from a self-contained

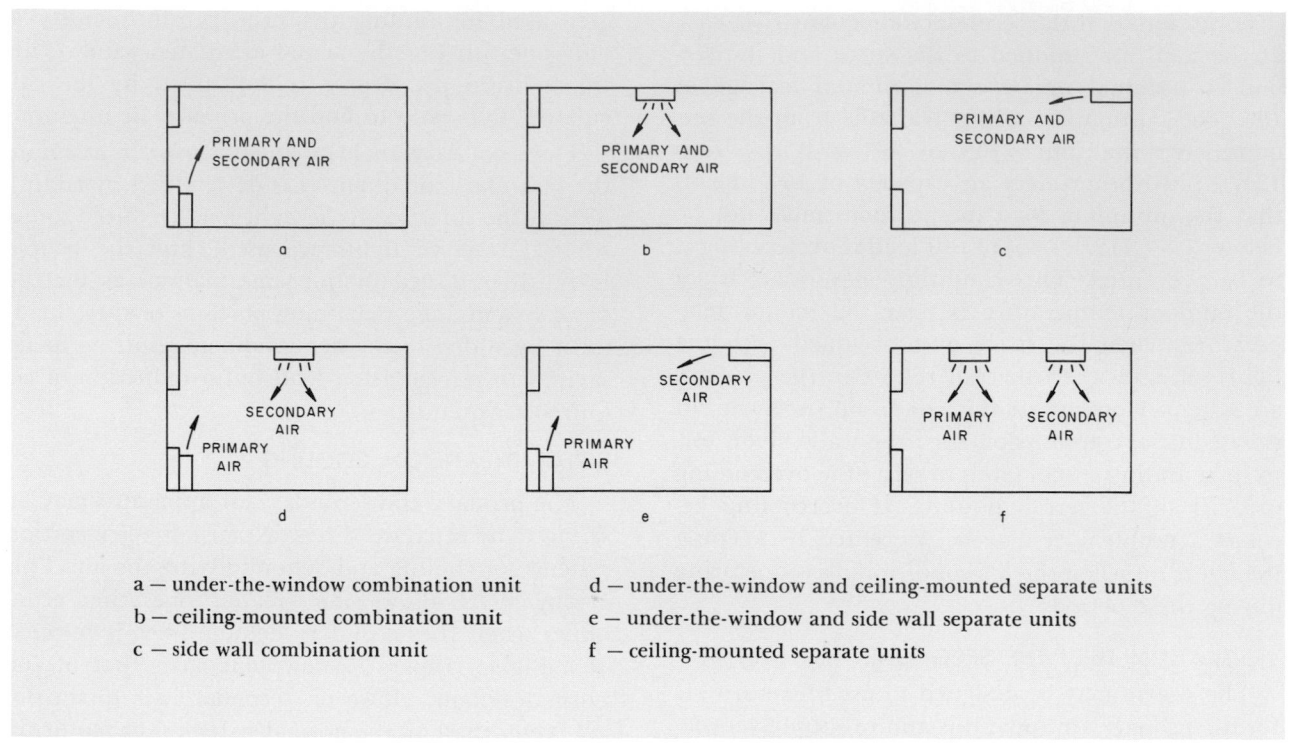

a — under-the-window combination unit   d — under-the-window and ceiling-mounted separate units
b — ceiling-mounted combination unit   e — under-the-window and side wall separate units
c — side wall combination unit   f — ceiling-mounted separate units

FIG. 32 — TERMINAL UNIT LOCATIONS

thermostat mounted on the unit or from pneumatic or electric thermostats mounted on a wall.

COMBINATION AIR TERMINAL UNITS

Mixing Unit

An air terminal may be used which mixes primary and secondary air before discharging into the room. A throttling damper is located in the secondary air plenum, and a balancing damper is located in the primary air duct. The mixing chamber is lined with sound attenuation material, and air is discharged to the room thru a single discharge. Care must be taken when designing this arrangement since air distribution is affected when the secondary air is throttled because of a difference in the outlet velocity. The unit can be located to distribute air from the ceiling, side wall or under the window.

DESIGN SUMMARY

The following is a summary of design guides for the Dual Conduit System:

1. Primary air is supplied to each space proportionally to the transmission per degree for the space.

2. The function of the primary air is to offset the transmission loads and the latent loads.

3. The A/T ratio for calculating primary air quantities is usually between 0.5 and 2.0.

4. Maximum suggested primary air temperature at the terminal unit is usually 125 F because the duct heat losses become excessive with higher temperatures.

5. Summer design primary air temperature generally used is 56 F, based on a 48 F adp and an 8 degree rise for the fan and duct heat gain.

6. Secondary air is supplied to each space at a constant temperature and a varying volume.

7. The function of the secondary air is to offset the sensible heat loads of the sun, lights and people.

8. Design secondary air temperature is generally 55 F, based on a 50 F adp and a 5 degree rise due to the fan and duct heat gain.

9. Secondary air temperature is maintained during intermediate and winter seasons by mixing the outdoor and return air.

10. Secondary air fan and the return air fan must be equipped with a method of volume control such as variable inlet vanes.

11. The secondary air fan should have a performance curve with a flat stable portion, and should operate within this portion.

12. Primary air dehumidifier should be the spray coil type to provide humidification and help with odor control.

13. A primary air reheater is selected to heat the total primary air to at least 15 degrees above the maximum supply air temperature.

14. Supply ductwork for both the primary and secondary air systems should be insulated.

15. The variable volume secondary air terminal unit must be capable of maintaining adequate air motion with a reduced quantity of conditioned air.

16. In northern climates where winter design conditions are below 20 F, the primary air should be distributed from under the window to offset downdrafts.

17. In single story buildings with the return plenum in the ceiling, only 33% of the roof transmission load should be used to calculate the primary air quantity.

Part 11
AIR-WATER SYSTEMS

CHAPTER 1. INDUCTION UNIT SYSTEM

The induction unit system is designed for use in perimeter rooms of multi-story, multi-room buildings such as office buildings, hotels, hospital patient rooms and apartments. Specifically, it is designed for buildings that have reversing sensible heat characteristics in which cooling may be required in one room and heating may be required in an adjacent room. In addition, it is especially adapted to handle the loads of modern skyscrapers with minimum space requirements for mechanical equipment.

This chapter includes System Description, System Features, Design Considerations, Engineering Procedure for designing the system, Controls and System Modifications.

SYSTEM FEATURES

Some of the features of the induction unit system are the following:

1. *Small Space Requirements* — The use of water to provide a major portion of the room cooling requirements reduces the air quantity distributed to each space when compared to the air quantity distributed in an all-air system. Thus, less space is required for both the air distribution system and the central air handling apparatus. At the same time adequate and constant room air circulation is maintained by the high induction of the secondary air from the room. In addition, the smaller primary air quantity can be distributed at high velocity without any increase in power requirements over systems using much more air with conventional distribution ductwork.

2. *Individual Room Control* — Zoning problems are solved since each room is a zone. Simultaneous heating or cooling is available in adjacent rooms when required.

3. *Winter Downdraft Eliminated* — The unit design permits under-the-window installation, eliminating downdrafts at the windows during severe winter weather.

4. *Minimum Service* — No individual fans or motors to create maintenance or servicing in the rooms. Most maintenance is centralized.

5. *Central Dehumidification* — Since all dehumidification occurs at the central apparatus, condensation on the room unit coil is eliminated. Thus, odor retention and corrosion problems at the unit are minimized.

6. *Quiet Operation* — All fans and other rotating equipment are remotely located.

SYSTEM DESCRIPTION

A typical induction unit system is shown in *Fig. 1.* Although the arrangement may differ for each application, this illustration includes the basic components common to most induction unit systems. The following description is for a nonchangeover system.

Outdoor air for ventilation is drawn into the central apparatus thru a louver, screen and damper. Return air may be introduced into the system if the total required primary air quantity is more than the minimum ventilation requirement. The preheater tempers the air in winter to increase the capacity of the air to absorb moisture and to prevent freezing air from entering the dehumidifier. The filters remove entrained dust and dirt particles from the air. The sprayed coil dehumidifier cools and dehumidifies the air during the warm weather; during cool weather the recirculation sprays may be used to add moisture to the air. The reheater heats the air to offset the building transmission losses. The high pressure fan delivers the conditioned air thru high velocity ducts to the induction units. A sound absorber on the leaving side of the fan is normally required to reduce the noise generated by the fan. Chilled water from a central refrigeration plant is circulated by the primary pump thru the dehumidifier coils in the apparatus. The secondary water pump circulates water to the induction unit coils.

The induction unit (*Fig. 2*) is supplied with high pressure primary air which is discharged within the unit thru nozzles. This air induces room air across the coil which is supplied with water from the secondary water pump. The induced air is heated or cooled depending on the temperature of the secondary water, and the mixture of primary air and induced air is discharged to the room.

The function of the primary air is to provide ventilation air, to offset the transmission loads, to provide dehumidification to offset the latent loads, and to provide the motivating force for induction and circulation of room air. The secondary water circuit functions to offset the heat gain from sun, lights and people. The primary air is tempered according to a reheat schedule to prevent the room temperature from falling below 72 F when there is a minimum load in the room.

On some applications it may be desirable to operate the system during the winter season with

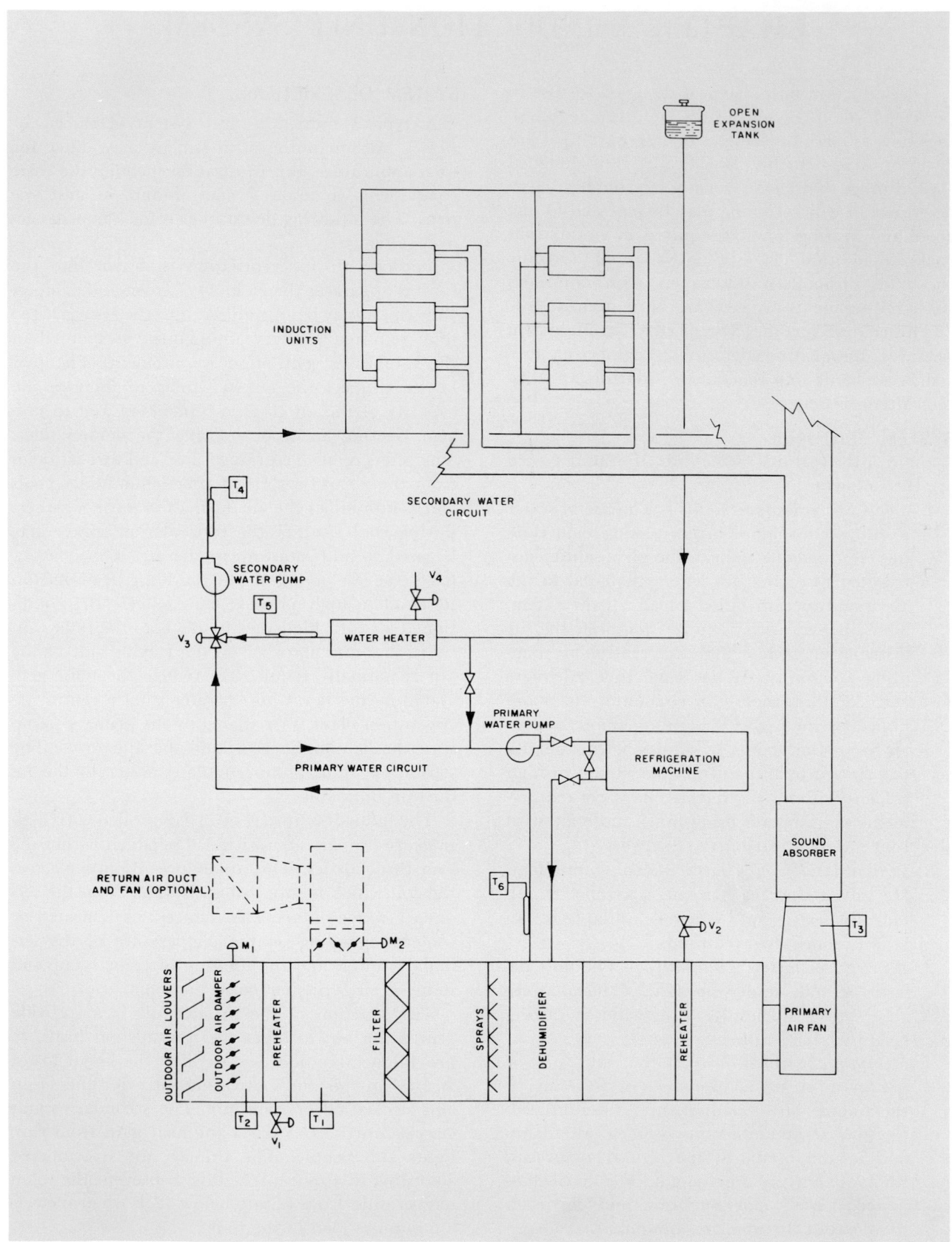

FIG. 1. — TYPICAL INDUCTION UNIT SYSTEM

hot water supplied to the coil and cold primary air. This is known as the change-over system which is explained under System Modifications.

SYSTEM DESIGN CONSIDERATIONS

AIR/TRANSMISSION RATIO

The design and the operation of an induction unit system is based on the Air/Transmission (A/T) ratio concept. It is important that this concept be fully understood.

Definition

The A/T ratio is the ratio of the unit primary air quantity (cfm) to the total room transmission per degree thru the exterior areas of the space served by the unit. Transmission per degree is determined by assuming a steady state heat flow; it is calculated for one degree of temperature difference across the outdoor walls, windows and roof. No credit is taken for storage effect since the effect of only the outdoor temperature is analyzed, regardless of the solar load. *Figure 3* illustrates an example of the calculation of the A/T ratio.

Function

The primary air cooling and heating capacity is varied to offset the effects of the transmission portion of the room load by reheat, scheduled in accordance with the outdoor dry-bulb temperature.

For each A/T ratio, there is a fixed reheat schedule which is calculated to prevent any room from having a temperature below 72 F with a minimum room load equivalent to 10 degrees multiplied by the transmission per degree for the room. When the load caused by sun, lights and people diminishes, the coil capacity is reduced to compensate for this lack of load. When these loads become a minimum, the design minimum room temperature is maintained by the controlled primary air temperature which offsets the transmission load.

Room humidity is not affected by the reheating of the primary air since this only adds sensible heat and the latent heat capacity remains unchanged.

Air Zoning

Units in spaces which have the same exposure or a similar loading may be grouped together to form a zone. All units within each zone must have the same A/T ratio so that the primary air may be reheated on a single schedule by an individual heater. The units must be located so they can be supplied from their heater with a minimum amount of duplication of ductwork. The purpose of air zoning is to provide a means of reducing the total amount of primary air.

Use

By surveying the values of the transmission per degree and the air quantities necessary to satisfy the room loads for a given building, a base A/T ratio can be found. This base A/T ratio may then be used to determine the primary air quantity for all other units in the zone and to establish the reheat schedule for the units. The primary air

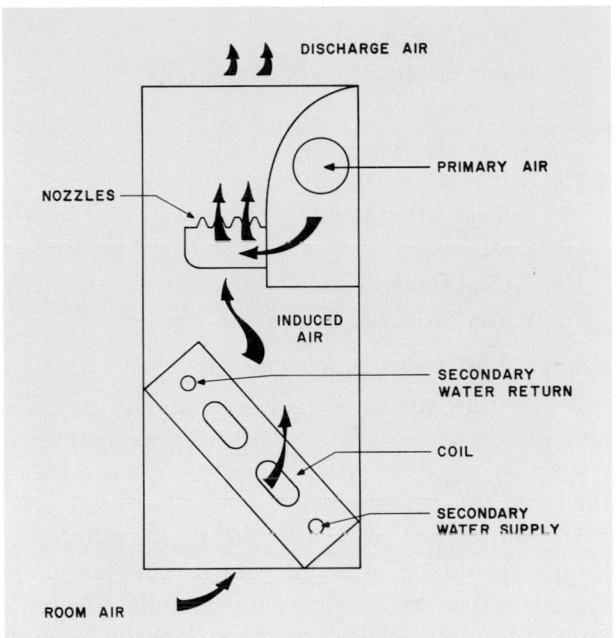

Fig. 2 — Typical Induction Unit

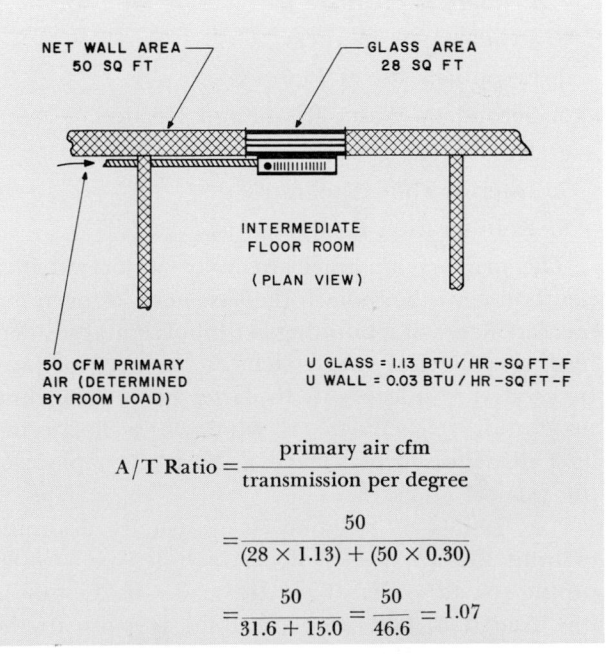

$$A/T \text{ Ratio} = \frac{\text{primary air cfm}}{\text{transmission per degree}}$$

$$= \frac{50}{(28 \times 1.13) + (50 \times 0.30)}$$

$$= \frac{50}{31.6 + 15.0} = \frac{50}{46.6} = 1.07$$

Fig. 3 — Calculation of A/T Ratio

quantity determined by using the base A/T ratio may be higher for some units than necessary if the units are selected to satisfy only the room sensible heat load.

ENGINEERING PROCEDURE

The following is an engineering procedure for designing an induction unit system:

1. Survey
2. Preliminary Layout
3. Room Cooling Load Calculations
4. Unit Selection
5. Room Heating Load Calculation
6. Apparatus Selection
7. Refrigeration Load
8. Duct Design
9. Piping Design
10. Water Heater Selection

SURVEY AND PRELIMINARY LAYOUT

An accurate survey of the load components, available space and services is a basic requirement for a system design. Refer to *Part 1* for a complete list of the items to be considered.

In conjunction with the survey a preliminary layout should be made. Consideration should be given to the arrangement and number of units around the building perimeter and to the location of the following items:

1. Primary Air Risers
2. Primary Air Apparatus
3. Primary Air Headers
4. Secondary Water Pump(s)
5. Secondary Water Risers and Headers
6. Return Air System (if used)
7. Interior Zone Apparatus
8. Refrigeration Equipment

The primary air apparatus may be located in a penthouse on the roof, in the basement, or on intermediate floors of a building, with horizontal headers feeding a vertical riser system at the exterior face. In modern buildings with large glass areas and little wall space between windows, a horizontal duct distribution may have to be used in place of the vertical risers.

For reasons of economy it is usually desirable to limit the number of floors included in a water piping system so that the static head plus the pumping head does not cause the total pressure in the system to exceed that allowable for standard weight piping and fittings.

The location of interior zone equipment should be considered with respect to the chilled water piping if the equipment is to obtain its refrigeration from the same central source as the induction unit system. Return air (if used) for the primary air apparatus may be taken thru the interior zone return air system for reasons of economy. Therefore, interior zone equipment should be located close to the primary air apparatus.

Location of the refrigeration machine in relation to its condenser water source (cooling tower, etc.) and the primary air apparatus (*Fig. 4*) may depend on these economic factors:

1. Insulated chilled water piping vs. condenser water piping costs.
2. Electric wiring vs. water piping costs.

However, when locating a given type of refrigeration equipment (centrifugal, absorption or reciprocating), there may be special engineering considerations involved such as structural reinforcement and vibration isolation on upper floors.

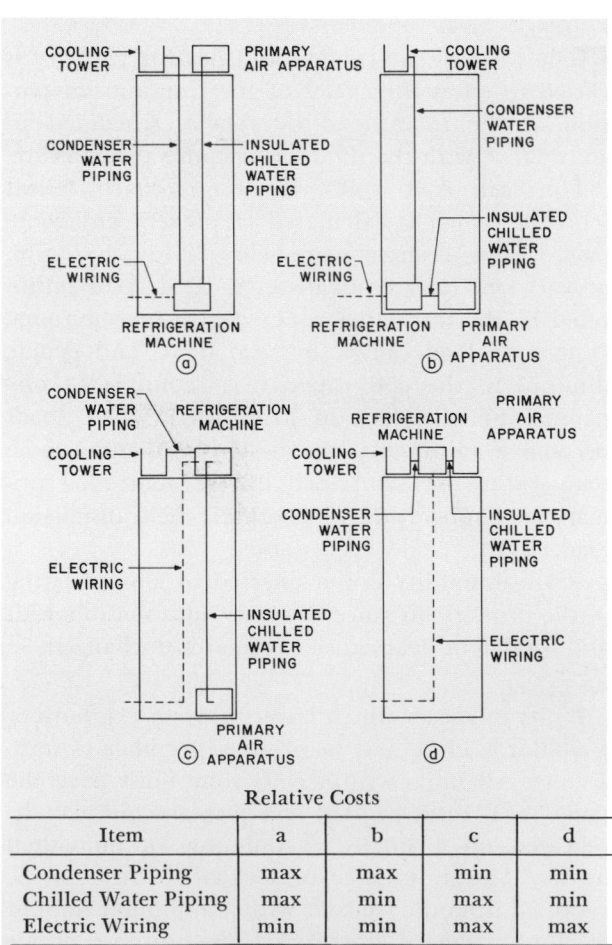

Relative Costs

Item	a	b	c	d
Condenser Piping	max	max	min	min
Chilled Water Piping	max	min	max	min
Electric Wiring	min	min	max	max

FIG. 4 — RELATIVE LOCATIONS
AND COSTS OF PIPING AND WIRING

BASIS OF TABLE 1

The data in *Table 1* is based on these conditions:

Outdoor design dry-bulb temperature	= 95 F
Room design dry-bulb temperature	= 75 F
Daily temperature range	= 20 deg

Yearly temperature range, 20° latitude = 75 deg
30° latitude = 85 deg
40° latitude = 100 deg
50° latitude = 115 deg

Light construction, wall — 40 lb/sq ft wall area
roof — 20 lb/sq ft roof area
room — 30 lb/sq ft floor area

TABLE 1—ROOM DESIGN LOAD FACTORS

EXPOSURE	NORTH				NORTHEAST*				NORTHEAST				EAST				SOUTHEAST			
LATITUDE (North)	20	30	40	50	20	30	40	50	20	30	40	50	20	30	40	50	20	30	40	50
LIGHT CONSTRUCTION																				
1. Design Sun Time — Month / Hour	June 5 p.m.				June 7 a.m.	July 7 a.m.			June 8 a.m.	July 8 a.m.			July 9 a.m.				Oct. 9 a.m.	Sept. 9 a.m.		
2. Outdoor Design Dry-bulb (F)	92	90	90	89	79	80	80	80	80	81	81	81	83	83	83	83	76	78	77	75
3. Solar Heat Gain Thru Glass Btu/(hr)(sq ft)																				
12-hour Operation	15	11	10	9					65	55	54	49	74	75	75	74	70	66	71	71
16-hour Operation	15	11	9	9					63	54	52	48	72	73	73	72	67	64	68	69
24-hour Operation	15	11	9	9	66	56	54	50	63	54	52	48	72	73	73	72	67	64	68	69
4. Equivalent Temp Diff (F)																				
Glass	17	15	15	14	4	5	5	5	5	6	6	6	8	8	8	8	1	3	2	0
Wall	16	14	14	13	13	12	12	11	17	15	15	14	32	32	32	32	19	20	21	19
Roof	49	47	45	42	3	4	4	4	2	3	3	3	4	4	4	4	-3	-1	-2	-4
MEDIUM CONSTRUCTION																				
1. Design Sun Time — Month / Hour	June 5 p.m.				June 7 a.m.	July 7 a.m.			June 8 a.m.	July 8 a.m.			July 9 a.m.				Oct. 9 a.m.	Sept. 9 a.m.		
2. Outdoor Design Dry-bulb (F)	92	90	90	89	79	80	80	80	80	81	81	81	83	83	83	83	76	78	77	75
3. Solar Heat Gain Thru Glass Btu/(hr)(sq ft)																				
12-hour Operation	14	11	9	9					55	47	46	42	64	64	64	64	63	60	63	64
16-hour Operation	13	10	9	8					53	45	43	40	61	62	62	61	59	56	60	60
24-hour Operation	13	10	9	8	52	44	43	39	49	42	41	37	57	58	58	57	55	52	55	56
4. Equivalent Temp Diff (F)																				
Glass	17	15	15	14	4	5	5	5	5	6	6	6	8	8	8	8	1	3	2	0
Wall	9	7	7	6	7	8	8	8	9	9	9	9	13	13	13	13	6	7	7	5
Roof	46	44	42	39	8	8	8	7	6	7	7	6	8	8	8	8	0	2	1	-2
HEAVY CONSTRUCTION																				
1. Design Sun Time — Month / Hour	June 5 p.m.				June 7 a.m.	July 7 a.m.			June 8 a.m.	July 8 a.m.			July 9 a.m.				Oct. 9 a.m.	Sept. 9 a.m.		
2. Outdoor Design Dry-bulb (F)	92	90	90	89	79	80	80	80	80	81	81	81	83	83	83	83	76	78	77	75
3. Solar Heat Gain Thru Glass Btu/(hr)(sq ft)																				
12-hour Operation	14	11	9	8					54	46	44	41	61	62	62	61	63	60	63	64
16-hour Operation	13	10	8	8					51	43	42	39	58	59	59	58	59	56	60	60
24-hour Operation	13	10	8	8	50	42	41	38	47	40	38	35	54	54	54	54	53	50	53	54
4. Equivalent Temp Diff (F)																				
Glass	17	15	15	14	4	5	5	5	5	6	6	6	8	8	8	8	1	3	2	0
Wall	7	5	5	4	10	10	10	10	11	11	11	11	14	14	14	14	8	10	9	7
Roof	44	42	41	37	14	14	13	12	11	11	11	10	13	13	12	11	4	6	4	9

*These factors are used for systems required to operate 24 hours continuously, as in hospitals, hotels and apartment houses.

BASIS OF TABLE 1 (Cont'd)

Medium construction,   wall — 100 lb/sq ft wall area
                         roof — 40 lb/sq ft roof area
                         room — 100 lb/sq ft floor area

Heavy construction,   wall — 140 lb/sq ft wall area
                      roof — 60 lb/sq ft roof area
                      room — 150 lb/sq ft floor area

Standard single-glazed double hung windows with venetian blinds.

TABLE 1—ROOM DESIGN LOAD FACTORS (Contd.)

EXPOSURE	SOUTH				SOUTHWEST				WEST				NORTHWEST			
LATITUDE (North)	20	30	40	50	20	30	40	50	20	30	40	50	20	30	40	50
LIGHT CONSTRUCTION																
1. Design Sun Time — Month / Hour	Nov. Noon	Oct. Noon			Oct. 3 p.m.	Sept. 3 p.m.			July 4 p.m.				June 5 p.m.	July 5 p.m.		
2. Outdoor Design Dry-bulb (F)	75	74	74	70	88	90	89	87	94	94	94	94	92	93	93	93
3. Solar Heat Gain Thru Glass Btu/(hr) (sq ft)																
12-hour Operation	71	73	81	84	78	74	79	80	75	75	75	75	69	59	57	52
16-hour Operation	70	72	80	82	77	73	78	79	74	75	75	74	69	59	57	52
24-hour Operation	70	72	80	82	77	73	78	79	74	75	75	74	69	59	57	52
4. Equivalent Temp Diff (F)																
Glass	0	−1	−1	−5	13	15	14	12	19	19	19	19	17	18	18	18
Wall	23	22	26	23	32	33	34	32	38	38	38	38	34	33	32	31
Roof	16	15	11	4	32	34	30	25	48	48	46	44	49	50	48	45
MEDIUM CONSTRUCTION																
1. Design Sun Time — Month / Hour	Nov. 2 p.m.	Oct. 2 p.m.			Oct. 3 p.m.	Sept. 3 p.m.			July 4 p.m.				June 5 p.m.	July 5 p.m.		
2. Outdoor Design Dry-bulb (F)	79	78	78	74	88	90	89	87	94	94	94	94	92	93	93	93
3. Solar Heat Gain Thru Glass Btu/(hr) (sq ft)																
12-hour Operation	61	63	70	72	69	66	70	70	66	66	66	66	60	51	50	46
16-hour Operation	55	56	63	65	63	60	63	64	60	61	61	60	56	48	46	43
24-hour Operation	55	56	63	65	63	60	63	64	60	61	61	60	56	48	46	43
4. Equivalent Temp Diff (F)																
Glass	4	3	3	−1	13	15	14	12	19	19	19	19	17	18	18	18
Wall	12	12	15	11	13	14	14	12	17	17	17	17	14	14	14	14
Roof	13	12	8	9	30	33	29	23	45	44	43	41	46	47	45	42
HEAVY CONSTRUCTION																
1. Design Sun Time — Month / Hour	Nov. 2 p.m.	Oct. 2 p.m.			Oct. 3 p.m.	Sept. 3 p.m.			July 4 p.m.				June 5 p.m.	July 5 p.m.		
2. Outdoor Design Dry-bulb (F)	79	78	78	74	88	90	89	87	94	94	94	94	92	93	93	93
3. Solar Heat Gain Thru Glass Btu/(hr) (sq ft)																
12-hour Operation	58	59	66	68	68	63	67	68	65	65	65	65	58	49	48	44
16-hour Operation	51	52	58	60	59	56	60	60	58	58	58	58	53	45	43	40
24-hour Operation	51	52	58	60	59	56	60	60	58	58	58	58	53	45	43	40
4. Equivalent Temp Diff (F)																
Glass	4	3	3	−1	13	15	14	12	19	19	19	19	17	18	18	18
Wall	−2	−5	−2	−5	7	8	8	6	15	15	15	15	12	12	12	12
Roof	12	11	6	−2	26	29	25	19	42	42	40	37	44	45	43	40

Example 1 — Typical Load Estimate

Given:

Typical floor plan (*Fig. 5*)
Wall U = 0.34 Btu/(hr)(sq ft)(deg F temp diff)
   Wt = 100 lb/sq ft (approx)
Windows — double hung wooden sash, single-glazed.
         venetian blinds, light color
Construction — medium, 100 lb/sq ft
Normal operation — 12 hours

Find:

For the numbered areas:
   Transmission per degree
   Room sensible heat gain
   Minimum ventilation requirements

Solution:

Fill out all appropriate columns using a form similar to that shown in *Fig. 6*.

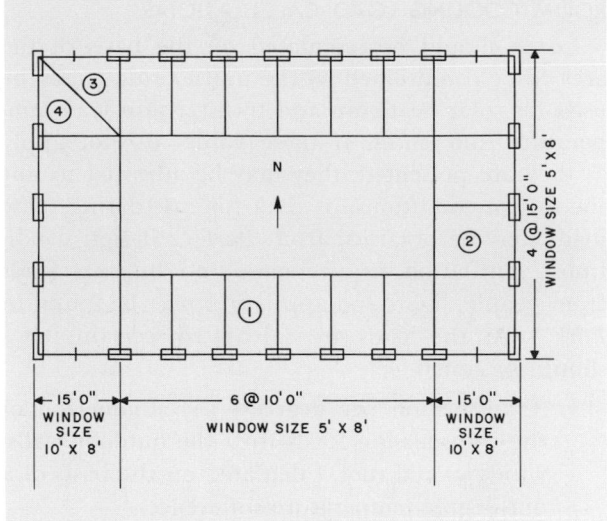

FIG. 5 — TYPICAL FLOOR PLAN

ITEM 1. "U" GLASS = 1.13				ITEM 2. "U" WALL = .34	ITEM 3. "U" ROOF = .18	ITEM 4. DESIGN ROOM DB = 78		ITEM 5. DESIGN OUTSIDE DB = 95		
PHYSICAL DATA PER UNIT FOR LOAD ESTIMATE										
6. EXPOSURE						SOUTH 1	EAST 2	NORTH 3	WEST 4	6
7. WIDTH OF OUTSIDE WALL PER UNIT						10'0"	15'0"	15'0"	15'0"	7
8. DISTANCE FROM OUTSIDE WALL SERVED BY UNIT						15'0"	15'0"	15'0"	15'0"	8
9. FLOOR TO FLOOR HEIGHT						10'0"	10'0"	10'0"	10'0"	9
10. GROSS WALL AREA (9 x 7)						100	150	150	150	10
11. NET SUN GLASS AREA (STEEL OR WOOD SASH)						40	40	80	40	11
12. NET SUN WALL AREA (10—11)						60	110	70	110	12
13. OTHER SUN GLASS AREA (CORNER ROOM)						-				13
14. OTHER WALL AREA (CORNER ROOM)										14
15. FLOOR AREA (7 x 8)						150	225	112	112	15
16. ROOF AREA (7 x 8)						150	225	112	112	16
17. LIGHTS (_4_ WATTS INPUT/SQ FT x 15)						600	900	450	450	17
18. HORSEPOWER OUTPUT TO ROOM										18
19. NUMBER OF PEOPLE PER UNIT						1	2	1	1	19
TRANSMISSION PER DEGREE (OUTSIDE EXPOSURE ONLY) BTU/HR/F										
20. ALL GLASS (11 + 13) x 1						45.2	45.2	90.4	45.2	20
21. WALL (12 x 2)						20.4	37.4	23.8	37.4	21
22. OTHER WALL (14 x 2)										22
23. INTERMEDIATE FLOOR TRANSMISSION PER DEGREE (20 + 21 + 22)						65.6	82.6	114.2	82.6	23
24. ROOF (16 x 3)						27.0	40.5	20.2	20.2	24
25. TOP FLOOR TRANSMISSION PER DEGREE (23 + 24)						92.6	123.1	134.4	102.8	25
LOAD ESTIMATE (ROOM SENSIBLE HEAT GAIN PER UNIT) BTU/HR										
26. SUN GLASS (11 x SOLAR FACTOR)	70	66	9	64		2800	2560	720	2640	26
27. OTHER SUN GLASS (13 x SOLAR FACTOR)										27
28. ALL GLASS TRANS (20 x TEMP DIFF AT DESIGN TIME)	0	16	12	5		0	226	1085	573	28
29. WALL TRANSMISSION (21 x EQUIV TEMP DIFF)	12	14	4	10		245	374	95	523	29
30. OTHER WALL TRANSMISSION (22 x EQUIV TEMP DIFF)										30
31. FLOOR, CEILING, OR PART. TRANS (SQ FT x TEMP DIFF x U)										31
32. LIGHTS (17 x 3.4)						2040	3060	1530	1530	32
33. HP (18 x _____ DIVERSITY FACTOR x 2545 ÷ EFFICIENCY)										33
34. PEOPLE (19 x _215_ BTU/HR SENSIBLE HEAT)						215	430	215	215	34
35. RSH GAIN, INTERMEDIATE FLOOR (TOTAL OF 26 THRU 34)						5300	6650	3645	5481	35
36. ROOF HEAT GAIN (24 x EQUIVALENT TEMP DIFF)	5	40	39	5		135	202	787	808	36
37. ROOM SENSIBLE HEAT GAIN, TOP FLOOR (35 + 36)						5435	6852	4432	6289	37
MINIMUM VENTILATION REQUIREMENTS PER UNIT · CFM										
38. NUMBER OF SMOKERS PER UNIT x 25 CFM/PERSON						25	25	25	25	38
39. NUMBER OF NON-SMOKERS PER UNIT x 15 CFM/PER PERSON							15			39
40. MINIMUM CFM BASED ON PEOPLE (38 + 39)						25	40	25	25	40
41. MINIMUM CFM BASED ON CFM/SQ FT (15 x _.25_ CFM PER SQ FT)						38	56	28	28	41
42. MINIMUM CFM PER ROOM (HOTEL) (HOSPITAL)										42
43. MINIMUM CFM PER UNIT (GREATEST CFM VALUE OF 40, 41, 42)						38	56	28	28	43

FIG. 6 — TYPICAL LOAD ESTIMATE

ROOM COOLING LOAD CALCULATIONS

Loads should be calculated on the basis of the area to be conditioned by the unit. *Table 1* may be used for solar heat gain and transmission load temperature differences. If these values do not apply as they are presented, they may be adjusted to suit the design conditions, or data for calculating these loads may be obtained from *Part 1*. Design conditions, ventilation requirements and internal loads from people, lights and appliances may be found in *Part 1*. As the loads are calculated, certain items should be noted.

1. Transmission per degree — the summation of the transmission loads thru the outdoor walls, windows and roof, calculated on the basis of a one-degree temperature difference.

2. Room sensible heat gain — the summation of solar heat, transmission, lights, people and appliance loads.

3. Minimum ventilation required — the largest of the air quantities calculated on a person or square foot basis.

The solar heat gain is usually the major load in the room and should be calculated with accuracy.

UNIT SELECTION

The induction units selected for a given space must be able to:

1. Supply a quantity of air to the space in a fixed proportion to the transmission per degree of the space. The outdoor air portion of the supply air must equal or exceed the ventilation requirement.

2. Produce a total cooling capacity that equals or exceeds calculated room sensible heat gain.

3. Operate at a nozzle pressure consistent with an acceptable sound level.

In addition to the above three conditions that must be met, there are three temperatures which must be known to select an induction unit.

1. Room temperature

2. Primary air temperature

3. Secondary water temperature

The room temperature is selected from the design conditions. This is the maximum room temperature which is acceptable at peak design load.

The primary air temperature depends on the selected apparatus dewpoint temperature. An apparatus dewpoint temperature may be selected for various moisture loads from *Table 2*. The primary

air temperature is usually 8 degrees higher than the apparatus dewpoint; this 8 degrees takes into account the effects of the bypass factor, fan motor heat, and duct heat gain. While this 8 degrees may not be uniform thruout the system because of duct lengths and air velocities in the ducts, it is a figure which can be used for most design purposes. For the south exposure, the peak loading usually occurs when the primary air is being reheated. Therefore, the primary air temperature for the south zone units is selected as being equal to the room temperature unless the south zone is supplied as a separate zone with its own reheater.

The secondary water temperature may be selected as low as 3 degrees below the room dewpoint temperature. Any temperature down to this minimum value allows a dry coil operation and does not require insulation of the water risers and runouts or installation of condensate drains.

The best design generally occurs when all the units are selected to be the same size and type and operate with the same air quantity. This also allows identical riser sizing and requires less installation and balancing time.

Make trial unit selections for the typical spaces. The units selected must have a total cooling capacity to meet the calculated room sensible heat load. The preferred unit selection should have a high capacity ratio (total sensible heat cooling capacity to primary air quantity). The primary air quantity must satisfy the minimum ventilation requirements. If the ventilation requirements are not met with the first trial selection, select a unit with a smaller capacity ratio. For each unit selected, the size, model, air quantity, total cooling capacity and nozzle pressure should be noted. The nozzle pressure must be below the acceptable limit consistent with the ambient sound level.

Calculate the A/T ratio for each selection. Units may be grouped in zones by exposure if desired. For each air zone, select the highest A/T ratio (representative of the majority of the spaces in that zone) as a base A/T ratio.

Find the maximum required primary air temperature by using *Table 3*. The temperature indicated under the column headed by the base A/T ratio and opposite the minimum design outdoor temperature is the maximum required primary air temperature. If this is above 140 F (generally accepted as an upper limit for the supply air temperature), then the base A/T ratio must be in-

TABLE 2—DEHUMIDIFIER APPARATUS DEWPOINT SELECTION GUIDE

Max. Design Room Temp and Percent RH	Normal Room Temp and Percent RH	Suggested Room Dewpoint	Design People Loading Sq Ft /Person	Dehumidifier Dewpoint*									
				8-Row Coil Primary Air (cfm/sq ft)					6-Row Coil Primary Air (cfm/sq ft)				
				.2	.3	.4	.5	.6	.2	.3	.4	.5	.6
80 F 45%	77 F 50%	56.6	125	50.2	52.5	53.6	54.3	54.8	49.5	52.0	53.4	54.3	54.8
			100	48.4	51.4	52.8	53.6	54.1	47.7	51.0	52.6	53.6	54.2
			75		49.5	51.4	52.5	53.2		48.6	51.0	52.4	53.2
			50			48.6	50.2	51.0			48.0	49.5	51.1
78 F 45%	75 F 50%	55.0	125	48.0	50.2	51.4	52.0	52.4	46.3	49.8	51.4	52.4	53.0
			100	46.2	49.2	50.6	51.6	52.0		48.4	50.3	51.4	52.1
			75		47.2	49.0	50.2	51.0		46.4	48.6	50.0	51.0
			50			46.0	47.8	49.0				47.4	49.0
76 F 45%	73 F 50%	53.5	125		47.8	49.2	50.0	50.5		47.3	49.0	50.0	50.5
			100		46.6	48.1	49.0	49.5		46.0	48.0	49.1	49.8
			75			46.8	48.0	48.8			46.6	48.0	48.8
			50					46.8					46.2
75 F 45%	72 F 50%	52.0	125		46.6	48.0	48.8	49.5		46.2	47.7	48.6	49.2
			100			46.9	48.1	49.0			46.6	47.7	48.5
			75				46.8	47.5				46.4	47.5
			50					45.5					46.0

*Apparatus dewpoints are based on:
8-row coil — 100% outside air bypass factor = .03
6-row coil — 0.1 cfm per sq ft ventilation air bypass factor = .1
Outside design conditions 95 F db, 75 F wb.

creased to reduce this temperature or the system may be designed as a change-over system as explained under *System Modifications*.

Calculate the design primary air quantity for each unit by multiplying the final selected base A/T ratio by the transmission per degree for each space making sure that the ventilation requirements are met.

Make the final unit selections using this design primary air quantity. The total cooling capacity of the units must meet the calculated room sensible heat load, and the nozzle pressure must be below the limit consistent with the ambient sound level.

In cases where the base A/T ratio is exceeded, there may be the possibility of overheating during low outdoor temperatures. To prevent this condition, the unit should be selected with sufficient coil capacity to handle the room load plus the excess reheat.

ROOM HEATING LOAD CALCULATION

If provision is to be made for gravity heating, the gravity heating load must be calculated for each typical unit. This load consists only of the trans-

mission heat loss but, on tall buildings, may include infiltration.

Calculate the water temperature required to meet the gravity heating load by using the ratings for the units selected. If the required water temperature is above a practical limit of 190 F, then the units may be increased in size or, more practically, the primary air fan can be operated during the periods of low outdoor temperatures occurring when gravity heating is required.

If the system is operated as a change-over system with hot water supplied to the units during periods of low outdoor temperatures, the total room heating load must be calculated. This load includes the transmission heat loss plus the amount of heat necessary to raise primary air to room temperature.

This may be required where the system operates for 12 to 16 hours daily and where a means must be provided to warm up the building after an overnight or weekend shutdown.

Calculate the water temperature required to meet the total room heating load for each typical space. The highest temperature required is the design temperature for the water heater.

TABLE 3—SCHEDULE OF PRIMARY AIR TEMPERATURES

OUTDOOR DRY-BULB TEMP. (F)	PRIMARY AIR TEMPERATURE (F) A/T Ratio												
	0.2	0.4	0.6	0.8	1.0	1.2	1.4	1.6	2.0	2.5	3.0	3.5	4.0
100	56	56	56	56	56	56	56	56	56	58	61	63	64
95	56	56	56	56	56	56	56	56	57	60	62	64	66
90	56	56	56	56	56	56	56	56	59	62	64	66	67
85	56	56	56	56	56	56	57	59	62	64	66	67	68
80	56	56	56	56	56	58	60	62	64	66	67	69	70
75	56	56	56	57	60	63	64	65	67	68	69	70	71
70	56	56	60	63	65	67	68	69	70	71	71	72	72
65	58	65	68	70	71	71	72	72	72	73	73	73	74
60	80	78	77	76	76	76	75	75	75	75	75	75	75
55	102	90	85	82	81	80	79	78	78	77	76	76	76
50	125	103	93	89	86	84	82	81	80	79	78	78	78
45	147	116	102	95	91	88	86	85	83	81	80	79	79
40		128	110	101	96	92	90	88	85	83	82	81	80
35		140	119	108	101	97	93	91	88	85	83	82	81
30			127	114	106	101	97	94	90	87	85	84	83
25			136	121	111	105	101	98	93	90	87	85	84
20			145	127	117	110	105	101	96	92	89	87	85
15				134	122	114	109	104	99	94	91	88	87
10				140	127	118	112	107	101	96	92	90	88
5				147	132	123	116	111	104	98	94	91	89
0 ●					137	127	120	114	106	100	96	93	91
− 5					143	131	124	117	109	102	98	94	92
−10						136	127	121	112	105	100	96	93
−15						140	131	124	114	107	101	97	95
−20							135	127	117	109	103	99	96

NOTE: These temperatures are required at the units, and thermostat settings must be adjusted to allow for duct heat gains or losses. The temperatures are based on:
1. Minimum average load in the space equivalent to 10 degrees multiplied by the transmission per degree.
2. Preventing the room temperature from dropping below 72 F. This schedule compensates for radiation and convection effect of the cold outdoor wall.
3. 140 F as the recommended upper limit of reheat.

APPARATUS SELECTION

The primary air apparatus consists normally of a supply fan, reheater, dehumidifier, air filter, preheater and outdoor air intake louver, screen and damper. A return air fan and damper may be used if required.

The equipment is selected for the sum of the air quantities supplied to the units.

The supply fan is generally a high pressure fan picked to handle the design air quantity at a calculated static pressure. The total static pressure required is generally from 5-8 in. wg. To provide the quietest operation, the fan should be picked near its maximum efficiency.

The reheater is selected to heat the design air quantity from 40 F to the temperature indicated by the reheat schedule plus an allowance of 15-20 degrees for duct heat loss and to provide a quick warm-up.

The preheater is selected to heat the design air quantity to approximately 55 F.

The dehumidifier may be either a dry coil or sprayed coil type; most installations are designed using the sprayed coil. The sprays serve the following functions:

1. Winter humidification.
2. Additional air cleaning and odor control by washing.
3. Evaporative cooling during marginal weather operation.

The dehumidifier is generally selected to cool the design air quantity to an apparatus dewpoint of 48 F unless conditions indicate that a lower temperature should be used.

The dehumidifier load is found from the formula:

$$\text{Load} = cfm_{da} \times 4.45 \times (1 - \text{BF})(h_{ea} - h_{adp})$$

where:

cfm_{da} = dehumidified air quantity
h_{ea} = entering air enthalpy
h_{adp} = apparatus dewpoint enthalpy
BF = dehumidifier bypass factor

The required apparatus dewpoint can be checked on an individual room basis by the formula:

$$W_{adp} = \frac{W_{rm} - (W_{ea} \times BF) - \dfrac{RLH}{.68 \times cfm_{da}}}{1 - BF}$$

where:

W_{adp} = apparatus dewpoint specific humidity (gr/lb)
W_{rm} = room specific humidity (gr/lb)
W_{ea} = air entering dehumidifier specific humidity (gr/lb)
RLH = room latent heat load
cfm_{da} = dehumidified air quantity
BF = dehumidifier bypass factor.

From the psychrometric chart the apparatus dewpoint temperature is found equal to the saturation temperature corresponding to W_{adp}.

Example 2 — Apparatus Dewpoint Calculation

Given:

Room specific humidity	= 64 gr/lb
Specific humidity, air entering dehumidifier	= 118 gr/lb
Dehumidifier bypass factor	= .05
Room latent heat	= 235 Btu/hr
Room air quantity	= 40 cfm

Find:

Apparatus dewpoint

Solution:

$$W_{adp} = \frac{W_{rm} - (W_{ea} \times BF) - \dfrac{RLH}{.68 \times cfm_{da}}}{1 - BF}$$

$$= \frac{64 - (118 \times .05) - \dfrac{235}{.68 \times 40}}{1 - .05}$$

$$= \frac{64 - 5.9 - 8.6}{.95} = \frac{49.5}{.95} = 52.2 \text{ gr/lb}$$

From the psychrometric chart at a specific humidity of 52.2 gr/lb, the apparatus dewpoint is found as 49.5 F.

The air filter is selected to handle the design air quantity at a high efficiency.

The outdoor air louver, screen and damper are selected for the design air quantity and a face velocity between 500 and 800 fpm. The higher values are normally used when there is a return air duct system without a fan.

DUCT DESIGN

The duct distribution system is made up of headers and risers to supply the induction units with a constant volume of air. High velocities are generally used, up to 3000 fpm in the headers and 4000-5000 fpm in the risers. Because the air distribution system is subject to high static pressures, tightness is essential. Therefore, rigid spiral conduit is generally used in place of conventional ductwork. Welded fittings are used for elbows and take-offs. Ducts must be carefully sealed to prevent air leakage.

The static regain method of duct sizing is recommended for this system. Details of duct design may be found in *Part 2*.

The duct distribution system generally includes a sound absorbing section at the fan discharge to attenuate the sound level of the high pressure fan. The attenuation required must be calculated and depends on the sound generated by the fan, the natural attenuation of the ducts, and the sound generated by other sources in the duct system.

The headers installed in unconditioned spaces should be insulated and vapor-sealed to prevent excessive heat gain and sweating. The risers should be insulated to reduce the temperature loss in winter to a minimum of 20% of the difference between the room temperature and the desired primary air temperature.

Since the risers are normally within the conditioned space, there is no need to use a vapor barrier or to seal the surface of the insulation in any way.

REFRIGERATION LOAD

Design Peak Building Load

When calculating the refrigeration load, the building as a whole should be considered. From a design viewpoint, the maximum demand for refrigeration is assumed to occur at the time of instantaneous building peak. The calculation of this load bears no relation to the room load calculated for unit selection as the design peak building load is not the sum of the individual room peak loads.

The time of day at which the building peak occurs depends on the relative amounts of east, south and west exposures. Where these exposures are of about the same magnitude, the building peak usually occurs in the afternoon when the sun is on the west side and the outdoor wet-bulb is high.

The refrigeration load is determined as follows:

1. Calculate the room sensible heat for the entire building at the time of peak load.

2. Add the total heat load of the primary air. The outdoor air is taken from outdoor conditions at the time of peak load to the required dewpoint temperature leaving the dehumidifier. Return air (when used) is taken from room conditions to the required leaving air dewpoint temperature.

3. Subtract the credit for primary air sensible cooling between room temperature and primary air temperature at the unit.

The room sensible heat includes the solar, transmission, lights and people sensible load, assuming peak zone rooms are at the design temperature; all other rooms are at their thermostat setting, usually taken as 3 degrees below design.

The refrigeration load can be reduced at peak loading by operating the equipment longer and taking advantage of storage and precooling. Refer to *Part 1* for explanations of storage and precooling.

Example 3 — Refrigeration Load Calculation

Given:

A typical building oriented with the largest exposures facing east and west.

Time of estimate — 4 p.m. during July

Outdoor dry-bulb temp	= 95 F
Outdoor wet-bulb temp	= 75 F
Indoor design dry-bulb temp	= 78 F
Indoor design relative humidity	= 45%
Apparatus dewpoint	= 48 F
Latitude	= 40° N
Daily range of temperature	= 14 deg
Length of operation	= 16 hours

Ordinary glass double hung, wood sash, light colored venetian blinds

Wall construction, U = .34, wt = 100 lb/sq ft

Roof construction, U = .18, wt = 40 lb/sq ft

Room temperatures, W exposure = 78 F

N, E, S exposures = 75 F

Air quantities

Exposure	Outdoor Air	Return Air
West	5060 cfm	4716 cfm
East	5300	4936
North	1720	1604
South	2160	2022
	14240 cfm	13278 cfm

Find:

Refrigeration load

Solution:

BUILDING SENSIBLE HEAT LOAD

Solar Gain — Glass

	Sq Ft		Heat Gain		Storage Factor		Shade Factor		Area Factor		Btu/hr
W Glass,	5100	×	164	×	.66	×	.56	×	1/.85	=	364,000
E Glass,	5100	×	164	×	.16	×	.56	×	1/.85	=	88,200
N Glass,	2270	×	15	×	.88	×	.56	×	1/.85	=	19,750
S Glass,	2270	×	69	×	.45	×	.56	×	1/.85	=	46,500

Solar and Transmission Gain — Walls and Roof

	Sq Ft		Trans Factor		Temp Diff		Btu/hr
W wall,	9800	×	.34	×	(12 + 5)	=	56,600
E wall,	9800	×	.34	×	(18 + 8)	=	86,600
N wall,	5180	×	.34	×	(4 + 8)	=	21,150
S wall,	5180	×	.34	×	(16 + 8)	=	42,300
Roof,	6336	×	.18	×	(38 + 8)	=	52,500

Transmission Gain — Glass

	Sq Ft		Trans Factor		Temp Diff		Btu/hr
W Glass,	5100	×	1.13	×	16	=	92,000
E-N-S Glass,	9640	×	1.13	×	19	=	206,000

Internal Heat Gain

	Heat Gain		Storage Factor		Factor Div		Btu/hr
People,	600	×	215	×	.89	×	.9 =	103,200
Lights,	130,000	×	3.4	×	.89	×	.85 =	334,000
						Subtotal		1,512,800

Storage

	Sq Ft		Storage Factor		Temp Diff		Btu/hr
W zone,	19,000	×	1.25	×	(−3)	=	−71,200
				Building sensible heat	=	1,441,600	

PRIMARY AIR LOAD

	Cfm		Conv Factor		Enthalpy Diff		Contact Factor		Btu/hr
Outdoor Air										
	14,240	×	4.45	×	(38.62 − 19.22)		(1 − .05)	=	1,168,000
Return Air, W zone										
	4,716	×	4.45	×	(28.79 − 19.22)		(1 − .05)	=	190,000
Return Air, E-N-S zones										
	8,562	×	4.45	×	(28.11 − 19.22)		(1 − .05)	=	321,000
					Primary air load				1,679,000
					Subtotal			=	3,120,600

CREDIT FOR PRIMARY AIR COOLING

	Cfm		Conv Factor		Temp Diff		Btu/hr
W zone,	9,776	×	1.08	×	(78 − 56)	=	−232,500
N-S-E zones,	17,742	×	1.08	×	(75 − 56)	=	−364,000
				Credit subtotal		=	−596,500
				Net refrigeration load		=	2,524,100

Notes: 1. All the values for the heat gain calculations may be obtained from *Part 1*.

2. Judgment must be used in estimating the diversity factors.

PIPING DESIGN

Water and steam piping arrangements and sizing may be found in *Part 3*.

The water distribution system *(Fig. 7)* consists of two interconnected circuits, the primary water and the secondary water circuits.

The primary water piping connects the dehumidifier in the primary air apparatus, the refrigeration machine, and the primary chilled water pump.

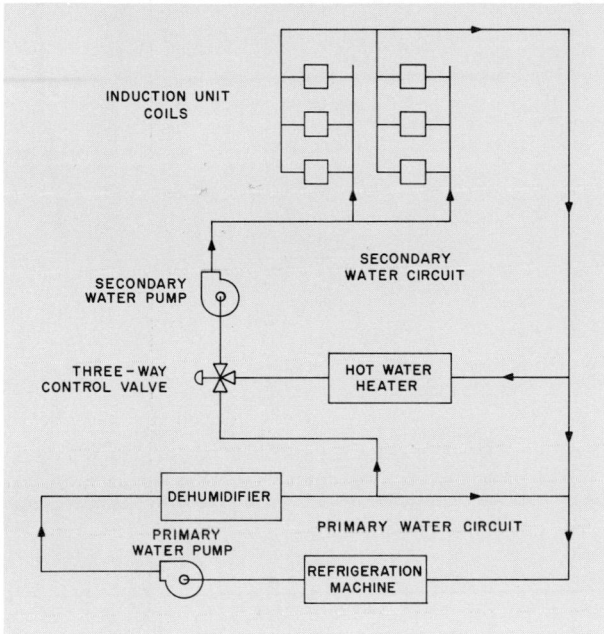

FIG. 7 — WATER DISTRIBUTION SYSTEM

The secondary water piping connects the induction unit coils, the secondary water pump, the three-way control valve, and the hot water heater. One side of the three-way valve is connected to the primary water circuit.

The secondary water piping is usually designed as a complete reverse return system, but may also be designed with reverse return headers and direct return risers where it is more convenient or more economical.

Proper allowances must be made for the expansion of all piping. Horizontal runouts from risers to units should have a minimum length of two feet to absorb the vertical expansion of the riser.

The primary chilled water pump should be selected for the total water quantity required by the dehumidifier(s). The pump head is the sum of the friction losses for the cooler, dehumidifier and primary circuit piping with maximum flow.

The secondary chilled water pump should be selected for the total water quantity required by the induction units multiplied by a diversity factor (if applicable). Diversity which is explained in *Part 3* may be used when more than one exposure is served by a common secondary water circuit and automatic modulating valves are used to control flow. Diversity cannot be used when there are no automatic modulating control valves used in the system. The pump head is determined by the total pressure drop thru the piping system, induction unit, valves, strainers and other accessories using the water quantity required by the pump.

The secondary water system should have an open type expansion tank to permit air venting, to provide for expansion with a rising temperature, and to provide a static head on the suction of the secondary pump.

When units use water throttling as a control means and the valves are partially closed, the water flow in the risers is reduced, as is the pressure drop. This tends to increase the pressure difference across the valves. To assure satisfactory control when many of the units are throttled, the system should be so designed that the pump head does not exceed the maximum pressure recommended by the valve manufacturer for valve close-off. It is desirable to select a pump with a flat head characteristic so excessive pressures do not result during reduced flow.

Air vents should be provided at system high points which cannot vent to the expansion tank. Pitch all piping upward to avoid air pockets so that air is carried along and vented.

All chilled water piping including valves and fittings other than the secondary water supply and return risers and runouts should be covered with not less than an equivalent of one-inch wool felt insulation with sealed canvas covering and an adequate vapor seal. If the design secondary water temperature is no more than 3 degrees below the design room dewpoint, it is not necessary to insulate the supply and return riser group. However, when omitting insulation from these risers, they must be sealed from unconditioned spaces, (basement or attic). If the 3 degree limitation is exceeded, supply and return risers should have a minimum of ½ inch of insulation with an adequate vapor seal. Under this condition the runouts to the units may be insulated with any insulation that provides a vapor seal.

Condensate drain piping from the induction units is usually not required if the secondary water temperature is maintained above a minimum of 3 degrees below the room dewpoint. When there is an unusually high latent load in a space such as a hotel or motel room with an adjoining shower bath or a room adjacent to a kitchen, condensate drains may be required. If there is any indication that the room dewpoint cannot be maintained by the primary air, drains should be installed. Size the drains as recommended in *Part 3*.

WATER HEATER

The water heater should be selected to have a capacity equal to the sum of the following three items:

1. The calculated transmission load of the zone or building.
2. Twenty percent of the transmission load to allow for quick warm-up.
3. The primary air load, calculated as the heat required to raise the temperature of the primary air from approximately 40 F to room temperature.

The water temperature leaving the heater is determined from the unit selections by using the highest water temperature required for the units in the zone served by the heater.

Example 4 — Water Heater Selection

Given:

Room temperature	= 75 F
Outdoor temperature	= 0 F
Secondary water quantity	= 570 gpm
Required hot water temperature	= 131 F
Same building as used in *Example 2*	

Find:

Total heat load for selecting water heater
Duty specifications for water heater

Solution:

TRANSMISSION LOSSES

	Sq Ft		Temp Diff		Trans Factor		Btu/hr
Roof	6,336	×	(75 − 0)	×	.18	=	85,500
Windows	14,720	×	(75 − 0)	×	1.13	=	1,247,000
Walls	29,970	×	(75 − 0)	×	.34	=	764,000
					Subtotal	=	2,096,500
20% for warm-up						=	419,500

PRIMARY AIR LOAD

Cfm		Conv Factor		Temp Diff		
27,518	×	1.08	×	(75 − 40)	=	1,040,000
				Total Heat Load	=	3,556,000

$$\text{Temperature rise} = \frac{\text{total heat load}}{500 \times \text{gpm}}$$

$$= \frac{3,556,000}{500 \times 570} = 12.5 \text{ deg}$$

Select a water heater to heat 570 gpm from 118.5 F to 131.0 F with a water pressure drop not to exceed 5 psi and with a fouling factor allowance of 0.001.

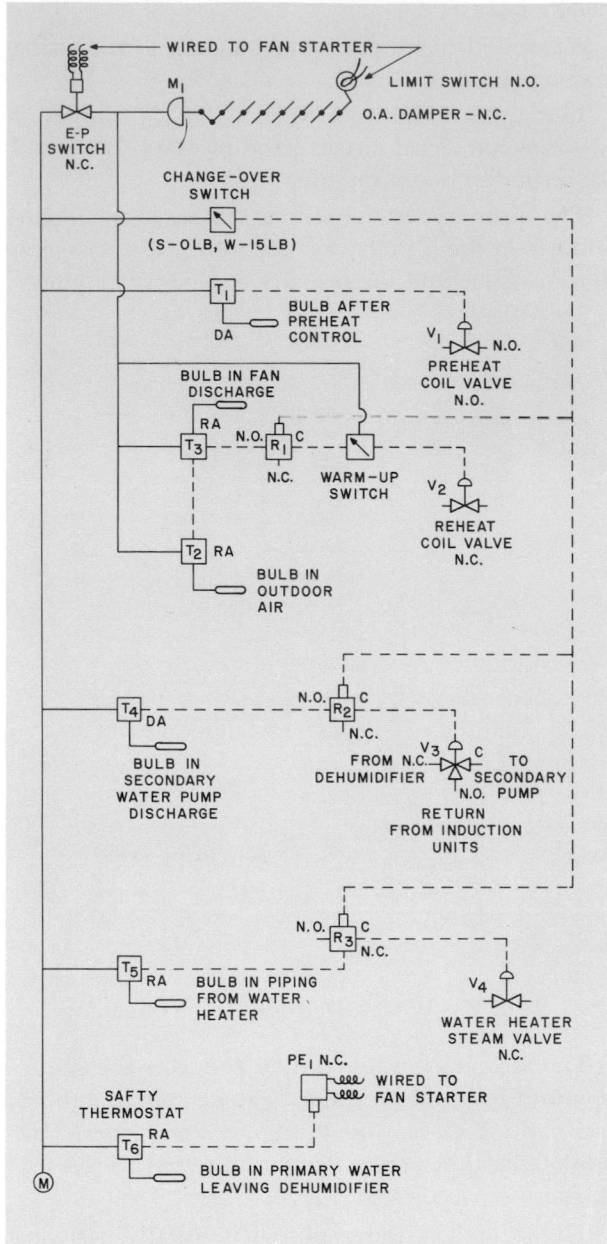

FIG. 8 — INDUCTION UNIT SYSTEM CONTROL

CONTROLS

The following description gives a recommended control sequence for a typical system. *Figure 8* is a schematic control diagram for the basic components of the system shown in *Fig. 1*. It is based on a system for an office building on 12-hour operation with 100% outdoor air utilizing the primary air dehumidifier to cool the secondary water during the winter season. Provision is made to change over the system to hot water when needed for winter heating or gravity heating.

PRIMARY AIR DAMPER

A momentary contact push button in the primary air fan starter energizes the electric-pneumatic switch which in turn causes damper motor M_1 to open the normally closed outdoor air damper. When the damper has traveled to a predetermined position, it energizes a limit switch to start the fan motor. With this arrangement the fan operates only when the dampers are open, thus preventing damage to the apparatus casing because of a vacuum.

PRIMARY AIR PREHEATER

A direct acting thermostat T_1 located after the preheater controls the preheater steam valve. The thermostat is set to control at 50 F since a lower setting may cause the dehumidifier sprays to freeze.

PRIMARY AIR REHEATER

The reheater is controlled according to a reheat schedule from *Table 3*. A master outdoor thermostat T_2 located in the outdoor air intake resets the control point of the submaster fan discharge air thermostat T_3 which controls the reheat valve. A manual switch is provided so the valve can be operated in a full open position if a quick warm-up is needed at start-up on cold mornings. When actuated by the change-over switch, relay R_1 allows valve V_2 to assume its normally closed position.

SECONDARY CHILLED WATER

A three-way mixing valve V_3 is controlled by direct-acting thermostat T_4 which has its thermal bulb located in the secondary chilled water line. The thermostat regulates valve V_3 to mix the proper amounts of chilled water and return water, satisfying the thermostat setting. When actuated by the change-over switch, relay R_2 allows valve V_3 to assume the position of full water flow thru the heater.

WATER HEATER

The water heater is normally inoperative but, when relay R_3 is actuated by the change-over switch, thermostat T_5 controls valve V_4 supplying steam to the water heater.

ROOM TEMPERATURE CONTROL

The room control is generally automatic, usually either pneumatic or self-contained.

When pneumatic control is used, a normally open control valve is provided. The control thermostat operation is direct-acting with hot water in the circuit, and is reverse-acting with cold water in the circuit. The reversing of the action of the thermostat is accomplished by varying the main air pressure to the thermostat. For gravity heating the control air pressure is bled to zero and the control valves assume their normally open position.

The self-contained control may be either a water control valve or a coil face and bypass damper. The control thermostats reverse their action depending on the water temperature available.

SAFETY THERMOSTAT

Safety thermostat T_6 located in the water line leaving the dehumidifier shuts down the primary air fan when the chilled water temperature drops to 35 F. This may occur if the preheater fails during outdoor temperatures below freezing.

SYSTEM MODIFICATIONS

This section points out certain variations that may be incorporated in the induction unit system. In this section are also included the calculation of the off-season cooling requirements, some sources of the off-season cooling, an explanation of the change-over system and the use of return air.

OFF-SEASON COOLING

As the outdoor temperature decreases, a point is reached where the main refrigeration system may be shut down and other sources used to cool the secondary water.

The total net cooling load on the secondary water is determined at the outdoor temperature (when the main refrigeration system is shut down), by making a block estimate for the exterior zone as is done to determine the design refrigeration load in summer.

When calculating the solar heat gain and light load, the storage load factors for 24-hour operation should be used, regardless of the length of time the system is operated.

Example 5 shows the calculations for the total net secondary coil load for summer operation and for two different periods of off-season operation.

Example 5 — Off-Season Refrigeration Requirement

Given:
  Same building as in *Example 2*

Find:
  Off-season refrigeration loads

Solution:

	July, 4 P. M.	October, 2 P. M.	April, 4 P. M.
Outdoor Temperature	94	48	48
Room Temp — Peak zone	W 78	S 78	W 78
— Other zones	E-N-S 75	E-W-N 75	E-N-S 75
Primary Air Temperature	56	88	88
Hours of operation	16	24 (equiv.)	24 (equiv.)

Solar Gain — Glass

	Sq Ft	Shade Factor	Area Factor	Heat Gain	Storage Factor	Btu/hr	Heat Gain	Storage Factor	Btu/hr	Heat Gain	Storage Factor	Btu/hr
West	5100	× .56	× 1/.85	× 164	× .66	364,000	× 122	× .36	147,500	× 162	× .66	359,000
East	5100	× .56	× 1/.85	× 164	× .16	88,200	× 122	× .20	82,000	× 162	× .16	87,000
North	2270	× .56	× 1/.85	× 15	× .88	19,750	× 7	× .85	8,920	× 11	× .88	14,500
South	2270	× .56	× 1/.85	× 69	× .45	46,500	× 162	× .89	168,000	× 102	× .45	69,000

Transmission Gain — Walls and Roof

	Sq Ft	Trans Factor	Temp Diff	Btu/hr	Temp Diff	Btu/hr	Temp Diff	Btu/hr
West	9800	× .34	× 17	56,600	× −32	−106,500	× −30	−100,000
East	9800	× .34	× 26	86,600	× −20	− 66,600	× −21	− 70,000
North	5180	× .34	× 12	21,150	× −37	− 65,100	× −35	− 61,700
South	5180	× .34	× 24	42,300	× −16	− 28,200	× −17	− 29,900
Roof	6336	× .18	× 46	52,500	× −20	− 22,800	× − 3	− 3,420

Transmission Gain — Glass

	Sq Ft	Trans Factor	Temp Diff	Btu/hr	Temp Diff	Btu/hr	Temp Diff	Btu/hr
West	5100	× 1.13	× 16	92,000	× −27	−155,500	× −30	−173,000
South	2270	× 1.13	× 19	48,000	× −30	− 77,000	× −27	− 69,300
Other	7370	× 1.13	× 19	158,000	× −27	−225,000	× −27	−225,000

Internal Heat Gain

	Heat Gain	Div Factor	Storage Factor	Btu/hr	Storage Factor	Btu/hr	Storage Factor	Btu/hr
600 people	× 215	× .9	× .89	103,200	× .83	96,300	× .87	101,000
130,000 watts	× 3.4	× .85	× .89	334,000	× .83	312,000	× .87	327,000
SUBTOTAL				1,512,800		68,020		225,180

Storage — Temp Swing

	Sq Ft	Temp Diff	Storage Factor	Btu/hr	Storage Factor	Btu/hr	Storage Factor	Btu/hr
W Zone	19,000	× 3	× −1.25	− 71,200			× −1.4	− 80,000
S Zone	9,500	× 3			× −1.4	− 40,000		
BUILDING SENSIBLE HEAT				1,441,600		28,020		145,180

Primary Air Load

	Cfm	Conv Factor	Temp Diff	Btu/hr	Temp Diff	Btu/hr	Temp Diff	Btu/hr
West	9776	× 1.08	× (56 − 78)	−232,500	× (88 − 75)	137,300	× (88 − 78)	105,500
South	4182	× 1.08	× (56 − 75)	− 85,800	× (88 − 78)	45,200	× (88 − 75)	58,700
Other	13,560	× 1.08	× (56 − 75)	−278,200	× (88 − 75)	190,200	× (88 − 75)	190,200
NET REFRIGERATION LOAD				845,100*		400,720		499,580

*Does not include outdoor air load.

NOTES: 1. Temperature differences for walls are corrected for solar radiation and for outdoor temperature.
Daily range is 14 degrees.
2. All values for heat gain calculations may be obtained in *Part 1*.

Example 5 shows that the off-season load is still substantial even though it has been greatly reduced from the summer peak.

SOURCES OF OFF-SEASON COOLING

Since a year-round source of cooling is required and since it is desirable to shut down the main refrigeration system during the winter months, an alternate economical means of cooling the secondary water must be provided.

One method is by using the outdoor air as a source of cooling in the primary air apparatus. When the outdoor air has enough capacity to cool the secondary water, the main refrigeration machine can be shut down and the secondary water circulated thru the coils of the primary air dehumidifier. By means of evaporative cooling, a considerable amount of heat can be removed from the secondary water and added to the primary air.

Although a considerable amount of cooling of the secondary water can be obtained from the primary air dehumidifier, it may not be sufficient to handle the entire off-season cooling load. Under such circumstances the interior zone dehumidifier becomes a supplementary cooling source which may be combined with the primary air dehumidifier to provide sufficient capacity.

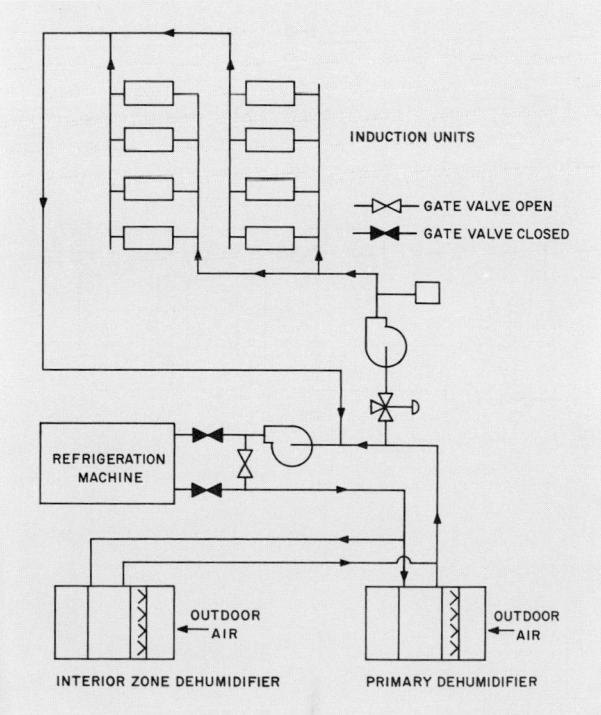

FIG. 9 — OFF-SEASON COOLING USING PRIMARY AIR COILS PLUS INTERIOR ZONE DEHUMIDIFIER COILS

Both interior zone and primary air dehumidifiers are available in most buildings, and can usually be used without any addition in first cost. The chilled water piping is the same as is used in summer operation, except that the refrigeration machine must be bypassed and the primary water quantity is made equal to the secondary water quantity. *Figure 9* shows the chilled water piping when using the dehumidifiers for the off-season cooling source.

Since water may be circulated thru the dehumidifiers when the outdoor temperature is below freezing, it is necessary to protect the coil from freezing.

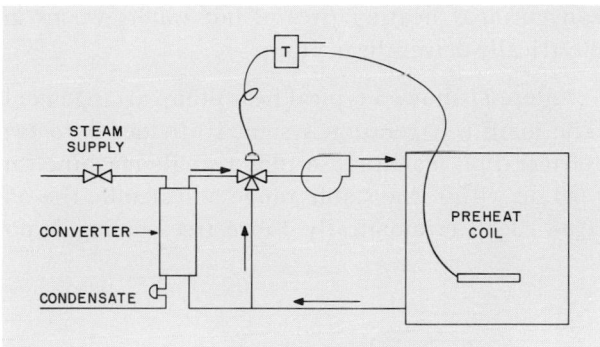

FIG. 10 — HOT WATER PREHEAT COIL

The most common method of freeze-up protection is the use of a conventional nonfreeze steam coil as a preheater. Normal precautions should be used to insure even temperature distribution, good steam distribution, and proper condensate removal. Another method involves the use of a hot water preheat coil (*Fig. 10*). The water circuit should be protected with an antifreeze solution. The water coil provides a more accurate control, and eliminates most of the stratification problems.

Air wet-bulb temperatures entering the dehumidifier coils must be maintained high enough to prevent freezing of sprays and dehumidifier; they must be low enough to provide cooling of the water supplied to the induction units. The entering air dry-bulb temperature is usually controlled at 50 F. This provides a wet-bulb temperature above freezing, and with the sprays providing evaporative cooling the temperature of the air entering the coils is generally low enough to provide the necessary water cooling.

If the interior zone dehumidifier is used as a cooling source, a reheater should be supplied to prevent overcooling of the interior spaces.

Another economical method of handling the off-season load is by using a supplementary refrigeration system as a heat pump. The heat removed from the secondary water is used to reheat the primary air by means of a condenser water reheat coil.

The heat pump unit may be a small package water chiller with only the capacity required to supplement the cooling available from the primary air dehumidifier, or it may be a centrifugal machine large enough to handle a good share of the peak summer load.

When selecting the most economical heat pump arrangement, several factors must be considered, the most important of which is the relative cost of conventional heating (steam, hot water) versus an electrically-driven heat pump.

Figure 11 shows a typical heat pump arrangement. The main refrigeration system is divided into two refrigeration machines so that a single machine can operate within the stable range and handle the off-season load economically. Since free cooling is not

desired, the dehumidifier is bypassed for winter operation.

When it is more economical to use steam for reheating the primary air, the heat pump should be used only to supplement the free cooling available from the dehumidifier. This type of arrangement is presented in *Fig. 12 and 13*. *Figure 12* shows the cooler of a heat pump unit piped in parallel with the cooler of a main refrigeration machine. *Figure 13* shows two coolers piped in series with only partial chilled water flow thru a heat pump cooler. In both drawings the piping is arranged so that during winter operation the water is cooled as much as possible in the dehumidifier before entering the heat pump. With this arrangement the refrigeration machine for the heat pump need be large enough to handle only that portion of the off-season load which cannot be handled with dehumidifier.

The heat pump arrangement may require an additional investment in equipment. However, since the refrigeration machine for the heat pump

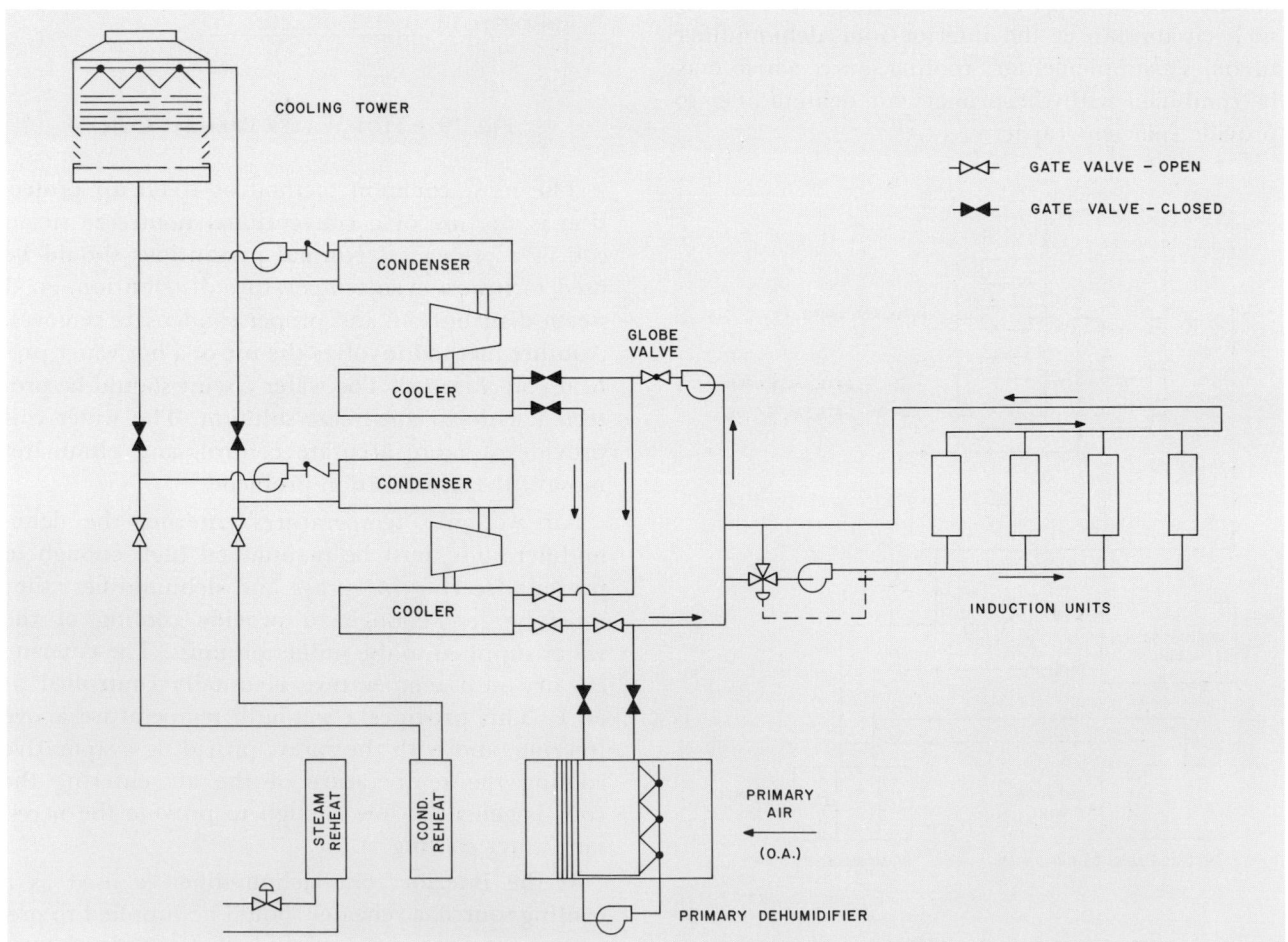

FIG. 11 — OFF-SEASON COOLING USING A HEAT PUMP SYSTEM

forms an integral part of the refrigeration system for summer operation, the increase in cost is not as much as it may appear to be at first glance.

CHANGE-OVER SYSTEM

The change-over system differs from the previously described nonchange-over system in that heating during low outdoor temperatures is accomplished by circulating hot water thru the secondary water circuit rather than by supplying warm primary air to the rooms. Cooling is available during the winter cycle by supplying cool primary air.

There are several conditions when the induction system may be designed as a change-over system.

1. When the air quantities must be increased to keep the primary air temperature below the suggested 140 F maximum of the nonchange-over system.

2. When a satisfactory change-over point above 35 F may be obtained with no increase in primary air.

3. When there is no means available for chilling the secondary water during the winter season.

The system is generally changed over from cold secondary water to warm secondary water when the transmission heat loss thru the outdoor exposures plus the cooling capacity of the primary air offsets the sun, lights and people heat loads of the room. The following emperical formula is offered to approximate this temperature.

$$t_{co} = t_{rm} - \frac{S + L + P - [1.08 \times cfm_{pa}\,(t_{rm} - t_{pa})]}{\text{transmission per degree}}$$

where:

t_{co} = change-over temperature

t_{rm} = room temperature at time of change-over (normally 76 F)

t_{pa} = primary air temperature at the unit after the system is changed over (normally 48 F)

cfm_{pa} = primary air quantity

S = net solar heat gain (Btu/hr)

L = heat gain from lights (Btu/hr)

P = sensible heat gain from people (Btu/hr)

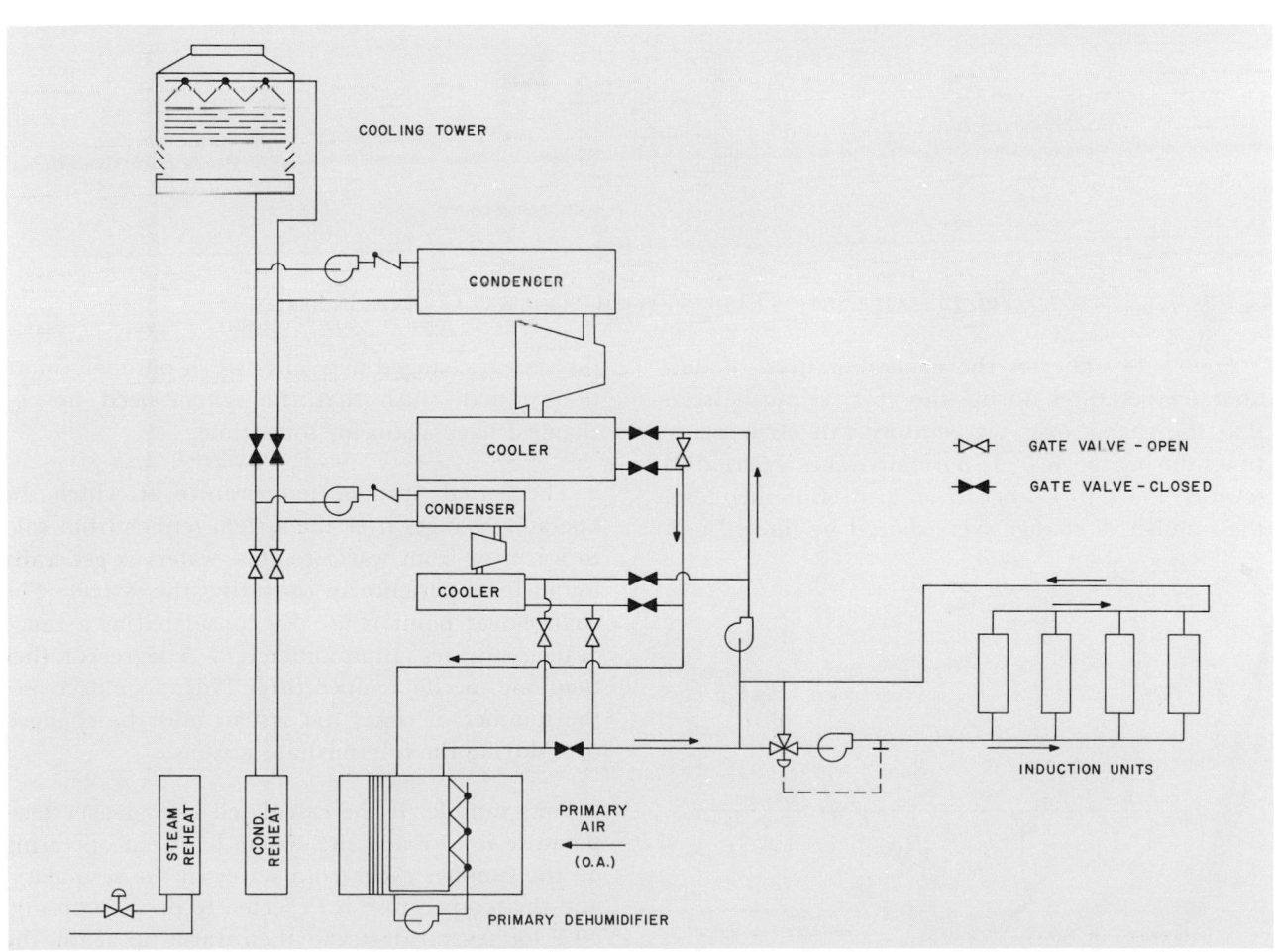

FIG. 12 — OFF-SEASON COOLING WITH AUXILIARY COOLERS IN PARALLEL

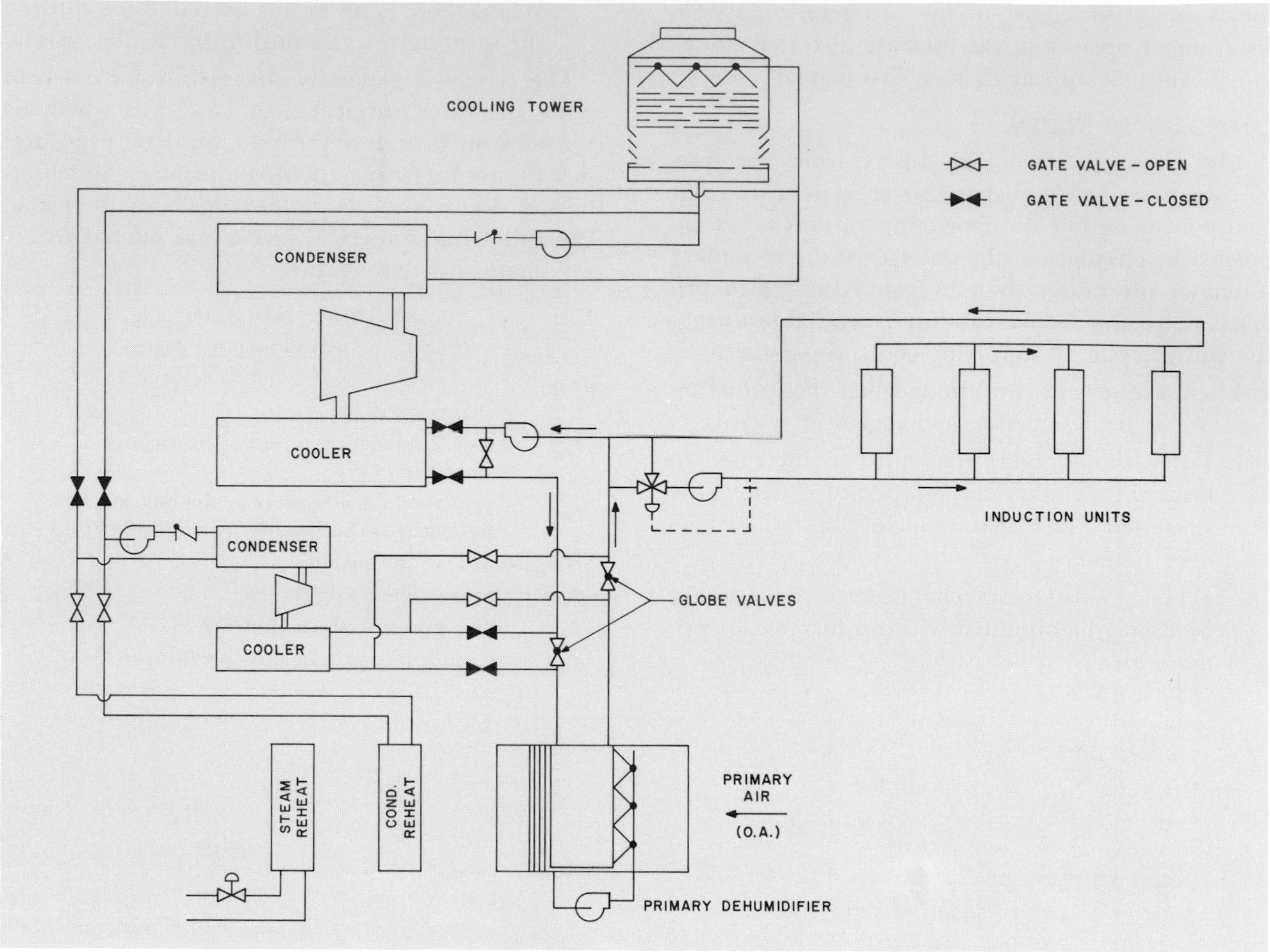

COOLING TOWER
GATE VALVE - OPEN
GATE VALVE - CLOSED
CONDENSER
COOLER
CONDENSER
COOLER
INDUCTION UNITS
GLOBE VALVES
STEAM REHEAT
COND. REHEAT
PRIMARY AIR
(O.A.)
PRIMARY DEHUMIDIFIER

Fig. 13 — Off-Season Cooling With Auxiliary Coolers In Series

Figure 14 indicates the general pattern of outdoor temperatures during the year. It can be seen that the change-over temperature can occur many times during the year. It normally takes a period of several hours to change over a system; therefore, the number of change-overs should be limited and the system changed over only when outdoor conditions remain such that the system need not be changed back again for some time.

AVERAGE TEMPERATURE CURVE
100
ACTUAL OUTDOOR TEMP CURVE
OUTDOOR AIR TEMPERATURE
CHANGE-OVER TEMP
CHANGE-OVER TEMP
0
SUMMER
AUTUMN
WINTER
SPRING
SUMMER

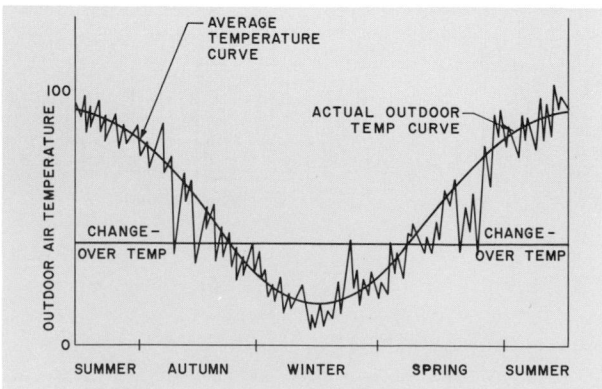

Fig. 14 — Outdoor Temperature Pattern

The actual outdoor temperature at which the operator changes over the system (either from cold to warm or from warm to cold water) is generally found by experience in operating the system. The change-over point is usually considered as a range of temperatures (approximately ± 5 degrees) rather than one specific temperature. This range decreases the number of times the system must be changed over during the intermediate seasons.

For example, if the calculated change-over temperature is 45 F and the system has been operating on the summer cycle (cold water in the secondary) and the temperature is expected to drop to around 40 F for several days and then warm up again, the system should not be changed over. But if the forecast predicts a temperature down to 30 F with high temperatures of 50 F during this period, then the system should be changed over. The reverse applies when the system is operated on the winter cycle.

Figure 15 shows graphically a temperature schedule for an induction system. This indicates the relative temperatures of the primary air and secondary water thruout the year, and also the change-over temperature range. The solid arrows show the temperature variation when changing over from summer cycle to winter cycle, and the open arrows show the temperature variation when going from winter cycle to summer cycle.

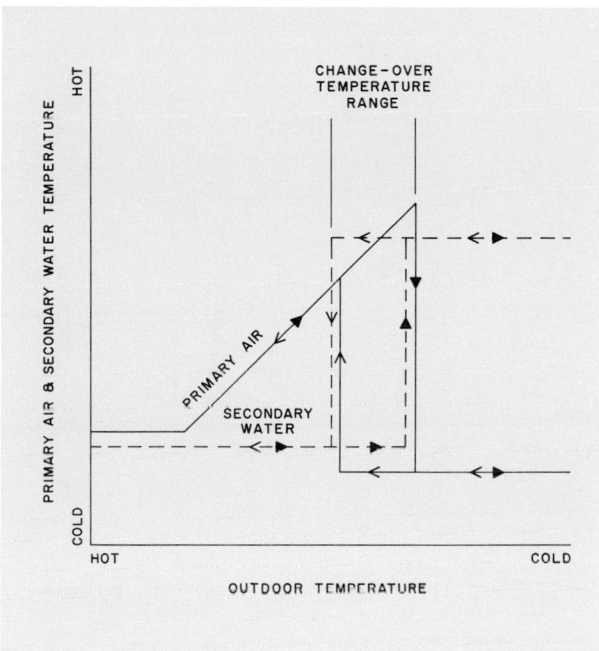

Fig. 15 — Temperature Schedule

In the calculation of the primary air quantity for the units of a change-over system, a deviation from the base A/T ratio may be used. This deviation is an experience factor which allows a closer matching of the unit capacity to the total cooling load of the space. The deviation allows a minimum temperature swing in the room. The maximum deviation (down to 0.7 multiplier) may be taken for buildings with heavy construction and small glass areas. For buildings with all glass or curtain wall construction, a deviation should not be used since there is a quicker response to the effects of outdoor temperature changes.

After the final unit selections are made, the water temperatures required by the units for winter heating are calculated when the units are operated with

primary air and when the units are used as gravity convectors. A lower room is usually used for the gravity heating requirements.

The reheat coil is selected with an entering temperature of 40 F. The leaving temperature is based on the primary air temperature specified by the reheat schedule (*Table 3*) at the calculated change-over temperature. The leaving air temperature used for the selection should be 25 degrees higher to allow for duct heat losses. This also provides a reserve capacity for a quick warm-up of the building and for any required adjustment of the reheat schedule.

With a change-over system, insulation is generally not required on the air risers, provided all openings in the floors are sealed to prevent stack effect and to retard circulation within the furred space. It is sometimes desirable to insulate the last two sections of the riser because the air quantities are small and the velocities are low; these low velocities allow a significant heat loss from these sections. As an alternate to insulating the risers, the air quantity of the units on the next to last floor may be selected on the basis of the base A/T ratio plus 10%, and the last units on the basis of the base A/T ratio plus 20%.

RETURN AIR

Room air may be returned to the primary air apparatus when the unit selections indicate that the primary air quantity exceeds the minimum ventilation requirements and when space is readily available for return air ducts.

When return air is used, it is possible to reduce the outdoor air quantity at peak loads only to 0.1 cfm per square foot, providing that the total primary air supplied to the units exceeds a minimum of 0.4 cfm per square foot. This provides a means to reduce the refrigeration requirement at peak loads. When the refrigeration machine can handle the load, the outdoor air quantity should be again operated at the design quantity.

When a return air system is used, the air is generally returned to the primary air apparatus by a return air fan. This fan operates at a static pressure to overcome the resistance of the return air system. This fan may also be used as an exhaust fan when the primary air system is operated with more than the minimum outdoor air quantity.

CHAPTER 2. PRIMARY AIR FAN-COIL SYSTEM

The primary air fan-coil system is in many ways similar to an induction unit system; the essential difference is the substitution of a fan-coil unit for the induction unit. The most suitable applications for the system are multi-room buildings such as hotels, hospitals and apartment houses, where the units need not be operated as convectors in winter.

This is a basic fan-coil system to which is added a second source of heating or cooling and positive ventilation. Its over-all performance is comparable to that of a change-over induction unit system. When performance is of more concern than first cost, this system may be considered. However, because of its first cost, an evaluation of an induction unit system may be advantageous before making a system choice.

Fan-coil units may be located along the perimeter of a building with the primary air supplied directly to the units (*Fig. 16a*) or from a corridor duct

directly into the room (*Fig. 16b*). Where the climate permits, the units may be suspended from the ceiling with the primary air supplied from a corridor duct (*Fig. 16c*). The latter arrangement may be less costly than that with the units along the perimeter of the building because of the more compact nature of the ductwork and piping layout.

This chapter includes System Description, System Features, Controls and Engineering Procedure for designing a complete primary air fan-coil system.

SYSTEM FEATURES

The primary air fan-coil system has the following features:

1. *Simultaneous Heating and Cooling* — The system provides two sources of capacity during the summer and winter seasons. In winter or below the change-over point, hot water is supplied to the room units and cool air is supplied from the primary air system. During the summer or above the change-over point, cold water is supplied to the room units and the primary air is heated according to a reheat schedule.

2. *Individual Room Temperature Control* — The system is ideally adapted to individual room temperature control because each unit has an integral cooling and heating coil designed for chilled and hot water.

3. *Confined Room Air Circulation* — Each unit recirculates room air only. Recirculation of air between rooms is kept at a minimum.

4. *Positive Ventilation At All Times* — A constant supply of outdoor air is delivered to each fan-coil unit after being properly conditioned, filtered, humidified or dehumidified, and heated or cooled in the central apparatus.

5. *Under-The-Window Air Distribution* — Under-the-window, upward air distribution is available and is superior to other types for small rooms, particularly in areas with low winter outdoor design temperatures.

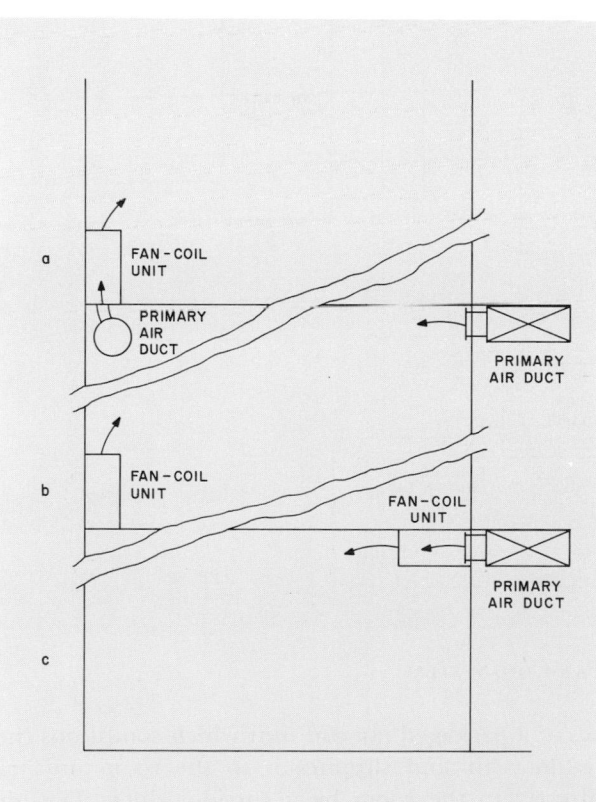

FIG. 16 — PRIMARY AIR FAN-COIL ARRANGEMENTS

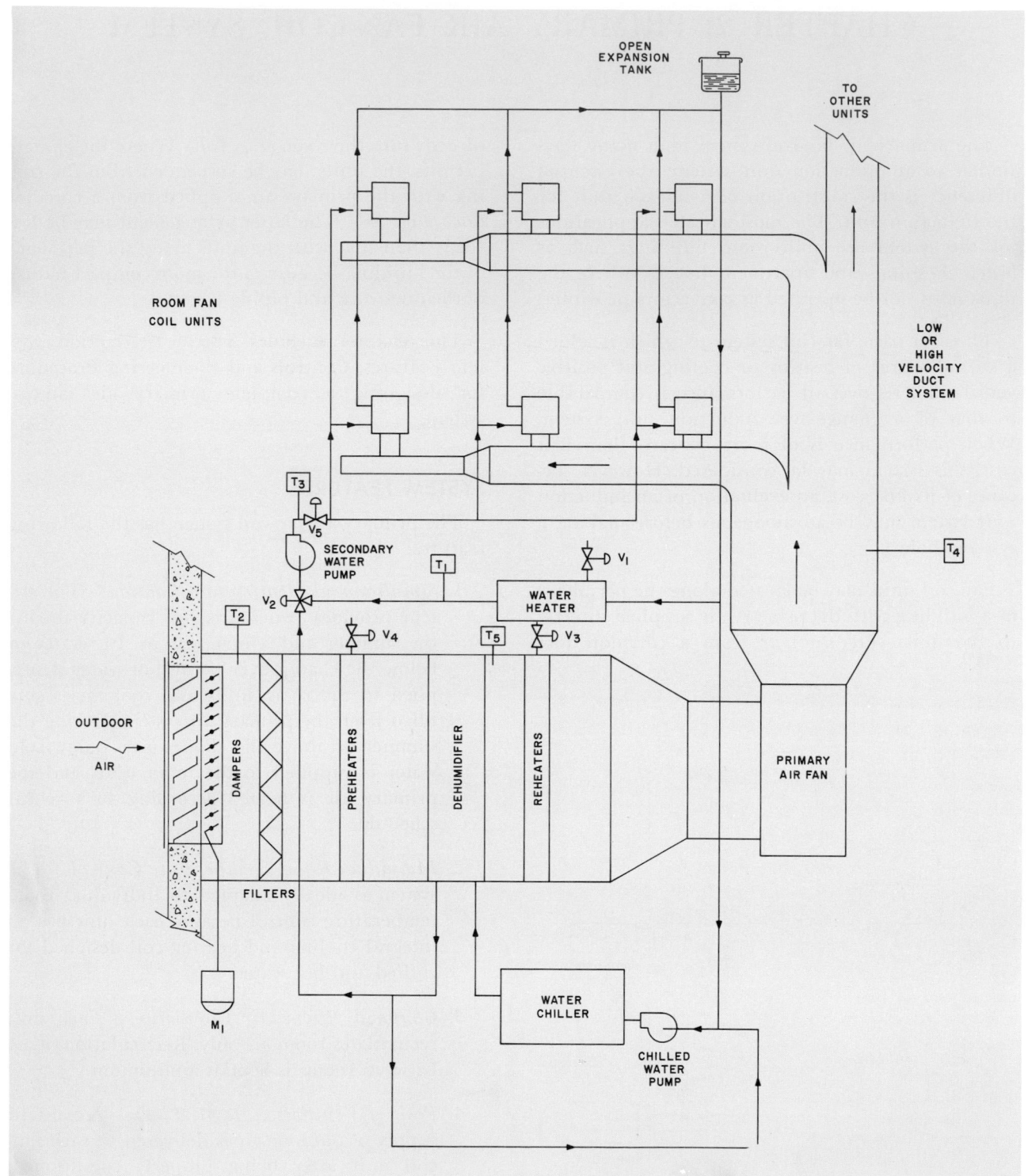

FIG. 17 — PRIMARY AIR FAN-COIL SYSTEM

SYSTEM DESCRIPTION

Figure 17 is a sketch of the system.

CENTRAL APPARATUS

The central apparatus is either a built-up apparatus or a packaged fan-coil unit which conditions the
outdoor air and supplies it to the room unit or
directly to the room by a corridor duct. The air
distribution system may be either low or high
velocity. A low velocity system is normally used if

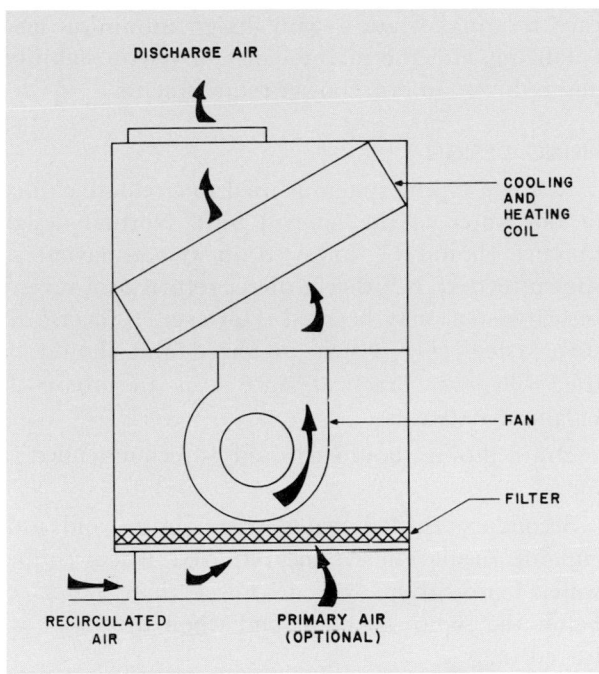

FIG. 18 — TYPICAL FAN-COIL UNIT

the primary air is discharged from a corridor duct directly into the room or supplied to units suspended from the ceiling. With space available a low velocity system results in the greatest economy of owning and operating costs.

The apparatus contains filters to cleanse the air, preheaters (when required) to temper the air, and a humidifier or dehumidifier to add humidification or remove excess moisture from the warm humid air. It also contains reheaters to heat the air from a predetermined schedule as the outdoor temperature falls to the change-over temperature. The primary air is held at a constant minimum temperature when the outdoor temperature is below the change-over temperature. When the primary air is supplied directly to the room, its minimum temperature is maintained sufficiently high to prevent drafts. Outdoor air to the apparatus is admitted thru a louver and screen.

Chilled water from a central refrigeration plant is circulated thru the dehumidifier coils in the central apparatus, and then mixes with recirculated water from the secondary water circuit to maintain a constant water temperature to the fan-coil units.

FAN-COIL UNIT

Figure 18 illustrates the basic elements of the fan-coil unit, including a recirculated air inlet, primary air inlet (optional), filter, fan, cooling and heating coil, and discharge air outlet.

The unit is supplied with cold or hot water depending on the outdoor temperature.

Room temperatures are maintained by thermostatically controlling the water flow.

ENGINEERING PROCEDURE

The following procedure is offered to assure a practical operating air conditioning system. A survey and preliminary layout are required as outlined in *Part 1*. Room loads and minimum ventilation air quantities are also determined from *Part 1*.

ROOM COOLING LOAD

Calculate the sensible and latent heat loads for all typical exposures: east, west, north, south, and any space that has unusual loads. It may be necessary to allow some flexibility in these calculations to allow for future partition changes, depending on the type of application. In most multi-room applications 8 to 16 room load calculations required.

ROOM HEATING LOAD

Calculate the room heating loads. They include the heating requirements to offset transmission and infiltration and also sufficient heat to temper the primary air from the temperature entering the room to the room winter design temperature.

PRIMARY AIR QUANTITY

Determine the ventilation air required for each unit from *Part 1*.

The primary air quantity should be determined in accordance with the A/T ratio concept as explained in *Chapter 1*. For each unit, calculate an A/T ratio which is the ratio of the unit ventilation air quantity to the total transmission per degree thru the outside exposed areas of the space served by the unit. Select the highest calculated A/T ratio as a base A/T ratio. Calculate the design primary air quantity for each unit by multiplying the base A/T ratio by the transmission per degree for the space served by each unit. The total primary air quantity for the system equals the sum of the primary air quantities required for each unit.

UNIT SELECTION

Select room units to satisfy these requirements:
1. Maximum room sensible load with a credit for the primary air cooling.
2. Maximum room and primary air heating load.

Unit selections can be made from a manufacturer's catalog.

The unit is normally adequate for zone depths of approximately 20 feet. Vertical air distribution from the perimeter unit spreads out in blanketing the exterior wall and travels along the ceiling for

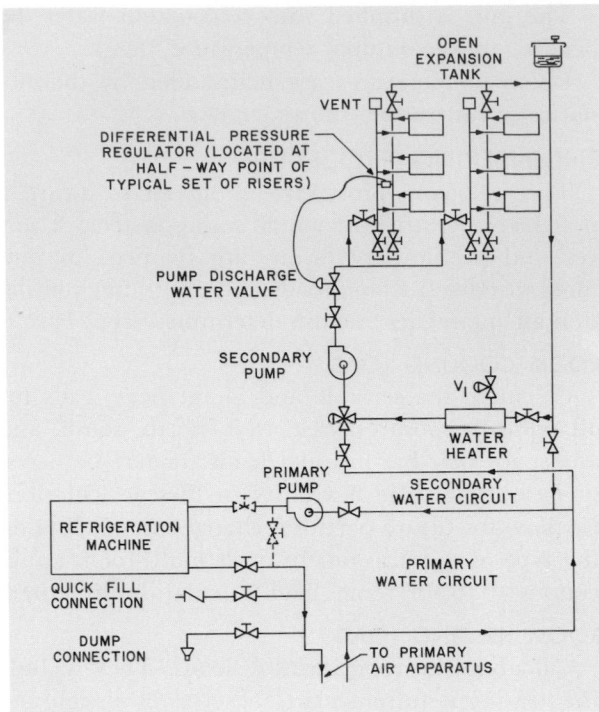

FIG. 19 — SCHEMATIC OF WATER PIPING, PRIMARY AIR
FAN-COIL SYSTEM

a distance of 15 to 20 feet before falling toward the floor in return air circulation.

The secondary water temperature should be selected to provide the required sensible heat capacity of the unit. In some cases the water temperature may be low enough to provide some latent heat removal. This may allow a higher apparatus dewpoint selection for the dehumidifier.

The water flow rate is dependent on the unit selection and cooling load, but should not be below the minimum flow which maintains turbulent conditions. Turbulent conditions for a 3/8, 1/2 and 5/8 inch OD tube is approximately 0.5, 0.7 and 0.9 gpm respectively.

The same water rate is used for heating as is used for cooling. The hot water temperature is calculated for each unit selection and the maximum temperature is used as the design.

DUCT DESIGN

High or low pressure ductwork can be used for the primary air system. Refer to *Part 2* for the design and sizing of the ductwork.

Although other methods of duct sizing such as equal friction or velocity reduction may be used, the static regain method is preferred. A system designed for static regain is nearly self-balancing because it is designed for the same static pressure at

each terminal. Static regain design minimizes field balancing, aids the maintenance of system stability, and reduces fan horsepower requirements.

PIPING DESIGN

A single piping system is used to circulate chilled or hot water to the fan-coil unit. Normal design practice should be followed in system layout as shown in *Part 3*. Either a direct return or a reverse return system may be used. However, a reverse return system *(Fig. 19)* is preferred and should be used whenever practical since it is an inherently balanced system.

Drain piping should be sized as recommended in *Part 3*.

Secondary chilled water riser piping and unit run-out insulation is not required when chilled water temperatures are no lower than 3 degrees below the room dewpoint and when the risers are furred in.

CENTRAL APPARATUS

Select the central air handling apparatus for the total primary air quantity.

The dehumidifier load is determined from the formula:

$$\text{Load} = cfm_{da} \times 4.45 \times (1 - \text{BF})\,(h_{ea} - h_{adp})$$

where:
cfm_{da} = dehumidifier air quantity
h_{ea} = entering air enthalpy
h_{adp} = apparatus dewpoint enthalpy
BF = dehumidifier bypass factor

The required apparatus dewpoint can be determined on an individual room basis by using the formula:

$$W_{adp} = \frac{W_{rm} - (W_{ea} \times \text{BF}) - \dfrac{\text{RLH}}{.68 \times cfm_{da}}}{1 - \text{BF}}$$

where:
W_{adp} = apparatus dewpoint specific humidity (gr/lb)
W_{rm} = room specific humidity (gr/lb)
W_{ea} = air entering dehumidifier specific humidity (gr/lb)
RLH = room latent heat load
cfm_{da} = dehumidified air quantity supplied to room
BF = dehumidifier bypass factor

The selected apparatus dewpoint should be representative of the majority of the spaces.

If this selected apparatus dewpoint is lower than approximately 48 F, an adjustment may be necessary in the selected secondary water temperature

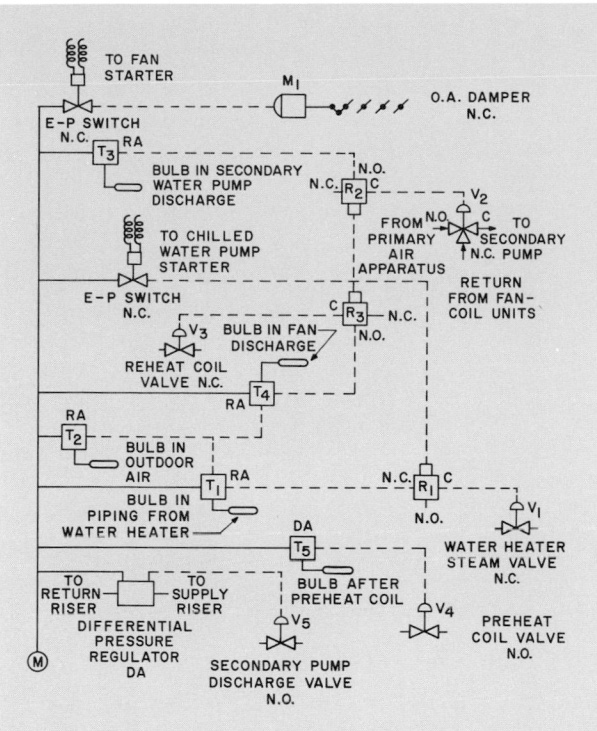

FIG. 20 — TYPICAL CONTROL DIAGRAM PNEUMATIC
PRIMARY AIR FAN-COIL SYSTEM

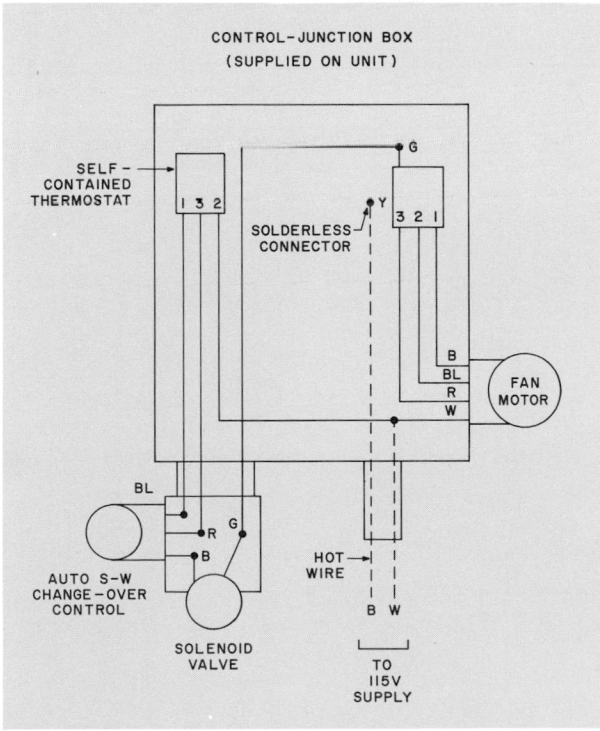

FIG. 21 — CONTROL PACKAGE, FAN-COIL UNIT,
MANUAL THREE-SPEED FAN CONTROL WITH
AUTOMATIC ON-OFF WATERFLOW

so the fan-coil units can accommodate part of the latent load. A few degrees drop in the secondary water temperature can provide a considerable amount of latent heat removal in the fan-coil unit.

Select the reheat coil to heat the primary air quantity from 40 F to a temperature based on the primary air temperature required at the change-over temperature (*Table 2, Chapter 1*). The leaving air temperature used for the selection should be increased by 25 degrees to provide for duct heat losses and reserve capacity for quick warm-up of the building. Change-over temperature and the reheat schedule are determined as in *Chapter 1*.

The preheater coil is selected to heat the primary air from the minimum outdoor design temperature to about 50-55 F.

The fan is selected for the primary air quantity and a static pressure sufficient to overcome the resistance in the apparatus and ductwork.

The filter is selected for the design air quantity and should have a good efficiency of about 85-95% based on the weight method of testing filters.

REFRIGERATION LOAD

The refrigeration load is equal to the sum of the peak building (or block estimate) sensible heat load and the dehumidifier load, less a credit for the primary air cooling of the conditioned spaces.

WATER HEATER

The water heater is selected as in *Chapter 1*.

CONTROLS

A basic control arrangement for the fan-coil unit, primary air apparatus and secondary water circuit is illustrated in *Fig. 20 and 21*. The controls are similar to those required for an induction system except at the fan-coil unit.

UNIT CONTROL

The fan-coil unit capacity is controlled by varying the water flow to the coil within the unit.

The room thermostat should be located on the room wall and not at the unit when primary air is admitted directly to the unit.

SECONDARY WATER CIRCUIT AND PRIMARY AIR APPARATUS CONTROLS

The secondary water circuit and primary air apparatus controls are similar to those used with the induction unit system and the control sequence is the same as described in *Chapter 1*.

Part 12
WATER AND DX SYSTEMS ————————
CHAPTER 1. FAN-COIL UNIT SYSTEM

The all-water fan-coil unit system is well suited for many applications. It is particularly applicable to a multi-room building where ductwork costs may be prohibitive. Where the climate permits, it is equally applicable to the relatively low temperature hot water characteristics of central heat pump systems, using an interior zone load as a heat source. A fan-coil unit system is not recommended for applications having high latent heat loads.

The system is used in many applications such as hotels, motels, hospitals, apartments, office buildings, professional buildings and clinics. Units may be located under the window, over closets, in dropped ceilings or furred down spaces.

This chapter discusses the fan-coil unit system covering the System Description, System Features, Engineering Procedure and a control arrangement.

TYPES OF SYSTEMS

All-water fan-coil systems may be classified in two major groups:

1. Single (2-pipe) piping system in which a single supply of water (cold or hot depending on the season) is available at each fan-coil unit and a single return piping system is utilized.
2. Multi-piping system in which a double supply of water (cold and hot) is available at each fan-coil unit and a single (3-pipe) or double (4-pipe) return piping system is utilized.

SINGLE PIPING SYSTEM

SYSTEM DESCRIPTION

This system (*Fig. 1*) consists of central water heating and cooling equipment, fan-coil units, controls, interconnecting piping and wiring as required.

The fan-coil unit system is designed to provide individual space temperature control without utilizing central station air handling equipment or ductwork. *Figure 2* illustrates the basic elements of the unit which include the air inlet, filter, fan and cooling and heating coil. The unit may be under-the-window or ceiling mounted.

Either a mixture of outdoor and return air or return air alone is supplied to the unit. Filters clean the air. The coil cools and dehumidifies during the summer and heats during the winter.

In *Fig. 1* the outdoor air for the under-the-window unit is shown delivered directly to the unit by low pressure ductwork. Such a method is preferable to ventilation air introduced thru a wall opening (*Fig. 2*). The use of a wall opening to introduce outdoor air to the fan-coil unit is not generally recommended for multi-story buildings. Stack and wind effects may affect adversely the performance of the units. In some instances air obtained by infiltration may be sufficient for ventilation, or air obtained from an interior zone system may be utilized for ventilation. Other methods of introducing outside air are discussed in *Part 11, Primary Air Fan-Coil System.*

Temperature control is maintained in two ways:
1. Fan speed adjustment or on-off fan control.
2. Water flow modulation or on-off water flow.

Electric reheat is sometimes installed at each unit to improve the operation of the fan-coil unit system during the intermediate season. Motels particularly are a good application for electric reheat.

Chilled water from a central refrigeration plant is circulated thru the unit coils to remove excess moisture and cool the air during the summer. Hot water from a water heater is supplied to the same unit coils during the winter.

SYSTEM FEATURES

Features of the fan-coil unit system are:
1. *Individual Room Temperature Control* — The system is adaptable to such a control because each unit has an integral cooling and heating coil designed for chilled and hot water.
2. *Confined Room Air Circulation* — Each unit recirculates only room air. Therefore, recirculation of air between rooms is minimized.
3. *Economy of Operation* — Outdoor air is available during marginal weather for free cooling. An outdoor-room air proportioning damper at the unit provides control.

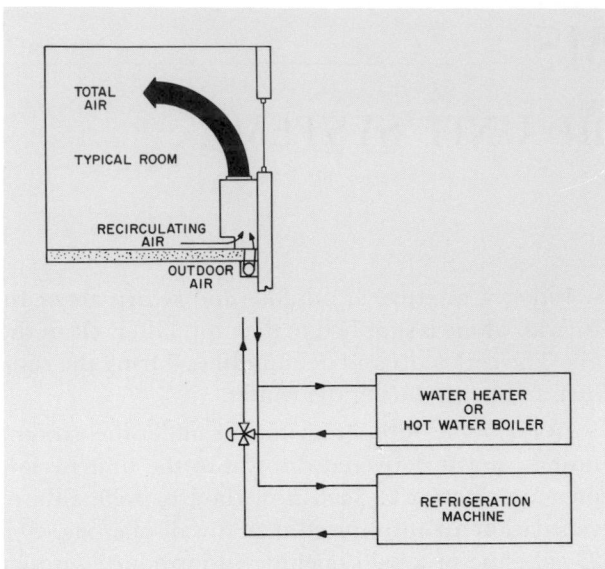

FIG. 1 — TWO-PIPE FAN-COIL UNIT SYSTEM

4. *Minimum Ductwork, Cutting and Patching* — Design costs of ductwork are minimized since supply and return air ductwork are not normally required. Cutting and patching in old buildings is normally confined to requirements for water piping and ventilation.

5. *Under-the-Window Air Distribution* — Under-the-window upward air distribution is superior to other methods for small rooms, particularly those having a heating or year-round load.

ENGINEERING PROCEDURE

The following is a design procedure for a fan-coil unit air conditioning system. As in all design work a survey and preliminary layout *(Part 1)* are required. Room loads and minimum ventilation air quantities are determined by methods outlined in *Part 1*.

Room Cooling Load

Calculate the load for all typical spaces on east, west, north and south exposures and for any space that has an unusual load. It may be necessary to provide flexibility in these calculations to allow for future partition changes, depending on the type of application. Eight to sixteen room load calculations may be required for most multi-room applications. The computations should include both sensible and latent load requirements.

To determine the total sensible and cooling capacity required of the unit, ventilation air and room loads must be combined.

Room Heating Load

Calculate the room heating load in the normal manner. The computations should include the heating requirements to offset transmission and infiltration and sufficient heat to temper the outdoor air from the temperature at which it enters the unit to the room design temperature (depending on the type of system).

Unit Selection

Select the room units to satisfy the following requirements:

1. Maximum room and ventilation air cooling load, both sensible and total.
2. Maximum room and ventilation air heating load.
3. External resistance imposed upon the unit by additional required ductwork. The external resistance should be considered relative to its effect on the air volume and the cooling and heating capacity of the unit.

The unit is normally adequate for zone depths of approximately 20 feet. Vertical air distribution of the perimeter unit spreads out in blanketing the exterior wall and travels along the ceiling for a distance of 15 to 20 feet before falling toward the floor in return air circulation.

Select the fan-coil units so that the smallest units are used coincident with the highest water temperature. Water temperatures are usually selected be-

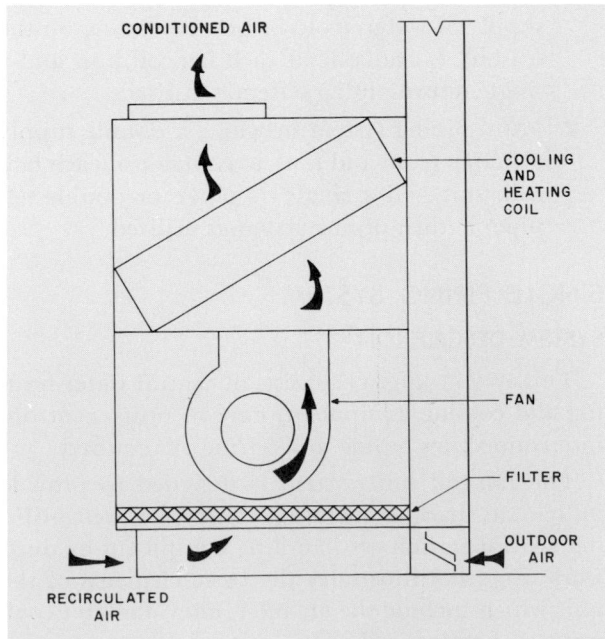

FIG. 2 — TYPICAL FAN-COIL UNIT

tween 45 F and 50 F. A few trial unit selections give an indication if the water temperature is acceptable.

The same water flow rate is used for heating as is used for cooling. System heating capacity is obtained by an adjustment of entering water temperature.

Convector heating during overnight, weekend or holiday periods is impractical; only one room heating load calculation with the unit fan operating is required.

Piping Design

Design the piping system in the normal manner. A single piping system is used to circulate chilled or hot water to the unit coil. Good design practice should be followed in system layout (Part 3). A reverse return system (Fig. 3) is recommended and should be used whenever it is adaptable to the building layout; this provides an inherently balanced system. Otherwise a direct return system may be used.

In buildings with units on more than one exposure a diversity factor should be applied as outlined in Part 3. In addition to resulting in smaller piping, headers and circulating pumps, diversity aids in minimizing sound problems when the system is operating at low water quantities. Diversity is not applied to the risers.

If more than one zone of water piping is desired, secondary circuits can be arranged accordingly. However, a three- or four-pipe system should be considered before designing a zoned single piping system.

The design water flow rate selected for the units can influence the total system cost. The lower the flow rate, the lower the first cost of the system piping and pump. However a check should be made to

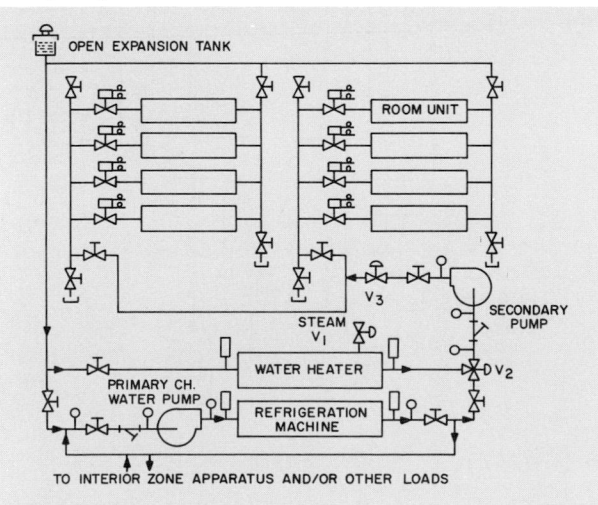

FIG. 3 — SCHEMATIC WATER PIPING, TWO-PIPE FAN-COIL UNIT SYSTEM

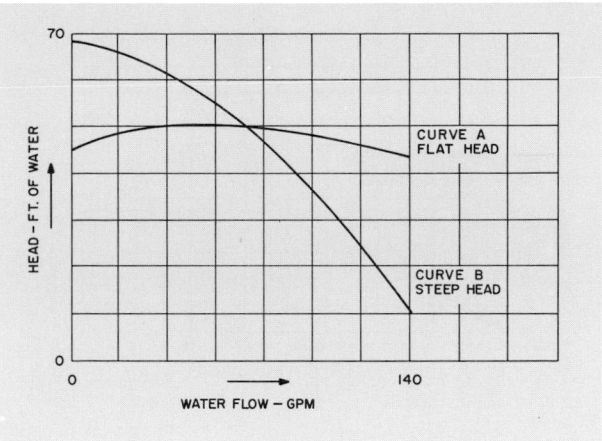

FIG. 4 — PUMP HEAD CHARACTERISTICS

assure turbulent flow in the unit coil. The minimum flow per coil circuit to maintain turbulent flow conditions for a ⅜, ½ or ⅝ inch OD tube is approximately 0.5, 0.7 or 0.9 gpm respectively.

The secondary water pump is operated continuously and should be selected for an extremely flat head characteristic (Curve A, Fig. 4). Do not select a pump that has a steep head characteristic (Curve B, Fig. 4).

Before the final secondary water pump selection is made, a check should be made to determine if head pressure control is required. When the water control valve (a three-way solenoid valve) in the closed position permits full water flow bypass around the coil, head pressure control is not required. If the water velocity in the runout is less than 10 fps, head pressure control is not required. Head pressure control requirements are determined by plotting the head curve of a trial pump selection (Curve A, Fig. 5).

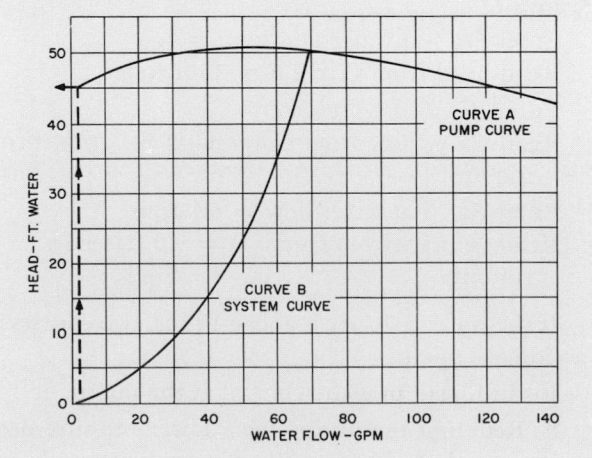

FIG. 5 — PUMP SELECTION

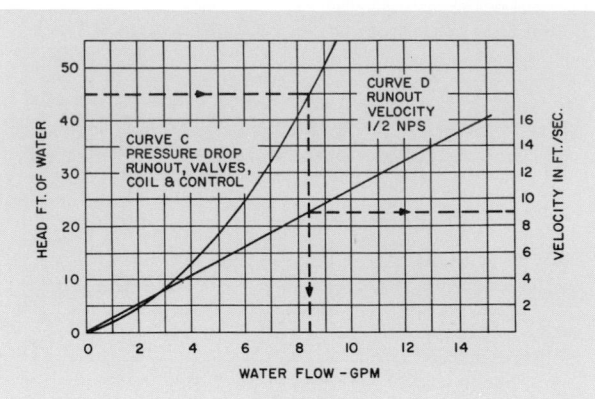

FIG. 6 — RUNOUT CHARACTERISTICS

The system pressure drop curve is added at its various flow rates (Curve B, *Fig. 5*). The intersection of Curves A and B is the operating point of the pump at full load.

The sum of the pressure drops of the unit runout, fittings, control devices and unit coil are plotted in curve form (Curve C, *Fig. 6*). This is the pressure drop of the system at various water flow rates from the supply riser thru the unit and back to the return riser. The velocity of the water flowing in the runout is also plotted for various flow rates (Curve D, *Fig. 6*). *Example 1* illustrates the use of *Fig. 5 and 6* in determining the water velocity in the runout when one unit is operating.

Example 1 — Runout Water Velocity

Given:
  Figures 5 and 6
  Unit design water quantity — 1 gpm

Find:
  Water velocity in runout

Solution:
  Enter *Fig 5* at water flow gpm = 1.
  Read head from Curve A = 45 ft water.

  Enter *Fig. 6* at 45 ft water.
  Read water flow from Curve C = 8.4 gpm thru
    the unit.

  Enter *Fig. 6* at water flow = 8.4 gpm.
  Read velocity from Curve D = 9.0 ft/sec in
    runout.

If the water velocity is above 10 fps, two courses of action are open:

1. Add head pressure control at the pump, or
2. Redesign the piping for a lower pressure drop so that the pump with a lower head at shutoff can be selected.

Optimum design of drain piping considers the type of ventilation air supply provided. When the fan-coil unit system handles the entire dehumidification of outdoor and recirculated air, there is a substantial amount of condensate from the units. However if the system draws ventilation air from an interior zone (conditioned by a separate system), very little condensate collects. The amount of condensate is, therefore, a consideration when sizing drain piping.

Vaporproof insulation is required for water piping and runouts.

Refrigeration Load

The refrigeration load is established by the peak building load (or block estimate) of the areas served by the system. Such an estimate includes all the applicable items to be found on an Air Conditioning Load Estimate form shown in *Part 1*. Provide 5% allowance for overcooling in the block load estimate.

CONTROLS

A typical control arrangement for the fan-coil unit and secondary water coil circuit is shown in *Fig. 7 and 8* respectively. *Figure 7* is a schematic of a control

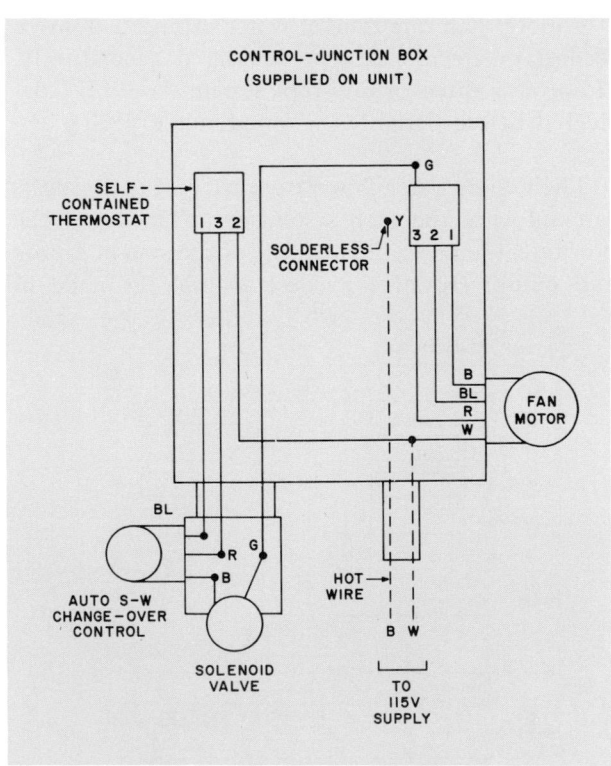

FIG. 7 — CONTROL PACKAGE, MANUAL THREE-SPEED
FAN CONTROL WITH AUTOMATIC ON-OFF
WATER FLOW

package consisting of a solenoid valve operated from a room thermostat in conjunction with a manual multi-speed fan switch.

Unit Control

Sensible and latent cooling capacity when performed by a single element cannot be adjusted to the wide range of coil sensible heat ratios which occur in comfort cooling. Maintenance within acceptable limits of room relative humidity is specified in order to provide desired dry-bulb temperature. Room conditions may be controlled by fan speed adjustment or on-off fan control. It may also be controlled by a combination of fan speed adjustment and on-off water flow to the heat transfer coil within the unit (or water flow modulation).

Basic control at the unit is accomplished by a multi-speed fan switch. Control of the water flow thru the unit coil may be manual, electric or pneumatic.

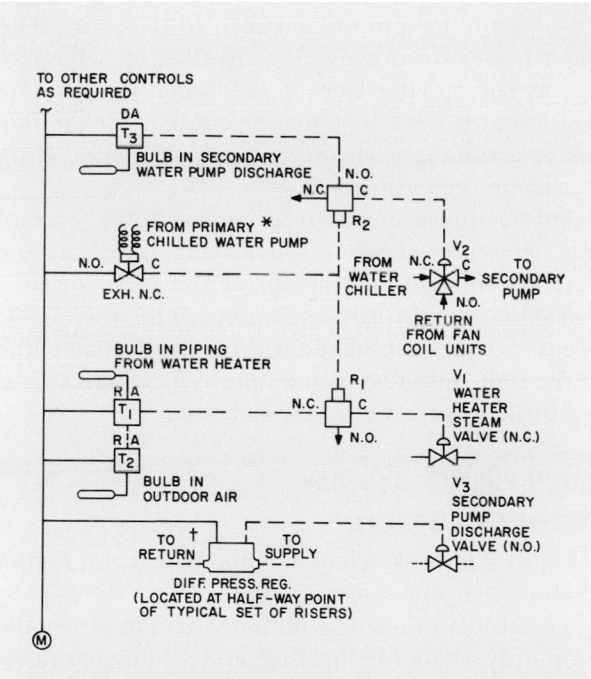

*Use a summer-winter switch (S—0 lb, W—15 lb) if there is more than secondary water load on the primary water circuit.
†Head pressure control required only if water velocity in unit runout exceeds 10 ft/sec.

FIG. 8 — CONTROL DIAGRAM, PNEUMATIC, TWO-PIPE
FAN-COIL UNIT SYSTEM

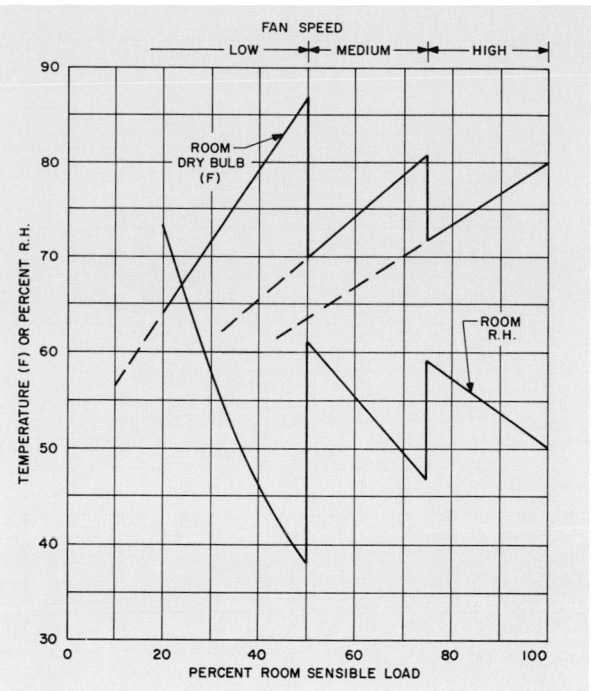

FIG. 9 — ROOM TEMPERATURE AND HUMIDITY
CONDITIONS, MANUAL THREE-SPEED FAN CONTROL

Continuous design flow of chilled water thru the cooling coil provides better maintenance of room humidity levels. However even with multi-speed fan control for loads below that equivalent to the minimum fan speed, the room temperature cannot be maintained and the fan must be shut down. Typical room temperature and humidity conditions for a fan-coil unit with manual, three-speed fan control are illustrated in *Fig. 9*.

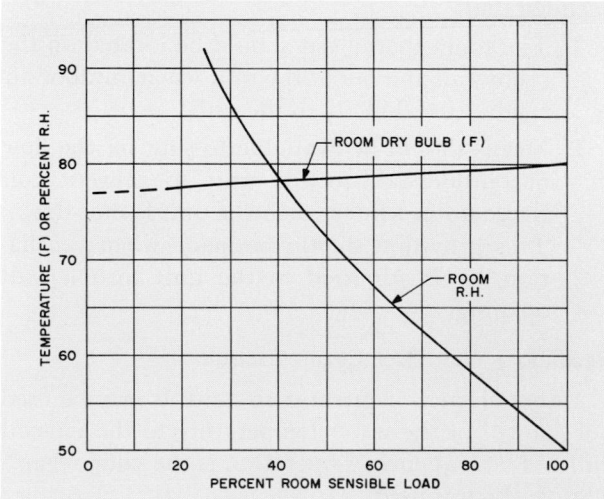

FIG. 10 — ROOM TEMPERATURE AND HUMIDITY
CONDITIONS, AUTOMATIC MODULATED WATER CONTROL

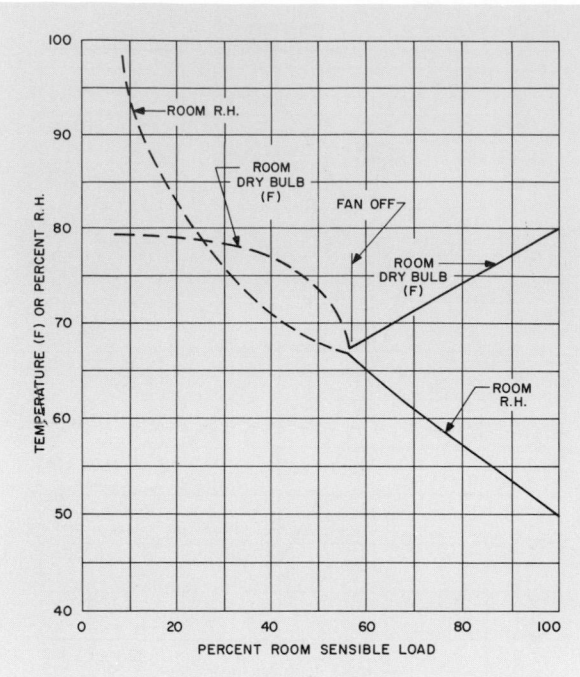

Fig. 11 — Room Temperature and Humidity
Conditions, Manual On-Off Fan Control

Modulated control of chilled water thru the cooling coil allows the room humidity level to rise as the room sensible load falls. Typical room temperature and humidity conditions for such a control are shown in *Fig. 10.*

Typical room temperature and humidity conditions for manual, on-off fan control are illustrated in *Fig. 11.*

When selecting a control system, it is recommended that:

1. The room thermostat should be located on the room wall and not at the unit when outdoor air is admitted directly to the unit.
2. Means should be available to shut off the flow of outdoor air to the unit to prevent coil freeze-up in winter when the unit is shut down. This provision should be made when ventilation air is admitted to the unit thru a wall opening.

Secondary Water Coil Circuit Controls

Either electric or pneumatic controls may be used for control of the water temperature to the fan-coil units. The sequence of operation is the same regardless of which is used.

Summer Operation

Outdoor air damper control may be either manual or automatic when ventilation is obtained thru an outside wall opening at the unit. If automatic control is used, the damper may be operated from the unit fan motor control to open when the fan is started. If outdoor air is supplied by a separate fan system thru low pressure ductwork, the fan may be started manually. With the primary chilled water pump running, the secondary water thermostat at the secondary water pump discharge controls a three-way mixing valve *(Fig. 8).* It maintains a constant secondary water temperature to the fan-coil units.

The room or unit thermostat *(Fig. 7)* controls the solenoid valve in the water line to the unit coil. A manual multi-speed fan switch is adjusted by the room occupant as required.

Winter Operation

The outdoor air damper operates as in summer. With no need for chilled water the primary chilled water pump is shut down. The three-way mixing valve is positioned to channel the secondary water from the fan-coil units thru the water heater, to the secondary water pump, and then to the units. The water heater steam valve is controlled by a thermostat at the leaving side of the water heater. The thermostat is reset by a master outdoor air thermostat to maintain a scheduled hot water temperature relative to the outdoors.

Automatic summer-winter change-over control *(Fig. 7)* senses the hot water temperature at the unit and reverses the action of the room or unit thermostat for winter operation. The thermostat opens or closes the solenoid valve in the water line to the unit coil. The manual multi-speed fan switch is adjusted by the occupant as desired.

MULTI-PIPING SYSTEM
SYSTEM DESCRIPTION

Figure 12 is a sketch of a 3-pipe system and *Figure 13* shows a 4-pipe system.

The ability of a single piping system to satisfy the constantly changing heating and cooling requirements of individual perimeter spaces increases with the number of zones provided. A multi-piping system provides hot and chilled water at each fan-coil unit the year round. In effect each unit then is a separate zone and functions independently. Its control valve selects either hot or chilled water for the unit coil depending on whether the module it serves requires heating or cooling. Simple, nonreversing thermostats may be utilized and change-over controls are not required. Design water quantities for the heating load may be minimized.

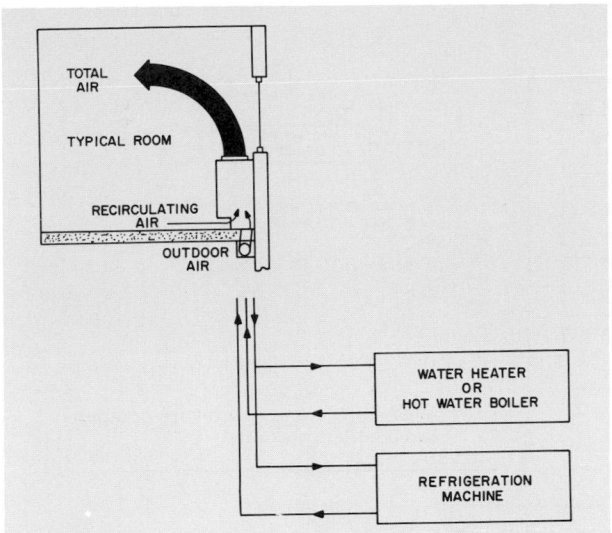

Fig. 12 — Three-Pipe Fan-Coil Unit System

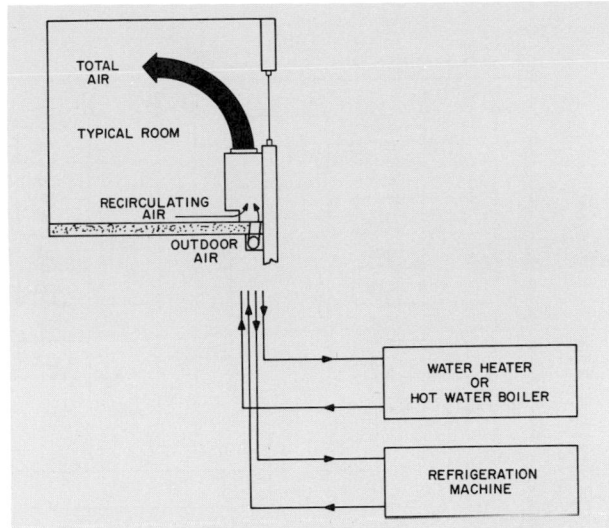

Fig. 13 — Four-Pipe Fan-Coil Unit System

Certain aspects of the three-pipe system should be investigated before deciding on a system design since three-pipe systems are covered by one or more patents. Although the system permits substantial first cost savings by using a common return, the operating costs are somewhat higher than those of a four-pipe system because of the mixing in the common return of water from the units operating on heating with water from the units operating on cooling.

The four-pipe system provides isolation of the hot and cold water circuits to minimize hydraulic problems. A single or split coil may be used at the unit. A split coil simplifies unit piping.

Two methods of operation are commonly used for multi-piping systems. The first method gives complete year-round temperature control in the room. Hot and cold water are necessary at the room unit at all times of the year. If the room temperature is too cold, hot water flows thru the fan-coil unit. If the room temperature is too warm, cold water flows thru the fan-coil unit. The second method of operation has hot and cold water available at the room unit only during the intermediate seasons. It is used when economy of operation is paramount and adequate room temperature control is desired. The sequence of equipment operation is shown in *Chart 1*. The temperatures at which the cooling and heating equipment is started and stopped vary from building to building.

SYSTEM FEATURES

In addition to the features of the fan-coil unit system as itemized under Single Piping System, additional features of the multi-piping system are:

1. *Quick Response to Thermostat Settings* — A change in thermostat setting is immediately apparent because hot and cold water are available at each unit. Quick response is psychologically advantageous.

2. *Elimination of Zoning by Exposure* — Multiple pumps, zoned piping and allied controls are eliminated because each space is a zone in itself as compared to the single piping system.

3. *Elimination of Operational Difficulties of Change-over* — When the system is properly designed, change-over is not required and operation is simplified.

4. *Year-round Room Temperature Control* — Tenant complaints during the intermediate season are eliminated because of the availability of both heating and cooling.

CHART 1 — OPERATING SCHEDULE

OPERATING SCHEDULE

OUTDOOR TEMPERATURE	REFRIGERATION EQUIPMENT	BOILER
95 — 70	ON	OFF
70 — 50	ON	ON
50 — 0	OFF	ON

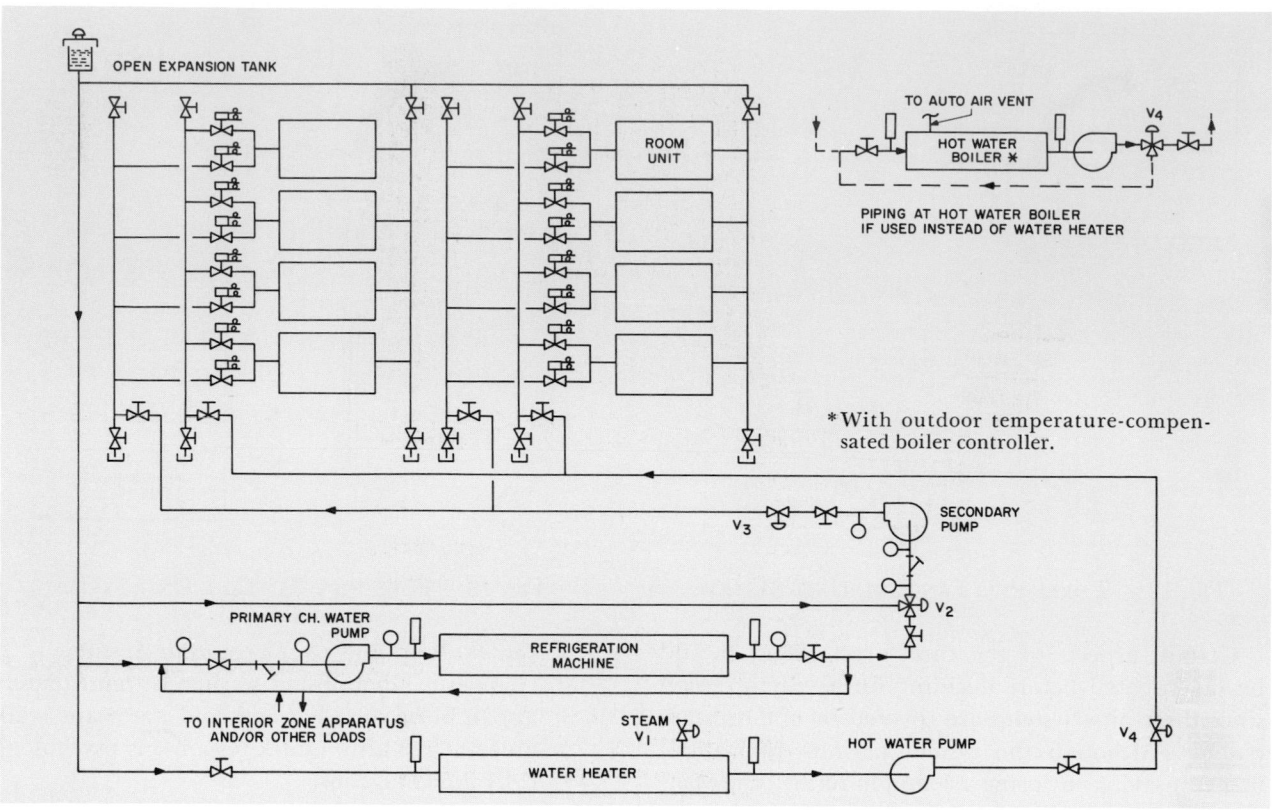

FIG. 14 — SCHEMATIC WATER PIPING, THREE-PIPE FAN-COIL UNIT SYSTEM

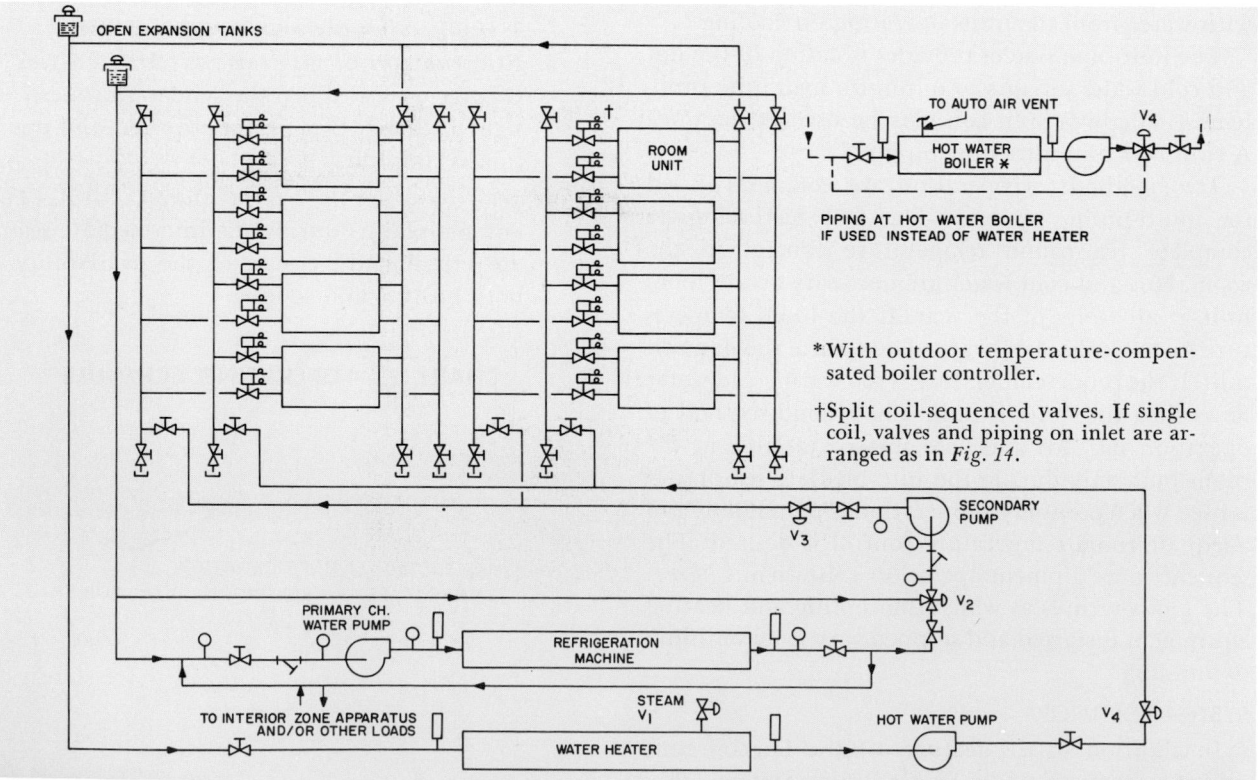

FIG. 15 — SCHEMATIC WATER PIPING, FOUR-PIPE FAN-COIL UNIT SYSTEM

ENGINEERING PROCEDURE

The design procedure for a multi-piping system is similar to the design of a single piping system. The few differences in design procedure are created by the piping system and involve piping design, controls and hot water quantities.

Unit Selection

Water quantities for the design heating load should be selected to give as high a rise as possible while matching the required heating load coincident with turbulent flow.

Piping Design

A multi-piping system is used to circulate chilled and hot water to the fan-coil unit.

The cold water supply and the common return line on a three-pipe system are sized first. Both headers are designed using a diversity factor. Several pumping arrangements can be used with this system. With a separate hot water pump (Fig. 14), the hot water piping is designed in accordance with normal practice.

A four-pipe system completely isolates the cold and hot water systems (Fig. 15) so that the piping for each system may be designed independently using a diversity factor on the chilled water supply and return headers.

The secondary and hot water pumps are selected for an extremely flat head characteristic (Fig. 4). Head pressure control requirements, if any, should be determined as in Example 1.

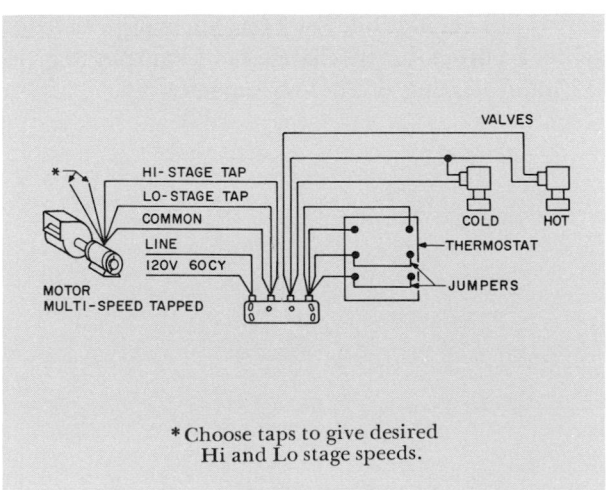

*Choose taps to give desired Hi and Lo stage speeds.

FIG. 16 — WIRING DIAGRAM, THREE-PIPE AND FOUR-PIPE FAN-COIL UNIT SYSTEM USING MULTI-SPEED MOTOR

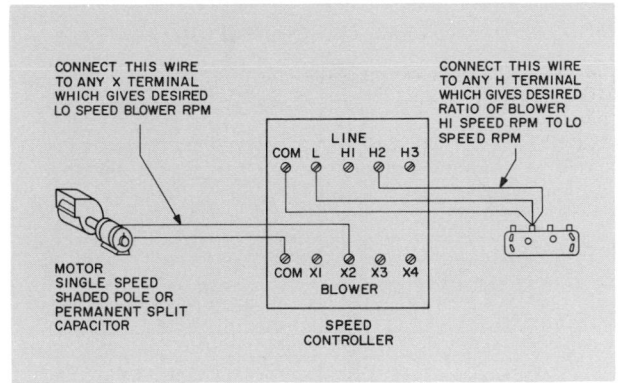

FIG. 17 — WIRING DIAGRAM, SINGLE SPEED MOTOR

Heating Load

The heating load is established from a summation of the room heating loads plus 20% for quick warm-up.

Heating equipment can be either a steam converter or a hot water boiler. The design water temperature is that which is determined from the unit selection. The water quantity is that determined from the sum of the individual room requirements.

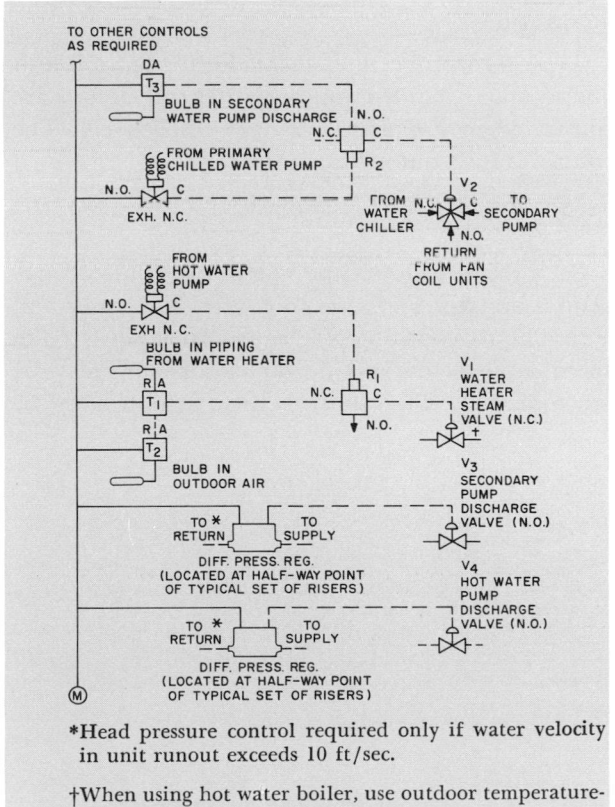

*Head pressure control required only if water velocity in unit runout exceeds 10 ft/sec.

†When using hot water boiler, use outdoor temperature-compensated boiler controller.

FIG. 18 — CONTROL DIAGRAM, PNEUMATIC, THREE-PIPE OR FOUR-PIPE FAN-COIL UNIT SYSTEM

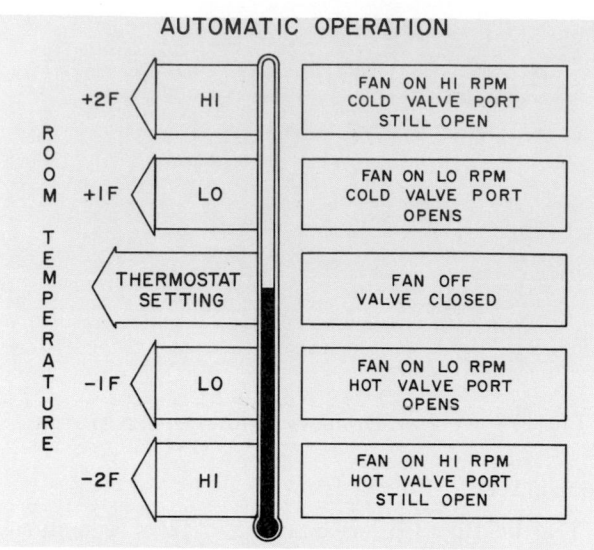

FIG. 19 — UNIT CONTROL OPERATION

CONTROLS

Control arrangements are many and varied. Pneumatic, electric or manual are used. Unit controls are often electric and may be automatic or partially automatic.

Typical control arrangements for the fan-coil unit and secondary chilled and hot water coil circuits are illustrated in *Fig. 16, 17 and 18* respectively. The control is fully automatic at the unit.

Unit Control

On a three- or four-pipe system two control valve actions are required, one to control hot water and a second to control cold water. In addition, fan speed adjustment and/or on-off air flow may be accomplished. The control valve action may be modulating or two-position. Manual control valves should not be used with any multi-piping system.

The sequence of operation for fully automatic control is shown graphically in *Fig. 19*.

Secondary Chilled Water and Hot Water Controls

Either electric or pneumatic controls may be used for control of the water temperature to the fan-coil units. The sequence of operation is the same regardless of which is used.

With the primary chilled water pump running, the chilled water thermostat at the secondary chilled water pump discharge controls a three-way mixing valve *(Fig. 18)*. It maintains a constant chilled water temperature to the fan-coil units. With the hot water pump running, the hot water thermostat at the leaving side of the water heater controls the water heater steam valve. The thermostat is adjusted by a master outdoor air thermostat to maintain a scheduled hot water temperature relative to the outdoors.

Referring to *Fig. 19,* on a rise in room temperature the unit fan is started on low speed and the cold water solenoid valve is opened. On a further rise in room temperature, the fan is automatically changed to high speed. As the room temperature approaches the thermostat setting, the fan returns to the low speed setting; when the thermostat is satisfied, the fan and cold water solenoid are shut off. On a drop in temperature the fan is started on low speed and the hot water solenoid valve is opened. On a further drop in room temperature the fan is automatically changed to high speed.

This control arrangement offers the advantage of longer operation at minimum sound level with close temperature regulation. For a few hours of peak load the fan is automatically changed to high rpm to give maximum heating or cooling capacity.

CHAPTER 2. DX SYSTEMS

DX systems are confined to the smaller and intermediate tonnage air conditioning and refrigeration applications. Packaged centrifugal and absorption machine liquid chillers are used for the higher tonnages.

DX systems for air conditioning or liquid chilling are those employing field-fabricated refrigerant piping. Condensing units and remote condenser (reciprocating) liquid chilling packages may be utilized in such systems. The piping interconnects reciprocating compressors, condensers and cooling coils or liquid chillers.

Refrigerant piping becomes economically less attractive as the amount of piping and distance between compressor, evaporator and condenser increases. The use of a cooling tower and/or reciprocating liquid chiller may then be more economical.

This chapter includes System Description, System Features and Engineering Procedure.

SYSTEM DESCRIPTION

Field-fabricated refrigerant piping is required for an air conditioning system using direct expansion coils and for a liquid chiller with a remote condenser. It is also required between the coils, the reciprocating compressor and the condenser when using direct expansion coils in either a built-up apparatus, fan-coil equipment or both. The minimum piping required for a liquid chiller (unless close-coupled) is that needed to connect the compressor and the condenser.

DIRECT EXPANSION AIR COOLING

Figure 20 is an isometric of a typical basic refrigeration system serving fan-coil equipment and utilizing direct expansion coils. Two compressors in parallel using an evaporative condenser and a subcooling coil provide refrigeration for direct expansion coils in each of three fan-coil units. A liquid suction interchanger in the liquid line from the subcooling coil increases the refrigerant cycle efficiency.

Capacity control at the compressors maintains a relatively stable suction pressure at the direct expansion coils. Solenoid valves in the liquid line to each coil are de-energized to shut off the flow of liquid refrigerant when its respective fan-coil unit is shut down.

LIQUID CHILLING

Figure 21 is an isometric of a typical DX system for water chilling. A packaged water chiller utilizes an evaporative condenser for condensing. The water chiller illustrated is a dry expansion cooler and is dual circuited. Chilled water temperature is controlled by compressor capacity and cooler circuit control.

Although dry expansion coolers are used on most applications, flooded coolers are available *(Fig. 22).* Dry expansion coolers are desirable because of:

1. Low first cost
2. Smaller space requirements
3. Minimum refrigerant requirements
4. Minimum freeze-up possibilities
5. Minimum oil return problems

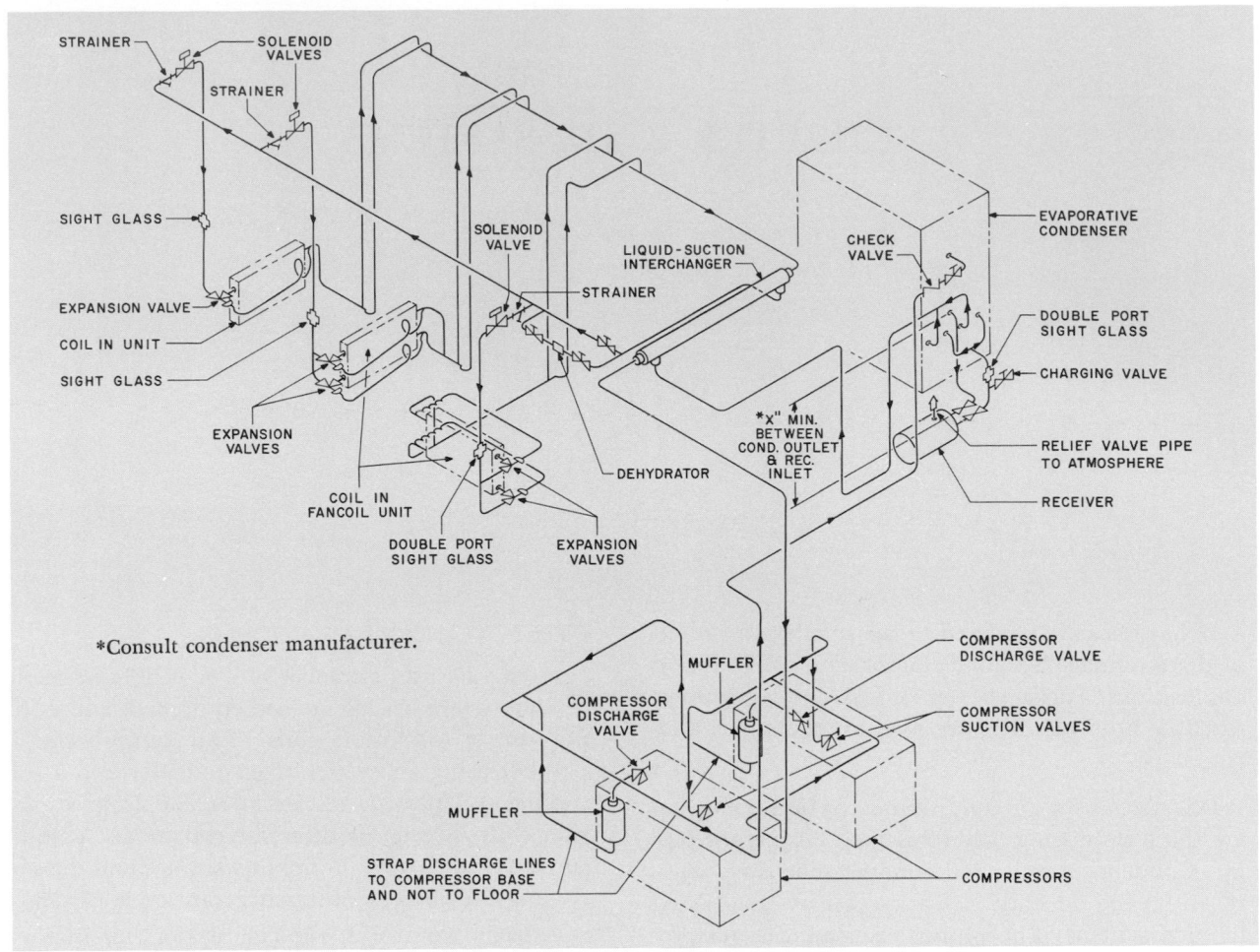

FIG. 20 — REFRIGERATION SYSTEM SUPPLYING DIRECT EXPANSION COILS

SYSTEM FEATURES

Features of the DX system are:

1. *Flexibility* — Physical layout and matching of equipment is more flexible.

2. *Low First Cost* — Reciprocating machines provide the lowest capital expenditure for refrigeration applications below approximately 100 tons.

3. *Suitability* — The trend towards air-cooled condensers makes the system suitable for many refrigeration applications. In addition the system can be used with an existing air handling apparatus which includes direct expansion cooling coils.

ENGINEERING PROCEDURE

The following procedure is offered to assure a practical basic refrigeration system. As in all design work a survey and preliminary layout are required as covered in *Part 1*. Refrigeration load is determined by using the method applicable to the particular type of air conditioning system *(Part 1)*.

The utilization of a single compressor and condenser is desirable when designing a DX system because of lower first cost, smaller space requirements and minimized hydraulic problems.

However multiple compressors and condensers are commonly used. Factory packages consisting of liquid chiller, multiple compressors and condensers are satisfactory because of low mass production costs and single controlled design.

When a single compressor or condenser of adequate size is unavailable to match load requirements, it may be advisable to design several systems each having its own evaporator, compressor and condenser. This is done with the understanding that the load may be readily divided into smaller portions which can be physically and more conveniently located apart.

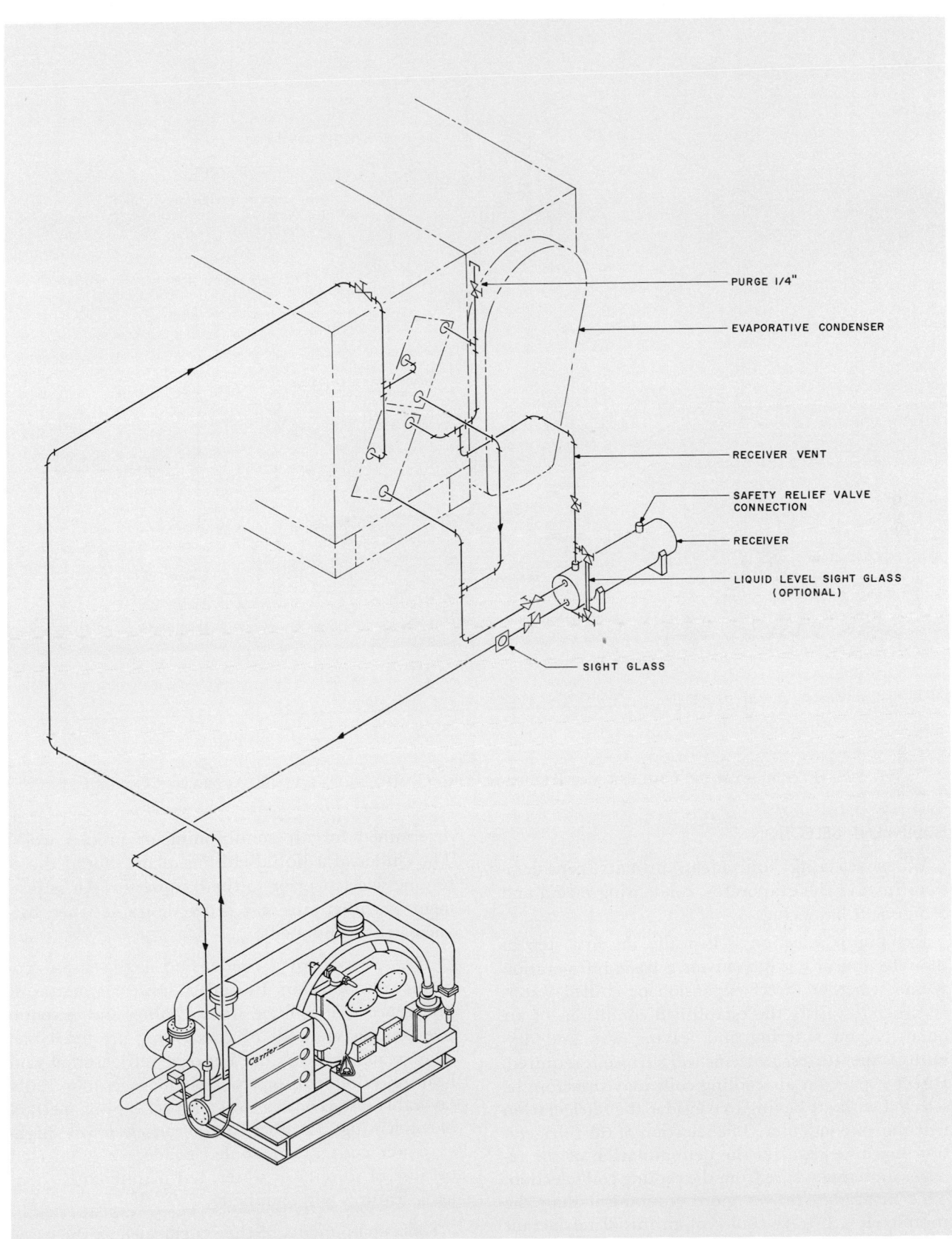

PURGE 1/4"

EVAPORATIVE CONDENSER

RECEIVER VENT

SAFETY RELIEF VALVE
CONNECTION

RECEIVER

LIQUID LEVEL SIGHT GLASS
(OPTIONAL)

SIGHT GLASS

FIG. 21 — DRY EXPANSION RECIPROCATING LIQUID CHILLING PACKAGE WITH EVAPORATIVE CONDENSER

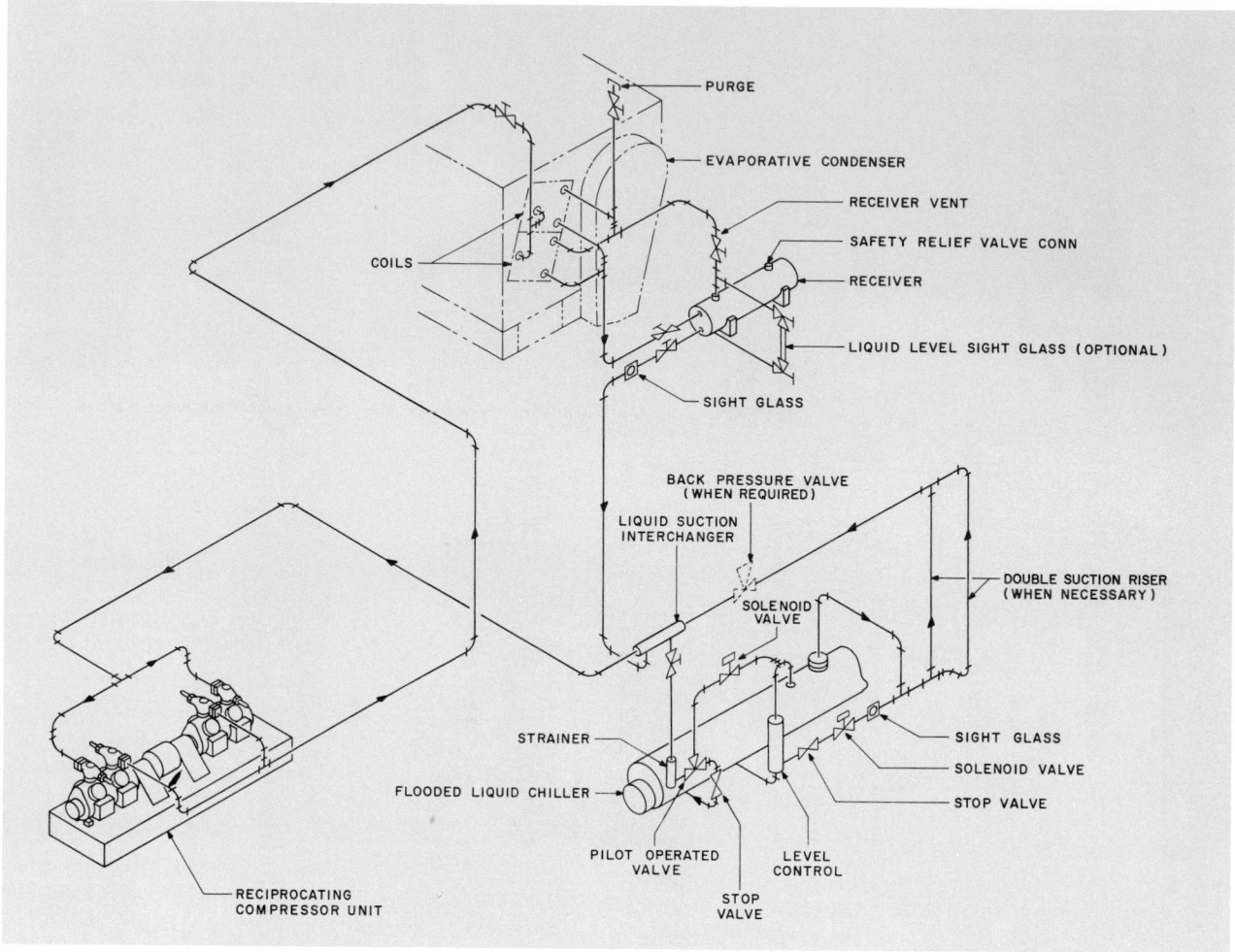

Fig. 22 — Flooded Liquid Chiller and Reciprocating Compressors with Evaporative Condenser

EQUIPMENT SELECTION

When selecting equipment, primary considerations involve the evaporator, condensing media and component balancing.

Evaporator selection is logically the first step in the selection of equipment for a basic refrigeration system, whether direct expansion or chilled water. In order to satisfy the established conditions of air quantity and entering and leaving wet- and dry-bulb temperatures, a specific coil surface is required. Direct expansion air cooling coils may therefore be selected without having to consider the refrigeration machine or condenser. The selection of the refrigeration machine requires the determination of the refrigerant temperature from the cooling coil selection. This invariably proves more economical than the completely arbitrary choice of an initial refrigerant temperature.

Chilled water quantity and temperature are pre-determined for air conditioning or process work. The chiller of a liquid chilling plant should therefore be selected prior to the compressor. An adjustment of chiller size may prove desirable when balancing the components.

The selection of a condensing media is an economic consideration. In some cases it is a matter of preference. Air cooling, water cooling and evaporative cooling methods of condensing are available. The greater the distance between refrigeration and heat rejection equipment, the more economically favorable is the cooling tower (water cooling method of condensing). In other circumstances, despite higher power costs an air-cooled condenser is selected because of scarcity of water, less maintenance, non-icing features and simplicity.

Some components of the system such as the evaporator can be selected individually based on operating requirements and without particular regard to

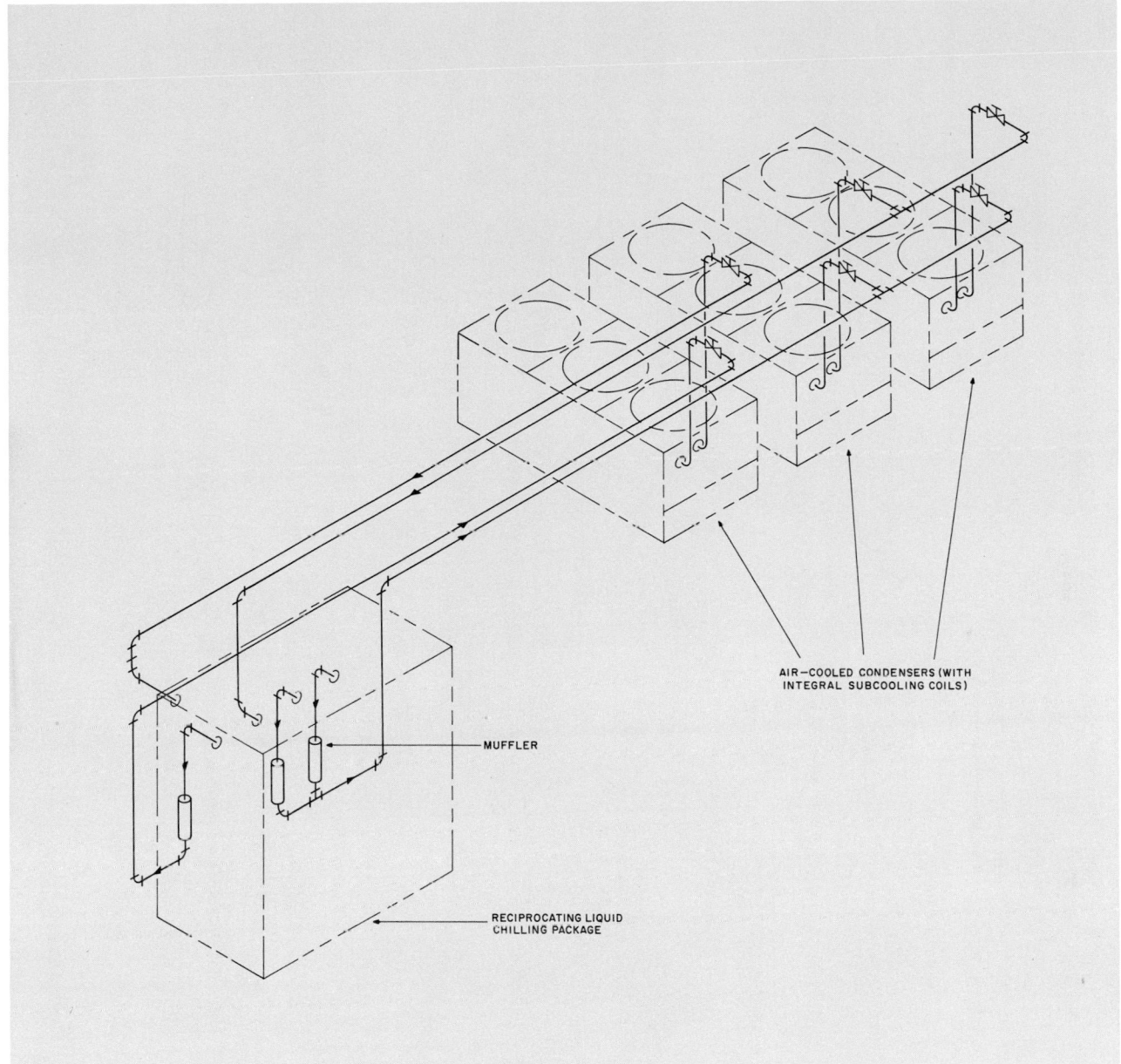

FIG. 23 — RECIPROCATING LIQUID CHILLING PACKAGE WITH AIR-COOLED CONDENSERS, THREE COMPRESSORS

other components in the system. However most components should be selected so that when operated together as a system, performance requirements are satisfied.

Balancing of the components of a basic refrigeration system is covered in *Part 7*. In general more than one combination of components meets the performance required of the system. Examination of several such combinations should be made to determine the optimum design.

The selection and balancing of system components influences the initial operating cost of the system. If optimum first cost is desired, compressor design, suction and head pressures are secondary; if optimum operating cost is desired, pressures are of primary importance.

Accessories such as liquid subcooling equipment and liquid suction interchangers should be considered when balancing components of the systems. These accessories are covered in *Parts 3, 4 and 7*.

Utilization of liquid subcooling increases the capacity of both the compressor and the condenser

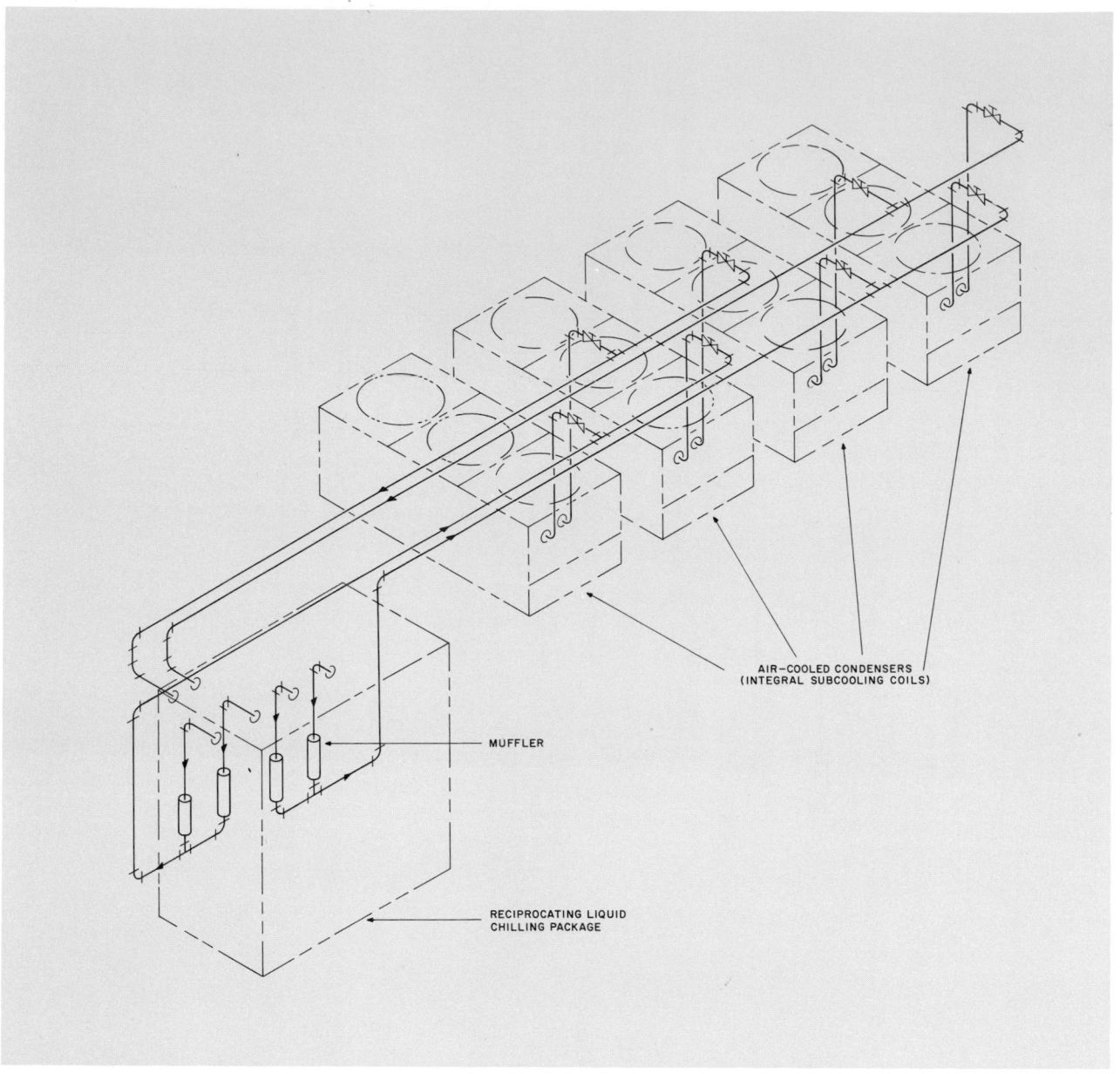

FIG. 24 — RECIPROCATING LIQUID CHILLING PACKAGE WITH AIR-COOLED CONDENSERS, FOUR COMPRESSORS

by the same amount. The evaporator is not affected.

The use of liquid subcooling:

1. Provides an increase in system capacity.

2. Subcools the liquid to offset the effect of moderate lifts.

3. Reduces horsepower per ton of refrigeration.

When the balanced capacity of a combination of compressor and evaporative condenser is slightly undersize, it is normally more economical to add a subcooling coil than to select the next larger com-

bination. The capacity of the condenser and compressor is increased by the same amount because with subcooled liquid each pound of refrigerant evaporated does more work. While increasing the capacity of the system, subcooling does not require any additional horsepower. Therefore the horsepower per ton of refrigeration is reduced.

Compressor ratings for Refrigerants 12 and 500 are generally based on 65 F actual suction gas temperatures. When this suction gas temperature is not obtained at the compressor, its rating must be low-

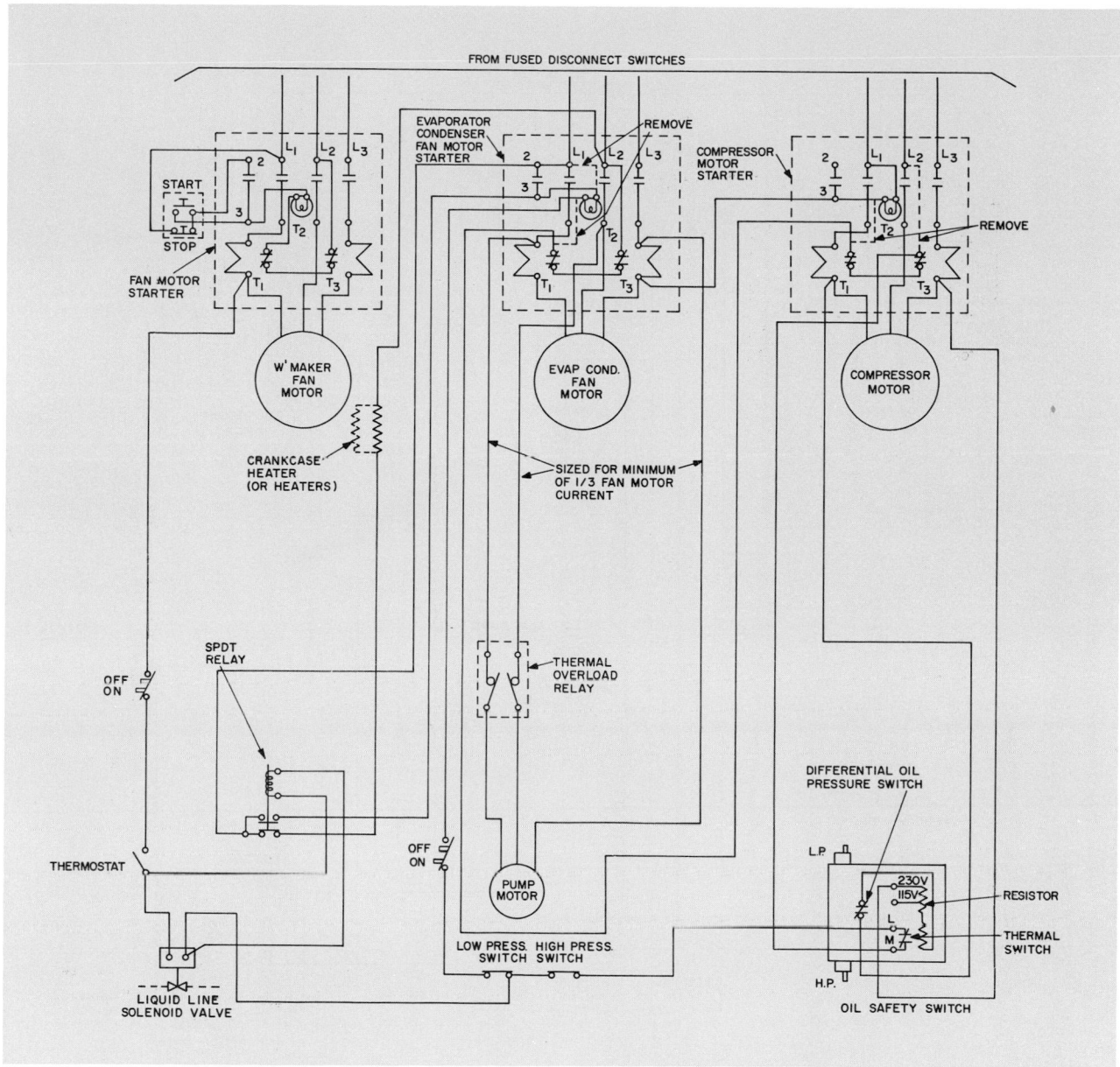

FIG. 25 — TYPICAL WIRING DIAGRAM FOR EVAPORATIVE CONDENSER INSTALLATION, 208–220 VOLTS, WITH SINGLE PUMPOUT COMPRESSOR CONTROL AND CRANKCASE HEATER

ered by an appropriate multiplier. To develop the full rating the required superheat over and above that available at the evaporator outlet may be obtained by means of a liquid suction interchanger. The effect of a liquid suction interchanger on the refrigeration cycle is discussed in *Part 4*.

PIPING DESIGN

Piping design *(Part 3)* should be followed when multiple or single compressors, condensers and evaporators are selected for a system.

Remote condenser type reciprocating liquid chiller packages require field-fabricated refrigerant piping between the compressors and condensers. Piping design and condenser selection should be made as recommended in *Part 3*.

Figures 23 and 24 illustrate a layout for a reciprocating water chiller package utilizing air-cooled condensers. In *Fig. 23* the water chiller package consists of three compressors and a dual circuited cooler. Two compressors, two condensers and one cooler

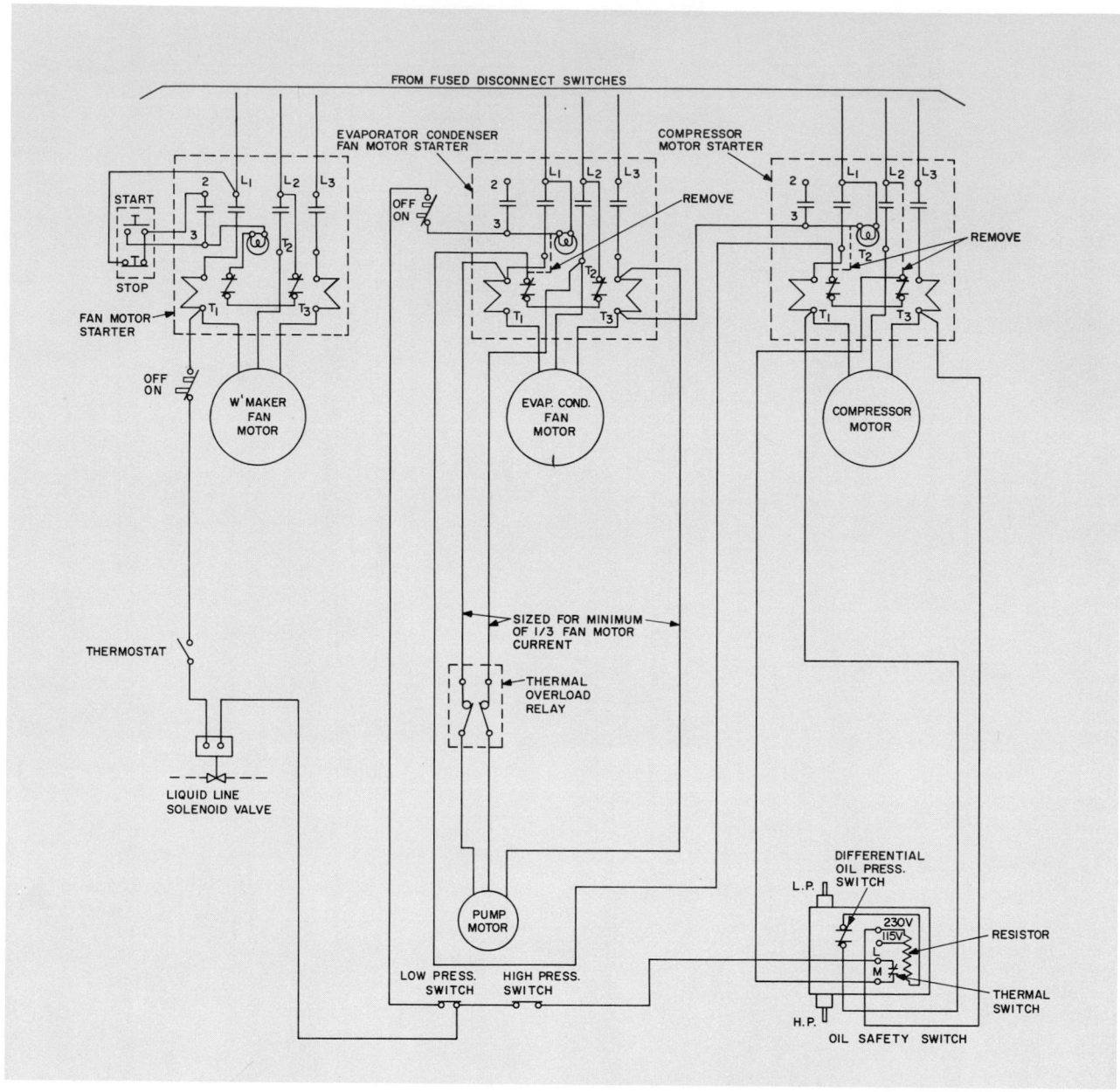

FIG. 26 — TYPICAL WIRING DIAGRAM FOR EVAPORATIVE CONDENSER INSTALLATION, 208–220 VOLTS, WITH
PUMPDOWN COMPRESSOR CONTROL

circuit are joined while the remaining compressor, condenser and cooler circuit are similarly connected. The condensers utilized in this instance are air-cooled with integral subcooling coils. Receivers are not used with this type of condenser because of the liquid seal maintained and the storage capacity afforded by the subcooling coil. The liquid refrigerant is piped directly from the liquid connection at the condenser to the chiller.

In *Fig. 24* the water chiller package consists of four compressors, four condensers and a dual cir-

cuited cooler. Two compressors, two condensers and one cooler circuit are connected while the remaining compressors, condensers and cooler circuit are similarly joined.

CONTROLS

A basic electric control arrangement is shown in *Fig. 25 and 26*. An electric control is particularly suited to this type of application because of the number of interconnections required between the controls and motor starters.

Single pumpout compressor control with crankcase heater is shown. It minimizes the accumulation of liquid refrigerant in the compressor crankcase during system shutdown. Single pumpout and automatic pumpdown control is covered in *Parts 3 and 7*.

The liquid line solenoid valve and crankcase heaters are interconnected with the fan starter and room thermostat. When the fan is running and the thermostat demands cooling, the liquid line solenoid valve opens, the compressor crankcase heaters are de-energized, and the single-pole double throw (SPDT) relay starts the evaporative condenser fan and pump motors. The compressor starts and continues to run if within a short period of time the oil pressure has built up sufficiently. Oil pressure is sensed at the differential oil pressure switch. The low and high pressure switches can shut down the compressor and evaporative condenser.

INDEX